2023 IEEE 5th Eurasia Conference on Biomedical Engineering, Healthcare and Sustainability

2023 IEEE 5th Eurasia Conference on Biomedical Engineering, Healthcare and Sustainability

Editors

Teen-Hang Meen
Kuei-Shu Hsu
Cheng-Fu Yang

MDPI

Basel • Beijing • Wuhan • Barcelona • Belgrade • Novi Sad • Cluj • Manchester

Editors
Teen-Hang Meen
National Formosa University
Yunlin
Taiwan

Kuei-Shu Hsu
Chia Nan University of
Pharmacy and Science
Tainan City
Taiwan

Cheng-Fu Yang
National University of
Kaohsiung
Kaohsiung
Taiwan

Editorial Office
MDPI AG
Grosspeteranlage 5
4052 Basel, Switzerland

This is a reprint of articles from the Proceedings published online in the open access journal *Engineering Proceedings* (ISSN 2673-4591) (available at: https://www.mdpi.com/2673-4591/55/1).

For citation purposes, cite each article independently as indicated on the article page online and as indicated below:

Lastname, A.A.; Lastname, B.B. Article Title. *Journal Name* **Year**, *Volume Number*, Page Range.

ISBN 978-3-7258-1441-1 (Hbk)
ISBN 978-3-7258-1442-8 (PDF)
doi.org/10.3390/books978-3-7258-1442-8

Contents

engineering proceedings

MDPI

Editorial

Preface of the Fifth IEEE Eurasia Conference on Biomedical Engineering, Healthcare and Sustainability, 2023 (IEEE ECBIOS 2023) †

Teen-Hang Meen [1], Kuei-Shu Hsu [2,*] and Cheng-Fu Yang [3,4,*]

1 Department of Electronic Engineering, National Formosa University, Yunlin 632, Taiwan; thmeen@gs.nfu.edu.tw
2 Department of Recreation and Health Care Management, Chia Nan University of Pharmacy & Science, Tainan City 71710, Taiwan
3 Department of Chemical and Materials Engineering, National University of Kaohsiung, Kaohsiung 811, Taiwan
4 Department of Aeronautical Engineering, Chaoyang University of Technology, Taichung 413, Taiwan
* Correspondence: kshsu888@mail.cnu.edu.tw (K.-S.H.); cfyang@nuk.edu.tw (C.-F.Y.)
† All papers published in this volume are presented at the IEEE 5th Eurasia Conference on Biomedical Engineering, Healthcare and Sustainability, Tainan, Taiwan, 2–4 June 2023.

Citation: Meen, T.-H.; Hsu, K.-S.; Yang, C.-F. Preface of the Fifth IEEE Eurasia Conference on Biomedical Engineering, Healthcare and Sustainability, 2023 (IEEE ECBIOS 2023). *Eng. Proc.* **2023**, *55*, 2. https://doi.org/10.3390/engproc2023055002

Published: 21 November 2023

This volume represents the proceedings of the fifth IEEE Eurasia Conference on Biomedical Engineering, Healthcare and Sustainability 2023 (IEEE ECBIOS 2023). This conference was held at Chia Nan University of Pharmacy & Science, Tainan, Taiwan, on 2–4 June 2023; it was co-organized by Chia Nan University of Pharmacy and Science, the Institute of Electrical and Electronics Engineers (IEEE), and the International Institute of Knowledge Innovation and Invention (IIKII). The conference provided a unified communication platform for researchers in the topics of biomedical engineering, healthcare and sustainability. The healthcare domain is currently experiencing a sector-wide transformation thanks to advances in computing, networking technologies, big data, and artificial intelligence. Healthcare is not only changing from reactive and hospital-centered to preventive and personalized medicine, but from disease-focused to well-being-centered treatment. Healthcare systems, as well as fundamental medicine research, are becoming smarter due to developments in biomedical engineering. Furthermore, with cutting-edge sensors and computer technologies, healthcare services and technologies are being used with better efficiency, higher quality and lower costs. However, these innovations often do not bring sustainability, health and happiness to all people. Science and technology are complemented by arts, humanities, social sciences and indigenous know-how and wisdom, therefore increasing the benefits for the needy in all regions and classes of people. We need an ethically aligned and driven healthcare system which is also sustainability. In this regard, this conference has promoted the interdisciplinary collaboration of science and engineering technology specialists in the academic and industrial fields, as well as fostering international networking. During the conference, there were extensive presentations and discussions in which attendants participated in various activities and gathered together in diverse groups across disciplines to generate new ideas, collaborations and business opportunities.

IEEE ECBIOS 2023 was held in a hybrid form of on-site and online presentations. Figure 1 depicts a group photo of the conference opening. The first keynote speech, entitled "Regulatory functions of glycosylation in cancer and neuroinflammation and the possibility of application", was delivered by Dr. Jian-Guo Gu, a Professor at Tohoku Medical and Pharmaceutical University, Japan. He divulged that glycosylation plays numerous roles in protein folding, targeting, recognition and other functions, and showed that the changes in glycan structures are associated with many physiological and pathological events, including cell adhesion, migration, growth, differentiation and tumor invasion. He mainly focused on N-glycans remodeled by several glycosyltransferases in cell growth, adhesion and the process of epithelial-to-mesenchymal transition (EMT) and cancer multidrug resistance (MDR) to address the potential roles of N-glycans in cancers. These results demonstrated that N-glycans can serve as an on/off switch to regulate cell adhesion and growth and provide new insights into the molecular mechanisms of cancer metastasis and chemoresistance. In the speech, the importance of core fucosylated N-glycans was also pointed out, addressing the molecular mechanisms of core fucosylation in several diseases, including hepatocellular carcinoma, pancreatic carcinoma and neuroinflammation. Possibilities for clinical applications were discussed.

Figure 1. Group photo at the opening ceremony of IEEE ECBIOS 2023.

The second keynote speech, entitled "Developing AI-based brain-computer interfaces in immersive VR environments", was presented by Dr. Po-Lei Lee, a Professor at the Department of Electrical Engineering, National Central University, Taiwan. The motor imagery (MI)-based brain–computer interface (BCI) is an intuitive interface that enables users to communicate with external environments through their minds. However, contemporary MI-BCI systems ask naïve subjects to perform unfamiliar MI tasks with simple textual instruction or visual/auditory cues. Dr. Lee stated that the unclear instructions for MI execution not only result in large inter-subject variability in the measured EEG patterns, but also cause difficulties in grouping cross-subject data for big data training. Dr. Lee introduced the design of a BCI training method in a virtual reality (VR) environment in which a head-mounted device (HMD) is used for action observations (Aos) with MI (i.e., AO+MI) in VR environments. EEG signals recorded in the AO+MI task are used to train an initial model, which is then continually improved using EEG data in subsequent BCI training sessions. In this experiment, five healthy subjects and each test subject participated in three tasks: an AO+MI task, an MI task, and an MI task with visual feedback (MI-FB) three times. By adopting a transformer-based spatial–temporal network (TSTN), users' MI intentions were decoded. In contrast to other convolutional neural network (CNN) or recurrent neural network (RNN) approaches, the TSTN extracted spatial and temporal features and applied attention mechanisms in spatial and temporal dimensions to perceive the global dependencies. The mean detection accuracies of TSTN were 0.63, 0.68, 0.75, and 0.77 in the MI, first MIFB, second MI-FB, and third MI-FB sessions, respectively. It was demonstrated that the AO+MI approach made it easier for subjects to conform to their imagery actions, and the BCI performance was improved with the continual learning in the MI-FB training process.

The third keynote speech was presented on "Optimizing Alternative Methods for Skin Sensitization Prediction: An Integrated Approach Using PaDEL and Machine Learning with the OECD QSAR Toolbox" by Dr. Kuan-Han Lee, a Distinguished Professor in the Department of Pharmacy, Chia Nan University of Pharmacy and Science, Taiwan. He emphasized that skin sensitization is critical in ensuring the safety and efficacy of chemicals used in various industries. The OECD QSAR Toolbox is an alternative tool for predicting skin sensitization with promising results; however, its current hit rate of 70–85% leaves room for improvement. He proposed the integration of machine learning techniques and PaDEL to analyze the molecular descriptors and fingerprints of the dataset to establish a skin sensitization prediction model using machine learning techniques such as grouping, combination selection and rolling corrections. In this speech, he explained that by enhancing the accuracy of the QSAR Toolbox, the resulting workflow was fed back into the QSAR Toolbox. A summary of the superior detection range and inaccurate aspects of allergens in OECD QSAR was provided to optimize the model and improve the prediction accuracy. The explained approach resulted in a higher hit rate than the present rate of 70–85% in the automated workflow of the QSAR Toolbox. By integrating PaDEL and machine learning techniques, a skin sensitization prediction model was constructed with enhanced accuracy. In his conclusion, the potential of integrating PaDEL and machine learning techniques with the OECD QSAR Toolbox was suggested to improve the accuracy of skin sensitization predictions. This technique is believed to promote animal welfare and the development of alternative methods and provide a valuable reference for future research on alternative skin sensitization tools. It is also expected to advance animal welfare and the safety of chemicals used in various industries.

In addition to the keynote speeches, IEEE ECBIOS 2023 provided six Regular Sessions and two Invited Sessions, covering various related topics of biomedical engineering, healthcare, and sustainability. Figures 2 and 3 show several on-site and online oral presentation sessions.

Figure 2. Presentation at an on-site session of IEEE ECBIOS 2023.

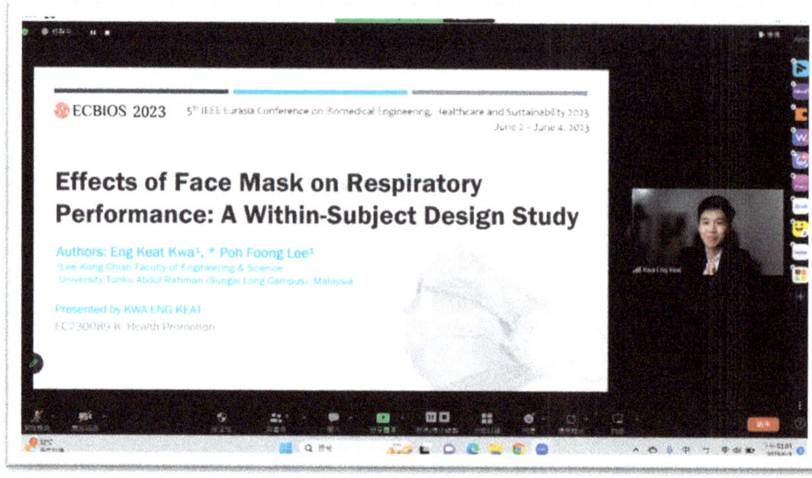

Figure 3. Online presentation of IEEE ECBIOS 2023.

Many substantial results were shared again at IEEE ECBIOS 2023 by enthusiastic participants. In total, 105 excellent papers in relevant engineering fields were selected through peer review for the publication of the IEEE ECBIOS 2023 proceedings. The proceedings are expected to help accelerate interdisciplinary collaboration and international networking within science and engineering technology specialists in the academic and industrial fields.

Author Contributions: All authors contributed equally to this paper. All authors have read and agreed to the published version of the manuscript.

Funding: This editorial received no external funding.

Institutional Review Board Statement: Not applicable.

Informed Consent Statement: Not applicable.

Data Availability Statement: Data is available at https://www.mdpi.com/2673-4591/55/1.

Conflicts of Interest: The authors declare no conflict of interest.

engineering proceedings

MDPI

Editorial

Statement of Peer Review †

Teen-Hang Meen [1], Kuei-Shu Hsu [2,*] and Cheng-Fu Yang [3,4,*]

1 Department of Electronic Engineering, National Formosa University, Yunlin 632, Taiwan; thmeen@gs.nfu.edu.tw
2 Department of Recreation and Health Care Management, Chia Nan University of Pharmacy & Science, Tainan City 71710, Taiwan
3 Department of Chemical and Materials Engineering, National University of Kaohsiung, Kaohsiung 811, Taiwan
4 Department of Aeronautical Engineering, Chaoyang University of Technology, Taichung 413, Taiwan
* Correspondence: kshsu888@mail.cnu.edu.tw (K.-S.H.); cfyang@nuk.edu.tw (C.-F.Y.)

† All papers published in this volume are presented at the IEEE 5th Eurasia Conference on Biomedical Engineering, Healthcare and Sustainability, Tainan, Taiwan, 2–4 June 2023.

In submitting the conference proceedings of the 5th IEEE Eurasia Conference on Biomedical Engineering, Healthcare and Sustainability 2023 (IEEE ECBIOS 2023) to *Engineering Proceedings*, the volume editors certify to the publisher that all papers published in this volume have been subjected to peer review administered by the volume editors. Reviews were conducted by expert referees to the professional and scientific standards expected of a *Proceedings* journal.

- Type of peer review: 2–3 reviews that were single-blind;
- Conference submission management system: https://2023.ecbios.asia/; (accessed on 1 January 2023)
- Number of submissions sent for review: 196;
- Number of submissions accepted: 105;
- Acceptance rate: 0.536;
- Average number of reviews per paper: 2;
- The total number of reviewers involved: 50;
- Any additional information on the review process: please refer to Figure 1.

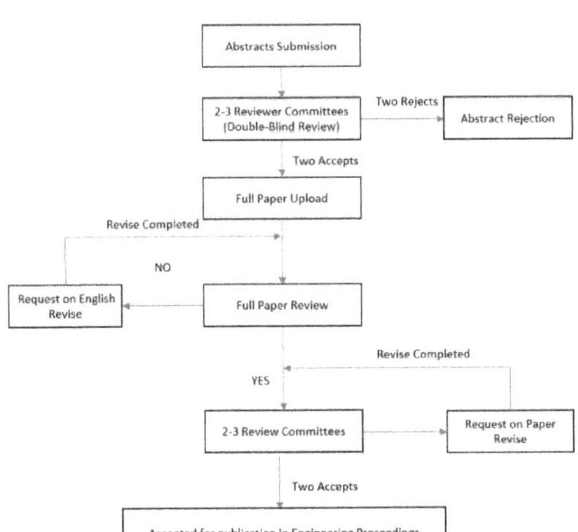

Figure 1. Review process for conference proceedings of the 5th IEEE Eurasia Conference on Biomedical Engineering, Healthcare and Sustainability 2023 (IEEE ECBIOS 2023).

Citation: Meen, T.-H.; Hsu, K.-S.; Yang, C.-F. Statement of Peer Review. *Eng. Proc.* **2023**, *55*, 1. https://doi.org/10.3390/engproc2023055001

Published: 21 November 2023

Author Contributions: All authors contributed equally to this paper. All authors have read and agreed to the published version of the manuscript.

Funding: This statement of peer review received no external funding.

Institutional Review Board Statement: Not applicable.

Informed Consent Statement: Not applicable.

Data Availability Statement: Data is available at https://www.mdpi.com/2673-4591/55/1.

Conflicts of Interest: The authors declare no conflict of interest.

engineering proceedings MDPI

Proceeding Paper

Anomaly Detection on Network Traffic for the Healthcare Internet of Things †

Hsiao-Ching Huang, I-Hsien Liu, Meng-Huan Lee and Jung-Shian Li *

Department of Electrical Engineering, Institute of Computer and Communication Engineering, National Cheng Kung University, Tainan 70101, Taiwan; hchuang@cans.ee.ncku.edu.tw (H.-C.H.); ihliu@cans.ee.ncku.edu.tw (I.-H.L.); mhlee@cans.ee.ncku.edu.tw (M.-H.L.)

* Correspondence: jsli@cans.ee.ncku.edu.tw

† Presented at the IEEE 5th Eurasia Conference on Biomedical Engineering, Healthcare and Sustainability, Tainan, Taiwan, 2–4 June 2023.

Abstract: The Internet of Things (IoT) has revolutionized technologies in society, including in households, offices, factories, and health centers. Among these, the Healthcare Internet of Things (HIoT) significantly transforms medical assistance for patients. By using wearable devices with remote network connections, caregivers monitor patients' physiological data to gain valuable insights into their health conditions. Despite the many benefits of the HIoT, several security vulnerabilities still exist. Hackers can exploit the internet connection to steal or modify credential information regarding patients, violating the integrity and confidentiality of the security policy. Moreover, they can launch cyberattacks on hospitals or critical life-support systems, further endangering patients' lives. Consequently, it is crucial to implement robust cybersecurity measures to enhance the security of healthcare services. Therefore, we proposed an anomaly detection method based on network traffic for the HIoT, adopting Markov models. Owing to their simplicity, interpretability, and well-developed theory, the Markov models have been applied to network traffic prediction and modeling, serving as a viable approach to cater to our needs. We evaluated the proposed method using the public dataset ToN_IoT and analyzed the results.

Keywords: healthcare internet of things; anomaly detection; network traffic; Markov models

Citation: Huang, H.-C.; Liu, I.-H.; Lee, M.-H.; Li, J.-S. Anomaly Detection on Network Traffic for the Healthcare Internet of Things. *Eng. Proc.* 2023, 55, 3. https://doi.org/10.3390/engproc2023055003

Academic Editors: Teen-Hang Meen, Kuei-Shu Hsu and Cheng-Fu Yang

Published: 22 November 2023

1. Introduction

The Healthcare Internet of Things (HIoT) redefines health services for patients. Due to the exponential growth in the network traffic generated by all connected devices in the HIoT, monitoring the network's performance and overcoming its inefficiencies pose a challenge. Network traffic prediction is one of the subfields of Network Traffic Monitoring and Analysis (NTMA) [1], which focuses on analyzing past characteristics of network traffic to predict future trends. This serves as a solution to be addressed, particularly in anomaly detection. Anomaly detection is crucial to cybersecurity and is further integrated into an intrusion detection system (IDS). An anomaly-based IDS, in comparison to its signature-based counterpart, characterizes the normal behavior of a system to differentiate attack traffic, whereas its counterpart searches for features that directly match the attack traffic from its pre-built database. Before the advent of the HIoT and the proliferation of IoT applications, most of the prediction and anomaly detection methods only considered univariate time series. However, network traffic consists of different attributes and statistical contexts, such as packet counts, interarrival time, protocol type, and connection status. Focusing solely on univariate time series may overlook the underlying correlation between the different attributes [2]. Therefore, in this study, we adopted a multivariate analysis to detect anomalies in network traffic in the HIoT environment.

2. Related Works

Anomaly detection methods have been a subject of ongoing research in many fields of study. Recent studies mostly focus on various machine learning and deep learning algorithms for anomaly detection. Wu et al. [3] applied the graph neural network (GNN) for anomaly detection in Industrial Internet of Things (IIoT) scenarios, specifically to the studies on smart factories, smart transportation, and smart energy. They provided a comprehensive investigation of different types of anomalies such as point anomalies, contextual anomalies, and collective anomalies. Chen et al. [4] proposed a transformer-based framework, called GTA, to learn graph structures for multivariate time series' anomaly detection in IoT sensor data. Park et al. [5] addressed the issue of data imbalance in an AI-based network intrusion detection system (NIDS) using a generative adversarial network (GAN) by generating synthetic data for minor attacks, along with an autoencoder-driven model for detection. Furthermore, Liu et al. [6] combined an attention-mechanism-based convolutional neural network long short-term memory (AMCNN-LSTM) model with federated learning to detect edge device failures. Qi et al. [7] introduced a novel approach called MDS_AD, which employed locality-sensitive hashing (LSH), isolation forest, and principal component analysis (PCA) to detect point and group anomalies considering multi-aspect data. Regarding studies of Markov models and network traffic analysis, Aceto et al. [8] applied high-order Markov chains and hidden Markov models (HMM) to predict mobile-app traffic. Liu et al. [9] introduced tensor operations to a multivariate, multi-order Markov chain for network traffic's multi-modal prediction.

3. Methodology

The multivariate high-order Markov chain with Hellinger distance (MHMC-HD) was proposed for detecting anomalies in network traffic for HIoT scenarios in this study.

3.1. Problem Formulation

Let $X = \{X_1, X_2, X_3, \ldots, X_{t-1}, X_t, X_{t+1}, \ldots\}$ be a set of consecutive random variables that describe the state of each network traffic flow at time t. The finite state set is denoted as

$$S \equiv \{1, 2, 3, \ldots, I\} \tag{1}$$

where I represents the total number of states. Similar to network traffic prediction, our objective is to obtain the transition probability of the state at the next time step, given the states at k's preceding ones.

3.2. MHMC

In a classical first-order Markov chain, the current state is determined solely by the previous state.

$$\begin{aligned} &\mathrm{P}(X_{t+1} = j | X_t = i, X_{t-1} = i_{t-1}, \ldots X_0 = i_0) \\ &= \mathrm{P}(X_{t+1} = j | X_t = i) \\ &= p_{i,j} \end{aligned} \tag{2}$$

where state $j, i, i_0, \ldots i_{t-1} \in S$. The assumption of temporal homogeneity is made, meaning that the transition probability does not depend on time t. Hence, the transition probability matrix can be expressed as follows:

$$p_{i,j} = \mathrm{P}(X_{t+1} = j | X_t = i) \tag{3}$$

$$P' = (p_{i,j}) \tag{4}$$

where $P \in \mathbb{R}^{I \times I}, \sum_{i=0}^{I} p_{i,j} = 1$.

Compared to the classical Markov chain, a k-order Markov chain not only depends on the previous state but also takes into consideration the k's preceding states.

$$\begin{aligned} &\mathrm{P}(X_{t+1} = j | X_t = i, X_{t-1} = i_{t-1}, \dots X_0 = i_0) \\ &= \mathrm{P}(X_{t+1} = j | X_t = i, X_{t-1} = i_{t-1}, \dots, X_{t-k+1} = i_{t-k+1}) \\ &= p_{i_{t-k+1}, \dots, i_{t-1}, i, j} \end{aligned} \tag{5}$$

where $i_{t-k+1}, \dots, i_{t-1}, i, j \in S$. Furthermore, for a network traffic flow, multiple attributes can be obtained, such as interarrival time, packet length, and others. If we apply separate univariate Markov chains to different attributes, the hidden correlation between these attributes may be neglected. Therefore, it is important to retain the underlying correlation between the different variables. In an m-variate k-order Markov chain, the state space can be defined as follows:

$$S' \equiv \{(1, 1, \dots, 1), (1, 1, \dots, 1, 2), \dots, (I_1, I_2 \dots, I_m)\}. \tag{6}$$

Thus, the transition probability can be represented as follows:

$$\begin{aligned} &p_{i_{t+1,1}, i_{t+1,2}, \dots, i_{t+1,m}, i_{t,1}, i_{t,2}, \dots, i_{t,m}, \dots, i_{t-k+1,1}, i_{t-k+1,2}, \dots, i_{t-k+1,m}} \\ &= \mathrm{P}(X_{t+1,1}, X_{t+1,2}, \dots, X_{t+1,m} = i_{t+1,1}, i_{t+1,2}, \dots, i_{t+1,m} | \\ &X_{t,1}, X_{t,2}, \dots, X_{t,m} = i_{t,1}, i_{t,2}, \dots, i_{t,m}, \dots, X_{t-k+1,1}, X_{t-k+1,2}, \\ &\dots, X_{t-k+1,m} = i_{t-k+1,1}, i_{t-k+1,2}, \dots, i_{t-k+1,m}) \end{aligned} \tag{7}$$

where

$$i_{t+1,1}, i_{t+1,2}, \dots, i_{t+1,m}, i_{t,1}, i_{t,2}, \dots, i_{t,m}, \dots, i_{t-k+1,1}, i_{t-k+1,2}, \dots, i_{t-k+1,m} \in S'$$

and the transition matrix can then be converted to a tensor

$$P'' \in \mathbb{R}^{I_1, I_2 \dots, I_m \times \dots \times I_1, I_2 \dots I_m}.$$

3.3. Maximum Likelihood Estimation (MLE)

After acquiring the network traffic flows, the unknown transition probability tensor is estimated based on these observations. The MLE is a common technique used for this purpose. For a classical first-order Markov chain, the transition probability matrix can be constructed as follows:

$$\hat{p_{i,j}} = \frac{n_{ij}}{\sum_k n_{ik}} \tag{8}$$

$$n_{ij} = \sum_{t=0}^{N_t} 1_{\{X_t = i\}} 1_{\{X_{t+1} = j\}} \tag{9}$$

where $\hat{p_{i,j}}$ and n_{ij} are the estimated transition probability and the count of transitions from state i to state j, respectively. N_t denotes the total number of time steps in the training data, and $1_{\{\cdot\}}$ is an indicator function.

3.4. MHMC with Hellinger Distance (MHMC-HD)

Initially, in the approach of this study, we applied the Hellinger distance to determine whether testing data samples exhibited a similar underlying probability distribution to the training data. The Hellinger distance is a measure to quantify the dissimilarity between two discrete probability distributions. According to [8], we considered two hypotheses, $\mathcal{H}_0$ and $\mathcal{H}_1$, to investigate whether the two datasets were represented by the same Markov model or by different ones. For a first-order Markov chain, the two probability distributions were described through $S \times S$ matrices $\mathbf{\Pi}^x$ and $\mathbf{\Pi}^y$, where each Π was one-to-one mapped to the corresponding transition matrix, with S denoting the finite state set.

Similar to estimating unknown transition matrices, the matrix Π was obtained using MLE.

$$\hat{\pi} = \frac{n_{ij}}{n} \tag{10}$$

Subsequently, we applied the Hellinger distance to measure the dissimilarity between the two matrices $\mathbf{\Pi}^x$ and $\mathbf{\Pi}^y$.

$$H(\hat{\mathbf{\Pi}}^x, \hat{\mathbf{\Pi}}^y) = \frac{1}{\sqrt{2}} \sqrt{\sum_{i=1}^{S} \sum_{j=1}^{S} \left(\sqrt{\hat{\pi}^x_{i,j}} - \sqrt{\hat{\pi}^y_{i,j}} \right)^2} \tag{11}$$

Given a threshold γ, if $H(\hat{\mathbf{\Pi}}^x, \hat{\mathbf{\Pi}}^y) > \gamma$, the two datasets belonged to different probability distributions ($\mathcal{H}_0$), whereas $H(\hat{\mathbf{\Pi}}^x, \hat{\mathbf{\Pi}}^y) < \gamma$ suggested that the two datasets used the same Markov model ($\mathcal{H}_1$). Following that, we applied MHMC to assess the probability of the generation of each testing data considering k.

4. Experiment and Result

4.1. Network Traffic Data

To evaluate our approach for anomaly detection on HIoT network traffic, we used the ToN_IoT datasets. The datasets comprised IoT/IIoT telemetry data from sensors, operating system data from Windows and Linux systems, as well as network traffic data collected during normal operations and under various attack interferences [10–12]. The network traffic datasets were derived from pcap and log files with Zeek logs. Among the various attributes in the dataset, we specifically selected source payload (src_bytes), destination payload (dest_bytes), and connection state (conn_state) to evaluate the proposed MHMC-HD approach. The testing data included different types of attack techniques, including DoS, injecting, ransomware, password attacks, and more [13–15].

4.2. Performance Metrics

We adopted the following common performance metrics to evaluate the results using the confusion matrix [16]. The confusion matrix provided an overview of the outcomes of predictive analytics and classification studies, presenting four different cases. Each case represented the number of testing data samples falling into one of the following categories as shown in Table 1: True Positive (TP), False Positive (FP), False Negative (FN), and True Negative (TN).

Table 1. Confusion Matrix.

Confusion Matrix		Actual Condition	
		Positive	**Negative**
Predicted Condition	Positive	TP	FP
	Negative	FN	TN

- Precision: The number of correctly detected anomaly samples over the total number of samples predicted as anomalies.

$$\text{Precision} = \frac{\text{TP}}{\text{TP} + \text{FP}} \tag{12}$$

- Recall: The number of correctly detected anomaly samples over the total number of actual anomaly samples.

$$\text{Recall} = \frac{\text{TP}}{\text{TP} + \text{FN}} \tag{13}$$

- F1-score: The harmonic mean of precision and recall providing a balance measure for the model's performance.

$$F1 = 2\frac{(\text{Precision})(\text{Recall})}{\text{Precision} + \text{Recall}} \tag{14}$$

- True Negative Rate (TNR): A metric for evaluating the false alarm rate. The number of correctly predicted normal traffic samples over the total number of normal traffic samples, as follows:

$$\text{TNR} = \frac{\text{TN}}{\text{TN} + \text{FP}}. \tag{15}$$

When evaluating the testing dataset consisting of 80,000 flows, we implemented the MHMC-HD with a threshold of $\gamma = 0.5$ for the Hellinger distance measure and an order of $k = 4$ for the MHMC. Next, we compared the MHMC-HD with three other approaches, including a four-order MHMC, the Hellinger distance measure without MHMC, and an ML-based long short-term memory (LSTM) with an autoencoder (AE). The results are presented in Table 2 and Figure 1. It was found that the MHMC_HD performed the best in terms of precision, F1, and TNR, while the LSTM-AE achieved the highest score in the recall metric.

Table 2. Comparison of the results with different methods.

Methods	Precision	Recall	F1	TNR
Four-order MHMC	0.9815	0.9951	0.9882	0.9436
Hellinger Distance	0.8894	0.9875	0.9359	0.6316
LSTM + AE	0.8104	**1**	0.8995	0.3
MHMC-HD	**0.9940**	0.9908	**0.9924**	**0.9821**

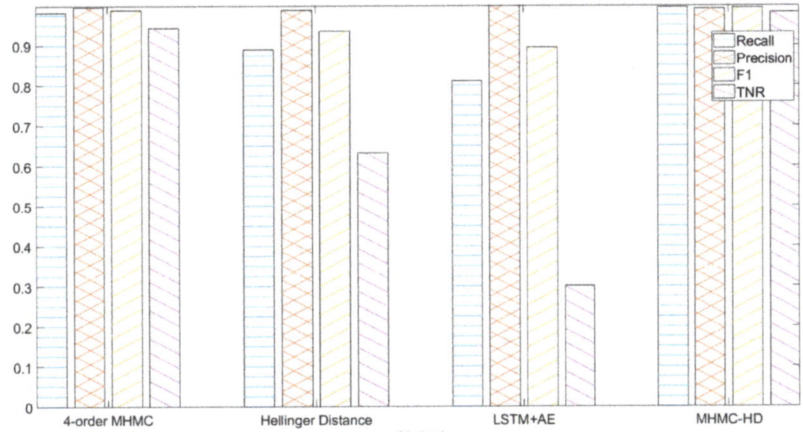

Figure 1. Comparison of the results with different methods.

To investigate the influence of a different order (k) on the MHMC-HD in terms of its performance in anomaly detection, we conducted experiments from order one to ten. The results are presented in Table 3 and Figure 2. The recall and F1 scores were improved at a higher order but the TNR metric dropped as the order increased. In Figure 3, a significant improvement of the TNR is shown, after integrating the Hellinger distance measure into the MHMC. This improvement indicated a reduction in the false alarm rate when implementing anomaly detection in IoT network traffic. Overall, the results suggested that the enhancement of the recall and F1 scores through the implementation of a higher order in the MHMC-HD undermined the TNR metric. Moreover, the integration of the Hellinger distance effectively reduced the false alarm rate in anomaly detection in IoT traffic.

Table 3. Evaluation of the MHMC-HD with different orders of *k*.

Order k	Precision	Recall	F1	TNR
1	**0.9979**	0.5164	0.6806	**0.9968**
2	0.9968	0.9387	0.9669	0.9910
3	0.9952	0.9816	0.9884	0.9858
4	0.9940	0.9908	**0.9924**	0.9821
5	0.9933	0.9950	0.9942	0.9799
6	0.9929	0.9968	0.9949	0.9786
7	0.9926	0.9979	0.9952	0.9775
8	0.9923	0.9993	0.9958	0.9766
9	0.9919	0.9996	0.9957	0.9756
10	0.9917	**0.9997**	0.9957	0.9748

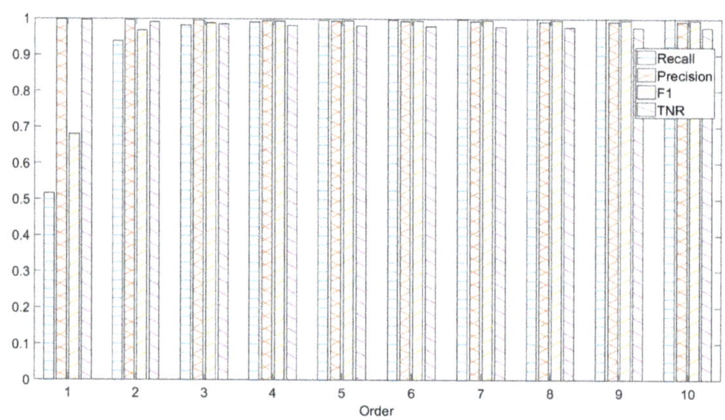

Figure 2. Evaluation of *k*-order Markov chain with Hellinger distance concerning *k*.

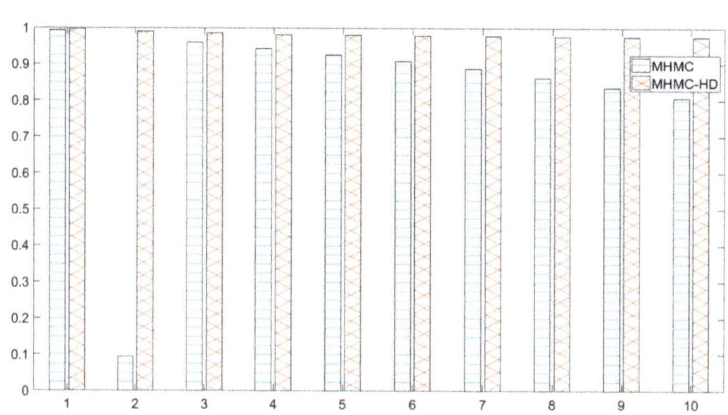

Figure 3. Comparison of the TNR metric depending on the existence of the Hellinger distance.

5. Conclusions

We developed the MHMC-HD method to detect anomalies specifically for attacking traffic flows on the IoT's network traffic. The impact of a higher order on the performance of the MHMC-HD was investigated. The results showed a considerable improvement in anomaly detection. The results also demonstrated that integrating the Hellinger distance into the MHMC produced a low false alarm rate, thereby enhancing the reliability of anomaly detection in HIoT network traffic analysis.

Author Contributions: Writing—original draft, H.-C.H.; Validation, M.-H.L.; Writing—review & editing, I.-H.L.; Supervision, J.-S.L. All authors have read and agreed to the published version of the manuscript.

Funding: This work was supported by the National Science and Technology Council (NSTC) in Taiwan under contract numbers 111-2221-E-006-079- and 112-2634-F-006-001-MBK.

Institutional Review Board Statement: Not applicable.

Informed Consent Statement: Not applicable.

Data Availability Statement: The public data used in this work is mentioned and cited in the text. Link to this dataset: https://research.unsw.edu.au/projects/toniot-datasets (accessed on 1 May 2023).

Conflicts of Interest: The authors declare no conflict of interest.

References

1. Lohrasbinasab, I.; Shahraki, A.; Taherkordi, A.; Jurcut, A.D. From statistical- to machine learning-based network traffic prediction. *Trans. Emerg. Telecommun. Technol.* **2021**, *33*, e4394. [CrossRef]
2. Sha, W.; Zhu, Y.; Chen, M.; Huang, T. Statistical Learning for Anomaly Detection in Cloud Server Systems: A Multi-Order Markov Chain Framework. *IEEE Trans. Cloud Comput.* **2018**, *6*, 401–413. [CrossRef]
3. Wu, Y.; Dai, H.-N.; Tang, H. Graph Neural Networks for Anomaly Detection in Industrial Internet of Things. *IEEE Internet Things J.* **2022**, *9*, 9214–9231. [CrossRef]
4. Chen, Z.; Chen, D.; Zhang, X.; Yuan, Z.; Cheng, X. Learning Graph Structures with Transformer for Multivariate Time-Series Anomaly Detection in IoT. *IEEE Internet Things J.* **2022**, *9*, 9179–9189. [CrossRef]
5. Park, C.; Lee, J.; Kim, Y.; Park, J.-G.; Kim, H.; Hong, D. An Enhanced AI-Based Network Intrusion Detection System Using Generative Adversarial Networks. *IEEE Internet Things J.* **2023**, *10*, 2330–2345. [CrossRef]
6. Liu, Y.; Garg, S.; Nie, J.; Zhang, Y.; Xiong, Z.; Kang, J.; Hossain, M.S. Deep Anomaly Detection for Time-Series Data in Industrial IoT: A Communication-Efficient On-Device Federated Learning Approach. *IEEE Internet Things J.* **2021**, *8*, 6348–6358. [CrossRef]
7. Qi, L.; Yang, Y.; Zhou, X.; Rafique, W.; Ma, J. Fast Anomaly Identification Based on Multiaspect Data Streams for Intelligent Intrusion Detection Toward Secure Industry 4.0. *IEEE Trans. Ind. Inform.* **2022**, *18*, 6503–6511. [CrossRef]
8. Aceto, G.; Bovenzi, G.; Ciuonzo, D.; Montieri, A.; Persico, V.; Pescapé, A. Characterization and Prediction of Mobile-App Traffic Using Markov Modeling. *IEEE Trans. Netw. Serv. Manag.* **2021**, *18*, 907–925. [CrossRef]
9. Liu, H.; Yang, L.T.; Chen, J.; Ye, M.; Ding, J.; Kuang, L. Multivariate Multi-Order Markov Multi-Modal Prediction with Its Applications in Network Traffic Management. *IEEE Trans. Netw. Serv. Manag.* **2019**, *16*, 828–841. [CrossRef]
10. Moustafa, N. A new distributed architecture for evaluating AI-based security systems at the edge: Network TON_IoT datasets. *Sustain. Cities Soc.* **2021**, *72*, 102994. [CrossRef]
11. Ashraf, J.; Keshk, M.; Moustafa, N.; Abdel-Basset, M.; Khurshid, H.; Bakhshi, A.D.; Mostafa, R.R. IoTBoT-IDS: A novel statistical learning-enabled botnet detection framework for protecting networks of smart cities. *Sustain. Cities Soc.* **2021**, *72*, 103041. [CrossRef]
12. Booij, T.M.; Chiscop, I.; Meeuwissen, E.; Moustafa, N.; Hartog, F.T.H.D. ToN_IoT: The Role of Heterogeneity and the Need for Standardization of Features and Attack Types in IoT Network Intrusion Data Sets. *IEEE Internet Things J.* **2022**, *9*, 485–496. [CrossRef]
13. Alsaedi, A.; Moustafa, N.; Tari, Z.; Mahmood, A.; Anwar, A. TON_IoT Telemetry Dataset: A New Generation Dataset of IoT and IIoT for Data-Driven Intrusion Detection Systems. *IEEE Access* **2020**, *8*, 165130–165150. [CrossRef]
14. Moustafa, N.; Keshky, M.; Debiez, E.; Janicke, H. Federated TON_IoT Windows Datasets for Evaluating AI-Based Security Applications. In Proceedings of the 2020 IEEE 19th International Conference on Trust, Security and Privacy in Computing and Communications (TrustCom), Guangzhou, China, 29 December 2020–1 January 2021.

15. Moustafa, N.; Ahmed, M.; Ahmed, S. Data Analytics-Enabled Intrusion Detection: Evaluations of ToN_IoT Linux Datasets. In Proceedings of the 2020 IEEE 19th International Conference on Trust, Security and Privacy in Computing and Communications (TrustCom), Guangzhou, China, 29 December 2020–1 January 2021.
16. Fathi-Kazerooni, S.; Rojas-Cessa, R. Countering Machine-Learning Classification of Applications by Equalizing Network Traffic Statistics. *IEEE Trans. Netw. Sci. Eng.* **2021**, *8*, 3392–3403. [CrossRef]

engineering proceedings

MDPI

Proceeding Paper

A Compartment Pharmacokinetics Model of THC and Its Metabolites after Smoking †

Thanachok Mahahong and Teerapol Saleewong *

Department of Mathematics, Faculty of Science, King Mongkut's University of Technology Thonburi, Bangkok 10140, Thailand; thanachok.maha@mail.kmutt.ac.th
* Correspondence: teerapol.sal@kmutt.ac.th
† Presented at the IEEE 5th Eurasia Conference on Biomedical Engineering, Healthcare and Sustainability, Tainan, Taiwan, 2–4 June 2023.

Abstract: This article presents the application of the compartment model to investigate the pharmacokinetics of delta9-tetrahydrocannabinol (THC), one of the cannabinoids found in cannabis, after smoking cannabis. The behavior of THC and its metabolite concentrations in the body following smoking were investigated, and the result offered a guideline for determining the appropriate duration between each smoking session to prevent intoxication and potential harm to the body. The compartment model was transformed into ordinary differential equations (ODEs) to describe the rate of change in the concentration of THC and its metabolites in each compartment, employing the law of mass action. For simulating the solution curve, the exact solutions of the ODE system and an actual data sample were processed using Microsoft Excel to obtain the optimal rate constants using curve fitting and generating a simulated curve that closely matched the actual data. The findings in this study indicated that the proposed model effectively described the concentration behavior of THC and its metabolites in plasma and other tissues. Therefore, the model will serve as a valuable tool for the determination of the appropriate duration between each smoking session to prevent harm to the body.

Keywords: compartment model; delta9-tetrahydrocannabinol (THC); pharmacokinetics; smoking

Citation: Mahahong, T.; Saleewong, T. A Compartment Pharmacokinetics Model of THC and Its Metabolites after Smoking. *Eng. Proc.* **2023**, *55*, 4. https://doi.org/10.3390/engproc2023055004

Academic Editors: Teen-Hang Meen, Kuei-Shu Hsu and Cheng-Fu Yang

Published: 22 November 2023

1. Introduction

Currently, many countries around the world have legalized or adopted a liberal approach to the use of cannabis for medical, consumption, cosmetic, and recreational purposes. Cannabis consists of approximately 500 chemical components, with at least 100 of them being unique to the cannabis plant and known as "cannabinoids" [1]. Among these cannabinoids, delta9-tetrahydrocannabinol (THC) is the most well-known, possessing psychoactive properties and the potential to become addictive. Nevertheless, THC exhibits several clinically beneficial pharmacological effects, such as for chronic pain, nausea, vomiting, and stimulating appetite [1]. Due to its psychoactive and addictive nature, inappropriate doses and durations of THC consumption, particularly through smoking without medical supervision, can lead to psychoactive effects and harm the body. Therefore, studying the pharmacokinetics of THC allows for a greater understanding of the behavior of THC and its metabolites in the body and a determination of the appropriate dosage and duration for each administration of THC.

Pharmacokinetics, which encompasses the processes of drug absorption, metabolism, distribution, and excretion [2], plays a crucial role in understanding drug behavior and establishing the relationship between drug absorption, distribution, and elimination rates within the body. It assists in developing appropriate therapeutic responses. Many different pharmacokinetics study approaches have been used so far. Compartmental modeling has been frequently employed in pharmacokinetics research since 1932 [3]. A compartment model serves as a mathematical representation of either the entire body or a specific portion, dividing it into compartments, and is useful for investigating the kinetics of physiological or pharmacological processes [4]. This modeling technique aids in comprehending the underlying biological processes that dictate the kinetic behavior of drugs or chemicals administered to the body.

In pharmacokinetics, THC rapidly enters the lungs and subsequently enters the bloodstream or systemic circulation, from where it is distributed to various tissues, including the effect site [1]. Additionally, THC undergoes metabolism in the liver by enzymes, resulting in the formation of two major metabolites, namely THC-OH and THCCOOH. THC-OH is the primary active metabolite, while THC-COOH is the secondary inactive metabolite. THC is predominantly eliminated from the body through metabolic processes, with THC and its metabolites being excreted in the feces and urine as well. There are few current studies on the pharmacokinetics of THC using the compartment model because medical cannabis is still in its early stages of research in various fields [5,6].

In this study, we examined the pharmacokinetics of THC and its metabolites after smoking, using the compartment model. The objectives of this research are to investigate the concentration–time relationship and to provide a guideline for determining the optimal dosage and duration of each smoking session. The result is expected to help mitigate intoxication and potential harm to the body. The proposed compartment model was accurate and found to be a valuable tool for understanding the pharmacokinetics of THC and its metabolites after smoking it. Moreover, a guideline to determine the optimal quantity and duration of each smoking session can be used to minimize potential harm to the body.

2. Methods

2.1. Compartmental Model

We constructed a compartment model to describe the pharmacokinetics of THC and its metabolites after smoking. The model was created in three phases: the pharmacokinetics of THC, THC-OH, and THCCOOH.

For the pharmacokinetics of THC and THC-OH, we divided the body into three physiologically significant compartments. The first compartment was the central compartment, representing plasma or systemic circulation. The second compartment was a rapidly equilibrating tissue compartment, including organs such as the heart, liver, lungs, and kidneys. The third compartment was a slowly equilibrating tissue compartment, encompassing tissues like muscle and bone. Additionally, we defined effective compartments for THC and THC-OH to describe their concentrations in target sites. As for the inactive metabolite THCCOOH, we considered and included only the central compartment. This compartmental model is illustrated in Figure 1.

Using the compartment model in Figure 1, we investigated the case in which the redistribution rates of THC and THC-OH from the rapidly equilibrating tissue compartment and slowly equilibrating tissue compartment to the central compartment were equal or $k_{31} = k_{21}$ and $k_{64} = k_{54}$. By applying the law of mass action, we derived the following ordinary differential equations (ODEs).

$$\begin{aligned} \frac{dC_1(t)}{dt} &= -(k_{12}+k_{13}+k_{14}+k_{17}+k_{1e})C_1(t)+k_{21}C_2(t)+k_{21}C_3(t); & C_1(0) &= C_0, \\ \frac{dC_2(t)}{dt} &= k_{12}C_1(t)-k_{21}C_2(t); & C_2(0) &= 0, \\ \frac{dC_3(t)}{dt} &= k_{13}C_1(t)-k_{21}C_3(t); & C_3(0) &= 0, \end{aligned} \tag{1}$$

$$\begin{aligned} \frac{dC_4(t)}{dt} &= k_{14}C_1(t)-(k_{40}+k_{45}+k_{46}+k_{47}+k_{4e})C_4(t)+k_{54}C_5(t)+k_{54}C_6(t); & C_4(0) &= 0, \\ \frac{dC_5(t)}{dt} &= k_{45}C_4(t)-k_{54}C_5(t); & C_5(0) &= 0, \\ \frac{dC_6(t)}{dt} &= k_{46}C_4(t)-k_{54}C_6(t); & C_6(0) &= 0, \end{aligned} \tag{2}$$

$$\frac{dC_7(t)}{dt} = k_{17}C_1(t)+k_{47}C_4(t)-k_{70}C_7(t); \qquad C_7(0) = 0, \tag{3}$$

$$\frac{dC_{e1}(t)}{dt} = k_{1e}C_1(t) - k_{e10}C_{e1}(t); \qquad C_{e1}(0) = 0, \quad (4)$$

$$\frac{dC_{e2}(t)}{dt} = k_{4e}C_4(t) - k_{e20}C_{e2}(t); \qquad C_{e2}(0) = 0, \quad (5)$$

where C_0 is the initial concentration of THC.

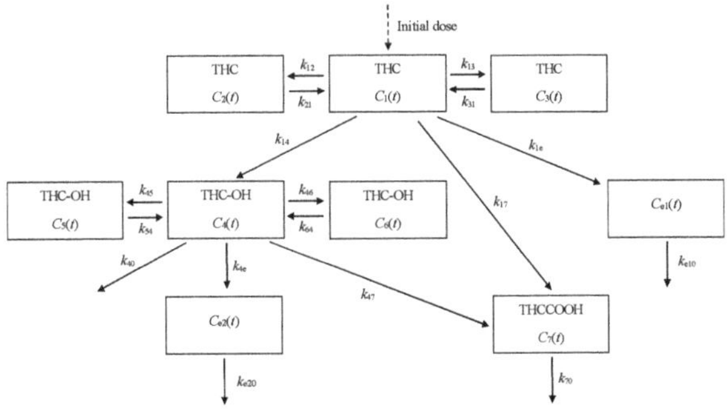

Figure 1. A compartment pharmacokinetics model of THC and its metabolites after smoking cannabis. $C_1(t)$: concentration of THC in the central compartment. $C_2(t)$: concentration of THC in the rapidly equilibrating tissue compartment. $C_3(t)$: concentration of THC in the slowly equilibrating tissue compartment. $C_{e1}(t)$: concentration of THC in the effect compartment. $C_4(t)$: concentration of THC-OH in the central compartment. $C_5(t)$: concentration of THC-OH in the rapidly equilibrating tissue compartment. $C_6(t)$: concentration of THC-OH in the slowly equilibrating tissue compartment. $C_{e2}(t)$: concentration of THC-OH in the effect compartment. $C_7(t)$: concentration of THCCOOH in the central compartment. k_{12}, k_{21}, k_{13}, k_{31}, k_{14}, k_{17}, and k_{1e}: distribution rate constants of THC in each compartment. k_{45}, k_{54}, k_{46}, k_{64}, k_{47}, and: distribution rate constants of THC-OH in each compartment. k_{40} and k_{70}: elimination rate constants of THC-OH and THCCOOH from the central compartment. k_{e10} and k_{e20}: elimination rate constants of THC and THC-OH from the effect compartment.

2.2. Data

We determined the concentration of THC and its metabolites, THC-OH and THC-COOH, in the plasma after smoking cigarettes containing 150 μg THC/kg body weight in 10 male volunteers [7]. These concentrations were quantified by using a gas chromatography/mass spectrometer (GC/MS), and the results are presented in Table 1.

Table 1. Mean plasma concentration of THC and its metabolites after smoking.

Time after Smoking (min)	THC Concentration (ng/mL)	THC-OH Concentration (ng/mL)	THCCOOH Concentration (ng/mL)
5	85.08	5.89	6.98
10	78.43	18.74	11.40
20	57.83	18.94	18.25
30	41.40	18.54	28.69
50	27.29	12.73	35.60
65	16.97	9.58	24.49
120	9.76	6.65	15.14
180	4.96	3.33	12.48
240	3.66	2.76	7.23
300	2.22	1.85	4.10

3. Results

The exact solutions of the system (Equations (1)–(5)) demonstrated the simulation results of the concentration–time relationship of THC and its metabolites after smoking cigarettes that contain 150 μg THC/kg body weight.

3.1. Exact Solutions

From Equations (1)–(5), we obtained the exact solutions as follows.

$$C_1(t) = \frac{C_0(k_{21}+\lambda_1)e^{\lambda_1 t}}{\lambda_1-\lambda_2} + \frac{C_0(k_{21}+\lambda_2)e^{\lambda_2 t}}{\lambda_2-\lambda_1}, \quad (6)$$

$$C_2(t) = \frac{k_{12}C_0e^{\lambda_1 t}}{\lambda_1-\lambda_2} + \frac{k_{12}C_0e^{\lambda_2 t}}{\lambda_2-\lambda_1}, \quad (7)$$

$$C_3(t) = \frac{k_{13}C_0e^{\lambda_1 t}}{\lambda_1-\lambda_2} + \frac{k_{13}C_0e^{\lambda_2 t}}{\lambda_2-\lambda_1}, \quad (8)$$

$$C_{e1}(t) = -\frac{k_{1e}(k_{21}+\lambda_1)e^{-k_{e10}t}}{k_{e10}} + \frac{k_{1e}C_1(t)}{k_{e10}}, \quad (9)$$

$$C_4(t) = \frac{k_{14}k_{54}C_1(t)}{\lambda_4\lambda_5} + \frac{k_{14}C_1(t)(\lambda_4+k_{54})e^{\lambda_4 t}}{\lambda_4(\lambda_4-\lambda_5)} + \frac{k_{14}C_1(t)(\lambda_5+k_{54})e^{\lambda_5 t}}{\lambda_5(\lambda_5-\lambda_4)}, \quad (10)$$

$$C_5(t) = \frac{k_{14}k_{54}C_1(t)}{\lambda_4\lambda_5} + \frac{k_{14}k_{45}C_1(t)(\lambda_4+k_{54})e^{\lambda_4 t}}{\lambda_4(\lambda_4-\lambda_5)} + \frac{k_{14}k_{45}C_1(t)(\lambda_5+k_{54})e^{\lambda_5 t}}{\lambda_5(\lambda_5-\lambda_4)}, \quad (11)$$

$$C_6(t) = \frac{k_{14}k_{54}C_1(t)}{\lambda_4\lambda_5} + \frac{k_{14}k_{46}C_1(t)(\lambda_4+k_{54})e^{\lambda_4 t}}{\lambda_4(\lambda_4-\lambda_5)} + \frac{k_{14}k_{46}C_1(t)(\lambda_5+k_{54})e^{\lambda_5 t}}{\lambda_5(\lambda_5-\lambda_4)}, \quad (12)$$

$$C_7(t) = -\frac{1}{k_{70}}(k_{17}C_0e^{-k_{70}t} - k_{17}C_1(t) - k_{47}C_4(t)), \quad (13)$$

$$C_{e2}(t) = \frac{k_{4e}C_4(t)}{k_{e20}} \quad (14)$$

where

$$\lambda_1 = -\frac{1}{2}(k_{12}+k_{13}+k_{14}+k_{17}+k_{1e}+k_{21}) + \frac{1}{2}\sqrt{(k_{12}+k_{13}+k_{14}+k_{17}+k_{1e}+k_{21})^2 - 4(k_{14}k_{21}+k_{17}k_{21}+k_{1e}k_{21})},$$

$$\lambda_2 = -\frac{1}{2}(k_{12}+k_{13}+k_{14}+k_{17}+k_{1e}+k_{21}) - \frac{1}{2}\sqrt{(k_{12}+k_{13}+k_{14}+k_{17}+k_{1e}+k_{21})^2 - 4(k_{14}k_{21}+k_{17}k_{21}+k_{1e}k_{21})},$$

$$\lambda_4 = -\frac{1}{2}(k_{40}+k_{45}+k_{46}+k_{47}+k_{4e}+k_{54}) + \frac{1}{2}\sqrt{(k_{40}+k_{45}+k_{46}+k_{47}+k_{4e}+k_{54})^2 - 4(k_{40}k_{54}+k_{47}k_{54}+k_{4e}k_{54})},$$

$$\lambda_5 = -\frac{1}{2}(k_{40}+k_{45}+k_{46}+k_{47}+k_{4e}+k_{54}) - \frac{1}{2}\sqrt{(k_{40}+k_{45}+k_{46}+k_{47}+k_{4e}+k_{54})^2 - 4(k_{40}k_{54}+k_{47}k_{54}+k_{4e}k_{54})}.$$

3.2. Simulation Results

3.2.1. THC Concentration

For the simulation results of the THC concentration in plasma after smoking cigarettes containing 150 μg THC/kg body weight, we used the solution of Equation (6) and the optimal parameters of the rate constants shown in Table 2, which were obtained by curve fitting the exact solution and the actual data of the THC concentration. Consequently, the simulation results are displayed in Table 3.

Table 2. Optimal values of rate constants for pharmacokinetics of THC after smoking cigarettes that contain 150 μg THC/kg body weight.

Parameters	Values	Parameters	Values
C_0	85.08 ng/mL	k_{14}	0.0174 min^{-1}
k_{12}	0.0091 min^{-1}	k_{17}	0.0000 min^{-1}
k_{21}	0.0067 min^{-1}	k_{e10}	0.0657 min^{-1}
k_{13}	0.0010 min^{-1}	λ_1	−0.0039 min^{-1}
k_{1e}	0.0003 min^{-1}	λ_2	−0.0306 min^{-1}

Table 3. Actual and simulated THC concentration in plasma after smoking cigarettes that contain 150 μg THC/kg body weight.

Time after Smoking (min)	Actual THC Concentration (ng/mL)	Simulated THC Concentration (ng/mL)
5	85.08	85.05
10	78.43	74.12
20	57.83	56.57
30	41.40	43.58
50	27.29	26.76
65	16.97	19.26
120	9.76	8.01
180	4.96	4.91
240	3.66	3.67
300	2.22	2.87

3.2.2. THC-OH and THCCOOH Concentration

For the simulation results of the THC-OH and THCCOOH concentrations in plasma after smoking cigarettes containing 150 μg THC/kg body weight, we used Equations (10) and (13), respectively, along with the optimal parameters of rate constants shown in Table 4. These parameters were obtained through curve fitting the exact solutions and the actual data of the THC-OH and THCCOOH concentrations. The simulation results are presented in Table 5.

Table 4. Optimal values of rate constants for pharmacokinetics of THC-OH and THCCOOH after smoking cigarettes that contain 150 μg THC/kg body weight.

Parameters	Values	Parameters	Values
k_{40}	0.0144 min^{-1}	k_{47}	0.0048 min^{-1}
k_{45}	0.0013 min^{-1}	k_{70}	0.0028 min^{-1}
k_{54}	0.0112 min^{-1}	k_{e20}	0.0010 min^{-1}
k_{46}	0.0009 min^{-1}	λ_4	−0.0092 min^{-1}
k_{4e}	0.0004 min^{-1}	λ_5	−0.0237 min^{-1}

Table 5. Actual and simulated THC-OH and THCCOOH concentration in plasma after smoking cigarettes that contain 150 µg THC/kg body weight.

Time after Smoking (min)	Actual THC-OH (ng/mL)	Simulated THC-OH (ng/mL)	Actual THCCOOH (ng/mL)	Simulated THCCOOH (ng/mL)
5	5.89	7.02	6.98	11.95
10	18.74	11.60	11.40	19.76
20	18.94	16.00	18.25	27.24
30	18.54	16.77	28.69	28.56
50	12.73	14.32	35.60	24.38
65	9.58	11.83	24.49	20.13
120	6.65	6.15	15.14	10.48
180	3.33	4.09	12.48	6.96
240	2.76	3.15	7.23	5.36
300	1.85	2.50	4.10	4.26

4. Discussion and Conclusions

In this study, we created a compartment model for investigating the pharmacokinetics of THC and its metabolites after smoking. The compartment model transformed ordinary differential equations (ODEs) using the law of mass action (Equations (1)–(5)) to describe the change in the rate constants of THC and its metabolites' concentration in each compartment. We obtained the solutions for the system as shown in Equations (6)–(14).

To find the simulation results, the exact solutions and a sample of actual data on the concentration of THC and its metabolites were processed using Microsoft Excel. The rate constants were fitted by curve fitting to closely match the actual data, and a simulation graph was drawn. First, we considered the THC concentration in plasma after smoking cigarettes that contained 150 µg THC/kg body weight. Using Equation (6) with the optimal rate constants (Table 2), a rapid decrease in the THC concentration in the plasma from the initial concentration of 85.08 ng/mL during the 0–180-min interval could be explained. The concentration gradually decreased during the 180–300-min interval, as depicted in Figure 2. When comparing the simulated model with the actual data, an error was identified between the two with a root mean squared error (RMSE) of 1.8434 and an R-squared value of 0.9962. In addition, we applied the obtained optimal rate constants (Table 2) to examine the concentration of THC in other compartments, namely, $C_2(t)$, $C_3(t)$, and $C_{e1}(t)$. The simulation results are displayed in Figure 3.

We simulated the concentration of two metabolites of THC, namely THC-OH and THCCOOH, in plasma after smoking 150 µg of THC using Equation (10) for the THC-OH concentration and Equation (13) for the THCCOOH concentration with the optimal rate constants (Table 4). We observed a steady increase in the THC-OH concentration during the 0–30 min interval, reaching a maximum concentration of 16.77 ng/mL at 30 min. Subsequently, the concentration gradually decreased during the 30–300 min interval, as shown in Figure 4. When comparing the simulation results with the actual data, we found an error between the model simulation and the actual data, with an RMSE of 2.7022 and an R-squared value of 0.8295. For the THCCOOH concentration in plasma, we observed a rapid increase during the 0–30 min interval, with a maximum concentration of 28.56 ng/mL at 30 min. Afterwards, the concentration gradually decreased during the 30–300 min interval. Comparing the simulation results with the actual data, we found an error between the model simulation and the actual data, with an RMSE of 6.1314 and an R-squared value of 0.6062. Moreover, we applied the obtained optimal rate constants (Table 4) to examine the concentration of THC-OH in other compartments, namely, $C_5(t)$, $C_6(t)$, and$C_{e2}(t)$. The simulation results are displayed in Figure 5.

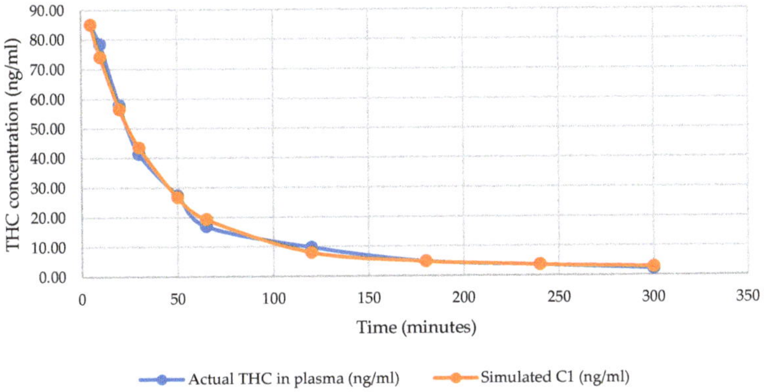

Figure 2. Graph of actual and simulated THC concentration in plasma after smoking cigarettes that contain 150 µg THC/kg body weight.

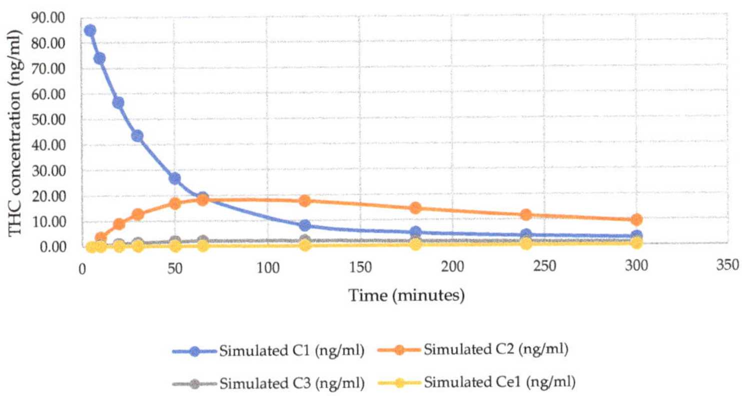

Figure 3. Graph of simulated THC concentration after smoking cigarettes which contain 150 µg THC/kg body weight in each compartment.

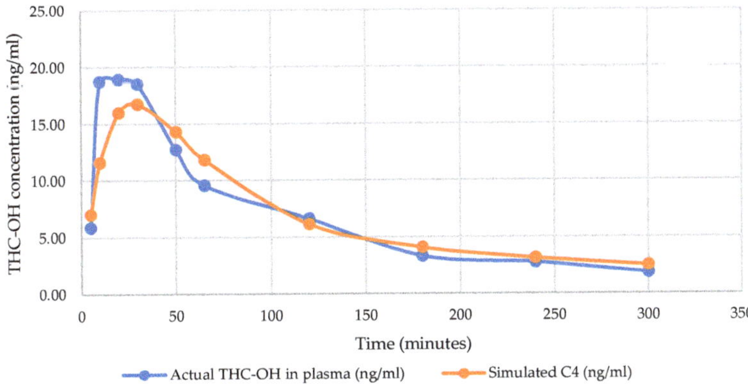

Figure 4. Graph of actual and simulated THC-OH concentration in plasma after smoking cigarettes that contain 150 µg THC/kg body weight.

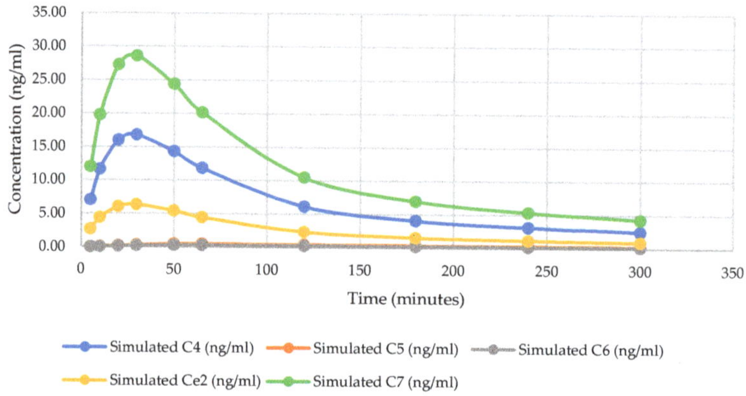

Figure 5. Graph of simulated THC-OH and THCCOOH concentration after smoking cigarettes that contain 150 μg THC/kg body weight.

The proposed compartment model accurately captured the pharmacokinetics of THC and its metabolites in plasma after smoking, matching the actual data. In addition, it predicted the pharmacokinetics of THC and its metabolites in other tissues. Therefore, the model was a useful tool in determining the optimal dosage and timing for each smoking session of THC-containing products, avoiding intoxication, and reducing potential harm. In this study, we used an initial dosage of 150 μg THC/kg body weight. The result of this study suggested that smokers should wait for a minimum of 300 min between smoking sessions to prevent the excessive accumulation of THC and its metabolite in plasma and other tissues.

Author Contributions: Conceptualization, T.M. and T.S.; methodology, T.M. and T.S.; software, T.M.; validation, T.M. and T.S.; formal analysis, T.M. and T.S.; investigation, T.M. and T.S.; resources, T.M.; data curation, T.M. and T.S.; writing—original draft preparation, T.M.; writing—review and editing, T.M. and T.S.; supervision; T.S. All authors have read and agreed to the published version of the manuscript.

Funding: This research received no external funding.

Institutional Review Board Statement: Not applicable.

Informed Consent Statement: Not applicable.

Data Availability Statement: Data is contained within the proceeding paper.

Acknowledgments: The authors would like to sincerely thank the Science Achievement Scholarship of Thailand and the Department of Mathematics, Faculty of Science, King Mongkut's University of Technology Thonburi for their support.

Conflicts of Interest: The authors declare no conflict of interest.

References

1. Martin, W.A. *Primer to Medicinal Cannabis: An Introductory Text to the Therapeutic Use of Cannabis*; Bedrocan International: Veendam, The Netherlands, 2018; pp. 1–57.
2. William, J.S.; William, E.W.; Joseph, T.D.; Robert, A.B.; Jane, M.P. Introduction to Pharmacokinetics and Pharmacodynamics. In *Concepts in Clinical Pharmacokinetics*, 6th ed.; American Society of Health-System Pharmacists: Bethesda, MD, USA, 2014; p. 2.
3. Widmark, E.M.P. *Principles and Applications of Medico Legal Alcohol Determination*; Biomedical Publications in English; Biomedical Publications: Seal Beach, CA, USA, 1981.
4. Khanday, M.A.; Rafiq, A.; Nazir, K. Mathematical models for drug diffusion through the compartments of blood and tissue medium. *Alex. J. Med.* **2017**, 53, 245–249. [CrossRef]
5. Rakesh, A. *Application of Modeling-Based Approaches to Study the Pharmacokinetics and Pharmacodynamics of Delta-9-Tetrahydrocannabinol (THC) and Its Active Metabolite*; The University of Iowa: Iowa City, IA, USA, 2017.

6. Sempio, C.; Huestis, M.A.; Mikulich-Gilbertson, S.K.; Klawitter, J.; Christians, U.; Henthorn, T.K. Population pharmacokinetic modeling of plasma Delta9-tetrahydrocannabinol and an active and inactive metabolite following controlled smoked cannabis administration. *Br. J. Clin. Pharmacol.* **2020**, *86*, 611–619. [CrossRef] [PubMed]
7. McBurney, L.J.; Bobbie, B.A.; Sepp, L.A. GC/MS and EMIT Analyses for Delta9-Tetrahydrocannabinol Metabolites in Plasma and Urine of Human Subjects. *J. Anal. Toxicol.* **1986**, *10*, 56–64. [CrossRef] [PubMed]

engineering proceedings

MDPI

Proceeding Paper

Comparative Analysis of Digital Contact-Tracing Technologies for Informing Public Health Policies †

Tsz Ho Kwan [1,2]

1 S.H. Ho Research Centre for Infectious Diseases, The Chinese University of Hong Kong, Shatin, Hong Kong; thkwan@cuhk.edu.hk
2 Jockey Club School of Public Health and Primary Care, The Chinese University of Hong Kong, Shatin, Hong Kong

† Presented at the IEEE 5th Eurasia Conference on Biomedical Engineering, Healthcare and Sustainability, Tainan, Taiwan, 2–4 June 2023.

Abstract: Contact tracing is the cornerstone of epidemic control of infectious diseases, especially in the era of COVID-19. This labor-intensive task calls for the use of digital technology to help identify individuals who have potentially been exposed to the infection to deliver necessary interventions and treatment. Mobile applications based on different technologies and system architectures have been developed and widely used in concert with public health policies and regulations. Three main types of digital contact-tracing technologies, namely Bluetooth low energy (BLE), location tracking, and check-in, were adopted in contact-tracing apps and implemented with a centralized or decentralized system architecture to protect privacy and facilitate spatiotemporal co-occurrence matching. Passive data collection methods, including BLE and location tracking, could be contrasted with check-in-based apps, which require users' effort to record their whereabouts. Persuasive computing with a centralized system for collecting contact-tracing data from users might require additional legislation for authorization and privacy protection. Technology options should therefore be taken into account when designing enforceable public health policies on the use of contact-tracing apps. At the same time, public health policies also inform the information system's design. This paper aims to delineate and contrast current technologies and system architectures used for developing contact-tracing apps and examine the intertwined relationship between the design and implementation of public health policies and the design of digital contact-tracing systems.

Keywords: requirements engineering; system analysis and design; technical planning; software development management; computer architecture; COVID-19

Citation: Kwan, T.H. Comparative Analysis of Digital Contact-Tracing Technologies for Informing Public Health Policies. *Eng. Proc.* **2023**, *55*, 5. https://doi.org/10.3390/engproc2023055005

Academic Editors: Teen-Hang Meen, Kuei-Shu Hsu and Cheng-Fu Yang

Published: 22 November 2023

1. Introduction

To control the Coronavirus disease 2019 (COVID-19) pandemic, an unprecedented global effort has been in place since its first detection in Wuhan, China. As it is transmitted primarily through respiratory droplets and contact routes, minimizing person-to-person contact in social settings is an effective and important means of reducing transmissions [1]. Collectively referred to as "non-pharmaceutical interventions", strategies involving non-biomedical products to help prevent the spread of infections, such as social distancing, hand washing, wearing facemasks, restricting mass gatherings, lockdown, and contact tracing, have been implemented in different places worldwide, especially when effective vaccines were not available [2]. When there are no lockdowns or restrictions on domestic travel, citizens may be free to visit multiple places for work, study, daily activities, and even entertainment. During the incubation period, which can be as long as 18 days [3], an infected individual may already have visited plenty of places; hence, direct contacts might have been exposed to the infection, so early identification, quarantine, and testing of these individuals would be necessary to contain further transmission [4]. Labor-intensive manual efforts by healthcare workers were required to help consolidate one's travel and contact history. However, not everyone, both the patients and the potentially exposed individuals, can recall his or her whereabouts with spatiotemporal niceties. It could also be difficult for people to keep a diary of their daily itinerary. Tools for assisting contact tracing are therefore needed. To this end, health authorities in various jurisdictions have developed and promoted the use of contact-tracing apps based on different technical frameworks. Although app-based contact tracing provides an additional reduction in transmission, as these apps often collect data relating to users and several apps require registration with phone numbers, controversies about privacy and data security have never ended. In this paper, technologies used by contact-tracing apps were evaluated and juxtaposed with a comparative analysis framework.

2. Related Work

Previous works have been extensively discussed the privacy issues relating to contact-tracing apps, with a particular focus on applicable privacy laws, data collection statements, data sharing policies, and compulsory use. Data relevant to one's whereabouts over an extended period would play an important role in epidemic control, but there is a tradeoff between the extent of privacy sacrificed by individuals and the effectiveness of contact tracing [5]. Legal tools to protect users' privacy have always been in place in most jurisdictions. With the introduction of contact-tracing apps, the compatibility of the two required investigation. Existing data protection laws in the European Union and the United States were evaluated and contrasted to predict the challenges faced by implementing a contact-tracing app [6]. In Australia, in order to protect data collection and use from contact-tracing apps, new legislations were introduced to provide trust to app users so as to increase uptake and maximize its effectiveness in controlling the spread of infection [7]. Du et al. identified data protection laws in different jurisdictions and argued that global pandemic control could be jeopardized by the limited interoperability between jurisdictions as constrained by domestic data protection legislation [8]. Even if one jurisdiction could successfully contain the epidemic, with the extraneous introduction of pathogens should international travel resume, the risk of local transmission could re-emerge. This phenomenon is similar to dengue virus transmission in a non-endemic region surrounded by endemic places [9]. Although the perceived severity of infection might not be associated with the contact-tracing app uptake [10], it could be difficult to encourage a population with a low perceived risk to continue to comply with non-pharmaceutical interventions as they might consider it no longer necessary, aside from concerns about privacy [11]. Barriers to adoption in the general public population were identified in different countries. Difficult-to-read data collection policies could be a reason why the general public is reluctant to use these apps [12]. Digital illiteracy was often cited as a hindrance in using general and contact tracing apps [11,13]. People with lower incomes might not have an updated mobile phone that is compatible with the latest version of the operating system to support contact-tracing apps [14]. A higher uptake rate was shown in Germany in the older population, as well as among those who had attained a higher level of education and had higher household incomes [15]. On the other hand, with an aspiration to "return to normal" and to stop local transmissions, people were willing to use contact-tracing apps [11]. Monetary incentives could be a booster for one-third of respondents who did not use a contact-tracing app [13]. Adopters' satisfaction in using contact-tracing apps is the next paramount issue to consider for their continued use. However, most apps received negative reviews and ratings in app stores [16,17], often accompanied by complaints about battery consumption and the perceived effectiveness of the app [17]. These highlighted technical challenges and logistical considerations when implementing policies on the use of contact-tracing apps [18], both of which will be further discussed in this paper.

3. Digital Contact-Tracing Technologies

While several communication technologies could be used for digital contact tracing [19], three major ones were used in the actual implementation of contact-tracing apps [20], which are therefore included in this paper: Bluetooth low energy (BLE), location tracking, and check-in using quick response (QR) codes or barcodes (Figure 1).

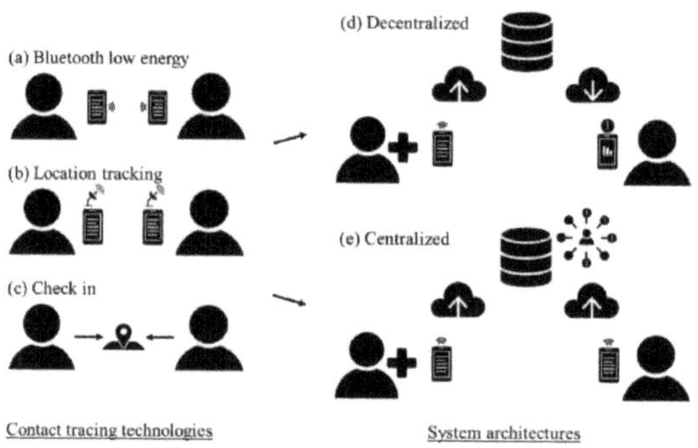

Figure 1. Overview of technologies and system architectures used in the development of contact-tracing apps.

3.1. BLE

Being the most prevalent technology adopted in the implementation of contact-tracing apps [21], devices with BLE broadcast their identifiers to nearby counterparts, who would log all received anonymized identifiers for the later detection of close contact with a patient. To protect privacy, the identifiers would be re-generated randomly from time to time, while a list of all identifiers ever generated in the device would be stored. Based on BLE, a privacy-preserving exposure notification system was thus jointly developed by Apple and Google and released to health authorities for developing their contact-tracing apps [22]. Other frameworks with different implementations, such as DP-3T and BlueTrace, were also developed [22].

With the strength of signals measured in the Received Signal Strength Indicator, the distance between devices could be estimated, although the precision could be low [23]. Bringing two pieces of information—contact history and signal strength—together, contact detection algorithms could be developed to define contact and recommend follow-up actions, which is the main purpose of contact tracing. However, such definitions varied across jurisdictions. In Taiwan, contact was defined as being within 2 m for at least 2 min [24]. In Australia and Ireland, at least 15 min of contact were needed, but the former requires close contact of less than 1.5 m, while the latter requires contact of under 2 m [25]. In Europe, BLE contact-tracing apps in Switzerland, Germany, and Italy adopted the definition of close contact with a distance shorter than 2 m for 10 (Germany) or 15 (Switzerland and Italy) min [26]. However, the correlation between signal strength and physical distance was far from accurate, even after calibrating the device models [26]. The algorithms adopted in Swiss and German contact-tracing apps did not result in any exposure notifications, although the devices were positioned in line with the definitions. It was also identified that there was a 50:50 chance to have triggered a correct exposure notification, similar to a coin flip event.

3.2. Location Tracking

A few contact-tracing apps opted for location-based technologies, such as a global positioning system (GPS), to identify people who had visited venues where a patient had also been by constantly collecting the user's location data. However, a GPS's error in measuring an exact position could be as wide as 7 to 13 m in urban environments [27]. Rather than detecting exact physical contact with another individual (or device), location-based apps aimed to identify occasions where the patients and other users coincided in terms of time and space. They also could not differentiate people on different floors, nor could they tell if the people were separated by a wall partition [28]. The draining of the mobile phone battery was also another issue identified in the pilot study. This can be alleviated by reducing the scanning frequency, at the expense of lower accuracy [29].

3.3. Check-In

A more accurate location history could be logged and stored on the device when an app user actively checks in and out by scanning a QR code or barcode posted at the entrance of a building or part of a building such as restaurants. In effect, this serves a similar purpose to recording one's whereabouts in a diary. Yet, the app facilitates the logging process and allows for comparison with patients' location histories. Similar to location-based apps, so long as an individual has been to the same place at the same or similar time, an exposure notification would be issued. This is the primary technology adopted in the contact-tracing app in Hong Kong, where the detection rules for venues and taxis were different: direct contact was defined as being in the same venue for at least 1 s, whereas indirect contact in a taxi was defined as taking the same taxi within 24 h after the patient left [30]. The implementation in the United Kingdom (UK) adopted a wider definition—a notice would be sent if the user were present at the same venue on the same date as the patient did [31].

3.4. Comparison of Digital Contact-Tracing Technologies

BLE and location-based apps are passive, which means that they send and collect data without users' extra effort, whereas check-in-based app users should actively scan the QR codes or barcodes whenever they arrive at a venue. While the former are hassle-free solutions for forgetful users, the latter requires additional reminders or policies to encourage users to scan the codes when entering and to complete the log when leaving. In an ideal scenario where all app users check in and out appropriately with a high uptake rate, check-in-based apps offer a greater granularity of contact history compared with the low accuracy of location or distance estimation with other technologies. However, BLE is advantageous in identifying close contacts, the signal of which is stronger regardless of the definition of closeness. The most location-based and check-in-based apps can do is identify people who have been to the same place or part of a building for the latter one. It does not imply that the app users who have visited the same location have a physical encounter with each other. Therefore, data collected using different technologies also carry distinct implications for the policy evaluation process. It is almost certain that contacts detected with BLE could be defined as close contacts due to its technological constraints, whereas check-in apps rely on spatiotemporal co-occurrence, so linkages between individuals would require the design of an appropriate algorithm [32].

4. System Architectures of Contact-Tracing Systems

Logging data per se is insufficient for identifying potentially infected individuals. An appropriate system architecture should be designed for comparing contact histories between patients and other app users. Intuitively, the more data a third party possesses, the more concerns about an individual's privacy will arise. In view of potential trust and legal issues, different types of system architectures were adopted [33]. A centralized approach was developed in which the centralized server collects data uploaded constantly by app users and performs matching once a positive user is identified for notifying potential contacts. This often requires the registration of personal information, including, but not limited to, mobile phone numbers, so that the health authority or the delegated agency can contact exposed individuals for necessary follow-up actions. If no data, no matter whether it is personal or not, were transferred to a third party or a centralized server, privacy could be protected while necessary notifications could still be received by the user. This decentralized approach could be achieved by constantly cross-checking with locally stored data after downloading patients' contact histories, which were stored on a centralized server, without uploading their data.

The system architecture is fundamental to contact-tracing app design and implementation. When choosing a system architecture, relevant legislation on the power of data collection and data privacy and protection in the jurisdiction should be considered. A decentralized approach does not require the collection of personal data, which could be more trusted and acceptable by potential users and further reinforced by additional protection by enacting new laws [7]. Constantly uploading data to a centralized server could drain device batteries as well. From the point of view of the service provider, a larger server storage capacity and processing power would be required, incurring extra financial resources for deployment. The data collected in the centralized approach could, on the other hand, be used to monitor outbreak clusters in real time and in retrospect [34].

5. Contact-Tracing Apps in Practice

For a public health policy or regulation to be justly enforceable with the use of contact-tracing apps, additional system requirements in their design and implementation are always necessary. In Hong Kong, where a check-in-based contact-tracing app is in use in line with the dynamic COVID-zero policy implemented following China [35], a compulsory testing notice could be issued under local regulation if a patient had visited certain venues [36]. In April 2021, after the identification of two cases with unknown sources, such a notice was issued to anyone who had stayed in a one-million-square-feet four-story shopping mall for more than two hours in the preceding two weeks, although the two cases only visited two floors [37]. In contrast, in July, only selected floors and shops in a shopping mall that was visited by a patient were included in the compulsory testing notice [38]. It may seem unnecessary for people who did not have close contact with patients to be tested. This highlights the granularity of check-in-based contact-tracing technologies, which limit how a reasonably enforceable policy can be established. To affect the least irrelevant people, BLE could be used to help identify true potentially exposed individuals, even if a weaker signal threshold was set to determine close contacts. If the other two technologies were used, the best they could do is limit the location down to the unit of a building or part of a building, so more people would need to be tested. Thus, more resources would be required to test a larger amount of people.

While the use of the contact-tracing app was not mandatory, it serves the purpose of notifying people who should be compulsorily tested. There is a legal risk for non-users who were present in the specified place but did not undergo testing, even if they were unaware of the notice [39]. On the other hand, it could be labor-intensive or even impractical for law enforcement units to identify and track down all visitors by checking CCTVs. The uptake of these voluntary measures depended on one's perception of their appropriateness as an antiepidemic policy [40], privacy concerns, and trust in the government's management of the digital contact-tracing app [41]. Therefore, minimizing data collection with a decentralized approach and transparent data management policies would be some of the important elements to be taken into consideration when planning and designing a successful digital contact-tracing app.

6. Conclusions

In this paper, intertwining technological issues and public health policies are discussed, highlighting the decisions on technology and system architecture adopted that confine the scope of policy and regulation enforcement. When designing and developing contact-tracing apps, input from relevant public health policymakers is warranted to ensure the policies are enforceable and compatible with the system. With different system designs, the resources required for their implementation and subsequent policy execution could vary. Bearing in mind that the uptake of contact-tracing apps is paramount for maximizing their effectiveness in identifying patients to curb transmissions, trust building with app users and appropriate, proportionate complementary measures are vital.

Funding: This research received no external funding.

Data Availability Statement: No new data were created or analyzed in this study. Data sharing is not applicable to this article.

Acknowledgments: The icons in Figure 1 are adopted from Megan Chown's Glyphs Collection on thenounproject.com (accessed on 11 March 2022).

Conflicts of Interest: The author declares no conflict of interest.

References

1. Wong, N.S.; Lee, S.S.; Kwan, T.H.; Yeoh, E.K. Settings of virus exposure and their implications in the propagation of transmission networks in a COVID-19 outbreak. *Lancet Reg. Health West Pac.* **2020**, *4*, 100052. [CrossRef] [PubMed]
2. Flaxman, S.; Mishra, S.; Gandy, A.; Unwin, H.J.T.; Mellan, T.A.; Coupland, H.; Whittaker, C.; Zhu, H.; Berah, T.; Eaton, J.W.; et al. Estimating the effects of non-pharmaceutical interventions on COVID-19 in Europe. *Nature* **2020**, *584*, 257–261. [CrossRef] [PubMed]
3. Xin, H.; Wong, J.Y.; Murphy, C.; Yeung, A.; Ali, S.T.; Wu, P.; Cowling, B.J. The incubation period distribution of coronavirus disease 2019 (COVID-19): A systematic review and meta-analysis. *Clin. Infect. Dis.* **2021**, *73*, 2344–2352. [CrossRef] [PubMed]
4. Kucharski, A.J.; Klepac, P.; Conlan, A.J.K.; Kissler, S.M.; Tang, M.L.; Fry, H.; Gog, J.R.; Edmunds, W.J.; Emery, J.C.; Medley, G.; et al. Effectiveness of isolation, testing, contact tracing, and physical distancing on reducing transmission of SARS-CoV-2 in different settings: A mathematical modelling study. *Lancet Infect. Dis.* **2020**, *20*, 1151–1160. [CrossRef]
5. Seto, E.; Challa, P.; Ware, P. Adoption of COVID-19 contact tracing apps: A balance between privacy and effectiveness. *J. Med. Internet Res.* **2021**, *23*, e25726. [CrossRef]
6. Bradford, L.; Aboy, M.; Liddell, K. COVID-19 contact tracing apps: A stress test for privacy, the GDPR, and data protection regimes. *J. Law Biosci.* **2020**, *7*, lsaa034. [CrossRef]
7. Greenleaf, G.; Kemp, K. Australia's 'COVIDSafe' law for contact tracing: An experiment in surveillance and trust. *Int. Data Privacy Law* **2021**, *11*, 257–275. [CrossRef]
8. Du, L.; Raposo, V.L.; Wang, M. COVID-19 contact tracing apps: A technologic tower of babel and the gap for international pandemic control. *JMIR Mhealth Uhealth* **2020**, *8*, e23194. [CrossRef]
9. Kwan, T.H.; Lee, S.S.; Chan, D.P.C.; Cheung, M.; Kam, K.M. Assessing the risk of dengue virus transmission in a non-endemic city surrounded by endemic and hyperendemic areas. *Int. J. Infect. Dis.* **2017**, *55*, 99–101. [CrossRef]
10. Walrave, M.; Waeterloos, C.; Ponnet, K. Adoption of a contact tracing app for containing COVID-19: A health belief model approach. *JMIR Public Health Surveill.* **2020**, *6*, e20572. [CrossRef]
11. Smoll, N.R.; Walker, J.; Khandaker, G. The barriers and enablers to downloading the COVIDSafe app—A topic modelling analysis. *Aust. N. Z. J. Public Health* **2021**, *45*, 344–347. [CrossRef] [PubMed]
12. Zhang, M.; Chow, A.; Smith, H. COVID-19 Contact-Tracing Apps: Analysis of the readability of privacy policies. *J. Med. Internet Res.* **2020**, *22*, e21572. [CrossRef] [PubMed]
13. Lang, R.; Benham, J.L.; Atabati, O.; Hollis, A.; Tombe, T.; Shaffer, B.; Burns, K.K.; MacKean, G.; Léveillé, T.; McCormack, B.; et al. Attitudes, behaviours and barriers to public health measures for COVID-19: A survey to inform public health messaging. *BMC Public Health* **2021**, *21*, 765. [CrossRef] [PubMed]
14. Blom, A.G.; Wenz, A.; Cornesse, C.; Rettig, T.; Fikel, M.; Friedel, S.; Möhring, K.; Naumann, E.; Reifenscheid, M.; Krieger, U. Barriers to the large-scale adoption of a COVID-19 contact tracing app in Germany: Survey study. *J. Med. Internet Res.* **2021**, *23*, e23362. [CrossRef] [PubMed]
15. Munzert, S.; Selb, P.; Gohdes, A.; Stoetzer, L.F.; Lowe, W. Tracking and promoting the usage of a COVID-19 contact tracing app. *Nat. Hum. Behav.* **2021**, *5*, 247–255. [CrossRef]
16. Bano, M.; Zowghi, D.; Arora, C. Requirements, politics, or individualism: What drives the success of COVID-19 contact-tracing apps? *IEEE Softw.* **2021**, *38*, 7–12. [CrossRef]
17. Garousi, V.; Cutting, D.; Felderer, M. Mining user reviews of COVID contact-tracing apps: An exploratory analysis of nine European apps. *J. Syst. Softw.* **2022**, *184*, 111136. [CrossRef]
18. Amann, J.; Sleigh, J.; Vayena, E. Digital contact-tracing during the COVID-19 pandemic: An analysis of newspaper coverage in Germany, Austria, and Switzerland. *PLoS ONE* **2021**, *16*, e0246524. [CrossRef]
19. Nguyen, C.T.; Saputra, Y.M.; Van Huynh, N.; Nguyen, N.-T.; Khoa, T.V.; Tuan, B.M.; Nguyen, D.N.; Hoang, D.T.; Vu, T.X.; Dutkiewicz, E.; et al. A comprehensive survey of enabling and emerging technologies for social distancing-part i: Fundamentals and enabling technologies. *IEEE Access* **2020**, *8*, 153479–153507. [CrossRef]
20. Min-Allah, N.; Alahmed, B.A.; Albreek, E.M.; Alghamdi, L.S.; Alawad, D.A.; Alharbi, A.S.; Al-Akkas, N.; Musleh, D.A.A.; Alrashed, S. A survey of COVID-19 contact-tracing apps. *Comput. Biol. Med.* **2021**, *137*, 104787. [CrossRef]
21. O'Neill, P.H.; Ryan-Mosley, T.; Johnson, B. A Flood of Coronavirus Apps Are Tracking Us. Now It's Time to Keep Track of Them. MIT Technology Review. Available online: https://www.technologyreview.com/2020/05/07/1000961/launching-mittr-covid-tracing-tracker/ (accessed on 11 March 2022).

22. Ahmed, N.; Michelin, R.A.; Xue, W.; Ruj, S.; Malaney, R.; Kanhere, S.S.; Seneviratne, A.; Hu, W.; Janicke, H.; Jha, S.K. A survey of COVID-19 contact tracing apps. *IEEE Access* **2020**, *8*, 134577–134601. [CrossRef]
23. Bertuletti, S.; Cereatti, A.; Caldara, M.; Galizzi, M.; Della Croce, U. Indoor distance estimated from Bluetooth Low Energy signal strength: Comparison of regression models. In Proceedings of the 2016 IEEE Sensors Applications Symposium, Catania, Italy, 20–22 April 2016; pp. 1–5. [CrossRef]
24. Liu, L. Download Taiwan Social Distancing App to Win Prizes. Taiwan News. Available online: https://www.taiwannews.com.tw/en/news/4203422 (accessed on 11 March 2022).
25. Leslie, M. COVID-19 fight enlists digital technology: Contact tracing apps. *Engineering* **2020**, *6*, 1064–1066. [CrossRef] [PubMed]
26. Leith, D.J.; Farrell, S. Measurement-based evaluation of Google/Apple Exposure Notification API for proximity detection in a light-rail tram. *PLoS ONE* **2020**, *15*, e0239943. [CrossRef] [PubMed]
27. Merry, K.; Bettinger, P. Smartphone GPS accuracy study in an urban environment. *PLoS ONE* **2019**, *14*, e0219890. [CrossRef] [PubMed]
28. Poolian, A.; Armoogum, S.; Armoogum, V.; Suddul, G. On the implementation of contact tracing via GPS. Presented at the Zoom. In Proceedings of the 2022 IEEE Zooming Innovation in Consumer Technologies Conference (ZINC), Novi Sad, Serbia, 25–26 May 2022. Available online: https://ieeexplore.ieee.org/document/9840554 (accessed on 11 March 2022).
29. Shubina, V.; Ometov, A.; Basiri, A.; Lohan, E.S. Effectiveness modelling of digital contact-tracing solutions for tackling the COVID-19 pandemic. *J. Navigat.* **2021**, *74*, 853–886. [CrossRef]
30. Office of the Government Chief Information Officer. Technical Specifications for LeaveHomeSafe Mobile App. Available online: https://www.leavehomesafe.gov.hk/_files/docs/Technical_Specifications.pdf (accessed on 11 March 2022).
31. National Health Service. How's It Decided If I Get an Alert That I've Been at the Same Veenue as People with Coronavirus (COVID-19)? Available online: https://faq.covid19.nhs.uk/article/KA-01213/en-us (accessed on 11 March 2022).
32. Kwan, T.H.; Wong, N.S.; Yeoh, E.K.; Lee, S.S. Mining relationships between transmission clusters from contact tracing data: An application for investigating COVID-19 outbreak. *J. Am. Med. Inform. Assoc.* **2021**, *28*, 2385–2392. [CrossRef]
33. White, L.; van Basshuysen, P. Privacy versus public health? A reassessment of centralised and decentralised digital contact tracing. *Sci. Eng. Ethics* **2021**, *27*, 23. [CrossRef]
34. Kwan, T.H.; Wong, N.S.; Yeoh, E.K.; Lee, S.S. Shifts of SARS-CoV-2 exposure settings in the transmission clusters of 2 epidemic waves in Hong Kong. *Int. J. Env. Health Res.* **2022**, *33*, 911–923. [CrossRef]
35. Li, H.; He, J.; Chen, J.; Pan, S.; Feng, J.; Liu, S. The governance of imported 2019-nCov infections: What can be learned from China's experience? *Glob. Health Res. Policy* **2022**, *7*, 8. [CrossRef]
36. Prevention and Control of Disease (Compulsory Testing for Certain Persons) Regulation 2021 (HK). Available online: https://www.elegislation.gov.hk/hk/cap599J (accessed on 11 March 2022).
37. The Government of the Hong Kong Special Administrative Region. Government Gazettes Compulsory Testing Notice. Available online: https://www.info.gov.hk/gia/general/202104/02/P2021040200786.htm (accessed on 11 March 2022).
38. The Government of the Hong Kong Special Administrative Region. Government Gazettes Compulsory Testing Notice. Available online: https://www.info.gov.hk/gia/general/202107/21/P2021072100893.htm (accessed on 11 March 2022).
39. Kwan, T.H. Enforcement of the use of digital contact-tracing apps in a common law jurisdiction. *Healthcare* **2022**, *10*, 1613. [CrossRef]
40. Kwan, T.H.; Wong, N.S.; Chan, C.P.; Yeoh, E.K.; Wong, S.Y.; Lee, S.S. Mass screening of SARS-CoV-2 with rapid antigen tests in a receding omicron wave: Population-based survey for epidemiologic evaluation. *JMIR Public Health Surveill.* **2022**, *8*, e40175. [CrossRef] [PubMed]
41. Fitriani, W.R.; Handayani, P.W.; Hidayanto, A.N. Challenges in coronavirus contact-tracing application implementation in Indonesia: Users' perspective. In Proceedings of the 2022 International Conference on Information Management and Technology (ICIMTech), Semarang, Indonesia, 11–12 August 2022. Available online: https://ieeexplore.ieee.org/document/9915106 (accessed on 11 March 2022).

engineering proceedings

MDPI

Proceeding Paper

Classification of Breast Cancer Using Radiological Society of North America Data by EfficientNet †

Hoang Nhut Huynh [1], Ngoc An Dang Nguyen [1], Anh Tu Tran [2], Van Chinh Nguyen [3] and Trung Nghia Tran [1,*]

1 Laboratory of Laser Technology, Faculty of Applied Science, Ho Chi Minh City University of Technology (HCMUT), VNU-HCM, 268 Ly Thuong Kiet Street, District 10, Ho Chi Minh City 72506, Vietnam; hhnhut@hcmut.edu.vn (H.N.H.); dnnan.sdh19@hcmut.edu.vn (N.A.D.N.)

2 Laboratory of General Physics, Faculty of Applied Science, Ho Chi Minh City University of Technology (HCMUT), VNU-HCM, 268 Ly Thuong Kiet Street, District 10, Ho Chi Minh City 72506, Vietnam; tranatu@hcmut.edu.vn

3 Faculty of Nursing & Medical Technology, University of Medicine and Pharmacy at HCMC (UMP), 217 Hong Bang Street, District 10, Ho Chi Minh City 72714, Vietnam; nvchinh@ump.edu.vn

* Correspondence: ttnghia@hcmut.edu.vn

† Presented at the IEEE 5th Eurasia Conference on Biomedical Engineering, Healthcare and Sustainability, Tainan, Taiwan, 2–4 June 2023.

Abstract: Breast cancer is a common cancer that affects women all over the world. Therefore, detection at an early stage is crucial for reducing the mortality rate linked to this disease. Mammography is the primary screening method for breast cancer. However, it has drawbacks, including high rates of false-positive and negative results, inter-observer variability, and limited sensitivity with dense breast tissue. To solve such problems, breast cancer was analyzed and classified using mammography images and deep learning models from the Radiological Society of North America (RSNA) database. This database contains processed and raw images from the RSNA that consist of annotated malignancies and clinical data. Using deep learning models based on convolutional neural network (CNN) models such as visual geometry group (VGG), Googlenet, EfficientNet, and Residual Networks, mammograms were classified into cancer or non-cancer categories. In this study, a novel architecture was proposed by combining CNNs and attention mechanisms, which extracted and highlighted the relevant features. A dataset of 8000 patients with 47,000 photographs was used to train and assess the model via 5-fold cross-validation. The results outperformed prior methods using the same database and reached an average accuracy rate of 95%. The results showed that mammography with deep learning methods considerably improved breast cancer detection and diagnosis.

Keywords: breast cancer; EfficientNet; RSNA database; 5-fold cross-validation; mammography

Citation: Huynh, H.N.; Nguyen, N.A.D.; Tran, A.T.; Nguyen, V.C.; Tran, T.N. Classification of Breast Cancer Using Radiological Society of North America Data by EfficientNet. *Eng. Proc.* **2023**, *55*, 6. https://doi.org/10.3390/engproc2023055006

Academic Editors: Teen-Hang Meen, Kuei-Shu Hsu and Cheng-Fu Yang

Published: 27 November 2023

1. Introduction

Following cardiovascular diseases, breast cancer is the second-most prevalent cause of mortality for women. Early breast cancer detection and diagnosis increase survival rates and reduce treatment costs. To identify breast tissue abnormalities before clinical symptoms start, mammography is the most widely used screening procedure. Nevertheless, mammography interpretation is subjective and complex, contingent on the radiologist's experience and expertise. In addition, the quality and appearance of mammography images vary depending on factors such as acquisition parameters, compression levels, and breast density. Therefore, objective and automated methods are required for analyzing and classifying mammography images.

Deep learning belongs to the discipline of machine learning, which deals with massive amounts of data and complex tasks. Deep learning has recently been used to analyze medical images, most notably in identifying and diagnosing breast cancer. As a result, significant advancements have been achieved in detecting and diagnosing breast cancer and medical imaging processes. However, the generalizability and validation of existing deep-learning methods for mammography are limited by the use of small or private datasets. Various deep learning models, including convolutional neural networks (CNNs), have been used in many studies to categorize mammograms into benign or malignant tumors and find problematic areas of interest (ROIs) in mammograms [1]. The breast dataset and the Digital Database for Screening Mammography (CBIS-DDSM) are two examples of publicly accessible datasets used for research for model training and assessment [2].

Using the mammography database of the Radiological Society of North America (RSNA), numerous images were acquired for this study, which was conducted to develop a novel deep learning method for evaluating and diagnosing breast cancer. The dataset contains mammograms of breast cancer, which are essential for identifying early diagnosis and treatment. False positives such as anxiety and unnecessary biopsies often negatively affect patients. Thus, we (1) preprocessed the mammography images to enhance their quality and reduce their variability, (2) extracted features from the mammography images using a CNN-based deep learning model, and (3) classified the mammography images into benign or malignant categories using a fully connected neural network (FCN). A subset of test data containing 2725 mammograms from 2725 women with 272 cancers was used to evaluate the proposed method. A comparison of the proposed method with previous methods for mammography analysis was conducted.

2. Materials and Methods

2.1. Dataset

With the RSNA Mammography Data, a deep learning approach was built and evaluated for breast cancer detection and diagnosis [3]. The RSNA Mammography Data are a large-scale collection of screening mammograms from Australia and the U.S. with detailed labels, radiologists' evaluations, and follow-up pathology results for suspected malignancies. The RSNA is a nonprofit organization that represents 31 radiology institutions in 145 different nations. Through education, research, and technological innovation programs, the RSNA strives to improve the quality of health care for patients [4]. The RSNA Mammography Data cover various mammography images with variable quality and appearance due to different factors such as acquisition parameters, compression levels, and breast density. The Breast Imaging Reporting and Data System (BI-RADS) lexicon is the main communication tool in mammography reports in most nations with established breast cancer screening programs [5]. The BI-RADS lexicon categorizes breast imaging findings into seven categories, from BI-RADS 0 (incomplete) to BI-RADS 6 (known biopsy-proven malignancy) [4]. The RSNA Mammography Data also provide information on breast density, lesion type, lesion size, lesion location, and histologic diagnosis. Figure 1 shows an example of images from the dataset.

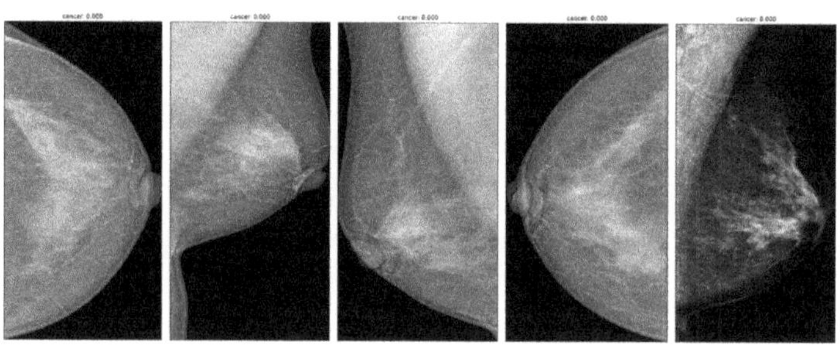

Figure 1. Example of mammography images from RSNA Mammography Data.

RSNA Mammography Data are valuable for constructing and evaluating artificial intelligence (AI) algorithms for the detection and diagnosis of breast cancer. It is also used to evaluate the performance of AI models on various patient types. In this study, the dataset was divided into a training set (80% of the images), a validation set (10%), and a test set (10%). The images were stored in DICOM format with a 1024 × 1024-pixel resolution. The images were labeled as 'benign' or 'malignant' according to the BI-RADS assessment category, a standard methodology for reporting mammography findings and recommendations [4]. Table 1 depicts the label distribution of the given dataset. The training set was augmented with techniques such as rotation, scaling, rotating, and cropping to increase its diversity and size.

Table 1. Label distribution of the RSNA Mammography Data.

Set	Original Data	Augment Data
Training	43,765	86,000
Validation	5741	10,000
Testing	5740	10,000
Total	54,706	106,000

2.2. Models

Four distinct deep CNN models were applied for breast cancer analysis and classification using RSNA mammography data. The models included VGG, Googlenet, EfficientNet, and Residual Networks.

VGG is a convolutional neural network architecture that consists of multiple blocks of convolutional layers followed by max-pooling layers [6]. As the network expands, the number of filters that are used in the convolutional layers has a higher priority and increases. VGG is known for its simplicity and high performance on image classification tasks. GoogLeNet is a CNN architecture based on the Inception module, which allows the network to choose between multiple convolutional filter sizes in each block [7]. GoogLeNet also uses skip connections to connect activations of a layer to further layers to avoid vanishing gradients and improve feature extraction. GoogLeNet won the ImageNet 2014 challenge with its high accuracy and low complexity. Residual Networks (ResNets) are a type of CNN that uses residual blocks, which consist of skip connections that add the input of a layer to its output [8]. This eases the training of very deep networks by avoiding vanishing or exploding gradients. ResNets have high accuracy on image classification and object detection tasks by increasing the network depth. EfficientNet has a CNN architecture and scaling method that uniformly scales all dimensions of depth, width, and resolution using a compound coefficient [9]. This allows the network to adapt to different resource constraints and improve the efficiency and performance of the model. EfficientNet uses inverted bottleneck residual blocks with squeeze-and-excitation modules and a swish activation function. EfficientNet achieves state-of-the-art accuracy on several image classification benchmarks with an order of magnitude fewer parameters than other models.

3. Results

We compared the performance of deep convolutional neural networks for breast mass classification into benign or malignant categories, using two different training strategies: (a) training from scratch and (b) fine-tuning. The performance of four CNN models was evaluated on the RSNA dataset. The dataset consists of 54,706 screening mammograms from 8000 women with annotations of breast cancer diagnosis within one year of the screening. The dataset was divided into training, validation, and test sets with 80, 10, and 10% of the total images, respectively. The task was to predict the probability of breast cancer for each image.

In the training-from-scratch strategy, the model weights were randomly initialized and updated using only the target dataset. In the fine-tuning strategy, the model weights were initialized from a pre-trained model on ImageNet and updated using the source and target datasets. The fine-tuning strategy leveraged the knowledge learned from a large generic dataset and applied it to a specific task or dataset. Binary cross-entropy was used as the loss function, and the area under the receiver operating characteristic curve (AUC) was used as the evaluation metric. Data augmentation techniques such as random cropping, flipping, and rotation were used to increase the diversity of the training data. The models were implemented using PyTorch and trained on an NVIDIA Tesla V100 GPU.

For the training-from-scratch strategy, the model parameters were randomly initialized and trained for 50 epochs using the Adam optimizer with a learning rate of 0.001 and a batch size of 32. The binary cross-entropy loss was used as the objective function and applied early stopping based on the validation loss. For the fine-tuning strategy, the model parameters were initialized with a pre-trained model on ImageNet and trained for 20 epochs using Adam optimizer with a learning rate of 0.0001 and a batch size of 32. The binary cross-entropy loss function was used as the objective function and applied to early stopping based on the validation loss.

The results of the developed models on the test set are shown in Table 2. The fine-tuning strategy achieved a higher AUC than the training from scratch strategy for all models, indicating that the fine-tuning strategy improved the performance of CNNs for breast cancer detection. Among the four models, EfficientNet showed the highest AUC of 92% with the fine-tuning strategy, followed by Residual Networks with an AUC of 90%, Googlenet with an AUC of 88%, and VGG with an AUC of 86%. Such results indicated that EfficientNet learned more discriminative features for breast cancer detection than the other models. The EfficientNet had fewer parameters and a faster inference time than the other models, enabling its efficiency and scalability, as shown in Figure 2.

Table 2. Accuracy and AUC of different models on test dataset.

Model	Accuracy	AUC	Accuracy	AUC
	Training from scratch strategy		Fine-tuning strategy	
VGG	0.83	0.91	0.82	0.86
Googlenet	0.85	0.93	0.85	0.88
EfficientNet	0.89	0.95	0.89	0.92
Residual Networks	0.87	0.94	0.87	0.90

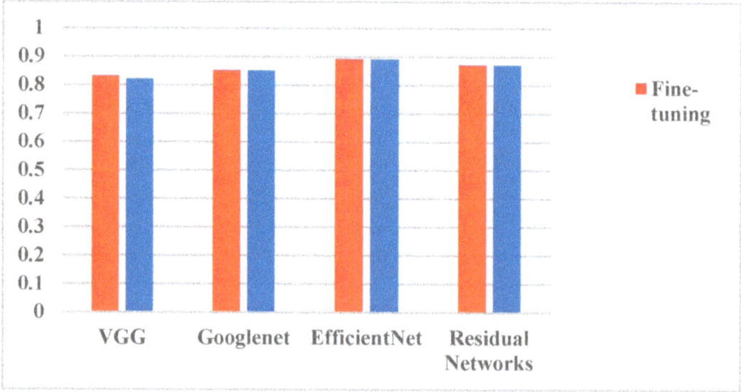

Figure 2. AUC scores of four models on test set.

The accuracy of the developed models on the test set is shown in Table 2. As a result, fine-tuning a pre-trained model improved the performance and generalization ability of the developed models by leveraging the knowledge learned from a large-scale dataset such as ImageNet. Among the four models, EfficientNet achieved the highest accuracy of 95.6% using the fine-tuning strategy, followed by Residual Networks with an accuracy of 94.8%, Googlenet with an accuracy of 93.2%, and VGG with an accuracy of 91.6%. Using the training from scratch strategy, EfficientNet also achieved the highest accuracy of 89.2%, followed by Residual Networks with an accuracy of 87.4%, Googlnet with an accuracy of 85.6%, and VGG with an accuracy of 83.2%, as shown in Figure 3.

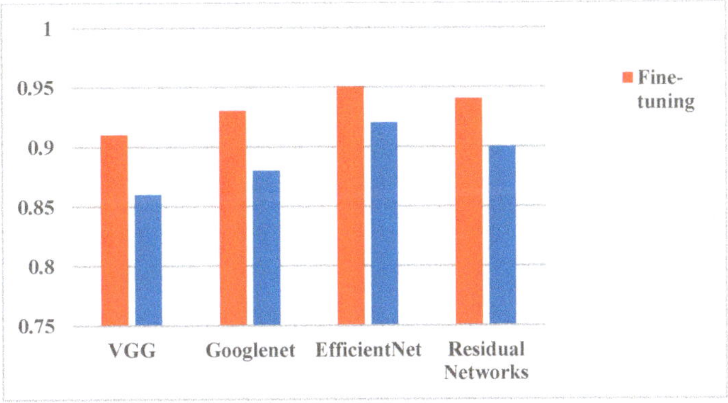

Figure 3. Accuracy of four models on test data set.

4. Discussion

We developed and investigated four deep CNN models to analyze and classify breast cancer using RSNA mammography data. EfficientNet outperformed the other models regarding AUC, parameters, and inference time. The proposed method was compared with existing ones on the same dataset to understand their implications and limitations. The effectiveness of deep CNNs for breast cancer detection using mammography data was also confirmed and validated. The CNN architecture significantly impacted the model's efficacy and efficiency.

EfficientNet showed the best results among the four models, with the least number of parameters and the fastest inference time. This suggested that EfficientNet was an appropriate architecture for breast cancer detection as it learned more relevant features for the task with fewer computational resources. One probable reason why EfficientNet performed better than the other models was using a compound scaling method that jointly scaled up the network's depth, width, and resolution with a fixed scaling coefficient. This allowed the network to capture more fine-grained details and contextual information from the images, which were beneficial for distinguishing benign and malignant lesions.

Moreover, EfficientNet also used depthwise separable convolutions and squeeze-and-excitation modules to improve the efficiency and performance of the network. These techniques reduced the redundancy and complexity of the network and enhanced feature representation and selection. The result of this study demonstrated that deep CNNs achieved high accuracy and reliability for breast cancer detection using mammography data, potentially improving the screening process and reducing false positives and negatives. It was proved in this study that EfficientNet was a promising architecture for breast cancer detection, as it achieved better results with less computational cost than other models. It facilitated the deployment and scalability of deep CNNs in resource-limited settings for further improving the performance of deep CNNs for breast cancer detection by exploring different data augmentation techniques, loss functions, optimization methods, and hyper-parameters.

5. Conclusions

We proposed deep-learning approaches for breast cancer detection using the RSNA dataset. The novel proposed architecture combined CNNs and attention mechanisms to extract and highlight the relevant features from the mammograms. In addition, data augmentation, transfer learning, and ensemble methods were applied to improve the performance and robustness of the developed model. The results showed that the developed model achieved a high accuracy of 95.6% on the test set, outperforming state-of-the-art methods and human radiologists. The results also demonstrated the potential of deep learning methods for breast cancer screening and diagnosis and the value of large-scale and diverse datasets for training and evaluating such models. The proposed approaches can contribute to reducing the mortality and morbidity of breast cancer by providing accurate and timely detection of malignant lesions.

Author Contributions: Conceptualization, H.N.H.; methodology, T.N.T.; software, H.N.H.; validation, A.T.T.; formal analysis, T.N.T.; investigation, N.A.D.N.; resources, V.C.N.; data curation, T.N.T. and V.C.N.; writing—original draft preparation, T.N.T.; writing—review and editing, H.N.H.; visualization, H.N.H.; supervision, H.N.H.; project administration, H.N.H. and T.N.T. All authors have read and agreed to the published version of the manuscript.

Funding: This research received no external funding.

Institutional Review Board Statement: Not applicable.

Informed Consent Statement: Not applicable.

Data Availability Statement: Data underlying the results presented in this paper are not publicly available but may be obtained from the authors upon reasonable request.

Acknowledgments: We acknowledge Ho Chi Minh City University of Technology (HCMUT) VNU-HCM for supporting this study.

Conflicts of Interest: The authors declare no conflict of interest.

References

1. Tsochatzidis, L.; Costaridou, L.; Pratikakis, I. Deep learning for breast cancer diagnosis from mammograms—A comparative study. *J. Imaging* **2019**, *5*, 37. [CrossRef] [PubMed]
2. Kumar, A.; Singh, S.K. Evaluation of deep learning models for detecting breast cancer using mammograms. In *Advances in Computer Vision and Information Technology*; Springer: Cham, Switzerland, 2020; pp. 117–126.
3. Kaggle. RSNA Breast Cancer Detection. Available online: https://www.kaggle.com/competitions/rsna-breast-cancer-detection (accessed on 8 February 2023).
4. Pesce, K.; Orruma, M.B.; Hadad, C.; Bermúdez Cano, Y.; Secco, R.; Cernadas, A. BI-RADS Terminology for Mammography Reports: What Residents Need to Know. *Radiographics* **2019**, *39*, 319–320. [CrossRef] [PubMed]
5. Halling-Brown, M.D.; Warren, L.M.; Ward, D.; Lewis, E.; Mackenzie, A.; Wallis, M.G.; Wilkinson, L.S.; Given-Wilson, R.M.; McAvinchey, R.; Young, K.C. Optimam mammography image database: A large-scale resource of mammography images and clinical data. *Radiol. Artif. Intell.* **2020**, *3*, e200103. [CrossRef] [PubMed]
6. Simonyan, K.; Zisserman, A. Very deep convolutional networks for large-scale image recognition. *arXiv* **2014**, arXiv:1409.1556.
7. Szegedy, C.; Liu, W.; Jia, Y.; Sermanet, P.; Reed, S.; Anguelov, D.; Erhan, D.; Vanhoucke, V.; Rabinovich, A. Going deeper with convolutions. In Proceedings of the IEEE Conference on Computer Vision and Pattern Recognition, Boston, MA, USA, 7–12 June 2015.
8. Zhang, K.; Sun, M.; Han, T.X.; Yuan, X.; Guo, L.; Liu, T. Residual networks of residual networks: Multilevel residual networks. *IEEE Trans. Circuits Syst. Video Technol.* **2017**, *28*, 1303–1314. [CrossRef]
9. Tan, M.; Le, Q. Efficientnet: Rethinking model scaling for convolutional neural networks. In Proceedings of the International Conference on Machine Learning, Long Beach, CA, USA, 10–15 June 2019.

engineering proceedings

MDPI

Proceeding Paper

Analysis of Outpatient Prescription Trends of Non-Narcotic Analgesics and Proton Pump Inhibitors (PPIs) in Taiwan from 2010 to 2021: A Nationwide Population-Based Study †

Jack C. Yue [1,2], I-Wei Lu [2,*], Jung-San Chang [3] and Chien-Nan Lin [2]

1 Department of Statistics, National Chengchi University, Taipei 11605, Taiwan; csyue@nccu.edu.tw
2 Center for Fundamental Science, Kaohsiung Medical University, Kaohsiung 80708, Taiwan; lincn@kmu.edu.tw
3 Department of Renal Care, College of Medicine, Kaohsiung Medical University, Kaohsiung 80708, Taiwan; cjs@kmu.edu.tw
* Correspondence: seaniwei@gmail.com or iwlu@kmu.edu.tw

† Presented at the IEEE 5th Eurasia Conference on Biomedical Engineering, Healthcare and Sustainability, Tainan, Taiwan, 2–4 June 2023.

Abstract: The purpose of this study is to describe the prevalence, trends, and associations of the use of outpatient prescription non-narcotic analgesics and proton pump inhibitors (PPIs) in Taiwan from 2010 to 2021. We identified the most commonly prescribed analgesics and PPIs, including paracetamol (acetaminophen) and nonsteroidal anti-inflammatory drugs (NSAIDs), in outpatient departments according to the data from Taiwan National Health Insurance Research Database and Taiwan's NHI Annual Statistical Report. This study shows that the traditional non-selective NSAIDs take the largest proportion of usage, representing over 60% of the annual usage of non-narcotic analgesics. Acetaminophen is the second and takes over 30% of annual usage. The selective COX-2 (coxibs) takes less than 10% of usage, but its use is growing rapidly. This study also shows that the annual usage of PPIs is growing rapidly and is not decreasing with the use of selective COX-2. The findings in this study can help us to focus on the most commonly prescribed drugs to investigate the prescription pattern of NSAIDs and PPIs, especially the combined use of selective COX-2 and PPIs, in future study.

Keywords: non-narcotic analgesics; acetaminophen; nonsteroidal anti-inflammatory drugs (NSAIDs); selective COX-2; proton pump inhibitors (PPIs)

Citation: Yue, J.C.; Lu, I.-W.; Chang, J.-S.; Lin, C.-N. Analysis of Outpatient Prescription Trends of Non-Narcotic Analgesics and Proton Pump Inhibitors (PPIs) in Taiwan from 2010 to 2021: A Nationwide Population-Based Study. *Eng. Proc.* **2023**, *55*, 7. https://doi.org/10.3390/engproc2023055007

Academic Editors: Teen-Hang Meen, Kuei-Shu Hsu and Cheng-Fu Yang

Published: 27 November 2023

1. Introduction

Non-narcotic analgesics, also known as non-opioid analgesics or non-addictive analgesics, are widely used in outpatient prescriptions as a preferred, safe, and effective first-line therapy for mild-to-moderate acute and chronic pain. Non-narcotic analgesics encompass paracetamol (acetaminophen) and nonsteroidal anti-inflammatory drugs (NSAIDs). In addition to pain relief and fever reduction, NSAIDs also have anti-inflammatory effects; therefore, the outpatient use of NSAIDs is growing rapidly [1]. However, NSAIDs have the risk of adverse gastrointestinal effects (ulceration and bleeding), and so they are often combined with the use of gastrointestinal protection drugs, such as proton pump inhibitors (PPIs) [2]. This not only places a greater burden on Taiwan's National Health Insurance (NHI) drug expenditures, but also induces several different forms of kidney injury, including hemodynamically mediated acute kidney injury (AKI) and acute interstitial nephritis (AIN) [3]. Since drug expenditures account for 25 percent of total Taiwan's NHI expenditures, and as the treatment expenses of acute kidney failure and chronic kidney disease have been the largest proportion of Taiwan's NHI annual outpatient expenditures over a long period of time, it is important to understand the prescription patterns of non-narcotic analgesics and gastrointestinal protection drugs to prevent these nephrotoxic drugs from causing kidney damage to patients [4].

We want to investigate the following questions: (1) What were the trends in outpatient prescription of non-narcotic analgesics and PPIs under Taiwan's NHI from 2010 to 2021? (2) Does the usage of PPIs decrease with the use of the selective COX-2? (3) What are the most commonly prescribed analgesics and PPIs according to annual usage in recent years? (4) Are these analgesics used in combination with the PPIs? (5) What are their prescription patterns? We conduct a two-step study to answer these research questions. In study 1, we list all the non-narcotic analgesics and PPIs. We rate the risk of adverse gastrointestinal effects of each non-narcotic analgesic and calculate the annual usage of each type of analgesics and PPIs using data from Taiwan's NHI Annual Statistical Report from 2010 to 2021. We describe the prevalence and trends of the use of non-narcotic analgesics and PPIs. Then, we identify the most commonly prescribed analgesics and PPIs in recent years. This study mainly focuses on study 1 and tries to answer the research question (1), (2), and (3). Working with study 2, we use the list of the most commonly prescribed analgesics and PPIs to search the outpatient prescriptions in Taiwan's NHI Research Database. We will collect the prescription data for these drugs, including the international classification of disease (ICD), the dosage, the demographic characteristics of the patients, and the cities in which a hospitals are. We will use these data to analyze the prescription patterns of the most commonly prescribed analgesics and how they are used in combination with PPIs? Study 2 will use the result of study 1 to answer the research question (4) and (5).

2. Materials and Methods

2.1. Data Source

We conducted a literature review in order to list all the commonly used non-narcotic analgesics included in Taiwan's NHI drug reimbursement. These drugs can be searched in Taiwan's NHI drug database [5] and can be used in outpatient prescriptions. We used the nationwide drug usage data in Taiwan's NHI Annual Statistical Report [6] from 2010 to 2021 to calculate the annual usage of each type of analgesics and PPIs.

2.2. Definitions of Variables

Non-narcotic analgesics encompass paracetamol (acetaminophen) and non-steroidal anti-inflammatory drugs (NSAIDs). All these drugs have both analgesic and antipyretic properties, and they are widely used to manage mild-to-moderate pain conditions. However, they also have the disadvantages of the potential for gastrointestinal (GI) and renal toxicity. Paracetamol (acetaminophen) is considered to be a safe analgesic/antipyretic compound and its side effects are less severe than those of the NSAIDs. Accordingly, it is widely accepted that paracetamol is particularly suitable for patients at high risk of developing gastrointestinal (GI) ulcers or bleeds. At high doses, it is believed that paracetamol may induce upper GI symptoms such as abdominal pain/discomfort, nausea or vomiting [7]. Therefore, we identify the risk of gastrointestinal (GI) toxicity of paracetamol as low.

NSAIDs are categorized as traditional 'non-selective COX-1/COX-2' NSAIDs and selective cyclooxygenase COX-2 inhibitors (also known as coxibs). Mucosal injury in the gastrointestinal (GI) tract is the most common adverse effect associated with NSAIDs. Patients who receive long-term treatment with a traditional non-selective NSAID have an approximate five-fold higher risk of peptic ulcer disease compared to non-users. The increased risk of gastrointestinal complications associated with non-selective NSAIDs promotes the development of newer COX-2 inhibitors (coxibs); patients receiving selective COX-2 inhibitors have significantly lower risk of GI toxicity than those receiving non-selective NSAIDs. Due to the high risks of GI side effects of traditional NSAIDs, patients at increased risk of GI complications should receive either a non-selective NSAID with proton pump inhibitors (PPIs), or a COX-2-selective inhibitor alone [1,8]. Therefore, we identify the risk of GI toxicity of the traditional non-selective NSAIDs as high, and that of the selective COX-2 inhibitors as low.

In Table 1, we list all the non-narcotic analgesics, including paracetamol (acetaminophen), traditional non-selective NSAIDs, and selective COX-2 inhibitors (coxibs) [9,10], and also list all the proton pump inhibitors (PPIs), that are approved by the United States Food and Drug Administration (FDA) and are included in Taiwan's NHI drug reimbursement [11,12]. The drugs are organized by name alphabetically:

Table 1. Non-narcotic analgesics and PPIs under Taiwan's NHI drug reimbursement.

Drug Category: Paracetamol (GI toxicity: Low)
Paracetamol (acetaminophen)
Drug Category: Traditional non-selective NSAIDs (GI toxicity: High)
Aspirin (Acetylsalicylic acid) Diclofenac Diflunisal Etodolac Fenoprofen Flurbiprofen Ibuprofen Indomethacin Ketoprofen Ketorolac Mefenamic Meloxicam Nabumetone Naproxen Piroxicam Sulindac Tiaprofenic Tolmetin
Drug Category: Selective COX-2 (coxibs) (GI toxicity: Low)
Celecoxib Etoricoxib Nimesulide
Drug Category: Proton pump inhibitors (PPIs)
Dexlansoprazole Esomeprazole Lansoprazole Omeprazole Pantoprazole Rabeprazole

3. Results

We used the drugs in Table 1 and the nationwide drug usage data in Taiwan's NHI Annual Statistical Report for 12 years from 2010 to 2021 to summarize the annual usage amount of one paracetamol variety, 18 non-selective NSAIDs, three selective COX-2 drugs, and six PPIs in Table 2. We compared the results by category and by each drug to show the prescription trends of each category of the non-narcotic analgesics and PPIs, and to identify the most commonly prescribed drugs in each category.

Table 2. The annual usage amount of non-narcotic analgesics and PPIs.

YEAR	2010	2011	2012	2013	2014	2015
Paracetamol (acetaminophen)	289,968,123.0	318,500,285.0	337,740,212.0	341,250,942.0	353,803,635.0	363,656,649.0
Traditional non-selective NSAIDs						
Diflunisal	242,486.0	256,434.0	245,354.0	217,710.0	231,813.0	179,761.0

Table 2. *Cont.*

YEAR	2010	2011	2012	2013	2014	2015
Tolmetin	630,951.0	753,216.0	1,297,261.0	1,304,548.0	1,419,728.0	962,601.0
Ketoprofen	3,503,597.0	3,427,570.0	3,521,078.0	3,226,268.0	2,821,006.0	1,860,493.0
Fenoprofen	5284.0	69,385.0	211,378.0	296,596.0	109,203.0	301,229.0
Ketorolac	8,698,936.0	8,542,896.0	9,852,274.0	9,751,331.0	9,676,173.0	9,221,865.0
Aspirin (Acetylsalicylic acid)	298,507,629.0	312,403,169.0	330,143,889.0	345,834,212.0	355,183,608.0	358,988,218.0
Indomethacin	4,913,995.0	4,989,848.0	5,400,171.0	5,272,001.0	5,080,380.0	4,969,189.0
Diclofenac	105,388,188.0	113,914,625.0	114,958,443.0	114,371,503.0	117,778,230.0	119,748,550.0
Ibuprofen	26,112,685.0	29,869,023.0	30,398,472.0	31,111,930.0	33,478,919.0	34,317,810.0
Naproxen	10,022,990.0	10,738,575.0	10,753,129.0	10,496,405.0	10,723,367.0	10,121,624.0
Mefenamic	31,710,604.0	36,115,006.0	36,890,593.0	37,485,769.0	38,383,954.0	38,143,173.0
Piroxicam	4,366,846.0	4,326,158.0	4,403,368.0	4,911,987.0	5,051,388.0	5,090,472.0
Sulindac	10,355,898.0	9,220,923.0	8,781,799.0	8,135,472.0	8,078,662.0	7,484,095.0
Flurbiprofen	36,953,085.0	41,426,770.0	41,636,956.0	41,106,175.0	41,761,267.0	40,911,453.0
Tiaprofenic	5,923,437.0	4,446,153.0	3,821,656.0	3,282,326.0	3,335,603.0	2,904,869.0
Meloxicam	26,131,337.0	25,862,349.0	25,322,478.0	24,300,303.0	23,651,596.0	22,060,609.0
Etodolac	11,888,349.0	11,367,502.0	11,325,817.0	11,110,919.0	10,639,812.0	10,509,084.0
Nabumetone	3,736,036.0	3,243,771.0	3,015,628.0	2,880,562.0	2,300,341.0	1,829,965.0
Total—non-selective NSAIDs	589,092,333.0	620,973,373.0	641,979,744.0	655,096,017.0	669,705,050.0	669,605,060.0
			selective COX-2 (coxibs)			
Celecoxib	29,342,734.0	30,049,890.0	32,085,398.0	34,482,671.0	37,819,244.0	41,604,467.0
Nimesulide	2,678,711.0	2,649,918.0	2,740,999.0	2,928,377.0	2,845,256.0	1,908,762.0
Etoricoxib	10,362,270.0	15,424,968.0	19,954,160.0	240,80,649.0	25,918,132.0	27,480,648.0
Total—selective COX-2	42,383,715.0	48,124,776.0	54,780,557.0	61,491,697.0	66,582,632.0	70,993,877.0
			Proton pump inhibitors (PPIs)			
Dexlansoprazole	0.0	0.0	0.0	0.0	1,195,419.0	6,285,019.0
Lansoprazole	15,937,975.0	16,700,483.0	21,500,786.0	21,592,943.0	22,158,816.0	18,213,277.0
Omeprazole	4,801,435.0	4,432,011.0	5,813,645.0	6,935,420.0	5,862,678.0	0.0
Esomeprazole	17,467,967.0	19,753,245.0	19,159,927.0	22,363,013.0	23,658,524.0	30,698,636.0
Pantoprazole	10,244,121.0	9,830,389.0	10,931,944.0	10,313,028.0	11,492,542.0	11,764,842.0
Rabeprazole	7,120,164.0	6,839,929.0	8,096,476.0	7,597,643.0	7,252,430.0	8,447,124.0
Total—PPIs	55,571,662.0	57,556,057.0	65,502,778.0	68,802,047.0	71,620,409.0	75,408,898.0
YEAR	**2016**	**2017**	**2018**	**2019**	**2020**	**2021**
Paracetamol (acetaminophen)	379,597,405.0	393,951,533.0	410,070,873.0	431,226,591.3	431,238,313.6	444,156,324.0
			Traditional non-selective NSAIDs			
Diflunisal	15,925.0	946.0	0.0	42.0	126.0	742.0
Tolmetin	40,118.0	38,568.0	40,849.0	30,057.0	26,687.0	29,282.0

Table 2. *Cont.*

YEAR	2016	2017	2018	2019	2020	2021
Ketoprofen	1,291,454.0	1,111,806.0	1,105,834.0	1,077,942.1	849,578.2	778,493.0
Fenoprofen	962,147.0	539,040.0	768,356.0	735,468.5	714,498.2	666,893.0
Ketorolac	9,574,316.0	9,878,009.0	10,056,409.0	9,716,596.9	9,031,225.5	8,835,210.0
Aspirin (Acetylsalicylic acid)	366,376,105.0	375,643,630.0	383,120,666.0	387,843,093.0	396,552,486.6	405,604,122.0
Indomethacin	4,699,619.0	4,500,193.0	4,311,930.0	4,051,962.9	3,872,131.3	3,665,919.0
Diclofenac	124,424,299.0	126,483,717.0	129,178,384.0	134,009,228.4	123,874,738.5	117,569,088.0
Ibuprofen	37,103,722.0	36,957,953.0	39,371,690.0	42,481,771.0	34,680,307.2	31,025,793.0
Naproxen	9,726,423.0	9,893,139.0	9,715,407.0	9,650,615.5	9,314,947.9	8,955,645.0
Mefenamic	39,116,748.0	38,837,548.0	40,038,274.0	40,749,495.9	35,870,159.3	32,708,550.0
Piroxicam	4,939,335.0	4,919,879.0	4,742,708.0	4,363,130.7	4,082,355.6	3,844,189.0
Sulindac	6,866,559.0	4,751,456.0	2,982,862.0	2,701,063.6	2,369,758.9	2,131,301.0
Flurbiprofen	40,789,198.0	40,477,203.0	39,277,977.0	37,531,871.0	32,659,331.5	30,632,617.0
Tiaprofenic	2,784,405.0	2,915,730.0	2,503,034.0	2,382,682.1	2,334,178.0	2,280,914.0
Meloxicam	19,968,809.0	18,737,852.0	17,643,748.0	16,743,526.7	15,806,439.2	15,187,917.0
Etodolac	9,774,633.0	9,479,509.0	9,406,691.0	8,920,161.2	8,193,847.7	7,423,861.0
Nabumetone	1,627,681.0	1,452,938.0	1,238,375.0	1,291,593.0	1,005,009.0	844,861.0
Total—non-selective NSAIDs	680,081,496.0	686619116.0	695503194.0	704,280,301.5	681,237,805.6	672,185,397.0
			selective COX-2 (coxibs)			
Celecoxib	47,641,788.0	54,574,797.0	65,058,810.0	73,424,446.8	77,509,160.3	79,913,423.0
Nimesulide	1,356,054.0	1,258,767.0	543,465.0	796,130.5	871,588.5	914,066.0
Etoricoxib	26,845,756.0	28,292,589.0	30,736,150.0	30,914,130.0	34,270,097.4	34,938,118.0
Total—selective COX-2	75,843,598.0	84,126,153.0	96,338,425.0	105,134,707.3	112,650,846.2	115,765,607.0
			Proton pump inhibitors (PPIs)			
Dexlansoprazole	9,213,852.0	11,516,407.0	12,539,314.0	15,043,897.5	15,068,329.3	13,804,902.0
Lansoprazole	19,129,981.0	21,591,598.0	22,822,871.0	26,078,750.0	26,921,078.8	26,332,941.0
Omeprazole	0.0	0.0	0.0	0.0	0.0	0.0
Esomeprazole	31,523,402.0	36,632,282.0	39,599,272.0	44,737,220.1	48,726,306.5	54,194,445.0
Pantoprazole	13,299,493.0	15,487,044.0	16,468,282.0	18,291,923.5	18,253,321.6	18,980,534.0
Rabeprazole	9,600,093.0	10,877,557.0	11,962,574.0	14,235,057.2	14,163,626.5	14,020,734.0
Total—PPIs	82,766,821.0	96,104,888.0	103,392,313.0	118,386,848.3	123,132,662.7	127,333,556.0

3.1. Annual Usage by Category

We compared annual usage by each category to find the prescription trends shown in Figure 1. Figure 1 shows that the non-selective NSAIDs are the largest in usage, acetaminophen is the second most used, and selective COX-2 (coxibs) have the smallest use in non-narcotic analgesics. Figure 1 also shows that all the analgesics and PPIs are increasing in amount each year; only the non-selective NSAIDs decreased from 2020 to 2021. We illustrated the annual usage of the non-narcotic analgesics in percentage terms in Figure 2, which shows the portions of the three categories clearly.

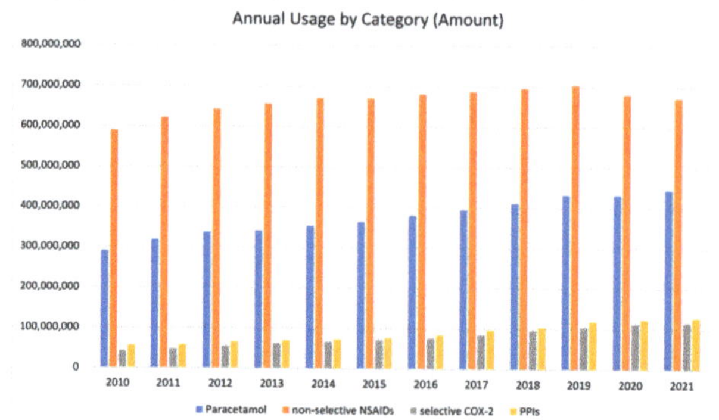

Figure 1. Annual usage by each category in amount from 2010 to 2021.

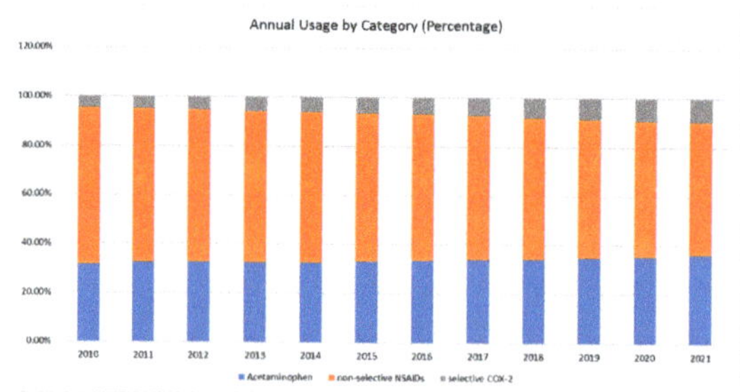

Figure 2. Annual usage of the non-narcotic analgesics in percentage from 2010 to 2021.

3.2. Annual Growth Rate by Category

We calculated the annual growth rate of usage by each category, and we also listed the growth rate of the annual usage of all drugs as an overall trend to see how the usage of non-narcotic analgesics and PPIs varies compared with the overall trend of all drug usage. The result is illustrated in Figure 3. We could see drug usage increasing every year; only the non-selective NSAIDs decreased in 2020 and 2021. Figure 4 shows the trends in a line graph. We see that the annual growth rates of selective COX-2 and PPIs are greater than those the overall drug usage, the growth rate of acetaminophen is closer to the overall trend, and the growth rate of non-selective NSAIDs is lower than the overall growth rate.

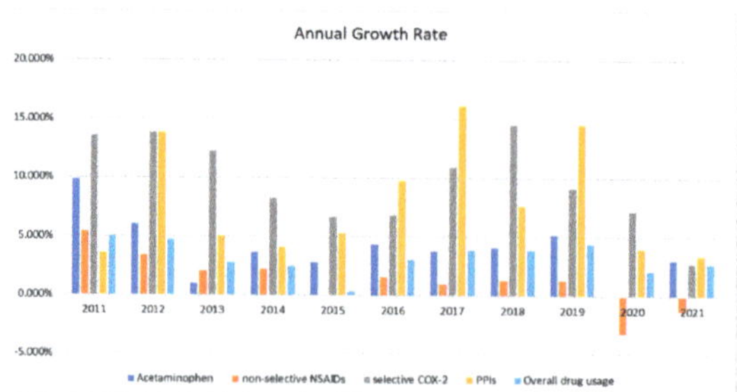

Figure 3. The annual growth rate of usage by each category and overall drug usage.

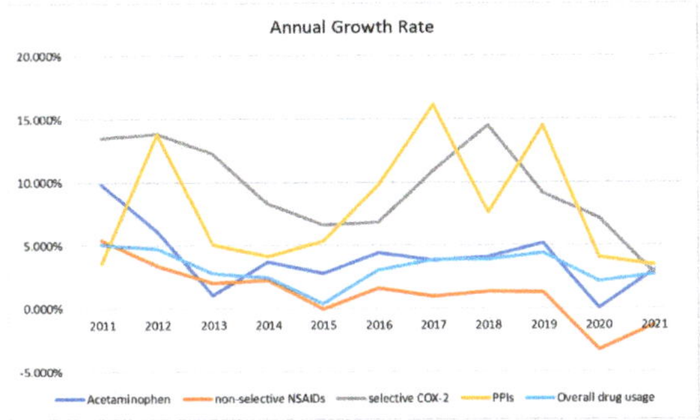

Figure 4. The annual growth rate of usage by each category and overall drug usage by line graph.

3.3. Annual Use Percentage of Each Drug

We calculated the use percentage of each drug in three categories of the traditional non-selective NSAIDs, the selective COX-2, and the PPIs. We found there are some drugs in a dominant position in the traditional non-selective NSAIDs. Thus, we illustrated the six most dominant non-selective NSAIDs, covering over 90% of the usage in 18 drugs, in Figure 5. Aspirin (acetylsalicylic acid) takes the largest proportion, with over 50% of annual usage. Diclofenac is the second largest at about 18% of usage. Flurbiprofen, mefenamic acid, ibuprofen, and meloxicam follow, with use rates of around 5% or less.

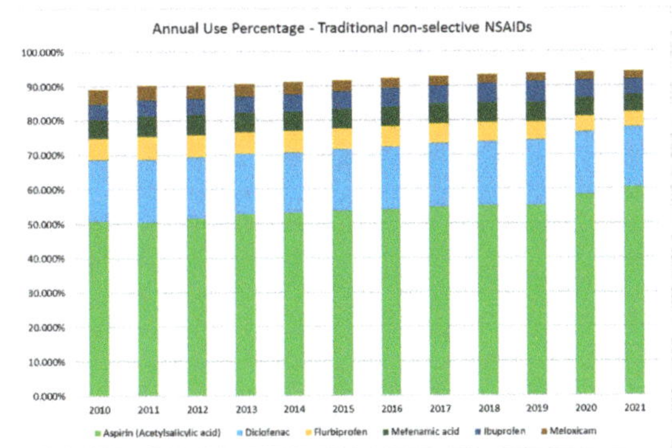

Figure 5. The annual use percentage of six traditional non-selective NSAIDs.

In Figure 6, there are three drugs in the selective COX-2 (coxibs). Celecoxib takes over 60% of annual selective COX-2 usage. Etoricoxib is the second largest, taking over 20% of usage. Nimesulide is the third, with usage of less than one percent in recent years.

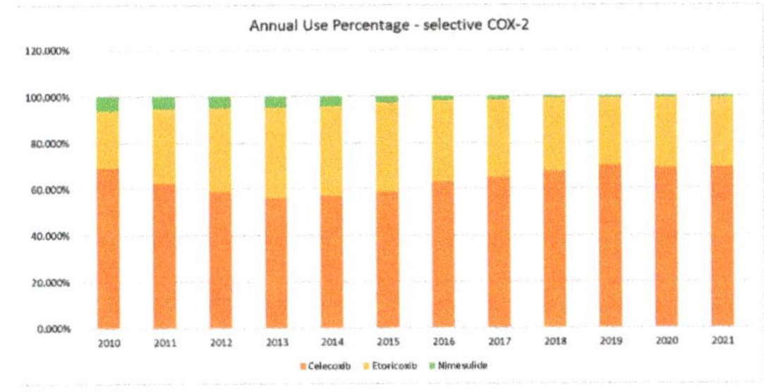

Figure 6. The annual use percentage of three selective COX-2 (coxibs).

Six PPIs are illustrated in Figure 7. Esomeprazole takes the largest proportion of the usage of PPIs. The usage of Omeprazole in Taiwan's NHI Annual Statistical Report has been aggregated with Esomeprazole since 2015. The proportion of Esomeprazole/Omeprazole accounts for around 40% of the annual PPIs usage. Lansoprazole is the second largest, taking over 20% of usage. Pantoprazole and rabeprazole are the third and the fourth, with under 20% use. It is remarkable that the usage of dexlansoprazole is increasing rapidly. The usage proportion of dexlansoprazole has taken over 10% of usage since 2016. There is only one drug, paracetamol (acetaminophen), in the paracetamol category, and so there is no need to figure the graph.

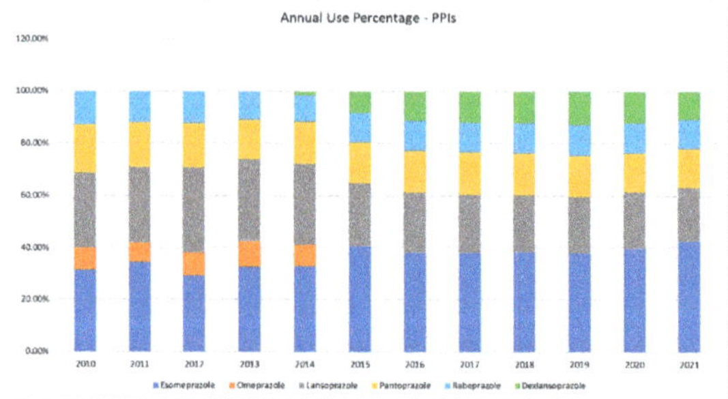

Figure 7. The annual use percentage of six PPIs.

4. Discussion

Figures 1 and 2 help us to answer the question of the prevalence and the trend of non-narcotic analgesics. The traditional non-selective NSAIDs take the largest proportion of usage over 60%, which is twice as many as the acetaminophen. The selective COX-2 (coxibs) takes the smallest proportion of the non-narcotic analgesics usage, but grows rapidly. In Figure 4, we can see the selective COX-2 has the highest annual growth rate. The annual growth rate of acetaminophen is also greater than the annual growth of overall drug usage. The traditional non-selective NSAIDs show a lower annual growth rate. Their use had been decreasing since 2020. The reason for this could be that the selective COX-2 are taking the place of the traditional non-selective NSAIDs. Although NSAIDs are widely used, acetaminophen still takes an important place—one-third of non-narcotic analgesic usage. It is noteworthy that the PPIs also have a high growth rate of annual usage over the growing trend of overall drug usage. This means the usage of PPIs did not decrease with the use of the selective COX-2 (coxibs). The result implies that it is necessary to further investigate the prescription pattern of the NSAIDs and PPIs, especially the combined use of selective COX-2 (coxibs) and PPIs.

Figures 5–7 help us to answer the question of what the most commonly prescribed non-narcotic analgesics and PPIs in Taiwan's NHI are. Aspirin (Acetylsalicylic acid), diclofenac, flurbiprofen, and mefenamic acid are the four most commonly prescribed traditional non-selective NSAIDs. They take over 80% of the annual usage. Aspirin takes the largest proportion, at over 50%. Celecoxib and etoricoxib are the two most commonly prescribed selective COX-2 (coxibs). They take 90% of the selective COX-2 usage. Esomeprazole/omeprazole, lansoprazole, and pantoprazole are the four most commonly prescribed PPIs. They take over 80% of the annual PPI usage.

This study gives us two important findings. The usage of PPIs is not decreasing with the use of the selective COX-2 (coxibs); on the contrary it is increasing rapidly. It is necessary to investigate the prescription pattern of the NSAIDs and PPIs, especially the combined use of selective COX-2 (coxibs) and PPIs. We identify the most commonly prescribed NSAIDs and PPIs in Taiwan's NHI. We can focus on these drugs to investigate the prescription pattern of the NSAIDs and PPIs using Taiwan National Health Insurance Research Database in future research.

Author Contributions: Writing—original draft, I.-W.L.; Validation, J.C.Y.; Writing—review & editing, I.-W.L.; Supervision, J.-S.C.; Project Administration, C.-N.L. All authors have read and agreed to the published version of the manuscript.

Funding: This research was supported by a grant from the Clinical Trial Center of Kaohsiung Medical University Hospital (grant no. H111095).

Institutional Review Board Statement: Not applicable.

Informed Consent Statement: Not applicable.

Data Availability Statement: The data presented in this study are openly available in Taiwan's NHI Annual Statistical Report (https://www.nhi.gov.tw/Content_List.aspx?n=5AA7CAFFF61CB16D&topn=5FE8C9FEAE863B46, accessed on 1 July 2023).

Acknowledgments: We thank the Center for Medical Informatics and Statistics of Kaohsiung Medical University for providing administrative support.

Conflicts of Interest: The authors declare no conflict of interest.

References

1. Sjøgren, P.; Elsner, F.; Kaasa, S. Non-opioid analgesics. In *Oxford Textbook of Palliative Medicine*; Cherny, N., Fallon, M., Kaasa, S., Portenoy, R.K., Currow, D.C., Eds.; Oxford University Press: Oxford, UK, 2015.
2. Meng, Q.; Zhang, Z.; Li, F.; Li, J.; Wang, N.; Guo, Z.; Wang, J.; Ye, X.; Li, Y. The prescription patterns and safety profiles of oral non-steroidal anti-inflammatory drugs in China: An 8-year real-life analysis. *Ann. Palliat. Med.* **2021**, *10*, 2224–2237. [CrossRef] [PubMed]
3. Zhang, X.; Donnan, P.T.; Bell, S.; Guthrie, B. Non-steroidal anti-inflammatory drug induced acute kidney injury in the community dwelling general population and people with chronic kidney disease: Systematic review and meta-analysis. *BMC Nephrol.* **2017**, *18*, 256. [CrossRef] [PubMed]
4. Lee, P.-C.; Wang, J.T.-H.; Chen, T.-Y.; Peng, C.-H. *Digital Health Care in Taiwan*; Springer Nature: Berlin, Germany, 2022.
5. Taiwan's NHI Drug Database. Available online: https://www.nhi.gov.tw/QueryN_New/QueryN/Query1 (accessed on 1 July 2023).
6. Taiwan's NHI Annual Statistical Report. Available online: https://www.nhi.gov.tw/Content_List.aspx?n=5AA7CAFFF61CB16D&topn=5FE8C9FEAE863B46 (accessed on 1 July 2023).
7. Bannwarth, B. Gastrointestinal safety of paracetamol: Is there any cause for concern? *Expert Opin. Drug Saf.* **2004**, *3*, 269–272. [CrossRef] [PubMed]
8. Tai, F.W.D.; McAlindon, M.E. Non-steroidal anti-inflammatory drugs and the gastrointestinal tract. *Clin. Med.* **2021**, *21*, 131. [CrossRef]
9. Bonnesen, K.; Schmidt, M. Recategorization of non-aspirin nonsteroidal anti-inflammatory drugs according to clinical relevance: Abandoning the traditional NSAID terminology. *Can. J. Cardiol.* **2021**, *37*, 1705–1707. [CrossRef] [PubMed]
10. Ghlichloo, I.; Gerriets, V. *Nonsteroidal Anti-Inflammatory Drugs (NSAIDs)*; StatPearls Publishing: Treasure Island, FL, USA, 2022.
11. Ahmed, A.; Clarke, J.O. *Proton Pump Inhibitors (PPI)*; StatPearls Publishing: Treasure Island, FL, USA, 2022.
12. Strand, D.S.; Kim, D.; Peura, D.A. 25 years of proton pump inhibitors: A comprehensive review. *Gut Liver* **2017**, *11*, 27. [CrossRef]

engineering proceedings

Proceeding Paper

Shanghai Smart Health Station Research Application in Fusion of Sports and Medicine Community Health Service †

Si-Jing Tu [1,2], Chen Jin [2], Xiao-He Wang [2] and Bi-Yan Wang [1,*]

1 School of Public Health and Management, Guangxi University of Chinese Medicine, 13 Wuhe Ave., Nanning 530200, China
2 School of Public Health, Hangzhou Normal University, 2318 Yuhangtang Rd., Hangzhou 311121, China; chenjin2020@stu.hznu.edu.cn (C.J.); xhewang@163.com (X.-H.W.)
* Correspondence: forgive_the_sin@foxmail.com
† Presented at the IEEE 5th Eurasia Conference on Biomedical Engineering, Healthcare and Sustainability, Tainan, Taiwan, 2–4 June 2023.

Abstract: The "fusion of sports and medicine (FSM)" is the key to promoting national fitness and health in deep fusion. However, there are disadvantages such as isolation of information networks and separation of health data in the progress of FSM. A smart health station (SHS) is a new health service mode of FSM to provide a place for community residents for health self-assessment and self-management. SHS forms an integrated health information management system based on the Internet of Things, mobile internet, and big data. It integrates community health services, community physical fitness surveillance, and other resources. In order to put forward health management and improve people's ability to manage their health, Shanghai has promoted the construction of SHS. The construction of SHS in Shanghai provides strong support and guarantees disease prevention and treatment in the community. "Early prevention, early diagnosis, and early treatment" are also achieved to eliminate the disease. In this study, we introduce the background, service, character, and management of SHS construction in Shanghai, and the application of health service data in health education, chronic disease management, and intervention guidance.

Keywords: health station; health service; fusion of sports and medicine

Citation: Tu, S.-J.; Jin, C.; Wang, X.-H.; Wang, B.-Y. Shanghai Smart Health Station Research Application in Fusion of Sports and Medicine Community Health Service. *Eng. Proc.* **2023**, *55*, 8. https://doi.org/10.3390/engproc2023055008

Academic Editors: Teen-Hang Meen, Kuei-Shu Hsu and Cheng-Fu Yang

Published: 27 November 2023

1. Introduction

To promote the development of a "fusion of sports and medicine (FSM)" and the deep integration of public fitness and public health, and better meet the diverse needs of people for fitness and health, the Shanghai Smart Health Station (SHS), an intelligent service mode with Shanghai regional characteristics, was launched [1]. As a practical project of the Shanghai government, the first 85 SHSs in Shanghai opened to the public in 2019, and 219 SHSs were built and used in every community and sub-districts by 2020. SHS (also previously named the health hut) has become an important place for residents to obtain health management resources, and plays an extremely important role in chronic disease management services. With communities and sub-districts as the main body and the support of the Shanghai "Healthcare Cloud" application, SHS realizes the self-examination, self-assessment, and self-management of residents' health and provides community residents with fitness testing, basic health testing, scientific fitness guidance, chronic disease sports intervention, sports rehabilitation training, health knowledge popularization, and other "one-stop" sports healthcare services [2]. Residents can set up a personal health account in "Healthcare Cloud" to master their health conditions in real time.

2. Function of SHS

The highlight of SHS includes a unique identity by integrating health indicators and shared health data (within health facilities). The health data are fully used to increase the residents' self-perception and experience of health management. With population characteristics, disease characteristics, behavior patterns, and other dimensions as the core, SHS monitors residents' health behavior, early disease, and risk factors through self-examination and self-evaluation using the internet information method, and integrates physical examination. Under the guidance of family-doctor teams or professionally trained volunteers, the risk factors affecting health status are assessed, and residents in need are guided for a timely connection with family doctors, specialist doctors, and social sports instructors. This improves the identification, filing, management, and control rates of key diseases through the development of individual exercise programs and the physical quality of chronic diseases.

2.1. *Self-Examination and Self-Evaluation*

SHS contains various health data collection devices for residents to perform health self-examination and self-assessment by wearable devices, sports equipment, and self-rating scales for basic physical conditions, physical monitoring, and lifestyle. By docking with the "Healthcare Cloud" application, residents' health data are uploaded and analyzed in real time. Health records can be dynamically updated to form a health database of Shanghai and establish a public health management system integrated with health management and medical treatment. Through the self-examination and evaluation function, SHS managed to realize the primary screening of residents' health conditions, promote the risk warning of key diseases, and move forward the community health management threshold.

2.2. *Risk Self-Assessment*

SHS plays an extremely important role in the analysis of health risk factors and the initial screening of diseases. According to the characteristics of residents such as gender, age, health status, and lifestyle, and under the guidance of professional medical staff, targeted health scales including TCM healthcare and chronic disease risk are selected for evaluation in SHS. It mainly evaluates their health status, factors affecting health hazards, and physical functions such as TCM constitution identification, normal constitution or biased constitution, and the degree of cognitive impairment of the elderly. Through self-assessment, SHS helps residents scientifically understand health risk factors, correctly accept health intervention measures, and actively participate in health self-management. This is a "sports + medical" characteristic health service.

2.3. *Health Guidance*

The corresponding health assessment report is issued after receiving the health assessment at the SHS. If the unhealthy lifestyle causes changes in health status and does not develop into early stages of diseases, SHS distributes health education materials such as health education prescriptions on exercise, diet, psychology, and traditional Chinese medicine. Residents need to follow the advice from SHS to establish a healthy lifestyle, change unhealthy habits harmful to health, and reduce the harm of health risk factors and the occurrence of diseases. In case of abnormal health conditions and early symptoms of the disease during initial screening, SHS guides residents to professional institutions for disease screening or treatment and provides community health service information.

2.4. *Service Process*

Residents only need to present their ID cards or social security cards to complete the intelligent identification, automatically create or call the residents' health account, and can independently choose to obtain 22 self-help health and physical fitness tests and 15 health scale self-assessment services, and can directly share the test data to the residents' health files in real time through the "Health Cloud" app to achieve family doctor contracting, chronic disease management, appointment registration, and many other "through train" services. Based on the combination of online and offline methods, the Smart Health Station provides health prescriptions and exercise advice for different groups of people based on self-help physical tests, links medical resources with residents' health needs, and provides targeted health guidance and advice, while guiding residents to refer to family doctors and contracting services or other professional institutions based on the results of health self-examinations and self-assessments. This will strengthen community residents' active participation in their own health management, enrich ways for community residents to carry out self-management and gradually improve their health literacy. The service flow of the Smart Health Station is outlined in Figure 1.

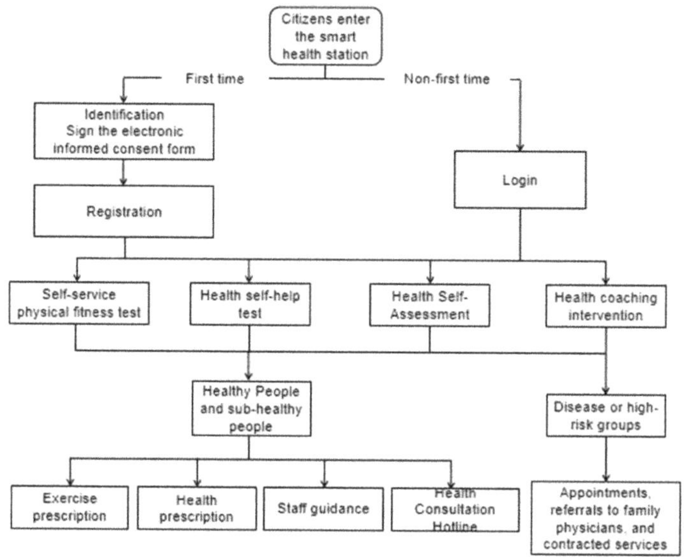

Figure 1. The service flow of the Smart Health Station.

3. Data Application

Supported by the "healthcare cloud" information system, personal health information is uploaded to the health accounts in the healthcare cloud account after SHS service [3]. The changing trend of health information such as blood pressure and blood sugar levels of patients with chronic disease is recorded and analyzed in the information system. With the updated information, family doctors can adjust health management programs, daily monitoring and management, health consultation, follow-up, and other services. Social sports instructors can improve residents' fitness programs according to changes in exercise prescriptions from family doctors.

3.1. Health Monitoring

In combination with their health requirements, residents visit the nearest SHS, perform smart identity identification with their ID card or social security card, automatically create or connect to their health accounts, and independently choose 11 items of the self-service health testing, 11 items of the self-service physical testing, and 15 items of the health scale self-evaluation services. The intelligent equipment of the SHS provides residents with a health self-examination service, completing the update of personal electronic health data information and long-term and continuous health condition indicator monitoring. Under the prompt of intelligent analysis of big data in the information system, it initially screens out high-risk groups such as chronic diseases and common diseases. The SHS effectively helps patients achieve the self-management of chronic diseases and provides diabetes monitoring and management services, including diagnosis reports, examination reports, follow-up reports, blood glucose monitoring records, and closed-loop management of diabetes.

3.2. Health Analysis

Relying on SHS to collect residents' health information, the normal range, different ages, gender, and health index to the health of the residents are provided through long-term, continuous monitoring and data analysis. Scientific monitoring achieves long-term continuous personal health indicators to monitor and analysis of risk factors that affect health. It also screens for the beginning of early disease symptoms and gives the family doctor the role of an "assistant". High blood pressure, diabetes, chronic diseases, and common diseases of high-risk groups are updated with the information of the health records. The family doctor signs for the residents' health files after finding residents' health abnormalities for timely medical treatment to reduce the disease damage to health.

3.3. Personalized Intervention

Residents choose different health self-test programs and scale assessment services. Compared with the normal range, a preliminary assessment of personal health is performed to understand the health index or disease risk score threshold of obesity, blood pressure, blood sugar, and cognitive impairment. The self-test system issues specific scientific health prescriptions for dietary nutrition and emotional nursing to help residents manage their health under the guidance of medical staff or community volunteers. If a resident's health index or disease risk score reaches the risk value and is listed as a high-risk group, the family doctor team of the health service center of the community or sub-district provides professional help and guidance to timely regulate the diagnosis and treatment in medical institutions to slow down the malignant development of the disease and the occurrence of complications. If diagnosed with chronic diseases such as hypertension and diabetes, they are brought to standardized management by family doctors to take comprehensive prevention and treatment measures provided by health management services. Such a "family doctor + social sports instructor + exercise and fitness guidance" mode provides residents with "chronic disease treatment, preventive treatment of disease" health guidance services.

Author Contributions: Conceptualization, S.-J.T. and B.-Y.W.; writing—review and editing, C.J.; supervision, X.-H.W. All authors have read and agreed to the published version of the manuscript.

Funding: This research was funded by the Basic Scientific Research Competence Improvement Project of Young and Middle-Aged Teachers at Guangxi universities (2023KY0279), and the Youth Fund Project of the Guangxi University of Chinese Medicine (2022QN001).

Institutional Review Board Statement: Not applicable.

Informed Consent Statement: Not applicable.

Data Availability Statement: Data are contained within the article.

Conflicts of Interest: The authors declare no conflict of interest.

References

1. Tu, S.J.; Jin, C.; Chen, B.T.; Xu, A.Y.; Luo, C.; Wang, X.H. Study on the Fusion of Sports and Medicine in China from 2012 to 2021: A Bibliometric Analysis via CiteSpace. *Front. Public Health* **2022**, *10*, 939557. [CrossRef] [PubMed]
2. Chen, C.; Gong, X.; Wang, X.; Cao, C.; Sun, X.; Zhang, Y.; Jing, L.; Chen, N.; Liu, Q.; Wang, H.; et al. Evaluation of the implementation and effect of the healthcare cloud information platform for diabetes self-management: A case study in Shanghai. *Int. J. Health Plan. Manag.* **2019**, *34*, 986–997. [CrossRef] [PubMed]
3. Poorejbari, S.; Mansoor, W. Smart healthcare systems on improving the efficiency of healthcare services. In Proceedings of the 2019 2nd International Conference on Signal Processing and Information Security (ICSPIS), Dubai, United Arab Emirates, 30–31 October 2019; pp. 1–4.

engineering proceedings

MDPI

Proceeding Paper

Defect Detection by Analyzing Thermal Infrared Images Covered with Shadows with a Hybrid Approach Driven by Local and Global Intensity Fitting Energy †

Yishuo Huang *, Chia-Chien Hung and Chih-Hung Chiang

Department of Civil and Construction Engineering, Chaoyang University of Technology, Taichung 413310, Taiwan; lovedog50107@gmail.com (C.-C.H.); chiangc@cyut.edu.tw (C.-H.C.)
* Correspondence: yishuo@cyut.edu.tw
† Presented at the IEEE 5th Eurasia Conference on Biomedical Engineering, Healthcare and Sustainability, Tainan, Taiwan, 2–4 June 2023.

Abstract: Defect detection using thermal infrared images is used in nondestructive evaluation and testing because it is easy to use. Thermal infrared images recorded the surface temperatures of the target with a thermal infrared camera. Image segmentation is a technique to group those pixels with similar surface temperatures to form thermal patterns. Defects can be identified by the segmented patterns having different surface temperatures in their neighborhoods. In this study, a hybrid approach combining fitting energy is proposed for describing the contamination illustrated in the recorded surface temperatures and regional constants averaging the surface temperatures of the segmented regions. The numerical implementation is completed by applying the level set functions on an iteration scheme. The initial level sets evolve till a convergence can be reached. The processed results demonstrate that the hybrid approach can be used for defect detection.

Keywords: image segmentation; fitting energy; regional constant; level sets

Citation: Huang, Y.; Hung, C.-C.; Chiang, C.-H. Defect Detection by Analyzing Thermal Infrared Images Covered with Shadows with a Hybrid Approach Driven by Local and Global Intensity Fitting Energy. *Eng. Proc.* **2023**, *55*, 9. https://doi.org/10.3390/engproc2023055009

Academic Editors: Teen-Hang Meen, Kuei-Shu Hsu and Cheng-Fu Yang

Published: 28 November 2023

1. Introduction

Defect detection is an essential issue in nondestructive testing (NTD). Thermal infrared images are widely employed by identifying the recorded surface temperatures presented in the given thermal infrared images. Defects are usually identified by finding the differences in surface temperatures. However, the recorded surface temperatures presented in the thermal infrared images can be contaminated by sunlight refraction and environmental deficiencies like shadows. Those contaminations make the pixels of the thermal infrared image contain not only the surface temperature but also extra information related to those contaminations.

Removing the intensity inhomogeneity presented in the given thermal infrared images is an important issue in analyzing thermal infrared images. Huang et al. modeled the shadow effects in a multiplicative way, and the shadow effects can be approximated by implementing level sets and iteration schemes [1]. Traditionally, Li et al. introduced the local fitting energy to model the intensity inhomogeneity in their segmentation algorithm [2]. Zhang et al. employed regional standards and regional constants to limit the segmentation regions [3]. Zhang et al. proposed a hybrid approach combining the fitting energy and regional constants to segment the given images [4].

Image segmentation grouping those pixels with similar surface temperatures is employed to identify the defects by finding the differences in the surface temperature. The authors used their algorithms to approximate the intensity inhomogeneity and simultaneously segment the given images. Huang et al. successfully used Zhang's regional standard deviations and regional constants to locate those potential defects [1]. They employed a hybrid approach combining the local fitting energy and regional parameters (like standard deviations and regional constants) to approximate intensity inhomogeneity. The processed results demonstrated that the proposed hybrid approach can be used for defect detection. The paper is organized as follows. Section 2 introduces the hybrid system, Section 3 presents a series of thermal infrared images of the side wall of the Administration Building, Chaoyang University of Technology, Taichung, Taiwan, as the test target to verify the robustness of the proposed approach, and Section 4 provides the discussions and conclusions.

2. Hybrid System

2.1. Principal Component Analysis

Principal component analysis (PCA) is widely used to analyze a series of images by projecting the original data onto a low-dimension space. The projected data can inherit the major properties from the original data. In doing so, the template image extracted from the projected data can be analyzed instead of analyzing each given image. Suppose a given data matrix M can be decomposed as a low-dimension space L and a sparse space S. Then, M can be given as follows.

$$\mathrm{M} = \mathrm{L} + \mathrm{S} \tag{1}$$

Equation (1) needs to satisfy the given condition as follows.

$$\min\|\mathrm{M} - \mathrm{L}\| \ \mathrm{st.rank}(\mathrm{L}) \leq \mathrm{k} \tag{2}$$

where $\mathrm{k} < \mathrm{rank}(\mathrm{M})$. The singular value decomposition (SVD) is applied to find the optimal approximation.

2.2. Hybrid Systems

The proposed hybrid system contains the local fitting energy and regional parameters (including regional standard deviations and regional constants). The local fitting energy assumes that the intensity inhomogeneity illustrated in the given image can be approximated by the local fitting energy [2]. Hence, the local fitting energy can be given as follows [2].

$$E_1(\Phi, F, C) = \sum_{i=1}^{4} \iint K_\sigma(x-y)|I(y) - f_i(x)|^2 M_i(y)dydx \tag{3}$$

where $K_\sigma(x-y)$ is Gaussian filter with the parameter σ, I is the given image, f_i is the local fitting energy and $\mathrm{M_i}$ are the combinations of two-level set functions. $\mathrm{M_i}$ is presented as follows.

$$\begin{aligned} M_1(\phi_1, \phi_2) &= H(\phi_1)H(\phi_2) \\ M_2(\phi_1, \phi_2) &= H(\phi_1)(1 - H(\phi_2)) \\ M_3(\phi_1, \phi_2) &= (1 - H(\phi_1))H(\phi_2) \\ M_3(\phi_1, \phi_2) &= (1 - H(\phi_1))(1 - H(\phi_2)) \end{aligned} \tag{4}$$

where ϕ_1, ϕ_2 are level set functions, and H indicates the Heaviside function, shown as follows [5].

$$H_\varepsilon(x) = \frac{1}{2}\left[1 + \frac{2}{\pi} tan^{-1}\left(\frac{x}{\varepsilon}\right)\right] \tag{5}$$

where $\varepsilon > 0$.

The image model containing the intensity inhomogeneity is incorporated in a multiplicative way and regional constants. The image model is illustrated as follows [3].

$$\mathrm{I} = \sum_{i=1}^{4} \mathrm{BC_i} \tag{6}$$

where B is the intensity inhomogeneity, and C_i is the regional constant. The regional constants are the average values of the segmented regions. Zhang et al. introduced the standard deviations of the segmented regions into the segmentation algorithm, and the algorithm can be presented as follows [3].

$$E_2(\sigma, c, B) = \int \left(\sum_{i=1}^{4} \int K(y,x) \left(\log \sigma_i + \frac{(I - BC_i)^2}{2\sigma_i^2} \right) M_i(\Phi) dx \right) dy \tag{7}$$

where σ_i are the regional standard deviations of the segmented regions. The hybrid system is the linear combination of Equations (3) and (7) and is given as follows.

$$\begin{aligned} E = \omega E_1 + (1-\omega) E_2 & \\ & +\nu \int |\nabla H(\phi_1)| dx \\ & +\nu \int |\nabla H(\phi_2)| dx + \mu \int \tfrac{1}{2}(|\nabla \phi_1| - 1)^2 dx \\ & +\mu \int \tfrac{1}{2}(|\nabla \phi_2| - 1)^2 dx \end{aligned} \tag{8}$$

where ω, ν, and μ are positive constants.

Then, the local fitting energy can be obtained and given as follows.

$$f_i(x) = \frac{K_\sigma(x-y) \otimes (IM_i(\Phi))}{K_\sigma(x-y) \otimes M_i(\Phi)} \tag{9}$$

where $\otimes$ is the convolution operator. Similarly, the regional constants, regional standard deviations, and the intensity inhomogeneity illustrated in the image can be given as follows.

$$C_i = \frac{\int K_\sigma(y,x) \otimes (IBM_i(\Phi)) dxdy}{\int K_\sigma(y,x) \otimes (B^2 M_i(\Phi)) dxdy} \tag{10}$$

$$\sigma_i^2 = \frac{\int K(y,x) \otimes \left((I - BC_i)^2 M_i(\Phi) \right) dxdy}{\int K(y,x) \otimes M_i(\Phi) dxdy} \tag{11}$$

$$B = \frac{\sum_{i=1}^{4} \int K_\sigma(y,x) \otimes \left(IM_i(\Phi) \frac{C_i}{\sigma_i^2} \right) dxdy}{\sum_{i=1}^{4} \int K(y,x) \otimes \left(M_i(\Phi) \frac{C_i^2}{\sigma_i^2} \right) dxdy} \tag{12}$$

The iteration scheme is applied such that the level set functions can evolve till the convergence is reached. Firstly, several parameters are defined as follows.

$$\begin{aligned} e_1 &= \iint K_\sigma(x-y) |I(y) - f_1(x)|^2 dydx \\ e_2 &= \iint K_\sigma(x-y) |I(y) - f_2(x)|^2 dydx \\ e_3 &= \iint K_\sigma(x-y) |I(y) - f_3(x)|^2 dydx \\ e_4 &= \iint K_\sigma(x-y) |I(y) - f_4(x)|^2 dydx \\ F_1 &= \int K(y,x) \left(\log \sigma_1 + \tfrac{(I - B(x) C_1)^2}{2\sigma_1^2} \right) dxdy \\ F_2 &= \int K(y,x) \left(\log \sigma_2 + \tfrac{(I - B(x) C_2)^2}{2\sigma_2^2} \right) dxdy \\ F_3 &= \int K(y,x) \left(\log \sigma_3 + \tfrac{(I - B(x) C_3)^2}{2\sigma_3^2} \right) dxdy \\ F_4 &= \int K(y,x) \left(\log \sigma_4 + \tfrac{(I - B(x) C_4)^2}{2\sigma_4^2} \right) dxdy \end{aligned} \tag{13}$$

The level set functions are rewritten for the time parameter and can be presented as follows.

$$\frac{\partial\phi_1}{\partial t} = \omega\delta(\phi_1)H(\phi_2)(e_3 - e_1) + \omega\delta(\phi_1)(1 - H(\phi_2))(e_4 - e_2) + (1-\omega)\delta(\phi_1)H(\phi_2)(F_3 - F_1) + (1-\omega)\delta(\phi_1)(1 - H(\phi_2))(F_4 - F_2) + \nu\delta(\phi_1)div\left(\frac{\nabla\phi_1}{|\nabla\phi_1|}\right) + \mu\left(\nabla^2\phi_1 - div\left(\frac{\nabla\phi_1}{|\nabla\phi_1|}\right)\right) \quad (14)$$

$$\frac{\partial\phi_2}{\partial t} = \omega\delta(\phi_2)H(\phi_1)(e_2 - e_1) + \omega\delta(\phi_2)(1 - H(\phi_1))(e_4 - e_3) + (1-\omega)\delta(\phi_2)H(\phi_1)(F_2 - F_1) + (1-\omega)\delta(\phi_2)(1 - H(\phi_1))(F_4 - F_3) + \nu\delta(\phi_2)div\left(\frac{\nabla\phi_2}{|\nabla\phi_2|}\right) + \mu\left(\nabla^2\phi_2 - div\left(\frac{\nabla\phi_2}{|\nabla\phi_2|}\right)\right) \quad (15)$$

In doing so, the intensity inhomogeneity can be approximated. Furthermore, the image can be calibrated by removing the intensity inhomogeneity.

3. Materials for Evaluation

A series of thermal infrared images were taken on 30 January 2019. Those thermal infrared images were recorded by NEC InfReC R500Pro, and the image sizes of each recorded image are 480 by 640 pixels. The accuracy of the recorded surface temperature reached 0.01 °C. The test target is the side wall of the Administration Building, Chaoyang University of Technology, Taichung, Taiwan. There were 80 frames recorded by NEC InfReC R500Pro, and the highest surface temperature and the average surface temperature in each recorded thermal infrared image are presented in Figure 1. Figure 2 shows the images recorded by NEC InfReC R500Pro and a digital camera installed on the thermal infrared camera.

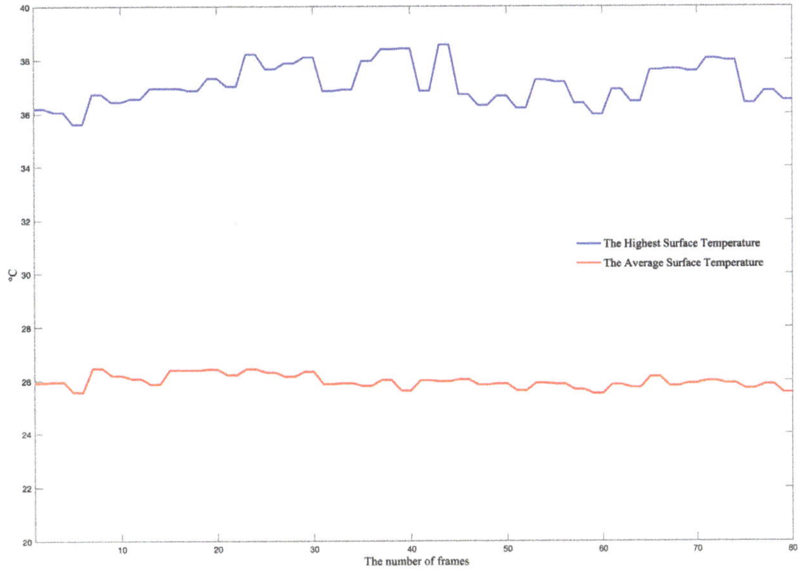

Figure 1. Surface temperature ranges of recorded thermal infrared images.

Then, PCA was applied to the recorded thermal infrared images, and the first image was extracted from the low-dimension space. The extracted image is illustrated in Figure 3. It is obvious that the information on the surface temperatures was lost because the projected data onto a low-dimension space could not keep the temperature information. However, the thermal patterns remained.

(**a**) (**b**)

Figure 2. (**a**). Thermal infrared image recorded by NEC InfReC R500Pro. (**b**). The corresponding image was recorded by a digital camera installed on NEC InfReC R500Pro.

Figure 3. The first image was extracted from the low-dimension space generated by using PCA.

The proposed hybrid system with settings $\omega = 0.1$, $\mu = 0.00001 \times 256 \times 256$, $\nu = 1$, and $\Delta t = 0.1$ was used on the results by employing PCA. The approximated intensity inhomogeneity is presented in Figure 4. The segmented results are shown in Figure 5. The calibrated image with removing intensity inhomogeneity is illustrated in Figure 6. The convergence is given in Figure 7.

Figure 4. An optimal approximation of intensity inhomogeneity after 1000 iterations is presented.

Figure 5. Segmented results by employing the hybrid system are presented.

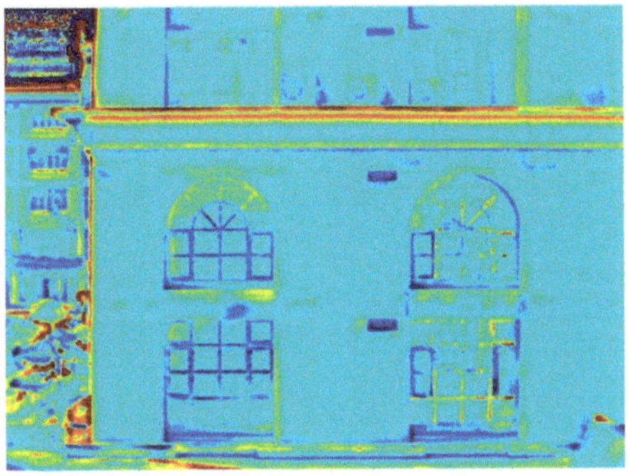

Figure 6. A calibrated image with removing the intensity inhomogeneity is shown.

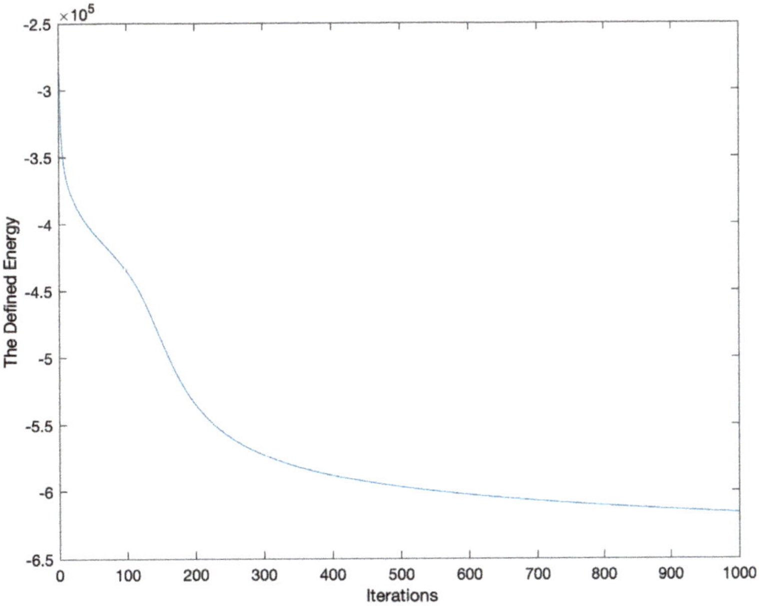

Figure 7. Convergence after 1000 iterations is illustrated.

4. Discussions and Conclusions

The proposed hybrid system can segment the given thermal infrared images such that the differences in the recorded surface temperatures can be identified. From Figure 5, it is evident that those segmented regions colored in yellow can be potential defects. Different NDT techniques can be employed to verify the results.

The proposed hybrid system employs the Gaussian function to assume that the intensity inhomogeneity is illustrated at location x and in its neighborhoods. The Gaussian function shows that for those neighborhoods, their influences decrease while their locations are far away from location x. With the specified σ, the Gaussian function with different window sizes is applied in the hybrid system. The window sizes are 5×5, 15×15, 25×25, and 35×35. The performance is presented in Table 1. The processing times were calculated by taking the averages after running the same program ten times. The processing time was increased with the window sizes. The estimated intensity inhomogeneity is presented in Figure 8a–d.

Table 1. Performances of Employing Different Window sizes.

Window Sizes	Processing Time (s)
5×5	256.37
15×15	334.43
25×25	457.79
35×35	610.10

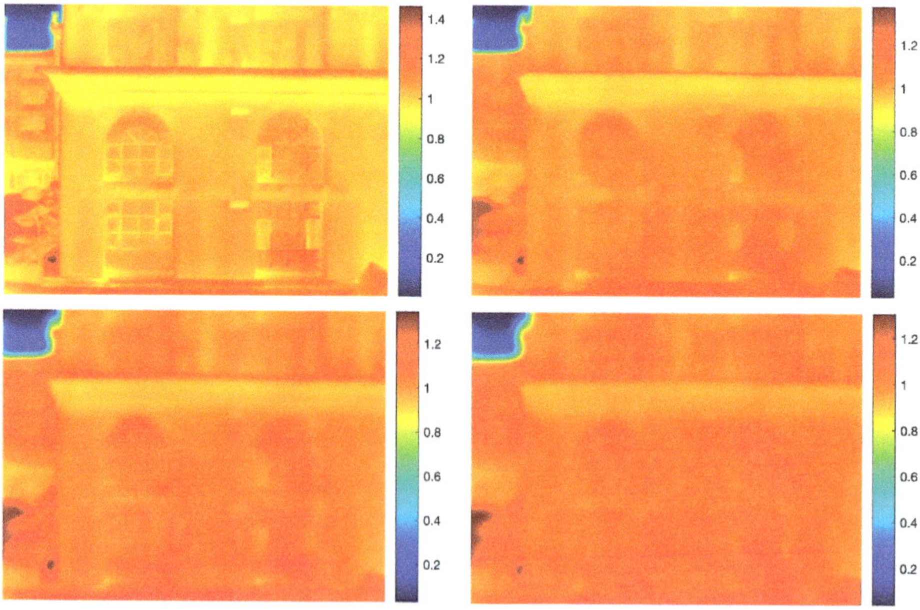

Figure 8. Estimated intensity inhomogeneity by employing 5×5 window sizes.

The intensity inhomogeneity does exist in the thermal infrared images, and its ranges are in the [0, 1.4]. The small window sizes, like 5×5, seem to have bigger ranges than those larger window sizes because the small window sizes give the influences from the neighborhoods such that the ranges are increased. Shadows illustrated in the thermal infrared images have intensity inhomogeneity, and the effects can be approximated. Furthermore, the intensity inhomogeneity can be removed. The estimated intensity inhomogeneity enhances the original image, while the intensity inhomogeneity is less than 1.0. Otherwise, the image is smoothed while the intensity inhomogeneity is larger than 1.0. As for that intensity inhomogeneity equal to 1.0, nothing can be done on the images.

Eventually, the conclusions are given as follows.

(1) Image segmentation can be employed to find potential defects by segmenting the given thermal infrared images.
(2) PCA can project the given data onto a low-dimension space such that the properties of the given data can be inherited from the images extracted from the low-dimension space.
(3) Intensity inhomogeneity does exist, and it needs to be estimated such that the thermal infrared images can be calibrated.

The proposed hybrid system seems to work well in analyzing thermal infrared images. Different methods to remove intensity inhomogeneity will be compared in the future.

Author Contributions: Conceptualizatopn, Y.H. and C.-H.C.; methodology, Y.H. and C.-C.H.; software Y.H.; formal analysis, Y.H. and C.-C.H.; writing original draft preparation, Y.H. All authors have read and agreed to the published version of the manuscript.

Funding: This research received no external funding.

Institutional Review Board Statement: Not applicable.

Informed Consent Statement: Not applicable.

Data Availability Statement: Data sharing is not applicable to this article.

Conflicts of Interest: The authors declare no conflict of interest.

References

1. Huang, Y.; Shih, P.; Hsu, K.T.; Chiang, C.H. To identify the defects illustrated on building facades by employing infrared thermography under shadow. *NDT E Int.* **2020**, *111*, 102240. [CrossRef]
2. Li, C.; Kao, C.Y.; Gore, J.C.; Ding, Z. Minimization of Region-Scalable Fitting Energy for Image Segmentation. *IEEE Trans. Image Process.* **2008**, *17*, 1940–1949. [PubMed]
3. Zhang, K.; Zhang, L.; Lam, K.M.; Zhang, D. A Level Set Approach to Image Segmentation with Intensity Inhomogeneity. *IEEE Trans. Cybern.* **2016**, *46*, 546–557. [CrossRef]
4. Zhang, W.; Wang, X.; You, W.; Chen, J.; Dai, P.; Zhang, P. RESLS: Region and edge synergetic level set framework for image segmentation. *IEEE Trans. Image Process.* **2019**, *29*, 57–71. [CrossRef]
5. Vese, L.A.; Chan, T.F. A Multiphase Level Set Framework for Image Segmentation Using the Mumford and Shah Model. *Int. J. Comput. Vis.* **2002**, *50*, 271–293. [CrossRef]

engineering proceedings

MDPI

Proceeding Paper

Willingness, Proficiency, or Supports? Challenges in Implementing Content and Language Integrated Learning for Taiwan K-12 Teachers †

Shing-Yun Jung, Shih-Wen Su, Xiaojing Yu, Shyan-Ming Yuan * and Chuen-Tsai Sun

Department of Computer Science, National Yang Ming Chiao Tung University, Hsinchu 30010, Taiwan; syjung@cs.nycu.edu.tw (S.-Y.J.); alvin.cs00@nycu.edu.tw (S.-W.S.); jing.cs09@nycu.edu.tw (X.Y.); ctsun@cs.nctu.edu.tw (C.-T.S.)

* Correspondence: smyuan@nycu.edu.tw

† Presented at the IEEE 5th Eurasia Conference on Biomedical Engineering, Healthcare and Sustainability, Tainan, Taiwan, 2–4 June 2023.

Abstract: After the Taiwanese government launched the Blueprint for Developing Taiwan into a Bilingual Nation by 2030, the Implementation Project of Bilingual Instruction in several domains of primary and junior high school education was promoted by the Taiwan Ministry of Education. Content and language integrated learning (CLIL) is a dual-objective strategy in which students simultaneously acquire language skills and subject knowledge. CLIL has been widely implemented and proven to be successful in European countries. This strategy will become the primary method of bilingual education for instructors in grades K-12. Other non-European countries, such as Latin America and Indonesia, however, found difficulties applying CLIL in class. The issues need to be identified to encourage researchers and practitioners to find solutions. The purpose of this paper is to identify the barriers to implementing CLIL from the perspective of K-12 teachers. We surveyed 102 K-12 teachers in Keelung using a questionnaire. The findings indicate that (1) 49.0% of teachers feel concerned if they are required to teach half of their content in English, (2) 64.8% of teachers are willing to improve their English proficiency in their spare time for implementing CLIL in class, and (3) only 36.3% of teachers are willing to adopt CLIL in class, while 50.0% of teachers are willing to implement CLIL if a teaching assistant system supports them with English course materials. Teaching assistant systems that provide English course materials might influence teachers' willingness to implement CLIL in class.

Keywords: content language integrated learning; bilingual education; K-12; technology enhanced

Citation: Jung, S.-Y.; Su, S.-W.; Yu, X.; Yuan, S.-M.; Sun, C.-T. Willingness, Proficiency, or Supports? Challenges in Implementing Content and Language Integrated Learning for Taiwan K-12 Teachers. *Eng. Proc.* **2023**, *55*, 10. https://doi.org/10.3390/engproc2023055010

Academic Editors: Teen-Hang Meen, Kuei-Shu Hsu and Cheng-Fu Yang

Published: 28 November 2023

1. Introduction

Taiwan has advanced English education from the fifth grade to the third grade in elementary schools since 2005. While this advancement has been in place for nearly two decades, its effectiveness has been less than anticipated [1]. The reasons for the lack of effectiveness may be because (1) Taiwan's English education is in an examination-led teaching atmosphere, and English for students is merely a subject at school rather than a useful tool after school, and (2) the non-English speaking environment in Taiwan prevents students from using English in their daily lives, reducing students' motivation to learn English.

In the past, it has been thought that learning a second or foreign language is best done in a way that emphasizes meaning rather than accuracy. Such an environment makes it hard to reach the target level of language, but it can be carried out through the creation of situations that encourage language practice and use in a meaningful way [2]. Perspectives from research in cognitive psychology, language learning theories, and second language acquisition have helped to support the development of content and language integrated learning programs that allow students to learn a new language naturally through meaningful communication. Content and language integrated learning (CLIL) is a dual-focused educational approach where students acquire language skills and subject knowledge at the same time [3]. The principal property of CLIL is that students "learn to use language and use language to learn" [4]. Chamot et al. [5] suggested that the CLIL environment enables non-English speaking students to learn English as a lingua franca and utilize English in an academic context within a globalized international setting. In recent years, scholars have suggested that the CLIL approach is about the aforementioned two objectives of learning and about learning the culture implied behind the language as well. Therefore, the CLIL approach has been presently developed into a "language, subject literacy and learning" teaching method [6].

In 2018, the Taiwan government launched the Blueprint for Developing Taiwan into a Bilingual Nation by 2030 to cultivate bilingual talents to increase Taiwan's global competitiveness. The Implementation Project of Bilingual Instruction in Some Domains of Primary and Junior High School Education was soon promoted by the Taiwan Ministry of Education [7]. Since CLIL has been widely implemented and proven to be successful in European countries [8,9], most Taiwanese primary and junior high schools have adopted CLIL to implement bilingual classes. CLIL tends to be the primary method of bilingual education for instructors in grades K-12. However, other non-European countries, such as Latin America [10] and Indonesia [11], have found difficulties applying CLIL in class.

Issues need to be identified to encourage researchers and practitioners to contribute more solutions. Thus, the purpose of this paper is to identify the barriers to implementing CLIL from the perspective of K-12 teachers. We surveyed 102 K-12 teachers in Keelung using a questionnaire. The results imply that the teaching assistant systems that provide English course materials might influence teachers' willingness to implement CLIL in class.

2. Related Work

2.1. CLIL in Indonesia

Isnaini et al. [11] studied the challenges in implementing CLIL for Indonesian primary school students from the head teacher's viewpoint. They suggested that the main challenges emerging during CLIL implementation include teachers' lack of knowledge, teaching experience, and sufficient preparation.

2.2. CLIL in Ecuador

On the other hand, Vega et al. [10] attempted to compare language learning outcomes between CLIL and English for specific purposes programs for Ecuadorean college students. Their study revealed no significant increase in language proficiency or difference in achievement between the two programs. They also pointed out that the students' initial language levels influence the learning effectiveness of the CLIL program. Due to the historical and cultural differences between Taiwan and other countries, the issues in implementing CLIL in Taiwan should be clearly identified. The bilingual education plan must be tailored to local needs to achieve tangible outcomes.

3. Material and Methods

We designed an online questionnaire containing potential challenges in implementing CLIL for K-12 teachers in Taiwan from three aspects: teachers' willingness, English proficiency, and support for CLIL curriculum development. The questions are listed below:

3.1. Willingness

- How willing are you to teach 50% of your subject in English?
- How concerned do you feel if you are now required to adopt CLIL in class?

3.2. English Proficiency

- How willing are you to spend your leisure time improving your English proficiency to smoothly adopt CLIL?

3.3. Support for CLIL Curriculum Development

- How willing would you be to implement CLIL in your class if a teaching assistant system could generate 50% of the English content, including materials and exercises when developing the CLIL curriculum?
- How willing would you be to implement CLIL in your class if a teaching assistant system could not only generate 50% of the English content, including materials and exercises, when developing the CLIL curriculum but also support students with self-study?

The participants were asked to answer the questions with a 5-point Likert scale, as shown in Table 1. In addition to the above questions, we also asked participants to leave comments on implementing CLIL in K 12 classes. We recruited 102 K-12 teachers in Keelung City as participants to answer the above questions anonymously. Figure 1 illustrates the age and gender distributions of the recruited participants. Female is the major subgroup, and almost half of the K-12 teachers are between the ages of 41 and 50.

Table 1. Five-point Likert scale in this study.

	1	**2**	**3**	**4**	**5**
Willing	Completely unwilling	Unwilling	Neutral	Willing	Completely willing
Concerned	Extremely concerned	Concerned	Neutral	Uncon-cerned	Extremely unconcerned

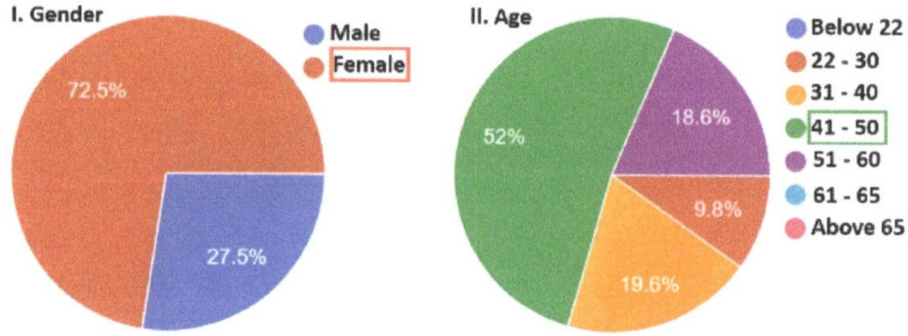

Figure 1. (**I**) and (**II**) are the gender and age distributions of the recruited participants, respectively.

4. Results and Discussion

The results related to teachers' willingness are shown in Figures 2 and 3. Teachers who are completely unwilling to teach their subjects with half the content in English make up 23.5%, the highest portion of all teachers, while only 36.3% of all teachers are willing to teach their subjects in English (Figure 2I). Figure 2II shows that teachers who feel extremely concerned about adopting CLIL in class make up 30.5%, the highest portion of all teachers. Figure 3 depicts the level of willingness of teachers between the ages of 41 and 50. Similar to the result shown in Figure 2II, teachers who feel extremely concerned when adopting CLIL in class make up 39.62%, the highest portion in this subgroup.

Figure 4 presents the results related to teachers' English proficiency. In total, 64.8% of teachers are willing to spend their leisure time improving their English proficiency to adopt CLIL. The results in Figures 2 and 3 imply that teachers may not be confident enough with their English proficiency. The lack of English proficiency appears to reduce teachers' willingness to adopt CLIL.

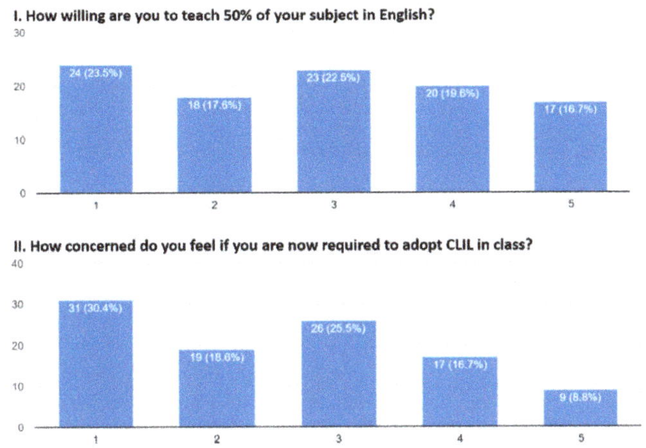

Figure 2. Results related to teachers' willingness.

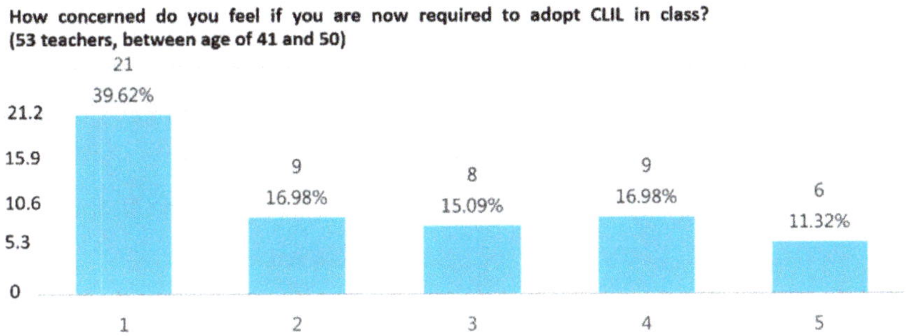

Figure 3. Results related to teachers' willingness in the subgroup of teachers between the ages of 41 and 50.

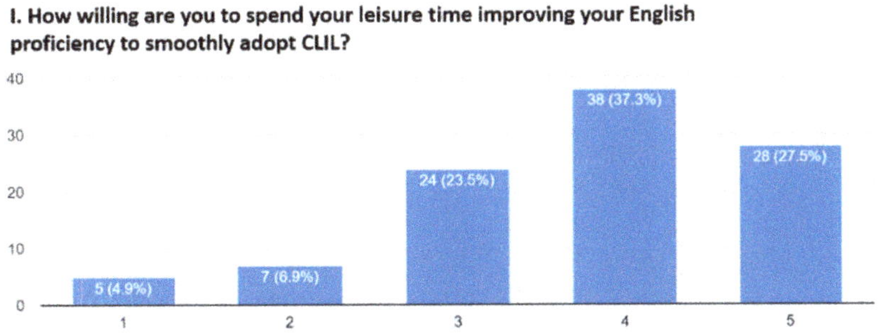

Figure 4. Results related to teachers' English proficiency.

The results related to the support for CLIL curriculum development are shown in Figure 5. Overall, 50% of teachers are willing to implement CLIL with a teaching assistant system providing English course materials (Figure 5I). A total of 53.9% of teachers are willing to implement CLIL with a teaching assistant system that not only provides English course materials but also helps students with self-study (Figure 5II). The teaching assistant system that generates English course materials seems to increase the teachers' willingness to deploy CLIL in class. Furthermore, the teaching assistant system supporting students with self-study boosts the teachers' willingness to deploy CLIL in class more. Figure 6 presents the female teachers' results related to support for CLIL curriculum development. The willingness to deploy CLIL in class accounts for 57.54% of all female teachers. The results in Figure 6 are almost in line with the results in Figure 5.

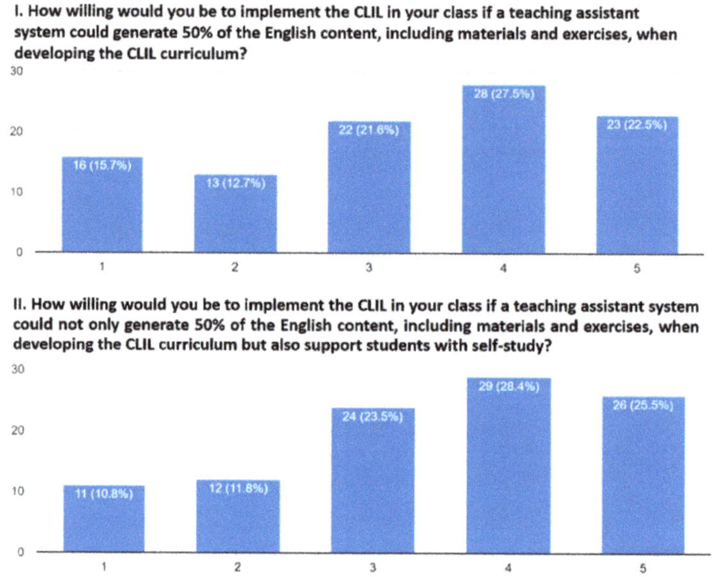

Figure 5. Results related to support for CLIL curriculum development.

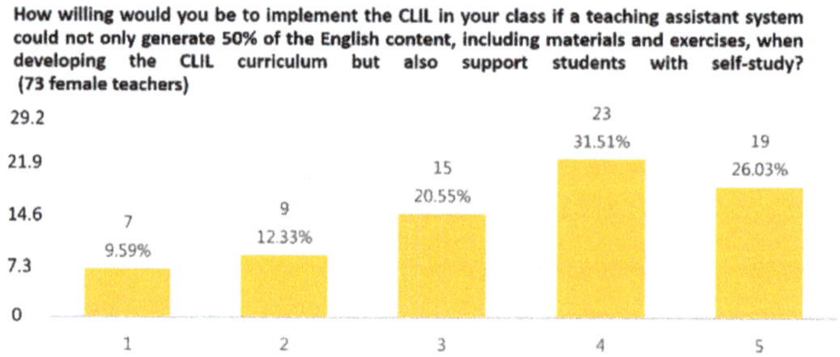

Figure 6. Results related to support for CLIL curriculum development in the subgroup of female teachers.

Table 2 describes the results of teachers' comments manually categorized by topics. In total, 31.37% of teachers feel concerned about their English proficiency regarding the implementation of CLIL. The results suggest that teachers' English proficiency is currently the major challenge in implementing CLIL in class. One possible explanation for the results is that the existing teacher training process does not contain training in bilingual teaching. English proficiency is not a requirement for K12 teachers. In addition, there is no shortcut to improving English proficiency to an appropriate level for implementing CLIL. Improving teachers' English proficiency is a long-run plan. However, short-term solutions are still in demand because teachers are the key players implementing CLIL in class.

Table 2. Teachers' comments on implementing CLIL categorized by topics.

Topic	Percent
Teachers' English proficiency	31.37%
Students' competency levels	19.60%
Bilingual education policy	12.75%
Curriculum development	8.82%
Appropriate subjects for CLIL	5.88%
Supporting measures	4.90%
Others	8.82%
No comments	7.84%

In addition to teachers' English proficiency, 19.60% of teachers are concerned about students' competency levels. Teachers' comments imply that students who already fail to understand Chinese lectures are likely to struggle even more with English lectures. Students' learning effectiveness might decrease because of the language barrier. However, students can still defeat their learning hardship as long as they are highly motivated to learn.

According to the results shown in Figures 5 and 6, a teaching assistant system that provides English course materials and helps students with self-study may be a short-term solution. Inspired by the concept of technology-enhanced CLIL [12], which was initially proposed to find useful discursive patterns that promote knowledge co-construction and language development, we identify the roles of technologies in facilitating CLIL implementation from teachers' perspectives, and this is shown in Figure 7. Technologies such as natural language processing and text-to-speech processing are helpful for teachers to develop bilingual curricula. Other technologies, such as augmented reality and chatbots, can be designed with proper human–computer interactions to help students with self-study. Integrating the technologies mentioned above may put the teaching assistant system into practice.

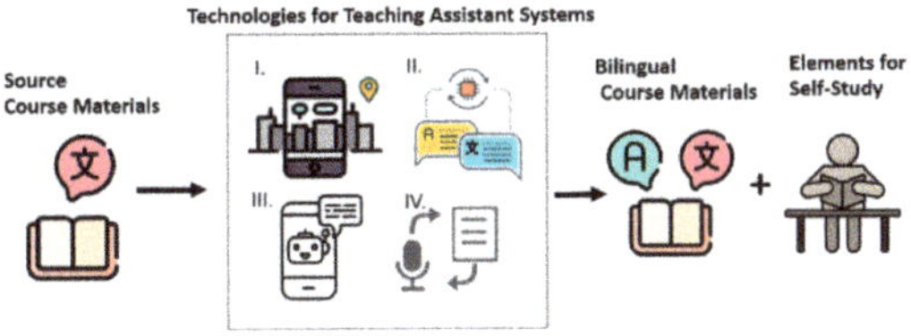

Figure 7. (**I–IV**) are augmented reality, natural language processing, chatbot, and text to speech plus speech recognition, respectively. These technologies may help teachers in implementing CLIL in class.

5. Conclusions

Implementing CLIL in class is currently imperative in Taiwan. This work discovers that teachers' English proficiency is the main challenge in implementing CLIL in class. Our findings also indicate that the teaching assistant system providing English teaching materials and helping students with self-study increases teachers' willingness to implement CLIL in class. Based on the results of this work, we also identify the roles of technologies in facilitating CLIL implementation from teachers' perspectives. Integrating proper technologies to develop teaching assistant systems may be a short-term solution to mitigate teachers' lack of confidence in their English proficiency.

Author Contributions: Conceptualization, S.-Y.J. and S.-M.Y.; methodology, S.-Y.J. and S.-M.Y.; validation, S.-Y.J., S.-W.S. and X.Y.; formal analysis, S.-Y.J. and S.-W.S.; investigation, X.Y.; resources, S.-M.Y. and C.-T.S.; data curation, S.-Y.J. and S.-W.S.; writing—original draft preparation, S.-Y.J., S.-W.S. and S.-M.Y.; writing—review and editing, S.-Y.J. and S.-M.Y.; visualization, S.-Y.J. and S.-W.S.; supervision, S.-M.Y. and C.-T.S.; project administration, S.-M.Y. and C.-T.S. All authors have read and agreed to the published version of the manuscript.

Funding: This research received no external funding.

Institutional Review Board Statement: Not applicable.

Informed Consent Statement: Not applicable.

Data Availability Statement: Data sharing is not applicable to this article.

Conflicts of Interest: The authors declare no conflict of interest.

References

1. Chang, W.-C. "Changes" and "No Changes" in Taiwan's English Education: Facing Challenges and Enhancing English Proficiency. *Second. Educ.* **2014**, *65*, 6–17. (In Traditional Chinese)
2. Burger, S.; Weinberg, A.; Wesche, M. Immersion studies at the University of Ottawa: From the 1980s to the present. *OLBI J.* **2013**, *6*. [CrossRef]
3. Coyle, D. Content and language integrated learning: Towards a connected research agenda for CLIL pedagogies. *Int. J. Biling. Educ. Biling.* **2007**, *10*, 543–562. [CrossRef]
4. Mehisto, P.; Marsh, D.; Frigols, M.J. *Uncovering CLIL: Content and Language Integrated Learning in Bilingual and Multilingual Education*; Macmillan Education: Oxford, UK, 2008.
5. Chamot, A.U. Developing self-regulated learning in the language classroom. In *New Perspectives on the Development of Communicative and Related Competence in Foreign Language Education*; De Gruyter Mouton: Berlin, Germany, 2018; pp. 41–52.
6. Coyle, D. The place of CLIL in (bilingual) education. *Theory Into Pract.* **2018**, *57*, 166–176. [CrossRef]
7. M. o. E. K-12 Education Administration. *Implementation Project of Bilingual Instruction in Some Domains of Primary and Junior High School Education*; M. o. E. K-12 Education Administration: Taichung, Taiwan, 2021.
8. Marsh, D. *CLIL/EMILE-The European Dimension: Actions, Trends and Foresight Potential*; University of Jyväskylä: Jyväskylä, Finland, 2002.
9. Merzlykin, O.V.; Topolova, I.Y.; Tron, V.V. Developing of key competencies by means of augmented reality at CLIL lessons. *High. Second. Sch. Pedagog.* **2019**, *51*, 58–73.
10. Vega, M.; Moscoso, M.D.L. Challenges in the Implementation of CLIL in Higher Education: From ESP to CLIL in the Tourism Classroom. *Lat. Am. J. Content Lang. Integr. Learn.* **2019**, *12*, 144–176. [CrossRef]
11. Isnaini, I.; Rohmah, T.; Saleh, M.; Faridi, A.; Fitriati, S.W. The Challenges in Implementing Content and Language Integrated Learning: The Head Teachers' Personal View of Teachers' Professional Development. In *International Conference on Science and Education and Technology (ISET 2019)*; Atlantis Press: Amsterdam, The Netherlands, 2020; pp. 536–539.
12. Zhao, K.; Zhou, J.; Zou, B. Developing subject knowledge co-construction and specific language use in a technology enhanced CLIL programme: Effectiveness and productive patterns. *Int. J. Biling. Educ. Biling.* **2021**, *25*, 2172–2185. [CrossRef]

engineering proceedings

MDPI

Proceeding Paper

A Case Study of Wind Farm Re-Powering †

Ming-Hong Chen * and Yan-Ting Lin

National Atomic Research Institute (NARI), Taoyuan City 325207, Taiwan; yantinglin@nari.org.tw
* Correspondence: minghongchen@nari.org.tw

† Presented at the IEEE 5th Eurasia Conference on Biomedical Engineering, Healthcare and Sustainability, Tainan, Taiwan, 2–4 June 2023.

Abstract: The Jhongtun wind farm in the area of Penghu Archipelago is studied. The first part of the Jhongtun wind farm has been operated for 19 years, and the second part has been operated for 15 years. It is about time to evaluate the feasibility of this wind farm's re-powering process to promote its power production. The evaluated AEPs are 82.65 GWh/year and 107.32 GWh/year. With the current deployment, roughly 12% of the power is provided by the wind farm with a 600 kW wind turbine, and its AEP is 11.46 GWh/year. The newly proposed 3 MW wind turbine has the potential to provide 100% of the power needed for the entire Penghu Archipelago.

Keywords: wind turbine; wind farm; re-powering

Citation: Chen, M.-H.; Lin, Y.-T. A Case Study of Wind Farm Re-Powering. *Eng. Proc.* **2023**, *55*, 11. https://doi.org/10.3390/engproc2023055011

Academic Editors: Teen-Hang Meen, Kuei-Shu Hsu and Cheng-Fu Yang

Published: 28 November 2023

1. Introduction

The research related to the exploitation of renewable energy (mainly wind and solar power) has been important in recent years. The traditional way of power production (mainly via fossil fuel) is not attractive due to its impact on the environment and the possibility of losing it entirely in the future. The goal of net zero emission and limiting the temperature change to 1.5 °C was declared in the Conference of the Parties 26 (COP26), who are also promoting the development of renewable energy.

Most of the onshore wind farms are developed by the Taiwan Power Company (TPC) and WPD [1]. The operation of TPC's wind farms has been ongoing for over 10 years. It is worth to conduct a comprehensive assessment on the benefit/cost of prolonging, decommissioning, and re-powering operational wind farms.

Besides the main island of Taiwan, the wind resources in the Penghu Archipelago are sufficient. About 10−12% of the needed energy was generated by wind turbine systems in the Penghu Archipelago in recent years. Capital investment is largely conducted in the Penghu Archipelago with the goal of 100% green energy. As the wind turbine systems have also been operating for more than 10 years in the Penghu Archipelago, re-powering to upgrade the power generation capacity is beneficial to meet the proposed goal.

The pre-process of a wind resources assessment (WRA) is an important step in acquiring wind power. Besides the technical consideration, the profitability of evaluating wind farms is vital for risk assessment planning. In recent years, larger wind turbines have been proposed, and the effects of complex terrain have to be considered for potential applications.

Wu et al. [2] evaluated the power generated in a wind farm with different heights and layouts by using the LES (large-eddy simulation) model. Eight layouts of turbine arrays were considered with 120 turbines in 30 rows. The aligned and staggered configurations were employed along the wake direction. Results showed that an obvious power reduction (45−65%) was observed for the first 12 turbines. More power was produced for the laterally staggered wind farm due to its better adaptability in the first 6 turbine rows. Meanwhile, the reduction of velocity and turbulence intensity of wake flow was observed for the vertically staggered wind farm configuration.

2. Method and Analysis

The effects of the characteristics of the targeted wind farm, compatibility with the considered wind turbines, price, and reliability of manufacturers need to be considered. The selection of wind turbines is based on the wind condition of the investigated wind farm. The effects of average wind speed and turbulence intensity (TI) must be considered in the assessment. For an area with a higher TI, a wind turbine with stronger robustness and a lower capacity factor is superior. In general, the same type of wind turbine is employed for an entire wind farm. In special cases, two or more different wind turbines are deployed in a wind farm with different hub heights for a maximum power output. It is also suggested that different wind turbines should belong to the same series and be made by the same manufacturer to be considerate of integration and management.

An onsite investigation must be conducted to assess the characteristics of a wind farm. The procedure of micro-siting is indicated in Figure 1. The feasibility of the selected wind turbine on a planned wind farm is evaluated. An adjustment is made for an optimization if needed. With the calculated net annual electricity production (AEP) and load on a wind turbine, the finalized plan of micro-siting can be proposed.

In this study, the evaluation model is developed by the software Wind Atlas Analysis and Application Program (WAsP) [3]. The data from the published wind atlas [4,5] are introduced for a comparison. The calculated AEP is then compared with the experimental value for validation of the proposed WAsP model. With the verified model, several scenarios are considered as possible deployments in the future for wind farm re-powering.

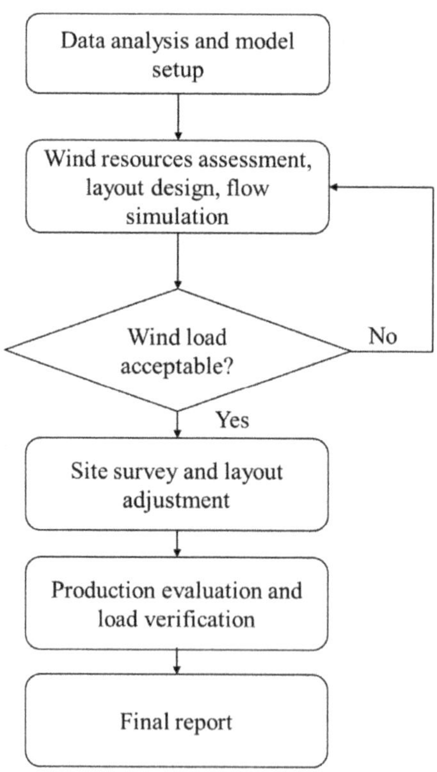

Figure 1. Procedure of micro-siting for a wind farm [6].

3. Results and Discussion

In this study, the Jhongtun wind farm is analyzed, as shown in Figures 2 and 3. The Jhongtun wind farm is the second wind farm in Taiwan produced by the TPC. There are eight wind turbines in the Jhongtun wind farm, and their generated power is integrated into the electricity network, providing 12% of the total demand of the Penghu Archipelago.

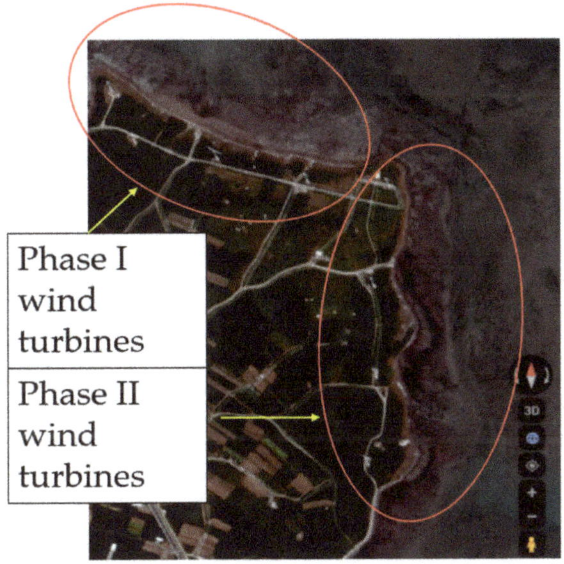

Figure 2. Aerial photography of the Jhongtun wind farm [7].

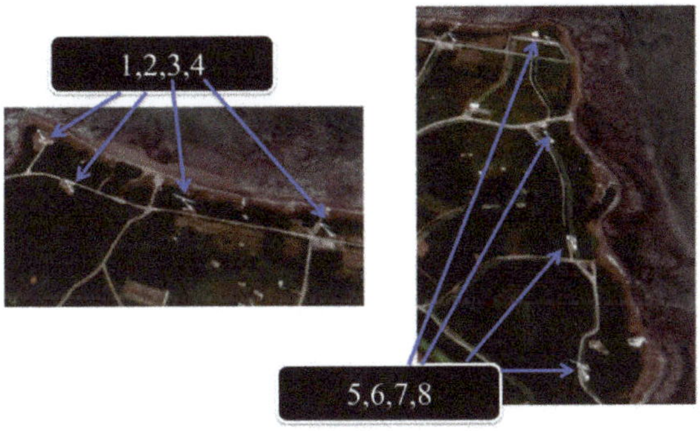

Figure 3. Plan of the Jhongtun wihd farm [7].

The first part of the Jhongtun wind farm was developed from May 2000 to December 2011 with four wind turbines produced by Enercon. The second part was developed from January 2004 to April 2005 with the same four wind turbines produced by Enercon. Their total capacity is 4.8 MW. The surrounding area of the Jhongtun wind farm was refurbished into a wind farm by the local government.

The hub height of the wind turbine installed in Jhongtun is 46 m. The statistical performance of the Jhongtun wind farm is shown in Figure 4.

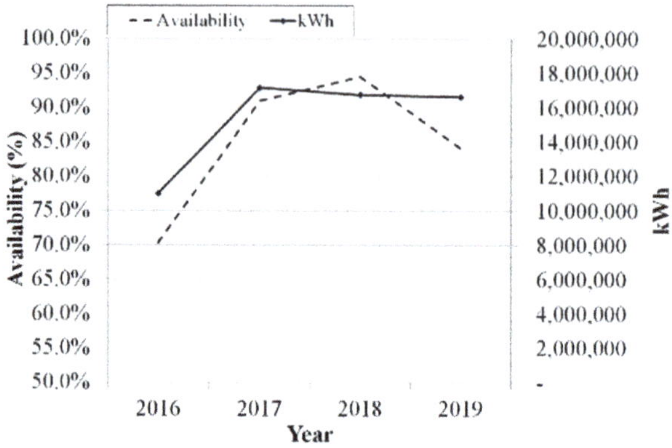

Figure 4. Four-year performance of the Jhongtun wind farm [8].

As shown in Figure 4, the overall availability is higher than 85% (except in the case of 2016), with the AEP being larger than 15 GWh/year. Compared to the other operating wind farms produced by the TPC, the performance of the Jhongtun wind farm has recently been good. However, the first part of the Jhongtun wind farm has been operated for 19 years, and the second for 15 years. It is necessary to evaluate the feasibility of the wind farm re-powering process to promote the power production of the Jhongtun wind farm.

The wind resources near the Jhongtun wind farm are the main topic of this study. The wind atlas with a high resolution is evaluated with the proposed WAsP model.

The statistical treatment result of hourly weather data is shown in Figure 5. The annual wind speed at the site is about 8.03 m/s. The contour plot assessed with the proposed WAsP model is shown in Figure 6. The calculated annual wind speed at the height of 80 m is about 9 m/s, and it is comparable with the data from reference [5].

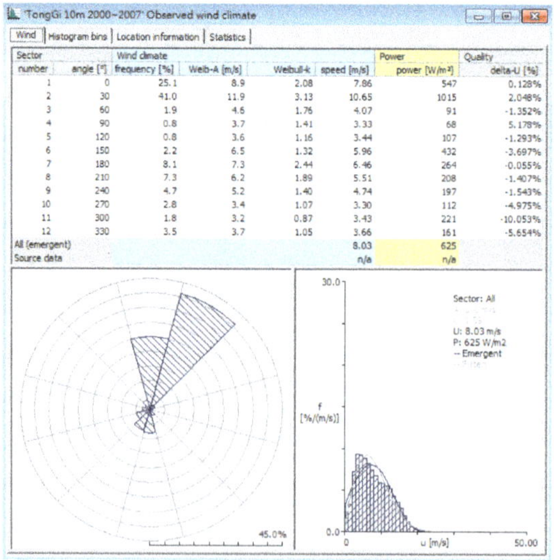

Sector		Wind climate				Power	Quality
number	angle [°]	frequency [%]	Weib-A [m/s]	Weibull-k	speed [m/s]	power [W/m²]	delta-U [%]
1	0	25.1	8.9	2.08	7.86	547	0.128%
2	30	41.0	11.9	3.13	10.65	1015	2.048%
3	60	1.9	4.6	1.76	4.07	91	-1.352%
4	90	0.8	3.7	1.41	3.33	68	5.178%
5	120	0.8	3.6	1.16	3.44	107	-1.293%
6	150	2.2	6.5	1.32	5.96	432	-3.697%
7	180	8.1	7.3	2.44	6.46	264	-0.055%
8	210	7.3	6.2	1.89	5.51	208	-1.407%
9	240	4.7	5.2	1.40	4.74	197	-1.543%
10	270	2.8	3.4	1.07	3.30	112	-4.975%
11	300	1.8	3.2	0.87	3.43	221	-10.053%
12	330	3.5	3.7	1.05	3.66	161	-5.654%
All (emergent)					8.03	625	
Source data					n/a	n/a	

Figure 5. Wind distribution of the Jhongtun wind farm as assessed with the WAsP model.

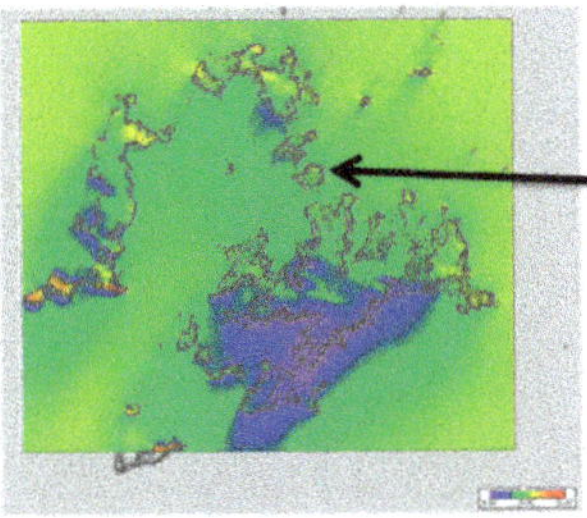

Figure 6. Atlas of the Jhongtun wind farm created with the WAsP model.

On the website of ref. [5], a simple and quick evaluation can be conducted. Results are shown in Table 1 with four sets of wind turbines. With the assumption that the new installation of the upgraded wind turbines have the same localization, the gross AEP is obtained by timing the AEP of a single wind turbine, as indicated in the last row of Table 1. The capacity of the considered wind turbines ranges from 850 to 2300 kW. The corresponding AEP of the re-powered wind farm ranged from 26 to 53 GWh/year. The effects of wake flow are not considered in such a simplified evaluation. For the scenario with different positions, heights, and numbers, the performance is evaluated with the proposed WAsP model.

Table 1. Summary of wind farm re-powering.

Specification	Vestas V60	Vestas V90	Vestas V80	Enercon E-70
V_hub (m/s)	7.7	8.1	7.85	7.79
Capacity (kW)	850	1800	2000	2300
Hub height (m)	60	80	67	64
AEP (kWh/y)	3,293,760	7,717,560	6,508,680	6,675,120
CF (%)	44.24	48.94	37.15	33.13
AEPX8(GWh/y)	26.4	61.7	52.1	53.4

The most feasible plan is to install a new and upgraded wind turbines at the Jhongtun wind farm in the same area via a preliminary discussion with the operator of the plant. The capacities of the considered wind turbine are 2 and 3 MW. The turbine characteristics are introduced into the WAsP model, and the results are shown in Figure 7.

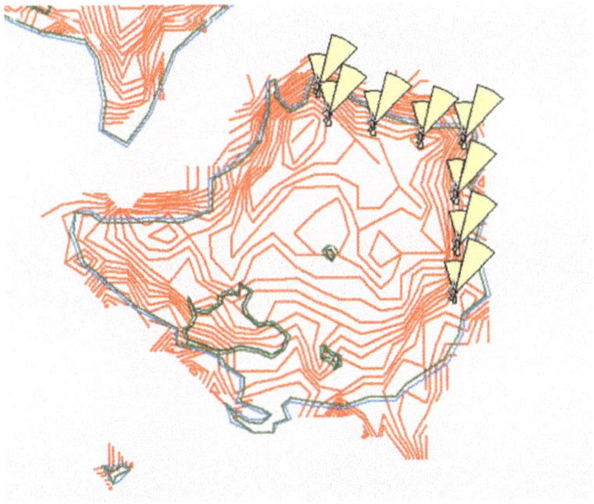

Figure 7. Re-powering of the Jhongtun wind farm with the WAsP model.

As shown in Figure 7, the effects of wake flow on the Jhongtun wind farm are insignificant. This also indicates that the original design of this wind farm is well organized with regard to the orientation and distance between turbines. Thus, installing a new and upgraded wind turbine in the same position of an existing one would also lead to good results. It is also expected that such a plan leads to a minimal impact on residents. The AEPs are re-calculated with new wind turbines via the proposed WAsP model, and the results are shown in Table 2.

Table 2. Summary of wind farm re-powering.

Turbine Capacity	Model	Hub Height (m)	AEP (GWh)
600 kW	Enercon E-40	46	11.46 (WAsP) 10.99 (Real, 2016)
2.0 MW	Vestas V90	80	82.65
3.0 MW	Vestas V90	80	107.32

As summarized in Table 2, the calculated AEP with the WAsP model is consistent with the true value of power production in 2016, validating the reliability of the proposed model. The wind turbines of Vestas V90 with a capacity of 2 and 3 MW are introduced into the evaluation. The hub heights for both wind turbines are the same and have different power curves. For a wind turbine of 3 MW, the rated power can be produced with the rated wind speed. The evaluated AEPs are 82.65 and 107.32 GWh/year, respectively.

When comparing the current deployment, roughly 12% of the area's power is provided by the wind farm with a 600 kW wind turbine with an AEP of 11.46 GWh/year; the newly proposed 3 MW wind turbine has the potential to provide 100% of the power needed for the entire Penghu Archipelago considering the yearly average power requirement.

In this study, a numerical model is proposed for the evaluation of the performance of a wind farm. The results are compared with a reference value to validate the proposed model. The AEP of the newly designed wind farm, with different scenarios, is evaluated and discussed. In the future, the economic assessment of the designed wind farm plans can be conducted as a comprehensive evaluation for the re-powering of a wind farm in question.

4. Conclusions

The pre-process of an assessment of wind resources is important in the process of acquiring wind power. Besides technical considerations, the profitability of evaluating a wind farm is also vital for risk assessment planning. In this study, the software Wind Atlas Analysis and Application Program is employed to develop a wind atlas of the investigated wind farm. The calculated AEP is then compared with the experimental value for validation of the proposed WAsP model. With the verified model, several scenarios are considered as possible deployments in the future for wind farm re-powering. Suggestions are provided based on the calculated results.

The Jhongtun wind farm is analyzed in the area of Penghu Archipelago. The Jhongtun wind farm is the second wind farm developed in Taiwan by the TPC. There are eight wind turbines installed in the Jhongtun wind farm, and the generated power is integrated into the electricity network, providing 12% of the total demand of the Penghu Archipelago area.

The overall availability is higher than 85% (except for the case in 2016), with an AEP larger than 15 GWh/year. Compared to the other operating wind farms produced by the TPC, the performance of the Jhongtun wind farm in recent years is good. However, the first part of the Jhongtun wind farm has been operated for 19 years, and the second for 15 years. The evaluation of the feasibility of the wind farm is needed for the re-powering process to promote the power production of the Jhongtun wind farm. The calculated annual wind speed at the height of 80 m is about 9 m/s, which is comparable with the reference data.

The most feasible plan is to install new and upgraded wind turbines in the same positions in the Jhongtun wind farm via a preliminary discussion with the operator of the plant. The capacities of the considered wind turbines are 2 and 3 MW, respectively. Results show that the original design of this wind farm is well organized in terms of the orientation and distance between turbines. Thus, installing new and upgraded wind turbines in the same positions also leads to good results. Such a plan leads to a minimal impact on residents. The calculated AEP with the WAsP model is consistent with the true value of power production in 2016, validating the reliability of the proposed model. The wind turbines produced by Vestas V90 with the capacity of 2 and 3 MW are introduced. The hub heights for both wind turbines are the same, with a different power curve. The evaluated AEPs are 82.65 and 107.32 GWh/year, respectively. When comparing the current deployment, roughly 12% of the power is provided by the wind farm with a 600 kW wind turbine and an AEP of 11.46 GWh/year. The newly proposed 3 MW wind turbine has the potential to provide 100% of the power needed for the entire Penghu Archipelago considering the yearly average power requirement.

In this study, a numerical model is proposed for the evaluation of the performance of a wind farm. The results are compared with a reference value to validate the proposed model. The AEP of the newly designed wind farm, with different scenarios, is evaluated and discussed. In the future, the economic assessment of the designed wind farm plans can be conducted as a comprehensive evaluation for the re-powering of a wind farm in question.

Author Contributions: Conceptualization, M.-H.C.; methodology, M.-H.C.; software, M.-H.C.; validation, Y.-T.L.; formal analysis, Y.-T.L.; investigation, Y.-T.L.; resources, Y.-T.L.; data curation, Y.-T.L.; writing—original draft preparation, M.-H.C.; writing—review and editing, M.-H.C.; visualization, M.-H.C.; supervision, M.-H.C.; project administration, M.-H.C.; funding acquisition, M.-H.C. All authors have read and agreed to the published version of the manuscript.

Funding: This research received no external funding.

Institutional Review Board Statement: Not applicable.

Informed Consent Statement: Not applicable.

Data Availability Statement: Data is unavailable due to privacy.

Conflicts of Interest: The authors declare no conflict of interest.

References

1. WPD. Available online: https://www.wpd.de/en/ (accessed on 26 October 2023).
2. Wu, Y.T.; Liao, T.L.; Chen, C.K.; Lin, C.Y.; Chen, P.W. Power output efficiency in large wind farms with different hub heights and configurations. *Renew. Energy* **2019**, *132*, 941–949. [CrossRef]
3. DTU Risoe. Wind Atlas Analysis and Application Program. 2012. Available online: http://www.wasp.dk/ (accessed on 12 October 2023).
4. Industrial Technology Research Institute. Thousand Wind Turbines Project. 2014. Available online: http://www.twtpo.org.tw/intro.aspx?id=462 (accessed on 26 October 2023).
5. MOEA. Wind power Information Integration Platform. 2020. Available online: https://pro.twtpo.org.tw/GIS/ (accessed on 26 October 2023).
6. Zhang, M.H. *Wind Resource Assessment and Micro-Siting*; Wiley Subscription Services, Inc.: Hoboken, NJ, USA, 2015.
7. Google Map. 2020. Available online: www.google.com.tw/maps/ (accessed on 18 October 2023).
8. TPC, Open Data Website. 2020. Available online: https://www.taipower.com.tw/tc/page.aspx?mid=96 (accessed on 21 October 2023).

engineering proceedings

Proceeding Paper

Using Machine Learning of Artificial Intelligence to Analyze Business Opportunities and Applications of the Massively Multiplayer Online Role-Playing Game Case in Metaverse †

Kuo-Hsien Lee [1,*], Wen-Hsien Tsai [1], Cheng-Tsu Huang [1], Jerry Tao [2], Hank Lee [3], Ching-Hui Chen [4], Li-Yun Lee [5] and Hsiao-Ting Tseng [6]

1 Department of Business Administration, National Central University, Taoyuan 320317, Taiwan; whtsai@mgt.ncu.edu.tw (W.-H.T.); chengtsu.huang@gmail.com (C.-T.H.)
2 AI Value Co., Ltd., Taipei City 114065, Taiwan; jerry0216@gmail.com
3 Samoi Techology Co., Ltd., Taipei City 10683, Taiwan; 1199et0309@gmail.com
4 Chief Strategic Planning & Promotion Division, Metal Iindustries Research & Development Center, Kaohsiung City 811225, Taiwan; fenychen@mail.mirdc.org.tw
5 Department of Tourism Recreation Management, Overseas Chinese University, Taichung City 40760, Taiwan; yun92388@ocu.edu.tw
6 Department of Information Management, National Central University, Taoyuan 320317, Taiwan; httseng@mgt.ncu.edu.tw
* Correspondence: durmaslee@gmail.com
† Presented at the IEEE 5th Eurasia Conference on Biomedical Engineering, Healthcare and Sustainability, Tainan, Taiwan, 2–4 June 2023.

Abstract: By using the machine learning of artificial intelligence to explore the application business opportunities of the Metaverse in the MMORPG (Massively Multiplayer Online Role-Playing Game) interactive game market, we study the supply and demand laws of buyers and sellers at the market economy level, future trends, and business opportunities. The feasibility of its new products and services is explored under a pragmatic, cooperative model of the game community platform "Key to the Desert" case for the application level and business opportunities of Taiwan's Metaverse markets. Online and offline integration (OMO; Online Merge Offline), precision marketing, and the customer management data platform (Customer Data Platform) are also explored in the application business opportunities of the Metaverse market. By combining the NFT (Non-Fungible Token) Monopoly game and MMORPG interactive games, we study the laws of supply and demand of buyers and sellers at the market economy level to provide third-party payment, electronic payment, mobile payment, and other transaction method certifications such as NFT (Non- Fungible Token). We also evaluation the future and security issues of cryptocurrency.

Keywords: Metaverse; artificial intelligence; machine learning; NFT; MMORPG

Citation: Lee, K.-H.; Tsai, W.-H.; Huang, C.-T.; Tao, J.; Lee, H.; Chen, C.-H.; Lee, L.-Y.; Tseng, H.-T. Using Machine Learning of Artificial Intelligence to Analyze Business Opportunities and Applications of the Massively Multiplayer Online Role-Playing Game Case in Metaverse. *Eng. Proc.* **2023**, *55*, 12. https://doi.org/10.3390/engproc2023055012

Academic Editors: Teen-Hang Meen, Kuei-Shu Hsu and Cheng-Fu Yang

Published: 28 November 2023

1. Introduction

By using AR (Augmented Reality) [1] technology and combining it with animation design (Figure 1) [1], or using AR alone, we can enhance the visual and auditory effect of a member card. Therefore, through NFT (Non-Fungible Token) supply and demand sales certification [2], products are integrated through virtual and real sales platforms. Online Merge Offline (OMO) [3] intelligently collides with AI (Artificial Intelligence) machine learning technology to match buyers and sellers on the cloud service platform [4], so that buyers and sellers are allowed to classify the big data provided by the cloud service for information docking, in a similar way to Uber's method [5]. To make matching transactions in the actual field, the transaction equilibrium points (E1, E2, and E3) must reach the supply and demand side mentioned in Figure 2.

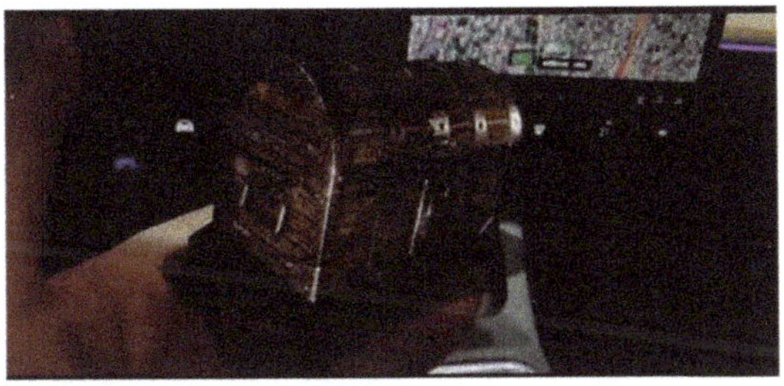

Figure 1. Member card with AR technology implemented (https://www.youtube.com/embed/48NEaMFm6Go?feature=oembed, accessed on 24 November 2023).

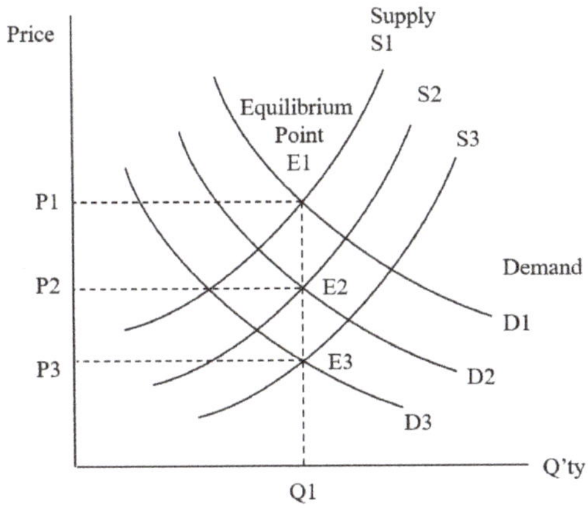

Figure 2. Equilibrium point in between supply and demand.

2. AI and Metaverse

2.1. NFT Monopoly Game and MMORPG

Figure 3 shows that with the collected data and information, it is possible to quickly find suitable buyers and sellers and conduct balanced supply and demand transactions through the calculation and classification of the cloud platform. Online and offline sales can also be carried out. The integrated model confirms the transaction amount and identity authentication of both parties. In addition to preventing transaction fraud, this does not need to be guaranteed by banks or financial institutions, reducing intermediate handling costs [6].

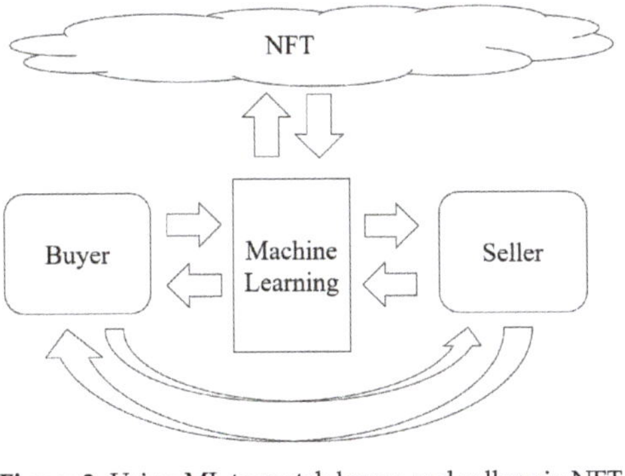

Figure 3. Using ML to match buyer and seller via NFT.

Next, we use CDPs (Customer Data Platform) for precise marketing [7] and customer management to explore the applications and business opportunities of the Taiwan Metaverse [8] market. Combining the NFT Monopoly game and MMORPG (Massively Multiplayer Online Role-Playing Game) [9] interaction games, new markets and future developments are created as shown in Figure 4.

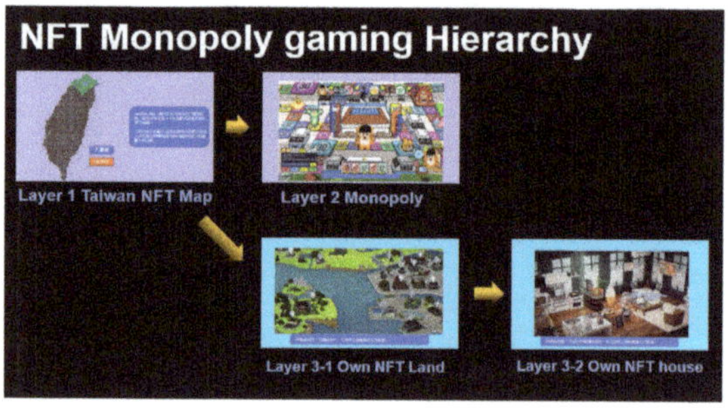

Figure 4. NFT Monopoly gaming hierarchy and MMORPG.

The NFT Monopoly game is a virtual online game that has existed for a long time. Being combined with MMORPG's massively multiplayer online role-playing interactive game and the recently rammed metaverse land purchase business opportunity, the Taiwan map is first matrixed. However, pixel bitmap files, which are restricted to digitizing and building roads and infrastructure in important cities, are started such as traffic signs, bridges, and other public buildings.

2.2. *Cryptocurrency*

Combined with the cold wallet conversion of virtual currency (cryptocurrency) [10] as shown in Figure 5, the fee-free electronic payment and the online and offline membership card system make it easier for consumers to conduct transactions. This is connected to the MMORPG cloud platform by the import of advanced technologies of AR (Augmented Reality), VR (Virtual Reality), MR (Mixed Reality), and XR (X-Reality; Cross Reality; Extended Reality) [11]. As the user experience is more appropriate, consumers are more able to contact the real objects and have a better user experience with the purchased products.

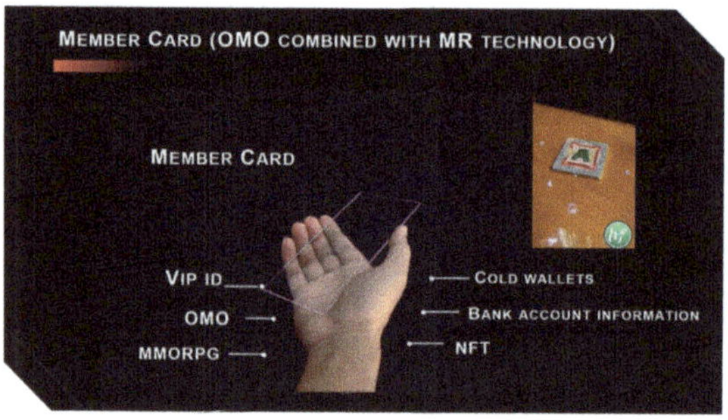

Figure 5. Cold wallet with AR, VR, MR, and XR technologies.

3. Circular Economy

The next step is to combine online and offline brand enterprises and use the OMO and CDP platforms to find the right TA (target audience; target customer; target group; target) [12] for the products and services through the AI algorithm of precise marketing. By lowering the advertising budget, consumers gather in the game, and the brand enterprise sells products and services to the customer group with a faster and more effective fraction, increasing click through rate (CTR) and conversion rate (CVR). The return on advertising spend (ROAS) can then be calculated [13].

Therefore, the NFT Monopoly game continues to create a positive cycle between consumers, brand business owners, and game platforms and continuously increases the CTR and CVR of online games. Thus, buyers and sellers can obtain maximum benefits, and supply and demand reaches the equilibrium trading point as shown in Figure 6.

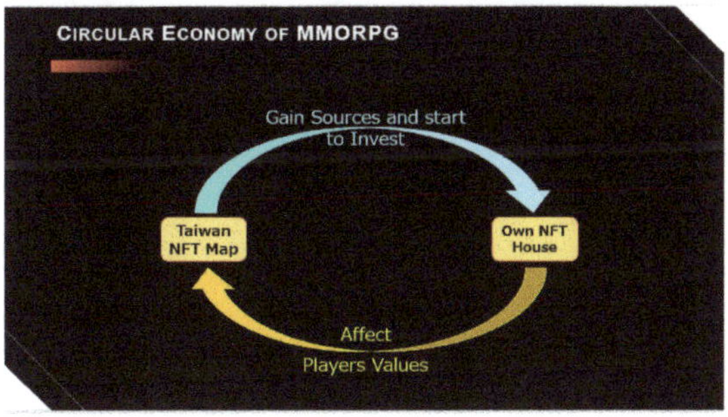

Figure 6. Circular economy via MMORPG.

4. Case Study

This chapter discusses the new product and service "Key to the Desert" game community platform as a case study and analyzes OMO plus CDP's precision marketing.

4.1. *"Key to the Desert" Game Community Platform*

"Key to the Desert" is a social platform that leaps from Web 2.0 to Web 3.0 [14]. From the Metaverse's original concept, it covers life, work, entertainment, and community. Through consensus gamification, online and offline behaviors are linked to each other to achieve virtual reality. It includes the applications and connections of new technologies to achieve new customers' expectations [15] in terms of hardware and services.

4.2. *OMO and CDP*

CDP is a collection platform for customers and consumers. It requires the traces of consumers and customers, such as how long consumers and customers have seen the products for and how soon to take action to buy them. It is based on the internal information of the enterprises, plus external public information to draw a more accurate consumer profile, and it is then used as shown in Table 1. This information is used for the accurate marketing of automatic classification.

Table 1. Example of CDP case.

Consumer: Mr. Lee
Sex: Male Age: 25 Address: No. 10, Section 4, Zhongxiao East Road, Taipei City Purchase time: 20 July 2021 Where to buy: 7–11 Purchased items: milk powder Purchase amount: 120 NTD Mobile: 0922XXXXX5

With dynamic data and static data, precise customer marketing can be achieved by combining these with CRM (Customer Relationship Management) [16]. In our current data-based time, it has becomes more difficult for enterprises to operate without CDP. The cross-use of CRM, DMP (Data Management Platform) [17], and CDP can maximize the use of marketing resources for enterprises as shown in Figure 7.

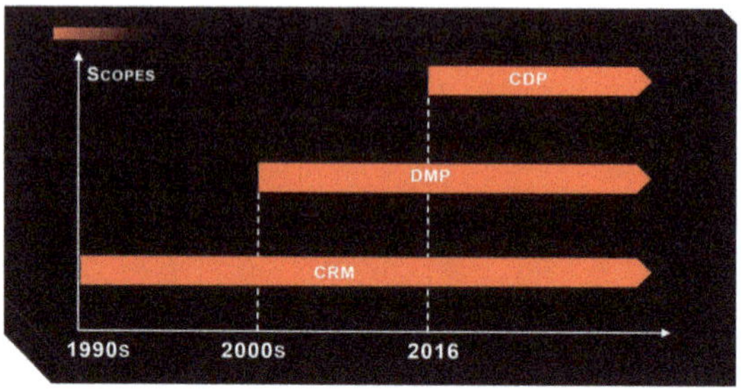

Figure 7. History of CRM, DMP, and CDP.

Because the data of the customer journey span many different companies and different behavior trajectories and also use a variety of different systems, it is important to track and label the customer's behavior completely. The platform has a function built for this purpose. It combines AI automatic labeling and the automatic analysis system so that the functions of the CDP platform can enhance the ability of accurate prediction, thereby making the marketing delivery more efficient (Figure 8).

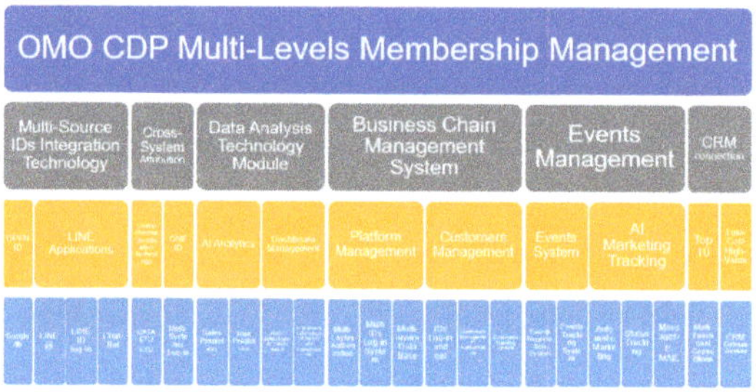

Figure 8. OMO CDP is used to classify the data collected from customers and consumers to transit more useful information in cloud computing.

The OMO CDP platform uses advanced AR and MR visual technologies to make users feel and experience better, and so they are more willing to join and use the functions of the platform and download the APP (Applications) [18]. With the data, the ecosystem manufacturers that cooperate with the platform also have more complete customer behavior data [19] and predictive analysis to understand the best customer experiences and data integration that is closer to customer needs. This is also the unique function of this platform system.

4.3. Next Steps

The ultimate goal of the OMO CDP platform is to create a data platform with a 360-degree view of customers, including dynamic and static data, and maintain dynamic and real-time updates. Here, the protection of customer privacy data becomes extremely important. This system uses an API (Application Interfaces) [20] encryption technology method to maintain an anti-hacking personal information protection system, which creates a high-security database for enterprises to prevent hackers access to private information.

To implement the OMO CDP platform into MMORPG and create new businesses and opportunities, we need to consider the current and future government act and relative regulations and social rules. Therefore, the next step in further study is to discuss and analyze the challenges of different peoples, countries, and communities.

Author Contributions: Conceptualization, H.L. and J.T.; methodology, K.-H.L.; software, J.T.; validation, C.-T.H. and H.-T.T.; formal analysis, C.-H.C.; investigation, L.-Y.L.; resources, H.L.; data curation, K.-H.L.; writing—original draft preparation, K.-H.L.; writing—review and editing, W.-H.T.; visualization, J.T.; supervision, W.-H.T.; project administration, K.-H.L. All authors have read and agreed to the published version of the manuscript.

Funding: This research received no external funding.

Institutional Review Board Statement: Not applicable.

Informed Consent Statement: Not applicable.

Data Availability Statement: No new data were created or analyzed in this study.

Conflicts of Interest: The authors declare no conflict of interest. AI Value Co., Ltd. and Samoi Techology Co., Ltd. have no conflict of interest with this paper.

References

1. Rauschnabel, P.A.; Babin, B.J.; tom Dieck, M.C.; Krey, N.; Jung, T. What is augmented reality marketing? Its definition, complexity, and future. *J. Bus. Res.* **2022**, *142*, 1140–1150. [CrossRef]
2. Dowling, M. Is non-fungible token pricing driven by cryptocurrencies? *Financ. Res. Lett.* **2021**, *44*, 102097. [CrossRef]
3. Liu, K.-H.; Cao, K.-D.; Wang, S.-M. Mini-Program Design and Evaluation to Create Online-Merge-Offline Service for Retailers. In Proceedings of the 2020 2nd International Conference on Big Data Engineering, Shanghai China, 29–31 May 2020.
4. Sun, S.; Cao, Z.; Zhu, H.; Zhao, J. A survey of optimization methods from a machine learning perspective. *IEEE Trans. Cybern.* **2019**, *50*, 3668–3681. [CrossRef] [PubMed]
5. Grabher, G.; van Tuijl, E. *Uber*-production: From global networks to digital platforms. *Environ. Plan. A Econ. Space* **2020**, *52*, 1005–1016. [CrossRef]
6. Clemons, E.K.; Madhani, N. Regulation of digital businesses with natural monopolies or third-party payment business models: Antitrust lessons from the analysis of Google. *J. Manag. Inf. Syst.* **2010**, *27*, 43–80. [CrossRef]
7. Zhang, H.-N.; Dwivedi, A.D. Precise Marketing Data Mining Method of E-Commerce Platform Based on Association Rules. *Mob. Netw. Appl.* **2022**, *27*, 2400–2408. [CrossRef]
8. Dionisio, J.D.N.; Burns, W.G., III; Gilbert, R. 3D virtual worlds and the Metaverse: Current status and future possibilities. *ACM Comput. Surv. (CSUR)* **2013**, *45*, 1–38. [CrossRef]
9. Cole, H.; Griffiths, M.D. Social interactions in massively multiplayer online role-playing gamers. *CyberPsychology Behav.* **2007**, *10*, 575–583. [CrossRef] [PubMed]
10. Hileman, G.; Rauchs, M. *Global Cryptocurrency Benchmarking Study*; Cambridge Centre For Alternative Finance: Cambridge, UK, 2017; pp. 33–113.
11. Ko, S.H.; Rogers, J. Functional Materials and Devices for XR (VR/AR/MR) Applications. *Adv. Funct. Mater.* **2021**, *31*, 2106546. [CrossRef]
12. Lefebvre, R.C. The new technology: The consumer as participant rather than target audience. *Soc. Mark. Q.* **2007**, *13*, 31–42. [CrossRef]
13. Lewis, R.A.; Rao, J.M. The unfavorable economics of measuring the returns to advertising. *Q. J. Econ.* **2015**, *130*, 1941–1973. [CrossRef]
14. Hendler, J. Web 3.0 Emerging. *Computer* **2009**, *42*, 111–113. [CrossRef]
15. Hoyer, W.D.; Kroschke, M.; Schmitt, B.; Kraume, K.; Shankar, V. Transforming the customer experience through new technologies. *J. Interact. Mark.* **2020**, *51*, 57–71. [CrossRef]
16. Winer, R.S. A framework for customer relationship management. *Calif. Manag. Rev.* **2001**, *43*, 89–105. [CrossRef]
17. Elmeleegy, H.; Li, Y.; Qi, Y.; Wilmot, P.; Wu, M.; Kolay, S.; Dasdan, A.; Chen, S. Overview of turn data management platform for digital advertising. *Proc. VLDB Endow.* **2013**, *6*, 1138–1149. [CrossRef]
18. Alanzi, T. A review of mobile applications available in the app and google play stores used during the COVID-19 outbreak. *J. Multidiscip. Healthc.* **2021**, *14*, 45–57. [CrossRef] [PubMed]
19. Rita, P.; Oliveira, T.; Farisa, A. The impact of e-service quality and customer satisfaction on customer behavior in online shopping. *Heliyon* **2019**, *5*, e02690. [CrossRef] [PubMed]
20. Roldan, J.J.; Pena-Tapia, E.; Garcia-Aunon, P.; Del Cerro, J.; Barrientos, A. Bringing adaptive and immersive interfaces to real-world multi-robot scenarios: Application to surveillance and intervention in infrastructures. *IEEE Access* **2019**, *7*, 86319–86335. [CrossRef]

engineering proceedings

Proceeding Paper

Prevention of Burnout Syndrome in Social Workers to Increase Professional Self-Efficacy †

Alida Samusevica and Santa Striguna *

Faculty of Pedagogy and Social Work, Liepaja University, LV-3401 Liepaja, Latvia; alida.samusevica@liepu.lv
* Correspondence: santa.striguna@liepu.lv

† Presented at the IEEE 5th Eurasia Conference on Biomedical Engineering, Healthcare and Sustainability, Tainan, Taiwan, 2–4 June 2023.

Abstract: The reality of the professional burnout of social workers has never been more topical than in today's transformative, dynamic, and anxiety-filled society. Thus, this study aims to stress the necessity and importance of prevention of social workers' burnout syndrome based on empirical research data on the topicality of the problem and the application of diverse and contradictory solutions. Although social work is purposefully structured, mentally rewarding, rejuvenating, and inspiring, it is also tedious, demanding, and emotionally draining at times, as it requires inexhaustible intellectual and emotional resources from social work professionals, consistently qualified professional activities, and effective cooperation with the client. The research data obtained from surveying social workers confirm the topicality of the problem and the need for preventive activities for the promotion and preservation of the mental health of professionals. The study result offers theory-based recommendations and suggestions for the reduction of burnout syndrome symptoms and prevention strategies when implementing professional self-efficacy of social workers in the era of constantly pulsating changes and challenges.

Keywords: burnout syndrome; social worker; prevention; professional activity; self-efficacy

Citation: Samusevica, A.; Striguna, S. Prevention of Burnout Syndrome in Social Workers to Increase Professional Self-Efficacy. *Eng. Proc.* **2023**, *55*, 13. https://doi.org/10.3390/engproc2023055013

Academic Editors: Teen-Hang Meen, Kuei-Shu Hsu and Cheng-Fu Yang

Published: 28 November 2023

1. Introduction

The topicality of the research problem analyzed in this study is the burnout of social workers and the risks related to it [1]. It must be noted that professional burnout and its prevention remain highly topical, especially in the context of the COVID-19 pandemic [2]. The war in Ukraine and the increased number of refugees in several European countries require complex solutions to life contradictions in social work practice that are vital to the existence of those involved in society [3].

The occupational vulnerability of social workers, including the work environment, is a well-founded concern. Research conducted in Latvia [3] on the reduction of stress in the work environment of social workers supported by their employers revealed that almost half of the social work professionals (42%) indicated that the support provided by employers was insufficient and highlighted the need for employers to prioritize stress reduction in workplaces. It is also important to educate employers about burnout risks and their prevention so that timely burnout and appropriate action can be conducted for employees' professional self-efficacy.

The social and practical significance of the problem was reinforced by the fact that the World Health Organization (2019) classified burnout as a professional disease and included it in the International Classification of Diseases. The survey conducted by APA outlines another important trend, which is that people do not want to talk to management about burnout syndrome. Stereotypes that have come along for years must also be taken into account. People are afraid to go to a psychiatrist to explore their health problems. Only 41% of social workers are willing to talk about the signs of burnout to their employees.

Professionals need to be educated on this matter to be aware that by approaching specialists on time, the problem can be solved without experiencing the worst consequences of burnout syndrome. However, 47% of social workers do not want their employers to know about the signs of burnout because they are afraid of condemnation and termination of their work contracts [4].

Therefore, it is necessary to stress the necessity and importance of prevention of social workers' burnout syndrome based on the analysis of the literature and relevant research data [4], empirical research data on the problem, and the application of diverse contradictory solutions in everyday practice. Two research questions are put forward while implementing the research goal.

1. What factors contribute to the professional burnout of social workers?
2. What burnout syndrome prevention aspects should be stressed in the professionalization of social workers?

2. Research Methodology

In this study, the content analysis of theoretical sources of the social work literature was implemented (22 units), along with the thematic analysis of empirical research data. The analyses allowed focusing on the content of information to build the knowledge of professionals and the public about the social and psychological reality of social work and promote the understanding of the defined problems. The result was summarized for a productive view of the research questions raised concerning the findings on professional burnout of social workers.

The research methods were selected based on the advantages of theoretical content analysis and thematic analysis of the topic so that the findings obtained as a result of these methods could be transparently structured and easier to perceive for the general public. The causes and consequences of the problem were identified. Combining theory and practice, understanding the theoretical conclusions, and delving into the result of qualitative and quantitative content analysis, we determined the content of data interpretation and the way of expression.

3. Interpretation of Burnout Syndrome

One of the early researchers of burnout syndrome, Herbert, defined burnout syndrome as an individual's state of fatigue or dissatisfaction in a professional relationship that does not allow the expected result to be achieved [5]. The term "burnout syndrome" describes the symptoms of exhaustion and mental weakness, going into their causes and the signs of behavioral manifestations, including the following: (1) the person's awareness about the inability to prove him/herself, obsessively demonstrating his/her value, (2) the inability to overcome professional difficulties and to switch to other activities, and (3) the neglect of needs, inadequate sleep, eating disorders, and lack of social interactions. Disadvantageous influences result in the displacement of existing conflicts, rejection of problems, personality feeling threatened, panic and nervousness, the uncertainty of values, and social alienation, which pose a significant threat not only to a person but also to the quality of his/her professional activities. Thus, "burnout syndrome can cause mental and physical collapse and requires complex medical assistance" [5]. Lee and Ashforth [6], referring to the research studies of Maslach [7,8] on burnout syndrome and evaluation criteria related to it, pointed out that burnout syndrome could be described in three dimensions:

Emotional exhaustion (lack of energy or feeling of emptiness);
Depersonalization (negative attitude or increased distancing from work);
Reduced personal achievement and professional effectiveness (characterized by negative assessment of work achievements).

Pines et al. [9] extended the definition of burnout syndrome, stating that it is not only a state of emotional exhaustion but also a state of physical and mental exhaustion in long-lasting, emotionally demanding situations often associated with the work environment [10].

In turn, Maslach, Jackson, and Leiter [10] have noted that it is not the employees' fault that they experience burnout, and the strong influence of the work environment either contributes to professional burnout or its prevention. According to the basic theoretical principles of burnout, it is considered the main mediator between chronic work stressors and different outcomes of professional activities and attitudes. Among them, the intention of employees to change their place of employment was also proposed in several studies [11] that confirmed the main result of burnout.

4. Burnout Syndrome in the Work Environment

Today's employment places are characterized by a faster pace of work, increased self-fulfillment expectations, increased reliance on interpersonal coordination to accomplish tasks, and increased changes resulting in job insecurity. Against this background, the impact of psychosocial job conditions on mental health, as well as on cardiovascular and musculoskeletal disorders, was discussed [12]. It was pointed out that anxiety development might be facilitated by a prolonged stress situation caused by overload, as well as by previously experienced alerts. Anxiety can arise from unclear goals and career orientation, low self-esteem, and self-reliance, which are personality components that significantly contribute to or hinder the effectiveness of the professional activity and the self-efficacy of the social worker.

A social worker is a facilitator of change and is largely linked to the goals of individual clients, engaging directly or indirectly in various psychosocial risk situations and promoting changes through direct interaction with a person/client. In the long term, the multifaceted and often unsolvable challenges of diverse social workers' activity have negative consequences if the professional activity is influenced by various burnout factors. Burnout syndrome impairs both personal and social functioning. Interventions for reducing burnout and promoting involvement can take place at the organizational and personal levels [10], as professional burnout is an unfavorable response to the stress at work, including psychological, psychophysiological, and behavioral components.

Dr. Cirule, an occupational health therapist in Latvia, has indicated, based on the Health Examination Cabinet Regulation No. 219, that several psycho-emotional factors can be identified: night shifts, increased responsibility, frequent making of decisions with a significant impact on human life or health or people's quality of life, and responsibility for great material values [13]. All of the above-mentioned factors contribute to high psycho-emotional stress at work, including psycho-emotional overload.

Purviņa noted that, in many ways, burnout could be similar to depression [14]. The doctor states that burnout is a more moderate form of depression, not a categorically different condition. This phenomenon must be regarded as a significant mental health condition and, consequently, as a major obstacle to the ability to carry out one's work. Purviņa pointed out that burnout syndrome and depression had common and distinctive features. The common features include visible weakness, depressiveness, and reduced ability to work, while the distinctive features include isolation, especially from work, and negative thoughts about work.

5. Pre-Requisites for Burnout Syndrome and Intent to Terminate Employment

An in-depth content analysis of the scientific literature reveals the impact of the interaction between stress, work autonomy, and social support on job stress prediction. The Job-Demand-Control model [15] and the extended Job-Demand-Control-Support model [16] predict that workers in conditions of high demand, low control, and low support are at the highest risk of psychological impairments. However, the risk of psychological stress can be reduced by changing job-related factors. The Job Demands-Resources burnout model [17] describes that work autonomy and social support diminish the link between job-related stress and burnout. While several studies emphasized that social support and job autonomy provided a buffer between job-related stress and burnout [18], little attention has been paid to the impact of the interaction of job conditions to predict the intention to change a career. It is generally recognized that supportive job conditions help employees cope with job-related stress and, thus, cause employees' attachment to the current organization [19]. Based on this idea, Nissly, Mor, Barak, and Levin [20] suggested that social support cushions the negative impact of job and family conflicts among social workers regarding intentions to change their place of employment. Applying the job stress model and the burnout model, we hypothesize that social workers experience a higher level of burnout and desire to change employment when job stress is high, and job autonomy and social support are limited.

6. Empirical Data in Latvia

To determine the relevance and significance of burnout syndrome issues in Latvian society as a whole, we collected research data from the Burnout of Population: Health Monitoring in 2021 (retingi.lv, 2021). Evidence shows that burnout has been a significant problem recently, and the COVID-19 pandemic has even contributed to the onset of burnout. According to this research, more than half, or 58% of respondents, experienced burnout symptoms (23% often and 35% occasionally). However, according to the results of health monitoring obtained in cooperation with the public opinion research center SKDS, another 28% of respondents faced burnout rarely, 11% never faced burnout, but 3% found it "hard to say". This survey revealed that women were more likely to experience burnout: 27% of surveyed women experience burnout symptoms often, 37% experience symptoms sometimes, 25% experience symptoms rarely, 6% experience symptoms never, and 4% found it "hard to say". Meanwhile, 19% of men responded that they encountered burnout often, 32% encountered burnout sometimes, 31% encountered burnout rarely, while 15% encountered burnout 'never', but 3% found it "difficult to say". Regarding age groups, the most affected respondents by burnout were aged 25−34, with 32% experiencing burnout often, 40% experiencing burnout occasionally, 19% experiencing burnout rarely, 7% experiencing burnout never, and 2% finding it "difficult to say". Meanwhile, seniors aged between 64 and 75 were the least affected by burnout syndrome: 10% reported being affected by burnout often, 34% reported being affected by burnout sometimes, 32% reported being affected by burnout rarely, 18% reported being affected by burnout never, and 6% found "difficult to say".

In turn, in the research study on professional burnout of social workers, there were 235 participants/social work specialists, among whom six were men, but 226 were women (one person had no data). The average age of participants was 46.4 years old. The differences in terms of gender were not considered further because of the high prevalence of women.

Most of the respondents (95% of the study participants) indicated that they have been doing social work for more than 12 years, so despite the heavy workload and possible psychosocial and psycho-emotional risks, the social workers have carried out their work for a long time; this can be explained by the theory of Dolaard et al. [19], which stated that for employees, their sense of affection and support in stressful situations were also influenced by job conditions.

Survey data showed that 84 respondents sometimes felt depressed and weak, 80 respondents indicated that after work, they sometimes wanted to distance themselves from everyone and be alone for some time, 51 respondents had such feelings often, and 16 respondents felt this way every day. A total of 67 respondents indicated that they sometimes became insensitive towards clients, 60 respondents felt this very rarely, 52 respondents rarely, and 24 respondents never felt this way. A total of 65 respondents often experienced feelings that their work turns them into a cruel person, 59 respondents sometimes felt this, 47 respondents "rarely" felt this way, 11 often felt this way, 7 very often felt this way, and 1 felt this way every day.

A total of 80 respondents stated that they have performed significant things in life often, 57 respondents stated they have performed significant things in life sometimes, but 53 respondents stated they have performed significant things in life very often. A total of 88 respondents said that they often easily dealt with emotional problems, 74 respondents said that they sometimes dealt with emotional problems, 41 said that they very often dealt with emotional problems, and 3 said that they dealt with emotional problems every day. A total of 71 respondents indicated that indifferent attitudes appeared sometimes, 15 indicated that indifferent attitudes appeared very often, 3 indicated that indifferent attitudes appeared every day, and 25 indicated that indifferent attitudes never appeared. A total of 34 respondents indicated that it was common when work-related duties were delegated to them by their colleagues: 18 said it was very common, 52 said it was often, 5 said it was every day, but 21 said it was never.

Concerning psycho-emotional changes in their daily lives, 124 respondents indicated possible memory impairment, 69 had attention disorders, 53 were angry quickly, 39 were often nervous, 102 had difficulty falling asleep, 72 had health disorders, such as high blood pressure and palpitations, 52 had stomach disorders, and 19 often cried without a reason. In total, 57 had no time at all for their hobbies, and 87 suffered from emotional eating. A total of 18 respondents increased the number of cigarettes, 20 respondents started drinking alcohol more often, 17 respondents experienced other manifestations: anxiety, panic, headache, unclear skin reactions, a feeling of being used, resentment, weakness, tiredness, drowsiness, and thoughts about whether they were in the right place. Sometimes, they felt indifferent to problem-solving and different changes due to aging. To understand the meaning of changes in feelings, thinking, and health and regain persistent attention, looking at everyday activities from a different point of view or trying other approaches are required because attention is attracted by the different, but not by routine and uniformity. Memory also needs to be trained; thus, the need to learn something new or complementary from time to time is also necessary. When dividing attention to several things simultaneously in daily life situations, fatigue quickly appears; thus, switching from one thing to another becomes slower, and irritation arises about the work that has to be left uncompleted.

On the question about the workload, 99 respondents noted that their workload was satisfactory; for one respondent, the workload was too small, but it might be that the respondent did not work a full working day. A total of 61 respondents felt that their workload was too high. For 65 respondents, their high workload caused anxiety. Thus, the employee who has too much work has a feeling of insufficiency and guilt, and motivation to act disappears when attempting to achieve the goals and fulfill the expectations. Consequently, stress arises, and feelings of disappointment that the expectations supposedly placed on him/her are not fulfilled.

7. Conclusions

The analysis of theoretical insights allows for the argument that professional burnout (diagnosis) of social workers is not identifiable only from the psychological stress factors of the personality from the psychosocial point of view. Occupational burnout must be understood as a response to the diverse social, economic, and psychological–pedagogical sources of chronic stress in the work environment. These are formed within the specific relationship between the clients as users of assistance services and social workers.

A positive and supportive psychosocial work environment improves the mental and physical well-being of employees. Then, social workers and clients can contribute to the effectiveness of social work. In turn, as psychosocial risks result from the inadequate planning of work, authoritarian organizations, and management, as well as unacceptable social conditions at work, they lead to negative psychological, physical, and social consequences, such as stress, anxiety, burnout syndrome, and depression, which belongs to the second largest group of job-related health problems identified by the population.

Analyzing the research conducted in the world and Latvia on the psychosocial risks of burnout syndrome in the work environment, as well as the consequences of psychological stress in professional activities, it is found that the above factors are significant contributors to depersonalization in the daily professional life of social workers. The professional efficiency of social workers can rapidly and critically decrease, causing them to experience burnout syndrome in the workplace, which is promoted by negative interpersonal relationships and destructive interaction in extreme professional life situations.

To solve the burnout syndrome problem in the field of contemporary social work, preventive education of future social workers in the components of professional mastery during the study process at the university and learning the specific aspects of professional identity becomes the primary issue.

Professional practice in social work can be improved in a targeted and high-quality way by researching the quality and efficiency of social workers' professional activities. It is essential to carry out timely professional testing and self-reflection of professional activities, report its results, recognize the symptoms of burnout on time, and respond adequately; this develops and implements an organizational strategy/action plan for solving the problems of a particular employee. Qualitatively and purposefully planned prevention in social work by implementing timely identification of burnout syndrome of social workers, development of support strategies and their competent application in the work environment, as well as the regular implementation of supervision, can contribute to the positive well-being of social workers. The development of educational support programs is important for increasing the self-efficacy of personalized professional activities.

Author Contributions: Conceptualization, A.S. and S.S.; methodology, A.S.; software, S.S.; validation, A.S. and S.S.; formal analysis, A.S. and S.S.; data curation, S.S.; writing—original draft preparation, A.S. and S.S.; writing—review and editing, A.S. and S.S. All authors have read and agreed to the published version of the manuscript.

Funding: This research received no external funding.

Institutional Review Board Statement: Not applicable.

Informed Consent Statement: Not applicable.

Data Availability Statement: Data is unavailable due to ethical restrictions.

Conflicts of Interest: The authors declare no conflict of interest. The funders had no role in the design of this study, in the collection, analyses, or interpretation of data, in the writing of the manuscript, or in the decision to publish the results.

References

1. Gomez-Garcia, R.; Alonso-Sangregorio, M.; Liamazares-Sanchez, M.L. Burnout in social workers and socio-demographic factors. *J. Soc. Work.* **2019**, *20*, 463–482. [CrossRef]
2. Holmes, M.R.; Rentrope, C.R.; Korsch-Williams, A.; King, J.A. Impact of COVID-19 Pandemic on Posttraumatic Stress, Grief, Burnout, and Secondary Trauma of Social Workers in the United States. *Clin. Soc. Work. J.* **2021**, *49*, 495–504. [CrossRef] [PubMed]
3. Roga-Vailza, V.; Ozola, I.; Apine, E. *Sociālais darbs ar gadījumu. Prakse teorijā*; Latvijas Republikas Labklājības Ministrija: Riga, Latvia, 2021.
4. World Health Organization. Burn-Out "Occupational Phenomenon". International Classification of Diseases. Available online: https://www.who.int/mental_health/evidence/burn-out/en/ (accessed on 1 July 2022).
5. Freudenberger, H.; Richelson, G. *Burnout: The High Cost of High Achievement*; Anchor Press: New York, NY, USA, 1980.
6. Lee, R.T.; Ashforth, B.E. On the meaning of Maslach's three dimensions of burnout. *J. Appl. Psychol.* **1990**, *75*, 743–747. [CrossRef] [PubMed]
7. Maslach, C. Burned-out. *Hum. Behav.* **1976**, *5*, 16–22.
8. Maslach, C.; Jackson, S.E. The Measurement of Experienced Burnout. 1981. Available online: https://smlr.rutgers.edu/sites/default/files/Documents/Faculty-StaffDocs/TheMeasurementofExperiencedBurnout.pdf (accessed on 1 July 2022).
9. Pines, A.; Aronson, E.; Kafry, D. *Vom Überdruss zur Selbstentfaltung*; Klett-Cotta: Stuttgart, Germany, 1992.
10. Maslach, C.; Jackson, S.E.; Leiter, M.P. Maslach Burnout Inventory. In *Evaluating Stress: A Book of Resources*, 3rd ed.; Zalaquett, C.P., Wood, R.J., Eds.; CPP: Palo Alto, CA, USA, 1997; pp. 191–218.
11. Huang, I.; Chuang, C.J.; Lin, H. The role of burnout in the relationship between perceptions of organizational politics and turnover intentions. *Public Pers. Manag.* **2003**, *32*, 519–531. [CrossRef]
12. Andel, R.; Crowe, M.; Kareholt, I.; Wastesson, J.; Parker, M.G. Indicators of job strain at midlife and cognitive functioning in advanced old age. *J. Gerontol. B Psychol. Soc. Sci.* **2011**, *66*, 287–291. [CrossRef] [PubMed]
13. Cīrule, J. Work Environment Psycho-Emotional Risk Factors and Occupational Diseases. 2019. Available online: http://stradavesels.lv/Uploads/2019/11/21/02_Jolanta_Cirule.pdf (accessed on 1 July 2022).
14. Purviņa, S. Izdegšanas Sindroms. 2020. Available online: http://stradavesels.lv/Uploads/2020/02/27/02_Santa_Purvina.pdf (accessed on 15 October 2022).
15. Karasek, R.; Theorell, T. *Healthy Work*; Basic Books: New York, NY, USA, 1990.

16. Johnson, J.V.; Hall, E.M. Job strain, work place social support, and cardiovascular disease: A cross-sectional study of a random sample of the Swedish working population. *Am. J. Public Health* **1988**, *78*, 1336–1342. [CrossRef] [PubMed]
17. Demerouti, E.; Bakker, A.B.; Nachreiner, F.; Schaufeli, W.B. The job demands-resources model of burnout. *J. Appl. Psychol.* **2001**, *86*, 499–512. [CrossRef] [PubMed]
18. Bakker, A.B.; Demerouti, E.; Euwema, M.C. Job resources buffer the impact of job requirements on burnout. *J. Occup. Health Psychol.* **2005**, *10*, 170–180. [CrossRef] [PubMed]
19. Dollard, M.F.; Winefield, H.R.; Winefield, A.M.; DeJonge, J. Psychological job strain and productivity in human service workers: A test of the demand-control-support model. *J. Occup. Organ. Psychol.* **2000**, *73*, 501–510. [CrossRef]
20. Nissly, J.A.; Mor Barak, M.E.; Levin, A. Stress, social support, and workers' intentions to leave their jobs in public child welfare. *Adm. Soc. Work* **2005**, *29*, 79–100. [CrossRef]

engineering proceedings

Proceeding Paper

Effect of an Active Reach-Out Program on Hepatitis C Screening Test for Drug-Addicted Inmates in Southern Taiwan †

Hsiao-Fang Chen [1,2] and Jin-Yuan Chern [3,*]

1 Business and Operations Management, Chang Jung Christian University, Tainan 71101, Taiwan; v7633333@gmail.com
2 Tainan Hospital, Ministry of Health and Welfare, Tainan 70043, Taiwan
3 Department of Health Care Administration, Chang Jung Christian University, Tainan 71101, Taiwan
* Correspondence: chern@mail.cjcu.edu.tw
† Presented at the IEEE 5th Eurasia Conference on Biomedical Engineering, Healthcare and Sustainability, Tainan, Taiwan, 2–4 June 2023.

Abstract: Chronic hepatitis C is an important threat to the world's public health. In Taiwan, 2~4% of the population is infected with hepatitis C, and 10~15% of those cases will lead to liver cirrhosis. This study examined the effect of a comprehensive screening test project conducted by the Addiction Treatment Center in southern Taiwan. In collaboration with the Drug Addiction Treatment Center, 154 screening tests were completed. It is demonstrated that through active reach-out screening service with innovative process design, the vulnerable groups of people with a potentially high prevalence of HCV could be targeted and cured. Nevertheless, close surveillance and follow-up would be necessary to prevent the reoccurrence.

Keywords: hepatitis C treatment; drug treatment center; drug-addicted inmate

Citation: Chen, H.-F.; Chern, J.-Y. Effect of an Active Reach-Out Program on Hepatitis C Screening Test for Drug-Addicted Inmates in Southern Taiwan. *Eng. Proc.* **2023**, *55*, 14. https://doi.org/10.3390/engproc2023055014

Academic Editors: Teen-Hang Meen, Kuei-Shu Hsu and Cheng-Fu Yang

Published: 28 November 2023

1. Introduction

According to the statistics of the Ministry of Health and Welfare, Taiwan, cancer ranks first among the top ten causes of death, and liver cancer ranks second among the top ten cancers [1]. The statistical results show that hepatitis, liver cirrhosis, liver cancer, and other diseases are seriously threatening the lives and health of Taiwanese people.

Hepatitis C is contracted through blood transfusion and is mainly characterized by long-term chronic symptoms. It is estimated that the global prevalence of hepatitis C is about 2.8%, while the prevalence rate in Taiwan is 3.28% [2]. A further analysis shows that the prevalence rates could reach 15% or even higher in certain towns or villages in the southwest area.

In terms of the development process of hepatitis C virus infection, about 70% to 80% of the patients who suffer from acute hepatitis C will develop into chronic hepatitis C, and among the patients with chronic hepatitis C, nearly 20% may turn into liver cirrhosis in 20 years. Those cirrhotic hepatitis C patients have a 1–4% chance of developing hepatocellular carcinoma and a 4–5% chance of liver function decompensation each year, which may result in considerable subsequent treatment costs [3]. Based on the 2020 National Health Insurance Administration statistics for the medical expenditures of cancer-related inpatient/outpatient health services, the number of outpatient visits for cancer was 181,000, and the annual health expenditure was about NTW 22.33 billion, all indicating that the financial burden for cancer care becomes an important issue and cannot be underestimated.

In September 2015, the World Health Organization (WHO) convened the World Hepatitis Summit in Glasgow, Scotland, at which a "Glasgow Declaration on Viral Hepatitis" was drawn up on the prevention and treatment of hepatitis. It was declared that universal access to the prevention, diagnosis, care, and treatment of viral hepatitis is a fundamental human right, and, therefore, calls on governments to work with all stakeholders, including non-profit organizations, medical professionals, and pharmaceutical companies, to develop and implement comprehensive hepatitis funding programs dedicated to eliminating this significant public health threat. The Declaration clearly highlights the goal of eliminating the threat of hepatitis C by 2030 [4].

Taiwan is one of the few countries in the world where the prevalence of hepatitis C exceeds 3%, much higher than that of other Asian neighboring countries such as Japan (0.98%) and South Korea (1.2%). Therefore, it becomes an urgent issue for the authorities in Taiwan to launch necessary healthcare strategies for the prevention and treatment of hepatitis C. In order to achieve the goal of eliminating hepatitis C by 2025 [5–7], the National Hepatitis C Elimination Office was established in 2016 to initiate the "National Hepatitis C Elimination Policy Program 2018–2025 (referred to as the Policy Program)". It was expected that a target of 250,000 people treated with direct-acting antiviral agent (DAA) for hepatitis C would be reached by 2025, and a core strategy and policy direction for hepatitis C elimination would be generated [8].

Among all the drug addicts who suffer from hepatitis C, the inmates in drug rehabilitation institutions are usually underprivileged groups of people. Therefore, the new cases of hepatitis C infection in Taiwan are mainly intravenous drug addicts or people with a higher risk of infection, especially younger males. In view of this, the subject hospital (a government-affiliated region-level hospital in southern Taiwan) undertook the responsibility of caring for the vulnerable people in the rural areas. To put this in practice, they took the initiative to ask the authorities, the Central Health Insurance Administration, for assistance in diagnosis and treatment to ensure patient safety.

2. Literature Review

2.1. Introduction to Hepatitis C

Hepatitis C virus is mainly transmitted through blood or body fluids. The transfusion of blood or plasma products is a well-known transmission route. In the United States, the incidence of infection due to blood transfusion is about 6%, but it could reach up to 42% with intravenous drug use. Other possible transmission channels include sexual behavior (6%), family contact (3%), medical issues (2%), and dialysis (0.6%). In Taiwan, the prevalence of serum antibodies of hepatitis C among family members is about 5.4%, while the positive rate of spouses' antibodies is 21%. It is, thus, suspected that prolonged contact may be the main cause of mutual infection between spouses. Pregnant women with risk factors (such as HIV infection) will affect the virus concentration, making the newborns more susceptible to hepatitis C virus infection, and they become chronic carriers, as defined by a detectable existence of HCV RNA in the blood of the patient for more than six months.

About 20–40% of people who are infected with hepatitis C can clear the virus on their own at the initial stage, and the rest will develop into chronic hepatitis C, some of whom may gradually deteriorate into liver cirrhosis or liver cancer within 40 years [4]. Clinically, the production efficiency of the hepatitis C virus in the human body is very high, and the copied gene body can often undergo mutations. Therefore, the virus in patients is seldom of a single species but a collection of many different mixed virus groups. This phenomenon may be related to the immune response of virus escape from the host. About 80% of people infected with hepatitis C will develop a chronic infection, and the speed of disease progression is influenced by certain factors such as age and gender. For example, among women or young infected people, the disease progression is relatively slow. It usually takes about 30 years to develop into chronic hepatitis C. However, for those who are male, older, alcoholic, or co-infected with hepatitis B or HIV, progressing into cirrhosis may be completed within 20 years, and there is a 1–5% chance of developing liver cancer after the occurrence of cirrhosis [3].

Hepatitis C virus genotype (HCV Genotype) can be classified into 1a, 1b, 2, 3, 4, 5, 6 and other types. Each genotype of the hepatitis C virus needs a different treatment. The treatment courses and therapeutic effects are different as well. Before using drugs to treat hepatitis C, the degree of liver fibrosis (hepatic fibrosis) must be diagnosed with abdominal ultrasonography. Based on the Metavir score, the diagnosis falls into one of the following five stages: F0 (normal liver tissue, without fibrosis), F1 (mild fibrosis, fibrosis limited to the hepatic portal area, no septum), F2 (moderate fibrosis, a few fiber bundles protrude from the hepatic hilar area to the lobules, with a few septa), F3 (severe fibrosis, many fiber bundles protrude from the hepatic hilar area to the lobules, with many septa), and F4 (severe fibrosis, fibrous tissue coherent and formed a circle, which can be diagnosed as liver cirrhosis) [9,10].

If the hepatitis C (Anti-HCV) antibody is detected positive, it means that the patient has been infected with the hepatitis C virus. Sometimes, the virus may not be detected in the blood because the hepatitis C antibody in the blood may continue to be positive for many years. Anyway, when the hepatitis C (Anti-HCV) antibody is positive, the patient needs to be referred to a medical institution for further testing of hepatitis C virus ribonucleic acid (HCV-RNA) and hepatitis C virus genotype (HCV Genotype). A positive test result for hepatitis C virus ribonucleic acid (HCV-RNA) indicates that the virus has been detected in the blood. The value of this test can be used to assess whether treatment should be given and the possible effect of treatment [9,10].

2.2. Status Quo in Taiwan

The prevalence rate of anti-hepatitis C virus antibodies in Taiwan is about 3.28% [2]. It is, thus, estimated that about 740,000 people have been infected with the hepatitis C virus, of which 550,000 people might develop chronic hepatitis C. The prevalence rates in certain townships in southwestern Taiwan could even reach 15% or higher, such as 38% of adults in Ziguan District, 35% in Taoyuan District, 16% in Alian District (all in Kaohsiung City), and 23% in Baisha Township (Penghu County). Surprisingly, the Mashagou area (in Tainan City) even has a prevalence rate as high as 67% [2].

Genotype 1b is the most widespread infection of hepatitis C virus in Taiwan, accounting for 50–70% of cases. Patients with genotype 1b infection usually have a more severe disease course and are more likely to develop liver cancer. About 60.3% of HCC patients infected with the hepatitis C virus have genotype 1b, while only about 16.8% of HCC patients are infected with genotype 2a or 2c, indicating that genotype 1b is the major risk factor for liver cancer.

In addition, the distribution of hepatitis C virus genotypes differed significantly between urban and rural areas in Taiwan. In northern Taiwan, genotype 1b accounts for 58–73% of cases, and genotype 2a for 7.4–26%. Nevertheless, in southern Taiwan, genotype 1b accounts for 48–64.3%, and genotype 2a for 35.7–41.4%. Taking Kaohsiung as an example, the Kaohsiung metro area, Ziguan Township, and Taoyuan Township were mainly infected with genotype 1b, accounting for 47%, 61.9%, and 76.9%, respectively. Nevertheless, Mashagou was massively infected with genotype 2a, accounting for about 63.5% [10].

A study by Yu et al. (2015) showed that, during the period 2012–2013, nearly 45% of the patients with chronic hepatitis C in Taiwan did not receive any treatment. One of the main reasons was that patients were afraid of possible adverse side effects caused by the interferon during the treatment process (36.9%), and what is more, elderly patients over 65 years old were 2.3 times more likely to be afraid of side effects than the general patients. Other reasons include failure to meet health insurance reimbursement regulations (17.6%) and lack of awareness of treatment (11.3%). The study also showed that among those with hepatitis C, only 8.1% were willing to seek treatment from a doctor's office and achieved successful treatment.

The Central Health Insurance Administration of the Ministry of Health and Welfare allocated NTW 3.101 billion and NTW 4.5 billion in 2016 and 2017, respectively. The funds would be used to pay for the treatment of hepatitis C and management of new oral hepatitis C drugs through optimal distribution operations to help avoid possible consumption of medical services and social resources with the development of liver cirrhosis and liver cancer in the future. With the aid of the funds, the goal of comprehensively eradicating hepatitis C is also hoped to be achieved in the near future. According to the report by the National Office for Hepatitis C Treatment, the 12-week viral load detection efficacy of the new oral drug treatment for hepatitis C was demonstrated to be remarkable, reaching a high value of 97.1% [6].

The main purpose of hepatitis C treatment is to avoid liver-related complications, and the treatment goal is to eliminate the virus so that the amount of virus in the serum can continue to be below the detection limit, i.e., to achieve the so-called "sustained viral response" (SVR, sustained virologic response). For example, SVR12 indicates the quantity of hepatitis C virus in the serum tested 12 weeks after the end of the treatment course. If no virus load is detected, the drug treatment is successful. In practice, some patients who have achieved SVR will still have their viral load detected after several years of follow-up. In those cases, most are due to re-infection rather than relapse of the original hepatitis C [11].

2.3. Rehabilitation Institutions

In Taiwan, most of the inmates in governmental correction institutions were jailed because of drug use, with an average age of 35 to 40 years old. The infection of hepatitis C is mainly through the abuse of injection drugs or the use of virus-contaminated injection needles, acupuncture needles, tattoo needles, etc. In view of the fact that most drug addicts are infected with hepatitis B, C, and HIV due to drug use, the Center for Disease Control has actively promoted the Harm Reduction Program since August 2005 to provide clean needles for drug addicts. However, since drug addicts still share needles when conducting injections, the risk of contracting blood-borne infectious diseases cannot be avoided. Therefore, for those who have been cured of drug addiction in rehabilitation institutions, free medical resources are provided.

In line with the government's hepatitis C eradication policy, hepatitis C screening provides medically underprivileged prisoners with the opportunity to receive effective medical care. Tainan Prison Mingde Rehabilitation and Treatment Branch is the first public drug rehabilitation institution in the country. It is located in Mingde Minimum Security Prison, Shanshang District, Tainan City. It is a long and arduous road for drug-abuse inmates to rid themselves of their psychological craving and dependence on drug addiction. Because the prison is a closed environment, the inmates are unlike ordinary patients. When they feel uncomfortable, they become less willing to go to the hospital. In addition, they have seen the treatment experience of other inmates. Therefore, they are more psychologically prepared to accept those uncomfortable symptoms. After completing the course of treatment, liver C is almost healed. Consequently, if complete and continuous professional services could be provided for drug addicts in rehabilitation facilities, not only could a drug-free environment be created to help inmates develop a new healthy lifestyle, but active medical treatment may also help discover health problems for the inmates.

3. Materials and Methods

A professional staff representative took the initiative to contact the Hepatitis Prevention and Control Association first and then a public drug rehabilitation institution based in the southern district. Through a briefing to the Superintendent, a task force was organized to facilitate the project planning and monitor subsequent operation process, especially acquiring the consent of the prisoners for blood testing and follow-up consultation and treatment. The operation process is illustrated as follows.

3.1. Contact and Scheduling

Arranged a professional staff representative to contact the authority, asking about the inmate's willingness to undergo hepatitis C screening and obtaining consent, followed by scheduling of blood screening time.

In order to make the inmates feel at ease to accept the examination, a professional team comprising a hepatologist/gastroenterologist and nurses was arranged to give lectures to the inmates. Through the lectures, the inmates could understand the impact of the disease caused by hepatitis C and explained the overall operation process. After the inmates' concerns were relieved and consent forms collected, a blood drawing timetable was scheduled to test for hepatitis C antibodies. If the hepatitis C (Anti-HCV) antibody turned out positive, further testing of hepatitis C virus ribonucleic acid (HCV-RNA) and hepatitis C virus genotype (HCV Genotype) was conducted.

3.2. Fund-Raising and Arrangement

With funding support from the Tainan Liver Disease Prevention Foundation, free hepatitis C antibody screening and HCV RNA TYPE and viral load testing were made possible.

The average cost of screening tests is about NTW 2000 for each subject, which might be too expensive for the inmates to pay for. Therefore, the task force decided to raise funding from the Tainan City Liver Disease Prevention Foundation.

3.3. Treatment Strategy

In consideration of the inmate's privacy and patient's safety, a customized treatment strategy was designed and proposed for the Superintendent's approval.

Normally, the inmates will be in handcuffs and shackles when seeking medical services outside correctional institutions. To protect the inmate's privacy, the doctor's consultation room was moved from the original place located on the first floor to the basement, and an abdominal ultrasound machine was set up in the consultation room. After the doctor's consultation, the abdominal ultrasound examination could be performed in the same spot. Based on the diagnosis, a prescription was issued immediately so that each inmate could possibly return to the institution within 2 h. In addition, prescriptions would be delivered to the institution by an on-duty RN, and health education was offered if needed. Meanwhile, the institution counselor provided assistance in confirming whether the inmates took the medicine as prescribed. When the on-duty RN delivered the medicines to the institution, they would conduct blood-drawings for each inmate every week and check if any symptoms or side effects occurred.

3.4. Financial Support from Social Service Fund

In view of the fact that some inmates are in financial difficulties and unable to pay medical expenses, assistance in applying for social service funds was offered.

Basically, most inmates are economically disadvantaged. Therefore, it would be very difficult, if not impossible, for the inmates to participate in this project voluntarily. To counter this challenge, with the aid of the institutional counselor, inmates in financial need were listed and the social service fund of the hospital was used to pay for the medical expenses.

4. Results

Under the pandemic threat, two case managers were assigned to perform blood-drawing procedures in a public drug rehabilitation institution in the southern district. A total of 154 people were screened, of which 83 were positive for hepatitis C antibody screening, with a positive rate of 53.2%. There were 70 subjects with detectable viral load. It is obvious that the prevalence rate of hepatitis C in drug addicts is more than 18 times that of the general population, which is 2–3%.

For those with viral load, the age distribution shows that 21 subjects were 41–45 years old, accounting for 30%, followed by 51–55 years old (20 subjects, 28.5%), 46–50 years old (17 subjects, 30%), There were even five subjects (7.2%) who were 56–65 years old (Table 1).

After consulting with the Superintendent, a treatment strategy was initiated, and corresponding procedures were conducted as of March 2019. The Xinhua branch hospital was appointed as the facility where treatment and service would be delivered. Among the 70 subjects with detectable viral load, 50 were retained for the treatment plan with new oral hepatitis C drugs. After a three-month follow-up, the viral load was tested again, and it turned out that no viral load reaction was detected. Moreover, 2 out of the 50 subjects did not complete the course of treatment due to personal factors; therefore, the cure rate was 97.9% (Table 2).

Table 1. Age distribution of the subjects (n = 70).

Age Category	n	%
35–40	7	10%
41–45	21	30%
46–50	17	24.3%
51–55	20	28.5%
56–65	5	7.2%
subtotal	70	100%

Table 2. Viral load of the subjects (n = 50).

Viral Load IU/mL	# of Patients	%
1–1999	12	24%
2000–2999	10	20%
3000–3999	6	12%
4000–4999	2	4%
5000–5999	8	16%
6000–6999	7	14%
7000–7999	2	4%
8000–8999	1	2%
Above 9000	2	4%
subtotal	50	100%

Further analysis shows that among the 50 subjects who received new oral drug treatment for hepatitis C, 16 were with genotype 6a, accounting for 32%, followed by genotype 1a (9 subjects, 18%) and genotype 3a (7 subjects, 14%) (see Table 3). In terms of liver fibrosis, 35 patients (70%) fell into grade F0, followed by F1 (19 patients, 18%), F2 (5 patients, 10%), and F3 (1 patient, 2%) (see Table 4). As for the medication, most were given Epclusa (21 patients, 42%), followed by Zepatie (18 patients, 36%), Marviret (8 patients, 16%), and Zepatier + RBV (3 patients, 6%), respectively (see Table 5). In order to reduce the medical burden of individual cases, the social service fund assisted in paying a total of 24 person-times.

Table 3. Hepatitis C genotype of the subjects (n = 50).

Genotype	# of Patients	%
1a	9	18%
2a	5	10%
3a	7	14%
6a	16	32%
1b	7	14%
1a + 1b	2	4%
1a + RAS	3	6%
1a + 2	1	2%
subtotal	50	100%

Table 4. Liver fibrosis of the subjects (n = 50).

Liver Fibrosis	# of Patients	%
F0	35	70%
F1	9	18%
F2	5	10%
F3	1	2%
subtotal	50	100%

Table 5. Hepatitis C medication of the subjects (n = 50).

Medication	# of Patients	%
Zepatier	18	36%
Marviret	8	16%
Epclusa	21	42%
Zepatier + RBV	3	6%
subtotal	50	100%

5. Discussion and Implications

According to a study by the Ministry of Justice, the recidivism rate of domestic drug users is as high as 80%, and in clinical experience, the success rate of life-long detoxification of heroin is only 10%. The harm caused by drugs is huge and long-term to individuals, families, and society. Most drug addicts will not seek medical advice, and, thus, experience delayed diagnosis and treatment. Significantly, many inmates entering drug rehabilitation institutions are relatively more underprivileged than the general public; therefore, they need more active medical support and assistance from medical institutions.

Through comprehensive active screening, inmates may have a better chance of receiving necessary screening tests and subsequent required medical treatment without strange vision. With a customized consultation process design, medical service providers could earn the trust of inmates in the drug rehabilitation institution and successfully complete liver C screening and treatment for the underprivileged who might otherwise delay the necessary treatment.

During this process, it was found that inmates in drug rehabilitation institutions infected with hepatitis C were mostly due to needle sharing. While the cure rate of hepatitis C is high, incidental contact with injecting drugs may easily lead to repeated infection of hepatitis C after the inmates leave the prison. It is suggested that special attention and tracking efforts must be paid to this group, supplemented with spiritual support to prevent drug addicts from going astray again.

Author Contributions: Conceptualization, H.-F.C. and J.-Y.C.; methodology, H.-F.C. and J.-Y.C.; software, H.-F.C.; validation, H.-F.C. and J.-Y.C.; formal analysis, H.-F.C.; investigation, H.-F.C.; resources, H.-F.C.; data curation, H.-F.C.; writing—original draft preparation, H.-F.C.; writing—review and editing, J.-Y.C.; visualization, H.-F.C. and J.-Y.C.; project administration, H.-F.C.; funding acquisition, H.-F.C. All authors have read and agreed to the published version of the manuscript.

Funding: This research received no external funding.

Institutional Review Board Statement: Not applicable.

Informed Consent Statement: Informed consent was obtained from all subjects involved in the study.

Data Availability Statement: Data are contained within the article.

Conflicts of Interest: The authors declare no conflict of interest.

References

1. Ministry of Health and Welfare. Cause of Death Statistics. 2022. Available online: https://www.mohw.gov.tw/np-126-2.html (accessed on 5 April 2022).
2. Yu, M.L.; Yeh, M.L.; Tsai, P.C.; Huang, C.I.; Huang, J.F.; Huang, C.F.; Hsieh, M.H.; Liang, P.C.; Lin, Y.H.; Hsieh, M.Y.; et al. Huge gap between clinica l efficacy and community effectiveness in the treatment of chronic hepatitis C: A nationwide survey in Taiwan. *Medicine* **2015**, *94*, 1–8. [CrossRef] [PubMed]
3. Maasoumy, B.; Wedemeyer, H. Natural history of acute and chronic hepatitis C. *Best Pract. Res. Clin. Gastroenterol.* **2012**, *26*, 401–412. [CrossRef] [PubMed]
4. WHO. Global Health Sectors Strategy on Viral Hepatitis 2016–2021: Towards Ending Viral Hepatits. Available online: https://www.who.int/hepatitis/strategy2016-2021/ghss-hep/en/ (accessed on 18 February 2021).
5. Chen, D.S. Taiwan commits to eliminating hepatitis C in 2025. *Lancet Infect. Dis.* **2019**, *19*, 466–467. [CrossRef] [PubMed]
6. Ministry of Health and Welfare. Taiwan Hepatitis C Policy Guideline 2018–2025. 2018. Available online: https://www.mohw.gov.tw/cp-4464-49019-1.html (accessed on 5 April 2022).
7. Wu, G.H.M.; Pwu, R.F.; Chen, S.C. Achieving hepatitis C elimination in Taiwan—Overcoming barriers by setting feasible strategies. *J. Formos. Med. Assoc.* **2018**, *117*, 1044–1045. [CrossRef] [PubMed]
8. Hammerstad, S.S.; Grock, S.F.; Lee, H.J.; Hasham, A.; Sundaram, N.; Tomer, Y. Diabetes and hepatitis C: A two-way association. *Front. Endocrinol.* **2015**, *6*, 134. [CrossRef] [PubMed]
9. Hsieh, P.; Kuo, H.; Cho, W.; Liao, Y.; Lin, C. The Genotype of Hepatitis C Virus Has Important Clinical and Therapeutic Implications. *J. Intern. Med. Taiwan* **2009**, *20*, 309–319.
10. Lee, C.; Hung, C.; Lu, S.; Changchien, C. Hepatitis C virus genotypes: Clinical relevance and therapeutic implications. *Chang. Gung Med. J.* **2008**, *31*, 16–25. [PubMed]
11. Cuypers, L.; Pérez, A.B.; Chueca, N.; Aldamiz-Echevarría, T.; Alados, J.C.; Martínez-Sapiña, A.M.; Merino, D.; Pineda, J.A.; Téllez, F.; Espinosa, N.; et al. Relapse or reinfection after failing hepatitis C direct acting antiviral treatment: Unravelled by phylogenetic analysis. *PLoS ONE* **2018**, *13*, e0201268. [CrossRef] [PubMed]

engineering proceedings

Proceeding Paper

Lightweight Network for Single Image Super-Resolution with Arbitrary Scale Factor †

Quang Truong Duy Dang, Kuan-Yu Huang and Pei-Yin Chen *

Department of Computer Science and Information Engineering, National Cheng Kung University, Tainan 70142, Taiwan; duyquang24251197@gmail.com (Q.T.D.D.); p78091523@gs.ncku.edu.tw (K.-Y.H.)
* Correspondence: pychen@mail.ncku.edu.tw

† Presented at the IEEE 5th Eurasia Conference on Biomedical Engineering, Healthcare and Sustainability, Tainan, Taiwan, 2–4 June 2023.

Abstract: The existing single image super-resolution (SISR) methods that consider integer scale factors (X2, X3, X4, and X8), have been developed well, but SISR methods with arbitrary scale factors (X1.3, X2.5, and X3.7) have gradually gained attention recently. Therefore, we proposed an efficient, lightweight model. In this study, there are two contributions as follows. (1) An efficient and lightweight network for SISR is combined with the up-scaled module, which determines its weights based on the size of the high-resolution (HR) image. (2) All scale factors are applied simultaneously using one model, which saves more storage and computational resources. Finally, we design various experiments to evaluate the proposed method based on multiple general datasets. The experimental results show that the proposed model is lightweight while the performance is relatively competitive.

Keywords: single image super-resolution (SISR); arbitrary scale factor; lightweight network

Citation: Dang, Q.T.D.; Huang, K.-Y.; Chen, P.-Y. Lightweight Network for Single Image Super-Resolution with Arbitrary Scale Factor. *Eng. Proc.* **2023**, *55*, 15. https://doi.org/10.3390/engproc2023055015

Academic Editors: Teen-Hang Meen, Kuei-Shu Hsu and Cheng-Fu Yang

Published: 28 November 2023

1. Introduction

In recent years, convolutional neural networks (CNN) have become one of the most ubiquitous machine learning solutions for computer vision tasks. CNN is used intentionally in most fields of image processing, and single image super-resolution (SISR) is one of them. SISR was known as the up-scaling method in the past, which indicates generating a high-resolution (HR) image from a single low-resolution (LR) image. The SISR techniques [1–6] based on CNN have been developed since SRCNN [1], but most of them only consider the integer scale factors (X2, X3, X4, and X8), as shown in Figure 1a. Since there are situations where users need to up-scale low-resolution (LR) images to customize the size instead of fixing the size in real-world scenarios, the SISR methods with arbitrary scale factors (X1.3, X2.5, and X3.7) become important. In addition, if we train a single specific model with every scale factor, it saves time and effort, as shown in Figure 1b. Putting all the reasons together, the researchers have come up with the idea to solve that problem.

Meta Super-Resolution (Meta-SR) [2] was proposed in 2019. Their Meta-Upscale Module is applied to up-scaling LR based on different scale factors. In contrast, the up-scaling module of the SISR methods with the integer scale factors is the deconvolution layer or sub-pixel layer at the end of networks. In particular, the sub-pixel layer [3] is widely used in SR works, such as the Residual Dense Network (RDN) [4] and residual channel attention network (RCAN) [5]. Meta-SR adopts RDN [4] as the backbone and proves that Meta-SR obtains high performance and deals with arbitrary scale factors for SISR. However, Meta-SR [2] has high complexity, and implementing it involves many challenges in terms of the hardware requirements, making it computationally expensive.

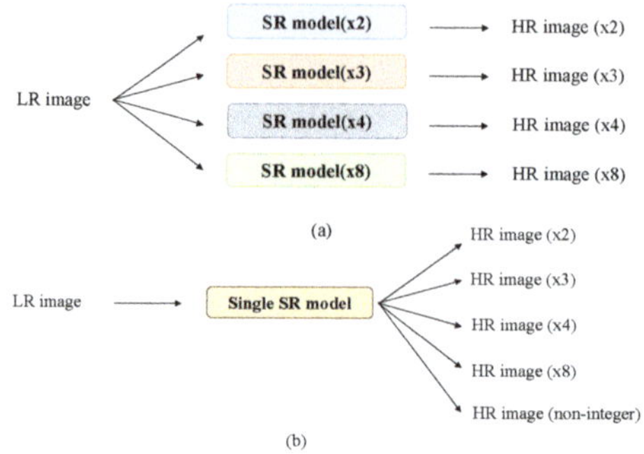

Figure 1. (**a**) Multiple SR model for different scale factors and (**b**) single SR model for the arbitrary scale factor.

Therefore, the proposed method is focused on constructing a lightweight model, which is more appropriate and likely to work in real-life scenarios. The proposed model is inspired by Meta-SR [2] and is called Light Arbitrary-SR (LAS), which is much lighter in weight than the original Meta-SR [2]. Compared to a similar study [6] of very deep super-resolution (VDSR) with arbitrary scale factors, the research result shows a better quality of HR images with lower usage of weights and computational cost.

2. Proposed Method

The proposed LAS is inspired by RCAN [5] and Meta-SR [2]. We found an efficient and lightweight network as the backbone based on RCAN [5] and combined it with the Meta-Upscale Module [2], as shown in Figure 2. One novelty in RCAN [5] is to establish a very deep network based on the residual in residual (RIR) structure. This network comprises several residual groups and long skip connections. Each group consists of multiple residual blocks and short skip connections.

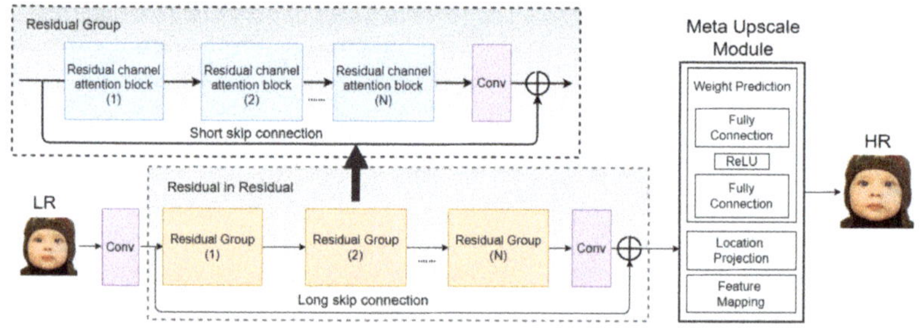

Figure 2. Architecture of the Meta-RCAN network.

Generally, RIR makes the main network concentrate on learning high-frequency information by allowing plentiful low-frequency information to be surpassed via numerous skip connections. The channel attention mechanism is also introduced to improve the representational ability of the network further. The dominant part of the network is the residual channel attention block (RCAB), which helps the network recognize informative components of the LR features efficiently. The RCAB, inspired by the success of channel attention (CA) and residual blocks (RB), helps the network learn and explore more information to improve the overall performance. RCAN is constructed using the foundation of RCAB and RIR structure.

However, a very deep RCAN brings about higher accuracy and superior results for the SR image. Thus, we aimed to build a low-complexity network for SR images with arbitrary scale factors, as the RCAN is still too complicated with a high computational cost, which makes it challenging to implement. In the original RCAN, 20 RCABs and 10 residual groups are set up, and the usage of total weights is about 16 M. To reduce the complexity and make it more appropriate for hardware implementation, we reduced it to around 90% of the entire implementation to only 3 or 6 RCABs and a single residual group.

Moreover, another highlight in the proposed LAS is the use of the Meta-Upscale Module, which has three core functions: Location Projection, Weight Prediction, and Feature Mapping. The block of Location Projection projects pixels of the HR image onto the LR image based on the scale factor and the kernel weights for each pixel on the HR image are predicted by the Weight Prediction Module. Lastly, the feature maps on the LR image and the predicted kernel weights are mapped back to the HR image using the Feature Mapping function to compute the value of the pixel of the HR image. We attempted to simplify the Meta-Upscale Module as well. Since the Weight Prediction function of the Meta-Upscale Module uses a network to predict the kernel weights using two fully connected layers, it consumes a lot of computational resources. We experimented with alleviating the neurons from 256 to 128 and then to 64 to observe their performance. Finally, the Meta-Upscale Module was simplified by reducing the number of neurons in the fully connected layer from 256 to 64. The proposed method is confirmed to be a lightweight SR method with arbitrary scale factors.

3. Experimental Result

To achieve a lightweight model of super-resolution with non-integer scale factors, we attempted to combine the Meta-Upscale Module and RCAN and simplified them. In the experiment, three versions were presented, LAS_A, LAS_B, LAS_C, and Meta-RCAN, based on a different setting. For LAS_A, we implemented three RCABs. A single residual block with a simplified Meta-Upscale Module reduced the number of neurons in the fully connected layer to 64. For LAS_B, there were six RCABs and a single residual block with a simplified Meta-Upscale Module reduced the number of neurons in the fully connected layer to 64. LAS_C contained six RCABs and one residual block with 256 fully connected layers in the Meta-Upscale Module. Lastly, Meta-RCAN indicates the use of a slightly simplified RCAN, which was set up with 16 RCABs and 10 residual groups. The setting of Meta-RCAN was adopted from the official source code in Ref. [2]. We re-trained the model and presented the test results, and we did not consider Meta-RCAN as one of the versions of LAS.

All the experiments were run in parallel on two GPUs (Nvidia GeForce GTX 1080 Ti). We used the Pytorch framework and Python 3 with CUDA (version 11.2.142). The training and testing required libraries, including Pytorch 0.5.0, Python 3.5 or higher, NumPy, ski-amge, imageio, and cv2. The training scale factors for the proposed methods varied from 1, 1.1, 1.2, 1.3, 1.4, ... to 4 with a stride of 0.1. The training dataset contained 800 images from the DIV2K [7] dataset. The test dataset was from on three datasets: Set5 [8], Set14 [9], and B100 [10]. For other details, the learning rate was decreased by half after every 200 epochs with an initialization of 10-4 for all the layers. The optimizer is Adam. For better convergence, the L1 loss function, instead of the L2, was used to train the network.

Since RCAN has a better representational ability of the model than RDN, the Meta-RCAN shows a similar value of the evaluated metric with around 40% lower parameters than Meta-RDN (Table 1). Moreover, LAS_C has more fully connected layers in the Meta-Upscale Module, so it has approximately 30% higher values of parameters than LAS_B. However, for the evaluated metric, LAS_B only has a slightly lower value than LAS_C. In the comparison of LAS_B and LAS_A, the more parameters there are, the better the quality of the image.

Table 1. Experimental results of the proposed method and comparison with other methods for the evaluated metric PSNR/SSIM.

		Bicubic	VDSR [6]	LAS_A	LAS_B	LAS_C	Meta-RCAN	Meta-RDN [2]
	Params	**N/A**	**665 K**	**411 K**	**634 K**	**967 K**	**12.7 M**	**22 M**
Set5	2	33.66/0.9299	37.53/0.9587	37.52/0.9583	37.67/0.9591	37.72/0.9593	38.22/0.9611	38.23/0.9610
	2.5	-	-	35.36/0.9395	35.60/0.9411	35.59/0.9410	36.20/0.9444	36.18/0.9441
	3	30.39/0.8682	33.66/0.9213	33.72/0.9209	34.03/0.9238	34.02/0.9240	34.73/0.9295	34.72/0.9296
	3.5	-	-	32.52/0.9019	32.81/0.9058	32.91/0.9071	33.56/0.9142	33.60/0.9146
	4	28.42/0.8104	31.35/0.8838	31.36/0.8807	31.72/0.8871	31.81/0.8886	32.52/0.8989	32.51/0.8986
Set14	2	30.24/0.8688	33.03/0.9124	33.05/0.9123	33.26/0.9149	33.27/0.9149	34.02/0.9206	34.03/0.9204
	2.5	-	-	31.17/0.8704	31.37/0.8740	31.37/0.8735	31.97/0.8819	31.89/0.8814
	3	27.55/0.7742	29.77/0.8314	29.83/0.8319	30.06/0.8364	30.06/0.8366	30.58/0.8463	30.58/0.8465
	3.5	-	-	28.86/0.7970	29.08/0.8023	29.11/0.8035	29.59/0.8139	29.60/0.8140
	4	26.00/0.7027	28.01/0.7674	28.05/0.7672	28.28/0.7735	28.31/0.7749	28.84/0.7872	28.86/0.7878
B100	2	29.56/0.8431	31.90/0.8960	31.81/0.8940	31.97/0.8966	31.99/0.8971	32.33/0.9008	32.36/0.9011
	2.5	-	-	29.95/0.8415	30.11/0.8450	30.12/0.8446	30.46/0.8509	30.48/0.8509
	3	27.21/0.7385	28.82/0.7976	28.75/0.7959	28.90/0.7999	28.92/0.8005	29.26/0.8079	29.28/0.8089
	3.5	-	-	27.89/0.7566	28.04/0.7617	28.06/0.7626	28.40/0.7718	28.42/0.7730
	4	25.96/0.6675	27.29/0.7251	27.22/0.7233	27.37/0.7289	27.40/0.7301	27.73/0.7409	27.76/0.7419

Compared to the lightweight VDSR, the evaluated metric of LAS_B is slightly higher, but it still requires relatively fewer parameters. For LAS_A, its parameters are 33% lower than those of VDSR and obtain almost similar evaluated metrics. The results of VDSR are obtained from the original data in Ref. [6], and the results of Meta-RDN [2] are obtained from the pre-trained model created by us.

We present the generated HR images made by LAS_B with several scale factors in Figure 3. The comparison of the proposed methods with others for practically generated x2.0, x3.0, and x4.0 HR images are provided in Figures 4–6, respectively. Finally, there is a trade-off between the performance, which is evaluated using PSNR and SSIM metrics, and the cost is assessed using the parameters. The proposed LAS_A with only 400 K parameters makes the proposed model reasonable and realistic for implementation into hardware devices. In particular, it considers non-integer scale factors.

Figure 3. Generated HR image of "ppt3" from Set14 dataset.

Figure 4. Visual comparison of the image "Monarch" from dataset Set14 with a scale factor of 2.

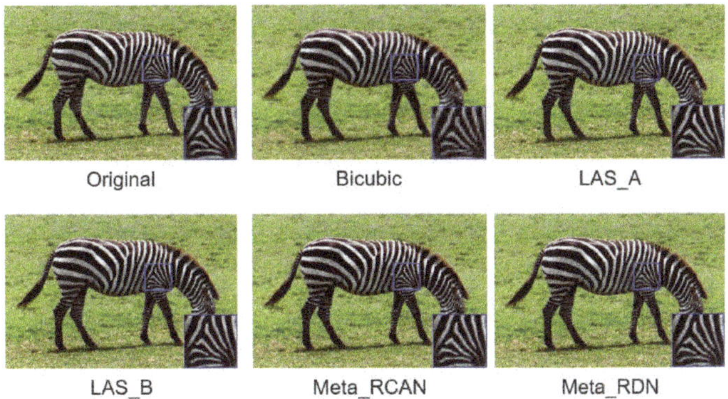

Figure 5. Visual comparison of the image "zebra" from dataset Set14 with a scale factor of 3.

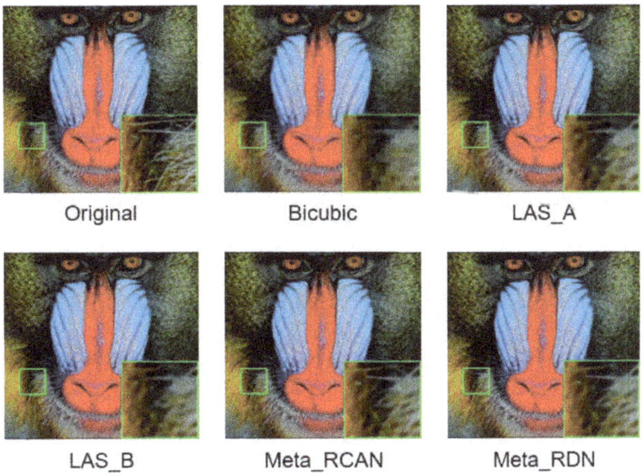

Figure 6. Visual comparison of the image "Baboon" from dataset Set14 with a scale factor of 4.

4. Conclusions

Super-resolution with non-integer scale factors is still a practical topic that has gradually gained attention in recent years. Meta-SR [2] is used for tackling this problem. A novel upscale module is proposed to proactively predict the kernel weights based on the corresponding scale factor. Based on this particular design, we need to train only a single model for all arbitrary scale factors. It saves time and effort compared to traditional training to train a specific model for each scale factor. However, it is still computationally expensive. Inspired by Ref. [2], we built a lightweight network, which is suitable for hardware applications. The main contribution of the proposed work is the creation of a single model for each arbitrary scale factor with a low computational cost. Its network is trained from scratch and only needs to be prepared once for all the scale factors.

Author Contributions: Conceptualization, Q.T.D.D. and K.-Y.H.; methodology, Q.T.D.D. and K.-Y.H.; software, Q.T.D.D.; validation, Q.T.D.D.; formal analysis, Q.T.D.D.; investigation, K.-Y.H.; resources, K.-Y.H.; data curation, K.-Y.H.; writing—original draft preparation, Q.T.D.D.; writing—review and editing, K.-Y.H.; visualization, K.-Y.H.; supervision, P.-Y.C.; project administration, P.-Y.C.; funding acquisition, P.-Y.C. All authors have read and agreed to the published version of the manuscript.

Funding: This work was supported in part by the National Science and Technology Council, R.O.C., under NSTC-110-2221-E-006-164-MY3, in part by National Academy of Marine Research, Taiwan, under NAMR-111001, and in part by Qualcomm through a Taiwan University Research Collaboration Project.

Institutional Review Board Statement: Not applicable.

Informed Consent Statement: Not applicable.

Data Availability Statement: Data are contained within the article.

Acknowledgments: The authors would like to thank anonymous editors for their valuable comments and suggestions to improve the quality of the paper.

Conflicts of Interest: The authors declare no conflict of interest.

References

1. Dong, C.; Loy, C.C.; He, K.; Tang, X. Image super-resolution using deep convolutional networks. *IEEE Trans. Pattern Anal. Mach. Intell.* **2015**, *38*, 295–307. [CrossRef] [PubMed]
2. Hu, X.; Mu, H.; Zhang, X.; Wang, Z.; Tan, T.; Sun, J. Meta-SR: A magnification-arbitrary network for super-resolution. In Proceedings of the IEEE/CVF Conference on Computer Vision and Pattern Recognition, Long Beach, CA, USA, 15–20 June 2019.
3. Shi, W.; Caballero, J.; Huszár, F.; Totz, J.; Aitken, A.P.; Bishop, R.; Rueckert, D.; Wang, Z. Real-time single image and video super-resolution using an efficient sub-pixel convolutional neural network. In Proceedings of the IEEE Conference on Computer Vision and Pattern Recognition, Las Vegas, NV, USA, 27–30 June 2016.
4. Zhang, Y.; Tian, Y.; Kong, Y.; Zhong, B.; Fu, Y. Residual dense network for image super-resolution. In Proceedings of the IEEE Conference on Computer Vision and Pattern Recognition, Salt Lake City, UT, USA, 18–23 June 2018.
5. Zhang, Y.; Li, K.; Li, K.; Wang, L.; Zhong, B.; Fu, Y. Image super-resolution using very deep residual channel attention networks. In Proceedings of the European Conference on Computer Vision (ECCV), Munich, Germany, 8–14 September 2018.
6. Kim, J.; Lee, J.K.; Lee, K.M. Accurate image super-resolution using very deep convolutional networks. In Proceedings of the IEEE Conference on Computer Vision and Pattern Recognition, Las Vegas, NV, USA, 27–30 June 2016.
7. Agustsson, E.; Timofte, R. NTIRE 2017 challenge on single image super-resolution: Dataset and study. In Proceedings of the 2017 IEEE Conference on Computer Vision and Pattern Recognition Workshops (CVPRW), Honolulu, HI, USA, 21–26 July 2017; pp. 1122–1131.
8. Bevilacqua, M.; Roumy, A.; Guillemot, C.; Alberi-Morel, M.-L. Low-complexity single-image super-resolution based on nonnegative neighbor embedding. In Proceedings of the British Machine Vision Conference 2012, Surrey, UK, 3–7 September 2012; pp. 1–10. [CrossRef]
9. Zeyde, R.; Elad, M.; Protter, M. On single image scale-up using sparse-representations. In Proceedings of the 7th international conference on Curves and Surfaces, Avignon, France, 24–30 June 2010; pp. 711–730.
10. Martin, D.; Fowlkes, C.; Tal, D.; Malik, J. A database of human segmented natural images and its application to evaluating segmentation algorithms and measuring ecological statistics. In Proceedings of the Eighth IEEE International Conference on Computer Vision, Vancouver, BC, Canada, 7–14 July 2001.

engineering proceedings

MDPI

Proceeding Paper

Neonatal Activity Monitoring by Camera-Based Multi-LSTM Network †

Imre Jánoki [1,*], Ádám Nagy [1], Péter Földesy [2], Ákos Zarándy [2], Máté Siket [3], Judit Varga [4] and Miklós Szabó [4]

1 Institute for Computer Science and Control, Pázmány Péter Catholic University, 1088 Budapest, Hungary; nagyadam@sztaki.hu
2 Institute for Computer Science and Control, 1111 Budapest, Hungary; foldesy.peter@sztaki.hu (P.F.); zarandy.akos@sztaki.hu (Á.Z.)
3 Institute for Computer Science and Control, Óbuda University, 1084 Budapest, Hungary; siket.mate@sztaki.hu
4 Budapest Division of Neonatology Istst Department of Pediatrics, Department of Obstetrics and Gynecology, Semmelweis University, 1082 Budapest, Hungary; varga.judit@med.semmelweis-univ.hu (J.V.); szabo.miklos@med.semmelweis-univ.hu (M.S.)
* Correspondence: janoki.imre.gergely@sztaki.hu
† Presented at the IEEE 5th Eurasia Conference on Biomedical Engineering, Healthcare and Sustainability, Tainan, Taiwan, 2–4 June 2023.

Abstract: The objective evaluation of an infant's activity and sleep pattern is critical in improving the comfort of the babies and ensuring the proper amount of quality sleep. The predefined behavioral states of an infant describe their consciousness and arousal level. The different states are characterized by different movements, body tone, eye movements and breath patterns. To recognize and adapt to these states is an essential part of development-friendly caring. It affects the neonate's sleep, influencing their brain development, while improving the bonding between mother and baby, and feeding is more successful during the state of quiet awakened. It can be a more difficult task to determine the level of arousal in premature neonates. In preterm clinics, the general practice is continuous observation, requiring the attention of the hospital staff. To create an automated, more objective system, helping the hospital staff and the parents, we developed a multi-RNN (multi-recurrent neural network) network-based solution to solve this classification problem, which works on a time-series-like feature set, extracted from cameras' video feeds. The set is composed of video actigraphy features, video-based respiration signal and additional descriptors. We separate infant caring from undisturbed presence based on our previous ensemble network solution. The network was trained and evaluated using our database of 402 h of footage, collected at the Neonatal Intensive Care Unit, Dept. of Neonatology of Pediatrics, Dept. of Obstetrics and Gynecology, Semmelweis University, Budapest, Hungary, with all-day recordings of 10 babies.

Keywords: actigraphy; sleep patter; motion estimation; breath rate; respiration rate; LSTM; non-contact; premature infant; neonatal

Citation: Jánoki, I.; Nagy, Á.; Földesy, P.; Zarándy, Á.; Siket, M.; Varga, J.; Szabó, M. Neonatal Activity Monitoring by Camera-Based Multi-LSTM Network. *Eng. Proc.* **2023**, *55*, 16. https://doi.org/10.3390/engproc2023055016

Academic Editors: Teen-Hang Meen, Kuei-Shu Hsu and Cheng-Fu Yang

Published: 28 November 2023

1. Introduction

The objective diagnosis of sleep patterns is essential to improve sleep comfort and to measure the effectiveness of interventions. Continuous observation and adjusting a caring schedule is important in development-friendly caring, but direct observations result in a burden on medical staff.

Polysomnography (PSG)—considered to be the "gold standard"—is complicated, needs hospital resources and causes discomfort for the infants. On the other hand, actigraphy is a cheap easy long-term measurement technique. Though its equivalency is up to debate, it is found to be an effective tool. Its video-based version is also an intensively investigated alternative [1]. This is a contact-free method, providing actigraphy using a camera, which quantifies body movements during long-term awake and sleep assessments, even in difficult situations, and, at the same time, it does not require wiring or attached sensors [2]. The classification of sleep–wake and intermediate stages, based on video observation with different solutions, are presented in [1,3,4].

Our aim is to quantitatively evaluate sleep–wake–caring states of infants, based on video cameras and recurrent neural networks (RNNs). In order to separate infant caring and presence we used our previous ensemble network solution [5]. This mentioned care state analyzer solution is also an integral part of the behavioral state classification algorithm presented here, and we refer to it as "Top-Level-Classification-Block" in [5]. This network is important because the determination of a behavioral state is only possible—or worthwhile—if there is a baby in the picture and there is no feeding, caring or other intervention happening. In these cases, the baby's behavioral state is trivial and cannot be classified, as described in [6].

It is important to note that, in this work, the output classes of the applied model are not the same as the sleep stages (e.g., NREM, REM) that you can read about in the mainstream literature. Instead, the output classes used here are based on the manual observations made by doctors and nurses, and they represent the activity phases of the observed infant [7]. This is a routinely employed classification in hospitals that doctors and nurses are trained for (e.g., in "Family and Infant Nerve Development Education" [6]). It also includes the state of sleep, but it is more about activity phases; henceforth we shall call them that. The literature defines five or six phases: quiet sleep, active sleep, intermediate, quiet alert, and active alert. In our case, medical advisors were most interested in how much the babies sleep, therefore, we merged some classes and simplified the classifications into the following three categories: sleep, intermediate, and alert. These three classes were satisfactory to estimate how much the babies sleep and to help move towards development-friendly caring.

2. Experimental Setup

The dataset used in the project was collected in cooperation with Semmelweis Medical University. We built a complete data collection system that monitored the infant with a conventional color camera (Basler acA2040-55uc, Basler AG, Ahrensburg, Germany) at 20 FPS, with a resolution of 500 × 500, that was able to record the physiological data of the babies in parallel and in sync. The recordings were made from several angles with zoom optics.

Later, it became necessary to observe the infants in closed incubators, even at night. We made a 3D printed capsule, which included both the camera and an infrared illuminator (Figure 1). Infrared illumination was only used for the night recordings. This solution always recorded the baby from the same fixed position on top of the incubator.

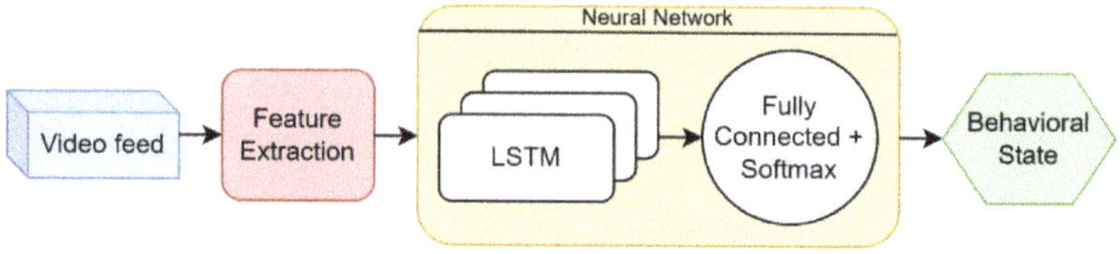

Figure 1. The data acquisition system that can be attached to the top of the incubator with suction cups. It includes a camera and an infrared illuminator.

Data storage, managing and annotation were carried out by custom software that we wrote. You can read more about this data collection and data managing system in [5]. A total of 402 h of data were collected in the neonatal intensive care unit from infants aged between 32 and 44 weeks. More details regarding the population of the recorded babies and their characteristics can be found in Table 1.

Table 1. Characteristics of the participants.

Subject	**1**	**2**	**3**	**4**	**5**	**6**	**7**	**8**
Recording time (hours)	96.7	5.5	39.4	27.4	51	105.5	50.1	36.4
Gender	F	M	M	F	F	M	F	F
Gestational age (weeks)	32	32 + 3	31 + 4	35 + 4	39	32	33	38 + 6
Birth weight (g)	2020	1840	1850	1870	3150	2120	2080	2840
Postnatal age (days)	4	4	10	8	4	7	2	7
Actual weight (g)	1900	1850	1680	1820	2905	2040	1960	3150
Length (cm)	46	44	-	45	57	45	44	48
Head circumference (cm)	32	29.5	-	32	34	30	32	33
Respiratory support	no	no	no	no	no	no	no	yes
Any drugs	no	no	no	no	yes	no	no	yes
Fitzpatrick scale	2	3	2	2	2	2	2	2

3. Methodology

We assumed that, if we extracted features from appropriate locations, in the absence of interventions or other disturbances, an RNN stack, which is designed for processing dynamic data, could solve the previously defined problem of activity phase classification.

Our solution for the activity phase classification algorithm can be divided into two main modules: **feature extraction** and **classification** (Figure 2). In addition, we extended the algorithm with a **scene analysis** module, which runs before the behavioral stage analysis and determines whether the current scene is suitable for the analysis. In the following subsections we introduce all modules separately.

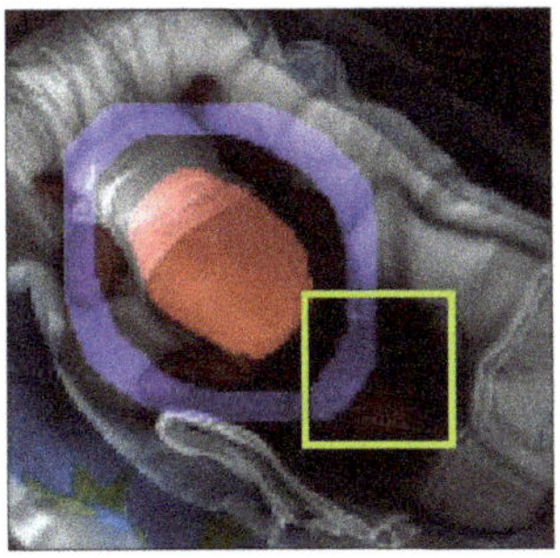

Figure 2. Summary of the presented behavioral state classification algorithm. The colors showing the different regions of interests: the head, the torso and the area around the torso.

3.1. Scene Analysis

First, we needed to determine that there was a baby in the picture, i.e., they had not been taken out of the incubator and there was no feeding or other care occurring. For this, we used the LSTM-based (long short-term memory) procedure that we introduced in [5] as "Top-Level-Classification-Block". This algorithm works very similarly to the behavioral state classification, that we explain in detail later, and it uses similar features calculated on the entire images. The vectors, constructed from extracted features, were processed by an RNN, which had a sequential input and a classification output. The network contained two fully connected (FC) layers with a Rectified Linear Unit (ReLU) activation function and a stateful LSTM cell. The final layer is a SoftMax one, which provides the classification probability-type output with the following classes:

- Baby is present and a respiration-like signal is detected.

- Baby is present and is showing intensive motion, interpreted as random and frequent self-motion.
- Caring or other intervention happens.
- No motion or minimal motion can be found in the incubator, but the baby is detected.
- The baby is not detected in the scene; empty incubator.
- Multiple subcategories incorporating unacceptable camera image quality and possible errors: low light conditions, blurry view, camera image is saturated, or consecutive frames do not differ from each other.

The top-level classification achieved 97.9% sensitivity and 97.5% specificity on the data set introduced in [5].

3.2. Feature Extraction

The feature extraction module started with the detection of different regions of interest (ROI). We started from the assumption that there were regions in the image that carried special information, separate from the other areas that could be useful to us in classifying activity phases (Figure 3). The examples are as follows:

- The whole picture;
- The baby's whole body;
- The area around the abdomen and torso;
- A ring around the baby's abdomen including the limbs;
- The infant's face.

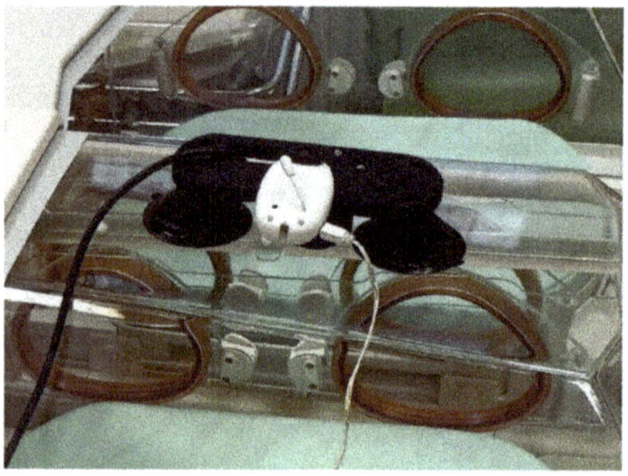

Figure 3. Regions that may be of key importance in determining activity phases. These are used as different ROIs in feature extraction.

Finding pixels for these regions was a segmentation task. Except for the baby's entire body, we segmented each area and collected movement, color and intensity information from them separately.

It was easy to conclude that these regions listed above carried specific information. The need to use the full picture was trivial. The abdominal region is the area where respiration can be observed, which is an important descriptor for determining activity phases [8–10]. The "ring around the baby's abdomen" was also important: this area was likely to include the majority of the limbs on the image, if the parameters were chosen correctly, and so we can extract information about the limb movements of the baby. The last region was the baby's face, giving us information about whether the baby's eyes are open or not, which is also an important descriptor.

The next step of this module was the feature extraction itself. To determine what features to use, we evaluated the ones we used in [5], which was in a similar domain. The selected features to extract were as follows:

- Image brightness;
- HSV (hue, saturation, value) image;

- Optical flow;
- Euclidean map;
- Respiration.

The listed features were calculated for all the ROIs that we defined as important regions above. The ROIs were stored as binary images (masks) that contained ones and zeros. The area designated by the ROI was obtained by multiplying the current image by the given ROI, using elementwise matrix multiplication. After the different features were calculated on the masked images, intra-frame statistics like the mean value and standard deviation were calculated by:

$$F_{mean} = \mu = \frac{1}{N}\sum_{i=1}^{N} I_t(i), \quad (1)$$

$$F_{std} = \sqrt{\frac{1}{N}\sum_{i=1}^{N} (I_t(i) - \mu)^2} \quad (2)$$

where N is the number of pixels, and $I_t(i)$ is the intensity value of the i-th pixel of the current frame. Inter-frame statistics like optical flow or the Euclidean map (F_{euc}) were also calculated:

$$f_x u + f_y v + f_t = 0;\ u = \frac{dx}{dt};\ v = \frac{dy}{dt}; \quad (3)$$

$$F_{euc} = \frac{dx}{dt}\frac{dx}{dt} + \frac{dy}{dt}\frac{dy}{dt}, \quad (4)$$

where f_x and f_y are the image gradients and f_t is the gradient along time, while u and v represents the displacements in the x and y directions during time dt.

Our basic assumption was that these intra- and inter-frame statistics, calculated for each of the regions, should be sufficient to classify the current activity state of infants. To analyze the data from a dynamic perspective, we had to transform them to another form which can be interpreted by RNNs. Since we did not need pixel-level interpretation, we converted them to 1D waveforms. This was usually carried out by calculating the average value for each feature.

3.3. Classification

As we classified using dynamic time-series data, it was evident to use recurrent neural networks, which have already been used successfully in many applications to solve time–domain classification problems like EEG classification [11–15]. LSTM and GRU (Gated Recurrent Unit) are such recurrent models. For our experiments, we implemented an LSTM stack and a GRU stack as well, extended with fully connected layers. Stacks constructed from these recurrent architectures were suitable for working on the multivariate input features. The expected outputs were the earlier mentioned activity phases, which were annotated by doctors on our video recordings.

Considering the parameters of the used RNN stack, we examined using a 2-layer LSTM, and a 2-layer GRU network with a cell state of 35 in both cases, and a hidden state of 35 lengths, respectively.

During the 100-epoch long training, we used the "CrossEntropy" loss function and "Adam-Optimizer". The database was split into three different sets: the training set, the test set and the validation set. We used Comet-ML (Comet.ml, NY, USA) for the visualization of loss, epoch loss, and for hyperparameter tuning.

4. Discussion

Our approach combines the activity classification with predictions of expected activity and with an observation of a broader view—of the incubators providing automatic statistics regarding procedure time, empty view, parents presence, etc. over an extended period of hospitalization.

Our system starts with the scene analysis, where we apply an LSTM-based method. This approach provided an impressive performance in the "scene-analyzing" step:

- Accuracy: 98.1%;
- Sensitivity: 97.5%;
- Specificity: 97.9%;
- Precision: 98.1%.

For the classification, two recurrent neural networks were tested. These were LSTM- and GRU-based. Our experiments showed that we could achieve slightly higher accuracy in this application by using GRU. These networks work on sequences of a certain fixed length (L) that heavily affects their performance and necessary hyperparameters. We tested a set of different input sequence lengths and found L = 200 to be the best.

The results with the better performing GRU stack and L = 200 are as follows:

- Accuracy: 82.60%;
- Sensitivity: 78.40%;
- Specificity: 87.67%;
- Precision: 88.46%.

The total number of trainable parameters of the GRU stack was 23,014.

This performance could be further increased by using additional features, for example we can specifically detect the eye and add its condition (opened or closed) as an additional feature. Besides adding eye detection, our ongoing work primarily focuses on increasing our database as sleep and wake scoring depends heavily on variables that are susceptible to the high heterogeneity of the subjects. We are also experimenting with measurements undertaken with different devices. Tuning the network hyperparameters and using different sensitivity thresholds is also still in progress.

5. Conclusions

We have shown that an RNN-based network may be suitable for classifying behavioral states using video recordings. We have shown a feature extraction method that is one possible way to create an input that can be used by these networks. We hypothesize that the presented automated behavioral state classification architecture may later be suitable to replace human observation to a certain level.

Author Contributions: Conceptualization, I.J. and Á.N.; methodology, I.J., Á.N., P.F. and M.S. (Máté Siket); software, I.J. and Á.N.; validation, J.V. and M.S. (Miklós Szabó); formal analysis, Á.N.; investigation, I.J. and Á.N.; resources, Á.Z. and M.S. (Miklós Szabó); data curation, I.J., Á.N., M.S. (Máté Siket) and J.V.; writing—original draft preparation, Á.N. and I.J.; writing—review and editing, I.J.; visualization, I.J. and Á.N.; supervision, P.F., Á.Z. and M.S. (Miklós Szabó); project administration, Á.Z. and M.S. (Miklós Szabó); funding acquisition, Á.Z. All authors have read and agreed to the published version of the manuscript.

Funding: This research was funded by the Hungarian grant NKFIH-1019658 and the Ministry of Innovation and Technology NRDI Office, Hungary within the framework of the Artificial Intelligence National Laboratory Program.

Institutional Review Board Statement: The study was conducted in accordance with the Declaration of Helsinki, and approved by the Ethics Committee of Semmelweis University (SE IRB number 265/2022).

Informed Consent Statement: Informed consent was obtained from all subjects involved in the study.

Data Availability Statement: The data presented in this study are available on request from the corresponding author. The data are not publicly available due to ethics and personal rights.

Conflicts of Interest: The authors declare no conflict of interest.

References

1. Long, X.; Espina, J.; Otte, R.A.; Wang, W.; Aarts, R.M.; Andriessen, P. Video-based actigraphy is an effective contact-free method of assessing sleep in preterm infants. *Acta Paediatr.* **2020**, *110*, 1815–1816. [CrossRef] [PubMed]
2. Yavuz-Kodat, E.; Reynaud, E.; Geoffray, M.-M.; Limousin, N.; Franco, P.; Bourgin, P.; Schroder, C.M. Validity of actigraphy compared to polysomnography for sleep assessment in children with autism spectrum disorder. *Front. Psychiatry* **2019**, *10*, 551. [CrossRef] [PubMed]
3. Liao, W.-H.; Yang, C.-M. Video-based activity and movement pattern analysis in overnight sleep studies. In Proceedings of the 2008 19th International Conference on Pattern Recognition, Tampa, FL, USA, 8–11 December 2008; pp. 1–4.
4. Unno, M.; Morisaki, T.; Kinoshita, M.; Saikusa, M.; Iwata, S.; Fukaya, S.; Yamashita, Y.; Nakayama, M.; Saitoh, S.; Iwata, O. Validation of actigraphy in hospitalised newborn infants using video polysomnography. *J. Sleep Res.* **2021**, *31*, e13437. [CrossRef] [PubMed]
5. Nagy, Á.; Földesy, P.; Jánoki, I.; Terbe, D.; Siket, M.; Szabó, M.; Varga, J.; Zarándy, Á. Continuous camera-based premature-infant monitoring algorithms for nicu. *Appl. Sci.* **2021**, *11*, 7215. [CrossRef]
6. Warren, I.; Mat-Ali, E.; Green, M.; Nyathi, D. Evaluation of the family and infant neurodevelopmental education (FINE) programme in the UK. *J. Neonatal Nurs.* **2019**, *25*, 93–98. [CrossRef]
7. Read, D.J.; Henderson-Smart, D.J. Regulation of breathing in the newborn during different behavioral states. *Annu. Rev. Physiol.* **1984**, *46*, 675–685. [CrossRef] [PubMed]
8. Maurya, L.; Kaur, P.; Chawla, D.; Mahapatra, P. Non-contact breathing rate monitoring in newborns: A review. *Comput. Biol. Med.* **2021**, *132*, 104321. [CrossRef] [PubMed]
9. Jorge, J.; Villarroel, M.; Chaichulee, S.; Guazzi, A.; Davis, S.; Green, G.; McCormick, K.; Tarassenko, L. Non-contact monitoring of respiration in the neonatal intensive care unit. In Proceedings of the 2017 12th IEEE International Conference on Automatic Face & Gesture Recognition, Washington, DC, USA, 30 May–3 June 2017; pp. 286–293.
10. Rossol, S.L.; Yang, J.K.; Toney-Noland, C.; Bergin, J.; Basavaraju, C.; Kumar, P.; Lee, H.C. Non-contact video-based neonatal respiratory monitoring. *Children* **2020**, *7*, 171. [CrossRef] [PubMed]
11. Wang, P.; Jiang, A.; Liu, X.; Shang, J.; Zhang, L. LSTM-based EEG classification in motor imagery tasks. *IEEE Trans. Neural Syst. Rehabil. Eng.* **2018**, *26*, 2086–2095. [CrossRef] [PubMed]
12. Zeng, H.; Yang, C.; Dai, G.; Qin, F.; Zhang, J.; Kong, W. EEG classification of driver mental states by deep learning. *Cogn. Neurodynamics* **2018**, *12*, 597–606. [CrossRef] [PubMed]
13. Hu, X.; Yuan, S.; Xu, F.; Leng, Y.; Yuan, K. Scalp EEG classification using deep bi-LSTM network for seizure detection. *Comput. Biol. Med.* **2020**, *124*, 103919. [CrossRef] [PubMed]
14. Rana, R. Gated recurrent unit (gru) for emotion classification from noisy speech. *arXiv* **2016**, arXiv:1612.07778v1.
15. Chen, J.X.; Jiang, D.M.; Zhang, Y.N. A hierarchical bidirectional gru model with attention for eeg-based emotion classification. *IEEE Access* **2019**, *7*, 118530–118540. [CrossRef]

engineering proceedings

MDPI

Proceeding Paper

Identification of Dominant Microbes and Functional Analysis of Sourdough Starters Made of Dried Longan and Raisin †

Yen-Chih Liu [1], Pei-Shan Wu [2], Shih-Hua Teng [3] and Ming-Jiuan Wu [2,*]

1 Department of Pharmacy, Chia-Nan University of Pharmacy and Science, Tainan 717301, Taiwan; rkmy070514@gmail.com

2 Department of Life and Health Science, Chia-Nan University of Pharmacy and Science, Tainan 717301, Taiwan; dc7575@gmail.com

3 Institute of Preventive Medicine, National Defense Medical Center, Taipei 114201, Taiwan; thomasteng6591@gmail.com

* Correspondence: imwu@gm.cnu.edu.tw; Tel.: +886-6-2664911 (ext. 2520)

† Presented at the IEEE 5th Eurasia Conference on Biomedical Engineering, Healthcare and Sustainability, Tainan, Taiwan, 2–4 June 2023.

Abstract: Baking bread is currently experiencing a profound transformation, propelled by an escalating interest in health and wellness, global cuisines, and sustainability efforts. The surface of fruits harbors various microbes, such as yeasts and lactic acid bacteria (LAB). In contrast to *Saccharomyces cerevisiae*, a commercial yeast, the microbial flora present on the surface of fruits allows bread with greater flavor, texture, and health benefits as seen in sourdough starters. In this research, sourdough starters were created using dried longan and raisin. The microbial flora of the sourdough starters and the fermentation profile were examined, and possible bacteriocin genes were identified. The dried longan starter exhibited bubbling, stickiness, and a drop in pH value from 5.80 to 4.45 at the third day of fermentation. Meanwhile, the raisin starter began bubbling on the sixth day with no change in fluid viscosity, and a small pH value decreased from 4.00 to 3.82. The dried longan starter contained *Weissella cibaria*, *W. paramesenteroides*, *S. cerevisiae*, *Pediococcus pentosaceus*, *Torulaspora delbrueckii* and *Leuconostoc citreum*. On the other hand, raisin starter only contained *Candida krusei* and *T. delbrueckii*. To detect antibacterial property in *W. paramesenteroides* and *P. pentosaceus*, four bacteriocins, Gassericin A (*gaaA*), Lactacin F (*lafA*), Plantaricin S (*pls*) and Pediocin (*ped*) were explored. Polymerase chain reaction (PCR) results revealed that *gaaA* and *lafA* were present in *W. paramesenteroides*, which suggested that this strain possessed antimicrobial properties.

Keywords: sourdough starters; *Weissella paramesenteroides*; bacteriocin gene

Citation: Liu, Y.-C.; Wu, P.-S.; Teng, S.-H.; Wu, M.-J. Identification of Dominant Microbes and Functional Analysis of Sourdough Starters Made of Dried Longan and Raisin. *Eng. Proc.* **2023**, *55*, 17. https://doi.org/10.3390/engproc2023055017

Academic Editors: Teen-Hang Meen, Kuei-Shu Hsu and Cheng-Fu Yang

Published: 29 November 2023

1. Introduction

Compared to breads made with commercial yeast, breads baked using wild yeasts and lactic acid bacteria (LAB) provide benefits such as glycemic response, satiety, and gastrointestinal well-being. Additionally, they allow a valuable natural method for enhancing the texture, flavor, and stability of bread [1]. Yeast plays a significant role in the production of beverages and bread thanks to its capability to produce carbon dioxide and ethanol as well as aroma compounds. Especially, LABs are especially crucial in fermentation as they produce desirable acids, flavor compounds, and peptides that inhibit the growth of undesirable microorganisms [2].

Fruits are commonly used ingredients for making sourdough starters, owing to the widespread presence of yeast on their surfaces. Longan (*Dimocarpus longan*) is one of the most important fruit crops in Tainan, Taiwan. In addition to being consumed fresh, approximately 50% of longan production is dried to extend its storage life and intensify its flavor and aroma. Raisins are popular ingredients for homemade sourdough starters. They naturally harbor wild yeast and a diverse microbial flora which contribute to vibrant and resilient fermentation. For this study, we chose dried longan and dark sultana raisins as the primary materials for creating sourdough starters. Bacteriocins are proteinaceous or peptidic toxins produced by bacteria to inhibit the growth of similar or closely related bacterial strains [3]. Bacteriocins have been identified in many strains of LAB, and their structural genes are identified using a PCR array of primers [4]. Based on the results, we investigated the fermentation profiles of sourdough starters made of dried longan and raisins. The dominant microbes and organic acids produced were also investigated.

2. Materials and Methods

2.1. Preparing Sourdough Starters with Dried Fruit

In total, 50 g of dried fruits, 25 g of sugar, and 100 g of sterile water were added to a beaker. The ingredients were mixed with a stirring bar until the sugar had completely dissolved. Finally, the beaker was covered with plastic wrap and placed at 25 °C.

2.2. Observation of Sourdough Starters

The fermentation of the sourdough starters was observed and recorded on the top and at the side for 7 days. pH values were recorded every day.

2.3. Organic Acid Analysis

The samples were collected on the seventh day of fermentation. After the samples were filtered, they were centrifuged at 12,000 rpm for 30 min. The supernatant was collected and filtered using a PVDF 0.45 μm filter. Samples were then sent to the Organic and Health Food Science and Technology Development Center (OHC) Lab for organic acid analysis following the method of the National Standards of the Republic of China, CNS 12635 [5].

2.4. Bacterial Identification

In this study, colonies were isolated using three media: Luria–Bertani agar (LB agar), De Man, Rogosa, and Sharpe agar (MRS agar), and potato–dextrose agar (PDA). The samples were diluted using sterile water and spread-plated on three types of media sequentially. LB plates were grown in 37 °C, PDA at 25 °C, and MRS plates at 30 °C under anaerobic conditions.

2.5. Identification of Bacteriocin Genes

Genomic DNA was isolated from *W. paramesenteroides* and *P. pentosaceus* using Chelex100 chelating resin [6]. Four specific primer sequences for *gaaA*, *lafA*, *pls*, and *ped* were listed in Table 1 [7,8]. The PCR process was as follows: an initial denaturation step at 95 °C for 5 min, followed by 34 cycles of denaturation at 95 °C for 1 min, annealing at 57 °C for 45 s, and extension at 72 °C for 1 min, using the GeneAmp PCR System 9700. After the final extension at 72 °C for 5 min, the PCR products were analyzed on 1.5% agarose gels, and stained with ethidium bromide.

Table 1. Primer for bacteriocin genes.

Primer	Sequence	Amplicon
gaaA	GAACAGGTGCACTAATCGGT CAGCTAAGTTAGAAGGGGCT	800 bp
Laf A	AGTCGTTGTTGGTGGAAGAAAT TCTTATCTTGCCAAAACCACCT	184 bp
pls	GCCTTACCAGCGTAATGCCC CTGGTGATGCAATCGTTAGTTT	320 bp
ped	AAAATATCTAACTAATACTTG TAAAAAGATATTTGA CCAAAA	711 bp

2.6. Fermentation Efficacy Test

A total of 25 g of the sourdough starter filtrate was thoroughly blended with 50 g of flour within a beaker. The initial and 41 h fermented dough heights were then meticulously measured and recorded.

3. Results

3.1. Observation of Sourdough Starters Made of Dried Longan and Raisin

The procedures outlined in Sections 2.1 and 2.2 were performed for the preparation and observation of the sourdough starters. On the third day, noticeable bubbles were discerned within the sourdough starter containing dried longan (Figure 1), whereas a similar bubbling phenomenon manifested on the fifth day within the raisin-infused starter (Figure 2). Additionally, it is noteworthy that the dried longan exhibited a sticky texture starting from the third day of the fermentation process.

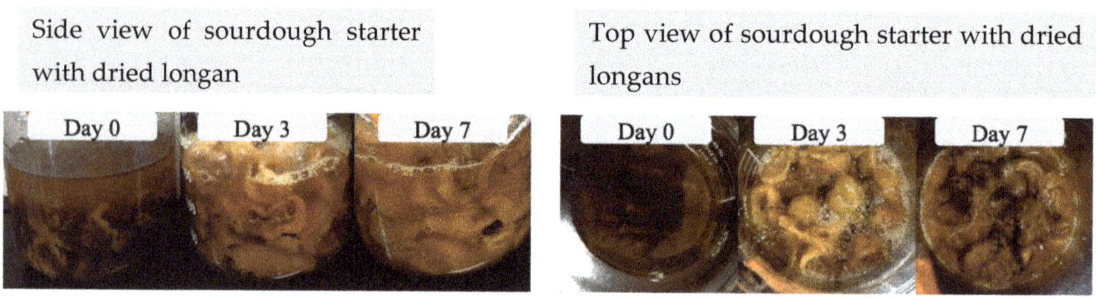

(a) (b)

Figure 1. Side view (**a**) and top view (**b**) of sourdough starter with dried longan during a 7-day fermentation.

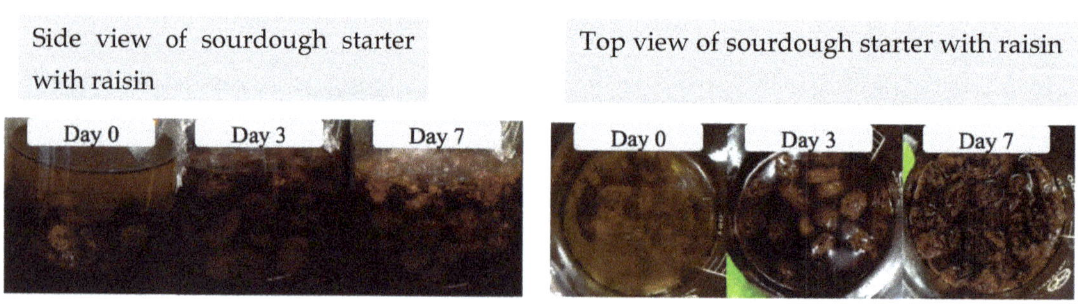

(a) (b)

Figure 2. Side view (**a**) and top view (**b**) of sourdough starter with raisin during a 7-day fermentation.

3.2. Change in pH of Sourdough Starters

pH values were monitored throughout the fermentation process. The pH of the sourdough starter with dried longan decreases from 5.80 to 4.45 during fermentation, particularly within the first three days (Figure 3). On the other hand, the pH of the sourdough starter with raisins decreased from 4.00 to 3.82 (Figure 3). There was a rapid drop in pH in the sourdough starter with dried longan compared to that with raisin.

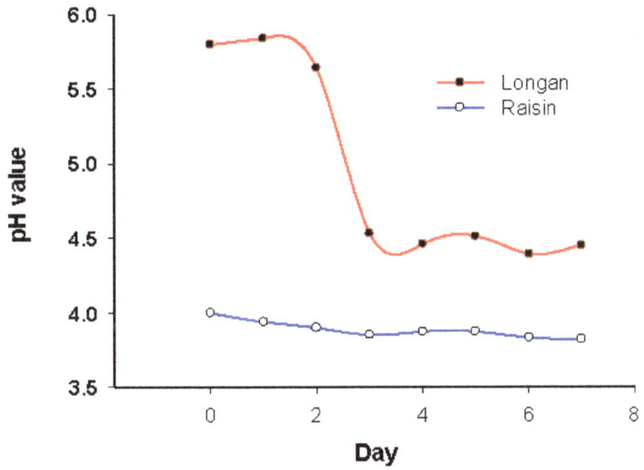

Figure 3. pH change during fermentation.

3.3. Organic Acid

The result of the high-performance liquid chromatography (HPLC) analysis revealed that within the sourdough starter infused with dried longan, lactic acid emerged as the predominant metabolite in terms of concentration, accompanied by notable levels of acetic acid and citric acid. Conversely, in the sourdough starter incorporating raisin, acetic acid took precedence as the primary metabolite, followed by lactic acid in the second position (Table 2).

Table 2. Organic acids in the sourdough starter with dried longan and raisin.

Organic Acid	Longan (mg/100 mL)	Raisin (mg/100 mL)
Acetic acid	192.14	167.42
Lactic acid	260.42	89.55
Citric acid	134.44	67.12
Malic acid	93.35	66.07
Succinic acid	14.50	41.25
Tartaric acid	5.44	22.29
Oxalic acid	25.19	15.59

3.4. Bacterial Identification

Two sourdough starters were plated onto Luria–Bertani (LB) agar, and ten colonies from each sourdough starter were isolated and transferred to de Man, Rogosa, and Sharpe (MRS) agar or potato–dextrose agar (PDA) for further analysis. From the isolated colonies of the sourdough starter featuring dried longan, a total of four distinct yeast strains were identified, encompassing *S. cerevisiae*, *Zygosaccharomyces* (or *Lachancea*), and two strains of *Torulaspora delbrueckii*. Further, the analysis yielded the isolation of six LAB strains, namely *Weissella paramesenteroides*, *Leuconostoc citreum*, two strains of *Pediococcus pentosaceus*, and two strains of *W. cibaria*. In contrast, the sourdough starter incorporating raisin was home to two yeast species, specifically four strains of *Candida krusei* and an additional four strains of *T. delbrueckii* (see Table 3). These findings underscore the discernible differences in microbial profiles between the two sourdough starters. Notably, the sourdough starter containing dried longan displayed a higher diversity of yeast and LAB strains compared to the counterpart with raisin.

Table 3. Species of strains found in sourdough starters with dried longan and raisin.

Possible Species	Classification
Dried Longan	
Torulaspora delbrueckii	Yeast
Zygosaccharomyces or Lachancea	Yeast
Saccharomyces cerevisiae	Yeast
Weissella cibaria	LAB
Leuconostoc citreum	LAB
Weissella paramesenteroides	LAB
Pediococcus pentosaceus	LAB
Raisin	
Torulaspora delbrueckii	Yeast
Candida krusei	Yeast

3.5. Identification of Bacteriocin Genes

The evaluation of potential antibacterial bacteriocins encompassed four types: Gassericin A (*gaaA*), Lactacin F (*lafA*), Plantaricin S (*pls*), and Pediocin (*ped*). Employing PCR, this analysis was conducted on both *W. paramesenteroides* and *P. pentosaceus*. Notably, the detection of the 800 bp *gaaA* gene and the 184 bp *lafA* gene within *W. paramesenteroides* provided tangible evidence of the strain's inherent antibacterial properties (Figure 4).

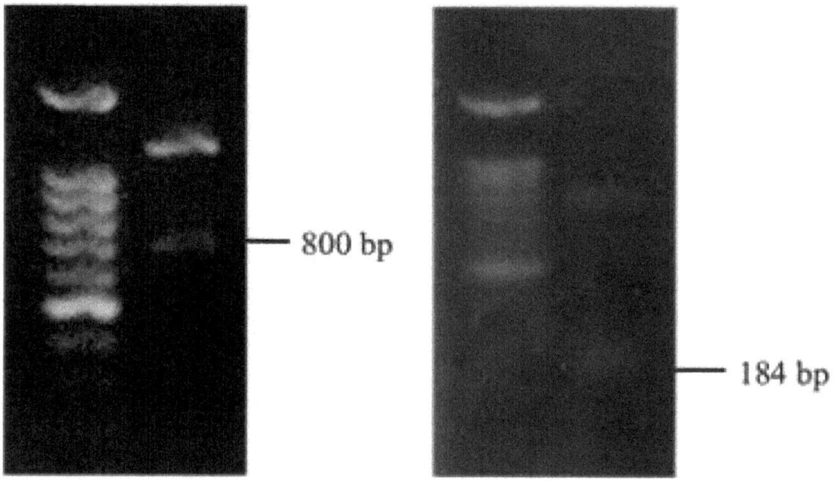

Figure 4. PCR analysis of bacteriocin genes in *W. paramesenteroides*.

3.6. Fermentation Efficacy

Fermentation efficacy was assessed for the two distinct sourdough starters. Over a 41 h fermentation period, the volume of the sourdough starter containing dried longan underwent a doubling effect. In stark contrast, the volume of the sourdough starter enriched with raisins remained unaltered (see Figure 5). These findings underscore that the sourdough starter featuring dried longan underwent a more pronounced fermentation compared to its raisin-containing counterpart.

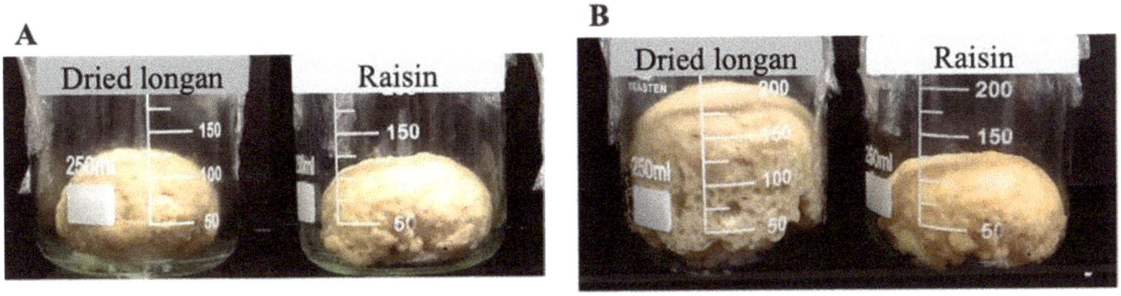

Figure 5. Fermentation efficacy test of two sourdough starters. (**A**) at the beginning of fermentation, (**B**) after 41 h of fermentation.

4. Conclusions

This study delved into the fermentation dynamics of sourdough starters infused with dried longan and raisin. On the third day, the sourdough starter containing dried longan exhibited bubbling along with an augmented stickiness. Simultaneously, a notable decline in pH was observed. Conversely, the sourdough starter with raisin displayed bubbling on the fifth day, without any significant alteration in viscosity. These outcomes underscore the dissimilarities in the activity of microbial flora between the two sourdough starters, a disparity attributed to distinct metabolite production.

The sourdough starter enriched with dried longan harbored a more abundant microbial diversity in comparison to its raisin counterpart. This divergence in microbial flora and metabolite synthesis among the two sourdough starters underscores their potential uniqueness and individuality. Recognizing these distinctions and their potential implications serves as a foundation for further investigations into refining the fermentation process and enhancing the development of superior sourdough starters for baking applications.

Author Contributions: Conceptualization, M.-J.W.; methodology, Y.-C.L. and S.-H.T.; data curation, P.-S.W.; writing—original draft preparation, Y.-C.L.; writing—M.-J.W.; funding acquisition, M.-J.W. All authors have read and agreed to the published version of the manuscript.

Funding: This research was supported by MOE STEM Project-Talent Cultivation Program in Dietary Supplements, Fermentation Industry and Food Safety, and MOST-110-2320-B-041-002-MY3.

Institutional Review Board Statement: Not applicable.

Informed Consent Statement: Not applicable.

Data Availability Statement: Data sharing is not applicable to this article.

Conflicts of Interest: The authors declare no conflict of interest.

References

1. Ribet, L.; Dessalles, R.; Lesens, C.; Brusselaers, N.; Durand-Dubief, M. Nutritional benefits of sourdoughs: A systematic review. *Adv. Nutr.* **2023**, *14*, 22–29. [CrossRef] [PubMed]
2. Faria-Oliveira, F.; Diniz, R.; Godoy Santos, F.; Mezadri, H.; Castro, I.; Brandão, R. *The Role of Yeast and Lactic Acid Bacteria in the Production of Fermented Beverages in South America*; IntechOpen: London, UK, 2015; pp. 107–135.
3. Negash, A.W.; Tsehai, B.A. Current Applications of Bacteriocin. *Int. J. Microbiol.* **2020**, *2020*, 4374891. [CrossRef] [PubMed]
4. Macwana, S.J.; Muriana, P.M. A 'bacteriocin PCR array' for identification of bacteriocin-related structural genes in lactic acid bacteria. *J. Microbiol. Methods* **2012**, *88*, 197–204. [CrossRef] [PubMed]
5. *CNS 12635*; Method of Test for Fruit and Vegetable Juices and Drinks—Determination of Organic Acids. CNS: Taipei, Taiwan, 2004.
6. Walsh, P.S.; Metzger, D.A.; Higushi, R. Chelex 100 as a medium for simple extraction of DNA for PCR-based typing from forensic material. *Biotechniques* **2013**, *54*, 134–139. [CrossRef] [PubMed]

7. Hasannejad Bibalan, M.; Eshaghi, M.; Rohani, M.; Pourshafie, M.R.; Talebi, M. Determination of Bacteriocin Genes and Antibacterial Activity of Lactobacillus Strains Isolated from Fecal of Healthy Individuals. *Int. J. Mol. Cell. Med.* **2017**, *6*, 50–55. [PubMed]
8. Zommiti, M.; Bouffartigues, E.; Maillot, O.; Barreau, M.; Szunerits, S.; Sebei, K.; Feuilloley, M.; Connil, N.; Ferchichi, M. In vitro Assessment of the Probiotic Properties and Bacteriocinogenic Potential of Pediococcus pentosaceus MZF16 Isolated From Artisanal Tunisian Meat "Dried Ossban". *Front. Microbiol.* **2018**, *9*, 2607. [CrossRef] [PubMed]

engineering proceedings

Proceeding Paper

Effectiveness of Small Amount of Surface Penetrant against Chloride Ion Penetration †

Rinchen Gyeltshen * and Shinichi Miyazato

Kanazawa Institute of Technology, Nonoichi 921-8501, Japan; koho@kanazawa-it.ac.jp
* Correspondence: rinche77@gmail.com

† Presented at the IEEE 5th Eurasia Conference on Biomedical Engineering, Healthcare and Sustainability, Tainan, Taiwan, 2–4 June 2023.

Abstract: To increase the durability of concrete structures, surface coating is widely used as a preventive maintenance strategy against de-icing salts. We investigated the effectiveness of a small amount of surface penetrant for chloride-induced corrosion on concrete structures exposed to low NaCl concentrations. The diffusion coefficient of mortar specimens with and without coating was determined using the electric migration test. The results indicated that even a small amount of Silane-based penetrant was effective against chloride ion penetration.

Keywords: surface penetrant; chloride ion penetration; apparent diffusion coefficient

Citation: Gyeltshen, R.; Miyazato, S. Effectiveness of Small Amount of Surface Penetrant against Chloride Ion Penetration. *Eng. Proc.* **2023**, *55*, 18. https://doi.org/10.3390/engproc2023055018

Academic Editors: Teen-Hang Meen, Kuei-Shu Hsu and Cheng-Fu Yang

Published: 29 November 2023

1. Introduction

Concrete structures play a vital role in modern infrastructure, from highways and bridges to buildings and parking structures. These structures, however, are often exposed to harsh weather conditions, particularly during the winter when de-icing salts are often used to ensure safe traffic on roads and walkways. The de-icing salts, particularly chloride-based compounds, harm the durability and condition of concrete with chloride attack and eventual deterioration. Therefore, to avoid premature deterioration, the use of surface treatment is required, and related research has been conducted to secure excellent workability and effective preventive measures [1].

The application of surface coatings to concrete structures considerably increased concrete's resistance to freezing, thawing, and chloride ion penetration [2]. Swamy and Tanikawa investigated the effect of surface penetrant coating on concrete durability and concluded that applying a surface coating to concrete is an efficient strategy for protecting new and old concrete structures as the coating makes concrete impervious [3]. The application of silicate-type surface penetrants significantly slows the initiation of corrosion in concrete structures located in coastal areas [4]. Ibrahim et al. showed that Silane with a topcoat had the best performance in terms of the effectiveness of acrylic and Silane coatings [5].

In Bhutan, light deicer products are often sprayed during the winter months to prevent vehicles from sliding due to ice formation on the road surface. The use of surface penetrants on concrete structures prevents chloride ion penetration induced by the application of deicers. However, there is limited information about the effectiveness of the surface penetrants against chloride ion penetration when concrete structures are exposed to normal and low sodium chloride (NaCl) concentrations.

2. Material and Methods

2.1. Properties of Fine Aggregate and Binding Materials

Fine aggregate/land sand was used for preparing mortar specimens that had a particle size of less than 4.75 mm and a density of 2.57 g/m^3 following the ASTMC33 standard as stated in Table 1. In this experiment, Ordinary Portland Cement (OPC) of a density of

3.16 g/m^3 meeting the ASTM C150 requirement, which is widely used in the construction of various concrete structures around the world, was adopted.

Table 1. Proportion of mortar mixture.

W/C	S/C	Unit Weight (kg/m^3)		
		W	C *	S **
0.50	2.5	12.296	24.591	61.5

* Ordinary Portland Cement (density of 3.16g/m^3); ** Land Sand (density of 2.57g/m^3).

2.2. Research Methodology

The proportion of the mixture of mortar specimens (20 cm in length × 10 cm in diameter) is shown in Table 1. The specimens were cured in water for 28 days at 20 °C. After that, the test specimens (3 cm in length × 10 cm in diameter) were obtained from prepared mortar specimens. The specimens were coated with the surface penetrant as shown in Table 2 for 7 days at 20 °C and a relative humidity of 60%. Then, the diffusion coefficient of the coated specimens for normal and low concentrations of NaCl was calculated based on JSCE G 571. Figure 1 shows an overview of the electric migration test method used. For testing, test specimens of 3 cm in thickness with and without coating were exposed to 3, 1, and 0.5% NaCl at a constant 15 V during the test duration. A steady state was assumed to be reached once the rate of chloride concentration in the anode became constant. The flux of chloride ions in the pore solution was determined as shown in Figure 2.

Table 2. Types of surface penetrant.

Case	Classification-Type	Application Quantity (g/m^2)
Case No.	No penetrant	0
Case A	Silane-type	200
Case B		100
Case C		50

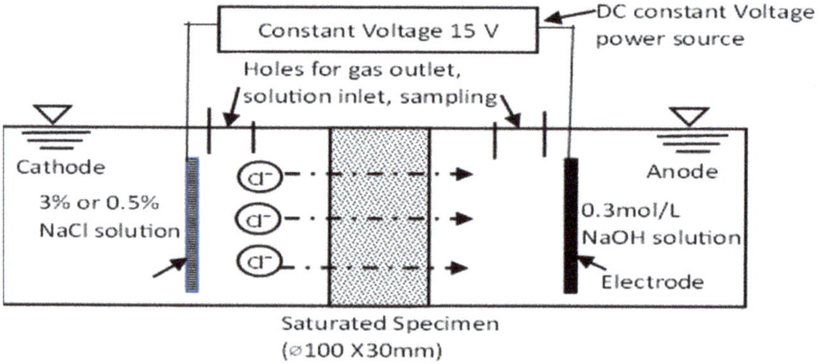

Figure 1. Overview of the electric migration test.

The effective diffusion coefficient of the chloride ion (Table 3) was calculated from the flux using the Nernst–Planck Equation. The effective diffusion coefficient was converted to the apparent diffusion coefficient (shown in Table 3 and Figure 3) using Equation (1).

$$D_{ae} = K_1 \cdot K_2 \cdot D_e \quad (1)$$

(D_{ae}: apparent diffusion coefficient, D_e: effective diffusion coefficient obtained in the migration test (cm^2/year), K_1: coefficient reflecting the equilibrium concentration of chloride ions on the cathode side and at the concrete surface, and K_2: coefficient reflecting the effect of immobilization of chloride ions in the hydrated cement system).

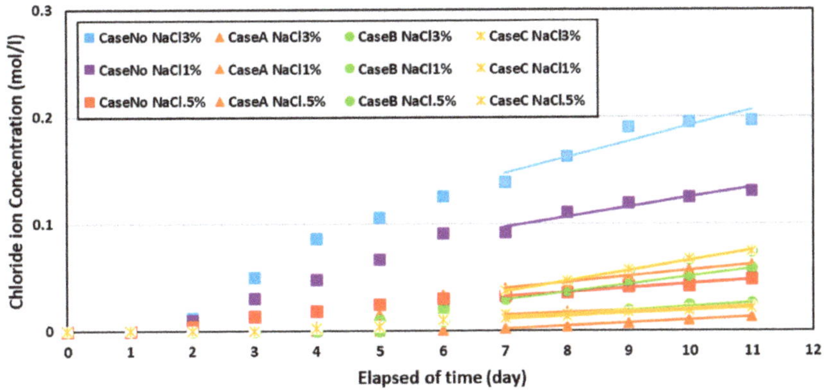

Figure 2. Change in chloride ion concentrations on the anode side.

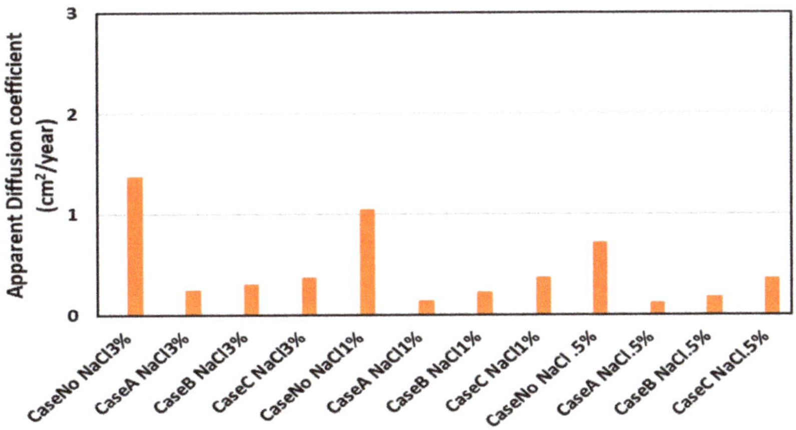

Figure 3. Apparent chloride ion diffusion coefficient of all experiment specimens.

When Ordinary Portland Cement was used, $(K_1 \cdot K_2) = 0.2 \exp(1.8\ (W/C))$ $(0.3 \leq W/C < 0.55)$ [6]. As shown in Table 3, the diffusion coefficient (D_s) of surface penetrants was estimated using Equation (2) [7]. The impregnation depth surface penetrant was 3 mm for a full coating, 2.3 mm for a half coating, and 1.7 mm for a quarter coating.

$$\frac{C'}{\sqrt{D'}} = \frac{C_c}{\sqrt{D_c}} + \frac{C_s}{\sqrt{D_s}} \tag{2}$$

(C': thickness of coated specimens (mm), D': diffusion coefficient of coated specimens (cm^2/year), D_c: diffusion coefficient of uncoated specimens (cm^2/year), C_c: thickness of uncoated specimens (mm), C_s: penetrating depths (mm), D_s: diffusion coefficient of penetrant (cm^2/year), and S: converted cover depth (mm)).

$$S = C'\left[\frac{\sqrt{D_c}}{\sqrt{D'}} - 1\right] = C_s\left[\frac{\sqrt{D_c}}{\sqrt{D_s}} - 1\right] \tag{3}$$

The converted cover depth [7] in Table 4 was calculated from the diffusion coefficient (D_s) of the surface penetrant, the apparent diffusion coefficient (D_c) of uncoated mortar specimens, and the apparent diffusion coefficient (D′) of coated mortar specimens using Equation (3) [7].

Table 3. Result of migration test.

Cases	Effective Diffusion (cm^2/Year)	Apparent Diffusion (cm^2/Year)	Diffusion Coefficient (D_s) Penetrant (cm^2/Year)
Case No NaCl 3%	2.67	1.36	
Case A NaCl 3%	0.47	0.24	0.0062
Case B NaCl 3%	0.59	0.30	0.0055
Case C NaCl 3%	0.71	0.36	0.0025
Case No NaCl 1%	2.03	1.04	
Case A NaCl 1%	0.26	0.13	0.0029
Case B NaCl 1%	0.43	0.22	0.0040
Case C NaCl 1%	0.71	0.36	0.0035
Case No NaCl 0.5%	1.40	0.71	
Case A NaCl 0.5%	0.21	0.11	0.0025
Case B NaCl 0.5%	0.34	0.17	0.0034
Case C NaCl 0.5%	0.69	0.35	0.0059

Table 4. Equivalent cover depth.

Case	Converted Cover Depth (mm)
Case A NaCl 3%	41.50
Case B NaCl 3%	39.94
Case C NaCl 3%	28.08
Case A NaCl 1%	48.84
Case B NaCl 1%	37.27
Case C NaCl 1%	18.84
Case A NaCl 0.5%	47.27
Case B NaCl 0.5%	36.25
Case C NaCl 0.5%	12.73

3. Result and Discussion

Figures 2 and 3 show that the chloride ion concentration and the apparent diffusion coefficient of the mortar specimens were higher for the specimens exposed to the high NaCl concentration and lower for the specimens exposed to the low NaCl concentration. Also, the concrete surface coated with a Silane-based surface penetrant significantly reduced the chloride ion concentration on the anode side and the apparent diffusion coefficient. As shown in Table 4 and Figure 4, the equivalent cover depth of full and half coatings increased significantly when exposed to NaCl concentrations of 0.5 to 3%, whereas quarter coatings showed an increase in cover depth. The converted cover depth remains roughly the same when the application amount is increased from half to full coating regardless of specimens exposed to 0.5% to 3% NaCl solution; however, quarter application shows a less increase in equivalent cover depth as illustrated in Table 4 and Figure 5. It indicated that quarterly-coated concrete structures with a surface penetrant did not allow chloride ion penetration when exposed to low NaCl concentrations.

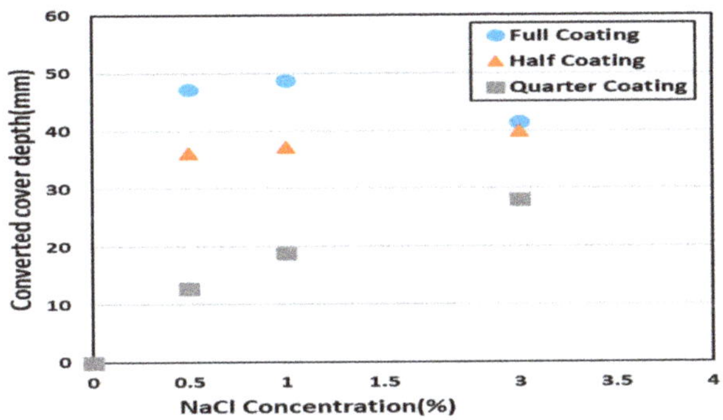

Figure 4. Equivalent cover depth over increase in NaCl concentration.

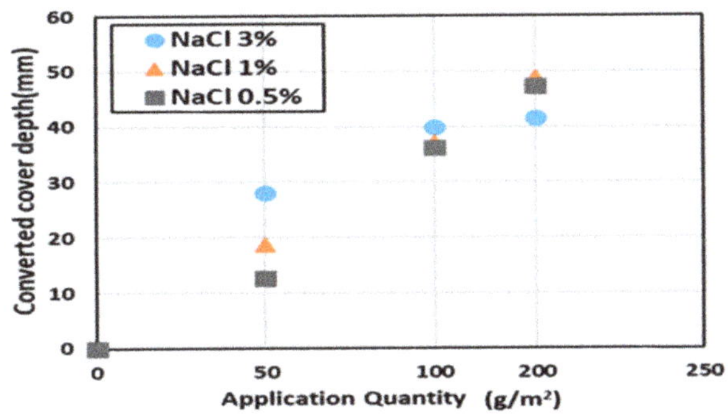

Figure 5. Equivalent cover depth over application rate of coating.

4. Simulation of Corrosion Initiation Time

To evaluate the delay in corrosion initiation time when concrete structures were coated with the Silane-based surface penetrant, the concrete cover depth was set to 5 cm, and the surface chloride ion concentration was fixed to be 1.80 kg/m^3, assuming that the amount of application of deicers was relatively low. The converted cover depth shown in Table 4 and the apparent diffusion coefficient of the uncoated specimens shown in Figure 3 were calculated following Fick's law of diffusion (Equation (4)) to estimate the time for the total chloride ion concentration around the steel bars to reach a threshold value of 1.2 kg/m^3 [8].

$$C_{(x,t)} = C_0\left\{1 - \mathrm{erf}\left(\frac{x}{2\sqrt{D_{ae} \cdot t}}\right)\right\} \quad (4)$$

($C_{(x,t)}$: chloride ion concentration at depth x (cm) and time t (years) (kg/m^3), C_0: chloride ion concentration at the concrete surface (kg/m^3), D_{ae}: apparent chloride ion diffusion coefficient (cm^2/year), and erf error function).

As shown in Figure 6, the time required for corrosion without a surface penetrant was calculated as 49 years for specimens exposed to 3% NaCl, 64 years for 1% NaCl, and 96 years for 0.5% NaCl. However, when specimens were coated with a quarter amount of Silane-based surface penetrant, the time for corrosion was delayed to 120 years for 3% NaCl, 160 years for 1% NaCl, and 230 years for 0.5% NaCl concentrations. The performance deterioration of surface penetrants was not considered in this calculation. However, it was obvious that using only a small quantity of the Silane-based penetrant on concrete structures significantly delayed the onset of rebar corrosion when concrete structures were exposed to normal to low NaCl concentrations.

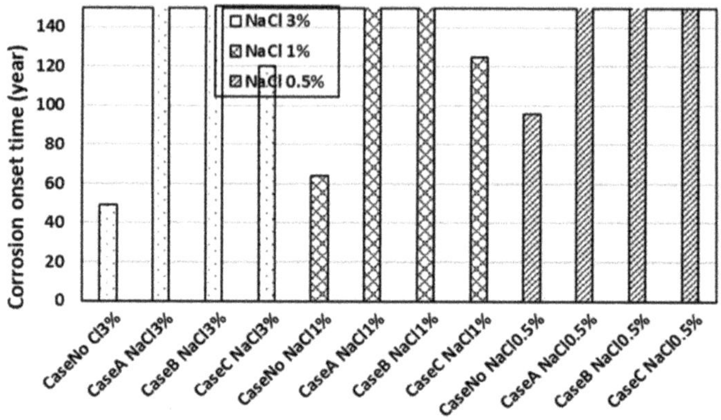

Figure 6. Onset corrosion time of all experiment specimens.

5. Conclusions

In this study, it was found that the chloride ion concentration and the apparent diffusion coefficient of the mortar specimens were higher than those of the specimens exposed to high and low NaCl concentrations. The concrete surface coated with the Silane-based surface penetrant significantly reduced the apparent diffusion coefficient of the chloride ion on the anode side. Even a small amount of the surface penetrant was effective in delaying the corrosion initiation of concrete structures exposed to any concentration of de-icing salts.

Author Contributions: Conceptualization, R.G. and S.M.; Methodology, R.G. and S.M.; Experiment, R.G.; Analysis, R.G.; Writing-Original draft preparation, R.G.; Writing-Review and editing, S.M.; Guidance and supervision, S.M.; All authors have read and agreed to the published version of the manuscript.

Funding: This research received no external funding.

Institutional Review Board Statement: Not applicable.

Informed Consent Statement: Not applicable.

Data Availability Statement: The current data set analysis is available and can be shared upon request.

Acknowledgments: The authors highly acknowledge the support of JICA and (Japan International Cooperation Agency) and the Royal Government of Bhutan throughout the study.

Conflicts of Interest: The authors declare no conflict of interest.

References

1. Hanaoka, D.; Amino, T.; Habuchi, T.; Miyazato, S.; Tabata, S. A study on prediction method of chloride ion penetration into concrete with surface penetrants. In *Life-Cycle of Structural Systems: Design, Assessment, Maintenance and Management, Procceedings of the 4th International Symposium on Life-Cycle Civil Engineering, IALCCE 2014, Tokyo, Japan, 16–19 November 2014*; Taylor & Francis: Abingdon, UK, 2014; pp. 950–955. [CrossRef]
2. Moon, H.Y.; Shin, D.G.; Choi, D.S. Evaluation of the durability of mortar and concrete applied with inorganic coating material and surface treatment system. *Constr. Build. Mater.* **2007**, *21*, 362–369. [CrossRef]
3. Swamy, R.N.; Tanikawa, S. Surface coatings to preserve concrete durability. In *Protection of Concrete*; Dhir, R.K., Green, J.W., Spon, F.N.E., Eds.; CRC Press: London, UK, 1990; Volume 1, pp. 149–165.
4. Kuroiwa, D.; Miyazato, S. Proposal of Estimation Method for Apparent Chloride Ion Diffusion Coefficient At Reform Part By Silicate Type Surface Penetrant and Trial Calculation for Corrosion Occurrence Time. *J. Jpn. Soc. Civ. Eng. Ser. E2 (Mater. Concr. Struct.)* **2015**, *71*, 124–134. [CrossRef]
5. Ibrahim, M.; Al-Gahtani, A.S.; Maslehuddin, M.; Almusallam, A.A. Effectiveness of concrete surface treatment materials in reducing chloride-induced reinforcement corrosion. *Constr. Build. Mater.* **1997**, *11*, 443–451. [CrossRef]
6. Japan Society of Civil Engineers. *Standard Specifications for Concrete Structures—2007 "Materials and Construction"*; Japan Society of Civil Engineers: Tokyo, Japan, 2010.

7. Yoshitani, T.; Miyazato, S. Estimation of Chloride ion Diffusion coefficient of concrete surface layer modified by Silane-based surface impregnation. In Proceedings of the 70th Annual Conference of the Japan Society of Engineers, Okayama, Japan, 16–17 September 2015; pp. 1173–1174.
8. Uomoto, T.; Ishibashi, T.; Nobuta, Y.; Satoh, T.; Kawano, H.; Takewaka, K.; Uji, K. Standard Specifications for Concrete Structures-2007 by Japan Society of Civil Engineers. *Concr. J.* **2008**, *46*, 3–14. [CrossRef]

engineering proceedings

Proceeding Paper

ISO 27001 Information Security Survey of Medical Service Organizations †

Hung-Hsiou Hsu [1,*] and Jyun-Rong Shih [2]

1 Department of Multimedia and Game Development, Chia Nan University of Pharmacy & Science, Tainan 717301, Taiwan
2 XieTong B&B, Hualien 970342, Taiwan; sniperkiller73@gmail.com
* Correspondence: hhhsu127@mail.cnu.edu.tw
† Presented at the IEEE 5th Eurasia Conference on Biomedical Engineering, Healthcare and Sustainability, Tainan, Taiwan, 2–4 June 2023.

Abstract: The differences between medical institutions in the security management of information systems were investigated by comparing the differences and the means used by personnel in different units in public and private hospitals. Personnel responsible for information security require the protocol of relevant units to solve information security issues. Based on ISO 27001 as a reference standard, a questionnaire survey was conducted to investigate the need for information security management in medical institutions. The information system security in each unit of medical institutions needs to pursue the goal of more perfection for a fully optimized information system. To help medical institution personnel understand the importance of information security and allow appropriate decision making, the results of this study can be used as a reference.

Keywords: ISO 27001; information security; medical information systems

Citation: Hsu, H.-H.; Shih, J.-R. ISO 27001 Information Security Survey of Medical Service Organizations. *Eng. Proc.* **2023**, *55*, 19. https://doi.org/10.3390/engproc2023055019

Academic Editors: Teen-Hang Meen, Kuei-Shu Hsu and Cheng-Fu Yang

Published: 29 November 2023

1. Introduction

In medicine and health care, the rapid development of the information industry has shed the security problems in their information systems such as the leakage of medical information. Several medical staff have even been caught selling patients' private information illegally. Thus, equipment and system software in the medical and health sector for information security are frequently updated. However, the new system has too many functions to be used adequately, leading to many potential security risks. The risk of hacking and cyber attacks also exists, threatening the information security of medical institutions. Related incidents increase every year, which makes people aware of the damage caused by such incidents.

The information network allows users to access relevant information and data quickly under relevant instructions. Due to the openness of the information network and system, they become useful but dangerous. Use of the information system by an unauthorized person damages the management and financials. Therefore, the development of a system for information security that does not sacrifice convenience and usability is required for securing confidentiality, integrity, and availability. The system also protects information assets and increases overall competitiveness. Information systems can be accessed by users and administrators, but information can be retrieved by inappropriate means. Hackers can damage the system and steal information using point-of-entry attacks, backdoors, Trojan horses, and viruses. Therefore, building a defensive information system security is necessary to provide complete and uninterrupted operation, including resistance, detection, and recovery [1].

Information security in medical institutions differs from that in other industry sectors. As medical care is related to patient safety, information security must be considered more seriously. If the system provides false information, this harms patients as well as the hospital's credit and reputation. In addition, patients' medical records and personal information may be leaked due to inadequate information security protection. If the patient's private information is used by a third party illegally, the patient's rights will be seriously compromised. Therefore, special attention must be paid to the security of patient information. The design of systems must meet the requirements of information security. Therefore, it is necessary to build an information security system that is accurate and effective in security.

The degree of informatization of an organization is closely related to the information strategy, information department structure, information system architecture, and information application areas. Information security strategy is influenced by the differentiation of the internal environment of the organization, so the characteristics of the organization and the degree of informatization lead to different combinations of information risk due to the types of organizations [2]. The different information risks require different information security strategies. The growth of organizational information can be measured in four aspects: information system architecture, information department strategy, information department organization, and information application area [3]. Although there is no research on the relationship between information security and the degree of informatization, it was found that information security is affected by information threats to personal computers and internal servers (mainframe) and improper operation of networked architectural systems [4,5].

To maintain information security, it is necessary to understand "the prevention and detection of unauthorized situations by users of the processing system" [6]. Information security is defined broadly as the protection of confidentiality, integrity, and availability of data stored in a system. There are many norms set for information security, such as BS7799 [7] and ISO 27001 [8]. To regulate the security and confidentiality of information and to meet the privacy needs of individuals, the US federal government passed the Health Insurance Portability and Accountability Act (HIPAA) in 1996, which specified the security mechanisms that must be in place for information systems. Its contents include the following four categories: administrative procedures, physical safeguards, technical security services, and technical security mechanisms. Based on BS7799, ISO 27001, and HIPAA, we investigated the need for an effective management system for information security through a questionnaire survey to provide the basis for the construction of the system. To assess the information security of an organization, indicators of risk assessment and information protection capability need to be determined. For understanding of the risks in information security and the degree of informatization, an accurate assessment of the existing information protection capability is required. Therefore, information risks were assessed from four aspects: hardware, software, information, and network, in terms of the physical environment, personnel, and management in this study. A questionnaire survey was conducted to define measurement indexes for these seven risks.

2. Questionnaire Survey

Based on the theory of Icove et al. [9], information risk was analyzed to understand the level of protection of information using the protection measures of each risk. Siegel referred to ISO 17799 [10] and InfoSec international standards, while Richardson [11] referred to the Computer Security Institute (CSI) against network risk threats for analysis. Referring to CNS 27001 [12] of the Ministry of Economic Affairs, BS7799 of the UK, and the international information security standard ISO 27001, we created a questionnaire to understand preventative measures for information security. The questionnaire was revised by experts to improve its validity. The final questionnaire was designed to measure the level of knowledge of information security and the importance of information security in the organization.

The questionnaire consisted of three parts including basic information, the current situation of information security, and risk awareness of information security. In this study, the risk to information security was classified into physical and system aspects.

In the physical aspect, the risks were grouped as the physical, human, and managerial risks, while in the system aspect, the risks in hardware, software, data, and network were included. The questionnaire included questions on equipment, management, virus, training, troubleshooting, and accounts of the information security systems. Figure 1 illustrates the structure of the questionnaire. The questions for the seven were contained in the six categories.

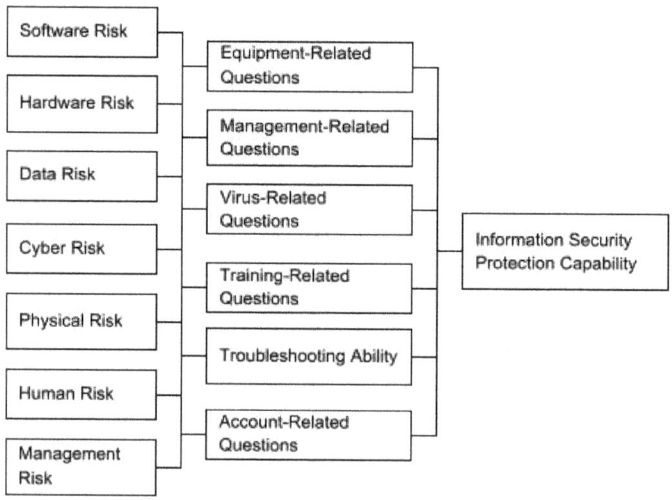

Figure 1. Structure of questionnaire in this study.

A five-point Likert scale was used in the questionnaire. Scores were given to strongly agree, agree, average, disagree, and strongly disagree with the organization's ability for information security, with one point representing no protection and five points representing sufficient protection. The reliability of the questionnaire was analyzed using Cronbach's alpha coefficient [13,14]. The internal consistency was also tested with Cronbach's alpha coefficient. The overall reliability was 0.949. The reliability of questions of software risk, hardware risk, data risk, network risk, physical risk, human risk, and managerial risk was 0.796, 0.639, 0.709, 0.790, 0.724, 0.751, and 0.813, all of which showed acceptable reliability and credibility.

A total of 150 questionnaires were distributed to the personnel in charge of information security in public and private hospitals in the southern region and military and security units in Taiwan, with 123 returned. The total number of valid questionnaires was 113. The respondents included administrative and information security staff, network administrators, and supervisors. The education of the respondents was at least college- and university-level (Table 1).

Table 1. Information on respondents of questionnaire survey.

Position	Number	Ratio (%)
Administration staff	60	53.1%
Information security personnel	31	27.4%
Network administrator	10	8.8%
Director	6	5.3%
Missing value	6	5.3%
Sum	113	100.0%

Some 72.6% of the respondents were male, and 27.4% were female. A majority of the respondents were executives (56.1%), followed by information security personnel (29.0%),

network administrators (9.3%), and supervisors (5.6%). Respondents with eight years of work experience accounted for 31.0%, followed by those with less than one year (29.2%), those with two to four years (25.7%), and those with five to seven years (14.2%) (Table 2).

Table 2. Working seniority scale table.

Work Experience	Number	Ratio (%)
Over 8 years	35	31.0%
1 year or less	33	29.2%
2~4 years	29	25.7%
5~7 years	16	14.2%
Missing value	0	0.0%
Sum	113	100.0%

As shown in Table 3, the respondents mainly served in military health units (46.9%), followed by private hospitals (37.2%), and then public hospitals (15.9%). Detailed information on the units was recorded in the interview.

Table 3. Unit type ratio table.

Institutions	Number	Ratio (%)
Military medical service	53	46.9%
Private hospital	42	37.2%
Public hospital	18	15.9%
Missing value	0	0.0%
Sum	113	100.0%

3. Results

The results of the survey are shown in Table 4. The highest protection ability of information security was observed for software risk, with a sum of scores of 436.835, while the lowest was for data risk, with a sum of scores of 400.882. For hardware risk, the ability showed the highest average score of 3.357, while the lowest score was observed for managerial risk, with an average score of 3.000. The highest average score was obtained for software risk, with a score of 4.196, while the lowest score was for data risk, with an average score of 3.912 (Tables 4 and 5).

Table 4 shows the standard deviation of the scores in this study. The smaller the standard deviation, the more concentrated the data are in this dimension, while the larger the standard deviation, the more scattered the data are in this dimension. It was found that the dimension of network risk was widely dispersed, while the data for the aspect of risk management were more concentrated. This result indicated that the respondents held diverse perspectives on cyber risks, which might be influenced by personal experiences, knowledge, or preferences. However, their views on risk management are more uniform, and may be influenced by shared norms or standards.

Table 4. Protection ability of information security for each risk.

	Sum of Scores	Average of Minimum Score	Average of Maximum Score	Standard Deviation
Software Risk	436.835	3.304	3.304	0.574
Hardware Risk	415.615	3.357	3.357	0.603
Data Risk	400.882	3.08	3.08	0.504

Table 4. Cont.

	Sum of Scores	Average of Minimum Score	Average of Maximum Score	Standard Deviation
Cyber Risk	415.355	3.223	3.223	0.625
Physical Risk	413.657	3.083	3.083	0.5
Human Risk	407.546	3.088	3.088	0.525
Managerial risk	407.173	3.000	3.000	0.425

Table 5. Summary of result of questionnaire survey.

		Director		Administration Staff		Information Security Personnel		Network Administrator		Total	
Institutions	Response	Number	Ratio (%)	Number	Ratio (%)	Number	Ratio (%)	Number	Ratio (%)	Number	Ratio (%)
Private Hospital	Very well understood	1	50.0%	1	11.1%	0	0.0%	0	0.0%	2	11.1%
	Fairly well understood	1	50.0%	4	44.4%	3	75.0%	3	100.0%	11	61.1%
	Somewhat	0	0.0%	4	44.4%	1	25.0%	0	0.0%	5	27.8%
Private Hospital	Very well understood	0	0.0%	2	9.5%	2	15.4%	2	33.3%	6	14.3%
	Fairly well understood	1	50.0%	8	38.1%	4	30.8%	1	16.7%	14	33.3%
	Somewhat	1	50.0%	6	28.6%	7	53.8%	3	50.0%	17	40.5%
	Do not know	0	0.0%	5	23.8%	0	0.0%	0	0.0%	5	11.9%
Public Hospital	Very well understood	0	0.0%	7	23.3%	1	7.1%	0	0.0%	8	17.0%
	Fairly well understood	1	50.0%	9	30.0%	7	50.0%	0	0.0%	17	36.2%
	Somewhat	1	50.0%	5	16.7%	6	42.9%	1	100.0%	13	27.7%
	Do not know	0	0.0%	9	30.0%	0	0.0%	0	0.0%	9	19.1%

4. Conclusions

Personnel in public hospitals understood the importance of all risks to information security better than those in other healthcare institutions. However, public hospitals do not invest more in information security than private hospitals, nor do they have a higher degree of information restriction than military health care units. Thus, public hospitals need to place more emphasis on the quality of personnel and management systems with various information security policies, rapid troubleshooting, reliable backup of important data, regular updates to key system security, training of personnel on information security, and control of information flow. The maintenance of information security must be conducted by constantly updating software and hardware equipment and maintaining the information security system. All information in the system must be restricted to authorized personnel and operators. The most effective way to maintain information security is to cultivate personnel and establish an effective managerial system.

Author Contributions: Conceptualization, H.-H.H. and J.-R.S.; methodology, H.-H.H. and J.-R.S.; software, J.-R.S.; validation, H.-H.H. and J.-R.S.; formal analysis, J.-R.S.; investigation, H.-H.H. and J.-R.S.; resources, H.-H.H. and J.-R.S.; data curation, J.-R.S.; writing—original draft preparation, H.-H.H. and J.-R.S.; writing—review and editing, H.-H.H.; visualization, H.-H.H. and J.-R.S.; supervision, H.-H.H.; project administration, H.-H.H.; funding acquisition, H.-H.H. All authors have read and agreed to the published version of the manuscript.

Funding: The researchers thank the Ministry of Economic Affairs for the research grant (112th Annual Funding Support for the Academic Promotion of Value-added Technology and Innovation Program in Local Industry).

Institutional Review Board Statement: Not applicable.

Informed Consent Statement: Not applicable.

Data Availability Statement: Data sharing is not applicable to this article.

Conflicts of Interest: The authors declare no conflict of interest. XieTong B&B has no conflict of interest with the paper.

References

1. Ellison, R.J.; Linger, R.C.; Longstaff, T.; Mead, N.R. Survivable network system analysis: A case study. *IEEE Softw.* **1999**, *16*, 70–77. [CrossRef]
2. Kumar, V.; Telang, R.; Mukhopadhyay, T. Enterprise Information Security: Who Should Manage it and How? In Proceedings of the 5th Annual Workshop on the Economics of Information Security (WEIS 2006), Cambridge, UK, 26–28 June 2006.
3. Earl, M.J. *Management Strategies for Information Technology*; Prentice-Hall, Inc.: Hoboken, NJ, USA, 1989.
4. Loch, K.D.; Carr, H.H.; Warkentin, M.E. Threats to information systems: Today's reality, yesterday's understanding. *MIS Q.* **1992**, *16*, 173–186. [CrossRef]
5. Ryan, S.D.; Bordoloi, B. Evaluating security threats in mainframe and client/server environments. *Inf. Manag.* **1997**, *32*, 137–146. [CrossRef]
6. Hartel, P.H.; Junger, M.; Wieringa, R.J. *Cyber-Crime Science= Crime Science + Information Security*; Technical Report TR-CTIT-10-34; CTIT, University of Twente: Enschede, The Netherlands, 2010.
7. *BS 7799-2*; Information Security Management—Part 2: Specification for Information Security. British Standards Institution: London, UK, 1999.
8. *ISO 27001*; Information Security Management Standard. ISO: Geneva, Switzerland, 2005.
9. Icove, D.; Seger, K.; VonStorch, W.R. *Computer Crime: A Crimefighter's Handbook*; O'Reilly & Associates, Inc.: Sebastopol, CA, USA, 1995.
10. *ISO/IEC 17799*; Information Technology-Code of Practice for Information Security Management. ISO: Geneva, Switzerland, 2005.
11. Gordon, L.A.; Loeb, M.P.; Lucyshyn, W.; Richardson, R. 2005 CSI/FBI computer crime and security survey. *Comput. Secur. J.* **2005**, *21*, 1.
12. *CNS 27001*; Information Security, Cybersecurity and Privacy Protection—Information Security Management Systems—Requirements. CNS: Taipei, Taiwan, 2023.
13. Kankanhalli, A.; Teo, H.H.; Tan, B.C.; Wei, K.K. An integrative study of information systems security effectiveness. *Int. J. Inf. Manag.* **2003**, *23*, 139–154. [CrossRef]
14. Shah, N.R.; Aragones, A.; Schaefer, E.W.; Stevens, D.; Gourevitch, M.N.; Glasgow, R.E. Validation of the Spanish Translation of the Patient Assessment of Chronic Illness Care (PACIC) Survey. *Prev. Chronic Dis.* **2008**, *5*, A113.

engineering proceedings

Proceeding Paper

Estimation and Implication of Time-Varying Reproduction Numbers during the COVID-19 Pandemic in the UK †

Jiangjiang Yan [1], Ruochen Huang [2,*] and Wuliang Yin [1]

1 Department of Electrical and Electronic Engineering, University of Manchester, Manchester M13 9PL, UK; jiangjiang.yan@postgrad.manchester.ac.uk (J.Y.); wuliang.yin@manchester.ac.uk (W.Y.)
2 College of Electrical Engineering and Automation, Fuzhou University, Fuzhou 350108, China
* Correspondence: ruochen_huang@fzu.edu.cn
† Presented at the IEEE 5th Eurasia Conference on Biomedical Engineering, Healthcare and Sustainability, Tainan, Taiwan, 2–4 June 2023.

Abstract: Infectious illness prevention and control is an important part of public health management. The early monitoring and numerical modelling of incidence data can help with the efficient prevention and control of infectious disease prevalence. The reproduction number R, as an essential index to understand the dynamics of COVID-19, can be predicted by using confirmed new incidence cases and serial interval data in the datasets provided by UK government. In this paper, an extended model is proposed to account for variable reporting rates instead of 1 for the estimation of the R number. The impact of using various modelling parameters is also evaluated, which provides insight into how to select a set of suitable parameters in modelling. The variation of the estimation of the R number by incorporating variable reporting rates can be observed and assessed.

Keywords: reproduction number; reporting rates; incidence cases; serial interval distribution; COVID-19

Citation: Yan, J.; Huang, R.; Yin, W. Estimation and Implication of Time-Varying Reproduction Numbers during the COVID-19 Pandemic in the UK. *Eng. Proc.* **2023**, *55*, 20. https://doi.org/10.3390/engproc2023055020

Academic Editors: Teen-Hang Meen, Kuei-Shu Hsu and Cheng-Fu Yang

Published: 29 November 2023

1. Introduction

Infectious illness prevention and control are an important part of public health and safety, as they have an impact on human health and social stability. Large-scale epidemics are likely to emerge if infectious disease propagation is not successfully controlled. During COVID-19, the importance of epidemic monitoring, modelling, and understanding, as well as early warning and policy guidance, is strongly emphasised.

Simultaneously, with the rapid advancement of information technology and significant rise in computing power, the data analytics (DA)-based modelling and monitoring of infectious illnesses have become widely accessible for public health practitioners. In this context, data analytics refers to a set of procedures for converting data into interpretable and actionable information sets [1] and for producing reliable and innovative insights into the current state of the world [2]. The discovery and tracking of epidemic diseases in the healthcare business is one of the most notable examples of data analytics.

Modelling the transmission of infectious illnesses can be performed in a variety of ways. They are largely divided into two categories: deterministic and statistical techniques. Statistical approaches use existing data to develop estimators using probability theory, whereas deterministic (mathematical) approaches focus on the construction of a mechanistic model describing disease transmission. Comprehensive evaluations of the methods can be found in works such as [3–7]. Dietz and Becker summarised multiple modelling strategies for estimating the propagation of infectious disorders in [3,4]. Perone presented hybrid models, such as the neural network autoregression (NNAR) model and the autoregressive moving average (ARIMA) model, that perform better in estimating the disease trend than single models, such as the neural network autoregression (NNAR) model and the autoregressive moving average (ARIMA) model [5]. Fraser suggested the home reproduction number [6], a method for predicting the reproduction number based on the household model. Using this strategy, it provides advice on the measures and control of families during epidemics. Godio et al., who combined the generalised SEIR epidemiological model with swarm intelligence, linked the model to reality movement in Italy and demonstrated that the stochastic approach is better for trend estimation [8].

The estimation of the reproductive number R is particularly important among the investigations. R is the average number of secondary cases induced by a single infected person [9]. R is used to measure the transmission capability of pathogens during epidemics; therefore, its estimation is critical in the prevention and control of infectious illnesses. R monitoring offers feedback on the success of control measures and the need to reinforce control [10]. It is a useful index for understanding the dynamics of infectious illnesses. Pharmaceutical interventions (PIs), such as vaccination, and non-pharmaceutical interventions (NPIs), such as lockdowns, are two types of public health control strategies that try to prevent and/or regulate transmission in the community. The purpose of intervention activities is to lower R below 1 and to be as close to 0 as possible, thereby bringing an epidemic under control. Various elements, such as the daily confirmed data, influence the accuracy of R estimation. An estimation mistake would result from the data's inaccuracy. However, there are only a few approaches that may be used to fix this type of problem.

To solve this problem, an extended model for R estimate based on variable reporting rates is proposed in this study. The proposed model can be used to more precisely analyse the evolution of the disease. In the theoretical demonstration of the model, varying reporting rates are introduced, followed by simulation outcomes due to various modelling settings. Finally, the utilisation of the reporting rate is assessed based on R number estimation.

2. The Theoretical Model for Estimating Time-Varying Instantaneous Reproduction Numbers

The term "epidemic model" describes a mathematical model used to describe infectious illnesses quantitatively. It uses quantitative research methodologies to study infectious diseases primarily in terms of growth and transmission characteristics in order to develop a model that can accurately depict the transmission dynamic characteristics of infectious diseases. Infectious disease modelling is primarily used to investigate disease transmission mechanisms, forecast future outbreaks, and assess the effectiveness of epidemic management efforts.

Researchers at Imperial College developed a novel framework and software to estimate time-varying reproduction counts during epidemics [9]. Because it provides a framework for estimating time-varying instantaneous reproduction numbers from incidence time series during epidemics, this line of models is referred to as the ETVR model in the study. Epidemiologists and public health organisations can use the ETVR model to alter public health responses in real time [11].

The calculation of the R number, according to the ETVR model, is dependent on accurate incident data (I_a^t) and the serial interval distribution (S^t). The serial interval distribution is used to approximate the infectivity profile of the infectious disease. It is common for infectious cases to be over- or under-reported. To address this problem, it is necessary to implement a reporting rate for the estimation of the R number. From here, both the reporting rate (λ^t) and the reported daily confirmed cases (I_{rp}^t) can be used to present the actual daily confirmed cases at timestep t, as expressed below:

$$I_a^t = \lambda^t I_{rp}^t \tag{1}$$

The reporting rate is considered to be constant at timestep t, but it can vary over the course of the period. The impact of different reporting rates on the R number estimation will be examined in this study.

The total infectivity of all infected individuals over a time period t can be calculated by combining the serial interval distribution and the actual confirmed cases:

$$R^t = \frac{I_a^t}{\Gamma^t} = \frac{\lambda^t I_{rp}^t}{\sum_{n=1}^t S^n I_a^{t-n}} \tag{2}$$

Then, by the definition of the reproduction number R [11], it can be obtained as

$$E\left(I_a^t\right) = R^t \sum_{n=1}^t S^n I_a^{t-n} \tag{3}$$

The time window is used to stabilise the change in the R number, producing less statistical noise, because the R number is particularly sensitive to small time steps, making it difficult to analyse the disease's transmissibility. Furthermore, it is expected that R remains constant during the sliding time window.

As a result, the R number can be more correctly calculated using this extended model based on the ETVR model and reporting rates. It can promote the use of surveillance data, analyse infectious disease incidence time series quickly, and quantify temporal changes in the transmission intensity of future epidemics.

3. The Impacts of Modelling Parameters on the Estimation of R

The reproduction number R is used to quantify transmissibility during epidemics. It varies over time due to various factors such as changes in underlying transmission mechanisms (e.g., due to seasonality or new variants), changes in contact patterns (changes in social interaction and hygienic habits), and the impact of control measures (lockdowns, etc.).

COVID-19 data were obtained from Our World in Data (https://ourworldindata.org/covid-cases, accessed on 10 November 2021), an organisation established by the Global Change Data Lab (GCDL), the Oxford Martin School, and the University of Oxford, for use in the simulation. Because of varied configuration factors, the R estimation differs. In the following sections, the effects of various modelling parameters are assessed.

3.1. Effect of Incubation Period in Serial Interval Distributions

The infectivity profile is generated using a serial interval distribution, which is an approximation of the generation time. As a result, it has a considerable impact on the R number estimation. Three serial interval distributions (S1, S2, and S3) are used to analyse their impact on the R number, as indicated in the Appendix A. In secondary cases, the symptom onset window was set to one day. Three alternative serial interval distributions represent varying levels of uncertainty about the onset of symptom within a single day.

The simulation results are shown in Figure 1 for three serial interval profiles. From S1 to S3, three serial interval profiles show an increase in generation time. When R > 1, as shown in Figure 1, an increase in generation time results in a general rise in estimated R for a given incidence profile. This is reasonable because as the generation time increases, the time delay between the infector and the infectee increases, necessitating a bigger R to produce the same number of cases down the line.

3.2. Effect of Time Window

The size of the time window used to estimate R is expected to have an impact on the results. Smaller R values result in a faster detection of transmission changes but higher statistical noise; larger R values result in more smoothing and lower statistical noise. Figure 2 depicts this observation.

Small windows, on the other hand, can result in highly volatile estimates with large credible intervals, whereas longer windows can result in smoothed estimates with narrower credible intervals. This is illustrated in Figure 2, where the R's standard deviation reduces as the temporal window length increases. In practice, the time window can be determined by the desired coefficient of variation or confidence level. A bigger time frame size is recommended for a low coefficient of variation.

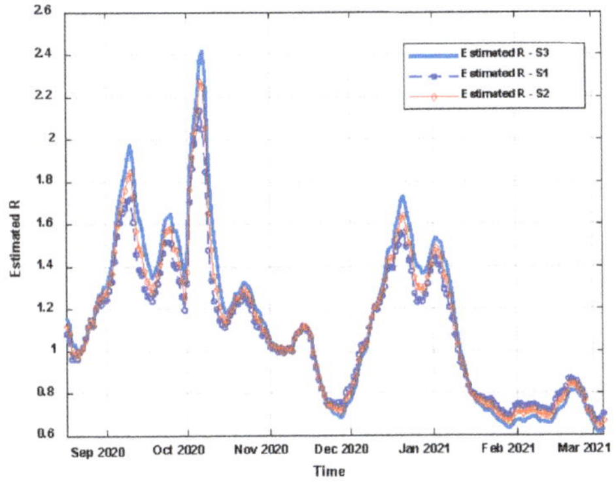

Figure 1. Three serial interval distributions and their R numbers, August 2020–March 2021.

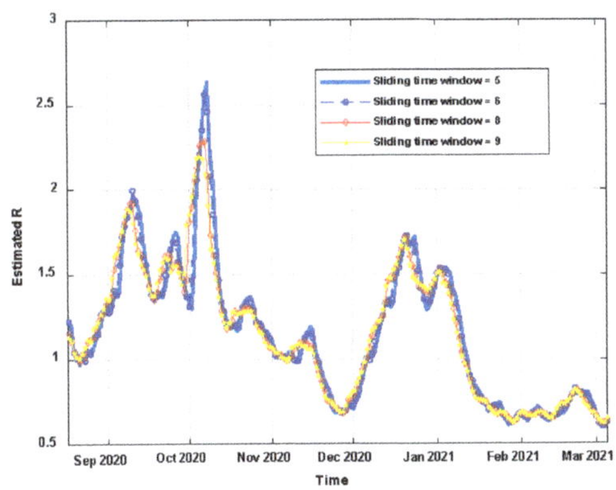

Figure 2. Estimates of R for varying time window size.

4. The Estimation of R Using the Extended Model with Variable Reporting Rates

The measurement of the R number is an important indicator for quantitatively describing infectious diseases since it could effectively reflect the transmission dynamic aspects of infectious diseases. In most research works, all cases are presumed to have been discovered and reported. However, in fact, this is not the case. As a result of including and considering the reporting rate to the gathered data, the incident data are then corrected due to over-reporting by scaling with the support of the reporting rate under 1. Because the reporting rate might be either constant or variable, both instances are covered here.

4.1. *Effect of Constant Reporting Rate Due to Over-Reporting*

Due to over-reporting, in this section, two over-reporting situations are simulated by keeping the reporting rate at 0.7 and 0.6, respectively. Figure 3 shows the simulated results for the estimated R.

It can be noted in Figure 3 that constant data scaling (constant over-reporting rate) has no effect on the prediction of the R number. This can be understood in the sense that the R number is a dimensionless ratio; it is not related to the overall number of incidence cases, but rather to how the cases increase or decrease, and thus is a reflection of pathogen transmission capacity during epidemics, which is the average number of secondary cases caused by an infected individual. This phenomenon is termed as the invariability of R with incidence data scaling.

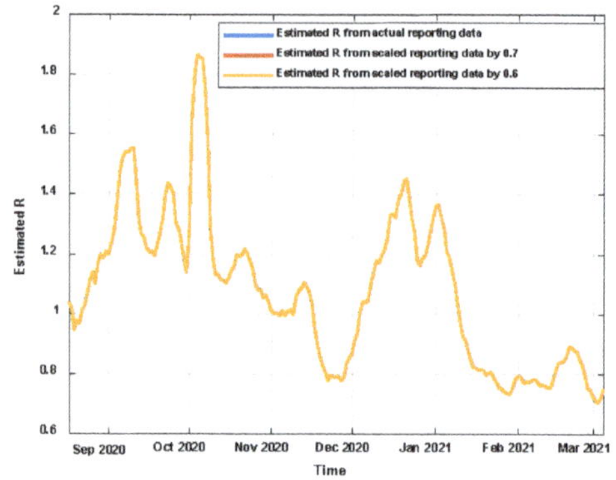

Figure 3. The results of the estimated R under constant over-reporting rates.

4.2. Effect of Variable Reporting Rates Due to Over-Reporting

Due to over-reporting, the gathered data can be adjusted by estimating the R number using the reporting rate. It cannot, however, be steady for the entire period. Herein, we present one situation whereby the reporting rate moves from a lower value to a higher value.

The new incidence data in this case are a mix of the new incidence data before the transition (scaled by 0.6) and the new incidence data after the transition (scaled by 0.7). Figure 4 shows the simulation result of the estimated R under these conditions. It appears that an erroneous R prediction occurs during the transition from a low reporting rate to a higher reporting rate (0.6 to 0.7). This can be explained by the fact that the model perceives an increase in cases as an increase in disease transmission capability because it is unaware of the variation in reporting rates.

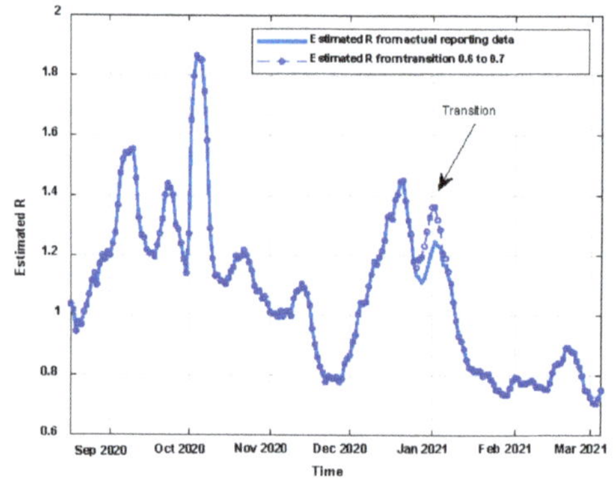

Figure 4. R numbers from the transition from a low reporting rate to a higher reporting rate (0.6 to 0.7).

5. Conclusions

For the estimation of R, we propose an extended model based on the ETVR model and reporting rate. The incidence statistics can be rectified using the reporting rate, resulting in a more accurate reproduction number for assessing infectious disease transmission, e.g., the COVID-19 pandemic.

In this paper, the mathematical manipulation for the proposed ETVR model is demonstrated. Several impacting factors were evaluated during simulations. The effect of serial interval distributions on the R number is found to be low, making their selection straightforward. The incubation period distribution has a considerable impact, with a longer incubation period leading to a higher R estimation and vice versa. Furthermore, the size of the time window matters, i.e., smaller windows resulting in highly variable estimates with broad credible intervals.

Further, according to the simulation results from the extended model with variable reporting rates, it reflects that, as long as the fraction of asymptomatic cases and the reporting rate remain constant throughout time, the reporting rate has no significant impact on the estimation of the R number. However, when the reporting rate is variable, it has an effect on the R estimate, especially during transition times between different reporting rates.

Author Contributions: Conceptualization, R.H. and W.Y.; methodology, J.Y., R.H., and W.Y.; software, J.Y. and R.H.; validation, J.Y.; writing—original draft preparation, R.H.; writing—review and editing, W.Y.; supervision, W.Y.; funding acquisition, W.Y. All authors have read and agreed to the published version of the manuscript.

Funding: This work was supported by UK Engineering and Physical Sciences Research Council (EPSRC) with the grant number: EP/P027237/1 (title: Real-time In-line Microstructural Engineering (RIME)).

Institutional Review Board Statement: Not applicable.

Informed Consent Statement: Not applicable.

Data Availability Statement: The data presented in this study are available on request from the corresponding author.

Conflicts of Interest: The authors declare no conflict of interest.

Appendix A

Table A1. Three serial interval distributions.

S1				S2				S3			
EL	**ER**	**SL**	**SR**	**EL**	**ER**	**SL**	**SR**	**EL**	**SR**	**SL**	**SR**
0	1	3	4	0	1	4	5	0	1	5	6
0	1	4	5	0	1	5	6	0	1	6	7
0	1	5	6	0	1	6	7	0	1	7	8
0	1	5	6	0	1	6	7	0	1	7	8
0	1	6	7	0	1	7	8	0	1	8	9
0	1	6	7	0	1	7	8	0	1	8	9
0	1	6	7	0	1	7	8	0	1	8	9
0	1	6	7	0	1	7	8	0	1	8	9
0	1	7	8	0	1	8	9	0	1	9	10
0	1	7	8	0	1	8	9	0	1	9	10
0	1	7	8	0	1	8	9	0	1	9	10
0	1	8	9	0	1	9	10	0	1	10	11
0	1	8	9	0	1	9	10	0	1	10	11
0	1	9	10	0	1	10	11	0	1	11	12
0	1	10	11	0	1	11	12	0	1	12	13
0	1	11	12	0	1	12	13	0	1	13	14

Three serial interval distributions are shown here: S1, S2, and S3. EL, ER, SL, and SR are the four columns that make up the serial interval data. The lower bound of the infector's symptom onset date is EL (given as an integer). The upper bound of the infector's symptom onset date is ER (given as an integer). ER should be more than or equal to EL. The lower bound of the infected person's symptom onset date is SL (given as an integer). SR is the upper limit of the infected person's symptom onset date (given as an integer). SR should be more than or equal to SL.

References

1. Bates, D.W.; Saria, S.; Ohno-Machado, L.; Shah, A.; Escobar, G. Big data in health care: Using analytics to identify and manage high-risk and high-cost patients. *Health Aff.* **2014**, *33*, 1123–1131. [CrossRef] [PubMed]
2. Singh, M.; Bhatia, V.; Bhatia, R. Big data analytics: Solution to healthcare. In Proceedings of the 2017 International Conference on Intelligent Communication and Computational Techniques (ICCT), Jaipur, India, 22–23 December 2017; pp. 239–241.
3. Dietz, K. The estimation of the basic reproduction number for infectious diseases. *Stat. Methods Med. Res.* **1993**, *2*, 23–41. [CrossRef]
4. Becker, N.G. *Analysis of Infectious Disease Data*, 1st ed.; Chapman & Hall/CRC Press: New York, NY, USA, 1989.
5. Fraser, C. Estimating individual and household reproduction numbers in an emerging epidemic. *PLoS ONE* **2007**, *2*, e758. [CrossRef]
6. Perone, G. Comparison of ARIMA, ETS, NNAR, TBATS and hybrid models to forecast the second wave of COVID-19 hospitalizations in Italy. *Eur. J. Health Econ.* **2021**, *23*, 917–940. [CrossRef] [PubMed]
7. Parshani, R.; Carmi, S.; Havlin, S. Epidemic threshold for the susceptible-infectious-susceptible model on random networks. *Phys. Rev. Lett.* **2010**, *104*, 258–701. [CrossRef]
8. Godio, A.; Pace, F.; Vergnano, A. SEIR modeling of the Italian epidemic of SARS-CoV-2 using computational swarm intelligence. *Int. J. Environ. Res. Public Health* **2020**, *17*, 3535. [CrossRef] [PubMed]
9. Cori, A.; Ferguson, N.M.; Fraser, C.; Cauchemez, S. A new framework and software to estimate time-varying reproduction numbers during epidemics. *Am. J. Epidemiol.* **2013**, *178*, 1505–1512. [CrossRef] [PubMed]
10. White, L.F.; Moser, C.B.; Thompson, R.N.; Marcello, P. Statistical estimation of the reproductive number from case notification data. *Am. J. Epidemiol.* **2020**, *190*, 611–620. [CrossRef] [PubMed]
11. Thompson, R.N.; Stockwin, J.E.; van Gaalen, R.D.; Polonsky, J.A.; Kamvar, Z.N.; Demarsh, P.A.; Dahlqwist, E.; Li, S.; Miguel, E.; Jombart, T.; et al. Improved inference of time-varying reproduction numbers during infectious disease outbreaks. *Epidemics* **2019**, *29*, 100356. [CrossRef] [PubMed]

engineering proceedings

Proceeding Paper

Effects of Face Masks on Respiratory Performance: A Within-Subject Design Study †

Eng Keat Kwa and Poh Foong Lee *

Lee Kong Chian Faculty of Engineering and Science, University Tunku Abdul Rahman (Sungai Long Campus), Kajang 43000, Malaysia; kwaek6644@1utar.my
* Correspondence: leepf@utar.edu.my
† Presented at the IEEE 5th Eurasia Conference on Biomedical Engineering, Healthcare and Sustainability, Tainan, Taiwan, 2–4 June 2023.

Abstract: There is an ongoing debate about whether wearing a facemask impacts respiratory performance, which is especially crucial in the prevention of COVID-19 transmission. This pre- and post-intervention within-subject design study aimed to quantify the impact of wearing a medical facemask during deep breathing on respiratory functions and compare it to deep breathing without a facemask. A total of 100 samples (n = 100) were obtained from a single healthy young male adult (age = 24 years) who underwent pulmonary function measurement before and after 5 min of deep breathing twice a day (morning and night) for 25 days without a facemask, followed by wearing a 4-ply medical face mask for the following 25 days. Significant improvements in all parameters (mean ± SD), including tidal volume (38.04 ± 46.97 mL, $p < 0.005$), vital capacity (34.08 ± 105.36 mL, $p = 0.027$), forced vital capacity (0.11 ± 0.11 L, $p < 0.005$), forced expiratory volume in 1 s (0.13 ± 0.13 L, $p < 0.005$), peak expiratory flow (0.36 ± 0.74 L/s, $p < 0.005$), and forced expiratory flow at 25–75% of forced vital capacity (0.04 ± 0.55 L/s, $p < 0.005$), were found without the facemask, whereas significant an improvement in forced expiratory volume in 1 s (0.05 ± 0.18 L, $p = 0.049$), a significant reduction in vital capacity (−29.98 ± 103.39 mL, $p = 0.046$), and no other significant changes were observed with the medical facemask. It was suggested that face masks exert breathing resistance but do not affect deep breathing performance. These results provide further knowledge of the effect of a facemask during deep breathing on respiratory performance.

Keywords: deep breathing; COVID-19; respiratory performance; mask; tidal volume

Citation: Kwa, E.K.; Lee, P.F. Effects of Face Masks on Respiratory Performance: A Within-Subject Design Study. *Eng. Proc.* **2023**, *55*, 21. https://doi.org/10.3390/engproc2023055021

Academic Editors: Teen-Hang Meen, Kuei-Shu Hsu and Cheng-Fu Yang

Published: 29 November 2023

1. Introduction

The mass outbreak of coronavirus disease 2019 (COVID-19) brought an epidemic to the majority of countries in the world. To prevent the further spread of the disease, many nations implemented lockdown strategies that extensively influenced daily life, communication, working, and economic systems [1]. Inevitably, it gave rise to the concern of issues related to lower physical health [2]. One of the most significant approaches, i.e., breathing exercises, has been proposed to improve undesirable conditions.

Deep breathing is involuntary respiration involving deep inhalation and exhalation in each regular respiratory cycle [3]. It has a long historical tradition in yoga, tai chi, and qigong [4,5]. Deep breathing has been well known for its ability to improve vagal tone and suppress the sympathetic nervous system, which aids in the regulation of system function and the alteration in parasympathetic activity [6], promoting individual physiological moderation and reduced stress levels [7]. In terms of respiratory moderation, sustained deep inhalation and exhalation continue to stimulate the dynamicity of lung volume and airflow [8]. This also suggests that the importance of pulmonary improvement on physical health and wellness. A pilot study showed that the pulmonary function of the healthy participants improved, particularly peak expiratory flow rate (PEFR) and forced expiratory volume (FEV1), after 6 weeks of yogic breathing [4]. Additionally, a research result showed that 2 weeks of deep breathing significantly increased tidal volume (TV), forced vital capacity (FVC), and FEV1 among healthy participants [9]. Pramanik et al. [10] presented similar outcomes on FVC and FEV1 using slow-paced breathing among the healthy participants, but they employed longer periods of about 2 months.

However, concern has been raised about whether facemask application affects the effectiveness of deep breathing exercises and pulmonary functions. During the global pandemic, there was elevated demand for individual protective equipment, such as medical face masks, as recommended by the World Health Organization [11], which were viable in suppressing infectious virus transmission, alongside practicing social distancing. The present recommendation impacts most people in the world, including healthcare professionals, who are consistently required to wear face masks.

In particular, no study to date has examined the quantitative impact of the deep breathing exercise on pulmonary functions in the presence of a medical face mask. Such data may facilitate the assessment of relaxation exercises while wearing a medical face mask. Thus, this pre-and post-intervention within-subject design study was carried out to determine if wearing a medical face mask during deep breathing exercises affected respiratory performance and compare the results to deep breathing without wearing any face mask.

2. Methodology

2.1. Subject Selection

The subject of this study was one of the authors. No additional subject was recruited or involved in this study. The healthy 24-year-old male subject met the following inclusion criteria: (1) performing deep breathing for more than 5 min, (2) owning a telecommunication device for file storage, (3) being free from any medical condition, (4) not receiving or practicing deep breathing before, and (5) being a nonsmoker. Despite it being a single-subject study, it adhered to the Declaration of Helsinki from the World Medical Association, as this study involved human samples.

2.2. Study Design and Procedures

The impact of audio-guided deep breathing on the spirometry measurements was explored in two different settings: without a face mask and with a 4-ply medical face mask. The present research employed a pre- and post-intervention within-subject design study that involved the baseline measures, followed by audio-guided deep breathing without a face mask and spirometry measurement twice a day, in the morning and at night, for 25 days. For the next 25 days, the same spirometry measurements were taken before and after the audio-guided deep breathing, where the subject was required to wear a 4-ply medical face mask (Medicos, Ultra Soft Series) throughout the deep breathing process.

The experiment was conducted in a room with ambient lighting, a minimal level of noise, and adequate ventilation. Prior to the experiment, the subject was instructed to rest for 5 min to achieve physiological stability. The baseline spirometry measurements of the subject were acquired and recorded. Subsequently, the subject performed deep breathing, following the audio for 5 min using standard wired earphones, without wearing any face mask (Figure 1a).

Following the deep breathing session, the spirometry measurements of the subject were again measured. The same procedure was repeated for the next consecutive 25 days: once in the morning and once at night. The subject then continued the same procedure by wearing a standard 4-ply medical face mask for the remaining 25 days (Figure 1b). The audio-guided deep breathing was repeated with consistent support and guidance for each session throughout the 30-day period of study. The procedures are detailed in Table 1.

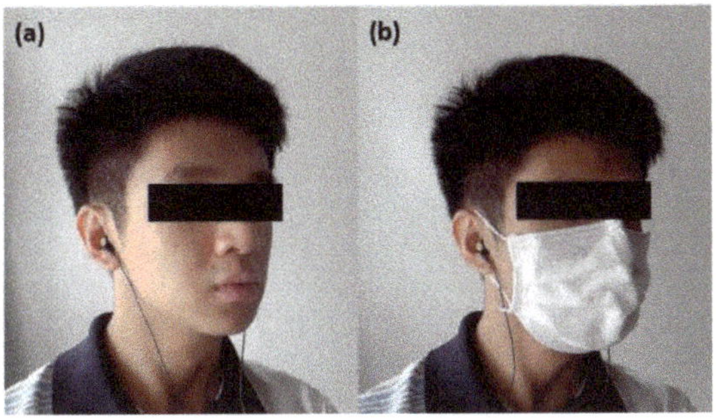

Figure 1. (**a**) Without a face mask and (**b**) with a standard 4-ply medical face mask (both with earphones).

Table 1. Measurement method used in this study.

	Baseline Measurement	**Intervention**	**Follow-Up Measurement**
Duration	Day 1 to day 25 (Without face mask) Day 26 to day 50 (With a standard 4-ply medical face mask)		
Outcomes [a]	TV, VC, FVC, FEV1, PEF, FEF25-75 (daily)		TV, VC, FVC, FEV1, PEF, FEF25-75 (daily)

[a]—TV—tidal volume; VC—vital capacity; FVC—forced vital capacity; FEV1—forced expiratory volume in 1 s; PEF—peak expiratory flow; FEF25-75—forced expiratory flow at 25–75% of forced vital capacity.

2.3. Development of Audio-Guided Deep Breathing

Figure 2 illustrates the audio-guided deep breathing method used in this study. Audacity®® Cross-Platform Sound Editor, version 2.4.2, was used to develop the audio clip. The audio clip was created at six breaths per minute with a total duration of 5 min. Cheng et al. [12] claimed that 5 min of deep breathing was the most optimal duration compared to the other durations proposed. Furthermore, the respiratory rate was proposed by Noble and Hochman [8], in which study it induced the coherent and resonance impacts of the neuro-mechanical interaction. The bird-chirping sound and the stream-flowing sound were developed and employed in this study. Each natural sound was adopted from royalty-free media clips [13,14] and selected according to the method in Ref. [15]. During the audio-guided deep breathing, the subject was required to inhale slowly and deeply in response to the bird-chirping sound and breathe out slowly and deeply in response to the stream-flowing sound while preserving the breathing rate. The subject needed to pay attention to his breathing and the flow through it while adhering to the natural sound guidance.

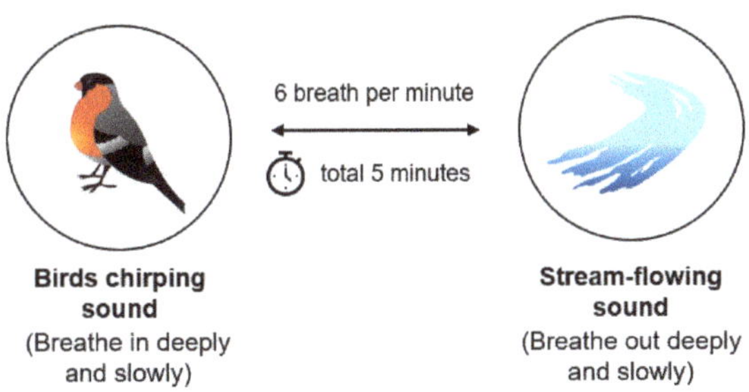

Figure 2. Audio-guided deep breathing.

2.4. Pulmonary Function Tests

The subject's spirometry was measured using two different autonomic digital spirometers: model SP80B (Contec Medical Systems Co., Ltd., Qinhuangdao, China) and model SY-8888 (Hangzhou Panhong Sports Goods Co., Ltd., Hangzhou, China), specifically designed for VC. For the measurement of TV, the subject exhaled in a one-way valve through a reusable mouthpiece attached to the spirometer during a single normal breath. To measure FVC and the forced expiratory volume in FEV1, PEF, and FEF25-75, the subject needed to exhale forcefully and speedily after a deep inhalation. Conversely, to measure VC, the subject exhaled in a similar setting but after a single deep inhalation. Following the reading displayed on the LCD screen, data from the three repeated measurements were used to obtain the best value. The attachable mouthpieces were removed and carefully sanitized according to the protocol before being reattached to the devices, maintaining appropriate hygiene before their subsequent usage.

2.5. Statistical Analysis

The data analysis was performed using IBM®® SPSS®® Statistics version 25. A normality test was conducted based on the statistical hypothesis to determine whether the data were normally distributed. The Shapiro–Wilk test was selected to identify the parametricity of the data, since this measure was used to appropriately analyze both small and large data samples at a significance level of $p < 0.05$. Using the mean difference of the spirometry before and after audio-guided deep breathing, a paired t-test was performed to test for the two-tailed test at a significance level of $p < 0.05$. As an alternative to the paired *t*-test, the Wilcoxon signed rank test was applied when the normality assumption of the data was disobeyed.

3. Discussions

We investigated whether deep breathing while wearing a face mask impacted the pulmonary functions associated with deep breathing with or without a face mask. The results are presented in Table 2. The results of this study revealed that all pulmonary parameters, including TV, VC, FVC, FEV1, PEF, and FEF25-75, significantly improved without a medical face mask. These findings were consistent with those of previous studies. A previous study demonstrated that 5 min of deep breathing with six breaths per minute significantly improved VC among healthy volunteers [16]. In this study, a healthy group using audio-guided deep breathing was tested for an identical breathing rate, and the results showed significant improvement in FVC, FEF25-75, TV, VC, and FEV1. During deep inspiration, the alveoli continued to expand beyond the regular stretch reflex, the intrapulmonary pressure was elevated, and the air occupied the alveoli sacs. The intrapulmonary pressure decreased and the alveoli sacs flattened at the beginning of the deep expiration [17]. Deep breathing improved ventilation via the strengthening of respiratory sinus arrhythmia when sympathetic activity decreased [18].

On the other hand, wearing a medical face mask impacted the respiratory function variables. One unanticipated result was a significant improvement in FEV1, whereas a significant reduction in VC was observed. Further, the results demonstrated that no significant changes were observed in TV, FVC, PEF, and FEF25-75 after audio-guided deep breathing, even though it marginally improved the functions. Previous studies showed that medical face masks imposed higher airway resistance on healthy populations in both resting and progressive exercising conditions [19,20]. There was also early evidence that lung function variables related to breathing resistance had similar impacts to that of upper airway obstruction caused by mouth resistance [21]. Since the average humidity and temperature of the exhaled air were approximately 31 °C to 35 °C and 66% to 76%, respectively [22], when carrying out deep breathing, the facial temperature, deformation, and humidity developed within the facemask increased breathing resistance [23]. Accordingly, the respiratory resistance altered the period required for breathing to achieve adequate ventilation, resulting in reduced ventilation function.

Table 2. Pair differences in respiratory function parameters before and after wearing a 4-ply medical face mask or no face mask.

Before–After	Paired Differences					*t*	*df*	Sig. (Two-Tailed) [b]
	M	SD	Standard Error	95% CI				
				Lower	Upper			
Without face mask								
TV (mL)	−38.04	46.965	6.709	−51.531	−24.551	−5.670	48	0.000 **
VC (mL)	−34.08	105.361	14.900	−64.023	−4.137	−2.287	49	0.027 *
FVC [a] (L)	−0.11	0.107	0.015	-	-	-	-	0.000 **
FEV1 (L)	−0.13	0.131	0.019	−0.170	−0.949	−7.120	49	0.000 **
PEF (L/s)	−0.36	0.736	0.106	−0.576	−0.149	−3.412	47	0.001 **
FEF25-75 (L/s)	−0.04	0.552	0.080	−0.593	−0.272	−5.422	47	0.000 **
With a 4-ply medical face mask								
TV (mL)	−2.96	34.187	4.835	−12.676	6.756	−0.612	49	0.543
VC (mL)	29.98	103.389	14.621	0.597	59.363	2.050	49	0.046 *
FVC (L)	−0.04	0.142	0.021	−0.081	0.002	−1.938	47	0.059
FEV1 (L)	−0.05	0.179	0.026	−0.104	0.000	−2.019	47	0.049 *
PEF (L/s)	−0.13	0.622	0.090	−0.312	0.049	−1.465	47	0.150
FEF25-75 (L/s)	−0.02	0.393	0.056	−0.133	0.092	−0.371	48	0.712

[a]—Wilcoxon signed-rank test; [b]—Significant value; * $p < 0.05$; ** $p < 0.005$.

However, these findings did not support the pulmonary implications of face masks during constant and maximum physical exercise. Fikenzer et al. [24] reported that pulmonary functions were significantly decreased in the healthy group in both rest and maximal exercise settings. Lässing et al. [19] also found that there was a significant reduction in pulmonary functions among healthy male subjects during continuous exercise. Combining all the pieces of evidence, it was found that the increment of the pulmonary functions with a medical face mask in this study was compromised by the deep breathing exercise. These findings had important implications for understanding the impacts of deep breathing exercises, despite the presence of a medical face mask.

The results need to be assessed in future studies using diverse samples or randomness to improve the external validity, since this study only involved a single male subject with repeated measures [25]. In addition, only medical facemasks were tested, and other types of facemasks, such as fabric and N95, still need to be tested. Different facemasks may lead to different psychological outcomes [26]. Despite such limitations, the findings of this study offered insights into the pulmonary impacts of deep breathing with and without facemasks.

4. Conclusions

Wearing a facemask during deep breathing appeared to impact respiratory performance with a decrease in TV, FVC, PEF, and FEF25-75. This implied that wearing a facemask was inadequate for physiological implications. In contrast, significant improvements were observed in all respiratory parameters of TV, VC, FVC, FEV1, PEF, and FEF25-75 in deep breathing without a facemask. Further investigations are needed with larger sample sizes and more diverse populations and the type of facemasks to generate more general implications.

Author Contributions: Conceptualization, P.F.L.; methodology, P.F.L.; formal analysis, E.K.K.; investigation, E.K.K.; writing—original draft preparation, E.K.K.; writing—review and editing, E.K.K. and P.F.L.; visualization, E.K.K.; supervision, P.F.L.; funding acquisition, P.F.L. All authors have read and agreed to the published version of the manuscript.

Funding: The authors would like to acknowledge the support of grants from the Fundamental Research Grant Scheme (FRGS/1/2021/ICT03/UTAR/02/2) and the Tunku Abdul Rahman University Research Fund (UTARRF/2020-C2/L05). The funders do not have any responsibility for data study, data collection and analysis or preparation of the manuscript.

Institutional Review Board Statement: The study was conducted in accordance with the Declaration of Helsinki, and the protocol was approved by the Ethics Committee of University.

Informed Consent Statement: Informed consent was obtained from all subjects involved in the study.

Data Availability Statement: No new data were created or analyzed in this study. Data sharing is not applicable to this article.

Conflicts of Interest: The authors declare no conflict of interest.

References

1. Talevi, D.; Socci, V.; Carai, M.; Carnaghi, G.; Faleri, S.; Trebbi, E.; di Bernardo, A.; Capelli, F.; Pacitti, F. Mental health outcomes of the COVID-19 pandemic. *Riv. Psichiatr.* **2023**, *55*, 137–144.
2. Sahu, A.; Naqvi, W.M. Quarantine exercises in the time of covid-19- a review. *J. Evol. Med. Dent. Sci.* **2020**, *9*, 1922–1927. [CrossRef]
3. Ma, X.; Yue, Z.-Q.; Gong, Z.-Q.; Zhang, H.; Duan, N.-Y.; Shi, Y.-T.; Wei, G.-X.; Li, Y.-F. The effect of diaphragmatic breathing on attention, negative affect and stress in healthy adults. *Front. Psychol.* **2017**, *8*, 874. [CrossRef] [PubMed]
4. Kupershmidt, S.; Barnable, T. Definition of a yoga breathing (pranayama) protocol that improves lung function. *Holist. Nurs. Pract.* **2019**, *33*, 197–203. [CrossRef] [PubMed]
5. Yeung, A.; Chan, J.S.M.; Cheung, J.C.; Zou, L. Qigong and tai-chi for mood regulation. *Focus* **2018**, *16*, 40–47. [CrossRef] [PubMed]
6. Breit, S.; Kupferberg, A.; Rogler, G.; Hasler, G. Vagus nerve as modulator of the brain–gut axis in psychiatric and inflammatory disorders. *Front. Psychiatry* **2018**, *9*, 44. [CrossRef] [PubMed]
7. Cheng, K.S.; Croarkin, P.E.; Lee, P.F. Heart rate variability of various video-aided mindful deep breathing durations and its impact on depression, anxiety, and stress symptom severity. *Mindfulness* **2019**, *10*, 2082–2094. [CrossRef]
8. Noble, D.J.; Hochman, S. Hypothesis: Pulmonary afferent activity patterns during slow, deep breathing contribute to the neural induction of physiological relaxation. *Front. Physiol.* **2019**, *10*, 1176. [CrossRef] [PubMed]
9. Awan, W.; Abid, N.; Rao, A.; Babur, M.; Ansari, M. Effect of deep breathing exercises in healthy smokers: A pilot study. *J. Pak. Med. Assoc.* **2020**, *70*, 1209–1213. [CrossRef]
10. Pramanik, T.; Shrestha, P.; Tako, R.; Bajracharya, S.; Chalise, A.; Pandit, R. Transient effect of slow pace breathing exercise on blood pressure, heart rate and pulmonary function tests. *Nepal Med. Coll. J.* **2016**, *18*, 48–50.
11. World Health Organization. Coronavirus Disease (COVID-19) Advice for the Public: When and How to Use Masks. December 2021. Available online: https://www.who.int/emergencies/diseases/novel-coronavirus-2019/advice-for-public/when-and-how-to-use-masks (accessed on 17 March 2023).
12. Cheng, K.S.; Chang, Y.F.; Han, R.P.S.; Lee, P.F. Enhanced conflict monitoring via a short-duration, video-assisted deep breathing in healthy young adults: An event-related potential approach through the Go/NoGo paradigm. *PeerJ* **2017**, *2017*, 10. [CrossRef] [PubMed]
13. Copyright Free Sound Effect Water Sound Small Stream Flowing Royalty Free Royalty Free. 26 October 2014. Available online: https://www.youtube.com/watch?v=JR3b3eY3408&ab_channel=copyrighyfreesound-effect (accessed on 26 June 2022).
14. Free Royalty Music. No Copyright Music] Forest Birdsong. Relaxing Nature Sounds [Video File]. 3 May 2019. Available online: https://www.youtube.com/watch?v=dDvgdixdD2w&ab_channel=FreeRoyaltyMusic (accessed on 26 June 2022).
15. Kow, F.P.; Adlina, B.; Sivasangari, S.; Punithavathi, N.; Ng, K.K.; Ang, A.H.; Ong, L.M. The impact of music guided deep breathing exercise on blood pressure control—A participant blinded randomised controlled study. *Med. J. Malays.* **2018**, *73*, 233–238.
16. GSivakumar; Prabhu, K.; Baliga, R.; Pai, M.K.; Manjunatha, S. Acute effects of deep breathing for a short duration (2–10 minutes) on pulmonary functions in healthy young volunteers. *Indian J. Physiol. Pharmacol.* **2011**, *55*, 154–159.
17. Keerthi, G.S.; Krishna, B.H.; Suresh, M.; Reddy, N.M. Effect of slow deep breathing (6 Breaths/Min) on pulmonary function in healthy volunteers. *Int. J. Med. Res. Health Sci.* **2013**, *2*, 597.
18. Nariyani, P.; Vyas, H. Immediate effect of deep breathing exercise on healthy subjects. *Int. J. Physiother. Res.* **2017**, *5*, 2420–2423. [CrossRef]
19. Lässing, J.; Falz, R.; Pökel, C.; Fikenzer, S.; Laufs, U.; Schulze, A.; Hölldobler, N.; Rüdrich, P.; Busse, M. Effects of surgical face masks on cardiopulmonary parameters during steady state exercise. *Sci. Rep.* **2020**, *10*, 22363. [CrossRef] [PubMed]
20. Mapelli, M.; Salvioni, E.; De Martino, F.; Mattavelli, I.; Gugliandolo, P.; Vignati, C.; Farina, S.; Palermo, P.; Campodonico, J.; Maragna, R.; et al. 'You can leave your mask on': Effects on cardiopulmonary parameters of different airway protective masks at rest and during maximal exercise. *Eur. Respir. J.* **2021**, *58*, 2004473. [CrossRef]

21. Melissant, C.; Lammers, J.; Demedts, M. Relationship between external resistances, lung function changes and maximal exercise capacity. *Eur. Respir. J.* **1998**, *11*, 1369–1375. [CrossRef]
22. Mapelli, M.; Salvioni, E.; De Martino, F.; Mattavelli, I.; Gugliandolo, P.; Vignati, C.; Farina, S.; Palermo, P.; Campodonico, J.; Maragna, R.; et al. Measurement of temperature and relative humidity in exhaled breath. *Sens. Actuators B Chem.* **2020**, *304*, 127371.
23. Mansour, E.; Vishinkin, R.; Rihet, S.; Saliba, W.; Fish, F.; Sarfati, P.; Haick, H. Effect of surgical masks on cardiopulmonary function in healthy young subjects: A crossover study. *Front. Physiol.* **2021**, *12*, S127371.
24. Fikenzer, S.; Uhe, T.; Lavall, D.; Rudolph, U.; Falz, R.; Busse, M.; Hepp, P.; Laufs, U. Effects of surgical and FFP2/N95 face masks on cardiopulmonary exercise capacity. *Clin. Res. Cardiol.* **2020**, *109*, 1522–1530. [CrossRef] [PubMed]
25. McEwan, B. Sampling and validity. *Ann. Int. Commun. Assoc.* **2020**, *44*, 235–247. [CrossRef]
26. Zheng, C.; Poon, E.T.-C.; Wan, K.; Dai, Z.; Wong, S.H.-S. Effects of wearing a mask during exercise on physiological and psychological outcomes in healthy individuals: A systematic review and meta-analysis. *Sports Med.* **2023**, *53*, 125–150. [CrossRef]

engineering proceedings

MDPI

Proceeding Paper

Development of Electric Power Generator by Using Hydrogen †

Ning Zhu [1,*], Yuta Takeuchi [1], Katsuhiro Amano [1] and Kazuhito Fukuda [2]

1 Shizuoka Institute of Science and Technology, Fukuroi 437-0032, Japan; 2221019.ty@sist.ac.jp (Y.T.); amano.katsuhiro@sist.ac.jp (K.A.)
2 Daytona Corporation, Morimachi 437-0226, Japan; fukuda.103@daytona-mc.jp
* Correspondence: zhu.ning@sist.ac.jp
† Presented at the IEEE 5th Eurasia Conference on Biomedical Engineering, Healthcare and Sustainability, Tainan, Taiwan, 2–4 June 2023.

Abstract: In this research, we developed a hydrogen (H_2) electric generator in an H_2 generation system based on chemical reactions. In the experiment, we tested the performance of the H_2 electric generator and measured the amount of H_2 generated. The maximum output was 700 W and the thermal efficiency was 18.2%. The theoretical value and measured value were almost the same, and the maximum error was 4%.

Keywords: hydrogen; electric power generator; intake manifold; on-site hydrogen generation

Citation: Zhu, N.; Takeuchi, Y.; Amano, K.; Fukuda, K. Development of Electric Power Generator by Using Hydrogen. *Eng. Proc.* **2023**, *55*, 22. https://doi.org/10.3390/engproc2023055022

Academic Editors: Teen-Hang Meen, Kuei-Shu Hsu and Cheng-Fu Yang

Published: 29 November 2023

1. Introduction

Recently, the impact of global warming has become serious due to an increase in greenhouse gases emitted from industrial activities [1]. The decarbonization of mobility and power generation systems is considered as a countermeasure. One of the representative technologies is the use of internal combustion engines (ICEs) using H_2 as fuel [2]. H_2 is attracting much attention as an alternative fuel for automobiles. However, there are few practical examples of electric generators that use H_2 as fuel. The purpose of this research was to develop a small H_2 electric generator that operates stably with a maximum output of 1 kW by improving the fuel supply part of the conventional gasoline electric generator, and establishing an H_2 generation system using chemical reactions. In the previous research [3], a surge tank was introduced to homogenize the mixture of H_2 and air. As a result of measuring the maximum output of the electric generator, power was supplied stably at 800 W. In this research, the effect of the shape of the intake manifold on electric generator performance was confirmed. In addition, the H_2 generation system using a chemical reaction between an aqueous solution of sodium borohydride ($NaBH_4aq$) and aqueous solution of citric acid ($C_6H_8O_7aq$) was adopted as an on-site H_2 generation method, and theoretical and actual values were compared.

2. Hydrogen Engine Generator Performance Test

2.1. Combustion Characteristics of Hydrogen

The combustion reaction of H_2 is shown in Equation (1). H_2 combines with oxygen (O_2) at a high temperature to produce water and thermal energy. H_2 is environmentally friendly because it does not emit carbon dioxide (CO_2). However, it also has the disadvantage of producing harmful nitrogen oxides (NOxs) at high temperatures.

$$2H_2 + O_2 = 2H_2O + 248 \text{ kJ/mol} \quad (1)$$

The main combustion characteristics of H_2 are early ignition and lean burn. Table 1 shows a comparison of the fuel properties of methane (CH_4) and H_2 [4]. The minimum ignition energy of H_2 is about 1/10 that of CH_4, indicating that even a small spark burns it. In addition, the flammability range of H_2 is 4 to 75, which indicates lean burn. In terms of flame propagation, H_2 has a six times faster speed than CH_4. H_2 is a material with excellent ignition and combustion properties. It is also susceptible to abnormal combustion such as backfire requiring a preventative measure. The basic experimental conditions are listed in Table 1.

Table 1. Fuel properties of CH_4 and H_2 [4].

	CH_4	H_2
Molecular weight (g/mol)	16	2
Density (kg/m^3)	0.651	0.084
Diffusion coefficient (m^2/s)	2.1×10^{-5}	6.7×10^{-5}
Thermal conductivity (W/m·K)	0.03	0.17
Minimum ignition energy (mJ)	0.28	0.02
Flammable range (Vol.%)	5~15	4~75
Flame propagation speed (m/s)	0.4	2.7

2.2. Experimental Method and Conditions

Figure 1 shows the H_2 electric generator test equipment. The electric generator in the experiment adopted a forced air-cooled 4-cycle gasoline overhead valve inverter with a total displacement of 79 cm^3, a rated output of 1.9 kW, and a compression ratio of 9.4. In the fuel supply system, a supply port injection was used. The H_2 supply port was attached to the bottom of the intake so that it could be mixed with the inflowing air. The surge tank was a box-type one with a capacity of 1670 cm^3 to temporarily store air and make the flow rate uniform.

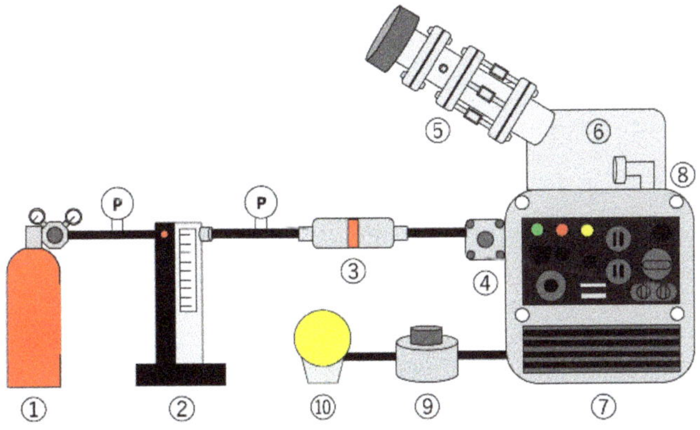

Figure 1. H_2 electric generator test equipment. ① H_2 cylinder; ② H_2 flowmeter; ③ backfire prevention valve; ④ regulator; ⑤ air flowmeter; ⑥ surge tank; ⑦ electric generator; ⑧ intake manifold; ⑨ converter; ⑩ light bulb.

Table 2 shows the experimental conditions. We designed and manufactured three types of intake manifolds to confirm the effect of intake air volume and air–fuel ratio (AFR). Figure 2 shows its appearance. In the experiment, the output was changed from 100 W to 1 kW using a converter. The engine speed was constant at each output, and partial load operation was performed. Then, we measured the intake air volume of H_2 at each output to evaluate the performance.

Table 2. Dimensions of intake manifolds.

Dimension	Type1	Type2	Type3
Inner diameter (mm)	21	23	23
Outer diameter (mm)	23	25	25
Length (upper part) (mm)	33	66	82
Length (lower part) (mm)	95	95	95

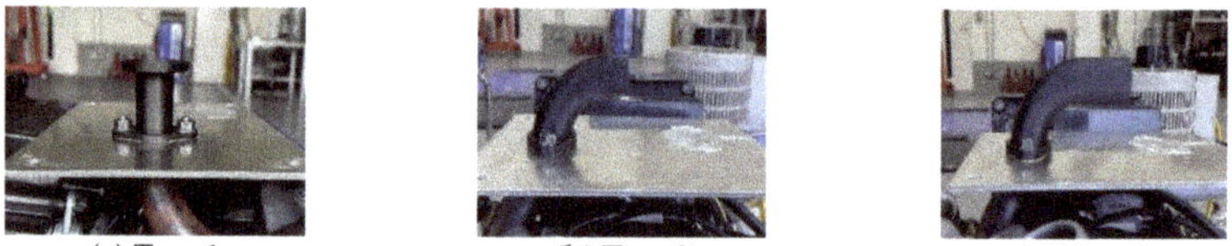

(**a**) Type1 (**b**) Type2 (**c**) Type3

Figure 2. Appearance of intake manifolds.

3. Hydrogen Generation

3.1. Principle of Hydrogen Generation

H_2 can be produced from various resources, such as the reforming method, which extracts H_2 from generated gas by burning fossil fuels, and the electrolysis method, which extracts H_2 by splitting water with electricity. H_2 is classified into three types, gray H_2, green H_2, and blue H_2, depending on the manufacturing method [5]. Gray and blue H_2 are produced using fossil fuels, and green H_2 is produced using renewable energy. Considering the global environment, the latter method is considered ideal. H_2 is known to be a clean energy that does not emit CO_2. However, there are issues with using it as fuel. One of the typical physical properties of hydrogen is its low density per volume. A common solution to these issues is pressurized gas and liquefied storage. These methods are not widely used due to high handling risks and costs.

We proposed an H_2 onsite generation system using sodium borohydride ($NaBH_4$) to solve the above problems. Table 3 shows the specifications of the samples and their appearances. $NaBH_4$ is a powdery white solid crystal and is stable at normal temperature and pressure. The mass density of H_2 is 10.6 wt.%, which is higher than high-pressure gas and liquified H_2. $NaBH_4$ reacts with water to generate H_2 which is accelerated under certain temperature or acidic conditions. In this research, as shown in Figure 3, we used this property and adopted a production method by a chemical reaction with $NaBH_4$ using a $C_6H_8O_7$ [6]. The chemical reaction formula is shown in Equation (2).

$$3NaBH_4 + 2H_2O + C_6H_8O_7 + 9H_2O \rightarrow 12H_2 + Na_3C_6H_5O_7 + 3B(OH)_3 \quad (2)$$

Table 3. Specifications of samples.

Chemical Formula	$NaBH_4$	$C_6H_8O_7$
Shape	White solid crystal	White solid crystal
Molecular weight	37.83	192.12
Density (g/cm^3)	1.074	1.665
Melting point (deg.)	400	153
Boiling point (deg.)	500	175
Solubility (g)	55/H_2O 100 (25 °C)	73/H_2O 100 (20 °C)

(**a**) $NaBH_4$ (**b**) $C_6H_8O_7$

Figure 3. Appearance of samples.

3.2. Experimental Method and Conditions

Figure 4 shows the H_2 generation system. In this system, $NaBH_4$ and $C_6H_8O_7$ were pumped up and reacted in a reactor to generate H_2. This reaction produces boric acid and sodium citrate in addition to H_2. These by-products were diverted in the flow path and sent to the waste tank. The H_2 generated is at a high temperature due to the exothermic reaction and contains water vapor that must be removed. The H_2 flow quantity was measured using a flow meter. Therefore, the temperature of H_2 was lowered by cooling water and measured after being passed through a desiccant.

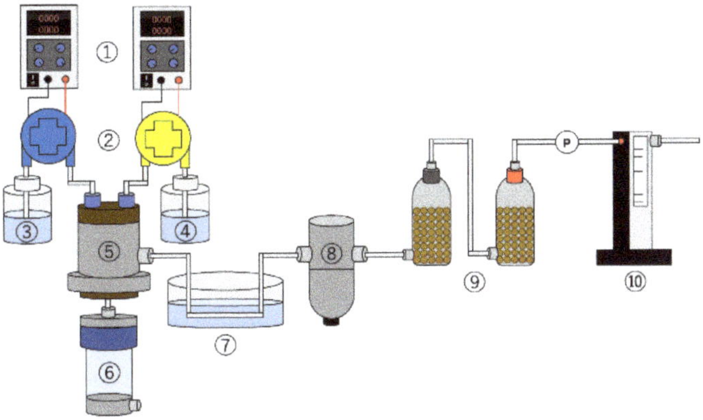

Figure 4. H_2 generation system. ① Power supply unit; ② pump; ③ $C_6H_8O_7$aq; ④ $NaBH_4$aq; ⑤ reactor; ⑥ waste tank; ⑦ cooling water; ⑧ separator; ⑨ desiccant; ⑩ flowmeter.

The concentration of the two aqueous solutions used in the experiment was an important factor, and precipitation and sticking occurred if the concentration was not appropriate. The results of previous research [7] indicated that the $NaBH_4$aq was 33.3 wt.% and the $C_6H_8O_7$aq was 27.0 wt.%. Also, the input ratio of the aqueous solution was 5:6, and the amount was derived from the following formula. Assuming a molecular weight of 37.83 for $NaBH_4$ and standard conditions (0 °C, 1 atm, and 22.4 L), we calculated the amount of aqueous solution required to generate H_2 (25 °C, 1 atm, and 10 L). Equation (2) shows that 4 mol of H_2 is generated from 1 mol of $NaBH_4$, and the amount of $NaBH_4$ can be obtained from Equation (3).

$$\mathrm{NaBH_4} = 37.83 \times \frac{1}{4} \times \frac{1}{22.4} \times \frac{273}{298} = 3.87\ \mathrm{g} \tag{3}$$

From the results of Equation (3), the amount of $NaBH_4$aq with a concentration of 33.3 wt.% is given by Equation (4).

$$\mathrm{NaBH_4aq} = 3.87 \times \frac{100.0}{33.3} = 11.6\ \mathrm{g} \tag{4}$$

Since the input ratio of $NaBH_4$ and $C_6H_8O_7$ is 5:6, the amount of $C_6H_8O_7$ is given by Equation (5).

$$C_6H_8O_7 = 3.87 \times \frac{6}{5} = 4.64 \text{ g} \quad (5)$$

From the results of Equation (5), the amount of $C_6H_8O_7aq$ with a concentration of 27.0 wt.% is given by Equation (6).

$$C_6H_8O_7aq = 4.64 \times \frac{100.0}{27.0} = 17.2 \text{ g} \quad (6)$$

4. Results and Discussion

4.1. Influence of Intake Manifold Shape on Intake Air Volume

Figure 5 shows a comparison of intake air volume. It increased for all types to produce 600 W. When the output reached 700 W, the intake air volume decreased, and the operation became unstable. ICE used in the experiment was naturally aspirated and the air could not be adjusted. Therefore, the required air volume was obtained during high output operation. The condition with the highest value was Type1, and the intake air volume increased by about 15% compared to Type2. As a result, the straight and short intake pipe was more susceptible to the pulsation effect. Therefore, the intake air volume increased.

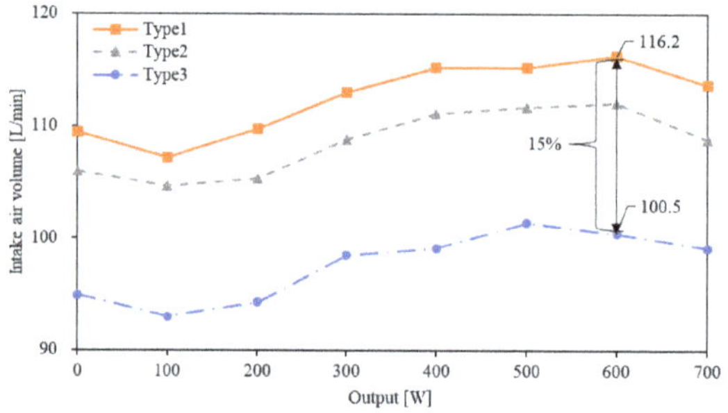

Figure 5. Comparison of intake air volume.

4.2. Influence of Intake Manifold Shape on AFR

Figure 6 shows a comparison of the AFR that is derived from the H_2 flow rate and intake air volume and is expressed in Equation (7).

$$\lambda = \frac{\rho_{Air} \times Q_{Air}}{\rho_{H_2} \times Q_{H_2}} \quad (7)$$

(λ: AFR; ρ_{Air}: air density (kg/m^3); ρ_{H_2}: H_2 density (kg/m^3); Q_{Air}: intake air volume (L/min); Q_{H_2}: H_2 flow rate (L/min)).

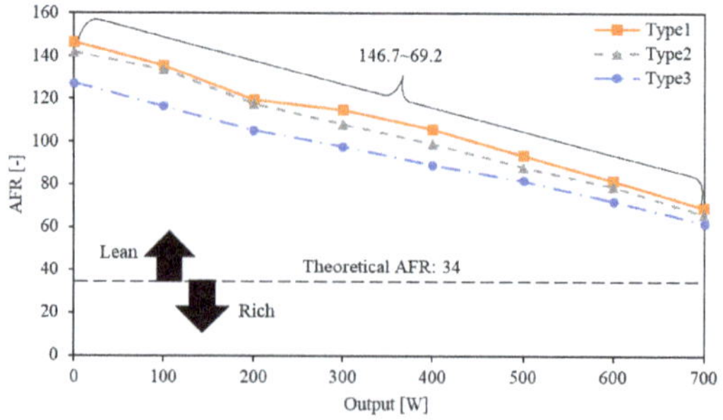

Figure 6. Comparison of AFR.

The theoretical AFR of the H_2 ICE was 34:1, so it tended to lean burn overall. As the power increased, the AFR gradually decreased, reaching approximately 70 at 700 W for Type1. The output did not increase as the intake air volume was greatly reduced in the high-output range. Therefore, the actual AFR fell below the theoretical AFR, and the operation became unstable.

4.3. Influence of Intake Manifold Shape on Thermal Efficiency

Figure 7 shows a comparison of thermal efficiency. It is derived from output and H_2 consumption and is shown in Equation (8).

$$\eta = \frac{3600W}{BH} \times 100 \tag{8}$$

(η: thermal efficiency (%); W: output (kW); B: H_2 consumption (kg/h); H: low heating value (kg/kJ)).

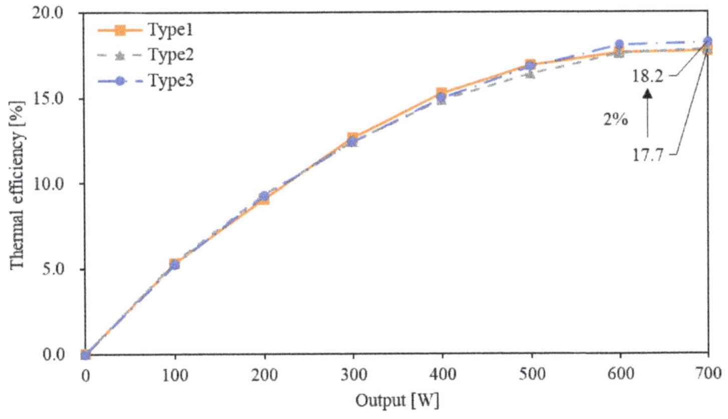

Figure 7. Comparison of thermal efficiency.

The maximum thermal efficiency was 18.2% under Type3. Comparing the maximum thermal efficiency of each type, the difference was about 2%. Therefore, there was no significant correlation between the shape of the intake manifold and the thermal efficiency.

4.4. Measurement of Hydrogen Generation

The amount of aqueous solution required to generate 10 L/min of H_2 was determined as 11.6 g ($NaBH_4$aq) and 17.2 g ($C_6H_8O_7$aq) (Equations (5) and (6)). Also, it must be mixed in a 5:6 ratio. The amount of aqueous solution varied with the voltage of the pump and had to be adjusted to achieve a 5:6 ratio. We explored the relationship between voltage and input amount prior to the performance of the pump. Table 4 shows the relationship between the aqueous solution and the pump voltage required to generate 10 to 20 L/min of H_2.

Figure 8 shows a comparison of theoretical and measured values of H_2. The theoretical value and measured value were almost the same, and the maximum error was about 4%. This result indicated that on-site power generation by combining an H_2 generator and an H_2 engine was possible. In the future, we will experiment with the two devices.

Table 4. Relationship of aqueous solution and pump voltage.

H_2 Flow	Voltage ($NaBH_4$)	$NaBH_4$	Voltage ($C_6H_8O_7$)	$C_6H_8O_7$
L/min	V	L/min	V	L/min
10	8.05	0.0116	8.95	0.0172
11	8.65	0.0128	9.68	0.0189
12	9.25	0.0139	10.42	0.0206
13	9.90	0.0151	11.15	0.0224
14	10.50	0.0162	11.90	0.0241
15	11.10	0.0174	12.55	0.0258
16	11.70	0.0186	13.25	0.0275
17	12.30	0.0197	14.02	0.0292
18	12.85	0.0209	14.75	0.0310
19	13.60	0.0220	15.50	0.0327
20	13.99	0.0232	16.20	0.0344

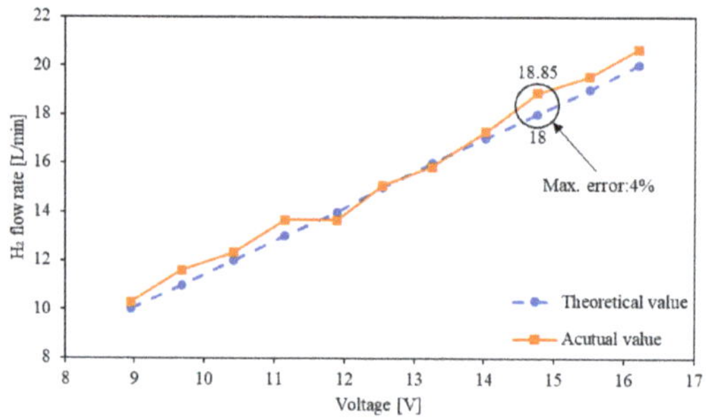

Figure 8. Comparison of theoretical and actual value.

5. Conclusions

In this research, the influence of the shape of the intake manifold on performance and the amount of H_2 generated were researched. As a result, the following conclusions were obtained. The maximum output of the H_2 electric generator was 700 W, which was less than half of the rating. It was found that the intake shape showed little effect on the thermal efficiency, and the maximum value was 18.2% for Type3. Type1 showed the highest intake air volume, improving by 15% compared to Type3. The amount of H_2 generated was the almost same as the theoretical value, and the maximum error was 4%.

Author Contributions: Conceptualization, K.F.; methodology, K.A.; Experiment, Y.T.; data curation, Y.T.; writing, Y.T.; writing—review and editing, N.Z. All authors have read and agreed to the published version of the manuscript.

Funding: This research received funding from the funder of Daytona Corporation.

Institutional Review Board Statement: Not applicable.

Informed Consent Statement: Not applicable.

Data Availability Statement: Data are contained within the article.

Acknowledgments: The authors want to thank Chenyu Huang for his kind cooperation.

Conflicts of Interest: The authors declare no conflict of interest. The funding sponsors had no role in the design of the study; in the collection, analyses, or interpretation of data; in the writing of the manuscript, and in the decision to publish the results.

References

1. Sixth Assessment Report. Available online: https://www.ipcc.ch/report/ar6/syr/(04302023) (accessed on 1 April 2023).
2. Yamane, K. Technology and Future of Reciprocating Hydrogen Internal Combustion Engine. *Hydrog. Energy Syst.* **2006**, *31*, 15–19.
3. Fang, J.S. Development of Small Hydrogen Engine Generator. Master's Thesis, Shizuoka Institute of Science and Technology, Fukuroi, Japan, 15 February 2022.
4. Sato, Y. Safety Engineering. *J. Inst. Electr. Install. Eng. Jpn.* **2005**, *44*, 1–30.
5. How to Make Hydrogen. Available online: https://www.enecho.meti.go.jp/about/special/johoteikyo/suiso_tukurikata.html(05052023) (accessed on 1 April 2023).
6. Hoshi, N. Verification of Sodium Borohydride as Fuel for Fuel Cell Vehicles. *Trans. Inst. Electr. Eng. Jpn. D* **2011**, *132*, 30–40.
7. Ishizuka, K. Research on Small Reactor for Emergency Hydrogen Generator Using Sodium Borohydride and Citric Acid. Graduation Thesis of Master Course, Tokyo University of Science, Tokyo, Japan, 15 February 2022.

engineering proceedings

MDPI

Proceeding Paper

The Development of a High-Concentration Oxygenated Water Generator Based on Nanobubbles and Its Application †

Ning Zhu [1,*], Minyu Li [2] and Kohei Shibata [3]

1. Department of Mechanical Engineering, Faculty of Science and Engineering, Shizuoka Institute of Science and Technology, Fukuroi 437-0032, Japan
2. Department of Complex Systems Science, Graduate School of Informatics, Nagoya University, Nagoya 464-8601, Japan; li.minyu.b0@s.mail.nagoya-u.ac.jp
3. Major of System Engineering, Graduation School, Shizuoka Institute of Science and Technology, Fukuroi 437-0032, Japan; 2321010.sk@sist.ac.jp

* Correspondence: zhu.ning@sist.ac.jp

† Presented at the IEEE 5th Eurasia Conference on Biomedical Engineering, Healthcare and Sustainability, Tainan, Taiwan, 2–4 June 2023.

Abstract: Water with a high concentration of oxygen is needed for the aquaculture industry in Japan. In the current study, the pressurized dissolution method was employed to generate high-concentration oxygenated water (HCOW) by producing oxygen nanobubbles in the water. In order to investigate factors such as temperature, geometric conditions, and their influence on the oxygen concentration, a special nanobubble generator was improved by changing the number and the diameter of the holes of the perforated plate in this study. Then, an experimental system where oxygen and water were separately introduced inside the proposed nanobubble generator was designed. The dissolved oxygen concentration was measured under different conditions. Finally, the produced HCOW was used to cultivate a mini-sunflower. Through a series of experiments, it was found that with the improved perforated plate, the dissolved oxygen concentration was increased and the nanobubble generator reached the saturation state quickly, while the mini-sunflower cultivated with the HCOW appeared to grow larger than that with tap water.

Keywords: high-concentration oxygenated water; nanobubble; pressurized dissolution method; perforated plate

Citation: Zhu, N.; Li, M.; Shibata, K. The Development of a High-Concentration Oxygenated Water Generator Based on Nanobubbles and Its Application. *Eng. Proc.* **2023**, *55*, 23. https://doi.org/10.3390/engproc2023055023

Academic Editors: Teen-Hang Meen, Kuei-Shu Hsu and Cheng-Fu Yang

Published: 29 November 2023

1. Introduction

Oxygen dissolved in water above the saturation concentration is called high-concentration oxygenated water (HCOW). Recently, the growth-promoting effect of leafy vegetable leaves in hydroponic cultivation by using high-concentration oxygen water was reported [1]. In the aquaculture industry, high-concentration oxygenated water is employed to enhance the growth of fish, while its sterilization ability has also been verified. In order to produce high-concentration oxygenated water, special oxygen-dissolving equipment is used based on the pressurized dissolution method (PDM) [2]. The principle of PDM is as follows.

As shown in Figure 1, oxygen gas is introduced into a cylindrical container, inside which a perforated plate is mounted. In the space between the top cover of the container and the perforated plate, the pressure increases to form a higher-pressure atmosphere. When water is introduced into the container at the same time, more oxygen dissolves into the water according to Henry's law. After the oxygenated water passes through the perforated plate, it is sheared and dropped into a lower-pressure space where the extra oxygen is released and oxygen nanobubbles are generated. Finally, the water with a high concentration of oxygen is pumped out of the container.

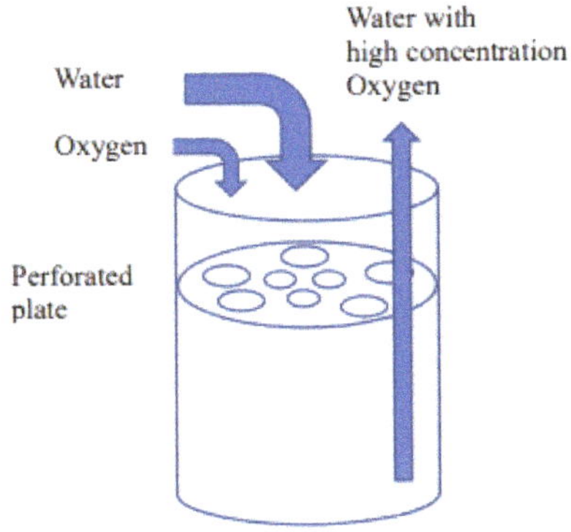

Figure 1. Pressurized dissolution method.

The pressure inside the container and the geometric conditions influence the concentration of the oxygenated water. In previous studies [3], the geometric structure of the perforated plate was improved to increase the dissolved oxygen concentration by changing the number and the diameter of the holes. Through a series of experiments, it was found that with the improved perforated plate, the dissolved oxygen concentration was increased, and the nanobubble generator reached the saturation state quickly.

Therefore, in this study, we designed and fabricated a new perforated plate with the number and diameter of the holes changed. An experiment system was established to verify the effect of HCOW when the new perforated plate was used. Finally, mini-sunflowers were cultivated with HCOW. In a series of experiments, it was found that the improved perforated plate increased the dissolved oxygen concentration and fish grew.

2. Methods

2.1. Experiment

The experiment system for the proposed oxygen-dissolving equipment is demonstrated in Figure 2. This experimental system consists of an oxygenated water generator, oxygen tank, and water pump water tank. Oxygen flows into the oxygenated water generator through a pressure regulator from the oxygen tank, and water is pumped into the generator at the same time. The oxygen concentration of the HCOW in the water tank is measured with an oxygen meter. The basic experimental conditions are listed in Table 1.

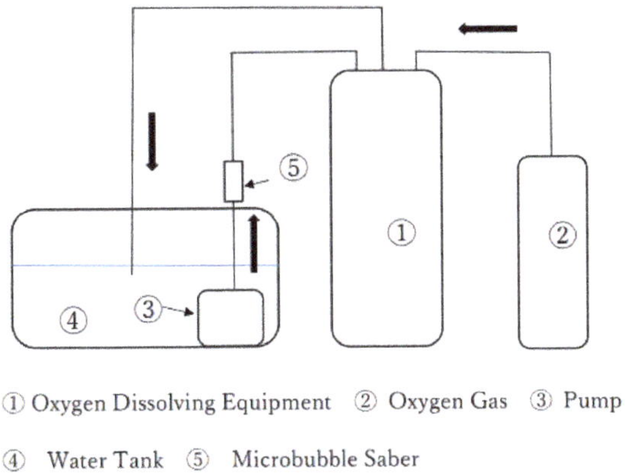

Figure 2. Experiment system for oxygen-dissolving generator.

Table 1. Basic experimental conditions.

Water quantity	54 L
Flow quantity of oxygen	1 L/min
Flow quantity of water	9 L/min
Oxygen supply pressure	0.7 MPa
Oxygen inlet pressure	0.2 MPa

2.2. *Perforated Plates*

To increase the oxygen concentration, two new perforated plates were designed and produced as shown in Figure 3. The diameter of the hole in both perforated plates was 3 mm, and the number of holes for each perforated plate was different. The new type 1 had 13 holes, and new type 2 had 77 (Table 2). We determined how much the oxygen concentration increased compared to the old perforated plate with the diameter of each hole being 5 mm and the number of holes being 13.

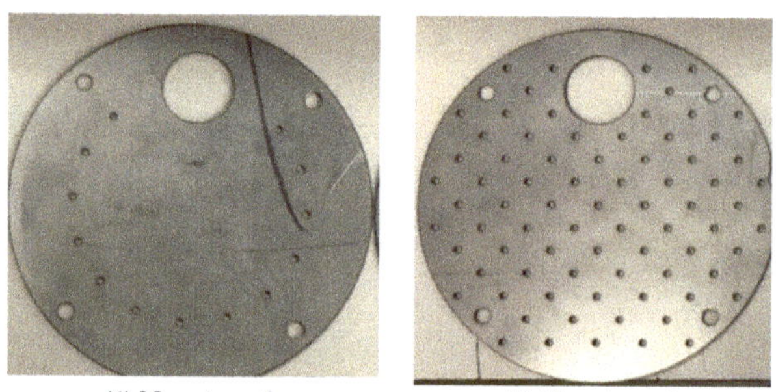

(**1**) New type 1. (**2**) New type 2.

Figure 3. Effect of geometric conditions of perforated plate (**1**) New type 1 with 13 holes; (**2**) New type 2 with 77 holes.

Table 2. Dimensions of the new perforated plate.

	New Type 1	**New Type 2**
Diameter [mm]	155	155
Thick [mm]	2	2
Number of holes	77	13
Hole diameter [mm]	3	3
Material	SUS304	SUS304

2.3. *Cultivation of Mini-Sunflowers*

An experiment on mini-sunflowers' growth using high-concentration oxygen water and tap water was conducted. Throughout the experiment lasting for 3 months, their growth states were observed by taking photos, and ultimately, the length and weight of their roots were measured.

3. Results and Discussions

3.1. *The Effect of the Geometric Conditions of the Perforated Plate*

The effect of the geometric conditions of the proposed perforated plate on oxygen concentration is demonstrated in Figure 4. Compared with the old type (the number of holes was 13 with a diameter of 5 mm), for the new type, the oxygen concentration reached saturation and its saturated oxygen concentration was higher than that of the old type. When the hole number was 77, the oxygen concentration showed a maximum value of

47 mg/L. The increased number of holes and the decreased diameter of the perforated plate sheared the falling water more effectively and increased the oxygen concentration.

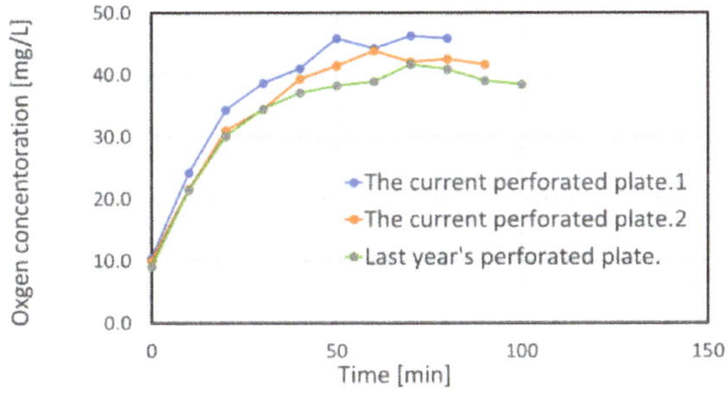

Figure 4. Effect of geometric conditions on oxygen.

3.2. Cultivation of the Mini-Sunflowers

The shape of the mini-sunflowers is illustrated in Figure 5, and the length and weight of them are also shown in Table 3. Essentially, mini-sunflower (a) was cultivated with tap water and mini-sunflower (b) was cultivated with HCOW daily. Figure 5 shows that mini-sunflower (a), cultivated with the high-oxygen-concentration water, had longer and thicker roots than (b). The weight of the mini-sunflowers cultivated with HOCW was 0.239 g, which was larger than those cultivated with tap water. With HOCW, the microbiota in the soil was activated, and the growth of the mini-sunflower enhanced.

(**a**) With tap water. (**b**) With HCOW.

Figure 5. Cultivation of mini-sunflowers with tap water (**a**) and HCOW (**b**).

Table 3. Comparison of growth of mini-sunflowers' cultivated with HCOW and tap water.

	Root Length [mm]	**Mass [g]**
HCOW	145	0.239
Tap Water	115	0.195

4. Conclusions

When new types of perforated plates were used, with the increased hole number and decreased diameter, the oxygen concentration increased and the maximum oxygen concentration reached 47 mg/L. By using HCOW, the growth of the mini-sunflower was enhanced significantly.

Author Contributions: Conceptualization, N.Z.; methodology, M.L.; experiment, K.S.; data curation, M.L; writing—N.Z.; writing—review and editing, N.Z. All authors have read and agreed to the published version of the manuscript.

Funding: This research received no external funding.

Institutional Review Board Statement: Not applicable.

Informed Consent Statement: Not applicable.

Data Availability Statement: Data are contained within the article.

Acknowledgments: The authors would like to thank Eiji Takeuchi for his kind suggestion.

Conflicts of Interest: The authors declare no conflict of interest.

References

1. Tanaka, Y.; Tamaki, H.; Tanaka, K.; Tozawa, E.; Matsuzawa, H.; Toyama, T.; Kamagata, Y.; Mori, K. Duckweed-Microbe Co-Cultivation Method for Isolating a Wide Variety of Microbes Including Taxonomically Novel Microbes. *Microbes Environ.* **2018**, *33*, 402–405. [CrossRef] [PubMed]
2. Ushikubo, F.Y.; Oshita, S.; Furukawa, T.; Makino, Y.; Kawagoe, Y.; Furihata, K.; Shiina, T. Properties of micro/nano-bubble water and its possible effect on physiological activity. *J. Agric. Mech.* **2008**, *70*, 15–16.
3. Zhu, N. Development of High Concentration Oxygen Water Generator based on Nanobubble. In Proceedings of the 2022 JSME-IIP/ASME-ISPS Joint International Conference on Micromechatronics for Information and Precision Equipment, Nagoya, Japan, 28–31 August 2021.

engineering proceedings

MDPI

Proceeding Paper

Performance Assessment of Food Safety and Hygiene in Physical Stores Engaged in Online Food Business in Taiwan †

Kuan-Chuan Tao [1], Jia-Jia Wu [1], Kai-Ling Chan [1] and Mei-Chin Mong [2,]*

[1] Leisure Services Management, Chaoyang University, Taichung 413310, Taiwan; willytao0218@cyut.edu.tw (K.-C.T.); wu12241224@gmail.com (J.-J.W.); kailing3528@gmail.com (K.-L.C.)
[2] Food Nutrition and Health Biotechnology, Asia University, Taichung 413305, Taiwan
* Correspondence: mmong@asia.edu.tw

† Presented at the IEEE 5th Eurasia Conference on Biomedical Engineering, Healthcare and Sustainability, Tainan, Taiwan, 2–4 June 2023.

Abstract: To examine the safety of food purchased online, a food safety project was conducted by the Office of Food and Drug Safety (OFDS) of Taichung City. Through collaboration with eight universities, the performance of food safety and hygiene in the physical stores engaged in online food businesses was assessed from October 2021 to January 2022. A total of 80 physical stores were involved in the present project. The results indicate that the most common nonconformities were as follows: (1) the food service standard form contract for distance transactions was not correctly displayed on websites (71%), (2) the food-related items were not sorted, were not stored on pallet boards or racks, and were not kept clean (46%), (3) the food preparation area and sales area did not have appropriate and adequate equipment for hand washing and drying (33%), and (4) the food safety and hygiene self-evaluation checklist was not prepared (33%). More efforts regarding food safety and hygiene are needed to improve and secure the knowledge of, attitude towards, and skills for good food safety and hygiene practices of online food businesses.

Keywords: food safety and hygiene; online food business; Taiwan

Citation: Tao, K.-C.; Wu, J.-J.; Chan, K.-L.; Mong, M.-C. Performance Assessment of Food Safety and Hygiene in Physical Stores Engaged in Online Food Business in Taiwan. *Eng. Proc.* **2023**, *55*, 24. https://doi.org/10.3390/engproc2023055024

Academic Editors: Teen-Hang Meen, Kuei-Shu Hsu and Cheng-Fu Yang

Published: 29 November 2023

1. Introduction

Since the rise of the internet in the 1990s, the global retail electronic commerce (e-commerce) market size was estimated to be USD 5.2 trillion in 2021 and is expected to reach USD 8.1 trillion by 2026 [1]. According to the Organization for Economic Cooperation and Development (OECD), e-commerce is defined as "the sale or purchase of goods or services, conducted over computer networks by methods specifically designed to receive or place order" [2]. Living in the digital or information age, there is no doubt that more and more consumers will be directly involved in e-commerce transactions. Moreover, the outbreak of the coronavirus disease 2019 (COVID-19) pandemic fueled the growth of e-commerce, particularly online food product shopping [3–5]. After the World Health Organization (WHO) announced that COVID-19 was a pandemic on March 11, 2020 [6], several public health approaches were implemented in countries to prevent its spreading. For example, to maintain social distancing and avoid crowds and close contact, it is recommended to wear a face mask properly and wash hands frequently and thoroughly with soap and water or an alcohol-based hand rub. Lockdowns encompass stay-at-home orders, quarantines, and travel restrictions [7,8]. Such approaches prohibit eating food together and indoor dining in restaurants, which has further pushed consumers toward online food shopping [3,9,10]. An analysis of the data from the agri-food e-commerce platform revealed that an additional confirmed case of COVID-19 raised sales by 5.7% and the number of customers by 4.9% [10]. Therefore, it is important to focus on the food safety of ordering and delivering food online to consumers. In addition, food safety must be resolved from a food safety engineering perspective to ensure the success of the food safety management system [11]. In response to the concerns and issues regarding the safety of food purchased online, the Office of Food and Drug Safety (OFDS) of Taichung City actively carried out a food safety project in collaboration with academic scholars of food-related fields from eight universities to evaluate the performance of food safety and hygiene self-management in physical stores engaged in online food businesses. The overall objective of the food safety project was to strengthen the ability of physical stores engaged in online food businesses in Taichung City to self-manage food safety and hygiene.

2. Materials and Methods

2.1. Expert Team and Qualified Food Safety Evaluatiors

The OFDS has relied on the expertise of academic scholars and industry specialists to complete food safety-related projects through the collaboration of industry–government–university in recent years. In the present food safety project, Chaoyang University of Technology was the project coordinator to monitor and control the process and efficiency of the project. The other partner universities were Hungkuang University, Taipei Medical University, Tunghai University, Asia University, Central Taiwan University of Science and Technology, National Chunghsing University, and Overseas Chinese University. Nine academic scholars from the eight universities formed an expert evaluation team. The project content was divided into different tasks according to the expertise of each academic scholar.

Before the project execution phase, the expert evaluation team members discussed the implementation details and time scheduling of the project, the applicable food safety-related laws and regulations, the food safety and hygiene evaluation checklist, the evaluation standards, and the key evaluation points. An industry specialist from the Food Industry Research and Development Institute was invited to join the meeting. The expert evaluation team carefully reviewed every question on the food safety and hygiene checklist for evaluation based on the "Regulations on Good Hygiene Practice (GHP) for Food" [12]. To increase the efficiency of project execution, more qualified evaluators were recruited to join the project. However, all potential evaluators had to attend a consensus meeting and 3 h lecture training for online food safety. The lecturers were experts who had practical work experience. After the meeting, the participants took an examination, and had to obtain a score of at least 80 points. The list of potential evaluators was sent to the OFDS to acquire the final approval for performing the on-site evaluations.

2.2. Food Safety Education and Regulations

Before running the on-site performance assessment, two workshops were held online for food businesses. Each workshop had 4 main topics: (1) the explanation of the food safety project (1 h); (2) the explanation of GHP regulations (1 h); (3) the registration platform of food business, the foodservice standard form contract for distance transactions, and what must be displayed on the web page (1 h); and (4) the regulations on imported food and relevant product inspection and correct food labeling in standard Chinese (1 h). The lecturers were food safety experts.

2.3. On-Site Food Safety Evaluations

The on-site assessment of food safety and hygiene in physical stores engaged in online food business started on 8 October 2021 and ended on 13 January 2022. The list of stores was provided by the OFDS. At least one evaluator and one OFDS officer made up the evaluation team to perform every on-site evaluation together. In addition to checking whether the stores were in keeping with the regulatory standards of GHP, the completeness of information for the registration platform of food business, the conformance levels of webpage content to the foodservice standard form contract for distance transaction, the completion of imported food and relevant products inspection procedures, and the accuracy of food labeling in standard Chinese were also checked.

2.4. Data Analysis

The data were analyzed and integrated after collecting all evaluation records. The major nonconformities found from the on-site food safety and hygiene evaluation were then identified. Practical suggestions regarding the nonconformities were proposed for reference of the OFDS in making related policies.

3. Results

Thirty-one qualified evaluators, including food safety experts, academic scholars, and OFDS retirees, were approved by the OFDS to perform the on-site evaluations. Before the on-site evaluations, more than 100 online food business workers attended the 2 workshops to obtain related information. A total of 80 physical stores, including 17 food manufacturing companies, 30 restaurants, and 33 online-only food stores participated in the present project (Table 1). Most physical stores were in the downtown area (71%), and those in mountain and coast areas were 16% and 13%, respectively.

Table 1. Distribution of physical stores engaged in online food businesses participated in the Food Safety project of Taichung City from October 2021 to January 2022.

District		Food-Manufacturing Company	Restaurant	Online-Only Food Store	Total
Downtown area	Central		4		4
	East	1			1
	South	1		2	3
	West	5		2	7
	North	5		4	9
	Beitun	1	2	5	8
	Xitun	6	3	5	14
	Nantun	7		4	11
Mountain area	Fengyuan			1	1
	Houli		1		1
	Xinshe	1			1
	Daya		1	3	4
	Wuri	1		2	3
	Dali	1		1	2
	Wufeng		1		1
Coast area	Dadu			1	1
	Shalu			2	2
	Wuqi		1		1
	Qingshui		2	1	3
	Dajia	1	1		2
	Daan		1		1
Total		30	17	33	80

The performance assessment results revealed that the top three nonconformities were as follows: (1) the foodservice standard form contract for distance transaction was not correctly displayed; (2) the food-related items were not sorted and stored on pallet boards or racks, and not kept clean; and (3) the food preparation area, as well as sales area, did not have appropriate and adequate equipment for hand washing and drying, and the food safety and hygiene self-evaluation checklist was not prepared. The other nonconformities found are shown in Table 2.

Table 2. Top 10 major nonconformities found in physical stores engaged in online food businesses inspected for food safety and hygiene in Taichung City from October 2021 to January 2022.

No.	Nonconformities	Unqualified Rate
1	The foodservice standard form contract for distance transaction was not correctly displayed.	71%
2	The food-related items were not sorted, were not stored on pallet boards or racks, and not kept clean.	46%
3	The food preparation area, as well as sales area, did not have appropriate and adequate equipment for hand washing and drying.	33%
4	The food safety self-evaluation checklist was not prepared.	33%
5	There were no assigned sanitation personnel to keep daily records regarding to GHP.	31%
6	The food products sold online did not have required food labeling.	28%
7	There was mold, dust, or peeling on walls, pillars, floors, or ceilings of workplace.	26%
8	The raw materials' semi-finished products and food packaging supplies were not properly stored or clearly marked with date.	23%
9	The workers did not attend sanitation workshop or keep the records.	20%
10	The food labeling as well as nutrition facts did not comply with relevant regulations.	20%

4. Conclusions and Suggestions

More and more people shop and sell foods and relevant products online with the development of e-commerce, especially after the COVID-19 pandemic. Online food shopping has gradually become the culture of the new normal for shopping for food from restaurants, food manufacturing companies, or online retailers in our daily life. There are increasing concerns about online food safety, such as online food labeling, food traceability, food hygiene practices, and food transportation, [13–15]. Food safety is the most critical issue related to food supply, which needs to be solved from the perspective of food safety engineering, including food safety detection and food safety management systems [11]. Food safety and hygiene performance assessment, therefore, could serve as an important measure for health protection [16].

The results of the current food safety project indicated that most physical stores engaged in online food businesses did not follow "the items that must and must not be stated in the foodservice standard form contract for distance transactions", amended and declared in 2017 by the Ministry of Health and Welfare (MOHW) of Taiwan for the contents displayed on the websites [17]. The incomplete and inaccurate information on the websites may mislead consumers and cause potential food safety hazards. Incorrect food labeling and nutrition fact information were also observed in several foods sold online. In the case of potential allergens, the customers may have severe allergic reactions if the allergens were not listed. The other major nonconformities were linked to the GHP standards, for instance, poor storage, lack of environmental hygiene, and poor self-management of food safety and hygiene practices. The training and knowledge about GHP seemed to be inadequate for the workers from the inspected physical stores engaged in online food businesses. It is interesting that new online stores selling many things besides food were not familiar with the food business registration system and thus did not apply for the food business registration number.

Based on the above findings, it was found that there is an urgent need for the authorities to perform on-site food safety and hygiene evaluations for online food businesses to increase the understanding and adherence of physical stores engaged in online food businesses to online business laws and regulations, to improve and strengthen the training and knowledge about GHP through continuous learning in sanitation workshops, and to raise the awareness of customers on how to shop safely for online food shopping.

Author Contributions: Conceptualization, K.-C.T. and M.-C.M.; methodology, K.-C.T.; software, J.-J.W. and K.-L.C.; validation, K.-C.T., K.-L.C. and M.-C.M.; formal analysis, K.-C.T., J.-J.W., K.-L.C. and M.-C.M.; investigation, K.-C.T. and M.-C.M.; resources, K.-C.T., J.-J.W. and K.-L.C.; data curation, K.-C.T., J.-J.W., K.-L.C. and M.-C.M.; writing—original draft preparation, K.-C.T. and M.-C.M.; writing—review and editing, M.-C.M.; visualization, M.-C.M.; supervision, M.-C.M.; project administration, K.-C.T.; funding acquisition, K.-C.T. All authors have read and agreed to the published version of the manuscript.

Funding: This project was funded by the Office of Food and Drug Safety of Taichung City, grant number Q110B12.

Institutional Review Board Statement: Not applicable.

Informed Consent Statement: Informed consent was obtained from all subjects involved in the study.

Data Availability Statement: Data are contained within the article.

Acknowledgments: The authors want to thank the administrative support of the Chaoyang University of Technology.

Conflicts of Interest: The authors declare no conflict of interest.

References

1. Chevalier, S. Retail e-Commerce Sales Worldwide from 2014 to 2026 (in Billion U.S. Dollars). Available online: https://www.statista.com/statistics/379046/worldwide-retail-e-commerce-sales/ (accessed on 10 October 2022).
2. Glossary of Statistic Terms—Electronic Commerce, the Organization for Economic Cooperation and Development (OECD). Available online: https://stats.oecd.org/glossary/detail.asp?ID=4721 (accessed on 10 October 2022).
3. Lee, S.; Ham, S. Food service industry in the ear of COVID-19: Trends and research implications. *Nutr. Res. Pract.* **2021**, *15* (Suppl. 1), S22–S31. [CrossRef] [PubMed]
4. Din, A.U.; Han, H.; Ariza-Montes, A.; Vega-Munoz, A.; Raposo, A.; Mohapatra, S. The impact of COVID-19 on the food supply chain and the role of e-commerce for food purchasing. *Sustainability* **2022**, *14*, 3074. [CrossRef]
5. Tyrvainen, O.; Harjaluoto, H. Online grocery shopping before and during the COVID-19 pandemic: A meta-analytical review. *Telemat. Inform.* **2022**, *71*, 101839. [CrossRef] [PubMed]
6. WHO Director-General's Opening Remarks at the Media Briefing on COVID-19—11 March 2020, World Health Organization (WHO). Available online: https://www.who.int/director-general/speeches/detail/who-director-general-s-opening-remarks-at-the-media-briefing-on-covid-19---11-march-2020 (accessed on 10 October 2022).
7. Advice for the Public: Coronavirus Disease (COVID-19), World Health Organization (WHO). Available online: https://www.who.int/emergencies/diseases/novel-coronavirus-2019/advice-for-public (accessed on 10 October 2022).
8. COVID-19 Lockdowns, Wikipedia. Available online: https://en.wikipedia.org/wiki/COVID-19_lockdowns (accessed on 1 November 2022).
9. Gavilan, D.; Balderas-Cejudo, A.; Fernandez-Lores, S.; Martinez-Navarro, G. Innovation in online food delivery: Learning from COVID-19. *Int. J. Gastron. Food Sci.* **2021**, *24*, 100330. [CrossRef] [PubMed]
10. Chang, H.H.; Meyerhoefer, C.D. COVID-19 and the demand for online food shopping services: Empirical evidence from Taiwan. *Am. J. Agric. Econ.* **2020**, *103*, 448–465. [CrossRef]
11. López-Gómez, A.; Fernández, P.S.; Palop, A.; Periago, P.M.; Martinez-López, A.; Marin-Iniesta, F.; Barbosa-Cánovas, G.V. Food Safety Engineering: An Emergent Perspective. *Food Eng. Rev.* **2009**, *1*, 84–104. [CrossRef]
12. Regulation on Good Hygiene Practice for Food, Taiwan Food and Drug Administration. Available online: https://www.fda.gov.tw/eng/lawContent.aspx?cid=16&id=2870 (accessed on 10 October 2022).
13. Food and Agriculture Organization. Thinking about the Future of Food Safety—A Foresight Report, Rome. 2022. Available online: https://www.fao.org/3/cb8667en/cb8667en.pdf (accessed on 10 October 2022).
14. Yiannas, F. FDA Works to Strengthen Food Safety Protections as Consumers Increasingly Order Foods Online. Available online: https://www.fda.gov/news-events/fda-voices/fda-works-strengthen-food-safety-protections-consumers-increasingly-order-foods-online (accessed on 10 October 2022).
15. Aprilianti, I.; Felippa, A. Promoting Food Safety in Indonesia's Online Food Delivery Services. Available online: https://hdl.handle.net/10419/249408 (accessed on 10 October 2022).

16. Barnes, J.; Whiley, H.; Ross, K.; Smith, J. Defining food safety inspection. *Int. J. Environ. Res. Public Health* **2022**, *19*, 789. [CrossRef] [PubMed]
17. The Items That Must and Must Not Be Stated in the Foodservice Standard-Form Contract for Distance Transactions, the Ministry of Health and Welfare of Taiwan. Available online: https://www.fda.gov.tw/tc/siteContent.aspx?sid=4024 (accessed on 10 October 2022).

engineering proceedings

Proceeding Paper

Prediction and Property Characterization of Injection Machining Parts Using Analysis for Relationship between Melt-Filling Pressure Measurement and Calculation Viscosity Index †

Chin-Chun Lee [1], Hsin-Shu Peng [1,*], Po-Wei Huang [2], Kai-Fu Liew [3], Dian-Ru Wu [1] and Wei-Jie Su [3]

1 Department of Mechanical and Computer-Aided Engineering, Feng Chia University, Taichung 40724, Taiwan; qinjun199706@gmail.com (C.-C.L.); dianru890321@gmail.com (D.-R.W.)
2 Intelligent Mold Design and Molding Center, Feng Chia University, Taichung 40724, Taiwan; bowei8915@gmail.com
3 Department of Ph.D. Program of Mechanical and Aerospace Engineering, Feng Chia University, Taichung 40724, Taiwan; liew5927@gmail.com (K.-F.L.); mopackd221511@gmail.com (W.-J.S.)
* Correspondence: hspeng@fcu.edu.tw
† Presented at the IEEE 5th Eurasia Conference on Biomedical Engineering, Healthcare and Sustainability, Tainan, Taiwan, 2–4 June 2023.

Abstract: We examine the effects of different process parameter settings on variations in melt-filling pressure, viscosity index, and part weight based on the use of real-time pressure sensor readings of the inside of the injection melt flow path. In the experiment, widely used polypropylene materials and round-shaped molded parts are selected as the molding and study object. At the same time, pressure sensors are used at different positions to perform data acquisition to enable research into the variation process of the melt-filling pressure, allowing for a viscosity index equation to simultaneously calculate viscosity indexes and observe weight variations in the samples. The obtained data will be a basis for setting up smart manufacturing in the future. The research showed the following: (i) the installation of the real-time pressure sensing modules allowed the variation process of the melt-filling pressure to be monitored along a path from the injection barrel to the mold cavity; (ii) the viscosity index was subject to changing the melt temperature, the injection speed, and the V/P switch-over point of the screw; and (iii) the melt temperature change had a considerable impact on part weight and changes in the injection speed had a relatively significant impact on viscosity variation.

Keywords: scientific injection molding; sensor for thermoplastics; melting filling pressure real-time mechanism design; viscosity index; weight analysis

Citation: Lee, C.-C.; Peng, H.-S.; Huang, P.-W.; Liew, K.-F.; Wu, D.-R.; Su, W.-J. Prediction and Property Characterization of Injection Machining Parts Using Analysis for Relationship between Melt-Filling Pressure Measurement and Calculation Viscosity Index. *Eng. Proc.* **2023**, *55*, 25. https://doi.org/10.3390/engproc2023055025

Academic Editors: Teen-Hang Meen, Kuei-Shu Hsu and Cheng-Fu Yang

Published: 29 November 2023

1. Introduction

Injection molding has seven stages: plasticization, clamping, injection, packaging, cooling, demolding, and ejection. Pressure during injection and packing may have a significant impact on the quality of products [1–10]. Kazmer et al. [1] mentioned that the variation of screw position and injection pressure is the factor to reduce the uncertainty of product quality. In addition to the screw position, injection speed is also important as a processing parameter of injection machines. Besides polymer materials, the relationship between pressure, volume, and temperature (P-V-T) influences the weight of the product significantly and is important in injection molding. By using the relationship between temperature and pressure to maintain a specific volume, the product is kept at a stable weight. Schreiber et al. [2] used the obtained feedback cavity pressure data and the P-V-T relationship of polymers for multiple cycles to control the product quality. Currently, there are many methods of controlling injection molding machines. For example, Chen et al. [3–5] found that there was a strong relationship between product weight and machine clamping force as a criterion for adjusting the V/P switching position. The feedback data of cavity pressure with the P-V-T relationship is used to control the injection machine. Agrawal et al. [6] established a master curve based on the cavity pressure curve to ensure the maintenance of injection speed. Based on the above, this study aimed to create a real-time melt-filling pressure sensing system for measuring the variation process of the melt-filling pressure in the barrel, nozzle, and mold cavity of an injection molding system. Polypropylene was used as the material to be molded and the subject of this study. The viscosity index of the polypropylene was calculated by integrating the melt-filling pressure for time to investigate the effects of different process parameters on variations of the pressure and viscosity index of the injected melt and the variation of part weight.

2. Experimental Works

2.1. Sample and Equipment

To sense in real time and observe the melt-filling pressure of the PP, a round-shaped sample and an injection mold were designed and used, producing a round-shaped sample 100 mm in diameter and 4 mm thick. Injection molding machine 60-TX of the Chuan-Lih-Fa Machinery Works Co. Ltd., Tainan, Taiwan, with 60 tons mold clamp force was used as it is a curve-elbow toggle mechanism injection molding machine with a rate of injection of 115 cm^3/s. The maximum injection pressure was 170 bar.

2.2. Materials

Polypropylene (PP) was used for this experiment. PP was a material that was commonly used in the experiment. The model of the PP was PP-6331, which was manufactured by LCY GROUP (Taipei, Taiwan).

2.3. Test Method

2.3.1. Flow of Investigations

First, the data were collected in real time with a data acquisition system and analyzed to observe changes in raw plastic material pressure during plasticization under various process parameter configurations. The correlated molding conditions and parameters for the samples are listed in Table 1 and include injection speed, melt temperature, and V/P switch-over position. In this study, we used a real-time measurement system for the observation.

Table 1. Configuration of the injection-molded PP.

Exp.	Melt Temperature (°C)	Injection Speed (%)	V/P Switch-Over Position (mm)
1	190	50	10
2	210	50	10
3	230	50	10
4	210	30	10
5	210	50	10
6	210	70	10
7	210	50	8
8	210	50	10
9	210	50	12

The standard molding criteria were established based on the given parameters, and variations in individual parameters and calculated viscosity indices were explored to examine their impact on the final weight of the products using injection molding. The changes in barrel pressure during the injection stage were correlated with changes in part weight using a high-precision weighing scale. Figure 1 shows the research flow chart.

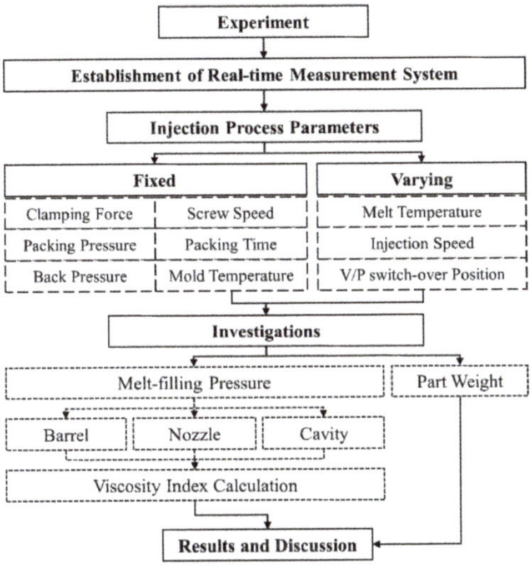

Figure 1. Research flow chart.

2.3.2. Melt-Filling Pressure of the Real-Time Mechanism

The injection barrel melt-filling pressure experiment was conducted using barrel/nozzle pressure sensors manufactured by American DYNISCO Corporation and mold cavity pressure sensors manufactured by Japanese FUTABA Corporation. The sensors were installed close to the injection-flow path and gate position (Figure 2) to detect a melt-filling pressure of the barrel–nozzle–cavity for PP. The sensors sensed the maximum peak of the pressure during filling and stabilized the condition during injection to change the material flow to be easier to measure.

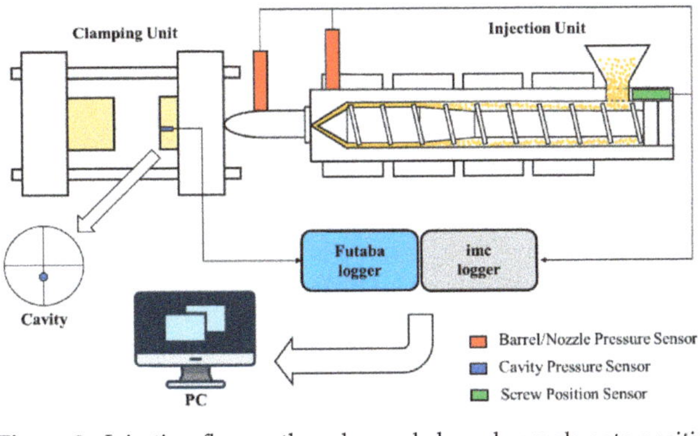

Figure 2. Injection-flow path and round-shaped sample-gate position of a barrel–nozzle–cavity pressure sensor.

Figure 3 shows the curves of the melt-filling pressure level in different sensing positions and the reaction trend of the screw position. In this figure, area (I) is for mold closing, area (II) is for injection and packing, area (III) is for plasticizing and cooling, and area (IV) is for mold opening. Figure 4 shows pressure profiles amid the filling of the injection molding were measured with a sensor, and the results are presented in [10–12]. The sensing system was also used to measure the mold cavity melt-filling pressure in this system.

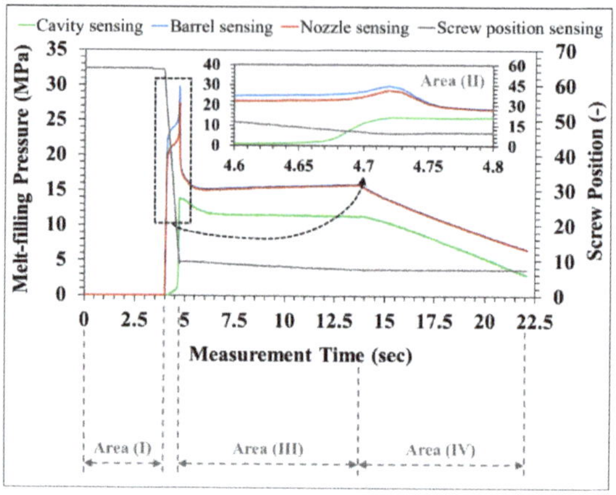

Figure 3. Real-time melt-filling pressure curves at different sensing positions and molding stages.

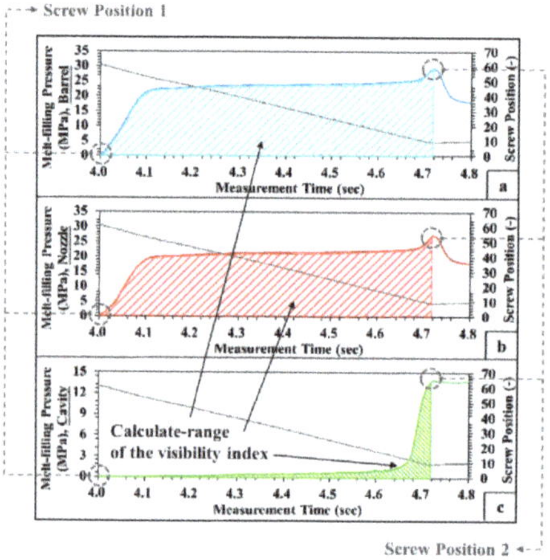

Figure 4. Calculation of the viscosity index: (a) barrel position; (b) nozzle position; (c) cavity position.

The viscosity index changes in the material during the melting process were calculated using Equation (1) and applying maximum pressure to the pressure peaks. The viscosity of the material was determined by considering the pressures in the barrel and nozzle cavity with the screw position displacement. The equation used to obtain the viscosity index includes the "*screw position*$_2$" term, which refers to the point where the V/P switch-over point [10,11].

$$VI_{Injection} = \int_{Screw\ Position_1}^{Screw\ Position_2} P_{Melt}(t)dt \tag{1}$$

3. Results and Discussion

3.1. Effect of Melt-Filling Pressure and Viscosity

- Effect of temperature of melt.

Figure 5 presents the variation trends of the melt-filling pressures at different sensing positions and different melt temperatures, the melt-filling process, and various trends of the corresponding viscosity indices of the melt. It was found that as the melt temperature setting was increased, the change in phase and temperature rise of the PP after it was plasticized, heated, and melted in the barrel increased the fluidity of the melt. The melt-filling pressure changed only slightly along the melt flow path, or more specifically, between the nozzle and barrel. In contrast, when the melt temperature was lowered, flow resistance, as well as the viscosity of the material, increased during the melt-filling process due to the melt-filling pressures in the barrel and the nozzle. The corresponding viscosity indices became relatively high. In addition, after the melt was injected into the mold cavity, the melt-filling pressure in it was not instituted until the mold cavity was almost full. Therefore, when the temperature and, hence, the fluidity of the melt was increased, it was relatively easy for the melt to flow into and fill the mold cavity, thus inducing the mold cavity pressure. A relatively high melt temperature also resulted in a rise in melt-filling pressure due to the limited mold cavity volume and the material temperature.

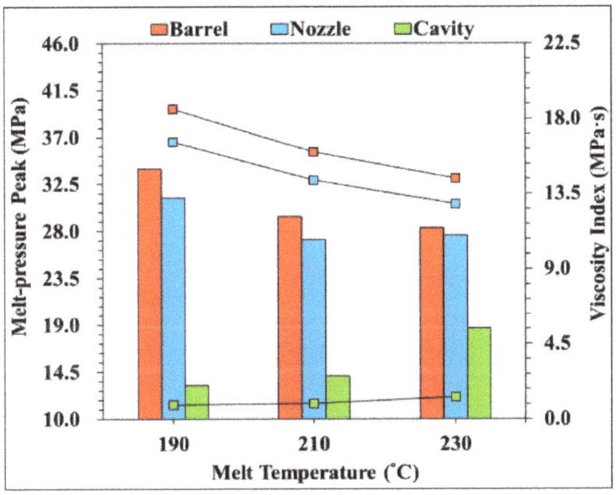

Figure 5. Influence of the temperature of the melt on barrel–nozzle–cavity pressure peak and viscosity index for PP melt filling.

- Effect of injection speed

Figure 6 depicts the variation trends in the melt-filling pressures and corresponding viscosity indices at different injection speeds over the filling stage, recorded from different sensing positions. It was found that by raising the injection speed, i.e., by raising the speed of advance of the screw mechanism in the injection barrel, the melt-filling pressure was also increased. Moreover, a relatively high injection speed caused the melt to apply a relatively high pressure when the mold cavity was full. In other words, melt-filling pressure in the mold cavity rose with injection speed. Alternatively, a relatively low injection speed led to a relatively large area under the integral curve for the melt-filling pressure with respect to time during the filling period, meaning the melt was subjected to relatively high flow resistance and had a relatively high viscosity index. When the injection speed was increased, flow resistance was experienced, and the viscosity of the melt was lowered as the area under the integral curve for the melt-filling pressure with respect to time during the filling period was reduced.

- Effect of V/P switch-over position

Figure 7 presents the variation trends of the melt-filling pressures and corresponding viscosity indices resulting from different V/P switch-over position settings recorded from different sensing positions during the filling process. A change in the position resulted in a greater change in the trends of the pressures in the barrel, the mold cavity, and the nozzle, which caused a change in the melt temperature or the injection speed. This was because a change in the position affected the material volume with which the mold cavity was filled. More specifically, if the V/P switch-over position is set to occur too late, the mold cavity, even when full, keeps being filled with the melt before the screw mechanism in the barrel unit reaches the position corresponding to the V/P switch-over position. In that case, the melt keeps pushing the wall of the mold cavity because of the limited volume of the mold cavity. Therefore, the melt-filling pressures in the barrel and the nozzle are affected by the reaction force acting on the melt in the overfilled cavity. The area under the integral curve for the melt-filling pressure with respect to time during the filling period increases, resulting in relatively high viscosity indices. A properly set V/P switch-over position, however, leads to a relatively small difference between the pressure peaks during the melt-filling process, as reflected by the variation trends of the viscosity indices.

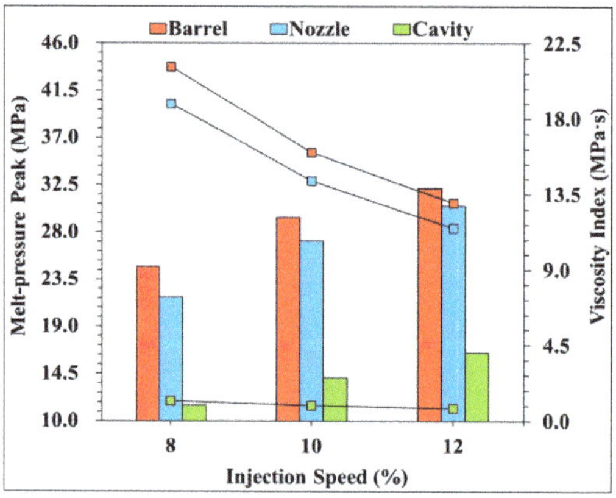

Figure 6. Effect of injection speed on barrel–nozzle–cavity pressure peak and viscosity index for PP melt filling.

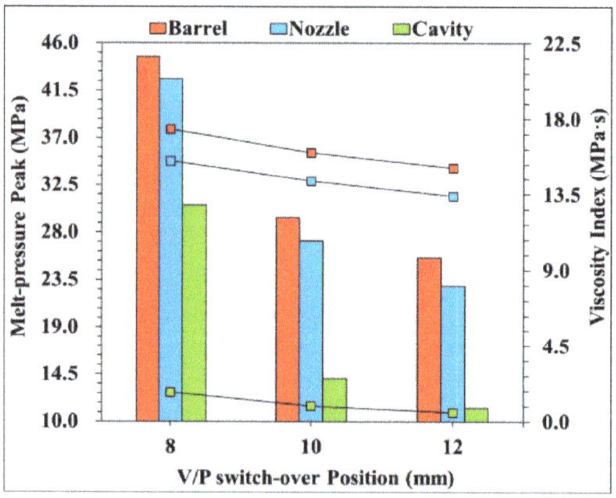

Figure 7. Effect of V/P switch-over position on barrel–nozzle–cavity pressure peak and viscosity index for PP melt filling.

3.2. Trend of Viscosity Index and Final Weight

By changing the injection conditions, i.e., by setting the injection speed, melt temperature, and V/P switch-over position differently, the melt-filling pressures at different sensing positions changed during the melt-filling process depending on changes in the melt fluidity, the limited volume of the mold cavity, whether the mold cavity was filled sooner than expected, and whether pressure was exerted on the melt in the mold cavity. This implies that a change in the injection conditions has a huge impact on the melt-filling pressure and indirectly changes the viscosity of the melt. Therefore, to fully understand the variation trends of the viscosity and pressure of the melt in the filling stage and the product quality by the solidification of the melt filling the mold cavity, it is necessary to measure the weight of injection-molded products and analyze the relationship of the part weight with viscosity. As there is relatively little literature on the effect of the pressure variation or viscosity index variation in the barrel of an injection unit on part weight, the correlation between viscosity index variation and part weight was observed by changing the injection molding conditions and then measuring the barrel pressure and calculating the viscosity index in this study.

- Variation in the melt temperature

Figure 8 describes that the change in the melt temperature affected the viscosity and part weight. An increase in the melt temperature lead to a lower viscosity index, lower resistance against melt flow, a lower melt density, a higher specific volume of the material, and lower weight. This indicates that a higher melt temperature leads to lower resistance against melt flow, greater ease with which the screw can push the melt into the mold cavity, and, therefore, a lower required molding pressure, as reflected by part weight.

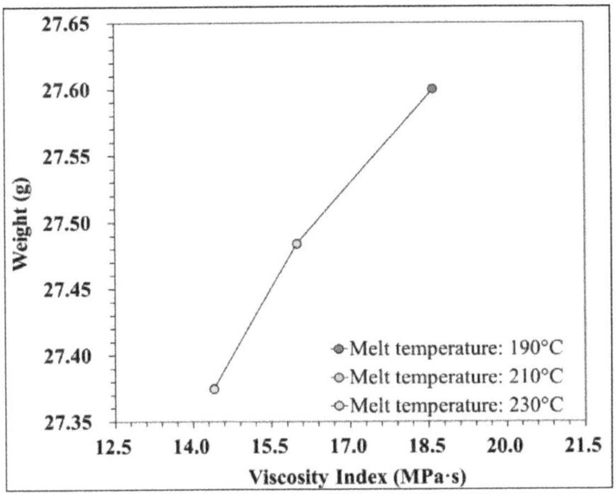

Figure 8. Variation in the melt temperature on viscosity index obtained from the barrel and injection-molded part weight.

- Variation in the injection speed

Figure 9 illustrates the change in the injection speed on the viscosity and part weight. A decrease in flow resistance during the melt-filling process and a lower melt viscosity were observed with an increase in injection speed. The mold cavity was filled up sooner with the melt, with the material in the mold cavity having a lower density. An increase in the injection speed resulted in a lower viscosity index and lower product weight. The weight reduction trend in the injection-molded product is reflected by the establishment of pressure in the mold cavity, which is directly affected by the speed at which the screw reaches the V/P switch-over position. A faster filling rate of the melt in the mold cavity is achieved as the screw approaches the V/P switch-over position earlier.

- Variation in the V/P switch-over position

Figure 10 presents the impact of changes in the V/P switch-over position on the viscosity index and product weight. When the position occurred too early, the mold cavity was unable to be filled with the melt, and the weight of the molded product was therefore affected, i.e., reduced. Conversely, when the position took place too late, the mold cavity was filled with excessive melt such that both material density and product weight were increased.

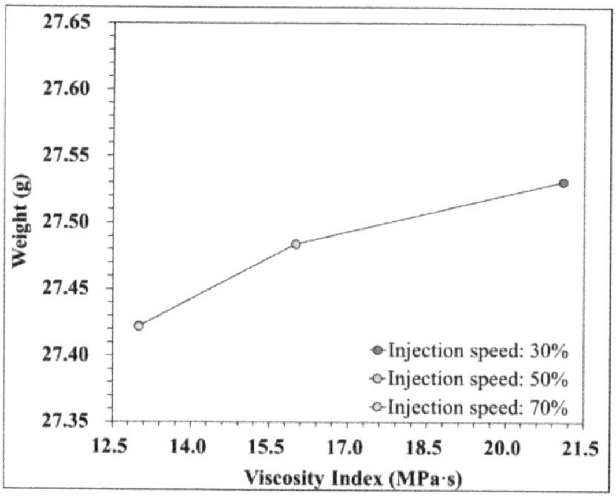

Figure 9. Variation in the injection speed on viscosity index obtained from the barrel and final injection-molded part weight.

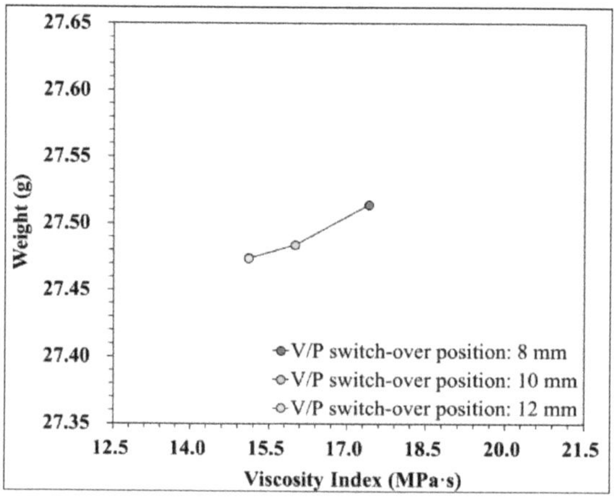

Figure 10. V/P switch-over position on viscosity index obtained from the barrel and injection-molded part weight.

4. Conclusions

The variation process of the melt-filling pressure in the injection barrel, nozzle, and mold cavity of an injection molding system in response to different process conditions (i.e., injection speed, melt temperature, and V/P switch-over position) was analyzed. The viscosity index of the melt was calculated by integrating pressure to time to find and observe correlations between the changes in each of the aforesaid process conditions and part weight. A real-time pressure sensing system was also created for this study. The creation of the real-time pressure sensing system allowed the variation process of the melt-filling pressure to be monitored along a path from the injection barrel to the mold cavity. The results of the research are as follows:

(1) The melt-filling pressures in the mold cavity, injection barrel, and nozzle changed with the process parameters. The injection speed had a relatively significant impact on the variation in the barrel pressure.

(2) Research into the correlation between the viscosity index variation in the injection barrel and part weight in response to different process parameters shows that changes

in the melt temperature had a relatively significant impact on part weight and that changes in the injection speed had a relatively significant impact on viscosity variation.

Author Contributions: Conceptualization, C.-C.L., H.-S.P., P.-W.H., K.-F.L. and W.-J.S.; data curation, C.-C.L., K.-F.L. and W.-J.S.; investigation, C.-C.L., K.-F.L., H.-S.P. and P.-W.H.; methodology, H.-S.P. and P.-W.H.; project administration, H.-S.P.; supervision, H.-S.P.; validation, P.-W.H. and W.-J.S.; writing–original draft, K.-F.L., D.-R.W. and C.-C.L.; writing–review & editing, H.-S.P., P.-W.H. and C.-C.L. All authors have read and agreed to the published version of the manuscript.

Funding: This research received no external funding.

Institutional Review Board Statement: Not applicable.

Informed Consent Statement: Not applicable.

Data Availability Statement: The data presented in this study are available upon request from the corresponding author.

Acknowledgments: Thanks to the Ministry of Science and Technology of Taiwan for funding (grant number MOST-110-2622-E-035-008) and its partner (CLF injection machines; Y.M. Molding Industry Mfg. Inc.).

Conflicts of Interest: The authors declare no conflict of interest.

References

1. Kazmer, D.O.; Velusamy, S.; Westerdale, S.; Johnston, S.; Gao, R.X. A comparison of seven filling to packing switchover methods for injection molding. *Polym. Eng. Sci.* **2010**, *50*, 2031–2043. [CrossRef]
2. Michaeli, W.; Schreiber, A. Online control of the injection molding process based on process variables. *Adv. Polym. Technol.* **2009**, *28*, 65–76. [CrossRef]
3. Huang, M.S.; Lin, C.Y. A Novel Clamping Force Searching Method Based on Sensing Tie-Bar Elongation for Injection Molding. *Int. J. Heat Mass Transf.* **2017**, *109*, 223–230. [CrossRef]
4. Chen, J.Y.; Liu, C.Y.; Huang, M.S. Enhancement of Injection Molding Consistency by Adjusting Velocity/Pressure Switching Time Based on Clamping Force. *Int. Polym. Process.* **2019**, *34*, 564–572. [CrossRef]
5. Chen, Z.; Turng, L.S. Injection molding quality control by integrating weight feedback into a cascade closed-loop control system. *Polym. Eng. Sci.* **2007**, *47*, 852–862. [CrossRef]
6. Agrawal, A.R.; Pandelidis, I.O.; Pecht, M. Injection-molding process control—A review. *Polym. Eng. Sci.* **1987**, *27*, 1345–1357. [CrossRef]
7. Zhang, S.; Dubay, R.; Charest, M. A principal component analysis model-based predictive controller for controlling part warpage in plastic injection molding. *Expert Syst. Appl.* **2015**, *42*, 2919–2927. [CrossRef]
8. Heinzler, F.A.; Mistier, M.; Wortberg, J. Quality Improvement by Enhanced Pressure Controlled Injection Molding. *SPE ANTEC Tech. Pap.* **2014**, *2*, 1694–1699.
9. Hopmann, C.; Abel, D.; Heinisch, J.; Stemmler, S. Self-Optimizing Injection Molding Based on Iterative Learning Cavity Pressure Control. *Prod. Eng.* **2017**, *11*, 97–106. [CrossRef]
10. Chen, Y.-S.; Wu, K.-T.; Tsai, M.-H.; Hwang, S.-J.; Lee, H.-H.; Peng, H.-S.; Chu, H.-Y. Adaptive process control of the changeover point for injection molding process. *J. Low Freq. Noise Vib. Act. Control* **2021**, *40*, 383–394. [CrossRef]
11. Chen, J.-Y.; Yang, K.-J.; Huang, M.-S. Online quality monitoring of molten resin in injection molding. *Int. J. Heat Mass Transf.* **2018**, *122*, 681–693. [CrossRef]
12. Schiffers, R.; Kruppa, S.; Moser, S. The right changeover point for each shot. *J. Kunststoffe* **2014**, *11*, 26–29.

engineering proceedings

Proceeding Paper

Innovative Recall Farm Situation with Sharing Experiential Learning by VR Technology †

Chi-Hui Chiang

Department of Information Management, Chia Nan University of Pharmacy & Science, Tainan 717, Taiwan; cscott@mail.cnu.edu.tw

† Presented at the 5th IEEE Eurasia Conference on Biomedical Engineering, Healthcare and Sustainability 2023, Tainan, Taiwan, 2–4 June 2023.

Abstract: In the early days of southern Taiwan, agriculture was important for living. Farmers used various agricultural implements for the cultivation and planting of rice, vegetables, and fruits for food and clothing. With the progress of the social environment, they gradually forgot the traditional agricultural implements that they depended on for a living. Recently, the government has promoted agricultural experience activities, and citizens' health concept has been improved. Growing vegetables and fruits on farms leads to the concept of shared experience. In particular, technological equipment has become popular. For leisure activities, the public has gradually accepted novel technologies (such as virtual reality games) and integrated them into their usual leisure and entertainment. For this reason, we investigate the degree of virtual interaction with the recall farm as a development basis and use the experiential learning theory to encourage the Silvers to learn the innovative recall farm situation to bring about the effect of leisure and entertainment. The results of this study show that the Silvers experience early agricultural situations and farming implements for entertainment purposes and listen to the agricultural stories to reminisce.

Keywords: Silvers; recall farm situation; sharing experiential learning

Citation: Chiang, C.-H. Innovative Recall Farm Situation with Sharing Experiential Learning by VR Technology. *Eng. Proc.* **2023**, *55*, 26. https://doi.org/10.3390/engproc2023055026

Academic Editors: Teen-Hang Meen, Kuei-Shu Hsu and Cheng-Fu Yang

Published: 29 November 2023

1. Introduction

Recently, the serious aging of Taiwan's agricultural labor force shows that the food produced by agriculture is a necessity for people's daily life, and that the tools used in agriculture are more closely related to it. The issue of population aging in Taiwan has been paid attention to by the government, academia, and several industries, and citizens are paying more and more attention to their health. In addition to the adjustment of dietary concepts, it is important to remain active and improve the function of the body, especially in the development of science and technology. The effects of these devices are divided into two categories: one is to increase the effect of leisure activities, and the other is to strengthen the effect of physical health care. As the government promotes agricultural experience activities, the public's health concept grows, and the growing of vegetables and fruits on farms leads to the concept of shared experience. Reference [1] mentioned that agricultural implements have preservation value, continue the use of traditional agricultural implements invented by ancestors, and have educational significance, with the purpose of inheriting historical relics. Ang [2] wrote that the Minnan language (also known as Taiwanese) accent is a characteristic of southern dialects, and Kaohsiung is a representative area of agriculture in which the Minnan language is used as a common dialect in everyday dialogue. This shows that Kaohsiung is representative of agriculture and the Minnan language. For this reason, we consider agriculture as representative of Taiwan's cultural history. Based on the concept of cultural creativity and technology, we can inherit the footprint of agriculture and present the past with an innovative model.

With the maturity of information technology, industries has developed many devices to assist the Silvers in participating in leisure activities, such as rehabilitation aids machines, fitness machines, leisure, and entertainment devices. Somatosensory technology is used especially with the emergence of virtual reality devices. The perceived enjoyment of virtual situations causes a pleasant experience [3]. With software development, the Silvers can easily participant in health-related activities and promote their health at their leisure. Many Silvers are attracted by these new technologies and their applications after learning how they function, and join the user base [4]. Lee, Lee [5] stated that the vividness and interactivity of virtual reality have an impact on users' perceived media richness, which further affects information-sharing behavior. Experiential learning is used in the field of health, and its benefits are considered to be beneficial to the physical health of participants. Participants from different backgrounds who participate in health activities can feel the benefits [6]. Bontchev, Vassileva [7] mentioned that the application of new entertainment games to the learning process helps users' psychological cognitive ability, and is mainly based on Kolb's experiential learning theory of new play styles [8]. Chiu [9] showed that the intensity of physical movements, hobbies, mood relaxation, and social skills of the Silvers after horticultural treatment was better than before. This promotes more gardening activities to help the Silvers be healthy. This study is performed to present the implementation of agricultural activities in a virtual situation, and it is expected that the research results will be of positive help to the Silvers.

To this end, we use tools used in agricultural practice in Kaohsiung and collect pictures of agricultural tools through field surveys, local story situations, vocabulary, and perform a semantic collection. The Minnan language's vocabulary pronunciation data of agricultural tools are collected from field surveys, and the method of using agricultural tools is gathered. Through the presentation of the virtual memory of agriculture, the entertainment is provided in combination with community activities. The purpose of the study is to respond to the needs of the Silvers' society, propose effective innovative technologies, bridge the common needs of technology research and development and the Silvers' society and economy, and enrich the research inspiration by participating in the pulse of the Silvers' society.

2. Literature Review

2.1. *Discussion on Kaohsiung Agricultural Appliances*

Taiwan was founded on agriculture. In order to solve the problem of people's livelihood, citizens went to the fields to cultivate rice, vegetables, and fruits. There still are all kinds of farming tools on their lands. Li (2019) mentioned that the preservation of early agricultural tools and their uses had been lost in the records due to changes in times. The older generation of citizens look at pictures of old agricultural tools and descriptions and realize the wisdom of their ancestors. However, their function and names are forgotten due to age, and even the memory of them is vague. If traditional agricultural tools disappear in the torrent of the times, it would be a great loss for the Taiwanese people.

To have the Silvers learn to evoke the memory of agricultural tools through shared experience learning, we use an innovative design concept to introduce the names of agricultural tools to the native Minnan speaker and deepen the Silvers' memory of agricultural tools. The process of planting and farming and how to use these tools were recorded in the Minnan language as relevant types of agricultural tools in a virtual environment. During the experience, the names of the Minnan language's agricultural tools and how to use them can be heard, and the semantics of the records are recorded. The pronunciation of the word is based on the International Phonetic Alphabet (IPA) pinyin, which is different to the Taiwanese Minnan language Romaji Pinyin symbol (referred to as "Tailuo") and phonetic symbols [1].

Chen-Su-Yun [10] studied the vocabulary of agricultural tools to inherit the local culture and divided the vocabulary corpus of agricultural tools into six categories, "farming tools for soil preparation", "tools for sowing and transplanting", "tools for weeding and insect trapping", "tools for irrigation and fertilization", and "tools for harvesting into warehouses", and "carrying animal equipment". Based on the classification methods of Chen-Su-Yun [10] and Li [1], we divide agricultural tools into soil preparation tools, planting tools, management tools, and harvesting tools to analyze the methods of using agricultural tools and categorize their vocabulary.

2.2. Application of Virtual Reality Situations

The applications of virtual situations include providing information on people, things, and other scenes of the real environment. By using special head-mounted displays (such as the HTC VIVE Cosmos), design results are presented through virtual situations (such as VR, MR, and XR) to provide users with a unique vision. These situations integrate many stories with a variety of sensory experiences. In order to allow users to interact with the combined real and virtual environments, the situations provide the operation of physical objects in the real world. It presents and feeds them back to users through virtual digital technology for an interactive experience. In the practical application of virtual situations in the past, Asai [11] pointed out that augmented reality superimposes virtual objects on real scenes to provide learners with a new learning model. Shahmoradi [12] found that virtual reality games have a high potential for rehabilitation and improve the performance of rehabilitated people. As such, it may be helpful for the Silvers to use virtual reality [12,13]. In order to study the characteristics of the virtual environment and its appropriate interface, a 3D presentation system was used for evaluation.

The results show that laptops are suitable as tools for the operation and display of virtual environments. The virtual environment provides the connection between reality and virtual objects and combines the functions of multimedia with the scene. The above application examples show that the design and development of this study based on the virtual memory of agriculture effectively present the guiding actions of the Silvers' memory of agriculture and provide the fun method of learning the Minnan language.

2.3. Applications of Shared Experiential Learning

The concept of sharing is mainly the behavior of "SHARING". In the early days, sharing appears in human society for family and friends. With technological progress, the sharing model began to emerge. The sharing of past objects between strangers and others is realized in the concept model, providing various new types of sharing, such as labor sharing and item sharing [4]. Experiential learning refers to the process by which a person constructs knowledge, acquires skills, enhances self-worth directly through experience [14], and shares with others to acquire knowledge or skills through experiential learning. The experiential learning proposed by Kolb [8] pertains to the practical application of experiential learning theory. The four stages are continuous and may occur at any time. It is worth noting that the generation of any experience in the "experiential learning circle" affects the future. In shared experiential learning, the role of the leader is to stimulate learners' learning motivation and attitude so that learners can actively digest external knowledge and internalize it into internal reference resources.

Shin and Biocca [15] clarified that in an immersive experience involving news stories in virtual reality, as users experience VR, emotional and behavioral factors are the main influencing factors to understand the content. The shared experience is more important than knowledge enhancement and ability acquisition after activities. One's memory is suitable for the methods that need to be observed and explored. When the Silvers can learn about the experience of farms, the content of the experience learning needs to be diverse through specific observation, operation, feeling, and comparison. In this way, users are willing to trust new technology, trust each other, and form a tacit understanding. At this time, users open their hearts and participate in activities for learning [16].

3. Research Methods

The study is based on the shared experience learning theory. For focusing on the design and the virtual recall of farms, we present the early Kaohsiung agricultural situation through virtual reality to form a cross-disciplinary research structure. Carbonell, Sánchez-Esguevillas [17] mentioned that context construction is presented in the form of storytelling, providing a rich story context and touching pictures. Based on the agricultural story context and constructs of the scene, the Silvers are eager to immerse themselves in the situation of the recall farm.

Ang [2] mentioned that the Minnan language in Kaohsiung is a local feature. In the past research, few virtual situations used the Minnan language as the main language for the design of works. In this study, the Minnan language is used for the Silvers to experience and learn about the names of agricultural implements and how to pronounce and listen. The plot of the recall farm's story helps increase speaking ability and helps the Silvers with listening, particularly to stories. Past research has shown that as the Silvers grow older, their speaking and hearing ability, and the memories of when they were young degenerate. Therefore, the Silvers are often heard talking about their youthful experiences. This study also aims to recall the memories of the Silvers. Shennong (also referred to as the Farmer God) is the protagonist used to tell the agricultural story in Taiwan (Figure 1) so that the Silvers who participate in the experience can learn how agricultural tools are used.

(a) (b) (c)

Figure 1. (**a**) Shennong, who tells the story of farm tools; (**b**) plow; (**c**) wheelbarrow.

Agricultural Story: *"In the blazing morning sun, the farmer pushed his wheelbarrow to the edge of the field. The wheelbarrow was filled with carefully prepared compost, ready to provide the nutrients required by the land. He first started plowing the field with his plow. The plow was inserted into the soil, and he used the plow handle to guide the tool, turning the soil via the plowshare to make it loose for creating ridges and sowing. Then, he transported the compost to this newly loosened land with his wheelbarrow, carefully depositing the compost from the wheelbarrow bit by bit into these plow grooves. This rich compost could provide nutrients to the new seeds, helping them grow robustly".*

The purpose of the study is to explore the influence of the integration of sharing experienced learning into the virtual recall of farms on the Silvers' learning motivation of the Minnan language and experience of Kaohsiung's agricultural situation (Figure 2). The experimental interview method was used for discussion, and the elderly in Kaohsiung area were taken as the research object. An innovative virtual recall farm situation operation was carried out through sharing experience learning.

In the study, 10 Silvers were interviewed, and the virtual situations of the recall of farms were designed for sharing. The Silvers could recall farms in different places via internet connection (Figure 3). In the situation, the Silvers can see other Silvers through the connection with the guidance of the Shennong avatar to allow them to quickly adapt to the virtual reality and reduce the feeling of discomfort.

Figure 2. Zhuangjia ancient house situation.

Figure 3. Architecture of silvers' sharing experience.

4. Conclusions

The use of experiential activities can make the Silvers immersed in the situation, and the guidance of Shennong's virtual characters serves as an effective learning method. The Silvers still have memories of Kaohsiung's agricultural situation. Over time, most agricultural tools no longer exist, and even their appearance is somewhat forgotten. After experiencing the farm environment, most of the Silvers can clearly explain how to use various agricultural tools. The research results show that the experience can make a deep impression on the Silvers, and they recall the local agricultural situation in the past. They can learn about the changes in the agricultural environment experience.

In an interview, the Silvers said that their favorite situations involved experiencing the old house of the Zhuangjia and the farm tool museum. They liked the content with operability, interaction, and high autonomy. In the design of similar interactive situations in the future, it is worth listing it as a design experience reference. Based on the face-to-face feedback from the interviewed Silvers, it is found that the Silvers are interested in virtual reality for shared experiential learning because the virtual character Shennong interacts with the situation with the Minnan language. Shennong's avatar guides the silvers to experience important scenes in the situation, which produces a good learning effect for the users.

Funding: The funding for this study was provided by the National Science and Technology Council, Taiwan, under the grant number MOST 110-2637-H-041-001.

Institutional Review Board Statement: This study was approved by the Human Research Ethics Committee of National Cheng Kung University, case number NCKU HREC-E-110-027-2, and adhered to the relevant ethical standards.

Informed Consent Statement: Informed consent was obtained from all participants involved in the study. They were fully informed about the nature, purpose, potential risks, and benefits of the research before participating, and their participation was voluntary.

Data Availability Statement: The data resulting from this study are available for interested researchers as a reference for subsequent extended studies.

Conflicts of Interest: There are no conflict of interest associated with this study.

References

1. Li, S.-C. *Agricultural Vocabulary in South Min*; National Kaohsiung Normal University: Kaohsiung, Taiwan, 2019.
2. Ang, U. *Kaohsiung County Minnan Language Dialect*; Kaohsiung County Government: Kaohsiung, Taiwan, 1997.
3. Manis, K.T.; Choi, D. The virtual reality hardware acceptance model (VR-HAM): Extending and individuating the technology acceptance model (TAM) for virtual reality hardware. *J. Bus. Res.* **2019**, *100*, 503–513. [CrossRef]
4. Schor, J. Debating the Sharing Economy. Available online: https://greattransition.org/publication/debating-the-sharing-economy (accessed on 1 October 2014).
5. Lee, S.A.; Lee, M.; Jeong, M. The role of virtual reality on information sharing and seeking behaviors. *J. Hosp. Tour. Manag.* **2021**, *46*, 215–223. [CrossRef]
6. Grace, S.; Stockhausen, L.; Patton, N.; Innes, E. Experiential learning in nursing and allied health education: Do we need a national framework to guide ethical practice? *Nurse Educ. Pract.* **2019**, *34*, 56–62. [CrossRef] [PubMed]
7. Bontchev, B.; Vassileva, D.; Aleksieva-Petrova, A.; Petrov, M. Playing styles based on experiential learning theory. *Comput. Hum. Behav.* **2018**, *85*, 319–328. [CrossRef]
8. Kolb, D.A. *Experiential Learning: Experience as the Source of Learning and Development*; Prentice Hall: Englewood Cliffs, NJ, USA, 1984.
9. Chiu, C.-Y. *To Verify the Effectiveness of Horticultural Therapy for the Health Status and Well-Being of Dementia Elderly*; National Taipei University of Nursing and Health Sciences Taipei: Taipei, Taiwan, 2018.
10. Chen, S.-Y. *Terminology of Traditional Farm Tools in Taiwanese Min-nan*; National Taiwan Normal University: Taipei, Taiwan, 2014.
11. Asai, K.; Kobayashi, H.; Kondo, T. Augmented instructions—A fusion of augmented reality and printed learning materials. In Proceedings of the Fifth IEEE International Conference on Advanced Learning Technologies (ICALT'05), Kaohsiung, Taiwan, 5–8 July 2005; pp. 213–215.
12. Shahmoradi, L.; Almasi, S.; Ahmadi, H.; Bashiri, A.; Azadi, T.; Mirbagherie, A.; Ansari, N.N.; Honarpishe, R. Virtual Reality Games for Rehabilitation of Upper Extremities in Stroke Patients. *J. Bodyw. Mov. Ther.* **2020**, *26*, 113–122. [CrossRef] [PubMed]
13. Liaw, S.Y.; Wu, L.T.; Soh, S.L.H.; Ringsted, C.; Lau, T.C.; Lim, W.S. Virtual Reality Simulation in Interprofessional Round Training for Health Care Students: A Qualitative Evaluation Study. *Clin. Simul. Nurs.* **2020**, *45*, 42–46. [CrossRef]
14. Xie, Z.-M. An Alternative Learning Way—Experience Learning. *Teacher World* **2003**, *127*, 6–13.
15. Shin, D.; Biocca, F. Exploring immersive experience in journalism. *New Media Soc.* **2017**, *20*, 2800–2823. [CrossRef]
16. Liou, H.-L. Action Research in Using Experiential Learning Theory. In *The Teaching of Social Studies*; National Taipei University of Education: Taipei, Taiwan, 2012.
17. Carbonell, J.; Sánchez-Esguevillas, A.; Carro, B. From data analysis to storytelling in scenario building. A semiotic approach to purpose-dependent writing of stories. *Futures* **2017**, *88*, 15–29. [CrossRef]

engineering proceedings

Proceeding Paper

Engagement with Music Technology in Special Educational Settings for Children with Disabilities †

Liza Lee [1] and Han-Ju Ho [2,*]

1 Department of Early Childhood Development & Education, Chaoyang University of Technology, Taichung 413310, Taiwan; lylee@gm.cyut.edu.tw
2 Department of Child Care and Education, Hungkuang University, Taichung 433304, Taiwan
* Correspondence: qqtomato18@gmail.com
† Presented at the IEEE 5th Eurasia Conference on Biomedical Engineering, Healthcare and Sustainability, Tainan, Taiwan, 2–4 June 2023.

Abstract: In line with the broader learning of integrating technology into science, there is growing demand for educational learning to apply electronic technologies in educational practice. Despite a handful of accounts indicating the advantages of this equipment and these tools, objective empirical research on their role in learning and their application in the learning process is lacking. Over the past ten years, studies have proven the Holistic Music Educational Approach for Young Children (HMEAYC) to be an effective method to assist in the learning and development of children with special needs. HMEAYC and technology integrated into instruction are both learning modes that expect learners to achieve learning orientation through musical expression. This study aimed to measure children with specific needs' language comprehension and self-conduct abilities when HMEAYC was combined with technology integrated into instruction in curriculum design. The research, conducted at an early-intervention center in central Taiwan, included 389 children aged between 4 and 6 years old with specific needs. These participants' traits of learning and developing more slowly than typical children of the same age were referred to as "specific needs" for this study. They all participated in one 40 min session per week for a total of 16 weeks. The research results showed that HMEAYC combined with music technology could enhance the development of language comprehension and the self-conduct of children with specific needs.

Keywords: music technology; Holistic Music Educational Approach for Young Children (HMEAYC); children with specific needs; language comprehension

Citation: Lee, L.; Ho, H.-J. Engagement with Music Technology in Special Educational Settings for Children with Disabilities. *Eng. Proc.* **2023**, *55*, 27. https://doi.org/10.3390/engproc2023055027

Academic Editors: Teen-Hang Meen, Kuei-Shu Hsu and Cheng-Fu Yang

Published: 29 November 2023

1. Introduction

Over the past ten years, research has proven the Holistic Music Educational Approach for Young Children (HMEAYC) to be an effective method for assisting in the learning and development of children with special needs [1–4]. The musical activity evidence base assists with early language and behavior development. Music can enhance learning motivation, which helps induce a different learning trigger whereby music has a positive effect on growth, making learning lively and dynamic. This article begins with an overview of HMEAYC and music technology in children with specific needs' education today, its history, and its current status. The applications of science and technology, multi-sensory equipment, and traditional instruments are mentioned. Technology equipment in music courses designed for assisted instruction is established. Research on young children and music technology is conducted, with references to work performed on integrating technology into science.

1.1. Child Music Course Models in Taiwan

For over 20 years in Taiwan, HMEAYC has commonly been accepted as an excellent practice for music education in the area. As a model, HMEAYC encourages training and experience in early childhood music [1] and shows that music is an excellent medium for young children's learning development [2,3]. Previous studies have shown positive effects in implementation experiments on the development of young children with developmental delays; the interaction of preschool children with developmental delays with musical technology instruments, whereby they learn the effectiveness of comprehension, physical movement, and social skills, can be improved by musical performances. Music addresses difficulties in children with disabilities, such as emotional, health, social interaction, intellectual, language, and creative [4]. Singing, rhymes, musical games, and instruments are all effective ways for young children to learn [5–8]. A significant finding is that HMEAYC positively affects young children with developmental delays.

Reviewing the research in the literature means finding positive results in learning efficiency [9], including improvements in language communication and physical and emotional abilities in children with special needs [5,10]. In addition to improvements in children with developmental delays, positive outcomes have been found in learning effectiveness through HMEAYC for all young children [11]. It is confirmed by research studies conducted on young children with developmental delays to enhance language comprehension and self-control. In other studies, the focus is on establishing particular musical elements, such as rhythmic and movement performance [12–14] and music image technology [10], to meet rehabilitation and treatment needs. Another evidence-based finding regarding the issue of children with developmental delays was that music could best enable children to process information visually. On a pragmatic note, the special preschool needs environment represents an obtainable source field, making opportunities to integrate music technology into education for young children with special needs easily accessible.

With the growth of research emphasizing the importance of early education for children with specific needs, educators face challenges in incorporating teaching practices. These practices focus less on the information technology integrated into instruction. HMEAYC and technology integrated into instruction are both learning modes that expect learners to achieve learning orientation through musical expression. This study aimed to measure children with specific needs' language comprehension and self-conduct abilities when HMEAYC was combined with technology integrated into instruction in curriculum design.

1.2. Current Study on Young Children with Developmental Delays in Music Technology

HMEAYC is a decades-old innovative education model that combines modern science and technology, multi-sensory equipment, and traditional instruments with creative music [9,15]. HMEAYC is a positive-enhancement course model. Its effects on the learning levels of young children with developmental delays were investigated by monitoring their participation in musical activities. Children with developmental delays improved their learning process via rhythmic and movement performance, demonstrating that music with technology effectively promotes learning [16–20].

2. Method

This study was designed with a pretest–post-test control group quasi-experimental research method. The research, conducted at an early-intervention center in central Taiwan, included 389 children aged between 4 and 6 years old with specific needs. These participants' traits of learning and developing more slowly than typical children of the same age were referred to as "specific needs" for this study. All the young children with developmental delays from one group were used in the experimental group, and all participants from the other group were determined to be the control group. The music image technology embedded into the HMEAYC course was utilized with the experimental group for 40 min weekly over 16 weeks.

The study was conducted in an educational institution in central Taiwan in a non-profit early-intervention center that frequently received young children with progression through predictable developmental phases of varying degrees, such as developmental delay. Gay et al. proposed that different student groups be selected as the experimental and control groups to reduce the interaction between the groups during the experimental intervention process and increase internal validity. All children between 4 and 6 years with developmental delays in the group in which the first author worked as a music teacher were chosen as the experimental group. All the participants in the other group were selected as the control group. In the research design, to examine participants' experiences in the experimental and control group through quantitative data, a structured assessment form was used.

3. Findings and Results

This study is based on the HMEAYC and technology integrated into instruction learning modes. Seven indices were selected for overall model evaluation. In the chi-square test, the p-value = 0.000. In this study, $\chi 2$ and the df ratio of the model are smaller than 5 (3.96). The GFI and AGFI of this model are 0.95, respectively. In this study, the RMSEA is 0.087. The CFI acceptable standard should be >0.9, and CFI for this study model is 0.95. The above results indicate that this study model is acceptable (Figure 1 and Table 1).

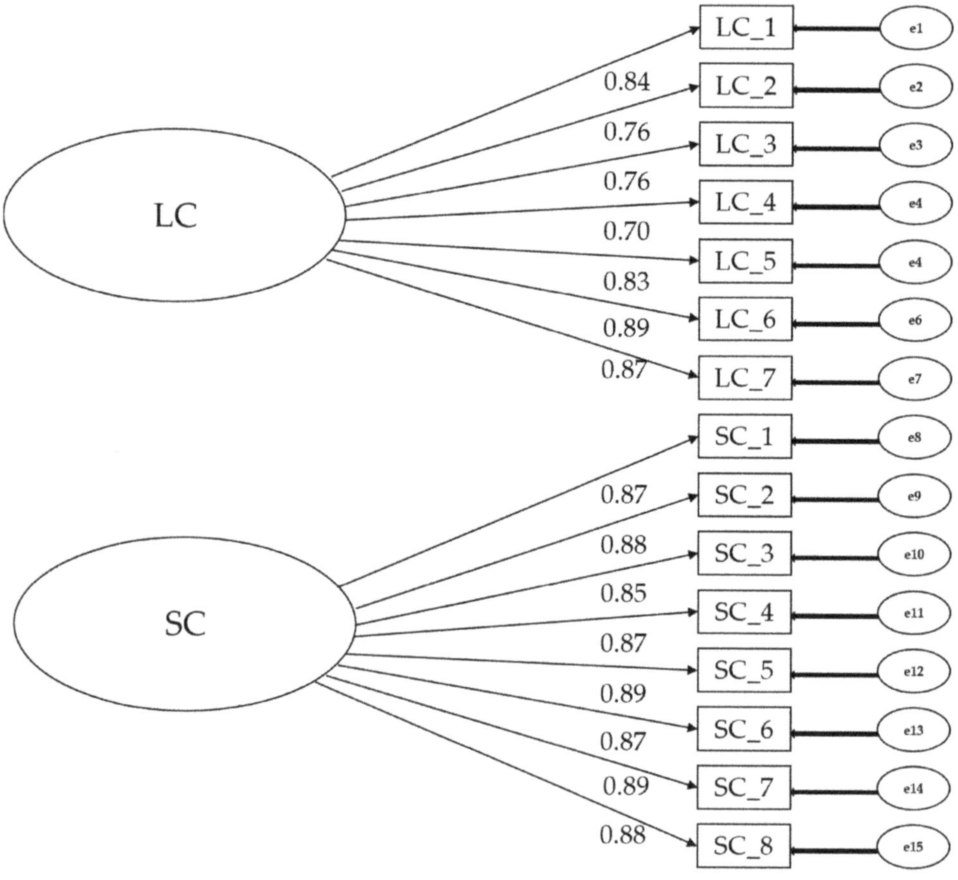

Figure 1. Covariance model of the effect of music image technology.

Table 1. Empirical results of research hypotheses on language comprehension (LC).

	Path Relationship	Path Coefficients	Support for Hypothesis
1.	LC→ look at each other properly when talking.	0.84	Yes
2.	LC→ listens carefully when others speak to him.	0.76	Yes
3.	LC→ could adequately express their inner thoughts.	0.76	Yes
4.	LC→ could respond to the content of the conversation.	0.70	Yes
5.	LC→ could actively participate in group activities.	0.83	Yes
6.	LC→ could relay the matters explained by teachers.	0.89	Yes
7.	LC→ could refuse the improper request of others.	0.87	Yes
8.	SC→ could maintain the cleanliness of personal appearance.	0.87	Yes
9.	SC→ will dress according to the weather changes.	0.88	Yes
10.	SC→ could adequately manage their belongings.	0.85	Yes
11.	SC→ when the requirements are unmet immediately, appropriate measures will be taken to deal with them.	0.87	Yes
12.	SC→ could cooperate with the daily routine of kindergarten.	0.89	Yes
13.	SC→ ability to work independently.	0.87	Yes
14.	SC→ try to do it by yourself before asking for help.	0.89	Yes
15.	SC→ can deal with the disruptive behavior of others.	0.88	Yes

The study results are divided into two sections. Language comprehension and self-conduct are learning path coefficients, and music technology's effects account for 72% of their variance. The estimated reliability of the music technology's effects on learning in children with specific needs is 0.72. The squared multiple correlations can be interpreted as follows.

- First, in Hypothesis 1, the path value, 0.84 ($p < 0.05$), reaches a significant level, indicating support for the hypothesis that participants have more positive scores for "look at each other properly when talking."
- The path value of Hypothesis 2, 0.76 ($p < 0.05$), does reach a significant level, and it gives support to the hypothesis that a participants have more positive "listens carefully when others speak to him" scores.
- In Hypothesis 3, the path value, 0.76 ($p < 0.05$), reaches a significant level. The hypothesis "could adequately express their inner thoughts" is supported, with more participants receiving positive scores in both learning modes.
- In Hypothesis 4, the path value, 0.70 ($p < 0.05$), does reach a significant level, and it gives support to the hypothesis of more positive "could respond to the content of the conversation" scores.
- In Hypothesis 5, the path value, 0.83 ($p < 0.05$), does reach a significant level, and it gives support to the hypothesis of more positive "could actively participate in group activities" scores.
- The path value of Hypothesis 6, 0.89 ($p < 0.05$), does reach a significant level and gives support to the hypothesis of more positive "could relay the matters explained by teachers" scores.
- In Hypothesis 7, the path value, 0.87 ($p < 0.05$), reaches a significant level, indicating support for the hypothesis that participants have more positive "could refuse the improper request of others" scores.

- Then, in Hypothesis 8, the path value, 0.87 ($p < 0.05$), reaches a significant level, indicating support for the hypothesis that participants receive more positive scores for "could maintain the cleanliness of personal appearance."
- The path value of Hypothesis 9, 0.88 ($p < 0.05$), does reach a significant level and gives support to the hypothesis of more positive "will dress according to the weather changes" scores.
- In Hypothesis 10, the path value, 0.85 ($p < 0.05$), reaches a significant level.
- In Hypothesis 11, the path value, 0.87 ($p < 0.05$), does reach a considerable level, and it gives support to the hypothesis of more positive "when the needs are unmet immediately, appropriate measures will be taken to deal with them" scores. The assumption "could adequately manage their belongings," is supported, with more participants receiving positive scores and benefitting from both learning modes.
- In Hypothesis 12, the path value, 0.89 ($p < 0.05$), does reach a significant level, and it gives support to the hypothesis of more positive "could cooperate with the daily routine of kindergarten" scores.

4. Conclusions and Recommendations

This study demonstrates that the test depends on an observed variable called earning, which could be interpreted as an underlying ability (language comprehension and self-conduct) that is not directly observed. According to the model, performance in the two tests depends on this ability. The research results showed that HMEAYC combined with music technology could enhance the development of language comprehension and the self-conduct of children with specific needs.

5. Discussion and Conclusions

This research found that music image technology effectively promoted participants' language comprehension. The results show that the model fit was adequate after adding error covariances between the pre- and post-tests in the hypothesis model for participants. The study suggests that music image technology significantly increased scores for young children with developmental delays in HMEAYC.

This study proved that there was an increase in the experimental group's language comprehension in the music image technology post-test scores. The post-test results were higher for young children with developmental delays in the control group than the pretest results. When the groups were compared, more significant improvements were observed in language comprehension in the experimental group due to the experimental intervention (the music image technology). In the end, all measurements showed a significant relationship with comprehension. The results of this study demonstrated that more activity using music technology related to educational practice is needed. At the same time, attention is given to methods and strategies to make more music technology accessible to various learners [2,7,17].

As a result of the experiments, we concluded that the significant relationships between language comprehension and self-control in the pretest and post-test are consistent with previous studies. From a language comprehension perspective, it was established that music image technology [4,5,9,12,14,18] was influential in language comprehension and promoted higher scores in the experimental group than the control group with statistical significance. This study's findings resemble previous experimental participants' findings regarding music technology in the language learning literature.

When participants understand the group rule, they demonstrate group-normative behavior in musical activities. As can be seen from this study, music is an excellent medium for young children's learning development [1,4,12,13]. The participants showed that the music image technology led to a positive change in learning behaviors, increased course engagement, and positive changes in interactive communication [10,11,20]. Within the limitations of these results, music image technology could likely contribute to evidence of language comprehension development in young children with developmental delays.

Author Contributions: L.L. was in charge of the project administration, supervision, methodology survey questionnaire, and data collection. Corresponding author H.-J.H. handled the writing, paper structure, and result analysis. All authors have read and agreed to the published version of the manuscript.

Funding: This research received no external funding.

Institutional Review Board Statement: Not applicable.

Informed Consent Statement: Informed consent was obtained from all subjects involved in the study.

Data Availability Statement: Not applicable.

Conflicts of Interest: The authors declare no conflict of interest.

References

1. Lee, L. Music Activities for Children with Disabilities: An Example from Taiwan. In *Exceptional Music Pedagogy for Children with Exceptionalities: International Perspectives*; Oxford University Press: Oxford, UK, 2016; pp. 131–153.
2. Lee, L.; Li, T.Y. The Impact of Music Activities in a Multi-Sensory Room for Children with Multiple Disabilities on Developing Positive Emotions: A Case Study. *J. Eur. Teach. Educ. Netw.* **2016**, *11*, 1–12.
3. Lee, L.; Chen, J.H.S.; Hogenes, M. *Music Activities for Teaching Young Children with Special Needs*; ISME: Glasgow, UK, 2016.
4. Lee, L.; Liu, Y.S. Training effects and intelligent evaluated pattern of the holistic music educational approach for children with developmental delay. *Int. J. Environ. Res. Public Health* **2021**, *18*, 10064. [CrossRef]
5. Kraus, N.; Chandrasekaran, B. Music training for the development of auditory skills. *Nat. Rev. Neurosci.* **2010**, *11*, 599–605. [CrossRef] [PubMed]
6. Israel, H.F. Language learning enhanced by music and song. *Lit. Inf. Comput. Educ. J. (LICEJ)* **2013**, *2*, 1360–1365. [CrossRef]
7. Strait, D.L.; Kraus, N. Biological impact of auditory expertise across the life span: Musicians as a model of auditory learning. *Hear. Res.* **2014**, *308*, 109–121. [CrossRef] [PubMed]
8. Miendlarzewska, E.A.; Trost, W.J. How musical training affects cognitive development: Rhythm, reward and other modulating variables. *Front. Neurosci.* **2014**, *7*, 279. [CrossRef] [PubMed]
9. Ventura, M.D. Creating inspiring learning environments by means of digital technologies: A case study of the effectiveness of WhatsApp in music education. In *E-Learning, E-Education, and Online Training*; Springer: Cham, Switzerland, 2017; pp. 36–45.
10. Lee, L.; Ho, H. Exploring Young Children's Communication Development through the Soundbeam Trigger Modes in the 'Holistic Music Educational Approach for Young Children' Programme. *Malays. J. Music.* **2018**, *7*, 1–19. [CrossRef]
11. Lee, L. Investigating the Impact of Music Activities Incorporating Soundbeam Technology on Children with Multiple Disabilities. *J. Eur. Teach. Educ. Netw.* **2015**, *10*, 1–12.
12. Kwak, E.E. Effect of rhythmic auditory stimulation on gait performance in children with spastic cerebral palsy. *J. Music Ther.* **2007**, *44*, 198–216. [CrossRef] [PubMed]
13. Lagasse, B. Music, Neuroscience, and Communication: Introduction to the Focus Area. *Music. Ther. Perspect.* **2017**, *35*, 105–106. [CrossRef]
14. Thaut, M.H.; McIntosh, G.C.; Hoemberg, V. Neurobiological foundations of neurologic music therapy: Rhythmic entrainment and the motor system. *Front. Psychol.* **2015**, *5*, 1185. [CrossRef] [PubMed]
15. Lee, L. *Theory & Practice of Music Educational Therapy for Young Children with Disabilities: A Report of the Industry-University Collaboration Research at Taichung Early Intervention Center*; Taiwan Fund for Children and Families: Taichung, Taiwan, 2012.
16. Frid, E. Accessible digital musical instruments—A review of musical interfaces in inclusive music practice. *Multimodal Technol. Interact.* **2019**, *3*, 57. [CrossRef]
17. Johnels, L.; Vehmas, S.; Wilder, J. Musical interaction with children and young people with severe or profound intellectual and multiple disabilities: A scoping review. *Int. J. Dev. Disabil.* **2021**, *69*, 487–504. [CrossRef] [PubMed]
18. Tierney, A.; Kraus, N. Music training for the development of reading skills. *Prog. Brain Res.* **2013**, *207*, 209–241. [PubMed]
19. Schon, D.; Boyer, M.; Moreno, S.; Besson, M.; Peretz, I.; Kolinsky, R. Songs as an aid for language acquisition. *Cognition* **2008**, *106*, 975–983. [CrossRef] [PubMed]
20. Costa-Giomi, E. The long-term effects of childhood music instruction on intelligence and general cognitive abilities. *Update Appl. Res. Music. Educ.* **2015**, *33*, 20–26. [CrossRef]

engineering proceedings

MDPI

Proceeding Paper

Research on User Experience and Needs of Virtual Reality Learning Home Appliances [†]

Yu-Hsuan Huang [1], Chien-Hsiung Chen [1] and Yo-Wen Liang [2,*]

[1] Department of Design, National Taiwan University of Science and Technology, Taipei 106335, Taiwan; sherry952036@gmail.com (Y.-H.H.); cchen@mail.ntust.edu.tw (C.-H.C.)
[2] Department of Industrial Design, National Taipei University of Technology, Taipei 106344, Taiwan
* Correspondence: yowenliangphd@gmail.com
† Presented at the IEEE 5th Eurasia Conference on Biomedical Engineering, Healthcare and Sustainability, Tainan, Taiwan, 2–4 June 2023.

Abstract: The surge in e-commerce and the COVID-19 pandemic has dramatically driven the demand for home appliances. However, this surge has also resulted in a proportional increase in electrical accidents. Furthermore, the current product manual does not provide users with intuitive and timely information. Virtual reality (VR) can deliver effective learning information with its exceptional simulation capabilities and immersive content. Thus, we explored the application of VR to enhance home appliances' display and user experience for user demand to deliver the manuals on the appliances and promote safe uses. A purposive sampling technique was employed, and 12 target users participated in the in-depth interviews. In the study, the VR application of the home appliance product experience and learning showed positive outcomes. The results can be used as a reference for future design. According to the users' feedback, we summarized and proposed eight future design directions for enhancing the virtual home appliance experience.

Keywords: virtual reality; home appliance; user experience; service design

Citation: Huang, Y.-H.; Chen, C.-H.; Liang, Y.-W. Research on User Experience and Needs of Virtual Reality Learning Home Appliances. *Eng. Proc.* **2023**, *55*, 28. https://doi.org/10.3390/engproc2023055028

Academic Editors: Teen-Hang Meen, Kuei-Shu Hsu and Cheng-Fu Yang

Published: 29 November 2023

1. Introduction

The rapid growth of e-commerce and the global pandemic's impact has boosted the stay-at-home economy, which has increased the demand for home appliances. Unfortunately, this increased demand has been accompanied by accidents stemming from improper operation and a lack of knowledge of products, especially for those that require high temperatures and complex procedures. In addition, product instructions do not convey enough information for users to learn how to use the home appliances intuitively and efficiently. Virtual reality (VR) has a high degree of simulation capability and immersion characteristics that can block external interference. Thus, it has better efficiency for good learning results. The purpose of this study was to explore the possibility of applying VR to use home appliances by understanding the needs and ideas of users to improve user understanding and safety in using home appliances.

2. Literature Review

2.1. VR Application

Nowadays, VR devices provide a highly realistic and more immersive experience. From a design point of view, VR can easily switch between physical and virtual spaces, becoming a convenient tool and saving the unnecessary waste of resources and time consumption with diverse innovations and applications [1,2]. Future VR technology is committed to simulating more realistic situations with the characteristics of the natural world to 3D virtual objects based on user behaviors and operations similar to the real world. This improves the quality of virtual situations [3–5]. However, VR technology in recent years has been mainly used in entertainment and games and has yet to be applied to solve problems in life.

2.2. Home Appliances and Hazards

According to the global home appliance market analysis in 2020, the total output value of the year was USD 448 billion, and more than 10% of households owned one or more home appliances [6]. At the same time, a home appliance safety research report conducted by London Economics from March to November 2021 pointed out that about 750,000 accidental injuries were caused by unsafe home appliances every year [7]. Common home appliances, such as ovens, microwave ovens, and induction cookers, generate high temperatures and heat, which may cause a fire if improperly used [8]. However, products with complex operation functions cause incorrect use behaviors. To prevent disasters, users must clearly understand the correct usage and equipment maintenance before purchasing the product [7]. However, most current product information provides cumbersome manuals and videos. If the product operation is complicated, the communication is limited [9,10].

2.3. User Experience

In operating a product or service, the user's subjective experience is affected by the difference in experience among individuals [11]. Although most products are purchased with manuals to assist in learning how to use them. Therefore, to communicate well, it is necessary to use the user-centered concept and understand the target group's use behavior and pattern before designing [12]. Sharp, Preece, and Rogers [13] pointed out that user experience goals included satisfying, enjoyable, fun, entertaining, helpful, motivating, aesthetically pleasing, supportive of creativity, rewarding, and emotionally fulfilling. In addition, Chauvergne, Hachet, and Prouzeau [14] developed a framework for the user guidance phase in VR, including onboarding context, degree of flexibility, degree of coverage, inputs onboarding, degree of interactivity, type of assistance, degree of feedback, and instruction modalities.

3. Methods

3.1. Participants

To ensure the authenticity and effectiveness of the research as well as the availability of data, we considered the age, gender, education level, and monthly income of the participants before conducting the interviews. They were selected by using a purposeful sampling technique. A total of 12 adult participants with home appliance experience and purchasing power were invited for the in-depth interviews.

3.2. Interview

The content of the interview was divided into the following three parts. To make it easier for the participants to recall, an oven with high temperature and an espresso machine with highly complex were chosen for the user experience as they were easily found at home.

3.2.1. Descriptive Statistics of Participants

The following questions were included in the interview to understand the personal background of the participants.

- What should I call you?
- What is your gender?
- What is your age?
- Your education level?
- What is your monthly income?

3.2.2. Home Appliances Using and Learning Experience

To understand the participants' experience in using home appliances, the following questions were asked during the interview.

- Did you encounter any difficulties when you first used the new oven? (Figure 1)

Figure 1. Examples of oven and espresso machine.

- Did you encounter any difficulties when you first used the new espresso machine? (Figure 1)
- Have you used any home appliances in the past that made you feel bad? (Open-ended/Psychological factors)
- Have you ever used other household electrical appliances with complex operation and high risk? (Open-ended/External factors)
- How do you learn to use a new home appliance? (See product manuals, teaching videos, others...)
- Learning by watching product manuals, what do you think are the advantages and disadvantages of this method?
- Learning by watching teaching videos, what do you think are the advantages and disadvantages of this method?
- The existing product introduction is almost dominated by manuals and videos. What is your opinion on this? How do you feel?
- Do you think the teaching method of watching the product operation on-site will improve your learning effect?
- When you get a new product, do you read about product hazards?

3.2.3. Home Appliances Virtual Experience Demands

To find out what demands the participants had for home appliances and their experience with VR devices, the following questions were asked during the interview.

- Have you ever used VR-related equipment?
- If there is a virtual home appliance product experience today, what services would you expect it to provide? (Open discussion based on the following items)
    - Display: product function introduction, display product structure, power knowledge, precautions, repeated viewing...
    - Auxiliary: visual product teaching, text description, audio commentary...
    - Experience: practical operation experience, error reminder, situational simulation...
- For the above question, what items do you think are particularly important?
- What kind of reasons will increase your willingness to try a VR virtual home appliance experience?

4. Results

In terms of a user-centered perspective, we explored users' experience in using and learning home appliances, as well as users' needs and opinions on the experience of using

VR in home appliances. The results of the interview with 12 participants were analyzed as follows.

4.1. Descriptive Statistics of Participants

The basic information of the participants is summarized in Table 1. In-depth interviews were followed by asking questions on the information as shown in Figure 2.

Table 1. Basic information of participants.

No.	Gender	Age	Income (monthly)	Date	Time	Place
V1	male	23	30,000–40,000 NTD	2023/05/22	30 min	research room
V2	female	23	20,000–30,000 NTD	2023/05/22	31 min	research room
V3	male	25	30,000–40,000 NTD	2023/05/22	45 min	research room
V4	male	38	60,000–70,000 NTD	2023/05/22	43 min	research room
V5	male	24	20,000–30,000 NTD	2023/05/23	32 min	research room
V6	female	37	40,000–50,000 NTD	2023/05/23	42 min	research room
V7	male	23	30,000–40,000 NTD	2023/05/23	38 min	research room
V8	female	24	30,000–40,000 NTD	2023/05/24	42 min	research room
V9	female	23	20,000–30,000 NTD	2023/05/24	59 min	research room
V10	male	25	30,000–40,000 NTD	2023/05/25	30 min	research room
V11	female	29	30,000–40,000 NTD	2023/05/25	47 min	research room
V12	female	32	40,000–50,000 NTD	2023/05/25	40 min	research room

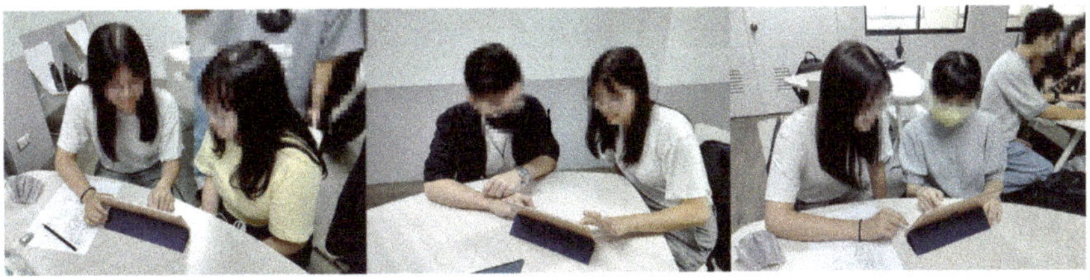

Figure 2. In-depth interviews.

4.2. Causes of Difficulties in Using Home Appliances

According to the interview with the participants on the user experience of the oven and espresso machine, users had trouble because they did not learn how to use the functions correctly and follow the steps to operate. This result was consistent with other literature. In the case of ovens, the product did not show the recommended temperatures in the manuals. In addition, for preheating the oven, temperature control was necessary, such as upper heating, lower heating, and upper and lower heating. It was difficult to use the product at a high temperature and heat risk. Users were worried and afraid when they did not understand how to operate the oven appropriately. For the espresso machine, the participants said it had more functions than other appliances. The operation included pre-work, use, and cleaning, which required the correct order of operation. More espresso machines are designed with complex user interfaces, including operation buttons, which need more time to understand the operation.

4.3. Limitations of Home Appliance Experience and Learning

According to the feedback from the participants, it was found that various home appliances were difficult to learn proper operations for due to different shapes, multiple ways of use, and complex functions. Most users read product manuals and watch teaching videos as the primary way of experiencing and learning. However, the existing methods were not effective and immersive. Regarding product manuals, most participants said that the content was too complicated to read. It took a long time for them to read it, and the manual was easy to lose. Several participants believed that the manuals were official information and were reliable (Table 2). Online teaching videos were more intuitive than manuals and easier to understand the operation method. However, most online teaching videos were about the evaluation of the products and not many videos were provided by the official. The quality of the videos often varied depending on the photographers' editing style, camera angle, and video script. With such a difference, the user must understand by themselves. Therefore, there were many disadvantages to using written and video manuals (Table 3).

Table 2. Results of user interviews on product manuals.

Comments	Participants Feedback
Negative Comments	The disadvantage of the manual is that if the product itself is complicated, it takes time to understand, and I prefer to read fewer words. (V3)
	Time is precious, and the manuals are very thick, so it will take time to find important information. (V4)
	He will make people instinctively feel that it should be complicated and it will feel so boring. (V5)
	I am a more visual person. (V6)
	I will not have the patience to watch it. (V8)
	For the details, the manual usually cannot present the parts clearly. (V10)
	Some manuals like to use irregular nouns, have a high repetition rate, do not speak directly, or have a long list of information without clear descriptions, making it difficult to read. (V12)
Positive Comments	Because it is an official message, it should be very reliable. In addition, the process will be very clear. (V7)
	I prioritize reading the manual, which should be trusted, because if you read it, you can see how he designed it, and you will know how to use it. There is a set of logic behind it. (V11)

Table 3. Results of user interviews on teaching video.

Comments	Participants Feedback
Negative Comments	The video is better than the product manual, but sometimes it does not even bring the scene to the scene. I watched it 2 or 3 times and still need to learn how to do it. (V2)
	Sometimes if I want to find out what a function is, and then I do not talk about it from the end to the end of the teaching video, it feels like a waste of time. (V3)
	Because I have become a passive recipient of information, the accuracy of the information will be low. (V7)
	The button design of the new product is different, but some brands will not update the teaching video, and it will always be the first or second generation, and I will not trust that video. (V8)
	I would not say I like teaching videos very much because I think it is a waste of time, the speech is very long, and there is no way you can scan it like pictures or text. (V11)
	It can be messy because of the photographer's style or because the script did not get it so well. (V12)
Positive Comments	The video is great, and I find it intuitive. (V1)
	Someone may tell you how to use it, and it feels more vivid. (V5)

4.4. Dangers of Home Appliances

In addition to the oven and espresso machine mentioned in the interview, 12 participants also pointed out that they had used high-risk household appliances in their experience. Participants (V2, V5, V6, V8, V11, and V12) were not sure which materials were microwavable and not, which might cause dangerous accidents. Air fryers (V1 and V6), vacuum cleaners (V2 and V4), and other products also had complex structures and were operated at a high temperature. However, 9 out of 12 participants said that they did not pay attention to the precautions related to danger because the appliances were easy to contact in our daily lives and subconsciously mistakenly thought that they were familiar with them. They did not like to spend too much time learning about home appliance products and ignoring important information. Only three participants paid attention to dangers at the beginning and tried to judge whether the product was dangerous. However, they worried because of the fear of product damage, not the safety and other reasons.

4.5. User Expectation for Virtual Home Appliance Experience

Compared with the written product manuals and teaching videos, users generally believed that a more specific visual representation improved the learning effect via an immersive product experience and display. After the interviews, the users' needs and expectations for VR for home appliances were compiled according to their descriptions, as shown in Table 4.

Table 4. Results of virtual home appliance experience interview.

Needs and Expectations	Participants Feedback
360° Product experience	I can be more specific about the product I am about to use. (V6)
	If I can watch the product in 3D from the front or the side of the product, it will improve the learning effect and experience. (V7)
Entertaining Teaching	If there is an element of entertainment, some interactivity, I want to use it before I buy it. (V1)
	I hope it does not look like a commercial film and it can simulate a situational case, like playing a game. (V6)
Deepen the Impression of Use	Because I have personally experienced it from the first-person perspective, it will be easier to deepen the impression of using it. (V4)
	Some words are forgotten after you read them, but the virtual experience can make me more aware of how to use this thing. (V9)
Remote Experience Anytime and Anywhere	If I suddenly want to buy in the middle of the night and want to place an order, the virtual experience is very convenient. (V11)
Increase Product Life	If you have been misusing it for a long time, the consumption rate of the product will be faster. If there is a reminder of the error in the virtual experience, the error can be corrected in time. (V6)
Full Communication of Official Content	The physical store is full of people, so he may miss something, but if it is a virtual experience, it is less likely to miss. (V7)
Safe to Try Out New Features	If I do not know how to use home appliances, I will be afraid to use them, but I can try them confidently in VR. (V2)

5. Discussion

According to the results, several emphasized suggestions for the user demands in VR were made for experiencing and learning home appliances.

5.1. Easy-to-Learn Methods

Even with the complicated content and difficulty, the manual has credibility as it was provided officially, and the operation process and introduction of the product were more precise in the manual. Therefore, the official initiative must provide a virtual home appliance experience. Users are less likely to feel fatigued in understanding the product's online teaching videos. However, due to the editing style and camera angle limitation, the visual teaching also shows problems. Therefore, the virtual experience is more acceptable as it can provide a 360-degree viewing angle.

5.2. Appropriate Reminders

Dangerous accidents occur with familiar and easily accessible household appliances. Since most users learn less about how to use home appliances, extra attention needs to be paid to hazard-related precautions. The participants said that if the product had a chance of causing harm, they were willing to learn how to prevent it via appropriate learning. Thus, reminders are required and must be short to the point or playful to grab the user's attention.

6. Conclusions

Considering the research results of this study and the user experience revealed in the literature, the following eight design directions for virtual home appliance experience were proposed.

1. Visual 3D function instruction of the product is required by viewing the product according to the user's needs. Users are allowed to move and watch freely on the product's front, side, or back.
2. Via block-based product teaching, product functions and operations need to be taught step by step. It needs to be formulated for users of different levels so that users can be more familiar with the product.
3. Precautions and knowledge need to be supplied for users to have more information. Rather than writing information in the product manual, precautions must be included in the product introduction for users to read it interestingly and shortly.
4. Errors must be shown instead of showing error reminders. Error reminders must be displayed in erroneous product operations. The error reminders can correct the users' wrong operation, as frequent mistakes easily cause negative emotions such as frustration.
5. To correct usage behavior, repeated learning is required. This also helps household appliances prolong their lifespan. For new functions and different operation methods of the product, users need to watch it repeatedly to become familiar with it.
6. Audio commentary is required to provide a product introduction. In the product introduction, sensory aids such as vision, hearing, or touch can be introduced, allowing users to choose them according to their preferences.
7. Actual cases need to be included in virtual teaching. As virtual teaching needs to be as close as possible to the actual behavior and situation, the gap needs to be narrowed between the virtual environment and actual situations to ensure no problems when the user operates.
8. Entertainment elements are necessary to effectively arouse users' interest and increase user interaction and participation in the product introduction, avoiding explanations only.

Author Contributions: Conceptualization, research, and interview, Y.-H.H.; methodology, Y.-H.H. and C.-H.C.; consultancy and suggestion of VR skills, Y.-W.L.; writing—original draft preparation, Y.-H.H.; writing—review and editing, C.-H.C. and Y.-W.L. All authors have read and agreed to the published version of the manuscript.

Funding: This research is partially supported by the "NSTC Research Program" and "Moe Teaching Practice Research Program" sponsored by the National Science and Technology Council and Ministry of Education, Taiwan, R.O.C. under Grant no. NSTC 111-2221-E-027-061 and MOE PHA1110385.

Institutional Review Board Statement: Not applicable.

Informed Consent Statement: Not applicable.

Data Availability Statement: All the relevant data have been included in the paper.

Conflicts of Interest: The authors declare no conflict of interest.

References

1. Karsakov, A.; Bilyatdinova, A.; Hoekstra, A.G. 3D virtual environment for project-based learning. In Proceedings of the 2014 IEEE 8th International Conference on Application of Information and Communication Technologies, Astana, Kazakhstan, 15–17 October 2014; pp. 1–5.

2. Liang, Y.W.; Lee, A.S.; Liu, S.F. Exploring the Design-oriented Demands of VR via ZMET-QFD Model for Industrial Design Education and Students' Learning. In Proceedings of the 2015 International Conference on Applied System Innovation, No.987, Osaka, Japan, 21–26 May 2015.
3. Liang, Y.W.; Lee, A.S.; Liu, S.F. A study on design-oriented demands of VR via ZMET-QFD model for industrial design education and students' learning. *Eurasia J. Math. Sci. Technol. Educ.* **2016**, *12*, 1205–1219.
4. Widder, M.; Gorsky, P. How students use a software application for visualizing 3D geometric objects to solve problems. *J. Comput. Math. Sci. Teach.* **2013**, *32*, 89–120.
5. Lombart, C.; Millan, E.; Normand, J.M.; Verhulst, A.; Labbé-Pinlon, B.; Moreau, G. Effects of physical, non-immersive virtual, and immersive virtual store environments on consumers' perceptions and purchase behavior. *Comput. Hum. Behav.* **2020**, *110*, 106374. [CrossRef]
6. Peysakhovich, R. 61 Home Appliances Industry Statistics. Onedesk. Available online: https://www.getonedesk.com/blog/home-appliances-industry-statistics (accessed on 12 August 2022).
7. London Economics. Safety of Smart Domestic Appliances: A Review of the Opportunities for Smart Technology to Enhance Product Safety. Available online: https://www.gov.uk/government/publications/safety-of-smart-domestic-appliances (accessed on 1 March 2023).
8. Taipei City Government Fire Department. Pay Attention to the Hidden Crisis Decryption when Using Home Appliances. Available online: https://www.119.gov.taipei/ (accessed on 25 June 2019). (In Chinese)
9. Brigham, T.J. Feast for the Eyes: An Introduction to Data *Visualization. Med. Ref. Serv. Q.* **2016**, *35*, 215–223. [CrossRef] [PubMed]
10. Germanakos, P.; Constantinides, M.; Kasinidou, M.; Samaras, G. A Metacognitive Perspective of Infovis in Education; Paper presentation. In Proceedings of the 27th Conference on User Modeling, Adaptation and Personalization, Larnaca, Cyprus, 9–12 June 2019; pp. 331–336. [CrossRef]
11. Lu, C.C.; Chang, Y.N. Dynamic Museum Exhibits and the User Experience. *Museol. Q.* **2015**, *29*, 89–102. (In Chinese)
12. Norman, D.A. *The Design of Everyday Things (Revised & Expanded Edition)*; Yuan-Liou Publishing: Taipei, Taiwan, 2014.
13. Sharp, H.; Preece, J.; Rogers, Y. *Interaction Design: Beyond Human-Computer Interaction*; Wiley Publishing: Hoboken, NJ, USA, 2007.
14. Chauvergne, E.; Hachet, M.; Prouzeau, A. User Onboarding in Virtual Reality: An Investigation of Current Practices. In Proceedings of the Extended Abstracts of the 2023 CHI Conference on Human Factors in Computing Systems, No. 711 (1–15), Hamburg, Germany, 19 April 2023. [CrossRef]

engineering proceedings

Proceeding Paper

Near-Infrared Wavelength Selection and Optimizing Detector Location for Apple Quality Assessment Using Molecular Optical Simulation Environment (MOSE) Software †

Quy Tan Ha [1], Thao Nguyen Dang Thi [1], Ngoc Tuyet Le Nguyen [1], Hoang Nhut Huynh [1], Anh Tu Tran [2], Hong Duyen Trinh Tran [1] and Trung Nghia Tran [1,*]

1 Laboratory of Laser Technology, Faculty of Applied Science, Ho Chi Minh City University of Technology (HCMUT), VNUHCM, 268 Ly Thuong Kiet Street, District 10, Ho Chi Minh City 72506, Vietnam; hqtan@hcmut.edu.vn (Q.T.H.); nguyen.dang_8600@hcmut.edu.vn (T.N.D.T.); lnntuyet.sdh20@hcmut.edu.vn (N.T.L.N.); hhnhut@hcmut.edu.vn (H.N.H.); tt_hd2005@hcmut.edu.vn (H.D.T.T.)

2 Laboratory of General Physics, Faculty of Applied Science, Ho Chi Minh City University of Technology (HCMUT), VNUHCM, 268 Ly Thuong Kiet Street, District 10, Ho Chi Minh City 72506, Vietnam; tranatu@hcmut.edu.vn

* Correspondence: ttnghia@hcmut.edu.vn

† Presented at the IEEE 5th Eurasia Conference on Biomedical Engineering, Healthcare and Sustainability, Tainan, Taiwan, 2–4 June 2023.

Abstract: As an alternate non-destructive analytical modality for monitoring from pre-harvest to post-storage, optical imaging with near-infrared wavelength is used to forecast the quality of numerous fruits. In the near-infrared spectrum, bio-chemicals are identified and measured with light by penetrating deeply into food components. In addition, apples and other fruits with a high water content benefit from water absorption capabilities. The optical approaches are efficient, inexpensive, and environmentally beneficial. This study is performed to examine the setup of reflection imaging to pick the near-infrared wavelength and optimize the distance between the detector and the light source. Molecular Optical Simulation Environment (MOSE) and Monte Carlo multi-layered programs (MCML) were used to simulate the light propagation in a model of apple tissue to select the appropriate wavelength for evaluating food quality in experiments and optimize the position of the reflected signal receiver. As a consequence, the 700–900 nm wavelength has great promise for use in assessing food quality, particularly apple quality. One centimeter is the optimal distance between the detector and the light source. The data may be used to organize an experiment and create an evaluation tool for determining the quality of fruits using optical methods, particularly apples.

Keywords: optical imaging; light propagation; near-infrared; Monte Carlo; MOSE; MCML

Citation: Ha, Q.T.; Thi, T.N.D.; Le Nguyen, N.T.; Huynh, H.N.; Tran, A.T.; Tran, H.D.T.; Tran, T.N. Near-Infrared Wavelength Selection and Optimizing Detector Location for Apple Quality Assessment Using Molecular Optical Simulation Environment (MOSE) Software. *Eng. Proc.* **2023**, *55*, 29. https://doi.org/10.3390/engproc2023055029

Academic Editors: Teen-Hang Meen, Kuei-Shu Hsu and Cheng-Fu Yang

Published: 29 November 2023

1. Introduction

Methods of nondestructive food quality evaluation based on optical features have been an increasing trend in fruit quality testing. As a result of the complexity of light propagation in multilayered food tissue, the optical assessment of food quality confronts several problems. The light propagation model using the Monte Carlo simulation (MC) approach is a typical and effective tool for gaining a deeper comprehension of the interaction between tissue and light. Moreover, in recent years, the modeling of light propagation in foods such as apples, onions, and citrus fruits has been investigated more extensively to anticipate their optical characteristics [1–3]. It provides guidelines for improving systems and creating noninvasive optical technologies for assessing food quality.

Light transmission through opaque tissues, such as fruit, is complex and is related to absorption and scattering of photons. Therefore, an understanding of optical characteristics is necessary to better comprehend the light-tissue interaction or to enhance nondestructive assessment methods. This research is carried out for the propagation of light in apples between 500 and 1000 nm with the Monte Carlo method. In this investigation, the simulation model is spherical. The light source is positioned at a certain distance from the surface of the sample. The purpose of this research is to analyze backscatter pictures using a simulated Charge-Couple Device (CCD) to determine the optimal probe location.

2. Materials and Methods

2.1. Monte Carlo Method

Radiative Transfer Equation describes the transmission of light in opaque tissues such as fruit (RTE). The RTE equation is used to represent the propagation of light in turbid media and light's wave qualities, such as diffraction and interference. There are several ways to solve RTE equations, including analytical and numerical approaches. However, the MC simulation approach is considered the standard for resolving this radiation transport issue due to its simplicity. Moreover, determining the appropriate inaccuracy to decrease computational effort is simple [4,5].

The MC approach enables simulating the transport of radiation in an opaque model based on the likelihood of a mean free route whose direction changes owing to Fresnel reflection scattering, and absorption. The probability relies on optical characteristics such as the scattering coefficient μ_s the absorption coefficient μ_a, and refractive index n, the anisotropic scattering coefficient g. The chemical makeup of opaque tissue is primarily responsible for its ability to absorb light. On the other hand, light scattering in tissue is dependent on the structural characteristics of the propagation medium (density, particle size) and the cellular structure [6].

2.2. Simulation Model

To run the simulation, the model needs geometrical parameters and optical qualities. In this study, the Monte Carlo Multi-Layered (MCML) program and the Molecular Optical Simulation Environment (MOSE) tool are used to simulate the Granny Smith apple [7,8]. The radius of the simulated apple sample model is 2.5 cm. The optical characteristics in Table 1 are derived from the research conducted by Cen et al. [9]. In addition, the Granny Smith sample's refractive index n and anisotropic scattering coefficient g are 1.35 and 0.80, respectively [6]. At MOSE, the acquisition of the reflected signal is mimicked by the default setup of the Charge Coupled Device (CCD). The resolution of the acquisition is 512 by 512 pixels.

Table 1. Inputs for MCML and MOSE simulation.

Parameter	Value
Photon	1,000,000
Resolution of radial distance *dr*	0.01 cm
Resolution of tissue depth *dz*	0.01 cm
Number of grids for radial distance r	500
Number of grids for tissue depth z	500
Refractive index for medium above/below (air)	1.0
Radius of the light beam ($1/e^2$ radius)	0.15 cm

3. Results and Discussion

3.1. Penetration Depth

Extremely dense materials prevent the transmission of light. Instead, the photon packet is separated into tissue absorption, diffuse reflection, and specular reflectance [10]. In the simulation model, photons penetrate to a maximum depth of about 3.0 cm at all wavelengths, as seen in Figure 1.

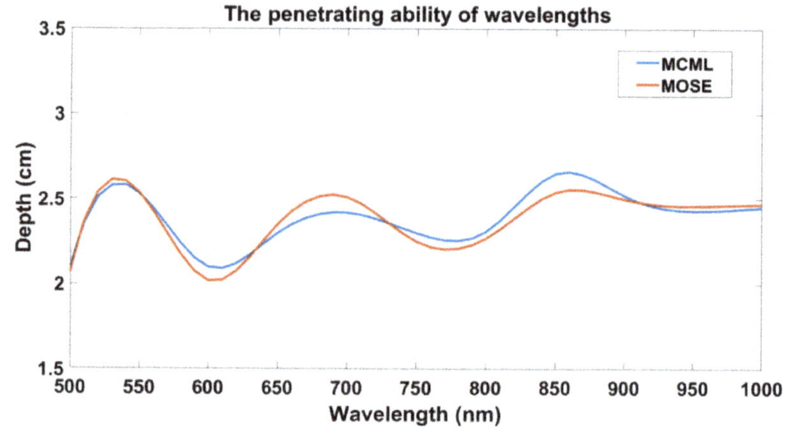

Figure 1. Penetrating ability of wavelengths.

As illustrated in Figure 1, 500–1000 nm wavelengths may profoundly enter apple tissue via absorption and scattering. However, it is difficult to transmit these wavelengths. As a result of the characteristics of the apple's optical properties, a high incident light intensity in transmittance is necessary for light to pass through the whole fruit. Due to the high power consumption in transmittance mode, the light may harm the apple's surface and affect the spectral properties. Several studies have referenced this material [3,10,11]. In the light-based approach for non-destructive quality evaluation, the reflection mode has more promise than the transmission mode.

3.2. *Diffuse Reflection*

Figure 2 shows the diffuse reflectance spectra of a simulated apple model throughout the wavelength of 500–1000 nm. The diffuse reflection is weakest at 650 nm wavelength and brightest at 750 nm. The optical properties of the simulated apple model at various wavelengths are characterized by the variation in diffuse reflectance at different wavelengths. As a result of the interior composition of apple tissue, high internal absorption occurs at 600 nm, absorbing the majority of photons, while mild absorption occurs at 750 nm.

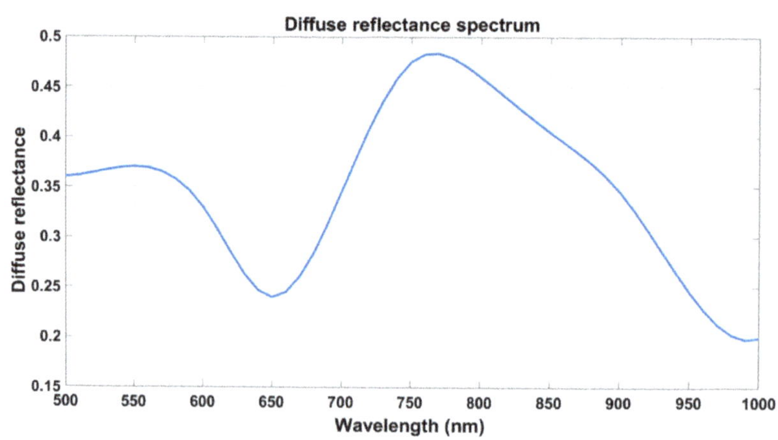

Figure 2. Diffuse reflectance spectrum.

3.3. Optimizing Source-Detector Distance

The distance between the source and detector is optimized with a wavelength of 750 nm. As demonstrated in Figure 2, this wavelength has more benefits than other wavelengths. Therefore, it gives additional information for evaluating apple quality. The diffuse reflection of radius r at 750 nm as simulated by the MCML software (version 1.2.2) is seen in Figure 3.

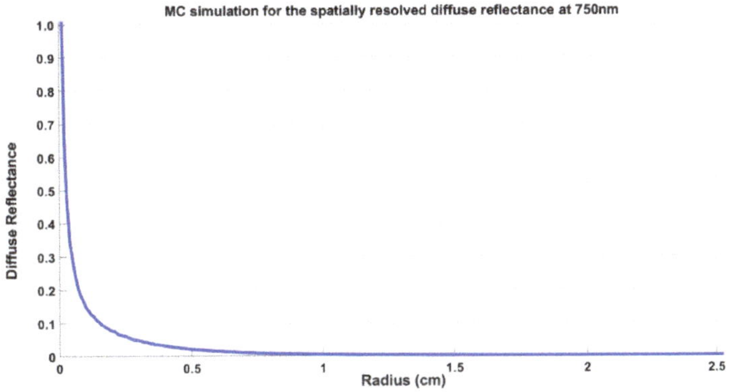

Figure 3. Diffuse reflectance at 750 nm.

Attenuation of the obtained diffusely reflected signal diminishes with increasing distance, with the strongest signal received at a distance of 0.0 cm (or close to the surface). In practice, it is altered by the light source diffusing straight into the detector. The acquisition of a diffusely reflected signal is deemed appropriate within a radius of 1.0 cm. After a radius of 1.0 cm, the reflected signal begins to degrade and provides insufficient data for assessment.

3.4. Back-Scattering Image by MOSE

Figure 4 depicts the simulated back-scattering picture and intensity profile generated by the MOSE program. As illustrated in Figure 5, the program is produced when the light source is positioned in the middle of the CCD. In this arrangement, the light source and CCD detector are simulated to be 1.0 cm from the surface of the apple. However, this setup is not working.

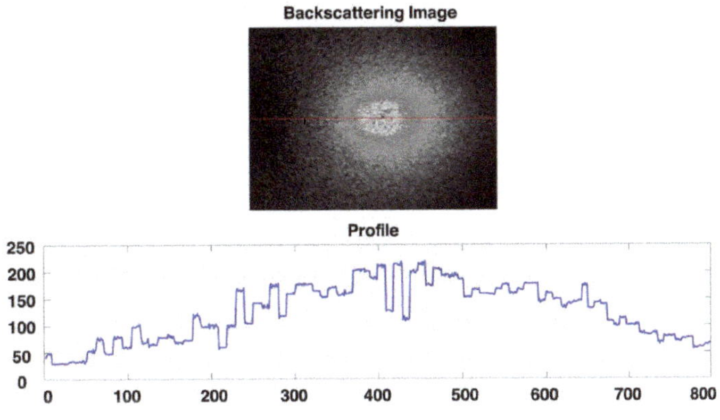

Figure 4. Simulated back-scattering image and its profile along its center.

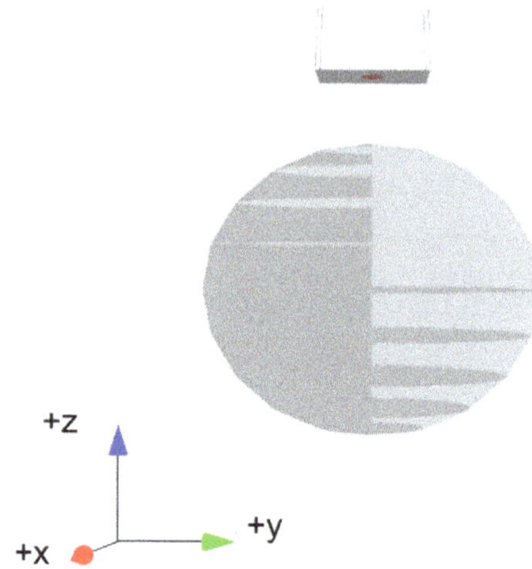

Figure 5. Light source (red dot) is placed at the center of the CCD detector.

Figures 6–8 illustrate the outcome of varying the distance between the light source and the CCD detector while maintaining the same arrangement as Figure 5.

When the distance of the light source from the center of the CCD detector is set as 0.9 cm, Figure 6 depicts the simulated back-scattering picture and its profile along its center.

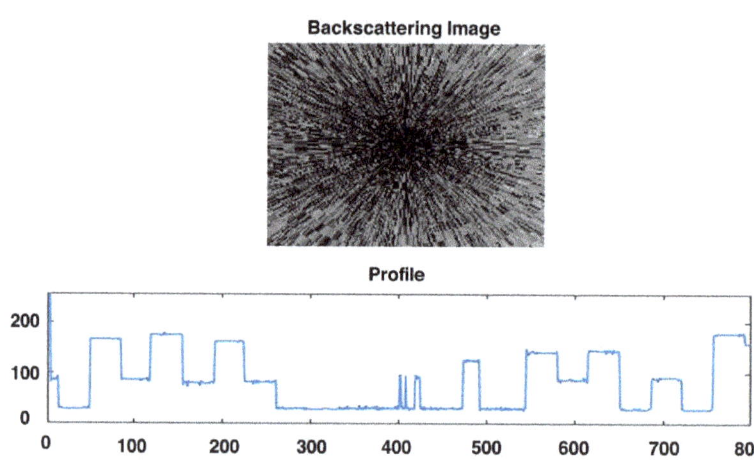

Figure 6. Simulated back-scattering image and its profile along its center (source-detector distance is 0.9 cm).

When the light source is situated 2.0 cm from the center of the CCD detector, Figure 7 depicts the simulated back-scattering picture and its profile along its center.

Figure 8 depicts the simulated back-scattering picture and its profile along its center when the light source is situated 3.0 cm from the CCD detector's center. As seen in Figures 7 and 8, a design with a source-detector distance of less than 1.0 cm provides a greater signal-to-noise ratio and more information.

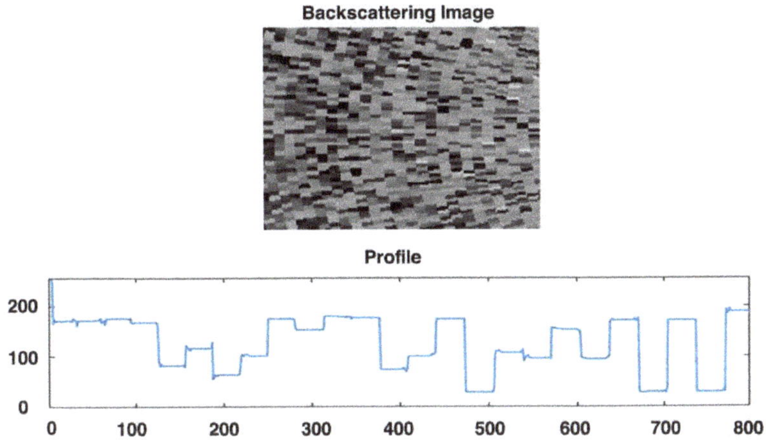

Figure 7. Simulated back-scattering image and its profile along its center (source-detector distance is 2.0 cm).

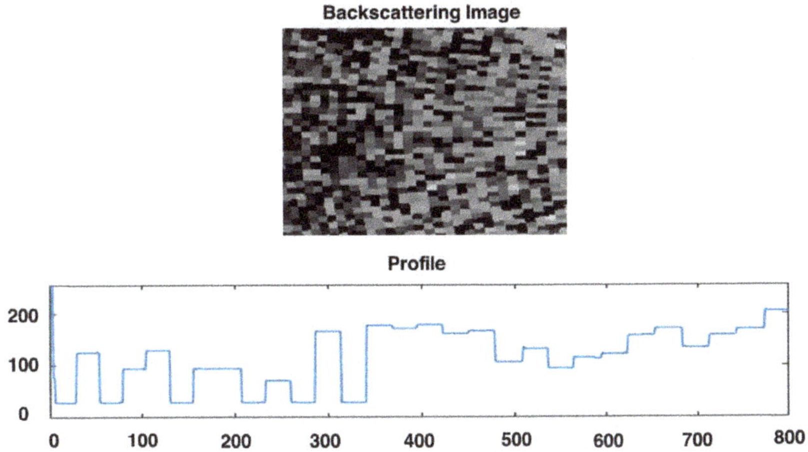

Figure 8. Simulated back-scattering image and its profile along its center (source-detector distance is 3.0 cm).

4. Conclusions

We study light propagation through a model of apple tissue using the Monte Carlo simulation approach and the MCML and MOSE programs. In the MOSE model, the Granny Smith apple was represented as a 3D spherical model. The findings of the simulation demonstrate that light cannot flow through the model. The result demonstrates that the reflection mode is superior to the transmission mode in the visual evaluation of food quality. The diffuse reflectance spectral band has the highest reflectivity at 750 nm. Therefore, it is recommended to utilize it to build a nondestructive optical method for assessing food quality. The optimal positioning of the detector probe is less than 1.0 cm from the light source. The arrangement with a source-detector distance of less than one centimeter may provide more data and a greater signal-to-noise ratio.

Author Contributions: Conceptualization, Q.T.H.; methodology, T.N.D.T.; software, N.T.L.N.; validation, H.N.H.; formal analysis, A.T.T.; investigation, H.D.T.T.; resources, T.N.T.; data curation, H.D.T.T. and Q.T.H.; writing—original draft preparation, T.N.T.; writing—review and editing, Q.T.H.; visualization, H.N.H.; supervision, T.N.T.; project administration, H.D.T.T. and T.N.T. All authors have read and agreed to the published version of the manuscript.

Funding: This research is funded by The Murata Science Foundation under grant number 23VH06. We acknowledge Ho Chi Minh City University of Technology (HCMUT), VNU-HCM for supporting this study.

Institutional Review Board Statement: Not applicable.

Informed Consent Statement: Not applicable.

Data Availability Statement: Data underlying the results presented in this paper are not publicly available but may be obtained from the authors upon reasonable request.

Conflicts of Interest: The authors declare no conflict of interest.

References

1. Wu, A.; Zhu, J.; Tao, Z. Light transmission analysis of laser scattering imaging and Monte Carlo simulation in apple issue. *IOP Conf. Ser. Mater. Sci. Eng.* **2018**, *392*, 052026. [CrossRef]
2. Chugunov, S.; Li, C. Monte Carlo simulation of light propagation in healthy and diseased onion bulbs with multiple layers. *Comput. Electron. Agric.* **2015**, *117*, 91–101. [CrossRef]
3. Sun, C.; Aernouts, B.; Van Beer, R.; Saeys, W. Simulation of light propagation in citrus fruit using monte carlo multi-layered (MCML) method. *J. Food Eng.* **2021**, *291*, 110225. [CrossRef]
4. Watté, R.; Aernouts, B.; Saeys, W. *Monte Carlo Modeling of Light Transfer in Food;* Lu, R., Ed.; CRC Press: Boca Raton, FL, USA, 2016; pp. 79–110.
5. Periyasamy, V.; Pramanik, M. Advances in Monte Carlo Simulation for Light Propagation in Tissue. *IEEE Rev. Biomed. Eng.* **2017**, *10*, 122–135. [CrossRef] [PubMed]
6. Askoura, M.; Vaudelle, F.; L'Huillier, J.-P. Numerical Study of Light Transport in Apple Models Based on Monte Carlo Simulations. *Photonics* **2015**, *3*, 2. [CrossRef]
7. Wang, L.; Jacques, S.L.; Zheng, L. MCML—Monte Carlo modeling of light transport in multi-layered tissues. *Comput. Methods Programs Biomed.* **1995**, *47*, 131–146. [CrossRef] [PubMed]
8. Ren, S.; Chen, X.; Wang, H.; Qu, X.; Wang, G.; Liang, J.; Tian, J. Molecular Optical Simulation Environment (MOSE): A Platform for the Simulation of Light Propagation in Turbid Media. *PLoS ONE* **2013**, *8*, e61304. [CrossRef] [PubMed]
9. Cen, H.; Lu, R.; Mendoza, F.; Beaudry, R.M. Relationship of the optical absorption and scattering properties with mechanical and structural properties of apple tissue. *Postharvest Biol. Technol.* **2013**, *85*, 30–38. [CrossRef]
10. Qin, J.; Lu, R. *Monte Carlo Simulation of Light Propagation in Apples*; American Society of Agricultural and Biological Engineers: St. Joseph, MI, USA, 2007.
11. Qin, J.; Lu, R. Monte Carlo simulation for quantification of light transport features in apples. *Comput. Electron. Agric.* **2009**, *68*, 44–51. [CrossRef]

engineering proceedings

MDPI

Proceeding Paper

Systematic Review of Fuzzy Scales for Multiple Criteria Decision-Making Issues during COVID-19 †

Venkateswarlu Nalluri, Yi-Yun Wang and Long-Sheng Chen *

Department of Information Management, Chaoyang University of Technology, Taichung 413310, Taiwan; nallurivenkey7@gmail.com (V.N.); winona25544@gmail.com (Y.-Y.W.)

* Correspondence: lschen@cyut.edu.tw

† Presented at the IEEE 5th Eurasia Conference on Biomedical Engineering, Healthcare and Sustainability, Tainan, Taiwan, 2–4 June 2023.

Abstract: The COVID-19 epidemic, which can be compared to the economic catastrophe of World War II, slowed down business activities and had a significant impact on all aspects of business operations. Fuzzy scales are popular MCDM (multi-criteria decision-making) methods in modeling COVID-19 problems owing to the multi-dimensionality and complexity of health and socio-economic systems. This study aims to examine 104 works that used MCDM approaches with fuzzy scales in various COVID-19 pandemic issues and were published in top peer-reviewed journals indexed in Web of Science and Scopus. This study presents a systematic review of (1) the prevalence of fuzzy scales in scientific research for multiple criteria decision-making during COVID-19; (2) bibliometric analysis was used to identify the most important articles, authors, journals, themes, and countries; and (3) the impact of fuzzy scales on spreading established fields of research in new directions was iew of Fuzzy Scales for Multiple Criteriaconsidered. Furthermore, it addresses pertinent filed criticism, validating certain claims and dispelling others. Finally, the present study result helps regulators, academic scholars, and policy-makers to understand the current perspective and trends on multiple criteria decision-making with fuzzy scales during COVID-19 and understand the relevant areas that require further investigation.

Keywords: fuzzy theory; trends and characteristics; multiple criteria decision making; COVID-19; systematic review

Citation: Nalluri, V.; Wang, Y.-Y.; Chen, L.-S. Systematic Review of Fuzzy Scales for Multiple Criteria Decision-Making Issues during COVID-19. *Eng. Proc.* **2023**, *55*, 30. https://doi.org/10.3390/engproc2023055030

Academic Editors: Teen-Hang Meen, Kuei-Shu Hsu and Cheng-Fu Yang

Published: 29 November 2023

1. Introduction

One of the most significant pandemics in the last two generations has affected the world from December 2019 to the present day. Every day, countless individuals lose their lives, and thousands of people become infected with this new coronavirus, which is characterized by its highly contagious nature [1]. Humans are typically affected by respiratory infections, which range in severity from the common cold to serious conditions like Middle East respiratory syndrome (MERS) or severe acute respiratory syndrome (SARS) [2]. The actual pandemic (COVID-19), which was introduced by the recently identified coronavirus SARS-CoV-2, has a higher mortality and contagiousness ratio than its predecessors [3]. The rapid spread of SARS-CoV-2, which has resulted in 102 million cases and 2.2 million fatalities worldwide, was made possible by ignorance and the virus' unknown nature.

Governments, nations, and societies have learned from each of these waves and improved their response mechanisms. However, given the virus' unpredictable behavior, somatological diversity, and the recent emergence of new virus variants [2], we continue to face numerous new challenges to the global healthcare system's viability. Numerous operational, logistical, organizational, and moral–ethical standards have been established in this area as a result of COVID-19 and among its waves for management, healthcare professionals, and associates. Similarly, it is more than likely that a fourth wave could occur, given the difficulties encountered in the distribution and application of the COVID-19 vaccine and the accumulated total of active cases. This is more likely to result in a saturation of the healthcare system. The management of specialized training for medical personnel must, therefore, be improved in light of current conditions, and buildings must be transformed into patient accommodations.

Decision-making models are currently used in patient emergencies to deal with this crisis by providing crucial data regarding the prediction of spreading, the evaluation of various factors, determining the weights of criteria, and restructuring decision-making [4]. In this regard, fuzzy-based decision models, based on the theory put forth in the early 1800s, have significant application in decision-making, as they are thought to be a more flexible and dynamic approach and, at the same time, a procedure that is more sensitive to real-life scenarios.

This broad shift in perspective has created the opportunity for future research questions, investigations of previously overlooked positive phenomena, and novel approaches to well-established areas of research. Accordingly, during recent decades, this movement has triggered a growth in research in various fields of study. As a result, this movement has made ripples with regard to the MCDM (multi-criteria decision-making) methods. Thus, we attempt to assess the situation and pose the following questions: what are the notable characteristics of the existing research inspired by this viewpoint?; how have researchers across the MCDM with fuzzy scales engaged and interpreted this viewpoint?; and what are amongst the most significant contributions that have originated?

2. Literature Review

Numerous fuzzy MCDM (FMCDM) methods have been proposed in the literature over the years, differing in terms of questions, theories, and results. Various techniques were not applied to other problems because they were developed for a specific problem. A variety of FMCDM techniques have recently been developed to select the best compromise options. The motivation for developing the FMCDM approaches came from a variety of real-world issues that call for the consideration of multiple criteria, as well as from the desire of practitioners to improve decision-making processes as a result of recent advancements in computer technology, scientific computing, and mathematical optimization [5]. All methods aim to improve decision-making by making it more informed and formalized. MCDM and FMCDM are classified into various fields and approaches in previous studies [6,7]. The MCDM approach can be divided into two groups [4,8]: traditional MCDM and FMCDM.

Since the MCDM problems are diverse, many different techniques have been suggested as solutions. Complete aggregation methods were the first ones including SAW (Simple Additive Weighting) with two stages in weighting [9], TOPSIS (Technique for Order Performance by Similarity to Ideal Solution) [10], WASPAS (Weighted Aggregated Sum Product Assessment) [11], VIKOR (VlseKriterijumska Optimizacija Kompromisno Resenje) [12], MAUA (Multi-Attribute Utility Analysis) [13], MOORA (Multi-Objective Optimization based on Ratio Analysis) [14], COPRAS (Complex Proportional Assessment) [15], ARAS-F (Fuzzy Additive Ratio Assessment Method), ARAS-G (Additive Ratio Assessment Method with Gray values), MULTIMOORA (Multiobjective Optimization Ratio Analysis Plus Full Multiplication Form) [16], COPRAS-G (Complex Proportional Assessment of alternatives to Grey Relations), and ARAS (Additive Ratio Assessment) [1]. Case studies of incomplete aggregation techniques, including NAIADE (Novel Approach to Imprecise Assessment and Decision Environments) [3], PROMETHEE (Preference Ranking Organisation Method for Enrichment Evaluations) [17], and ELECTRE (Elimination and Choice Translating Reality Method), involve comparing alternatives in a pair-wise manner [18]. Additionally, pair-wise comparisons have been used in the ANP (Analytical Network Process) and AHP (Analytic Hierarchy Process) [4]. Fuzzy multi-attribute decision-making (FMADM) and fuzzy multi-objective decision-making (FMODM) are the two subcategories of FMCDM [6]. Reference [2] investigated the development of MADM between 1900 and 2021.

The FMADM approach has a finite and implicit goal, whereas the FMODM approach has an infinite and explicit goal. The decision-objective-makers in FMADM are unified under the decision-utility-maker, which is a super function that depends on the selection criteria. Assigned fuzzy weights in FMODM reflect the relative importance of the decision-maker's goals, such as optimal resource utilization, quality improvement, and remaining explicit. The ability of FMCDM models to take into account various selection criteria is their most significant advantage. Fuzzy sets were used in the MCDM field [19]. The intersection of all the fuzzy goals is what Ref. [3] refers to as the fuzzy set of a decision. Reference [20] compiled a list of MADM problem applications for the fuzzy set theory.

The MCDM problems are greatly impacted by aggregation techniques, and grouping operators have been widely used in MCDM. Reference [19] created many functions for evaluating each alternative's suitability in light of a given set of MCDM criteria in a fuzzy environment. Comparatively speaking, however, very few studies have concentrated on the MCDM issues in a hesitant fuzzy environment. Additionally, hesitation and uncertainty are typically viewed as unavoidable issues when making decisions. Numerous tools have been developed in the literature to more accurately express the evaluation data of decision makers, including fuzzy sets, fuzzy multi-sets, intuitionistic fuzzy sets, linguistic fuzzy sets, interval-valued fuzzy sets, and type-2 fuzzy sets.

3. Methodology

To address the MCDM issues and avoid the inherent narrative reviews, we conducted a systematic review of the literature on the topic of fuzzy scales [1]. A systematic literature review is a type of secondary study defined by Ref. [13] as a method of locating, assessing, and analyzing the evidence that is currently available and relevant to a specific research issue or field of study. This type of analysis points us in the direction of a transparent and repeatable process of selecting, analyzing, and reporting previously conducted research into a specific subject [20]. A literature review, in terms of research methodology, combines qualitative and quantitative analysis to investigate a specific issue of subjective interest [21]. Furthermore, few studies have created frameworks for conducting a systematic review. The first stage of the review is a systematic literature review, as recommended by Ref. [4]. Then, data collection and screening procedures were devised under the guidelines adapted from Ref. [9]. Furthermore, bibliometric analysis follows from Ref. [22]. This systematic review was conducted for two primary reasons. Firstly, researchers reviewing healthcare studies [15], supply chain management [18], and sustainability in emerging situations have frequently used this method in the context of MCDM with fuzzy scales during COVID-19 and the effectiveness of fuzzy scales' perspectives and research trend studies [11]. The second concern is the different approaches for conducting a systematic literature review that considers how fuzzy theory affects MCDM problems [7]. The four main stages of the systematic review process are shown and described as follows:

(1) Systematic review of the literature (Section II);
(2) Data collection: Scopus and Web of Science (WoS) were utilized to collect the necessary data (Section A);
(3) Data screening: using a quantitative approach, formal aspects of the collected data were evaluated and examined (Section A);
(4) Applied techniques: bibliometric analysis and a review of the academic studies of MCDM issues were conducted during COVID-19 (Section IV).

Data Collection

The most complete scientific databases, namely Web of Science and Scopus, were used to gather the data (WoS), enabling the execution of trustworthy bibliometric studies [14]. The broad keywords MCDM issues, the COVID-19 pandemic, the fuzzy scales, and various country or sector names were used in both databases as research criteria to photograph fuzzy scales for MCDM issues affecting the scientific community. While authors, titles, and keywords were the research criteria in WoS, they were authors, abstracts, and keywords in

Scopus. The first step in the review procedure was defining the unit of analysis. The sole research article was taken into account for this review as a whole. As a result, the article was the sole source of results in both databases. Finally, we chose only English-language articles. As it is the language that is most frequently used in both databases and widely regarded as the international academic language, English was chosen as the study's main language [8]. There were no time restrictions. In August 2022, the queries on the various databases were performed. In comparison to WoS, which produced 43 recognized results, using Scopus, we produced a total of 61 results. Duplicates from both databases were subsequently deleted. As a result, in total, 104 papers were reviewed and listed in Figure 1.

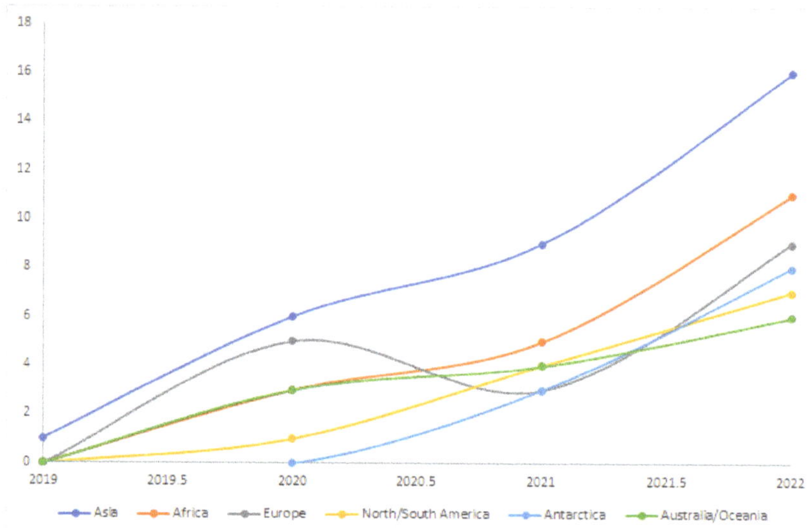

Figure 1. Recently published articles on this topic.

4. Result and Discussions

This section explains the prevalence and bibliometric analysis results.

4.1. Prevalence

Table 1 shows the proof that specific nations and areas frequently serve as hubs within continents. There is evidence that certain countries and regions tend to be at the center of this growth, even though the current systematic review confirms the use of fuzzy scales for MCDM issues during the COVID-19 pandemic. For instance, 54% of the publications using samples from (k = 67 articles) for the MCDM with fuzzy scale were in Asia. Only 5% of the articles were published in Europe, 8% were published in Australia/Oceania, and 33% were published in Africa. Antarctica and North/South America displayed comparable trends. The majority of articles in Asia were from China and India. In Southern Africa, Cabo Verde produced 50% of the articles, with the remainder distributed throughout Europe. However, there has been an increase in interest in several new countries and regions with regard to using fuzzy scales to address MCDM issues. In the dataset, participant samples only ever represented 84 of the world's countries once, and of these 14, 6 were only ever used in multinational studies' empirical research.

Table 1. Articles published statistical results.

Region	Country	Number of Articles
Asia	China, India	67
Africa	Southern Africa, Cabo Verde	24
Europe	United Kingdom	8
North/South America	USA	6
Australia/Oceania	Australia, Samoa	19

4.2. Bibliometric Analysis

Figure 2 shows that numerous research methodologies have been used to conduct studies in the field of MCDM. The 124 publications were manually grouped into 4 different study types to analyze the various research approaches that have been used in this field: quantitative, qualitative, mixed, and other. Quantitative studies emphasize the gathering of quantifiable data and the application of mathematical, statistical, or computational techniques. In the context of this article, qualitative studies are a type of research that focuses on gathering information through conversation and open-ended questions. Studies that combine quantitative and qualitative methods are called mixed studies. The other category includes case studies and other types of reports.

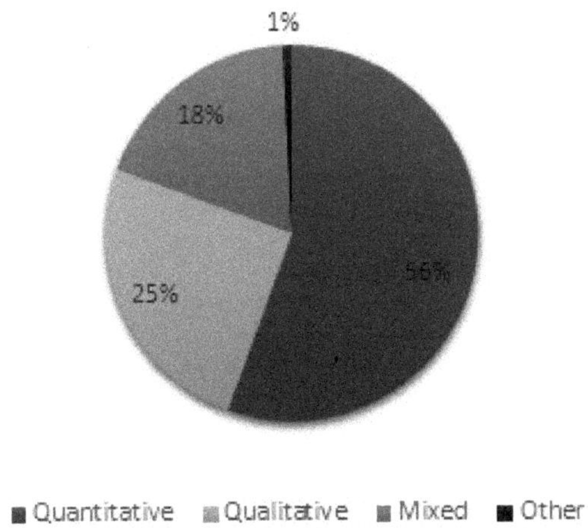

Figure 2. Sample statistics of 124 articles based on this study.

Figure 2 exhibits the percentage of various study types that were used in the fuzzy scale to address the study of MCDM issues. Only 25% of the 124 articles consisted of mixed-approach studies. Instead, 56% were qualitative and 18% were quantitative. Only a few articles used a combination of methods to develop their approaches. According to current qualitative and quantitative theories, the question "what" has been answered fairly effectively, but "how" is still a mystery. It is also clear that there are not many review articles available that offer current information on MCDM issues.

5. Conclusions and Future Directions

This study conducted a systematic literature review to understand the importance and characteristics of fuzzy scales among MCDM studies during the COVID-19 pandemic. We determined the current trends through prevalence and Bibliometric analysis. However, the current trends confirmed that fuzzy scales are key to investigating MCDM issues. The conducted studies originated from Asian countries. In general, Asian countries and businesses were affected by the COVID-19 pandemic [5]. This is the positive outcome of this study's results. This study is contributing to the theory by identifying the current trends in and usefulness of the fuzzy scales while addressing the MCDM issues among businesses and societies.

However, there are limitations to this study. Multi-criteria methodologies are still developing and exponentially spreading in the international scientific community, despite being a relatively new research method, to simplify complex decision-making situations. They have been successfully used as ideal decision-maker methodologies to handle complex situations, as demonstrated in prior studies [6]. Commercial and energy fields have emerged in engineering, in addition to gaining importance in other industries. The applications of FMCDA and FMCDM that have been reviewed confirm their enormous value in situations like the COVID-19 pandemic. The future of FMCDA must be made clear, and the sooner we succeed in the vast array of fuzzy sets [23], obtaining an improvement in the social health of the population, the earlier we can realize the significance of fMCDA decision sciences within the various scientific fields. This is a crucial fact to consider, especially for emergency first aid teams, who also need to concentrate on other pertinent issues [11].

In addition, future works need to consider fuzzy multi-criteria decision analysis techniques that are used to investigate various factors, including isolation planning, the location of quarantine centers, safe nursing homes, safe homes, safe masks, an epidemic-controlling model, and an intensive care unit beds augmentation model for COVID-19 hospitals, and ensure that a large number of patients receive the care they need. Better staffing for COVID-19 management can be ensured through studies involving various lockdown models or policing the individuals who have produced COVID-19 antibodies. Understanding the degree of community spread will be made easier with thorough research and data analytics. Additionally, new SARS-CoV-2 virus variants are discovered every week. Therefore, the decision criteria must either ensure that the current testing, treatment, and vaccines are still effective or consider alternative approaches to combating the virus' effects. The application of the fuzzy-based MCDM method can also be modified based on each patient's unique circumstances.

Author Contributions: Conceptualization, V.N.; methodology, V.N. and Y.-Y.W.; software, V.N. and Y.-Y.W.; validation, V.N.; formal analysis, V.N. and L.-S.C.; writing—original draft preparation, V.N.; writing—review and editing, L.-S.C.; visualization, V.N. and Y.-Y.W.; supervision, L.-S.C.; project administration, L.-S.C.; funding acquisition, L.-S.C. All authors have read and agreed to the published version of the manuscript.

Funding: This work was supported in part by the National Science and Technology Council, Taiwan (Grant No. MOST 111-2410-H-324-006).

Institutional Review Board Statement: Not applicable.

Informed Consent Statement: Not applicable.

Data Availability Statement: The used secondary data that were used for analysis in this study are available from the corresponding author upon request.

Conflicts of Interest: The authors declare no conflict of interest.

References

1. Mishra, A.R.; Rani, P.; Krishankumar, R.; Ravichandran, K.S.; Kar, S. An extended fuzzy decision-making framework using hesitant fuzzy sets for the drug selection to treat the mild symptoms of Coronavirus Disease 2019 (COVID-19). *Appl. Soft. Comput.* **2021**, *103*, 107155. [CrossRef] [PubMed]
2. Paul, A.; Shukla, N.; Paul, S.K.; Trianni, A. Sustainable supply chain management and multi-criteria decision-making methods: A systematic review. *Sustainability* **2021**, *13*, 7104. [CrossRef]
3. Maêda, N.; Rodrigues, M.V.G.; Ângelo, M.; Moreira, L.; Gomes, C.F.S.; dos Santos, M. Bibliometric Studies on Multi-Criteria Decision Analysis (MCDA) Applied in Personnel Selection. In *Modern Management Based on Big Data II and Machine Learning and Intelligent Systems III: Proceedings of MMBD 2021 and MLIS 2021*; IOS Press BV: Amsterdam, The Netherlands, 2021; Volume 341, p. 119.
4. Gao, Z.; Jiang, Y.; He, J.; Wu, J.; Xu, J.; Christakos, G. An AHP-based regional COVID-19 vulnerability model and its application in China. *Model. Earth Syst. Environ.* **2022**, *8*, 2525–2538. [CrossRef] [PubMed]

5. Nalluri, V.; Huynh-Cam, T.; Sama, H.; Chen, L. Decision-making model for the effective e-services adoption in the Indian educational organizations. *Decis. Sci. Lett.* **2023**, *12*, 211–224. [CrossRef]
6. Hou, L.X.; Mao, L.X.; Liu, H.C.; Zhang, L. Decades on emergency decision-making: A bibliometric analysis and literature review. *Complex Intell. Syst.* **2021**, *7*, 2819–2832. [CrossRef]
7. Erol, I.; Ar, I.M.; Peker, I. Scrutinizing blockchain applicability in sustainable supply chains through an integrated fuzzy multi-criteria decision making framework. *Appl. Soft Comput.* **2022**, *116*, 108331. [CrossRef]
8. Ortiz-Barrios, M.; Borrego-Areyanes, A.A.; Gómez-Villar, I.D.; De Felice, F.; Petrillo, A.; Gul, M.; Yucesan, M. A multiple criteria decision-making approach for increasing the preparedness level of sales departments against COVID-19 and future pandemics: A real-world case. *Int. J. Disaster Risk Reduct.* **2021**, *62*, 102411. [CrossRef]
9. Sama, H.R.; Chen, L.S.; Nalluri, V.; Chendragiri, M. Enhancing service quality of rural public transport during the COVID-19 pandemic: A novel fuzzy approach. *Public Transp.* **2023**, *15*, 479–501. [CrossRef]
10. Ocampo, L.; Tanaid, R.A.; Tiu, A.M.; Selerio Jr, E.; Yamagishi, K. Classifying the degree of exposure of customers to COVID-19 in the restaurant industry: A novel intuitionistic fuzzy set extension of the TOPSIS-Sort. *Appl. Soft Comput.* **2021**, *113*, 107906. [CrossRef]
11. Rani, P.; Mishra, A.R. Interval-valued fermatean fuzzy sets with multi-criteria weighted aggregated sum product assessment-based decision analysis framework. *Neural Comput. Appl.* **2022**, *34*, 8051–8067. [CrossRef]
12. Ayyildiz, E.; Taskin, A. A novel spherical fuzzy AHP-VIKOR methodology to determine serving petrol station selection during COVID-19 lockdown: A pilot study for İstanbul. *Socio-Econ. Plan. Sci.* **2022**, *83*, 101345. [CrossRef] [PubMed]
13. Saad Menezes, M.C.; Santinelli Pestana, D.V.; Ferreira, J.C.; Ribeiro de Carvalho, C.R.; Felix, M.C.; Marcilio, I.O.; HCFMUSP COVID-19 Study Group. Distinct Outcomes in COVID-19 Patients with Positive or Negative RT-PCR Test. *Viruses* **2022**, *14*, 175. [CrossRef] [PubMed]
14. Özkan, B.; Özceylan, E.; Kabak, M.; Dikmen, A.U. Evaluation of criteria and COVID-19 patients for intensive care unit admission in the era of pandemic: A multi-criteria decision making approach. *Comput. Methods Programs Biomed.* **2021**, *209*, 106348. [CrossRef] [PubMed]
15. Hezer, S.; Gelmez, E.; Özceylan, E. Comparative analysis of TOPSIS, VIKOR and COPRAS methods for the COVID-19 Regional Safety Assessment. *J. Infect. Public Health* **2021**, *14*, 775–786. [CrossRef]
16. Farhadinia, B. A Divergence-Based Medical Decision-Making Process of COVID-19 Diagnosis. *Math. Probl. Eng.* **2022**, *2022*, 7685033. [CrossRef]
17. Moreira, M.Â.L.; Gomes, C.F.S.; Dos Santos, M.; da Silva Júnior, A.C.; de Araújo Costa, I.P. Sensitivity Analysis by the PROMETHEE-GAIA method: Algorithms evaluation for COVID-19 prediction. *Procedia Comput. Sci.* **2022**, *199*, 431–438. [CrossRef]
18. Forestal, R.L.; Pi, S.M. A hybrid approach based on ELECTRE III-genetic algorithm and TOPSIS method for selection of optimal COVID-19 vaccines. *J. Multi-Criteria Decis. Anal.* **2022**, *29*, 80–91. [CrossRef]
19. Karuppiah, K.; Sankaranarayanan, B.; Ali, S.M.; Paul, S.K. Key challenges to sustainable humanitarian supply chains: Lessons from the COVID-19 pandemic. *Sustainability* **2021**, *13*, 5850. [CrossRef]
20. Ciampi, F.; Giannozzi, A.; Marzi, G.; Altman, E.I. Rethinking SME default prediction: A systematic literature review and future perspectives. *Scientometrics* **2021**, *126*, 2141–2188. [CrossRef]
21. Gul, M.; Guneri, A.F. Hospital location selection: A systematic literature review on methodologies and applications. *Math. Probl. Eng.* **2021**, *2021*, 6682958. [CrossRef]
22. Dohale, V.; Akarte, M.; Gunasekaran, A.; Verma, P. Exploring the role of artificial intelligence in building production resilience: Learnings from the COVID-19 pandemic. *Int. J. Prod. Res.* **2022**, *61*, 1–22. [CrossRef]
23. Zhong, M.; Lin, M. Bibliometric analysis for economy in COVID-19 pandemic. *Heliyon* **2022**, *8*, e10757. [CrossRef] [PubMed]

engineering proceedings

Proceeding Paper

A Comparison of Online and Offline Digital Gameplay Activities in Promoting Computational Thinking in K-12 Education †

Li-Xian Chen [1], Shih-Wen Su [2], Yen-Yun Chen [3], Chia-Hung Liao [2] and Shyan-Ming Yuan [2,*]

1 School of Big Data, Fuzhou University of International Studies and Trade, Fuzhou 350202, China; lixian.cs98g@g2.nctu.edu.tw
2 Department of Computer Science, National Yang-Ming Chiao Tung University, Hsinchu 30010, Taiwan; alvin.cs00@nycu.edu.tw (S.-W.S.); aiallen.cs07g@nctu.edu.tw (C.-H.L.)
3 Institute of Network Engineering, National Yang-Ming Chiao Tung University, Hsinchu 30010, Taiwan; yychen.c@nycu.edu.tw
* Correspondence: smyuan@nycu.edu.tw
† Presented at the IEEE 5th Eurasia Conference on Biomedical Engineering, Healthcare and Sustainability, Tainan, Taiwan, 2–4 June 2023.

Abstract: Due to the COVID-19 pandemic, courses for all ages in many countries moved online. The ability to use information technology fluently for learning has become a vital issue. Whether students can develop computational thinking (CT) literacy in online or offline learning environments is investigated. Two web games, Rummikub and Robozzle, are applied to teach computational thinking through Google Meet, Google Classroom, and offline live teaching. The results indicate that offline training activities are more appropriate for middle school students.

Keywords: computational thinking; digital games; digital literacy; game-based learning; K-12 education

Citation: Chen, L.-X.; Su, S.-W.; Chen, Y.-Y.; Liao, C.-H.; Yuan, S.-M. A Comparison of Online and Offline Digital Gameplay Activities in Promoting Computational Thinking in K-12 Education. *Eng. Proc.* **2023**, 55, 31. https://doi.org/10.3390/engproc2023055031

Academic Editors: Teen-Hang Meen, Kuei-Shu Hsu and Cheng-Fu Yang

Published: 29 November 2023

1. Introduction

With the rapid changes brought on by the COVID-19 epidemic situation, both students' learning methods and teachers' teaching ways faced unprecedented challenges and changes [1]. In response to various policies put in place during the epidemic, hybrid, online, and physical teaching methods have been alternately implemented. These conditions test whether teachers are flexible enough to change what and how they teach and whether students can learn the content. This process of change and adjustment requires using information technology to solve problems, conduct tests, and enhance interactions. The competencies required in this process are also known as computational thinking and digital literacy, which are skills that must be acquired through training [2]. With the increasing reliance of computer technology in the age of AI and big data, students must develop CT skills. Despite the increasing literature on CT, there is still a considerable amount of discussion and development regarding the cultivation of CT in education [3].

Digital games can support the development of a variety of computational thinking skills. Game-based learning environments enable students to understand complex systems independently or collaboratively, reflect on the solutions they find critically, and use analytical thinking to identify logical problem-solving strategies [4]. Gee [5] indicated that the characteristics of games, such as providing immediate feedback, analyzing the course of self-activity, involving students in trial and error, motivating the implementation of reasoning strategies, and providing algorithmic structures, are essential for developing students' problem-solving skills. For example, students who regularly play "information/logic games" demonstrate higher levels of algorithmic thinking, cooperation, and problem-solving than those who do not [4]. Pattern recognition is one of the four cornerstones of computational thinking. It involves finding similarities or patterns between minor, disaggregated problems that can help in solving complex problems more effectively. The goal of the game Rummikub is to place tiles in sets of either a group (i.e., include at least three tiles of the same number in a different color) or a run (i.e., include at least three tiles with consecutive numbers in the same color) [6]. The combinations in each round differ depending on how the tiles were placed in the previous round. Players must figure out the release pattern and constantly think about how to place tiles to conform to the set rules and figure out how to play (Figure 1).

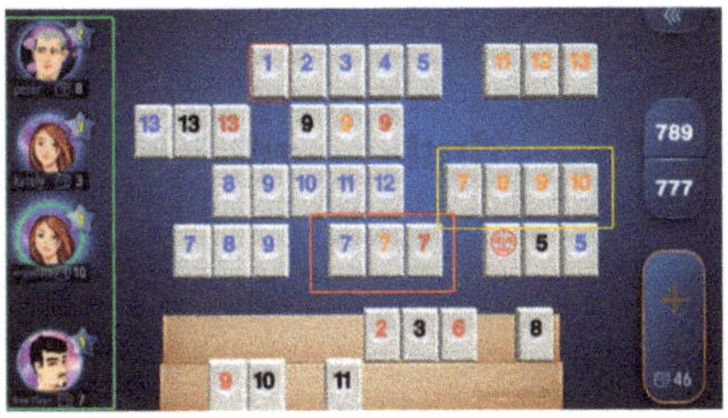

Figure 1. Pattern recognition and group interaction in Rummikub.

Algorithmic thinking is another derivative of computer science and coding. Its core concept is automating the problem-solving process by creating a series of systematic, logical steps. For example, Google applies the PageRank algorithm, which assigns a webpage's importance based on the number of sites linking to it, to bring important search results to the top of the page [7]. Another example is long division, which follows the standard division algorithm by dividing multi-digit integers to calculate the quotient. Robozzle is a puzzle game that requires the player to program a robot through a maze to collect all the stars on the board (Figure 2). Students have to think about how to use the existing instructions and a limited number of moves to design a path so that the robot can collect all the stars during the gameplay [8]. The process mentioned above needs algorithmic thinking to design a good solution.

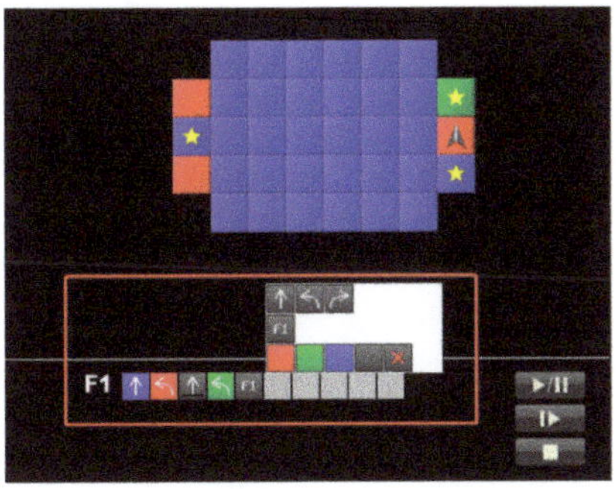

Figure 2. Planning algorithm in Robozzle.

Due to the epidemic's impact, the mode of teaching has changed from offline to online. Without the regular classroom environment and teachers' direct regulation, the effect on students' learning has been a significant concern [9]. This study aimed to explore and compare different teaching modes (online and offline) to train students' computational thinking ability by asking them to solve problems logically and go step by step when they encounter problems.

2. Literature Review

2.1. Online and Offline Learning

In 2020, teaching globally was switched to emergency online teaching [9]. Students need basic digital literacy to take advantage of online learning opportunities. Digital literacy, including using digital technologies to search, process, manipulate and create data, online communication, and collaboration skills are also perceived as a form of computational thinking [10]. Before the COVID-19 pandemic in 2020, many researchers employed offline activities (e.g., unplugged activities and board games) to investigate computational thinking concepts and found that they could serve as a platform with which to promote students' CT learning [11]. Others also used online resources to cultivate CT abilities, such as programming or game-based learning [12]. The method of course delivery (i.e., online or offline) affects student performance, satisfaction, and understanding [13]. However, the benefits of online instruction for students are expected to be highly heterogeneous, and there will be learning boundaries and occasionally even no learning outcomes during online learning [14]. There is a lack of investigations that compare the effectiveness of CT gameplay activities with online and offline learning approaches.

2.2. Digital Game-Based Learning and CT

Computational thinking enables people to understand and process complex problems and formulate possible solutions [15,16]. Both computers and humans can understand these solutions. Computational thinking has four cornerstones: decomposition, pattern recognition, abstraction, and algorithm thinking [16]. For example, imitating codes with similar functions allows students to discover small pieces of code, and adjusting a suitable algorithm can speed up problem-solving and reduce the computing power need for computers in CS course teaching. However, many Asia students are accustomed to accepting and following teachers' orders for learning and lack the ability and opportunity to manage their own learning time independently [17]. It is difficult for teachers to adequately describe the search for rules and conceptualization in the classroom. Nowadays, teachers and researchers are applying digital games as a potential tool for teaching computational thinking in the classroom [18]. Students can be guided to find solutions, cultivate their CT abilities, and enhance their learning performance in a digital game-based learning (DGBL) environment [19]. For example, students in the UK use a free online learning resource, a gaming platform, that explains all aspects of CT skills [15]. In addition, adopting digital games in CT training activities can promote students' creativity and stimulate their motivation [17]. Learners with different CT abilities will develop different strategies in the game to complete the game levels. Digital games facilitate students' ability to apply logical thinking, memory, visualization, and problem-solving skills in real life [18]. Therefore, we used two games, Rummikub and Robozzle, to compare the effectiveness of online and offline activities in developing CT.

3. Methods

3.1. Study Design and Participants

Our primary study goal was to determine the influence of online and offline activities on CT abilities among a group of seventh-grade students playing two web-based games. Study participants volunteered in a two-day computational thinking training camp in July 2021. Students in the online camp joined the activities via the Google Meet and Google Classroom applications and the offline group attended a face-to-face camp at the school. The camp activities were divided into three parts. The first part included a Bebras pre-test and game teaching for two hours. In this stage, students were guided on how to play Robozzle and Rummikub. Then, students had one-day of gameplay time with an accompanying teacher and instructions. A game competition was held after self-gameplay. The goals were to solve the most Robozzle puzzles and get the highest Rummikub scores. The last part included the Bebras post-test to test students' skills (Figure 3).

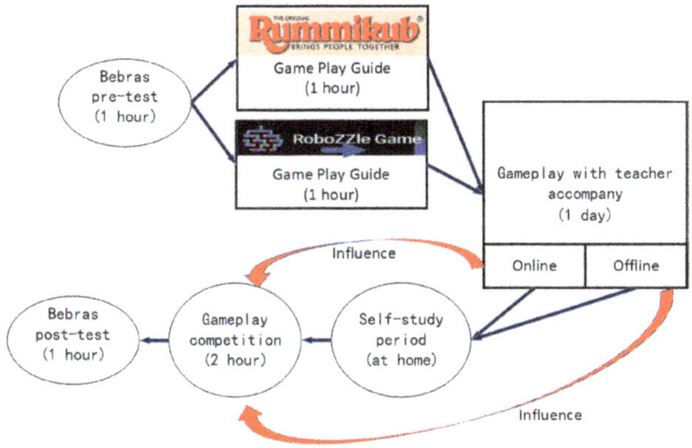

Figure 3. Online and offline training camp process.

We used the Bebras Challenge [20] to test students' computational thinking abilities in this study. The participant sample comprised 66 seventh-grade students from secondary high schools in Northern Taiwan. A total of 26 students participated in the online camp, while 40 participated in the offline camp. Half the people in each camp played Rummikub and half played Robozzle. Seven questions were selected from the seventh-grade question datasets, including easy, medium, and hard levels. The score range for the pre- and post-tests was 0–7.

3.2. *Exploration and Competiton Strategy*

Camp activities were held during holidays, so students participated in computational thinking training activities with their parents' consent or through voluntary registration. Self-exploration and competition strategies were applied to enhance learning motivation and encourage engagement in the activities. During the self-exploration stage, students could use the taught game skills to explore gaming environments and find any game tips. Students in the online group could use Google Meet to discuss solutions with other participants, while students in the offline group could only complete the paperwork specified by the activity. In addition, the competitive mode was applied to prompt the middle school students to devote more time and energy to mastering one job [21]. Students played against the computer in single-player mode for two games. Their ranking in the game was recorded, and the highest-ranked student was rewarded.

4. Results and Discussion

Students' CT ability was improved after the game-based learning camp activities (t = 2.96, $p = 0.032$). The post-test scores (M = 3.273, SD = 1.750) were significantly higher than the pre-test scores (M = 2.773, SD = 1.537) (Tables 1 and 2). To determine whether the online teaching method differs from the offline teaching method, we looked at descriptive statistics for both the online and offline groups and found that the mean of both groups improved (Table 3). After analyzing students' progress (differences in pre- and post-test scores) by using a paired-sample t-test (assuming unequal variance), we found no significant difference between the online and offline groups (t = 0.14, $p = 0.886$) (Table 4).

Table 1. Descriptive statistics of pre-test and post-test CT scores of all participants.

	N = 66	
	Mean	**S.D.**
Pre-test	2.773	1.537
Post-test	3.273	1.750

Table 2. Paired-sample *t*-test of pre-test and post-test scores of all participants (n = 66).

	df	t	*p*-Value
Overall	65	2.96 *	0.032

* $p < 0.05$.

Table 3. Descriptive statistics of pre-test and post-test scores of online and offline groups.

	Min	Max	Mean	S.D.
Online (N = 26)				
Pre-test	0	7	3.077	1.787
Post-test	0	7	3.538	1.985
Offline (N = 40)				
Pre-test	0	5	2.575	1.338
Post-test	0	6	3.1	1.582

Table 4. Paired-sample *t*-test of the pre-and post-test scores of online and offline groups (n = 66).

	df	t	*p*-Value
Overall	62	0.14	0.886

Next, we attempted to determine if any differences existed between the two games. We divided students into groups according to the type of game played and the method of teaching activities. Table 5. shows all students' progress and regression in performance by subtracting the post- and pre-test scores.

Table 5. Descriptive statistics of progress in CT performance in playing Rummikub and Robozzle.

	N	Min	Max	Mean	S.D.
Rummikub					
Online	13	−5	4	0.69	1.702
Offline	20	−4	5	1.15	2.300
Robozzle					
Online	13	−2	2	0.23	1.363
Offline	20	−3	3	−0.01	1.586

A paired-sample *t*-test was applied to detect differences between the same game with different teaching methods (Table 6), and the results show a significant difference when Rummikub was played in the offline mode. The progress in offline activity performance (M = 1.15, S.D. = 2.330) was better than that in online activity performance (M = 0.69, S.D. = 1.702). However, we found no significant difference between the online and offline modes for students playing Robozzle. Comparing the Rummikub and Robozzle games, the former involves group interactions and emphasizes peer interaction and communication in the learning process. Therefore, we can infer that if learning is carried out in groups or classes, the offline mode will be more effective than the online mode.

Table 6. Paired-sample *t*-test of pre- and post-test scores for different game-based learning environments.

	N	df	t	*p* Value
Rummikub				
Online	13	12	−0.61	0.553
Offline	20	19	−2.24	0.037 *
Robozzle				
Online	13	12	−1.47	0.168
Offline	20	19	0.28	0.781

* $p < 0.05$.

5. Conclusions

Comparing students' performance in the digital game camps with online and offline CT abilities, we found that students in both the online and offline groups played games on their computers, so the results were not significantly different. However, when analyzing Rummikub and Robozzle separately, the CT training outcomes for online and offline activities differed.

During the Rummikub multiplayer game, players must take turns taking cards and figuring out how to win the game. The opponents are nearby in the offline activity, so the students playing the game can further judge how to play their card by observing their opponent's voice, expressions, and even body language. In the online group, the cameras were set above the neck of the students, and they could only look at the computer screen, so students in this group had far less information than in the offline group. Furthermore, in offline activities, students can hear the voices of competitors from other groups, while in online games, students can only hear the voices of the people from their groups, so offline activities have a stronger sense of immersion than online activities. In addition, the player does not need to interact with other players, and the winner of the competition is determined by students' tanking in the Robozzle single-player game. Individual gameplay did not affect CT ability regardless of whether the students participated in online or offline activities.

The study results show that different teaching methods and gameplay methods will cause differences in students' CT ability performance. Therefore, when planning for online-game-based learning or developing an online learning platform, it is necessary to consider the gameplay type so that it is possible to produce learning outcomes and promote computational thinking abilities.

Author Contributions: Conceptualization, Y.-Y.C. and S.-M.Y.; methodology, L.-X.C. and Y.-Y.C.; validation, L.-X.C., S.-W.S. and Y.-Y.C.; formal analysis, Y.-Y.C. and C.-H.L.; data curation, L.-X.C. and Y.-Y.C.; writing—original draft preparation, L.-X.C. and S.-W.S.; writing—review and editing, L.-X.C. and C.-H.L.; visualization, L.-X.C. and Y.-Y.C.; supervision, S.-M.Y. All authors have read and agreed to the published version of the manuscript.

Funding: This work was partially funded by National Science and Technology Council Taiwan (grant number: 108-2511-H-009-009-MY3) and the High-level Talent Research Project at Fuzhou University of International Studies and Trade (grant no. FWKQJ201909).

Institutional Review Board Statement: Not applicable.

Informed Consent Statement: Not applicable.

Data Availability Statement: Data are contained within the article.

Acknowledgments: The authors wish to thank the blind reviewers for their insightful and constructive comments.

Conflicts of Interest: The authors declare no conflict of interest.

References

1. Hofer, S.I.; Nistor, N.; Scheibenzuber, C. Online teaching and learning in higher education: Lessons learned in crisis situations. *Comput. Hum. Behav.* **2021**, *121*, 106789. [CrossRef] [PubMed]
2. Su, S.W.; Jung, S.Y.; Yu, X.; Yuan, S.M.; Sun, C.T. Modify, Decompose and Reassemble: Learner-Centered Constructive Teaching Strategy for Introductory Programming Course in College. In Proceedings of the 2022 IEEE 5th Eurasian Conference on Educational Innovation (ECEI), Taipei, Taiwan, 10–12 February 2022; pp. 197–200. [CrossRef]
3. Saxena, A.; Lo, C.K.; Hew, K.F.; Wong, G.K.W. Designing unplugged and plugged activities to cultivate computational thinking: An exploratory study in early childhood education. *Asia-Pac. Educ. Res.* **2020**, *29*, 55–66. [CrossRef]
4. Durak, H.Y.; Yilmaz, F.G.K.; Yilmaz, R. Examining the Relationship between Digital Game Preferences and Computational Thinking Skills. *Contemp. Educ. Technol.* **2017**, *8*, 359–369.
5. Gee, J.P. Learning by design: Games as learning machines. *Digit. Educ. Rev.* **2004**, *8*, 15–23.
6. van Rijn, J.N.; Takes, F.W.; Vis, J.K. The complexity of Rummikub problems. *arXiv* **2016**, arXiv:1604.07553.

7. O'Hara, I. Feedback Loops: Algorithmic Authority, Emergent Biases, and Implications for Information Literacy. *Pa. Libr. Res. Pract.* **2021**, *9*, 8–15. [CrossRef]
8. Lindberg, R.S.; Laine, T.H.; Haaranen, L. Gamifying programming education in K-12: A review of programming curricula in seven countries and programming games. *Br. J. Educ. Technol.* **2019**, *50*, 1979–1995. [CrossRef]
9. Li, D. The Shift to Online Classes during the COVID-19 Pandemic: Benefits, Challenges, and Required Improvements from the Students' Perspective. *Electr. J. E-Learn.* **2022**, *20*, 1–18. [CrossRef]
10. Carretero, S.; Vuorikari, R.; Punie, Y. *DigComp 2.1: The Digital Competence Framework for Citizens with Eight Proficiency Levels and Examples of Use EUR*; Scientific and Technical Research Series; Publications Office of the European Union: Seville, Spain, 2017; Volume 28558.
11. Looi, C.-K.; How, M.-L.; Longkai, W.; Seow, P.; Liu, L. Analysis of linkages between an unplugged activity and the development of computational thinking. *Comput. Sci. Educ.* **2018**, *28*, 255–279. [CrossRef]
12. Lye, S.Y.; Koh, J.H.L. Review on teaching and learning of computational thinking through programming: What is next for K-12? *Comput. Hum. Behav.* **2014**, *41*, 51–61. [CrossRef]
13. Singh, S.; Rylander, D.H.; Mims, T.C. Efficiency of online vs. offline learning: A comparison of inputs and outcomes. *Int. J. Bus. Humanit. Technol.* **2012**, *2*, 93–98.
14. Hofer, S.I.; Holzberger, D.; Reiss, K. Evaluating school inspection effectiveness: A systematic research synthesis on 30 years of international research. *Stud. Educ. Eval.* **2020**, *65*, 100864. [CrossRef]
15. BBC. Introduction to Computational Thinking. Available online: https://www.bbc.co.uk/education/guides/zp92mp3/revision/1 (accessed on 1 December 2022).
16. Wing, J.M. Computational thinking. *Commun. ACM* **2006**, *49*, 33–35. [CrossRef]
17. Chen, C.; Stevenson, H.W. Motivation and mathematics achievement: A comparative study of Asian-American, Caucasian-American, and East Asian high school students. *Child Dev.* **1995**, *66*, 1215–1234. [CrossRef]
18. Ch'ng, S.I.; Low, Y.C.; Lee, Y.L.; Chia, W.C.; Yeong, L.S. Video games: A potential vehicle for teaching computational thinking. In *Computational Thinking Education*; Springer: Singapore, 2019; pp. 247–260.
19. Sun, C.-T.; Chen, L.-X.; Chu, H.-M. Associations among scaffold presentation, reward mechanisms and problem-solving behaviors in game play. *Comput. Educ.* **2018**, *119*, 95–111. [CrossRef]
20. Hsu, T.-C.; Chang, S.-C.; Hung, Y.-T. How to learn and how to teach computational thinking: Suggestions based on a review of the literature. *Comput. Educ.* **2018**, *126*, 296–310. [CrossRef]
21. Kim, B.; Park, H.; Baek, Y. Not just fun, but serious strategies: Using meta-cognitive strategies in game-based learning. *Comput. Educ.* **2009**, *52*, 800–810. [CrossRef]

engineering proceedings

MDPI

Proceeding Paper

Impact of Generative Artificial Intelligence on Footwear Design Concept and Ideation †

Shih-Hung Cheng

Department of Design, National Taiwan University of Science, Taipei City 106335, Taiwan; achille@mail.ntust.edu.tw

† Presented at the IEEE 5th Eurasia Conference on Biomedical Engineering, Healthcare and Sustainability, Tainan, Taiwan, 2–4 June 2023.

Abstract: The impact of Generative artificial intelligence (GAI) on footwear design creativity and feasibility was investigated in this study. Using a text-to-image GAI tool called Midjourney, 17 prompts were tested to generate footwear concepts. Ten distinct outcomes were selected from the results and evaluated by seven experts in footwear design. Prompt implementation correlated weakly positively with design creativity and feasibility. A set of prompts in the Japanese style showed significantly higher creativity due to explicit style descriptions. Sharp generative design features had higher creativity but lower feasibility. Concepts generated without subcategories induced lower feasibility and prompt implementation. These findings offer insights into GAI's role in footwear design innovation.

Keywords: AI-generated content; footwear; design ideation; expert evaluation

1. Introduction

Artificial intelligence (AI) has been rapidly adopted by the industry, revolutionizing various aspects of our lives. Generative artificial intelligence (GAI) has especially shown extensive and profound advancements and significantly impacted the global economy [1]. GAI generates new content of text, images, or audio, based on training data. Prominent examples of GAI include GPT-4 and Midjourney, both of which are widely adopted, impacting work and communication practices [2]. The development of AI traces back to 1956 when John McCarthy coined the term "artificial intelligence." In 1997, the supercomputer "Deep Blue" achieved a significant milestone by defeating the world champion chess player and establishing a clear benchmark in AI capabilities. More recently, in 2022, Midjourney, Inc. (San Francisco, CA, USA) released the Midjourney Discord bot. The refinement of large language models, such as ChatGPT, has been developed via various methods, including supervised learning, machine learning, and reinforcement learning. Moreover, latent text-to-image diffusion models such as Stable Diffusion and Midjourney have been widely applied in visual GAI software. Due to its capacity to process and analyze vast amounts of data quickly and accurately, AI is already being employed in diverse applications, employing natural language processing and autonomous vehicles. Its prevalence is expected to continue soaring in the coming years [3]. Machine learning is one of the key techniques used in AI. By learning from patterns and features in previous data, AI combines vast amounts of information with fast and iterative processing and intelligent algorithms to generate new content. In machine learning, AI is trained to identify patterns and make predictions based on existing data. By automating tasks that are currently handled by humans, AI allows for more time and resources for other activities that require human attention. In other words, its potential to enhance efficiency and productivity leads to cost savings, quicker delivery times, and improved overall performance [3]. This presents both opportunities and challenges across a wide range of professions and research areas, including education [4], healthcare [5], marketing [6], and transportation [7]. The application of GAI has been extensively explored in various design research, such as architectural

Citation: Cheng, S.-H. Impact of Generative Artificial Intelligence on Footwear Design Concept and Ideation. *Eng. Proc.* **2023**, *55*, 32. https://doi.org/10.3390/engproc2023055032

Academic Editors: Teen-Hang Meen, Kuei-Shu Hsu and Cheng-Fu Yang

Published: 30 November 2023

design [8], fashion design [9], household appliance design [10], and industrial design [11]. In particular, Taiwan has gained recognition for its advanced footwear manufacturing technology and thriving footwear industry cluster.

This study aims to delve into the utilization of GAI to harness the accumulated industrial and technological knowledge and its influence on practical footwear design and the innovation process and investigate how GAI effectively integrates the industrial innovation process for footwear design. To examine the potential advantages and disadvantages of such integration, the creative concept development process of footwear design was incorporated into Midjourney V4, a text-to-image GAI. Midjourney V4 was selected based on its specific ability to generate visual outputs from textual input. Furthermore, to gain valuable insights into the impact of integrating GAI into the design process, seven experts in footwear design were invited to evaluate the results from the incorporation of Midjourney V4. The evaluation focused on assessing the creativity and feasibility of the generated designs. By involving the experts in the assessment, a comprehensive understanding of the implications and potential benefits of applying text-to-image GAI was obtained in footwear design. In this study, the various advantages that text-to-image GAI offers were explored in footwear design. Additionally, the influence of GAI on the overall design thinking process was assessed with a particular emphasis on creative concept development. By investigating these aspects, the result of this study suggests novel possibilities and challenges in the application of GAI in the footwear industry.

2. Literature Review

2.1. Defamiliarization

In the research of GAI, designers have challenges of the authority of a technocratic elite by employing the technique of defamiliarization. This approach allows for the present design or artistic works that highlight the disparities between machinic decision-making and human intuition [12]. The concept of defamiliarization was introduced by Shklovsky, who suggested that artistic or literary works can portray familiar objects or situations in an unfamiliar manner, thus prolonging the perceptive process and offering a fresh perspective [13]. Shklovsky's examples were derived from literature showing how Tolstoy depicted ordinary things by describing them in intricate detail rather than using conventional names or signifiers. By avoiding precise nouns or conventional representations, these objects were presented as something distinct from well-known practices, prompting readers to think inductively rather than simply matching ideas with conventional concepts and archetypal images [14]. In particular, art removes objects from automatic perception to estrange common objects and present them in an unfamiliar light. This process slows down the reader's perceptual experience and deautomates their perception and restoring awareness of perpetual change [15].

2.2. Design Thinking and Black Box

Design solutions are proposed to address specific problems. In the exploration of potential solutions, creativity is consistently recognized as a crucial element. However, design experts tend to rely heavily on their intuition during the generation phase [16]. This intuitive and enigmatic aspect of the design process often leads to a misconception of design expertise [17]. As a result, significant research has been devoted to demystifying creative design [16]. The black box metaphor is frequently employed to illustrate the process of creative development because the inputs and outputs of the design process are sometimes unobservable. Furthermore, the black box metaphor implies that failures in the design and creativity process remain hidden [17]. By providing clear descriptions of design methods and processes, it becomes possible to unveil designers' private thinking and clarify the overall process. This enables other stakeholders to participate more consciously and rationally. These characteristics have a profound influence on designers, allowing them to adapt their working methods and thinking styles, with the Double Diamond model serving as a valuable reference [18].

The Double Diamond model, introduced by the British Design Council, represents a design process that incorporates both divergent and convergent thinking and forms cycles in an iterative development process. Divergent thinking enables designers to explore an issue more extensively or deeply, while convergent thinking involves taking focused action. The Double Diamond model consists of four distinct phases that shape the iterative process, namely Discover, Define, Develop, and Deliver [19,20]. The first phase, Discover, marks the initial stage of divergent thinking in the Double Diamond model. It allows designers to deeply understand the problem by engaging with stakeholders affected by the issues. The insights gathered from the discovery phase then assist designers in defining the challenge from a fresh perspective, completing the first diamond. Moving on to the second diamond, the Development phase encourages designers to seek inspiration and explore diverse possibilities. In this phase, different solutions and ideas are generated. Lastly, the Delivery phase prototypes and tests various solutions for designers to identify and reject those that are not suitable or effective [19]. Experienced designers often apply additional constraints during the solution generation phase to refine the solution and facilitate the generation of viable concepts. Throughout such a design process, designers modify goals and adjust constraints for a deeper understanding of the problem and progress in defining the solution. Despite these changes, designers endeavor to maintain their major solution concept for as long as possible. The purpose of these adjustments is to overcome challenges that emerge during the design process [16].

3. Method

3.1. Concept Ideation with Midjourney

The impact of GAI was explored on the application of both conventional and unfamiliar categories in the image generation process using Midjourney. Specifically, it investigated how GAI influenced the generation of different types of footwear designs. It was assumed that the existing subcategories of footwear were considered as known archetypal images while defining unfamiliar shoe styles, such as "protective devices for the feet", which required a careful redefinition of component attributes. A lack of archetypal definitions increases conceptual creativity and flexibility but entails greater risks. On the other hand, explicit subcategory definitions may reduce the risk in design thinking but potentially limit creativity and innovativeness. To test these hypotheses, seventeen sets of footwear design proposals were generated using prompts with obviously different word counts and image styles. After decreasing homogeneity and similarity, ten sets of outcomes were selected for expert evaluation. Table 1 provides a summary of ten prompts for Midjourney V4. Then, all ten sets of results generated by Midjourney are shown in Figures 1–5.

Table 1. Ten sets of prompts for Midjourney V4.

No.	Subcategories (Image Style)	Word Count of Prompts
1	Soccer shoe (Chelsea style)	10
2	Women's sporty sneaker (Futuristic style)	15
3	Football shoe (Mutation style)	19
4	Police tactical boots (Geometric Style)	38
5	Running shoe (Geometric Style)	73
6	Lady running shoes (Futuristic style)	34
7	Children's shoes (Vivian Westwood style)	59
8	Children's rain boots (Vivian Westwood style)	54
9	High-heeled shoes (Japanese kimono-style)	45
10	Protective devices for the feet (No style)	220

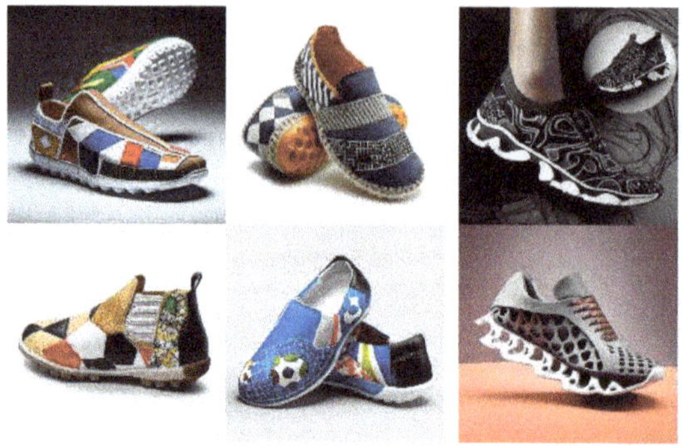

Figure 1. Generative results of soccer shoes with Chelsea style (**left**) and women sporty sneakers with futuristic style (**right**).

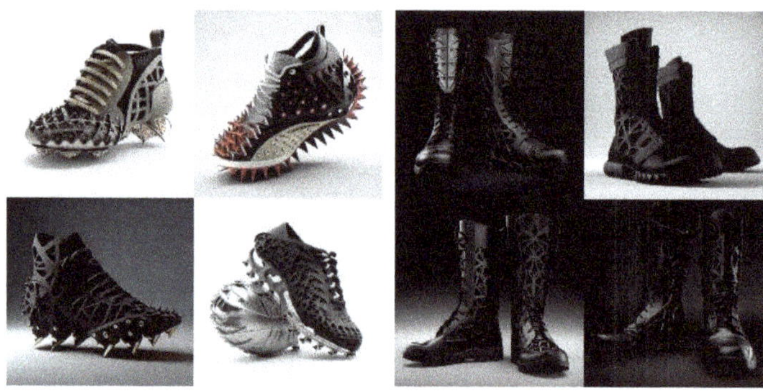

Figure 2. Generative results of football shoes with mutation style (**left**) and police tactical boots with geometric style (**right**).

Figure 3. Generative results of running shoes with geometric style (**left**) and lady running shoes with futuristic style (**right**).

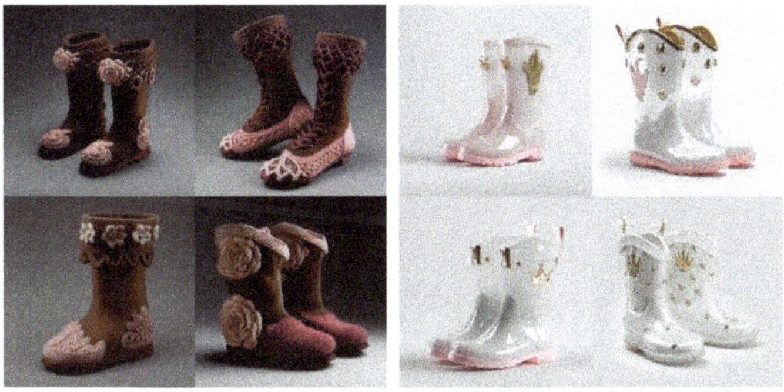

Figure 4. Generative results of children's shoes with Vivian Westwood style (**left**) and children's rain boots with Vivian Westwood style (**right**).

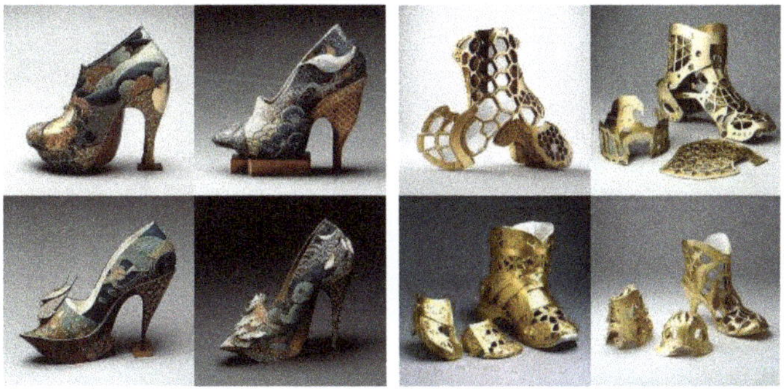

Figure 5. Generative results of high-heeled shoes with Japanese kimono-style (**left**) and protective devices for the feet without definition of style (**right**).

3.2. Expert Evaluation

Seven experts in footwear design evaluated and examined the ten sets of conceptual designs using explanations in English and their Chinese translations. The 11-point Likert scale (ranging from 0 to 10, where 0 represented the least and 10 represented the most) was applied to subjectively assess the results in terms of prompt implementation, design creativity, and practical feasibility. Additionally, these ten experts described three aspects of the outcomes qualitatively to articulate professional insights. Table 2 presents the profile of the design experts who participated in the evaluation process.

Table 2. Profile of participating design experts.

No.	Gender	Age	Years of Experience
E1	Woman	34	10
E2	Woman	25	1
E3	Woman	37	12
E4	Man	35	6
E5	Woman	30	7.5
E6	Woman	23	1
E7	Woman	27	1

4. Results

4.1. Quantitative Analysis

In the descriptive statistical analysis, the sixth and ninth sets of generative results received higher scores in all three aspects. The tenth set received the lowest scores in prompt implementation. The seventh set scored the lowest in creativity, and the third set scored the lowest in feasibility (Figure 6). The results of ANOVA indicated significant differences in prompt implementation (F [9, 69] = 2.165, $p = 0.037$), creativity (F [9, 69] = 4.017, $p = 0.000$), and feasibility (F [9, 69] = 3.598, $p = 0.001$). The post hoc test result (Duncan) for prompt implementation revealed a significant difference, with the 9th set (7.86) scoring significantly higher than the 3rd set (5.29), 7th set (5.00), 4th set (4.71), 5th set (4.71), and 10th set (4.57). In creativity, the post hoc test (Duncan) indicated that both the 3rd set (7.71) and 9th set (7.71) scored significantly higher than the 4th set (4.43) and 7th set (3.71). For feasibility, the 8th set (8.00), 6th set (7.43), and 4th set (7.29) scored significantly higher than the 10th set (3.86) and 3rd set (3.57).

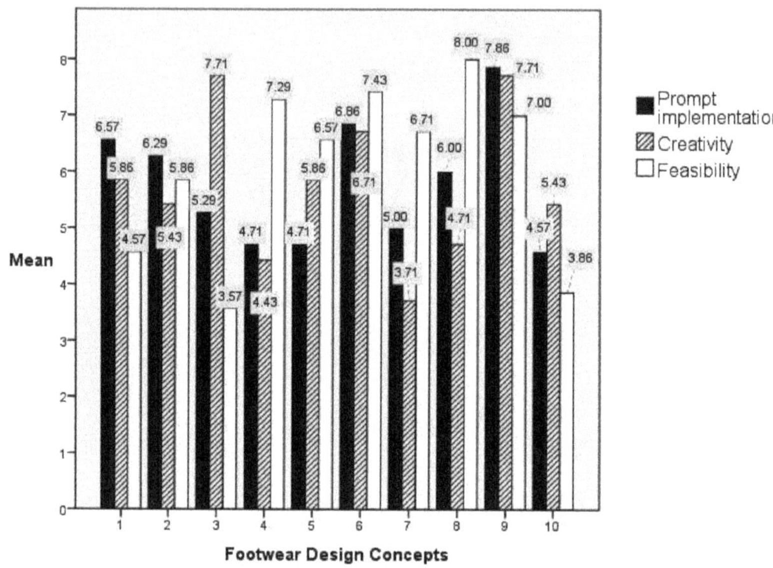

Figure 6. Average evaluation scores of concept ideas by design experts.

4.2. Qualitative Analysis

Qualitative assessments were conducted for prompt implementation, design creativity, and practical feasibility by the experts. The summary of the qualitative assessment regarding the implementation of prompts is presented in Tables 3–5.

Table 3. Key assessment for prompt implementation.

Expert	Ideation	Comments
E1	9th	I found this set of results to be highly artistic, with almost all the prompts successfully fulfilled, except for the generation of fish scales.
E4	9th	The Japanese style is evident in the presentation.
E5	9th	The implementation rate of prompts is not high, and many conditions interfere with each other.
E3	10th	The implementation rate of prompts is not high, and many conditions interfere with each other.
E5	10th	The settings and conditions in the prompts are specific and detailed, but GAI lacks integration capability, essentially only fulfilling the first two prompts.

Table 4. Key assessment for creativity.

Expert	Ideation	Comments
E1	9th	GAI doesn't excel as much in the creativity of footwear design, perhaps because it analyzes big data using the shoe's original stereotypes and archetypes to determine the shape the shoe should be.
E3	9th	GAI can combine difficult-to-imagine style features to generate designs rapidly.
E2	3rd	I find these computation results very innovative and creative! However, the design of this footwear makes it challenging to convey the elegance of women's high-top shoes.
E2	4th	In terms of creativity, I believe these results are rather conservative, resembling the boots available in the market. Therefore, from a development standpoint, they seem quite reasonable.
E6	4th	These are relatively common among AI-generated results, but I believe they would be more widely accepted. Moreover, the development process should be easier compared to other results.
E4	7th	The prompts are more specific, resulting in a lower level of creativity.

Table 5. Key assessment for feasibility.

Expert	Ideation	Comments
E6	8th	I personally find it quite attractive, and the development process shouldn't be difficult either. This design has a certain charm that makes you want to buy it and keep it as a collectible.
E4	8th	The shiny polyethylene and transparent materials are presented reasonably.
E2	6th	Most sock-style sneakers tend to enhance the stability of the heel. In this set of results, I noticed that both the upper and the heel counter are supported by TPU or leather materials.
E5	10th	Regarding setting prompts, I recommend using no more than three design conditions for design generation and then using them as inspiration for the design.
E1	3rd	It has a visual impact but lacks consideration for wearability.
E4	3rd	It is challenging for GAI to simultaneously assess the practical feasibility of both materials and craftsmanship in the development process.

5. Conclusions

The application of GAI in footwear design was evaluated from three perspectives. The results of the quantitative analysis showed that footwear designs referencing particular cultural styles demonstrated creativity and prompt implementation, especially in the case of the ninth set of results. However, deliberate avoidance of the specification of subcategories for footwear and refraining from incorporating specific cultural styles hindered the high integration of design definitions as suggested by prompts. In design creativity, the results exhibited a negative correlation between creativity and feasibility. For instance, the fourth, seventh, and eighth sets lacked creativity due to their excessive feasibility and similarity to existing products. Moreover, the majority of low feasibility resulted from shattered designs or sharp features that conflicted with comfort requirements, resulting in relatively low feasibility. Expert evaluations also indicated that GAI lacked integration, particularly when design conditions interfered with each other. Additionally, GAI's prediction of material wearability during usage still requires improvement. These are areas where GAI can potentially enhance its design applications in the future.

Funding: This research was funded by the National Science and Technology Council of Taiwan grant number NSTC 111-2410-H-011-042.

Institutional Review Board Statement: The study was conducted in accordance with the Declaration of Helsinki, and approved by the Research Ethics Committee for Human Subject Protection, National Yang Ming Chiao Tung University (protocol code NYCU-REC-111-056E and date of approval: 19 July 2022).

Informed Consent Statement: Informed consent was obtained from all subjects involved in the study.

Data Availability Statement: Data are contained within the article.

Conflicts of Interest: The author declares no conflict of interest. The funders had no role in the design of the study; in the collection, analyses, or interpretation of data; in the writing of the manuscript; or in the decision to publish the results.

References

1. Generative AI Could Raise Global GDP by 7%. Available online: https://www.goldmansachs.com/insights/pages/generative-ai-could-raise-global-gdp-by-7-percent.html (accessed on 19 July 2023).
2. Feuerriegel, S.; Hartmann, J.; Janiesch, C.; Zschech, P. Generative AI. *Bus. Inf. Syst. Eng.* **2023**, 1–16. [CrossRef]
3. Boymamatovich, S.M. Exploring the Benefits and Future of Artificial Intelligence. *Cent. Asian J. Theor. Appl. Sci.* **2023**, *4*, 108–113.
4. Hwang, G.J.; Chen, N.S. Editorial Position Paper: Exploring the Potential of Generative Artificial Intelligence in Education: Applications, Challenges, and Future Research Directions. *Educ. Technol. Soc.* **2023**, *26*, i–xviii.
5. Huang, K.; Fu, T.; Gao, W.; Zhao, Y.; Roohani, Y.; Leskovec, J.; Coley, C.W.; Xiao, C.; Zitnik, M. Artificial intelligence foundation for therapeutic science. *Nat. Chem. Biol.* **2022**, *18*, 1033–1036. [CrossRef]
6. Nesterenko, V.; Olefirenko, O. The impact of AI development on the development of marketing communications. *Mark. Manag. Innov.* **2023**, *1*, 169–181. [CrossRef]
7. Wang, X.; Wang, D.; Chen, L.; Lin, Y. Building Transportation Foundation Model via Generative Graph Transformer. *arXiv* **2023**, arXiv:2305.14826, 2023.
8. Jaruga-Rozdolska, A. Artificial intelligence as part of future practices in the architect's work: MidJourney generative tool as part of a process of creating an architectural form. *Architectus* **2022**, *3*, 95–104.
9. Yan, H.; Zhang, H.; Liu, L.; Zhou, D.; Xu, X.; Zhang, Z.; Yan, S. Toward intelligent design: An ai-based fashion designer using generative adversarial networks aided by sketch and rendering generators. *IEEE Trans. Multimed.* **2023**, *25*, 2323–2338. [CrossRef]
10. Fang, Y.M. The Role of Generative AI in Industrial Design: Enhancing the Design Process and Learning. In Proceedings of the IEEE conference proceedings: 2023 9th International Conference on Applied System Innovation (ICASI), Chiba, Japan, 21–25 April 2023.
11. Vartiainen, H.; Tedre, M. Using artificial intelligence in craft education: Crafting with text-to-image generative models. *Digit. Creat.* **2023**, *34*, 1–21. [CrossRef]
12. Curry, D. Artistic defamiliarization in the Age of Algorithmic prediction. *Leonardo* **2023**, *56*, 177–182. [CrossRef]
13. Gunn, D.P. Making art strange: A commentary on defamiliarization. *Ga. Rev.* **1984**, *38*, 25–33.
14. Shklovsky, V. Art as technique. In *Contemporary Literary Criticism: Modernism through Poststructuralism*; Con Davis, R., Ed.; Longman Press: New York, NY, USA; London, UK, 1986.
15. Bell, G.; Blythe, M.; Sengers, P. Making by making strange: Defamiliarization and the design of domestic technologies. *ACM Trans. Comput.-Hum. Interact. (TOCHI)* **2005**, *12*, 149–173. [CrossRef]
16. Cross, N. Research in Design Thinking. In *Research in Design Thinking*; Cross, N., Dorst, K., Roozenburg, N., Eds.; Delft University Press: Delft, The Netherlands, 1992; pp. 3–10.
17. Gulari, M.N. Metaphors in design: How we think of design expertise. *J. Res. Pract.* **2015**, *11*, M8.
18. Kochanowska, M.; Gagliardi, W.R. The double diamond model: In pursuit of simplicity and flexibility. In *Perspectives on Design II: Research, Education and Practice*; Raposo, D., Neves, J., Silva, J., Eds.; Springer: Cham, Switzerland, 2022; pp. 19–32.
19. The Double Diamond. Available online: https://www.designcouncil.org.uk/our-resources/the-double-diamond/ (accessed on 19 July 2023).
20. Design Counci, "History of the Double Diamond". Available online: https://www.designcouncil.org.uk/our-resources/the-double-diamond/history-of-the-double-diamond/ (accessed on 19 July 2023).

engineering proceedings

Proceeding Paper

Following the Leader's Innovation or Leading Follower's Innovation: Taiwan's Banking Industry Analysis †

Jen-Ying Shih * and Tsai-Hsiu Lin

Graduate Institute of Global Business and Strategy, National Taiwan Normal University, Taipei 10610, Taiwan; b60202brian@gmail.com

* Correspondence: jyshih@ntnu.edu.tw

† Presented at the IEEE 5th Eurasia Conference on Biomedical Engineering, Healthcare and Sustainability, Tainan, Taiwan, 2–4 June 2023.

Abstract: In response to the rapid development of the global economy and FinTech innovation, Taiwan's government has continuously promoted digital transformation policies for the financial industry and encouraged the industry to strengthen infrastructure constructions and upgrade innovative financial services and patents. Novel ideas are proposed to create a new situation and make banks invincible but it is questionable if such innovations necessarily bring competitiveness or business performance. To answer this question, we analyzed market competition using the PR test and data envelopment analysis (DEA) to study the operational efficiency of banks. K-means were also used to segment banks into three groups, Leader, Chaser, and Laggard. The data included financial (business performance) and patent data (innovation) of Taiwan's banks from 2013 to 2020. The research results revealed that the market competition in the Leader group was the most intense. Quasi-public banks were relatively inefficient in creating revenue, while private banks were more efficient. The Chaser group showed the most apparent changes in operational efficiency from 2013 to 2020. The banks in the Laggard group needed to strengthen the relative efficiency of revenue, and financial innovation needed to be individualized in this competitive market group.

Keywords: financial innovation; market competition; relative efficiency; DEA

Citation: Shih, J.-Y.; Lin, T.-H. Following the Leader's Innovation or Leading Follower's Innovation: Taiwan's Banking Industry Analysis. *Eng. Proc.* **2023**, *55*, 33. https://doi.org/10.3390/engproc2023055033

Academic Editors: Teen-Hang Meen, Kuei-Shu Hsu and Cheng-Fu Yang

Published: 30 November 2023

1. Introduction

Is innovation always good for competitiveness or business performance? Two cases were studied to answer this question. The Shanghai Commercial and Savings Bank (SCSB) was the first bank in Taiwan to offer 24 h service of automatic teller machines (ATM) and cooperate with convenience stores for the service. SCSB was undoubtedly a pioneer in the banking industry at that time. However, under fierce competition, the SCSB's ATM occupancy rate of convenience stores in recent years has decreased due to large consumer banks' marketing. E. Sun Bank once wanted to launch electronic money but as various online payments in Taiwan have become popular, they gave up the project. After adjusting the innovation strategy to the target market, they introduced mobile payment in the e-commerce market combined with cross-border transactions as the services are used by people. Although it is not the most advanced technology, it meets the expectations of consumers.

The above two cases showed that the use of the newest innovative technology for services may not be enough to outperform the competitors. The service needs to meet the demand of the target customer to obtain a competitive edge. Novel ideas sometimes create a situation for banks to remain competitive but it is unclear when individual banks face an unsolvable situation and what they need to do for it. It is also questionable how much effort needs to be put into the development of the service and how long it would take.

In response to the rapid development of the global economy and financial technology, Taiwan's government has continuously promoted digital transformation in the financial industry and encouraged the industry to build infrastructures and provide innovative financial services. However, the banks always keep an eye on what competitors are doing. Innovation cannot be achieved by following what others do. The authority is vigorously promoting financial innovation and digital transformation to enhance the global competitiveness of Taiwan's financial industry.

Therefore, it is required to provide useful information in developing innovation strategies considering any difference in the intensity of market competition of banks according to their size and the promotion of financial patents to enhance competitiveness. In this study, we analyzed the degree of competition in Taiwan's financial market using the data envelopment analysis (DEA) and the PR test for the financial data of Taiwan's banking industry and the government's financial data [1]. The efficiency of the operation of banks of different sizes was evaluated.

The organization of this article is as follows. After introducing the research background and purpose in Section 1, Section 2 reviews the related literature regarding financial innovation, market competition, and business performance evaluation. Section 3 describes the data set and analytical models. Finally, the research results and conclusion are presented in Sections 4 and 5.

2. Literature Review

2.1. Financial Innovation

There is no consistent standard for financial innovation. Most research extended Schumpeter's viewpoint on innovation [2] to define financial innovation. Reference [2] claimed that innovation was the driving force for the advancement of society. Innovation was to constantly break the old framework through a cycle of continuous reorganization among new products, methods, markets, resources, and organizations. Reference [3] mentioned that financial innovation was to provide new products, new transaction models, new organizational behaviors, or new services to meet the needs of financial market participants. Financial innovation included novel financial technologies or ubiquitous equipment such as automatic teller machines (ATMs). Any change in financial-related participants, products, services, and processes can be called innovation. It is a financial phenomenon that reflects market changes in satisfying consumer demand or increasing suppliers' profits.

Compared with innovation, the patent system has cons and pros. References [4,5] mentioned that because financial innovation was relatively easy to imitate for a new financial product, competitors could launch similar ones in the market to obtain short-term benefits. It is difficult for financial institutions to use this innovation to obtain higher returns. Reference [6] argued that the emergence of the patent system encouraged and protected the benefits brought by innovation. The patent system is open and transparent, and provides better rewards for innovations that are easy to copy but costly to develop. However, the innovation is incremental or requires multiple different combinations of technology. Benefits and protection of the patent may suffer losses due to early disclosure of information.

Taiwan's financial regulatory authorities vigorously promote the transformation of banks and financial patents. More researchers have paid attention to the quantitative data of financial patents for their analytical value. Taiwan's central bank pointed out that if financial institutions implement the patent system, the system becomes more suitable for financial innovation.

Reference [7] studied consumer and corporate finance banks from 1994 to 2007 in Taiwan's financial environment. The innovation of technological services by corporate finance banks has not been effective for financial performance. However, the innovation deployed by consumer finance banks has effectively increased operating income and market share. Reference [8] showed that the banking industry gradually reduced its attention to financial innovation because the intensity of market competition increased, which let the degree of market competition and financial innovation show a negative correlation from

2003 to 2011 in Taiwan. It coincided with Schumpeter's theory. The financial innovation of the leader can create profits but competition and imitation may weaken leaders' willingness to re-innovate. Reference [9] suggested that banks appropriately increased employee salaries to stimulate innovation activities and create more profits. Then, better profits could allow banks to have more resources for financial innovation. Positive feedback could be formed within the enterprise.

In terms of the external environment, it was mentioned that financial innovation improved the efficiency of the public's use of financial services and the enhancement of this momentum stimulated the invention and popularization of new technologies [10]. Conversely, changes in demand caused by technological progress and economic growth catalyzed financial innovation. Therefore, financial innovation, technological progress, and economic growth have a positive relationship. Based on the above arguments, investment in financial innovation is beneficial to the public in terms of convenience and efficiency. However, the revenue and competitiveness of banks are related to how many resources a bank needs to invest, which is the main concern of this study.

2.2. Market Competition and PR Test

For the impact of innovation, the degree of overall market competition is considered first. Two major theories were reviewed in this study to discuss innovation and market competition. One is the "Schumpeter Theory" which explains that innovative products disturb the equilibrium of a competitive market and allow gaining a short-lived monopoly [2]. In this theory, the innovator tries their best to prolong the monopoly, while the opponents try to break the innovator's monopoly through improvement or imitation. Such a competitive relationship weakens the innovator's willingness to re-innovate. The other is the "competition avoidance hypothesis". Although competition causes companies lagging far behind to stop innovating, it still helps companies compete with each other and accelerate innovation because companies hope to dominate the market and lead competitors by a large margin through innovation to avoid competition [11].

The PR test [1] was used to observe the degree of market competition in the non-structural analysis by calculating the relevant degree of market competition by the H statistical value [9]. $H \leq 0$ means monopoly, $0 < H < 1$ means monopolistic competition, and $H = 1$ means perfect competition. The PR test has been widely used to evaluate competitive behavior [12]. Reference [13] used linear regression to obtain the value of H statistics. However, it also has limitations and inconveniences. For example, to use the PR test method, it must be confirmed that the overall market is in a long-term equilibrium because a whole set of sample data is what is the only way to properly calculate H statistical value. Therefore, it is more suitable to calculate the degree of market competition within a certain range but, in this case, the individual competitiveness of a single case is not examined. The degree of market competition in Asian countries from 1994 to 2008 showed perfect or exclusive competition in the banking industry [14].

2.3. Business Performance Evaluation with DEA

The traditional profit sources of the banking industry are interest income and non-interest income. Reference [15] proposed that banks with more deposits have higher risks and therefore need higher net interest margins. Therefore, the correlation between deposit scale and net interest profit margin is positive. However, smaller banks have higher profit margins than large banks. In the non-interest income of Taiwan's banking industry, the higher the non-interest income, the lower the relative risk value [16].

For the evaluation of business performance with innovation, indicators of financial risks are used. For example, the current Basel III regulations provide a risk reference indicator for financial institutions. Reference [17] studied the relevant regulations of Basel III and pointed out that if the regulations of Basel III comply with the prediction of the bank's Z-Score and return on asset (ROA), the bank's default risk is reduced during the crisis. In recent years, Taiwan has referred to the capital adequacy ratio defined by Basel

III when measuring Domestic Systemically Important Banks (D-SIBs). Therefore, for the operating performance and profitability of various banks, the capital adequacy ratio needs to be considered as a relevant research factor.

Pareto proposed the idea of the non-dominance solution. Although apples and oranges cannot be compared, two oranges and three apples are better than one orange and two apples, This non-overriding solution is also called Pareto optimality. DEA adopts the concept of Pareto optimality as a mathematical model derived from the concept of technical efficiency evaluation. Reference [18] constructed a Charnes, Cooper, and Rhodes (CCR) model with various inputs and outputs assuming a fixed scale return (Constant Return to Scale (CRS)). The technical efficiency value model called the Banker, Chames, and Cooper (BCC) model was proposed assuming Variable Return to Scale (VRS) [19]. Most relevant studies were based on the expansion of the two models. Most of the DEA analysis tools provide improvement targets among variables. The most important information of the DEA model is to provide the relative efficiency of input and output factors and the most favorable results under the objective environment. The DEA model is widely used for research on the innovation and technology of the financial industry. For example, the factors affecting the technical efficiency of the US banking industry were studied using the model [20], and the result showed that the scale of the bank was positively correlated with the efficiency value.

Regardless of the businesses of the company, the core value is judged by the public based on its operating performance. Innovation is important in a competitive market. Market competitiveness and business performance are indicators of the innovation of new products, services, and models. Most of the previous studies analyzed the technical efficiency and performance of each bank in the market or its scale. The size of the banks varies all the time. Big banks are the leaders in the industry. To prevent the followers from having the opportunity to surpass, they use various innovative methods by cooperating with the government regarding government policies or securing unique customer bases. When discussing the technological innovation of the banking industry, it is appropriate to study different groups separately to observe what role banks of different sizes need to play for the competition.

3. Research Method

3.1. Data Set

We obtained the financial data from Taiwan banking companies from 2013 to 2020 in the Taiwan Economic Journal (TEJ) database and the patent information on the banking industry from the Intellectual Property Bureau of the Ministry of Economic Affairs. Because Taiwan's banking industry is subject to government supervision, the Bank of Taiwan (BOT) is responsible for cooperating with the government, and when observing the total assets of 34 banks in 2013, the assets of BOT far exceeded those of others. Therefore, we excluded BOT. A total of 33 banking companies were selected in this study.

3.2. Segment of Banks by Scale

Based on total assets and net income in 2013, we clustered 33 banks into three groups using K-means as shown in Table 1. The banks in the leader group had a large asset size and net income.

Table 1. Three bank groups in 2013.

	Laggard	Leader	Chaser
Number of banks	19	7	7
Average Total Assets (NT$ 1 m)	391,794	2,376,754	1,441,082
Average Net Income (NT$ 1 m)	8258	38,405	26,389

Data source: TEJ database.

3.3. *Financial Patents*

The number of financial patent applications in Taiwan's banking industry was scarce before 2016, but with the development and promotion of financial technology, it has increased since 2017. The number of patents of 33 banks has increased from 158 in 2017 to more than 500 pieces per year in 2020 (Figure 1). For the types of patent applications (code of the international patent classification (IPC)), Figure 2 shows the bar chart of the patents applied for by the 33 banks. G06Q40 (finance and insurance) is the largest, followed by G06Q20 (payment solutions), G06Q10 (administrative management) and G06Q30 (e-commerce). Most patents filed by Taiwan's banking industry focused on the application of financial technology.

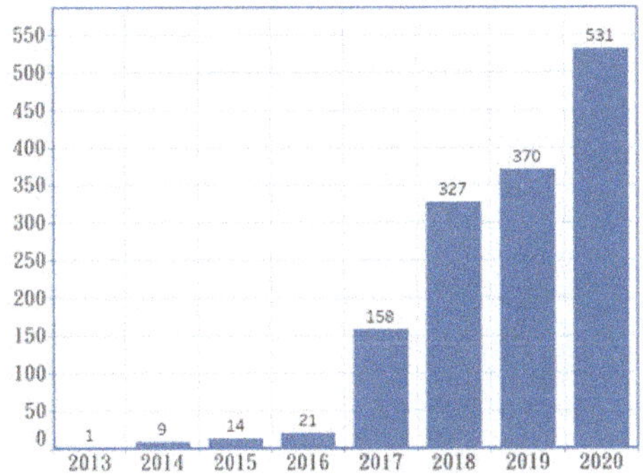

Figure 1. Number of patents publicly announced by 33 banks.

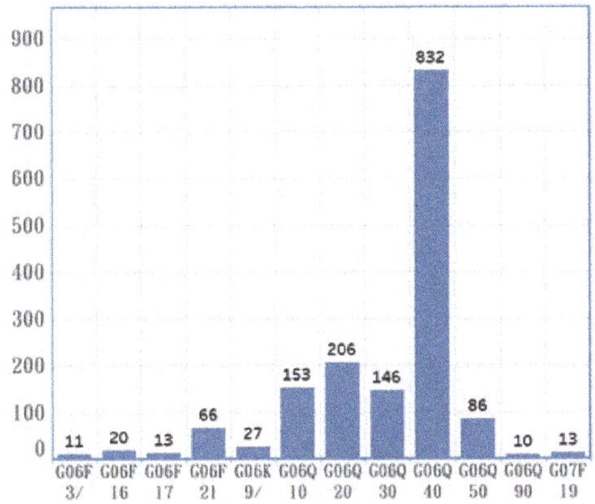

Figure 2. Bar chart of financial patents according to the IPC classification code of 33 banks from 2013 to 2020 (https://topic.tipo.gov.tw/patents-tw/sp-ipcq-full-101.html (accessed on 30 March 2023)).

3.4. *Market Competition by PR Test*

To construct a PR Model, we set Net Revenue as a target variable and labor price (PL), capital price (PC), and price of fund (PF) as input variables. The definition of the PR Model is shown in Table 2.

Table 2. Operational definition of variables in the PR model.

Variable	Variable Role	Variable Description
lnTR	Target Variable	*ln* (Net Revenue)
lnPL	PL	*ln* (Employee salary and benefits/(Total Assets-Total Fixed Assets))
lnPC	PC	*ln* (the other administration and operating expense/Total Fixed Assets)
lnPF	PF	*ln* (Interest expense/Deposit amount)
lnTA	Control Variable	*ln* (Total Assets)
lnBIS	Control Variable	*ln* (Capital adequacy ratio)
lnPT	Control Variable	*ln* (Effective number of patents)
lnLR	Control Variable	*ln* (Overdue loan ratio)

The relationship between market competition and innovation is presented in the PR Model (Equation (1)). H statistics are determined by Equation (2).

$$lnTR = \alpha + \beta_1 lnPL + \beta_2 lnPC + \beta_3 lnPF + \gamma_1 lnTA + \gamma_2 lnBIS + \gamma_3 lnPT + \gamma_4 lnLR, \quad (1)$$

$$H\ statistics = \beta_1 + \beta_2 + \beta_3, \quad (2)$$

where β_1, β_2, and β_3 are estimated from Equation (1).

3.5. Input and Output of DEA Model

In the analysis of relative cost efficiency, the input and output are as follows.

- Input: operating expenses, net self-owned capital, and total deposits.
- Output: total risk assets and total loans.

Output is at a fixed level, and input is relative efficiency.
The relative revenue efficiency analysis factors are as follows.

- Input: total risky assets, total loans, interest expenses.
- Output: fee income, interest income.

When the input is assumed to be fixed, the relative efficiency of output is obtained. Input-oriented efficiency (cost-relative efficiency) of the CCR model is represented by Equations (3)–(5), while the output-oriented efficiency (revenue-relative efficiency) is referred to Ref. [21].

$$Min\ H_k = \theta - \varepsilon\left(\sum_{i=1}^{m} S_i^- + \sum_{r=1}^{s} S_r^+\right), \quad (3)$$

$$s.t. \sum_{j=1}^{n} \lambda_j X_{ij} + S_i^- = \theta X_{ik}, \quad (4)$$

$$\sum_{j=1}^{n} \lambda_j Y_{rj} - S_r^+ = Y_{rk}, \quad (5)$$

$$\lambda_j,\ S_i^-,\ S_r^+ \geq 0,\ i = 1 \ldots m,\ j = 1 \ldots n,\ r = 1 \ldots s$$

H_k = relative efficiency value of the kth DMU.
$\theta = 1$ represents efficient.
λ_j= relative efficiency value of the jth DMU.
X_{ij} = the ith input of the jth DMU.
Y_{rj} = the rth output of the jth DMU.
S_i^- = the ith input slack.
S_r^+ = the rth output surplus.
n = the number of DMU.
m = the number of inputs.
s = the number of outputs.

4. Result and Discussion

4.1. PR Model Analysis

Table 3 shows the results of three bank groups during 2013–2016 and 2017–2020. The PR model adopted a regression method. The explanatory power of each group's model was above 80%. The degree of market competition of the Leader group was 0.9771 from 2013 to 2016, and 1.033 from 2017 to 2020, indicating a state of perfect competition. The degree of the Chaser group was 0.4319 from 2013 to 2016 and increased to 0.5456 from 2017 to 2020. The degree of the Laggard group was 0.1468 from 2013 to 2016 and 0.0942 from 2017 to 2020, which indicated that the degree of market competition was monopolistic. However, the PR model only calculated the H statistics within a group. Thus, it did not show the changes in the individual competitiveness and technology gap of a single bank. Therefore, the DEA model was used to explore the changes and relationships between the technical efficiency and business performance of each bank.

Table 3. Market competition analysis using H statistics.

Date Range	2013–2016 (Leader)	2017–2020 (Leader)	2013–2016 (Chaser)	2017–2020 (Chaser)	2013–2016 (Laggard)	2017–2020 (Laggard)
lnPC	0.3248 ***	0.1688 ***	−0.0487	−0.2494 *	−0.0034	0.0599 *
lnPF	−0.1129	0.1193 *	0.2049 *	0.022	−0.2136 ***	−0.0948
lnPL	0.7652 ***	0.7449 ***	0.2757 *	0.7731 *	0.3637 ***	0.1292
lnTA	0.8153 ***	0.78 ***	0.4459 ***	1.052 ***	0.9048 ***	0.9554 ***
lnBIS	0.156	0.5622 *	1.172 ***	0.8665 *	0.986 ***	1.1506 ***
lnPT	0.0125	0.002	−0.1798 *	−0.0692 *	0	−0.0167
lnLR	−0.001	0.1104 ***	−0.2355 ***	−0.22	−0.0385	−0.0338
H statistics	0.9771	1.033	0.4319	0.5456	0.1468	0.0942
R-square	0.9712	0.9848	0.8964	0.8452	0.9181	0.9287
Adjusted R-square	0.9611	0.9795	0.8601	0.791	0.9097	0.9214
Number of observations	28	28	28	28	76	76
Market Competition	High	High	Medium	Medium	Low	Low
Change in Competition	Increase		Increase		Decrease	

*** $p < 0.01$. * $0.01 < p < 0.1$.

4.2. CCR Model Results of Leader Group

Using the annual data from 2013 to 2020, the CCR model was used to obtain the input-oriented and the output-oriented efficiency (Figure 3). The relative cost efficiency and revenue efficiency of most banks in the Leader Group were above 75%. The relative cost efficiency of Cathay United Bank (Tick No: 5835) in 2020 was lower than 75%. The relative revenue efficiency of private banks was better than that of quasi-public banks. In 2013 and 2020, the Land Bank of Taiwan (Tick No: 5857) showed the best relative cost efficiency. In 2020, the gap between the relative cost efficiency of all banks and the Land Bank increased significantly. Most banks showed stable or slightly declined relative revenue efficiency. Only Taiwan Cooperative Bank (Tick No: 5854) increased efficiency by 5.50% from 79.20 to 84.70%.

Figure 4 shows that, since 2017, the number of financial patents increased significantly. In terms of the number of effective patents in the three bank groups, the Leader group and Chaser group accounted for 95%. We analyzed the patents of the two groups. According to the statistics, from 2017 to 2020, the number of patents by quasi-public banks was 981, and that by private banks was 334 (Figure 5). Table 4 shows the correlation coefficient between the number of effective patents and financial performance variables.

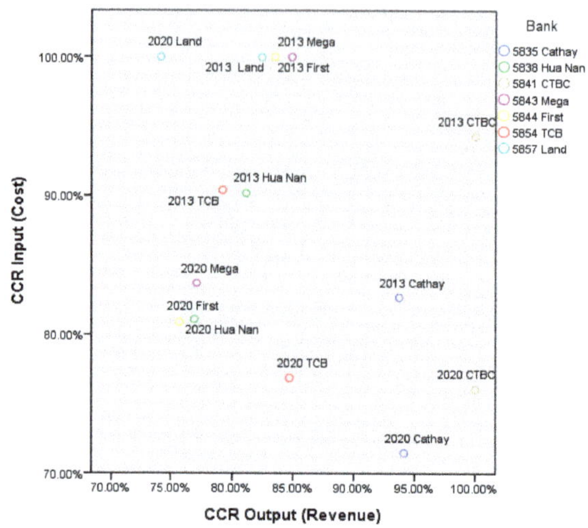

Figure 3. Scatter plot of CCR Model for Leader Group based on cost efficiency and revenue efficiency in 2013 and 2020.

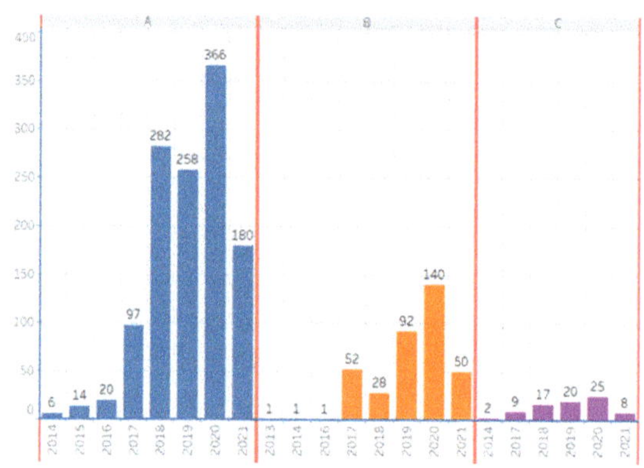

Figure 4. Number of effective patents of the three bank groups. (ps. Blue bars belong to Leader group; orange bars belong to Chaser group; purple bars belong to Lagger group).

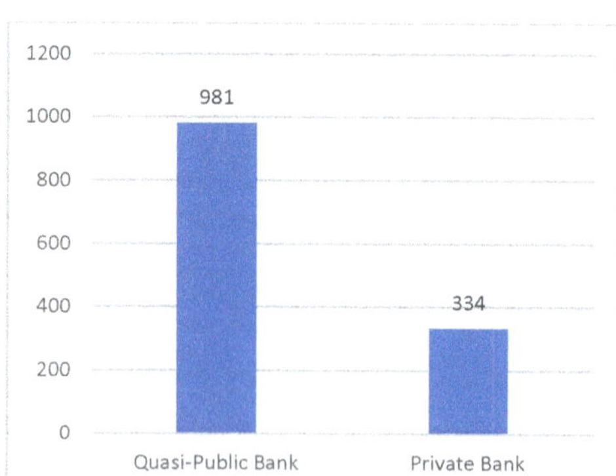

Figure 5. Statistics on number of invention and new model patents from 2017 to 2020. (Seven quasi-public banks).

Table 4. Correlation coefficient table between the number of effective patents and financial performance variables.

	Private Banks	Quasi-Public Banks
net self-owned capital	0.59	0.49
total deposits	0.66	0.5
total loans	0.66	0.46
fee income	0.59	0.55
interest income	0.57	0.31

5. Conclusions

Using financial innovation and efficiency analysis, we studied the current market competition in Taiwan's banking industry and the differences in the efficiency of their operations in different groups of banks. It was reviewed whether individual banks had to strengthen their investment in financial innovation or just follow in the footsteps of leaders. The analysis result of effective patents (Figure 5 and Table 4) showed that the correlation between the number of effective patents and financial performance variables of quasi-public banks was less significant than that of private banks. In cost and revenue efficiency (Figure 3), quasi-public banks showed lower values than private banks. The number of inventions and patents of quasi-public banks was about three times higher than that of private banks (Figure 5). The degree of market competition in the Lagger group showed monopolistic characteristics. They showed decreased efficiencies and less investment in patents. This unwillingness to invest in increasing patents and strengthening market competitiveness coincided with the "escape from competition hypothesis". The banks were losing their motivation to invest in innovation. The Leader group was in the phase of perfect competition. In the market competition of the Chaser group, the changes in the competition degree were rapid in the two periods. During the eight years, two banks had caught up with the leader group in terms of the size.

Author Contributions: Conceptualization, J.-Y.S. and T.-H.L.; methodology, J.-Y.S. and T.-H.L.; software, T.-H.L.; validation, J.-Y.S. and T.-H.L.; formal analysis, J.-Y.S. and T.-H.L.; data curation, T.-H.L.; writing—original draft preparation, J.-Y.S.; writing—review and editing, J.-Y.S.; project administration, J.-Y.S. All authors have read and agreed to the published version of the manuscript.

Funding: This research received no external funding.

Institutional Review Board Statement: Not applicable.

Informed Consent Statement: Not applicable.

Data Availability Statement: The data presented in this study are available on request from the corresponding author.

Conflicts of Interest: The authors declare no conflict of interest.

References

1. Panzar, J.C.; Rosse, J.N. Testing for "monopoly" equilibrium. *J. Ind. Econ.* **1987**, *35*, 443–456. [CrossRef]
2. Schumpeter, J.A. *Capitalism, Socialism and Democracy*; Routledge: London, UK, 2013.
3. Frame, W.S.; White, L.J. Empirical studies of financial innovation: Lots of talk, little action? *J. Econ. Lit.* **2004**, *42*, 116–144. [CrossRef]
4. Lerner, J. Where does State Street lead? A first look at finance patents, 1971 to 2000. *J. Financ.* **2002**, *57*, 901–930. [CrossRef]
5. Tufano, P. Financial innovation and first-mover advantages. *J. Financ. Econ.* **1989**, *25*, 213–240. [CrossRef]
6. Hall, B.H. *Business Method Patents, Innovation, and Policy*; National Bureau of Economic Research: Cambridge, MA, USA, 2003.
7. Yao, G.-Z. *A Study of Technology Service Innovation and Performance in Taiwanese Banking Industry*; National Taiwan University: Taipei, Taiwan, 2008.
8. Lu, P.-S. *Market Concentration, Competition and Financial Innovation in Taiwanese Banking Services Industry*; National Taipei University of Technology: Taipei, Taiwan, 2013.
9. Central Bank Quarterly. Available online: https://www.cbc.gov.tw/tw/public/Attachment/48201659571.pdf (accessed on 21 July 2023).

10. Perez, C. Technological revolutions and techno-economic paradigms. *Camb. J. Econ.* **2010**, *34*, 185–202. [CrossRef]
11. Aghion, P.; Bloom, N.; Blundell, R.; Griffith, R.; Howitt, P. Competition and innovation: An inverted-U relationship. *Q. J. Econ.* **2005**, *120*, 701–728.
12. Bikker, J.A.; Shaffer, S.; Spierdijk, L. Assessing competition with the Panzar-Rosse model: The role of scale, costs, and equilibrium. *Rev. Econ. Stat.* **2012**, *94*, 1025–1044. [CrossRef]
13. Shaffer, S. Patterns of competition in banking. *J. Econ. Bus.* **2004**, *56*, 287–313. [CrossRef]
14. Huang, T.-H.; Chiang, D.-L.; Chang, Y.-K. Re-visit the Market Competition of the Banking Industries in Eight East Asian Countries. *Manag. Rev.* **2015**, *34*, 59–76.
15. Ho, T.S.; Saunders, A. The determinants of bank interest margins: Theory and empirical evidence. *J. Financ. Quant. Anal.* **1981**, *16*, 581–600. [CrossRef]
16. Tseng, J.-L.; Chen, C.-J. The Risk Effects of Taiwan Listed and OTC-Listed Banks' Nontraditional Activities Revenues and Off-Balance-Sheet Activities Volumes. *Taiwan Acad. Manag. J.* **2012**, *12*, 139–157.
17. Giordana, G.A.; Schumacher, I. An empirical study on the impact of Basel III standards on banks' default risk: The case of Luxembourg. *J. Risk Financ. Manag.* **2017**, *10*, 8. [CrossRef]
18. Charnes, A.; Cooper, W.W.; Rhodes, E. Measuring the efficiency of decision-making units. *Eur. J. Oper. Res.* **1979**, *3*, 339. [CrossRef]
19. Banker, R.D.; Charnes, A.; Cooper, W.W. Some models for estimating technical and scale inefficiencies in data envelopment analysis. *Manag. Sci.* **1984**, *30*, 1078–1092. [CrossRef]
20. Aly, H.Y.; Grabowski, R.; Pasurka, C.; Rangan, N. Technical, scale, and allocative efficiencies in US banking: An empirical investigation. *Rev. Econ. Stat.* **1990**, *72*, 211–218. [CrossRef]
21. Kao, C.; Hwang, S.N.; Sueyoshi, T. *Management Performance Evaluation: Data Envelopment Analysis*; Hwatai Publishing Co.: Taipei, Taiwan, 2003; Volume 18.

engineering proceedings

Proceeding Paper

Trends of Adaptive/Personalized Learning and Intelligent Tutoring Systems in Mathematics: A Review of Academic Publications from 2010 to 2022 †

Thanyaluck Ingkavara [1], Wararat Wongkia [1] and Patcharin Panjaburee [2,3,*]

1 Institute for Innovative Learning, Mahidol University, Salaya 73170, Thailand; ing.thanyaluck@gmail.com (T.I.); wararat.won@mahidol.edu (W.W.)
2 Faculty of Education, Khon Kaen University, Khon Kaen 40002, Thailand
3 Digital Education and Learning Engineering Association, Nonthaburi 11110, Thailand
* Correspondence: patchapan@kku.ac.th
† Presented at the IEEE 5th Eurasia Conference on Biomedical Engineering, Healthcare and Sustainability, Tainan, Taiwan, 2–4 June 2023.

Abstract: Technology is used in research to satisfy the needs of the public. As there is a growing trend for adaptive/personalized learning, intelligent tutoring systems are used in various studies to overcome personal limitations. Thus, we reviewed the trends and characteristics of adaptive/personalized learning and intelligent tutoring systems in mathematics learning. Fifty-four relevant research articles (from 2010 to 2022) were selected and analyzed to investigate system parameters, roles of the systems in learning, mathematics contents, and learning outcomes. The analysis results showed the trends in research issues and challenges to be solved.

Keywords: personalized learning; adaptive learning; intelligent tutoring system; mathematics education; application in subject areas

Citation: Ingkavara, T.; Wongkia, W.; Panjaburee, P. Trends of Adaptive/Personalized Learning and Intelligent Tutoring Systems in Mathematics: A Review of Academic Publications from 2010 to 2022. *Eng. Proc.* **2023**, *55*, 34. https://doi.org/10.3390/engproc2023055034

Academic Editors: Teen-Hang Meen, Kuei-Shu Hsu and Cheng-Fu Yang

Published: 30 November 2023

1. Introduction

Abstract concepts in mathematics are always referred to for understanding why learners have difficulty in learning mathematics. Many scholars have proposed methods to help students learn mathematics with appropriate technology. Emerging adaptive/personalized learning and intelligent tutoring systems have been developed to support learners in acquiring mathematics skills more efficiently. According to the definition from the United States National Education Technology Plan 2017, personalized learning represents instructions in which the pace of learning and the instructional approach are optimized for the needs of each learner. Learning objectives, instructional approaches, and instructional content (and its sequencing) vary depending on learner needs. Learning activities need to be meaningful and relevant to learners, driven by their interests, and often self-initiated [1].

The activities are for learners who struggle with learning mathematics. In particular, an adaptive learning feature was considered regarding the state of technology-enhanced mathematics learning by Plass and Pawar. A learning system was used to investigate a learner's specific needs for appropriate adjustments to enhance learning outcomes [2]. The characteristics of personal learning showed that implementing technology-assisted learners was effective for learning mathematics. With technological development, an "Intelligent Tutoring System" has been proposed. For example, intelligent tutoring systems (ITSs) are computer programs that incorporate AI technology to provide tutors with what, who, and how to teach [3].

These three definitions focused on using technology to develop learning aids to assist learners and construct a learning environment in the classroom. Learning behavior needs to be understood for learner-centered teaching. Therefore, an application to the classroom is required. Based on students' characteristics, personalized learning systems in mathematics education can be developed [4–8]. Accordingly, we researched articles on personalized learning, adaptive learning, and intelligent tutoring systems in mathematics learning to investigate the system parameters used to develop the systems, the roles of the system in providing support, the contents and educational levels, and how to measure the learning outcomes in mathematics education using data from 2010 to 2022.

2. Data Collection and Process

2.1. Resources

The selected articles were found using the keywords "personalized learning in mathematics", "adaptive learning in mathematics", and "intelligent tutoring system in mathematics" in the Web of Science database. The publication years were set from 2010 to 2022. The results showed 197 related articles. A total of 54 articles not in the first quartile in education were excluded. In addition, articles related to pedagogical content knowledge, professional development, non-technological personalized learning, non-mathematics content, and STEM were excluded. Lastly, 54 articles were selected for this study.

2.2. Selected Articles

Figure 1 shows an overall increase in the number of relevant articles with decreases in 2011–2015 and 2016–2019. From 2020 to 2022, the number of articles increased significantly. The direction of researched showed a great leap forward. A previous study [9] in 2011 implemented Cognitive Tutor Authoring Tools (CTATs) for high school learners to study algebra. At that time, the system used embedded adaptive features (i.e., feedback and on-demand tips) to support learning by analyzing a learner's behavior, such as marking erroneous text/answers with red and providing clues after a third mistake. In other words, the system understood personal abilities based on a framework to recommend appropriate tasks, materials, and tools to assist the learner in reaching the learning goals. Digital games were designed to support mathematics learning and adaptive/personalized learning using learning concepts through playing [10–12].

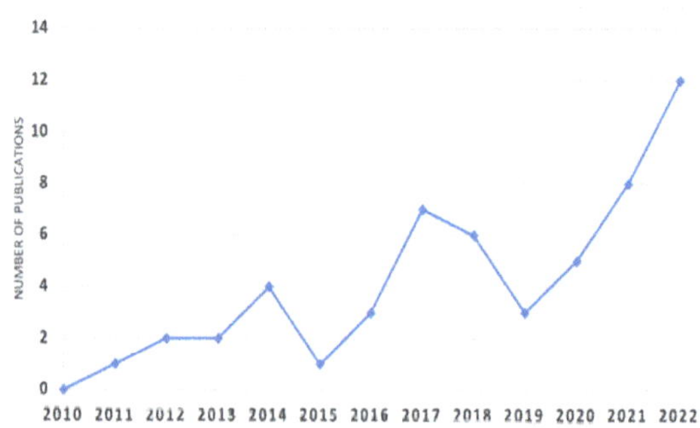

Figure 1. Number of publications from 2010 to 2022.

2.3. Coding Scheme

In this study, five categories of coding schemes were used.

1. Code for the system parameters: The code for system parameters refers to how technologies support mathematics learning, the learning activity environment, the assessment process, teacher–learner interactions, and the learning environment.
2. Code for system roles: The code for system roles is about how learners acquire knowledge while learning with the systems. In this study, the regulation proposed by Lai and Hwang [5] was used for accessible material, learning with the material, conducting assessments, and learning with full online support. Full online support

refers to systems that offer learning materials, allow learners to use the system, evaluate their learning abilities, and promote teacher–learner interaction.

3. Code for mathematics content: The code for mathematics content is used to categorize the systems based on the particular mathematics content designed for learners to learn. Likewise, this study classified rational numbers and fractions, algebra, calculus, geometry, probability, arithmetic operations, decimal numbers, modeling, arithmetic mean, mixed contents in mathematics, and non-specified content.
4. Code for learners: The code for learners is used to investigate the learners' levels of education. Therefore, we categorized it as kindergarten, elementary school, junior and senior high school, higher education, teachers, and non-specified educational level.
5. Code for learning outcomes: This code was used for three themes—cognitive, affective, and technical–behavioral correlation, referred to in Ref. [5].

3. Results

3.1. System Parameters

For 54 reviewed articles, 43.08% showed a feature that supported mathematics learning. There were five top-trend traits, including feedback, adaptive activity, on-demand hints, tutoring, and interactive tools. Other characteristics were content sequencing concerns, i.e., mastery approach, consecutive questions, or content. The support features for learning management were explored by 22.64% of the articles, including environmental structuring, individual learning paths, learning status, reviewing graded work, learning time, and student preferences. Similarly, assessment support features were used to identify a collection of log files to monitor learning paths and to serve formative and summative assessments in 14.78%. Additionally, the teacher–learner-interaction-supported feature was investigated in terms of collaborative activity, seeking assistance, and use of a chat room or online viewing mode and forum in 13.84%. These features were typically developed to allow learners and teachers to interact simultaneously. In 5.66% of articles, features that contributed to an excellent learning environment were explored using a combination of gamification (i.e., score, time) and emotional support (i.e., animated characters, avatars).

3.2. System Roles

Twenty-two articles (40.74%) showed that learners only participated in the systems by accessing the provided materials. However, accessible material did not mean handouts and assignments but activities for learning, practicing, and relearning by themselves. In total, 38.89% of learners used the provided materials. Full online learning support was important. In 20.37% of articles, the support was considered significant. For instance, an intelligent tutoring system was designed to help learners with intelligent tutoring systems (ITSs) combined with cognitive tutor authoring tools (CTATs) individually and collaboratively [13].

3.3. Mathematics Content

Rational numbers and fractions content was considered the most, in 20.37%. Non-specified content was the second most important as explored in 18.52% [14,15]. Arithmetic operations were also important in mathematics learning systems. Algebra-related content was the third most investigated, in 16.67%. The frequency of content was considered differently depending on geometry, decimals, calculus, and probability. Rational numbers and fractions, arithmetic operations, and algebra were considered appropriate to be implemented in the adaptive/personalized or intellectual tutoring systems at various difficulty levels. However, higher level concepts were considered in few studies, which needs further study.

3.4. Learner Level

Adaptive/personalized and intellectual tutoring systems in mathematics learning in the articles were explored for an elementary level in 51.72% (30 of 58 articles). At the junior and senior high school levels, 22.41% of the articles explored these systems. For higher education, 7 of 58 articles investigated the use of these systems. In Ref. [16], higher education, elementary level, and junior and senior high level were evenly explored for the use of the systems. Mathematics teachers' roles were regarded as important in 8.62%. However, the learner level was not specified in two articles. Instead, they explained the system design rather than the implementation. Lastly, in one study, a system for pupils at the kindergarten level was developed. The results showed that adaptive/personalized and intellectual tutoring systems could be used for mathematics learning in any age range.

3.5. Learning Outcomes

In 46.15%, learners' cognitive abilities were measured. In 32.05%, the technological–behavioral correlation was considered, while in 21.79%, learning effectiveness was determined. Cognitive abilities were focused on in all studies. In the cognitive category, learning achievement was studied in 23 articles, whereas low- or high-order thinking skills were explored in 7 articles. In terms of technical–behavioral correlation, in 14 articles the correlation between learning performance and the usability of systems was considered. In seven articles, the cause and effect of the use of the systems were explored. Learners' behaviors were investigated in four articles. An adaptive digital game called the Number Navigation Game (NNG) was used to enhance flexibility and adaptivity in mathematical thinking [10]. In six articles, self-efficacy was measured. Attitude, perception, self-regulation, and mathematics anxiety were examined in a few studies.

4. Discussions and Conclusions

The results of the research on articles from 2010 to 2022 showed what needs to be studied further in mathematics education. Elementary-level mathematics contents were usually taught at the elementary stage including arithmetic operations, rational numbers, and fractions as fundamental concepts, which are a foundation of in-depth knowledge. Understanding rational numbers was crucial for learning primary school mathematics as pointed out by Refs. [17,18], where students' arithmetic achievement was predicted. It was essential for mathematical proficiency to be developed every day in life. A few existing practical tools were used to support learning mathematical concepts [17]. For system development, feedback and adaptive features were considered top priorities in developing systems for learning mathematics. The lack of fundamental concepts impacted the learners' hierarchical mathematics learning. Thus, feedback and adaptive features were critical to a supported mathematics learning system. Reference [19] developed an adaptive software based on the concept of rational numbers and fractions, namely Woot Math Adaptive Learning (WMAL), a minor revision in 2020, for students to learn mathematics by providing automatic feedback. However, the system provided alternative features such as an assessment process, self-sequencing, collaborative working, or learning stimuli to increase learners' opportunities to participate in after-school or private learning [15,20].

Most learning systems were designed to provide learning materials. At the same time, several systems combined accessible learning materials and an evaluation of learning. In Ref. [12], a system was developed to embed adaptive learning based on an assessment approach. It was possible to provide feedback and support to learners individually. In an adaptive game [18], the learning materials and assessments were provided for personal users. Full online support systems were explored in a few studies. In the future, such systems are required to focus on the learning process, assessment, and the interaction between teachers and students. These features can be used to increase learners' engagement and their use of an adaptive/personalized or intellectual tutoring mathematics learning environment. The MathE platform was developed as an interactive environment where learners followed the individual learning process, increased their learning engagement, and improved learning outcomes [4]. It was an online collaborative system that consisted of three sections: learners' assessment, a library (learning tools), and a community for practice. These supports could provide a favorable environment for learners. Learning outcomes were in the cognitive domain. Technical–behavioral correlation was investigated in several studies as adaptive/personalized learning or intelligent tutoring systems responded directly to personal needs and abilities and positively impacted learning. Thus, efficacy studies on the impact of use, the relation between parameters, or even the expected behaviors/skills were carried out [13,21].

The findings in this study showed that the use of adaptive/personalized learning or intelligent tutoring systems in learning mathematics enhanced learners' abilities or learning process in any mathematics content at different learner levels. The characteristics of the systems support the learning process and help learners acquire and improve their mathematics knowledge/skills. Their cognition ability could be improved at the right pace. Therefore, the development of adaptive/personalized learning or intelligent tutoring systems is necessary to consider in the future to teach at various levels and provide appropriate mathematics content. The features of the system need to be adaptive depending on learners' preferences. In addition, assessment is required to process learning achievements and provide on-demand hints/suggestions. The collaboration in learning mathematics through the system allows learner and learner or learner and teacher to work online/learn together in real-time, which requires further research in mathematics education.

Author Contributions: Conceptualization, T.I. and W.W.; methodology, P.P. and T.I.; validation, P.P. and W.W.; formal analysis, T.I.; investigation, W.W.; resources, T.I.; data curation, T.I.; writing—original draft preparation, T.I.; writing—review and editing, P.P. and W.W.; visualization, T.I.; supervision, P.P. and W.W.; project administration, T.I. All authors have read and agreed to the published version of the manuscript.

Funding: This research was funded by the NATIONAL RESEARCH COUNCIL OF THAILAND (NRCT), grant number N41A640223.

Institutional Review Board Statement: This study was conducted in accordance with the Declaration of Helsinki and approved by the Institutional Review Board of INSTITUTE FOR POPULATION AND SOCIAL RESEARCH, MAHIDOL UNIVERSITY (IPSR-IRB-2022-113 and 26 May 2022).

Informed Consent Statement: Informed consent was obtained from all subjects involved in the study.

Data Availability Statement: The data presented in this study are available on request from the corresponding author.

Conflicts of Interest: The authors declare no conflict of interest. The funder had no role in the design of the study; in the collection, analyses, or interpretation of data; in the writing of the manuscript; or in the decision to publish the results.

References

1. U.S. Department of Education. Reimagining the Role of Technology in Education: 2017 National Education Technology Plan Update. 2017. Available online: https://tech.ed.gov/files/2017/01/NETP17.pdf (accessed on 25 May 2023).
2. Plass, J.L.; Pawar, S. Toward a taxonomy of adaptivity for learning. *J. Res. Technol. Educ.* **2020**, *52*, 275–300. [CrossRef]
3. Nwana, H. Intelligent tutoring systems: An overview. *Artif. Intell. Rev.* **1990**, *4*, 251–277. [CrossRef]
4. Azevedo, B.F.; Pereira, A.I.; Fernandes, F.P.; Pacheco, M.F. Mathematics learning and assessment using MathE platform: A case study. *Educ. Inf. Technol.* **2022**, *27*, 1747–1769. [CrossRef] [PubMed]
5. Lai, C.L.; Hwang, G.J. Strategies for enhancing self-regulation in e-learning: A review of selected journal publications from 2010 to 2020. *Interact. Learn. Environ.* **2021**, *31*, 3757–3779. [CrossRef]
6. Panjaburee, P.; Hwang, G.J.; Triampo, W.; Shih, B.Y. A multi-expert approach for developing testing and diagnostic systems based on the concept-effect model. *Comput. Educ.* **2010**, *55*, 527–540. [CrossRef]
7. Hwang, G.J.; Panjaburee, P.; Triampo, W.; Shih, B.Y. A group decision approach to developing concept-effect models for diagnosing student learning problems in mathematics. *Br. J. Educ. Technol.* **2013**, *44*, 453–468. [CrossRef]
8. Panjaburee, P.; Komalawardhana, N.; Ingkavara, T. Acceptance of personalized e-learning systems: A case study of concept-effect relationship approach on science, technology, and mathematics courses. *J. Comput. Educ.* **2022**, *9*, 681–705. [CrossRef]

9. Mullins, D.; Rummel, N.; Spada, H. Are two heads always better than one? Differential effects of collaboration on students' computer-supported learning in mathematics. *Int. J. Comput. Support. Collab. Learn.* **2011**, *6*, 421–443. [CrossRef]
10. Debeer, D.; Vanbecelaere, S.; Van Den Noortgate, W.; Reynvoet, B.; Depaepe, F. The effect of adaptivity in digital learning technologies. Modelling learning efficiency using data from an educational game. *Br. J. Educ. Technol.* **2021**, *52*, 1881–1897. [CrossRef]
11. Bhatia, P.; Le Diagon, S.; Langlois, E.; William, M.; Prado, J.; Gardes, M. Impact of a game-based intervention on fraction learning for fifth-grade students: A pre-registered randomized controlled study. *J. Comput. Assist. Learn.* **2022**, *39*, 49–62. [CrossRef]
12. Chu, H.C.; Chen, J.M.; Kuo, F.R.; Yang, S.M. Development of an Adaptive Game-Based Diagnostic and Remedial Learning System Based on the Concept-Effect Model for Improving Learning Achievements in Mathematics. *J. Educ. Technol. Soc.* **2021**, *24*, 36–53.
13. Olsen, J.K.; Rummel, N.; Aleven, V. It is not either or: An initial investigation into combining collaborative and individual learning using an ITS. *Int. J. Comput. Support. Collab. Learn.* **2019**, *14*, 353–381. [CrossRef]
14. Jacobs, J.; Scornavacco, K.; Harty, C.; Suresh, A.; Lai, V.; Sumner, T. Promoting rich discussions in mathematics classrooms: Using personalized, automated feedback to support reflection and instructional change. *Teach. Teach. Educ.* **2022**, *112*, 103631. [CrossRef]
15. Kooken, J.W.; Zaini, R.; Arroyo, I. Simulating the dynamics of self-regulation, emotion, grit, and student performance in cyber-learning environments. *Metacogn. Learn.* **2021**, *16*, 367–405. [CrossRef] [PubMed]
16. Xie, H.; Chu, H.C.; Hwang, G.J.; Wang, C.C. Trends and development in technology-enhanced adaptive/personalized learning: A systematic review of journal publications from 2007 to 2017. *Comput. Educ.* **2019**, *140*, 103599. [CrossRef]
17. Brezovszky, B.; McMullen, J.; Veermans, K.; Hannula-Sormunen, M.M.; Rodríguez-Aflecht, G.; Pongsakdi, N.; Laakkonen, E.; Lehtinen, E. Effects of a mathematics game-based learning environment on primary school students' adaptive number knowledge. *Comput. Educ.* **2019**, *128*, 63–74. [CrossRef]
18. Kärki, T.; McMullen, J.; Lehtinen, E. Improving rational number knowledge using the NanoRoboMath digital game. *Educ. Stud. Math.* **2022**, *110*, 101–123. [CrossRef]
19. Bush, J.B. Software-based intervention with digital manipulatives to support student conceptual understandings of fractions. *Br. J. Educ. Technol.* **2021**, *52*, 2299–2318. [CrossRef]
20. Craig, S.D.; Hu, X.; Graesser, A.C.; Bargagliotti, A.E.; Sterbinsky, A.; Cheney, K.R.; Okwumabua, T. The impact of a technology-based mathematics after-school program using ALEKS on student's knowledge and behaviors. *Comput. Educ.* **2013**, *68*, 495–504. [CrossRef]
21. Wang, S.; Christensen, C.; Cui, W.; Tong, R.; Yarnall, L.; Shear, L.; Feng, M. When adaptive learning is effective learning: Comparison of an adaptive learning system to teacher-led instruction. *Interact. Learn. Environ.* **2020**, *31*, 793–803. [CrossRef]

engineering proceedings

MDPI

Proceeding Paper

Performance Evaluation of Lithium Nitrite-Based Gel against Corrosion of Rebar with Partial Short Cover Depth in Chloride Environment †

Muhammad Afaq Khalid [1,*], Shinichi Miyazato [1], Hibiki Mizuguchi [1] and Katsuichi Miyaguchi [2]

1 Department of Environmental and Civil Engineering, Kanazawa Institute of Technology, Nonoichi 921-8501, Japan; miyazato@neptune.kanazawa-it.ac.jp (S.M.); c6302296@st.kanazawa-it.ac.jp (H.M.)
2 SHO-BOND Corporation, Tokyo 103-0015, Japan; miyaguchi-ka@sho-bond.co.jp
* Correspondence: c7000029@planet.kanazawa-it.ac.jp
† Presented at the IEEE 5th Eurasia Conference on Biomedical Engineering, Healthcare and Sustainability, Tainan, Taiwan, 2–4 June 2023.

Abstract: Surface coatings are widely used for preventive maintenance, as they prolong the durability of concrete structures. Partial short cover depth may result due to poor construction quality and uneven formwork. In this study, lithium nitrite-based gel was used as an anodic corrosion inhibitor coating material to investigate its influence on the corrosion of rebar embedded in a partial short cover depth in the presence of chloride. The specimens were cured at a temperature of 20 °C and a relative humidity of higher than 95%. (a) Macrocell corrosion current density, (b) microcell corrosion current, and (c) polarization curves were measured, and the results were compared with those of uncoated specimens after aging for 28 days. The results revealed that the corrosion rate was reduced considerably due to the application of the coating with the gel type.

Keywords: chloride attack; corrosion current densities; lithium nitrite-based gel; partial short cover depth; rebar corrosion

Citation: Khalid, M.A.; Miyazato, S.; Mizuguchi, H.; Miyaguchi, K. Performance Evaluation of Lithium Nitrite-Based Gel against Corrosion of Rebar with Partial Short Cover Depth in Chloride Environment. *Eng. Proc.* **2023**, *55*, 35. https://doi.org/10.3390/engproc2023055035

Academic Editors: Teen-Hang Meen, Kuei-Shu Hsu and Cheng-Fu Yang

Published: 30 November 2023

1. Introduction

The corrosion of steel in concrete depends on cover depth, chloride environment, coarse and fine aggregate, and the water–cement (W/C) ratio of the concrete. The rebar is highly alkaline due to the large amount of calcium hydroxide contained in cement. Calcium hydroxide protects the rebar from external effects due to a protective coating effect of the passive film. This passive film is destroyed when the amount of chloride ions present in the concrete exceeds a certain level or threshold for corrosion. Sea sand, chloride ions from seawater, airborne chloride, and anti-freezing agents are the main sources of chlorides. A partial short cover depth results from the dealignment of spacers or poor construction quality. A rebar with an adequate cover depth significantly decreases the corrosion density by 50% compared to that with a low W/C mortar subjected to an aggressive chloride environment [1]. The gel contains lithium nitrite as an anodic corrosion inhibitor. Nitrite ions (NO_2^-) diffuse through the concrete surface and react with ferrous ions (Fe^{2+}) to form a passive film [2]. Lithium-ion gel has been used for the repair of ASR deterioration in the concrete and also as a rust inhibitor in solution form. The gel-type penetrant effectively delays the corrosion rate of rebar in concrete [3,4]. However, its influence on the corrosion of rebar embedded in partial short cover depth has not yet been investigated. Thus, we explored the effect of the nitrite-based gel-type surface penetrant against the corrosion of rebar with partial short cover depth in the presence of chlorides using quantitative electrochemical methods.

2. Materials and Methods

2.1. Specimen

The specimen is shown in Figure 1. For the measurement of corrosion current densities in the rebars, six (06) (D10, 295) steel elements [5] 25 mm in length were connected through the high-insulating epoxy resin to form a 180 mm long bar. The cover depth was 20 mm and 7.5 mm. Ordinary Portland Cement (OPC) was used with W/C = 0.50 for both the 20 mm and 7.5 mm cover depth. Chloride ions of 2.5 kg/m^3 were added to the mortar of the 7.5 mm cover depth portion only. The chloride ions were added by mixing sodium chloride (NaCl) in water at the kneading time.

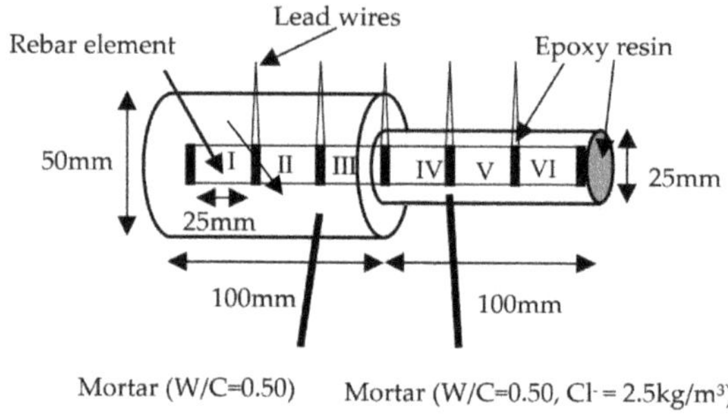

Figure 1. Partial short cover depth specimen with special divided bar.

The purpose of adding chloride ions was to simulate the penetration of chloride ions in short cover depth and to achieve a potential difference that enhances macrocell corrosion [6]. The mortar mix proportions are shown in Table 1. After casting, initial curing was completed in wet conditions at 20 °C and RH > 95% for 27 days, followed by 7 days of drying at 20 °C and RH = 60% to ensure that the surface moisture was less than 6%, as recommended by the manufacturer for coating. The gel-type coating was applied to the specimens at 1 kg/m^2 by using the brush. The antirust agent-mixed gel was an aqueous solution of lithium nitrite that was moderately thickened using a thickener. The amount of nitrite ions in the gel was 360 kg/m^3. The specimens were dried in dry conditions for 7 days and then cured in wet conditions for 27 days.

Table 1. Mortar mixture proportions.

W/C	S/C	Unit Weight (kg/m^3)		
		W	C	S
0.50	2.5	276	553	1384

2.2. Electrochemical Measurement

The current flowing between the steel elements was referred to as macrocell corrosion current density. The positive macrocell was anodic and the negative macrocell was cathodic in the macrocell corrosion current. However, the current flowing in the individual steel element was referred to as microcell corrosion current density. The summation of the anodic macrocell of an individual steel element and the microcell for that element was defined as total corrosion current density [7]. The polarization resistance was determined by the AC impedance method. Polarization curves were drawn in the three-electrode system. Ag/Agcl was used as a reference electrode. Potential at the rate of 1 mV/s was applied and the corresponding current was measured.

3. Results and Discussions

3.1. Corrosion Current Density

The graphs in Figure 2a,b show the variation in macrocell, microcell, and total corrosion current densities for all the steel elements along the rebar with cover depth and amount of chloride ion concentration for coated and uncoated specimens at the age of 28 days after coating. Figure 3 shows the result of the comparison of the average of the maximum total corrosion current densities of specimens that were coated and uncoated. The corrosion current densities for the coated specimen were lower than for the uncoated specimen. The average total corrosion current density for the specimen coated with the gel type was less than 0.1 $\mu A/cm^2$, which was a "Negligible corrosion level" as per RILEM recommendations [8].

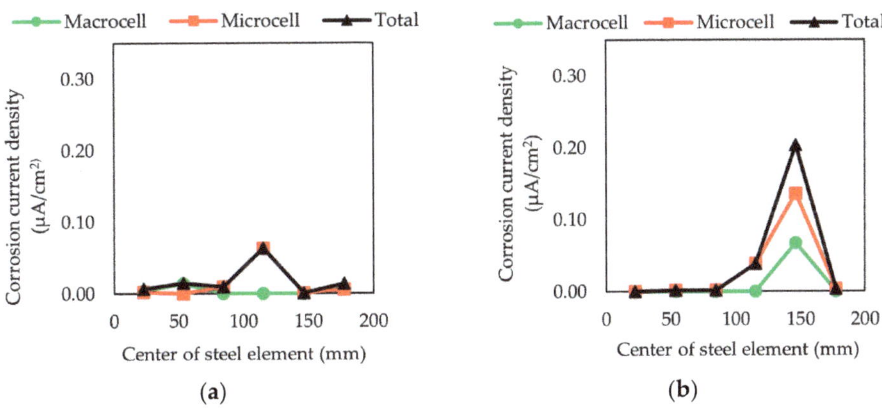

Figure 2. Showing comparison of macrocell, microcell, and total corrosion current densities of coated and uncoated specimens. (**a**) Variation in corrosion current densities along the length of rebar for specimen coated with nitrite-based gel; (**b**) variation in corrosion current densities along the length of rebar for the uncoated specimen.

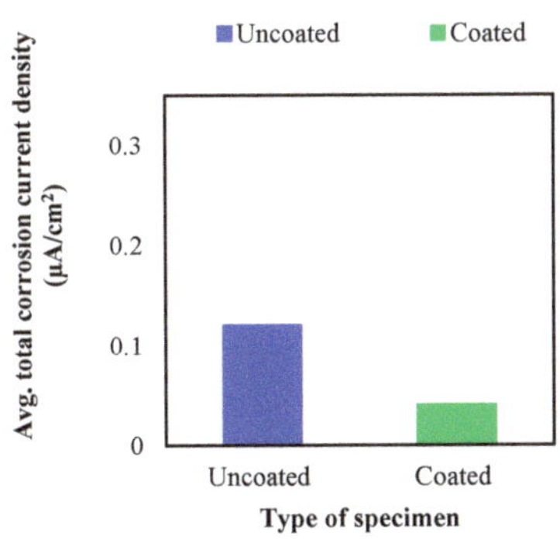

Figure 3. Comparison of avg. total corrosion current densities of coated and uncoated specimens.

3.2. Anodic Polarization Curves

Figure 4 shows the comparison of the anodic polarization curves at the age of 28 days after coating. The graph depicts that the current density at any potential difference is lower for coated specimens than the uncoated specimens. This confirmed the existence of lithium nitrite as an anodic corrosion inhibitor which reacted with the Fe^{2+} to regenerate the passive film and delay the corrosion reaction at the anode.

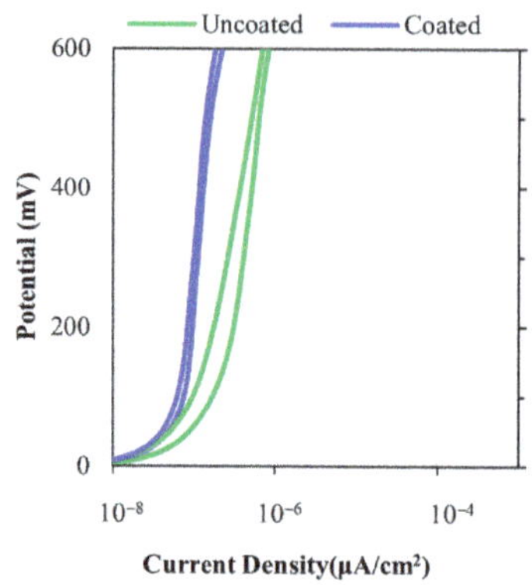

Figure 4. Anodic polarization curve of the coated specimen with anti-rust lithium nitrite gel type and uncoated specimen.

4. Conclusions

The performance of lithium nitrite-based gel as an anodic corrosion inhibitor was evaluated for the corrosion of rebar embedded in specimens with a partial short cover depth by using quantitative electrochemical measurements. The total corrosion current density for the coated specimen with the nitrite-based gel was less than that for the uncoated specimen for the partial short cover depth specimen. The magnitude of the current for the gel type was less than that of the uncoated specimen in the anodic polarization curves, indicating that the gel-type penetrant regenerated the passive film and delayed the initiation of corrosion. The application of the gel-type penetrant for partial short cover depth saves labor by delaying the corrosion that results in the spalling of concrete, thereby decreasing maintenance costs.

Author Contributions: Conceptualization, M.A.K. and S.M.; methodology, M.A.K. and H.M.; Experimentation, M.A.K. and H.M.; Results preparations, M.A.K. and H.M.; writing—original draft preparation, M.A.K.; writing—review and editing, M.A.K., S.M. and K.M.; supervision, S.M. and K.M.; Materials for the experiment, K.M. All authors have read and agreed to the published version of the manuscript.

Funding: This research received no external funding.

Institutional Review Board Statement: Not applicable.

Informed Consent Statement: Not applicable.

Data Availability Statement: Data are contained within the article.

Acknowledgments: The authors highly acknowledge the support of the Japan International Cooperation Agency (JICA) throughout the study.

Conflicts of Interest: Author Katsuichi Miyaguchi is working in the company SHO-BOND Corporation. The remaining authors declare that the research was conducted in the absence of any commercial or financial relationships that could be construed as a potential conflict of interest.

References

1. Muhammad, A.K.; Miyazato, S.; Sugawara, N. Influence of Cover depth and mortar quality on Rebar corrosion under aggressive chloride environment. In Proceedings of the 4th Asian Concrete Federation Symposium on Emerging Technologies for Structural Longevity (ACF2023_ETSL), Shenzhen, China, 11–13 March 2023.
2. Lee, H.S.; An, J.M.; Shin, S.W. Study on the evaluation of corrosion protection effect with LiNO2, inhibitor. *J. Korea inst. Struct. Maint. Insp.* **2002**, *6*, 339–342.

3. Miyaguchi, K.; Takai, K.; Miyazato, S. Corrosion inhibition effects of reinforcing steel bars of reinforced concrete structures subjected to salt damage by Gel coating material containing Anti-corrosive. In Proceedings of the 77th Annual Lecture of Civil Engineering Society National Conference, Kyoto University Yoshida campus, Kyoto, Japan, 15–16 September 2022; Volume 548. (In Japanese)
4. Miyaguchi, K.; Takai, K.; Miyazato, S. Corrosion prevention effect of Reinforced concrete Using Gel composed Inhibitor against chloride attack, carbonation and composite deterioration. *Proc. Concr. Struct. Scenar. JSMS* **2022**, *22*, 1–6. Available online: https://cir.nii.ac.jp/crid/1520576137321497472 (accessed on 30 November 2023). (In Japanese).
5. Miyazato, S.; Otsuki, N. Steel corrosion induced by chloride or carbonation in mortar with bending cracks or joints. *J. Adv. Concr. Technol.* **2010**, *8*, 135–144. [CrossRef]
6. Hansson, C.M.; Poursaee, A.; Laurent, A. Macrocell and microcell corrosion of steel in ordinary Portland cement and high-performance concretes. *Cem. Concr. Res.* **2006**, *36*, 2098–2102. [CrossRef]
7. Miyazato, S.; Otsuki, N. Measurement Method for Macrocell Corrosion in Concrete Specimen using a Segmented Steel Bar. *J. Adv. Concr. Technol.* **2022**, *20*, 222–235. [CrossRef]
8. Andrade, C.; Alonso, C. Test methods for on-site corrosion rate measurement of steel reinforcement in concrete by means of the polarization resistance method. *Mater. Struct.* **2004**, *37*, 623–643. [CrossRef]

engineering proceedings

MDPI

Proceeding Paper

An Innovative Design for Drawlooms with an Open-Type Heald †

Jian-Liang Lin

Visitor Services Division, National Science and Technology Museum, Kaohsiung 811, Taiwan; golduser007@gmail.com

† Presented at the IEEE 5th Eurasia Conference on Biomedical Engineering, Healthcare and Sustainability, Tainan, Taiwan, 2–4 June 2023.

Abstract: An innovative and simple design for a weaving loom was proposed to simplify the pattern design and the decoding process of weaving. Using the structural analysis of the existing looms and based on the concepts of weaving, a creative open-type rigid heald was developed to replace the traditional one and generate warp shedding. Corresponding to the raster graphics of the designed pattern, the heald shape had convex–concave features to control warping in shedding. The pattern design was integrated into the heald design using mechanical coding.

Keywords: open-type rigid heald; loom; raster graphic

1. Introduction

Textile is essential in culture with weaving being an ancient technique. Bone needles were found at the cave site of the Paleolithic period in China showing the earliest known origin of weaving, far before the time of pottery making [1]. The oldest textiles found in Peru date back to between 10,100 and 9080 BC [2]. Textiles were found in the birthplaces of civilizations around the world, including manual rope weaving, handmade instruments, and the design and application of textile machinery. The existing textile teaching focuses on knitting as knitting tools are cheap and easy to use and carry. For weaving, a mechanical device is necessary but expensive. A simple loom for plain weaving costs several thousand NTD, and a loom used for pattern weaving (Figure 1) costs more than ten thousand NTD. The device is heavy to carry, the procedure is cumbersome, and the pre-operation is time-consuming. The above problems make it difficult to promote the weaving science popularization activities.

Figure 1. Modern manual loom.

Therefore, there is a need to develop an innovative drawloom with creative flat healding to increase the popularity of learning weaving. The new device has a simple structure, low cost, and simplified pre-operation procedures. In addition, the bitmap of the

Citation: Lin, J.-L. An Innovative Design for Drawlooms with an Open-Type Heald. *Eng. Proc.* **2023**, *55*, 36. https://doi.org/10.3390/engproc2023055036

Academic Editors: Teen-Hang Meen, Kuei-Shu Hsu and Cheng-Fu Yang

Published: 30 November 2023

woven pattern is used to design the flat healding to control the warps. The relationship between the weaving pattern and warp motions are coded by weaving program. By integrating the pattern design and the healding design, the popularization of weaving education is achieved.

2. Basic Concepts of Weaving and Drawlooms

The procedure for the typical weaving process includes three steps: shedding, picking, and beating up. In shedding, the operator controls the up–down motions of heddles to pull or press the warps and then divides them into upper and lower layers to create a shed for the warp layer. Picking is passing the beater with a weft through the shed from one side to the other side. The last stage of beating up involves pushing the new weft to the cloth using a reed or beater to evenly arrange the wefts. By repeating the motion of beating up, the horizontal wefts are arranged tightly to generate a tight cloth. When continuous weaving is carried out in traditional weaving, two synchronous actions of "warp control" and "cloth control" are necessary and generated by scrolling two beams as shown in Figure 2. The warp beam delivers the warps and the piece beam receives the cloth. The warps between these two shafts can maintain the required tensions.

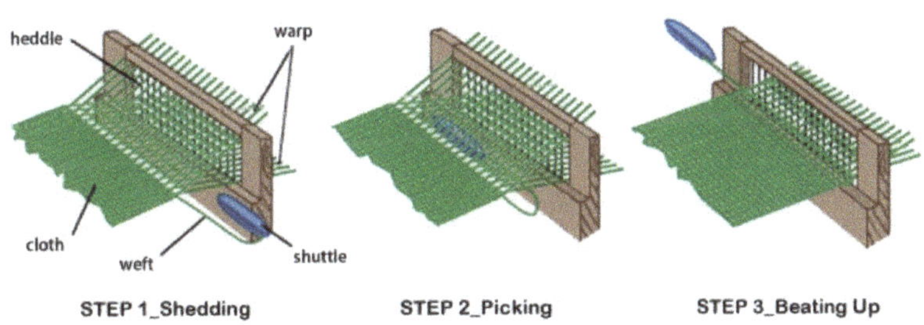

Figure 2. Manual procedure of a common drawloom.

The repetitive patterns in weaving are essential in textile machinery. The pattern design corresponds to its weaving procedure, i.e., the pattern cards, jacquard design, or weaving plan. In ancient Chinese weaving, the pattern cards were called "Yi Jiang Tu (意匠圖)" or "Hua Ben (花本)" (Figure 3). The principle was to convert the pattern of the jacquard into the regular action program of each warp in the weaving process (that is, "moving up and down" or "not moving"), and the actions of the warps depended on whether the warp passes through the harness eyes or the cord eyes. This is called "sleying" or "healding." Under the combination design of the healding arrangement and the number of harnesses, all of the warps' actions in a specific sequence were controlled by operating certain harnesses to complete the function of shedding.

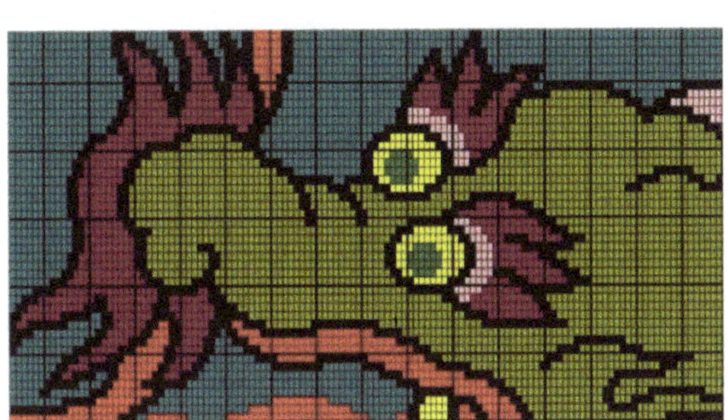

Figure 3. Pattern cards (Yi Jiang Tu).

The up-down action of the warps is called "lifting." In the design of machinery, it is common to lift harnesses using a direct manual or mechanism linkage method. The simple looms are manually operated and usually found in the traditional culture. These ancient looms consisted of parts that were not mechanical devices. The looms that can weave fabrics with complicated patterns were operated by mechanisms, and floor-type looms were common.

For the jacquard machine, the structure of multiple treadles and harnesses is designed to be a pattern storage method for fabric production. The information of the pattern is stored in several healds. In the weaving process, the healds are lifted in sequence according to information on the fabric [3]. Aside from the complexity of its pattern, the relationship between warp and weft is described as "passing through" or "not passing through" depending on the "lifting" and "not lifting" of the warp in weaving. The concept is equivalent to the symbols "0 and 1" in computer coding [4]. If the color yarn and weave changes are added, the pattern is designed as a "color matching pattern" [5], so the weaving sequence of warps is a loom-specific program translated based on the textile patterns. Therefore, the "lifting" and "healding" processes realize the textile pattern, and the pattern card is a decoder.

However, it is difficult for beginners to learn weaving techniques and understand pattern design using pattern cards. Integrating pattern design with weaving in the teaching plan is difficult. If the pattern design is not taught in weaving lessons, learners cannot learn how to translate programming language into the drawloom. With logical thinking and image coding concepts, the most critical concept of weaving can be skipped in weaving teaching.

3. Structural Analysis of Looms

Weaving is an ancient technique of making clothes. The drawlooms have developed into several types in the development of the textile industry. In this article, three representative ancient looms and one common modern loom are introduced.

The slanting loom and the drawloom (Figures 4 and 5) were introduced in the ancient Chinese Book *Tian Gong Kai Wu* (天工開物) by Song Ying-Xing (宋應星). The slanting loom is more common than the drawloom, and the loom is divided into three types, including "Yao Jia (腰機)," "Bu Ji (布機)," and "Wo Ji (臥機)" according to the operating angles of the looms and the fixed method of warps. The warp motions of a slanting loom are usually operated by stepping on the treads. A treadle connects to one heald by a scale link and strings. Such a mechanism is called a heddle-raising device. Based on previous research [6,7], the slanting looms fall into four categories: two treadle, single heddle rack (TTSH); single treadle, single heddle rack (STSH); single treadle, half heddle rack (STHH); and two treadle, two heddle rack (TTTH). By analyzing the topological structure, the heddle-raising device was found to be a planar or a spatial mechanism with four, five, or six links. After operating the treadle to finish the shedding and picking, the loom with one heddle must be weighed on the scale link to make the heddle recover its initial weight. Mostly, a slanting loom has two healds to control the odd number of warps and the even number of warps for a plain weave.

The structure of the drawloom is more complex than the slanting loom to generate sophisticated patterns on clothes. It results in a diversity of weaving patterns and the elegance of clothes. In ancient China, the drawloom called "Da Hua Lou Ti Hua Ji (大花樓提花機)" (Figure 4) was used to weave the emperor's court dress. The weaving work of a drawloom is time-consuming and complicated. Two people simultaneously operate a drawloom. One weaves and the other controls the warp weaving procedure, i.e., the program of the pattern design. Owing to the complicated weaving pattern and cloth structure, the drawloom needs several healds to control the sequential motion in each row weft of the pattern design. In the structure design and layout of the loom, it is hard to increase the number of essential healds. Therefore, the drawloom uses two mechanisms to control healds. One is the foot-raising heald mechanism, and the other is the foot-falling heald mechanism. They are called "Wen Zong (紋綜)" and "Di Zong (地綜)" and are used to make repeated patterns in the ground-woven clothes. According to the reconstructions [6,8], the foot-raising heald mechanism can be a planar or spatial mechanism with five or seven links. The foot-falling device is a planar or spatial mechanism with four, five, or six links. However, this is not enough to fulfill the pattern weaving of the drawloom. The embossed pattern is arranged by the "Hua Ben (花本)", i.e., the pattern cards. By applying the knots to mark the weaving sequence, a person operates the weaver and controls the warps for weaving the embossed patterns. Therefore, pattern weaving requires a special design for the textile machines in ancient China. According to the functions, the pattern weaving of the drawloom includes five subsystems: a foot-raising heald mechanism, foot-falling heald mechanism, hand-raising warp mechanism, weft pressing mechanism, and fabric reeling mechanism. Although the design of the loom is excellent, it needs to be operated by a human, not by an automated machine.

Figure 4. Slanting loom in ancient China.

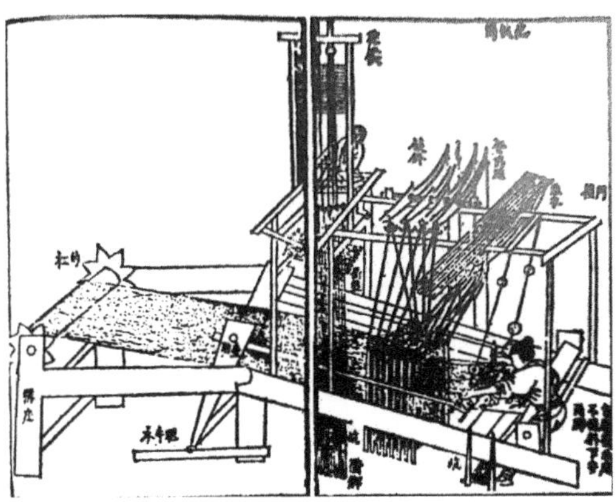

Figure 5. Drawlooms in ancient China.

In the West, a Jacquard loom with similar functions was invented by Joseph Marie Jacquard in 1807 (Figure 6). One weaver can operate the Jacquard loom. The healds are also used to control the warps' motions based on the repeated pattern design of the cloth. A punch card made of a piece of wood with several circle holes is used to represent a row of weft as shown in Figure 6b. A skilled worker transfers the pattern design into punched cards. These punch cards are stitched together in a continuous belt and fed into the loom. A pattern design must be generated by several pattern cards. The function of the holes on the pattern card is to move the iron stick with a circle and spring at its end. An iron hook passes through the stick's circle and controls a warp. Therefore, for a punch card, the stick can pass through the hole without affecting the hook's position. Then, the hook can be pulled on to draw the warp. Otherwise, for a position on the punch card without a hole, the stick is pressed, and the hook is moved out of its original position. Then, the hook cannot be pulled on, and the corresponding warp is not drawn. With the punch card, the spring can recover the stick's position. Owing to such a design, the Jacquard loom does not need a person to draw the warps to operate the foot-raising heald mechanism, foot-falling heald mechanism, or hand-raising warp mechanism. This machine simplifies the process of manufacturing woven fabrics with complex patterns, such as brocade, damask, and matelassé [9].

(a) Machine (b) Punch cards

Figure 6. Jacquard machine.

4. Innovative Design of Drawlooms

Regardless of the ancient looms or modern desktop looms, the price of the devices is high. Therefore, an innovative drawloom was developed by using the simplest structure (the three-links mechanism). The loom has similar functions as basic weaving with convenient functions for fabricating quickly. Figure 7 shows the innovative and simple drawloom, which consists of a front plate, a rear plate, and two side plates with a slide. The maximum size of the existing design was about 32 cm in length and 15 cm in width. The device includes a flat heald to replace the traditional heald or reed but maintains the function of generating a shed. To control the motions of the warps, a V-shaped notch replaces the cord eyes. The concave and convex shape of the end plate allows the warp to be in the right position. The left and right side panels have a chute design to adjust the size of the machine and maintain the warp tension according to the user's hand length or fabric length.

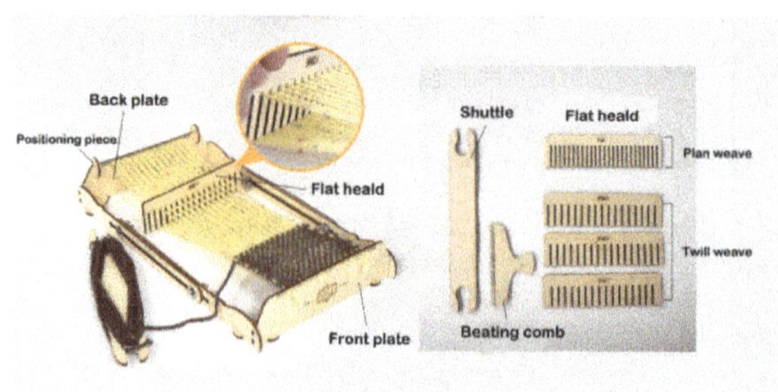

Figure 7. Innovative design of manual drawloom.

According to the description shown in the figures, the heald is lifted to control the movement of the warp, and the warp action in a lifting sequence is determined by the heald frame. The steel sheet is matched with the heald eye and the rigid-body heald frame; the rigid-body heald frame is presented as a closed one-piece flat plate with a chute and a steel sheet with a heald eye since the drawing is a process of connecting the pattern and lifting the heald. The innovative feature of the design of this teaching aid was to replace the heald frame of the traditional weaving machine with flat heddles. The shape of the heald determined the movement of the warp, and the "heddle" was transformed into the shape of the heald, which uses a binary code of 0 and 1 for bump sorting. The shape of the flat heald is a flat comb, and the concave and convex order controlled the up and down position of the warp according to the pattern design. The edge of the protruding part of the heald is in the shape of an inverted V, which is convenient for clamping and pressing the warp. The depth of the concave part determines the stroke of the warp that was displaced up and down. During the process of combing and combing the yarn by the heald, the starting position is confirmed through the positioning piece of the rear end plate to avoid the problem of jumping wires during the combing and combing process of the warp yarn and improve the accuracy of the yarn opening operation. The shuttle has the same function as the shuttle to wrap the weft yarn and guide the weft yarn during the weft insertion process. The beating-up comb is the same design as the beating-up knife of the ground loom and the reed of the typical drawloom to complete the beating-up process.

5. Bitmap Concept of Pattern Design

Learning Weaving for beginners requires considerable time. Thus, a method to simplify the complicated pre-work of weaving and the concept of pattern design is presented in this article. An innovative drawloom with a simple structure is presented under the concepts of three weaving procedures. This innovative loom is used on a desktop and without the warp control and cloth control. Hence, it weaves clothes of a limited size based on the length and weight of the device and cannot weave continuously. To simplify the patterns converted into warp action procedures, the image of the weaving pattern was created into a bitmap so that abecedarians could code the weaving procedure with intuitive concepts. They did not execute the healding process, the most complicated step in the weaving technique. The lifting action generated a shed based on the designed sequence of warps.

The pattern coding procedure for the simplified drawloom is as follows: (1) The number of rows and columns is determined to generate a grid; each column corresponds to each weft yarn and each row corresponds to each warp yarn. (2) The warp and weft are expressed in multiple selected colors. (3) By filling in the selected color of the warp and weft in the squares of the grid, the pattern is produced as a bitmap of a graph. (4) For a specific row, the color change of the row determines the upper and lower positions of each warp in the picking process, which correspond to pressing or not pressing the warp, and confirm whether the warp needs to be moved. If the field is the color of the weft yarn, the warp yarn is pressed down, and the inverted V shape must be filled. If it is displayed as the color of the warp yarn, the warp yarn does not move, and the horizontal line figure must be filled in. (5) The inverted V shapes and horizontal lines are connected to each column to generate the concave–convex sequence for the side of flat heald corresponding to the designed weft yarn. (6) The bitmap of the design pattern in turn is created to generate the heald geometry for each row (weft yarn). In the example shown in Figure 8, the warp is white, the weft is blue, and the warp is pressed down by the flat heald to generate a shed so that the weft is on the top after picking. According to the blue and white sequence of the bitmap, the side geometry of the flat heald in the red line corresponds to the action procedure of the warps as a code. Moreover, the combination of the flat healds is regarded as a program translating the pattern.

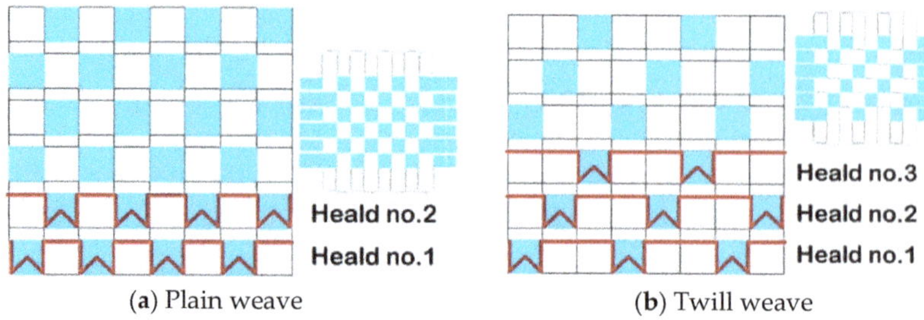

Figure 8. Bitmap concept of pattern design for new drawloom.

6. Conclusions

Textiles are an element of civilization, representing humanities, history, craftsmanship, and machinery. It is worthy of promotion as a popular science education project. In this study, the basic principle of weaving and the existing drawlooms were analyzed for the looms in ancient China and the ancient West. Considering the concept of simplified structure and weaving process and lower cost, an innovative simple drawloom with flat healds was developed. The weaving pattern was directly drawn in the form of a bitmap. The shape design of the heald was used to control the upper and lower sequence of the warp yarns so that the weaving pattern could be converted into the weaving coding more intuitively. Through the combination of the design method and the new drawloom, the complicated process of "sleying" or "healding" was simplified. In this way, the open-type flat heald was invented. The result was successful in creative weaving. Learning weaving techniques and design became much easier. With the integration of this innovative design, the popularization of science activities for textiles has proliferated. Also, the programming of weaving was improved.

Funding: This research was funded by the National Science and Technology Council (Taipei, Taiwan), grant number MOST 111-2410-H-359-002.

Institutional Review Board Statement: Not applicable.

Informed Consent Statement: Not applicable.

Data Availability Statement: Data sharing is not applicable to this article.

Acknowledgments: The author is grateful to the National Science and Technology Council (Taipei, Taiwan) and the National Science and Technology Museum for the support for this work.

Conflicts of Interest: The author declares no conflict of interest.

References

1. Soffer, O. Recovering perishable technologies through use wear on tools—Preliminary evidence for Upper Paleolithic weaving and net making. *Curr. Anthropol.* **2004**, *45*, 407–413. [CrossRef]
2. Jolie, E.A.; Edward; Lynch, T.F.; Geib, P.R. Cordage, Textiles, and the Late Pleistocene Peopling of the Andes. *Curr. Anthropol.* **2011**, *52*, 285–296. [CrossRef]
3. Luo, Q. Gu Dai Duo Zong Duo Nie Ti Hua Ji Jie Guo Ji Zhuang Zao Xing Shi Tan Tao. *Si Chou* **2011**, *48*, 45–47. (In Chinese)
4. Long, B.; Zhao, F. Zhong Guo Gu Dai Zao Qi Ti Hua Zhi Ji De He Xin: Duo Zong Ti Hua Zhuang Zhi. *Si Chou* **2020**, *57*, 72–77. (In Chinese)
5. Miao, Q.J.; Jiang, X.X. *Zhi Wu Yu Jie Gou*; Donghua University Press: Shanghai, China, 2007. (In Chinese)
6. Hsiao, K.H.; Yan, H.S. *Mechanism in Ancient Chinese Books with Illustrations*; Springer Nature: Berlin, Germany, 2014.
7. Hsiao, K.H.; Yan, H.S. Structural Synthesis of Ancient Chinese Foot-operated Slanting Loom. *Proc. Inst. Mech. Eng. Part C J. Process Mech. Eng.* **2011**, *225*, 2685–2699. [CrossRef]

8. Hsiao, K.H.; Yan, H.S. Structural Synthesis of Ancient Chinese Drawloom for Pattern-weaving. *Trans. Can. Soc. Mech. Eng.* **2011**, *35*, 291–308. [CrossRef]
9. Victoria and Albert Museum, Youtube Movie: How Was It Made? Jacquard Weaving. Available online: https://www.youtube.com/watch?v=K6NgMNvK52A&ab_channel=VictoriaandAlbertMuseumaccessed (accessed on 1 July 2023).

engineering proceedings

Proceeding Paper

Catalytic Reduction of Acetophenone Promoted with Quinuclidinol-Based Quaternary Ammonium Ionic Liquid as a Sustainable Solvent †

Gaojin Zhou [1], Hangping Chen [2], Ruoheng Li [1], Yu Cao [3] and Shun Yao [1,*]

1 School of Chemical Engineering, Sichuan University, Chengdu 610065, China; 2022223070115@stu.scu.edu.cn (G.Z.); pharmposter2012@163.com (R.L.)
2 College of Pharmacy, Jinan University, Guangzhou 510632, China; chenhp2013@126.com
3 College of Life Science & Biotechnology, Mianyang Normal University, Mianyang 621000, China; agnes_cy2004@163.com
* Correspondence: cusack@scu.edu.cn

† Presented at the IEEE 5th Eurasia Conference on Biomedical Engineering, Healthcare and Sustainability, Tainan, Taiwan, 2–4 June 2023.

Abstract: As a new sustainable catalyst and green solvent, ionic liquid has important applications in catalytic reduction. In this study, four quinuclidinol-based quaternary ammonium ionic liquids (ILs)were at first prepared using simple procedures, and then they were used as catalyst for acetophenone reduction. The main catalytic conditions were studied and screened through a comparison among the reaction results. It showed that the highest yield of 1-phenylethanol was 88.6% when [MenQu]Cl was used as a catalyst; the ideal solvent was ethanol; the amount of catalyst was 20 mol%, and the reaction temperature was 25 °C. More substrates were employed to validate the universality of the developed method using such an IL. Moreover, potential reaction mechanisms and an IL recovery method were also suggested, which laid the foundation for its further applications in the field of sustainable chemistry and cleaner industrial production.

Keywords: ionic liquid; catalysis; reduction; acetophenone; mechanisms

Citation: Zhou, G.; Chen, H.; Li, R.; Cao, Y.; Yao, S. Catalytic Reduction of Acetophenone Promoted with Quinuclidinol-Based Quaternary Ammonium Ionic Liquid as a Sustainable Solvent. *Eng. Proc.* **2023**, *55*, 37. https://doi.org/10.3390/engproc2023055037

Academic Editors: Teen-Hang Meen, Kuei-Shu Hsu and Cheng-Fu Yang

Published: 30 November 2023

1. Introduction

A cleaner and friendly technological route is the focus of both academic and industrial circles. As one of the new sustainable means to achieve an intensive development mode of economy and society, the use of ionic liquid (IL) is becoming more and more attractive. Ionic liquid is a kind of room-temperature-molten salt composed of organic cations and inorganic anions. It has many characteristics, such as its low vapor pressure, non-volatility, easy recovery, good stability, and so on. It has gradually become a good substitute for traditional organic solvents [1]. The application of this green solvent for catalytic reduction is an important sustainable way to solve the shortcomings of the existing reaction system. It has been widely used in organic synthesis because of advantages such as its high activity, strong catalytic effect, broad solubility, friendliness, adjustable function, flexible applied forms, low production cost, etc. Ionic liquids can be used as catalysts for olefine reduction, nitro reduction, carbonyl reduction, and other reactions and have a great application prospect in many reaction systems [2–4], especially for those used for the preparation of products with great value in the medical and healthcare fields.

Acetophenone, also known as acetylbenzene, is the simplest aromatic ketone, and its aromatic nucleus is directly connected to a carbonyl group. Acetophenone exists in the essential oils of some plants in a free state and has an aroma like hawthorn, so it can be used to prepare spices and as a raw material for pharmaceutical and other organic synthesis. Acetophenone can be reduced to form 1-phenylethanol (also known as α-phenylethylalcohol). This kind of product is a colorless liquid with a light gardenia flavor, which also has a wide range of applications in the pharmaceutical field. Benzeneethanol can be used as an analgesic, antibacterial, antiviral, and anti-tumor agent, among other things, and it can also be used as an intermediate in the preparation of some drugs, such as phenylethanolamine, phenylethanolic acid, etc. Traditional acetophenone reduction systems have shortcomings such as a low catalytic efficiency, high reagent costs, unfriendly conditions, and poor sustainability, which urgently need improvement [5].

At present, the applied ILs for reduction reactions include [Bmim][Br], [Bmim][BF_4], [Bmim][Al_mCl_n], [DMEA][Lac], [HBth]HSO_4, etc. These green sustainable solvents have been found to contribute to the formation of intermediates, reducing the activation energy and enhancing system miscibility. Overall, the current IL types are very limited, mainly consisting of the imidazolium type, and there is still room for further improvement in the yield of reduction products (usually below 80%). For instance, electroreduction with sodium borohydrate occurs in ionic liquids [Bmim][Br]/water mixtures. When a certain amount of ionic liquid is added to pure water as a reaction solvent, the substrate and sodium borohydrate dissolved in the water are changed from a two-phase system to a homogeneous system, with a final product yield of 75% [6]. In order to expand the types of IL used for reduction and explore potential reaction mechanisms, this study prepared a unique type of IL using quinolinol and menthol as the main raw materials, exploring its potential application possibilities in the catalytic field.

2. Experiment

2.1. Reagents and Materials

Acetophenone and isopropanol were provided byKelong Chemical Plant (Chengdu, China). R-3-quinuclidinol, menthol, chloroacetic acid, o-hydroxyacetophenone, o-methylacetophenone, p-methylacetophenone, p-nitroacetophenone, parabromoacetophenone, potassium hexafluorophosphate, potassium trifluoromethanesulfonate, sodium tetrafluoroborate, and p-bromophenone were purchased from Adamas Reagent Company (Shanghai, China). Silica thin-layer plates were sourced from Ocean Chemicals (Qingdao, China). N-hexane was purchased from Thermo Fisher Scientific (Waltham, MA, USA). Except for the chromatographic-grade isopropanol and n-hexane used for a quantitative analysis, all the reagents and solvents were of an analytical pure grade and were used without further purification if not stated otherwise.

2.2. Instruments

The quantitative determination of 1-phenylethanol was performed usingLC3000 high-performance liquid chromatography (Innovation Tongheng Technology Co., Ltd., Beijing, China) equipped with a Welchrom-C_{18} chromatographic column (4.6 mm × 250 mm, 5 μm) with methanol-water (60:40, v/v) and detected at 254 nm. When the enantiomers of the 1-phenylethanol were analyzed, the Chiralcel OD-H column (250 × 4.6 mm, 5 μm; DAICEL Inc., Tokyo, Japan) was used with an n-Hexane and iso-propanol mixture (95:5, v/v) in an isocratic mode at 0.9 mL/min [7].

2.3. Synthesis of Quinuclidinol-Based Quaternyl Ammonium Ionic Liquids

Taking the IL of [MenQu]Clas an example (see Figure 1), firstly, menthol and chloroacetic acid were combined to obtain the product of menthol chloroacetate rapidly. Secondly, 30 mL ethyl acetate was poured into a single neck bottom flask of 100 mL; then 5 mmol (0.635 g) of R-3-quinuclidinol was added, and ultrasonic oscillation was applied to assist its dissolution. After thorough dissolution, the round bottom bottle was put into an oil bath at 40 °C for heat preservation, and magnetic stirring was turned on. In addition, 5 mmol of menthol chloroacetate(1.164 g) was dissolved in 10 mL of ethyl acetate and added in the reactive system using the drop funnel with a speed of one to two drops per second. After adding the menthol chloroacetate solution, the drop funnel was replaced with a spherical condensate tube, and the condensate water was turned on; meanwhile, the oil bath's temperature was raised to 60 °C to continue the reaction. During the reaction, white solids continued to precipitate. After 5 h of reaction, the heating and stirring power was turned off. After the reactant was slightly cold, a great amount of white precipitates were obtained. After filtration, the filter cake was washed with an appropriate volume of ethyl acetate for three times to remove unreacted components and other impurities. After drying, the white solid was placed in the air-dryer (50 °C) for 24 h and then transferred to the vacuum-dryer for preservation. Three other ionic liquids with the same cation could be obtained via an anion exchange with corresponding salts, and then Cl^- was changed to be BF_4^-, PF_6^-, and $CF_3SO_3^-$. All the IL products were checked and confirmed using a spectral analysis.

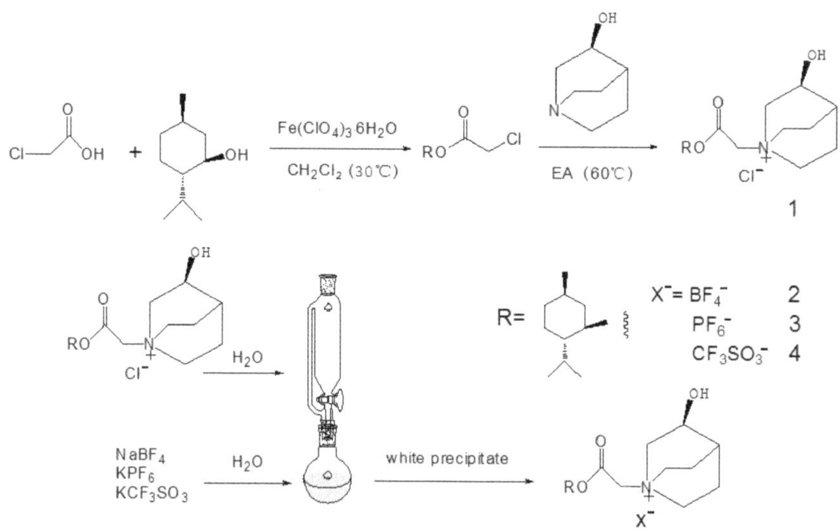

Figure 1. Schematic diagram of the experimental process of quinuclidinol-based quaternyl ammonium ILs (1: [MenQu]Cl; 2: [MenQu]BF_4; 3: [MenQu]PF_6; and 4: [MenQu]CF_3SO_3).

2.4. Reduction of Acetophenone

At room temperature, 5 mL of solvent (water/acetone/methanol, etc.) was added to the 25 mL round bottom flask; then, 1 mmol (0.1202 g) of acetophenone and 20 mol% of quinuclidinol-based quaternyl ammonium IL were also added to the flask with thorough mixing. Then, 0.0567 g (1.5 mmol) of sodium tetrafluoroborate was slowly added to the reactive system under continuous magnetic stirring. The reaction process was detected using thin-layer chromatography with chloroform-methanol as the developing reagent, and the position and size of the reactants and product spots were observed under a UV lamp with a wavelength of 254 nm. When the spots of the reactants completely disappeared, 5 mL of saturated ammonium chloride solution was added in the reactive system to quench the reduction reaction; then, 8 mL of ethyl acetate was applied to extract the product of 1-phenylethanol from the reaction solution three times, which could be obtained with liquid–liquid separation and concentration under vacuum. The yield was calculated using the results through liquid chromatography.

3. Results and Discussion

3.1. Investigation on Reaction Conditions

3.1.1. Effect of the IL Anions on the Reaction Results

The applied frequency of the Cl^-, BF_4^-, PF_6^-, and $CF_3SO_3^-$ of the IL is high for the catalysis in the current study, and the performance difference among them was investigated first. The experimental results are included in Table 1, and the reaction route is shown in Figure 2. After comparing the yield and reaction time of the reduction process, it was found that the IL with the best performance was [MenQu]Cl, and the yield was 77.7% (that of the other three was below 70%); moreover, the reaction duration was less than 3 h. The reason is that the polarity of ionic liquids depends on their anions when their cation remains unchanged; meanwhile, the reactive system is more inclined to form homogenization in the IL with Cl^- than in the ILs with BF_4^-, PF_6^-, and $CF_3SO_3^-$, so their reaction efficiency differed [8,9].

Table 1. Yields of catalytic reactions with four ILs with different anions.

Experimental Group	Catalyst	IL Content	Reaction Time/h	Yield (%)
1	[MenQu]Cl	20 mol%	2 h	77.7
2	[MenQu]BF_4	20 mol%	2 h	63.7
3	[MenQu]PF_6	20 mol%	3 h	66.3
4	[MenQu]CF_3SO_3	20 mol%	3 h	60.2

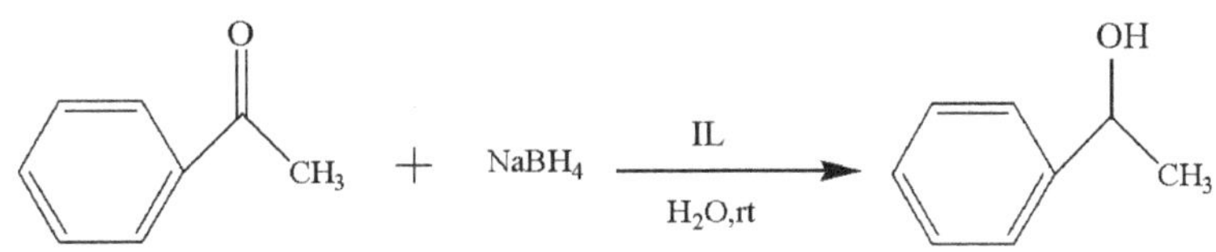

Figure 2. Reaction scheme for the investigation of IL anions.

3.1.2. Effect of Different Solvents on the Reaction Results

The experimental results of solvent replacement are summarized in Table 2, and the reaction route is shown in Figure 3. After screening the solvents, the best candidate is found to be EtOH, which can result in the highest yield (88.6%) within the shortest duration (1 h). This may be related to the viscosity, polarity, and solubility of these solvents, and the reaction with the solvent system with a good miscibility with acetophenone is faster [10]. To make the reaction more efficient, ethanol was selected as the ideal solvent for the reaction in subsequent experiments.

Table 2. Results of catalytic reactions in different solvents.

Experimental Group	Catalyst	Solvent	Reaction Time/h	Yield (%)
1	[MenQu]Cl	isopropanol	4.5 h	70.0
2	[MenQu]Cl	THF	20 h	54.29
3	[MenQu]Cl	EtOH	1 h	88.6
4	[MenQu]Cl	MeOH	1 h	79.2
5	[MenQu]Cl	n-Hexanol	24 h	-
6	[MenQu]Cl	DMF	4 h	73.0
7	[MenQu]Cl	glycol	9 h	87.8
8	[MenQu]Cl	glycerol	9 h	67.9
9	[MenQu]Cl	n-Butanol	24 h	-
10	[MenQu]Cl	H_2O	2 h	77.7

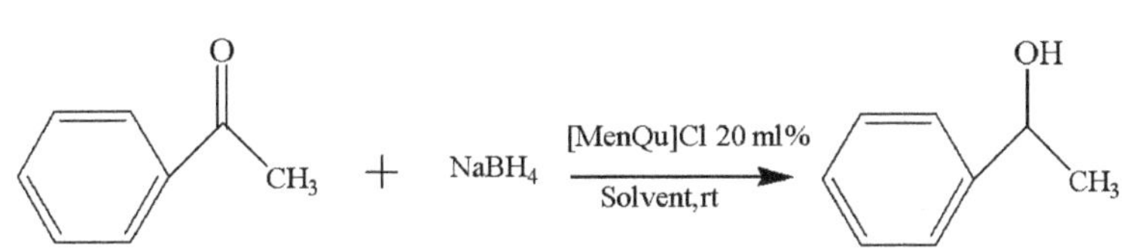

Figure 3. Reaction scheme for the investigation on different solvents.

3.1.3. Effect of IL Dosage on the Reaction Results

Through the screening experiment of catalyst dosage, it is found that the amount of IL catalyst has little effect on the experimental results; the specific results are included in

Table 3, and the reaction route is shown in Figure 4. This suggests that a small amount of IL can achieve the desired catalytic effect and that its activity is very high, while a large amount of IL can actually cause an increase in system viscosity, as well as a difficult recovery and unnecessary waste; at the same time, the probability of side reactions may also increase as a result [11]. Finally, the 20 mol% [MenQu]Cl was selected to explore the subsequent reaction conditions.

Table 3. Effect of IL dosage on yield of reaction results.

Experimental Group	Catalyst	IL Content (mol%)	Reaction Time (h)	Yield (%)
1	[MenQu]Cl	10	1	78.9
2	[MenQu]Cl	20	1	88.6
3	[MenQu]Cl	40	1	82.5
4	[MenQu]Cl	60	1	81.6
5	[MenQu]Cl	100	1	80.7

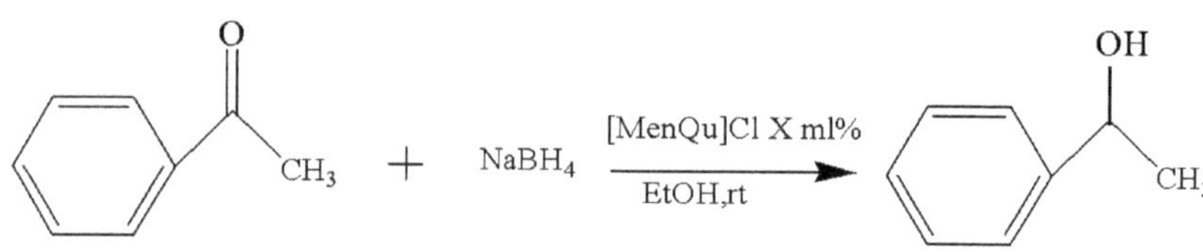

Figure 4. Reaction scheme for the investigation on IL dosage.

3.1.4. Effect of Temperature on the Reaction

Based on the reaction route in Figure 5, the data in Table 4 show that the change of temperature has a less obvious effect on the yield of the reaction. Considering that the freezing point of ethanol is much lower than 0 °C, here, we have investigated the reaction results below the freezing point. It is found that mild conditions under room temperature can achieve the satisfied results. However, with the decrease in temperature, the reaction time will be prolonged; meanwhile, it still maintains a yield near 70% at −20 °C. Previous studies also indicate that the temperature may affect the reaction efficiency without noticeably changing the reaction yield, but it can have an impact on enantiomers [12,13]. According to the experimental screening results, the subsequent reaction at a room temperature of 25 °C was selected.

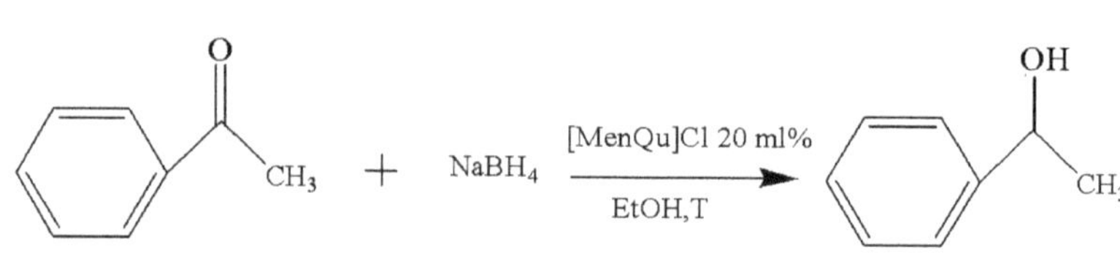

Figure 5. Reaction scheme for the investigation on temperature.

Table 4. Effect of temperature on reaction results.

Experimental Group	Catalyst	Reaction Temperature (°C)	Reaction Time (h)	Yield (%)
1	[MenQu]Cl	25	1	88.6
2	[MenQu]Cl	10	1.5	73.1
3	[MenQu]Cl	0	2	71.9
4	[MenQu]Cl	−10	4	70.3
5	[MenQu]Cl	−20	10	68.9

3.2. Expansion of Reaction Substrate

With the reaction route in Figure 6, it is found from the results in Table 5 that the developed IL-based catalytic method is applicable to other substrates with similar structures of acetophenone. Here, o-hydroxyacetophenone, o-methylacetophenone, p-methylacetophenone, p-nitroacetophenone, and parabromoacetophenone were selected to make comparison with acetophenone. It can be found that different R groups have some influence on the reaction carbonyl group and that the reaction efficiency of the electron-absorbing group is higher [12]. Therefore, the best catalytic result is still obtained with the [MenQu]Cl-promoted reduction of p-nitroacetophenone within 0.5 h, and the reaction speed is high.

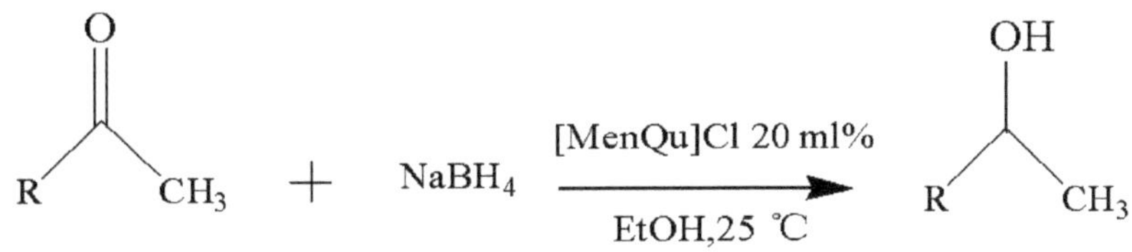

Figure 6. Reaction scheme for the investigation on different substrates.

Table 5. Reaction results of different substrates.

Experimental Group	Reaction Substrate	Reaction Temperature (°C)	Reaction Time (h)	Yield (%)
1	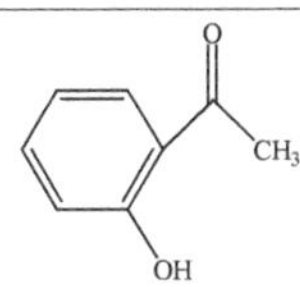	25	2	88.5
2	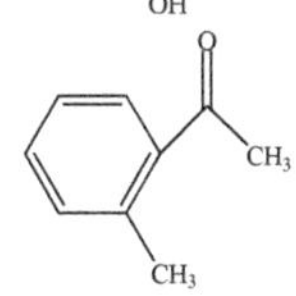	25	4	91.3
3	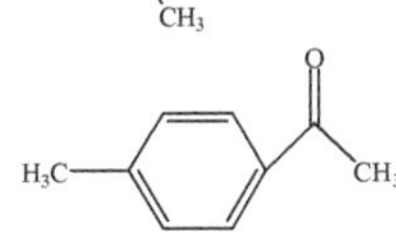	25	4	84.2
4	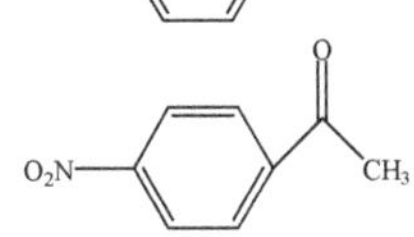	25	0.5	94.8
5	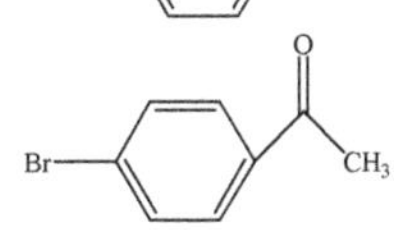	25	1	86.1

3.3. Potential Reaction Mechanism

Based on the above results and the existing literature [5,6], the following potential mechanism is suggested. As shown in Figure 7a, sodium borohydride first forms a reductive borane complex with the ionic liquid, and then a hydrogen of borane is removed, attacking the carbonyl's carbon-positive ion, finally forming a new chiral center. If the system where the N^+ center is located has a stabilizing effect on the carbonyl group of acetophenone, the transition state is easier to form and more stable. Furthermore, through the steric hindrance in the menthol plane and the steric hindrance outside the ring of quinuclidinol, the spatial steric hindrance of menthol is promoted; the anions of BH_4^- from the back of the bridge ring approach the center of N^+, so acetophenone is close to the anions of BH_4^- from the back. Therefore, the main product should be 1-phenylethanol with α-OH, which has been confirmed with the enantioseparation results using achiral chromatographic column under the analytical conditions in Section 2.2.

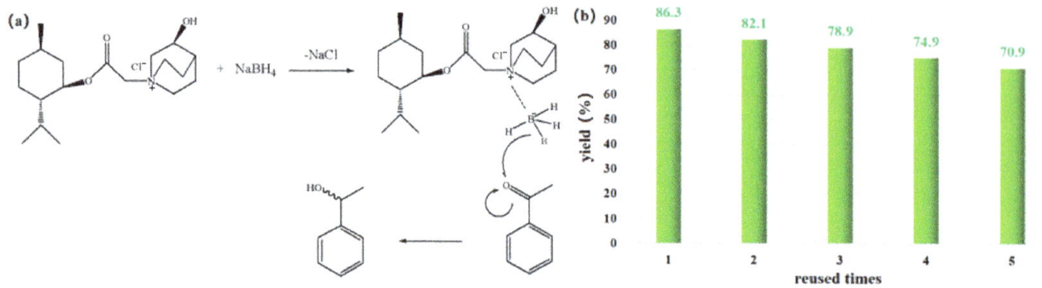

Figure 7. (**a**) Possible reduction mechanisms catalyzed via IL and (**b**) yield in the IL reused process.

3.4. Recovery and Reuse of IL

As a sustainable solvent, a good recyclability is required for [MenQu]Cl. After the reaction was completed, ethyl acetate was used to extract the product, and then isochoric n-butanol was further applied to extract and recover the ionic liquid from the residual solution two times. After back-washing with redistilled water, the n-butanol extract was concentrated and dried under vacuum and then used for the next parallel reuse experiment. The results of five recycling experiments are shown in Figure 7b (when the reused time = 0, yield = 88.6%), indicating that the above recovery method can maintain an acceptable IL catalytic activity over a certain number of cycles. But, when the color of the recovered [MenQu]Cl was found to have significantly deepened, a more complex regeneration way was needed, and the experimental results suggested that the D101 macroporous resin can achieve a satisfactory performance with 95% of ethanol as the eluent reagent.

4. Conclusions

The newly synthesized quinuclidinol-based quaternary ammonium ionic liquids were applied to the hydrogenation and reduction of chiral ketones. The optimum catalytic conditions were determined by screening the catalyst type, reaction solvent, catalyst dosage, reaction temperature, and so on. Through the investigation on their effects, the results indicate that the catalytic activity of different ILs from high to low follows the order of [MenQu]Cl > [MenQu]BF_4 > [MenQu]PF_6 > [MenQu]CF_3SO_3. The reaction can be completed under mild conditions around room temperature, and a low IL dosage is enough for its effective effect. The most effective catalytic conditions were as follows: [MenQu]Cl as a catalyst, ethanol as a solvent, a catalyst dosage of 20 mol%, and a reaction temperature of 25 °C. The yield of 1-phenylethanol was 88.6%. It was found in the substrate expansion experiments that similar substrates can be also suitable for such an IL catalyst, which means that it has good universality. On the basis of the above results and current studies, possible experimental mechanisms were suggested, which provided a meaningful reference for the subsequent structure optimization of the catalyst. Finally, the IL was well-reused through simple recovery procedures, and it is necessary for sustainable chemistry and cleaner production processes.

Author Contributions: Conceptualization, G.Z. and R.L.; methodology, G.Z.; software, H.C.; validation, Y.C.; formal analysis, R.L.; data curation, H.C.; writing—original draft preparation, G.Z.; writing—review and editing, S.Y.; visualization, Y.C. All authors have read and agreed to the published version of the manuscript.

Funding: This research was funded by Sichuan Science and Technology Program (No. 2021YFG0276).

Institutional Review Board Statement: Not applicable.

Informed Consent Statement: Not applicable.

Data Availability Statement: Data are contained within the article.

Acknowledgments: All the authors' affiliations provided the convenience for the related studies, respectively. A special thanks goes to the Engineering Experimental Teaching Center, in the School of Chemical Engineering at Sichuan University, for related measurements and characterizations, including FT-IR and NMR, etc.

Conflicts of Interest: The authors declare no conflict of interest.

References

1. Pendleton, J.N.; Gilmore, B.F. The antimicrobial potential of ionic liquids: A source of chemical diversity for infection and biofilm control. *Inter. J. Antimicrob. Agents* **2015**, *46*, 131–139. [CrossRef] [PubMed]
2. Zhao, Y.; Han, B.; Liu, Z. Ionic-liquid-catalyzed approaches under metal-free conditions. *Acc. Chem. Res.* **2021**, *54*, 3172–3190. [CrossRef] [PubMed]
3. Chinnappan, A.; Tamboli, A.H.; Chung, W.J.; Kim, H. Green synthesis, characterization and catalytic efficiency of hypercross-linked porous polymeric ionic liquid networks towards 4-nitrophenol reduction. *Chem. Eng. J.* **2016**, *285*, 554–561. [CrossRef]
4. Javle, B.R.; Kinage, A.K. Chiral amino-acid-amide based ionic liquids as a stereoselective organocatalyst in asymmetric transfer hydrogenation of acetophenone at room-temperature. *Chemistryselect* **2018**, *3*, 2365–6549. [CrossRef]
5. Singh, S.K.; Savoy, A.W. Ionic liquids synthesis and applications: An overview. *J. Mol. Liq.* **2020**, *297*, 112038. [CrossRef]
6. Su, Z.; Zhang, L.; Zhang, Y.P. Ionic liquid/water mixed solvent promotes the reduction of aldosterone. *J. Lanzhou Univ. Technol.* **2008**, *34*, 74–76.
7. Özlem, A.; Emine, B.; Ülkü, M. Determination of effective diffusion coefficient of acetophenone in κ-carrageenan and asymmetric bioreduction in packed bed reactor. *J. Mol. Catal. B Enzym.* **2011**, *72*, 46–52.
8. Yu, B.; Zhou, F.; Liu, G.; Liang, Y.; Huck, W.T.S.; Liu, W. The electrolyte switchable solubility of multi-walled carbon nanotube/ionic liquid (MWCNT/IL) hybrids. *Chem. Commun.* **2006**, *22*, 2356–2358. [CrossRef]
9. Ogoshi, T.; Onodera, T.; Yamagishi, T.-A.; Nakamoto, Y. Green polymerization of phenol in ionic liquids. *Macromolecules* **2008**, *41*, 8533–8536. [CrossRef]
10. Bertero, N.M.; Trasarti, A.F.; Apesteguía, C.R.; Marchi, A.J. Solvent effect in the liquid-phase hydrogenation of acetophenone over Ni/SiO_2: A comprehensive study of the phenomenon. *Appl. Catal. A Gen.* **2011**, *394*, 228–238. [CrossRef]
11. Akopyan, A.V.; Eseva, E.; Polikarpova, P.; Kedalo, A.; Vutolkina, A.; Glotov, A.P. Deep Oxidative desulfurization of fuels in the presence of brönsted acidic polyoxometalate-based ionic liquids. *Molecules* **2020**, *25*, 536. [CrossRef] [PubMed]
12. Xu, J.; Wei, T.; Zhang, Q. Effect of Temperature on the Enantioselectivity in the Oxazaborolidine-Catalyzed Asymmetric Reduction of Ketones. Noncatalytic Borane Reduction, a Non-neglectable Factor in the Reduction System. *J. Org. Chem.* **2003**, *68*, 10146–10151. [CrossRef]
13. Sun, W.; Xia, C.G.; Zhao, P.Q. Chiral salen-Co(II) complex catalyzed asymmetric reduction of acetophenone with sodium borohydride. *Acta Chim. Sin.* **2001**, *59*, 976–978.

engineering proceedings

MDPI

Proceeding Paper

Application of Geospatial Data to Architectural Design Education †

Kenji Sato

Department of Architecture, Faculty of Science and Technology, Shizuoka Institute of Science and Technology, 2200-2 Toyosawa, Fukuroi City, Shizuoka Prefecture 437-8555, Japan; sato.kenji@sist.ac.jp or ishikawa.haruno@sist.ac.jp

† Presented at the IEEE 5th Eurasia Conference on Biomedical Engineering, Healthcare and Sustainability, Tainan, Taiwan, 2–4 June 2023.

Abstract: The objective of this study is to incorporate varied geospatial data into CAD/BIM systems, which are now essential design tools for architects. Those geospatial data include digital information of topography, water features, roads and streets, existing buildings, and so on. Those data have been released by the government as public information, and their full utilization in actual design projects would be highly beneficial.

Keywords: geospatial data; architectural design; CAD; BIM; GIS

Citation: Sato, K. Application of Geospatial Data to Architectural Design Education. *Eng. Proc.* **2023**, *55*, 38. https://doi.org/10.3390/engproc2023055038

Academic Editors: Teen-Hang Meen, Kuei-Shu Hsu and Cheng-Fu Yang

Published: 30 November 2023

1. Background

In the current field of Architectural Design Education, a course of learning CAD (Computer Aided Design) is incorporated in the curriculum, and students are encouraged to use CAD for their design studies. Given the fact that the CAD system has widely been used in design firms, students need to improve their skills in using CAD in their school days. To meet with the desire of architects for the novelty in designs, the software as a design tool has evolved gradually since 1980s. Nowadays, architects can easily handle the complexity of geometric form such as solid objects and free curved surfaces.

Recently its evolution is observed in the design system called BIM (Building Information Modelling). There, such architectural components as column, wall, floor, roof, etc., are assembled within a virtual 3-dimensional space. They are 3D solid objects, and each object knows how to behave itself when it meets with the other components.

On the other hand, in the field of geography, valuable geospatial data have been prepared and provided by the government, so as to meet varied public needs. The extensive use of the Satellite Positioning and Navigating System has backed up this trend. In Japan, the Geospatial Information Authority (hereinafter GSI) started the distribution of "Fundamental Geospatial Data" in 2007 under the law of the "Basic Act on the Advancement of Utilizing Geospatial Information". Those data include topographic contour lines, water features, railways, streets and sidewalks, footprints of existing building, and so on. This set of information will give a tremendous advantage to architectural design studies. They are freely downloadable and cover the entire land space of Japan. Those data can be read and visualized on the software called GIS (Geographic Information System).

However, at this point, architects are not able to easily handle those valuable data in their design practices. Those geographic data are not transferrable among the systems of CAD, BIM, and GIS. From an architect's viewpoint, this prevents the full utilization of data related to an existing urban context for a specific design project.

For architectural design practices, those design tools need to be integrated so that a seamless dataflow between the systems is enabled. To examine the portability of data, a piece of land, approximately the surface area of 1000 m × 750 m adjacent to the campus of the Shizuoka Institute of Science and Technology, is chosen as a case study site.

In this study, AutoCAD 2024, Revit 2024 and QGIS 3.26 are used as basic design tools. Additionally, a series of Python and Lisp scripts are written and utilized for the data transfer from GIS to CAD. In the Python codes, the GeoPandas library is used for data analysis and to coordinate conversion between the global polar and the local orthogonal systems.

2. Site

The site for the case study is shown on the map and in the aerial photo below (Figures 1 and 2), which were visualized on QGIS (https://qgis.org). The specified land is located on a hilly terrain, and several settlements extend along the valleys. Geospatial data within the boundary, as shown in a red rectangle, are to be transferred to CAD systems. Both the map and the aerial photo are raster images on GIS at this stage, and vector data of varied features are needed for the development on CAD systems. The size of the bounding rectangle is 1000 m × 750 m. On QGIS, the CRS (Coordinate Reference System) of the image is set to Japan Plane Rectangular CS VIII, which covers the area over the Shizuoka Prefecture.

Figure 1. Site for case study/1000 m × 750 m.

Figure 2. Aerial photo and the bounding rectangle.

3. Fundamental Geospatial Data

Anyone can freely access and download the "Fundamental Geospatial Data" on the website of the Geospatial Information Authority (GSI). On the download page, "Fundamental Geospatial Data" or "Digital Elevation Model" can be chosen. Both data are used for this study. It is noteworthy that those digital data cover the entire land space of Japan.

Downloaded data are stored in a GML file format, which is a kind of text data containing the coordinates of longitude, latitude and altitude. This needs to be converted into Shape file format, which is commonly used in GIS systems. For this purpose, a short Python script is written using GeoPandas library. Pandas is a library for handling 2-dimensional tables like spread sheets, and GeoPandas is its extension to geospatial data. GeoPandas can read GML files and transform them into Shape files, while conducting coordinate conversions.

The image below (Figure 3) shows the Fundamental Geospatial Data of the area over Fukuroi city on QGIS. This set of geospatial data includes topographic contour lines, water features, railways, roads and streets, and existing buildings.

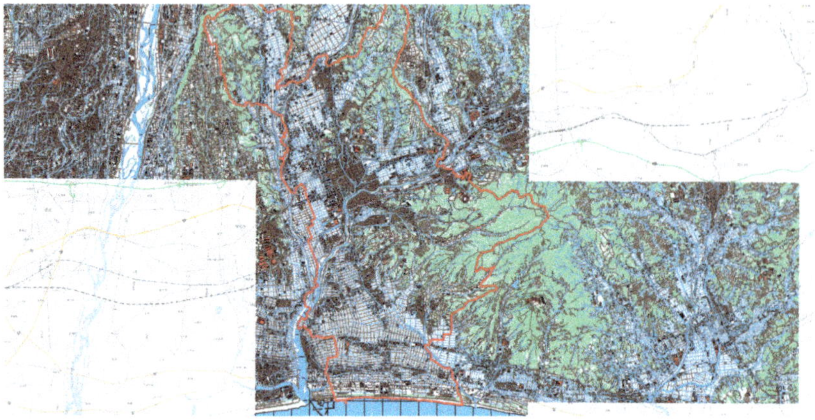

Figure 3. Fundamental Geospatial Data/Visualization on QGIS.

Now let us zoom into the specified land of 1000 m × 750 m for this case study (Figure 4). In this image, the map underlay is not displayed. The geospatial data of contour lines, water features, roads and buildings are all vectors consisting of coordinates of longitude and latitude. Converting those coordinate values into the local orthogonal system leads to the utilization of data in CAD and BIM systems. For this operation, we are reminded that the geospatial data are basically 2-dimensional. For instance, as for the contour data, the altitude is given as an attribute to the 2-dimensional point data.

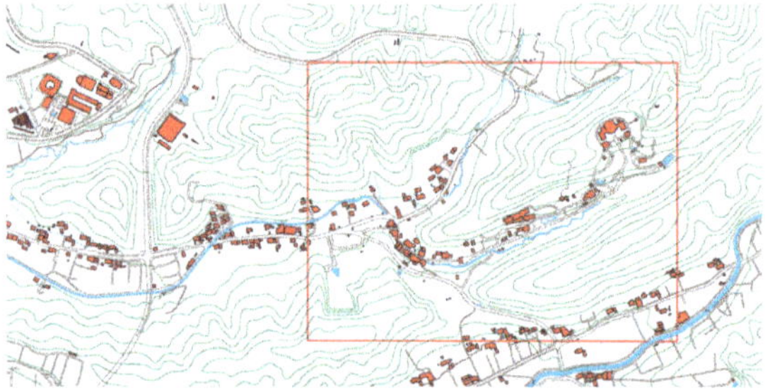

Figure 4. Zooming into the specified site.

In the following study, these contour data are not used because their vertical interval is set to 10 m. Instead, more precise information of the altitude is gained from the "Digital Elevation Model", and contour lines are calculated based on the altitude data on each point of the 5 m × 5 m mesh grid over the land. The accuracy of the data is reported to be less than 30 cm, and this figure is more than necessary for preliminary studies of a specific architectural design project.

4. Digital Elevation Model

The DEM (Digital Elevation Model) data are also downloadable from the website of GSI. The downloaded dataset is also stored in GML file format. However, it contains a list of three-dimensional point data and, therefore, GeoPandas cannot read the data. For reading and visualizing the data, one idea is to use a utility program called the Fundamental Geospatial Data Reader, which is provided by GSI. With that program, file conversion from GML format to a simple XYZ text format is possible. Another idea is to write a Python script to read GML files and output a simple CSV file, which contains a list of coordinates consisting of longitude, latitude, and altitude of points. The original dataset is huge, and another script for cropping data within the specified boundary is also needed.

So, data created in this way should be structured grid-data; however, the original data contain missing points over the water features. The dataset is discrete, and needs to be structured. For this, using Scipy library on Python, linear interpolation to the data is applied and a structured-grid dataset is achieved. Then, on the dense grid data, the calculation of contour lines is enabled with the help of ContourPy library.

5. Data Transfer to AutoCAD

In order to read geospatial data in AutoCAD [1], an intermediate data file is prepared. That file stores point data of the serially described x, y, and z coordinate values. To make this intermediate data file, a Python script is written. That script mainly works on two points: (1) the conversion of a coordinate reference system (from polar to orthogonal), and (2) the Boolean operation between objects ("features", in geospatial terms).

For CRS (Coordinate Reference System) conversion, GeoPandas library is used. More specifically, it works on the conversion from JGD2011 to JGD2011/Japan Plane Rectangular CS VIII. JGD2011 is the most common CRS in publications of GSI.

For cropping varied features within the rectangular boundary, Boolean operations are applied, which include such operations as union, subtract, and intersect between 2-dimensional geometric objects. A geospatial dataset comprises basically three object types: Point, LineString and Polygon. To work on those Boolean operations in Python codes, a library called Shapely is used.

Once this intermediate data file is created, then an application program to read and draw data in AutoCAD is needed. Historically, in AutoCAD, Lisp interpreter works as a man–machine interface for the system. Therefore, a Lisp script is written for reading geospatial data described in the intermediate files. There, the data types of Point, LineString, and Polygon are replaced with Point, Open Polyline and Closed Polyline of AutoCAD.

Thus, the geospatial data are imported to AutoCAD (Figures 5 and 6), which includes contour lines (3D Polyline), water features (2D Closed Polyline), roads (2D Polyline), and building footprints (2D Closed Polyline).

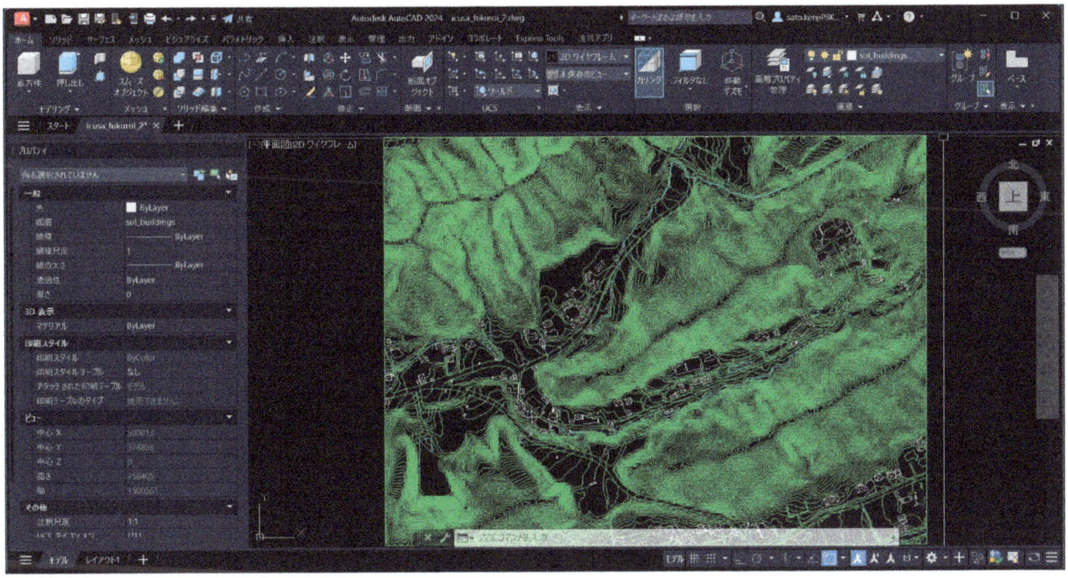

Figure 5. Visualization of data on AutoCAD/plan view.

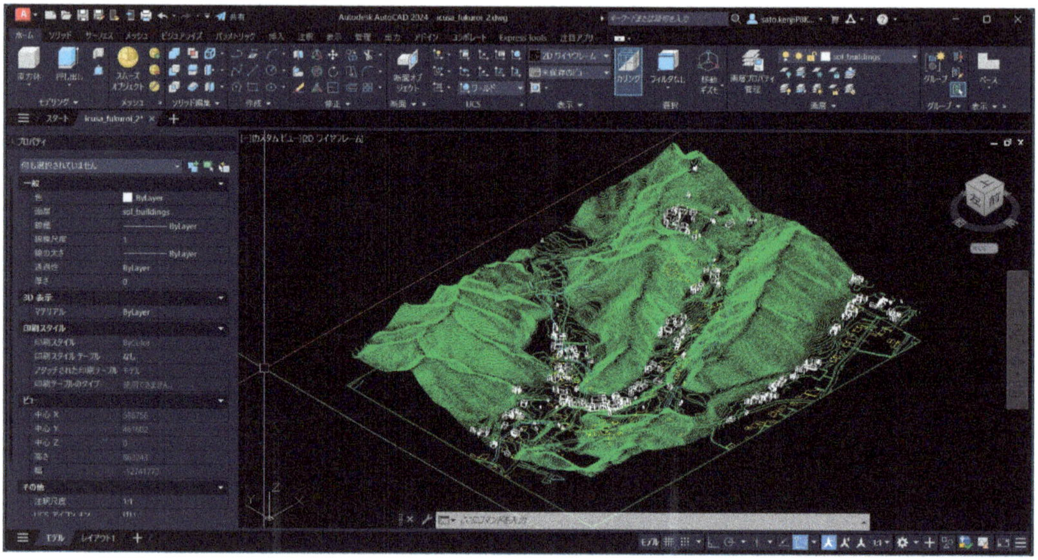

Figure 6. Visualization of data on AutoCAD/isometric view.

By extruding the drawn footprints, 3D building objects are created as 3D Solid. For this, building elevations need to be calculated beforehand. Since we already know the altitudes of every 5 m grid over the land, it is easy to calculate the elevation of each building. However, as for the height of the building, there is no information in the current Fundamental Geospatial Data. The profile of each building in the drawing does not represent its actual height. Knowing this limitation, the drawing still seems highly effective for architectural studies.

6. Development on Revit

On Revit, the topography of the land is generated by importing the grid data of altitude. The generated topographic object is called topo-solid in Revit (Figure 7). Regarding the roads and the water features, sub-regions on the topo-solid need to be specified.

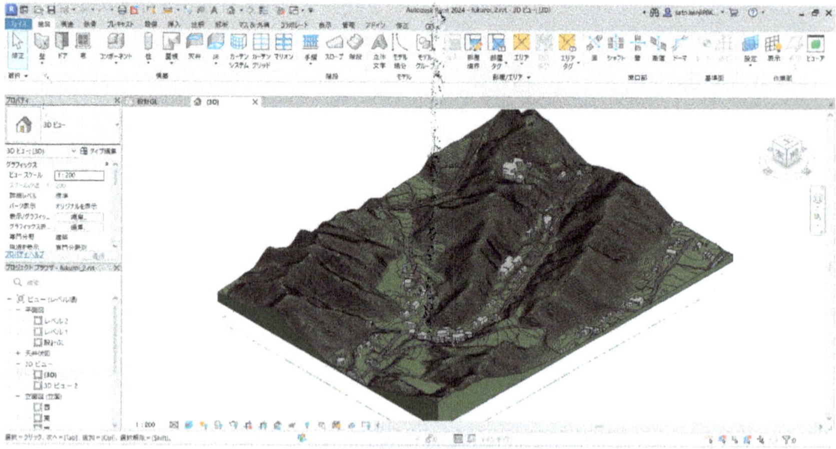

Figure 7. An environmental model of Revit/Topo-solid and sub-regions.

To create sub-regions, firstly AutoCAD data are imported, and the boundaries of each sub-region are traced. In Revit, this operation is called "sketching". In fact, this sketching is a time-consuming task, and another solution needs to be developed in the future. As for the buildings, they are imported from the AutoCAD data and converted into Mass objects of Revit.

On the surface of the topo-solid, you can easily place such landscape elements as natural trees, human profiles, and so on. Based on this environmental model, a specific site for a project can be set up as a sub-region on a topo-solid. Students will develop their design exercise on that project site.

7. Conclusions

The objective of this study has been to create an urban-scale model on CAD/BIM systems for architectural design on a specific site. By utilizing geospatial data, which are now available through public services, architects can extend their capability of handling the complexity of the real city.

The photograph below (Figure 8) is an example of a site model, which was made by students at a design studio for a certain project. That model is made with paper and styrofoam. Architects make such urban-scale models at the very beginning of the preliminary study for a project.

Figure 8. An example of a site model.

The aim of the study as so far carried out is to replace this real model with a virtual model on the computer. At this moment, the situation of data sharing between the CAD, BIM and GIS systems is inadequate. To read geospatial data, which basically comprise 2-dimensional geometric objects stored in the form of Shape file, and also to visualize the data on CAD/BIM systems, it is necessary to write a series of custom add-in programs at this stage. CAD, BIM and GIS systems are essentially the tools for manipulation of geometric objects and, therefore, integration of those tools is expected to be achieved soon.

Another subject to be resolved is related to the height, the shape, and the structure of existing buildings. At this moment, 2-dimensional information of existing buildings is provided in the "Fundamental Geospatial Data". When three-dimensional data of features are made available, our virtual model will become much closer to the real urban environment. It will definitely contribute to the design studies by architects.

Funding: This research received no external funding.

Institutional Review Board Statement: Not applicable.

Informed Consent Statement: Not applicable.

Data Availability Statement: The original GIS data were downloaded from the website of GSI https://www.gsi.go.jp/kiban (accessed on 15 May 2023). In this study, the following open-source libraries were used: Pandas (https://pandas.pydata.org/), GeoPandas (https://geopandas.org/en/stable/), Shapely (https://pypi.org/project/shapely/), scipy.spatial (https://scipy.org/), ContourPy (https://pypi.org/project/contourpy/).

Conflicts of Interest: The author declares no conflict of interest.

Reference

1. Sato, K. *Learning AutoCAD for Architectural 2D and 3D Drawings*; Gakugei Shuppansha: Kyoto, Japan, 2021; Chapters 15 and 16; ISBN 978-4-7615-3270-3.

engineering proceedings

Proceeding Paper

Using Machine Learning to Identify Product Styles †

Hung-Hsiang Wang and Yen-Ling Chen *

Department of Industrial Design, National Taipei University of Technology, Taipei 106344, Taiwan; wanghh@mail.ntut.edu.tw
* Correspondence: lillian95085@gmail.com
† Presented at the IEEE 5th Eurasia Conference on Biomedical Engineering, Healthcare and Sustainability, Tainan, Taiwan, 2–4 June 2023.

Abstract: The Waikato Environment for Knowledge Analysis (WEKA), a machine learning tool, was used to develop a model to identify product styles, and the style of classic chairs was determined using the model. Data used to develop the model consisted of 100 images of four styles of chairs such as Windsor, Shaker, Thonet, and Ming. After pre-processing the images using the image filters of WEKA, the images were used to train the model to classify chair styles. The accuracy of the model ranged from 96 to 98%. This validated the performance of the proposed method in classifying the styles of chairs, which helps the design of new chairs.

Keywords: product style; image recognition; machine learning

1. Introduction

In the field of industrial design, the style of a product may vary depending on the design concept, material, and function [1]. For the ever-changing needs of consumers, designers are looking for inspiration for better product designs. The design industry has been inspired by the art and styles of the past [2]. Product style is important in the communication of consumers with designers, allowing consumers to understand the function of the product and the concept and meaning of the product design [3]. Therefore, designers need to choose the most appropriate and accurate product style to meet the consumer's needs. Product style is defined by the visual impressions affected by shape, line, and decoration [4]. Designers categorize styles based on experience by referring to relevant image recognition theories. The following are the relevant theories of image recognition.

- Feature-matching theory: Things or shapes have attributes or features that must be analyzed while considering the quality and quantity of the attributes [5].
- Shape recognition: If the pattern is identified for each component, the recognition of the whole pattern is required with generalization, which is a bottom-up processing of the feature comparison theory. If the pattern is identified by the overall pattern, each component can be identified. This process is called a top-down process of the template comparison theory [6].

Manual classification is not rational or sufficiently scientific to classify product styles. Therefore, machine learning (ML) is used for classification models in design. ML imitates the central nervous system of humans to learn multi-level concepts and has been applied to image, motion, and speech recognition [7]. Image processing with ML is used to identify colors and textures in images to determine features [8]. Those features are used with hue saturation value (HSV) extraction and the gray level co-occurrence matrix (GLCM) calculation [9]. In the manufacturing industry, the inspection of parts is important. Therefore, deep learning (DL) is used to learn the correct object image and identify defective products [10]. Davis stated that it was difficult to define the boundaries of images but that ML could be used to build a database for the classification of images [11].

Citation: Wang, H.-H.; Chen, Y.-L. Using Machine Learning to Identify Product Styles. *Eng. Proc.* **2023**, *55*, 39. https://doi.org/10.3390/engproc2023055039

Academic Editors: Teen-Hang Meen, Kuei-Shu Hsu and Cheng-Fu Yang

Published: 1 December 2023

The recognition of the product style belongs to image recognition and concept classification [3]. Thus, ML can be used as a tool for the development of products and designs. In industrial design, the product style is determined according to the design concept, material, and function [1]. Thus, it is necessary for designers to accurately identify the product. The purpose of this study is to investigate whether Waikato Environment for Knowledge Analysis (WEKA) can be used as an effective method for the categorization of the product style. In this study, we used the designs of chairs to investigate the use of the WEKA model for the extraction of the features of the images. The model was trained to recognize and classify product styles. The result helps develop new products and their styles that are preferred by consumers.

2. Method

2.1. Chairs

There are a variety of design styles for chairs according to history, region, designer, or thought. Whether it is modern or classical; Chinese or European; or made of solid wood, steel, bamboo, or man-made materials, each style represents the history and design philosophy [4]. Therefore, the most representative and influential chairs were selected in this study. In the history of furniture, the Windsor chair represents the origin of the chair. For hundreds of years, throughout Europe and the United States, it has been used, and this has greatly contributed to the popularization of the chair [12]. Mr. Sanshiro Ikeda of Matsumoto Folk Art Furniture, who made the Windsor chair in Japan, said, "The Windsor chair is the ultimate chair [13]. This shows that the style of the Windsor chair really influenced the development of the chair in the future".

After the Second World War, Wegner's "The Chair", with its organic curves from the back to the arms and legs made of teak wood, became a model of Danish organic design [14]. Hans Wegner designed more than 500 chairs in his lifetime, the most classic being the Chinese Chair designed in 1944. His design concept was inspired by the ancient tradition of Corinthian design, and he developed this, bringing the essence of the Chinese Ming-style circle chair [15]. The smooth lines show a deep Chinese flavor and foundation of the Ming-style chair. Shimazaki said that in the history of chairs, whenever a new material was developed, an epoch-making chair was introduced [16]. Therefore, in the case of chairs made of wood, the wood-bending technology developed by Thonet allows for a wider range of chair designs. It is even possible to produce a curved surface that conforms to the human body, and mass production is also possible. Thonet's wood-bending technology developed the furniture manufacturing industry and influenced the furniture of future generations. Therefore, we selected the Windsor chair, the Thonet chair, the Ming chair, and the Shaker chair for this study.

2.2. Research Methodology

To use the WEKA model to identify the product style, appropriate images needed to be chosen for the model to learn. A total of 100 pictures of each style of the four chairs were collected as shown in Table 1. Categories were defined (Table 2) for the chairs that did not belong to these four styles to help the model detect the style of chairs accurately. The final database was imported into the data for learning. The styles of the chairs were classified into five categories (Figure 1).

In the first phase of this study, we identified the filter to extract the image features and the classifier to classify the recognized image features for ML. In the second phase, we investigated whether the proposed WEKA model classified the styles of chairs accurately. Table 3 shows the result of the classification of the chair style with different filters. The best accuracy was obtained with JpegCoefficientFilter. Compared with the other filters, the accuracy (95%) of JpegCoefficientFilter was higher with a lower absolute difference, which means that the stability of this filter was higher. The filter captured features by calculating quantile coefficients in the image that were not visible to humans. The absolute difference was the other way to verify the feature extraction.

Table 1. Four types of chair styles.

Images				
Chair	Shaker	Ming	Thonet	Windsor

Table 2. Chairs not classified as these four styles.

Figures	
Chair	Other

Selected attribute			
Name: class		Type: Nominal	
Missing: 0 (0%)	Distinct: 5	Unique: 0 (0%)	
No.	**Label**	**Count**	**Weight**
1	Ming	100	100
2	Other	110	110
3	Shock	110	110
4	Thonet	110	110
5	Windsor	110	110

Figure 1. Five categories for machine learning training.

Table 4 shows that the correctness of each classifier was similar, but the performance of SMO was significantly lower than the other classifiers in terms of failure rate and absolute difference. This indicated that SMO was more stable in classification.

Table 3. Accuracy of four methods with different image filters.

Filter	Function	Accuracy	Absolute Dispersion
BinaryPatterns PyramidFilter	Extraction of rotation-invariant numerical histograms of local binary patterns from images.	90%	25%
EdgeHistogramFilter	A filter for extracting MPEG7 boundary histogram features from pictures.	85%	37%
JpegCoefficientFilter	A batch filter for extracting JPEG coefficients from images	95%	11%
PHOGFilter	A filter for extracting the directional gradient histogram value PHOG from the image.	90%	26%

Table 4. Accuracy rates of the five methods with different classifiers.

Classifier	Success Rate	Failure Rate	Absolute Dispersion
SMO	98%	0.073	8%
J48	96%	0.097	16%
RandomForest	99%	0.096	19%
RandomComitee	99%	0.097	16%
RandomSubspace	99%	0.097	18%

The JpegCoefficientFilter filter and the SMO classifier were tested in the first phase for image recognition and the most stability. Then, they were tested in the second phase to classify the chair styles. The accuracy of the training was 100%, which implied that the WEKA model accurately learned the images (Figure 2).

```
Correctly Classified Instances         620              100      %
Incorrectly Classified Instances         0                0      %
Kappa statistic                          1
Mean absolute error                      0.0625
Root mean squared error                  0.1749
Relative absolute error                 97.1721 %
Root relative squared error             97.8816 %
Total Number of Instances              620

=== Detailed Accuracy By Class ===

                 TP Rate  FP Rate  Precision  Recall   F-Measure  MCC      ROC Area  PRC Area  Class
                 1.000    0.000    1.000      1.000    1.000      1.000    1.000     1.000     Ming
                 1.000    0.000    1.000      1.000    1.000      1.000    1.000     1.000     Other
                 1.000    0.000    1.000      1.000    1.000      1.000    1.000     1.000     Shock
                 1.000    0.000    1.000      1.000    1.000      1.000    1.000     1.000     Thonet
                 1.000    0.000    1.000      1.000    1.000      1.000    1.000     1.000     Windsor
```

Figure 2. Numerical result of classification of five categories.

The chairs in Figure 3 were used to train the model for the prediction and classification of chair styles. The results are shown in Figures 3–5. Chairs 5, 6, 7, and 8 in Table 2 were classified as Ming style. Chair 9 was classified as a new style. This prediction was accurate because the WEKA model accurately classified chairs. The WEKA model can be used to classify other product styles without manual classification and the influence of subjectivity and uncertainty.

Figure 3. Random chair images.

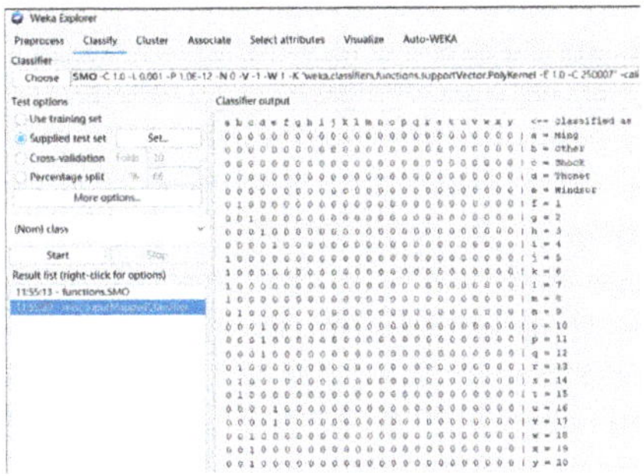

Figure 4. Classification matrix.

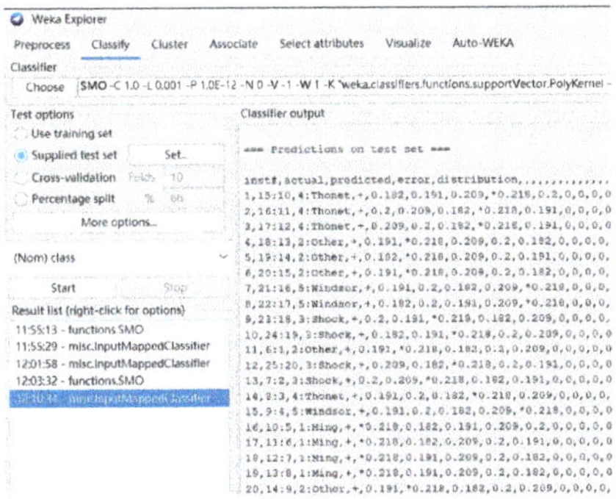

Figure 5. Predicted data by WEKA model.

3. Conclusions

We demonstrated the effectiveness of the Weka model in classifying the four styles of 100 chairs. The results indicated that industrial designers could use the proposed model to classify and identify product styles. New applications of ML could be developed for product design based on the result of this study. Future research is required to classify and identify product styles using AI technology.

Author Contributions: Conceptualization, H.-H.W.; method, H.-H.W.; resources, Y.-L.C.; data curation, Y.-L.C.; valida-tion, H.-H.W.; writing—original draft preparation, Y.-L.C.; writing—review and editing, H.-H.W.; experiments conducted, Y.-L.C.; supervision, H.-H.W. All authors have read and agreed to the published version of the manuscript.

Funding: The authors gratefully acknowledge the financial support of NSTC, Taiwan, through grant 111-2410-H-027-019-MY2.

Institutional Review Board Statement: Not applicable.

Informed Consent Statement: Informed consent was obtained from all subjects involved in the study.

Data Availability Statement: Data are unavailable due to privacy reasons.

Conflicts of Interest: The authors declare no conflict of interest.

References

1. Tsai, H.-C. An Analysis of Study on the Features Design from Hans J. Wegner in Denmark. Master's Thesis, National Taiwan Normal University, Taipei, Taiwan, 2011.
2. Bhaskaran, L. *Design of THE Times: Using Key Movements and Styles for Contemporary Design Paperback*; RotoVision: Brighton, UK, 2005.
3. Chan, C. Exploring individual style in design. *Environ. Plan B Plan. Des.* **1992**, *19*, 503–523. [CrossRef]
4. Qiu, M.; Cheng, Y.; Chen, P. *Taiwanese Wood Furniture New Craftmanship*; Jackwood Culture: Taipei, Taiwan, 2007.
5. Cheng, C. *Cognitive Psychology—Theory and Practice*; Laureate Publishing: Slaton, TX, USA, 1993.
6. Pahl, G.; Beitz, W. *Engineering Design: A Systematic Approach*; The Design Council: London, UK, 1988.
7. Kuo, P. *Image Super Resolution Based on Deep Learning*; National Cheng Kung University: Tainan, Taiwan, 2015.
8. Kim, D.; Shin, K.; Woo, J. Displacement Measurement of Steel Pipe Support Using Image Processing Technology. *J. Image Graph.* **2020**, *8*, 80–84. [CrossRef]
9. Vimina, E.R.; Poulose, K. Content Based Image Retrieval Using Low Level Features of Automatically Extracted Regions of Interest. *J. Image Graph.* **2013**, *1*. [CrossRef]
10. Syakirin, N.; Fauadi, M.; Awang, N. Some Technique for an Image of Defect in Inspection Process Based on Image Processing. *J. Image Graph.* **2016**, *4*, 55–58. [CrossRef]
11. Haffey, M.K.D.; Dufty, A.H.B. Knowledge discovery and data control mining in design environments. In *From Knowledge Intensive CAD to Knowledge Intensive Engineering*; Cugini, U., Wozny, M., Eds.; ITIFIP; Springer: Boston, MA, USA, 2000; Volume 79, p. 59.
12. Kane Watanabe, W. *Western Furniture Integration*; Kodansha Ltd.: Tokyo, Japan, 1980.
13. Nishikawa, R. *Illustrated Classic Chairs*; Taiwan Tohan: Taipei, Taiwan, 2015; pp. 93, 224.
14. Hauffe, T. An Analysis of Study on the Features Design from Hans J. Wegner in Denmark. In *Design: A Concise History*; Laurence King: London, UK, 1998; pp. 126–127.
15. Fang, H.; Hu, J.; Peng, L. *History of Modern Furniture in the World*; Central Compiler: Beijing, China, 2005; pp. 270–275.
16. Shimazaki, S. *Japanese Chairs*; Seibundo Shinkosha: Tokyo, Japan, 2007.

engineering proceedings

MDPI

Proceeding Paper

Effect of Learning Style on Non-Programmed Computational Thinking Activities †

Yen-Yun Chen [1], Shih-Wen Su [2], Li-Xian Chen [3], Chia-Hung Liao [2] and Shyan-Ming Yuan [2,*]

1 Institute of Network Engineering, National Yang Ming Chiao Tung University, Hsinchu 30010, Taiwan; yychen.c@nycu.edu.tw
2 Department of Computer Science, National Yang Ming Chiao Tung University, Hsinchu 30010, Taiwan; alvin.cs00@nycu.edu.tw (S.-W.S.); aiallen.cs07g@nctu.edu.tw (C.-H.L.)
3 School of Big Data, Fuzhou University of International Studies and Trade, Fuzhou 350202, China; lixian.cs98g@g2.nctu.edu.tw
* Correspondence: smyuan@nycu.edu.tw
† Presented at the IEEE 5th Eurasia Conference on Biomedical Engineering, Healthcare and Sustainability, Tainan, Taiwan, 2–4 June 2023.

Abstract: As information explosion and algorithms are proliferated in the digital age, computational thinking becomes more critical. Nevertheless, most activities that promote computational thinking are always presented in a programming or block language, which tends to horrify students. This study aimed to develop a series of non-programmed, computational-thinking training activities to demonstrate if they improve students' computational thinking skills without writing code. In the meantime, it is checked whether students' learning styles influence their outcomes. After an investigation consisting of a 16 h gamified learning session, we observed that students' computational thinking skills improved overall. Furthermore, the initial learning style of the students influenced the effect of different dimensions of computational thinking in the learning procedures.

Keywords: computational thinking; learning style; game-based learning; digital learning

Citation: Chen, Y.-Y.; Su, S.-W.; Chen, L.-X.; Liao, C.-H.; Yuan, S.-M. Effect of Learning Style on Non-Programmed Computational Thinking Activities. *Eng. Proc.* **2023**, *55*, 40. https://doi.org/10.3390/engproc2023055040

Academic Editors: Teen-Hang Meen, Kuei-Shu Hsu and Cheng-Fu Yang

Published: 1 December 2023

1. Introduction

In the digital age, encouraging students to use computational thinking to solve, analyze, criticize, judge, and summarize is the focus of AI and Big Data generation education. As the community increasingly depends on computer technology, developing computing thinking becomes essential in long-term teaching principles, and different countries have developed separate national education strategies to meet these challenges [1]. For example, the UK has incorporated computational thinking education into the regular curriculum, meaning primary and secondary school students must learn how to program. Moreover, states and provinces in the United States, such as Massachusetts, have developed exclusive curricula and standards to evaluate skills for computational thinking [2].

With the development of computational thinking training, approaches to training computational thinking have expanded into non-programming (e.g., digital games) and unplugged activities in addition to general programming training [3,4]. Asbell-Clarke et al. exploited Zoombinis, a popular Computational Thinking (CT) learning game for age eight to adults, to train students in grades three to eight [5] and proved that well-crafted learning games could be effective in promoting student learning. Hooshya et al. also obtained similar results when they developed an adaptive educational game, AutoThinking, to train computational thinking skills and conceptual knowledge in fifth graders [6]. In gamified and non-programmed environments, students have demonstrated higher levels of interest, satisfaction, and technology acceptance in learning computational thinking [7,8].

The non-programmed activities' goal is to enable students become accustomed to thinking in the mode of computer scientists, including through programming and related classroom activities, learning to discover problems, disassemble problems, conceptualize and model, and find feasible computing solutions. However, students' acceptance and management of the learning materials vary according to each student's thinking ability. Students' habits of absorbing and managing material affect their learning styles [9], which is a preferred way of learning that affects how an individual receives stimuli, remembers, thinks, and solves problems. Students who have different learning styles have different problem-solving abilities [10]. To cultivate students' computational thinking literacy, we must consider their different learning styles.

In this study, we aim to prove that the latest training improves CT literacy and to investigate the relationship between learning style and computational thinking ability.

2. Literature Review

2.1. Learning Style

A learning style is a preferred way of learning that affects how an individual receives stimuli, remembers, thinks, and resolves problems. Since students with different learning styles solve problems differently, how they solve them influences how well they absorb information [11]. Many study results have shown how to advance the teaching of various subjects, including mathematics and science [9]. In this study, we applied the Felder–Silverman learning style model from North Carolina State University to group our students into four dimensions of learning style (Figure 1) [12].

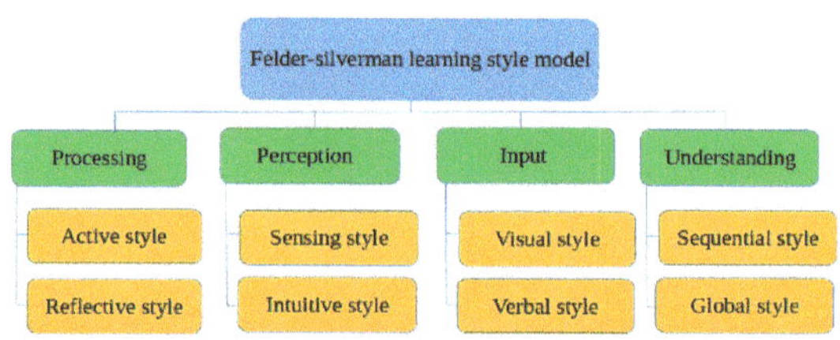

Figure 1. Felder–Silverman learning style model.

(1) Active/reflective continuum: First, there is the active/reflective continuum, which determines the method of processing information preferred by learners. Learners with an active learning style show action-oriented learning characteristics and are accustomed to understanding how things work through trying. Reversely, learners with a reflective learning style are likely to be introverted but good at thinking and prefer to work alone.

(2) Sensing/Intuitive continuum: The second dimension determines how learners prefer to take in the information, which is divided into the sensing and intuitive continuums. A learner with the sensing style is good at memorizing concrete facts rather than abstractive theories and generally performs with patience and caution. Conversely, a learner with the intuitive style is excellent at figuring out innovative concepts. Most of them find it hard to endure to learn with many memorable and tedious calculations.

(3) Visual/verbal continuum: The visual/verbal continuum is the third dimension determining how learners prefer information to be presented. Students who prefer the visual style perform better on graphics or video scaffolds. On the other hand, students who prefer the verbal style benefit from the written or spoken narration. However, there is no noticeable difference when studying because the information is generally present in visual and written narrations simultaneously.

(4) Sequential/global continuum: As the last dimension, the sequential/global continuum indicates learners' preferences for organizing information. A learner with a sequential-style preference is good at linear inference, which means learning step by step. A learner who prefers the global learning method is used to mastering the whole cycle of the learning content first. Once the global-style learners understand the gen-

eral direction of the whole, they can quickly solve complex problems or summarize things innovatively.

Even though the model presents learning styles in four pairs of opposing styles, the learning styles possessed by learners are not strictly dichotomous but rather entail a number of heavily weighted items. It means students may have both active and reflective learning styles, but one's learning attitude may be more active. Therefore, each learner obtains four tendencies in four dimensions after completing the measurement.

2.2. Computational Thinking

The concept of computational thinking (CT) first appeared in Jeanette Wing's study in 2006. In Wing's definition, CT is the concept of representing the formulation and solution of a problem in a way that computers can understand and implement [13]. Over the past decades, increasing attention has been paid to applied CT in education, and various definitions of CT have emerged from different disciplines [14].

We applied the classification of CT from the Google educational resource website to design our curriculum [15]. The four indicators in the definition are decomposition, pattern recognition, abstraction, and algorithm design separately. First, decomposition relates to breaking down problems into a series of minor ones. Pattern recognition refers to the ability to generate predictions and test models by observing each small problem and considering how similar problems have been solved. Another indicator is an abstraction, which focuses on the essential details while ignoring redundant information. In the process of abstraction, laws or principles will be discovered that underlie the patterns. Lastly, algorithm design relates to the ability to develop the instructions to solve similar problems and repeat the process.

3. Method

3.1. Experimental Design

The research designed a 16 h course separated into four days to examine whether non-programmed gamified activities could cultivate students' computational thinking. The Google educational resource website [15] (Computational Thinking for Educators) suggests four concrete elements in computational thinking: problem decomposition, pattern recognition, abstraction, and algorithm design. In the courses, we applied different types of games to cultivate these computational literacies, e.g., Little Alchemy, Rummikub, Gartic.io, and Robozzle.

In the experiment, we designed three parts of the process in each of the four competence courses. The first part focused on game teaching for an hour. The second part was a self-exploring period that allowed for students to play the game and find winning tips in two hours. The last part was the game competition and post-test the next day.

3.2. Self-Exploring Design

After game teaching, a self-exploring period was arranged for students to explore the tips of the games taught. During the self-exploring period, the experiment requested students to finish their study sheet as a scaffold when learning the games. There were also two trainers for 40 students to help if students encountered any problems that were difficult to resolve, ensuring that self-exploration was smooth.

3.3. Competition Design

The courses designed the game competition to increase students' motivation when studying alone. After the self-exploring, students played the game in the single-mode in all the games to battle with the computer. Meanwhile, the trainers recorded their game ranking and rewarded the class's highest-ranking students.

3.4. Assessment Design

In order to evaluate students' original degree of CT competence and its promotion after the courses, the study used Bebras Challenge [16], an international challenge that tests informatics and computational thinking, to examine the results. The researchers selected questions from the Bebras Australia Computational Thinking Challenges of 2017, 2020, and 2021 and translated them into the junior high school level. There were 15 questions in total as a formal contest with 5 questions for each level (easy, medium, hard).

4. Results and Discussion

4.1. Dataset

The sample for this study consisted of 352 seventh-grade students from a secondary school in Taichung. In the experiment, all the students participated in the gamified courses corresponding to problem decomposition, pattern recognition, abstraction, and algorithm design, which are four kinds of core competencies designed by experts. Since the course lasted four days, several subjects could not participate in the entire process, so 307 samples were obtained after deducting the subjects absent in the middle.

The pretest assessment consisted of fifteen questions worth 8 points, while the post-test consisted of six questions on each dimension, totaling twenty-four. The score range was 0 to 120.

4.2. Data Results

Among the 175 males (56.8%) and 133 females (43.2%) in the course, the mean post-test score was higher in both genders than the pretest score (Table 1). In this study, gender was used as an independent variable, and an independent-sample one-way analysis of covariance (ANCOVA) was performed on the post-test scores, controlling for pretest to eliminate student differences. According to the analysis, CT learning outcomes were not different between genders, F (1, 304) = 0.902, p = 0.343 (Table 2).

Table 1. Descriptive statistics of pretest and posttest.

	Gender	N	Min	Max	Mean	Standard Deviation
Pretest	Overall	307	0	112	59.01	23.34
	Male	174	8	112	59.84	23.69
	Female	133	0	112	57.92	22.91
Posttest	Overall	307	20	120	61.09	18.47
	Male	174	20	120	62.14	19.48
	Female	133	20	105	59.70	17.03

Table 2. ANCOVA of gender on post-test.

Source	Type III Sum of Squares	df	Mean Square	F	Sig.
Between groups	227.50	1	227.5	0.902	0.343
Within groups	76,710.906	304	252.239		
Total	1,248,500	307			

Previously, this course has been tested for its learning effect in a short-term version, so we are confident that this experiment improves students' computational thinking skills, and the research results support this hypothesis (Table 3).

Table 3. Paired *t*-test of pretest and post-test of overall.

	N	df	t stat	t critical	*p*-Value
Overall	307	306	1.724	0.089	0.043 *

* $p < 0.05$.

Since we want to examine whether students' original learning styles influence their computational thinking literacy improvement, we first observed the frequency distribution table of their learning styles (Table 4). According to the results, the student's preference is average in the first, second, and fourth dimensions. Nevertheless, more students prefer the visual learning style to the verbal one.

Table 4. Frequency distribution table of Soloman's learning style.

	Preference	N	Percentage
1st dimension	Active	190	61.9%
	Reflective	117	38.1%
2nd dimension	Sensing	160	52.1%
	Intuitive	147	47.9%
3rd dimension	Visual	261	85%
	Verbal	46	15%
4th dimension	Sequential	138	45%
	Global	169	55%

We implemented four UNIANOVA tests to detect the difference between four learning style dimensions with four competence improvements. In each ANOVA test, the interaction effect between dimensions was examined to guarantee independence, and all interaction effects had no significance.

The competencies' advances in composition and abstraction were not affected differently depending on the type of learning style. In contrast, pattern recognition and algorithm design differed significantly in how they were taught in terms of the second and first dimensions (Tables 5 and 6), in which pattern recognition was more advanced among intuitive learners than sensing learners, and algorithm design displayed more remarkable improvements among reflective learners than active learners.

Table 5. UNIANOVA of pattern recognition improvement.

Source	Type III Sum of Squares	df	Mean Square	F	Sig.
Between groups (2nd dimension)	227.655	1	227.655	4.179	0.042 *
Within groups	16,616.597	289	54.481		
Total	18,156.052	307			

* $p < 0.05$.

Table 6. UNIANOVA of algorithm design improvement.

Source	Type III Sum of Squares	df	Mean Square	F	Sig.
Between groups (1st dimension)	305.671	1	305.671	5.48	0.020 *
Within groups	16,120.267	289	55.779		
Total	16,978.184	307			

* $p < 0.05$.

5. Conclusions

The experimental results showed no significant difference between male and female groups. Gender was not a significant factor that caused apparent differences when conducting the computational thinking training courses. Despite this, students' computational thinking skills improved after the courses, and their original learning styles significantly influenced their computational thinking literacy growth.

We found that intuitive learners showed more significant gains in pattern recognition competence training than sensing learners. Sensing learners preferred to learn concrete and factual information and details, facts, and figures and disliked surprises and complications. Conversely, intuitive learners preferred abstract and original information, preferring to discover relationships and possibilities. To learn pattern recognition, learners must experiment with several combinations to find the best fit. It may not appeal to intuitive learners, which explains why intuitive learners performed better than sensing learners.

Aside from that, the analysis also found that reflective learners performed better in algorithm design training than active learners. Students in algorithm design training solved puzzles independently without a need to discuss or compete with others, which required more patience to brainstorm and try. In contrast to reflective learners, active learners processed information by attempting it rather than negotiating and explaining it. It may be a crucial difference that separates reflective learners from active learners.

As discussed above, this study proved that the gamification training method is effective for secondary school students. However, students' learning styles influence the course improvement. In order to ensure that students with distinct learning preferences can participate in the curriculum to their strengths, learning styles and a variety of teaching activities (e.g., teamwork) need to be considered for each core competency in the future.

Author Contributions: Conceptualization, Y.-Y.C. and S.-M.Y.; methodology, Y.-Y.C. and S.-M.Y.; validation, Y.-Y.C.; S.-W.S. and C.-H.L.; formal analysis, Y.-Y.C. and S.-W.S.; data curation, Y.-Y.C. and S.-W.S.; writing—original draft preparation, Y.-Y.C. and S.-W.S.; writing—review and editing, Y.-Y.C. and S.-M.Y.; visualization, L.-X.C. and C.-H.L.; project administration, S.-M.Y. All authors have read and agreed to the published version of the manuscript.

Funding: This work was partially funded by National Science and Technology Council Taiwan (grant number: 108-2511-H-009-009-MY3).

Institutional Review Board Statement: Not applicable.

Informed Consent Statement: Not applicable.

Data Availability Statement: The data presented in this study are available upon request from the corresponding author.

Acknowledgments: The authors wish to thank the blind reviewers for their insightful and constructive comments.

Conflicts of Interest: The authors declare no conflict of interest.

References

1. Angeli, C.; Voogt, J.; Fluck, A.; Webb, M.; Cox, M.; Malyn-Smith, J.; Zagami, J. A K-6 computational thinking curriculum framework: Implications for teacher knowledge. *J. Educ. Technol. Soc.* **2016**, *19*, 47–57.
2. Masami, I.; Roberto, A.; Maitree, N. Develop Computational Thinking on AI and Big Data Era for Digital Society-Recommendations from APEC InMside I Project. 2018. Available online: https://www.criced.tsukuba.ac.jp/math/apec/2021/pdf/InMsideFinal.pdf (accessed on 10 December 2022).
3. Liao, C.H.; Yan, J.Y.; Yuan, S.M.; Chen, L.X. Research on cultivating the computational thinking literacy of secondary school students through non-programming activities. In Proceedings of the 4th Eurasian Conference on Educational Innovation 2021, Taitung, Taiwan, 5–7 February 2021.
4. Chen, L.X.; Su, S.W.; Chen YYLiao, C.H.; Yuan, S.M. Cultivating The Computational Thinking Literacy Through Online and Offline Game-based Activities. In Proceedings of the 5th IEEE Eurasian Conference on Educational Innovation, Taipei, Taiwan, 10–12 February 2022.

5. Asbell-Clarke, J.; Rowe, E.; Almeda, V.; Edwards, T.; Bardar, E.; Gasca, S.; Baker, R.S.; Scruggs, R. The development of students' computational thinking practices in elementary-and middle-school classes using the learning game, Zoombinis. *Comput. Hum. Behav.* **2021**, *115*, 106587. [CrossRef]
6. Hooshyar, D.; Pedaste, M.; Yang, Y.; Malva, L.; Hwang, G.J.; Wang, M.; Lim, H.; Delev, D. From gaming to computational thinking: An adaptive educational computer game-based learning approach. *J. Educ. Comput. Res.* **2021**, *59*, 383–409. [CrossRef]
7. Sun, C.-T.; Chen, L.-X.; Chu, H.-M. Associations among scaffold presentation, reward mechanisms and problem-solving behaviors in game play. *Comput. Educ.* **2018**, *119*, 95–111. [CrossRef]
8. Soflano, M.; Connolly, T.M.; Hainey, T. Learning style analysis in adaptive GBL application to teach SQL. *Comput. Educ.* **2015**, *86*, 105–119. [CrossRef]
9. Veronica, A.R.; Siswono TY, E.; Wiryanto, W. Primary School Students' Computational Thinking in Solving Mathematics Problems Based on Learning Style. *Eduma Math. Educ. Learn. Teach.* **2022**, *11*, 84–96. [CrossRef]
10. Hung, Y.H.; Chang, R.I.; Lin, C.F. Hybrid learning style identification and developing adaptive problem-solving learning activities. *Comput. Hum. Behav.* **2016**, *55*, 552–561. [CrossRef]
11. Fadly, W. Profile of students analytical thinking skills in learning style for completing substance pressure problems. *Jambura Phys. J.* **2021**, *3*, 1–15. [CrossRef]
12. North Carolina State University. Teaching and Learning STEM. Available online: https://www.engr.ncsu.edu/stem-resources/ (accessed on 16 July 2022).
13. Wing, J.M. Computational thinking. *Commun. ACM* **2006**, *49*, 33–35. [CrossRef]
14. Shute, V.J.; Sun, C.; Asbell-Clarke, J. Demystifying computational thinking. *Educ. Res. Rev.* **2017**, *22*, 142–158. [CrossRef]
15. Exploring Computational Thinking. Available online: https://ai.googleblog.com/2010/10/exploring-computational-thinking.html (accessed on 15 July 2022).
16. Bebras Offical. What is Bebras. Available online: https://www.bebras.org/?q=goodtask (accessed on 15 July 2022).

engineering proceedings

Proceeding Paper

Necessity of Notification System Application According to Elementary School Teacher's Environmental Behavior †

Haruno Ishikawa

Department of Architecture, Faculty of Science and Technology, Shizuoka Institute of Science and Technology, 2200-2 Toyosawa, Fukuroi City 437-0032, Shizuoka Prefecture, Japan; ishikawa.haruno@sist.ac.jp

† Presented at the IEEE 5th Eurasia Conference on Biomedical Engineering, Healthcare and Sustainability, Tainan, Taiwan, 2–4 June 2023.

Abstract: In rural regions of Japan, specifically in Shizuoka, the majority of elementary school classrooms lack ventilation systems, and the operation is manually conducted by teachers and students. Instead of relying on the implementation of high-performance hardware solutions, the aim is to strive for a harmonious coexistence of COVID-19 mitigation and Zero Energy Building (ZEB) realization through appropriate information dissemination and proactive environmental behavior. This paper investigates the environmental behavior of educators and its indicators, and assuming homeroom teachers in X City implement classroom ventilation based on threshold value notifications, it is demonstrated that a reduction of up to 20% in the current air conditioning heat load can be achieved.

Keywords: elementary schools; environmental actions; ventilation in classrooms; ZEB; carbon neutrality

Citation: Ishikawa, H. Necessity of Notification System Application According to Elementary School Teacher's Environmental Behavior. *Eng. Proc.* **2023**, *55*, 41. https://doi.org/10.3390/engproc2023055041

Academic Editors: Teen-Hang Meen, Kuei-Shu Hsu and Cheng-Fu Yang

Published: 1 December 2023

1. Introduction

Japan aims to achieve carbon neutrality by 2050 and reduce CO_2 emissions by 46% by 2030. In the construction sector, it is planned that the majority of new public facilities will achieve Zero Energy Building (ZEB) status by 2030. School facilities, which account for 40% of public facilities, need to implement measures at an early stage.

Therefore, around 2018, when the government announced the promotion of ZEB in public facilities, the existing school buildings in Shizuoka Prefecture, a warm region, had extremely low insulation performance and minimal building facilities in classrooms. The building facilities in classrooms include lighting, air conditioning, and ventilation systems.

Regarding ventilation systems, due to the lack of installed ventilation facilities in classrooms in Shizuoka Prefecture, the preservation of the learning environment in each classroom relies on the operation of the homeroom teachers [1,2]. For COVID-19 prevention measures, manual window opening by teachers is the main method of ventilation.

In this paper, we investigate the operational conditions of teachers and their environmental behavior in primary school classrooms during the summer, with a focus on COVID-19 countermeasures. We aim to grasp the actual situation and demonstrate the potential reduction in heat load achievable through systemization.

In the future, to maintain the classroom environment in X City's elementary and middle schools while minimizing energy consumption, the introduction of an announcement system in addition to the efforts of teachers is deemed effective.

In this paper, we aim to grasp the operational status and environmental behavior of teachers focusing on COVID-19 measures in summer elementary school classrooms. We calculate the potential reduction in energy consumption by introducing an unmanned environmental notification system that allows users to engage in environmental behavior to achieve a comfortable classroom environment.

In the future, to suppress energy consumption while maintaining classroom environments in elementary and junior high schools in X City, unmanned environmental notification system applications that facilitate appropriate environmental behavior by homeroom teachers will be effective.

2. Methodology

Shizuoka Prefecture is located on the Pacific side of mainland Honshu and is known for its mild climate. Under the Building Energy Efficiency Law, Japan is divided into eight regions for the calculation of target buildings based on regional characteristics. Due to the large area of Shizuoka Prefecture, the cities and towns within the prefecture are divided into five to seven regional classifications.

Therefore, in any region of Shizuoka Prefecture, the transitional period is long, and the duration of heating and cooling through facilities is short. It has been found that public elementary and junior high schools often fail to meet the standards for room temperature set by the government, even when air conditioning is used for both cooling and heating during the summer and winter, according to existing surveys. Classroom environmental maintenance, including ventilation, has primarily been carried out by classroom teachers.

In this study, the target municipality was X City, Shizuoka Prefecture, and a questionnaire survey was conducted targeting teachers who are homeroom teachers in elementary schools. X City has a total of 13 public elementary schools, with 223 regular classrooms and 35 special needs classrooms as the subjects.

The survey was conducted during the summer period from 29 September 2022 to 31 October 2022, for a duration of one month, using the web-based survey platform Forms.

The questionnaire items include demographic information, as well as the following four items: (1) Psychological aspects of students and teachers during the summer as perceived by teachers; (2) Concerns and actions related to air conditioning operation; (3) Concerns and actions related to ventilation; and (4) Actual window opening time and extent.

3. Simulation and Results

A total of 159 responses were obtained from teachers in 10 elementary schools (X City), including 140 homeroom teachers and 19 special education teachers. The effective response rate for homeroom teachers was 62.7%. Responses were received from individuals spanning the age range of 20s to 60s, with relatively equal participation across all age groups. The ratio of male to female respondents was 7:9.

Figure 1 presents the attributes. The distribution of homeroom teachers (n = 140) based on the grade level they teach and the floor level of their classrooms is depicted in Figure 1a. When classified into first floor, intermediate floors, and top floor categories, lower grade levels were allocated to lower floors, while higher grade levels were positioned on upper floors, with the top floor classrooms accounting for 41% of the total.

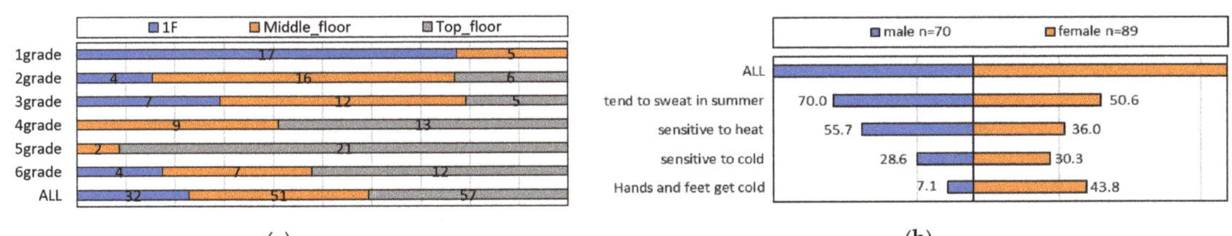

Figure 1. The attributes. (**a**) The grades of the teachers and the floor of the classrooms; (**b**) The teacher's own thermal sensation.

The thermal sensation responses provided by the participating educators are illustrated in Figure 1b. Approximately 50% of male respondents and 40% of female respondents indicated being either "cold hands and feet" or "sweaty".

3.1. Results of Psychometric Survey of Students and Teachers

Figure 2 depicts the thermal sensations and satisfaction levels of the children, as perceived by the faculty members, as well as the thermal sensation and satisfaction levels reported by the faculty members themselves. There were no significant differences observed between the faculty's evaluation of the children's condition and their own evaluation. Regarding thermal sensation, approximately 75% of both the children and teachers reported feeling "hot", while 20% of the respondents, including both children and teachers, reported feeling "very hot". As for satisfaction levels, around 50% indicated being "satisfied", while approximately 35% expressed some level of dissatisfaction.

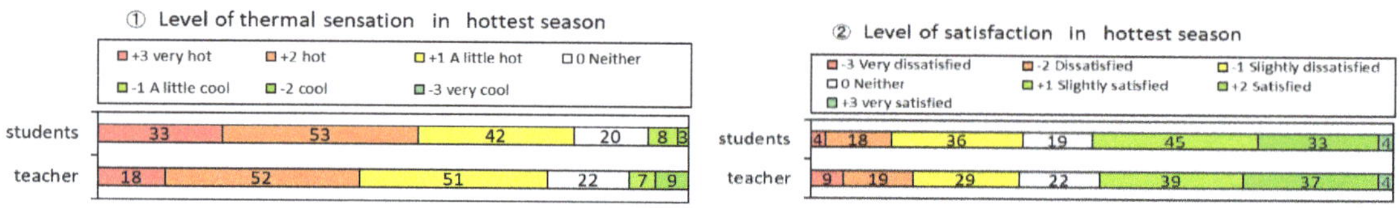

Figure 2. The thermal sensations and satisfaction levels of the students and the classroom teachers.

Figure 3 illustrates the operating hours and non-operating hours of the air conditioning system, as reported by the faculty members. During instructional periods when students are present or during lunchtime, the operating rate exceeds 90%. However, there is variation in the usage of the system during non-instructional periods when students are not present, such as during cleaning time or after-school hours when teachers utilize the facility. Multiple responses indicated that determining the appropriate operation during these periods is challenging for the faculty members, and it was also difficult to make definitive judgments based on the written responses.

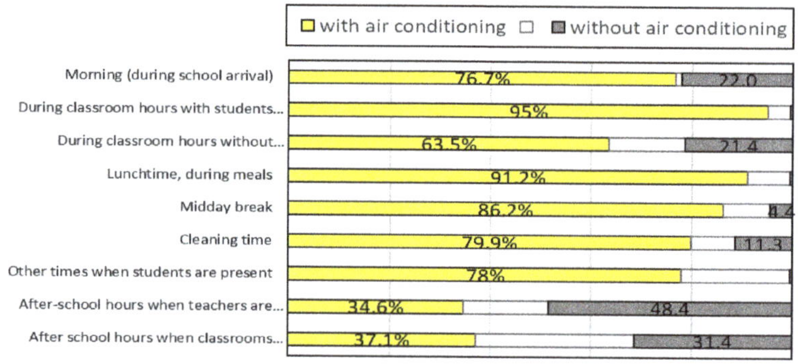

Figure 3. Air conditioning operating hours and non-operating hours.

Figure 4 displays the actual temperature settings adjusted by the faculty members using the remote control, as well as their desired temperature settings. According to the air conditioning system operation guidelines provided by the Education Committee of X City, a recommended temperature range of 26 to 28 degrees is suggested for remote control settings. Among the faculty members, the temperature setting of 26 degrees accounted for 48% of the total responses. On the other hand, the desired temperature settings indicated by the faculty members were predominantly in the range of 24 to 26 degrees, representing three-fourths of the total responses. This reveals that the faculty members have not lowered the temperature setting to match their desired settings.

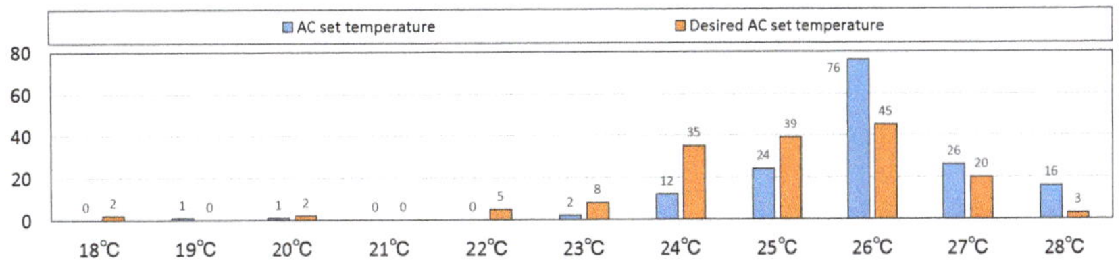

Figure 4. Teacher's air conditioning remote control set temperature and desired set temperature.

3.2. Points of Teachers' Environmental Behavior

Figure 5a presents the evaluation of environmental factors as perceived by the faculty members, categorized by assessment items. They rated the environmental factors during classes on a 6-point scale.

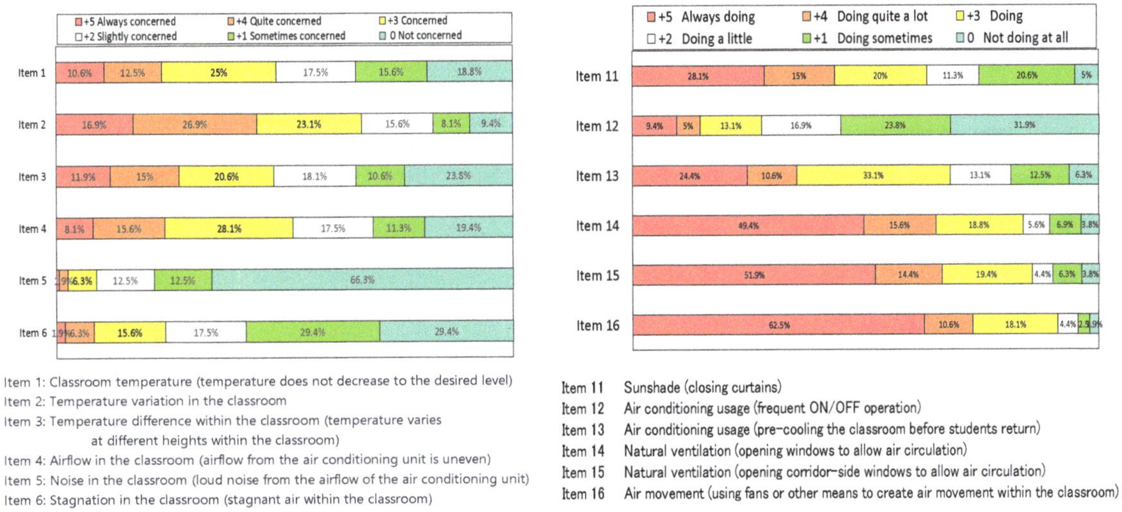

(a) (b)

Figure 5. (**a**) Classroom environment items of concern during class. (**b**) Teacher environmental behavior.

Regarding the temperature distribution in the classroom (item 2), approximately two-thirds of the faculty members responded with ratings indicating concern (ratings +5, +4, or +3). Furthermore, for the following three items: (1) inability of the classroom temperature to reach the desired level, (2) temperature differences at different heights within the classroom, and (3) uneven distribution of air conditioning airflow, each of them garnered concern ratings (ratings +5, +4, or +3) from approximately half of the respondents. On the other hand, approximately two-thirds of the faculty members reported not being concerned about noise in the classroom (specifically, noise generated by the air conditioning system airflow).

Figure 5b presents the evaluation of the environmental behaviors carried out by the faculty members, categorized by assessment items. These behaviors were also rated on a 6-point scale.

Regarding classroom ventilation, approximately 80% of the faculty members responded with ratings indicating that they consciously engage in actions such as "Opening the outer windows" (item 13) and "Opening the hallway windows" (item 15), as well as "Stirring the indoor air with a fan" (item 16). Regarding sunshade measures, the most commonly reported behavior was "Closing the curtains" (item 11), accounting for 63% of responses. As for air conditioning usage, measures such as "Pre-cooling the room in accordance with the children's usage" (item 15) were reported by 54% of the faculty members.

3.3. Window Opening Amount and Opening Time during Cooling

The relationship between the open area during class and the open area during cleaning is shown in Figure 6. The horizontal axis represents the classroom window open area during class, with an average of 2.80 square meters (SD ± 2.90). This average corresponds to an air exchange rate of 5.8 exchanges per hour when a natural wind speed of 0.5 m/s is introduced. Conversely, at the ventilation rate specified by the school environmental hygiene standards of 2.2 exchanges per hour, the open area is 0.476 square meters when the wind speed is 0.5 m/s. The vertical axis represents the difference in open area between cleaning and class hours.

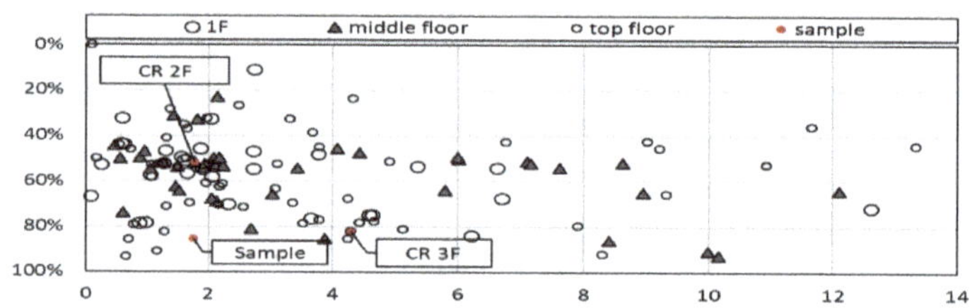

Figure 6. Opening area during class and during ventilation.

It can be observed that the open area during cleaning significantly differs from that during class hours, as X City recommends window opening as a COVID-19 countermeasure in each elementary school. According to the calculations based on teacher reports for the second-floor classroom (Q Elementary School), the open area during class hours was 0.48 square meters. However, the average CO_2 concentration measured in the classroom was 1023 ppm, exceeding 1000 ppm in approximately 60% of the time when students were present. On the other hand, for the third-floor classroom (Q Elementary School), the calculated open area during class hours based on teacher reports was 4.42 square meters. The measured CO_2 concentration in the classroom had an average of 591 ppm, and it did not exceed 1000 ppm during student occupancy. The significant variation in open area reported by teachers indicates that in some cases, excessive ventilation may occur due to the discretion of individual teachers.

Figure 7 presents the calculated classroom opening areas for indoor and outdoor spaces based on teacher reports, along with actual measurement examples. The average values reported by teachers indicate an average opening area of 1.32 square meters for outdoor spaces and 1.93 square meters for indoor spaces, but with significant variation. In the case of classrooms where "sufficient ventilation was achieved", the opening area towards the outdoor environment accounted for only 14% of the total.

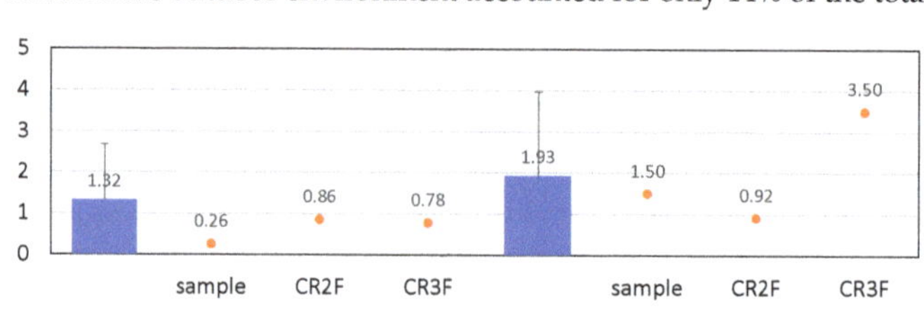

Figure 7. Average window opening areas for indoor and outdoor spaces.

Figure 8 presents the results of thermal load calculations [3] per classroom for four different cases (Case 1 to Case 4) conducted in Q Elementary School, categorized into upper and intermediate floors. Cases 1 and 3 represent scenarios where ventilation is ensured, while Cases 2 and 4 are calculated based on the average values derived from the teacher-reported data in Figure 8. For Cases 1 and 2, the room temperature was set at 28 degrees Celsius, while for Cases 3 and 4, it was set at 26 degrees Celsius.

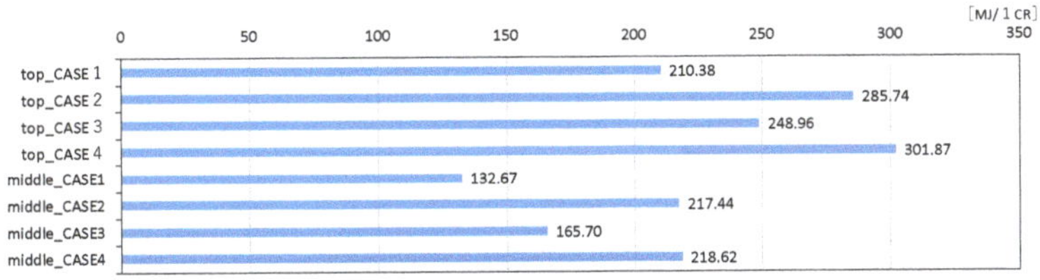

Figure 8. The results of thermal load calculations per classroom for four different cases.

By transitioning to "cases with sufficient ventilation," the thermal load of the classrooms can be reduced by 26% for the upper floor with a 28-degree setting, and by 12% for a 26-degree setting. Similarly, for the intermediate floor, the reduction amounts to 38% for a 28-degree setting and 23% for a 26-degree setting.

The thermal load reduction for Q Elementary School was calculated for each case. In the scenario where sufficient ventilation is ensured (Case 3), a 31% reduction in thermal load is achievable compared with the current state of all regular classrooms (Case 4) at a 28-degree setting, and a 20% reduction at a 26-degree setting. When the calculation is extended to the entire Q Elementary School with a 26-degree setting, the thermal load on the hottest day, which was initially 7151.48 MJ, can be reduced to 5722.83 MJ.

4. Conclusions

In this paper, the following findings were obtained:

- The designated air conditioning temperature in X City is 26 to 28 degrees Celsius, but the majority of teachers operate it at 26 degrees Celsius and express a preference for temperatures ranging from 24 to 26 degrees Celsius.
- The homeroom teacher is particularly attentive to ventilation among the classroom environmental conditions and takes actions such as opening windows and using fans. However, these environmental practices are being carried out without any scientific basis.
- Natural ventilation during air conditioning shows significant variation in terms of window opening size and duration, as determined by individual teacher judgment, leading to instances of excessive ventilation compared to the required amount.
- By implementing an announcement system that encourages appropriate environmental behaviors, it is estimated that classroom thermal load can be reduced by approximately 20%, resulting in a potential reduction of 1400 to 2000 MJ in thermal load on the hottest days at the school level.

Funding: Part of this work was supported by JSPS KAKENHI Grant Number 21K04384.

Institutional Review Board Statement: Not applicable.

Informed Consent Statement: Not applicable.

Data Availability Statement: Data sharing is not applicable to this article.

Acknowledgments: I express our sincere gratitude to the X City Board of Education and all the staff members at Q Elementary School for their invaluable cooperation in conducting this study.

Conflicts of Interest: The author declares no conflict of interest.

References

1. X City Board of Education. *X City Elementary and Junior High School Air Conditioner Operation Guidelines*; Ministry of Education, Culture, Sports, Science and Technology: Tokyo, Japan, 2020.
2. Ministry of Education, Culture, Sports, Science and Technology. *Hygiene Management Manual for New Coronavirus Infectious Diseases in Schools "New School Lifestyles"*; Version 9; Ministry of Education, Culture, Sports, Science and Technology: Tokyo, Japan, 2023.
3. The Japan Society of Air Conditioning and Sanitary Engineers, HASPEE (Heating, Air conditioning and Sanitary Engineers' Program for Experience and Education)1.0.11, SHASE-G_1001-2022, 4 August 2022. Available online: http://www.shasej.org/tosho/haspee/download.html (accessed on 10 July 2023).

engineering proceedings

MDPI

Proceeding Paper

Study on Low-Frequency Acoustic Environment of Hospital Buildings by Acoustic Simulation Software †

Chiung Yao Chen and Xiao Lin *

Department of Architecture, Chaoyang University of Technology, Taichung 413310, Taiwan; chychen@cyut.edu.tw
* Correspondence: lin2291702@gmail.com

† Presented at the IEEE 5th Eurasia Conference on Biomedical Engineering, Healthcare and Sustainability, Tainan, Taiwan, 2–4 June 2023.

Abstract: The noisy environment of hospitals has always been a problem for medical workers and patients as people in hospitals need a quiet environment more than others. Thus, the study of the noisy environment of hospital buildings has been carried out for a long time. This study aims to find out the existing problems and explore solutions for improvement through research. The purpose is to investigate and analyze the lower frequency range of current noise (25–200 Hz) in four central hospitals (single building or group of buildings using a single lobby), following the research of "Characterizing the subjective noise in hospital lobbies". In this research, the acoustic environment condition in the lobby after the road noise spread into the hospital and its environmental noise problems were analyzed by the commercial acoustic simulation software SoundPLAN for scene restoration and simulation. Focusing on reconstructing the noisy condition of the hospital lobby and comparing the measurements with those of our previous study, acoustic experimental simulations are conducted to acquire the indoor frequency spectrum. The experimental data are also compared with the live measurement data to verify the reliability of SoundPlAN. Finally, the experimental result is analyzed with the psychological experimental noisiness values in our previous study to identify the sources of the main noise existing in the lobbies of the four hospitals.

Keywords: hospital-type building lobby; low-frequency noise (25–200 Hz); SoundPLAN acoustic simulation software; correlation

Citation: Chen, C.Y.; Lin, X. Study on Low-Frequency Acoustic Environment of Hospital Buildings by Acoustic Simulation Software. *Eng. Proc.* **2023**, *55*, 42. https://doi.org/10.3390/engproc2023055042

Academic Editors: Teen-Hang Meen, Kuei-Shu Hsu and Cheng-Fu Yang

Published: 1 December 2023

1. Introduction

Many researchers have raised the noise problems of the hospital, indicated improvement directions, and proposed related solutions. Nevertheless, in a noisy environment, one phenomenon is generally ignored: low-frequency noise. When people are forced to adapt to this noise, they would insensibly become irritable, and even irrational [1]. Prolonged exposure to low-frequency noise is likely to result in neurological weakness, insomnia, headaches, and other neurological dysfunctions, and even affect the fetus in the womb. Compared with low-frequency noise, high-frequency noise decreases rapidly as the distance increases or when encountering obstacles. As low-frequency noise decreases slowly and its sound waves are longer, they can easily pass through obstacles, run over long distances, and penetrate walls and ears. Low-frequency noise, in general, has a common nature. Their waves are formed by vibration being a form of energy transmission. A hospital requires a favorable rehabilitation environment and a working environment. Topf [2] Noise in hospitals is not conducive to patients' recovery, and they need an environment that is quieter than the environment. They are usually experienced in daily life. Noise in hospitals also distracts hospital staff and increases the risk of medical incidents. Torija [3,4] suggested that low-frequency noise (or other acoustic features that are usually less important) is less dominant in the subjective response to other acoustic features (e.g., mid- and high-frequency noise) in noise annoyance, and is a more critical factor under indoor conditions. Low-frequency noise may be more dominant than the other frequency bands under indoor conditions Tomei [5] Increased noise will damage the human auditory system and may also induce hypertension or other irreversible cardiovascular diseases. Using the SoundPLAN acoustic simulation software, the propagation paths of low-frequency noise and the coverage of low-frequency noise are investigated in the acoustic environment of the four hospital halls. Then, the measured data in the field are compared to verify their reliability and to improve the strategies and recommendations.

2. Purpose and Objects of Research

2.1. Purpose of Research

The following are the objectives of this study.

- To study the types of low-frequency noise in hospitals;
- To investigate the impact of low-frequency noise on people's lives and their psychological perception of it;
- To continue the investigation of the current situation of noise in the lobby of a general hospital in central Taiwan;
- To implement noise measurements in four hospitals in central Taiwan;
- To use the acoustic simulation software SoundPLAN to restore the acoustic environment scene during the measurement by Chen [6,7] for analysis and research.

2.2. Objectives of Research

In this study, low-frequency noise was measured and simulated at four hospitals (Table 1) in Taichung City in order to design and improve the physical environment.

Table 1. Hospitals and the area of their lobbies.

Number	Name	Area/m^2
a.	Taichung Hospital, Ministry of Health and Welfare	1050
b.	Jen-Ai Hospital Dali	1175
c.	Ching-Chyuan Hospital Daya	531
d.	Cheng Ching Hospital Chung Kang Branch	480

3. Experimental Sites

Torija [3] suggested that low-frequency noise (or other usually less important acoustic features) may be more important than other acoustic features (such as MF and HF content) that are less dominant such as indoor conditions. Low-frequency noise is relatively more dominant indoors and behind noise barriers and relatively more significant, as it may lead to irritation, excessive fatigue, inattention, sleep disorders, and troubles. Through animal experiments, Spreng [8] found that the cellular cortisol of the tested animals would be affected by noise and cause changes, and it would cause small changes in the heart and adrenal glands.

Paunović [9] explored traffic noise-induced changes in blood pressure, heart and hemodynamics in young adults. Their study included a total of 130 people (42 men, 88 women) with an average age of 24 years old. The results of the study pointed out that 89 decibels of traffic noise will produce blood changes, including increased blood vessel resistance and reduced cardiac blood elimination. These changes occur when noise is exposed The interplay of effects results in an increase in systolic and diastolic blood pressure.

3.1. Floor Plan Acquisition

Four central hospitals near the road were selected to study the low-frequency noise. The criteria for measuring and setting the measuring points are selected following Chen's research [6].

The measurement was conducted on normal working days from 9:00 a.m. to 5:00 p.m. and from Monday to Friday. The reason was to avoid the special impact noise when the hospital opens in the morning such as the noise of opening the iron gate, air conditioning, mechanical equipment trial operation, and so on. Noise at night occurs from 9 p.m. to 5 a.m. The measuring points need to meet the following conditions.

1. The height between the measuring point and the ground must be 1.2–1.6 m;
2. The distance between the measuring point and each radiation surface in space must be greater than 1.0 m;
3. The distance between each measuring point must be greater than 1.5 m;
4. The distance between the measuring point and the noise source in space must be greater than 1.5 m.

The floor plan is shown in Figure 1.

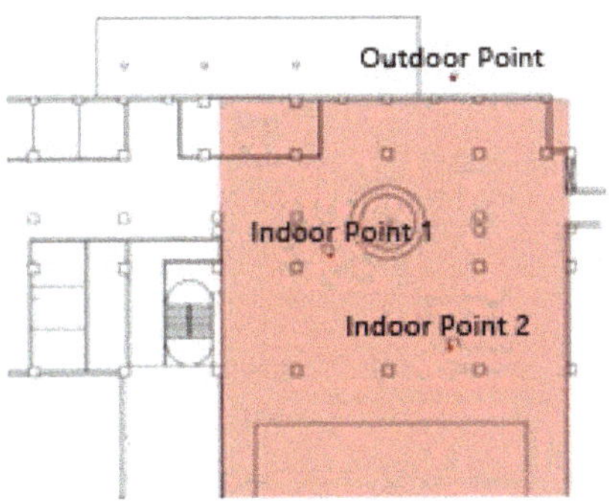

(**a**) Taichung Hospital, Ministry of Health and Welfare

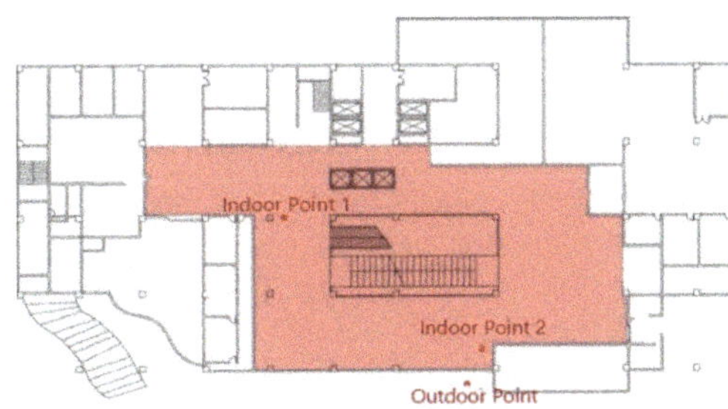

(**b**) Jen-Ai Hospital Dali

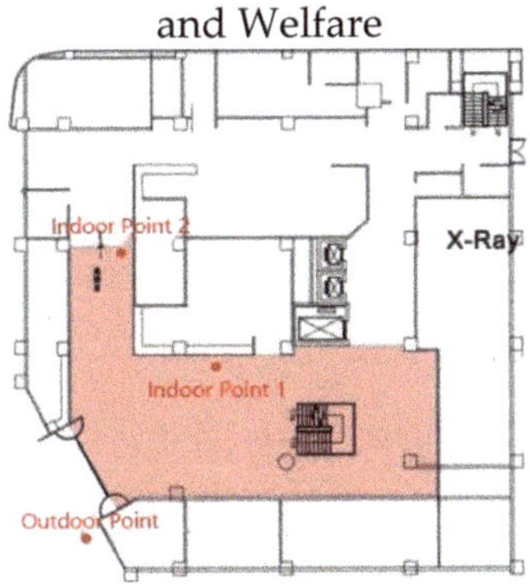

(**c**) Ching-Chyuan Hospital Daya

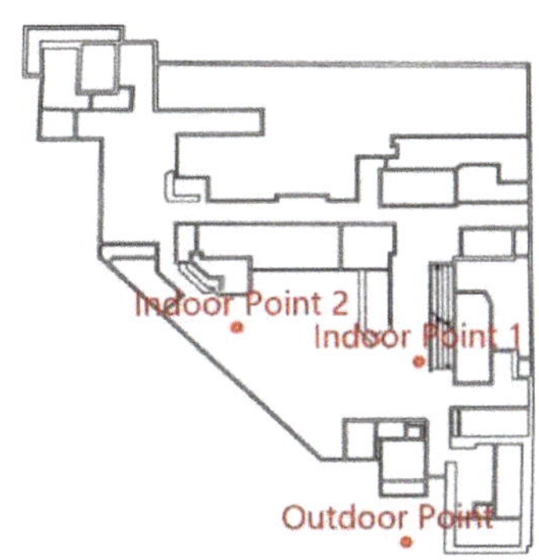

(**d**) Cheng Ching Hospital Chung Kang Branch

Figure 1. Floor plane of hospital lobbies for the research [6].

3.2. Data Acquisition

Chen [6] performed a psychological quantitative survey on the audience of hospital users using the Likert Scale. It is a psychological reaction scale and is often used for questionnaire surveys. The Likert scale has two extreme quantitative methods to measure positive or negative responses. The subject scored the noise in the hospital hall at five levels of very agree, agree, ordinary, disagree, and very disagree (Table 2).

Table 2. Summary of noise incident questionnaire data.

Number	a	b	c	d
Reason for staying	3.16	3.41	3.06	3.45
Means of transportation	2.06	1.66	2.11	2.10
Home environment	2.41	2.49	2.21	2.35
Noise severity	2.78	3.16	2.85	3.02
Conversational voice	3.08	3.06	2.95	3.17
Children's frolicking	3.08	3.52	3.31	3.23
Ambulance alarm	2.78	3.15	2.85	2.70
Medical equipment	2.57	2.48	2.40	2.37
Mechanical equipment	2.57	2.69	2.40	2.28
Broadcast	2.65	2.82	2.89	2.65
Call name	2.57	2.93	2.82	2.77
Footsteps	2.60	2.94	2.70	2.53
Opening/closing sound	2.72	2.57	2.63	2.52
Car sound	2.65	2.71	2.85	2.98
Construction operation	2.83	3.34	3.37	3.42
TV set	2.58	2.42	2.80	2.53
Traffic noise in the location	2.58	2.45	2.60	2.97
Ambulance sound in the location	2.50	2.71	2.63	2.78
Counterspeech in the location	2.52	2.65	2.33	2.68
Elevator/escalator in the location	2.52	2.62	2.25	2.39
Shuttle footsteps in the location	2.60	2.89	2.52	2.55
Sound of medical instruments in the location	2.70	2.78	2.38	2.36
Emergency room sound in the location	2.78	2.43	2.55	2.42

After collecting all data, the psychological noise parameters of the subjects in four hospitals were obtained as shown in Table 3.

Table 3. Psychological noise parameter table.

Name	Psychological Noise Value
Hospital a	3.393
Hospital b	3.417
Hospital c	2.600
Hospital d	3.617

The collated parameters were used in the later stage of the experiment with the simulated indoor spectrum.

4. Methods

A SoundPlAN model was used for data acquisition, measurement, and simulation.

4.1. Model

Based on the floor plans of four hospitals obtained by Chen [6], AutoCAD was used to restore the plans on an equal scale. The floor plans were imported into SoundPlAN to establish a model as shown in Figure 2.

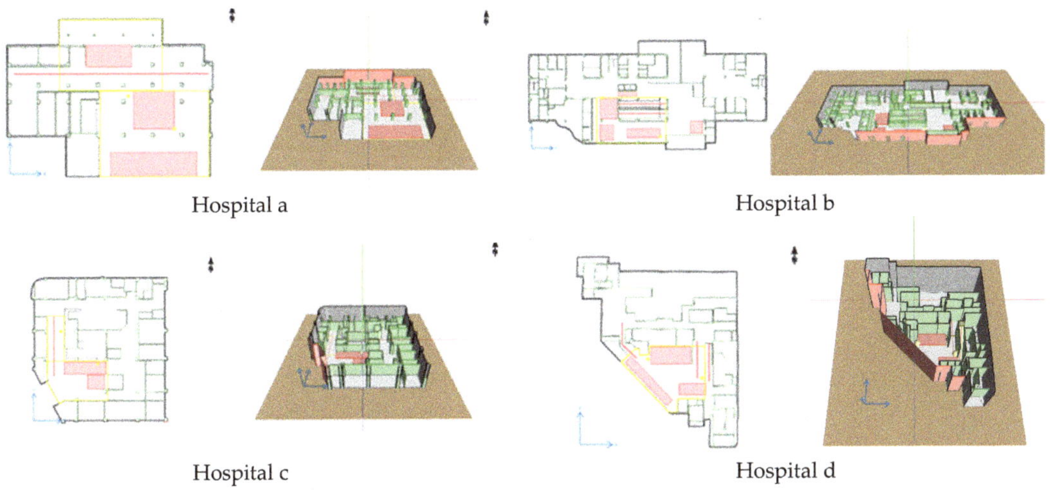

Figure 2. Four hospital plans and 3D models in SoundPLAN.

4.2. Sound Absorption and Penetration

According to the field investigation and collection of data on the penetration coefficient (ISO 12354) [10], a sound absorption coefficient (ISO 354) [11] of 20 cm reinforced concrete walls was obtained as shown in Table 4.

Table 4. The 20 cm RC wall penetration and absorption coefficient spectrum.

Name/Band Hz/dB	Penetration Coefficient	Sound Absorption Coefficient
25	20.1	0.1
31	21.2	0.1
40	22.8	0.1
50	23.2	0.1
63	24.4	0.1
80	25.5	0.1
100	27.7	0.1
125	30.4	0.1
160	32.7	0.1
200	34.3	0.1
250	36.3	0.1
315	38.4	0.1
400	40.4	0.1
500	42.4	0.1
630	44.4	0.1
800	46.45	0.1
1000	48.4	0.1
1250	50.4	0.1
1600	52.4	0.1
2000	54.5	0.1
2500	56.5	0.1
3150	58.4	0.1
4000	60.5	0.1

5. Process and Results

Due to the shortage of outdoor noise measurement data in Chen [6], the field measurement was obtained as the spectrum of outdoor noise sources. The measurement principle followed by this outdoor noise spectrum collection is as follows.

The measurement was carried out during the normal working day from Monday to Friday every 30 min. The reason was to avoid the special noise that occurred when the hospital opened in the morning such as opening the iron gate, air conditioning, mechanical equipment trial operation, and other noises. In case of sudden noise (except the noise brought by the hospital such as the siren of an ambulance), the measurement was terminated. The measurement time was 15 min and was repeated to ensure accuracy. The measuring points meet the previous conditions.

Precision Sound 140 Sound Analyzer was used in this research. The obtained measurement data are shown in Table 5.

Table 5. The 25–4 kHz frequency spectrum of outdoor noise in four hospitals.

Name/Band Hz	Hospital a/dB	Hospital b/dB	Hospital c/dB	Hospital d/dB
25	68.30	69.10	67.00	64.94
31	68.75	68.10	66.50	67.69
40	72.55	64.40	70.80	64.34
50	72.65	65.30	68.00	64.69
63	69.15	65.60	68.00	64.74
80	68.45	65.00	68.40	60.74
100	66.65	62.60	67.40	59.34
125	65.45	62.40	68.70	61.44
160	65.65	61.20	67.50	59.04
200	65.70	60.20	64.70	55.69
250	64.20	59.40	63.90	58.39
315	62.30	57.60	61.90	60.69
400	62.35	59.20	61.20	56.19
500	62.10	57.90	61.00	64.19
630	61.65	56.70	61.50	54.89
800	60.65	56.10	63.30	50.54
1000	60.45	56.90	64.50	52.19
1250	59.95	57.50	58.70	53.09
1600	58.55	56.30	57.60	52.84
2000	57.60	53.70	56.00	49.59
2500	56.05	52.60	54.70	48.74
3150	55.05	51.40	53.50	48.89
4000	52.15	48.60	51.30	42.94
Total	79.12	73.90	78.40	72.90

Data acquisition: field measurement.

After the establishment of the model, the input of various values and the setting of the analytical program, the indoor acoustic environment spectrum of the four hospitals is finally obtained as shown in Table 6. The indoor sound pressure distribution diagram is shown in Figure 3.

Table 6. Indoor noise spectrum of four hospitals after simulation.

Name/Band Hz	Hospital a/dB	Hospital b/dB	Hospital c/dB	Hospital d/dB
25	68.30	64.50	59.60	66.4
31.5	68.75	69.1	59.1	69.8
40	72.55	68.1	58	65.95
50	72.65	64.4	57.1	66.4
63	69.15	65.3	58.4	66.05
80	68.45	65.6	54.6	62.25
100	66.65	65	55.5	61.25
125	65.45	62.6	54.5	63.5
160	65.65	62.4	52.5	63.5

Table 6. *Cont.*

Name/Band Hz	Hospital a/dB	Hospital b/dB	Hospital c/dB	Hospital d/dB
200	65.7	61.2	54.2	58.15
250	64.2	60.2	55.6	59.75
315	62.3	59.5	55.9	61.55
400	62.35	57.7	57.8	57.8
500	62.1	59.3	57.5	64.85
630	61.65	58	57.4	56.8
800	60.65	56.8	57	53.4
1000	60.45	56.2	55.5	53.65
1250	59.95	57	57.1	53.9
1600	58.55	57.6	52.4	53.6
2000	57.6	56.4	50.1	55.15
2500	56.05	53.8	49.4	49.7
3150	55.05	52.6	55.9	49.6
4000	52.15	51.6	57.7	44.2
Total	78.82	74.9	69.2	74.22

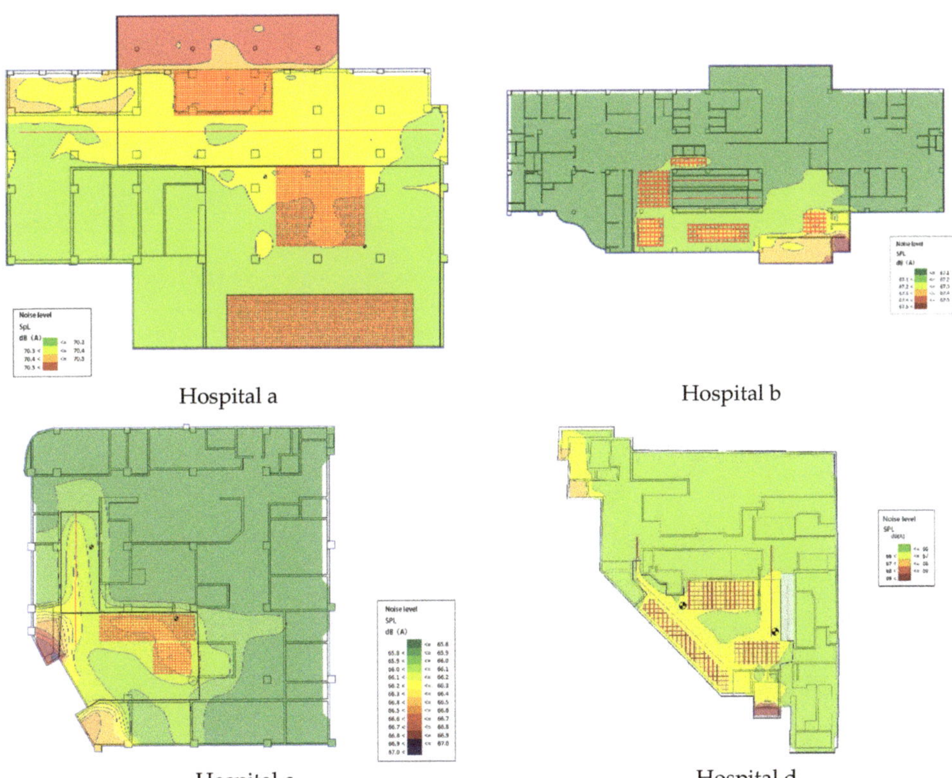

Figure 3. Sound pressure distribution in the four hospitals.

The sound pressure distribution diagram shows the acoustic environment of the hospital lobby. The red areas included the medication collection area, the entrance lobby, and the rest and waiting area which were crowded. The second red area is for the sound sources in the corridors and escalators are linear sources. The sound level was the most prominent in the Daya Ching Chuen Hospital, presumably due to its proximity to the road.

The correlation analysis result of the indoor spectrum in SoundPlaN and the psychological noise value is shown in Table 7.

Table 7. Correlation analysis result.

Name/Band Hz	Correlation Coefficient
25	0.853337
31	0.928248
40	0.944554
50	0.833262
63	0.915337
80	0.969916
100	0.975709
125	0.933431
160	0.954515
200	0.877396
250	0.873824
315	0.834558
400	0.457622
500	0.490722
630	0.485701
800	0.244667
1000	0.450634
1250	0.274137
1600	0.882458
2000	0.984785
2500	0.782741
3150	−0.3497
4000	−0.515241

The experimental results show that the frequency band of noise was mainly concentrated in 25–315 Hz and 1600–2500 Hz and present the direction of looking for noise source mainly in low frequency (Figure 4). It is shown that the main source of low-frequency noise is in the halls of the four tested hospitals.

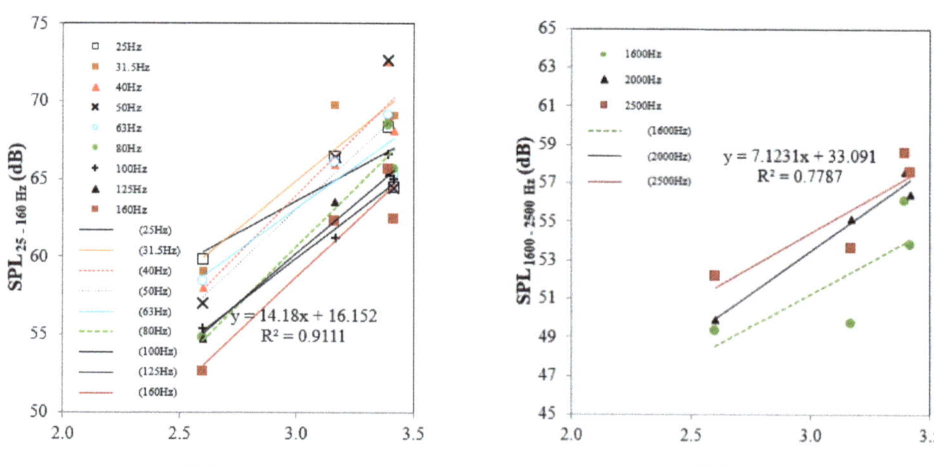

Figure 4. Correlation between subjective noisiness and sound pressure levels of simulated noise at the range 25–160 Hz and 1600–2500 Hz.

6. Conclusions

In this study, the acoustic environment of the lobby of the four hospitals was investigated and compared to those of Chen [6], and the correlation between the measured data and the psychological experimental data was compared.

According to the experimental results, the noise affecting the four hospitals is mainly low-frequency noise. The psychological experiment data shows that the highest correlation frequency bands are 25–315 Hz and 1600–2500 Hz. Therefore, the noise frequency band with the highest perception by people in the lobby contains the middle and low followed by the high-frequency band. The acoustic simulation software (SoundPLAN) was used to restore the measurement by placing noise sources, installing noise barriers, and placing measurement points and surfaces. The experiment results present that the four hospitals had a wider distribution of indoor sound levels, and the most significant noise level was found at Daya Ching Chuen Hospital, presumably because it is closer to the road than the other three hospitals.

Author Contributions: Conceptualization, C.Y.C. and X.L.; methodology, C.Y.C.; software, X.L.; validation, C.Y.C. and X.L.; formal analysis, X.L.; investigation, X.L.; resources, C.Y.C.; data curation, X.L.; writing—original draft preparation, X.L.; writing—review and editing, X.L.; visualization, X.L.; supervision, C.Y.C.; project administration, C.Y.C.; funding acquisition, X.L. All authors have read and agreed to the published version of the manuscript.

Funding: This research received no external funding.

Institutional Review Board Statement: Not applicable.

Informed Consent Statement: Not applicable.

Data Availability Statement: Due to restrictions such as privacy or ethics, upon request provide data.

Conflicts of Interest: The authors declare no conflict of interest.

References

1. Kknews. Chronic Damage to Human Caused by Low-Frequency Noise Can not be Ignored. 2018. Available online: https://kknews.cc/zh-tw/health/9vzkgpl.html (accessed on 30 November 2023).
2. Topf, M. Noise induced occupational stress and health in critical care nurses. *Hosp. Top* **1988**, *66*, 30–34. [CrossRef] [PubMed]
3. Torija, A.J.; Flindell, I.H. Differences in subjective loudness and annoyance depending on the road traffic noise spectrum. *J. Acoust. Soc. Am.* **2014**, *135*, 1–4. [CrossRef] [PubMed]
4. Torija, A.J.; Flindell, I.H. The subjective effect of low frequency content in road traffic noise. *J. Acoust. Soc. Am.* **2015**, *137*, 189–198. [CrossRef] [PubMed]
5. Tomei, F.; Fantini, S.; Tomao, E.; Baccolo, T.P.; Rosati, M.V. Hypertension and chronic exposure to noise. *Arch. Environ. Health* **2014**, *55*, 319–325. [CrossRef] [PubMed]
6. Chen, C.Y. Characterizing Subjective Noisiness in Hospital Lobbies. *Arch. Acoust.* **2015**, *40*, 235–246. [CrossRef]
7. Chen, C.Y. Monitoring time-varying noise levels and perceived noisiness in hospital lobbies. *Build. Acoust.* **2021**, *28*, 35–55. [CrossRef]
8. Spreng, M. Possible health effects of noise induced cortisol increase. *Noise Health* **2000**, *2*, 59. [PubMed]
9. Paunović, K.; Stojanov, V.; Jakovljević, B.; Belojević, G. Thoracic bioelectrical impedance assessment of the hemodynamic reactions to recorded road-traffic noise in young adults. *Environ. Res.* **2014**, *129*, 52–58. [CrossRef] [PubMed]
10. ISO12354-1. 2017. Available online: https://www.iso.org/standard/70242.html (accessed on 30 November 2023).
11. ISO354. 2003. Available online: https://www.iso.org/standard/34545.html (accessed on 30 November 2023).

engineering proceedings

Proceeding Paper

Architectural Design Guidelines Based on Computer Core Evaluation and Processing Paradigms †

Jose L. Reategui

Facultad de Arquitectura, Universidad Peruana de Ciencias Aplicadas, Lima 15023, Peru; pcarjrea@upc.edu.pe

† Presented at the IEEE 5th Eurasia Conference on Biomedical Engineering, Healthcare and Sustainability, Tainan, Taiwan, 2–4 June 2023.

Abstract: The architectural design process may utilize decision-making algorithms; usually, this takes a long time when using analog methods. Today, sequential processing (CPU) or parallel processing (GPU) works with architectural design in different ways, such as Graphical User Interfaces using CPU or ray-tracing rendering using GPU. Architectural design paradigms have not yet fully adapted to the processing computer core paradigm. Thus, this research proposes design guidelines to take the computing core paradigm and transport it to architectural design with a practical case that tests the core's performance with a genetic algorithm (form-finding process). The genetic algorithm was tested on a computer based on a central processing unit (CPU), another on a graphics processing unit (GPU), and the last one on quantum computers (Quantum Gates). The guidelines operate these cores under the cloud computing paradigm, supported by new components made in Rhinoceros–Grasshopper 3D. The discussion shows the integration of the computation core paradigm with the current digital architecture design process.

Keywords: cloud computing; quantum computing; parametric design; genetic algorithm; architectural design

Citation: Reategui, J.L. Architectural Design Guidelines Based on Computer Core Evaluation and Processing Paradigms. *Eng. Proc.* **2023**, *55*, 43. https://doi.org/10.3390/engproc2023055043

Academic Editors: Teen-Hang Meen, Kuei-Shu Hsu and Cheng-Fu Yang

Published: 4 December 2023

1. Introduction

The architectural design process shows a relationship between the design procedure and the calculation with or without computational tools. This research seeks a comparison criterion to adapt the decision-making process in architectural design based on the advantages of three different computing paradigms. The first is sequential computing with CPU, the second is parallel processing using GPU, and the third is quantum processing using quantum processors (QPU).

A genetic algorithm (GA) implemented in different computing paradigms proposes how the architectural design process can improve considering the computer core processing. The genetic algorithm requires an intensive calculation process. The CPU processor works in Amazon Web Services (AWS) based in Intel, a GPU processor based in Nvidia was used in AWS, and quantum computing works with AWS Bracket with a Rigetti server. A graphic user interface (GUI) created in Rhinoceros–Grasshopper 3D allows connecting an API connection located in AWS to access different computing processors.

Usually, studies on architectural design are associated with division and task integration into the abstraction process. Likewise, the abstraction process reduces the time process, which is possible with parametric design technologies. In this way, Rhinoceros-Grasshopper incorporation in this core computation evaluation is crucial.

The specific information technology architecture based on CPU, GPU, and QPU where the GA is implemented is shown in Figure 1.

(Figure 1) shows the information technology (IT) architecture. Grasshopper and Rhinoceros 3D in the right-side work as the frontend, which means a GUI; for this purpose, there are two main areas: Grasshopper with the new component that works with send, receive, and transform data, and Rhinoceros 3D with the visual geometry. Another area is on the top-left side, which consists of the EC2, a virtual machine that allows us to use a CPU processor (based on Intel) and a GPU processor (based on Nvidia). The last area, on the bottom-left side, is the Rigetti server, where the quantum algorithm runs, but to use this service we have to connect through AWS Bracket.

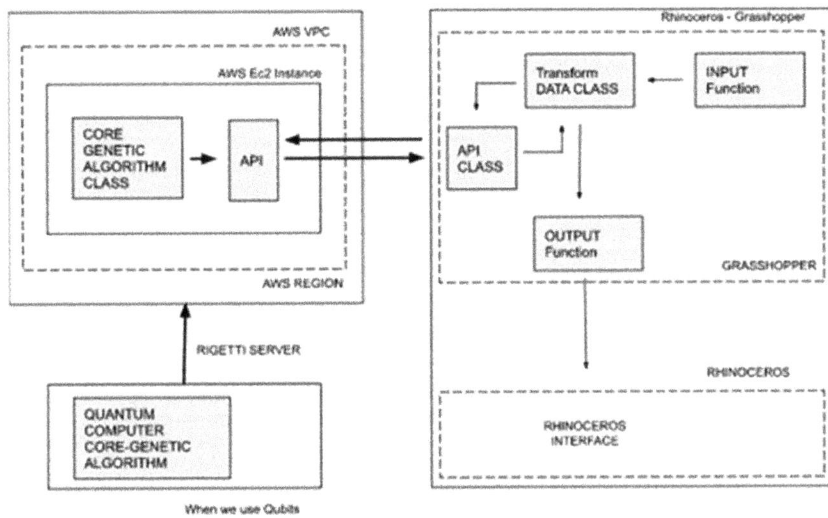

Figure 1. Information technology architecture: backend and frontend.

The design process shown in this research has two main branches: the digital morphogenesis process (A) and the analog morphogenesis process (B). In this way, a process guided by the simulation of natural evolution was used (A.1).

1.1. *Digital Morphogenesis*

1.1.1. Guided by Simulation of Natural Evolution through Genetic Algorithms

In these processes, the algorithm produces a population that has the parameters as genes. Once the selection and mutation mechanisms work, the best individuals can reach the goal; in this way, analyzing the performance of each individual is necessary through a mathematical procedure [1].

1.1.2. Guided by Emerging Patterns through Autonomous Agents

In this process, a population of agents is created, and each agent can communicate with others through basic state rules and constant analysis of their environment [2].

1.1.3. Guided by Parameters and Geometry Transformation Processes through Visual Programming

The process uses computational thinking to create shape relationships and data transformations, usually via visual programming.

1.1.4. Guided by Animation Sequences through Dynamic Patterns

In these processes, design decisions are usually made based on the achievement of frames in animated sequences where time is a variable to make decisions.

1.1.5. Guided by Materials' Behavior and Computational Physics

In these processes, the behavior of materials is usually analyzed with variables and physical laws such as gravity or Hooke's law to then make design decisions based on these observations.

1.1.6. Guided by Manual Control of Digital Curve Points

In these processes, the designer determines the position of control points that model a shape in space; Bezier or Nurbs curves are often used for these processes.

1.2. Morphogenesis with Analog Resources

1.2.1. Guided by Physical Model Transform to Digital Model

The process can transform the physical model made by hand or another method in bits through points digitalization or 3D scanning.

1.2.2. From Free-Hand Drawing to Digital Model

The process uses free-hand drawings to transform into pictures using a scanning process or a digital camera.

2. Information Technology Architecture

2.1. Genetic Algorithm Architecture

The genetic algorithm was made with the following parts:

1. Generating a population method:

The population is generated randomly with individuals that have binary-encoded genes. These genes represent parameters in binary format.

2. Iterating with a stopping method:

The selection and evolution mechanism iteration reaches a parameter combination to target approximate results and finally stop.

3. Individual evolving method:

The evolution and selection mechanisms have a closer relation. The best parent genes (parameter value) are determined according to a selection rule for evolution.

4. Selecting the parent method:

The performance rule defines which individuals are better to adapt to environmental conditions.

5. Applying the crossover and mutation method to parents:

This operation uses a gene selection rule that determines, through operations in pairs, which genes will be taken for the next generation and, in some cases, gene mutation.

2.2. Computation Processors in the Cloud

Amazon Web Services allows the use of different computation technologies through servers such as AWS EC2 service (Intel i7 CPUs, the NVIDIA T4 Tensor Core GPUs) and AWS Bracket using Aspen-10-25Q Rigetti processors used in this research.

1. AWS EC2 based on CPU

In this system architecture, connectivity was developed through a Rest API created with the Express system, and then the genetic algorithm was implemented with NodeJs that, through the Google V8 engine, can make use of the CPU (Figure 2).

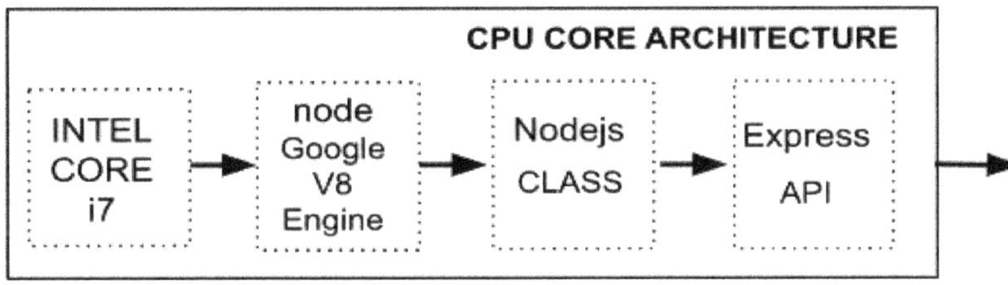

Figure 2. AWS EC2 instance based on Intel Core i7.

2. AWS EC2 based on GPU

In this architecture, the node-rapids technology connects the Nvidia GPU with NodeJs through CUDA; this technology is exposed by using Rest API (Figure 3).

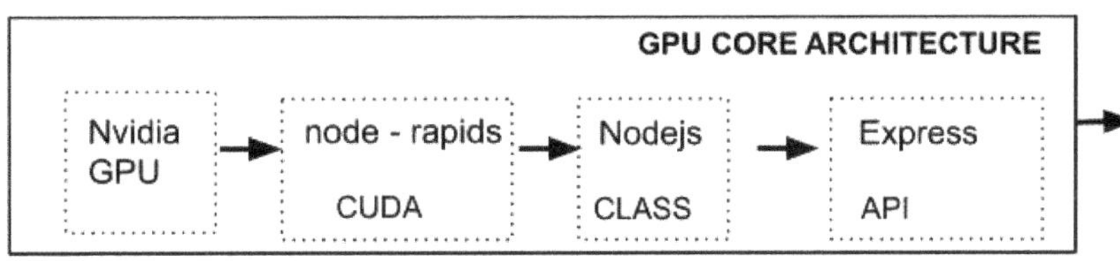

Figure 3. AWS EC2 instance based on Nvidia GPU.

3. AWS BRACKET based on Aspen-10-25Q Rigetti processors

This system architecture works with two cloud computing servers. First, Amazon Web Services Bracket allows the use of Python-based Jupyter Notebooks to control load execution and find out the status of computing tasks; second, the Rigetti Aspen 10 can access this from AWS Bracket (Figure 4).

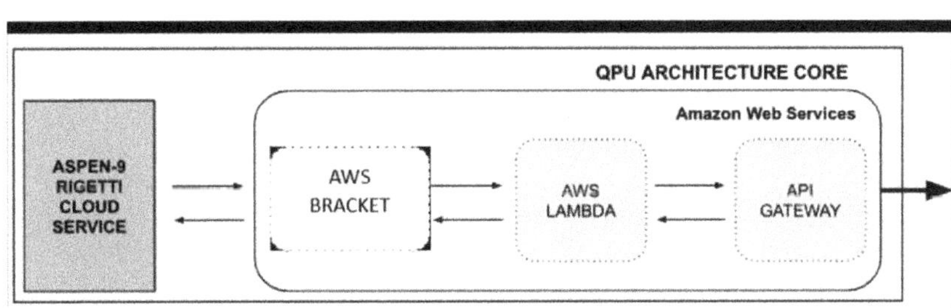

Figure 4. AWS Bracket is used to control a Rigetti Quantum core.

2.3. Frontend Anatomy in Grasshopper–Rhinoceros 3D

A new component was developed to connect the server with Grasshopper 3D. This component considers the sending and receiving values in JSON (JavaScript Object Notation) format, and their subsequent transformation into inputs to be used within parametric design processes.

Figure 5 shows the Grasshopper 3D functions. The API function allows inserting data from the user into the Grasshopper 3D component; after that, the transform class uses this information to obtain data from the server with the API function. The API function can obtain information from the server and transform the server response into data to use. The output function takes the information from the transform function and gives the information that the Grasshopper 3D user needs; after that, this information is shown in Rhinoceros 3D.

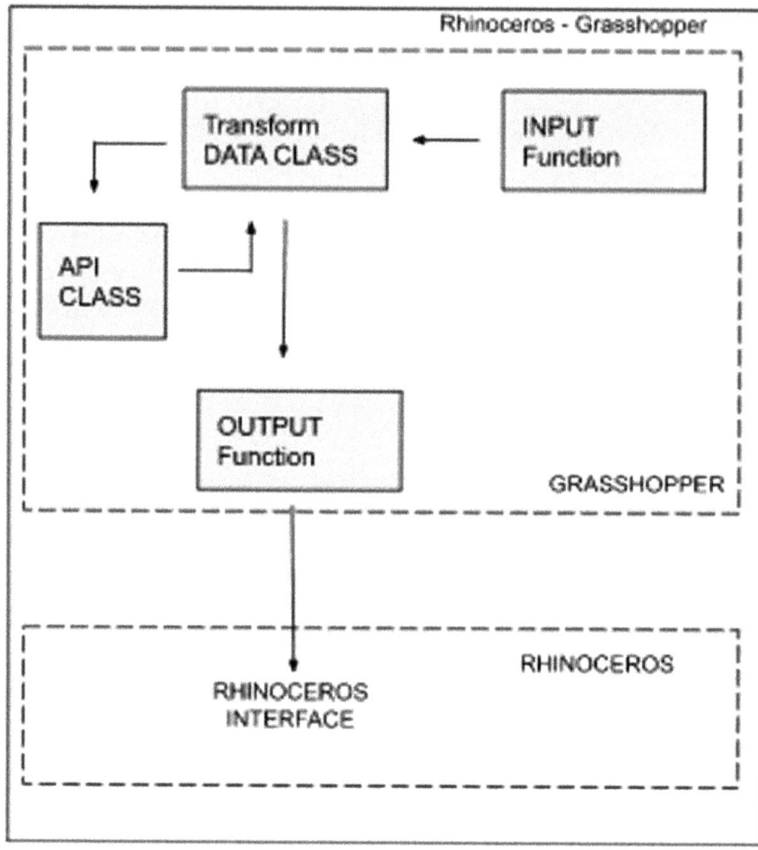

Figure 5. Frontend architecture is based on Rhinoceros 3D–Grasshopper.

3. Results

Architectural design currently does not differentiate between which computing tasks to direct its processes towards. The different morphogenesis processes require different computing capacities related to the following tasks:

A.1 Searches for the best combination and the best route.

A.2 Determination of states based on neighboring states.

A.3 Re-calculation based on a time frame.

A.4 Re-calculation based on picture frames.

A.5 Subdivision to generate analysis mesh.

A.6 Analysis via raytracing.

A.7 Statistical analysis of point concurrence.

In this research, the comparison of the "Search for the best combination" task was carried out using a genetic algorithm implemented in the CPU, GPU, and QPU architectures.

For the first case (CPU), sequential tasks were developed based on the use of a server with the Intel Core i7 processor (Figure 6).

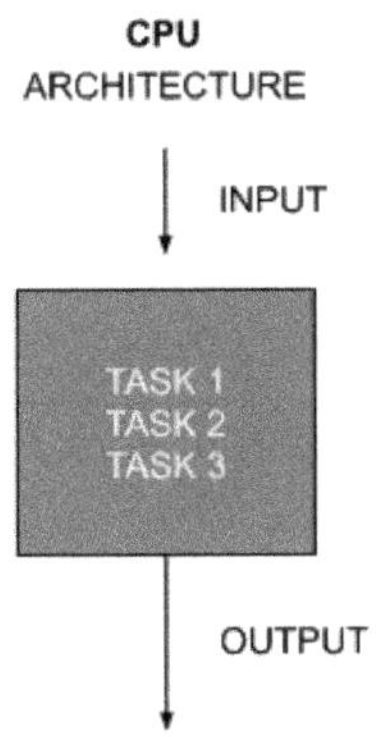

Figure 6. CPU task-solving model.

For the second case (GPU), the combination tasks were distributed in the different cores of the GPU to obtain information about the performance of the offspring (Figure 7).

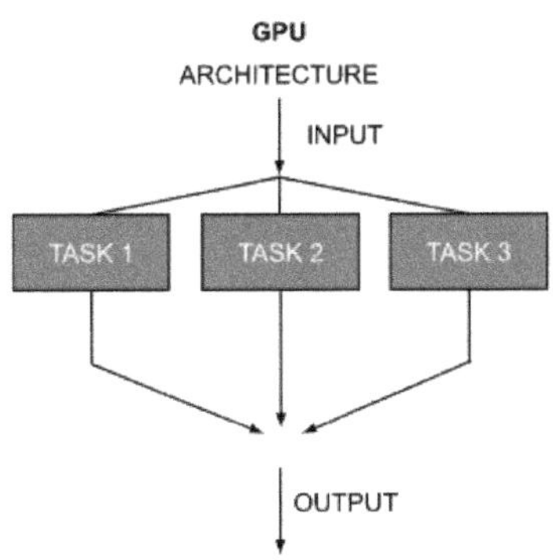

Figure 7. GPU task-solving model.

For the third case (QPU), the tasks were executed at the same time thanks to the quantum entanglement and superposition mechanisms (Figure 8).

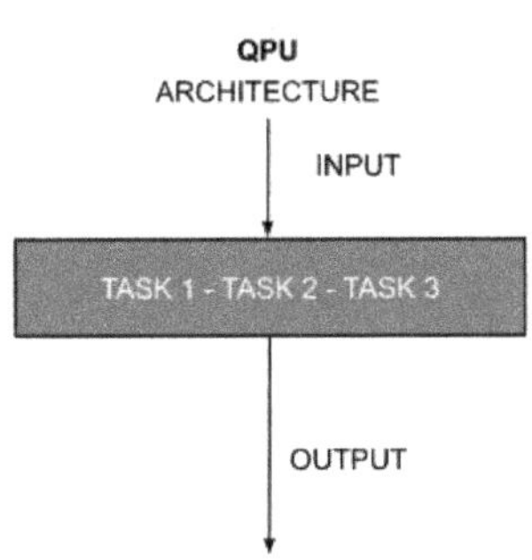

Figure 8. QPU task-solving model.

The results between the different processors show that the quantum computing processor performed better in finding the best combination of genetic algorithm genes to achieve a performance goal (Figure 9).

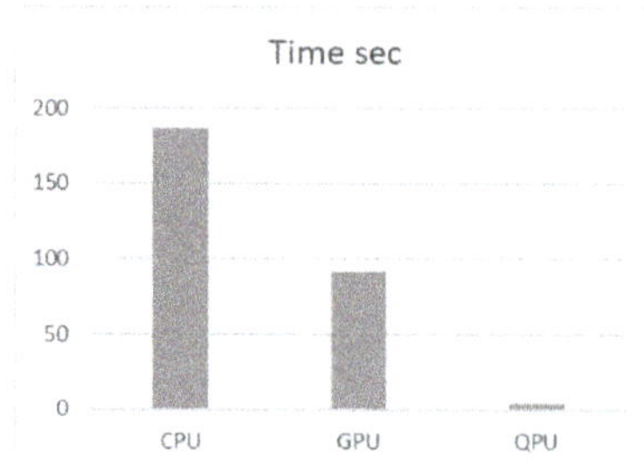

Figure 9. Time evaluation between CPU-GPU-QPU.

The objective of implementing a genetic algorithm is to show how the different processing cores behave to determine strategies in architectural design processes. In terms of processing speed, the QPU had better performance than the GPU, while the GPU was superior to the CPU. With this observation, the following steps are proposed.

1. Define the type of problem to solve according to the morphogenesis process to take.
2. Analyze the compatibility of the problem with the type of processing paradigm to be used, considering the following.
    - If it is a general-purpose task and requires design platforms and low-intensity computational capacity, it is recommended that the CPU is used; preferably, the tasks should be executed sequentially and interdependently (Figure 10).
    - If the task can be divided as in the determination of the state in autonomous agents, the GPU should be used. Tasks are not sequential and interdependent [3] (Figure 11).
    - If the task can be divided and consists of reaching an objective solution in multiple ways, as in the case of searching for the best route to reach an objective, it is recommended to use the QPU. It seeks to take advantage of superposition and quantum entanglement to execute tasks with Qubits (Figure 12).

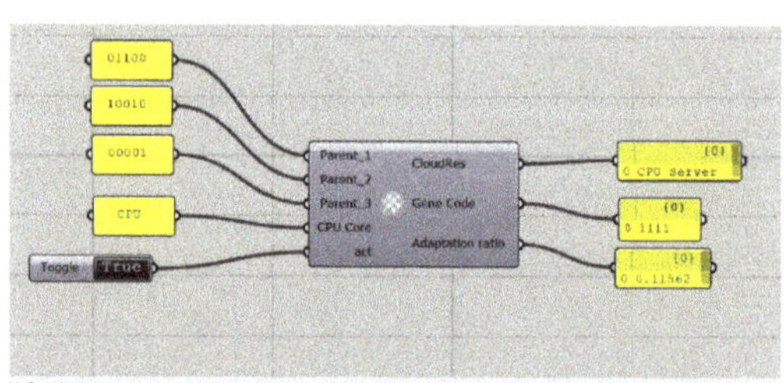

Figure 10. Grasshopper component to connect AWS Instance based on CPU core.

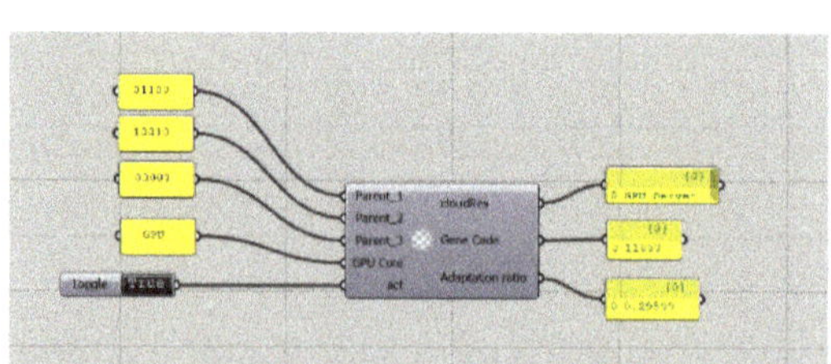

Figure 11. Grasshopper component to connect AWS Instance based on GPU core.

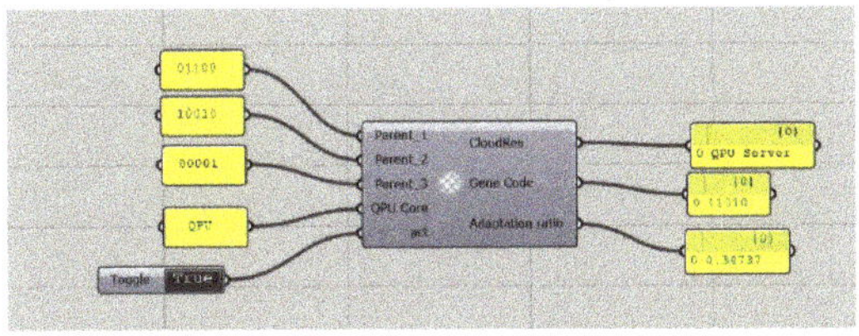

Figure 12. Grasshopper component to connect AWS Bracket based on QPU core.

3. Define the design problem to be solved according to a computational design process; the use of a visual programming system such as Rhinoceros 3D–Grasshopper, or Revit-Dynamo is recommended, or if possible, the use of explicit programming.
4. The indicators for this guideline are computing execution time and decision-making time in the architectural design.

4. Future Work

This research proposes guidelines based on the evaluation of processing paradigms of the CPU, GPU, and QPU. The implementation of a customizable genetic algorithm through a component created ad hoc in the Rhinoceros 3D–Grasshopper computer system was used as an example of implementation and demonstration.

The main selection criteria for the processing core were the computing tasks' divisibility, the task's flexibility, and the need for intensive computing. Future research will expand the range of implementation examples focused on the creation and performance analysis of machine learning engines; in this way, the focus will be on neural networks applied in computing core paradigms and architectural design processes.

Funding: This research was supported by the Dirección de Investigación de la Universidad Peruana de Ciencias Aplicadas.

Institutional Review Board Statement: Not applicable.

Informed Consent Statement: Not applicable.

Data Availability Statement: Data are contained within the article.

Conflicts of Interest: The author declares no conflict of interest.

References

1. Horn, J.; Nafpliotis, N.; Goldberg, D.E. A niched Pareto genetic algorithm for multiobjective optimization. In Proceedings of the First IEEE Conference on Evolutionary Computation, IEEE World Congress on Computational Intelligence, Orlando, FL, USA, 27–29 June 1994; pp. 82–83.
2. Kennedy, J.; Eberhart, R. Particle swarm optimization. In Proceedings of the ICNN'95-International Conference on Neural Networks, Perth, WA, Australia, 27 November–1 December 1995; pp. 5–6.
3. Gregg, C.; Hazelwood, K. Where is the data? Why you cannot debate CPU vs. GPU performance without the answer. In Proceedings of the (IEEE ISPASS) IEEE International Symposium on Performance Analysis of Systems and Software, Austin, TX, USA, 10–12 April 2011; p. 143.

engineering proceedings

Proceeding Paper

Regression Analysis to Investigate Contributions of Process Parameters on Transesterification of Waste Soybean Oil †

Masjudin and Wei-Chin Chang *

Department of Mechanical Engineering, Southern Taiwan University of Science & Technology, Tainan 71005, Taiwan; da51y205@stust.edu.tw
* Correspondence: wcchang@stust.edu.tw

† Presented at the IEEE 5th Eurasia Conference on Biomedical Engineering, Healthcare and Sustainability, Tainan, Taiwan, 2–4 June 2023.

Abstract: Four process parameters were investigated via regression analysis for their effects on biodiesel production in the transesterification of biodiesel from waste soybean oil (WSO). The maximum biodiesel conversion rate reached 98.56% at a methanol/oil ratio of 6:1, a reaction time of 60 min, a reaction temperature of 60 °C, and a catalyst loading of 0.5 wt.%. The functional group was dominated by an ester with the wavenumber 1743.78, the group attribution C=O, a strong absorption intensity, and a stretching vibration. The major methyl ester compounds of the biodiesel yielded from the waste soybean oil were dominated by linoleic acid (37.53%), oleic acid (24.69%), and palmitic acid (12.95%). The regression analysis result showed that the parameters influenced the biodiesel yield, and that the reaction temperature contributed to the reaction more than the others.

Keywords: biodiesel production; waste soybean oil; transesterification process; methyl ester; regression analysis

1. Introduction

Biodiesel is one of the most promising fuels because its raw materials are abundant. Even though the cost of producing biodiesel is more than that of fossil fuels, many countries are producing biodiesel commercially to meet their energy needs because biodiesel assures energy security, environmental friendliness, and socio-economic stability. With similar properties to diesel oil, biodiesel can be used for any engine because of its excellent lubricity. It is a renewable resource (vegetable oil or animal fat) and is biodegradable and environmentally friendly. To produce biodiesel from vegetable oil, transesterification is process most used to form a mixture of fatty acid methyl esters (FAMEs). Oils change their chemical and physical properties if they are cooked many times. Waste or used cooking oil can be used to produce biodiesel. The use of used cooking oil as a raw material (feedstock) may increase the production of biodiesel and reduce environmental pollution [1]. Soybean oil is the second most consumed vegetable oil in the world; hence, used soybean oil can be used in many countries for the production of biodiesel [2–5].

Producing biodiesel from waste soybean oil via transesterification requires a catalyst. There are many kinds of catalysts, including base and acid catalysts, that can increase biodiesel production in the transesterification reaction. Catalysts must be chosen considering environmental aspects and costs. Potassium hydroxide, one of the base catalysts, is used in the transesterification reaction and has shown a conversion yield of ethyl or methyl ester of over 90% [1,6]. The transesterification reaction is shown in Equation (1).

Citation: Masjudin; Chang, W.-C. Regression Analysis to Investigate Contributions of Process Parameters on Transesterification of Waste Soybean Oil. *Eng. Proc.* **2023**, *55*, 44. https://doi.org/10.3390/engproc2023055044

Academic Editors: Teen-Hang Meen, Kuei-Shu Hsu and Cheng-Fu Yang

Published: 4 December 2023

$$\begin{array}{lcccl} CH_2-OCOR^1 & & & CH_2OH & \\ \quad | & & Catalyst & | & R^1COOCH_3 \\ CH-OCOR^2 & +\ 3CH_3OH & \rightleftharpoons & CHOH & +\ R^2COOCH_3 \\ \quad | & & KOH & | & R^3COOCH_3 \\ CH_2-OCOR^3 & & & CH_2OH & \\ \text{Triglyceride} & +\ \text{Methanol} & \rightleftharpoons & \text{Glycerol} & +\ \text{Methyl Ester} \end{array} \quad (1)$$

Triglycerides are heavy molecular fats and oils saponified (hydrolyzed) in a solution to produce soap and glycerol. The glycerol is insoluble in the transesterification product and easily separated with methyl ester because of its higher density. In transesterification reactions, the soap is formed because free fatty acids (FFAs) from the oil react with the catalyst. It is a reactant in the transesterification containing water, and the reaction increases the quantity of FFAs in hydrolysis. This increases the formation of soap. The process of soap formation with a high FFA content is expressed as follows:

$$\begin{array}{lcccccl} CH_2-OCOR^1 & & & & CH_2OH & & \\ \quad | & & & & | & & R^1COOCK \\ CH-OCOR^2 & + & 3KOH & \rightarrow & CHOH & + & R^2COOCK \\ \quad | & & & & | & & R^3COOCK \\ CH_2-OCOR^3 & & & & CH_2OH & & \end{array} \quad (2)$$
$$\text{Triglyceride} + \text{Potassium Hydroxide} \rightarrow \text{Glycerol} + \text{Potassium Carboxylate}$$

In transesterification, the most active catalysts are methoxide radicals. The activity of a particular catalyst depends on the number of methoxide radicals. These are used in the reaction, as each mole of triglyceride reacts with 3 moles of primary alcohol, and yields 3 moles of alkyl ester (biodiesel) and 1 mole of glycerol (by-product). Dissolving potassium in methanol produces the methoxide ion according to the following reaction:

$$OH^- + 3CH_3OH \rightarrow 3CH_3O^- + H_2O \quad (3)$$

Biodiesel purification using a conventional catalyst requires high energy and large water consumption, which causes environmental issues and affects the quality of the biodiesel. To increase the biodiesel quality, a purification process for removing impurities such as catalysts, methanol, glycerol, glyceride, water, soap, and other materials is necessary. There are two often used methods, namely wet washing and dry washing. The wet washing method is the best choice so far in purifying biodiesel after the transesterification reaction.

2. Materials and Methods

A transesterification method was used to obtain methyl ester biodiesel. In the transesterification and purification, the operating conditions and homogeneous mixture were controlled. The required raw materials were waste soybean oil, potassium hydroxide, and methanol. Soybean oil was purchased from a local supermarket in Tainan, Taiwan. It was used for frying fish or meat, and then became waste soybean oil as a feedstock. Methanol was chosen because it was more suitable than other alcohols for transesterification when considering the price and physical and chemical properties. Potassium hydroxide (KOH) or sodium hydroxide (NaOH) were recommended to produce methyl ester. However, considering the cost, KOH was chosen as it is cheaper than NaOH. The operating conditions, regarding the oil/methanol molar ratios, catalyst loading, reaction time, and reaction temperature, are presented in Table 1.

Table 1. Operating conditions used to convert triglycerides from waste soybean oil to methyl ester biodiesel.

Variable Conditions	Reaction Temperature (°C)	Methanol/Oil Ratio (mol/mol)	Reaction Time (min.)	Catalyst Loading (%)
Reaction temperature	30, 40, 50, 60, 70, 80, 90, 110	6:1	60	0.5
Methanol/oil molar ratio	60	3:1, 4:1, 5:1, 6:1, 8:1	60	0.5
Reaction time	60	6:1	30, 50, 60, 90, 120, 180	0.5
Catalyst loading	60	6:1	60	0.2, 0.5, 1, 2

First, waste soybean oil was heated up to 40 °C and then filtered to remove dirt, charred food, and non-oil materials. Then, the oil was heated up to above 100 °C to remove the water content. The catalyst was added to the methanol and stirred until all KOH dissolved, and then it was mixed into the heated oil. The mixture was heated on a hot plate stirrer. After heating, the separation of methyl ester and glycerin was performed for 60 min in a separatory funnel. The glycerin waste was drawn off from the bottom of the settling vessel via gravity. The experimental process of biodiesel production is shown in Figures 1 and 2.

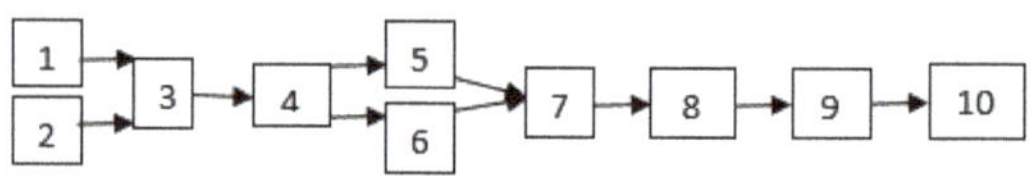

Figure 1. Biodiesel production process; 1. Catalyst and methanol, 2. WSO, 3. Transesterification tank, 4. Separatory funnel, 5. Glycerol, 6. Biodiesel result of transesterification, 7. Purification/washing, 8. Heating, 9. Methyl ester biodiesel, 10. Physical and chemical property test.

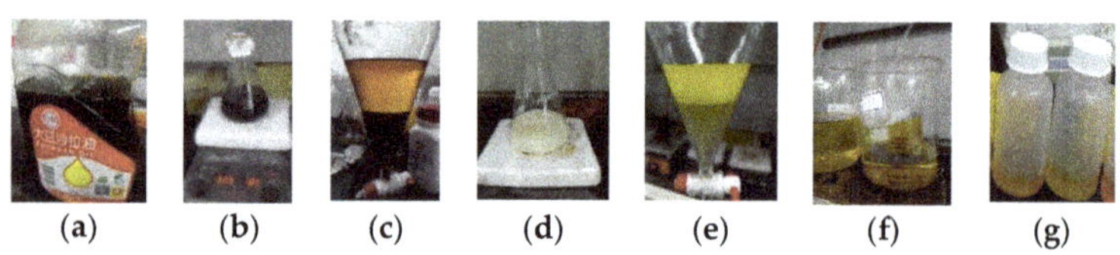

Figure 2. Process of biodiesel production: (**a**) waste soybean oil (WSO); (**b**) transesterification reaction; (**c**) separation process between biodiesel and glycerol; (**d**) the washing process; (**e**) separation process between biodiesel and water; (**f**) the heating process; (**g**) pure biodiesel.

2.1. *Analysis of Physicochemical Properties*

The properties of biodiesel depend on the type and composition of the oil and the transesterification operating conditions. FT-IR spectroscopy was used to identify the presence of functional groups in the soybean oil, waste soybean oil, and biodiesel samples. Equation (4) was applied to calculate the yield of the obtained biodiesel.

$$\text{Yield of biodiesel (\%)} = [(\text{biodiesel (wt.)})/(\text{WCO (wt.)})] \times 100\% \quad (4)$$

2.2. *Regression Model*

Regression analysis was used to analyze the experimental results. The variables are shown in Table 1. The F-test was conducted to determine the effects of the independent variable (X) on the dependent variable (Y) by comparing F-values at a significance level of 5%. By finding a partial coefficient, the fitness of the model was tested. The effect was identified by calculating the t-value and comparing it at a significance level of 5%. In the regression model, the parameters (Table 1) of eight reaction temperatures, five methanol/ oil ratios, six reaction times, and four catalyst loadings were included as independent variables, while biodiesel production (%) was included as a dependent variable. In the regression analysis, each independent variable's effect on the dependent variable was identified, and function coefficients and polynomial functions were found. The regression equation was a polynomial function that depended on the degree of freedom. The polynomial model equation was as follows:

$$f(X) = Y = a_nX^n + a_{n-1}X^{n-1} + a_{n-2}X^{n-2} + ... + a_2X^2 + a_1X + a_0 \quad (5)$$

By entering all the variable conditions, the regression equation for eight reaction temperatures was derived as follows:

$$\begin{aligned} Y = {} & a_{11}X_1 + a_{01} + a_{12}X_2 + a_{02} + a_{13}X_3 + a_{03} + a_{14}X_4 + a_{04} + a_{15}X_5 + a_{05} + a_{16}X_6 \\ & + a_{06} + a_{17}X_7 + a_{07} + a_{18}X_8 + a_{08} \end{aligned} \quad (6)$$

$$\begin{aligned} Y = {} & a_{21}X_1^2 + a_{11}X_1 + a_{01} + a_{22}X_2^2 + a_{12}X_2 + a_{02} + a_{23}X_3^2 + a_{13}X_3 + a_{03} + a_{24}X_4^2 \\ & + a_{14}X_4 + a_{04} + a_{25}X_5^2 + a_{15}X_5 + a_{05} + a_{26}X_6^2 + a_{16}X_6 + a_{06} + a_{27}X_7^2 \\ & + a_{17}X_7 + a_{07} + a_{28}X_8^2 + a_{18}X_8 + a_{08} \end{aligned} \quad (7)$$

$$\begin{aligned} Y = {} & a_{31}X_1^3 + a_{21}X_1^2 + a_{11}X_1 + a_{01} + a_{32}X_2^3 + a_{22}X_2^2 + a_{12}X_2 + a_{02} + a_{33}X_3^3 + a_{23}X_3^2 \\ & + a_{13}X_3 + a_{03} + a_{34}X_4^3 + a_{24}X_4^2 + a_{14}X_4 + a_{04} + a_{35}X_5^3 + a_{25}X_5^2 \\ & + a_{15}X_5 + a_{05} + a_{36}X_6^3 + a_{26}X_6^2 + a_{16}X_6 + a_{06} + a_{37}X_7^3 + a_{27}X_7^2 \\ & + a_{17}X_7 + a_{07} + a_{38}X_8^3 + a_{28}X_8^2 + a_{18}X_8 + a_{08} \end{aligned} \quad (8)$$

where Y is for the biodiesel conversion, a_n, $a_{(n-1)}$, $a_{(n-2)}$, . . ., a_2, a_1, a_0 are function coefficients, n is the degree of freedom of the polynomial function, and X is the independent variable.

3. Results and Discussion

3.1. Regression Analysis

Figure 3 presents the estimation curves, which resemble quadratic and cubic curves. The coefficients of determination (R^2) of the linear, quadratic, and cubic curves were 0.209, 0.972, and 0.982 (with the variation in reaction temperature), 0.405, 0.985, and 0.991 (with the variation in the methanol/oil ratio), 0.150, 0.450, and 0.729 (with the variation in reaction time), and 0.839, 0.960, and 1 (with the variation in catalyst loading). The cubic curve was similar to the observation data (R^2 = 1) compared with the quadratic estimation curve for all the variations.

The parameters of the reaction temperature (quadratic = 0.00 and cubic = 0.001), methanol/oil ratio (quadratic = 0.015 and cubic = 0.009), and catalyst loading (Sig. = 0.000) had a significant effect on biodiesel conversion. On the other hand, the reaction time did not influence the biodiesel conversion significantly. The effect of the reaction temperature (X_T), methanol/oil ratio (X_{mo}), reaction time (X_t), and catalyst loading (X_c) on biodiesel conversion Y((Y_T), (Y_{mo}), (Y_t), (Y_c) and (Y_T)), respectively, was estimated using the following regression models:

$$Y_T = 87.335 + 0.243X_T + 0.00062X_T^2 - 2.494X_T^3 \quad (9)$$

$$Y_{mo} = 67.106 + 7.407X_{mo} - 0.06215X_{mo}^3 \quad (10)$$

$$Y_t = 80.888 + 0.529X_t - 0.0054X_t^2 + 0.0000155X_t^3 \quad (11)$$

$$Y_c = 89.912 + 36.096X_c - 43.625X_c^2 + 12.073X_c^3 \quad (12)$$

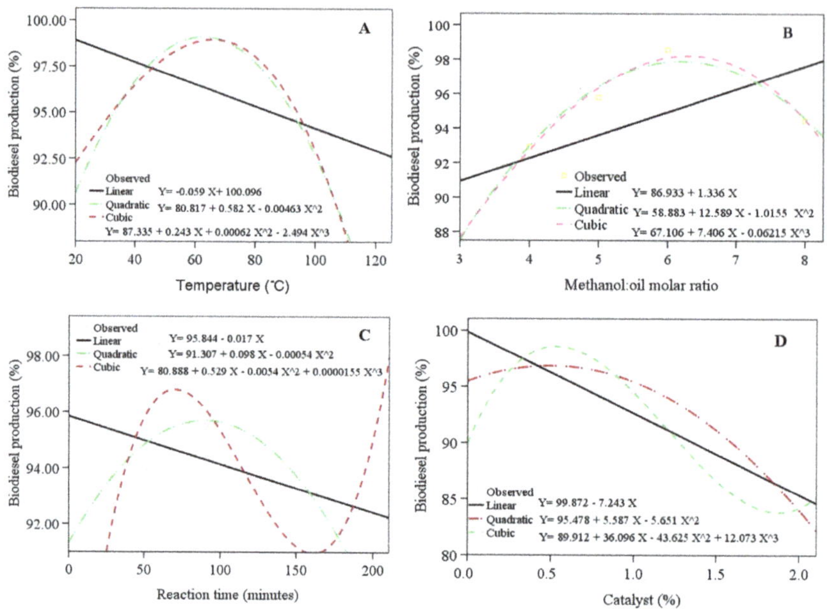

Figure 3. The biodiesel production with the variations in the reaction temperature (**A**), methanol/oil ratio (**B**), reaction time (**C**), catalyst loading (**D**), the observation data compared with the estimation curves.

3.2. Chemical Analysis of Biodiesel Using Fourier Transform Infrared (FTIR) Spectrum

Figure 4 displays the FTIR spectra results for the main functional group of soybean oil, waste soybean oil, and biodiesel. Figure 5 shows the FTIR spectra of biodiesel under different reaction times and catalyst loadings, respectively. The spectra of the samples from soybean oil, waste soybean oil, and biodiesel were obtained at the ranges of 4000 to 400 cm^{-1}.

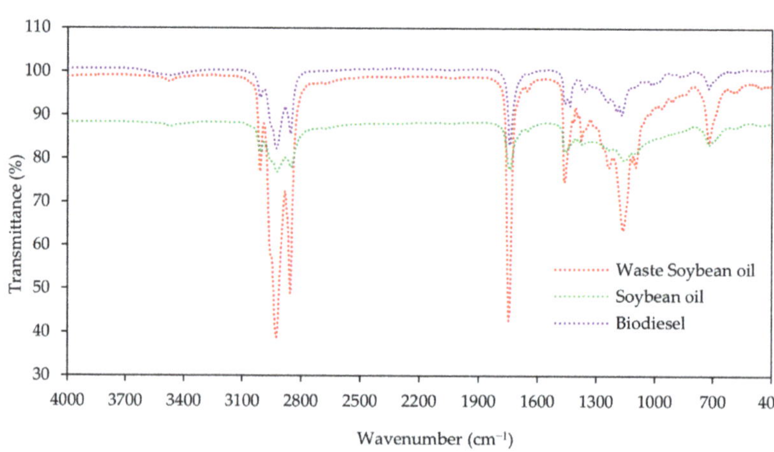

Figure 4. Spectra at 4000–400 cm^{-1} were obtained for the soybean oil, waste soybean oil, and biodiesel using the FTIR technique.

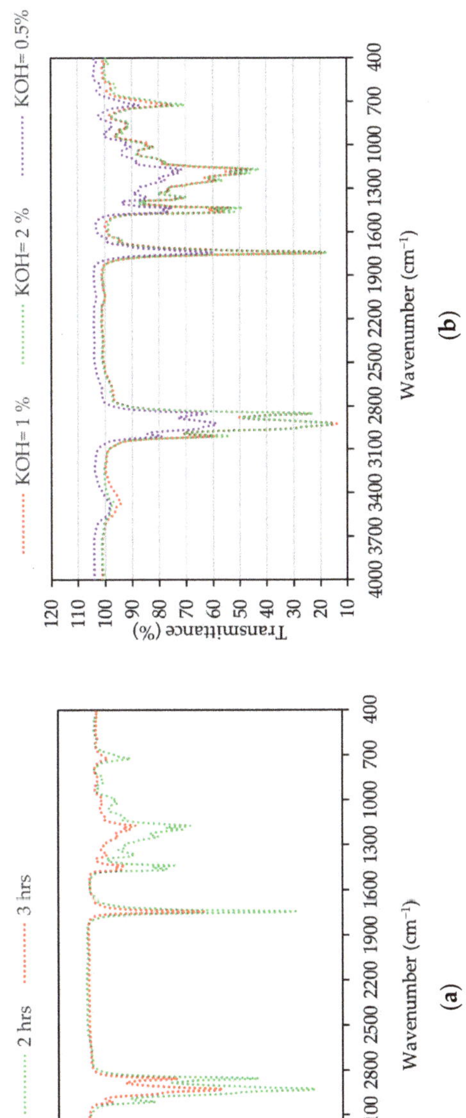

Figure 5. Biodiesel: (**a**) FTIR spectra at the reaction time of 2 and 3 hours; (**b**) FTIR spectra under the different catalyst loadings.

There were no significant peaks in the regions of 1700−2800 cm^{-1} and 3000−4000 cm^{-1} for the soybean oil, waste soybean oil, and biodiesel for all conditions. There were strong peaks between 2800 cm^{-1} and 3000 cm^{-1}, corresponding to the C-H vibration. The main characteristic peak for biodiesel was obtained at the spectrum of 1245 cm^{-1} (C-O), corresponding to the proportion of fatty acid methyl ester. Other peaks for biodiesel were identified at the spectra of 1744 cm^{-1} (C=O) and 1057 cm^{-1} (C-O).

3.3. Gas Chromatography-Mass Spectrum (GC-MS) of Biodiesel Yield from Waste Soybean Oil

GC-MS was applied to determine the substance compositions or quality of the biodiesel in terms of the presence of free fatty acid methyl ester. Figure 6 shows the chromatography of biodiesel used to identify the compositions present in FAME. Table 2 shows data from the GC-MS results, indicating the peak number, methyl ester names, molecular weight, and approximate area values for each of the methyl ester compositions. The dominant fatty acids were linoleic acid (37.53%), followed by oleic acid (24.69%) and palmitic acid (12.95%).

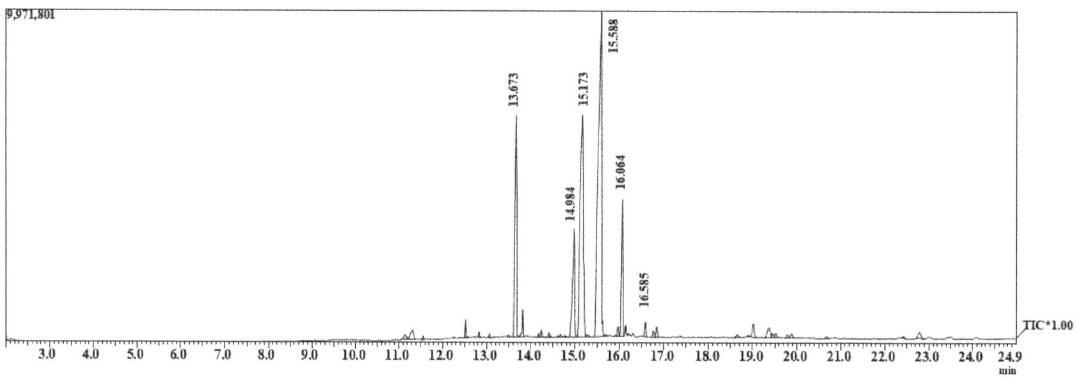

Figure 6. GC-MS of biodiesel made of waste soybean oil.

Table 2. Contents of methyl ester in the biodiesel yielded from waste soybean oil using GC-MS.

Peak no.	Molecular Formula	Molecular Weight	Formal Name	Common Name	Approximate Area (%)	MW (g/mol)
4	C15H30O2	242	Methyl Tetradecanoate acid	Myristic	0.45	1.089
8	C17H34O2	270	Hexadecenoic acid	Palmitic	12.95	34.965
18	C19H36O2	296	9-Octadecenoic acid	oleic	24.69	73.082
20	C19H36O2	294	9, 12-octadecadienoic acid	linoleic	37.53	110.338
29	C21H42O2	354	Docosanoic acid	Behenic	0.92	3.257
			Total		96.5	282.995

4. Conclusions

The transesterification process produced biodiesel by reacting triglycerides with used soybean oil with methanol and a KOH catalyst. Through the analysis of regression with four process parameters, the maximum methyl ester biodiesel produced from waste soybean oil was predicted at the operating conditions of 6:1 (methanol/ oil ratio), 0.5% (KOH catalyst), 60 min (reaction time), 60 °C (reaction temperature), and 600 rpm (stirrer speed). The reaction temperature, methanol to oil ratio, reaction time, and catalyst loading to the biodiesel conversion formed a cubic curve, which was fitted in the regression analysis. The FTIR spectra of the ester (C=O) were found at 1743, 1746, and 1743 cm^{-1} for the soybean oil, waste soybean oil, and biodiesel, respectively. The prominent characteristic peaks for biodiesel were identified at the spectra of 1245 cm^{-1} (C-O) and 1057 cm^{-1} (C-O). The GC-MS result showed that the major methyl ester components of the biodiesel yielded from waste soybean oil were linoleic acid (37.53%), followed by oleic acid (24.69%) and palmitic acid (12.95%).

Author Contributions: Conceptualization, M.; methodology, M. and W.-C.C.; software, M.; validation, M. and W.-C.C.; experiment data capture, M.; formal analysis, M.; writing—original draft preparation, M.; writing—review and editing, W.-C.C.; supervision, W.-C.C.; funding acquisition, W.-C.C. All authors have read and agreed to the published version of the manuscript.

Funding: This research was funded by National Science and Technology Council, Taiwan, grant number NSTC 111-2221-E-218-010.

Institutional Review Board Statement: Not applicable.

Informed Consent Statement: Not applicable.

Data Availability Statement: Data described in this study will be provided upon request.

Acknowledgments: The authors would like to express special thanks to the Instrument Equipment Division, Core Facility Center, NCKU, for providing FTIR and GC-MS to complete the important tests of biodiesel.

Conflicts of Interest: The authors declare no conflict of interest.

References

1. Ejikeme, P.M.; Anyaogu, I.D.; Ejikeme, C.L.; Nwafor, N.P.; Egbuonu, C.A.; Ukogu, K.; Ibemesi, J.A. Catalysis in Biodiesel Production by Transesterification Processes-An Insight. *J. Chem.* **2010**, *4*, 1120–1132. [CrossRef]
2. Rajesh, K.; Natarajan, M.P.; Devan, P.K.; Ponnuvel, S. Coconut fatty acid distillate as novel feedstock for biodiesel production and its characterization as a fuel for diesel engine. *Renew. Energy* **2021**, *164*, 1424–1435. [CrossRef]
3. Karthikeyan, M. Production of biodiesel from Cordiamyxa bio-oil using $BaMoO_4$ -Ce_2O_3 nanoparticles as an alternative fuel for diesel engine. *Mater. Lett.* **2019**, *243*, 199–201. [CrossRef]
4. Gónzalez, J.P.C.; Gutiérrez, P.E.Á.; Medina, M.A.; Zapata, B.Y.L.; Guerrero, G.V.R.; Valdés, L.G.V. Effects on biodiesel production caused by feed oil changes in a continuous stirred-tank reactor. *Appl. Sci.* **2020**, *10*, 992. [CrossRef]
5. García-Martín, J.F.; Barrios, C.C.; Alés-Álvarez, F.J.; Dominguez-Sáez, A.; lvarez-Mateos, P.A. Biodiesel production from waste cooking oil in an oscillatory flow reactor: Performance as a fuel on a TDI diesel engine. *Renew. Energy* **2018**, *125*, 546–556. [CrossRef]
6. Blinová, L.; Fiala, J.; Balog, K. Biodiesel Production from Waste Cooking Oil in Laboratory Scale. *Renew. Energy Environ. Technol.* **2014**, *448*, 1656–1659. [CrossRef]

engineering proceedings

MDPI

Proceeding Paper

Image Enhancement CNN Approach to COVID-19 Detection Using Chest X-ray Images †

Chamoda Tharindu Kumara [1], Sandunika Charuni Pushpakumari [1], Ashmini Jeewa Udhyani [1], Mohamed Aashiq [2,*], Hirshan Rajendran [1] and Chinthaka Wasantha Kumara [2]

1 Department of Electrical and Telecommunication Engineering, South Eastern University of Sri Lanka, Oluvil 32360, Sri Lanka; mrchamodatharindu@gmail.com (C.T.K.); sandunikacharuni9674@gmail.com (S.C.P.); hpajudhyani@seu.ac.lk (A.J.U.); rajehirshan@seu.ac.lk (H.R.)

2 Department of Computer Science and Engineering, South Eastern University of Sri Lanka, Oluvil 32360, Sri Lanka; chinthakawk@seu.ac.lk

* Correspondence: aashiqmnm@seu.ac.lk

† Presented at the IEEE 5th Eurasia Conference on Biomedical Engineering, Healthcare and Sustainability, Tainan, Taiwan, 2–4 June 2023.

Abstract: Coronavirus (COVID-19) is a fast-spreading virus-related disease. On 28 March 2022, Worldometer (COVID-19 live update) reported that there were about 482,338,923 COVID-19 cases and 6,149,387 fatalities worldwide. Moreover, there were about 416,884,712 recovered patients. The primary clinical mechanism currently utilized for COVID-19 identification is the Reverse Transcription–Polymerase Chain Reaction (RT-PCR). Hospitals only have small quantities of COVID-19 test kits available due to the daily increase in cases. As an alternative diagnosis possibility, an automatic detection system was implemented. A vigorous technique for the automatic COVID-19 identification is the deep learning approach. Chest X-ray (CXR) imaging is a modest tool that can be an alternate for diagnosing COVID-19-infected patients. With the use of deep learning, deep layer characteristics that are hidden from human sight may be observed using CXR imaging. One of the largest public databases, the "COVID-19 Radiography Database", comprises 21,164 CXR images and was taken from Kaggle. To achieve the best accuracy in this work, data cleansing and the balanced dataset approach were applied. The primary goal of data cleansing is to remove duplicate CXR images from the database. The accuracy of three distinct pre-trained Convolutional Neural Networks (CNNs) was compared and then analyzed (Xception, InceptionV3, and MobileNetV2). Among other models, Xception achieved the best testing accuracy of 94.13% with plain lung CXR pictures. The Gabor filtering image enhancement approach was also employed to identify COVID-19. Only for the MobileNetV2 model did enhance CXR images perform significantly better for classification than plain lung CXR images. This study attempts to enhance the system's accuracy to 100%, outperforming previous tests.

Keywords: COVID-19; Xception; InceptionV3; MobileNetV2; deep learning; CXR images; CNN; image enhancement

Citation: Kumara, C.T.; Pushpakumari, S.C.; Udhyani, A.J.; Aashiq, M.; Rajendran, H.; Kumara, C.W. Image Enhancement CNN Approach to COVID-19 Detection Using Chest X-ray Images. *Eng. Proc.* **2023**, *55*, 45. http://doi.org/10.3390/engproc2023055045

Academic Editors: Teen-Hang Meen, Kuei-Shu Hsu and Cheng-Fu Yang

Published: 4 December 2023

1. Introduction

COVID-19 is a highly transmittable disease [1]. A CXR image-based COVID-19 detection system is fast and widely available and has the ability to analyze multiple cases simultaneously. A CNN model was proposed to categorize COVID-19, viral pneumonia, lung opacity, and otherwise normal classes. In this paper, various previous works were focused on to formulate a deep learning technique to recognize COVID-19-infected individuals. Table 1 summarizes the previous related studies. Section 2 presents the methodology. Section 3 demonstrates the experiment's results and the related discussion. Section 4 has the conclusion.

Table 1. Database [2] and used dataset statistics.

Research Paper.	Dataset CXR Images	COVID-19	Number of Classes	Best Model Name	Best Accuracy (%)
[3]	3487	423	2 (COVID-19, Normal) 3 (COVID-19, Viral Pneumonia, Normal)	DenseNet201	99.70 97.94
[4]	7406	341	2 (COVID-19, Normal) 2 (COVID-19, Viral Pneumonia) 2 (COVID-19, Bacterial Pneumonia)	ResNet50 ResNet101 ResNet50	96.10 99.50 99.70
[5]	6432	576	3 (COVID-19, Pneumonia, Normal)	Xception	97.97
[6]	18,479	3616	3 (COVID-19, Lung opacity, Normal)	ChexNet	96.29
[7]	458	295	3 (COVID-19, Pneumonia, Normal)	SqueezeNet	99.27
[8]	1125	125	3 (COVID-19, Pneumonia, No Findings) 2 (COVID-19, No Findings)	DarkCovidNet	98.08 87.02
[9]	13,975	358	3 (COVID-19, Pneumonia, Normal)	COVID-Net	93.30
[10]	1251	284	4 (COVID-19, Viral Pneumonia, Bacterial Pneumonia, Normal) 3 (COVID-19, Pneumonia, Normal) 2 (COVID-19, Normal)	CoroNet	89.60 95.00 99.00
[11]	1125 9000	125 3000	3 (COVID-19, Pneumonia, No Findings) 2 (COVID-19, No Findings) 3 (COVID-19, Pneumonia, No Findings) 2 (COVID-19, No Findings)	Multiscale deep CNN	96.00 100.00 97.17 96.06
[12]	450	150	3 (COVID-19, Pneumonia, Normal)	QuNet	92.92
[13]	21,165	3616	4 (COVID-19, Viral Pneumonia, Normal, Lung Opacity)	Modified MobileNetV2	95.80

2. Methodology

2.1. Dataset

In this experimental analysis, four classes of the 'COVID-19 Radiography Database' were used [2]. Normal, viral pneumonia, lung opacity, and COVID-19 were the classes that related to CXR images. Among the most significant public datasets with 21,164 CXR images is the "COVID-19 Radiography Database". All data are CXR images, and each image is in PNG format with the resolution of 299 × 299 pixels. Table 2 displays the number of images in the dataset.

Table 2. Summary of the related work.

Medical Case	Number of CXR Images	Used Images Training	Validation	Testing
COVID-19 (positive)	3615	1000	145	200
Normal (healthy)	10,192	1000	145	200
Lung opacity	6012	1000	145	200
Viral pneumonia	1345	1000	145	200

2.2. Pre-Processing

Before applying CXR images as input to the systems, they were resized. Each CNN had a different set of input specifications. With CAD, the resizing procedure is integrated. Most CNNs were trained on 224 × 224 pixels image resolution [14]. Therefore, the size changed from 299 × 299 pixels to 224 × 224 pixels. An image data generator was used for data augmentation with rescaling at 1/255 and a rotation range value of 40. The range of height shift, width shift, shear, and zoom was the same as 0.2, and the horizontal flip was set as 'True'.

The classes in the database contained a various number of images. The class imbalance issue affects accuracy [10], and hence the dataset was balanced by using an exact number of images in the viral pneumonia class because it has the least number of images.

Many duplicate CXR images were available in the dataset, and the "dupeGuru" tool was used to identify the duplicates [15]. Some images had 100% similar copies, and some had more than 70% similarity. Figure 1 shows a set of sample duplicate images in the database. The CXR images with less than 70% similarity display some visible differences. Due to this, images with less than 70% similarity were selected to process. A new dataset was created by removing duplicates with more than 70% similarity. After the cleansing process, the "dupeGuru" tool was used several times to create a balanced dataset.

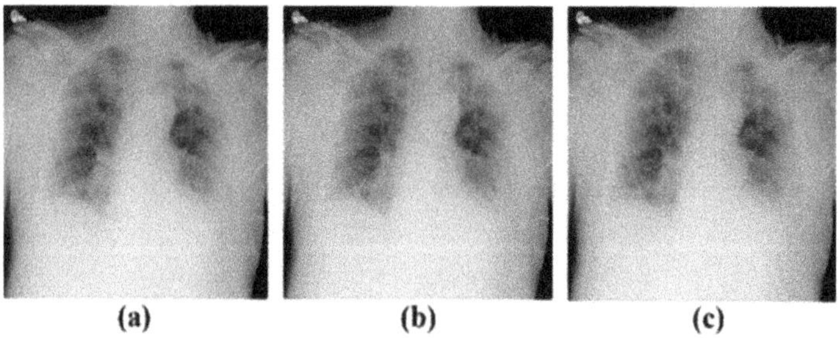

Figure 1. A sample of duplicate CXR images: (**a**) COVID-1, (**b**) COVID-140, and (**c**) COVID-591.

As a technique for the image enhancement process, Gabor filtering was used on the CXR image. A convolution filter called Gabor represents a term that combines sinusoidal and gaussian [16]. The sine element gives the directionality, whereas the Gaussian element yields the weights. The optimal parameter values were lambda = 0.785; Theta = 1.571; Gamma = 0.9; sigma = 1; phi = 0.8; ksize = 15 in Gabor filter. The variation between the original and enhanced images in the four classes is exhibited in Figure 2.

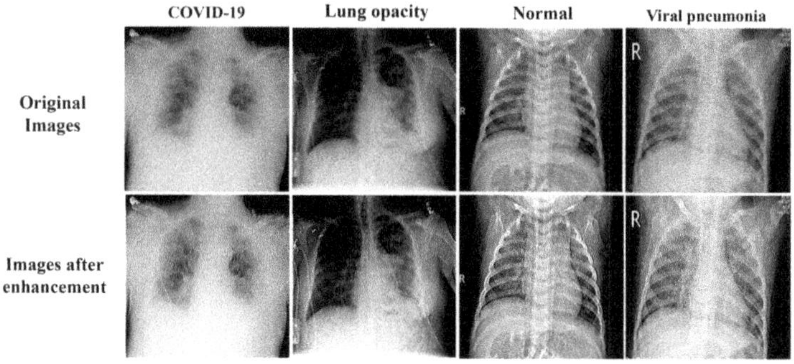

Figure 2. The input CXR images and the output CXR images after passing through the Gabor filter bank.

2.3. Proposed Method

Xception: It is an alteration of the Inception Net. In this model, the Inception Net changed in terms of the size of the parameter, but the parameters of this method are closer to the inception net. And also, the performance of the model was moderately better. This Xception model has 22.9 million parameters and is 88 MB in size [17].

InceptionV3: It consists of a pre-trained CNN architecture with 48 layers. It is an earlier version of the network that was trained using the ImageNet database. This model has 23.9 million parameters and is 92 MB in size [17].

MobileNetV2: It is a CNN network that contains 53 layers. It has several advantages such as a reduced network, fewer parameters, a smaller number of operations, excellent efficiency, and low power consumption. This model has 3.5 million parameters and is 14 MB in size [17].

Fine-tuning is a common technique for transfer learning [18]. It uses a model that has already been trained to complete a specific task and then fine-tunes it to carry out a second similar task [19]. Fine-tuning was used to match the pre-trained model with the created dataset. The proposed architecture layout is shown in Figure 3.

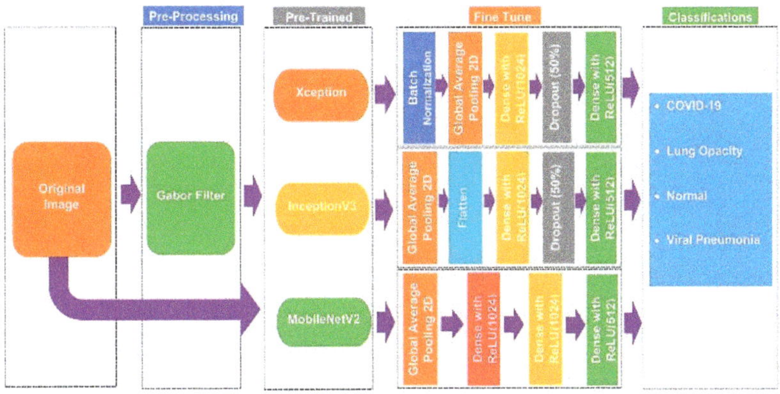

Figure 3. Proposed method architecture for multiclass classification.

2.4. *Specification of Model Training*

Many pre-trained CNN models were processed in this study. Pre-trained models were run on a Jupyter notebook with an Intel core i7 CPU, 2.8 GHz processor, and 16 GB Memory, as well as an NVIDIA GTX 1650 GeForce 4-GB Graphics Processor Unit (GPU) card running Windows 10 64-bit. Moreover, Google Colab was employed, which was linked to Python 3 Google Compute Engine (TPU) RAM: 35.25 GB Disk: 107.72 GB [20].

3. Result and Discussion

The experiments were performed to detect and categorize COVID-19, normal, lung-opacity, and pneumonia (viral) using the CXR images. The database contained cleansed data, and balanced CXR images were used during the process.

3.1. *Without the Enhancement Technique*

As the initial step, three different pre-trained CNN models were used to obtain training, testing, and validation accuracies and a confusion matrix. The CNN models were pre-trained with an adaptive moment estimation (ADAM) optimizer. Every experiment used an experimental parameter of 40, 0.0001, 100, and 20 for the batch size, learning rate, steps per epoch, and the total number of epochs. The last layer initiation function was softmax, while the loss function was categorical cross-entropy. In addition, here, we used 224 × 224 pixel-resized images for all model processing.

3.2. *With the Enhancement Technique*

Secondly, three pre-trained CNN models (Xception, InceptionV3, and MobileNetV2) were again processed with an image enhancement technique named Gabor filtering. The hyperparameters were selected as the same as in the previous process. In a classification problem, the confusion matrix can be used to summarize prediction results. Figure 4 shows the comparison of confusion matrix plots for pre-trained CNN models with and without enhancement.

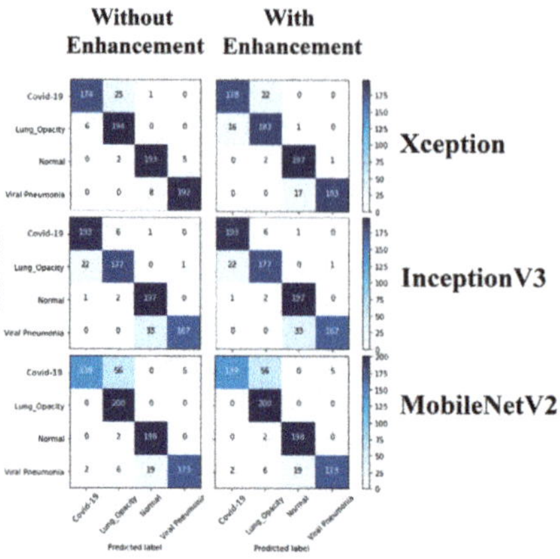

Figure 4. Confusion matrix for pre-trained models with (without) enhancement.

Table 3 shows a comparison between training, validation, and testing accuracies for Xception, InceptionV3, and MobileNetV2 models with and without using the enhancement technique for the dataset. A comparison of performance evaluation matrices of different pre-trained models with and without enhancement for the classification of the four classes is shown in Table 4.

Table 3. Accuracy of pre-trained models with and without enhancement.

Model Name	Enhancement Techniques Used	Accuracy (%)			Time Spent (min)
		Training	Validation	Testing	
MobileNetV2	Not used	83.67	83.28	88.75	50
	Gabor filter	93.05	87.76	89.25	43
InceptionV3	Not used	97.85	91.38	91.75	84
	Gabor filter	96.63	89.48	92.00	76
Xception	Not used	99.10	91.21	94.13	111
	Gabor filter	98.68	89.83	92.62	125

Table 4. Comparison of performance evaluation matrices with and without enhancement technique.

Class	Enhancement Techniques	Xception Precision	Recall	F1-Score	InceptionV3 Precision	Recall	F1-Score	MobileNetV2 Precision	Recall	F1-Score	Support
COVID-19	Not used	0.967	0.870	0.916	0.894	0.965	0.928	0.986	0.695	0.815	200
	Gabor Filter	0.918	0.890	0.904	0.865	0.960	0.910	0.855	0.945	0.898	200
Lung opacity	Not used	0.878	0.970	0.922	0.957	0.885	0.919	0.758	1.000	0.862	200
	Gabor Filter	0.884	0.915	0.899	0.994	0.865	0.925	0.983	0.845	0.909	200
Normal	Not used	0.955	0.965	0.960	0.853	0.985	0.914	0.912	0.990	0.950	200
	Gabor Filter	0.916	0.985	0.949	0.857	0.990	0.919	0.988	0.790	0.878	200
Viral pneumonia	Not used	0.975	0.960	0.967	0.994	0.835	0.908	0.972	0.865	0.915	200
	Gabor Filter	0.995	0.915	0.953	1.000	0.865	0.928	0.802	0.990	0.886	200
Accuracy	Not used			0.941			0.917			0.887	800
	Gabor Filter			0.926			0.920			0.892	800
Macro avg	Not used	0.944	0.941	0.941	0.924	0.917	0.917	0.907	0.887	0.886	800
	Gabor Filter	0.928	0.926	0.926	0.929	0.920	0.920	0.907	0.893	0.893	800
Weighted avg	Not used	0.944	0.941	0.941	0.924	0.917	0.917	0.907	0.887	0.886	800
	Gabor Filter	0.928	0.926	0.926	0.929	0.920	0.920	0.907	0.892	0.893	800

Training and validation loss values at each epoch were reduced in consecutive epochs. A comparison of training and validation losses in contrasting pre-trained models with and without containing the enhancement mechanism is shown in Figure 5.

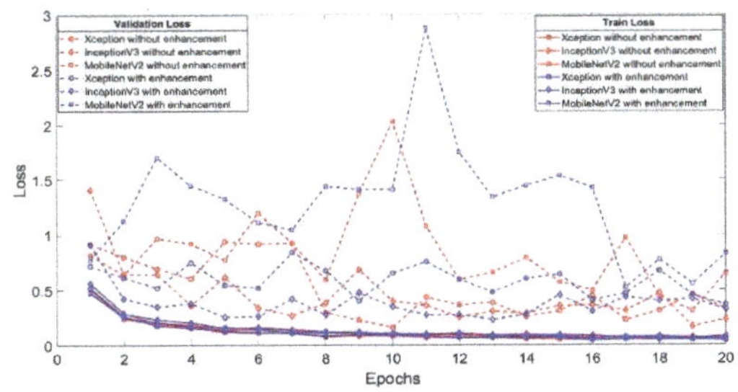

Figure 5. Comparison of three different pre-trained models with and without enhancement technique.

4. Conclusions

In this work, a deep learning-based method using CXR images to predict COVID-19 patients automatically was proposed. The Xception model showed the best testing accuracy of 94.13% without enhancement technique. Considering all of the models with Gabor filtering enhancement, the Xception model has the highest testing accuracy, but it is lower than the accuracy result without enhancement for the Xception model. Moreover, the Gabor filtering enhancement technique only positively affects MobileNetV2. As a result, training, validation, and testing accuracies increased when we used the Gabor filtering enhancement technique for MobileNetV2. In future work, the proposed methodology will be tested on different CNN models to increase the system's accuracy toward 100%, exceeding the previous studies.

Author Contributions: Conceptualization, M.A.; methodology, C.T.K. and S.C.P.; software, C.T.K. and S.C.P.; validation, C.T.K., S.C.P., A.J.U., M.A., H.R. and C.W.K.; formal analysis, C.T.K. and S.C.P.; investigation, C.T.K. and S.C.P.; resources, M.A., H.R. and C.W.K.; data curation, C.T.K., S.C.P. and M.A.; writing—original draft preparation, C.T.K., S.C.P. and A.J.U.; writing—review and editing, M.A., H.R. and C.W.K.; visualization, C.T.K. and S.C.P.; supervision, M.A., H.R. and C.W.K.; project administration, M.A, H.R. and C.W.K.; All authors have read and agreed to the published version of the manuscript.

Funding: This research received no external funding.

Institutional Review Board Statement: Not applicable.

Informed Consent Statement: Not applicable.

Data Availability Statement: The data used for this research are from Kaggle and available at https://www.kaggle.com/tawsifurrahman/covid19-radiography-database/activity (accessed on 25 May 2021).

Conflicts of Interest: The authors declare no conflict of interest.

References

1. COVID Live Update: 222,276,536 Cases and 4,593,898 Deaths from the Coronavirus—Worldometer. Available online: https://www.worldometers.info/coronavirus/?utmcampaign=homeAdvegas1? (accessed on 8 September 2021).
2. Kaggle. COVID-19 Radiography Dataset. Available online: https://www.kaggle.com/tawsifurrahman/covid19-radiography-database/activity (accessed on 25 May 2021).
3. Kadir, M.A.; Mahbub, Z.B.; Islam, K.R.; Khan, M.S.; Can, A.I. Help in Screening Viral and COVID-19 Pneumonia? *IEEE Access* **2020**, *8*, 132665–132676. [CrossRef]
4. Narin, A.; Kaya, C.; Pamuk, Z. Automatic detection of coronavirus disease (COVID-19) using X-ray images and deep convolutional neural networks. *Pattern Anal. Appl.* **2021**, *24*, 1207–1220. [CrossRef] [PubMed]
5. Jain, R.; Gupta, M.; Taneja, S.; Hemanth, D.J. Deep learning based detection and analysis of COVID-19 on chest X-ray images. *Appl. Intell.* **2021**, *51*, 1690–1700. [CrossRef] [PubMed]

6. Rahman, T.; Khandakar, A.; Qiblawey, Y.; Tahir, A.; Kiranyaz, S.; Kashem, S.B.A.; Islam, M.T.; Al Maadeed, S.; Zughaier, S.M.; Khan, M.S.; et al. Exploring the effect of image enhancement techniques on COVID-19 detection using chest X-ray images. *Comput. Biol. Med.* **2021**, *132*, 104319. [CrossRef] [PubMed]
7. Toğaçar, M.; Ergen, B.; Cömert, Z. COVID-19 detection using deep learning models to exploit Social Mimic Optimization and structured chest X-ray images using fuzzy color and stacking approaches. *Comput. Biol. Med.* **2020**, *121*, 103805. [CrossRef] [PubMed]
8. Ozturk, T.; Talo, M.; Yildirim, E.A.; Baloglu, U.B.; Yildirim, O.; Acharya, U.R. Automated detection of COVID-19 cases using deep neural networks with X-ray images. *Comput. Biol. Med.* **2020**, *121*, 103792. [CrossRef] [PubMed]
9. Wang, L.; Lin, Z.Q.; Wong, A. COVID-Net: A tailored deep convolutional neural network design for detection of COVID-19 cases from chest X-ray images. *Sci. Rep.* **2020**, *10*, 19549. [CrossRef] [PubMed]
10. Khan, A.I.; Shah, J.L.; Bhat, M.M. CoroNet: A deep neural network for detection and diagnosis of COVID-19 from chest X-ray images. *Comput. Methods Programs Biomed.* **2020**, *196*, 105581. [CrossRef] [PubMed]
11. Muralidharan, N.; Gupta, S.; Prusty, M.R.; Tripathy, R.K. Detection of COVID19 from X-ray images using multiscale Deep Convolutional Neural Network. *Appl. Soft Comput.* **2022**, *119*, 108610. [CrossRef] [PubMed]
12. Asghar, U.; Arif, M.; Ejaz, K.; Vicoveanu, D.; Izdrui, D.; Geman, O. An Improved COVID-19 Detection using GAN-Based Data Augmentation and Novel QuNet-Based Classification. *BioMed Res. Int.* **2022**, *2022*, 8925930. [CrossRef] [PubMed]
13. Sanida, T.; Sideris, A.; Tsiktsiris, D.; Dasygenis, M. Lightweight Neural Network for COVID-19 Detection from Chest X-ray Images Implemented on an Embedded System. *Technologies* **2022**, *10*, 37. [CrossRef]
14. SIIM-ISIC Melanoma Classification. Available online: https://kaggle.com/c/siim-isic-melanoma-classification (accessed on 15 March 2022).
15. Results—dupeGuru 4.0.3 Documentation. Available online: https://dupeguru.voltaicideas.net/help/en/results.html (accessed on 26 March 2022).
16. Shah, A. (Exploring Neurons) Through the Eyes of Gabor Filter. Medium, 17 June 2018. Available online: https://medium.com/@anuj-shah/through-the-eyes-of-gabor-filter-17d1fdb3ac97 (accessed on 21 March 2022).
17. Team, K. Keras Documentation: Keras Applications. Available online: https://keras.io/api/applications/ (accessed on 21 March 2022).
18. Marcelino, P. Transfer Learning from Pre-Trained Models. Medium, 23 October 2018. Available online: https://towardsdatascience.com/transfer-learning-from-pre-trained-models-f2393f124751 (accessed on 15 September 2021).
19. Team, K. Keras Documentation: Transfer Learning & Fine-Tuning. Available online: https://keras.io/guides/transfer-learning/ (accessed on 26 March 2022).
20. Google Colaboratory. Available online: https://colab.research.google.com/?utm-source=scs-index (accessed on 26 March 2022).

engineering proceedings

Proceeding Paper

CameraEEG: Synchronous Recording of Electroencephalogram and Video Data for Neuroergonomics Applications †

Doli Hazarika *, Srihari Madhavan and Cota Navin Gupta *

Neural Engineering Lab, Department of Biosciences and Bioengineering, Indian Institute of Technology Guwahati, Guwahati 781039, India; madhavan@alumni.iitg.ac.in

* Correspondence: dhazarika@iitg.ac.in (D.H.); cngupta@iitg.ac.in (C.N.G.)

† Presented at the IEEE 5th Eurasia Conference on Biomedical Engineering, Healthcare and Sustainability, Tainan, Taiwan, 2–4 June 2023.

Abstract: Lab-confined electroencephalogram experiments generally impel the subject's mobility. Hence, we provide a wearable solution enabling human brain activity while monitoring during everyday activities, especially for neuroergonomics. This paper introduces CameraEEG, a new Android application that allows for synchronized smartphone acquisition of electroencephalogram (EEG) and camera data. Using a button on the app, the subject can record the witnessed audio-visual events of interest. Android SDK version 28 and mBrainTrain's Smarting Mobi SDK were used to develop the app. The app can be used across all Android smartphones that have Android OS–Lollipop at least. In this paper, we used the app to record synchronized video and EEG data from four subjects during two tasks (namely, closed and open eyes), each under sitting conditions. We used the POz electrode data for analysis. There was a visible difference between the power spectrums of both the tasks, with the eyes-closed task reflecting an alpha band peak. Also, the obtained video and EEG data showed accurate synchronization. A download weblink for the .apk file along with a detailed help document for the developed app is provided for further testing.

Keywords: electroencephalogram; smartphone; Android application; EEG–video synchronization; Neuroergonomics

Citation: Hazarika, D.; Madhavan, S.; Gupta, C.N. CameraEEG: Synchronous Recording of Electroencephalogram and Video Data for Neuroergonomics Applications. *Eng. Proc.* **2023**, *55*, 46. https://doi.org/10.3390/engproc2023055046

Academic Editors: Teen-Hang Meen, Kuei-Shu Hsu and Cheng-Fu Yang

Published: 4 December 2023

1. Introduction

With the advancements in cognitive psychology, the present requirement is to investigate the neural responses that are naturally triggered by a dynamic environment in synchronicity with the participant's task [1]. Applications using restricted stimuli in natural settings have also been developed in order to examine the variations in brain responses [2]. Electroencephalogram (EEG) and smartphone integration have made brain activity measurement possible on the move [3]. Attention studies that gauge the neural activity corresponding to different contexts within a classroom also employ portable electroencephalogram systems [4]. Recently, a study on the neuronal dynamics of interactive synchrony among musicians reported using smartphone-based hyper-scanning [5]. The above investigations might have been effective in locating events of interest if the camera data were also added alongside the EEG.

Synchronized EEG and camera data streams seem to be of current interest for natural environment experiments. Hence, we set up to develop a smartphone Android app called "CameraEEG" that enables this. Since the app was built for natural environment cognitive experiments, the EEG and rear camera data stream are synchronized. We used a transparent mobile holder pouch for placing the smartphone to enable hands-free video mode recording. Eventually, we tested the app's performance with four subjects performing eyes-closed and eyes-open tasks while sitting. Offline analysis was carried out for the synchronization between the camera and electroencephalogram data streams. We then studied each task's power spectral density (PSD) from the EEG signals. We believe that the CameraEEG app is the first of its kind that enables synchronized video and brain signals to be recorded on an Android platform.

Wascher et al. reviewed mobileEEG in neuroergonomics to understand mental states in workplace environments [6]. Our CameraEEG app will play a vital role in such scenarios by providing the opportunity to obtain EEG as well as video data. A portable solution for examination of cognitive states in workplace and daily life situations is becoming increasingly important for the neuroergonomics community.

2. Materials and Methodology

2.1. Software Architecture of Android Application

Java Development Kit (JDK) was used for the app development on Android Studio (AS). The EEG device manufacturer's software development kit (SDK), version 2019 for data acquisition was also used, thus restricting the app to a particular EEG device, namely, mBrainTrain. However, the performance was robust. Two modules related to camera and EEG made up the app. An application programming interface (API) named camera2 was used from AS to modify the camera module. Built-in camera unit facilitates and stores image frames through a buffer. Google's "Camera2Basic" functions were used to create the camera preview of the app (https://github.com/googlearchive/android-Camera2Video, accessed on 1 December 2023). This function uses a background handler thread to parallelize camera preview as well as record the video on smartphone. The closing and opening overlay of the camera device object was hindered by a binary semaphore. Resolution of the recorded video through the back camera was kept at 480 p. The camera preview feed is automatically scaled to fit the surface by rescaling its dimensions through a modified SurfaceTexture class called "AutofitTexture".

2.2. Operation of Android Application

The steps for utilizing the app to record data are shown in Figure 1. First, we must connect the app to mBraintrain's smarting EEG device's Bluetooth. The app then activates the recording button (Figure 2), which records the synchronized EEG data as well as the video feed onto the smartphone as .bdf and .mp4 files, respectively.

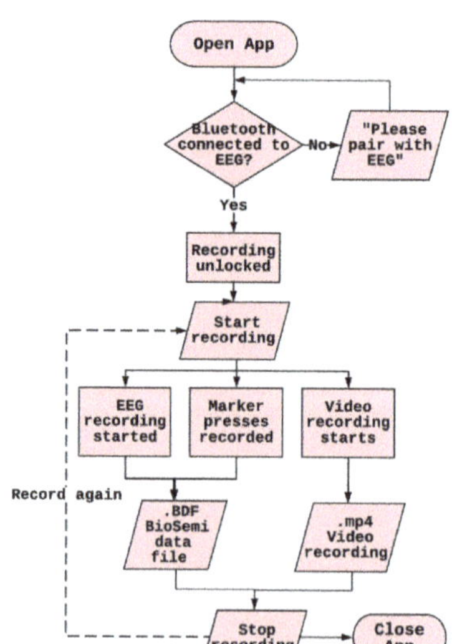

Figure 1. CameraEEG application's operational flow.

Figure 2. Graphical user interface of the app: (**a**) preview of video recording, (**b**) button for connecting/disconnecting to the EEG device, (**c**) button to mark events, (**d**) start/stop record button for video–EEG, (**e**) chronometer for tracking the recording time.

Stopping and re-recording can be carried out using the same button on the app (Figure 2d). An event marker button is also available for the subject to mark any unique events during the experiment (Figure 2c). The apps. apk file is available for download at the following link: https://github.com/NeuralLabIITGuwahati/CameraEEG, accessed on 1 December 2023.

2.3. Synchronized EEG and Video Data Acquisition

Consent was obtained from our institute's ethics committee at IIT Guwahati for the collection of data from human participants. Data were recorded from four subjects, three male and one female with a mean age of 25 and standard deviation 2.16 for this study. We used the developed app with an Easycap 24-channel headcap of mBraintrain smarting device (https://mbraintrain.com/smarting-wireless-eeg/, accessed on 1 December 2023). However, it is also compatible with 20-channel concealed-EEG (cEEGrid) electrodes. Before starting the synchronized EEG and camera recording, impedance levels of each electrode were checked using the SMARTING app. The Android smartphone was placed in a clear mobile holder pouch as in Figure 3, which permitted the recording of video stream with the rear camera and enabling touch-screen operation. The cell phone was not held continuously in hand by the subject but worn as in Figure 3, with the rear camera pointed away from the subject. A wearable experimental setup for the CameraEEG app is shown in Figure 3 and can include 20 cEEGrid electrode or headcap EEG setups.

Synchronization of video and EEG streams as well as data fidelity were checked when the subjects performed two tasks, namely, with their eyes open and closed. Each task lasted for about 5 min. The app had an event marker button to record situations of interest, but this was not used in this study.

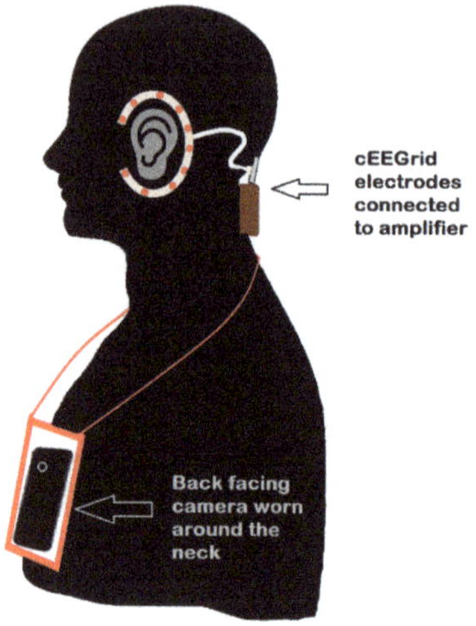

Figure 3. CameraEEG experimental setup with cEEGrid electrodes.

3. Results

The video and EEG from the synchronized recordings were analyzed for any information loss and mismatch. The EEG and video data streams matched with less than a ±5 ms inaccuracy for all four subjects when we evaluated the recorded sessions for synchronization.

The recorded EEG time series during the two tasks were partitioned into two segments using MATLAB (Version 2021a) and EEGLAB (Version 2021.1) for further analysis. Custom scripts were written to analyze the data from the two tasks offline. The POz channel from the 24-channel EEG electrode setup was employed in this study. Following this, the change in the alpha activity was obtained as reported in [7]. Power spectral analysis depicted a peak around ~10 Hz during the eyes-closed task (Figure 4) across all subjects, indicating, as expected, a higher-alpha power.

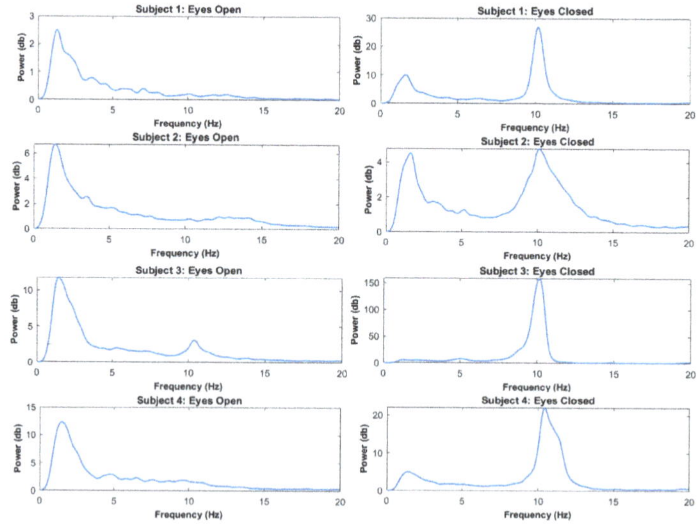

Figure 4. Comparison of eyes-open and eyes-closed multisubject EEG recordings across POz channels.

4. Discussion

The app's performance was deciphered from the data fidelity using the eyes-closed and eyes-open sessions, as performed in earlier works [8]. We observed an alpha peak in the eyes-closed sessions across all subjects, showing that the built app correctly acquired EEG data. Our CameraEEG app offers a straightforward integrated application to record unrestricted, natural electroencephalogram and video data for neuroergonomics applications. This data can also be utilized as a biomarker in intelligent building applications [9]. It also finds applications in understanding the mental implications for people living in urban environments [10,11]. Garza et al. found that high-beta power PSDs from prefrontal and frontal electrodes appeared during a museum walk compared with the baseline condition (staring at the wall) [12]. They also observed high-alpha power PSDs in central, parietal and occipital electrodes for the baseline condition. Hence, through this work, we demonstrate that good-quality synchronized video–electroencephalogram data can be acquired using an Android smartphone. This cost-effective solution also has numerous biomedical applications like monitoring epileptic, Alzheimer's and sleep disorder patients.

We show that a single Android app can record synchronized EEG and video streams, in contrast to other research works, which appear to use several apps [1,13]. This configuration makes it simple to re-engineer for different applications by adding extra modules, decreasing the smartphone's load compared with using multiple apps at once. However, our app is only compatible with mBrainTrain-based EEG devices, because it uses libraries from that specific EEG device manufacturer.

Although the recording sessions provided primary validation of the application's operation, longer-duration recordings are necessary to thoroughly assess the app's robustness. Running the CameraEEG application on Android phone models from later than 2020 (checked on the Redmi Note 7 pro with 4 GB RAM and Realme 5 pro with 6 GB RAM) showed very little heating. However, an older Android smartphone may become heated during extended recording sessions. Furthermore, since only 480 p resolution camera recordings were used, higher resolutions can degrade the smartphone's performance by causing stutters of the toolbar. However, an enhanced video quality may provide increased resolutions for observing specific natural environment stimulus. Going forward, future challenges for this work include exploring the possibilities of online EEG analysis on smartphones like artefact removal and classifying tasks.

Author Contributions: D.H. and S.M. contributed equally. Discussions between S.M. and C.N.G. led to this idea. S.M. developed the app. D.H. tested the app and carried out data analysis. The paper was written by all three authors. A major portion of this work was submitted for the M. Tech degree (Biotechnology) of S.M. at IIT Guwahati, Assam, India. All authors have read and agreed to the published version of the manuscript.

Funding: D.H. received funding from MHRD Doctoral fellowship and NEWGEN IEDC Grant, DST, Govt. of India. S.M. was funded by MHRD master's program fellowship. C.N.G's time was also funded by NEWGEN IEDC Grant, DST, Govt. of India.

Institutional Review Board Statement: The study was conducted in accordance with the Declaration of Helsinki, and approved by the Institutional Review Board (or Ethics Committee) of Indian Institute of Technology Guwahati, India (Date of approval: 27 March 2019).

Informed Consent Statement: Informed consent was obtained from all subjects involved in the study.

Data Availability Statement: Data can be made available upon request to the authors.

Acknowledgments: We also thank Dasari Shivakumar and Nanaki Singh for their time and help during experiments.

Conflicts of Interest: The authors declare no conflict of interest.

References

1. Hölle, D.; Blum, S.; Kissner, S.; Debener, S.; Bleichner, M.G. Real-Time Audio Processing of Real-Life Soundscapes for EEG Analysis: ERPs Based on Natural Sound Onsets. *Front. Neuroergonomics* **2022**, *3*, 793061. [CrossRef]
2. Liebherr, M.; Corcoran, A.W.; Alday, P.M.; Coussens, S.; Bellan, V.; Howlett, C.A.; Immink, M.A.; Kohler, M.; Schlesewsky, M. EEG and behavioral correlates of attentional processing while walking and navigating naturalistic environments. *Sci. Rep.* **2021**, *11*, 22325. [CrossRef] [PubMed]
3. Blum, S.; Debener, S.; Emkes, R.; Volkening, N.; Fudickar, S.; Bleichner, M.G. EEG Recording and Online Signal Processing on Android: A Multiapp Framework for Brain-Computer Interfaces on Smartphone. *BioMed Res. Int.* **2017**, *2017*, 3072870. [CrossRef] [PubMed]
4. Grammer, J.K.; Xu, K.; Lenartowicz, A. Effects of context on the neural correlates of attention in a college classroom. *NPJ Sci. Learn.* **2021**, *6*, 15. [CrossRef] [PubMed]
5. Zamm, A.; Palmer, C.; Bauer, A.-K.R.; Bleichner, M.G.; Demos, A.P.; Debener, S. Behavioral and Neural Dynamics of Interpersonal Synchrony Between Performing Musicians: A Wireless EEG Hyperscanning Study. *Front. Hum. Neurosci.* **2021**, *15*, 476. [CrossRef] [PubMed]
6. Wascher, E.; Reiser, J.; Rinkenauer, G.; Larrá, M.; Dreger, F.A.; Schneider, D.; Karthaus, M.; Getzmann, S.; Gutberlet, M.; Arnau, S. Neuroergonomics on the Go: An Evaluation of the Potential of Mobile EEG for Workplace Assessment and Design. *Hum. Factors* **2023**, *65*, 86–106. [CrossRef] [PubMed]
7. Hohaia, W.; Saurels, B.W.; Johnston, A.; Yarrow, K.; Arnold, D.H. Occipital alpha-band brain waves when the eyes are closed are shaped by ongoing visual processes. *Sci. Rep.* **2022**, *12*, 1194. [CrossRef] [PubMed]
8. Bateson, A.D.; Asghar, A.U.R. Development and Evaluation of a Smartphone-Based Electroencephalography (EEG) System. *IEEE Access* **2021**, *9*, 75650–75667. [CrossRef]
9. Guan, H.; Hu, S.; Lu, M.; He, M.; Zhang, X.; Liu, G. Analysis of human electroencephalogram features in different indoor environments. *Build. Environ.* **2020**, *186*, 107328. [CrossRef]
10. Smith, N.; Georgiou, M.; King, A.C.; Tieges, Z.; Webb, S.; Chastin, S. Urban blue spaces and human health: A systematic review and meta-analysis of quantitative studies. *Cities* **2021**, *119*, 103413. [CrossRef]
11. Norwood, M.F.; Lakhani, A.; Maujean, A.; Zeeman, H.; Creux, O.; Kendall, E. Brain activity, underlying mood and the environment: A systematic review. *J. Environ. Psychol.* **2019**, *65*, 101321. [CrossRef]
12. Cruz-Garza, J.G.; Brantley, J.A.; Nakagome, S.; Kontson, K.; Megjhani, M.; Robleto, D.; Contreras-Vidal, J.L. Deployment of Mobile EEG Technology in an Art Museum Setting: Evaluation of Signal Quality and Usability. *Front. Hum. Neurosci.* **2017**, *11*, 527. Available online: https://www.frontiersin.org/articles/10.3389/fnhum.2017.00527 (accessed on 24 February 2023). [CrossRef] [PubMed]
13. Scanlon, J.E.M.; Townsend, K.A.; Cormier, D.L.; Kuziek, J.W.P.; Mathewson, K.E. Taking off the training wheels: Measuring auditory P3 during outdoor cycling using an active wet EEG system. *Brain Res.* **2019**, *1716*, 50–61. [CrossRef] [PubMed]

engineering proceedings

MDPI

Proceeding Paper

Exploring Innovative Thinking of Bergson's Philosophy and Modern Art via Computer-Aided Design—A Case Study with Three Works as Examples †

Chung-Ho Tien [1], Xia-Na Ma [2] and Zi-Hui Sun [3,*]

1 College for Creative Studies, Dongguan City University, Dongguan 523419, China; wu06h94@yahoo.com.tw
2 Department of Marketing, Guangzhou College of Technology and Business, Guangzhou 510850, China; charlona123@163.com
3 Department of Digital Media Art Design, Guangdong Engineering Polytechnic College, Guangzhou 526100, China
* Correspondence: sunzihuiemail@foxmail.com

† Presented at the IEEE 5th Eurasia Conference on Biomedical Engineering, Healthcare and Sustainability, Tainan, Taiwan, 2–4 June 2023.

Abstract: The innovative thinking of artists highlights the influence of Bergson's philosophy on modern art, and the perception of the relationship of the inner essence of "mind and matter" through observing and experiencing daily life helps artists design works according to the artists' conception. The innovative thinking of learners is based on the creation of art, namely duration, movement, timeliness, and dynamics. We integrated the emotions of creators as the links of the creation of works with the "aesthetic duration" of the viewer, the "movement" rhythm of visual transformation of different "timeliness" provided by the workspace to evoke the "dynamics" of the viewers' internal thoughts. In this article, the creative thinking of three works, namely Growth, Fisherman, and Virtuality and Reality, was analyzed for learners to discuss how the creators designed their works to connect and reflect their creative thinking and creation with the help of emotion. The results of this analysis of creative thinking helped students understand the process of artistic creation and have the creative characteristics of Bergerson's philosophy. The emotional elements of the creator need to be integrated to evoke the deepest feelings and help viewers feel the beauty of works to the maximum.

Keywords: Bergson's philosophy; innovative thinking; aesthetic duration

Citation: Tien, C.-H.; Ma, X.-N.; Sun, Z.-H. Exploring Innovative Thinking of Bergson's Philosophy and Modern Art via Computer-Aided Design—A Case Study with Three Works as Examples. *Eng. Proc.* **2023**, *55*, 47. https://doi.org/10.3390/engproc2023055047

Academic Editors: Teen-Hang Meen, Kuei-Shu Hsu and Cheng-Fu Yang

Published: 4 December 2023

1. Introduction

A few years ago, the book titled "Bergson Style of Modern Painting—Creation and Realization" by You Zhaoliang, an associate professor at the Taipei University of Nursing and Health, was published, and it reflected Bergson's thoughts [1]. There are similar thoughts about past and present artistic creations. We analyzed Bergson's thoughts by connecting the innovative thinking of artistic creation with works of academic thought. The trend of thought of Henri Bergson (1896–1941), which was popular in France in the early 20th century, focused on the process of consciousness duration in the way of "intuition" and the analysis of the relationship between "mind and matter". Bergson's philosophy has a social status and cultural significance that cannot be ignored for modern art and exerts great impacts on modern art schools. He regarded the world as a process of constant change and creation. The world is full of vigor and vitality. The essence of life is "change". Through thinking in life, we can understand the meaning and new values of human life. When drawing, artists are stimulated by external objects, have a feeling, and then display their innovative thinking by drawing.

People usually observe things in an "intuitive" way, without any intention, then look at things "rationally" and take an objective view through many external factors to explore the truth through creative practice. In the history of Western art, from Impressionism to Fauvism, Henri Matisse (1869–1954), and from Cubism to Futurism, there are different ways of painting and thinking. This research was conducted to reveal the internal essence of life things, observe the paintings from different angles, and reflect on the inner feelings. We discussed the creative thinking of artistic creation with the relation between "mind and matter" in Bergson's philosophy in this article.

2. Literature Review

Rudolf Arnheim (1904–2007), an art and visual psychologist, mentioned that an artistic image is not only the product of perception but also the product of evocation, where the self-reproduction of the form is sought from a stimulus and food. We use the original material of the stimulus projected on the retina to organize and the established media to create an appropriate form to reproduce that perception [2]. The artist's innovative thinking has a feeling about the depicted objects. They take a step back when observing to make the vision broader, obtain a suitable vision, display the essence of the images, and reach the natural connotation. This is the artist's respect for the visual and aesthetic thinking.

Zhu Guangqian, a contemporary esthetician, said that "beauty is something in the objective aspect whose nature and form are suitable for the subjective ideology and can be blended together to become a complete image."; that is to say, only when people's subjective feelings, consciousness, and objects are combined to possess the unity of subjectivity and objectivity in the "state of consciousness"; that is, emotions and thoughts can be produced within it [3]. The beauty of art is the reaction triggered via a certain emotion, which depends on visual observation.

Li Zehou, a Chinese philosopher, mentioned that art must be created as a structural gestalt through materials to arouse similar reactions in the body and mind of the viewer, not just to only make the audience understand its meaning through the theme and content. The viewer only looks at the subject, but the expert directly understands the meaning of the work from the body structure, resonating the same feeling in the body and mind [3]. The mirror refraction effect is added to the work to explore the mutual refraction between people and things in the mirror. Through observing, the viewer's heart is captured and their feelings are affected.

The three Chinese and Western estheticians mentioned above put forward the same argument as Bergson's. If one wants to express the beauty of things, he/she needs to start from a subjective and objective perspective with feelings. Things change over time, which is not seen with eyes, but they are indeed changing. The creative thinking of the three works was used to verify the profound influence of Bergson's philosophy on artistic creation in this study.

3. Connection between Bergson's Philosophy and Modern Artistic Thinking

At the beginning of the Bergson trend, it exerted little influence on the art. Western art was guided by Paul Cézanne (1839–1906) in his later years. Later, with the continuation of various art schools, a Bergson style of "time revolution" with great changes and creativity enriched the entire modern art performance.

3.1. Duration of Fauvism

Matisse put forward the idea of "duration" based on Bergson's philosophy. Matisse was once a traditional painter and had studied the style of the impressionists. Later, he launched a revolution with a number impressionist artists. He abandoned the overly detailed realism and instead painted the impression of the object he grasped. Matisse said that no matter what color is used for harmony or contrast, it moves people. When he worked, he always paid attention to the first fresh and superficial feeling. In the past, he was satisfied when he achieved his goal. He secretly wondered whether he observed more deeply. Even if he was satisfied with his paintings, the picture gave him vague aspects. He recorded those fleeting feelings. Not only could he not show his current feelings, but the next day he no longer recognized what they were. The feeling no longer exists because the thoughts have changed over time. As a result, there is a need to rely on the precipitation of thoughts to sort it out.

Matthews also mentioned in his painting theory that he felt the cohesion to interpret his paintings. The "feeling condensation" is a dense feeling as time goes by. In addition to not making it fleeting, there is also a need to grasp this feeling and keep it in mind, just as the condensation of various instant impressions following precipitation. Matisse especially emphasized that "feeling is different", and distinguished "superficial" and "fleeting" feelings. In contrast, he put forward the relative vocabulary of "feeling cohesion" to judge. The "inner thinking" is guided by the "visual sense", and the deep feelings of the soul are revealed in the picture through creation.

3.2. Movement and Timeliness of Cubism

The works of "multiple viewpoints" of Pablo Ruiz Picasso (1881–1973) and Georges Braque (1882–1960), the great masters of Cubism, represent an artist's cognitive attitude and record the changing process of objects in time and space, which showed the spiritual connotation of Bergson's philosophy. Bergson said "I saw a static object, observed it in the same direction and angle, and at the same time. The "vision" I got is not what I just saw. The "vision" has changed. The external things change over time, let alone people have emotional thinking. I will not say much about what will change in it.". In the early development of Cubism, cubists tried to contact Bergson, who had a high status at that time. Although it was not accepted by Bergson, cubists were enthusiastic about Bergson's philosophy.

Cubists explored the moving process of object observation. They also introduced the factors of "movement" and "time" into "space". Bergson said, "When I see a moving object moving to a certain point, there is no doubt that I imagine it can stop. Even if it does not stop, I still tend to think that its process is like a temporary and infinite rest, because it takes at least a little time to think. It is my imagination that is resting, while the moving object is always moving".

Cubism established a new milestone for modern painting. As time goes on, artists must have a unique spiritual sentiment and reconstruct the "reproduction" of objects in their works, which is the innovative thinking that has been discussed in this paper.

3.3. Dynamics and Continuity of Futurism

In the Declaration of Futurism, Umberto Boccioni (1882–1916) advocated "showing universal dynamism, because everything is changing". Boccini deeply explored the "characteristics of perception" and combined the concept of "movement" and the "simultaneous juxtaposition" image of viewing objects. Human consciousness and the image are continuously changing, so there are no clear boundaries.

To some extent, the cubist thinking of "movement and timeliness" created futurism. Both of them expressed time and space in the same picture. Futurism projected real feelings into the picture, which made the images form a dynamic tension. Under the dynamic tension, the outline of the object becomes vaguer, which also induces the familiar dynamic sense. Futurism attempted to express the "continuous reality" of the object in creative thinking, as reality itself is dynamic and continuous.

At that time, there was a fashion in Europe for people to express their actions through snapshots and segmented images. Boccioni believed that this technique was just a vulgar and incompetent act. He believed that a symbol must be pursued and represented in a single form. Therefore, he especially pursued Bergson's "continuity", which replaced this concept of "segmentation" with "continuity". Anheim said "A person's observation of certain things at a specific moment was easily affected by what he has seen, thought about, or learned before. Experience may be beneficial or harmful to his current observation". Only after visual perception is internalized can works of art be truly successful.

4. Bergson's Philosophy and The Practice of Innovative Thinking in Three Works

In this study, Bergson's philosophical theory and the literature were analyzed through the above Chinese and Western estheticians' thoughts; the "mind and matter" were explored with the thinking method. Experience and data collection were implemented to assess the innovative thinking and material application.

In 2008, a series of works of "mirror image" were published. "Image" and "mirror refraction" were used to interweave images to present the inner image with rational expression. For the experience, Bergson said "Experience tells us that the life of the soul must be the combination and correlation of conscious life and physical life, and experience is like this.". In the creation of the "series of mirror images", we raised the question of "what is true" in an anti-traditional and anti-logical way of thinking. For example (Figure 1), the work of Growth has been presented from three viewing angles. It illustrates the growth and transformation of life and the unpredictable world through images. It is constantly changing, but its future is hard to predict.

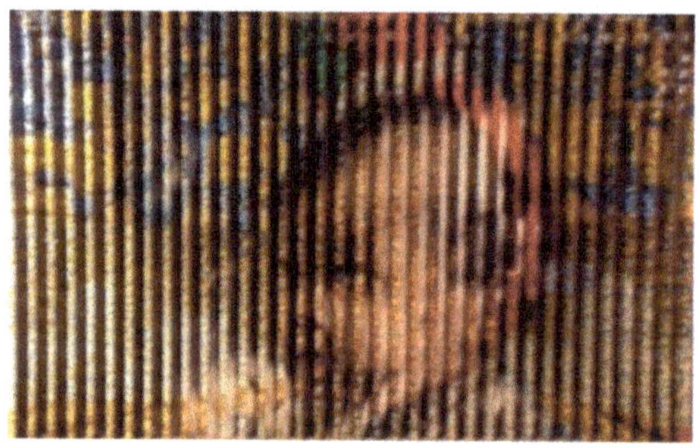

Figure 1. Tien–Chung Ho, Growth, 2008 (cardboard). Image output: 252 × 260 cm.

4.1. Growth

The front view of Growth (Figure 1) is a strip photo of a daughter when she was two years old. These photos comprise thousands of growth records from her birth until she was two years old. The growth of children is the ardent expectation of every family, and the care and love of parents are hidden in the stripes. However, the environment is changing.

On the right side of Growth (Figure 2), a black-and-white ultrasound picture of a fetus is shown. Thanks to modern technology, we can uncover a picture that cannot be seen via the naked eye. As the world changes, we cannot see it with our eyes, but it is transforming. On the left side of Growth (Figure 3), a normal image is presented. The strip disappears, making the picture appear clearer. When observing closely, the emotional cohesion of the creator and the visual impact of the viewer are uncovered.

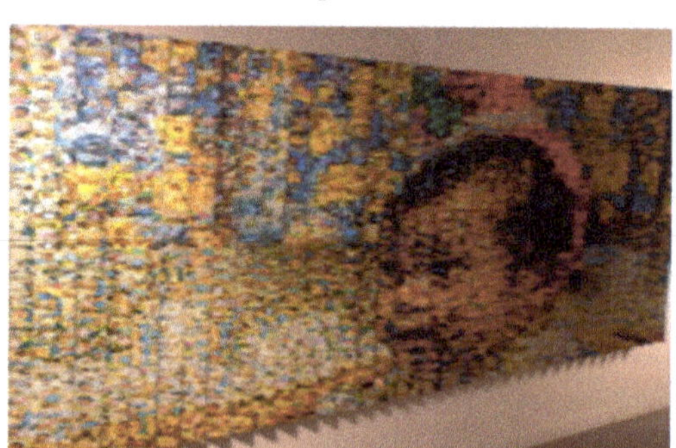

Figure 2. Growth—an angle of 45° on the right.

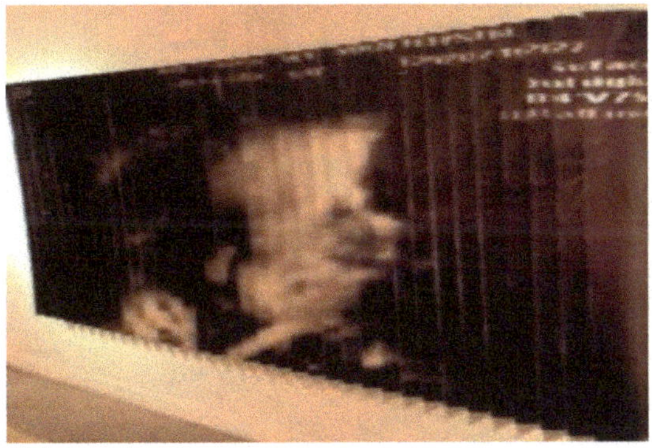

Figure 3. Growth—at an angle of 45° on the left.

4.2. Fisherman

Wu Kang divides memory into two forms in Bergson's philosophy, namely habitual memory (learning memory) and image recall memory (natural recall) [4]. The first memory is termed "habit". It has a fixed time pattern and a mechanical structure, depending on natural habits. Fisherman, which is the text, belongs to "habitual memory". "Memory of recall" was drawn from the text of "Fisherman". In countless memories, each memory denotes a moment of truth in life, occupying the time extension, as it is the result of numerous thoughts from childhood to the present.

Observed from the front, the Fisherman (Figure 4) is the interlaced picture composed of the text of the Fisherman and the black and white orchid shape similar to a billowing flower bud extending outward. The color image has been interwoven with the black-and-white image, showing the communication between the present and the past. The text "Fisherman" plays the key role of "traction" on time in the work, which illustrates a kind of concern for the father who has been fishing outside for many years.

Figure 4. Tien–Chung Ho, Fisherman, 2017 (cardboard, watercolor, and stainless steel). Output: 60 × 120 × 6 cm.

On the left side of Fisherman (Figure 5), it is a work of a seemingly ordinary orchid, which denotes the surging waves with the shape of the flower bud, constantly extending outwards. There is no so-called pure art as the art that contains the emotional content of people, and the construction and reconfirmation of human psychological feelings. On the right side of Fisherman (Figure 6), the orchids in the picture are removed from the color, showing that childhood memories have become fragments that are no longer clear. These works retrieve memories. Although the characters in the picture are childhood memories, many images in the past extend their activities and flow into the current reality perception through the structure of the human nervous system.

Figure 5. Fisherman—at an angle of 45 degrees on the right.

Figure 6. Fisherman—at an angle of 45 degrees on the left.

4.3. Virtuality and Reality

The factor of time is included in the painting of traditional space art to express the representation of objects via continuous planes. The binary contrast of "past and present", "static and moving", and "space and time" are incorporated into the picture, making the means of expression of art more diversified and richer.

The mirror stainless steel in the picture allows for the viewer's image and picture to be mutually interwoven into what Futurism calls "intuitive common feelings". "Virtuality and Reality" allows for different views of appreciation. Through the interaction between the work and the viewer, only when the viewer walks as they observe the work, "intuitive empathy" with the creator can be felt.

Observed from the front, "Virtuality and Reality" (Figure 7) depicts the interlaced and disordered picture between the virtual and the real. The viewer observes the picture from a multi-perspective perspective from left to right. Depending on people's keen observation, the emotion is conveyed through art. It is a kind of interaction between the means of artistic creation and attracting the viewer. On the left side of "Virtuality and Reality" (Figures 8 and 9), the scenery at first sight is an important source of creation. The deeper meaning can be retrieved to improve the perception of things that cannot be seen via the naked eye in your feelings and thoughts. Only by constructing and confirming human psychological feelings can you highlight the more profound spirit.

Figure 7. Tien–Chung Ho, Virtuality and Reality, 2017 (cardboard, watercolor, and stainless steel). Output: 60 × 120 × 6 cm.

Figure 8. Virtuality and Reality—at an angle of 45 degrees on the right.

Figure 9. Virtuality and Reality—at an angle of 45 degrees on the left.

5. Conclusions

The beginning of the 20th century was a major turning point in art. Whether it was Fauvism, Cubism, or Futurism, they were all inherited from Cezanne and started by criticizing Impressionism. They shared a common element, which was the study of the difference in "time". Artistic creation was defined as the "continuity" in Bergson's philosophy, constantly seeking innovation and change in the creation. The mission of innovative thinking is the common spiritual connotation of philosophy and art. The philosophical worldview under the Chinese cultural tradition is a "continuity of existence", which is different from the "fragmentation of existence". The concept of continuity of existence refers to the unity of the three aspects of "heaven and man, object and self, and person and self" [5]. The ancients said that "when you read a hundred times, you can see the meaning of the book". Only by diversified innovative thinking can the source of creation keep emerging.

Author Contributions: Conceptualization, C.-H.T. and Z.-H.S.; methodology, C.-H.T.; investigation, X.-N.M.; resources, X.-N.M.; writing—original draft preparation, C.-H.T.; writing—review and editing, C.-H.T. and Z.-H.S.; supervision, Z.-H.S.; funding acquisition, C.-H.T. All authors have read and agreed to the published version of the manuscript.

Funding: This research received no external funding.

Institutional Review Board Statement: Not applicable.

Informed Consent Statement: Not applicable.

Data Availability Statement: Data availability is not applicable to this article as no new data were created or analyzed in this study.

Conflicts of Interest: The authors have no conflict of interest to report.

References

1. Zhaoliang, Y. *The Bergson Trend of Modern Painting-Creation and Realization*; Nuan Nuan Bookstore & Cultural Business: New Taipei City, Taiwan, 2013; p. 87.
2. Arnheim, R. *Art and Visual Perception—A Psychology of the Creative Eye*; Li, C., Translator; Lion Book Company: Taipei, Taiwan, 1985; pp. 133–134.
3. Zehou, L. *Four Lectures on Aesthetics*; Sanmin Bookstore Co., Ltd.: Taipei, Taiwan, 2001; pp. 42–82.
4. Kang, W. *Bergson's Philosophy*; Taiwan Commercial Press: New Taipei City, Taiwan, 1980; p. 11.
5. Derong, P.; Xianzong, L. *Eastern and Western Philosophy and Ontological Interpretation: Mr. Cheng Zhongying's Essays on the 70th Birthday*; Kant Press: Taipei, Taiwan, 2005; p. 137.

engineering proceedings

MDPI

Proceeding Paper

Ancient Chinese Repeating Crossbow: Basic Research to Science Education in Museums †

Jian-Liang Lin [1], Li-Chun Lin [2] and Kuo-Hung Hsiao [2,*]

1 Technology Education Division, National Science and Technology Museum, Kaohsiung 80765, Taiwan; jllin927@mail.nstm.gov.tw
2 Exhibition Division, National Science and Technology Museum, Kaohsiung 80765, Taiwan; llc@mail.nstm.gov.tw
* Correspondence: khhsiao@mail.nstm.gov.tw

† Presented at the IEEE 5th Eurasia Conference on Biomedical Engineering, Healthcare and Sustainability, Tainan, Taiwan, 2–4 June 2023.

Abstract: The single-shot crossbow is an important ancient Chinese invention dating back to 600 B.C. It was a weapon using elastic force to shoot arrows at targets at a distance. It was composed of a stock, a bow, a string, and a trigger. By adding a magazine to a single-shot crossbow, the repeating crossbow was allowed to shoot repeatedly without hesitation by operating the operating handle. According to archaeological findings, the earliest repeating crossbow was excavated in Jiangling County, Hubei Province, China, and dated back to 400 B.C. Based on the restored design of ancient crossbows, we found six feasible designs of the earliest repeating crossbow. These designs adopted the mechanism of "six members and seven joints". One feasible design was selected to construct its model, using computer animation as a DIY teaching aid. The research results can be used for STEM education in museum exhibitions.

Keywords: repeating crossbow; reconstruction design method of ancient machinery; museum; science education; mechanism design

Citation: Lin, J.-L.; Lin, L.-C.; Hsiao, K.-H. Ancient Chinese Repeating Crossbow: Basic Research to Science Education in Museums. *Eng. Proc.* **2023**, *55*, 48. https://doi.org/10.3390/engproc2023055048

Academic Editors: Teen-Hang Meen, Kuei-Shu Hsu and Cheng-Fu Yang

Published: 5 December 2023

1. Introduction

Numerous arrowheads made of animal bones or stones have been excavated in many Neolithic sites. It is generally believed that prehistoric humans had developed hunting tools, from rock slings to bows. The elastic force of the limbs and the string was used to shoot targets over a long distance with a bow. With the gradual evolution of bows and arrows, a diverse range of projectile weapons have been developed [1,2]. Among these weapons, the development of single-shot crossbows was most eminent; by adding a stock and a trigger to the original bow, the single-shot crossbow was updated to a bow with delayed shooting. As shown in Figure 1, the trigger was operated with a cam mechanism using the input link, the connecting link, and the hook link [3,4]. To use a single-shot crossbow, the trigger was adjusted to the auto-locked state, then the crossbowman used arms or arms and legs to pull the string and hook it to the connecting link of the trigger. When the string was cocked tight, the crossbow was ready to shoot. When shooting, the crossbowman needed to tap the input link to loosen the string and eject the bolt. Single-shot crossbows overcame the three major shortcomings of original bows: manual cocking, the impossibility of long-time aiming, and a short shooting range. In 600 B.C., single-shot crossbows were widely used in the wars of ancient China. Since crossbows continued to be used until the early 20th century, they represented historic ancient weapons.

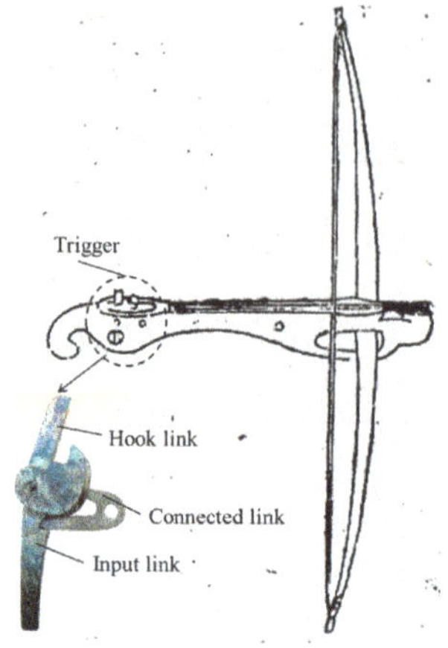

Figure 1. Single-shot crossbow [3,4].

The crossbow shooting was conducted in four steps: string cocking, bolt placement, string release, and bolt launching. Shooting was completed with a repeating crossbow by operating the input link [5]. According to archaeological findings, the earliest crossbow-enabled and repeated bolt launching in ancient China was excavated from an ancient tomb in Jiangling County of Hubei Province. Based on the location of the tomb and other relics found in the tomb, it was determined to be a tomb of the Chu State built in the Warring States period (475–221 B.C.). Since there is no historical record of this repeating crossbow, it is named the Chu State Repeating Crossbow after the excavated location. As shown in Figure 2, the magazine containing 20 bolts of the crossbow was affixed to the stock [6]. When the shooting was performed, the crossbowman held the input link ($K_I(4)$) and pushed it forward to hook the string to the hook link ($K_L(6)$). Then, the archer moved the input link backward until the percussion link ($K_{PL}(5)$) and touched the switch point. At this point, the two consecutive bolts were ready to launch. After a bolt was launched, the next bolt from the magazine was dropped down due to gravity and the next shooting was waited for. As the Chu State Repeating Crossbow used an ingenious mechanical structure, it is worthy of in-depth research and discussion. Based on the restored design of ancient crossbows, we formulated six feasible designs of the Chu State Repeating Crossbow. One feasible design was selected to construct its model using computer animation. The results have been already used for exhibition and science education in museums.

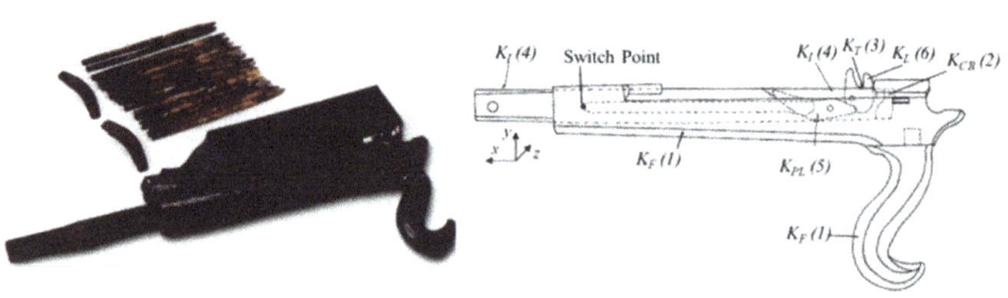

Figure 2. Chu State Repeating Crossbow [6].

2. Structure Analysis

A crossbow was assembled from multiple link members and various types of joints. It has the function of string cocking and bolt launching for the operation of the input link. The Chu State Repeating Crossbow comprised six link members: the stock (Member 1, K_F), the bow (Member 2, K_{CB}), the string (Member 3, K_T), the input link (Member 4, K_I), the percussion link (Member 5, K_{PL}), and the hook link (Member 6, K_L) [7,8]. A bamboo joint (J_{BB}) was used to connect the bow and the stock between the string. The bow, the string, and the hook link in the thread joint (J_T) between the input link and the stock of the sliding joint (J^{Px}), the percussion link, the switch point (the stock) as the cam joint (J_A) of the input link, the percussion link, the hook link with two revolved joints (J_{Rx}), and the cam joint (J_A) were all made of bamboo. The Chu State Repeating Crossbow consisted of six members and eight joints. The percussion link fixed the hook link so that the string could hook to the hook link; then, the string could be released when the percussion link touched the switch point by moving the input link to the extreme position. As shown in Figure 3 [7,8], the restored design of a "6-member and 8-joint" Chu State Repeating Crossbow was constructed based on the restored design.

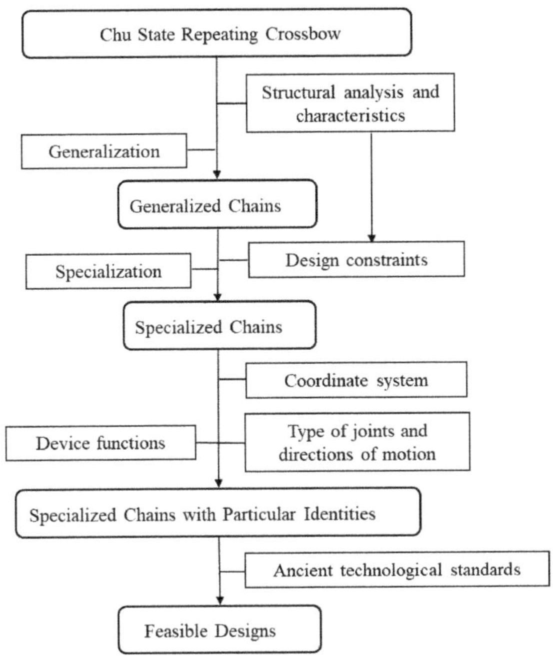

Figure 3. Restored design methods of ancient crossbows [7,8].

We analyzed the structure and summarized the structural characteristics, as follows:

- The Chu State Repeating Crossbow adopts a "6-member (Members 1–6) and 8-joint" mechanism.
- The stock (K_F) is a three-joint link.
- The bow (K_{CB}) is a two-joint link that connects the adjacent stock (K_F) with bamboo joints (J_{BB}).
- The string (K_T) is a two-joint link that connects the adjacent bow (K_{CB}) and hook link (K_L) with thread joints (J_T).
- The input link (K_I) connects the adjacent stock (K_F) with the sliding joint (J^{Px}), but is not adjacent to the string (K_T).
- The percussion link (K_{PL}) connects the adjacent stock (K_F) with a cam joint (J_A).

According to the structure, the diagrams of the "6-member and 8-joint" mechanism [9] were drawn as shown in Figure 4.

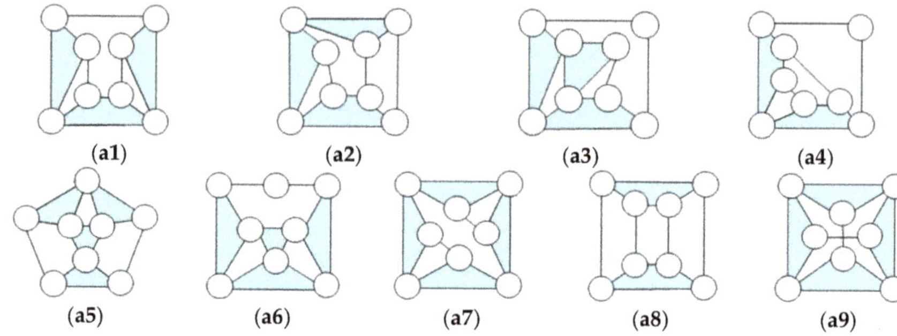

Figure 4. Diagrams of the "6-member and 8-joint" mechanism [9].

A pair of adjacent two-joint links must be used as the bow and the string, which was adjacent to the three-joint link used as the stock. Therefore, only (a3) and (a6) in Figure 4 met the requirements. All feasible specialized chains were obtained through the following steps.

Stock, bow, and string (K_F, K_{CB}, and K_T)

A three-joint link was used for the stock (K_F), and a pair of adjacent two-joint links was used as the bow (K_{CB}) and the string (K_T) to be connected to the stock with bamboo joints (J_{BB}) and the thread joint (J_T). Therefore, they were designed as follows. Regarding the generalized chain in Figure 4a3, the pattern of the stock, the bow, and the string was obtained as shown in Figure 5.

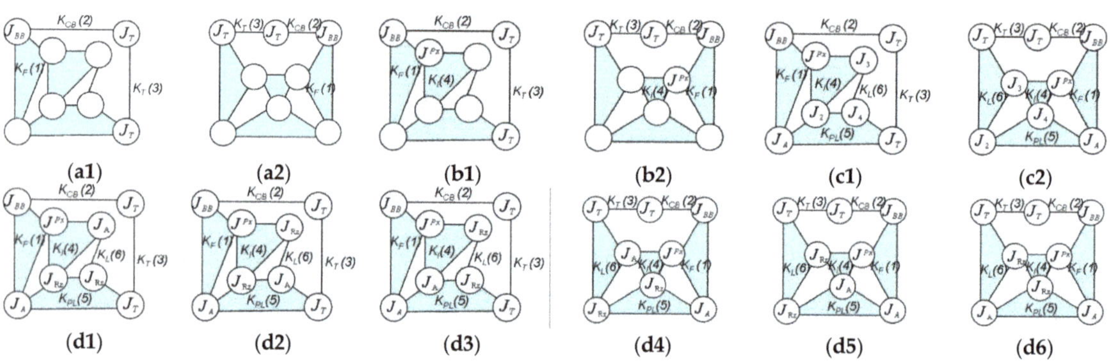

Figure 5. Specialization of the Chu State Repeating Crossbow.

Input link (K_I)

Since the input link (K_I) connected the adjacent stock (K_F) to the sliding joint (J^{Px}), it had not to be adjacent to the string (K_T). Two feasible designs of the specialized chains of the stock, bow, string, and input link were obtained, which are shown in Figure 5b1,b2.

Percussion link and hook link (K_{PL} and K_L)

Since the percussion link (K_{PL}) connected the adjacent stock (K_F) with the cam joint (J_A), only one pattern of the percussion link and the hook link was obtained, as shown in Figure 5c1. For the case shown in Figure 5b2, the pattern was obtained as shown in Figure 5c2.

Then, a rectangular coordinate system was defined as shown in Figure 2. Bolt shooting with the Chu State Repeating Crossbow was performed through the reciprocating motions of the input link. As the pattern of the original joints was uncertain with multiple possibilities, the uncertain joints were designed to have the specialized chains (Figure 5c1,c2) in six feasible patterns (Figure 5d1–d6). Considering the requirements for the movement and functions of the mechanism, all specialized chains with designated joints were visualized in the diagrams of the feasible mechanisms that conformed to the technical level of ancient craftsmanship. As shown in Figure 6a–f, these diagrams were transformed into computer graphics.

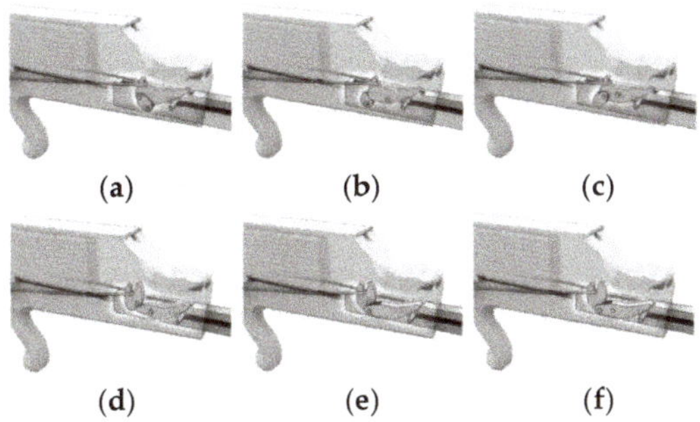

Figure 6. Diagrams of feasible designs for the Chu State Repeating Crossbow.

3. Museum Applications

The International Council of Museums (ICOM) defines a museum as "a non-profit, permanent institution in the service of society and its development, open to the public, which acquires, conserves, and researches the heritage of humanity and its environment for education, study, and enjoyment" [10]. Therefore, the museum collects, researches, exhibits, educates, and delivers information on the precious heritage of humanity to the public. The National Science and Technology Museum is the largest museum of applied sciences in Taiwan. Its main functions are research, collection, showing exhibitions of various science and technology themes, and the promotion of social scientific and technological education to introduce the impact of key scientific and technological development on human life. The Chu State Repeating Crossbow has an ingenious mechanical structure. It is the earliest repeating crossbow excavated and has a special historical position. Based on the restoration of the mechanism and the feasible design (Figure 6d), the prototype was manufactured using computer simulation, as shown in Figures 7 and 8. Figure 9 shows the design of the DIY teaching aid for the relevant course for the promotion of science education.

Figure 7. A prototype for the Chu State Repeating Crossbow.

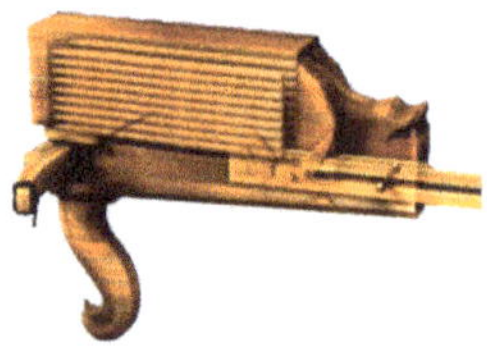

(**a**) Pushing input link forward

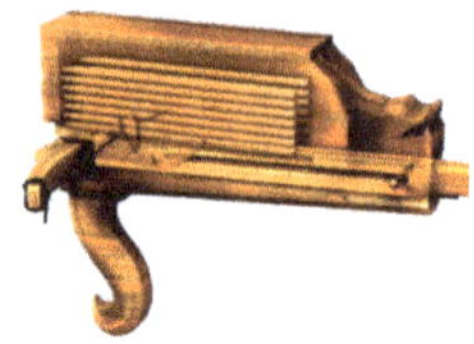

(**b**) Hooking string

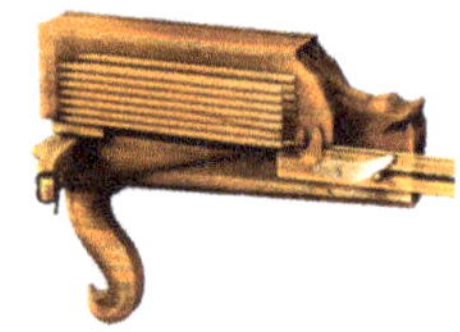

(**c**) Percussion link touching switch

(**d**) Loosing string

(**e**) Launching bolts

Figure 8. Computer animation of Chu State Repeating Crossbow.

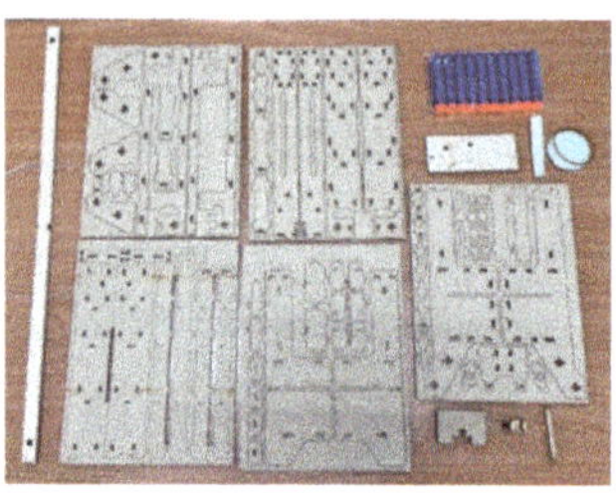

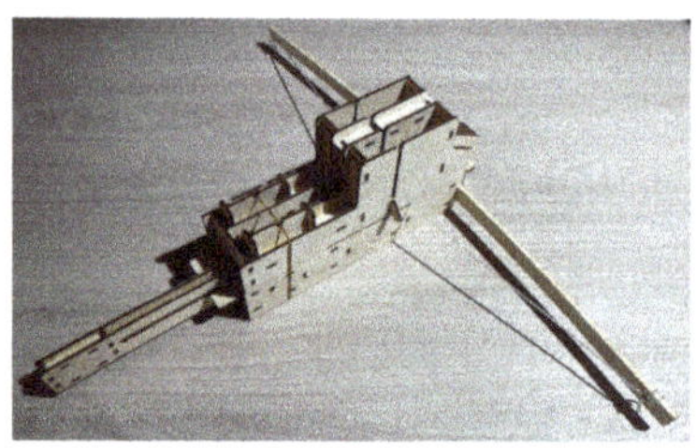

Figure 9. DIY teaching aid and course for the promotion of science education regarding the Chu State Repeating Crossbow.

4. Conclusions

We presented two types of ancient Chinese crossbows. The earliest use of single-shot crossbows dates back to the 6th century B.C., with numerous relics and historical records. The Chu State Repeating Crossbow appeared in the 4th century B.C., and was an ingenious invention with real excavated relics without any historical records. As it was the first crossbow used for repeated shooting, the demonstration of the ingenious design of mechanisms by the ancients is worthy of in-depth research and educational promotion in museums. In this study, we explained the process of the restoration of the design of the Chu State Repeating Crossbow based on the restored six feasible designs. One design was selected to construct its prototype using computer animation, and a DIY teaching aid was manufactured. The research results already have been used for the exhibition and promotion of science education in museums. People can learn about the ingeniously designed mechanical device with historical significance.

Author Contributions: The author' contributions are as follows: J.-L.L. was the lead author and provided the material of ancient Chinese crossbows. L.-C.L. directed the DIY teaching aid and course for the promotion of science education. K.-H.H. made contributions to presentation approaches regarding the mechanisms and provided reviews of the paper. All authors have read and agreed to the published version of the manuscript.

Funding: The authors are grateful to the National Science and Technology Council (Taipei, Taiwan) under grant NSTC 111-2410-H-359-001- for the financial support of this work.

Institutional Review Board Statement: Not applicable.

Informed Consent Statement: Not applicable.

Data Availability Statement: No new data were created.

Acknowledgments: Support from the Ancient Chinese Machinery Cultural Foundation (Tainan, Taiwan) is greatly appreciated.

Conflicts of Interest: The authors declare no competing interest.

References

1. Needham, J. *Science and Civilisation in China*; Cambridge University Press: Cambridge, UK, 1954; Volume 5, Part 6, pp. 101–145.
2. Zhang, C.H.; You, Z.H.; Wu, Z.Z.; Liu, Y.L. *History of Inventions in Chinese Mechanical Engineering*, 2nd ed.; Tsinghua University Press: Beijing, China, 2004; pp. 269–310. (In Chinese)
3. Mao, T.Y. *Treatise on Armament*, 1st ed.; Hainan Press: Hainan, China, 2001; p. 17. (In Chinese)
4. Xu, Z.Y. *Trigger Mechanism*, 1st ed.; Hebei Fine Arts Publishing House: Hebei, China, 2007; p. 38. (In Chinese)
5. Hsiao, K.H. Structural Synthesis of Ancient Chinese Original Crossbow, Trans. *Can. Soc. Mech. Eng.* **2013**, *37*, 259–271. [CrossRef]
6. Jingzhou Museum. *Important Archaeological Discoveries*; Cultural Relics Press: Beijing, China, 2009.
7. Hsiao, K.H.; Yan, H.S. Structural Synthesis of Ancient Chinese Chu State Repeating Crossbow. In *Advances in Reconfigurable Mechanisms and Robots I*, 1st ed.; Dai, J.S., Zoppi, M., Kong, X.W., Eds.; Springer: Berlin/Heidelberg, Germany, 2012; pp. 749–758.
8. Hsiao, K.H.; Yan, H.S. *Mechanisms in Ancient Chinese Books with Illustrations*, 1st ed.; Springer: Cham, Switzerland, 2014; pp. 230–234.
9. Yan, H.S. *Creative Design of Mechanical Devices*, 1st ed.; Springer: Singapore, 1998; p. 129.
10. Museum Definition. Available online: https://icom.museum/en/resources/standards-guidelines/museum-definition/ (accessed on 24 August 2022).

engineering proceedings

MDPI

Proceeding Paper

Handcrafting Objects made with Machine Learning: An Object Design Approach with Computer Vision †

Jose L. Reategui

Facultad de Arquitectura, Universidad Peruana de Ciencias Aplicadas, Lima 15023, Peru; pcarjrea@upc.edu.pe

† Presented at the IEEE 5th Eurasia Conference on Biomedical Engineering, Healthcare and Sustainability, Tainan, Taiwan, 2–4 June 2023.

Abstract: Many of today's computational design systems based on explicit or graphic programming software require designers to determine relationships for morphogenesis based on computational thinking supported by the abstraction process. This computational thinking process can reduce the ability to generate analogies in design development and adverse vision related to computational tools. It also reduces the innovation capacity of small companies that produce handicrafts and design teaching in a customized way. This research promotes a computational model based on machine learning combined with an analog creation process. Machine learning engines determine the objects similarity percentage between students' objects and master objects through a forecasting model. There is a proposal to combine parametric design systems, such as Grasshopper3D, with cloud computing and an edge computing device.

Keywords: edge computing; abstraction; machine learning; ceramics; SDG 1; SDG 9; cloud computing; IoT; digital craft

Citation: Reategui, J.L. Handcrafting Objects made with Machine Learning: An Object Design Approach with Computer Vision. *Eng. Proc.* **2023**, *55*, 49. https://doi.org/10.3390/engproc2023055049

Academic Editors: Teen-Hang Meen, Kuei-Shu Hsu and Cheng-Fu Yang

Published: 4 December 2023

1. Introduction

A digital process in design is supported by computational thinking; therefore, by a process of abstraction. Abstraction generates two main mental processes. The first is the reduction of the variables that are present in any phenomenon of reality. An example is the production of algorithms unlike the traditional design process based on analog objects. The second is a generalization with the aim of generalizing reduction using the software [1].

To transform an analogic process into an industrialized one depends on the automatization of the process that can reduce the skill to generate analogs and change the design teaching process for handcrafted objects.

Nowadays, computational thinking in digital design is often underestimated because there is the misconception in the current practice where computational design is equal to superficial toll knowledge without algorithms knowledge [2]. The actual design practice uses sketches, models, or another way to express ideas process, is closer to ancient practice of art, there, creative achievement was linked to in-depth knowledge of tools. Also, the computational thinking process is not part of the traditional practice with analogic tools because the implementation cost is huge, and it needs professionals with higher education training [3]. In this context, another creative practices such as craftsman process cannot incorporated digital technologies based on computational thinking very well where the craft master knowledge is the key piece to improve digital technologies.

There are contradictory points between the analog process of handicrafts and industrialized production based on digital technologies. For instance, the production speed of craft objects depends on the master craftsman time, which is finite. On the other hand, digital technology has the advantage of reaching a lot of people around the world in a few seconds. The handcrafted design can imprint the designer's personality on each object, which is something that the industrial process cannot do.

The challenge to be overcome, in many cases, is the belief that being creative with software is knowing the commands associated with the production of geometry and not with the production of new strategies in the use and customization of software in original ideas; the over-dependence on programming logic can reduce the intuition and sensitivity.

Figure 1 shows a handcraft master workshop with different objects made with clay. The picture on the bottom shows the modeling platform; in this case, it is the workshop floor.

Figure 1. Handcraft master workshop; this is an ancestral practice.

2. Crafting Inference Engine

2.1. Research Site

This research was developed in Lamas city, San Martin, Peru, in the Wayku indigenous community (Figure 2) and includes images taken in an artisan's workshop located in the community and images from a laboratory.

Figure 2. Pictures from the research place, an indigenous community.

2.2. Inference Engine Architecture

The system architecture was: Frontend in Rhinoceros 3D-Grasshopper (1), Backend in Amazon Web Services (2), Frontend with Nvidia Jetson Nano (3), AI in edge computing model (4), AI engine on cloud computing server (5) The five parts work in the Nvidia Jetson Nano (3 and 4), Amazon Web Services (2 and 5), and Rhinoceros 3D-Grasshopper (1) (Figure 3).

Figure 3. Main architecture scheme.

The data collection starts in the Nvidia Jetson Nano and the ML engine-computer vision, both works to send information to Amazon Web Services (AWS), which stores the data and runs another ML engine for prediction, after Rhinoceros3D-Grasshopper receives this prediction from AWS using an API Rest.

2.2.1. Frontend in Rhinoceros3D-Grasshopper

A connection system with AWS was made to transmit parametric design data and know the inference results from the Nvidia Jetson Nano board (Figure 4).

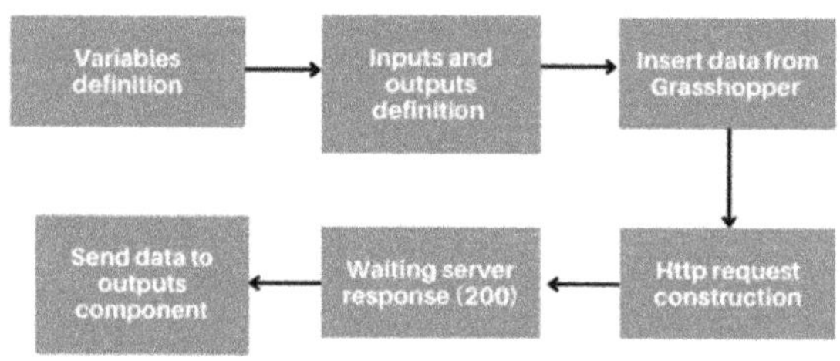

Figure 4. Data workflow to connect Grasshopper3D with AWS.

This means the new component receives data from variables such as parameters and produces results such as outputs. Once the user sends information through inputs, the component uses a method to obtain information from AWS; in this way, the JSON message format was used.

2.2.2. Backend in Amazon Web Services

The AWS backend was made to locate complementary microservices, such as databases, computing without servers, unstructured data storage, support connections to foreign AWS, and the forecasting engine model.

1. AWS API Gateway (API connections)
2. AWS DynamoDB (database)

AWS DynamoDB stores the data to be transferred from Grasshopper to Jetson, and vice versa. A database is used to query and write information at high speeds;

3. AWS S3 (unstructured data storage)

It is used to store images that come from the Nvidia Jetson Nano device and allows us to obtain this information not from the AWS application;

4. AWS Lambda (computation without servers)

To execute code without turning on a server, providing viability to the prototype, it is used to process images, save them, and save information in a database;

5. AWS IoT Core (to connect edge device to AWS)

It is used to transfer the edge device results to the cloud using MQTT and HTTP protocols;

6. AWS Forecast (ML engine to predict)

It is used with the DeepAR+ predictor that allows for a performance prediction based on time series; this time series was previously saved using edge computing.

APIs are built based on the information transfer needs between Rhinoceros 3D-Grasshopper and Nvidia Jetson Nano under the REST protocol.

2.2.3. Frontend with Nvidia Jetson Nano (Edge Computing)

This allows code execution for taking and sending images and also image processing with a neural network.

2.2.4. ML Engine in Edge Computing Device

1. Finished object.

An ML engine recognizes that a ceramic work looks so much like one completed well and another finished with deficiencies. This computer vision is a classification task built on the Nvidia board with the ResNet 18 algorithm. The objective is to identify how much the apprentice object resembles that of the master craftsman through a bank of photographs (see Figure 5).

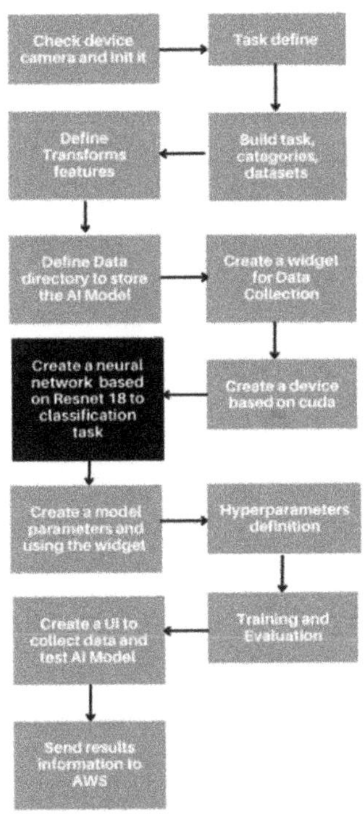

Figure 5. ML engine creation process for finished object.

2. Object making process.

To recognize the step in the object making process, a regression task was used with the goal of detecting the position of the last ceramic mass with a real process (Figure 6). It was also used to detect imperfections that can damage the final object (wrinkles and cracks).

In machine learning, the classification task is associated with engine training based on categorical values as labels, where an object can be categorized as one type or another. A regression task is also related to supervised learning, but the labels are not categorical and are values that change according to a trend; this ML engine can recognize a specific point in the image (Figure 7).

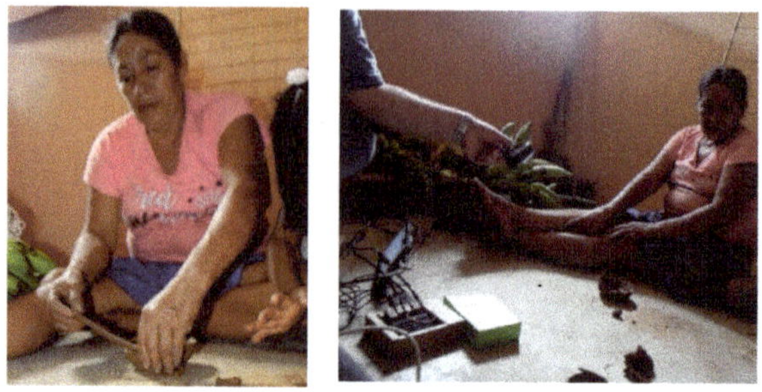

Figure 6. First AI version test in a handcraft master workshop.

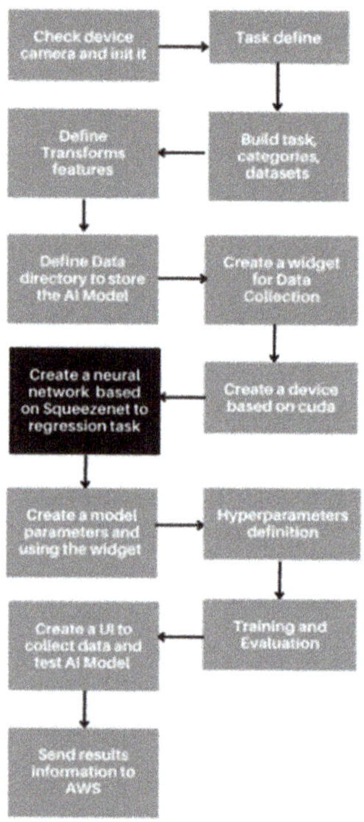

Figure 7. ML engine architecture to determine the completion percentage.

The reason for choosing an edge device was the need for ease-of-use in artisan communities, where they do not usually have a personal computer with which to carry out their activities; the other reason was the ability of the edge device to run high-intensity computing tasks.

Regarding the connectivity of the device, the workshop has a telephone signal which allows the Wi-Fi network to be generated from a mobile phone to speed up the training of neural networks. In the case of not having a connection that allows the use of a mobile phone, it is possible to use a GPRS/GSM device, which allows access to the telecommunications chip and, through it, the internet.

Electricity is important for the project; in this prototype, the artisans provided us with the electrical connections, and the use of batteries is suggested.

2.2.5. ML Engine on Cloud Computing Server

The methods are divided into two sets of microservices, to respond to (1) information sent from the edge computing device, and (2) information sent from Rhinoceros3D-Grasshopper. For both, the following procedure was used (Figure 8).

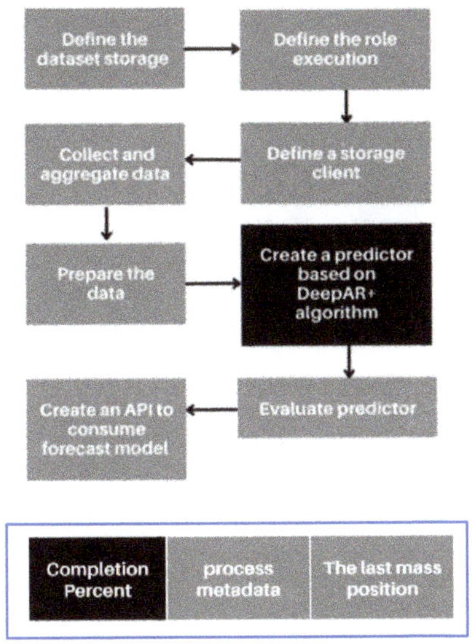

Figure 8. At the top, the forecast architecture process; at the bottom, the side temporal series schema.

After the ML forecasting was trained, it was important to connect this microservice with AWS.

2.3. Data Capture and Processing

For the collection of information, the creation process stages suggested by the master craftsman were taken. These stages were stored in the Jetson device, such as pictures to be consumed, and a bank of images was developed to have labels related to the position of the last clay mass (Figure 9). For this reason, it is possible to associate the advance percentage with the position of the mass. These images were taken from various angles and under different lighting conditions.

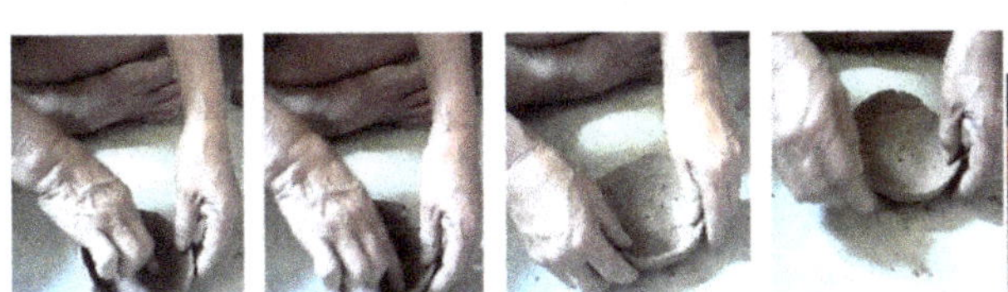

Figure 9. Handcraft master process for teaching.

2.4. System Features

The system uses two user interfaces—first, on the edge computing device, and second in Rhinoceros 3D through a Grasshopper3D component. The edge computing device shows that it looks very much like the object of the master (Figure 10). The Grasshopper interface shows the prediction vector made by the Amazon Web Services server (Figure 11).

Figure 10. The edge device allows the use two ML engines, classification and regression models.

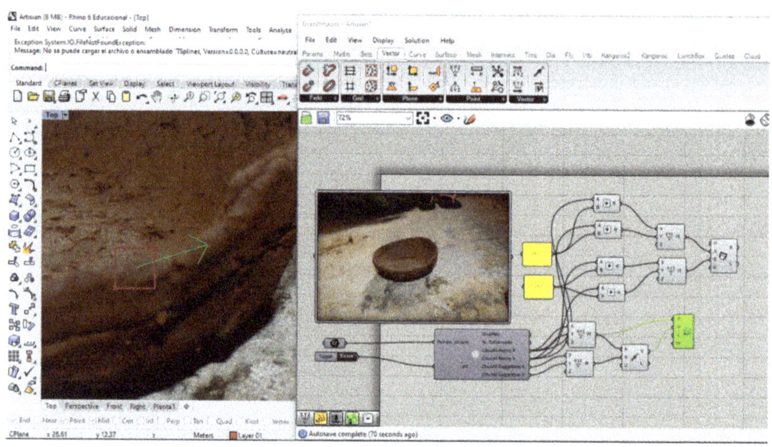

Figure 11. Progress suggestion vector in Grasshopper3D from Amazon Web Services.

Figure 12 shows the image analysis using the custom Grasshopper component, this sample image was attached to the Grasshopper route component, and after that, the custom component sends the image to AWS as a base 64 string to activate the forecasting machine learning engine on AWS, and the Nvidia Jetson to execute the second-, and third-machine learning (ML) engine. The custom component has two input parameters, the image route and the activation button. There are six output parameters; "cloudRes" shows the complete response from AWS; "%_ToComplete" shows the percentage for completing the craftsman object, this comes from Nvidia Jetson Nano with the ML classification Engine; "CloudAI-Recog-X" and "CloudAI-Recog-Y" parameters show the point at X and Y coordinates on the sample picture, this information comes from Nvidia with the ML regression engine; Nvidia engines use AWS to send data; "CloudAI-Suggestion-X" and "CloudAI-Suggestion-Y" come from a forecasting model located on AWS, and show the suggested vector that the apprentice needs to get (green arrow) (Figure 11).

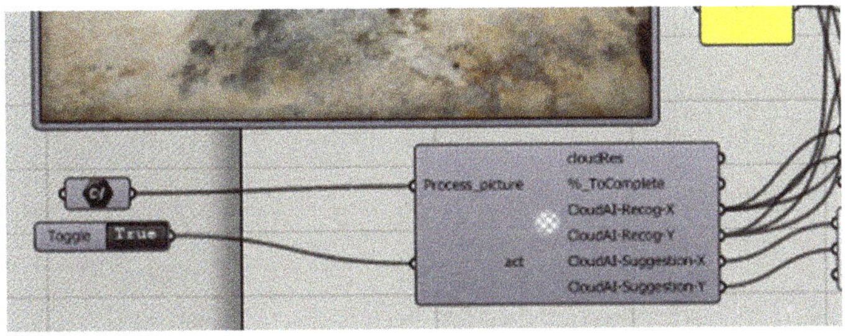

Figure 12. Custom Grasshopper component to communicate AWS with Rhinoceros3D.

3. Conclusions

Usually, teaching design is a process of accompanying a teacher and apprentices. This process is linked to the time and space in which the teacher can develop. In this research, it is proposed that the prediction system allows the teacher's design process to be captured as a continuous process of geometry development to increase the scope of the teaching process and provide better education with the teacher's guidance in the computer system.

Designers can reduce the time taken with the suggestions and avoid losing the analogies made by analog design. This means that the designer can learn to create forms without explicit programming and with an assistant who suggests how to continue the development of the physical object in a vector.

Analog design allows for the flexibility to incorporate different ideas into the design and make quick decisions. The system captures the imprint of the designer or master craftsman in the creation of physical objects to reduce the time it takes to learn the master design process.

4. Discussion and Future Work

Small craft workshops in Latin America sell their products to tourists. Their productive capacity and income are related to the time that the master craftsman can dedicate to the production of new pieces. To develop the skills of apprentices, years of accompaniment are needed for education, and understand that the digital resources as a means for digital craftsmanship to bring together visual thinking with manual dexterity and practical knowledge [4].

For the development and modernization of infrastructures, contemporary design tools would be incorporated into a traditional process to improve the workshop's production capacity. Likewise, as it is connected to a server, the work can be viewed by potential buyers on a website to evaluate which pieces are being developed as well as the connection with other artisan workshops around the world to transmit their imprint through the machine learning model.

How can the processes of machine learning be integrated into the manufacturing creation process?

The apprentice needs the master's guidance. This guidance is crucial to learning about the development of physical objects and is delivered through suggestions made by machine learning. In addition, the process of advancing the object and the detection of possible faults in its creation are delivered by the Nvidia Jetson Nano through the user interface.

This research proposes the complementation of the parametric–generative system to the analog process to use parameters and generative relations, such as application for computational thinking, in explicit or visual programming with machine learning suggestions made by an analog process.

Funding: This research was supported by the Dirección de Investigación de la Universidad Peruana de Ciencias Aplicadas, with the grant with code UPC-A-020-2021-2.

Institutional Review Board Statement: Not applicable.

Informed Consent Statement: Not applicable.

Data Availability Statement: All data are in the manuscript.

Conflicts of Interest: The author declares no conflict of interest.

References

1. Gardner, N.; Meng, L.; Haeusler, M. Computational pragmatism: Computational design as pragmatist tools for the age of the Anthropocene. In Proceedings of the 25th International Conference of the Association for Computer-Aided Architectural Design Research in Asia: Anthropocene, CAADRIA 2020, Bangkok, Thailand, 5–8 August 2020; pp. 487–496.
2. Senke, N. Digital minds, materials, and ethics: Linking computational thinking and digital craft. In Proceedings of the 19th International Conference of the Association for Computer-Aided Architectural Design Research in Asia: Rethinking Comprehensive Design: Speculative Counterculture, CAADRIA 2014, Kyoto, Japan, 14–17 May 2014; pp. 831–840.
3. Herrera, P. Artesanía en Latinoamérica: Experiencias en el contexto de la Fabricación Digital. In Proceedings of the XX Congreso de la Sociedad Ibero-americana de Gráfica Digital, SIGraDi 2016, Buenos Aires, Argentina, 9–11 November 2016; pp. 426–432.
4. Nitsche, M.; Zwaan, S. Teaching Digital Craft. In Proceedings of the 32th Annual ACM Conference on Human Factors in Computing Systems, CHI EA 2014, New York, NY, USA, 26 April–1 May 2014; pp. 831–840.

engineering proceedings

MDPI

Proceeding Paper

Cyber Security in a 5G-Based Smart Healthcare Network: A Base Station Case Study †

Meng-Huan Lee, I-Hsien Liu, Hsiao-Ching Huang and Jung-Shian Li *

Department of Electrical Engineering, Institute of Computer and Communication Engineering, National Cheng Kung University, Tainan 70101, Taiwan; mhlee@cans.ee.ncku.edu.tw (M.-H.L.); ihliu@cans.ee.ncku.edu.tw (I.-H.L.); hchuang@cans.ee.ncku.edu.tw (H.-C.H.)

* Correspondence: jsli@cans.ee.ncku.edu.tw

† Presented at the IEEE 5th Eurasia Conference on Biomedical Engineering, Healthcare and Sustainability, Tainan, Taiwan, 2–4 June 2023.

Abstract: 5G is transforming the healthcare industry by providing novel use cases and applications. With a lower latency, a higher bandwidth, and massive machine-type communication capabilities, 5G supports the smart healthcare system in expanding its applications. This rapid adoption of 5G technology increased the number of connected devices. However, the deployment of 5G networks also introduced security challenges, especially for sensitive patient data which must be secured. Base stations are a key component in connecting devices to the mobile network and need to be secured. As unauthorized base stations cause interference or can be used as fake base stations to attack devices in the network, it is crucial to address the security concerns associated with 5G-based smart healthcare networks. Thus, we studied the case of base station security to investigate different methods for detecting and mitigating the risks associated with unauthorized base stations. The results can be used to better secure 5G-based smart healthcare networks.

Keywords: 5G; smart healthcare; base station; cyber security

Citation: Lee, M.-H.; Liu, I.-H.; Huang, H.-C.; Li, J.-S. Cyber Security in a 5G-Based Smart Healthcare Network: A Base Station Case Study. *Eng. Proc.* **2023**, *55*, 50. https://doi.org/10.3390/engproc2023055050

Academic Editors: Teen-Hang Meen, Kuei-Shu Hsu and Cheng-Fu Yang

Published: 5 December 2023

1. Introduction

Amid the COVID-19 pandemic, the use and demand for telemedicine and remote healthcare have increased significantly. Thus, new approaches for smart healthcare are required for urgent application [1]. Smart healthcare integrates advanced technologies such as 5G, Internet of Things (IoT) devices, AI, and data analytics into the healthcare industry. It revolutionizes the delivery of healthcare services by enabling real-time monitoring, personalized treatment, and improved patient outcomes. Interconnected devices and networks are used to collect and analyze health data, facilitating remote monitoring, telemedicine consultations, and efficient data management. Wearable devices are used to monitor vital signs for early detection and personalized interventions [2]. Smart healthcare systems empower patients and provide access to medical records, personalized recommendations, and virtual consultations. They also optimize care resource utilization and enhance the efficiency of healthcare delivery.

Figure 1 illustrates an example architecture of a 5G-based smart healthcare network connecting different devices to healthcare services. With the help of 5G, these applications have become more reliable than before while serving more people in rural areas where healthcare services were not easily available. The rapid adoption of 5G and IoT technology increases the number of connected devices, including medical devices such as blood glucose meters and heart rate monitors. These devices collect sensitive patient data that must be protected from cyber-attacks. Base stations, which act as access points between these devices and the network, are particularly vulnerable to cyber-attacks, as they are exposed to public internet or radio attacks. Therefore, addressing cyber security concerns in 5G-based smart healthcare networks is of paramount importance. Therefore, we explored base station security and how fake base stations became a threat to a smart healthcare network. Then, possible measures were proposed to secure and detect such threats.

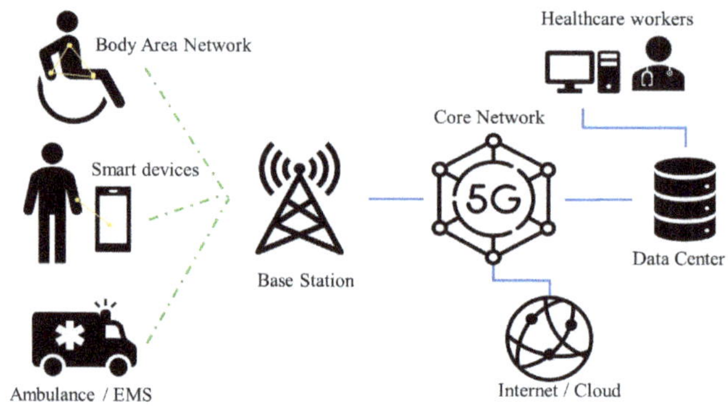

Figure 1. Example of a 5G-based smart healthcare network.

2. Research Background

We reviewed the related studies on 5G and 5G-based smart healthcare networks to define cyber security problems in their use.

2.1. 5G and 5G-Based Smart Healthcare Networks

5G is an advanced mobile network technology that provides a low latency, a high bandwidth, and massive machine-type communications serving large numbers of IoT applications. These applications range from cars and robots to sensors and, in the context of smart healthcare, medical equipment and wearable electronics. A 5G network mainly comprises user equipment (UE), base stations, and core network [3], as follows:

- The UE is a modem and stores a permanent identifier called International Mobile Subscriber Identity (IMSI) and a key used for mutual authentication between the user and the network.
- The base station (BS) acts as the access point for UEs to attach to the Radio Access Network (RAN), connecting it to the mobile network and, thus, to the Internet.
- The core network (CN) is the backbone of the mobile network; it performs all management tasks and traffic routing.

Reference [4] described the architecture of 5G and IoT-enabled smart healthcare considering specific key-enabling technologies such as small cells and the software-defined network (SDN). A taxonomy for 5G smart healthcare was also explained, and future research opportunities applied to IoT-based 5G smart healthcare were reviewed. In Taiwan, a 5G-based smart healthcare network has been trialed in hospitals with mobile network operators that deployed 5G private networks. In the network, medical robots were used to help doctors and nurses [5].

2.2. Cyber Security in a Smart Healthcare Network

The cyber security of a smart healthcare network is important in safely deploying its healthcare services. Securing sensitive data and personal smart healthcare devices is critical to the protection of the privacy of patients and medical personnel. A smart healthcare service is also a mission-critical application that involves human well-being, which requires high availability and reliability. Any security breach or service interruption threatens patients and working personnel. Ahad et al. presented a comprehensive review of 5G-based smart healthcare network security [6]. They investigated various technological features and services related to 5G smart healthcare security, including authentication, confidentiality availability, non-repudiation, and integrity. They also discussed many security threats in 5G smart healthcare and proposed available solutions for issues of 5G-based smart healthcare security. Algarni classified schemes for smart healthcare systems to explore important security issues and challenges and propose countermeasures and directions for future research [7].

3. Base Station Security

Base stations play an important role in a mobile network and are terminals for encryption and integrity protection. Therefore, they are accessible to anyone controlling them and implementing the software. We reviewed the security issues involved in the RAN and BSs.

3.1. Security of Base Station in a Mobile Network

The security of a BS and RAN is often underestimated in a mobile network compared to the UE and CN. Securing the BS helps protect the mobile network as a whole. From the perspective of security, the BS acts as the first line of defense against unauthorized access and malicious activities by enforcing authentication and encryption protocols, monitoring network traffic, and detecting and mitigating potential threats. For 5G to facilitate a higher demand and amount of devices, the number of base stations is much larger than in previous generations. A large number of base stations are now processing large amounts of data.

3.2. Fake Base Station

A fake base station (FBS), also known as a rogue base station (RBS), is a type of malicious device that impersonates a real base station. The main goal of the attacker is to trick UEs into connecting to unauthorized services and/or to steal the identity of the user by tracking and collecting IMSI information from the user's devices [8]. These "IMSI-Catchers" have been spotted since the use of 2G globally. There are reports of suspicious devices appearing in major cities and records of law enforcement using similar equipment to conduct surveillance on suspects [9–11]. There was a recent case of using a fake BS for phishing in Taiwan. The suspects were found to have purchased the fake BS in China to transmit phishing SMS messages to trick people into giving away their credit card information. For this, they downgraded the victims' smartphones to use 2G networks, so the victims were disconnected from the legitimate 4G/5G BS and became vulnerable to phishing messages [12]. Similar cases had been also found in China before and characterized to provide insights into the FBS spam ecosystem in China [13]. A typical procedure for using a fake BS is illustrated in Figure 2.

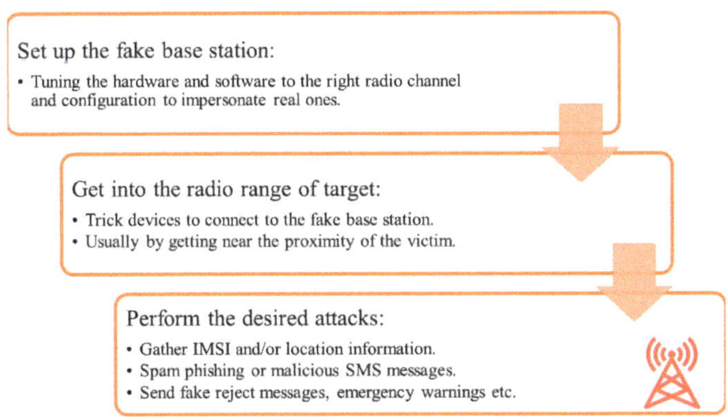

Figure 2. Typical procedure of cyber-attacks using an FBS.

In using FBSs for phishing to steal credit card information in a 4G/5G mobile network, phishing messages are usually modified to cheat unsuspected victims with fake hospitals, healthcare websites, or banks. For example, during the pandemic, a fake website of an economic relief package application was spread on social media. Advanced fake BSs are capable of performing Denial-of-Services (DoS) and Man-in-the-Middle (MitM) attacks. While the cost of deploying mobile networks becomes cheaper thanks to much research and development, so does the cost for malicious cyber-attacks to obtain the hardware and software for setting up FBSs. Thus, cyber-attacks using low-cost FBSs are increasing (Figure 3).

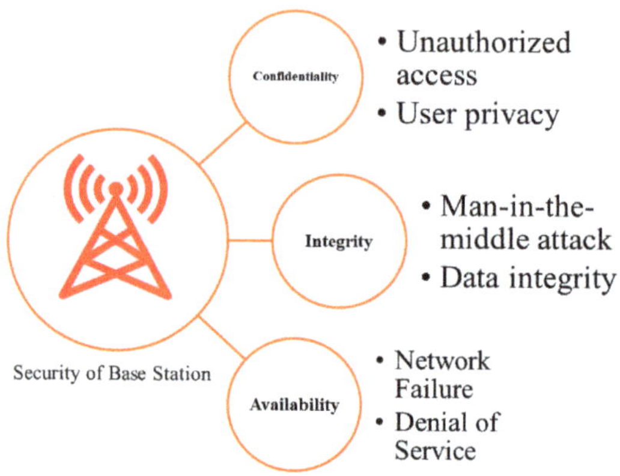

Figure 3. Security threat of a base station.

3.3. *Threat to a Smart Healthcare Network*

FBSs are a threat to smart healthcare networks. These threats were described in terms of confidentiality, integrity, and availability in a CIA triad model, as follows:

- Confidentiality: Malfunctioned access control and authentication allow unauthorized access to a BS, leading to malicious cyber-attacks on sensitive healthcare data. Poorly implemented encryption makes system information and user data vulnerable to cyber-attacks. As a result, patients' privacy is infringed by the theft of sensitive healthcare data.
- Integrity: An FBS is placed between the victims and the real BS and is open to MitM attacks. Information transmitted between the devices and the base station is intercepted and modified to steal smart healthcare data.
- Availability: An FBS causes a DoS, and sensors and devices in the healthcare network lose their connections to the real network, which paralyzes healthcare services and data.

4. Discussion

Potential solutions for security problems are necessary to protect base stations in 5G-based smart healthcare networks. The following needs to be considered for networks' security in this study.

Solution for Security

To prevent sensitive data from being exposed to a public mobile network, the use of a private network dedicated to smart healthcare is encouraged. In a private network of institutions, their connectivity can be more easily controlled, and the specific needs of a smart healthcare service also can be satisfied. The private network still is vulnerable to cyber-attacks using an FBS on the wireless interface. Several private networks share resources with public networks through network function virtualization (NFV). In this case, if the shared BS is attacked, the private network service is also affected [8]. Cryptography and blockchain technology can be used to protect the network. Novel approaches were already proposed for using those methods. Blockchain technology stores records and allows safe and transparent data sharing for patients and network users. The decentralized framework provided by the distributed ledger facilitates privacy needs for numerous applications in 5G. However, blockchain requires considerable computing power for small IoT devices [14]. Detecting suspicious BSs for potential threats is necessary to identify threats and take action before the damage occurs. To detect an FBS, network-, signal-, and location-based detection schemes were suggested for 5G smart healthcare networks [15,16].

5. Conclusions

We explored the cyber security of 5G-based smart healthcare networks with an emphasis on BSs. We reviewed the related literature to explore the increasing importance of the rapid expansion and importance of cyber security for 5G smart healthcare networks and the IoT devices connected to them. Through a detailed investigation of the FBS attacks in Taiwan, the security problems caused by FBSs were analyzed to describe the failure to secure BSs and 5G-based smart healthcare networks. Potential security measures and technology were also proposed to raise awareness of the importance of cyber security in 5G smart healthcare networks.

Smart healthcare and mobile network technology are constantly evolving, and so are cyber-attackers. Although the proposed solutions are considered to solve security problems, it is still necessary for healthcare service providers to be alerted to an increasing number of exploits and attacks. Developing safe and efficient authentication methods is key for the security of smart healthcare networks and connected devices. However, networks and devices have different computing powers. Therefore, an ideal algorithm is required for security by minimizing latency and computing costs. The large amounts of sensitive data on the network require improved privacy and proper encrypted and processed methods. Policies and security standards must be updated often to keep up with the constantly evolving security threats. Healthcare workers, patients, and administrators must be aware of the threats to smart healthcare technology.

Author Contributions: Validation, H.-C.H.; Writing—original draft, M.-H.L.; Writing—review & editing, I.-H.L.; Supervision, J.-S.L. All authors have read and agreed to the published version of the manuscript.

Funding: This work was supported by the National Science and Technology Council (NSTC) in Taiwan under contract numbers 111-2221-E-006-079- and 112-2634-F-006-001-MBK.

Institutional Review Board Statement: Not applicable.

Informed Consent Statement: Not applicable.

Data Availability Statement: Data available on request due to restrictions e.g., privacy or ethical. The data presented in this study are available on request from the corresponding author. The data are not publicly available due to the dataset containing base station signal records that do not comply with regulatory approvals.

Conflicts of Interest: The authors declare no conflict of interest.

References

1. Taiwo, O.; Ezugwu, A.E. Smart healthcare support for remote patient monitoring during COVID-19 quarantine. *Inform. Med. Unlocked* **2020**, *20*, 100428. [CrossRef] [PubMed]
2. Baker, S.B.; Xiang, W.; Atkinson, I. Internet of Things for Smart Healthcare: Technologies, Challenges, and Opportunities. *IEEE Access* **2017**, *5*, 26521–26544. [CrossRef]
3. 3GPP TS 23.501 System Architecture for the 5G System (5GS). Available online: https://www.etsi.org/deliver/etsi_ts/123500_123599/123501/16.06.00_60/ts_123501v160600p.pdf (accessed on 22 April 2023).
4. Ahad, A.; Tahir, M.; Yau, K.-L.A. 5G-Based Smart Healthcare Network: Architecture, Taxonomy, Challenges and Future Research Directions. *IEEE Access* **2019**, *7*, 100747–100762. [CrossRef]
5. Taipei City Government. 5G Private Network Intelligent Disinfection and Transportation Robot. Available online: https://smartcity.taipei/projdetail/175?lang=en (accessed on 12 May 2023).
6. Ahad, A.; Ali, Z.; Mateen, A.; Tahir, M.; Hannan, A.; Garcia, N.M.; Pires, I.M. A Comprehensive review on 5G-based Smart Healthcare Network Security: Taxonomy, Issues, Solutions and Future research directions. *Array* **2023**, *18*, 100290. [CrossRef]

7. Algarni, A. A Survey and Classification of Security and Privacy Research in Smart Healthcare Systems. *IEEE Access* **2019**, *7*, 101879–101894. [CrossRef]
8. Lee, M.-H.; Liu, I.-H.; Li, J.-S. Fake Base Station Threats in 5G Non-Public Networks. In Proceedings of the ICAROB 2023, Oita, Japan, 9–12 February 2023.
9. Cullen, C.; Bureau, B. Someone Is Spying on Cellphones in the Nation's Capital. CBC News. Available online: https://www.cbc.ca/news/politics/imsi-cellphones-spying-ottawa-1.4050049 (accessed on 7 March 2023).
10. Ramirez, A. *ICE Records Confirm That Immigration Enforcement Agencies Are Using Invasive Cell Phone Surveillance Devices*; American Civil Liberties Union: New York, NY, USA, 2020.
11. Chlosta, M.; Rupprecht, D.; Pöpper, C.; Holz, T. 5G SUCI-catchers: Still catching them all? In Proceedings of the 14th ACM Conference on Security and Privacy in Wireless and Mobile Networks (WiSec '21), New York, NY, USA, 28 June–2 July 2021.
12. Chiang, Y.; Su, S.; Ko, L. 12 Indicted over Credit Card Fraud Using Fake Base Stations. Focus Taiwan. Available online: https://fostaiwan.tw/society/202304100016 (accessed on 22 April 2023).
13. Zhang, Y.; Liu, B.; Lu, C.; Li, Z.; Duan, H.; Hao, S.; Liu, M.; Liu, Y.; Wang, D.; Li, Q. Lies in the Air: Characterizing Fake-base-station Spam Ecosystem in China. In Proceedings of the 2020 ACM SIGSAC Conference on Computer and Communications Security (CCS '20). Association for Computing Machinery, New York, NY, USA, 9–13 November 2020; pp. 521–534.
14. Tahir, M.; Habaebi, M.H.; Dabbagh, M.; Mughees, A.; Ahad, A.; Ahmed, K.I. A review on application of blockchain in 5G and beyond networks: Taxonomy, field-trials, challenges and opportunities. *IEEE Access* **2020**, *8*, 115876–115904. [CrossRef]
15. Park, S.; Shaik, A.; Borgaonkar, R.; Martin, A.; Seifert, J.P. White-Stingray: Evaluating IMSI Catchers Detection Applications. In Proceedings of the Workshop on Offensive Technologies (WOOT), Vancouver, BC, Canada, 14–15 August 2017.
16. Huang, K.-W.; Wang, H.-M. Identifying the Fake Base Station: A Location Based Approach. *IEEE Commun. Lett.* **2018**, *22*, 1604–1607. [CrossRef]

engineering proceedings

MDPI

Proceeding Paper

Application of Tabu Search for Job Shop Scheduling Based on Manufacturing Order Swapping †

Jung-Sing Jwo [1], Cheng-Hsiung Lee [1], Jian-Tan Chen [1], Ching-Sheng Lin [1,*], Chun-Yu Lin [2], Wen-Kai Cheng [2], Chun-Hao Chang [2] and Jen-Kai King [2]

1 Master Program of Digital Innovation, Tunghai University, Taichung 40704, Taiwan
2 Aerospace Industrial Development Corporation, Taichung 407803, Taiwan
* Correspondence: cslin612@thu.edu.tw

† Presented at the IEEE 5th Eurasia Conference on Biomedical Engineering, Healthcare and Sustainability, Tainan, Taiwan, 2–4 June 2023.

Abstract: Due to the characteristics of small-volume-large-variety production in the aircraft industry, determining how to intelligently take men, machines, materials, methods, and environments into account is critical for production, posing the challenge of scheduling the job shop. The existing scheduling system is slow to adapt when the requirements change and, most importantly, slow to consider the current situation of the production line. In this study, we apply the Tabu search algorithm to resolve the scheduling problem. This method swaps manufacturing orders with the consideration of the due date, working hours, and current delays via the finite capacity assumption. Meanwhile, we set up extreme bounds using the infinite capacity assumption for comparison. The case study is conducted in a work center that uses pliers to trim parts for use in the aircraft industry. The results show the closeness to the extreme bounds and indicate this method's practical feasibility for industrial applications.

Keywords: job shop scheduling; Tabu search; manufacturing orders; extreme bounds; aircraft industry

Citation: Jwo, J.-S.; Lee, C.-H.; Chen, J.-T.; Lin, C.-S.; Lin, C.-Y.; Cheng, W.-K.; Chang, C.-H.; King, J.-K. Application of Tabu Search for Job Shop Scheduling Based on Manufacturing Order Swapping. *Eng. Proc.* **2023**, *55*, 51. https://doi.org/10.3390/engproc2023055051

Academic Editors: Teen-Hang Meen, Kuei-Shu Hsu and Cheng-Fu Yang

Published: 5 December 2023

1. Introduction

With the rapid development of Internet of Things (IoT) technology, the concepts and implementations of smart manufacturing have been widely studied. Since the production of composite materials is a crucial manufacturing process in the aircraft industry, it is important to introduce smart manufacturing-related technologies into current manufacturing processes. Properly arranging the manufacturing orders (MOs) of the jobs is the key to increasing the efficiency of product life cycles and successfully moving toward smart manufacturing. The main characteristic of processes in the aircraft industry is small-volume-large-variety production, posing the challenge of scheduling the job shop.

Job shop scheduling is considered to be combinatorial optimization and belongs to the class of NP-hard problems [1]. Many different methods have been explored in relation to this research topic, including analytical algorithms and artificial intelligence-based methods [2]. Queuing theory and simulation models are popular in analytical algorithms. Artificial intelligence-based optimization techniques mainly include Tabu search and genetic algorithms. In this study, we apply the Tabu search algorithm to address the job shop scheduling by swapping MOs. Using the finite capacity assumption, we consider the due date, working hours, and current delays for each job. To evaluate the effectiveness of the proposed solution, we compute extreme bounds based on the infinite capacity assumption to determine how far the proposed method is from optimality. The empirical study is conducted in a work center which uses pliers to trim parts for use in the aircraft industry.

The rest of this paper is structured as follows. Section 2 reviews several prior studies. Section 3 discusses the proposed scheduling method. The experiments are described in Section 4 and the conclusions are provided in Section 5.

2. Literature Review

Prior studies related to this paper are discussed with regard to Tabu search-based approaches and AI-related models.

2.1. Tabu Search Overview

Tabu search was created by Glover in 1986 for optimization problems [3,4]. There are two properties, local search and the Tabu list, in the Tabu search that enhance its optimization ability. The local search technique includes movements which do not improve the performance if no better moves are found. These worse but acceptable moves help the optimization process escape from local optima by exploring the solution space more thoroughly. The Tabu list is a form of short-term memory that records the most recently visited moves and discourages the Tabu search from revisiting them to prevent cycles [5]. To generate more neighborhoods, a method to resolve the distributed permutation flow-shop scheduling problem (DPFSP) is proposed, which involves swapping the sub-sequences of jobs [6]. A parallel computing approach based on Tabu search is used to speed up the computation time taken to resolve the blocking job shop scheduling problem [7]. The experiment was conducted on a cluster-based supercomputer using 512 CPU cores.

2.2. Other AI Approaches

Due to the booming development of deep learning techniques, neural network-based approaches have also been investigated in relation to the scheduling problem. Actor–Critic is a model-free reinforcement learning technique in which the actor network is taught how to behave in given environments, whereas the critic network learns to give feedback to the actor network [8]. An actor–critic deep reinforcement learning model is proposed to solve the job shop scheduling problem, in which the actor network assigns a job and the critic network provides rewards according to the processing time of the job [9]. A Deep Q-Network (DQN) is adopted for semiconductor manufacturing scheduling by using an agent to perform job assignments for each work center [10]. Another DQN method based on edge computing is proposed to solve the scheduling problem for a smart semiconductor manufacturing factory [11]. Unlike the traditional DQN, which supports only one decision, this method outputs multiple dispatching rules for multiple edge devices. To better handle the highly dynamic and changeable conditions in the factory, a dueling double DQN is designed to deal with the adaptive job shop scheduling problem by training a CNN to approximate the state-action value function [12].

3. Scheduling by Manufacturing Order Swapping

The proposed Tabu search scheduling based on manufacturing order swapping is introduced to discuss the research problem, the concept of the proposed solution, and the objectives. The detailed algorithm used for Tabu search scheduling is also described in the second section.

3.1. Problem Definition

We focus on scheduling one of the work centers in the aircraft process. For each day, there are new MOs to be dispatched. The unprocessed MOs from the previous day, they are combined with new MOs for the current day for dispatch. The initial sorting of MOs each day is performed using First in First out (FIFO) and Earliest Due Date (EDD) processes. The notations are defined as follows.

WO: working hours per day for the current work center (CWC).
i: MO, i = 1, 2, . . .I.
PT(i): processing time (scaled by hours) for MO i.

LT(i): average lead time (scaled by days) for MO i after CWC.
SD(i): the scheduled day (noted by 1, 2, 3, . . .) for the MO i where 1 means today, 2 means tomorrow, and so on.
DD(i): due date for MO i.
BPDD(i): number of days the MO i before or past the due date at the CWC by computing (Today-DD(i)).
EDD_FCA(i): expected delay days for MO i under finite capacity assumption by computing BPDD (i) + SD(i) + LT(i) where a positive value represents a delay and a negative value is one that occurs before the start of the schedule.

Due to the specific production requirements of composite materials in the aerospace industry, there are three minimum optimization problems to overcome: (i) total expected early completion days (TEECD), (ii) total expected delay days (TEDD), and (iii) total number of delayed manufacturing orders (TNDMO). For each MO i, we calculate the EDD_FCA(i). The TEECD is defined as the total sum of those EDD_FCA which are smaller than 0, as in "(1)", whereas TEDD is defined as the total sum of those EDD_FCA which are greater than 0, as in "(2)". The TNDMO is the number of MOs which have EDD_FCA greater than 0, as in "(3)". Another notation is the total number of early completion manufacturing orders (TNECMO), as in "(4)". After each schedule, we assign SD(i) by accumulating PT(i) and considering WO. The estimated values of TEECD, TEDD, and TNDMO are calculated as well.

$$\text{TEECD} = \sum_{i=1}^{I} \text{EDD\_FCA}(i), \text{where EDD\_FCA}(i) < 0 \quad (1)$$

$$\text{TEDD} = \sum_{i=1}^{I} \text{EDD\_FCA}(i), \text{where EDD\_FCA}(i) \geq 0 \quad (2)$$

$$\text{TNDMO} = |\{i|\text{EDD\_FCA}(i) \geq 0 \forall i\}| \quad (3)$$

$$\text{TNECMO} = |\{i|\text{EDD\_FCA}(i) < 0 \forall i\}| \quad (4)$$

The workflow of the scheduling process is shown in Figure 1. For each daily scheduling, we set the initial sorting (FIFO or EDD) for MOs. The Tabu search is applied to the schedule to compute TEECD, TEDD, and TNDMO. These values are used to evaluate the effectiveness of the method. Those MOs that are unable to be processed are moved to the next day for further scheduling.

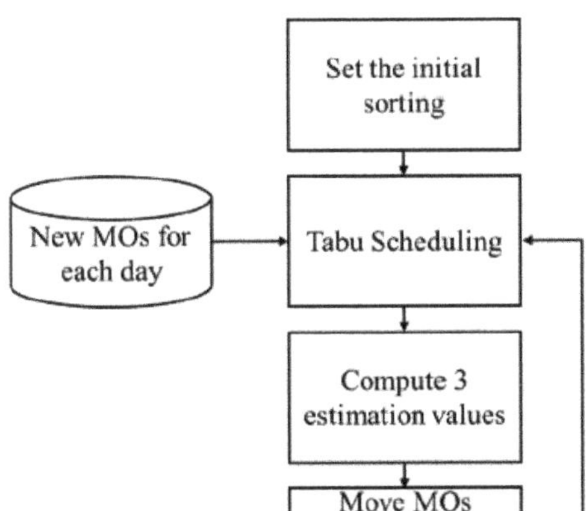

Figure 1. Process of scheduling.

3.2. Scheduling Algorithm

The concept of the Tabu search scheduling is to switch the MOs and obtain the best TEECD, TEDD, and TNDMO. To be specific, for each MO i with EDD_FCA$(i) \geq 0$, we switch i with all other MOs and re-calculate SD to obtain new EDD_FCA. With the new EDD_FCA, we compute new TEECD, TEDD, and TNDMO for comparison. The switch frequencies are TNDMO $\times$ (TNDMO $+$ TNECMO $-$ 1). The above process operates iteratively.

The details of the scheduling algorithm are shown in Algorithm 1. The input S_init is an initial sorting by FIFO or EDD. The variable assignment occurs from lines 1 to 5, where sBest denotes the overall best scheduling and iterBest denotes the best scheduling in each iteration. Each day, we perform an N iteration to find the best schedule. The function getNeighbors is used to find out all possible schedules by swapping the elements in setDelay with the elements in set_All and storing all possible schedulings into sNeighborhood (lines 7–10). Then, we loop over each scheduling in sNeighborhood to find the best scheduling for the given iteration (lines 11–16). The function checkMinOptimization checks whether the current scheduling sCandidate has lower TEECD, TEDD, and TNDMO than iterBest. If the current schedule already exists in the tabuList, we skip the comparison. In the next step, we use the same function to compare iterBest with sBest (lines 17–18). Lastly, iterBest is put into tabuList to avoid being trapped in the local optimum (lines 19–20).

Algorithm 1: Tabu Search Scheduling by Manufacturing Order Swapping.

Input: S_init which is an initial scheduling by FIFO or EDD

```
1:  sBest ← S_init
2:  iterBest ← S_init
3:  tabuList ← []
4:  tabuList.push(S_init)
5:  iteration ← N
6:  while iteration:
7:     set_Delay = { MO i | EDD_FCA(i) ≥ 0 ∀ i based on iterBest}
8:     set_All = {MO i based on iterBest}
9:     #Get all possible schedulings based on iterBest
10:    sNeighborhood ← getNeighbors(set_Delay, set_All)
11:    #Find the best scheduling in this iteration
12:    for (sCandidate in sNeighborhood)
13:         if (not tabuList.contains(sCandidate))
14:              iterBest ← checkMinOptimization(sCandidate, iterBest)
15:         end
16:    end
17:    #Is iterBest better than sBest?
18:    sBest ← checkMinOptimization(iterBest, sBest)
19:    #Put iterBest into tabuList to avoid trapped in the local optimum
20:    tabuList.push(iterBest)
21: End
```

4. Experimental Results

For each day, there are new manufacturing orders (NMO) sent to the work center: total manufacturing orders (TMO), finished manufacturing orders (FMO), finished but delayed manufacturing orders (FDMO), and total number of delayed manufacturing orders (TNDMO). The TMO is the NMO combined with the previous unprocessed MOs. The FMO represents the number of MOs which finish on that day, while FDMO indicates those FMO that pass the due dates. Based on the scheduling, TNDMO is used to represent the total number of MOs that pass the due dates.

To obtain evaluation metrics with which to assess the approach, we introduce three bounds including UBTEECD, LBTEDD and LBTNDMO. The assumption is based on infinite capacity, which implies all new MOs are finished on one day without carrying over to the next day. The UBTEECD is used to represent the maximum of total expected early completion days, LBTEDD represents the minimum of total expected delay days, and LBTNDMO represents the minimum total number of delayed manufacturing orders. As mentioned before, the minimum of TEECD is considered due to the special requirements of composite materials.

In this empirical study, five days of scheduling are presented. Since the total number of delayed manufacturing orders is the most critical factor in the factory, we start with the evaluation of TNDMO. Both FIFO and EDD are used for initial scheduling. The results are shown in Tables 1 and 2. The scheduling approach with FIFO initialization reaches the LBTNDMO every day, whereas EDD initialization is not promising. More detailed evaluations of TEECD, TEDD, and TNDMO with FIFO and EDD initializations are also conducted. Table 3 demonstrates the FIFO scheduling and the proposed scheduling with FIFO initialization. Directly applying FIFO for scheduling does not provide good results, while the proposed method achieves good performance in TEDD and TNDMO. In Table 4, we employ EDD and the proposed method with EDD initialization for comparison. Directly applying EDD for scheduling does not achieve good performance either, but the FIFO scheduling outperforms EDD scheduling. Similarly, the proposed method with FIFO initialization is better than EDD initialization. In conclusion, the proposed method works better in terms of TEDD and TNDMO, though there is still room to improve the values of the TEECD.

Table 1. Tabu search with initial scheduling via FIFO compared with the LB_{TNDMO}.

	Day1	**Day2**	**Day3**	**Day4**	**Day5**
NMO	369	66	94	123	121
TMO	369	393	402	508	594
FMO	42	85	17	35	56
FDMO	34	54	9	34	48
TNDMO	63	61	21	40	50
LB_{TNDMO}	63	61	21	40	50

Table 2. Tabu search with initial scheduling via EDD compared with the LB_{TNDMO}.

	Day1	**Day2**	**Day3**	**Day4**	**Day5**
NMO	369	66	94	123	121
TMO	369	397	451	557	655
FMO	38	40	17	23	48
FDMO	27	37	13	20	41
TNDMO	71	81	53	65	84
LB_{TNDMO}	63	68	45	60	84

Table 3. Comparison results between the FIFO, our method and bound (% is the comparison with the bound).

	Day1			Day2			Day3			Day4			Day5		
	TEECD	TEDD	TNDMO	TEECD	TEDD	TNDMO	TEECD	TEDD	TNDMO	TEECD	TEDD	TNDMO	TEECD	TEDD	TNDMO
FIFO	4524 (−12%)	1230 (18%)	86 (37%)	5114 (−11%)	1029 (17%)	67 (10%)	5869 (−12%)	572 (24%)	30 (43%)	7623 (−14%)	883 (17%)	43 (8%)	9144 (−13%)	876 (34%)	54 (8%)
Our method	4312 (−16%)	1079 (3%)	63 (0%)	4950 (−14%)	898 (2%)	61 (0%)	5509 (−18%)	475 (3%)	21 (0%)	7420 (−16%)	758 (1%)	40 (0%)	8737 (−17%)	655 (0%)	50 (0%)
Bound	5113	1043	63	5739	882	61	6682	463	21	8877	752	40	10510	653	50

Table 4. Comparison results between the EDD, our method and bound (% is the comparison with the bound).

	Day1			Day2			Day3			Day4			Day5		
	TEECD	TEDD	TNDMO	TEECD	TEDD	TNDMO	TEECD	TEDD	TNDMO	TEECD	TEDD	TNDMO	TEECD	TEDD	TNDMO
EDD	4193 (−18%)	1154 (11%)	95 (51%)	4704 (−17%)	1146 (13%)	97 (43%)	5595 (−19%)	751 (28%)	80 (78%)	7488 (−18%)	819 (18%)	90 (50%)	8877 (−17%)	939 (24%)	109 (30%)
Our method	4399 (−14%)	1131 (8%)	71 (13%)	4669 (−18%)	1085 (7%)	81 (19%)	5511 (−20%)	657 (12%)	53 (18%)	7428 (−18%)	860 (8%)	65 (5%)	8779 (−18%)	814 (7%)	84 (0%)
Bound	5113	1043	63	5668	1015	68	6893	585	45	9088	692	60	10721	760	84

5. Conclusions

We present a Tabu search schedule based on swapping the manufacturing orders. We conduct empirical studies in the aerospace industry with FIFO and EDD initializations. The proposed approach achieves good performance with regard to minimizing the total expected delay days and the total number of delayed manufacturing orders. In future work, we will focus on the improvement of the total expected early completion days and also expand our method to different work centers to evaluate its effectiveness and robustness.

Author Contributions: Supervision, J.-S.J.; methodology, J.-T.C., C.-S.L. and C.-H.L.; investigation, J.-S.J., J.-T.C., C.-Y.L., W.-K.C., C.-H.C. and J.-K.K.; writing—review and editing, C.-S.L.;. All authors have read and agreed to the published version of the manuscript.

Funding: This work is financially supported by the Ministry of Science and Technology (MOST) of Taiwan under Grant MOST 110-2622-E-029-015 - .

Institutional Review Board Statement: Not applicable.

Informed Consent Statement: Not applicable.

Data Availability Statement: Data are contained within the article.

Conflicts of Interest: The authors declare no conflict of interest.

References

1. Garey, M.R.; Johnson, D.S.; Sethi, R. The complexity of flowshop and jobshop scheduling. *Math. Oper. Res.* **1976**, *1*, 117–129. [CrossRef]
2. Zhang, Z.; Wang, W.; Zhong, S.; Hu, K. Flow shop scheduling with reinforcement learning. *Asia-Pac. J. Oper. Res.* **2013**, *30*, 1350014. [CrossRef]
3. Glover, F. Tabu search—Part I. *ORSA J. Comput.* **1989**, *1*, 190–206. [CrossRef]
4. Navarro, A.; Rudnick, H. Large-scale distribution planning—Part II: Macro-optimization with Voronoi's diagram and tabu search. *IEEE Trans. Power Syst.* **2009**, *24*, 752–758. [CrossRef]
5. Caldeira, R.H.; Gnanavelbabu, A.; Joseph Solomon, J. Solving the Flexible Job Shop Scheduling Problem Using a Hybrid Artificial Bee Colony Algorithm. In *Trends in Manufacturing and Engineering Management*; Springer: Berlin/Heidelberg, Germany, 2021; pp. 833–843.
6. Gao, J.; Chen, R.; Deng, W. An efficient tabu search algorithm for the distributed permutation flowshop scheduling problem. *Int. J. Prod. Res.* **2013**, *51*, 641–651. [CrossRef]
7. Dabah, A.; Bendjoudi, A.; AitZai, A.; Taboudjemat, N.N. Efficient parallel tabu search for the blocking job shop scheduling problem. *Soft Comput.* **2019**, *23*, 13283–13295. [CrossRef]
8. Konda, V.; Tsitsiklis, J. Actor-critic algorithms. In Proceedings of the Advances in Neural Information Processing Systems, Denver, CO, USA, 29 November–4 December 1999; Volume 12.
9. Liu, C.L.; Chang, C.C.; Tseng, C.J. Actor-critic deep reinforcement learning for solving job shop scheduling problems. *IEEE Access* **2020**, *8*, 71752–71762. [CrossRef]
10. Waschneck, B.; Reichstaller, A.; Belzner, L.; Altenmüller, T.; Bauernhansl, T.; Knapp, A.; Kyek, A. Deep reinforcement learning for semiconductor production scheduling. In Proceedings of the 2018 29th Annual SEMI Advanced Semiconductor Manufacturing Conference (ASMC), Saratoga Springs, NY, USA, 30 April–3 May 2018; pp. 301–306.
11. Lin, C.C.; Deng, D.J.; Chih, Y.L.; Chiu, H.T. Smart manufacturing scheduling with edge computing using multiclass deep Q network. *IEEE Trans. Ind. Inform.* **2019**, *15*, 4276–4284. [CrossRef]
12. Han, B.A.; Yang, J.J. Research on adaptive job shop scheduling problems based on dueling double DQN. *IEEE Access* **2020**, *8*, 186474–186495. [CrossRef]

engineering proceedings

MDPI

Proceeding Paper

Application of AIoT Image Sensor for Lifting Operation Safety Monitoring of Mobile Crane †

Wen-Der Yu [1], Hsien-Chou Liao [2], Jian-Wei Li [3], Zi-Yi Lim [3,*] and Wen-Ta Hsiao [1]

[1] Department of Construction Engineering, Chaoyang University of Technology, Taichung 413310, Taiwan; wenderyu@cyut.edu.tw (W.-D.Y.); wdshiau@cyut.edu.tw (W.-T.H.)
[2] Department of Computer Science and Information Engineering, Chaoyang University of Technology, Taichung 413310, Taiwan; hcliao@cyut.edu.tw
[3] Department of Information and Communication Engineering, Chaoyang University of Technology, Taichung 413310, Taiwan; lijw@cyut.edu.tw
* Correspondence: zylim@cyut.edu.tw

† Presented at the IEEE 5th Eurasia Conference on Biomedical Engineering, Healthcare and Sustainability, Tainan, Taiwan, 2–4 June 2023.

Abstract: The emerging advances in deep learning and computer vision have enabled traditional cloud-based decision-making through edge computing with artificial intelligent internet of things (AIoT) image sensors (AIoT-IS). As a result, the timeliness and security of image recognition can be obtained. This study aims to develop an AIoT-IS-based smart safety control device for construction sites ("SmartCon" hereafter) for the operations of mobile cranes. The research is carried out to include the definition of hazardous control areas and the identification of unsafe scenarios for crane operations, the development of an intelligent prototype system for the safety control of lifting operations, system application demonstrations, and the evaluation of effectiveness. It is intended to assist labor inspection agencies and industry practitioners with a tool for monitoring lifting operations. The results of two empirical field tests show that the proposed SmartCon improves the safety monitoring of the operation of the machine through the real-time response to the identified potential risks on construction sites.

Keywords: construction safety; AIoT image sensor; site management; YOLO; image recognition

Citation: Yu, W.-D.; Liao, H.-C.; Li, J.-W.; Lim, Z.-Y.; Hsiao, W.-T. Application of AIoT Image Sensor for Lifting Operation Safety Monitoring of Mobile Crane. *Eng. Proc.* **2023**, *55*, 52. https://doi.org/10.3390/engproc2023055052

Academic Editors: Teen-Hang Meen, Kuei-Shu Hsu and Cheng-Fu Yang

Published: 5 December 2023

1. Introduction

The construction industry is considered a key driver for economic development by the governments of most countries in the world. According to the latest statistics from the Industry and Service Census conducted by the Directorate General of Budget, Accounting and Statistics (DGBAS) of the Executive Yuan in Taiwan (R.O.C), the construction industry employs more than 913,000 workers, which accounts for 7.49% of the total employed population [1]. More than 2.49 million family members are supported by construction employees. About 3.77% of Taiwan's gross national product (GNP) is contributed by the construction industry. Dividing the industry's output value by the number of employees shows that the per-capita GNP of the construction industry is significantly lower than the national average. On the other hand, the income of 913,000 construction workers is the main source of income for their families. When occupational accidents occur, the families of wounded workers often face a significant economic impact. This, in turn, affects the whole society.

Construction accidents have long taken up the highest percentage of industrial and serious occupational accidents around the world. There are significant harmful impacts on industrial production and workers' lives. The main reason is that construction sites are open to risks.

According to the latest statistics from the DGBAS of the Executive Yuan in Taiwan in 2019 [2], there were 313 significant occupational casualties in Taiwan's entire industry, of which the construction industry accounted for 46.32% (a total of 145 casualties). Statistics from 2012 to 2020 showed the total number of fatal occupational accidents in the industry was 2879, and the construction industry accounted for 47.65% (a total of 1372 casualties), which was about half of that of all industries. These statistics highlight the severity of safety issues for construction workers.

Referring to Article 22 of "Taiwan's Enforcement Rules of Occupational Safety and Health Act", dangerous machinery includes fixed and mobile cranes. According to statistics collected by earlier research [3,4], there have been 12 to 19 major occupational accidents involving mobile cranes in Taiwan annually since 2002. Between that year to 2012, 104 people died of these accidents. Three people suffered from minor injuries, whereas ten people suffered from serious injuries. In the analysis of the causes of accidents, "management factors" constituted the most significant portion (66%), followed by "human factors" (22%), "equipment factors" (8.5%), and "environmental factors" (3%). Materials falling, cranes tipping over, being caught, being hit, and electric shock are the leading causes of accidents. Although the relevant safety regulations have been issued by the competent authorities [5], the primary approaches to safety management and disaster prevention still mainly rely on manual inspection by occupational safety management personnel. There are still problems in practice. Even with detailed laws and regulations governing occupational safety, reducing site risks and implementing immediate control and prevention remain challenging. A promising solution for construction site safety management is to use information and computer technology (ICT) technology to help site safety managers improve the deficiencies of manual safety management.

Deep neural networks (DNN) have advanced quickly during the past decade as an emerging artificial intelligence (AI) technology [6]. Machine learning (ML) can judge and evaluate a worker's safety on site in real time. Based on the above discussion, this research is carried out to apply AI technology based on DNN to the technology for construction site visual recognition and analysis to detect labor safety hazards of workers automatically during crane operations. The goal is to provide early warning of labor hazards to reduce or avoid the occurrence of construction accidents on site and achieve a safer working environment for construction workers.

The rest of this article is organized as follows. Section 2 sorts out some related works about construction safety. The methodology and framework of the proposed method are described in Section 3. The testing results are shown in Section 4. Lastly, Section 5 concludes this study.

2. Related Works

In recent years, various advanced technologies have been developed to assist in the safety control of construction sites and reduce the risks caused by lifting operations. For example, the Taipei City government in Taiwan continues to promote smart city-related plans, one of which is to use Bluetooth low-energy (BLE) technology to assist in the safety management of construction sites [7]. This technology installs Bluetooth transmitters (iBeacon) on the worker's helmet or personal protection equipment (PPE). When the Bluetooth receiver receives the signal from the transmitter, it accurately calculates the position of the personnel or equipment, thereby notifying the construction workers to avoid entering the safety control area. In addition, CTCI Inc. developed technology to calculate the construction site by recognizing the triangular cone of lifting operations. It sends a warning to notify the workers when they work under lifting objects or pass through the range of a construction area [8]. Japan's Taisei Corporation also uses the GPS positioning of construction machinery and construction workers to remind the workers not to enter the lifting area of construction machinery [9]. If workers are close to construction machinery, alarms are sent to smart watches.

In addition to general site safety control technology, the Swedish company Gigasense has developed a set of weight sensors installed on lifting equipment to monitor equipment in real time to ensure the balance state of lifting operations [10]. China Xiamen Baima Technology has also developed a remote monitoring system to monitor the lifting operations in a construction site [11]. The various sensors installed on the lifting equipment can obtain the data of the operation in real time and carry out calculations to identify the limitations of lifting operations such as anti-collision calculation, weight limit, torque limit, height limit, amplitude limit, angle limit, wind speed alarm, and main track anti-tilt.

Nath et al. proposed a construction site safety monitoring method in 2020 [12], which applied the convolutional neural network (CNN) to identify and monitor the protective equipment of workers. The method achieved 72.3% mAP (mean average precision) at an 11-FPS (frames per second) detection speed. However, the proposed method can only identify the correct wearing of labor personal protective equipment and no other automatic hazard identification functions.

Golcarenarenji et al. proposed a method called CraneNet in 2021 to detect people under crane lifting equipment [13]. In this method, a camera is hung on a hanging hook, and an image is transmitted to a small computer with NVIDIA Jetson Xavier through a wireless network. The ML technology based on CNN is applied (YOLOv4) to detect the position of personnel under crane lifting operations. On construction sites with complex environments that require real-time detection, this method achieves an accuracy of 92.59% at a detection speed of 19 FPS, which is close to the accuracy of human judgment. Although the accuracy is improved, the disadvantage of the CraneNet method is similar to the limitation of the method of Chen and Fang et al. [14,15]. Misrecognition was found due to the dynamic change of the background or the similarity of the object's color to the background. Therefore, there are still difficulties in application in construction site practice.

On the other hand, Liu et al. proposed a method in 2021 to detect whether people are dozing off to prevent accidents caused by operational errors [16]. They used a camera to detect a driver's face through DL technologies using CNN and LSTM and then extract the facial features to classify the driver's behavior such as talking, sleepy eyes, yawning, and napping. If any behavior that affects safety is detected, it warns the driver to avoid accidents.

The above summarizes the advanced technology and latest research literature on the application of image recognition and sensing technology to the safety control of construction lifting operations. This technology's technical viability, maturity, mobility, and economic viability must be considered to make the technology applicable to the actual construction site in this study.

3. Proposed Method

In this section, we describe the design of the proposed AIoT device, the smart construction site safety control device (SmartCon), for the lifting operations of mobile cranes on construction sites.

3.1. System Architecture

The SmartCon is an all-in-one and single independent device that can operate independently. The schematic design of the proposed SmartCon is shown in Figure 1.

To meet the practical needs of the construction site, multiple control devices can be moved and set up quickly at different locations according to the on-site environmental conditions to monitor multiple cranes and multiple lifting operations. The SmartCon uses a customized extendable stand to set up a waterproofed IP camera, network device, AI development board, alarm, battery, and other devices. All the devices and a battery must be set up in the case (except the IP camera) to make sure that the SmartCon is waterproof for all weather.

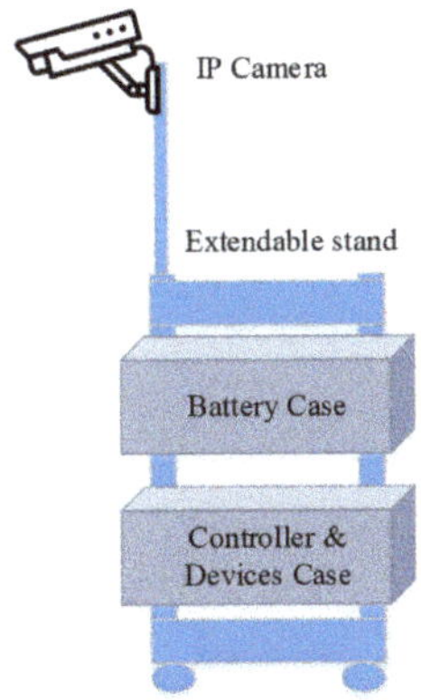

Figure 1. SmartCon all-in-one diagram.

The following SmartCon features, including portability, heat dissipation, and water resistance, are considered in designing the SmartCon.

1. Portability: the extendable stand includes all components, including the battery, controller, computing devices, and IP camera. Furthermore, the Jetson Nano computing device utilized in the design has a compact size, is lightweight, and has low power consumption, which increases the mobility of the SmartCon.
2. Heat dissipation: considering the SmartCon's outdoor operation requirement, the SmartCon may need to operate under the hot sun for a long time. If it is overheated, it damages the hardware. In this regard, we must consider isolating the external temperature and strengthening the internal heat dissipation. The internal heat dissipation is related to the device's portability. The smaller the size, the higher the portability. By contrast, the interior heat dissipation space of the device is smaller. Therefore, it is necessary to install some fans to ensure that the heat in the space can be dissipated.
3. Water resistance: considering the uncertainty of the weather at the construction site, the SmartCon may be used on rainy days. Thus, water resistance must be implemented to ensure that the device circuit will not be damaged.

The system architecture of the SmartCon is shown in Figure 2. The blue block is the hardware module in the device, including a 4G Wi-Fi router, Nvidia AI development board, RF receiver, alarm, and IP cam. The green blocks are the operator's smartphone device and RF remote control. Furthermore, the device is connected to Line Notify for alert notifications.

The detailed specifications of the above components are described as follows.

1. 4G Wi-Fi router: the SmartCon uses a 4G Wi-Fi Router to establish local networks for connecting devices. The IP camera uses PoE (power over ethernet) for the power supply and network connection. The router also connects the Jetson® development board with ethernet. Externally, the 4G network is used for external communication, e.g., Line notification.
2. Jetson development board and core: the SmartCon uses the Jetson Nano®, an embedded system developed by Nvidia®. The development board contains a high-performance computer. The hardware design is optimized for AI computing, and the product has the features of a small size and low power consumption. The small size reduces the weight and volume of the SmartCon, and the low power consumption reduces the battery capacity requirement. Since the battery capacity is proportional to the weight and volume of the battery, the physical volume and weight can be reduced by using the selected development board.

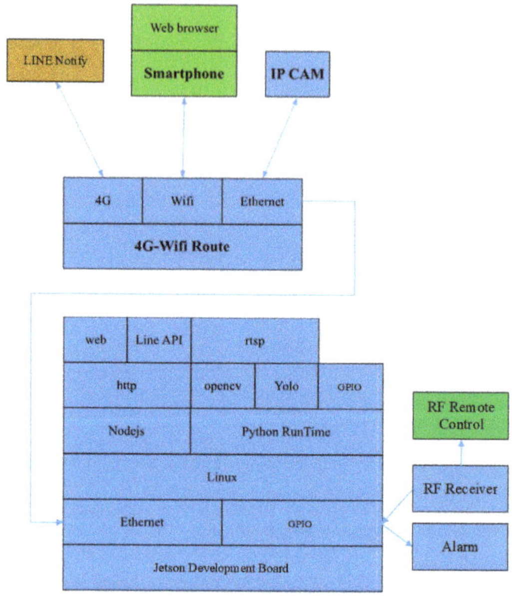

Figure 2. SmartCon system architecture.

The proposed core includes a network module, video streaming module, AI recognition module, GPIO control module, Line API, and electronic fence module. The video stream of the IP camera is communicated through the RTSP streaming protocol and then detected by the AI recognition module. In this study, YOLOv4 is used for object detection. YOLO is a neural network for object detection characterized by its light weight and high efficiency. YOLOv4 is implemented using the darknet architecture. After identifying the object, it determines whether it conforms to the set rules of the electronic fence. If the preset conditions of the alarm are met, the recording and alarm notification are performed, including Line notification and the activation of the alarm.

1. IP camera: the SmartCon uses a network camera with full HD (FHD) resolution, a PoE power supply, and a real-time streaming protocol (RTSP) as the video source. The characteristics of the IP camera in this study are as follows.
    - FHD resolution: the image resolution is proportional to the recognition rate, but inversely proportional to the recognition speed. Such cameras transmit lower-resolution images that need to be adjusted according to the requirements to achieve balance with speed.
    - PoE power supply: although this system provides Wi-Fi as the function of device connection, the wired network is still more stable than the wireless network. Users can connect to the internet without a network line through Wi-Fi, but it still requires a power line for the power supply. Through the PoE function, it only needs a single network line to power on, and connects to the internet simultaneously.
    - Real-time streaming protocol (RTSP): RTSP is an application layer communication protocol for multimedia streaming. It is used in multimedia transmission to establish a connection between a multimedia server and a client to monitor multimedia.
2. Smartphone: the web interface can be used to set and monitor the SmartCon. Users can use any smartphone with a web browser and connect to the 4G Wi-Fi Router. The web interface functions include video streaming monitoring, electronic fence settings, enabling and disabling controls, and more parameter settings.

3. Remote control: the SmartCon system provides a radio frequency (RF) remote controller, which is convenient for users to perform wireless control operations quickly. This RF communication refers to radio waves with frequencies ranging from 3 kHz to 300 GHz to transmit over long distances.
4. Alarm: to create a more significant warning impact in the construction site environment, the SmartCon employs a higher-power alarm as a warning device.
5. Line notification: when the danger alarm is triggered, the system transmits the time, place, and photo through the Line API to a Line account owned by the relevant personnel managing the construction site. This is convenient to understand the situation on the construction site quickly.

3.2. Deep Learning Module

As described in the previous system architecture, the SmartCon applied YOLO object recognition techniques for AI recognition module development, and the model training dataset uses the MS COCO dataset [17]. The MS COCO dataset is a massive object database with many images; more than 200,000 objects are labeled in the images. These labels are divided into 81 categories, with detailed information, including object location and image context.

3.3. Development Board

In order to reduce the cost of system implementation, we use the "Jetson Nano" for system development. "Jetson Nano" is the basic entry-level model of the Jetson embedded systems. Although the hardware capability is the lowest, it also has the lowest power consumption and weight, which reduces the battery capacity requirement and the weight of the host. Moreover, it optimizes the system environment and code as much as possible. The maximum hardware capacity is used to reduce hardware costs. As the functions increase and the system operation burden increases, higher-end hardware can be adopted in the future.

When image data are transmitted through the network, the image data are usually compressed through image coding to reduce the bandwidth required for transmission. Therefore, the receiving end must decode the information to obtain the original multimedia information. The CPU usually processes this decoding work, but the Jetson Nano® is equipped with a multimedia decoding chip, which effectively improves the speed of the system reading camera images and reduces the CPU load.

4. Implementation and Testing

This section presents the implementation of the SmartCon and the experimental results in real-world construction sites.

4.1. System Implementation

The hardware architecture of the SmartCon is shown in Figure 3. The device is a stainless steel waterproof case covering the battery and electronics. On one side of the stand, the tube can be derived to give the camera a better angle for monitoring. The size of the battery case is 44 cm × 30 cm × 24 cm (width × height × depth). The size of the second case is 44 cm × 30 cm × 24 cm (width × height × depth). The extendable stand can be extended to 148 cm. The total weight without a battery is about 10 kg, whereas it is about 12.8 kg with a battery.

The operation web interface of the proposed system is shown in Figure 4, which includes the following functions: user login, real-time screen, control area setting, fine-tuning button, undo setting, clear area, status, enabling/disabling controls, showing alarm status, modifying the location name, modifying the equipment name, setting Line notifications, system log, and shutdown.

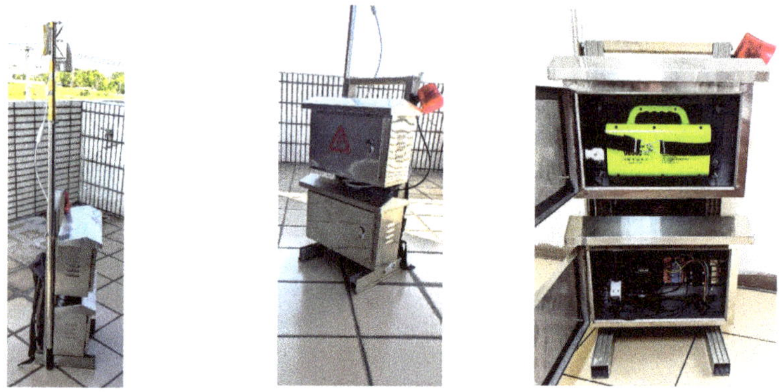

(**a**) (**b**) (**c**)

Figure 3. Photo of the SmartCon implementation. (**a**) Left-side view, (**b**) front view, (**c**) inside view.

Figure 4. Web interface of SmartCon.

Users need to log in with a password to access the web interface. The user can monitor the real-time IP camera image in the interface and set the range of the control area. After enabling the control state via a one-click button, the workers are not allowed to enter the preset control area. If any workers enter the control area, the alarm is set, and the Line notification is sent immediately.

4.2. AI Model Selection

The AI recognition module refers to the preset usage situation. We set the camera resolution to 1920×1080, and set the camera at a height of 10 m to overlook the ground for actual testing. During the test, a video is recorded in this environment. The video has 495 image frames recording one worker passing through the control area. Figure 5 is a screenshot of the two experimental site recognition results. The recognized object (person) is displayed on a frame line with a confidence value (the larger the value, the more similar the object, a worker, is).

(a) **(b)**

Figure 5. Experimental site recognition result. (**a**) Site 1, (**b**) site 2.

In this test, various YOLO models were used for the object recognition test, and the confidence value was set to be greater than 0.7 before the recognition result was passed. The recognition rate of the model for this environment was calculated through the inference results of each model.

The actual test data results in Table 1 show that the yolov4-608 (FP16) model obtained the highest recognition rate, and the subsequent system deployment is based on yolov4-608 (FP16).

Table 1. Evaluation of Recognition Precision and FPS.

Model Name: Version-Pixel Size	**Recognition Precision**	**Frame Per Second (FPS)**
yolov3-tiny-288	11/495 = 0.022	26.54
yolov3-tiny-416	41/495 = 0.082	20.40
yolov3-288	157/495 = 0.317	7.56
yolov3-416	266/495 = 0.537	4.70
yolov3-608	290/495 = 0.585	2.40
yolov3-spp-288	181/495 = 0.365	7.80
yolov3-spp-416	113/495 = 0.228	4.68
yolov3-spp-608	294/495 = 0.593	2.45
yolov4-tiny-288	28/495 = 0.050	26.72
yolov4-tiny-416	81/495 = 0.163	20.40
yolov4-288	246/495 = 0.496	7.56
yolov4-416	298/495 = 0.602	4.46
yolov4-608	**321/495 = 0.648**	**2.30**
yolov4-csp-256	42/495 = 0.084	11.80
yolov4-csp-512	269/495 = 0.543	4.00
yolov4x-mish-320	176/495 = 0.355	4.60
yolov4x-mish-640	291/495 = 0.587	1.44

4.3. Power Consumption

The developed equipment is easy to carry and is not troubled by the power supply. Thus, a large-capacity battery of 96,000 mAh is set up to power the system's operation. After the system starts to detect, the maximum wattage detected per hour is 23.5 W. We estimated that the battery could run the system for about 13 h. The battery can run the system for 13.06 h in theory. In the field test, the SmartCon could be used for about 10 h.

4.4. System Evaluation

Table 2 shows the evaluation results of the two sites. In this experiment, we calculate the accuracy with only the alarm triggered by the SmartCon when workers enter the control area. For the alarm trigger accuracy, it reaches over 98% in the threshold of confidence between 0.5 to 0.7. In this test, 0.5 is better for the alarm trigger accuracy. The Jetson Nano® development board has been optimized as systematically as possible to maximize its benefits. The average response time is 0.53 s in Jetson Nano®. Without cost limitations, it could recognize larger or more real-time images if upgraded to higher-end hardware (for example, Jetson AGX Xavier). The price will undoubtedly rise along with the upgrade.

Table 2. Evaluation of response time and alarm trigger accuracy.

Experimental Site	Threshold	Average Response Time (s)	Alarm Trigger Accuracy
Site 1	0.7	0.53	100.00%
Site 1	0.6	0.53	100.00%
Site 1	**0.5**	**0.53**	**100.00%**
Site 2	0.7	0.53	98.80%
Site 2	0.6	0.53	99.60%
Site 2	**0.5**	**0.53**	**100.00%**

4.5. Comparison

We also developed a previous device version with comparable features to the one developed in this study. To examine the differences between the previous and current versions, we compared the outcomes of this research and the previous systems. The comparisons are shown in Table 3. Figure 6 shows the previous edge device without computing. All of the image recognition must be recognized in the remote server (with GPU). The previous version device was developed with a server and client system (cloud-based) architecture, so the device is hard to install and store. Even if the detection speed is better than the current version, the connection response becomes slow when the internet connection is unstable.

Table 3. Comparison of Previous and New-version Devices.

Items	Previous Version	Current Version
System architecture	Server and client (compute in server)	AIoT (compute in edge, Jetson)
Hardware architecture	Tripod stands with a case, but needs to set up the server host	All-in-one
Ease of installation and storage	Hard	Easy
Detection speed	Normal	Slow
Connection response	Slow	Fast
Water resistance	Not Available	Yes
Heat dissipation	Not Available	Yes
Battery capacity	53,600 mAh	96,000 mAh

By contrast, the current version has advantages because it is developed with edge computing. The AI algorithm can be processed at the edge, so the overall device is designed as all-in-one, which is convenient for installation and storage. Table 3 shows that the new design is more portable and convenient with water resistance and heat dissipation, making the equipment more suitable for construction management environments.

Figure 6. Photo of the previous-version device.

5. Conclusions

The proposed SmartCon offers an effective tool for improving the safety monitoring of construction site operations by a real-time response to detecting potential risks, according to the findings of two empirical field tests.

This experiment was conducted on a real-world construction site, with Earthpower Construction providing a site for us to finish the experiment. The cost of hardware parts was about USD 1600 excluding assembly, software system installation, maintenance, and repair costs. In the future, the price of hardware will become lower. If the user uses a chip with better performance to achieve a more accurate or faster recognition effect, the cost of the hardware will increase to upgrade the level of the Jetson development board.

We hope to deploy the SmartCon to more construction sites to obtain more test results in future research. Furthermore, we also plan to complete the detection of helmets and vests to better ensure the safety of workers in the monitoring area to make the system perfect.

Author Contributions: Conceptualization, W.-D.Y. and H.-C.L.; methodology, H.-C.L. and J.-W.L.; software, J.-W.L.; validation, J.-W.L. and Z.-Y.L.; formal analysis, J.-W.L. and H.-C.L.; investigation, W.-D.Y.; resources, W.-D.Y.; data curation, Z.-Y.L. and W.-T.H.; writing—original draft preparation, Z.-Y.L.; writing—review and editing, W.-D.Y., H.-C.L., J.-W.L., Z.-Y.L. and W.-T.H.; visualization, W.-T.H.; supervision, W.-D.Y. and H.-C.L.; project administration, W.-D.Y.; funding acquisition, W.-D.Y. and H.-C.L. All authors have read and agreed to the published version of the manuscript.

Funding: This research project was funded by the Institute of Labor, Occupational Safety and Health, Ministry of Labor, Taiwan, under project No. 1110051; and by the Ministry of Science and Technology, Taiwan, under project No. MOST 111-2221-E-324-011-MY3.

Institutional Review Board Statement: Not applicable.

Informed Consent Statement: Not applicable.

Data Availability Statement: Data is unavailable due to privacy.

Acknowledgments: Sincere appreciation is given to the sponsor by the authors. The research team would also like to express our sincere appreciation to Earthpower Construction for providing experimental sites for this research.

Conflicts of Interest: The authors declare no conflict of interest.

References

1. Summary Analysis of the Preliminary Statistical Results of the 2018, Industrial and Commercial and Service Industry Census, Executive Yuan, Taiwan, 2018. Available online: https://www.stat.gov.tw/public/Attachment/8427181926BYT9Y7B1.pdf (accessed on 1 May 2022). (In Chinese)
2. TOSHA, 2010 Labor Inspection Annual Report, Occupational Safety and Health Administration, Ministry of Labor, Executive Yuan, Taiwan, 2021. Available online: https://www.osha.gov.tw/1106/1164/1165/1168/34345/ (accessed on 1 May 2022). (In Chinese)

3. TIOSH, Work Safety Alert Mobile Crane Safety, Institute of Labor, Occupational Safety and Health, Ministry of Labor, Taiwan, 2022. Available online: https://www.ilosh.gov.tw/menu/1169/1172/%25E7%25A7%25BB%25E5%258B%2595%25E5%25BC%258F%25E8%25B5%25B7%25E9%2587%258D%25E6%25A9%259F%25E5%25AE%2589%25E5%2585%25A8/ (accessed on 1 May 2022). (In Chinese)
4. The Number One Culprit of Occupational Accidents-Mobile Cranes, Occupational Safety and Health Newsletter No. 125, 2014. Available online: https://www.ilosh.gov.tw/media/2232/f1416792627939.pdf (accessed on 1 May 2022). (In Chinese)
5. TOSHA, Safety Rules for Lifting Equipment, Laws and Regulations Database of the Republic of China (Taiwan), 2022. Available online: https://law.moj.gov.tw/LawClass/LawAll.aspx?pcode=N0060013 (accessed on 1 May 2022). (In Chinese)
6. Russell, S.; Norvig, P. *Artificial Intelligence, a Modern Approach*; Prentice Hall: Upper Saddle River, NJ, USA, 2010; ISBN 0132071487.
7. Taipei City Government, "Smart Construction Site for Social Housing-Demonstration Project for Application of Internet of Things Safety Management", Taipei City, 2022. Available online: https://smartcity.taipei/projdetail/241 (accessed on 1 May 2022). (In Chinese)
8. CTCI, "Application of Safety and Health Technology in Engineering Industry", CTCI e-Newsletter, No.470, 2022. Available online: https://www.ctci.com/e-newsletter/CH/470/sustainable-future/article-01.html (accessed on 1 May 2022). (In Chinese)
9. Taisei Corporation, "Expanding the Functions of the Site Management System "T-iDigital Field" That Utilizes Digital Data," 2022. Available online: https://www.taisei.co.jp/about_us/wn/2020/201223_5058.html (accessed on 1 May 2022). (In Japanese)
10. Gigasense CSM Crane Safety Monitor, 2022. Available online: https://www.gigasense.se/product/crane-safety-monitor/ (accessed on 1 May 2022).
11. Baima Technology Presented at the "Crane Safety Monitoring and Management System", Baimatech, 2022. Available online: https://en.baimatech.com/baima-crane-safety-monitoring-management-system.html (accessed on 1 May 2022).
12. Nath, N.D.; Behzadan, A.H.; Paal, A.H. Deep Learning for Site Safety: Real-Time Detection of Personal Protective Equipment. *Autom. Constr.* **2020**, *112*, 103085. [CrossRef]
13. Golcarenarenji, G.; Martinez-Alpiste, I.; Wang, Q. Machine-Learning-Based Top-View Safety Monitoring of Ground Workforce on Complex Industrial Sites. *Neural Comput. Appl.* **2021**, *34*, 4207–4220. [CrossRef]
14. Fang, Y.; Cho, Y.K.; Chen, J. A framework for real-time pro-active safety assistance for mobile crane lifting operations. *Autom. Constr.* **2016**, *72*, 367–379. [CrossRef]
15. Chen, J.; Fang, Y.; Cho, Y.K. Real-Time 3D Crane Workspace Update Using a Hybrid Visualization Approach. *J. Comput. Civ. Eng.* **2017**, *31*. [CrossRef]
16. Liu, P.; Chi, H.-L.; Li, X.; Guo, J. Effects of dataset characteristics on the performance of fatigue detection for crane operators using hybrid deep neural networks. *Autom. Constr.* **2021**, *132*, 103901. [CrossRef]
17. Lin, T.-Y.; Maire, M.; Belongie, S.; Hays, J.; Perona, P.; Ramanan, D.; Dollár, P.; Zitnick, C.L. Microsoft COCO: Common Objects in Context. In *Lecture Notes in Computer Science (Including Subseries Lecture Notes in Artificial Intelligence and Lecture Notes in Bioinformatics)*; Springer: Berlin/Heidelberg, Germany, 2014; pp. 740–755.

engineering proceedings

Proceeding Paper

A Preliminary Study on Taiwanese Sexual Rights from Medical and Legal Perspective †

Chun-Ming Shih [1], Jiin-Chyuan Mark Lai [2,*] and Ming-Yuan Hsieh [1,3,*]

1 Graduate School of Human Sexuality, Shu-Te University, Kaohsiung 82445, Taiwan; cmshih0901@stu.edu.tw
2 Department of Applied Foreign Languages, TransWorld University, Douliu 640302, Taiwan
3 Department of International Business, National Taichung University of Education, Taichung 403514, Taiwan
* Correspondence: marklai07@gmail.com (J.-C.M.L.); s22871101@stu.edu.tw or capwisely@mail.ntcu.edu.tw (M.-Y.H.)

† Presented at the IEEE 5th Eurasia Conference on Biomedical Engineering, Healthcare and Sustainability, Tainan, Taiwan, 2–4 June 2023.

Abstract: Under the influence of sexually conservative Chinese culture, sex has always been an untouchable issue for a long time. As "Sex" has been a long-standing social taboo, sex-related knowledge and discourses have become scarce. The knowledge of the majority of Taiwanese on sex has been obtained from banned books and videos such as porn videos. This made people have the worst knowledge and information on it. No matter how open people are, the ability to feel sexuality rights among sexuality, sexuality education, sexology, trans/Gender studies, and LesBiGay due to the sexual legal norms (SLNs) in Taiwan is always lacking. Therefore, not only can young Taiwanese people enjoy the thrill of riding on the sexual waves, but as time passes, old Taiwanese people are also supposed to enjoy themselves without any pressure from society. Therefore, we developed sexual education on Taiwanese sexuality rights from a legal perspective by integrating the four essential SLNs: (1) the Sexual Harassment Prevention Act was instituted for establishing the sources of law in the various sexuality interactive behaviors; (2) the Act of Gender Equality in Employment was instituted for providing sources of law to promote male and female sexual employment; (3) the Communicable Disease Control Act was instituted for constructing the sources of law to prevent the STDs; and (4) the Gender Equity Education Act was instituted for instituting the sources of law. There is still unfair and unbalanced sexual violence in the current society. Thus, a future study is required to analyze and identify the legal determinants of sexual rights. For this, empirical suggestions were made to form more efficient SLNs for a harmonious and diverse Taiwanese society.

Keywords: Taiwanese sexuality development; sexuality rights; sexual legal norms (SLNs)

Citation: Shih, C.-M.; Lai, J.-C.M.; Hsieh, M.-Y. A Preliminary Study on Taiwanese Sexual Rights from Medical and Legal Perspective. *Eng. Proc.* **2023**, *55*, 53. https://doi.org/10.3390/engproc2023055053

Academic Editors: Teen-Hang Meen, Kuei-Shu Hsu and Cheng-Fu Yang

Published: 5 December 2023

1. Introduction

Since 1987, martial law has been suspended due to the most powerful rise of various social movements and the long-repressed surging Taiwanese social energy. This developed tendency of social change has been driven by the economic boom in previous decades which again has changed interpersonal relationships and connections. With the realization of the social circumstances, diversified sexuality activities and products are available and gradually overflowing (not referring to adult sex toys, but all kinds of commodities associated with sexual desire as their symbolic values). The sexy body and open sexual remarks have been regarded as a new social phenomenon to attract wide public attention. However, it produced various anxieties and concerns in society. Significantly, the popular but authoritative foreign sexology reports such as the Kinsey sexology report caught the public's attention. Furthermore, the academic achievements and discourses of professional sexologists have set up new institutes of sexology and gender research in higher education [1]. Under the influence of conservative Chinese culture, sex has always been taboo, so sex-related knowledge and discourses have become scarce. The sexual knowledge of the majority of Taiwanese people has been obtained from the circulated forbidden books and videos such as porn videos. This induced the worst sexual knowledge and information. Therefore, with the awareness and acceptance of sexuality, the sexual revolution has been paid for by the public which increased demand for sex information and knowledge. The popularization and dissemination of sexuality reports have become important in the sexuality revolution, and deepen and normalize the practices in the sexuality revolution. In 1992, the Teacher Zhang Publishing House boldly issued the Chinese translation of "New Jinsai Sexology Report", which sold 100,000 copies in a year. At the same time, sex-related topics, themes, and writings have been dealt with by social media including popular magazines and TV programs, and a lot of columns were written about sex. Taiwanese publishers have even conducted self-made questionnaires to obtain knowledge of sexology to impress people to quickly produce local sexuality reports, articles, and papers to meet the Taiwanese public sexuality demands [2].

Since the end of the 1980s, experts and scholars who have spoken and published publicly on sexuality-targeted issues, and health professionals or physicians (urologists and gynecologists) have often talked about sex-related issues from their professional standpoints and perspectives of disease and hygiene. Regarding the exploration and experimentation of sexuality, the sexuality anxiety and pleasure brought about by the erotic frenzy were not represented. "Sex Mood Workshops" sprang up like mushrooms in Taiwan at the end of the 1990s to discuss the experience of women's sexual desires and development and love trends. To open up in a positive, where positive concerned women's sexuality development, taboos needed to be removed for the feelings of the Taiwanese public defenders' uneasiness to deal with eroticism. People were driven to talk about sex but did not consider sexually transmitted diseases (STDs) and sexual legal norms (SLNs). These emotions did not confront the challenge and issue of the sexual revolution. STDs and SLNs were regarded as a threat, which did not coincide with the professional position of public health or medical treatment. As a result, most experts and professionals only describe sexuality as dangerous and terrible activities and behaviors and regard abstinence as the main resolution without talking about how to protect one's health and create one's pleasure in sexual activities and behaviors. Intimidating words are used for STD prevention and SLN development [3].

This social atmosphere continues to equate sex with danger, making women lack cultural enjoyment of eroticism and avoid practicing due to fear and guilt. Thus, there are still questions about developing and mastering the body and desires, how to face the rising culture of lust, what to use to participate in related activities, and how to actively and positively meet the sexuality revolution. The answers to these questions have instigated feminist thought that requires active thinking for freedom from chronic repression and has continued for a long time. To discuss it simply, the sexuality revolution and openness have certainly created a demand for sexuality information and knowledge and asked 'Does the existing information and knowledge meet the interests and needs of male and female subjects?' The sexuality experts, professionals, and researchers have no overlap on these issues. Although experts such as doctors join the debate, they no longer continue the dialogue. On the contrary, such dialogues in social media have gradually increased the public's interest in and reflection on sexuality issues for the male and the female.

At the beginning of 2000, a series of sexual movements opened up space for sexology development that caused erotic discourses and social movements for public health in social media. Concretely, new sexual realities brought about by rapid social changes have constantly forced public health and sex education professionals to consider how to apply their knowledge to the policy. Therefore, erotic discourses have confronted public health and medical professionals.

However, the sex movement has uncovered the blind spots of sexuality, sexuality education, sexology, trans/Gender studies, and LesBiGay, as well as the indifference to human sexuality rights in society. Therefore, academic conferences were held to have different analytical perspectives and propose different points of view to enrich the research and thinking of "sexuality" regarding sexology to combine gender studies and homosexual studies (lesbian and gay studies), and research in interdisciplinary perspectives. As a result, since the Fourth International Conference on Sexuality Education, Sexology, Trans/Gender Studies, and LesBiGay Studies (Four Sex Symposiums) was hosted in 1996, the topics have been officially discussed by the public health, medical, and sex education professionals. Sexology was defined by Taiwanese academic legitimacy to announce related professional discourses for the first time in Taiwan. This conceptual view emphasized sexual norms and normality. The Sex/Gender Research Office of Central University, known for promoting sexual liberation, has actively collaborated with the Gender and Space Research Office of National Taiwan University and the Gender and Society of Tsinghua University according to the Taiwanese past cross-examination experience and public consent.

Significantly, the developed mainstream research of sexuality, sexuality education, sexology, trans/Gender studies, and LesBiGay often fails to consider and discuss the related issues using the theoretical presuppositions in various research directions such as human culture, society, history, and rights. The majority of research topics regarding sexual preference have been shunned. The reason was that academic sexology focusing on indoctrination and a normal/abnormal distinction created more anxiety and reluctance to recognize and respect differences, and inhibited thought about erotic justice. How social restrictions for erotic justice are allocated on sexual issues regarding the topics is also lacking in pertinence to the ever-growing sexual issues.

With the awareness of human rights, most researchers and experts have paid more attention to sexuality rights and researched the scope and topics of sex research, the importance of sex research, the pragmatic presentation of sex, the possibility of sex research, the effect of conservative ethics and outdated laws on the research, academic autonomy, and the diversified philia and sexual preference link case related to the academic legitimacy of sexology and sexuality.

Foremost, the sex rights movement and sexology research are facing the changing social reality: various new legislations and enforcements (SLNs) in Taiwan still limit sex research in society with taboos (especially age and gender stereotypes). The new legislation (SLNs) includes the Sexual Harassment Prevention Act as the jurisdiction of the Ministry of Health and Welfare for establishing the sources of law in the various sexuality interactive behaviors (in 2009), the Act of Gender Equality in Employment as the jurisdiction of the Ministry of Labor for providing sources of law to promote male and female sexual employment (in 2016), The Communicable Disease Control Act as the jurisdiction of the Ministry of Health and Welfare for constructing the sources of law to prevent the STDs (in 2019), and the Gender Equity Education Act as the jurisdiction of the Ministry of Education for instituting the sources of law to develop sexual education (in 2022). The social space of sexuality imposes increasingly tight legal norms due to the rapid development of diversified sexuality. The circulation of various sexual speeches, articles, videos, and sexual information developed under the jurisdiction of the laws is inhibited. However, to "protect individuals from inappropriate information", new public opinions have been formed to "purify" the media, education, publishing, and academia. The development of these legalisms has further standardized the territory of sexology, and at the same time promoted the positive development of sexual rights [4].

Therefore, we preliminarily discussed the impact of these four new legislations and enforcements (SLNs) on Taiwanese sexuality rights development from a legal analytical perspective, as illustrated in Figure 1.

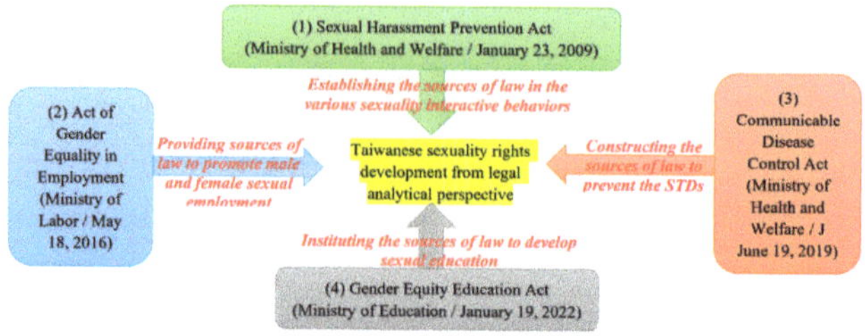

Figure 1. The four main Taiwanese SLNs.

2. Literature Review

To recognize the impact of the SLNs on the four legislations, the definition of sexual harassment and the handling and prevention of incidents need to be stipulated under the law. What is not regulated must be dealt with under other laws. Except for Articles 12, 24, and 25, the Gender Equality in Employment Act and the Gender Equality Education Act do not apply to the regulations of the present Act" [5]. In the Act of Gender Equality in Employment (2016), Article 1 reads, "The Act is enacted to protect gender equality in right-to-work, implement thoroughly the constitutional mandate of eliminating gender discrimination, and promote the spirit of substantial gender equality" [5]. In the Communicable Disease Control Act (2019), Article 1 is written as "To arrest the occurrence, infection, and spread of communicable diseases, this Act is specifically formulated"; Article 2: "Competent authorities" in this Act mean the Ministry of Health and Welfare at the central level; the municipality governments at the municipality level; and the county (city) governments at the county (city) level." [5]. In the Gender Equity Education Act (2022), Article 1 states "This Act is prescribed to promote substantive gender equality, eliminate gender discrimination, uphold human dignity, and improve and establish education resources and environment of gender equality" [5].

3. Evaluation Method

We comprehensively explored the impact of the sexual legal norms (SLNs) of the (1) Sexual Harassment Prevention Act, (2) Act of Gender Equality in Employment, (3) Communicable Disease Control Act, and (4) Gender Equity Education Act on the development of the Taiwanese sexuality rights from a legal perspective to view the Taiwanese sexual rights from the legal perspective as shown in Figure 2 [6–11].

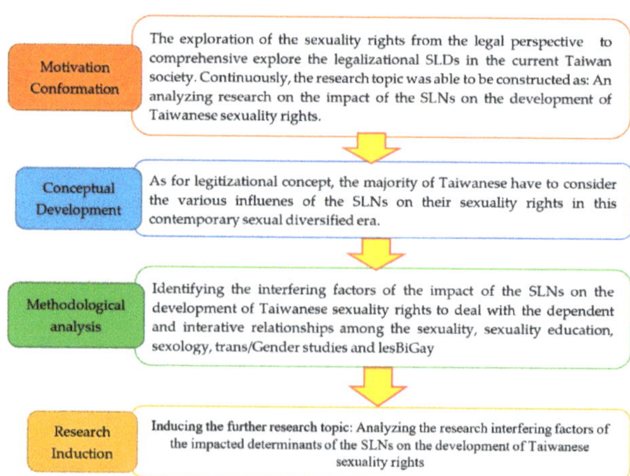

Figure 2. The research method.

4. Conclusions and Recommendations

No matter how open people are, the ability to enjoy sexuality rights for sexuality, sexuality education, sexology, trans/Gender studies, and LesBiGay is limited due to the SLNs in Taiwan. Therefore, young Taiwanese are on the verge of sexual waves, while old Taiwanese may enjoy their rights without any pressure from society. We integrated the four essential SLNs: (1) the Sexual Harassment Prevention Act was instituted for establishing the sources of law in the various sexually interactive behaviors; (2) Act of Gender Equality in Employment was instituted for providing sources of law to promote male and female sexual employment; (3) Communicable Disease Control Act was instituted for constructing the sources of law to prevent the STDs; and (4) the Gender Equity Education Act was instituted for instituting the sources of law to develop sexual education on the development of the Taiwanese sexuality rights from legal perspective. However, although sexuality rights have been positively developed owing to these essential SLNs, there is still unfair and unbalanced sexual violence in the current society [12–18]. Further study is necessary to analyze and identify the legal determinants of sexuality rights and propose effective suggestions to form efficient SLNs for a more harmonious and diverse Taiwanese society.

Author Contributions: Conceptualization, C.-M.S. and M.-Y.H.; methodology, M.-Y.H.; validation, J.-C.M.L.; formal analysis, M.-Y.H.; investigation, C.-M.S.; resources, J.-C.M.L.; writing—original draft preparation, M.-Y.H.; writing—review and editing, M.-Y.H.; visualization, C.-M.S.; supervision, M.-Y.H.; project administration, M.-Y.H.; funding acquisition, M.-Y.H. All authors have read and agreed to the published version of the manuscript.

Funding: This research was supplied by the research supporting research projects (MOST 110-2420-H-002-003-MY3-Y11209) of Ministry of Science and Technology and (NTCU112101) from the National Taichung University of Education.

Institutional Review Board Statement: This research did not execute questionnaires or interviews and, hence, this research did not require ethical approval.

Informed Consent Statement: Not applicable because this research did not involve any humans.

Data Availability Statement: No new data were created.

Conflicts of Interest: There are no conflict of interest in this research.

References

1. International Planned Parenthood Federation. Sexual Rights: An IPPF Declaration. Available online: https://www.ippf.org/sites/default/files/sexualrightsippfdeclaration_1.pdf (accessed on 31 October 2022).
2. Dewson, H.; Rix, K.J.; Le Gallez, I.; Choong, K.A. Sexual rights, mental disorder and intellectual disability: Practical implications for policy makers and practitioners. *BJPsych Adv.* **2018**, *24*, 386. [CrossRef]
3. Gomez, M.T.; Carlson, G.M.; Van Dooren, K. Practical approaches to supporting young women with intellectual disabilities and high support needs with their menstruation. *Health Care Women Int.* **2012**, *33*, 678. [CrossRef] [PubMed]
4. Pérez-Curiel, P.; Vicente, E.; Morán, M.L.; Gómez, L.E. The Right to Sexuality, Reproductive Health, and Found a Family for People with Intellectual Disability: A Systematic Review. *Int. J. Environ. Res. Public Health* **2023**, *20*, 1587. [CrossRef]
5. The Laws and Regulations Database of the Republic of China (Taiwan). Available online: https://law.moj.gov.tw/ENG/LawClass/LawAll.aspx?pcode=D0050074 (accessed on 1 March 2023).
6. Huang, C.C.; Chan, Y.-K.; Hsieh, M.Y. The Determinants of ESG for Community LOHASism Sustainable Development Strategy. *Sustainability* **2022**, *14*, 11429. [CrossRef]
7. Chan, Y.-K.; Hsieh, M.Y. An Empirical Study on Higher Education C-ESG Sustainable Development Strategy in Lower-Birth-Rate Era. *Sustainability* **2022**, *14*, 12629. [CrossRef]
8. Hsieh, M.Y. The Sustainable Development and Strategic Approaches for Contemporary Higher Education. *Sustainability* **2022**, *14*, 12925. [CrossRef]
9. Huang, C.-C.; Chan, Y.-K.; Hsieh, M.-Y. Preliminary Research on the Sustainable Determinants of Taiwanese Ecotourism with the International Standards. *Int. J. Environ. Res. Public Health* **2022**, *19*, 14489. [CrossRef] [PubMed]
10. Chan, Y.-K.; Hsieh, M.-Y.; Usak, M. A Concrete Study on Social-Media Connection of Global Literacy Abilities in MOOCs under the Dual Impacts of Lower Birth-Rate and COVID-19. *Sustainability* **2021**, *13*, 2203. [CrossRef]
11. Usak, M.; Hsieh, M.-Y.; Chan, Y.-K. A Concertizing Research on Making Higher Education Sustainability Count. *Sustainability* **2021**, *12*, 2724. [CrossRef]

12. Duran, M.; Usak, M.; Hsieh, M.-Y.; Uygun, H. A New Perspective on Pedagogical Content Knowledge: Intellectual and Emotional Characteristics of Science Teachers. *Rev. De Cercet. Si Interv. Soc. (RCIS) (Rev. Res. Soc. Interv.)* **2021**, *72*, 9–32. Available online: https://www.ceeol.com/search/article-detail?id=959021 (accessed on 21 March 2023). [CrossRef]
13. McCracken, H. The End of the Zero-Sum Game. 27 January 2011. Available online: http://technologizer.com/2011/01/27/the-end-of-the-zero-sum-game/ (accessed on 1 April 2023).
14. Greenwald, A.; Littman, M.L. Introduction to the special issue on learning and computational game theory. *Mach. Learn.* **2007**, *67*, 3. [CrossRef]
15. Wu, T.-L.; Hsieh, M.-Y.; Min, K.-W.; Yu, M.-T.; Ho, C.-T. Use of Sensor Technologies in Online Courses in Post-COVID-19 Era. *Sens. Mater.* **2021**, *33*, 2045–2062. [CrossRef]
16. Hsieh, Y.-M. Online learning era: Exploring the most decisive determinants of MOOCs in Taiwanese higher education. *Eurasia J. Math. Sci. Technol. Educ.* **2016**, *12*, 1163. [CrossRef]
17. Hsieh, Y.-M.; Usak, M. High Education Radical Transformation Era: How Teachers' Competency can Enhance the Students' Employability. *Rev. De Cercet. Si Interv. Soc. (RCIS)* **2020**, *68*, 95. Available online: https://www.ceeol.com/search/article-detail?id=848474 (accessed on 23 March 2023). [CrossRef]
18. Hsieh, Y.-M. Employing MCDM methodology to verify correlationship between social media and service quality in the dynamic m-commerce era. *J. Internet Technol.* **2018**, *19*, 225. Available online: https://jit.ndhu.edu.tw/article/view/1674 (accessed on 11 May 2023).

engineering proceedings

MDPI

Proceeding Paper

Prediction Model for Preoperative Diagnosis of Ovarian Cancer Using Tumor Markers, CBC, and LFT †

Sorawit Tongyib and Teerapol Saleewong *

Department of Mathematics, Faculty of Science, King Mongkut's University of Technology Thonburi, Bangkok 10140, Thailand; sorawit08@hotmail.com

* Correspondence: teerapol.sal@kmutt.ac.th

† Presented at the IEEE 5th Eurasia Conference on Biomedical Engineering, Healthcare and Sustainability, Tainan, Taiwan, 2–4 June 2023.

Abstract: The preoperative diagnosis of ovarian cancer (OC) was developed based on risk factor groups using secondary data. Binary and multiple logistic regression and its operating characteristic curve were used to analyze the data of risk factor groups for tumor markers, complete blood count (CBC), and liver function tests (LFT), respectively, and to explore potential predictors for each risk factor group. The data of 202 patients with ovarian cancer were analyzed in this research. As the tumor markers group, menopausal status, human epididymal protein 4, and cancer antigen 19-9 were included as the derivation of the preoperative diagnosis index. For the CBC group, menopausal status, lymphocyte count, and basophil cell ratio were used as predictors. Menopausal status, albumin, alkaline phosphatase, and indirect bilirubin were used as predictors for the LFT group. The area under the receiver operating characteristic curve (AUROC) for tumor markers, CBC, and LFT were 0.89 (95% CI, 0.845–0.935; sensitivity = 0.776, specificity = 0.919), 0.813 (95% CI, 0.755–0.871; sensitivity = 0.741, specificity = 0.767), and 0.81 (95% CI, 0.751–0.868; sensitivity = 0.664, specificity = 0.837), respectively.

Keywords: ovarian cancer; tumor marker; complete blood count; liver function tests; preoperative prediction

Citation: Tongyib, S.; Saleewong, T. Prediction Model for Preoperative Diagnosis of Ovarian Cancer Using Tumor Markers, CBC, and LFT. *Eng. Proc.* **2023**, *55*, 54. https://doi.org/10.3390/engproc2023055054

Academic Editors: Teen-Hang Meen, Kuei-Shu Hsu and Cheng-Fu Yang

Published: 5 December 2023

1. Introduction

Gynecological cancer is the most common cancer in women. However, its diagnosis is complicated because the cancer is found in the pelvis and diagnosed only by internal examination. Diagnosis may be delayed, which adversely affects the treatment of gynecological cancer because it is more effective at an early stage than at an advanced stage. Among gynecological cancers, ovarian cancer is the second most common, and the number one cause of death. The age-standardized incidence rates per 100,000 women were 7.1 and 5.8 for countries with high/very high Human Development Index (HDI) and low/medium HDI in 2020 [1]. Ovarian cancer does not show symptoms in its early stages often, causing patients to come to the doctor in the advanced stage. This results in a high mortality rate compared to other gynecological cancers. The important diagnostic method of ovarian cancer is to detect pelvic masses that are relatively hard and have a rough texture along with the presence of ascites. The stage of ovarian cancer is determined by the International Federation of Gynecology and Obstetrics (FIGO) system for describing how much cancer is in the body and determining how serious the cancer is. The initial treatment of patients at early stages is planned, while cytoreductive surgery is usually considered for patients at the advanced stage of ovarian cancer [2]. An important treatment can be decided based on a preoperative diagnosis for patients with high or low risk. Generally, patients at high risk must be referred to gynecologic oncologists for appropriate management.

Many indexes have been developed for preoperative diagnosis, such as the Risk of Malignancy Index (RMI) [3] and the Risk of Ovarian Malignancy Algorithm (ROMA) [4].

RMI is recommended by the American College of Obstetricians and Gynecologists (ACOG) as a tool for referring patients to gynecologic oncologists [2]. These indexes require the collection of demographic data, blood tests, morphological patterns, and biomarkers for prognostic purposes in the preoperative diagnosis. Therefore, several hospitals are unable to use these indexes to analyze patients due to a shortage of ultrasound specialists or gynecologists. Therefore, preoperative diagnosis indexes were developed based on tumor markers, complete blood count (CBC), and liver function tests (LFT), respectively, in women without pelvic or adnexal mass data.

2. Methods

This research was for developing preoperative diagnostic methods as a retrospective study using clinical data published by Mi et al. [5]. There were three groups of interesting risk factors for ovarian cancer: tumor markers, CBC, and LFT. The data of each group with different risk factors were analyzed to construct preoperative diagnosis indexes using multiple logistic regression (or binary logistic regression) with SPSS. The significant predictors were obtained based on the result of multiple logistic regression, and their predictive significance was assessed based on the diagnostic odds ratio and p-value. These significant predictors were then reanalyzed through regression until all variables became statistically significant to determine the logistic response function. After that, the Hosmer–Lemeshow goodness-of-fit statistics for logistic regression models was used to calibrate the models. The discriminative ability or predictive performance was assessed by calculating the area under the receiver operating characteristic curve (AUROC), which plotted the sensitivity against 1-specificity at various possible cut-off points. The optimal cut-off point was determined by considering the point on the receiver operating characteristic curve which was closest to the perfect cut-off point [6]. Furthermore, the sensitivity, specificity, positive predictive value, negative predictive value, and accuracy were calculated using a 2×2 table.

3. Results

The data were collected from the records of 202 patients to develop the preoperative diagnosis indexes [5]. The data contained information on the status, blood test results, and biomarkers of each patient, including 86 without ovarian cancer and 116 with ovarian cancer. The risk factors of ovarian cancer were divided into three groups: tumor markers, CBC, and LFT. Menopause was included in the analysis because postmenopausal patients are known to have a higher risk of developing epithelial cancer.

3.1. Logistic Regression Analysis: Tumor Marker Group

The preoperative diagnosis index based on tumor markers was created by using binary logistic regression analysis. The tumor markers consisted of the levels of tumor markers: human epididymis protein 4 (HE4) and carbohydrate antigen 19-9 (CA19-9). The details and parameters of tumor markers are presented in Table 1. The logistic response function of the tumor marker group was described as

$$P(\mathrm{OC}) = \frac{e^{\beta_0+\beta_1 Menopausal+\beta_2 HE4+\beta_3 CA19-9}}{1+e^{\beta_0+\beta_1 Menopausal+\beta_2 HE4+\beta_3 CA19-9}}, \quad (1)$$

where β_0 is the regression constant, $\beta_{1\leq i\leq 3}$ is the regression coefficient of each independent variable, and $0 \leq P(OC) \leq 1$; represents the probability of a patient with ovarian cancer.

Table 1. Parameters and their details for creation of preoperative diagnosis index of tumor marker group.

Parameter	Details
Menopausal status	$menopausal = \begin{cases} 1, postmenopausal \\ 0, premenopausal \end{cases}$
HE4	Level of human epididymis protein 4
CA19-9	Level of carbohydrate antigen 19-9

After conducting binary logistic regression based on tumor markers, the menopausal status, human epididymis protein 4 level, and carbohydrate antigen 19-9 level were included in the multiple logistic regression analysis. The model of the binary logistic regression was described as

$$P(\text{OC}) = \frac{e^{-3.867+0.623Menopausal+0.056HE4+0.008(CA19-9)}}{1+e^{-3.867+0.623Menopausal+0.056HE4+0.008(CA19-9)}}, \quad (2)$$

where the levels of HE4 and CA19-9 are measured in pmol/L and units/mL, respectively. The results of multiple logistic regression analysis are presented in Table 2. For discriminative ability, the area under the AUROC of the model of the tumor marker group was 0.89 (95% CI, 0.845–0.935), and the p-value of the Hosmer–Lemeshow goodness-of-fit test was 0.694. With the optimal cutoff point of 0.5, the sensitivity, specificity, positive predictive value, negative predictive value, and accuracy were 0.776, 0.919, 0.928, 0.752, and 0.837, respectively.

Table 2. Result of logistic regression analysis of preoperative diagnosis index of tumor marker group.

Parameter	Coefficient	OR	95% CL	p-Value
Menopausal status	0.623	1.865	0.682–5.104	0.225
HE4	0.056	1.058	1.030–1.086	<0.001
CA19-9	0.008	1.008	1.000–1.015	0.051
Constant	−3.867	0.021		<0.001

Abbreviations: OR = odds ratio; Cl = confident interval; Significant at $p < 0.05$.

3.2. Logistic Regression Analysis: CBC Group

For the binary logistic regression analysis for creating the preoperative diagnosis index of the CBC group, basophil count (BASO) and lymphocyte count (LYM) were included in the regression analysis. The logistic response function is as follows.

$$P(\text{OC}) = \frac{e^{\beta_0+\beta_1 BASO+\beta_2 LYM+\beta_3 Menopausal}}{1+e^{\beta_0+\beta_1 BASO+\beta_2 LYM+\beta_3 Menopausal}}, \quad (3)$$

where β_0 is the regression constant, $\beta_{1\leq i\leq 3}$ is the regression coefficient of each independent variable, and $0 \leq P(OC) \leq 1$; represents the probability of a patient having ovarian cancer. Table 3 presents information on the parameters of the index for the CBC group.

Table 3. Parameters and their details for creation of preoperative diagnosis index of CBC group.

Parameters	Details
Menopausal status	$menopausal = \begin{cases} 1, postmenopausal \\ 0, premenopausal \end{cases}$
BASO	Basophil cell ratio
LYM	Amount of lymphocyte

The model based on the result of binary logistics regression analysis included the data of BASO, LYM, and menopausal status where BASO and LYM were remeasured as their ratios to the number of white blood cells and $10^9/l$ of it. The detail of the parameters for multiple logistic regression analysis are shown in Table 4, and the logistic response function is as follows:

$$P(\mathrm{OC}) = \frac{e^{2.395-1.02BASO-1.283LYM+1.87Menopausal}}{1+e^{2.395-1.02BASO-1.283LYM+1.87Menopausal}}. \tag{4}$$

Table 4. Result of logistic regression analysis of preoperative diagnosis index of CBC group.

Parameter	Coefficient	OR	95% CL	p-Value
Menopausal status	1.870	6.48	0.682–14.018	<0.001
BASO	−1.020	0.361	0.143–0.911	0.031
LYM	−1.283	0.277	1.000–1.015	<0.001
Constant	2.395	10.698		<0.001

Significant at $p < 0.05$.

The AUROC of the model was 0.813 (95% CI, 0.755–0.871), and the model fitted the data well (p-value = 0.2 in the Hosmer–Lemeshow test). At the optimal cutoff point of 0.53, the sensitivity, specificity, positive predictive value, negative predictive value, and accuracy were 0.741, 0.767, 0.811, 0.688, and 0.753, respectively.

3.3. Logistic Regression Analysis: LFT Group

The data of the LFT group included albumin (ALB), alkaline phosphatase (ALP), and indirect bilirubin (IBIL). These parameters were used to create a preoperative diagnosis index by using binary logistic regression. The information for each parameter is shown in Table 5. The logistic function that describes the response of this case is expressed as

$$P(\mathrm{OC}) = \frac{e^{\beta_0+\beta_1 Menopausal+\beta_2 ALB+\beta_3 ALP+\beta_4 IBIL}}{1+e^{\beta_0+\beta_1 Menopausal+\beta_2 ALB+\beta_3 ALP+\beta_4 IBIL}}, \tag{5}$$

where β_0 is the regression constant, $\beta_{1\le i\le 4}$ is the regression coefficient of each independent variable, and $0 \le P(OC) \le 1$; represents the probability of a patient having ovarian cancer.

Table 5. Parameters and their details for creation of preoperative diagnosis index of CBC group.

Parameters	Details
Menopausal status	$menopausal = \begin{cases} 1, postmenopausal \\ 0, premenopausal \end{cases}$
ALB	The amount of albumin
ALP	The amount of alkaline phosphatase
IBIL	The amount of indirect bilirubin

In the binary logistics regression analysis, ALB, ALP, and IBIL, and menopausal status, the units of ALB, ALP, and IBIL are g/L, units/L, and μmol/L, respectively. The coefficient and detail of each parameter are shown in Table 6. The model is described as

$$P(\mathrm{OC}) = \frac{e^{2.606+1.919Menopausal-0.093ALB+0.02ALP-0.198IBIL}}{1+e^{2.606+1.919Menopausal-0.093ALB+0.02ALP-0.198IBIL}}. \tag{6}$$

Table 6. Result of logistic regression analysis of preoperative diagnosis index of LFT group.

Parameter	Coefficient	OR	95% CL	*p*-Value
Menopausal status	1.919	6.813	3.057–15.182	<0.001
ALB	−0.093	0.911	0.848–0.979	0.011
ALP	0.020	1.021	1.005–1.037	0.011
IBIL	−0.116	0.890	0.769–1.030	0.119
Constant	2.606	13.543		0.089

Significant at $p < 0.05$.

The AUROC of the receiver operating characteristic curve was 0.81 (95% CI, 0.751–0.868), indicating a good fit of the model to the data (*p*-value = 0.596). With an optimal cutoff point of 0.61, the sensitivity, specificity, positive predictive value, negative predictive value, and accuracy were 0.664, 0.837, 0.846, 0.649, and 0.738, respectively.

4. Discussion and Conclusions

In this study, three binary logistic regression models were developed for the preoperative diagnosis of ovarian cancer based on the data of tumor markers, CBC, and LFT. The model based on the data of tumor markers showed that HE4 and CA19-9 levels were significant predictive parameters for the prediction of ovarian cancer, while the HE4 level was presented the highest odds ratio. Menopausal status, BASO, and LYM were significant predictive parameters in the preoperative diagnosis based on the CBC data. For the preoperative diagnosis based on the data of LFT, menopausal status, ALB, ALP, and IBIL were the significant parameters. Menopausal status had the highest odds in the CBC and LFT groups. The discriminative performance of each model (Figure 1) was evaluated with the AUROC of the logistic regression models for the tumor marker, CBC, and LFT groups. The model of the tumor marker group had the largest AUROC, indicating better diagnostic performance than that of the CBC and LFT groups. The diagnostic indexes of the models for each group are compared in Table 7.

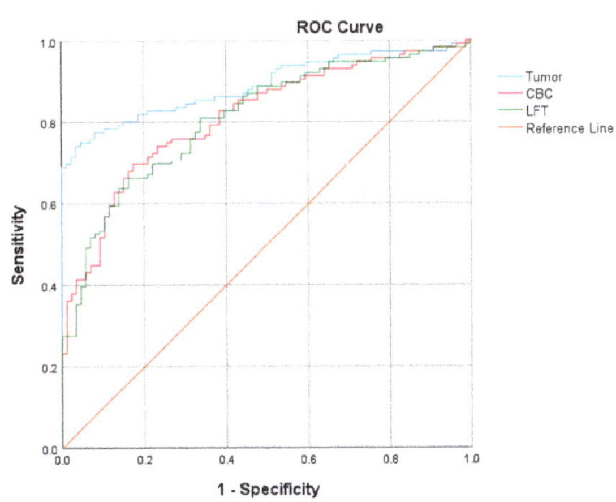

Figure 1. AUROC of logistic regression models of the tumor marker, CBC, and LFT groups.

Table 7. Comparison of AUROC curves logistic regression models of the tumor marker, CBC, and LFT groups.

Model	Hosmer and Lemeshow	AUROCC	95% CI
Tumor marker group	Appropriate	0.890	0.845–0.935
CBC group	Appropriate	0.813	0.755–0.871
LFT group	Appropriate	0.810	0.751–0.868

It was found that the likelihood of ovarian cancer in patients decreased as the levels of BASO and LYM increased. Therefore, treatments for increasing the level of BASO and LYM are beneficial for treatment. The indexes of the model of the LFT group indicated that increasing levels of ALB, IBIL, and DBIL decreased the likelihood of having the disease.

The developed preoperative diagnosis indexes of ovarian cancer without pelvic or adnexal mass data can be used as a reference for the evaluation of patients presenting with ovarian tumors by physicians or gynecologists and assist in management planning and patient prioritization for surgery, potentially reducing surgical risks.

Author Contributions: Conceptualization, S.T. and T.S.; methodology, S.T. and T.S.; software, S.T.; validation, S.T. and T.S.; formal analysis, S.T.; investigation, S.T. and T.S.; resources, S.T. and T.S.; data curation, S.T.; writing—original draft preparation, S.T.; writing—review and editing, S.T. and T.S.; visualization, S.T.; supervision, T.S. All authors have read and agreed to the published version of the manuscript.

Funding: This research received no external funding.

Institutional Review Board Statement: Not applicable.

Informed Consent Statement: Not applicable.

Data Availability Statement: Data is available in a publicly accessible repository. The data presented in this study and supporting information are openly available in Mendeley Data at https://doi.org/10.17632/th7fztbrv9.11 [5].

Acknowledgments: The authors sincerely thank for Science Achievement Scholarship of Thailand and the Department of Mathematics, Faculty of Science, King Mongkut's University of Technology Thonburi for supporting us.

Conflicts of Interest: The authors declare no conflict of interest.

References

1. Sung, H.; Ferlay, J.; Siegel, R.L.; Laversanne, M.; Soerjomataram, I.; Jemal, A. Global cancer statistics 2020: GLOBOCAN estimates of incidence and mortality worldwide for 36 cancers in 185 countries. *CA Cancer J. Clin.* **2021**, 209–249. [CrossRef] [PubMed]
2. Chirdchim, W.; Wanichsetakul, P.; Phinyo, P. Development and Validation of a Predictive Score for Preoperative Diagnosis of Early Stage Epithelial Ovarian Cancer. *Asian Pac. J. Cancer Prev.* **2019**, *20*, 1207–1213. [CrossRef] [PubMed]
3. Jacobs, I.; Oram, D.; Fairbanks, J.; Turner, J.; Frost, C.; Grudzinskas, J.G. A risk of malignancy index incorporating CA 125, ultrasound and menopausal status for the accurate preoperative diagnosis of ovarian cancer. *Br. J. Obstet. Gynaecol.* **1990**, *97*, 922–929. [CrossRef] [PubMed]
4. Moore, R.G.; McMeekin, D.S.; Brown, A.K.; DiSilvestro, P.; Miller, M.C.; Allard, W.J.; Gajewski, W.; Kurman, R.; Bast, R.C.; Skates, S.J. A novel multiple marker bioassay utilizing HE4 and CA125 for the prediction of ovarian cancer in patients with a pelvic mass. *Gynecol. Oncol.* **2009**, *112*, 40–46. [CrossRef] [PubMed]
5. Data for: Using Machine Learning to Predict Ovarian Cancer. Available online: https://data.mendeley.com/datasets/th7fztbrv9/11 (accessed on 23 October 2020).
6. Pepe, M.S. *The Statistical Evaluation of Medical Tests for Classification and Prediction*; Oxford University Press: Oxford, UK, 2003.

engineering proceedings

MDPI

Proceeding Paper

Application of Voronoi Diagram to School Districts in Shizuoka Prefecture †

Kenji Sato * and Haruno Ishikawa

Department of Architecture, Faculty of Science and Technology, Shizuoka Institute of Science and Technology, 2200-2 Toyosawa, Fukuroi City 437-8555, Shizuoka Prefecture, Japan; ishikawa.haruno@sist.ac.jp

* Correspondence: sato.kenji@sist.ac.jp

† Presented at the IEEE 5th Eurasia Conference on Biomedical Engineering, Healthcare and Sustainability, Tainan, Taiwan, 2–4 June 2023.

Abstract: Comparing the distribution of school districts in Shizuoka Prefecture with a Voronoi diagram generated for school locations, the application of Voronoi diagrams to the field of urban planning is pursued. The original statistics have been released by the government in the form of GIS data. Data processing and analysis are carried out utilizing open-source libraries.

Keywords: school district; Voronoi diagram; urban studies; GIS

Citation: Sato, K.; Ishikawa, H. Application of Voronoi Diagram to School Districts in Shizuoka Prefecture. *Eng. Proc.* **2023**, *55*, 55. http://doi.org/10.3390/engproc2023055055

Academic Editors: Teen-Hang Meen, Kuei-Shu Hsu and Cheng-Fu Yang

Published: 5 December 2023

1. Background

Japan is suffering from depopulation, and Shizuoka Prefecture is not an exception to this tendency. For the last 30 years, the total number of elementary schools in Shizuoka Prefecture has decreased by approximately 10%. The loss of a school causes significant damage to the community. However, the reorganization or the relocation of schools is underway in many local municipalities. This has led to an inevitable metamorphosis in urban structure.

Here, the distribution of school districts is chosen as a case study. A school district is designated by the municipality, and all school children in that zone commute to that school. And the cluster of the school district fills up the entire land space of the municipality. In this configuration, a school can be seen as an attracter in each territory.

A similar configuration of a point and its territory is drawn as a Voronoi diagram [1] (Figure 1). A Voronoi diagram is known as a method of spatial subdivision of a plane based on the notion of "distance" between the two points. A set of distributed points creates a cluster of polygons, and the boundary of each polygon represents the territory belonging to the specific point.

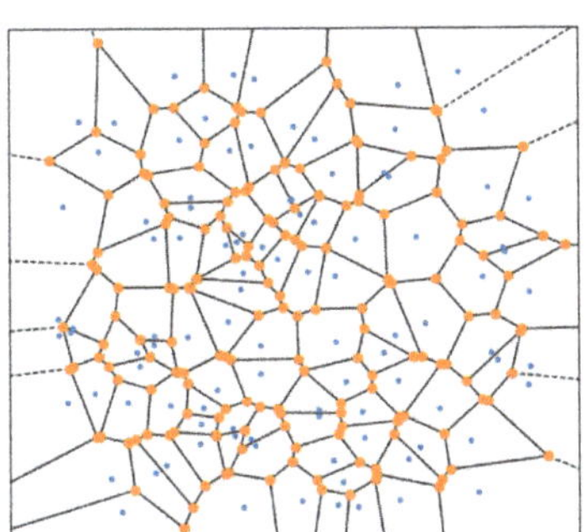

Figure 1. Voronoi diagram generated from 100 seeds.

The locations of elementary schools and each school district are accessible through the webpages of the Ministry of Land, Infrastructure, Transport and Tourism (hereinafter MLIT). Also, the population forecast until 2050 has been released by MLIT in the form of GIS data. By synthesizing these data, the distribution of the population in each school district is calculated and the future of elementary schools can be predicted. At the same time, drawing a Voronoi diagram based on the school locations and calculating the population in each Voronoi polygon reveal the effectiveness and the limitations of applying a Voronoi diagram as a method of urban planning.

2. Population Forecast until 2050

The GIS data of the population forecast until 2050 is downloadable from the website of MLIT [2]. The data include the population forecast on every 500 m × 500 m mesh grid over the entire land of Shizuoka Prefecture for each time span of 5 years until 2050. This forecast is carried out based on the national census of 2015, and it predicts the population by age and sex.

By writing a short Python script [3], the population forecast is summarized as shown in the bar chart below (Figure 2). Shizuoka Prefecture had a population of 3,615,586 in 2020, and it is predicted to decrease down to 2,791,692 in 2050. The decreasing rate is approximately 77% in 30 years. For this calculation, the Pandas and GeoPandas libraries are used to manipulate data tables.

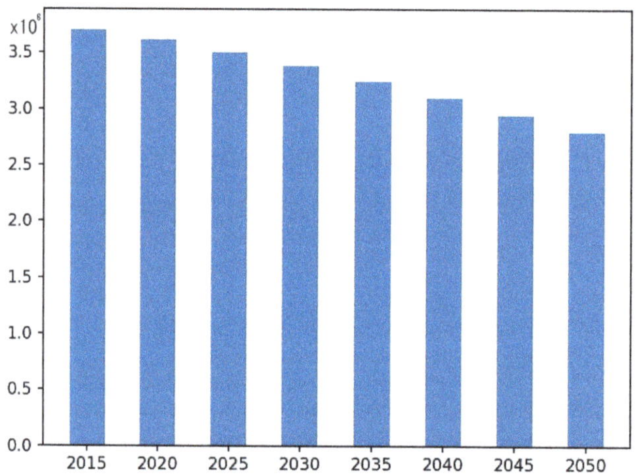

Figure 2. Population forecast in Shizuoka Prefecture until 2050.

The original dataset has population columns for every 5 years from 2015 to 2050 on every 500 m × 500 m mesh grid. It is visualized on a map using red-blue gradation (Figure 3). The reddish cells show dense areas with a population of around 10 to 40 per hectare. The most populated cell has a density of 128 per hectare, and is located in the southern part of Shizuoka City. On the contrary, in the mountain area, the cells are bluish and have a density of less than 5 per hectare.

The original data are described by age, and the population of school children, that is, between the ages 6 and 11, is calculated (Figure 4). In 2020, the total number of school children amounts to 181,434, while this figure decreases down to 122,458 in 2050. The decreasing rate is 67.4%. This figure is lower than the case of all ages, which seems to reflect the ongoing tendency toward an aging society.

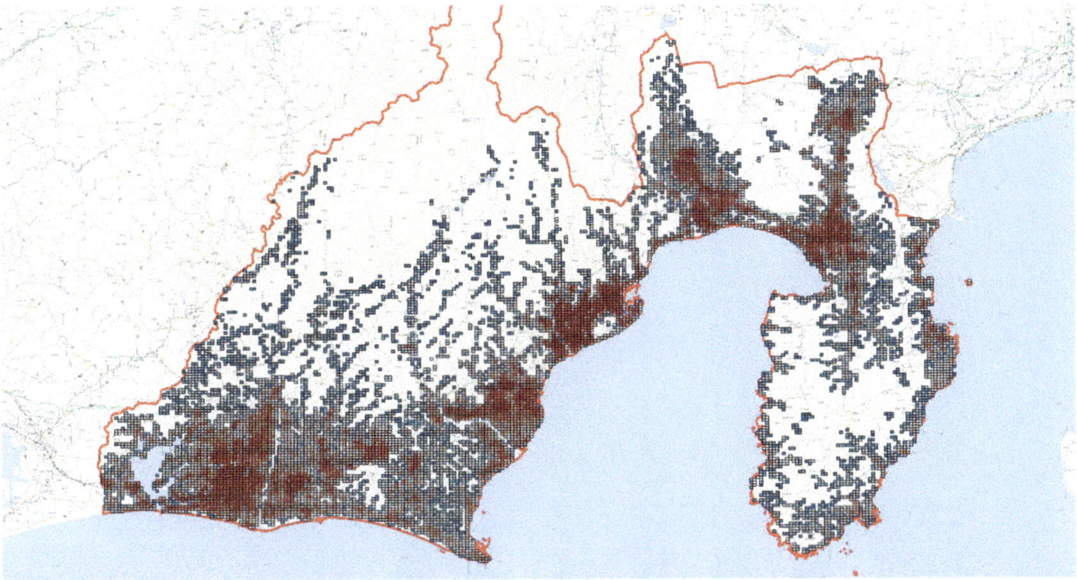

Figure 3. Population on 500 m × 500 m mesh grids in 2050, shown as red-blue gradation (red: dense, blue: sparse).

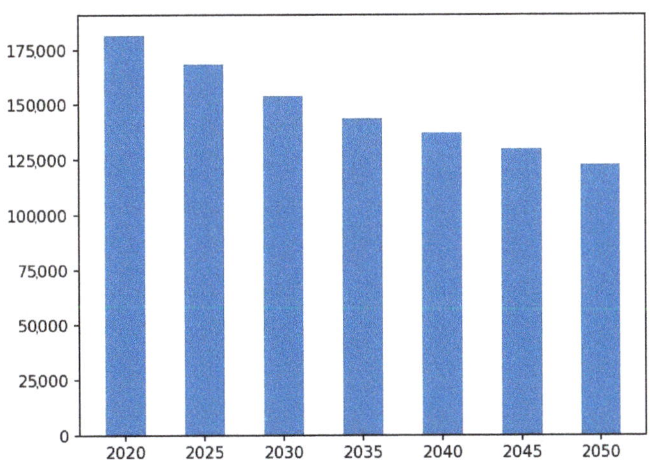

Figure 4. Forecast/number of school children.

3. Applying Population Forecast to School Districts

Combining the two datasets, the population forecast of school children on each 500 m × 500 m grid and the distribution of school districts, a diagram showing the chronological changes on each school district is obtained. The original GIS data of school districts are available on the website of MLIT. That dataset basically reflects the status as of 2021; however, some of the data have not yet been updated in some municipalities. For those cases, the data of school districts as of 2010 are combined with the dataset of 2021.

On the map below (Figure 5), altogether, 473 elementary schools within Shizuoka Prefecture are represented, whose locations are shown as blue points. Each school district is drawn as a closed polyline, and the inside of the region is filled with a red-colored gradation. This color represents the calculated population of school children in 2050 in each school district. The deeper the red, the lower the population becomes, while the whiter districts have more school children.

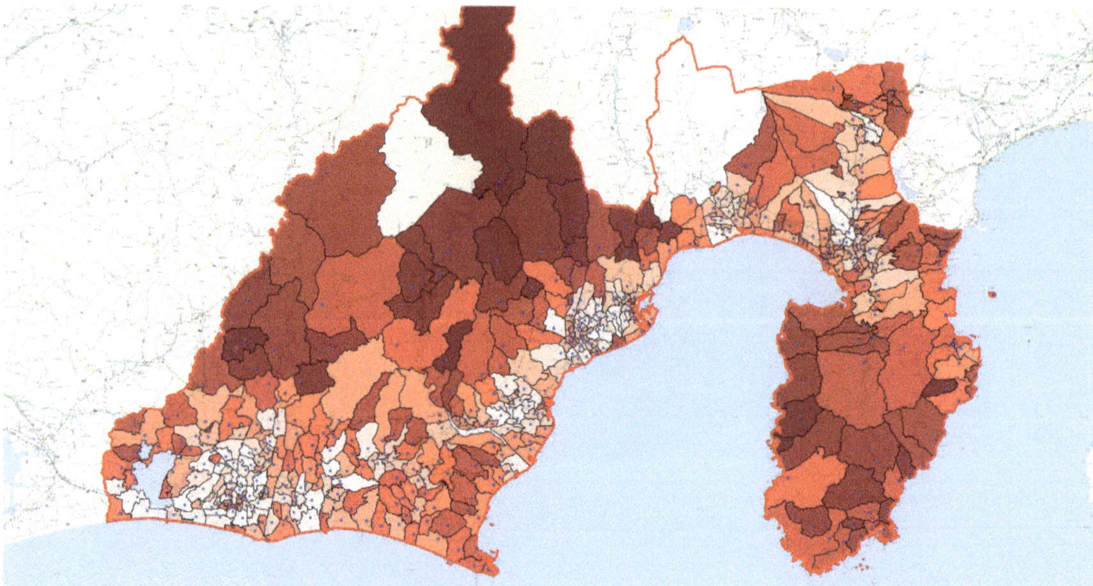

Figure 5. School districts and population (school children) in 2050.

Extracting the number of school children in 2050 in each school district and sorting the data in ascending order results in the following bar chart (Figure 6). There are 149 cases out of 473 samples that are predicted to have fewer than 120 school children in the year 2050. Suppose there are 20 students in each grade, and the school has 120 children altogether. If the number of children becomes smaller than this figure, it seems that it would be difficult to run that school any longer. This means that approximately 1/3 of the schools would need to be reorganized.

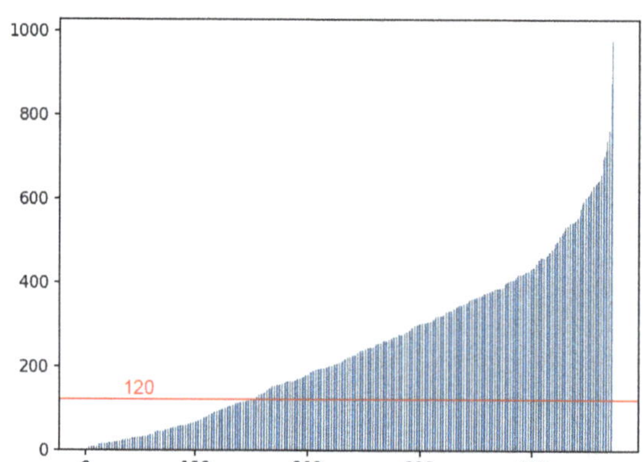

Figure 6. Number of school children in 2050 (473 samples).

The line graph below (Figure 7) shows the chronological changes in the decreasing rate of school children of the 473 samples in different colors.

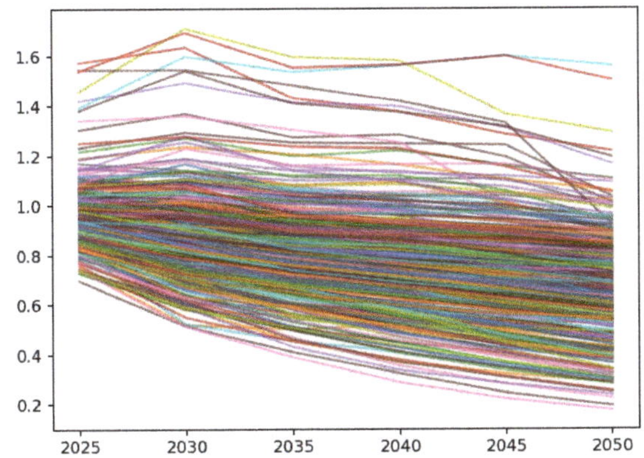

Figure 7. Decreasing rate of school children (473 samples).

4. Generating a Voronoi Diagram from the School Locations

Using QGIS [4], a visualizer of geospatial data, it is possible to create a Voronoi diagram from the dataset of school locations. Also, the geometric intersection of the Voronoi polygons and the bounding polygon of Shizuoka Prefecture are visualized using a Python script. For this operation, i.e., a 2-dimensional Boolean operation, a library called Shapely is used. The calculation of the number of school children on each Voronoi polygon is made through the same process as explained before in the case of school districts.

On the following map (Figure 8), the number of school children within each Voronoi polygon in the year 2050 is shown in red gradient colors. The deeper the red, the lower the population becomes. This is the same as in the case of Figure 5.

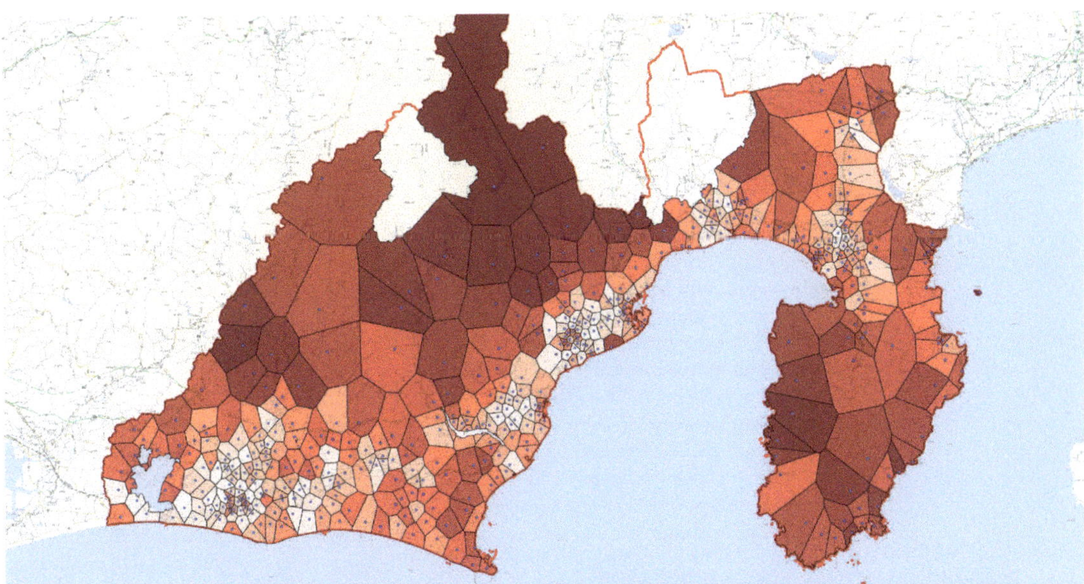

Figure 8. Voronoi diagram generated from the 473 location points of elementary schools.

The shape of each Voronoi polygon approximates the bounding lines of the corresponding school district. However, the actual school districts are determined based on their topographical and, sometimes, cultural characteristics. This causes a difference in shapes between the actual school districts and virtual Voronoi polygons.

To numerically compare the cases of actual school districts (Figure 5) and the Voronoi diagram (Figure 8), correlation coefficients are calculated. For this calculation, a Python script in the Pandas library is written. A correlation coefficient takes a value between −1.0 and +1.0, and the relationship between the two variables is stronger if the value becomes closer to +1.0.

Its result is shown in the bar chart below (Figure 9). Regarding the number of school children, the coefficients exceed 0.87 throughout each time span of 5 years. As to the coefficient related to the area of polygons, i.e., the square meterage of school districts and Voronoi polygons, the value is 0.81, and it seems appropriate to conclude that their relationship is relatively strong.

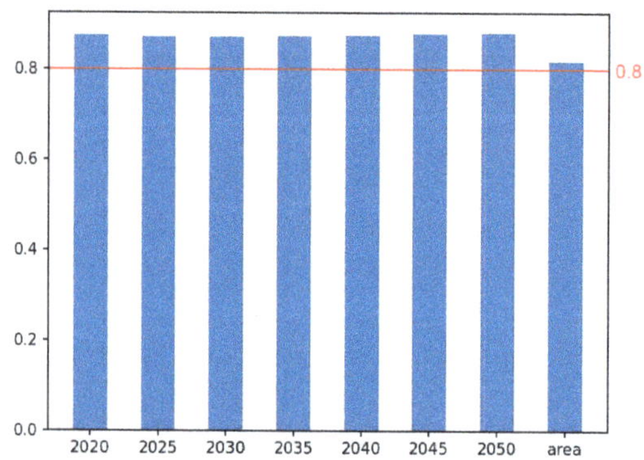

Figure 9. Correlation coefficients between school districts and Voronoi polygons.

5. Conclusions

Herein, the relationship of the actual school districts in Shizuoka Prefecture and the Voronoi regions generated from the school locations is determined and examined. They both fill up the entire land space, and geometrically, they both are configured as closed polylines on a map. Therefore, it seems quite natural that they show a strong correlation. However, when we apply the idea of the Voronoi diagram to various aspects in the field of actual city planning, its effectiveness and limitations need to be considered.

A Voronoi diagram theoretically subdivides the land into a cluster of polygons. However, it is mathematically drawn on an abstract plane, which is 2-dimensional and completely flat. On the contrary, in the real environment, mountains and rivers create natural topography, and human communities have been built on that basis. This seems to give the difference in the values of correlation coefficients.

On the other hand, there can be many scenes predicted where the method of the Voronoi diagram is applied to varied studies of urban planning. The operation of spatial subdivision forms the fundamental basis for planning. Its optimization has always been the central issue in urban planning.

A city can be viewed as a cluster of communities. Each community comprises a population of approximately 7000 to 10,000. Elementary schools have functioned as cultural and spiritual symbols of the community. The reorganization and consolidation of schools is inevitable under the current tendency of depopulation. The studies that have been conducted thus far can build a basis for planning the further reorganization of communities.

In this study, school districts are chosen as a case study. However, a real city can be conceived as a polymerized entity consisting of numerous types of community. The configuration of school districts is understood as one such example. In a real city, there exist multiple layers of territories with different sizes, which have been generated by public institutions like hospitals, post offices, museums, libraries, or even convenience stores. In the practice of urban planning, Voronoi diagrams can be an effective tool for architects and planners.

Author Contributions: Conceptualization, K.S.; methodology, K.S.; data curation, H.I.; analysis, K.S. and H.I.; writing—original draft preparation, K.S.; writing—review and editing, H.I. All authors have read and agreed to the published version of the manuscript.

Funding: This research received no external funding.

Institutional Review Board Statement: Not applicable.

Informed Consent Statement: Not applicable.

Data Availability Statement: All GIS data used in this study have been curated from the public archive of Ministry of Land, Infrastructure, Transport and Tourism. In this study, the following open-source libraries were used: Pandas (https://pandas.pydata.org/), GeoPandas (https://geopandas.org/en/stable/), Shapely (https://pypi.org/project/shapely/), scipy.spatial (https://scipy.org/), and matplotlib.pyplot (https://matplotlib.org/).

Conflicts of Interest: The authors declare no conflict of interest.

References

1. Sugihara, K. *Mathematical Models of Territories*; Kyoritsu Shuppan: Tokyo, Japan, 2009; Chapters 1–5; ISBN 978-4-320-01878-5.
2. The Original GIS Data Were Downloaded from the Website of MLIT. Available online: https://nlftp.mlit.go.jp (accessed on 31 May 2023).
3. Sato, K. *Learning AutoCAD for Architectural 2D and 3D Drawings*; Gakugei Shuppansha: Kyoto, Japan, 2021; Chapters 15 and 16; ISBN 978-4-7615-3270-3.
4. QGIS. Available online: https://qgis.org (accessed on 14 September 2022).

engineering proceedings

MDPI

Proceeding Paper

Influence of Interior Decorations on Indoor Air Quality in Fitness Centers †

Po-Yen Kuo *, Chiung-Yao Chen and Tzu-Yu Wu

Department of Architecture, Chaoyang University of Technology, Taichung 413310, Taiwan; chychen@cyut.edu.tw (C.-Y.C.); fsummer1027@gmail.com (T.-Y.W.)
* Correspondence: kuoboen@gmail.com
† Presented at the IEEE 5th Eurasia Conference on Biomedical Engineering, Healthcare and Sustainability, Tainan, Taiwan, 2–4 June 2023.

Abstract: In order to understand the air quality of fitness centers according to the Environmental Protection Administration's "Indoor Air Quality Standards", as well as to discuss how fitness center air-conditioning ventilation systems can effectively remove indoor air pollutants, this study focused on six Taichung fitness centers in a seven-sample air quality investigation, employing handheld precision instruments in space air testing and using linear regression analyses of the concentrations of indoor chemical pollutants, including factors such as temperature, relative humidity, CO_2, O_3, CO, CH_2O, TVOC, PM2.5, and PM10. The results of this research are as follows: (1) Quantity of indoor decorations: The decorative materials used in each of the sample spaces and their sources are not the same. Even with the same quantity of decorations, the concentrations of CH_2O and total volatile organic compounds (TVOCs) that escaped into the atmosphere were different across the samples, leading to a low correlation (R = 0.0316, R = −0.0976). Our findings on the influence of the fitness center's establishment date on the concentrations of formaldehyde (CH_2O) and total volatile organic compounds (TVOCs) that escaped into the indoor air indicate that this correlation is low and insignificant (R = −0.3598, R = −0.5523), but show that the indoor concentration of formaldehyde decreases with time. (2) Occupants' indoor activities: The CO_2 concentration generated by the static and dynamic activities of indoor occupants is not reflected in real time but will gradually accumulate, resulting in a moderate to low and insignificant correlation between the number of active occupants and the level of CO_2 (R = 0.4343). (3) The PM2.5 and PM10 sources of suspended particles are not only related to the external air and interior decoration materials, but also to coarse surfaces, which can easily attract dust accumulation. Therefore, materials made from fabric and artificial turf should be reduced in order to reduce dust accumulating on the materials' surfaces.

Keywords: indoor air quality; fitness center; ventilation; interior decorators

Citation: Kuo, P.-Y.; Chen, C.-Y.; Wu, T.-Y. Influence of Interior Decorations on Indoor Air Quality in Fitness Centers. *Eng. Proc.* **2023**, *55*, 56. https://doi.org/10.3390/engproc2023055056

Academic Editors: Teen-Hang Meen, Kuei-Shu Hsu and Cheng-Fu Yang

Published: 5 December 2023

1. Background and Purposes

The prevalence of national sports in Taiwan has influenced the increasing emergence of fitness centers in recent years. Expectedly, people have a preference for fitness centers that are convenient and meet their fitness/exercise/health requirements. To reduce the impact of the heavy weight of fitness center equipment on floors, many shock-absorbent mats are usually laid on the floors of fitness centers. Cabinets and walls also have many decorative materials on them. However, these interior decoration materials usually contain formaldehyde and other volatile organic matters. Hence, excessive decoration and improper construction methods in fitness centers lead to increased health risks and the possibility of chemical gas emissions. Moreover, the combination of closed interior spaces and large amounts of carbon dioxide produced after exercise can lead to the accumulation of indoor pollutants [1].

This study aimed to investigate the influence of decoration and usage patterns on indoor air quality in emerging fitness centers. The indoor air quality of six private fitness centers was measured in order to discuss the influence of indoor users, decorations, and air exchange rates on their indoor air quality. This study measured the air quality concentrations of all exercise periods to check whether they met the standards set by the Environmental Protection Administration and to establish their correlation with the measured air pollutant concentrations [2]. It also discusses the relationship between indoor air quality and various factors. The air exchange rates of the spaces that would meet the carbon dioxide concentration standard set by the Environmental Protection Administration were estimated using the mass balance equation, and are provided here for future design organizations or owners to estimate the pollution concentrations of their spaces in advance, so as to make provisions for good and healthy indoor air quality in these spaces. The purposes of this study are as follows:

(1) to investigate the influence of people's activities on the indoor air quality in fitness centers;
(2) to investigate the influence of interior decorations on the indoor air quality in fitness centers;
(3) to understand the rationality behind the air conditioning mechanisms currently operating in fitness centers;
(4) to evaluate the fitness centers' ventilation and propose the best operative method of ventilation according to indoor users' activities.

2. Theories and Methods

This study investigated the indoor air quality of fitness centers, mainly through onsite diachronic monitoring, supplemented by manual records of the number of indoor occupants. It focused on recording and discussing the indoor air quality in fitness centers, so as to understand the influence of indoor space utilization and indoor decorations on indoor air quality. This study can serve as a reference for important decisions on indoor air quality in the future.

2.1. Theory Regarding the Concentration of Carbon Dioxide (CO_2) Produced via the Metabolism of Indoor Occupants

The relationship between indoor air quality and the density of indoor occupants and the resulting different indoor air mixing efficiencies can be estimated using the calculation method of the "pollutant mass balance model" under the condition of an existing legal standard for the ventilation of an area [3–5]. This study explored the activities of dynamic and static people indoors, so the equation was modified as follows (Equation (1)):

$$C_i = C_o + \frac{S_i \times (\text{met}_{st} \times n + \text{met}_{dy} \times n) \times DF}{K \times Q_0} \quad (1)$$

Ci: indoor CO_2 concentration (ppm)

Co: outdoor CO_2 concentration (ppm)

Si: when the human metabolic rate reached 1 met, the amount of carbon dioxide produced was assumed as a fixed value of 21.6 (m^3/s)

met: metabolic equivalence, which was assumed as 1.2 for static people (st), and 6.0 for dynamic people (¬dy)

n: the number of indoor occupants

DF: people flow rate; the ratio of the maximum number of people in a day to the total number of people on that day for each sample

K: air mixing (0.5)

Q_0: external air volume (m^3/s)

2.2. Theory Regarding Indoor Ventilation

As an empirical coefficient, the air exchange rate is related to not only the nature of air-conditioning in the rooms but also many other factors, such as the room's size, height, location, and mode of air supply, as well as the degree of indoor air deterioration. The air exchange rate per unit hour was estimated using the ACH because it is related to the indoor pollution concentrations, so as to estimate the air exchange rate required to achieve a good air quality in each of the sample spaces (Equation (2)) [5], and serve as reference for owners and design organizations.

$$ACH = Q \times 60/V \tag{2}$$

Q: air flow (m^3/h)
V: indoor volume (m^3)

2.3. Defining the Quantity of Decorations

The "decoration quantity" of each space was measured according to the ISO 16000 standard and then defined based on the "Green Building Evaluation Manual—Basic Version (2015 edition)" (p. 100–104). Walls and ceilings decorated simply by painting, ceilings decorated with simple flat panels under fire-extinguishing pipelines, or ceilings decorated with simple lighting systems specified in the decoration quantity of a sample's basic structure were excluded from the decoration area in this study. Only the decoration area outside of that specified in the decoration quantity of each sample's basic structure was included in this study. However, among the study samples, only the ceiling of one sample was equipped with flat panels. Hence, this flat panel ceiling was included in the decoration calculation herein (Equation (3)) (Figure 1).

$$LF = \sum A_i/V \tag{3}$$

A_i: area of all indoor decoration materials and furniture (m^2)
V: indoor volume (m^3)

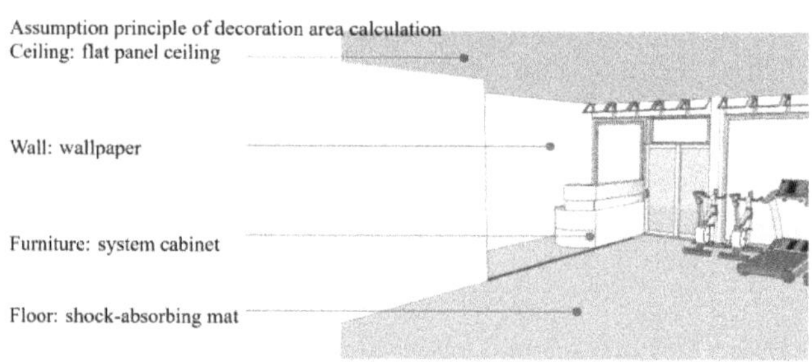

Figure 1. Calculation principle of the decoration area.

3. Measurement Process

3.1. Description of Measurement Method

In this study, portable precision instruments were used for sampling to investigate indoor air quality. These precision instruments can simultaneously detect humidity and six other items specified by the Environmental Protection Administration. They also display all detected data in real time. The concentrations of indoor pollutants specified by the Environmental Protection Administration, including CO, O_3, CH_2O, CO_2, VOC, PM10, and PM2.5, were measured, respectively. The instrument's specifications and accuracy are shown in Table 1. The detection points were set according to the "measures for the management of indoor air quality detection and measurement" of the Environmental Protection Administration. The instruments were placed in areas which could reflect the main of users' activities within their detection range and which tried to minimize the instruments' influence on the users within the tested area. They were positioned at a distance of at least 0.5 m from indoor hardware structures and facilities and at least 3 m from doors or elevators. The sufficiency of the sample size to represent the level of indoor air pollution in the detected area depended on the changes in the concentration levels of all pollutants. The floor area of each of the seven samples in this study was less than 500 m^2, and according to regulations, at least one sampling point should be selected for each sample. Because our instruments were limited and there was only one instrument of each kind, it was impossible to complete sampling in one day. Therefore, sampling was carried out separately, and the concentration levels were measured within the same week. The measurements were made during business hours, and this lasted for 3–10 h.

Table 1. Ranges and accuracy levels of the measuring instruments.

Measuring Instruments	Measured Gas	Measurement Range	Location
Portable air quality tester YesAir Datalog Session YA1807K00903	CO	0–50.00 ppm	Indoor
	O_3	0–01.00 ppm	
	CH_2O	0–05.00 ppm	
	CO_2	0–9999 ppm	
	VOC	0–30.00 ppm	
Direct-reading instrument for measuring suspended particles Aerocet- 831 Aerosol	Suspended particle size PM_1, $PM_{2.5}$, PM_4, PM_{10}	0–1000 μg/m^3	Indoor
Carbon dioxide tester TES-1370CO/CO_2	CO_2	0–6000 ppm	Outdoor

3.2. Basic Introduction of All Sample Spaces

According to the field investigation, the average area of all samples was 160 m^2. Their decoration situations were as follows: their walls and ceilings were mainly painted; their floors were mainly laid with shock-absorbent mats and artificial turf; their quantity of decorations varied from 0.03 to 0.42 m^2/m^3; their establishment dates were all within the last three years; and the measurement time was between 3 and 10 h (Table 2).

Table 2. Sample spaces overview.

Sample Code	Establishment Date (year)	Construction Area (m^2)	Interior Volume (m^3)	Decoration Rate (m^2/m^3)	Overview of Interior Decoration	Measurement Time
FL5	2017	198.35	595.05	0.42	Ceiling: basic paint Wall: basic paint, wall paper Floor: artificial turf, shock-absorbing mats	10:00–22:00
CY	2005	72.56	277.11	0.26	Ceiling: light steel joist ceiling Wall: basic paint, some wooden sideboards Floor: plastic flooring	17:00–20:00
FL4	2017	88.30	264.90	0.40	Ceiling: basic paint Wall: basic paint Floor: artificial turf, plastic flooring	10:00–12:00 14:00–15:00 19:00–20:00
IW	2018	145.00	580.00	0.28	Ceiling: basic paint Wall: basic paint, wall paper Floor: shock-absorbing mats	14:30–22:00

Table 2. *Cont.*

Sample Code	Establishment Date (year)	Construction Area (m^2)	Interior Volume (m^3)	Decoration Rate (m^2/m^3)	Overview of Interior Decoration	Measurement Time
HU	2019	294.31	1177.24	0.25	Ceiling: basic paint Wall: basic paint Floor: artificial turf, shock-absorbing mats	10:00–22:00
EF	2017	110.00	308.00	0.03	Ceiling: basic paint Wall: basic paint, some cabinets Floor: plastic flooring	10:00–12:00 14:00–20:00
PS	2018	211.80	868.38	0.24	Ceiling: basic paint Wall: basic paint Floor: shock-absorbing mats	10:00–12:30 14:30–21:00

3.3. Current Indoor Air Quality of All Samples

This study investigated indoor air quality from both physical and chemical perspectives by referring to the concentration levels specified in the 2012 "Indoor Air Quality Standards" from the Environmental Protection Administration. The diachronic measurement results are shown in Table 3. The average CO_2 concentration in CY was 2209 ppm, which was the highest, while the average CO_2 concentration in IW was 1492 ppm; both of these were higher than the standard of 1000 ppm recommended by the Environmental Protection Administration. The reasons for these excessive CO_2 concentrations were imperfect indoor ventilation and a large number of indoor occupants. The average concentrations of CO and O_3 in all sample areas were lower than the standards of 9.00 ppm and 0.06 ppm, respectively. The average CH_2O concentrations in all of the sample places were higher than the standard of 0.08 ppm, and the average VOC concentrations in CY, FL4, and IW were higher than the standard of 0.56 ppm, mainly because their decoration materials contained CH_2O and VOC and their indoor ventilation was perfect. The average concentrations of PM2.5 and PM10 in FL5, FL4, and EF were higher than the standards of 35 μg/m^3 and 75 μg/m^3. These excessive PM2.5 and PM10 concentrations in FL5 and FL4 were mainly attributed to improper use of interior floor materials and inadequate cleaning, and the EF concentration was mainly due to the external air.

Table 3. Diachronic measurement of the current indoor air quality of each sample.

Samples	CO ppm	O_3 ppm	CH_2O ppm	VOC ppm	$PM_{2.5}$ μg/m^3	PM_{10} μg/m^3
Standards	9.00	0.06	0.08	0.56	35	75
FL5	0.56	0	0.57 *	0.05	55.3 *	102.6 *
CY	3.51	0	1.36 *	0.65 *	6.2	26.0
FL4	0.03	0	0.51 *	0.66 *	35.9 *	82.3 *
IW	2.41	0	0.91 *	0.57 *	8.6	28.2
HU	0.79	0	0.62 *	0.20	21.2	34.6
EF	0.41	0	0.56 *	0.17	95.1 *	188.7 *
PS	0.07	0	0.37 *	0.09	25.0	61.6

Table 3. *Cont.*

Samples	CO_2 ppm	Number of Static People	Number of Dynamic People	Decoration Rate m^2/m^3	Indoor People Flow Rate
Standards	1000	-	-	-	-
FL5	776	6	6	0.42	0.63
CY	2209 *	3	16	0.26	1.00
FL4	519	1	4	0.40	0.26
IW	1492 *	6	11	0.28	0.89
HU	951	0	11	0.25	0.57
EF	563	10	8	0.03	0.94
PS	573	3	8	0.24	0.57

* indicates that the result is higher than the standard recommended by the Environmental Protection Administration.

3.4. Relationship between the Indoor Air Quality and Various Factors

This section discusses the CO_2 concentration of IW and the suspended particles of FL5, which both exceeded the standards of the Environmental Protection Administration. This discussion is based on the data presented in Table 3. According to our diachronic measurements, the CO concentrations in IW and EF5 were lower than the Taiwan Environmental Protection Administration standard of 9 ppm.

- Carbon dioxide (CO_2)

The number of users in a space influences its CO_2 concentration. In the FL5, CY, IW, and HU sample spaces, after 17:00, the CO_2 concentration exceeded the Environmental Protection Administration standard of 1000 ppm due to the increasing number of users after that time. Also, due to the awful indoor ventilation in the CY sample space, the concentration accumulated to 3680 ppm within a short time. In the other sample spaces (i.e., FL5, IW, and HU), due to the large number of people there at night and poor ventilation, the CO_2 accumulated and its concentration exceeded the standard (Figure 2).

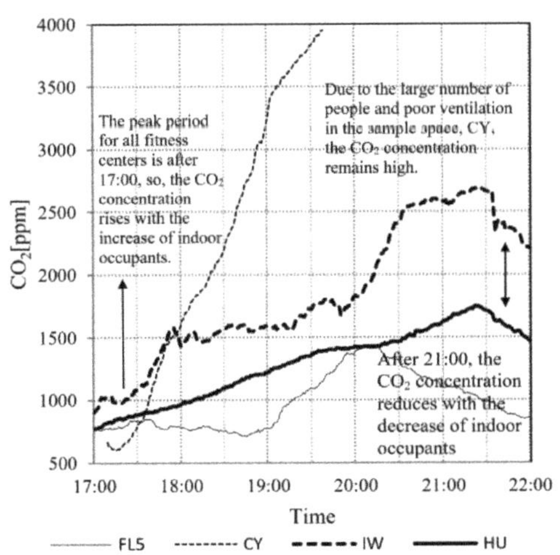

Figure 2. Diachronic monitoring of variations in the CO_2 concentration across the samples.

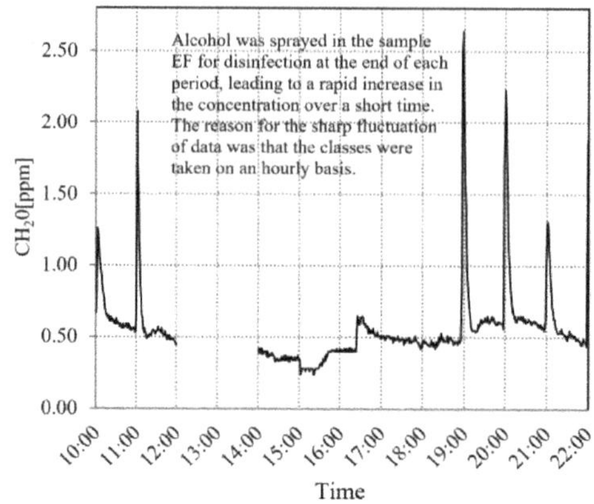

Figure 3. Diachronic monitoring of variations in the formaldehyde (CH_2O) concentration across the samples.

- Formaldehyde (CH_2O)

The CH_2O concentrations in all samples exceeded the standard of 0.08 ppm given by the Environmental Protection Administration, due to CH_2O in the decoration materials. The tested samples were all new fitness centers established within the last four years; therefore, the CH_2O in these spaces had not yet been dispersed, thus resulting in excessive CH_2O concentrations. For example, in sample EF (Figure 3), not only did the building materials release formaldehyde, but also, spray cleaners caused the CH_2O concentration to rise sharply within a short time. The reason for this sharp fluctuations of data in the EF sample within certain periods was that this fitness center's classes were taken on an hourly basis, and students entered and leaved the space on the hour.

- Total volatile organic compounds (TVOCs)

The total volatile organic compounds (TVOCs) were mainly found in building materials, furniture, and cleaning materials. Their concentrations in the CY and IW samples occasionally exceeded the Environmental Protection Administration standard of 0.56 ppm. The air conditioning in IW kept this sample under a constant temperature and humidity, but the total volatile organic compounds (TVOCs) in this space increased gradually, because the humidity rose slightly with increases in the number of indoor occupants (Figure 4). As mentioned in the literature review in Section 2, the diffusion coefficient of total volatile organic compounds increases under high humidity (Kun-Chih Huang: 'A study on the influence of indoor thermal environment variation on adsorption and reduction of formaldehyde from building materials', 2011). In sample CY, the awful indoor ventilation led to a serious accumulation of indoor pollutants.

- Suspended particles (PM2.5, PM10)

The suspended particles (PM2.5, PM10) mainly arose from the accumulation of indoor pollutants and the external air. The suspended particle (PM2.5, PM10) concentrations in the FL5, FL4, and EF samples exceeded the Environmental Protection Administration standards of 35 $\mu g/m^3$ and 75 $\mu g/m^3$, respectively. The excessive suspended particle concentrations in samples FL5 and FL4 can be attributed to the improper use of indoor floor materials (plastic turf that is prone to dirt accumulation was used), which caused the pollutants to release dust as people moved around; the concentrations rose as the number of indoor users increased (Figures 5 and 6). The excessive suspended particle concentrations in sample EF can be attributed to the external air. This sample space was naturally ventilated all day long, leading to an excessive suspended particle (PM2.5, PM10) concentration, because the polluted outdoor air entered the room.

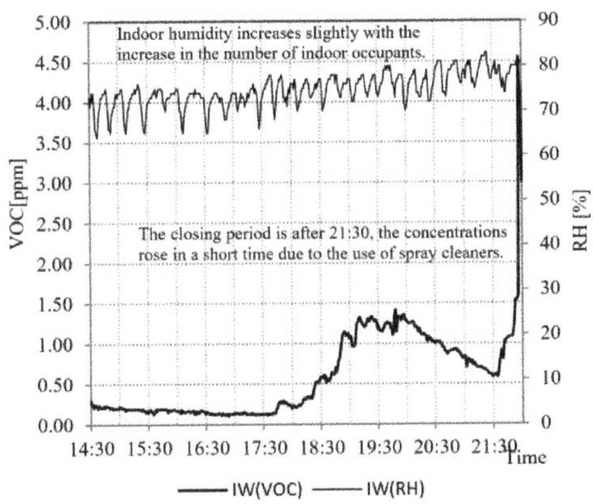

Figure 4. Diachronic monitoring of variations in the total volatile organic compound (TVOC) concentration and humidity (RH) in the IW sample.

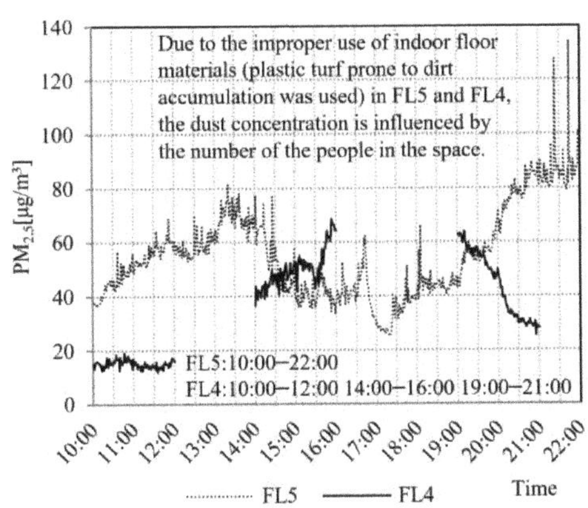

Figure 5. Diachronic monitoring of variations in the suspended particle (PM2.5) concentration across the FL5 and FL4 samples.

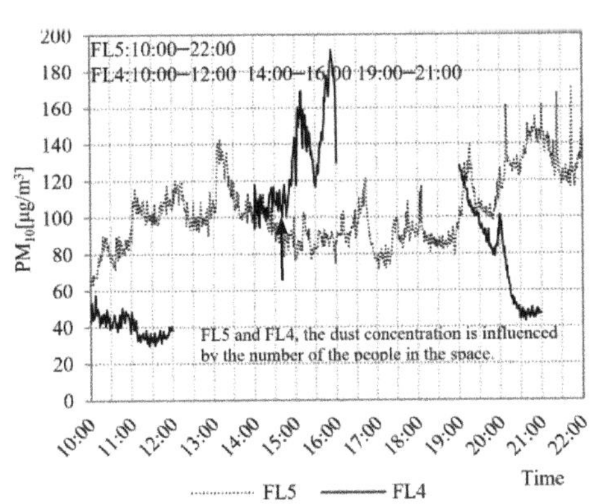

Figure 6. Diachronic monitoring of variations in the suspended particle (PM10) concentration across the FL5 and FL4 samples.

4. Analyses of the Measured Results

To understand the relationship between various indoor factors and the indoor air quality in fitness centers, our analyses involved a pairwise comparison with the factors identified in seven samples. This analysis was carried out to establish the correlations between indoor air quality and air pollutant concentrations and provide a basis for future design organizations or owners to estimate the pollution concentrations of their buildings in advance. The correlation analysis of various factors is shown in Table 4.

4.1. Summary of the Correlations between Various Pollutants and Indoor Air Quality

- Temperature (temp.): temperature was moderately correlated to various pollutant factors ($0.3 < R < 0.7$). Since the sample spaces were under constant temperature control via air conditioning systems, temperature was not the main factor influencing the pollution factors.
- Humidity (RH): humidity was moderately correlated to the VOC concentration ($0.3 < R < 0.7$), indicating that high humidity aids gas dispersion [6].
- CO: the indoor CO concentrations were lower than the corresponding Environmental Protection Administration standard, but the CH_2O, CO_2, and VOC were highly correlated ($R > 0.7$), indicating that they mutually influenced one another.
- O_3: the indoor O_3 concentrations were all 0 ppm, which did not correlate to any of the various pollutants.
- CO_2: due to the nature of air concentration accumulation, the CO_2 concentration generated through the activities of indoor occupants could not be measured in real time, resulting in a moderate correlation ($R = 0.4343$).
- Suspended particles: the suspended particles (PM2.5 and PM10) only had a moderate and positive correlation with the temperature ($0.3 < R < 0.7$), and had a low correlation with all other factors, indicating that external air mainly influenced the suspended particles.

Table 4. Results of the correlation (R) analysis for the concentrations of various factors in indoor air.

	Temp.	RH	CO	O_3	CH_2O	CO_2	VOC	$PM_{2.5}$	PM_{10}	People
Temp.	1.00	0.22	0.28	N/A	0.12	0.37	0.33	0.52	0.42	0.41
RH		1.00	0.19	N/A	0.16	0.21	0.37	0.22	0.21	0.29
CO			1.00	N/A	0.89	0.70	0.89	0.28	0.28	0.25
O_3				1.00	N/A	N/A	N/A	N/A	N/A	N/A
CH_2O					1.00	0.69	0.90	0.21	0.23	0.20
CO_2						1.00	0.62	0.41	0.39	0.43
VOC							1.00	0.33	0.29	0.22
$PM_{2.5}$								1.00	0.96	0.29
PM_{10}									1.00	0.26
People										1.00

4.2. Influence of Indoor Decorations on Indoor Air Quality

The CH_2O and VOC concentrations in the samples were different even when they had the same quantity of decorations, leading to a low correlation between the decoration quantity and the CH_2O and VOC concentrations ($R = 0.0316$, $R = -0.0976$) (Figures 7 and 8). The establishment date had a low and negative correlation with the CH_2O and VOC ($R = -003598$, $R = -0.5523$) (Figures 9 and 10); this is consistent with the concept that gas concentrations indoors decrease with time. Due to the awful environment in the CY sample space, the concentrations were several times higher than those of the other samples, so the CY data were excluded from this analysis.

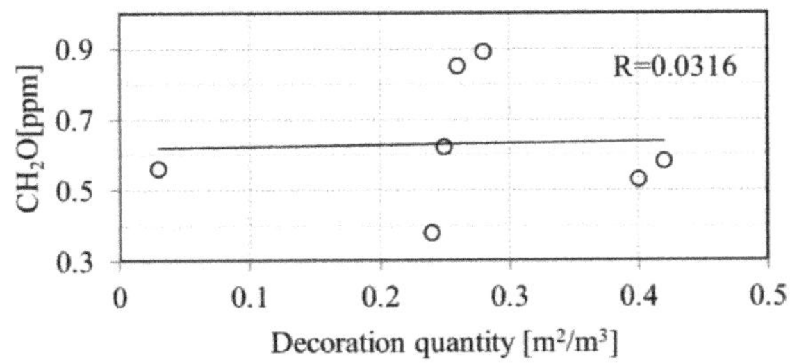

Figure 7. Correlation between the quantity of decorations and the CH_2O concentration.

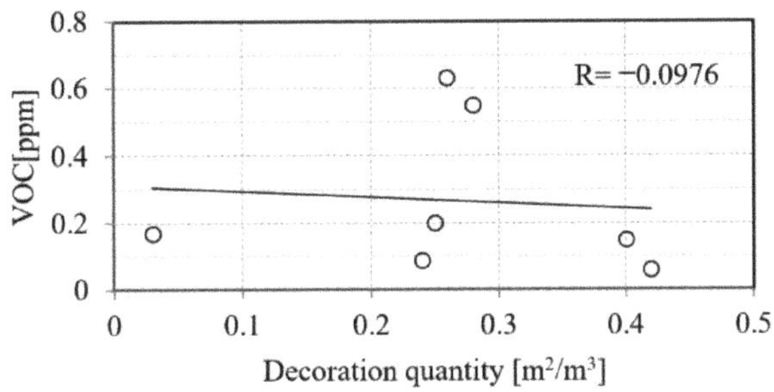

Figure 8. Correlation between the quantity of decorations and the VOC concentration.

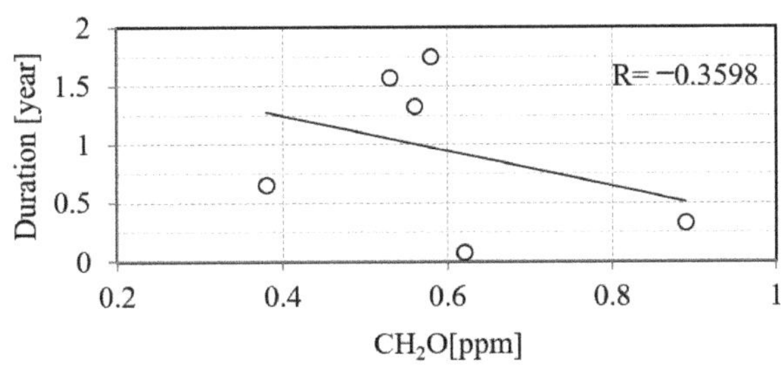

Figure 9. Correlation between the establishment date and the CH_2O concentration.

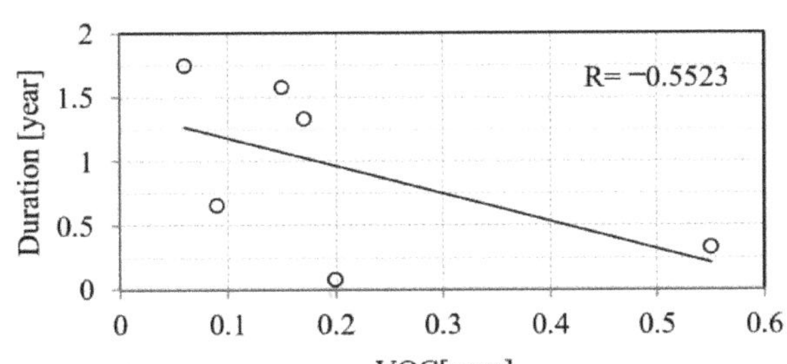

Figure 10. Correlation between the establishment date and the VOC concentration.

4.3. Air Exchange Volume Estimation for the Sample Spaces

Based on the measured data, the air exchange volume decreased during the periods when the average CO_2 concentrations in fitness centers exceeded the estimated standard. The reasons for these excess concentrations were the maximum number of indoor occupants allowable, and more importantly, the imperfect ventilation of the indoor air conditioning.

The air exchange volume during the periods when the concentrations exceeded the standard was modified and evaluated according to the calculation method of the "pollutant mass balance model" and the theory of indoor ventilation. The FL5 sample entered its peak period at 19:00, and its air exchange rate was only 0.79 h^{-1} at an indoor CO_2 concentration of 1104 ppm. To reduce the CO_2 concentration in FL5 to below the 1000 ppm standard set by the Environmental Protection Administration, the air exchange per hour (ACH) should be increased from 0.79 h^{-1} to 1.00 h^{-1}, which is equivalent to 53 min of continuous operation of an exhaust fan with an exhaust air rate of 680 m^3/h, so as to achieve good air quality. Due to the small space and large number of people, the indoor CO_2 concentration exceeded the standard when the CY sample opened at 17:00 and violently rose to 3680 ppm at 19:00, and the air exchange rate was only 0.34 h^{-1} at this concentration. To reduce the CO_2 concentration in CY to below the standard of 1000 ppm set by the Environmental Protection Administration, the ACH should be increased from 0.34 h^{-1} to 2.22 h^{-1}, as shown in Table 5, which is equivalent to 55 min of continuous operation of an exhaust fan with an exhaust air rate of 680 m^3/h, so as to achieve good air quality.

Table 5. Air exchange volume estimations before and after modification.

Samples	Indoor Volume (m^3)	Periods When Standards are Exceeded	Indoor CO_2 Concentration (ppm)	Outdoor CO_2 Concentration (ppm)	Dynamic and Static People	Original ACH (h^{-1})	Ideal ACH (h^{-1})
FL5	595.05	19:00–20:00	1104	612	14, 4	0.79	1.00
		20:00–21:00	1286	557	9, 3	0.24	0.40
CY	277.11	17:00–18:00	1003	547	12, 4	1.26	1.27
		18:00–19:00	2322	527	15, 4	0.47	1.78
		19:00–20:00	3680	514	19, 1	0.34	2.22
IW	580.00	17:00–18:00	1180	498	21, 8	0.89	1.21
		18:00–19:00	1544	504	9, 6	0.14	0.29
		19:00–20:00	1693	514	11, 3	0.13	0.31
		20:00–21:00	2330	505	15, 5	0.16	0.60
		21:00–22:00	2523	498	7, 7	0.05	0.22
HU	1177.24	18:00–19:00	1098	529	12, 0	0.14	0.17
		19:00–20:00	1350	526	8, 0	0.04	0.07
		20:00–21:00	1495	520	23, 0	0.30	0.61
		21:00–22:00	1638	524	24, 0	0.28	0.67

5. Conclusions

More and more people have a preference for fitness centers that are convenient and meet their fitness/exercise/health requirements. However, indoor fitness centers are often overdecorated with heavy sports activities and have an over-concentrated number of occupants, hence their concentrations of CO_2, CH_2O, and volatile organic compounds (VOCs) are high. Inadvertently, people will be affected by prolonged exposures to these poor quality environments even if they are in good health. This study examined the current air quality in Taiwan's fitness centers and proposes measures for the prevention of poor air quality and for proper operation management in such centers. These are to serve as references for owners and designers in the future, so that the concept of good and healthy air quality can be implemented and extended to all fields.

5.1. Influence of Indoor Decorations and Occupants' Activities on Fitness Center Air Quality

Indoor decoration quantity: The decoration materials used in all sample spaces and their sources are not the same. Even with the same decoration quantity, the concentrations of CH_2O and total volatile organic compounds (TVOCs) that escaped into the atmosphere were different in each of the samples, leading to a low correlation (R = 0.0316, R = 0.0976). Our findings on the influence of the establishment date on the concentrations of CH_2O and total volatile organic compounds (TVOCs) that escaped into the indoor air indicate that the correlation is low and insignificant (R = 0.1968, R = 0.4429), but show that the indoor concentration decreases with time. Indoor occupants' activities: The CO_2 concentration generated via the static and dynamic activities of indoor occupants can not be reflected in real time but will gradually accumulate.

5.2. Indoor Ventilation Predictions and Improvements

Indoor ventilation: The indoor volume, number of indoor occupants, and the various factors causing indoor pollution were not the same in all fitness centers. In planning, designing, and selecting ventilation facilities, attention should be paid to decoration materials and the ventilation effects of facilities. The suspended particles we recorded were not only from the external air but also from interior decoration materials, and rough and uneven surfaces that easily attracted dust accumulation and pollution; therefore, materials made from fabric and artificial turf should be reduced in order to achieve less dust accumulation on material surfaces. While most of the tested sample spaces were equipped with air purifiers, the effects of these were insignificant. Due to the large areas of fitness centers, the air purifiers could not purify the indoor air effectively, resulting in the accumulation and high concentration of indoor pollutants. Indoor ventilation improvements: CO_2 was 60% correlated to indoor CH_2O and total volatile organic compounds (TVOCs) (R = 0.6856, R = 0.6156), 40% correlated to suspended particles (PM2.5, PM10) (R = 0.3924, R = 0.413), and 70% correlated to CO (R = 0.7045), indicating that CO_2 is highly correlated to various indoor air pollution indicators. By establishing real-time indoor CO_2 measurement systems, fitness centers can reduce their indoor concentrations of CO_2 and other pollutants with effective ventilation.

Author Contributions: Conceptualization, P.-Y.K. and T.-Y.W.; methodology and formal analysis, P.-Y.K. and T.-Y.W.; investigation, P.-Y.K. and T.-Y.W.; writing, P.-Y.K. and T.-Y.W.; review and editing, P.-Y.K. and C.-Y.C. All authors have read and agreed to the published version of the manuscript.

Funding: This research received no external funding.

Institutional Review Board Statement: Not applicable.

Informed Consent Statement: Not applicable.

Data Availability Statement: The data supporting the findings of this study are available within the article.

Conflicts of Interest: The authors declare no conflict of interest.

References

1. Andrade, A. Indoor air quality of environments used for physical exercise and sports practice Systematic review. *J. Environ. Manag.* **2018**, *206*, 277–286. [CrossRef] [PubMed]
2. Cianfanelli, C. Environmental Quality in Sports Facilities: Perception and Indoor Air Quality. *J. Phys. Educ. Sports Manag.* **2016**, *3*, 57–77. [CrossRef]
3. *ANSI/ASHRAE Standard 62.2-2019*; Ventilation and Acceptable Indoor Air Quality in Residential Buildings. ASHARE: Peachtree Corners, GA, USA, 2019.
4. *ANSI/ASHRAE Standard 62.1-2022*; Ventilation and Acceptable Indoor Air Quality. ASHARE: Peachtree Corners, GA, USA, 2019.
5. Li, Y.-Y. *A Study on Indoor Air Quality Management Strategies in Office Space*; Department of Architecture, National Cheng Kung University: Tainan, Taiwan, 2004.
6. Huang, K.C. A study on the influence of indoor thermal environment variation on adsorption and reduction of formaldehyde from building materials. In Proceedings of the 23rd Second Architectural Research Achievement Presentation Conference, Tainan, Taiwan, November 2011.

engineering proceedings

Proceeding Paper

Fabrication of Titanium Oxide Thin-Film Electrodes with Photocatalytic Activities and an Evaluation of Their Photoelectrochemical Properties †

Naoya Wakatsuki [1,*] and Tomohiro Tojo [2,*]

1 Department of System Engineering, Shizuoka Institute of Science and Technology, 2200-2 Toyosawa, Fukuroi 437-8555, Japan

2 Department of Electrical and Electronic Engineering, Faculty of Science and Technology, Shizuoka Institute of Science and Technology, 2200-2 Toyosawa, Fukuroi 437-8555, Japan

* Correspondence: 2221036.wn@sist.ac.jp (N.W.); tojo.tomohiro@sist.ac.jp (T.T.)

† Presented at the IEEE 5th Eurasia Conference on Biomedical Engineering, Healthcare and Sustainability, Tainan, Taiwan, 2–4 June 2023.

Abstract: We synthesized hollow spherical titanium oxide particles, which are one of the structural features of fabricating a thin-film photoelectrode, using the particles and evaluated their properties. The XRD diffraction results confirmed the main phase peaks of the target rutile-type hollow spherical titanium oxide (HSTR) and bronze-type hollow spherical titanium oxide (TiO_2(B)). The calcium carbonate used in the core material was also removed. The photocatalytic reaction measurement result showed that the activity of TiO_2(B) in ultraviolet light of 365 nm was higher than that of TiO_2(B). As shown in the visible spectrum, the photo adsorption wavelength of HSTR was approxim ately 700 nm, whereas TiO_2(B) was generally absorbed around 400 nm. A relationship between an electric current peak and the square root of a scan potential speed suggested a reversible reaction system in light irradiation.

Keywords: environment energy; TiO_2; hollow spherical nanoparticles; photoelectrochemical measurement

Citation: Wakatsuki, N.; Tojo, T. Fabrication of Titanium Oxide Thin-Film Electrodes with Photocatalytic Activities and an Evaluation of Their Photoelectrochemical Properties. *Eng. Proc.* **2023**, *55*, 57. https://doi.org/10.3390/engproc2023055057

Academic Editors: Teen-Hang Meen, Kuei-Shu Hsu and Cheng-Fu Yang

Published: 6 December 2023

1. Introduction

Energy is indispensable for our lives; however, the biggest problems in energy are the depletion of energy resources and environmental problems. Energy production based on fossil fuels causes air pollution and global warming, and the limited number of reservoirs threatens the stability of future energy supplies. To solve this problem, the conventional energy production method needs to be changed. Thus, renewable energy has been attracting attention. Although solar cells are most popular in the production of renewable energy using sunlight as an energy source, they have problems such as manufacturing costs and the need for a large area of land to increase the amount of generated power.

Dye-sensitized solar cells (DSSCs) [1] and perovskite solar cells [2] have been extensively studied to solve the above-mentioned problems. As DSSCs generally adopt halides as electrolytes and perovskite solar cells employ lead halides as constituent elements, the fabrication of devices does not emit toxic materials that harm the environment and the human body. As shown in Figure 1a, a solar cell generates electric power that needs to be converted and stored in the battery through a DC–DC converter. The quantity of energy conversions increases the energy loss in the process of solar energy to electrical energy and electrical energy to chemical energy. In addition, since the power generation part and the storage part are separated, the whole system becomes large, which causes high operation costs.

To solve the above-mentioned problems, we constructed a photoactive electrode of solar cells without non-toxicity using titanium oxides. Titanium oxides undergo photoactivation, known as the Honda–Fujishima effect [3]. This is a photoelectronic reaction in which titanium oxide particles are irradiated with ultraviolet light and electrons and holes are excited, promoting a redox reaction. In addition, titanium oxides are used as an electrode material for lithium-ion batteries; therefore, it is also possible to occlude lithium ions. If the excited electrons produced from photoirradiation are consumed to store lithium ions into titanium oxide particles on a single transparent electrode, a conventional solar system can be replaced with an integrated single system of a solar cell and a lithium-ion battery (Figure 1b). By making the primary particle shape a hollow sphere, the specific surface area increases to promote the reaction near the particle interface and broaden the active wavelength without adding other elements. Therefore, titanium oxides can be used as an integrated electrode material to generate and store electricity, simplify the entire system, and decrease the loss in energy conversion. Furthermore, they have little effect on the environment and human health as they are used in cosmetics.

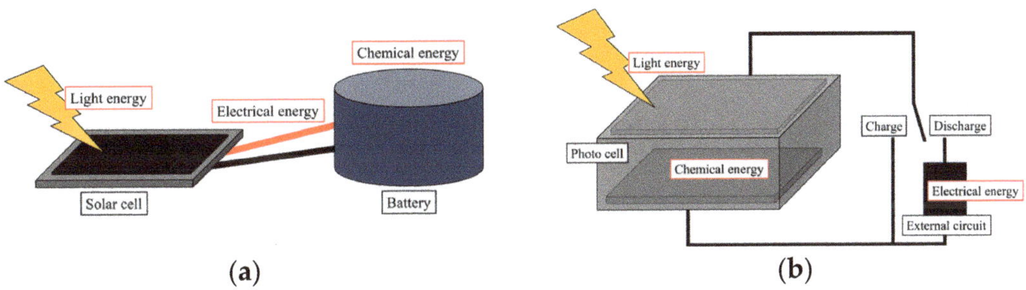

Figure 1. Difference between the conventional solar cell system (**a**) and the proposed photovoltaic battery system (**b**).

Purpose

Figure 2 shows that titanium oxide has various crystal structures and that the bandgap changes depending on the crystal structures [4]. In this study, we analyzed a photocatalytic effect on the absorption wavelength produced by the differential crystal structure only by controlling the primary particle shape without adding other elements. Furthermore, we constructed the integrated battery system by fabricating the photoactive thin-film electrode and evaluating its characteristics.

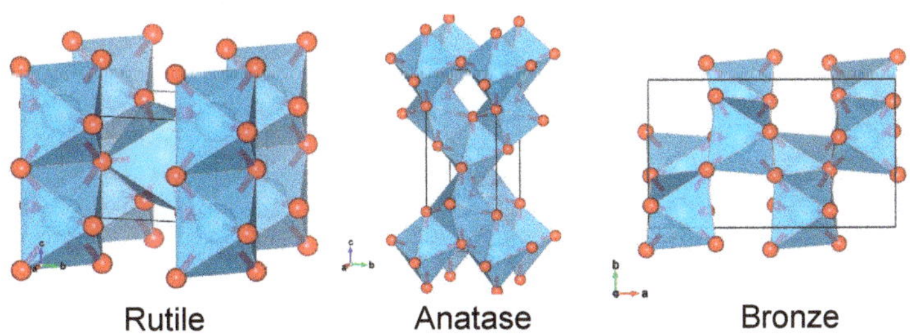

Figure 2. Crystal structures of polymorphic titanium oxide.

2. Material and Method

2.1. Sample Synthesis

We synthesized titanium oxides with hollow spherical primary particles [5]. First, we fabricated calcium carbonate-coated titanium alkoxide (TCC). Starting from titanium alkoxide (TTIP), it was hydrolyzed with distilled water and coated on calcium carbonate ($CaCO_3$) in ethanol. Next, amorphous titanium was synthesized as a precursor. $CaCO_3$ particles in the core material were removed by stirring TCC at 300 rpm in a nitric acid aqueous solution. Finally, hollow spherical titanium oxide ($HSTiO_2$) was fabricated. Hollow spherical rutile-type titanium oxide (HSTR) was obtained by calcining spherical amorphous titanium at 900 °C for 48 h under an air atmosphere, and hollow-spherical, anatase-type titanium oxide (HSTA) was obtained by calcining spherical amorphous titanium at 500 °C for 1 h in an air atmosphere (Figure 3).

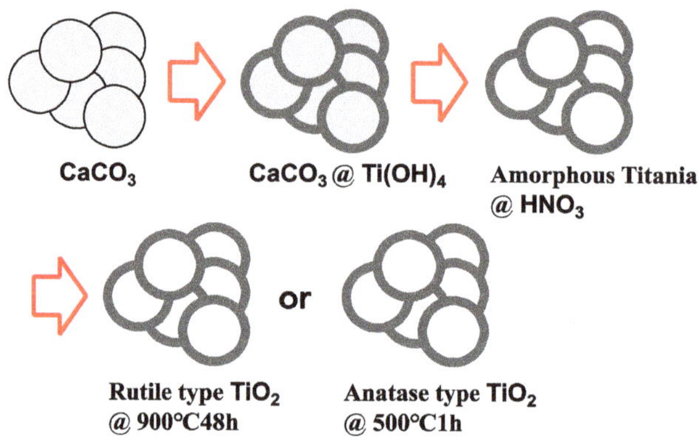

Figure 3. Synthesis method of polymorphic hollow spherical titanium oxide.

2.2. Sample Analysis

The synthesized samples were analyzed using powder X-ray diffraction (XRD, Rigaku (Akishima, Japan), SmartLab-SRD), a scanning electron microscope (SEM, Jeol (Akishima, Japan), JSM-IT100), and energy-dispersive X-ray spectroscopy (EDX). The optical properties of the samples were evaluated using ultraviolet–visible spectroscopy (UV-Vis., Jasco (Easton, MD, USA), V-650). In the structural analysis, an XRD apparatus was set to a Cu Kα ray, an accelerating voltage current of 40 kV-20 mA, and an angle range of 2θ = 10–90°. A scanning electron microscope with built-in EDX was used for shape observation and elemental analysis. The acceleration voltage was set to 15 kV, the shape observation was to a probe current (P.C.) of 40 mA, and the elemental analysis was set to P.C. 60 mA. The optical properties were evaluated using UV-Vis. with irradiation wavelengths at 200–900 nm.

2.3. Fabrication of Thin-Film Electrodes

An electrode was fabricated using the synthesized sample in this study. A mixed solution of polyvinylidene fluoride (PVdF) and HSTR was coated on an indium tin oxide substrate (ITO-PET) to a thickness of 200 µm using a bar coater. Table 1 shows a mixing ratio in weight of PVdF and HSTR.

Table 1. Mixing ratio of PVdF and HSTR.

PVdF [g]	HSTR [g]	wt. Ratio
0.05	non	1:0
0.05	0.05	1:1
0.05	0.02	1:0.4
0.05	0.01	1:0.2

2.4. Photoelectrochemical Properties

Using the prepared electrode as the working electrode, a photocatalytic reaction measurement and photoelectrochemical measurement were performed. Distilled water and absolute ethanol were used as solvents for photocatalytic reaction measurements, and a carbon rod was used as the counter electrode. The generated amount of hydrogen gas was estimated using the Faraday electrolysis formula (Equation (1)) by measuring the electrical current value while 365 nm light was irradiated for 60 min.

$$m = \frac{It}{F} \cdot \frac{M}{n}[\mathrm{g}] \tag{1}$$
$$(\text{Current}: I[\mathrm{A}], \text{Time}; t[\mathrm{s}], \text{Faraday constant}; F(9.6485 \times 10^4\ \mathrm{C/mol}),$$
$$\text{Formula weight of reactants } M[\mathrm{g/mol}], \text{Number of rection electrons}; n[-])$$

In photoelectrochemical measurements, a Pt wire electrode was used as the counter electrode, an Ag/AgCl electrode was used as the reference electrode, and a 1M Na_2SO_4 aqueous solution was used as the electrolyte. The conditions were in a potential range between −0.4 and +0.4 V vs. Ag/AgCl, a sweep rate between 5 and 50 mV/s, and an irradiation wavelength of 800 nm. Measurements were performed in UV-Vis. (Jasco, V-630BIO) using a halogen lamp as the light source.

3. Results and Discussion

3.1. Sample Analysis

The results of the sample analysis of HSTR are shown in Figure 4. Based on the XRD analysis result, we produced the rutile-type structure. The SEM observation result showed spherical particles about 0.2 μm in diameter. The EDX elemental mapping result indicated that the residual amount of calcium was negligible. TiO_2(B) analysis results are shown in Figure 5. XRD patterns in Figure 5a matched the crystal peak pattern of a previous report [6] for the titanium oxide bronze, indicating successful synthesis. However, as shown in Figure 5b, the desired hollow spherical shape was maintained, and the crystals grew into the needle-like shape that is characteristic of the bronze type.

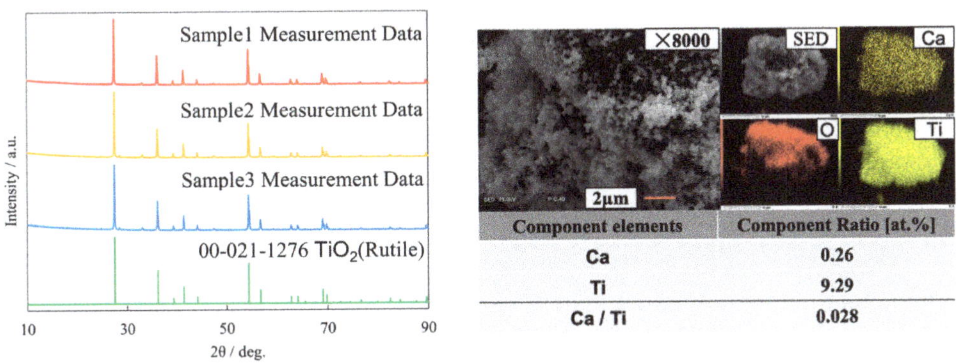

Component elements	Component Ratio [at.%]
Ca	0.26
Ti	9.29
Ca / Ti	0.028

(a) (b)

Figure 4. XRD analysis: (**a**) SEM image observation and EDX quantitative result (**b**) of HSTR.

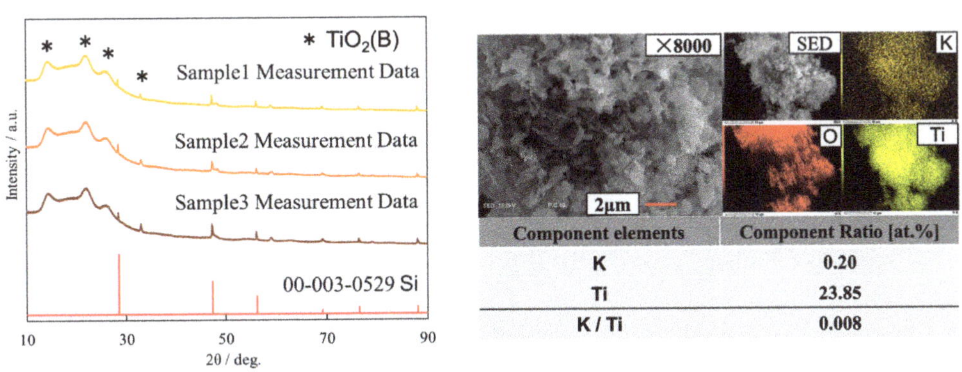

Component elements	Component Ratio [at.%]
K	0.20
Ti	23.85
K / Ti	0.008

(a) (b)

Figure 5. XRD patterns: (**a**) SEM image and EDX elemental mapping images (**b**) of TiO_2(B).

The results of the UV-visible measurements (Figure 6) presented a red shift of a photo adsorption wavelength in the rutile type due to a shaping effect on the hollow spheres, whereas the wavelength of rutile-type TiO_2 is generally observed at approximately 400 nm. Since bronze crystals grew in a needle-like (layered) shape, the peak was the same as the conventional peak.

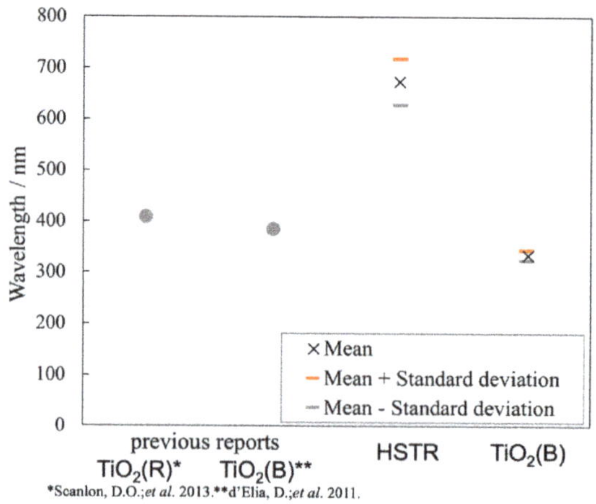

Figure 6. UV-visible absorption wavelength of HSTR, TiO_2(B) and previous reports [7,8].

3.2. *Photoelectrochemical Properties*

Figure 7 shows the results of a photocatalytic reaction measurement using HSTR, which maintained the desired hollow and spherical shape. It was confirmed that the catalytic reaction was carried out through light irradiation.

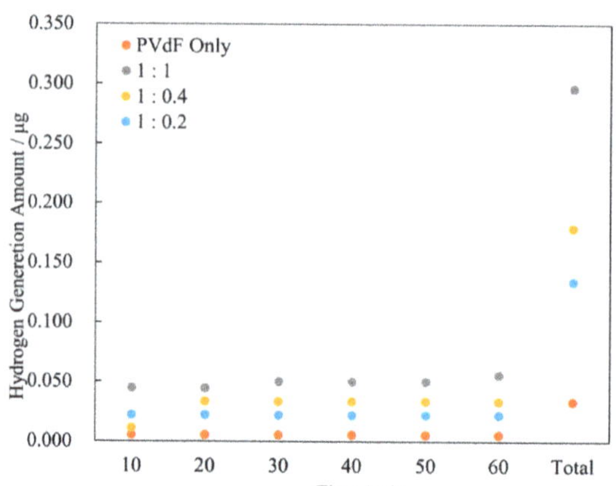

Figure 7. Emitting measurement of hydrogen gas during a photocatalytic reaction of thin-film photoelectrodes coated with HSTR.

Photoelectrochemical measurements using cyclic voltammetry revealed a current peak caused by redox reactions in the cyclic voltammogram (Figure 8a) when irradiated with light at a specific wavelength. In the absence of light irradiation, no peaks associated with redox reactions appeared during reductive and oxidative scans, and only peaks attributed to a series resistance appeared. Two current peaks were observed on the reduction side. However, Figure 8b showed that both peaks showed good linearity, suggesting that the reaction progressed at the electrode interface in two stages on the reduction side.

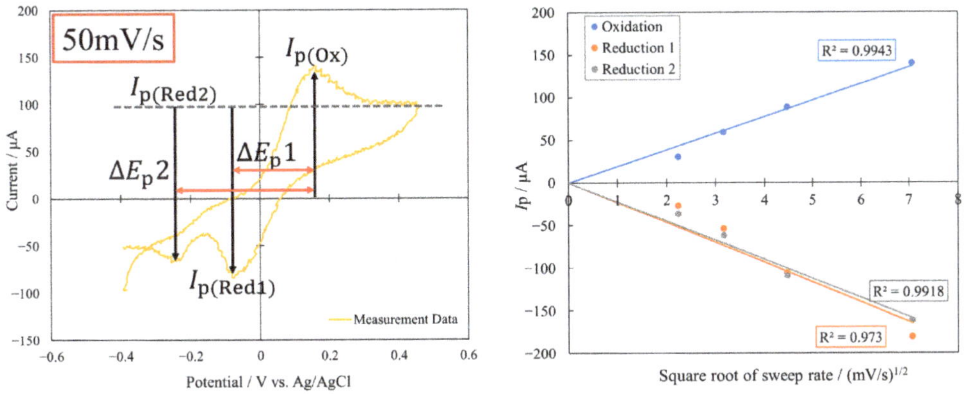

(a) (b)

Figure 8. Cyclic voltammetry measurement result of HSTR-coated, thin-film photoelectrode with UV irradiation (**a**). Relationship between peak current and the square root of sweep speed (**b**).

4. Conclusions

The primary particles of TiO_2(B) could not maintain hollow spheres and grew needle-like shapes. In addition, the light absorption wavelength corresponded to that of a previous report. On the other hand, the absorption wavelength of HSTR shifted from the previous report forward to the infrared wavelength side. This shift suggested that the shape affects the primary particles. Oxidation and reduction peaks were confirmed in a cyclic voltammogram, and current peak values changed linearly against the square root of the sweep speed. Therefore, the fabricated photoelectrodes had a reversible redox reaction and promoted the redox reaction in light irradiation.

Author Contributions: N.W.; research and writing—original draft preparation, T.T.; research assistance, writing—review and editing. All authors have read and agreed to the published version of the manuscript.

Funding: This research received no external funding.

Institutional Review Board Statement: Not applicable.

Informed Consent Statement: Not applicable.

Data Availability Statement: Raw data were generated at Shizuoka Institute of Science and Technology. Derived data supporting the findings of this study are available from the corresponding authors; N. Wakatsuki and/or T. Tojo on request.

Acknowledgments: This work was supported by the Advanced Instrumental Analysis Center of the Shizuoka Institute of Science and Technology.

Conflicts of Interest: The authors declare no conflict of interest.

References

1. O'Regan, B.; Grätzel, M. A low-cost, high-efficiency solar cell based on dye-sensitized colloidal TiO_2 films. *Nature* **1991**, *353*, 737–740. [CrossRef]
2. Kojima, A.; Teshima, K.; Shirai, Y.; Miyasaka, T. Organometal Halide Perovskites as Visible-Light Sensitizers for Photovoltaic Cells. *J. Am. Chem. Soc.* **2009**, *131*, 6050–6051. [CrossRef] [PubMed]
3. Fujishima, A.; Honda, K. Electrochemical Photolysis of Water at a Semiconductor Electrode. *Nature* **1972**, *238*, 37–38. [CrossRef] [PubMed]
4. Yang, G.; Abraham, C.; Ma, Y.; Lee, M.; Helfrick, E.; Oh, D.; Lee, D. Advances in Materials Design for All-Solid-state Batteries: From Bulk to Thin Films. *Appl. Sci.* **2020**, *10*, 4727. [CrossRef]
5. Huang, X.; Cui, S.; Wieboldt, R.C.; Hallac, P.B.; Fell, C.R.; Metz, B.; Jiang, J.; Chen, J. Hollow TiO_2 as an Anode for Lithium Ion Batteries: Synthesis and In Situ Visualization of State of Charge. *Adv. Electron. Mater.* **2015**, *1*, 1500256. [CrossRef]

6. Kolen'Ko, Y.V.; Kovnir, K.A.; Gavrilov, A.I.; Garshev, A.V.; Frantti, J.; Lebedev, O.I.; Churagulov, B.R.; Van Tendeloo, G.; Yoshimura, M. Hydrothermal synthesis and characterization of nanorods of various titanates and titanium dioxide. *J. Phys. Chem. B* **2006**, *110*, 4030–4038. [CrossRef] [PubMed]
7. Scanlon, D.O.; Dunnill, C.W.; Buckeridge, J.; Shevlin, S.A.; Logsdail, A.J.; Woodley, S.M.; Catlow, C.R.A.; Powell, M.J.; Palgrave, R.G.; Parkin, I.P.; et al. Band alignment of rutile and anatase TiO_2. *Nat. Mater.* **2013**, *12*, 798–801. [CrossRef] [PubMed]
8. d'Elia, D.; Beauger, C.; Hochepied, J.F.; Rigacci, A.; Berger, M.H.; Keller, N.; Keller-Spitzer, V.; Suzuki, Y.; Valmalette, J.C.; Benabdesselam, M.; et al. Impact of three different TiO_2 morphologies on hydrogen evolution by methanol-assisted water splitting: Nanoparticles, nanotubes and aerogels. *Int. J. Hydrogen Energy* **2011**, *36*, 14360–14373. [CrossRef]

engineering proceedings

MDPI

Proceeding Paper

AI-DAS: AI-Based Driving Assistance System for Scooters for Traffic Accident Avoidance †

Yen-Da Chen and Yu-Cheng Chang *

Department of Computer Information and Network Engineering, Lunghwa University of Science and Technology, Taoyuan 33306, Taiwan; ydchen@mail.lhu.edu.tw
* Correspondence: yucheng208@outlook.com

† Presented at the IEEE 5th Eurasia Conference on Biomedical Engineering, Healthcare and Sustainability, Tainan, Taiwan, 2–4 June 2023.

Abstract: Due to strong mobility, low costs, and easy parking, the scooter has become an important mode of transportation in Taiwan. The number of traffic accidents is also increasing because of drivers' bad driving habits. For road traffic injury prevention, the government has approved the road safety policy to promote awareness of road safety. Thus, we propose an AI-based driving assistance system (AI-DAS) for scooter riders to prevent traffic accidents. In the AI-DAS, Raspberry Pi controls the camera and sensors for the scooter. The AI-DAS detects pedestrians and monitors lanes not only to warn riders and pedestrians by sound and light but also to drastically reduce accidents and traffic jams. The AI-DAS is implemented using the python programming language and is tested on the Raspberry Pi hardware platform.

Keywords: scooter; lane line detection; traffic accident; Raspberry Pi; OpenCV

Citation: Chen, Y.-D.; Chang, Y.-C. AI-DAS: AI-Based Driving Assistance System for Scooters for Traffic Accident Avoidance. *Eng. Proc.* **2023**, *55*, 58. https://doi.org/10.3390/engproc2023055058

Academic Editors: Teen-Hang Meen, Kuei-Shu Hsu and Cheng-Fu Yang

Published: 6 December 2023

1. Introduction

Due to strong mobility, low costs, and easy parking, the scooter has become an important mode of transportation in Taiwan. The number of traffic accidents is also increasing year by year because of the bad driving habits of riders, especially young riders. Therefore, the number of people killed in traffic accidents is increasing in Taiwan.

Recently, there has been much research focusing on safety message dissemination [1,2] and safe driving assistance systems [3–5] to avoid traffic accidents. In Refs. [1,2], each vehicle broadcasts or rebroadcasts the safety message after receiving it. The safety message for car crash alerts or rescue operation notifications must be rapidly and successfully disseminated in risk areas to avoid potential traffic accidents. However, broadcasting storms [6] may happen if all vehicles' safety messages rebroadcast.

To prevent the broadcast storm problem, the safe driving assistance system [3–5] has been developed. In Ref. [4], each vehicle collects near-crash and crash data by video cameras to alert the driver and avoid potentially risky driving behaviors. In Ref. [3], LASER radio detection sensors and global positioning system (GPS) sensors are necessary to detect the vehicle and lane position. In Ref. [5], the obstacles in front of the car can be detected and the distance between the car and the nearest obstacles can be estimated. In Refs. [3–5], the obstacles and lane position can be seen. However, there no further operations are performed for the purpose of avoiding traffic accidents.

As a result, an AI-based driving assistance system (AI-DAS) is proposed not only to detect pedestrians and monitor lane lines but also to warn riders and pedestrians by sound and light for in order to avoid accidents. The rest of this paper is organized as follows. The hardware design of the AI-DAS is introduced in Section 2.1. The concepts of the lane position detection and accident alerting system are proposed in Section 2.2. Finally, Section 3 concludes the paper.

2. AI-Based Driving Assistance System (AI-DAS)

The functions of the AI-DAS are organized as follows.

Lane Line Detection Function

This function is used to detect the lane line by the output picture from the camera module.

Scooter Position Detection Function

It is used to recognize the scooter position between two neighboring lane lines.

Turn Signal Function

The turn signal function is the main contribution of the AI-DAS. It is a function designed to warn other drivers or pedestrians when the scooter begins to move out of its lane according to the recognition results from the scooter position detection function. If the scooter position is a little too far to the left/right, and the turn signal is not working, the turn signal function turns on automatically.

The hardware and software design of the AI-DAS are illustrated as follows.

2.1. Hardware Design of AI-DAS

The hardware design and the actual implementation of the AI-DAS are shown in Figures 1 and 2, respectively. The AI-DAS consists of a Raspberry Pi module, cellular phone module, and camera module, where the scooter position detection function, lane line detection function, and turn signal function are embedded in the Raspberry Pi module. The camera module shown in Figure 2a is fixed to the handle pipe to obtain pictures or video of the lane. In addition, the camera module communicates with the Raspberry Pi module using a USB port connection.

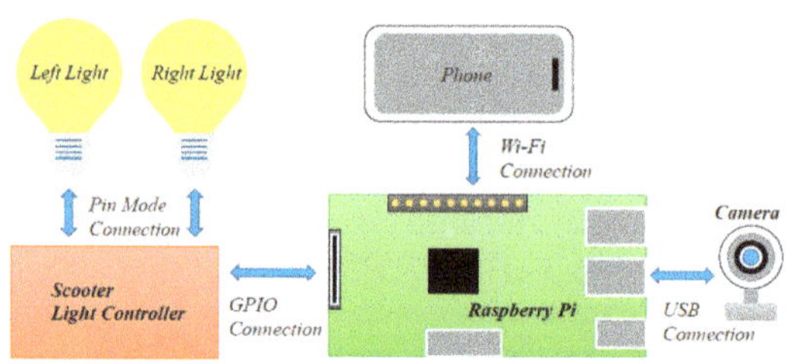

Figure 1. Hardware components of the AI-DAS.

(a) (b) (c)

Figure 2. Actual implementation of the AI-DAS in a (**a**) camera, (**b**) Raspberry Pi, and (**c**) cellular phone, respectively.

As shown in Figure 2b, the Raspberry Pi module is powered by the scooter and is placed in the storage compartment of the scooter. The cellular phone module is also fixed to the handle pipe. As shown in Figure 2c, it displays the operational results of the lane line detection and turn signal functions. The connection between the Raspberry Pi and cellular phone modules is established as follows.

The cellular phone is set up as a virtual access point (AP).

The Raspberry Pi connects to the virtual AP by Wi-Fi communication.

A virtual IP address is assigned to the Raspberry Pi.
An SSH connection app is installed on the cellular phone.
The Raspberry Pi is logged in with the virtual IP address by the SSH app.

2.2. Software Design of the AI-DAS

The software design of the AI-DAS consisting of the lane line detection, scooter position detection, and turn signal functions is described in this section.

Lane line detection function: The lane line detection function is composed of the following 5 stages.

Distortion Correction

Figure 3a shows the raw image from the camera module. In digital photography, the optical aberration in a camera deforms and twists physically upright lines in the resulting images and makes the images appear curvy. To make a distortion correction to raw images, we obtained the camera calibration matrices and distortion coefficients utilizing chessboard images.

(a) (b)

Figure 3. (**a**) Raw images of the lane lines. (**b**) Distortion correction of (**a**).

OpenCV (Ver. 2.x) [7] is open source and widely used for image processing applications. In this paper, three functions, findChessboardCorners (), calibrateCamera (), and undistort () are used for distortion correction. As a result, Figure 3a is modified to Figure 3b.

Thresholding

To make images easier to analyze, we convert the images from color to binary images using gradients and color transforms. After thresholding, Figure 3b becomes Figure 4a.

Color Conversion

Figure 4a shows that the lane is not easy to recognize. Therefore, before thresholding, we first transform RGB images into the HSV color space. cv2.cvtColor is used for conversion, and the thresholding and color conversion results of Figure 3a are shown in Figure 4b.

Perspective Transformation

For lane line detection, we transform images into bird's-eye-view image. cv2.getPerspectiveTransform() is used to obtain the transformation matrix, and cv2.warpPerspective() is used for bird's-eye-view images. The transformation results are shown in Figure 5a.

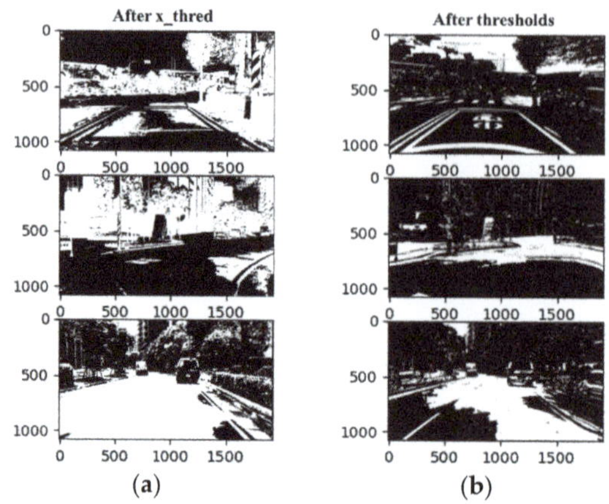

Figure 4. (**a**) Results of the thresholding. (**b**) Results of the color conversion.

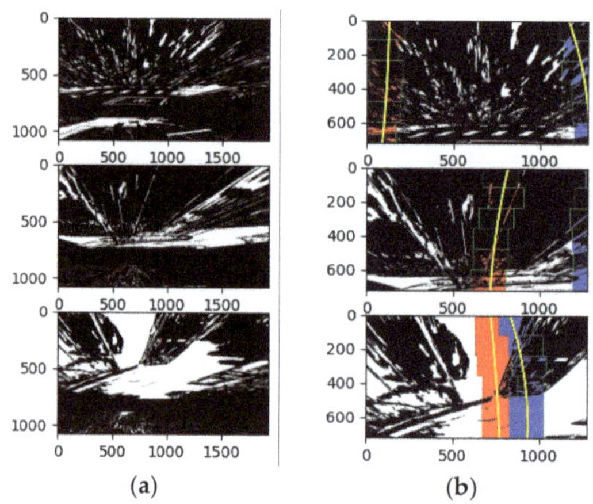

Figure 5. (**a**) Result of the perspective transformation. (**b**) Results of the lane line detection.

<u>Lane detection</u>

Since the pixels of the lane lines are concentrated within a certain range of the x-axis, the sliding window polynomial fitting approach [8,9] is used to detect the lane lines. The detection results are shown in Figure 5b. The left lane line is marked as a red line and the right lane line is marked as a blue line.

Scooter position detection: Figure 5b reveals that the position of the scooter is located in the middle of the picture. In addition, we obtain the center position of the lane through the center of the two-lane lines. Therefore, the offset of the scooter can be calculated by the position of the scooter and the center position of the lane.

Finally, the results of the lane detection are shown in Figure 6.

Turn signal function: The turn signal controller controls the left and right lights and is embedded in the scooter's computer. After receiving a turn instruction from the lane line detection module, the GPIO control pin of Raspberry Pi sends the signal to the turn signal controller of the scooter. Therefore, the scooter rider can avoid damage as much as possible.

Figure 6. Result of the lane detection.

3. Conclusions

The scooter is an important mode of transportation in Taiwan, and the number of traffic accidents is also increasing because of drivers' bad driving habits. Thus, we propose an AI-based driving assistance system (AI-DAS) for scooters in order to prevent road traffic injuries. In the AI-DAS, we detect lanes and control the turn signals not only to warn riders and pedestrians by sound and light but also to drastically reduce traffic accidents. In the future, the AI-DAS will include a module for the detection of moving cars and a module for the prediction of the direction of moving cars to further avoid traffic accidents.

Author Contributions: Conceptualization, Y.-D.C. and Y.-C.C.; methodology, Y.-C.C.; software, Y.-C.C.; validation, Y.-D.C. and Y.-C.C.; formal Analysis, Y.-C.C.; investigation, Y.-C.C.; resources, Y.-C.C.; data curation, Y.-C.C.; writing—original draft preparation, Y.-C.C.; writing—review and editing, Y.-D.C.; visualization, Y.-C.C.; supervision, Y.-D.C.; project administration, Y.-C.C.; funding acquisition, Y.-C.C. All authors have read and agreed to the published version of the manuscript.

Funding: This research received no external funding.

Institutional Review Board Statement: Not applicable.

Informed Consent Statement: Not applicable.

Data Availability Statement: Data are contained within the article.

Conflicts of Interest: The authors declare no conflict of interest.

References

1. Suthaputchakun, C.; Dianati, M.; Sun, Z. Trinary partitioned blackburstbased broadcast protocol for time-critical emergency message dissemination in vanets. *IEEE Trans. Veh. Technol.* **2014**, *63*, 2926–2940. [CrossRef]
2. Chen, Y.-D.; Shih, Y.-P.; Shih, K.-P. An emergency message dissemination protocol using n-way search with power control for vanets. In Proceedings of the 2015 IEEE International Conference on Communications (ICC), London, UK, 8–12 June 2015; pp. 3653–3658.
3. McCall, J.; Trivedi, M. Video-based Lane estimation and tracking for driver assistance: Survey, system, and evaluation. *IEEE Trans. Intell. Transp. Syst.* **2006**, *7*, 20–37. [CrossRef]
4. Li, L.; Zhu, X. Design concept and method of advanced driver assistance systems. In Proceedings of the 2013 Fifth International Conference on Measuring Technology and Mechatronics Automation, Hong Kong, China, 16–17 January 2013; pp. 434–437.
5. Fang, C.-Y.; Hsu, W.-H.; Ma, C.-W.; Chen, S.-W. A vision-based safety driver assistance system for motorcycles on a smartphone. In Proceedings of the 17th International IEEE Conference on Intelligent Transportation Systems (ITSC), Qingdao, China, 8–11 October 2014; pp. 328–333.
6. Tseng, Y.-C.; Ni, S.-Y.; Chen, Y.-S.; Sheu, J.-P. The broadcast storm problem in a mobile ad hoc network. In Proceedings of the ACM International Conference on Mobile Computing and Networking (MOBICOM), Seattle, WA, USA, 15–19 August 1999; pp. 153–167.
7. Intel Open-Source Computer Vision Library Reference Manual. 2001. Available online: https://www.cs.unc.edu/Research/stc/FAQs/OpenCV/OpenCVReferenceManual.pdf (accessed on 5 December 2023).

8. Dragojevi'c, M.; Stevi'c, S.; Kruni ́c, M.; Luki ́c, N. Advanced Lane finding prototype based on autoware platform. In Proceedings of the 2020 Zooming Innovation in Consumer Technologies Conference (ZINC), Novi Sad, Serbia, 26–27 May 2020; pp. 169–173.
9. Yılmaz, M.E.; Erkmen, B. Advanced Lane line detection using heterogeneous embedded computing. In Proceedings of the 2021 International Conference on Innovations in Intelligent SysTems and Applications (INISTA), Kocaeli, Turkey, 25–27 August 2021; pp. 1–5.

engineering proceedings

MDPI

Proceeding Paper

Water Treatment of Manganese Oxides and Organic Matter through Pre-Oxidation and Coagulation/Sedimentation †

Yi-Chang Wu [1,*], Chi-Chuan Kan [2], Shih-Chieh Lee [3] and Feng-Yu Yang [4]

1 Program of Biotechnology and Industry, Da-Yeh University, No. 168, University Rd., Dacun, Changhua 515006, Taiwan
2 Department of Tourism Management, Chia-Nan University of Pharmacy and Science, No. 60, Sec. 1, Erren Rd., Rende Dist., Tainan City 717301, Taiwan; cckanev@mail.cnu.edu.tw
3 Department of Medical Botanicals and Foods on Health Applications, Da-Yeh University, No. 168, University Rd., Dacun, Changhua 515006, Taiwan; slee@mail.dyu.edu.tw
4 Department of Marine Environment and Engineering, National Sun Yat-sen University, No. 70 Lien-hai Road, Kaohsiung 804201, Taiwan; loveccat100@gmail.com
* Correspondence: charleswu@mail.water.gov.tw
† Presented at the IEEE 5th Eurasia Conference on Biomedical Engineering, Healthcare and Sustainability, Tainan, Taiwan, 2–4 June 2023.

Abstract: We investigated the reducing rate of manganese oxides (MnOs) and organic matter in water by using pre-oxidation and coagulation/sedimentation methods with different chemicals. The reduction rate using NaOCl for organic matter was about 11%, while that of manganese was 12%. The reduction rate using chlorine dioxide (ClO_2) was only 7% for organic matter. However, the rate of manganese was 29% when using ClO_2. Potassium permanganate ($KMnO_4$) removed organic matter and MnOs more effectively, with a rate of 18 and 71%. Moreover, aluminum sulfate ($Al_2(SO_4)_3$), ferric chloride ($FeCl_3$), and polysilicate iron (PSI) worked more effectively, with a reduction rate of 99% and 55% for turbidity and organic matter.

Keywords: pre-chlorination; manganese; organic matter; turbidity

Citation: Wu, Y.-C.; Kan, C.-C.; Lee, S.-C.; Yang, F.-Y. Water Treatment of Manganese Oxides and Organic Matter through Pre-Oxidation and Coagulation/Sedimentation. *Eng. Proc.* **2023**, *55*, 59. https://doi.org/10.3390/engproc2023055059

Academic Editors: Teen-Hang Meen, Kuei-Shu Hsu and Cheng-Fu Yang

Published: 6 December 2023

1. Introduction

Drinking water in Taiwan is obtained from natural reservoirs. In summer, strong sunlight and abundant nutrients in water trigger frequent eutrophication, which complicates water purification processes. During the dry season, the storage of water in the reservoir decreases, and dissolved oxygen (DO) decreases, which forms an anaerobic environment at the bottom. The microbial reduction of iron and manganese oxides increases the concentration of Fe^{2+} and Mn^{2+} in water and turbidity, which makes the water body black and worsens the water quality. This also causes discoloration and a metallic taste in drinking water, and residual manganese ions promote the growth and reproduction of microorganisms. An increase in the number of microorganisms allows manganese oxides to be formed and creates clogging in the pipelines [1,2]. The forms of manganese oxides in water are diverse, including particulate manganese (>0.45 μm), colloidal manganese (<30,000 Da~0.20 μm), and dissolved manganese. Colloidal and particulate manganese can be removed using coagulation and filtration. Dissolved manganese requires oxidants (chlorine, ClO_2, and $KMnO_4$) to be removed from water by forming manganese dioxide, which coagulates and can be filtered out.

Natural organic matter is primarily derived from plants, animals, and their degradation products in aquatic, terrestrial, and marine environments [3]. Owing to the diversity of sources, the composition and concentration of natural organic matter in the environment vary depending on backgrounds, locations, climate, and human activities [4]. Natural organic matter is a complex multiphase mixture of organic substances with different molecular sizes and physical and chemical properties, because it contains various functional groups that include aromatic, aliphatic, phenolic, and quinone structures [5].

In the water-chlorination purification process, the phenols and carboxyls of organic matter in water react with chlorine to generate disinfection byproducts (DBPs) such as total trihalomethane, haloacetic acid, haloketone, and haloacetonitrile. The generation of DBPs is determined by the type of organic matter, temperature, inorganic salts, and pH value [6]. In general, pre-oxidation with coagulants for water purification is carried out to remove organic matter. For this, coagulants such as iron and aluminum salts are hydrolyzed into positively charged metal complex ions, which are then adsorbed onto the negatively charged colloidal surface. Consequently, large polymers containing organic matter are coagulated and deposited on the bottom. As the composition of natural organic matter in different water bodies varies, the removal efficiency of organic matter becomes different.

Therefore, the optimal conditions and strategies for removing manganese oxides and organic matter need to be determined. Thus, we performed related tests under different types and doses of oxidants and coagulants to improve the efficiency of water purification in this study. The result provides a basic reference for further investigation and the development of more effective water treatment procedures.

2. Materials and Methods

The experimental process of this study is shown in Figure 1. The water samples were obtained from reservoirs in south Taiwan. The concentrations of iron, manganese, and total organic carbon (TOC) and the collective property of water were analyzed. The characteristics of organic matter were also analyzed. In the pre-oxidation test, NaOCl, ClO_2, and $KMnO_4$ were used as oxidants to investigate the effect of oxidants on the TOC concentration in water. Changes in the organic matter composition in pre-oxidation were analyzed using the excitation–emission matrix (EEM). The rate of coagulation and sedimentation was measured using coagulants such as $Al_2(SO_4)_3$, $FeCl_3$, and PSI. The pH and turbidity of water samples were measured to identify suitable dosing modes of the coagulants. A simulated distribution system (SDS) was used for chlorination, as shown in Figure 2. The SDS simulated the generation of DBPs in the water pipe after the chlorination of treated water.

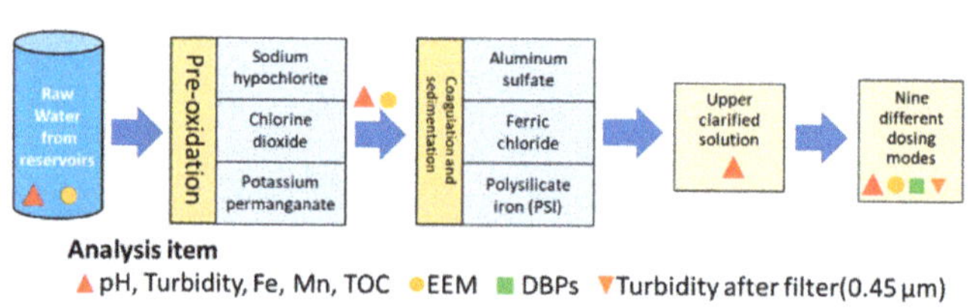

Figure 1. Experimental procedure of this study.

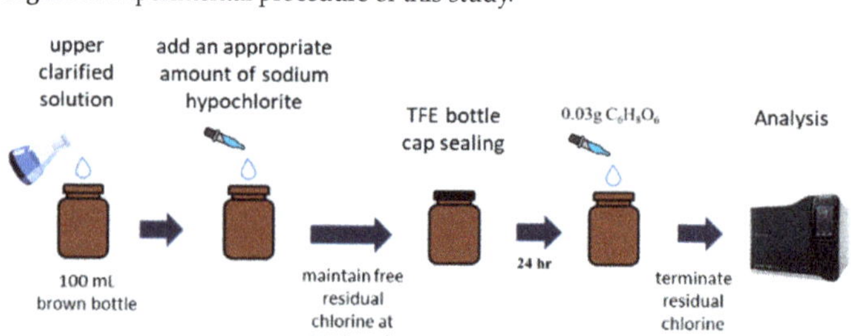

Figure 2. Illustration of SDS chlorination.

Water samples were collected in a 100 mL brown bottle, and an appropriate amount of NaOCl was added to it to control the residual chlorine between 0.5 and 1 mg/L. After the bottle was sealed, the samples were left for 24 h for enough of a reaction to be produced (in real situations, drinking water needs a 24 h wait after NaOCl has been added in a water purification station before reaching domestic homes). After 24 h, ascorbic acid was added to terminate the chlorine reaction.

3. Results

3.1. Effect of Oxidants on Organic Matter

3.1.1. Types and Doses

As the dose of the oxidant increased, the removal rate of TOC increased (Figure 3). However, an excessive amount of the oxidant decreased the removal rate of TOC. References [7,8] pointed out that certain oxidants and excessive oxidants ruptured algal cells, thus resulting in the release of intracellular organic matter. This increased the organic matter content in water, which was reproduced in the results of the experiment in this study. The water samples in this study were not filtered and directly oxidized. The dominant algae species in the water samples were green and blue-green algae.

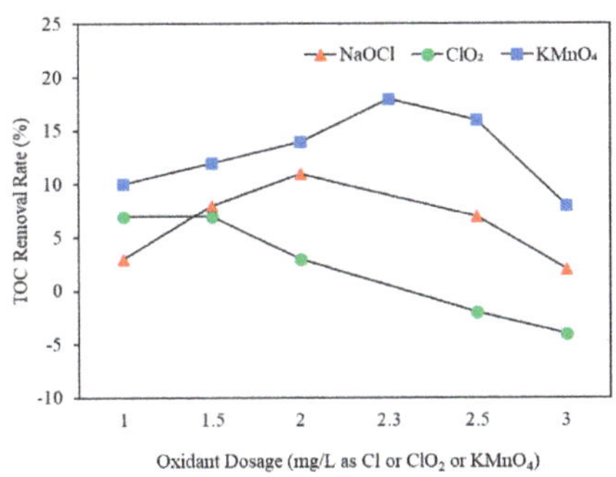

Figure 3. Effect of different oxidants and doses on TOC removal rate.

3.1.2. Effect of Oxidants on Coagulation and Sedimentation

The composition of particulate and dissolved organic matter in water was determined based on the characteristic interval of the EEM. In this study, the excitation and emission wavelength ranges were set to 200–400 and 250–500 nm, respectively. The spectra were classified into four areas, such as by-product-like soluble microbial (Area I, SMP), humic-acid-like soluble microbial (Area II, HA), fulvic-acid-like soluble microbial (Area III, FA), and aromatic proteins soluble microbial (Area IV, AP).

Figure 4 presents the optimal removal dose of the TOC for different oxidants. In this study, 2 mg/L of NaOCl, 1.5 mg/L of ClO_2, and 2.3 mg/L of $KMnO_4$ were added to the water sample for pre-oxidation, and the mixture was stirred for 30 s at 200 rpm. Changes in the organic matter composition before and after stirring were analyzed with EEM. Before oxidation, the spectral intensity of the raw water was the strongest in Area III, followed by Areas II, IV, and I. The organic matter composition in the water was primarily FA-like in composition. With the other three oxidants, the spectral intensities of Areas I, II, and III showed slight changes compared with those before oxidation. Meanwhile, the spectral intensity of Area IV changed significantly after NaOCl was added. Fluorescence data showed a significant weakening trend in Area IV. The decrease in the fluorescence intensity of aromatic proteins was attributed to the generation of DBPs [9]. The addition of ClO_2 and $KMnO_4$ also changed the spectral intensity slightly.

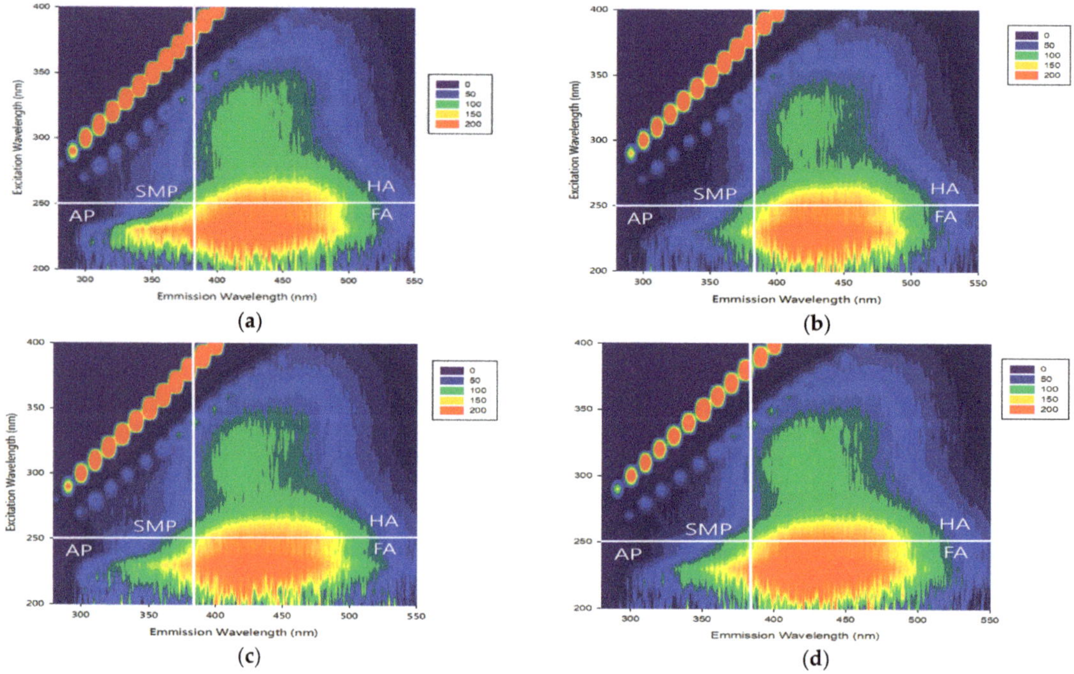

Figure 4. EEM spectra of organic matter oxidized by oxidants: (**a**) no oxidation in water sample; (**b**) 2 mg/L of NaOCl; (**c**) 1.5 mg/L of ClO_2; (**d**) 2.3 mg/L of $KMnO_4$.

3.2. Effect of Oxidants with Coagulants on Coagulation and Sedimentation

In this study, changes in the organic matter content in the water sample were caused by different doses of oxidants. The best treatment method, using oxidants for water purification, completely removed algal cells without rupture, which avoided an increase in dissolved organic matter in water and the generation of DBPs [10]. Therefore, the appropriate doses of oxidants and coagulants needed to be selected for coagulation and sedimentation tests. The selected doses of NaOCl, ClO_2, and $KMnO_4$ were 2, 1.5, and 2.3 mg/L in this study.

Figures 5–8 show the changes in pH and the content of organic matter with three oxidants and the different doses of $Al_2(SO_4)_3$ and $FeCl_3$. When the dose of $Al_2(SO_4)_3$ increased, the removal rate of TOC also increased, while the pH value decreased to 6.7. According to the on-site operation protocol of the water purification plant, the pH of the treated water is maintained above 7 to not exceed the regulatory standards for pH level due to the gradual acidification of drinking water during its transportation in the distribution system. When the dose of $FeCl_3$ increased, the removal rate of TOC kept increasing. The dose of PSI was set independently. From the result, it was found that the appropriate doses of $Al_2(SO_4)_3$, $FeCl_3$, and PSI were 8 mg/L (as 4.2 mg/L Al), 8 mg/L (as 2.8 mg/L Fe), and 3 mg/L (as 3 mg/L as Fe), respectively. In the test for coagulation and sedimentation, the doses of the three oxidants were used with the doses of the three coagulants to produce nine different dosing modes (Table 1). The result was analyzed to investigate the effects of different dosing modes on pH, the concentration of manganese and iron ions, the content of organic matter, and the generation of DBPs. The water samples in this study were not filtered and pH adjusted.

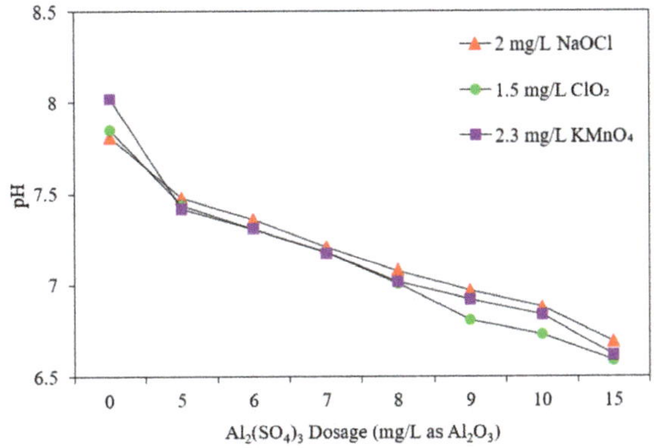

Figure 5. pH changes caused by three oxidants combined with different doses of $Al_2(SO_4)_3$.

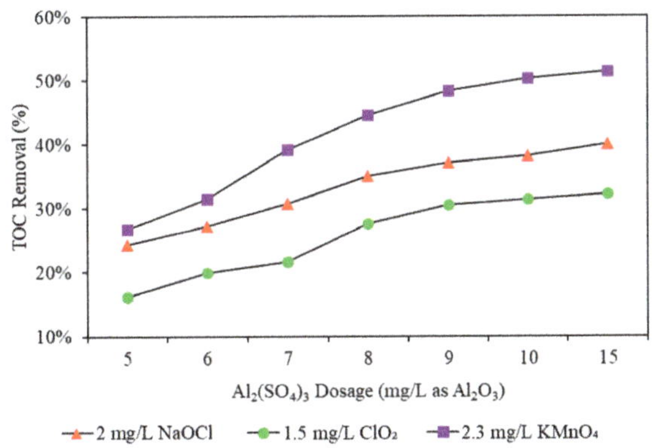

Figure 6. Changes in TOC removal rates caused by three oxidants combined with different doses of $Al_2(SO_4)_3$.

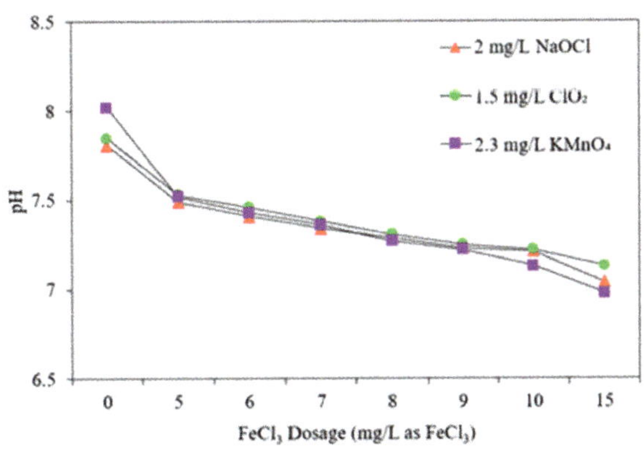

Figure 7. pH changes caused by three oxidants combined with different doses of $FeCl_3$.

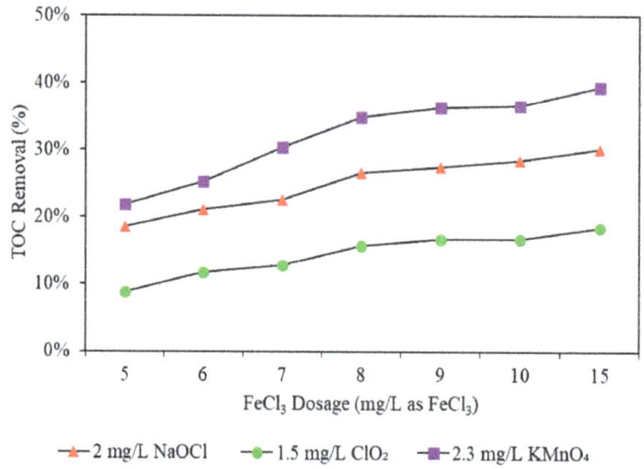

Figure 8. Changes in TOC removal rates caused by three oxidants combined with different doses of $FeCl_3$.

Table 1. Nine different dosing operation modes.

Mode	Dosing Operation
A	2 mg/L NaOCl + 8 mg/L $Al_2(SO_4)_3$
B	2 mg/L NaOCl + 8 mg/L $FeCl_3$
C	2 mg/L NaOCl + 3 mg/L PSI
D	1.5 mg/L ClO_2 + 8 mg/L $Al_2(SO_4)_3$
E	1.5 mg/L ClO_2 + 8 mg/L $FeCl_3$
F	1.5 mg/L ClO_2 + 3 mg/L PSI
G	2.3 mg/L $KMnO_4$ + 8 mg/L $Al_2(SO_4)_3$
H	2.3 mg/L $KMnO_4$ + 8 mg/L $FeCl_3$
I	2.3 mg/L $KMnO_4$ + 3 mg/L PSI

3.2.1. Manganese Removal

Table 2 shows the removal effects of manganese ions with different dosing modes. With no coagulant, manganese ions were removed only via pre-oxidation with NaOCl, and the removal rate was approximately 12%. When $Al_2(SO_4)_3$, $FeCl_3$, and PSI were added, the removal rate of manganese ions increased to 37–50% with coagulation and sedimentation but the concentration of manganese ions still exceeded 0.05 mg/L, which is a drinking water standard. The higher concentration over 0.05 mg/L was maintained because NaOCl could not effectively oxidize manganese ions in pre-oxidation, and a trace amount of manganese ions were not oxidized. Consequently, manganese removal in subsequent coagulation and sedimentation became less prominent. When ClO_2 was used for pre-oxidation, the removal rate of the manganese ions was approximately 30%. If three types of coagulants were added, then the removal rate of manganese ions in coagulation and sedimentation was 63–72%. Correspondingly, the concentration of residual manganese ions in water was lower than 0.05 mg/L. The high oxidizing ability of ClO_2 allowed a considerable removal of manganese ions, which resulted in a better manganese removal effect. When $KMnO_4$ was used for pre-oxidation, the removal rate of manganese ions reached 71%. With $KMnO_4$ and the three coagulants, the removal rate of manganese ions increased to 84–92%. Additionally, the residual manganese concentration was below the drinking water standard. $KMnO_4$ effectively and rapidly oxidized dissolved manganese ions in water to form manganese particles. Coagulation and sedimentation derived an effective removal effect. The experimental results showed that removing manganese ions from water required extensive oxidation in the pre-oxidation process and coagulation and sedimentation. The best oxidant for manganese removal was $KMnO_4$. Appropriate oxidants with coagulants for coagulation and sedimentation guarantee decreased residual manganese in water, which satisfies the quality standards of drinking water.

Table 2. Changes in manganese removal under nine dosing combinations.

Mode	Mn Removal Rate (%)	Residual Mn (mg/L)
Raw water	N/A	0.134
NaOCl	12%	0.118
A	37%	0.085
B	54%	0.062
C	51%	0.065
ClO_2	29%	0.095
D	65%	0.047
E	72%	0.038
F	69%	0.041
$KMnO_4$	71%	0.039
G	84%	0.022
H	92%	0.011
I	89%	0.015

3.2.2. Turbidity Decrease

Table 3 shows the decrease in turbidity with nine dosing modes. Without coagulants, the turbidity of the water sample was decreased only by pre-oxidation. When NaOCl was used as the oxidant, the decrease rate of turbidity was only 3%, and no significant improvement was indicated. With $Al_2(SO_4)_3$, $FeCl_3$, and PSI, the turbidity decreased by 98%. Using coagulants and filtration, turbidity was reduced to 0.46 nephelometric turbidity units (NTU). When ClO_2 was used for pre-oxidation, turbidity decreased to 20%. Pre-oxidation oxidized metal ions in water and produced metal particles that coagulated and sedimented. If three coagulants were added, turbidity decreased by 98–99% through coagulation and sedimentation. Oxidation by ClO_2 did not change the distribution of particle sizes but resulted in the micro-gelation of colloidal particles, thus facilitating particle removal [11]. The maximum decrease rate of turbidity via pre-oxidation using $KMnO_4$ was approximately 36%. Using $KMnO_4$ with the three coagulants, turbidity decreased by 99%, mainly owing to the adsorption of manganese dioxide. Reference [7] discovered that $KMnO_4$ facilitated the removal of organic and inorganic particles. Their experimental results showed that ClO_2 and $KMnO_4$ contributed to the removal of colloidal particles in water. $Al_2(SO_4)_3$ decreased turbidity most significantly. Meanwhile, using ClO_2 and $KMnO_4$ with $Al_2(SO_4)_3$ showed a better effect on turbidity decrease. Using filtration, turbidity could be maintained below the drinking water standard. Therefore, turbidity could be decreased through coagulation, sedimentation, and filtration.

Table 3. Changes in turbidity removal under nine dosing combinations.

Mode	Turbidity Removal Rate (%)	Residual Turbidity (NTU)
Raw water	N/A	213
Only NaOCl	3%	206
A	99%	2.86
B	98%	3.96
C	98%	4.22
Only ClO_2	20%	170
D	99%	1.43
E	98%	3.71
F	98%	3.34
Only $KMnO_4$	36%	136
G	100%	0.11
H	99%	2.92
I	99%	2.69

3.2.3. TOC Removal

Table 4 shows the removal rates of TOC in the water sample using oxidants and coagulants. The results showed that TOC was removed by $KMnO_4$ more than by NaOCl and ClO_2. Among the coagulants, $Al_2(SO_4)_3$ was the most effective in removing TOC, with a removal rate of up to 55%. No significant difference was found in the removal of TOC between $FeCl_3$ and PSI. As organic matter was composed of various large- and small-molecule organic components, the removal of different types of organic matter by different coagulants could not be determined based on the removal results of the TOC.

Table 4. Changes in TOC removal under nine dosing combinations.

Mode		TOC Removal Rate (%)	Residual TOC (mg/L)
Raw water		N/A	2.82
NaOCl	A	46%	1.51
	B	30%	1.98
	C	28%	2.03
ClO_2	D	38%	1.75
	E	23%	2.17
	F	24%	2.13
$KMnO_4$	G	55%	1.26
	H	40%	1.69
	I	37%	1.78

3.2.4. Effect of Organic Matter Composition on Its Removal

Table 5 shows the effect of oxidants and coagulants on the organic matter with different compositions. The spectral intensity of the raw water was the highest in Area III, and Areas II, IV, and I, in that order. Fluorescence intensities decreased according to the degree of coagulation and sedimentation. The fluorescence intensities in Areas II (HA) and III (FA) decreased most significantly. Thus, it was found that coagulation and sedimentation effectively removed organic matter from water, and in particular large-molecule organic matter. Higher removal rates were achieved with modes A, D, and G, in which $Al_2(SO_4)_3$ was used as a coagulant. Mode G showed the best removal efficiency of organic matter in Areas II (HA) and III (FA), where the fluorescence reduction rates were 60 and 56%, respectively. The fluorescence reduction rates using $FeCl_3$ and PSI were similar to $Al_2(SO_4)_3$. However, the fluorescence intensity in Area IV with mode E removed more organic matter. The high oxidizing ability of ClO_2 allowed the disintegration of large-molecule organic matter containing multiple benzene rings and phenolic groups in Areas II and III into aromatic organic matter with smaller molecular weights. In Ref. [12], the molecular weight changes in organic matter were observed after adding ClO_2. It was discovered that ClO_2 oxidized and disintegrated large-molecule organic matter into smaller molecules. The combination of $KMnO_4$ and $Al_2(SO_4)_3$ enabled the most significant decrease in the fluorescence intensity of water after sedimentation. The reduced form of $KMnO_4$ was adsorbed on the surface of algal cells and organic matter, thus increasing the weight. Using coagulants reduced the coagulation and sedimentation of organic matter. When the pH was greater than 7, the hydrolysis products of $Al_2(SO_4)_3$ became colloidal precipitates of $Al(OH)_{3(s)}$. At this point, the gel plume formed became large [13], and organic matter in the water was removed via sedimentation.

Table 5. Analysis results of raw water EEM under different dosing operation modes.

Organic Matter Type	Raw Water	Mode A		Mode B		Mode C	
	Spectral Intensity (E_0)	Spectral Intensity (E_0)	Removal Rate * (%)	Spectral Intensity (E_0)	Removal Rate (%)	Spectral Intensity (E_0)	Removal Rate (%)
SMP	51.172	34.842	32%	36.286	29%	34.971	32%
HA	75.366	34.634	54%	49.108	35%	48.607	36%
FA	128.23	63.388	51%	88.975	31%	88.355	31%
AP	64.506	49.504	23%	49.140	24%	46.599	28%
Organic Matter Type	**Mode D**		**Mode E**		**Mode F**		
	Spectral intensity (E_0)	Removal rate * (%)	Spectral intensity (E_0)	Removal Rate (%)	Spectral intensity (E_0)	Removal Rate (%)	
SMP	33.550	34%	42.301	17%	39.552	23%	
HA	35.454	53%	51.184	32%	51.952	31%	
FA	62.282	51%	95.687	25%	90.628	29%	
AP	43.605	32%	66.207	−3%	46.727	28%	
Organic Matter Type	**Mode G**		**Mode H**		**Mode I**		
	Spectral intensity (E_0)	Removal rate * (%)	Spectral intensity (E_0)	Removal Rate (%)	Spectral intensity (E_0)	Removal Rate (%)	
SMP	29.337	43%	32.914	36%	31.947	38%	
HA	29.959	60%	49.305	35%	47.358	37%	
FA	56.034	56%	90.962	29%	87.158	32%	
AP	39.681	38%	44.134	32%	39.896	38%	

* Removal rate (%) = $(E_0 - E_1)/E_0$.

3.2.5. DBPs

With coagulation and sedimentation in different dosage modes, the number of DBPs in water with the SDS chlorination experiment (24 h) was measured as shown in Table 6. When $KMnO_4$ and ClO_2 were used as the pre-oxidant, the concentration of DBPs was more effectively decreased than with NaOCl, and the decrease in haloacetic acid was significant. If $Al_2(SO_4)_3$ was used as the coagulant, the concentration of DBPs (total trihalomethane and haloacetic acid) was lowered compared to those with $FeCl_3$ and PSI. $Al_2(SO_4)_3$ had the best removal efficiency for organic matter in Areas II and III of the EEM. Since large-molecule organic matter such as HA and FA contained more benzene rings and phenolic groups, their disinfection by-product generation potential was higher than that of small-molecule organic matter [14,15].

Table 6. Generation of DBPs under different dosing operation modes.

Mode	Disinfection By-Product	
	TTHMs	HAAs
A	0.077	0.066
B	0.119	0.091
C	0.129	0.088
D	0.079	0.052
E	0.115	0.074
F	0.119	0.079
G	0.071	0.045
H	0.107	0.078
I	0.117	0.086

4. Conclusions

The experimental results on removing manganese ions from the water samples showed that NaOCl, ClO_2, and $KMnO_4$ were effective oxidants. $KMnO_4$ showed the highest manganese removal rate, followed by ClO_2. The best removal rate of TOC was observed when using $KMnO_4$. Among the three coagulants, $Al_2(SO_4)_3$ was better at removing TOC. Using ClO_2 and $KMnO_4$ instead of NaOCl as pre-oxidants reduced the generation of trihalomethane (THM) and haloacetic-acid (HAA). The amount of organic matter in Areas II (HA) and III (FA) in the water sample was considerably decreased through coagulation and sedimentation. A 0.45 μm nylon filter was used to simulate the filtration process in this study; filters fabricated using quartz sand, anthracite, or other materials may be used to construct a better water treatment process.

Author Contributions: Conceptualization, Y.-C.W., C.-C.K., S.-C.L.; methodology, Y.-C.W., C.-C.K., S.-C.L., F.-Y.Y.; software, Y.-C.W., F.-Y.Y.; validation, Y.-C.W., C.-C.K., F.-Y.Y.; formal analysis, Y.-C.W., F.-Y.Y.; investigation, Y.-C.W., C.-C.K., S.-C.L., F.-Y.Y.; resources, C.-C.K., S.-C.L.; data curation, Y.-C.W., C.-C.K., S.-C.L., F.-Y.Y.; writing—original draft preparation, Y.-C.W., C.-C.K., F.-Y.Y.; writing—review and editing, Y.-C.W., S.-C.L., F.-Y.Y.; visualization, Y.-C.W.; supervision, C.-C.K., S.-C.L.; project administration, C.-C.K., S.-C.L.; funding acquisition, C.-C.K. All authors have read and agreed to the published version of the manuscript.

Funding: This research was funded by the Taiwan Water Corporation, No.2-1, Sec. 2, Shuangshih Rd., North District, Taichung City 404403, Taiwan.

Institutional Review Board Statement: Not applicable.

Informed Consent Statement: Not applicable.

Data Availability Statement: Data are contained within the article.

Conflicts of Interest: The authors declare no conflict of interest.

References

1. Cerrato, J.M. Biogeochemical Cycling of Manganese in Drinking Water Systems. Ph.D. Thesis, Virginia Tech, Blacksburg, VA, USA, 2010.
2. Kohl, P.M.; Medlar, S.J. *Occurrence of Manganese in Drinking Water and Manganese Control*; AWWA: Denver, CO, USA, 2006.
3. Uyguner Demirel, C.; Bekbolet, M.; Swietlik, J. *Natural Organic Matter: Definitions and Characterization. Control of Disinfection By-Products in Drinking Water Systems*; Nova Science Publishers, Inc.: Hauppauge, NY, USA, 2007; pp. 253–277.
4. Dawson, J.J.C.; Malcolm, I.A.; Middlemas, S.J.; Tetzlaff, D.; Soulsby, C. Is the composition of dissolved organic carbon changing in upland acidic streams? *Environ. Sci. Technol.* **2009**, *43*, 7748–7753. [CrossRef] [PubMed]
5. Aliverti, N.; Callegari, A.; Capodaglio, A.G.; Sauvignet, P. Nom removal from freshwater supplies by advanced separation technology. In *Advanced Water Supply and Wastewater Treatment: A Road to Safer Society and Environment*; Springer: Dordrecht, Germany, 2011.
6. Delpla, I.; Jung, A.V.; Baures, E.; Clement, M.; Thomas, O. Impacts of climate change on surface water quality in relation to drinking water production. *Environ. Int.* **2009**, *35*, 1225–1233. [CrossRef] [PubMed]
7. Chen, J.J.; Yeh, H.H. The mechanisms of potassium permanganate on algae removal. *Water Res.* **2005**, *39*, 4420–4428. [CrossRef] [PubMed]
8. Huang, W.J.; Wang, C.C. Evaluation Metabolic Organic Products of Algae-*Chlorella* sp. and Assessing the Fate of Organic Compounds During Ozonation. *Hung Kuang Unversity J.* **2004**, *43*, 159–165.
9. Tian, C.; Guo, T.T.; Liu, R.P.; Jefferson, W.; Liu, H.J.; Qu, J.H. Formation of disinfection by-products by *Microcystis aeruginosa* intracellular organic matter: Comparison between chlorination and bromination. *Huan Jing Ke Xue* **2013**, *34*, 4282–4289. [PubMed]
10. Plummer, J.D.; Edzwald, J.K. Effects of chlorine and ozone on algal cell properties and removal of algae by coagulation. *Water Supply Res. Technol.-Aqua* **2002**, *51*, 307–318. [CrossRef]
11. Chen, S.H. The Effect of ClO_2 Pre-Oxidation on the Coagulation Efficiency of Drinking Water Treatment. Master's Thesis, Department of Environmental Engineering and Science, Chia-Nan University of Pharmacy and Science, Tainan, Taiwan, 2003.
12. Świetlik, J.; Raczyk-Stanisławiak, U.; Biłozor, S.; Ilecki, W.; Nawrocki, J. Adsorption of natural organic matter oxidized with ClO_2 on granular activated carbon. *Water Res.* **2002**, *36*, 2328–2336. [CrossRef] [PubMed]
13. Kan, C.C.; Huang, C.P.; Pan, J.R. Time requirement for rapid-mixing in coagulation. *Colloids Surf. A Physicochem. Eng. Asp.* **2002**, *203*, 1–9. [CrossRef]
14. Xu, X.; Kang, J.; Shen, J.; Zhao, S.; Wang, B.; Zhang, X.; Chen, Z. EEM–PARAFAC characterization of dissolved organic matter and its relationship with disinfection by-products formation potential in drinking water sources of northeastern China. *Sci. Total Environ.* **2021**, *774*, 145297. [CrossRef] [PubMed]
15. Pan, Y.J. The Relationship between Organic Fractions of NOM and Disinfection by-Products Formation Potential. Master's Thesis, Department of Environmental Engineering, National Cheng Kung University, Tainan, Taiwan, 2015.

engineering proceedings

Proceeding Paper

Development of a New Heat Source Based on Inducing Heat for Greenhouses †

Ning Zhu [1,*], Minyu Li [2], Ben Nanzai [1], Shigeru Kubono [3], Hiromi Fujimura [3] and Mitsuhiro Sakamoto [3]

1 Department of Material and Life, Faculty of Science and Engineering, Shizuoka Institute of Science and Technology, Fukuroi 437-0032, Japan; nanzai.ben@sist.ac.jp
2 Department of Complex Systems Science, Graduate School of Informatics, Nagoya University, Nagoya 464-8601, Japan; li.minyu.b0@s.mail.nagoya-u.ac.jp
3 TSK Corporation, Fukuroi 437-0032, Japan; shigeru@tskcorp.com (S.K.); fujimura.0409@uv.tnc.ne.jp (H.F.); sakamoto@tskcorp.com (M.S.)
* Correspondence: zhu.ning@sist.ac.jp
† Presented at the IEEE 5th Eurasia Conference on Biomedical Engineering, Healthcare and Sustainability, Tainan, Taiwan, 2–4 June 2023.

Abstract: In Fukuroi City Japan, greenhouses are usually used for crown musk melon (CMM) cultivation. When the temperature inside the greenhouse is lower than 25 °C, the CMM quality deteriorates. Hence, from late autumn until the middle of spring, oil is used to supply hot water to greenhouses from a boiler through a heat exchanger . However, since oil prices have recently soared, fuel expenses have drastically increased which heavily pressures the CMM business. In addition, with the promotion of the carbon-neutral policy, environmentally friendly heat sources are emphasized instead of fossil fuels. The TSK corporation has produced a new inducing heating (IH) source where heat is generated by rotating a plate on which a couple of permanent magnets are mounted using a motor. Since the rotation speed is easily controlled, the IH heat source capacity can be freely adjusted. To apply the IH source to the greenhouse, an experimental system was created in this study. During the experiment, water inside a vessel (maximum volume of 90 L) was heated to 70 °C by the IH system. Then, the heated water was circulated for heat dissipation through a heat exchanger by a pump. When the temperature of the water was lower than the target temperature, the IH system restarted to heat the water back up to 70 °C. Under several experimental conditions, the heating time, reheating time, and electric power were measured and evaluated. It was confirmed that the new IH heat source could possibly be applied to greenhouses.

Keywords: greenhouse heat source; IH; carbon-neutral; permanent magnet

Citation: Zhu, N.; Li, M.; Nanzai, B.; Kubono, S.; Fujimura, H.; Sakamoto, M. Development of a New Heat Source Based on Inducing Heat for Greenhouses. *Eng. Proc.* **2023**, *55*, 60. https://doi.org/10.3390/engproc2023055060

Academic Editors: Teen-Hang Meen, Kuei-Shu Hsu and Cheng-Fu Yang

Published: 6 December 2023

1. Introduction

Fukuroi greenhouse crown musk melons (CMMs) are a branded product of Fukuroi City. Producing CMMs requires strict temperature control. From mid-autumn to mid-spring, the temperature of the soil in greenhouses must be maintained above 25 °C. Usually, oil is used to produce warm water for greenhouses. However, fuel prices have continued to soar due to the impact of recent high crude oil prices and the weaker yen, which pressures farmers. In addition, the exhaust gas from heavy oil fuel contains greenhouse gases including carbon dioxide and NOx, which have a negative effect on the environment. Therefore, while the government promotes decarbonization policies, it is necessary to also introduce electric heat source technology in the greenhouses used for growing CMMs [1].

The TSK corporation, located in Fukuroi City, has developed MAGHEAT as a heat source, which uses a magnet for high-speed heating. MAGHEAT is a new type of heater that rotates a disk with a strong magnetic field bundled with permanent magnets to heat non-ferrous metals such as aluminum alloys and allow for self-heating by Joule heat. Compared to conventional heating methods, MAGHEAT is economical, responsive, controllable, and space-saving. In this study, we established the following research objectives based on the current use of hot water boilers, taking the greenhouses owned by members of the Fukuroi Greenhouse Crown Melon Association and exchanging information with representatives of the association and the TSK Corporation. We built a new hot water supply system for melon greenhouses using a new electric heat source instead of a hot water supply system (conventional heat source). A small pilot hot water supply system using MAGHEAT was created and a series of experiments were conducted to verify the possibility of using MAGHEAT.

2. Principle of MAGHEAT

MAGHEAT is an electromagnetic induction heating device (Figure 1). It consists of a motor, a MAGHEAT magnet board, and permanent magnet arrays. Electromagnetic induction is generated using the magnet board in which a magnet is embedded and an eddy current is generated inside the metal. The current (eddy current) is replaced by heat (Joules) due to the internal resistance of the metal. The quantity of the eddy current is supposed to be proportional to the motor rotation speed, which means that the heating time can be adjusted if the motor rotation speed is controlled.

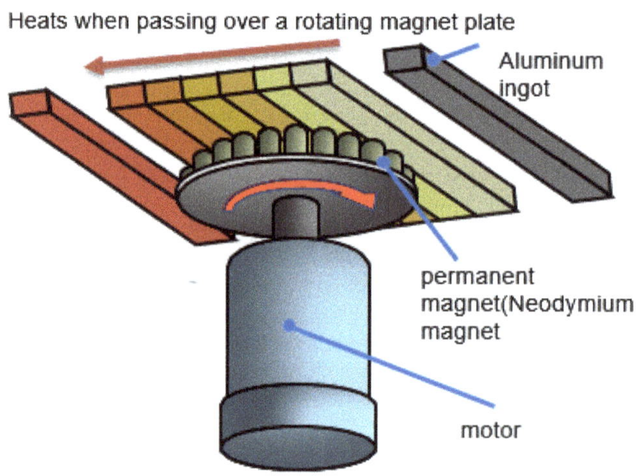

Figure 1. Structure of MAGHEAT.

3. Experiment

The experimental system using MAGHEAT as a heat source for heating water is shown in Figure 2. It consists of a vessel, MAGHEAT (with a motor with an output of 55 kW), water pump, heat exchanger, flowmeter, thermocouples, and PLC controller. During the experiment, first, water in the vessel was heated by MAGHEAT to 70 °C. Then, the power of the MAGHEAT was turned off while water pumps were powered on in order to circulate the water through the heat exchanger. When the temperature after cooling reached the target value, the MAGHEAT restarted to heat the cooled water back to 70 °C. This process was repeated several times. In the experiment, the heating time, cooling time, and electric energy were measured at different motor rotation speeds, target temperatures, and water quantities. The experimental conditions are shown in Table 1, and the MAGHEAT heating source is presented in Figure 3.

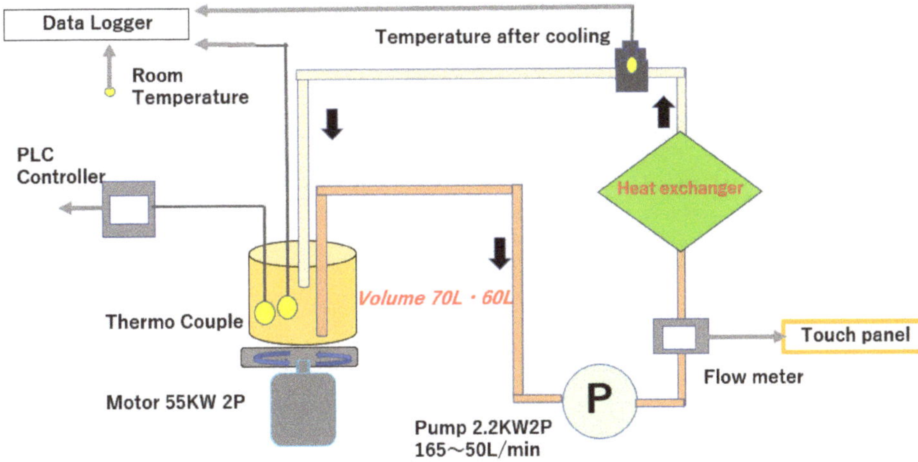

Figure 2. Experimental system using MAGHEAT as a heat source for heating water.

Table 1. Experimental conditions (Flow rate = 30 L/min).

Ex.No	1	2	3	4	5	6
Water Quantity (L)	70	70	70	70	90	90
Target temperature (°C)	60	60	50	40	60	60
Motor Rotation speed (rpm) 10^3	2	3	2	2	2	3

Figure 3. MAGHEAT heating source.

4. Results and Discussions

4.1. Heating and Cooling Process

The temperature changes in the heating and cooling process are shown in Figure 4. This result was obtained under the conditions of 2000 RPM and 70 L/min. At the beginning, the water in the vessel was heated by MAGHEAT to 70 °C, and then the MAGHEAT was stopped to begin cooling. When the temperature value after cooling reached 60 °C, the MAGHEAT was restarted to heat the cooled water back to 70 °C. This heating and cooling process was repeated 5 times. MAGHEAT efficiently started and stopped according to the target temperature, which showed that MAGHEAT could be used as the new heat source for the greenhouse. The first heating time was 604 s because water had to be heated from 18 °C. The average time for heating was 160 s, which was shorter than the first heating time. The accumulated consumed electric energy was 12 kWh.

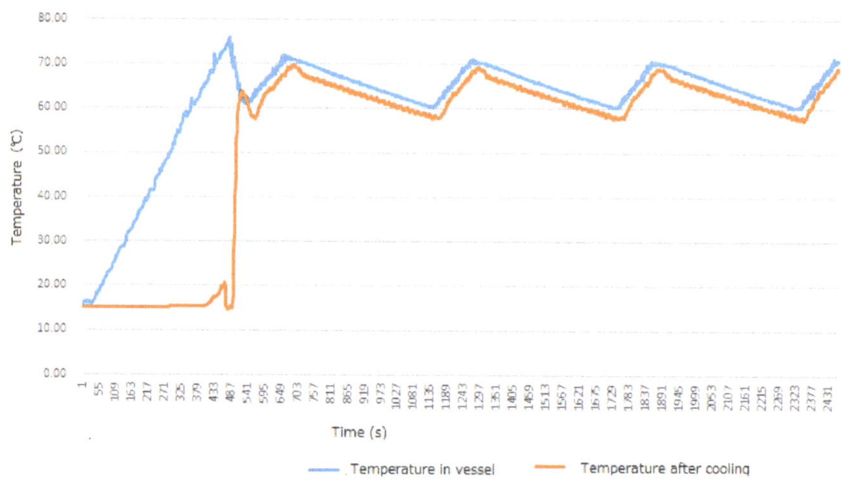

Figure 4. Temperature changes in the heating and cooling process.

4.2. Effect of Motor Rotation Speed on Heating Time

The effect of motor rotation speed on heating time is shown in Figure 5. The heating time was 604 s at 2000 RPM and 456 s at 3000 RPM. The average time for the rest of the heating process was 160 s at 2000 RPM and 120 s at 3000 RPM. The heating time was shorter than that at 3000 RPM. Higher motor rotation speed increased heating and shortened the heating time.

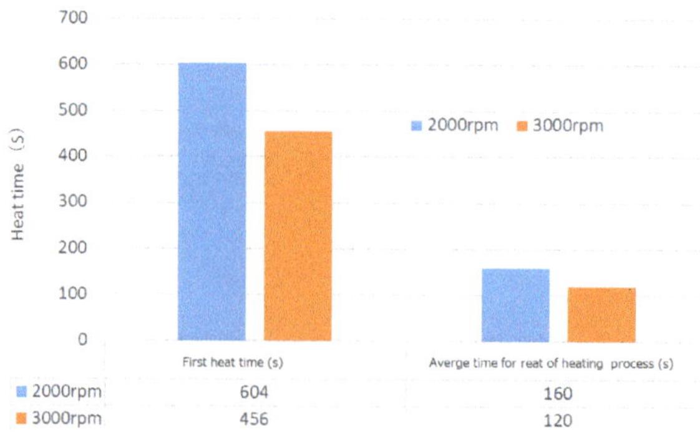

Figure 5. Effect of motor rotation speed on heating time.

4.3. Effect of Motor Rotation Speed on Heating Time

The effect of motor rotation speed on the accumulated electric energy is shown in Figure 6. At 2000 RPM, it was 12 kWh, while at 3000 RPM, it was 11.4 kWh. At 3000 RPM, the energy conversion efficiency became better.

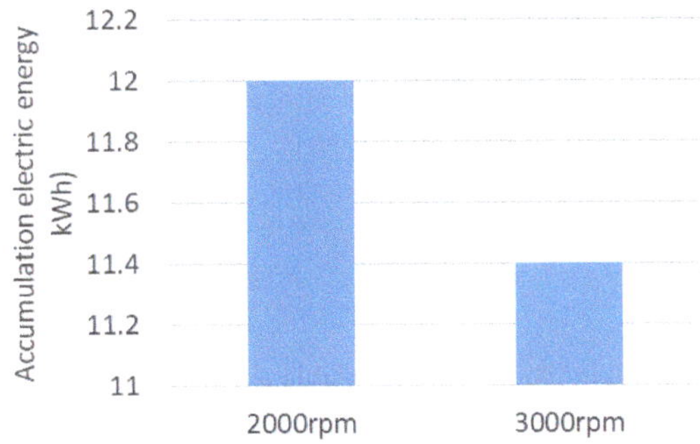

Figure 6. Effect of motor rotation speed on the accumulation of electric energy.

5. Conclusions

To validate the use of MAGHEAT as a new heat source for greenhouses, an experiment was conducted to investigate the effect of rotation speed on heating and accumulated electric energy. It was found that MAGHEAT can be the new heat source for controlling temperature changes and heating processes and applied to greenhouses. In addition, we found that a higher motor rotation speed shortened the heating time and improved energy conversion efficiency.

Author Contributions: Conceptualization, N.Z.; methodology, B.N.; experiment, H.F.; M.S.; data curation, M.L.; writing—S.K.; writing—review and editing, N.Z. All authors have read and agreed to the published version of the manuscript.

Funding: This study is supported by Fukuroi Industry Innovation Center District Research Grant which is offered by Fukuroi Municipal Government.

Institutional Review Board Statement: Not applicable.

Informed Consent Statement: Not applicable.

Data Availability Statement: The data presented in this study are available upon request from the corresponding author.

Acknowledgments: The author wanted to thank Kohei Shibata for his kind help.

Conflicts of Interest: Authors Shigeru Kubono, Hiromi Fujimura and Mitsuhiro Sakamoto were employed by TSK cooperation. The remaining authors declare that the research was conducted in the absence of any commercial or financial relationships that could be construed as a potential conflict of interest.

Reference

1. Katsurai, M. *Basic Energy Engineering*; Math and Science Engineering Press: Tokyo, Japan, 2002; pp. 73–89; ISBN 4-901683-04-7.

engineering proceedings

MDPI

Proceeding Paper

Applying the Engage, Explore, Explain, Elaborate, and Evaluate Procedure in STEAM Education for Primary Students: A Sample with the Topic "My Green Garden" †

Hang Nguyen Thi Thu *, Bich Tran Ngoc and Thanh-Binh Nguyen

Department of Primary Education, Thai Nguyen University of Education, Thai Nguyen City 24000, Vietnam; bichtn@tnue.edu.vn (B.T.N.); binhnt@tnue.edu.vn (T.-B.N.)

* Correspondence: hangntt@tnue.edu.vn

† Presented at the IEEE 5th Eurasia Conference on Biomedical Engineering, Healthcare and Sustainability, Tainan, Taiwan, 2–4 June 2023.

Abstract: STEM education has been a topic of interest in many countries around the world. Many STEM research studies have been conducted; however, most of them focus on middle-school students and high-school students. At the primary level, STEAM education is known as an advanced mode of STEM education, which is highly integrated between science, technology, engineering, arts, and mathematics. This study proposes a procedure of organizing primary-school students' learning activities in STEAM education as an approach to help students experience positive emotions and stimulate passion, thereby developing the capacity of problem solving and creativity for students—one of the common competencies emphasized in the General Education Program 2018 in Vietnam. The procedure of organizing STEAM education activities in this study is close to the 5E procedure, but it is flexibly applied to suit the psycho-physiological characteristics of primary-school students in Vietnam. This research was conducted under the project title "Developing STEM education in a number of schools in Lang Son province to meet the new general school program" according to the cooperation document between the University of Education—Thai Nguyen University with the Department of Education and Training of Lang Son Province, No. 77/HT-SGD and DTLS-ĐHSPTN, dated 10 January 2020. The trial of this study was conducted at eight primary schools in Lang Son, Vietnam, with the theme "My green garden" in teaching nature and society in grade 1. Through observing the students' participation in learning activities, evaluating the learning products made by those students, and collecting comments and feedback from teachers about the procedure of organizing the learning activities, it can be initially seen that the students actively participated in learning activities with joyful feelings, ready to cooperate with each other to complete assigned tasks, and that the students' creativity was clearly expressed through many unique ideas.

Keywords: STEAM; primary school; 5E procedure

Citation: Nguyen Thi Thu, H.; Tran Ngoc, B.; Nguyen, T.-B. Applying the Engage, Explore, Explain, Elaborate, and Evaluate Procedure in STEAM Education for Primary Students: A Sample with the Topic "My Green Garden". *Eng. Proc.* **2023**, *55*, 61. https://doi.org/10.3390/engproc2023055061

Academic Editors: Teen-Hang Meen, Kuei-Shu Hsu and Cheng-Fu Yang

Published: 6 December 2023

1. Introduction

1.1. Common Issues in STEM and STEAM Education

Entering the twenty-first century, STEM has emerged as an effective educational method, equipping learners with interdisciplinary knowledge and skills that are related to each other within the fields of science, technology, engineering, and math. Instead of approaching each subject separately, STEM connects the learning objects of these subjects based on situations associated with high practice [1] helping to solve complex problems relevant to the context of the changing world [2]. When researching STEM, many researchers have pointed out the outstanding advantages of this educational method, such as helping learners to develop problem-solving competency and increasing learners' creativity [3].

STEM education is understood as an interdisciplinary approach to the fields of science, technology, engineering, and mathematics, in which the learning content is linked to practice based on a series of teaching activities which are interactive and active. In the current trend of STEM development, STEAM is a newer integrated trend, linking artistic activities in correlated relationships with the fields of science, technology, engineering, and mathematics. STEAM's philosophy revolves around the following concepts: STEAM is the result of the combination of science and technology interpreted through technical and artistic elements, and this combination is based on relevant mathematical elements [2]. The artistic content integrated into STEAM education can include many factors, such as the aesthetics of the artistic products created by students; humanity in communication, behavior, and coordination with each other to solve learning problems associated with practice; and sociality integrated through activities. The letter "A" (art) in STEAM education is an important element and an indispensable part of STEAM education. It represents language arts, social studies, aesthetic arts, and music. STEAM education helps to promote learners' creative thinking through discovering art [4]. STEAM education can be considered as a transformation of STEM education, creating a way to organize students' learning activities based on highly practical problems through experience and discovery, and significantly contribute to the development of emotional values and problem-solving skills that are critical and flexible. One of the goals of STEAM education is to develop learners' ability to create and innovate [2]. In STEAM education, the difference between each student is the factor that promotes learning [5,6]; the learners' creativity is supported and encouraged [6–8]; their interdisciplinary thinking is enhanced [3] and learners are stimulated to apply science, technology, engineering, and math into their work and life [9].

In STEAM education, learners can interact more with each other; students participating in STEAM topics not only simply form their understanding of science, but above all, they can connect interdisciplinary knowledge to form and develop a cultural behavioral ability, thereby responding to the changes and developments within society [2,4,8].

1.2. STEAM Education for Primary-School Students in Vietnam

In primary teaching in Vietnam, STEM has only been approached by schools and teachers recently. In 2013, the Vietnamese Ministry of Education and Training launched interdisciplinary movements in schools, mainly middle and high schools. In December 2018, when the General Education Program was enacted, STEM was a term that was emphasized and interested in the curriculum from primary to junior high school and high school [10–12]. In the General Education Program, the following is emphasized: "Implementing integrated education, especially the integration of science, technology, engineering, and math (STEM education); integrated education in environmental protection; promoting the economical and efficient use of energy; preventing natural disasters; adapting to climate change; and meeting the requirements of sustainable development in society" [10]. In particular, at the primary level, STEAM has a lot of potential because the curricula in mathematics, nature and society, science, information technology, and technology were built in an open and practical way, creating favorable conditions and facilitating schools and teachers to build STEAM topics on the basis of integrating interdisciplinary activities between science, technology, engineering, arts, and math to solve problems associated with everyday life.

Lower primary students are aged 6 to 8 years old; this is the transitional period from preschool to primary. Due to the change in the learning environment, they usually have certain changes in terms of their psycho-physiology, with the following common features: (1) Body: their high-level nervous system is improving in terms of function, and their bodies are also on the rise in terms of height and weight. Most students in this age group enjoy active games and physical activities. (2) Consciousness: the students' thinking is gradually shifting from action visualization to figurative and abstract thinking; intellectual games are also gradually being interested and noticed. (3) Language: gradually improving, but the ability to attain information through images is faster and more accurate through texts; at this age, students are more interested in experiential learning and field learning than traditional classroom-learning activities. (4) Personality: their stance is not stable;

they do not have selective receptive skills, so they are easily affected and changed by their surrounding objects [13].

From a number of psycho-physiological characteristics of primary-school students, it can seen that, for them, learning activities based on manipulating objects through practice, experiments, and exploration play an important role. These learning activities not only attract students to participate, but also stimulate their creativity, curiosity, and problem-solving capacity. When students approach and learn through STEAM, they will have the opportunity to learn and explore science through light and highly attractive learning activities, thereby developing their cooperation, communication, and creative thinking abilities in solving practical problems.

1.3. The 5E Procedure in STEAM Education

5E stands for five words beginning with letter "E": engage, explore, explain, elaborate, and evaluate. This procedure is suitable and advantageous for teaching science and STEM or STEAM education [8,9]. It was invented in 1987 by Dr. Rodger W. Bybee and his colleagues in the Biological Sciences Curriculum Study (BSCS) in the US. This is a constructivism-based model in which students mobilize their existing knowledge and experiences to experience and form ideas and build new knowledge [7,14].The Bybee group's 5E model was inherited and developed from the learning cycle of Myron Atkin and Robert Karplus [6]. Based on this model, the learning process of biology is undertaken according to a unified logic. From those biology lessons, the 5E model is continued to be developed and applied in many other disciplines, including math, engineering, and technology. So far, there have been many studies in many different countries showing the effectiveness of the 5E model in STEM education [6,15].

In STEAM education, the 5E model can be understood in association with the five steps of the students' learning process (Figure 1) [15], in which the following takes place:

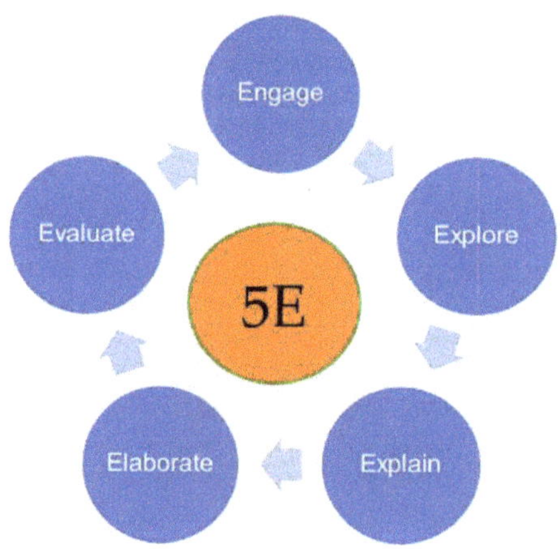

Figure 1. The 5E procedure.

Engage: This is the beginning stage of the 5E-based learning process. In this step, the teacher poses a situation, a question, or a problem to arouse students' curiosity so that they can connect and mobilize existing knowledge and learning experiences to solve the problem.

Explore: students are encouraged to participate in the experience to propose measures, come up with ideas and plan an investigation, and collect information to solve the problems raised in the previous step.

Explain: At this stage, students mobilize their own knowledge and skills to investigate, explore, and prove proposed ideas related to STEAM educational topics. During this stage, the teacher's engagement with questions and instructions can guide learners to understand more deeply the STEAM topics being studied.

Elaborate: At this stage, students share the results and knowledge they have gained. Through the presentation and sharing of results and the implementation of ideas, the students' knowledge is confirmed, and scientific knowledge is formed and developed.

Evaluate: This is the last stage in the 5E procedure. At this stage, students are encouraged to assess themselves, to re-evaluate the process of participating in the activity, and gain scientific knowledge about the STEAM topic; at the same time, the teacher can also evaluate them by observing group activities and assessing their learning products to acquire an overview of their progress.

2. Methodology

2.1. Target Group

The goal of this study is to find out how appropriate it is to use the 5E procedure in STEAM education for Lang Son primary students; what effects can be brought to students' teamworking ability and creativity; and the feasibility of applying the 5E procedure in STEAM education for different students in various places with different living standards and competences; or whether it can only apply to those who live in cities with high living standards, thereby pointing out recommendations for teachers when applying the procedure in organizing STEAM educational activities for primary-school students in Lang Son in particular and in Vietnam in general.

2.2. Methods of Inquiry

In order to achieve the goals set, the research team designed the theme My green garden for students in grade 1 (nature and society) following the 5E procedure, including 5 steps: (1) engage; (2) explore; (3) explain; (4) elaborate; and (5) evaluate. This topic was deployed to teach students in grade 1 on the basis of applying a combination of teaching methods, including group discussion and problem detecting and solving.

My green garden (grade 1) [10] was conducted in 2 periods; the main objectives of the topic were that, after the lesson, the students should be able to achieve the following:

- Draw and annotate the external parts (stem, roots, leaves, flowers, fruits) of a plant;
- Tell about the color, shape, and size characteristics of some common plants;
- Design and create a model of a green garden;
- Develop a love for nature and one's surroundings;
- Develop creative thinking.

The students' learning progressed according to the 5 steps of the 5E procedure as follows:

Step 1: The teacher asks students to move to the school garden and observe the trees in the school garden in groups of 4 to 6 students to answer the following questions: What are the names of the plants you observe? What are their sizes and colors? What external organs do those plants have? After that, the teacher asks the students to draw the observed plant and annotate its external parts on the observation sheet and to share it with their friends in the group.

Step 2: The teacher asks the students to come up with questions and answers on how those plants are organized and planted in the yard, and assigns the following task: With recycled or natural materials, design the model of a green garden.

Step 3: The teacher asks the students to work in groups to design and implement the concept of a green garden model. The teacher instructs the students to create models of green plants from clay, colored paper, glue, recycled materials, etc. Then, they are instructed to arrange the plant models into cardboard or foam, creating paths, decorative lights, entertainment, and relaxation corners on their green garden models.

Step 4: The teacher asks the students to report and share the results of their group work. Each group can introduce the plants that they have created; how they decorated it; and how they would feel if they could play in that garden.

Step 5: The teacher asks the students to self-assess what they have learned in the topic by stating the activities they have participated in and the activities they enjoyed most, coloring the appropriate emotional faces after learning the subject.

After organizing the teaching process according to the above 5 steps, the research team conducted a survey to evaluate the feasibility of the 5E model when applied to teaching STEAM in nature and society for lower-primary-school students, and answer the questions outlined in Section 2.1 of this article.

My Green Garden was taught experimentally in 12 classes of 6 schools in 3 regions with students from different living standards. Each class was divided into 4 to 6 groups. Those who participated in the observation of the STEAM lesson included 31 teachers of grade 1; 28 teachers of grade 2; 10 academic leaders; and 8 managers and members of the research team. They had their own observation sheets. At the end of the lesson, the research team took their ideas, interviewed them, and asked the students to carry out a self-assessment to evaluate their level of excitement as well as their average scores of their group work and products after participating in the STEAM lesson.

3. Research Findings

3.1. Information about the Experiment

Subjects: 358 students in grade 1 from 6 schools in 3 areas with different levels of living standards and educational quality within Lang Son province, a northern mountainous province of Vietnam where many ethnic minority groups live. Compared with the general educational level of students in Vietnam, those in Lang Son seem to have a lower quality of education. The living conditions of students in different areas are significantly different; most students are shy when participating in learning activities.

Area 1: 138 students (74 boys and 64 girls) from 4 classes in grade 1 at Hoang Van Thu primary school and Chi Lang primary school in Lang Son city. These schools have favorable conditions for educational development, located in the center of Lang Son city; the students' level of awareness is relatively equal, with good educational conditions and quality.

Area 2: 122 students (62 boys and 60 girls) from 4 classes in grade 1 at a primary school in the Tu Mich commune (Loc Binh district, Lang Son province) and another primary school in the Long Dong commune (Bac Son district, Lang Son province). These schools are of moderate conditions, located in districts which are about 100 km from the center of Lang Son city; the students' cognitive level is average; the facilities and the quality of the teachers may ensure minimum learning needs.

Area 3: 98 students (47 boys and 51 girls) from 4 classes in grade 1 at Chau Son primary school (Dinh Lap district, Lang Son province) and Thien Thuat primary school (Binh Gia district, Lang Son province). These schools are located in highland districts along the Vietnam–China border, in disadvantaged areas of Lang Son province, occupied by a more than 95% proportion of ethnic minorities (Tay, Nung, Dao, San Chi); the students have unequal cognitive levels, and there is a shortage of infrastructure conditions and facilities.

Trial teaching period: 3 February 2020 to 10 February 2020. At this time, students in grade 1 already had the ability to read short sentences, but their writing ability was limited; it took them 2 or 3 min to write a short sentence, which is one of the students' limitations that teachers need to consider to design appropriate learning activities for their STEAM education.

The trial teaching period was specifically implemented as follows:

On 3 February 2020 and 4 February 2020, the research team met the students and teachers of grade 1 in person to study and evaluate the students' cognitive levels and living conditions, and to observe the classroom and school garden to design a STEAM lesson and develop an appropriate teaching plan.

On 5 February 2020, the research team completed the design of a teaching plan for the topic *My green garden*.

On 6 February 2020 and 7 February 2020, the research team met and discussed with the teachers who were in charge of the experimental lessons about the teaching plan on the topic of *My Green Garden*, thereby consulting homeroom teachers and first-grade academic leaders from primary schools in 3 areas in Lang Son city.

On 10 and 11 February 2020, the research team met the students of the experimental classes from 3 areas in Lang Son city and asked them to prepare necessary supplies to study the theme My Green Garden, such as colored paper, cardboard, clay, foam board, and other recycled materials.

On 13 and 14 February 2020, the research team and teachers who were responsible for the experimental lesson prepared equipment to teach the theme My Green Garden, and designed observation sheets, evaluation sheets, and questionnaires, which would be used while the students were participating in activities to evaluate the effectiveness of STEAM topic teaching after the end of the lesson.

From 17 February to 19 February 2020, 12 teachers organized the teaching of STEAM to students in 3 different areas of Lang Son with the theme My green garden (time: 70 min) under the observation of a number of teachers, the first-grade academic leader, and the managers of the Lang Son Department of Training and Education.

3.2. The Results of the Experimental Lessons

Students: * Pre-lesson: In the preparation phase, prior to the participation of learning "My Green Garden", the students brought a number of materials in preparation for the lesson. When they were directly interviewed with some questions, such as "Are you curious about the learning content in today's topic?" and "What do you anticipate to be doing in this topic?", many students happily replied, "I am very curious and I am eager to learn; I really enjoy learning with these materials; I'm wondering what the teacher will organize for us to do with these materials". Based on the students' voluntary preparation for the learning materials of the topic, as well as their responses and enthusiastic attitude before the lesson, the authors noticed that the designed STEAM topic attracted and evoked their curiosity and excitement before they actually took part in the lesson.

During the lesson (following the 5E procedure):

The first step of the procedure aimed at raising the topic and encouraging students learn about plants in the school garden, thereby forming a visual of the plants' outer parts and a feeling of love for trees and the environment. In this step, the students were excited to work in groups and observe as assigned by the teachers; the process was full of fun; the students worked in groups with high cooperation and discipline; and there was no confusion, disturbance, or disorder. When the teacher asked the students to discuss and share the results of learning about the plants in the school garden, the students in groups boldly presented; all groups showed the signal that they wanted to present by raising their hands, and a group of students even picked up yellow leaves, fallen leaves, and dry branches of special shapes for teachers and other students to observe.

When it came to the step of designing the model under the theme of My Green Garden, all the students in the groups participated enthusiastically, expressing their emotions, and showing high concentration and great excitement.

The products that students made in their My Green Garden theme not only showed their creativity but also their artistic ideas. The products were made from many different materials: yogurt boxes, clay, foam, colored paper, plastic spoons, etc. (Figure 2). The plants created by the students came in a variety of different colors and shapes. They were held upright with clay, bottle caps, and candlestick glue. There were three groups that even created fruits and flowers with clay, colored paper, and white foam; there were green plants plugged inside plastic caps, imitating plant pots. There was a group that used colored paper to drop into a small plastic cap, presenting that they were the lotus (an aquatic plant with flowers and leaves floating on a water surface). Thus, the trees were not only freely shaped and recreated by students, but also decorated and arranged in different ways, forming a beautiful campus. There was a case where a group of students put small plastic spoons (used to eat yogurt) next to their created green trees. This made the team curiously ask them about the idea. The students excitedly replied, "It is a fountain and I want to have a beautiful fountain in my green garden". The students' response surprised the research team, because at first, everyone in the research group thought they were light poles. After the students' answer, it could be seen that their thinking about the issue in the lesson had been expanded a lot with creative ideas, arising from what they observed in practical life and expressed with their own moods and emotions.

Figure 2. Some of the students' work on the *My Green Garden* theme.

When presenting the product, all groups made a very clear presentation about the parts of a plant (roots, stems, and leaves) and clearly expressed their ideas of the model My Green Garden as their dream of a green world.

In general, when implementing their models, each group worked at a different speed, but they all managed to complete and demonstrate their ideas well. At the same time, by observing the presentation of the model product, the teachers could also clearly assess the goal that this STEAM topic was aimed at through product modeling, speech, and the students' collaborative activities while they participated in the lesson.

Besides observing to assess the students' interest and creative problem-solving abilities when participating in the 5E model lessons with a STEAM topic, the research team also used student self-assessment, peer assessment between groups, and teacher assessment forms. The contents of those forms are shown in Tables 1–3.

Table 1. Self-assessment form.

Full Name:					**Group:**				
1. Put an X under the box that expresses your feeling after learning the topic My Green Garden									
Really happy			Happy			Ok		Unhappy	
2. Put an X under the box that shows the mark you think you deserve when learning this lesson.									
10	9	8	7	6	5	4	3	2	1

Table 2. Peer-assessment form (in the same group).

Full Name:			**Group:**		
No.	**Full Name**	**Participating**	**Supporting**	**Being Creative**	**Total**
Maximum score		5	3	2	10
1	Tran Thu An				
2	Nguyen Tu				
3	...				

Table 3. Teacher assessment form.

No.	Criteria	Maximum Score	Score
1	Draw and annotate the parts of a plant observed	1	
2	Figure out the main characteristics including colors, shapes, and sizes of several plants in the school garden	1	
3	Present the idea of the model My Green Garden	2	
4	Prepare enough materials to make a model	1	
5	Complete the model of a green garden	2	
6	Feasibility and creativity of the model	2	
7	Clearly present about the group's model	1	
	Total	10	

To evaluate the results of the trial teaching on the theme My Green Garden for students in grade 1 in three different areas in Lang Son City, the research team synthesized two groups of results: (1) results on the emotions assessed by the students themselves (item 1, Table 1); as well as (2) the average three-point results (rounded to one decimal place) for each student using the self-assessment form (item 2, Table 1); the peer-assessment scores of students in the same group (Table 2); and the teachers' assessment scores for each student (Table 3). The evaluation results are ranked according to four levels: excellent completion (9.0–10 points); good completion (7.0–8.4 points); completion (5.0–6.9 points); no completion (under 5.0 points). The evaluation results are shown in Table 4 and Figures 3 and 4.

Table 4. The results of experimental teaching.

Emotion	Result		Average	Result	
	Number of Students	Percentage		Number of Students	Percentage
Really happy	269	75	8.5–10	221	61.8
Happy	63	17.6	7.0–8.4	105	29.3
OK	26	7.4	5.0–6.9	32	8.9
Unhappy	0	0	Under 5.0	0	0

Figure 3. Students' self-assessment results.

Besides evaluating the general result, the research team made a comparison among three areas of Lang Son city in terms of the interest and average scores after the experimental lessons. The results are shown in Figures 5 and 6.

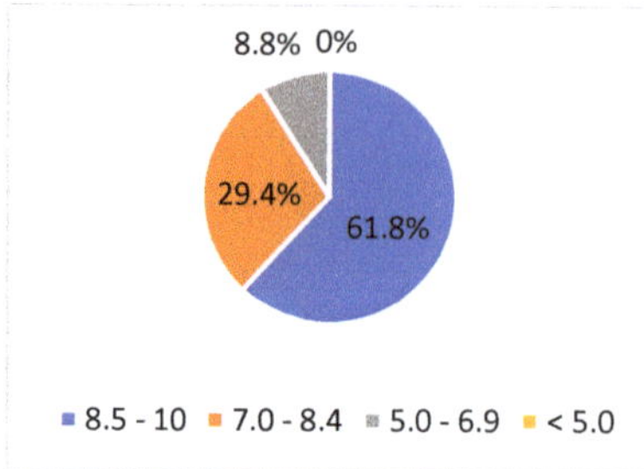

Figure 4. Students' average scores.

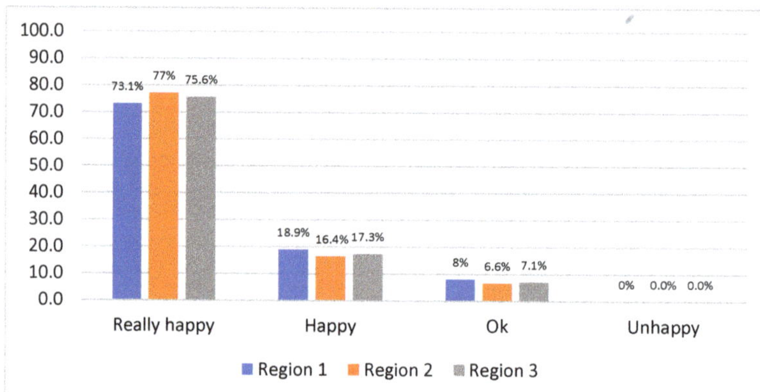

Figure 5. Results of evaluating students' emotions (in 3 different experimental areas).

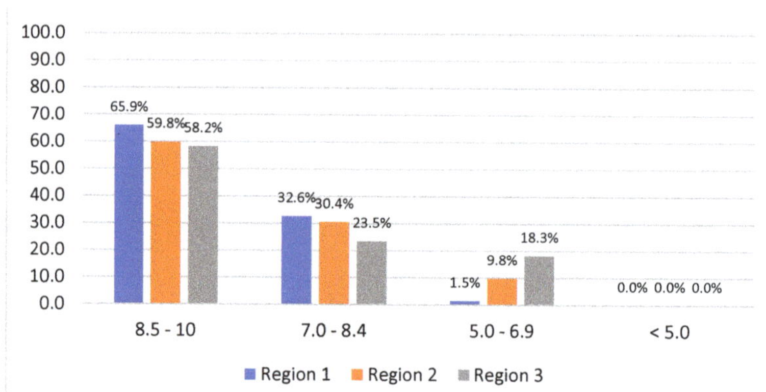

Figure 6. Results of students' average scores in 3 different experimental areas).

Looking at the results shown in Figure 5, we can see that there is no significant difference in the emotions of the students from the three different areas when participating in the lessons; the difference rate only ranges from 1% to 3%. The results in Figure 6 show that there is a slight difference in the students' average score while participating in the lessons. Students in area 1 had an excellent completion rate that was 7 to 8% higher than those of the other areas; the figure for good completion was 2% to 3% higher than those in sectors 2 and 3; and the figure for no completion was lower than those of the other two areas. We studied the criteria in the teacher evaluation form assessing group performance according to Table 3 that are related to this score. The results showed that the average score of students in area 2 and area 3 was lower than that of area 1 due to a number of reasons: students in areas 2 and 3 did not receive points, or even had deductions, in terms of presenting and expressing their ideas about the model of My Green Garden (due to limited Vietnamese language and not much flexibility or courage). In addition, we devised more questions for the managers and teachers at the pilot schools in areas 2 and 3, such as What is the biggest change you have seen in students in this class? The results showed that 31/31 teachers recognized that the students had a good spirit, a cheerful attitude, and an excitement during the activities; 30 out of 31 teachers appreciated that their students made significant changes when working in groups and presenting their products to the class, presenting more clearly and confidently; and 31 out of 31 teachers in area 2 and area 3 said that they were surprised by their students' green garden models.

Looking at Table 4 and Figures 3–7, it can be said that the students were really interested in participating in the activities under the STEAM topic (92.6% of students were really happy or happy). In addition, the students' average scores were rather high (61.8% of students achieved excellent completion; 29.3% of students achieved good completion; 8.9% of students achieved completion). No students were assessed with No completion.

Figure 7. Photos of students' learning activities.

Beside evaluating the students, the research team asked 77 teachers who observed the lessons in the 12 pilot classes for their opinions and evaluation with regard to the feasibility and possibility of applying the 5E procedure into STEAM education for grades 1, 2, and 3 in primary schools. The teachers' questionnaire was designed as follows (Table 5):

Table 5. Teachers' questionnaire.

Question	Level			
	(1)	(2)	(3)	(4)
How can applying the 5E procedure into STEAM education help students achieve a lesson's objectives?	Not good ☐	Normal ☐	Good ☐	Very good ☐
What is your opinion about the feasibility of applying the 5E procedure into organizing activities of STEAM education for lower primary grades?	Unfeasible ☐	Normal ☐	Feasible ☐	Highly feasible ☐

The results of the questionnaire are shown in Figure 8.

Looking at Figure 8, it can be seen that using the 5E procedure to organize STEAM educational activities would be effective and highly feasible; 100% of the teachers rated the effectiveness and feasibility at level 3 and 4, and no teachers chose level 1 or 2. Besides the questionnaire, the research team also interviewed teachers who observed the lessons about the students' level of positivity and thinking ability when participating in the STEAM educational theme compared with the traditional teaching organization. The in-depth interview questions were What expresses students' ability to communicate cooperatively when using the 5E process in STEAM education? What expresses students' creativity and hardworking? What expresses students' interest in participating in the lesson? What products were made after the lesson? Here are several comments made by the teachers: "In my opinion, students participated enthusiastically and more creatively than they did in the conventional classes." (N.T.V.A—Academic leader of grade 1); "I am surprised about the students' creativity expressed in the products" (T.H.T and H.T.T—teachers of grades 1 and 2); "They are wonderful!" (G.T. N—manager) (Figure 9).

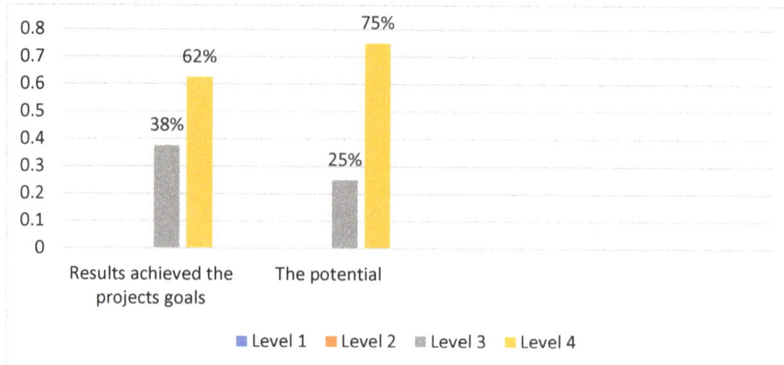

Figure 8. The results of teachers' questionnaire about the feasibility of the application of the 5E procedure into organizing STEAM educational topics.

Figure 9. Teachers expressing their opinions after the lesson.

In addition, the research team also surveyed the teachers about the difficulties they would face when implementing and expanding this STEAM model in primary schools. Many teachers expressed their concerns about a number of issues, such as the training and expansion of the 5E model in STEAM education for primary school teachers; training students to develop autonomy in preparing materials when learning STEAM; building and arranging school-year plans to set an appropriate amount of time for organizing STEAM educational topics; and managing large-sized classrooms. Also, many teachers proposed setting up Facebook pages; organizing online training sessions for primary school managers and teachers; and setting up groups of teachers in the same grade at a school to support each other to prepare materials to organize STEAM educational activities.

4. Conclusions

From the procedure of implementing the abovementioned research, some conclusions can be made as follows:

- The 5E procedure in STEAM education can be organized in five steps: engage, explore, explain, elaborate, evaluate;
- The use of the 5E procedure in organizing STEAM educational activities for primary students is reasonable and feasible. It can be adopted into the teaching process for primary students from different areas with a wide variety of levels and teaching conditions;

- The use of the 5E procedure in teaching STEAM educational topics not only evokes students' interests but also contributes towards motivating their creativity, developing their teamworking ability, and attaching educational contents with daily problem-solving skills;
- The 5E procedure can be used to organize STEAM educational activities for primary-school students to meet the goals of developing students' personalities and competencies under Vietnam's new General Training Program.

Author Contributions: Conceptualization, H.N.T.T.; methodology, H.N.T.T. and B.T.N.; software, H.N.T.T. and B.T.N.; validation, H.N.T.T., B.T.N. and T.-B.N.; formal analysis, H.N.T.T. and T.-B.N.; investigation, B.T.N.; resources, H.N.T.T.; data curation, H.N.T.T.; writing—original draft preparation, H.N.T.T.; writing—review and editing, H.N.T.T., B.T.N. and T.-B.N. All authors have read and agreed to the published version of the manuscript.

Funding: This research received no external funding.

Institutional Review Board Statement: Not applicable.

Informed Consent Statement: Not applicable.

Data Availability Statement: Data are contained within the article.

Conflicts of Interest: The authors declare no conflict of interest.

References

1. Hom, E.J. What Is STEM Education? 2015. Available online: http://www.livescience.com/43296-what-is-stem-education.html (accessed on 5 October 2020).
2. Yakman, G.; Lee, H. Exploring the exemplary STEAM education in the US as a practical educational framework for Korea. *J. Korean Assoc. Sci. Educ.* **2012**, *32*, 1072–1086. [CrossRef]
3. Morrison, J. *TIES STEM Education Monograph Series: Attributes of STEM Education*; TIES: Baltimore MD, USA, 2006.
4. Cho, B. The Effects of Creativity and Flow on Learning through the Steam Education on Elementary School Contexts. In Proceedings of the Korean Society for Educational Technology Academic Conference, November 2013; pp. 206–210. Available online: https://kiss.kstudy.com/Detail/Ar?key=3407442 (accessed on 5 October 2020).
5. Ah-Nam, L.; Osman, K. Developing 21st Century Skills through a Constructivist-Constructionist Learning Environment. *K-12 STEM Educ.* **2017**, *3*, 205–216.
6. Atkin, J.M.; Karplus, R. Discovery or invention? *Sci. Teach.* **1962**, *29*, 45–51.
7. Bybee, R.W. The BSCS 5E instructional model: Personal reflections and contemporary implications. *Sci. Child.* **2014**, *51*, 10–13. [CrossRef]
8. Bybee, R.W. Using the BSCS 5E instructional model to introduce STEM disciplines. *Sci. Child.* **2019**, *56*, 8–12. [CrossRef]
9. Kaniawati, D.S.; Kaniawati, I.; Suwarma, I.R. Implementation of STEM Education in Learning Cycle 5E to Improve Concept Understanding On Direct Current Concept. In Proceedings of the International Conference on Mathematics and Science Education, Malang, Indonesia, 29–30 August 2017; Atlantis Press: Amsterdam, The Netherlands, 2017.
10. Ministry of Education and Training. *General Education Program*; Ministry of Education and Training: Ha Noi, Vietnam, 2018.
11. Ministry of Education and Training. *Science Curriculum*; Ministry of Education and Training: Ha Noi, Vietnam, 2018.
12. Ministry of Education and Training. *Nature and Society Curriculum*; Ministry of Education and Training: Ha Noi, Vietnam, 2018.
13. Van Hue, B.; Mai, P.T.H.; Thuc, N.X. *Primary Psychology Curriculum*; Hanoi National University of Education Publisher: Ha Noi, Vietnam, 2018.
14. DeBoer, G.E. *A History of Ideas in Science Education: Implications for Practice*; Teachers College, Columbia University: New York, NY, USA, 1991.
15. Tezer, M.; Cumhur, M. Mathematics through the 5E Instructional Model and Mathematical Modelling: The Geometrical Objects. *Eurasia J. Math. Sci. Technol. Educ.* **2017**, *13*, 4789–4804. [CrossRef]

engineering proceedings

Proceeding Paper

Digital Game Approaches for Cultivating Computational Thinking Skills in College Students †

Li-Xian Chen [1,*], Shih-Wen Su [2], Chia-Hung Liao [2], Mei-Jin Hsiao [3] and Shyan-Ming Yuan [2]

1 School of Big Data, Fuzhou University of International Studies and Trade, Fuzhou 350202, China
2 Department of Computer Science, National Yang Ming Chiao Tung University, Hsinchu City 30010, Taiwan; alvin.cs00@nycu.edu.tw (S.-W.S.); aiallen.cs07g@nctu.edu.tw (C.-H.L.); smyuan@nycu.edu.tw (S.-M.Y.)
3 School of Art & Design, Fuzhou University of International Studies and Trade, Fuzhou 350202, China; meijins@ms69.hinet.net
* Correspondence: lixian.cs98g@g2.nctu.edu.tw

† Presented at the IEEE 5th Eurasia Conference on Biomedical Engineering, Healthcare and Sustainability, Tainan, Taiwan, 2–4 June 2023.

Abstract: Computational thinking (CT) has become one of the critical goals of teaching CS programming courses. Computational skills consist of skills taken in a computational form in learning programming and dealing with daily life. More research adopted games to teach CT skills. This paper investigated two games, Little Alchemy 2 and Dr. Sudoku, to promote CS students' CT skills and applied international Bebras tests to measure their CT skills. The results showed that CT skills in problem decomposition and pattern recognition could be improved via digital games. Thus, this study contributes to computing education using available digital games to promote CS college students' CT abilities.

Keywords: computational thinking; digital game-based learning; CS education; problem decomposition; pattern recognition

Citation: Chen, L.-X.; Su, S.-W.; Liao, C.-H.; Hsiao, M.-J.; Yuan, S.-M. Digital Game Approaches for Cultivating Computational Thinking Skills in College Students. *Eng. Proc.* **2023**, *55*, 62. https://doi.org/10.3390/engproc2023055062

Academic Editors: Teen-Hang Meen, Kuei-Shu Hsu and Cheng-Fu Yang

Published: 6 December 2023

1. Introduction

Computational thinking is an algorithmic skill used by computer and mathematical scientists and is one of the basic daily life skills that everyone needs [1,2]. Computational thinking applications are ubiquitous in life, and anyone can use computational thinking skills to solve problems [2]. CT skills include problem decomposition, pattern recognition, abstraction, and algorithm design as the concept of computational thinking [1,2]. For example, computer programmers who are programming games often break the code into small chunks to control a character's movement and sound effects. In a game-based learning environment, like Little Alchemy 2, students learn that mist can be decomposed into air and water (Figure 1). Problem decomposition is a useful problem-solving strategy. It can help us write a complex computer program, plan a holiday, or make a model plane. Using computational thinking skills can help people simplify complex things in different subject areas [3].

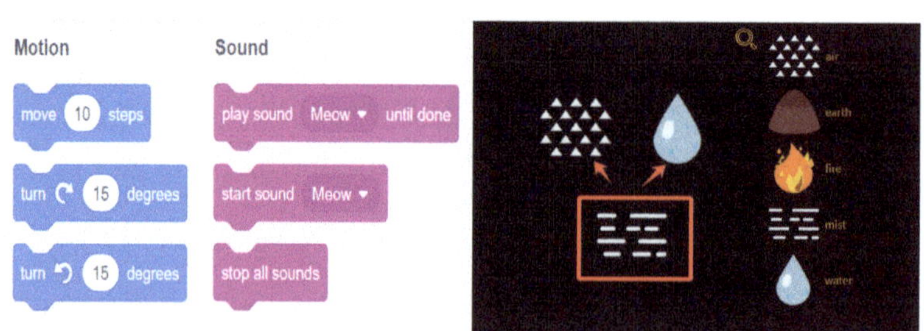

Figure 1. Problem decomposition in the programming and game environment.

The skills of CT enable students to decompose problems, find the rules, and reconstruct them into familiar problems when they encounter new difficulties [4]. In programming education, students imitate codes and learn algorithms to strengthen their mathematical modeling and calculation abilities. However, nearly half of college students have no experience with programming, and it is not accessible to "code" [5]. Teaching rule discovery and pattern recognition in programming courses is challenging [6]. As a result, there are considerable problems with CT training in the curriculum.

Many researchers move from programming to non-programming environments to cultivate computational thinking. Recent programming education has emphasized how computational thinking can be fostered using more game-based learning in programming education [5,7,8]. The digital game environment provides students with repetitive actions to find rules, plan resource arrangements, and optimal paths, and solve encountered problems at the current level [7–9]. Planhing is like computational thinking in a game environment, and following directions is like programming. The playing process reflects students' logic and ability to find rules and solve problems. Therefore, this study applied two digital games (Little Alchemy 2 and Dr. Sudoku) to explore the possibility of enhancing students' computational thinking skills.

2. Related Works

Computational thinking is valued in science, technology, engineering, and mathematics (STEM) education [10]. In the 21st century, people have more opportunities to apply information technology to integrate knowledge, skills, values, and attitudes. Computational thinking and programming are problem-solving methods used at different learning stages and disciplines. Whether programming education or courses that cultivate computational thinking skills, it enables students to discover helpful information, find rules, and use existing skills to solve problems in their daily lives [11,12]. Computational thinking and digital literacy are components of programming skills [5]. An attractive computer science education, such as building block programming and game-based learning environments, keeps students engaged in learning [1]. People have the opportunities to explore and make mistakes in a digital game-based learning environment, synthesize information, discover rules, and formulate strategies [11]. Game-based learning environments also stimulate students' imagination, creativity, and motivation to learn, thereby helping to improve CT ability. Therefore, digital games have been promoted as effective systems for teaching rules and knowledge [13,14]. Block-based game environments, like Scratch and Webduino Blockly, have been adopted as CT development tools [15]. Commercial or online web games enable students to engage in more natural playing behaviors that are not constrained by specific learning subjects. Sudoku can be used to teach collaboration and problem-solving abilities, which can be transferred to other learning and living conditions [16]. Little Alchemy 2 lets players combine icons representing concepts, objects, substances, and elements to discover more items. Effective combinations form a new project waiting to be combined with other things [17]. Sudoku has been identified as a tool for helping players to teach reasoning skills and logical thinking [11]. Using Dr. Sudoku and Little Alchemy 2, we explored whether games can improve students' problem decomposition and pattern recognition skills.

A proper evaluation of CT competencies can review the effectiveness of training activities. Bebras is an international information science challenge designed to increase learners' interest in informatics and information and communication technology as a basic science by solving problems on computer science topics. Questions about computers and various logical thinking skills (e.g., pattern recognition, problem decomposition, algorithm, abstraction) are used to increase interest and creativity in the computer science field. This study applied the Bebras challenge to investigate whether the game activity was effective in improving students' CT skills.

3. Methods

3.1. Participants and Procedures

In June 2021, two weeks before the end of the semester, we conducted a quasi-experimental study with 76 juniors studying at an applied university in Fuzhou, China. These students have rudimentary C language and network programming concepts. Two classes were divided into playing Little Alchemy 2 (32 students) and Dr. Sudoku (44 students). To reduce potential distractions, we did not tell the students the research purpose of playing the game, nor did we give them any other information about group conditions. Bebras' CT tests were taken before and after 20-minute game training individually for 30 min. Finally, the students had an open discussion about the gameplay process and filled out a gaming questionnaire. The experimental process is shown in Figure 2.

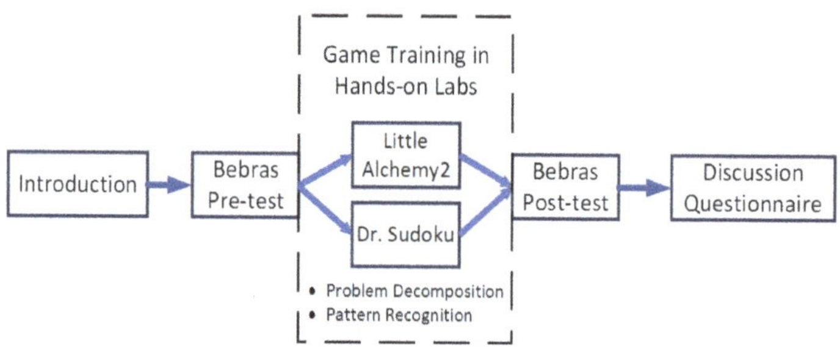

Figure 2. Game education process for training CT.

3.2. Training Environments

This study explored whether games enhance learners' problem-solving and pattern-recognition CT skills. Little Alchemy 2 was adopted to teach students problem decomposition skills by disassembling and synthesizing elements. According to the characteristics of two or more elements, such as air, fire, water, and earth, the game combines them into 720 items. Observing the crafted items, the player can break them into related constituent elements. Figure 3 shows that steam can be decomposed into water and lava. Then, a worksheet was provided for students to fill out. Another game, Dr. Sudoku, required players to fill in the missing 1 to 9 in the 9 × 9 cells. In this game, there are three functions (slips of different colors, display of all candidate numbers, and display of some candidate numbers) for students to solve the puzzle. Players use their logical reasoning abilities during the gameplay. For example, red boxes display that two numbers of 2 already exist in the two rows of green arrows, and students use the horizontal line division method to infer that the unfilled number in the middle block is 2 (Figure 4).

Bebras Challenge Test was taken to understand students' problem decomposition and pattern recognition skills before and after the CT game training. Considering that the subjects are college students with a fundamental programming foundation, and the difficulty of the competition is only up to 12-grade levels, eight questions, including four simple questions, three medium questions, and one difficult question, were chosen. The Bebra challenge questions from 2013 to 2016 were selected as pre-test and post-test to evaluate the two CT skills. After adding up the dimensions of each question and normalizing the aggregated scores to one hundred, we obtained the ability scores for the two dimensions (i.e., problem decomposition and pattern recognition) and the total computational thinking score. For example, the pre-test dimension contains six questions for problem decomposition and pattern recognition. The students correctly answered the first and second questions in the pre-test (q1: pattern recognition and problem decomposition, q3: pattern recognition). The total computational thinking skill score was 25, the problem decomposition skill score was 16.7 points, and the pattern recognition skill score was 33.3 points.

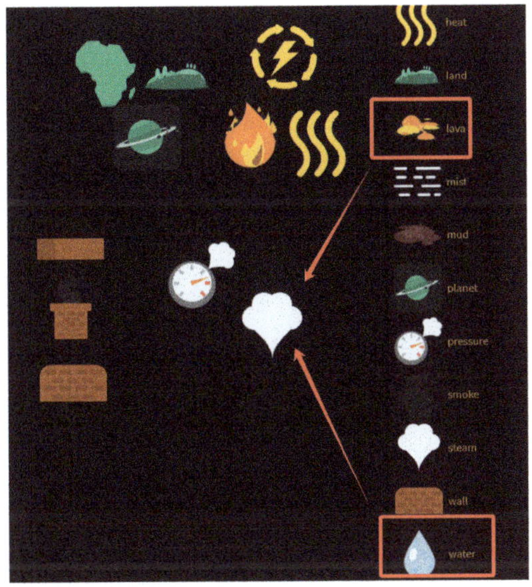

Figure 3. Little Alchemy 2 screen shot.

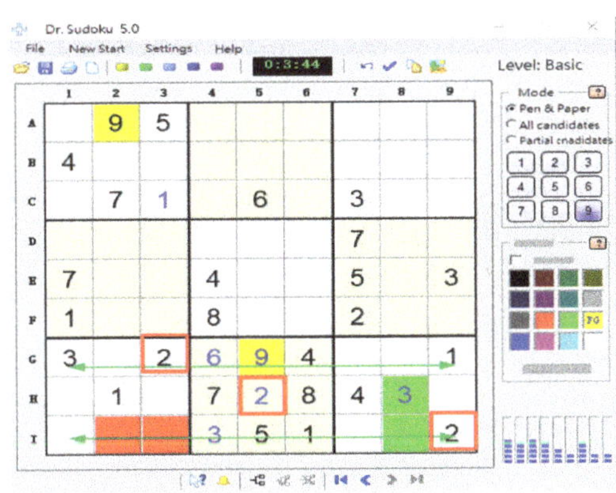

Figure 4. Dr. Sudoku screen shot.

4. Results and Discussion

The independent sample *t*-test showed no significant difference in pre-test scores between the two groups (t(56.92) = 1.64, $p = 0.106$, d = 0.396). The students' starting CT skills of the two groups showed no difference (Table 1).

Table 1. Independent sample *t*-test of pre-test in two groups.

	Mean (S.D.)		df	t	Effect Size (d)
	Little Alchemy 2 (N = 44)	Dr. Sudoku (N = 32)			
score	257.27 (76.72)	223.23 (97.43)	56.92	1.64	0.9

The paired sample *t*-test for paired data showed that the post-test total score of Little Alchemy 2 was significantly larger than the total pre-test score (t(31) = −4.12, p = 0.000, d = 0.858), and the total post-test score of Dr. Sudoku was significantly larger than total pre-test score (t(43) = −4.48, p = 0.000, d = 0.832). The average post-test scores of Little Alchemy 2 and Dr. Sudoku had improved, and the average improvement score (i.e., the post-test score minus the pre-test score) of Little Alchemy 2 was greater than that of Dr. Sudoku (78.85 > 57.43) (Table 2).

Table 2. Paired sample *t*-test for paired data of Little Alchemy 2 and Dr. Sudoku.

	Mean (S.D.)		df	t	Effect Size (d)
	Pre-Test	Post-Test			
Little Alchemy 2 (N = 32)					
Total score	223.22 (97.43)	302.08 (85.97)	31	−4.12 ***	0.858
Problem decomposition	50.00 (23.18)	76.04 (27.08)		−4.78 ***	1.033
Pattern recognition	60.93 (27.30)	75.00 (21.55)		−2.85 **	0.572
Dr. Sudoku (N = 44)					
Total score	257.27 (76.72)	314.69 (60.33)	43	−4.48 ***	0.832
Problem decomposition	63.25 (21.73)	81.81 (22.10)		−4.65 ***	0.847
Pattern recognition	69.31 (20.94)	72.27 (22.50)		−0.77	0.136

** $p < 0.01$, *** $p < 0.001$.

Next, we try to analyze whether these two games have an improved effect on problem decomposition ability and pattern recognition ability. Paired sample *t*-test results showed that pre-test and post-test scores of problem decomposition (t(31) = −4.78, p = 0.000, d = 1.033) and pre-test and post-test scores of pattern recognition (t(31) = −2.85, p = 0.008, d = 0.572) were all the statistically significant difference. After playing the Little Alchemy 2 game, the problem decomposition score progressed from 50.00 to 76.04, and the pattern recognition score improved from 60.93 to 75.00 (Table 2). We analyzed the participants' pre- and post-test performance and found that more than half of the students progressed their problem decomposition and pattern recognition scores after playing Little Alchemy 2 (Figure 5). Due to the lack of recognizable cues in this game-based learning environment, students were strongly driven to create new objects based on their exploration. This result shows that they use their world knowledge to drive their exploration.

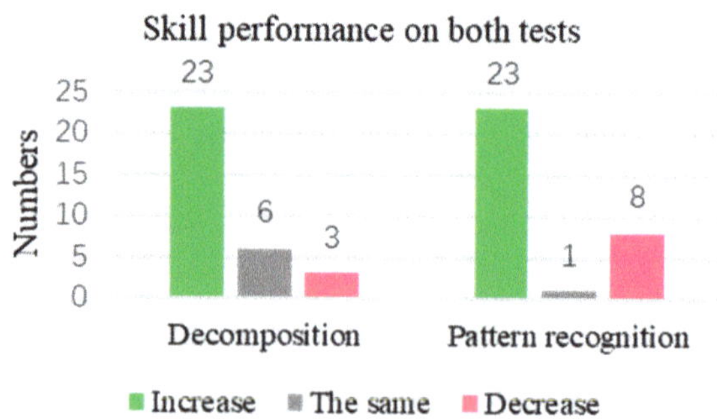

Figure 5. Skill performance after playing the Little Alchemy 2 game.

In the Dr. Sudoku game, problem decomposition pre-test and post-test scores (t(43) = −4.65, p = 0.000, d = 0.847) were statistically significant differences. Problem decomposition scores had progressed from 63.25 to 81.81 (Table 2). However, the pattern recognition score had no statistically significant difference, and more than half of students' pattern recognition scores decreased (Figure 6). This result suggested that filling in numbers from Dr. Sudoku to find the pattern was already familiar to college students, so there was no noticeable improvement.

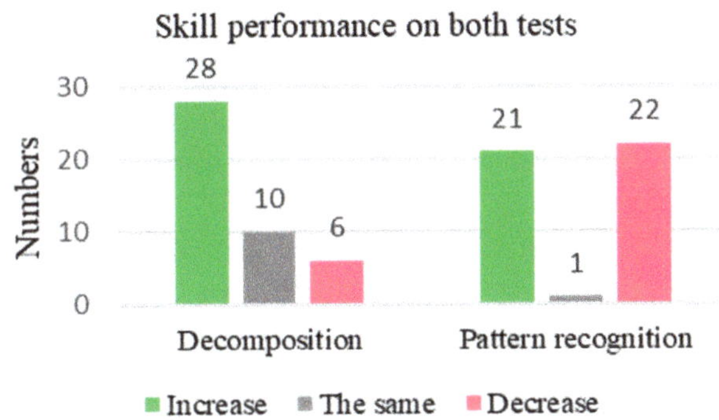

Figure 6. Skill performance after playing Dr. Sudoku game.

Although CS students already have basic programming concepts of C and Java, the problem decomposition and pattern recognition abilities of CT skills can be effectively enhanced via the digital game (Little Alchemy 2) with self-control and self-exploration. Therefore, non-programming digital games should be added to the CS course to teach students CT abilities.

5. Conclusions and Future Work

This paper investigated how the problem decomposition and pattern recognition of CT skills can be trained through specific digital games. This study found that utilizing non-programming digital games can promote computational thinking skills in college CS students. In addition, a digital game (Little Alchemy 2) with the function of free decomposition and integration can effectively improve the problem decomposition and pattern recognition ability. Since the experiment participants are only CS students, data can be collected from the School of Biology and Technology, the School of Foreign Language, and the School of Art and Design to analyze the benefits in the future.

Author Contributions: Conceptualization, L.-X.C. and C.-H.L.; methodology, L.-X.C. and S.-M.Y.; validation, L.-X.C.; S.-W.S. and M.-J.H.; formal analysis, L.-X.C. and S.-W.S.; writing—original draft preparation, L.-X.C. and S.-W.S.; writing—review and editing, L.-X.C. and C.-H.L.; visualization, L.-X.C. and M.-J.H.; project administration, S.-M.Y. All authors have read and agreed to the published version of the manuscript.

Funding: This work was partially funded by National Science and Technology Council Taiwan (grant number: 108-2511-H-009-009-MY3) and the High-level Talent Research Project at Fuzhou University of International Studies and Trade (grant no. FWKQJ201909).

Institutional Review Board Statement: Not applicable.

Informed Consent Statement: Not applicable.

Data Availability Statement: The data presented in this study are available upon request from the corresponding author.

Acknowledgments: The authors wish to thank the blind reviewers for their insightful and constructive comments.

Conflicts of Interest: The authors declare no conflict of interest.

References

1. Hsu, T.-C.; Chang, S.-C.; Hung, Y.-T. How to learn and how to teach computational thinking: Suggestions based on a review of the literature. *Comput. Educ.* **2018**, *126*, 296–310. [CrossRef]
2. Wing, J.M. Computational thinking. *Commun. ACM* **2006**, *49*, 33–35. [CrossRef]
3. BBC. Introduction to Computational Thinking. Available online: https://www.bbc.co.uk/bitesize/guides/zp92mp3/revision/1 (accessed on 11 January 2022).
4. Rijke, W.J.; Bollen, L.; Eysink, T.H.; Tolboom, J.L. Computational thinking in primary school: An examination of abstraction and decomposition in different age groups. *Inform. Educ.* **2018**, *17*, 77–92. [CrossRef]
5. Su, S.-W.; Jung, S.-Y.; Yu, X.; Yuan, S.-M.; Sun, C.-T. Modify, Decompose and Reassemble: Learner-Centered Constructive Teaching Strategy for Introductory Programming Course in College. In Proceedings of the 2022 IEEE 5th Eurasian Conference on Educational Innovation (ECEI), Taipei, Taiwan, 10–12 February 2022; pp. 197–200. [CrossRef]
6. Stachel, J.; Marghitu, D.; Brahim, T.B.; Sims, R.; Reynolds, L.; Czelusniak, V. Managing cognitive load in introductory programming courses: A cognitive aware scaffolding tool. *J. Integr. Des. Process Sci.* **2013**, *17*, 37–54. [CrossRef]
7. Liao, C.H.; Yan, J.Y.; Yuan, S.M.; Chen, L.X. Research on cultivating the computational thinking literacy of secondary school students through non-programming activities. In Proceedings of the 4th Eurasian Conference on Educational Innovation 2021, Singapore, 5–7 February 2021.
8. Shell, D.F.; Hazley, M.P.; Soh, L.-K.; Miller, L.D.; Chiriacescu, V.; Ingraham, E. Improving learning of computational thinking using computational creativity exercises in a college CSI computer science course for engineers. In Proceedings of the 2014 IEEE Frontiers in Education Conference (FIE) Proceedings, Madrid, Spain, 22–25 October 2014; IEEE: Piscataway, NJ, USA, 2014; pp. 1–7.
9. del Olmo-Muñoz, J.; Cózar-Gutiérrez, R.; González-Calero, J.A. Computational thinking through unplugged activities in early years of Primary Education. *Comput. Educ.* **2020**, *150*, 103832. [CrossRef]
10. Robins, A.; Rountree, J.; Rountree, N. Learning and teaching programming: A review and discussion. *Comput. Sci. Educ.* **2003**, *13*, 137–172. [CrossRef]
11. Sun, C.-T.; Chen, L.-X.; Chu, H.-M. Associations among scaffold presentation, reward mechanisms and problem-solving behaviors in game play. *Comput. Educ.* **2018**, *119*, 95–111. [CrossRef]
12. Deci, E.L.; Ryan, R.M. *Intrinsic Motivation and Self-Determination in Human Behavior*; Springer Science & Business Media: Berlin/Heidelberg, Germany, 2013.
13. Ch'ng, S.I.; Low, Y.C.; Lee, Y.L.; Chia, W.C.; Yeong, L.S. Video games: A potential vehicle for teaching computational thinking. In *Computational Thinking Education*; Springer: Singapore, 2019; pp. 247–260.
14. Gee, J.P. *Good Video Games+ Good Learning: Collected Essays on Video Games, Learning, and Literacy*; Peter Lang: Lausanne, Switzerland, 2007.
15. Zhao, W.; Shute, V.J. Can playing a video game foster computational thinking skills? *Comput. Educ.* **2019**, *141*, 103633. [CrossRef]
16. Emily, L. In Sudoku Club, Scotten School Students Love Thinking Logically. Available online: https://www.theunion.com/news/local-news/in-sudoku-club-scotten-school-students-love-thinking-logically/ (accessed on 12 January 2022).
17. Leino, O.T. STONE + LIFE = EGG—Little Alchemy as a limit-idea for thinking about knowledge and discovery in computer games. In Proceedings of the Philosophy of Computer Games Conference, Msida, Malta, 1–4 November 2016.

engineering proceedings

Proceeding Paper

Decision Support from Financial Disclosures with Deep Reinforcement Learning Considering Different Countries and Exchange Rates †

Yi-Hsin Cheng [1,*] and Hei-Chia Wang [1,2]

1 Institute of Information Management, National Cheng Kung University, Tainan 701, Taiwan; hcwang@mail.ncku.edu.tw
2 Center for Innovative FinTech Business Models, National Cheng Kung University, Tainan 701, Taiwan
* Correspondence: r78101011@gs.ncku.edu.tw

† Presented at the IEEE 5th Eurasia Conference on Biomedical Engineering, Healthcare and Sustainability, Tainan, Taiwan, 2–4 June 2023.

Abstract: The era of low-interest rates is coming. In addition to their basic salary, people hope to increase their income by doing part-time work, understanding how to use assets already on hand, and investing in assets to earn extra rewards. Goldman Sachs reports that over the past 140 years, the 10-year stock market return has averaged 9.2%. The investment firm also noted that the S&P 500 outperformed its 10-year historical average with an annual average return of 13.6% between 2010 and 2020. Nowadays, with increased computing power and advancements in artificial intelligence algorithms, the effective use of computing power for investment has become an important topic. In the investment process, venture capitalists form portfolios, a practice that improves investment efficiency and reduces risks in a relatively safe situation. Current investments are not limited to one country, allowing for investments in other countries. Given this situation, we must pay attention to Uncovered Equity Parity (UEP) conditions. Thus, we aim to find optimal dynamic trading strategies with Deep Reinforcement Learning, considering the aforementioned properties.

Keywords: decision support; deep reinforcement learning; Investment portfolio

Citation: Cheng, Y.-H.; Wang, H.-C. Decision Support from Financial Disclosures with Deep Reinforcement Learning Considering Different Countries and Exchange Rates. *Eng. Proc.* **2023**, *55*, 63. https://doi.org/10.3390/engproc2023055063

Academic Editors: Teen-Hang Meen, Kuei-Shu Hsu and Cheng-Fu Yang

Published: 7 December 2023

1. Introduction

The era of low-interest rates is coming. In addition to regular jobs, people hope to do part-time work to increase their income, understand how to use assets already on hand, and invest in them. Central Banks around the world are adopting negative interest rate policies. People who previously saved their money now tend to invest more [1]. Goldman Sachs reports that the 10-year stock market return averages 9.2%, with the S&P 500 outperforming its 10-year historical average from 2010 to 2020, achieving an annual average return of 13.6%.

However, while the S&P 500's Index achieved a high return, the average market return for the typical investor was only 5.19% during the same period. The main reason for this is that investors are sometimes illogical and irrational when investing [2]. We usually invest in various stocks, real estate, or gold. Different investments have different attributes and features. In the past, investors invested based on their experience, but were lacking a complete formula or exact rules. Nowadays, with increased computing power and developed artificial intelligence algorithms, the effective use of computing power for investment is an important topic.

People constantly adjust their various investment portfolio holdings, with past studies focusing on univariate time series analysis due to excessive noise in multivariate data causing inaccuracies [3]. However, daily stock prices are highly positively correlated with each other. As a result of recent advances in neural network research, particularly in

Long Short-Term Memory (LSTM) networks, multivariate models have been shown to outperform univariate networks [4].

In the context of global investing, it is essential to consider Uncovered Equity Parity (UEP), which suggests that a country's stock market often moves in the opposite direction to the exchange rate [5]. This means that the exchange rate can significantly impact investment performance, therefore making it crucial to incorporate this factor into portfolio design. Previous studies have shown that adding USD and gold to a portfolio can improve its hedging effect, with the optimal weighted stock-dollar-gold combination emerging as the best portfolio strategy, regardless of the reference return or risk [6].

To tackle the challenge of investing in different countries, we propose a deep reinforcement learning-based approach. Specifically, we use an A2C-based model to adjust the shares owned of each stock at each time step, with the goal of maximizing the portfolio return per period by selecting sequential trading directions for individual assets based on time-varying market characteristics.

2. Methodology

Stock returns become a time series forecasting problem, due to the daily changes in stock prices. Many researchers have undertaken extensive studies on this topic [3,7]. Recurrent Neural Networks (RNNs) are a type of neural network that contain feedback loops in their recurrent layers. Through feedback loops, RNNs can store information in "memory cells". Several studies have demonstrated that various forms of Recurrent Neural Networks (RNNs) perform better than conventional financial time series models across different markets and financial investment applications [8,9]. However, they perform poorly if learning has long-term time dependencies.

Reinforcement learning (RL) is a type of machine learning that allows an agent to learn by interacting with an environment and receiving feedback in the form of rewards or penalties. In contrast to supervised learning, where feedback provides explicit instruction on how to perform a task, RL learns through trial and error by receiving signals indicating whether its actions are positive or negative. Essentially, RL enables an agent to make optimal decisions based on past experiences and feedback from the environment, incorporating knowledge about when it acts and how its environment encourages rewards.

Recent research tries to find a learning method to optimize trading strategies when interacting with the dynamic real-world financial environment. Pendharkar and Cusatis [10] show two different RL methods: on-policy (SARSA) and off-policy (Q-learning). They also compared the performance of the two methods. Almahdi and Yang [11] considered real-world constraints and high transaction cost conditions to build a hybrid method that combines recurrent RL and particle swarm optimization.

These studies used different RL-based methods depending on the different problem settings they face. Advantage Actor Critic (A2C) is a typical actor-critic algorithm to improve policy gradient updates. In the basic Actor—Critic algorithm, the critic network estimates the value function, but there is a problem with the large variance of the policy gradient. The critic network in A2C not only estimates the value function of action, but also evaluates how the action could be performed better. To address the high variance of the policy gradient, A2C employs a baseline mechanism to make the model more robust.

In A2C, many agents work independently to pass average gradients to the global network.

The A2C's objective function $\nabla \mathcal{J}_\theta(\theta)$ can be written as follows:

$$\nabla \mathcal{J}_\theta(\theta) = \mathbb{E}\left[\sum_{t=1}^{T} \nabla_\theta \log \pi_\theta(a_t|s_t) A(s_t, a_t)\right] \tag{1}$$

$\pi_\theta(a_t|s_t)$ is the policy network at time step—t. $A(a_t|s_t)$ is the advantage function and can be written as:

$$A(s_t, a_t) = Q(s_t, a_t) - V(s_t) \tag{2}$$

or

$$A(s_t, a_t) = r(s_t, a_t, s_{t+1}) + \gamma V(s_t) - V(s_t) \tag{3}$$

Portfolio optimization involves adjusting the allocation of shares in a portfolio to achieve the objective of maximizing returns while minimizing risk. This can be achieved by considering various factors such as expected returns, volatility, and correlation between assets. In our study, we use the approach of changing the volume of shares in our portfolio to find the optimal allocation for maximizing returns while minimizing risk. We aim to strike a balance between generating high returns and managing the risk of losses, achieved through efficient diversification and asset allocation.

The Sharpe ratio is a commonly used metric for evaluating the risk-adjusted performance of an investment strategy. It measures the excess return of an asset or portfolio over the risk-free rate per unit of volatility or risk. A higher Sharpe ratio indicates better risk-adjusted performance. In our study, we use the Sharpe ratio to measure the performance of our proposed model in optimizing portfolio return with respect to risk. By evaluating the Sharpe ratio of our model, we can gain insights into its ability to generate higher returns while controlling risk, and compare its performance with other models or benchmark indices.

In this study, assumptions are made to ensure that individual investor trades do not have any impact on the market price of portfolio assets since trades are small. This means that if we buy or sell any number of shares, the stock price does not change. In the portfolio trading process, we have the freedom to buy and sell any number of shares at any time. To maximize the portfolio return per period, we formulate this process as a generalized Markov decision process (MDP), where the agent selects the sequential trading direction of individual assets in the portfolio based on time-varying market characteristics.

The state space of the agent in this system is where the agent observes the financial environment. The state at period, t, can be expressed as follows:

$$S_t = (X_t, cp_t, sh_t, ch_t, b_t) \tag{4}$$

(X_t: encoding features at time step—t
cp_t: each stock's close price at time step—t
sh_t: shares owned of each stock at time step—t
ch_t : exchange rate at time step—t
b_t: investment available balance at time step—t)

For each stock, our action space is defined as $\{-i \ldots, -1, 0, 1, \ldots, j\}$, where i represents the maximum number of stocks to buy and j is the maximum number of stocks to sell at each action step. For continuous actions, we normalized the action space to $[-1, 1]$. In each state, we first perform an action selection, followed by a sell operation, a buy, and a hold operation.

The encoding features consist of 5 features and OHCLV (Open, High, Close, Low, Volume) data for each stock. The five features are Simple Moving Average (SMA), Exponential Moving Average (EMA), Moving Average Convergence Divergence (MACD), Relative Strength Index (RSI), and On-Balance Volume (OBV). These five features are selected from widely used financial technical indicators to predict stock price trend. Table 1 shows the technical indicators used in the proposed model.

Table 1. Summary of five features.

Type	Financial Technical Indicators
Moving average	Simple Moving Average (SMA)
Moving average	Exponential Moving Average (EMA)
Trend	Moving Average Convergence Divergence (MACD)
Momentum	Relative Strength Index (RSI)
Volume	On-Balance Volume (OBV)

From the state processing network, the agent observes the financial environment and leverages its experience through the Actor—Critic network. After updating the average gradients to the global network, the agent performs the final action, as depicted in Figure 1.

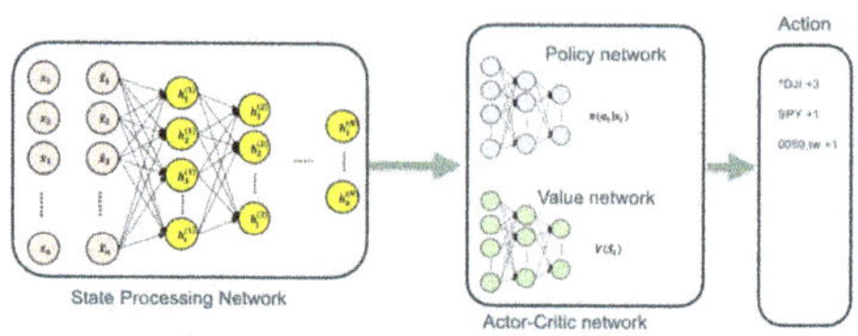

Figure 1. Flow diagram of the proposed model.

Our goal is to obtain the maximum return at the end of the investment process. For this goal, we define the reward function as recomputing the whole portfolio value in a new state—s using action—a.

$$R(s_t, a_t, s_{t+1}) = \left(b_{t+1} + cp_{t+1}s_{t+1}ch_{t+1}\right) - \left(b_t + cp_t s_t ch_t\right) \tag{5}$$

3. Experiment and Evaluation

To evaluate the effectiveness of the proposed A2C-based model for financial portfolio management, we conducted experiments on a real-world stock market dataset. The dataset includes historical stock prices of various assets, as well as economic indicators and news sentiment data that may affect the stock market performance.

We used Python, NumPy, pandas, and the scikit-learn machine learning library to implement and train the A2C model on this dataset.

To accelerate the computation process, we used a PC with an Intel four-core CPU 2.7 G, DDR4 16 G RAM, and the Ubuntu Desktop 20.04.5 LTS operating system. The experiment environment is further described in Table 2.

Table 2. Experimental environment.

Numerical and Machine Learning Package	
Python 3.9.2	scikit-learn
	imbalanced-learn
	NumPy
	SciPy
	pandas

To construct the portfolio, we selected Dow Jones Industrial Average (DJIA), SPDR S&P 500 ETF Trust (SPY), and Yuanta/P-shares Taiwan Top 50 ETF (0050.tw). In the portfolio, the first two stocks are in the US stock market and the last one is in the Taiwan stock market. We used daily data from 1 January 2010 to 1 January 2020 for training and data from 1 January 2020 to 7 June 2022 for validation (back-testing). Figure 2 shows the relative price of these three stocks from 1 January 2020 to 8 July 2022.

The close price of SPY is the highest at the end of this back-testing period, which is the reason why we choose SPY as the baseline. We would have obtained almost 30% if we bought SPY on 1 January 2020 and kept this stock until 7 June 2022.

The portfolio comprises SPY, DJIA, and 0050.tw, and the A2C model, which has been trained, is used by the agent to adjust the shares of each stock owned at each time step. The time step in this research is daily. Each day, the agent observes the financial environment and uses the trained A2C model to take action. Table 3 presents the experimental results of a real financial dataset, covering the period from 1 January 2020 to 7 June 2022.

Figure 2. Flow diagram of the A2C-based model.

Table 3. Performance of portfolio for stock trading.

	SPY	DJIA	0050.tw	Portfolio
Annual Return	0.11728	0.05538	0.12267	0.21484
Cumulative returns	0.30851	0.13962	0.30875	0.60423
Sharp Ratio	0.56783	0.33789	0.65887	0.81368

The portfolio outperforms other stocks that use a buy-and-hold strategy.

Figure 3 shows the cumulative returns of the portfolio and the baseline. The baseline is the buy-and-hold strategy, only investing in SPY stock.

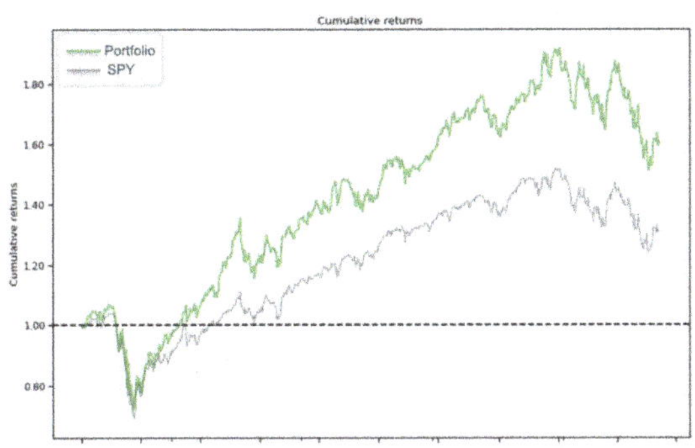

Figure 3. Returns and baseline of the portfolio.

4. Conclusions and Future Work

Investments in today's global economy are not limited to a single country. As a result, when constructing a portfolio, it is essential to consider the Unintended Effects Paradigm (UEP) that can affect the portfolio's performance. In this study, we propose an A2C-based model for financial portfolio management that takes into account different countries and exchange rates. Although exchange fees and stock transaction costs were not factored into the model, our findings indicate that investing in a portfolio yields better returns than investing in a single stock. Future research should focus on incorporating exchange fees and stock transaction costs into the model, and we will explore another deep reinforcement learning model that can be integrated into our proposed framework.

Author Contributions: Conceptualization, Y.-H.C. and H.-C.W.; methodology, Y.-H.C.; investigation, H.-C.W.; writing, Y.-H.C. All authors have read and agreed to the published version of the manuscript.

Funding: This research received no external funding.

Institutional Review Board Statement: Not applicable.

Informed Consent Statement: Not applicable.

Data Availability Statement: The authors declare that all data supporting the results of this research are available in this article.

Conflicts of Interest: The authors declare no conflict of interest.

References

1. Chandrasekhar, C.P. Negative interest rates: Symptom of crisis or instrument for recovery. *Econ. Political Wkly.* **2017**, *52*, 53–60.
2. Li, Y.-M.; Lin, L.F.; Hsieh, C.Y.; Huang, B.S. A social investing approach for portfolio recommendation. *Inf. Manag.* **2021**, *58*, 103536. [CrossRef]
3. Moghaddam, A.H.; Moghaddam, M.H.; Esfandyari, M. Stock market index prediction using artificial neural network. *J. Econ. Finance Adm. Sci.* **2016**, *21*, 89–93. [CrossRef]
4. Hochreiter, S.; Schmidhuber, J. Long Short-Term Memory. *Neural Comput.* **1997**, *9*, 1735–1780. [CrossRef] [PubMed]
5. Jung, J.; Jung, K.M. Stock market uncertainty and uncovered equity parity deviation: Evidence from Asia. *J. Asian Econ.* **2021**, *73*, 101271. [CrossRef]
6. Dong, X.Y.; Li, C.H.; Yoon, S.M. How can investors build a better portfolio in small open economies? Evidence from Asia's Four Little Dragons. *N. Am. J. Econ. Finance* **2021**, *58*, 19. [CrossRef]
7. Chiang, W.-C.; Enke, D.; Wu, T.; Wang, R. An adaptive stock index trading decision support system. *Expert Syst. Appl.* **2016**, *59*, 195–207. [CrossRef]
8. Bao, W.; Yue, J.; Rao, Y. A deep learning framework for financial time series using stacked autoencoders and long-short term memory. *PLoS ONE* **2017**, *12*, e0180944. [CrossRef]
9. Sermpinis, G.; Karathanasopoulos, A.; Rosillo, R.; de la Fuente, D. Neural networks in financial trading. *Ann. Oper. Res.* **2021**, *297*, 293–308. [CrossRef]
10. Pendharkar, P.C.; Cusatis, P. Trading financial indices with reinforcement learning agents. *Expert Syst. Appl.* **2018**, *103*, 1–13. [CrossRef]
11. Almahdi, S.; Yang, S.Y. A constrained portfolio trading system using particle swarm algorithm and recurrent reinforcement learning. *Expert Syst. Appl.* **2019**, *130*, 145–156. [CrossRef]

engineering proceedings

MDPI

Proceeding Paper

Risk Management in Latvia Municipal Social Services for Sustainable Well-Being of the Population †

Svetlana Lanka [1], Undine Vevere [2] and Santa Striguna [2,*]

1 Institute of Educational Sciences, Liepaja University, LV-3401 Liepaja, Latvia; svetlana.lanka@liepu.lv
2 Faculty of Pedagogy and Social Work, Liepaja University, LV-3401 Liepaja, Latvia; vevere.undine@gmail.com
* Correspondence: santa.striguna@liepu.lv

† Presented at the IEEE 5th Eurasia Conference on Biomedical Engineering, Healthcare and Sustainability, Tainan, Taiwan, 2–4 June 2023.

Abstract: The European Pillar of Social Rights, adopted in November 2017, sets out 20 principles with the explicit aim of promoting upward convergence to create better living and working conditions in Europe. This helps fight poverty in all its forms and ensures fair, adequate and sustainable social security systems. For several years now, the world and Latvia have been witnessing a change in political systems and ideologies, and rapid technological developments, especially in information technology. Today, we can see how the methods of administration in local government and the forms of customer service in the public sector are changing and what the benefits of change are for the population; however, the risks that can significantly affect the organization of work and communication with customers in local government institutions for the promotion of social welfare in a sustainable way are not fully identified. The aim of the research—based on the scientific literature, regulatory framework and the findings of the empirical study—is to develop a risk assessment, in accordance with the specifics of social services in Latvia, to promote the social well-being of the population with sustainability. Methods of research: theoretical research methods—literature research and analysis and regulatory analysis—and empirical research methods—survey, statistical analysis and expert interviews. Results of the research: Summarized conclusions on the need for risk management in municipal social services and a risk assessment developed in accordance with the specifics of social services. The obtained research results can become a basis for further research and risk management approval in Latvian municipal services.

Keywords: risk management; municipal social services; well-being; sustainability

Citation: Lanka, S.; Vevere, U.; Striguna, S. Risk Management in Latvia Municipal Social Services for Sustainable Well-Being of the Population. *Eng. Proc.* **2023**, *55*, 64. https://doi.org/10.3390/engproc2023055064

Academic Editors: Teen-Hang Meen, Kuei-Shu Hsu and Cheng-Fu Yang

Published: 6 December 2023

1. Introduction

Social work is a field that is exposed to uncertainty or risks on a daily basis because it is not always possible to balance positive and negative factors when working with people with a social problem. At the same time, social services in Latvia are currently institutions with a relatively large number of staff, so risk management in institutions may be incomplete or not yet relevant due to the fragmentation of regions and institutions. The need for risk assessment is also highlighted by the current geopolitical processes—in 2021, Latvia underwent an administrative territorial reform, which merged municipalities and significantly changed local government institutions, including social services [1]. This means that the reorganized institutions—the new social services—now have more than 50 employees under their management instead of the previous 30 employees (some municipal social services even had only 9–15 employees), which is reflected in the research part. The drastic change in the size of the team requires an assessment of risks that would not have been considered before or were very low, and it is also important to carry out a risk assessment specific to social services.

In a global context, risk management in social services is a long-standing issue. In 2014, Frederic G. Reamer, professor and social work researcher at the University of Chicago, highlighted the importance of risk management in social institutions, given the vital and important role of social institutions in crisis situations, where people need continuous social assistance, both in material terms and in the form of social services. The main challenges in social work are highlighted, namely, that the most significant risks in social services are related to privacy and confidentiality (particularly related to information technology and cyber security), maltreatment and service delivery, actions of practicing social workers and supervision of their work, consultation and referral, fraud and deception at work [2].

The lack of risk management in state and local government institutions was also highlighted in 2017 by researchers from the European University of Tirana in their article "Risk Management in Public Sector: A Literature Review", revealing the etiology of the problem, which is:

1. The mission and objectives of the public sector institution ignore other considerations, which is not a common phenomenon in the private sector;
2. Frequent management changes and the absence of leadership;
3. Managers who lack basic knowledge of risk management and business;
4. Insufficient budget and resources to develop and implement a risk strategy;
5. Lack of logical and detailed risk matrices in institutions;
6. Lack of understanding of the importance of risk management by the management and staff of the institution;
7. The developed risk management strategy is not applied on a day-to-day basis, so staff do not have practical skills in risk management [3].

It would not be acceptable for a social institution to suspend its work or restrict the provision of assistance or services to clients when internal or external risks occur, but, in the COVID-19 pandemic, which was an external risk, problems in risk management were more frequent and more pronounced. The absence of risk management was clearly visible during the COVID-19 pandemic in social care centers in Latvia and also in neighboring Estonia and Lithuania, demonstrating the "powerlessness" of institutional management against the consequences of the virus and the poor organization during the subsequent vaccination of clients [4]. Without anticipating the risks of the institution, it is impossible to prepare a plan of action in advance in case a risk occurs, which was also demonstrated by the simultaneous occurrence of several serious risks in social care centers both during the pandemic and in previous years (reputational, customer service, financial risks, etc.).

However, it should be noted that, at the global level, global crises that "paralyses" organizations are not only discussed in the context of the COVID-19 pandemic, but, increasingly, risks associated with environmental issues, geopolitical situations and societal processes, such as extreme weather events, natural disasters, biodiversity loss, man-made natural crises and protests and riots (by definition these are high risks by probability), are also being considered as high risks. As well, the use of biological weapons, economic conflicts between major powers, drinking water and food crises, extreme drought, including extreme heat waves, and social instability (high risks by impact) are being included [5]. It is important to take this into account when planning a sustainable risk management strategy, as these risks affect not only the well-being of society as a whole, but also organizations in both the private and public sectors.

As for social services, in the event of one of these risks (natural disasters, geopolitical conflicts, etc.), social workers should have a clear plan for how to continue their work providing adequate social assistance and social services to the population. The role of social institutions and social workers in times of crisis was analyzed in an article in the *Journal of Sociology and Social Work* by researchers of Iran. While such crises are not uncommon in Iran, the researchers point out that, with the rapidly changing climate, the risks are also becoming higher for temperate countries; heat waves, floods, prolonged internet and power outages and even attacks on the installations of sovereign states are becoming more frequent in European countries. The article highlights risk management, which allows one to anticipate these risks and to instruct employees when they occur, i.e., to develop a crisis scenario to prevent or mitigate disasters—the plan includes preventive measures and permitted and prohibited employee activities, as well as the necessary resources to cope with the crisis. Along with the planning process, practical training of social workers is also important, similar to what is carried out for police officers, firefighters and other professions that are fundamental in crisis situations [6].

Professor F. Reamer, on the other hand, highlights some of the most difficult risks to manage in both private companies and public institutions, namely, the information risks associated with an organization's crisis communication, public relations management, the media and customer behavior and opinion. Social work as a field has existed worldwide for almost a century and has therefore undergone many different transformations, keeping pace with innovation; today, social workers provide counseling in person, by telephone, online via internet platforms (video counseling) and through inquiries, emails and social networks. Cybertherapy and self-guided web-based interventions are described as some of the relatively newer social work methods which can be used by social workers when the client is more comfortable communicating online or when circumstances require them to do so [7]. These innovations also create new risks for institutions which may concern various ethical breaches in the social service itself. In the context of knowledge transfer in social work practice, from a risk management perspective, social workers and social services face the challenge of adopting and implementing a politically neutral position in their professional practice, focusing on individual cases and not focusing on structural causes in risk management.

Based on the facts above and the topicality of the problem, the research questions are:

1. What are the current risks in the social services of Latvian municipalities?
2. Do social services need to assess the risks in order to prevent or mitigate the potential consequences of the risks for the sustainable functioning of the institution?

2. Ethics and Risk Management in Social Service

However, a review of the literature suggests that risk management in the social sphere will be different from risk management in, for example, a private organization. Allan Barsky, professor of social work at Florida Atlantic University, discusses the risks in the risk management strategy itself in his article, stressing that the role of the institution is to avoid or mitigate potential risks and harms, but avoiding harm to the client, employee or institution may not achieve the ethical objectives of the institution. For example, social services often want to avoid actions that could have legal consequences (appealing against decisions of the institution, suing for 'bad practices' or actions of staff, etc.), but it remains an open question as to whether, in such cases, the principle of social justice towards clients is ensured and all the resources of the social service are used to deal with the situation. This brings up the concept of 'risk appetite', or the risks that an institution is willing to take and tolerate in order to achieve its goal as an organization, which is why risk management policies in social services are not straightforward to assess. The author concludes at the end of the article that risk management is essentially a teleological approach to ethics, focusing on the consequences of different actions, and that the key risk in risk management is when the employee and the institution carry out the less risky action rather than the right action [8].

Janet Seden, analyzing risks in social work, highlights social media crises, which often affect social services and have a negative impact on public confidence in social work as a field. Without taking measures to prevent or mitigate this risk, workers themselves may seek to avoid this risk by simply deciding not to perform their duties and shielding themselves from media attention. J. Seden compares social media crises to a 'moral panic', which is an intense public reaction to a sensational or scandalous news item in the media and which does not provide a fertile ground for social workers to continue their work. Moreover, it is concluded that the inability of a social institution to manage social media crises in the best possible way makes workers afraid to carry out their duties or makes them make mistakes, potentially leading to unprofessional social work or even inaction. [9].

It follows from the above that it is important for the risk assessment and later the risk management strategy or policy of a social service not only to be based on statistical data and the best practices of other institutions in risk management, but also to take into account the specificities and ethical aspects of the field of social work when developing sustainable risk management in social services with the aim of ensuring the well-being of the population. In practice, this would mean taking into account modern ethical guidelines, such as those proposed in the NASW Code of Ethics developed by the National Association of Social Workers (NASW). The document was last updated in February 2021, and most of the additions to the Code of Ethics relate specifically to the prevention of information technology and reputational risks, as well as a recommendation to assess the risks of each social institution in order to form a basis for implementation of an organizational policy or strategy to mitigate the impact of negative events on staff or the institution as a whole [10]. In Latvia, to address the risks mentioned above, the Sustainable Development Strategy 2030 includes a Sustainability Model as a response to global challenges, which is the basis for the research [11].

3. Method

The study was an analytical and research project aimed at identifying the need for risk assessment in Latvian municipal social services. The research methodology consisted of two parts: the first part used theoretical methods—analysis of scientific literature and analysis of normative acts—and the second part used empirical methods—questionnaires, statistical analysis of data and expert interviews—to assess the need for risk assessment in social services. The study concludes with the risk assessment itself and proposals for the implementation of risk management in municipal social services.

In the period from 1 March to 31 May 2021, a questionnaire survey was carried out with the heads of social services in Latvian municipalities on the existing risks in social services that the institutions face on a daily basis, as well as on the existing risk management and the methods used to prevent or mitigate the risks. A total of 76 respondents, or heads of social services, social administrations and social departments from all over Latvia, completed the questionnaires. The questionnaire was in electronic format and was sent to the respondent, completed and returned virtually via the internet. The questionnaire consisted of 10 closed and open-ended questions, including as the response type multiple-choice questions and a Likert scale, which is an ordinal scale aimed at the respondent to assess the extent to which he agrees or disagrees with a series of statements about the phenomenon or risk under study. The questionnaire selected different types of risk based on the results obtained from the analysis of literature sources, as well as following the best practice of the Exchequer of the Republic of Latvia in risk identification in institutions [12].

The questionnaires were collected manually, coded and statistically analyzed using Microsoft Excel software (version 2111) and an analysis package (Data Analysis) to perform a statistical analysis (Pearson analysis, linear regression analysis, simple analysis of variance and 2-group comparison tests). For the discussion with the experts, a semi-structured interview (the interviewer read out the prepared questions and wrote down the respondent's answers) with 6 questions based on the results of the questionnaire was developed to obtain an in-depth view of the social services managers on the current risk management and the need for a risk assessment in the institution. The expert interviews took place in October 2021, both face to face (the interviewer met the respondent in person) and via the online platform "Zoom"—managers of social services in Saldus, Sigulda, Tukums and Kuldīga municipalities were interviewed.

4. Findings

Social service managers were asked to rate the proposed risks on a Likert scale according to the level of risk (high, medium or low), thus revealing respondents' views on the risks that are relevant to social services as public institutions. In order to find out whether there were overall differences between respondents' opinions on the different risks, a simple analysis of variance (ANOVA: single factor) was carried out, which showed that there was a statistically significant difference between the groups ($p = 0.0003$; $p < 0.05$) (see Table 1). This means that the respondents assessed the risks differently, and the differences were significant enough to warrant further investigation into this issue.

Table 1. Risk assessment of social service managers by risk level.

Risk Level	Risk Type	Average (n)
Low	Risk of corruption	2.53
	Customer service risk	2.56
Medium	Strategic risks	2.97
	Reputation risk	2.91
	Cyber security risk	2.78
	Natural disaster/crisis risk	3.00
High	Information circulation risk	3.66
	Financial risk	3.34
	Work environment risk	3.03

After a simple analysis of variance and respondents assessment, it was possible to distinguish which risks are rated as high for social services and which are less relevant and rated as medium or low (see Table 1). The results showed that the lowest risks according to the social services managers are corruption and customer service risks, while the highest risks are information flow, financial and working environment risks.

According to the respondents' rating on a scale of 1 for "no risk" to 5 for "high risk", it was found that the social services representatives consider information circulation (average rating—3.66), financial (3.34) and working environment (3.03) risks to be high risks for the institution. The assessment is also in line with the analysis of the literature sources, especially as regards the risk of information circulation. The aspect of the working environment appears as a topical and high-level risk, which is relatively less frequently reflected in the literature review, but the results of the study show that this aspect needs more attention, specifically at the national level, and should be taken into account in the institution's risk assessment.

On the other hand, the risk of corruption (2.53) is assessed as a low-level risk, which can be explained by specific regulatory enactments that already require institutions to manage this risk [13]; thus, the risk has well-known mitigation and prevention measures that reduce the overall risk level. Similarly, the customer service risk is rated as low (2.56), which can be explained by the fact that social services inevitably have to deal with customer service risks (customer complaints, inaccuracies in documentation, etc.), so this risk occurs relatively frequently but its impact on the institution's performance is not as significant, and respondents rate other risks higher. The other risks rated as medium are: natural disasters/crises (3.0), cyber security (2.78), reputation (2.91) and strategic (2.97) risks, which is explained by the complexity of managing the risks in question, taking into account the theoretical analysis of the paper; for example, cyber security risks or natural disasters are not frequently experienced in Latvia.

A slightly different picture emerged when using the Pearson correlation method to find associations between the types of risks, their frequency of occurrence and some of the factors that, according to the theoretical analysis, could influence (most often increase) the risks, such as the communication of social services with clients, the number of employees in the institution, the institution's previous experience with negative incidents, etc. A statistically significant correlation (using Pearson correlation and linear regression analysis) was found between the number of staff and the frequency of current risks, as shown in Table 2.

Table 2. Relevance between current risks and the number of staff in social services.

Type of Risk	Relevance with the Number of Staff in the Institution (r Coefficient)	*p* (Confidence Level)
Information circulation risk	0.44	0.004
Customer service risk	−0.37	0.020
Corruption and interests conflict risk	0.11	0.305
Financial risk	−0.20	0.477
Nature disasters/crises risk	0.04	0.408
Strategic risk	0.04	0.385
Work environment risk	0.48	0.009
Reputational risk	0.19	0.610
Cyber security risk	0.1	0.421

When examining the relevance between the number of employees and the frequency of actual risks, positive or negative correlations were sought and found, such as a statistically significant positive correlation between the number of employees and information circulation risk ($r = 0.44$, $p = 0.004$), indicating a very strong relationship (at least 95% correlation), as well as with work environment risks ($r = 0.48$, $p = 0.009$). This suggests that the higher the number of employees in an institution, the higher the information flow and work environment risks. While information circulation as a high risk was already addressed in the theory part of the study, e.g., in the book by J. Seden [9], work environment risks, which include employee and customer safety violations, employee interpersonal relations (mobbing, bossing) and a negative psycho-emotional environment in the workplace, were relatively less frequently mentioned in the literature review. This aspect of the survey results leads to the conclusion that, in the opinion of social service managers, the working environment is a sufficiently important risk factor in Latvia and should be given increasing attention as a risk in the institution.

However, a statistically significant negative correlation ($r = -0.37$, $p = 0.020$) was found between the number of employees and customer service risks. This suggests that customer service risk is more acute in social services with fewer staff, which may be explained by the poor communication with the public and clients in municipal social services, which are probably smaller in terms of staff. This negative correlation led to an analysis of the methods of communication between social services and the public (see Table 3).

Table 3. Social services communication with the public by number of staff (averages).

Communication Channels	1–30 Employees	31–60 Employees	60 or More Employees	Average
Booklets, flyers	2.6	2	2	2.4
Social networks (Facebook, etc.)	3.5	5.3	6	4.1
Through employees	4.2	3.5	3.4	4.0
Newspapers (local, regional)	4.1	4.3	3.6	4.1
TV/radio	1.0	1.0	1.0	1.0
Institution/municipality website	5.6	5.0	5.0	5.4

Table 3 shows the types of communication with the public by different social services, i.e., respondents were asked to rank the communication tools used according to their frequency of use, assuming 6 to be "most frequently used communication tool" and 1 to be "least frequently used communication tool" (averages are shown in the table). The final result shows that social services communicate most frequently via the institution's or municipality's own website (average 5.4), social networks (4.1) and direct contact via staff (4.0), while radio, television (1.0) and printed information (leaflets, flyers) (2.4) are used relatively less frequently. There are differences between the communication styles of services with fewer employees and those with more employees, e.g., institutions with up to 30 employees mentioned their own or the municipality's website (5.6), and direct contact (4.2), as an important communication tool, but use social networks significantly less often (3.5), without creating a separate profile for the institution and without being committed to communicating with the public via this information channel. In contrast, in institutions with 30 or more employees, social networks are rated as frequently used and important (5.3 and 6), as is the institution's or municipality's website, suggesting that social services with a larger number of employees are more active online, which could also be explained by greater financial and human capital resources. Thus, the customer service risk discussed above, which appears in the results of the study as a more acute risk for social services with a small number of employees, could be due to the lack of public awareness of the tasks, objectives and assistance available to citizens, as the information channel of social networks is not fully exploited.

In order to further analyze the risk assessment, the question was raised as to whether there was a significant difference in the risk assessment between the answers of respondents whose institutions have been involved in a negative and public incident with social media, the municipal council or the supervising public authorities in the last five years and the answers of respondents whose institutions have not been involved in such incidents. The study found that almost half, or 46.9%, of the respondents answered in the affirmative, which leads to the conclusion that social services have to experience various negative events, which could potentially also turn into a crisis situation and have various impacts and consequences on the sustainable functioning of the institution in providing services to the population, frequently. In order to compare the responses of the two groups, respondents who have experienced negative incidents and respondents who have not experienced negative incidents in the last 5 years, a two-group comparison test (F-Test Two-Sample for Variance) was used, which concluded that the p-value for this sample variance is 0.4, which, at the significance level $p = 0.05$, indicates that the variance is homogeneous, and, therefore, a t-Test was performed accordingly (Two-Sample Assuming Equal Variances) for this particular sample to assess how different the responses of the two groups are.

It was found that at the significance level $p = 0.05$, there is a statistically significant difference between the responses of the above respondents, and the analysis carried out shows that respondents who have already experienced negative publicity or other negative experiences with other institutions or social media while managing a social service tend to rate risks at a higher risk level, thus giving more importance to the assessment of risks within the institution, than social service managers who have not experienced significant negative incidents so far (in the last 5 years).

The summary of the in-depth expert interview responses concludes that the risks with the highest level in social services are currently:

1. Compliance and customer service risk—due to the COVID-19 pandemic, there is prolonged staff sickness, staff turnover and staff shortages, making it difficult for services to ensure the availability of all services across the entire coverage area;
2. Information circulation risk—service managers stress the importance of communication with the public, which includes not only communication with clients, but also communication with local councilors and other related bodies;
3. Work environment risk—experts mention staff burnout as a major problem in social work, which is also linked to frequent staff changes and prolonged sickness and which affects the functioning of the social services.

In the conclusion of the expert interview, addressing the issue of the need for risk management and risk assessment in social services, it is found that risk assessment will become increasingly important in the future, as:

1. The number of staff in the institution will increase;
2. The area covered by the institution will increase;
3. There has already been an increase in certain risks;
4. In the past, risk management in social services was not an issue because institutions were small, fragmented and sometimes without competent management (before the administrative territorial reform [1]).

5. Risk Assessment

The analysis of the literature and documents shows that a risk management strategy appropriate to an organization includes both an assessment of the risks identified, identifying the level of risk acceptable to the organization, and an assessment of the possibility of the risk occurring and its impact on the organization, i.e., the institution has also developed a risk matrix. The risk matrix has been developed taking into account The State Audit Office of the Republic of Latvia Guidelines for the internal control system designed in 2018 [14] as well as the COSO 2020 Risk Management Guidelines (COSO Compliance Risk Management: Applying the COSO ERM Framework) [15] (see Table 4).

Table 4. Assessment of current risks in terms of likelihood and impact. Risk matrix.

Possibility and Impact Risk Matrix						
Possibility	5 Likely to occur				Information circulation Risk	
	4 Very likely			Compliance and customer service risk	Cyber security risk	
	3 Possible		Strategic risks	Work environment risk	Reputation risk	
	2 Unlikely			Financial risk	Risk of corruption	Risk of natural disasters/geopolitical crises
	1 Rare					
		1 Insignificant	2 Small	3 Serious	4 Strong	5 Catastrophic
				Impact		

Red—high risk, yellow—medium risk, green—low risk

The table shows that, depending on the possibility and impact, the low level of risk is a strategic risk, while the medium level of risk is much higher—compliance and customer service, work environment, reputational, financial, corruption and natural disaster risks—as the impact of these risks on the institution's operations can be severe. It is important to mention that, although the possibility of natural disasters/geopolitical crises in the Latvian context is relatively low (unlikely), as reflected in the survey and expert interviews, the impact on the institution when the risk occurs could be catastrophic, so these risks should definitely be considered in the risk assessment.

6. Results, Conclusions and Recommendations

The results show that almost half, or 43.8%, of social services have used a lawyer to deal with the consequences of various risks, while 53.1% have dealt with the problems within the institution, and 3.1% have not faced a situation where they had to deal with serious consequences of risks. The survey results show that 18.8% of social services already have their own lawyer to deal with the consequences of various risks on a regular basis. On a positive note, 81.2% of social services cooperate with their own or the municipality's public relations officer to inform the public about available social assistance and social services and to reduce negative publicity. However, the majority, or 89.1%, of social services do not have a separate risk management document, which is explained by the fact that the normative acts do not strictly prescribe the creation of such a document for the municipal units as it is prescribed for the state institutions for the development of the internal control system. Of respondents, 7.7% answered that the institution has developed a risk management document, while 3.2% chose another option, in which they noted that risk management in the social service is controlled by the municipal management which has developed an overall municipal risk management policy to ensure the maximum stability of the level of social well-being of the population in the long term.

In summary, it can be concluded that there are a number of high-level risks in the functioning of social institutions that could potentially lead to a crisis situation, and, therefore, risk management in the institution plays an important role in order to prevent these risks. However, a well-developed risk assessment does not necessarily imply good risk management in the institution, or, in the case of social services, risks in risk management itself. Foreign practice and research related to the functions of social services and communication with the public show the need for risk management in municipal social services, thereby sustainably mitigating the impact of various risks on the reputation of the institution and the social work sector as a whole. Risk management and risk assessment in social services will differ from the risk management strategy of the private sector as the introduction of risk management in social services should take into account the ethical dimension of the institution and the ethical norms with which social workers are confronted on a daily basis, thus integrating the principles of the International Code of Ethics for Social Workers into risk management in order to sustainably provide social services and social assistance based on societal needs and ethical principles.

The results of the survey of social services managers show that the most frequent risks in social services in the last five years are information circulation, customer service and work environment risks, while information circulation, work environment and financial risks are ranked as high risks by the heads of social services. The analysis of the survey results shows that there is a statically significant correlation between the number of staff and certain types of risks, i.e., there is a tendency that the higher the number of staff, the more frequent the information circulation and work environment risks, while, with lower staff numbers, the customer service risks become acute, which, in turn, can be explained by the communication style of the service with the public.

The empirical study concludes that risk management in the social services is particularly relevant at a time when, in the context of the administrative territorial reform of 1 July 2021, institutions have been reorganized, bringing together larger teams and covering a wider area, thus changing the various management processes of the institution. In order to facilitate the management of risks in social services based on the principles of sustainable development and performance, it can be generally concluded that risk assessment is feasible for social services and a sufficient number of risks can be identified, which means that risk management in services can be an additional tool in management processes and in ensuring optimal performance of the institution in providing sustainable and quality services to the community.

In the framework of the study, a risk assessment and a risk matrix were developed according to the specifics of Latvian municipal social services, based on the analysis of the literature and the data collected in the study, which reflect the current risks.

When planning and implementing risk management in social services, based on the results of the study, it is recommended to use the already developed guidelines and recommendations of state institutions, based on internationally recognized risk management models, principles and the key principles of the Sustainable Development Strategy. The risk assessment should be developed in a way that does not contradict the Code of Ethics developed by the National Association of Social Workers and/or the Code of Ethics for Social Workers developed within the institution, according to the specificities of the social service, as risk management in the social field involves different ethical considerations. In order for risk management to function fully in the social services once it has been introduced, the management of the institution should designate a risk management officer who regularly updates the identification of risks and reports to management on the urgency of risks and methods of mitigation or elimination.

The risk assessment developed in the study can serve as a basis for the development of a risk management policy or strategy for social services which includes regular risk identification and assessment, selection of risk management methods and a risk management monitoring system. The results of the study can be incorporated into the internal control system of social services as part of risk management, thereby facilitating the development and implementation of an internal control system in an institution that has not previously had such a system, thereby ensuring the sustainability of the institution's operations and the delivery of services to members of the public.

The study answers both research questions, showing the most common risks in social services and the need for risk assessment. In the context of knowledge transfer in social work practice, from a risk management perspective, it is necessary to focus on the structural causes of risk management.

Author Contributions: Conceptualization, S.L., U.V. and S.S.; methodology, U.V.; software, S.S.; validation, S.L. and U.V.; formal analysis, S.S. and U.V.; data curation, S.L. and U.V.; writing—original draft preparation, U.V. and S.S.; writing—review and editing, S.L. and S.S. All authors have read and agreed to the published version of the manuscript.

Funding: This research received no external funding.

Institutional Review Board Statement: Not applicable.

Informed Consent Statement: Not applicable.

Data Availability Statement: Data is unavailable due to ethical restrictions.

Conflicts of Interest: The authors declare no conflict of interest. The funders had no role in the design of this study, in the collection, analyses, or interpretation of data, in the writing of the manuscript, or in the decision to publish the results.

References

1. Law on Administrative Territories and Settlements. Available online: https://likumi.lv/ta/id/315654-administrativo-teritoriju-un-apdzivoto-vietu-likums (accessed on 13 January 2021).
2. Reamer, F. *Risk Management in Social Work: Preventing Professional Malpractice, Liability, and Disciplinary Action*; Columbia University Press: New York, NY, USA, 2014; pp. 152–180.
3. Remzi, A.; Besarta, V. Risk Management in the Public Sector: A Literature Review. *Eur. J. Multidiscip. Stud.* **2017**, *2*, 323–328. [CrossRef]
4. The Effects of COVID-19 in Social Care Homes in Lithuania and Estonia. Situation in Latvia. 2021. Available online: https://lr1.lsm.lv/lv/raksts/labriit/par-covid-19-sekam-socialas-aprupes-namos-lietuva-un-igaunija.-s.a143139/ (accessed on 10 October 2021).
5. The Global Risk Report. 2020. Available online: http://www3.weforum.org/docs/WEF_Global_Risk_Report_2020.pdf (accessed on 9 July 2021).
6. Iravani, M.; Parast, S. Examine the Role of Social Workers in Crisis Management. *J. Sociol. Soc. Work.* **2014**, *2*, 87–97.
7. Reamer, F. Social Work in a Digital Age: Ethical and Risk Management Challenges. *Soc. Work.* **2013**, *58*, 163–172. [CrossRef] [PubMed]
8. Risk of Risk Management. The New Social Worker. Available online: https://www.socialworker.com/feature-articles/ethics-articles/risks-of-risk-management/ (accessed on 5 October 2021).
9. Seden, J. *Social Work: Risks, Needs and Balanced Assessments*; The Open University: Milton Keynes, UK, 2016; pp. 5–15.
10. Highlighted Revisions to the Code of Ethics. Available online: https://www.socialworkers.org/About/Ethics/Code-of-Ethics/Highlighted-Revisions-to-the-Code-of-Ethics (accessed on 5 October 2021).
11. Sustainable Development Strategy of Latvia Until 2030. Available online: https://pkc.gov.lv/sites/default/files/inline-files/LIAS_2030_parluks_lv_0.pdf (accessed on 5 October 2021).
12. Comprehensive, Systematic and Coordinated Risk Management Approach. Available online: https://www.fm.gov.lv/lv/media/8439/download (accessed on 1 September 2021).
13. Regulations on the Basic Requirements of the Internal Control System for Prevention of Corruption and Conflict of Interest Risk in a Public Institution. Available online: https://likumi.lv/ta/id/294518-noteikumi-par-ieksejas-kontroles-sistemas-pamatprasibam-korupcijas-un-interestu-konflikta-riska-noversanai-publiskas-personas-institucija (accessed on 5 October 2021).
14. Guidelines for the Internal Control System. Available online: https://www.kase.gov.lv/sites/default/files/public/PD/Metodika/Rokasgr%25C4%2581matas/IKS_vadlinijas.pdf (accessed on 6 October 2021).
15. Compliance Risk Management: Applying the COSO ERM Framework. Available online: https://www.coso.org/Documents/Compliance-Risk-Management-Applying-the-COSO-ERM-Framework.pdf (accessed on 6 October 2021).

engineering proceedings

MDPI

Proceeding Paper

RG-Based (k, n)-Threshold Visual Cryptography with Abilities of OR and XOR Decryption †

Yu-Ru Lin and Justie Su-Tzu Juan *

Department of Computer Science and Information Engineering, National Chi Nan University, Nantou 545, Taiwan; s111321509@mail1.ncnu.edu.tw
* Correspondence: jsjuan@ncnu.edu.tw

† Presented at the IEEE 5th Eurasia Conference on Biomedical Engineering, Healthcare and Sustainability, Tainan, Taiwan, 2–4 June 2023.

Abstract: A (k, n) visual cryptography (VCS) is used to build a system for secret sharing. The system divides a secret image (S) into n shares and recovers S by stacking shares bigger than or equal to k, while shares below k provide no information about S. The fundamental idea of VCS is that, rather than relying on mathematical or cryptographic skills, human vision can be used to decrypt the secret image. Typically, a Boolean OR operation can be used to indicate the stacking action in a VCS. The reconstructed secret image gradually darkens as more shares are stacked. However, this intractable issue can be overcome by designing an XOR-based VCS that uses the Boolean XOR operator rather than the OR operation. This indicates that by using the XOR-based VCS, higher image quality can be attained. Because the XOR operation requires the use of additional equipment, scholars consider that when no equipment is available, the traditional OR operation can still be used to reveal the secret images. That is, the secret image can be decrypted without a computing device by stacking enough shares, and if a lightweight computing device is available, a better-quality image can be produced via an XOR operation. In 2015, an RG-based (k, n) VCS to restore the secret image by using an OR or XOR operation was proposed. In this study, we improve the scheme and design a new (k, n) VCS, called (k, n) 2D_VCS to encrypt a secret image into n shares. The secret image can be recovered when k or more shares are gathered and stacked (OR operation) together or when an XOR procedure is utilized. Both the theoretical proof and experimental results show that the quality of the restored image obtained by our method is better than that of the previous methods.

Keywords: secret sharing scheme; visual cryptography; (k, n)-threshold; XOR operation; random grids

Citation: Lin, Y.-R.; Juan, J.S.-T. RG-Based (k, n)-Threshold Visual Cryptography with Abilities of OR and XOR Decryption. *Eng. Proc.* **2023**, *55*, 65. https://doi.org/10.3390/engproc2023055065

Academic Editors: Teen-Hang Meen, Kuei-Shu Hsu and Cheng-Fu Yang

Published: 7 December 2023

1. Introduction

The Visual Cryptography Scheme (VCS) was suggested in 1995 for distributing secrets [1]. A binary secret image can be recovered by encrypting it into n random images (shares) and then stacking those n shares. They also developed a threshold k-out-of-n VCS method, (k, n)-threshold VCS. Encrypting a secret image in binary into n shares and then stacking any k or more than k shares ($k \leq n$) can reconstruct the secret image. Fewer than k shares cannot provide any clues about the binary secret image.

Visual secret sharing (VSS) by random grids (RG) was introduced by Kafri and Keren in 1987 [2]. It eliminates the secret pixel expansion issue, and this approach does not demand codebook designs, which is the disadvantage of Naor and Shamir's scheme [1]. Random grids mean that we obtain every pixel by randomly selecting between 1 (opaque or black) and 0 (transparent or white). In 2009, Shyu gave a scheme of (n, n) RG-based VCS [3]. Chen and Tsao constructed a (k, n) RG-based VCS [4]. Then, Guo, Liu, and Wu improved Chen and Tsao's scheme in 2013 [5]. The stacking operation in a VCS is denoted by a Boolean OR operation. However, the more shares of stacking, the darker the reconstructed image. An XOR-based VCS can solve this problem, as a better recovered image quality can be obtained when using an XOR operation instead of an OR operation. A new idea was suggested in Ref. [6]. In their method, OR or XOR is optionally used to restore the secret image when restoring the binary secret image. As a result, the secret image can be restored when computing resources are limited by stacking k or more shares (using an OR operation). When computing resources are available, a secret image of higher quality can be obtained (using an XOR operation). We improved their scheme by designing a new (k, n) VCS that can also extract binary secret images by OR or XOR operations and achieved a better quality of restored images than their scheme.

This paper is structured as follows. The related work is presented in the next section, the suggested scheme, (k, n) 2D_VCS is presented in Section 3, and an analysis of the (k, n) 2D_VCS is shown in Section 4. Sections 5 and 6 show the experimental results and conclusion, respectively.

2. Related Work

2.1. (2, 2) RG-Based VCS

In 1987, Kafri and Keren firstly proposed (2, 2) RG-based VCS [2], Algorithm 1, to encrypt a binary image S, and their algorithm output two shares B_1 and B_2, which looked like random grids. After superimposing them, the secret image S was reconstructed by using the Human Visual System (HVS) without specific computational and cryptographic knowledge.

Algorithm 1 KK [2]

Input:	Secret binary image S.
Output:	Two shares B_1, B_2.
Step 1.	Generate the first share B_1 randomly selecting 0 or 1 for each pixel of B_1.
Step 2.	Based on the pixels $S[i, j]$ of S and the pixel $B_1[i, j]$ of B_1, the pixel value $B_2[i, j]$ of share B_2 is calculated by $B_2[i,j] = \begin{cases} B_1[i,j], & if\ S[i,j] = 0 \\ \overline{B_1[i,j]}, & if\ S[i,j] = 1 \end{cases}$.

2.2. (k, n) RG-Based VCS

As mentioned above, a scheme of (k, n) RG-based VCS, Algorithm 2, was proposed in 2011 [4]. For comparison with the proposed scheme, we rewrite their algorithm as follows. Note that it is essentially the same as their original algorithm.

Algorithm 2 CT [4]

Input:	A secret binary image S.
Output:	n shares $(B_1, B_2, \ldots, B_n)$.
Step 1.	For each position $[i, j]$, repeat Steps 2–4:
Step 2.	Randomly select from $\{0, 1\}$ to generate n random pixels $b_1, b_2, \ldots, b_n$.
Step 3.	If $S[i, j] \neq b_1 \oplus b_2 \oplus \ldots \oplus b_k$, then $b_k = \overline{b_k}$.
Step 4.	Rearrange the above n bits $b_1, b_2, \ldots, b_n$ into $B_1[i, j], B_2[i, j], \ldots, B_n[i, j]$ randomly.
Step 5.	Output $(B_1, B_2, \ldots, B_n)$.

3. Proposed Scheme

A new (k, n) RG-based VCS used with two decryption methods (OR and XOR) for binary images is presented, called (k, n) 2D_VCS. Algorithm 2 only considers the case of stacking k shares and does not consider the case of stacking more than k shares. Furthermore, additional consideration is required to fully restore the original secret image when collecting all the shares. Therefore, we add two steps, resulting in the following algorithm.

Algorithm 3 differs only in Steps 4 and 5 from Algorithm 2. These two steps lead to important consequences. Step 4 of Algorithm 3 makes the reconstructed image clearer when staking t $(k < t \leq n)$ shares. Step 5 of Algorithm 3 makes it possible to fully recover the secret image after collecting all n shares. Since we do not change the first k shares constructed by Algorithm 2, when Algorithm 3 stacks k shares, the quality of the reconstructed image is similar to that by Algorithm 2. When stacking t shares, the probability of recovering the secret image is $1/C(n, t)$ $(n - k + 1)$, which makes the proposed scheme more clearly recover the secret using the XOR operator for tacking $k < t < n$ shares. Therefore, compared to earlier studies, the method has a higher quality of the recovered image and improved ability to recover the secret image with both XOR and OR operations. Figure 1 depicts the suggested scheme's schematic.

Algorithm 3 (k, n) 2D_VCS.

Input:	An $M \times N$ secret binary image S.
Output:	n shares $(B_1, B_2, \ldots, B_n)$.
Step 1.	For each position $[i, j]$, repeat Steps 2–6.
Step 2.	Randomly select from $\{0, 1\}$ to generate n random pixels $b_1, b_2, \ldots, b_n$.
Step 3.	If $S[i, j] \neq b_1 \oplus b_2 \oplus \ldots \oplus b_k$, then $b_k = \overline{b_k}$.
Step 4.	Select a number t randomly from $\{k + 1, k + 2, \ldots, n\}$. If $S[i, j] \neq b_1 \oplus b_2 \oplus \ldots \oplus b_t$, then $b_t = \overline{b_t}$.
Step 5.	If $S[i, j] \neq b_1 \oplus b_2 \oplus \ldots \oplus b_n$, then $b_n = \overline{b_n}$.
Step 6.	Randomly rearrange these n bits $b_1, b_2, \ldots, b_n$ into $B_1[i, j], B_2[i, j], \ldots, B_n[i, j]$.
Step 7.	Output $(B_1, B_2, \ldots, B_n)$

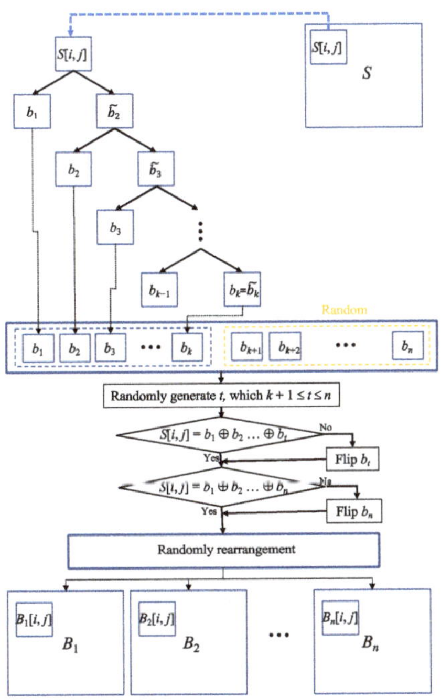

Figure 1. Schematic of the proposed scheme.

Therefore, we can directly stack enough shares ($\geq k$) to recover the binary secret image via HVS. Computers enable us to obtain more accurate secret images. However, when less than k shares are acquired, any clue about the binary secret image cannot be found.

4. Analysis

The proposed scheme's security is theoretically shown, and its performance is evaluated in terms of the visual aspect. First, we review useful tools and analyze the proposed scheme with these tools. We refer to definitions from previous works.

Definition 1. (Average light transmission [3]**).** *The probability that a given pixel, x, in the binary image X is transparent. Let Prob (x = 0) denote the light transmission of pixel x, which is shown by the symbol l[x]. The average light transmission of image X, with the size M × N, is expressed as*

$$LX \frac{1}{M \times N} \sum_{i=1}^{M} \sum_{j=1}^{N} l[X[i, j]] \tag{1}$$

So, l[x] = 0 for an opaque pixel x, l[x] = 1 for a transparent pixel x. Also, we say L[B] = 1/2 for a random grid B normally.

Definition 2. (Contrast [3]**)** *The visual quality will determine how effectively the reconstructed image is recognized by human eyes. Contrast α is expressed as*

$$\alpha = \frac{L[B[S(0)]] - L[B[S(1)]]}{1 + L[B[S(1)]]} \tag{2}$$

where S(0) (S(1), respectively) stands for all of S's transparent (opaque, respectively) pixels, and B[S(0)] (B[S(1)], respectively) for all of the encrypted pixels from S(0) (S(1), respectively).

Definition 3. (Visually recognizable) *When L[B[S(1)]] < L[B[S(0)]], the contrast of the restored image α > 0, and the binary secret image S is recognized. Moreover, if α = 1, the recovered image B is the same as the original image S.*

Definition 4. (Security).*When L[B[S(0)]] = L[B[S(1)]], the light transmission of the transparent part of the original image is the same as that of the opaque part of the original image. That means no clue about the original secret image S can be recognized from the reconstructed image B. Therefore, if α = 0 when less than k shares are collected, the approach is secured since no information of S is retrieved.*

Table 1 lists all possibilities for sharing pixels when $k = 3$ and $n = 4$ of the (k, n) 2D_VCS, which is used to show this scheme is secured and visually recognizable.

Table 1. All possibilities b_1, b_2, b_3, b_4 for the proposed (3, 4) RG-based VCS.

S[i, j]	b_1	b_2	b_3	b_4
0	0	0	0	0
	0	1	1	0
	1	0	1	0
	1	1	0	0
1	0	0	1	0
	0	1	0	0
	1	0	0	0
	1	1	1	0

Lemma 1. *Algorithm 3 (k, n) 2D_VCS gives a (k, n) RG-based VCS with two decryption methods (OR and XOR) when k = 3 and n = 4. That is, Algorithm 3 satisfies the (1) security and (2) visually recognizable conditions.*

Proof. The following is divided into two parts to prove: (1) We must prove that stacking less than $k = 3$ shares (including one share) fails to yield any information regarding the binary secret image. From Tables 1–4 are induced, where $b_{x \oplus y}$ means $b_x \oplus b_y$, and $b_{x \otimes y}$ means $b_x \otimes b_y$ for any integers $1 \leq x < y \leq 4$. Similar notations are used later for stacking more than two shares.

Table 2. Contrast of each bit for Table 1.

S[i, j]	b_1	b_2	b_3	b_4
0	0	0	0	0
	0	1	1	0
	1	0	1	0
	1	1	0	0
1	0	0	1	0
	0	1	0	0
	1	0	0	0
	1	1	1	0
$L[B[S(0)]]$	0.5	0.5	0.5	1
$L[B[S(1)]]$	0.5	0.5	0.5	1
α	0	0	0	0

Table 3. Contrast of stacking any two bits using XOR operator for Table 1.

S	$b_{1\oplus2}$	$b_{1\oplus3}$	$b_{1\oplus4}$	$b_{2\oplus3}$	$b_{2\oplus4}$	$b_{3\oplus4}$
0	0	0	0	0	0	0
	1	1	0	0	1	1
	1	0	1	1	0	1
	0	1	1	1	1	0
1	0	1	0	1	0	1
	1	0	0	1	1	0
	1	1	1	0	0	0
	0	0	1	0	1	1
$L[B[S(0)]]$	0.5	0.5	0.5	0.5	0.5	0.5
$L[B[S(1)]]$	0.5	0.5	0.5	0.5	0.5	0.5
α	0	0	0	0	0	0

Table 4. Contrast of stacking any two bits using OR operator for Table 1.

S	$B_{1\otimes2}$	$B_{1\otimes3}$	$B_{1\otimes4}$	$B_{2\otimes3}$	$B_{2\otimes4}$	$B_{3\otimes4}$
0	0	0	0	0	0	0
	1	1	0	1	1	1
	1	1	1	1	0	1
	1	1	1	1	1	0
1	0	1	0	1	0	1
	1	0	0	1	1	0
	1	1	1	0	0	0
	1	1	1	1	1	1
$L[B[S(0)]]$	0.25	0.25	0.5	0.25	0.5	0.5
$L[B[S(1)]]$	0.25	0.25	0.5	0.25	0.5	0.5
α	0	0	0	0	0	0

Tables 2–4 show that α is 0 for any possible case; therefore, the proposed (k, n) 2D_VCS is safe when $(k, n) = (3, 4)$. For proving Equation (2), we need to show that stacking $k = 3$ or 4 shares recovers the binary secret image S. Tables 5 and 6 show the correctness.

Table 5. Contrast of stacking any three or four bits using XOR operator for Table 1.

S	$b_{1\oplus2\oplus3}$	$b_{1\oplus2\oplus4}$	$b_{1\oplus3\oplus4}$	$b_{2\oplus3\oplus4}$	$b_{1\oplus2\oplus3\oplus4}$
0	0	0	0	0	0
	0	1	1	0	0
	0	1	0	1	0
	0	0	1	1	0
1	1	0	1	1	1
	1	1	0	1	1
	1	1	1	0	1
	1	0	0	0	1
$L[B[S(0)]]$	1	0.5	0.5	0.5	1
$L[B[S(1)]]$	0	0.5	0.5	0.5	0
Average α	0.1818				1

Table 6. Contrast of stacking any three or four bits using OR operator for Table 1.

S	$b_{1\otimes2\otimes3}$	$b_{1\otimes2\otimes4}$	$b_{1\otimes3\otimes4}$	$b_{2\otimes3\otimes4}$	$b_{1\otimes2\otimes3\otimes4}$
0	0	0	0	0	0
	1	1	1	1	1
	1	1	1	1	1
	1	1	1	1	1
1	1	0	1	1	1
	1	1	0	1	1
	1	1	1	0	1
	1	1	1	1	1
$L[B[S(0)]]$	0.25	0.25	0.25	0.25	0.25
$L[B[S(1)]]$	0	0.25	0.25	0.25	0
Average α	0.0526				0.25

According to Tables 5 and 6, the average α all > 0, which means that the proposed (3, 4) RG-based VCS is visually recognizable. Hence, the proof is complete.□

Then, the following theorem can be obtained. Due to the page limit, we skip the proof.

Theorem 1. *Algorithm 3 (k, n) 2D_VCS is a (k, n) RG-based VCS of two decryption methods (OR and XOR).*

Tables 7 and 8 show the theoretical contrast of the (k, n) 2D_VCS for stacking $2 \leq p \leq n$ shares for any $2 \leq k \leq n \leq 6$. We obtain these two tables by analyzing each case using a method similar to that in Lemma 1, which shows nothing is found when staking less than k shares, and the secret image S is revealed when staking more than or equal to k shares.

Table 7. Contrast of the proposed (*k*, *n*) RG-based VCS using OR operator.

(*k*, *n*)	$p = 2$	$p = 3$	$p = 4$	$p = 5$	$p = 6$
(2, 4)	0.0606	0.1579	0.375	-	-
(3, 4)	0	0.0526	0.25	-	-
(4, 4)	0	0	0.125	-	-
(2, 5)	0.0382	0.0786	0.1154	0.2083	-
(3, 5)	0	0.0204	0.0674	0.1875	-
(4, 5)	0	0	0.0227	0.125	-
(5, 5)	0	0	0	0.0625	-
(2, 6)	0.0260	0.0492	0.0621	0.0704	0.1094
(3, 6)	0	0.0106	0.0277	0.0495	0.1042
(4, 6)	0	0	0.0074	0.0294	0.0938
(5, 6)	0	0	0	0.0099	0.0625
(6, 6)	0	0	0	0	0.0313

Table 8. Contrast of the proposed (*k*, *n*) RG-based VCS using XOR operator.

(*k*, *n*)	$p = 2$	$p = 3$	$p = 4$	$p = 5$	$p = 6$
(2, 4)	0.1111	0.1818	1	-	-
(3, 4)	0	0.1818	1	-	-
(4, 4)	0	0	1	-	-
(2, 5)	0.0674	0.0440	0.0930	1	-
(3, 5)	0	0.0690	0.1429	1	-
(4, 5)	0	0	0.1429	1	-
(5, 5)	0	0	0	1	-
(2, 6)	0.0449	0.0167	0.0220	0.0571	1
(3, 6)	0	0.0333	0.0301	0.0769	1
(4, 6)	0	0	0.0455	0.1176	1
(5, 6)	0	0	0	0.1176	1
(6, 6)	0	0	0	0	1

5. Experimental Results

The results of the (*k*, *n*) 2D_VCS when (*k*, *n*) = (2, 4), (3, 4), and (3, 5) are obtained in this section. The 300 × 300 pixels images in those experiments are utilized for original secret images and shares. Figure 2 shows the result for the proposed (2, 4) 2D_VCS. Figure 2a shows the secret image *S*, and Figure 2b–e show the four random-noise-like shares B_1, B_2, B_3, and B_4. The recovered binary images based on the OR operator are shown in Figure 2f–h. The recovered binary images based on the XOR operator are shown in Figure 2i–k.

Similarly, the experimental findings for the suggested (3, 4) and (3, 5) 2D_VCS are shown in Figures 3 and 4. Both (a) are the original binary secret images. Figures 3b–e and 4b–f are random shares. Figures 3f–h and 4g–j show the recovered binary images based on the OR operator. Then, Figures 3i–k and 4k–n are the recovered binary images based on the XOR operator.

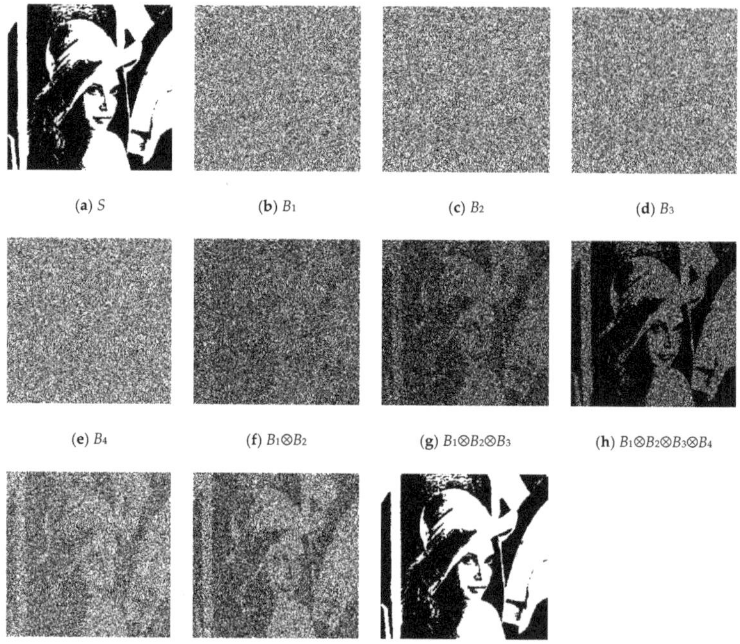

(**a**) S (**b**) B_1 (**c**) B_2 (**d**) B_3 (**e**) B_4 (**f**) $B_1 \otimes B_2$ (**g**) $B_1 \otimes B_2 \otimes B_3$ (**h**) $B_1 \otimes B_2 \otimes B_3 \otimes B_4$ (**i**) $B_1 \oplus B_2$ (**j**) $B_1 \oplus B_2 \oplus B_3$ (**k**) $B_1 \oplus B_2 \oplus B_3 \oplus B_4$

Figure 2. The experimental result of the proposed (2, 4) 2D_VCS.

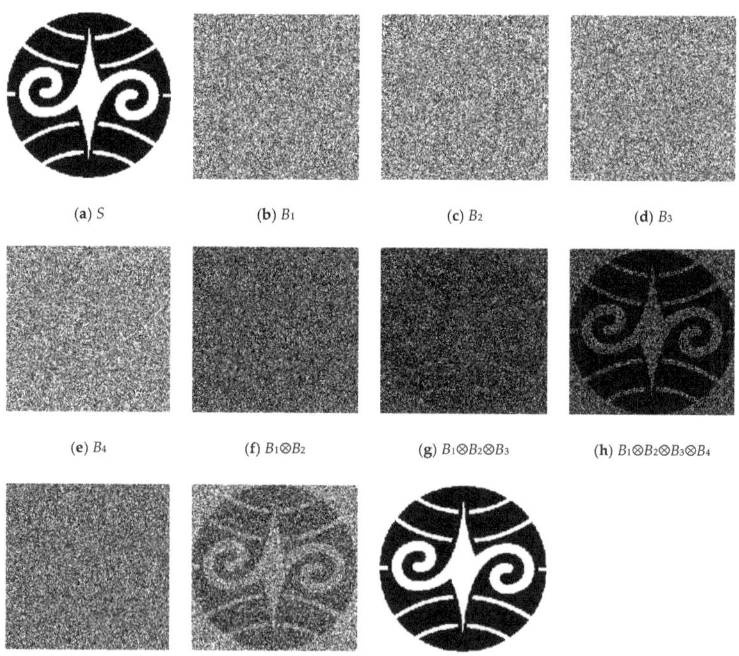

(**a**) S (**b**) B_1 (**c**) B_2 (**d**) B_3 (**e**) B_4 (**f**) $B_1 \otimes B_2$ (**g**) $B_1 \otimes B_2 \otimes B_3$ (**h**) $B_1 \otimes B_2 \otimes B_3 \otimes B_4$ (**i**) $B_1 \oplus B_2$ (**j**) $B_1 \oplus B_2 \oplus B_3$ (**k**) $B_1 \oplus B_2 \oplus B_3 \oplus B_4$

Figure 3. The experimental result of the proposed (3, 4) 2D_VCS.

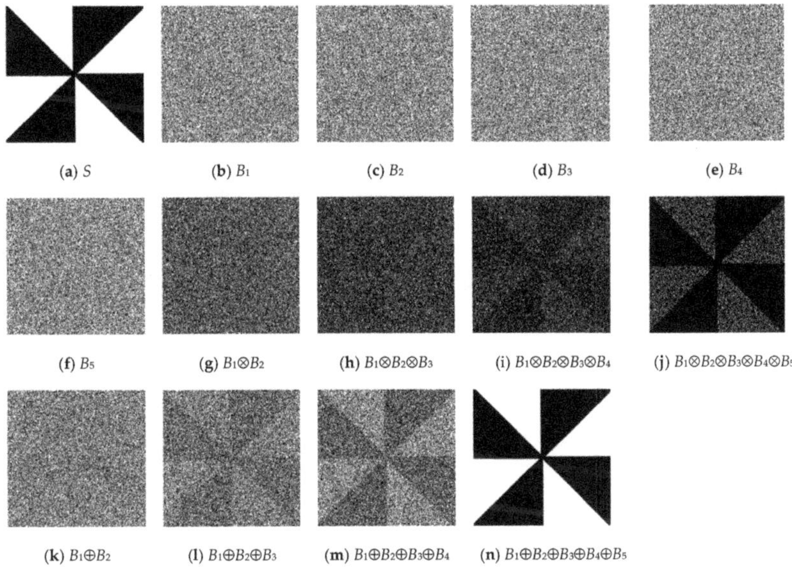

(a) S **(b)** B_1 **(c)** B_2 **(d)** B_3 **(e)** B_4 **(f)** B_5 **(g)** $B_1 \otimes B_2$ **(h)** $B_1 \otimes B_2 \otimes B_3$ **(i)** $B_1 \otimes B_2 \otimes B_3 \otimes B_4$ **(j)** $B_1 \otimes B_2 \otimes B_3 \otimes B_4 \otimes B_5$ **(k)** $B_1 \oplus B_2$ **(l)** $B_1 \oplus B_2 \oplus B_3$ **(m)** $B_1 \oplus B_2 \oplus B_3 \oplus B_4$ **(n)** $B_1 \oplus B_2 \oplus B_3 \oplus B_4 \oplus B_5$

Figure 4. The experimental result of the proposed (3, 5) 2D_VCS.

We present more results in Tables 9 and 10. Compared to Tables 7 and 8, the theoretical analysis with the proposed method is more accurate.

Table 9. Average contrast for five experiments of the proposed method using OR operator.

(k, n)	$p = 2$	$p = 3$	$p = 4$	$p = 5$	$p = 6$
(2, 4)	0.05976	0.15859	0.37504	-	-
(3, 4)	0	0.05356	0.25135	-	-
(4, 4)	0	0	0.12456	-	-
(2, 5)	0.03624	0.07584	0.11362	0.20624	-
(3, 5)	0	0.02107	0.06915	0.18865	-
(4, 5)	0	0	0.02155	0.12336	-
(5, 5)	0	0	0	0.06257	-
(2, 6)	0.02564	0.05044	0.06234	0.07023	0.1086
(3, 6)	0	0.01071	0.02679	0.04997	0.10429
(4, 6)	0	0	0.00747	0.02963	0.09385
(5, 6)	0	0	0	0.009	0.06202
(6, 6)	0	0	0	0	0.03117

Table 10. Average contrast for five experiments of the proposed method using XOR operator.

(k, n)	$p = 2$	$p = 3$	$p = 4$	$p = 5$	$p = 6$
(2, 4)	0.10901	0.18131	1	-	-
(3, 4)	0	0.18211	1	-	-
(4, 4)	0	0	1	-	-
(2, 5)	0.06857	0.04455	0.09344	1	-
(3, 5)	0	0.06945	0.14342	1	-

Table 10. *Cont.*

(*k*, *n*)	*p* = 2	*p* = 3	*p* = 4	*p* = 5	*p* = 6
(4, 5)	0	0	0.14142	1	-
(5, 5)	0	0	0	1	-
(2, 6)	0.04424	0.0168	0.02187	0.05863	1
(3, 6)	0	0.03139	0.0289	0.07784	1
(4, 6)	0	0	0.04554	0.11652	1
(5, 6)	0	0	0	0.12059	1
(6, 6)	0	0	0	0	1

6. Comparison and Conclusions

An RG-based (k, n) VCS restoring the secret image using OR or XOR operation was presented by Yan et al. [6]. We compared their scheme with the (k, n) 2D_VCS. The experimental outcome of Yan et al.'s scheme for the $(k, n) = (2, 4)$ is shown in Figure 5. The results in Figures 2 and 5 show the better visual quality of the proposed scheme, especially when staking three shares and using the XOR operator on them ((j) in both figures). Thus, an RG-based (k, n)-threshold visual cryptography with XOR and OR decryption capabilities, called (k, n) 2D_VCS, is proposed to restore the secret image in a human visual system (OR) and computer visual system (XOR). The results are better than all previous studies. Future work will be carried out to design other RG-based (k, n) threshold VCSs with better contrast of the restored image and demonstrate that the scheme is theoretically safe and correct.

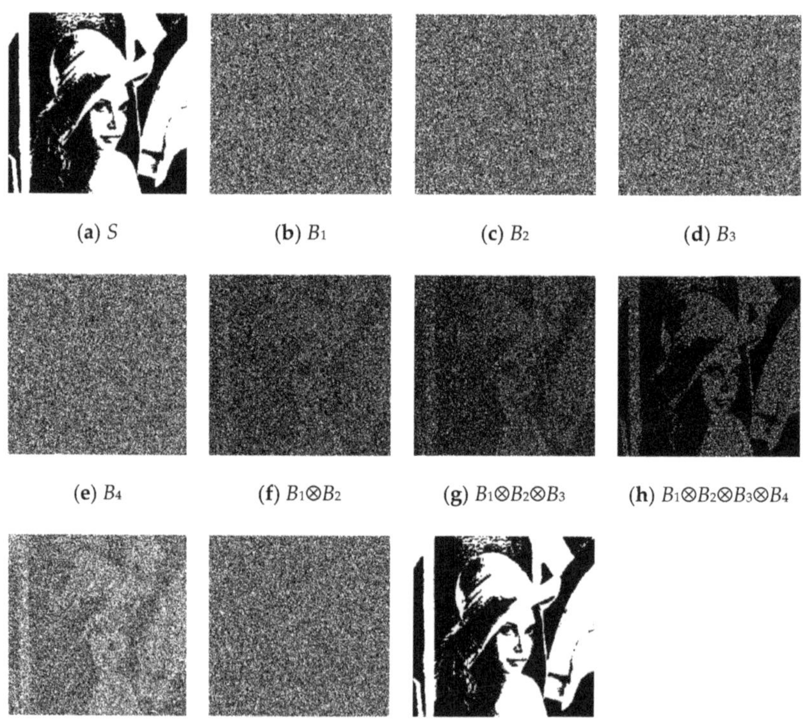

(**a**) S (**b**) B_1 (**c**) B_2 (**d**) B_3

(**e**) B_4 (**f**) $B_1 \otimes B_2$ (**g**) $B_1 \otimes B_2 \otimes B_3$ (**h**) $B_1 \otimes B_2 \otimes B_3 \otimes B_4$

(**i**) $B_1 \oplus B_2$ (**j**) $B_1 \oplus B_2 \oplus B_3$ (**k**) $B_1 \oplus B_2 \oplus B_3 \oplus B_4$

Figure 5. The experimental result used the (2, 4) RG-based VCS proposed in [6].

Author Contributions: Conceptualization, J.S.-T.J. and Y.-R.L.; methodology, J.S.-T.J.; software, Y.-R.L.; validation, J.S.-T.J. and Y.-R.L.; formal analysis, J.S.-T.J. and Y.-R.L.; investigation, J.S.-T.J. and Y.-R.L.; data curation, Y.-R.L.; writing—original draft preparation, Y.-R.L.; writing—review and editing, J.S.-T.J.; visualization, Y.-R.L.; supervision, J.S.-T.J.; project administration, J.S.-T.J.; funding acquisition, J.S.-T.J. and Y.-R.L. All authors have read and agreed to the published version of the manuscript.

Funding: This research was funded by Ministry of Science and Technology of the Republic of China grant number MOST 110-2221-E-260-003, and 110-2813-C-260-005-E. And The APC was funded by MOST 110-2221-E-260-003.

Institutional Review Board Statement: Not applicable.

Informed Consent Statement: Not applicable.

Data Availability Statement: Data sharing is not applicable to this article.

Conflicts of Interest: The authors declare no conflict of interest.

References

1. Naor, M.; Shamir, A. Visual cryptography. In *Workshop on the Theory and Application of Cryptographic Techniques*; Springer: Berlin/Heidelberg, Germany, 1994.
2. Kafri, O.; Keren, E. Encryption of pictures and shapes by random grids. *Opt. Lett.* **1987**, *12*, 377–379. [CrossRef] [PubMed]
3. Shyu, S.J. Image encryption by multiple random grids. *Pattern Recognit.* **2009**, *42*, 1582–1596. [CrossRef]
4. Chen, T.; Tsao, K. Threshold visual secret sharing by random grids. *J. Syst. Softw.* **2011**, *84*, 1197–1208. [CrossRef]
5. Guo, T.; Liu, F.; Wu, C. Threshold visual secret sharing by random grids with improved contrast. *J. Syst. Softw.* **2013**, *86*, 2094–2109. [CrossRef]
6. Yan, X.; Wang, S.; Niu, X.; Yang, C.N. Random grid-based visual secret sharing with multiple decryptions. *J. Vis. Commun. Image Represent.* **2015**, *26*, 94–104. [CrossRef]

engineering proceedings

MDPI

Proceeding Paper

Application of Typicality in Predicting Product Appearance [†]

Hung-Hsiang Wang, Yu-Yang Lin * and Hao-Ting Huang *

Department of Industrial Design, National Taipei University of Technology, Taipei 10608, Taiwan; wanghh@ntut.edu.tw
* Correspondence: t111588002@ntut.org.tw (Y.-Y.L.); t111588021@ntut.org.tw (H.-T.H.)

[†] Presented at the 5th IEEE Eurasia Conference on Biomedical Engineering, Healthcare and Sustainability 2023, Tainan, Taiwan, 2–4 June 2023.

Abstract: The applications of image classification are useful, especially in the prediction of object categories at a high accuracy rate. The technique helps industrial designers classify the existing products by product appearance and assess the extent to which newly designed products belong to their brand or product family. By doing this, designers classify products by typicality. Yet, the application of image classification of products and brand styles based on typicality is still limited. Thus, we presented a case study in which the extent to which a chair was recognized as a typical Apple product was predicted. In the technique, Waikato Environment for Knowledge Analysis (WEKA) was used. The results show that the industrial designer can use product typicality for the image classification of products to manage the style of a brand and its products.

Keywords: brand style; industrial design; image classification; deep learning

Citation: Wang, H.-H.; Lin, Y.-Y.; Huang, H.-T. Application of Typicality in Predicting Product Appearance. *Eng. Proc.* **2023**, *55*, 66. https://doi.org/10.3390/engproc2023055066

Academic Editors: Teen-Hang Meen, Kuei-Shu Hsu and Cheng-Fu Yang

Published: 7 December 2023

1. Introduction

Deep learning (DL) is a method of artificial intelligence (AI) used to process data in a brain-inspired way. The DL model identifies complex patterns in pictures, words, sounds, and other means for more accurate insights and predictions. At present, many applications of the DL model to industrial design concern data analysis and the integration of production information. However, few applications of image classification with deep learning in industrial design have been found at present.

Product development needs to design new products according to the brand strategy with new product appearances from the market demands, retaining the brand style elements. The products of Apple, in the past ten years, have maintained their unique style and thus established stable customer loyalty with innovative shapes and materials in different products. One of the most critical tasks of industrial design is to create novel products with the design elements of the original brand, as this is beneficial to judge the brand degree of products. Analysis of brand characteristics refers to finding consistent design characteristics and enabling the brand to be more stable in the market. However, with time, the epidemic factors and aesthetic preferences of different periods change rapidly. Therefore, regular data collection and analysis are needed to ensure the accuracy of the time effectiveness.

A complete product design needs to show a brand-friendly style in response to a variety of styling changes. Unique styling features enhance the product's recognizability and the loyalty of consumers. Product styling is the most direct way to attract consumers and the motivation for consumers to purchase. Based on the aesthetics of the product with appropriate novelty in appearance, the highest level of design can be achieved. Typicality and novelty are the key elements of the aesthetic design of a product in a brand [1]. The brand typicality presents a distinct difference between competitors [2]. Product differentiation is the key to the rise and fall of a brand. The visual typicality of a brand is used to express its goals and core values. We used the image classification of DL by training the prediction model according to the typicality of brand style as the classification benchmark, taking the Apple brand as the experimental object. The detrended fluctuation analysis (DFA) method was used to capture the brand features, apply the features to different types of products, and predict the brand style typicality.

Designers rely on their aesthetics and experience to design products, more or less with the subjective consciousness of miscellaneous designers to choose product directions. It is difficult to objectively verify whether its design direction conforms to the development direction proposed earlier. Using machine DL, we applied the product appearance of industrial design to a new application and explored brand typicality of product shape to create new product design strategies and improve design capability.

2. Related Work

2.1. Important Influence Elements of Product Appearance and Brand Style

A solid brand image improves brand value, identity, and brand characteristics in the market and encourages the interaction between brands and consumers. Therefore, brand style plays a vital role in the development of companies [3]. A design characteristic reflects the brand's essence, so design features need to explicitly and implicitly be embedded in product design so that customers can perceive and recognize the features [4]. Brand image is closely linked with product style and the typical DNA of brand style, shown in product design. Brand image intuitively affects consumers and brand awareness. Thus, it is necessary to study the typicality of products and brand styles in the industrial design of aesthetic concepts and styles [1,2,5]. Companies must also retain their brand style to meet the market trend and improve product development. Therefore, by adjusting the value between style typicality and novelty, the development of a new product design is conducted. Based on these, the concept of typicality is used as the criterion for the classification of brands and to predict the brand style in products' appearance.

2.2. Image Classification with DL

DL is a cutting-edge technique that has gained popularity for its ability to handle traditional computer images. This is because DL automatically learns valuable features for a specific task and eliminates the need for manual feature selection. The convolutional neural network (CNN), a type of DL architecture, can be used for product design using the mammalian visual system. It is commonly used for image recognition and segmentation due to its ability to consider the relationship between nearby pixels [6]. CNN layers are classified into convolutional, pooling, and fully connected layers. Waikato Environment for Knowledge Analysis (WEKA) is a collection of machine learning (ML) algorithms for data mining that is efficient and powerful and does not require separate coding. It contains tools for data preparation, classification, regression, clustering, association rule mining, and visualization.

In this study, WEKA was used to allow for the application of image classification of DL without coding. Texture and line make product appearance obvious and characteristic. Hence, image classification with DL was used to identify the features with specificity as the criterion of product brand classification for training a prediction model.

2.3. Apple Chair

Girelli Studio, an industrial product design company, designs products for companies venturing into the world of furniture. They designed a chair considering the concept of Apple AirPods Max as the style of Apple's brand. This example represents style typicality. No matter the shape or size of the product's appearance, as long as the design shows the style characteristics, it maintains style consistency. Image classification with DL is used to predict brand styles by training a model on a dataset of images of products associated with brand styles. Once the model is trained, it is used to predict the appearance of a new product based on its image. DL is used to improve the productivity of designing products.

Therefore, we used image classification with DL without coding so that all designers could learn and use it.

There were many published research results [7–9] showing examples of industrial design [10]. However, most of the DL studies showed industrial process development with data mining and data-driven design. However, there were not many studies on product appearance and design style [11,12]. Therefore, the relationship between ML and product appearance design was determined to enable a closer connection between product design and technology. In this study, we specified the typicality of products of a certain brand for training models to predict the product brand, demonstrated the image classification process of using WEKA for no-code designers, and summarized the opportunities to apply ML technology to product design and development.

3. Methods

Chairs were selected to predict the style, and the brand, Apple was used for analyzing the brand product style and design. This study was conducted in two stages. In the first stage, an appropriate brand-style training set was selected, and its features were quantified using WEKA's image filter. Model training was carried out for classification, and the parameters were adjusted to obtain the most efficient model sample. In the second stage, the known brand styles were used to redesign products in different categories, and the accuracy of the classification model was evaluated for the redesigned samples.

3.1. DFA to Extract Features

DFA was developed by Warell (2001) to explore design features in various products and explicit design clues for visual recognition [2]. In this study, the features of the product style of Apple were determined using DFA. The results showed that the product design features included rounded corners, symmetry, and simplicity. After obtaining the features, 150 training sets, including pictures of Apple products, were used to train the model for typical, novel, and non-Apple style features. WEKA ImageFilter was used to learn Apple brand product features in the model. After training, two senior designers selected 10 Apple brand-specific chairs based on rounded corners, symmetry, and simplicity and blended 20 non-Apple-style chairs into the test set.

3.2. Labeling Characteristics of Brand Style

The supervised ML method was used to train the model using the image label. The classification model was trained for WEKA classification, and the dominance of the brand style features in the image was rated according to the image level. According to the features of Apple's brand style [2], rounded outline, simplicity, and symmetry were chosen as the three most dominant adjectives. We collected 32 groups of pictures of Apple products from 2018 to 2022 (Figures 1 and 2). The product pictures were labeled using the three adjectives. The rules of extraction are shown in Figures 3–5. Pictures labeled for the three adjectives were considered with typical Apple brand styles. Pictures for two adjectives were considered novel Apple brand styles. Pictures for one adjective were considered non-Apple style. A total of 150 images with the features of Apple products were collected in the training set.

3.3. Model Accuracy

In this study, the rule for selecting chairs in the test set was Apple brand features. The test set consisted of 30 chairs, with 10 chairs having brand styles and the remaining 20 chairs having non-brand styles (Figures 6 and 7). The previous classification model was used to evaluate the accuracy of the model.

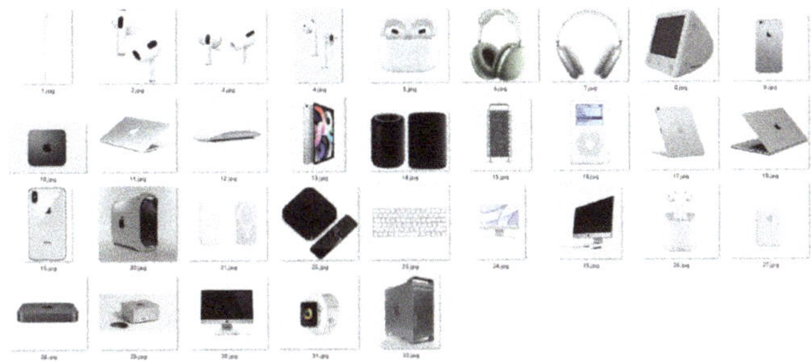

Figure 1. Apple product images.

Figure 2. Feature set of Apple products.

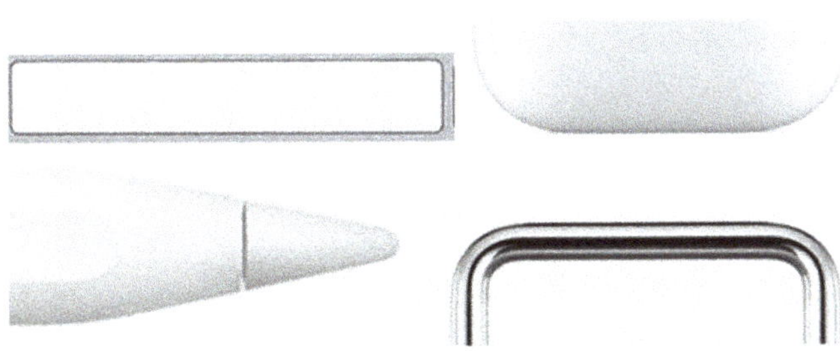

Figure 3. Typical brand style (rounded, symmetrical, and simple).

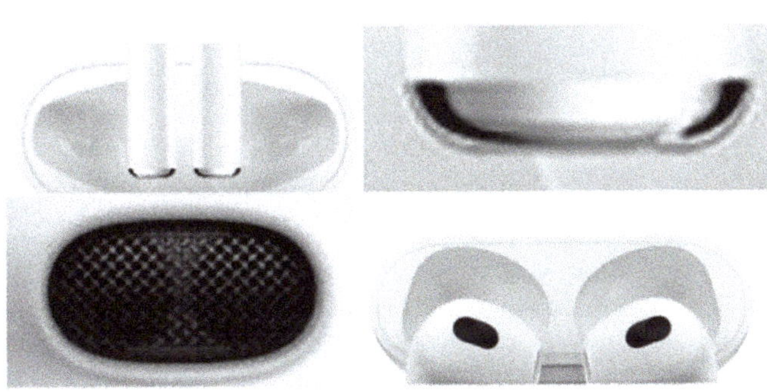

Figure 4. Novel brand style (rounded and symmetrical).

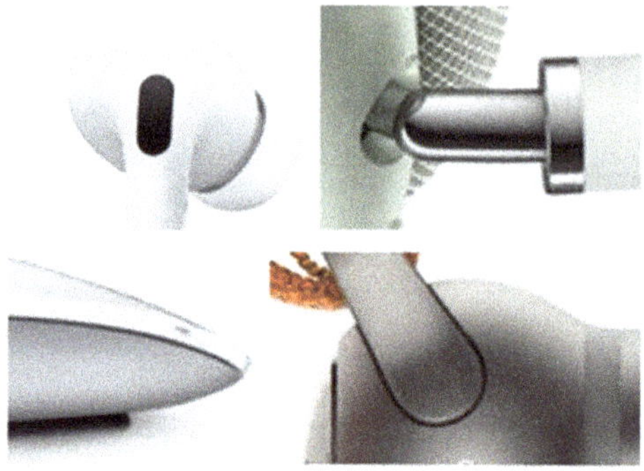

Figure 5. Non-Apple style (rounded corners).

Figure 6. Chairs with brand style (10 pics).

Figure 7. Chairs with non-Apple brand style (20 pics).

4. Results

4.1. Feature Extraction

An image filter was used to quantify image features, including geometric, frame, color, shape, histogram, and local binary pattern features. We used DFA to define brand features. The number of texture features was more than the number of color features. Thus, we used the texture image filter of WEKA and assigned features in a feature map to compare the performance to select the best prediction model. The texture filters available in WEKA included the PHOG-Filter, BinaryPatternsPyramid-Filter, EdgeHistogram-Filter, Gabor-Filter, and FCTH-Filter. The image training results showed that the PHOG-Filter, BinaryPatternsPyramid-Filter, and EdgeHistogram-Filter showed an accuracy of 100% in prediction after training with the training set. The FCTH filter with color and texture showed an accuracy of 86.333%. However, the Gabor-Filter performed the worst, with an accuracy of 43.333% (Table 1).

Most of the filters using texture units in feature extraction showed high accuracies. Thus, these filters were efficient in color extraction because they used the product edge lines of each image to convert them into quantified values. The PHOG-Filter is based on HOG values. In extraction, the image was considered colorless. The number and location of edge gradient directions were calculated so that features could be closely related to location. The process was performed on local cell units of the image to keep invariance to the geometric and optical deformation of the image. The BinaryPatternsPyramid-Filter was similar to the PHOG-Filter. In feature extraction, it divided image pixels into 16×16 cell units to compare them with those in the surrounding 8 units and save them as binary values. After normalizing the histogram, the quantified image feature was obtained. The EdgeHistogram-Filter extracts the features of MPEG7 edge histograms from images. The edges are generated according to color opposition or discontinuity. The Gabor-Filter is based on the Gabor wavelet transform texture value and analyzes specific frequency content in a particular direction in a local area around a point or area in an image. The FCTH-filter uses color and texture values to retrieve blurry color and texture features from images and unify color and histogram features in the highest 72-tuple pixels.

Table 1. Model accuracy using different image filters.

Filter Name	Class	Filter Function	Style Model Accuracy
PHOG-Filter	Texture	A filter for extracting the directional gradient histogram value PHOG from the image.	100%
BinaryPatternsPyramid-Filter	Texture	Extraction of rotation-invariant numerical histograms of local binary patterns from images.	100%
EdgeHistogram-Filter	Texture	A filter for extracting MPEG7 boundary histogram features from pictures.	100%
Gabor-Filter	Texture	Use Gabor wavelet transformation to extract texture features from pictures.	43.333%
FCTH-Filter	Color and Texture	A filter that extracts MPEG7 color layouts from images. It will cut the picture into 64 pieces and then calculate the average color in each piece.	86.666%

The least expected texture filter was the Gabor-Filter. Because the Gabor filter was similar to the filter of the human visual system, it could not perform the expected classification, and its accuracy was 43.333%. The other texture filters were effective in extracting quantified data for product features. Filters not based on histogram features were not efficient in the extraction.

4.2. Feature Extraction of Image Filters

According to the test results in this study, the classification model to predict the brand style was trained by ML and was used by designers. The application was conducted in three phases: image data collection, quantitative image collection, and classification and image prediction. In phase 1, designers collected products with style and appearance. In addition, DFA was used to define characteristic adjectives. In this phase, designers paid attention to the hierarchical arrangement of adjectives to define an appropriate number of adjectives. In phase 2, the adjectives in the previous phase were used to capture the product images in the image set, label brand typicality, and input the image into WEKA to filter the features. The user determined the product details in the feature in this phase by dividing the image into grade intervals. In this phase, attention had to be paid to selecting the characteristics of the image set to select appropriate filters. In phase 3, the features were input into the classification for the classification model, and the target design product was input into the test set to predict the typicality of the product style. The designer could modify the features. The classification model's performance depended on the adjustment of parameters and classification methods, and the accuracy in the prediction of the model was improved by adding training sets and cross-validation.

5. Conclusions

We introduced ML to use the product appearance to capture features and predict the target style after constructing the classification model. As a result, the method of the feature extraction of product style sand labeling rules for ML was established to guide and adjust models in a human–machine interface. The design process can be taught to students. The result of this study confirmed that DL could be applied to product development, and designers could use ML methods to modify product features according to brand typicality and optimize the design process. Many image filters and methods performed well in extracting the features from product images and training classification models. It is recommended to use more images with different angles, lightness, and positions to obtain better performance in the model.

Author Contributions: Conceptualization, H.-H.W.; methodology, H.-H.W.; resources, Y.-Y.L. and H.-T.H.; Data curation, Y.-Y.L. and H.-T.H.; validation, H.-H.W.; writing—original draft preparation, Y.-Y.L. and H.-T.H.; writing—review and editing, H.-H.W.; experiments conducted, Y.-Y.L. and H.-T.H.; supervision, H.-H.W. All authors have read and agreed to the published version of the manuscript.

Funding: This research was funded by NSTC, Taiwan, R.O.C., via grant 111-2410-H-027-019-MY2.

Institutional Review Board Statement: Not applicable.

Informed Consent Statement: Not applicable.

Data Availability Statement: Data are unavailable due to privacy reasons.

Acknowledgments: The authors gratefully acknowledge the financial support of NSTC, Taiwan, R.O.C., via grant 111-2410-H-027-019-MY2.

Conflicts of Interest: The authors declare no conflict of interest.

References

1. Hung, W.-K.; Chen, L.-L. Effects of novelty and its dimensions on aesthetic preference in product design. *Int. J. Des.* **2012**, *6*, 81–90.
2. Karjalainen, T.-M. It looks like a Toyota: Educational approaches to designing for visual brand recognition. *Int. J. Des.* **2007**, *1*, 67–81.
3. Wang, H.; Chen, J.; Hu, Y.; Ye, M. The consistency of product design and brand image. In Proceedings of the 2008 9th International Conference on Computer-Aided Industrial Design and Conceptual Design, Beijing, China, 22–25 November 2008; IEEE: Piscatawa, NJ, USA, 2008; pp. 1142–1144.
4. Andersson, T.; Warell, A.; Holmlid, S. Product gist: An approach to identifying form characteristics of the current product sign. In Proceedings of the Conference: Crafting the Future 2013, the 10th European Academy of design Conference, Gothenburg, Sweden, 17–19 April 2013.
5. Mayer, S.; Landwehr, J.R. Measuring design typicality–A comparison of objective and subjective approaches. In Proceedings of the Future Focused Thinking—DRS International Conference, Brighton, UK, 27–30 June 2016.
6. Yang, X.; Ye, Y.; Li, X.; Lau, R.Y.; Zhang, X.; Huang, X. Hyperspectral image classification with DL models. *IEEE Trans. Geosci. Remote Sens.* **2018**, *56*, 5408–5423. [CrossRef]
7. Yang, M.; Kumar, P.; Bhola, J.; Shabaz, M. Development of image recognition software based on artificial intelligence algorithm for the efficient sorting of apple fruit. *Int. J. Syst. Assur. Eng. Manag.* **2021**, *13*, 322–330. [CrossRef]
8. Hebbale, S.; Marndi, A.; Achyutha, P.N.; Manjula, G.; Mohan, B.; Jagadeesh, B. Automated medical image classification using DL. *Int. J. Health Sci.* **2022**, *6*, 1650–1667.
9. Zhou, J.; Li, Z.; Zhi, W.; Liang, B.; Moses, D.; Dawes, L. Using convolutional neural networks and transfer learning for bone age classification. In Proceedings of the 2017 International Conference on Digital Image Computing: Techniques and Applications (DICTA), Sydney, Australia, 29 November–1 December 2017; IEEE: Piscatawa, NJ, USA, 2017.
10. Bae, J.K.; Kim, J. Product development with data mining techniques: A case on design of digital camera. *Expert Syst. Appl.* **2011**, *38*, 9274–9280. [CrossRef]
11. Kim, J.; Lee, J.-K. Stochastic detection of interior design styles using a deep-learning model for reference images. *Appl. Sci.* **2020**, *10*, 7299. [CrossRef]
12. Wang, H.-H.; Chen, C.-P. Using t-SNE to Evaluate the Brand Style of New Mice Design. In Proceedings of the 2022 IEEE International Conference on Consumer Electronics-Taiwan, Taipei, Taiwan, 6–8 July 2022; IEEE: Piscatawa, NJ, USA, 2022.

engineering proceedings

MDPI

Proceeding Paper

Predicting Consumer Preferences by the Deformation Threshold of Product Appearance †

Hung-Hsiang Wang, Cheng-Kang Liu * and Shin-Bei Yu *

Department of Industrial Design, National Taipei University of Technology, Taipei 10608, Taiwan; wanghh@mail.ntut.edu.tw
* Correspondence: t111588029@ntut.org.tw (C.-K.L.); t111588027@ntut.org.tw (S.-B.Y.)

† Presented at the IEEE 5th Eurasia Conference on Biomedical Engineering, Healthcare and Sustainability, Tainan, Taiwan, 2–4 June 2023.

Abstract: Product appearance is essential in shaping consumers' preferences. Thus, we predicted consumer preference by establishing a feasible threshold for appearance innovation with a novel method. We used deep learning (DL)-based image classification and the Waikato Environment for Knowledge Analysis (WEKA) as the implementation tool. Advanced DL algorithms were adopted to classify images of disparate products according to their aesthetics. The results of the classification presented that the method could explain the level of innovation in appearance that consumers favored. The result can be used as a guideline for industrial designers in the creation of new products. Moreover, the method can be used to assess existing products that need enhancement to meet consumer preferences. The proposed method assists industrial designers in understanding consumer preferences and product innovation. Future research is required to ensure the successful design of new products.

Keywords: industrial design; product development; deep learning; image recognition; consumer preferences

Citation: Wang, H.-H.; Liu, C.-K.; Yu, S.-B. Predicting Consumer Preferences by the Deformation Threshold of Product Appearance. *Eng. Proc.* **2023**, *55*, 67. https://doi.org/10.3390/engproc2023055067

Academic Editors: Teen-Hang Meen, Kuei-Shu Hsu and Cheng-Fu Yang

Published: 8 December 2023

1. Introduction

With advancements in Artificial Intelligence (AI), Data Mining, and the Internet of Things (IoT), various approaches and conceptualizations in healthcare and pharmaceuticals [1], artistic creation [2], and agriculture have been proposed. However, such applications in industrial design and product design have been scarce. When it comes to designing a product's appearance, designers depend on their personal experience and individual aesthetic concepts, not on objective and quantified means. One major reason for this is the substantial time and effort required to collect and analyze data on market opportunities, competing products, and design strategies. Data associated with the appearance of products include images, which inherently challenges the quantified analysis. These challenges are barriers to designing product appearance strategically. Therefore, we used image classification and machine learning (ML) to find a way of innovative designs for predicting consumer preferences. The result helps designers find suitable designs for products and predict consumer preference for the designs.

2. Related Work

2.1. Typicality, Novelty, and Consumer Preference

Typicality refers to the appearance, shape, texture, and functional characteristics of product design. It signifies the wide recognition of people for a product or brand. For example, Apple Inc. has shown a unique round shape and polished appearance in product design, which has now become the trademark. Similarly, when one mentions "mammals", people are reminded of cats, dogs, pigs, or sheep rather than platypuses or dolphins. This signifies a stronger sense of typicality.

The aesthetics of a product are an important indicator for consumers to discern products. Brands nowadays have been striving to exhibit their distinctive brand style in their products' appearance. The design of vehicles is the most evident for typicality. Regardless of whether it is budget-friendly or luxurious, brand recognition with aesthetic design is the design goal of companies. The design of vehicles generally adheres to two design strategies: the single-driven and the market-driven approaches. The single-driven strategy is used for creating iconic product design elements and establishing a family look to attract target audiences. On the contrary, the market-driven strategy is used to develop diversified styles based on market demand to attract consumers. With the importance of aesthetic design, companies are emphasizing product appearance [3] and require innovations in design. Traditional and manual identification of styles is conducted by observing lines, curvatures, and other features that cannot match the speed of innovation and style shifts. For the innovation of product design, a balance between innovation, retaining brand style, and maintaining typicality is significant for designers and companies.

The concept of novelty corresponds to finely crafted design features. Consumers' judgment of a product's novelty is subjective. Hung and Chen [4] used 88 chairs in various shapes and applied them to the prototype production. They researched the semantic difference of bipolar adjectives to understand the impact on novelty and aesthetic preference. The results validated Berlyne's hypothesis [5] that novelty and aesthetic preference showed an inverted U-shaped relationship. Moderate novelty can enhance aesthetic preference.

2.2. Integrity of Specifications

Style encapsulates the character of a designer or brand. The brand style is one of the most applied design strategies in the market, and it serves as a crucial component in brand recognition. Chan [6] stated that style was discerned by recurring features, forms, and context. Car manufacturers developed a symbolic brand appearance in the market by repetitively employing similar design elements [7]. Chan [6] proposed a measurement method based on classification. However, measuring product styles was a challenge. Hyun put forth a quantitative method to evaluate the similarity of the styles of vehicles [8] using Fourier decomposition of the data of eye-tracking on shapes. Through the analysis of visual semantics, the significance of the design elements of vehicles was weighted to find the similarity of vehicle designs. Features are more restrictive than styles, concentrating more on the characteristics or symbolic implications of logos. Chan [6] claimed that designs with higher stimulation in shapes were more easily distinguishable than other designs. The measurement of features for the identification of features was categorized into a manual setting definition [9] and coding feature definition [10].

2.3. Design Freedom and Brand Recognition

Design freedom is the threshold with which designers deviate from existing designs for innovative designs. It pertains to the permissible variation that does not stray from the brand value and recognition. Hung and Chen's experiments affirmed that moderately innovative designs gained more popularity than excessively radical designs [4]. To investigate how to establish the range of design freedom, Burnap defined the distance across design and the difference between novel concept designs from existing ones. By incorporating customer feedback into the measurement function of design freedom, values of design freedom could be obtained for each design concept [11]. Brand recognition underscores the affiliation with the brand. Customer feedback is a key factor in purchasing decisions and affects the ranking of products [12]. In the automotive industry, the association between brand loyalty and consumer feedback is more prominent. A favorable market response can be secured if the design of the exterior of a vehicle is created within the appropriate range of design freedom. The front image of the vehicle is more significantly correlated with brand recognition than the side and rear views [13]. The front view unifies colors and the brand logo. Burnap confirmed that excessive design freedom would diminish brand recognition [11].

3. Methods

We used a deep learning (DL) method for image classification with machine learning (ML). The research was conducted in three phases: determining the reliability of ML in image classification, probing consumers' subjective perceptions of product innovation, and finding the feasibility of deploying ML in predicting consumer preferences and analyzing product innovation.

3.1. Reliability of ML in Image Classification

Liu [14] and Landwehr [9,15] researched the design innovation of vehicles using key features of product innovation using ML. In total, 28 frontal images of vehicles of four high-end brands (BMW, Mercedes Benz, LEXUS, and VOLVO) were collected in 2021 and 2022 for experiments. The sample images were collected from the online database of Consumer Reports issued by the American Consumer Union (CU), which is a non-profit and advertisement-free organization without any biased opinions. The database includes detailed information and images of consumer products as the source for collecting target images. The uniform image specification by Hyun [8] was employed to classify the front images of vehicles for the *X*-axis transformation, *Y*-axis transformation, and proportional transformation at different rates of 25, 75, and 100%. The morphed images were classified into training and testing sets as shown in Figure 1. The training group contained 24 deformed samples of models, while the test group consisted of four deformed samples of vehicle models.

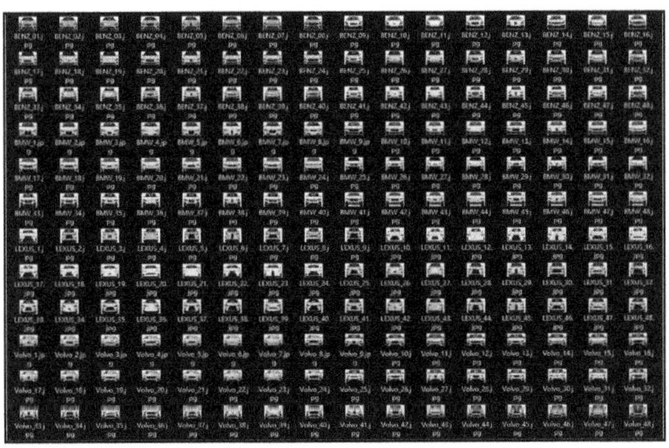

Figure 1. Classified image samples.

The training set samples were converted into files in Attribute-Relation File Format (ARFF) and fed into the Waikato Environment for Knowledge Analysis (WEKA). The "JpegCoefficientFilter filter" in the WEKA was used to analyze the data of the sample images. On the WEKA's "Classify page", sequential minimal optimization (SMO) was used to categorize the data. As displayed in Figure 2, the outcomes using WEKA for the training set showed an accuracy of 100%. The samples in the testing set were converted into ARFF files and fed into the model. As shown in Figure 3, an accuracy of 100% was obtained in classification. It was found that ML classified the images accurately.

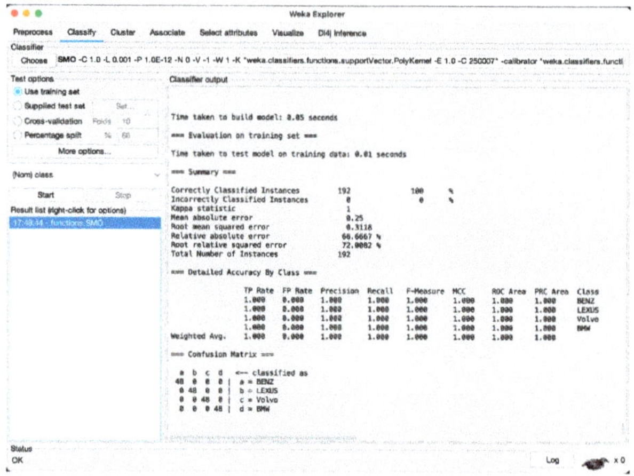

Figure 2. Training model established using WEKA.

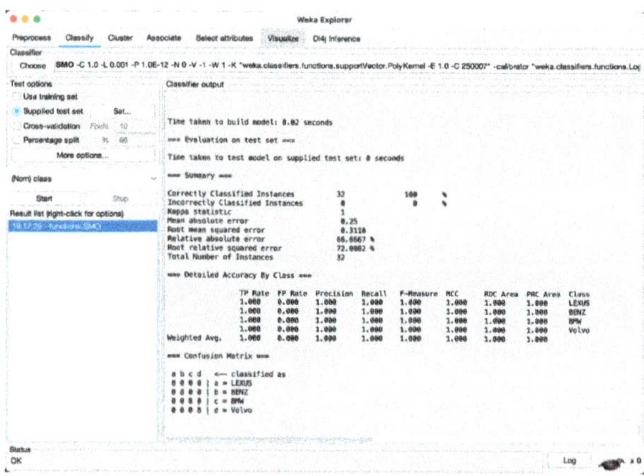

Figure 3. Classifying using WEKA training model.

3.2. Consumer Perceptions of Product Innovation

The relationship between consumer perceptions of innovation and consumer preference for a particular brand or product was found in this study. The samples were collected from the images of BMW M3 series vehicles as the experimental sample to evaluate "Typical-Novel" and "Dislike-Favored". To evaluate consumer perception of the vehicles, a questionnaire survey was conducted using the Semantic Differential Scale [16]. In the survey, the images of the BMW M3 model in the sixth generation were shown regarding research typicality and consumer preference. A total of 200 valid questionnaires were collected from 147 males and 53 females. In the data analysis, we obtained scores for consumer perception of the vehicle as depicted in Table 1. E36 scored the lowest (2.295). The key characteristics of the remaining five M3 models (E30, E46, E92, F80, and G80) were also scored in terms of innovation in the design of the front images. As shown in Figure 4, the order of consumer preference was E46, E30, E92, and F80. A substantial decline was found for G80 as it showed a significant decrease in favorability and innovation compared to the scores of other models: 10% (E46), 20% (E30), 40% (E92), and 50% (F80). To maintain the appropriate level of consumer preference, the innovation of the front images compared to the most typical M3 model E36 was required.

Table 1. Questionnaire survey results.

	E30	E36	E46	E92	F80	G80
Typical–Novel	2.34	2.295	2.59	3.08	3.42	3.885
Dislike–Favored	3.14	2.84	3.07	3.185	3.645	3.355
Innovation level	20%	0	10%	40%	50%	330%

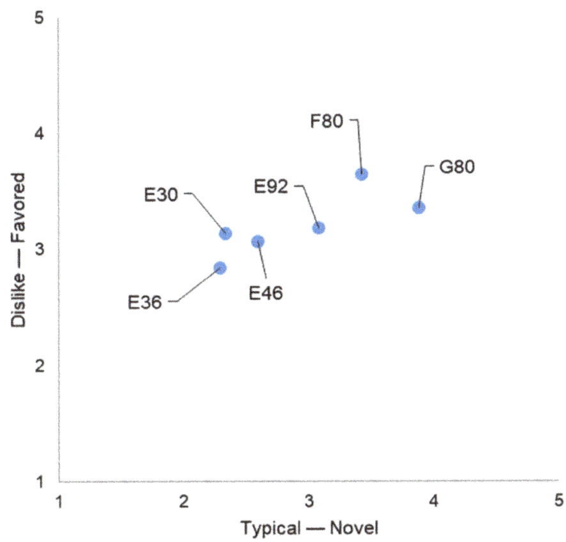

Figure 4. Scatter plot of typicality and consumer preference.

3.3. ML for Predicting Consumer Preferences and Analyzing Product Innovation

To investigate the capability of ML in predicting consumer preferences and identifying a suitable range of product innovation, a series of image samples was created based on the M3 and E36 models as they were regarded as typical models according to consumer perception. These samples were shown in three dimensions at different transformations of the images ranging from 10 to 100% and −10 to −100%. According to the rule of "the degree of innovation between 10 and 50%", they were inputted into WEKA for the ML analysis to explore the relationship between innovation and consumer preference as displayed in Table 2.

Table 2. Sample images for design innovation.

Unknow_1	**Unknow_2**	**Unknow_3**	**Unknow_4**	**Unknow_5**

Furthermore, the front grilles of the remaining five M3 models were transplanted onto the iconic M3 model, E36, to serve as a test to evaluate the accuracy of WEKA's prediction model in forecasting consumer preferences.

Subsequently, the grouped transformed image samples were utilized in WEKA to establish a prediction model capable of classifying the innovation level of BMW M3 models based on consumer preferences. These image samples yielded a prediction model with a classification accuracy of 84.127% after ten layers of Cross-validation, as depicted in Figure 5.

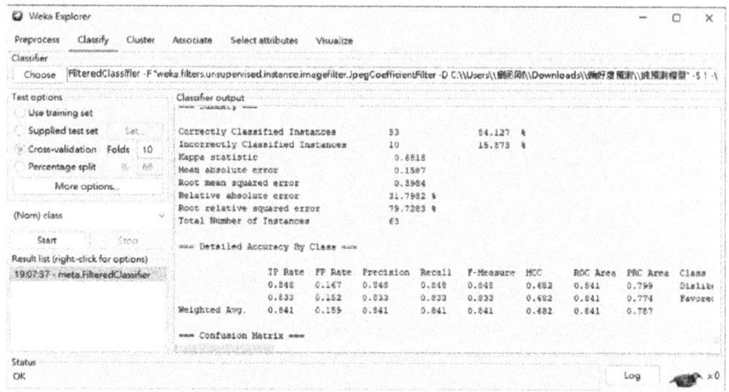

Figure 5. Prediction model of consumer preference for BMW M3 using WEKA.

Image samples of the front images of the M3 and E36 models were used in the unknow innovation level for testing the WEKA prediction model. The model analyzed these image samples and made predictions as illustrated in Figure 6. The model's prediction results were compared with the actual images as shown in Table 3. The accuracy of prediction was high. The result showed that ML was effective in predicting consumer preferences and determining the threshold of product design innovation.

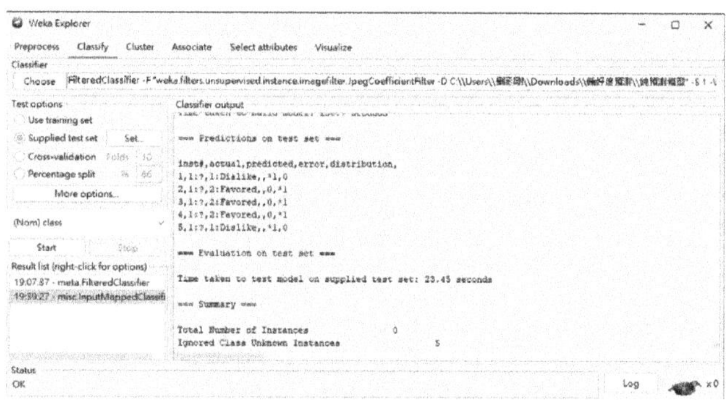

Figure 6. Prediction of consumer preference for unknow sample images using the model.

Table 3. Prediction results of WEKA prediction model.

	Unknow_1	Unknow_2	Unknow_3	Unknow_4	Unknow _5
Prediction Results	Dislike	Favored	Favored	Favored	Favored
Correct Answer	Dislike	Favored	Favored	Favored	Favored
Correct/ Incorrect	Correct	Correct	Correct	Correct	Correct

4. Conclusions

Using the images of the renowned M3 model of BMW, we investigated the performance of the WEKA prediction model with ML in predicting consumer perceptions and the threshold of design innovation. The result showed that the model was effective in predicting consumer perception and can be applied to other consumer-related parameters.

Author Contributions: Conceptualization, H.-H.W.; method, H.-H.W.; resources, C.-K.L. and S.-B.Y.; data curation, C.-K.L. and S.-B.Y.; validation, H.-H.W.; writing—original draft preparation, C.-K.L. and S.-B.Y.; writing—review and editing, H.-H.W.; experiments conducted, C.-K.L. and S.-B.Y.; supervision, H.-H.W. All authors have read and agreed to the published version of the manuscript.

Funding: This research was funded by NSTC, Taiwan, R.O.C., through grant 111-2410-H-027-019-MY2.

Institutional Review Board Statement: Not applicable.

Informed Consent Statement: Not applicable.

Data Availability Statement: Data are unavailable due to privacy reasons.

Acknowledgments: The authors gratefully acknowledge the financial support of NSTC, Taiwan, R.O.C., through grant 111-2410-H-027-019-MY2.

Conflicts of Interest: The authors declare no conflict of interest.

References

1. Shen, D.; Wu, G.; Suk, H.-I. Deep Learning in Medical Image Analysis. *Annu. Rev. Biomed. Eng.* **2017**, *19*, 221–248. [CrossRef] [PubMed]
2. Yi, R.; Liu, Y.-J.; Lai, Y.-K.; Rosin, P.L. APDrawingGAN: Generating Artistic Portrait Drawings from Face Photos with Hierarchical GANs. In Proceedings of the IEEE/CVF Conference on Computer Vision and Pattern Recognition (CVPR), Long Beach, CA, USA, 15–20 June 2019.
3. Orth, U.R.; Malkewitz, K. Holistic Package Design and Consumer Brand Impressions. *J. Mark.* **2008**, *72*, 64–81. [CrossRef]
4. Hung, W.; Chen, L. Effects of Novelty and Its Dimensions on Aesthetic Preference in Product Design. *Int. J. Des.* **2012**, *6*, 81–90.
5. Berlyne, D.E. Novelty, complexity, and hedonic value. *Percept. Psychophys.* **1970**, *8*, 279–286. [CrossRef]
6. Chan, C.-S. Can style be measured? *Des. Stud.* **2000**, *21*, 277–291. [CrossRef]
7. Karjalainen, T.-M. *Semantic Transformation in Design: Communicating Strategic Brand Identity Through Product Design References*; University of Art and Design Helsinki: Helsinki, Finland, 2004.
8. Hyun, K.H.; Lee, J.-H.; Kim, M.; Cho, S. Style synthesis and analysis of car designs for style quantification based on product appearance similarities. *Adv. Eng. Inform.* **2015**, *29*, 483–494. [CrossRef]
9. Mayer, S.; Landwehr, J.R. Objective measures of design typicality. *Des. Stud.* **2018**, *54*, 146–161. [CrossRef]
10. Monga, V.; Evans, B.L. Perceptual Image Hashing Via Feature Points: Performance Evaluation and Tradeoffs. *IEEE Trans. Image Process.* **2006**, *15*, 3452–3465. [CrossRef] [PubMed]
11. Burnap, A.; Hartley, J.; Pan, Y.; Gonzalez, R.; Papalambros, P.Y. Balancing Design Freedom and Brand Recognition in the Evolution of Automotive Brand Styling. In Proceedings of the 27th International Conference on Design Theory and Method, Boston, MA, USA, 2–5 August 2015; Volume 7.
12. General Motors. *General Motors Brand Equity Research*; Internal. General Motors: Warren, MI, USA, 2014.
13. Ranscombe, C.; Hicks, B.; Mullineux, G.; Singh, B. Visually decomposing vehicle images: Exploring the influence of different aesthetic features on consumer perception of brand. *Des. Stud.* **2012**, *33*, 319–341. [CrossRef]
14. Liu, Y.; Li, K.J.; Chen, H.; Balachander, S. The Effects of Products' Aesthetic Design on Demand and Marketing-Mix Effectiveness: The Role of Segment Prototypicality and Brand Consistency. *J. Mark.* **2017**, *81*, 83–102. [CrossRef]
15. Landwehr, J.R.; Labroo, A.A.; Herrmann, A. Gut Liking for the Ordinary: Incorporating Design Fluency Improves Automobile Sales Forecasts. *Mark. Sci.* **2011**, *30*, 416–429. [CrossRef]
16. Osgood, C.E.; Suci, G.J.; Tannenbaum, P.H. *The Measurement of Meaning*; University of Illinois Press: Urbana, IL, USA, 1957.

engineering proceedings

Proceeding Paper

Establishment of a Predictive Model for Cold Forging Force in Fastener Manufacturing Using Numerical Analysis †

Yi-Teng Wang [1] and Shao-Yi Hsia [2,*]

[1] 205 Arsenal of Armaments Bureau, Ministry of National Defense, Kaohsiung 806, Taiwan; frankwang1364a@yahoo.com.tw
[2] Department of Mechanical and Automation Engineering, Kao-Yuan University, Kaohsiung 821, Taiwan
* Correspondence: syhsia15@gmail.com
† Presented at the IEEE 5th Eurasia Conference on Biomedical Engineering, Healthcare and Sustainability, Tainan, Taiwan, 2–4 June 2023.

Abstract: Precise metal parts are manufactured using cold forged screw forming, but this method has drawbacks such as material damage and the possibility of imprecise parts. To ensure the accuracy and quality of parts, it is crucial to analyze and predict the forging force in this process. There are various methods to predict the forging force including laboratory testing, numerical simulation techniques, and simulation techniques. To provide forging information that cannot be measured by forging-forming machines, we simulated the nut-forming process based on actual manufacturing conditions using SolidWorks (v2020) with DEFORM-3D (v6.1). The results showed that the performance of the prediction of forging force specifications by the currently available nut-forming machines was improved by 8–20%. As the product becomes larger, the difference in forging force also increases proportionally. This research result, combined with appropriate experimental analysis, can be used as parameter controls for manufacturers in the future implementation of smart manufacturing.

Keywords: predictive model; cold forging force; fasteners; DEFORM-3D software

Citation: Wang, Y.-T.; Hsia, S.-Y. Establishment of a Predictive Model for Cold Forging Force in Fastener Manufacturing Using Numerical Analysis. *Eng. Proc.* **2023**, *55*, 68. https://doi.org/10.3390/engproc2023055068

Academic Editors: Teen-Hang Meen, Kuei-Shu Hsu and Cheng-Fu Yang

Published: 8 December 2023

1. Introduction

In nut production, the industry uses nut-forming machines. Due to the different design dimensions of nut-forming machines, machines with different forging forces are required. The size of nuts produced by this machine depends on the design of its forging force. In the past, the industry relied on accumulated practical experience and repeated testing results to design the forging force of the machines. The forging forces listed by different manufacturers for the same specification of forming machines are, accordingly, not consistent. For nut manufacturers, costs are an important factor. Thus, accurate specifications of nut forging forces provided by the machine manufacturer are important for nut manufacturers. Therefore, if the result of simulation and analysis using the software can be combined with the practical experience of designers for cross-data comparison and mutual verification, reasonable and effective design criteria can be obtained to reduce costs and enhance the overall competitiveness of the enterprise.

Matthew O'Connell et al. [1] developed the slab method with a finite element method DEFORM-2D to estimate the forming load of hexagonal head bolts without flash closed-die forging. The method was used to select suitable forming equipment to reduce energy consumption and increase the production rate. Kim et al. [2] used the FORMEX and DEFORM finite element software to design a process and establish a database for the process. They reduced the time for forming process sequencing and mold design during product production. Vazquez et al. [3] compared the use of tungsten carbide insert and segmented insert in cold forging dies and explored the influence of material changes on die life using the DEFORM finite element software. The results could be used to reduce the cost of forging dies by calculating the cold forging die life. Falk et al. [4] proposed a method related to the volume of forged parts using the DEFORM finite element software and provided and validated a simple and accurate estimation of the life of cold forging dies. Lee et al. [5] analyzed the axial extrusion process and designed parameters such as axial diameter, die corner radius, and friction coefficient with the finite element method. They investigated the individual effects. The research results showed that all design parameters including the axial diameter influenced the forming load, while increasing the die corner radius reduced the maximum forming load. The simulation results were consistent with the experimental results. MacCormack et al. [6] proposed multi-stage cold forging analysis using the finite element analysis software DFFORM for aerospace engineering key locks. The strain distribution, three stages of failure, and flow patterns were observed after product forming, allowing for the understanding of failure values and equivalent strain values, as well as identifying the locations of maximum failure values and maximum equivalent strain values.

Behrens et al. [7] performed a failure analysis of various production processes using the MSC.SuperForm2002 finite element analysis software. They investigated material fractures in cold forging and warm forging and showed that the generation of surface cracks and material damages was reduced by slightly modifying the geometric shape of critical components. Landre et al. [8] investigated the influence of three different preform shapes on the forming limits of the billet and predicted the strain locations of billet failure using finite element software to simulate the cylindrical compression process of 1040 carbon steel. The suitability of various criteria was compared using the fracture ratio. Kim et al. [9] conducted a multi-stage forming analysis of automotive steering joints using the DEFORM-3D finite element analysis software. A new process was proposed to improve product performance, increase production efficiency, shorten development time, and reduce costs for large-scale production. Hsia et al. [10] studied the extrusion processing of thin-layer aluminum products. In the simulation of the product, difficulties were found in welding when the material entered the welding chamber, which affected the subsequent forming analysis. By using the DEFORM-3D forming software, the development speed of the product was increased, development risks were reduced, and product quality was improved. Optimizing the extrusion forming of aluminum materials was important in extending the life of the mold. Hsia et al. [11] noted the scarcity of production techniques related to hobbing forming in previous studies. An example of a pointed tail tooth plate mold for iron sheets was investigated with the drawing software SolidWorks (v2012) and the DEFORM-3D (v6.1) finite element analysis software. They simulated plate hobbing to determine the conditions in hobbing forming. The consistency between the numerical analysis results and the actual finished products was determined to provide better quality and increase customer satisfaction.

Based on the previous results, we analyzed the forging force in the cold forming of screws. Cold forming of screws is a common metal processing method to manufacture various shaped parts. However, due to the complexity of cold forming of screws, there are unknown mechanical behaviors. Therefore, it is necessary to analyze the mechanical behavior of cold-forming screws to better understand their mechanical behavior. Firstly, we introduced the basic principles and mechanical behavior of cold-forming screws. Next, using the basic principles of simulation software, their functions and applications were reviewed. As there is limited research on thread rolling, we adopted examples of a plate with pointed teeth. We used SolidWorks to import the finite element analysis software DEFORM-3D and simulated plate thread rolling. The result was used to allow for the rapid and accurate determination of the conditions and directions of thread rolling. The DEFORM-3D simulation software was used to analyze the mechanical behavior of cold-forming screws to understand their mechanical behavior. The results of this study contribute to a better understanding of cold-forming screws and provides a reference for future research and applications.

2. Materials and Methods

2.1. Research Methods

In the design of the method in this study, the cold forging of external and external hexagon flange screws was considered. Finite element software was used to simulate and analyze the forming process to understand the corresponding stress and strain variations generated during the cold forging process. The simulation was conducted to verify the forging force generated during the production of screws and provide a reference for the improvement of the design. The mold design drawings provided by "Chung Hsing Ta Mold Co., Ltd. (Tainan, Taiwan)." were created using SolidWorks for three nut specifications: M18, M20, and M22. Since we focused on the magnitude of the forging force, the female die and the back punch were merged in the drawings. The workpiece material was made of low-carbon steel material C1010 in order to model it using the DEFORM-3D preprocessor. The material parameters were set using the built-in parameters of DEFORM-3D. The simulation was conducted in the 5 stages of forging. After analysis, post-processing was conducted to output the peak forging force and compare the result with those of various manufacturers.

2.2. Material Mechanical Properties

The material used in this study was low-carbon steel C1010. The mold used in the formation of hexagonal nuts is shown in Figure 1. It consisted of a front punch, back punch, die insert, and die casing. In practice, the front punch must be made of high-speed steel SKH55, the back punch must be made of SKH9, the die insert must be made of WC, and the die casing must be made of SKD61. However, since we focused on the design of the nut-forming machine, the mold materials were assumed to be rigid bodies in the simulation to prevent deformation caused by unnecessary idle energy and excessive forging force, which increased production costs.

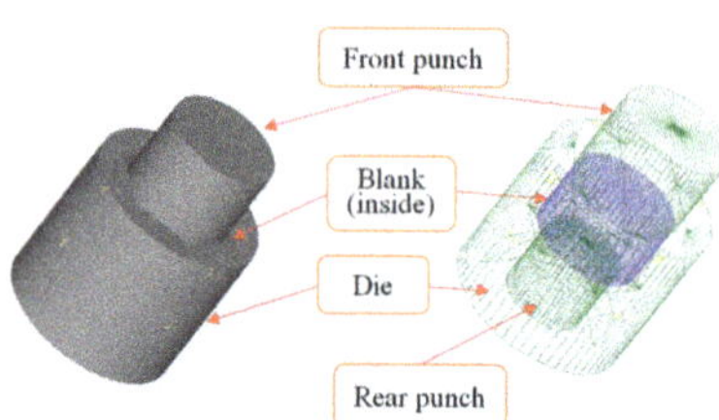

Figure 1. Configuration diagram showing the positions of the die, punch, and forging blank.

2.3. DEFORM Simulation Parameter Planning

The DEFORM-3D simulation software was used with the following conditions. During the forming process, except for the workpiece, all other mold components were regarded as rigid bodies, and the cold forging temperature was set to room temperature at 20 °C. The material selected was C1010, and the speed of the punch was set to 7.36 mm/s downward, with a step size of 0.1 mm per movement. The friction interface coefficient was set to 0.12 in the constant shear friction mode. Relevant simulation parameters are presented in Table 1. Figure 1 illustrates the configuration of the die, punch, and forging blank positions.

Table 1. Parameters for the forming simulation.

Mesh type	Tetrahedron
Material of blank	C1010
Blank/die property	Plastic/rigid

Table 1. *Cont.*

Temperature	20 °C
Speed of punch	7.36 mm/s
Quantity of mesh	90,000
Step distance	0.1 mm
Friction model	Constant shear friction
Coefficient of friction	0.12

2.4. Convergence Analysis

After establishing the workpiece and mold model and importing them into DEFORM-3D, element construction was performed. Finite element software was used for a numerical analysis method to divide the mesh for the structural analysis. As the number of meshes increases, the accuracy of the numerical analysis results increases, but the computational load also increases. With more data, the simulation calculation time also increases. Therefore, it is necessary to balance between achieving the accuracy of the simulation results and reducing the simulation calculation time by performing convergence analysis on the mesh. The convergence analysis method was used to compare the forging force in the Z-axis when the workpiece filled the mold with the difference in meshes. The difference in simulated load between the current and previous mesh numbers within 0.5% convergence was considered. Using the M18 workpiece material, the first forging consisted of 24 steps. The simulation was conducted with a mesh range of 10,000 to 100,000 in intervals of 10,000. The simulation results showed that convergence was reached when the mesh number reached 50,000. Based on actual factory interviews and considering current computer performance, a mesh number of 90,000 was used for the cold forging analysis to ensure the accuracy of the simulation results.

2.5. Calculation of Punch Speed

When simulating the 5-step forming of an M22 nut using DEFORM-3D, the speed of the front punch was initially set to 1 mm/s. After completing the entire process, 25 s were spent. The estimated production speed of the nut was 12 pcs/min, which was significantly different from the commonly set 100–140 pcs/min in manufacturer catalogs. In the second simulation, a speed of 350 mm/s was used, and 0.06 s were spent. The estimated production speed of the nut was 5000 pcs/min. In the third simulation, a speed of 8 mm/s was estimated, taking approximately 2.6 s. The estimated production speed of the nut was 118 pcs/min. If the simulated production rate was 100 pcs/min, and the average of the results from 3 simulations was used to calculate the velocity of the punch, then the velocity was 7.36 mm/s.

3. Results and Discussion

We estimated the forging force and related forming information for different workpiece dimensions using simulation methods. The data generated from the DEFORM-3D simulation were analyzed using linear regression to identify the forming trends. These results were then compared with the forging force data collected from manufacturer catalogs. The findings were expected to be used as a reference for forming machine manufacturers and nut manufacturers in terms of procurement costs and production management.

First, mold drawings provided by "Lian-Shyang industries Co., Ltd. (Tainan, Taiwan)" were created using SOLIDWORKS (v2020) and converted into STL files for simulation using DEFORM-3D (v6.1), as shown in Figures 2–4. However, there were differences between the simulated shape and the actual product (Figure 5).

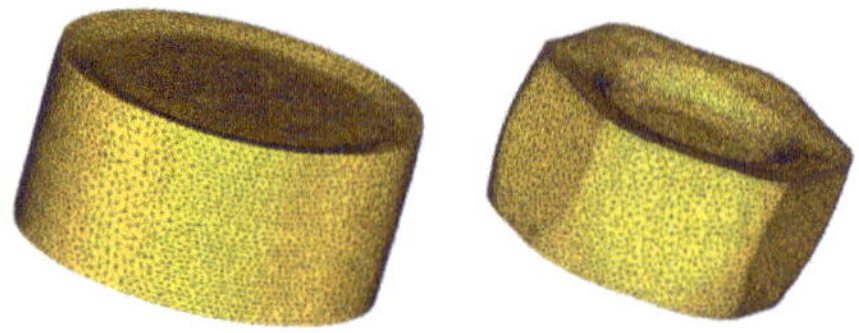

Figure 2. Completed blanks for the first pass (**left**) and the second pass (**right**).

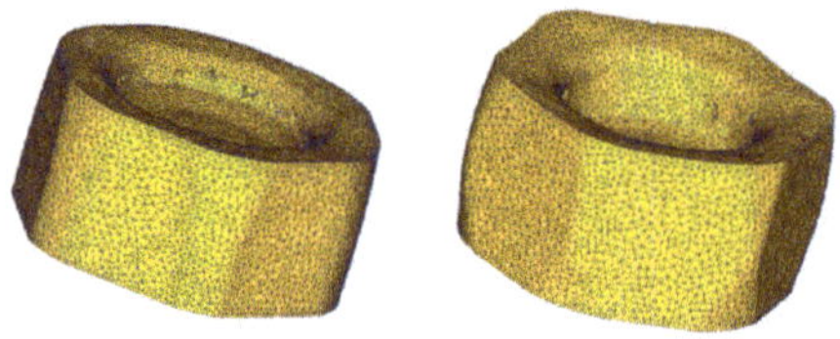

Figure 3. Completed blanks for the third pass (**left**) and the fourth pass (**right**).

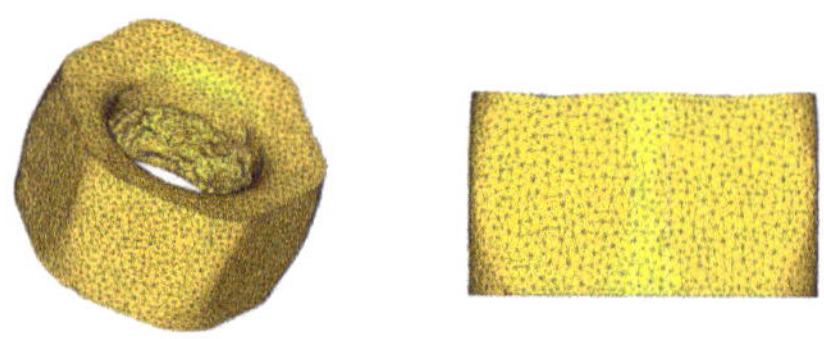

Figure 4. Completed blank for the fifth pass (**left**) and front view (**right**).

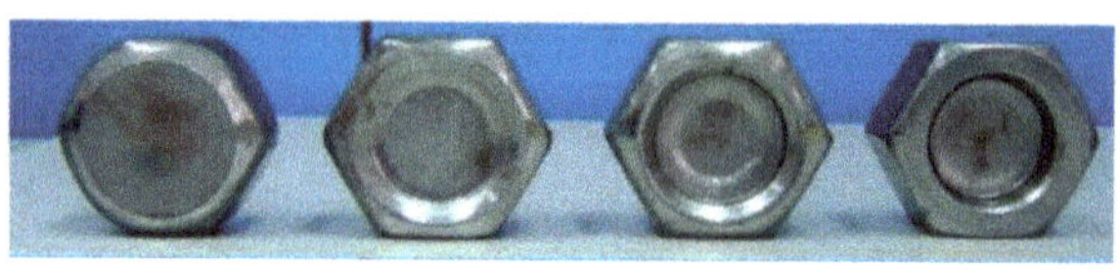

Figure 5. Actual stamped blank shape on-site.

The DEFORM-3D post-processing operation was used to extract the forging force data (Figure 6), and the sum of the peak forging forces for the punch was calculated as 183.3 tons using EXCEL. After identifying this error, the simulation was re-conducted, and the corrected workpiece profiles were obtained.

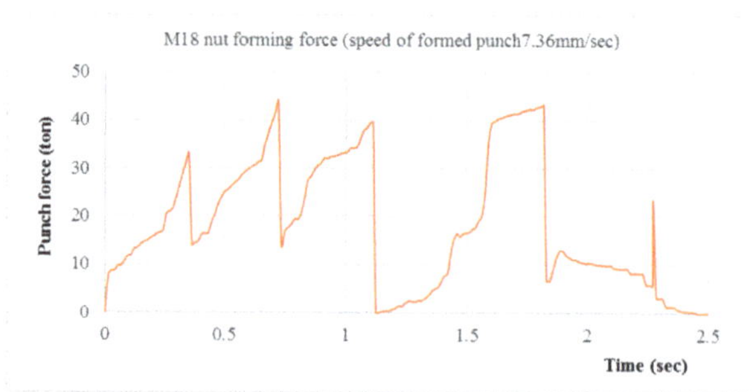

Figure 6. Distribution of the forging forces for the M18 nut.

Interviews were conducted with on-site engineers from "Lian-Shyang industries Co., Ltd. (Tainan, Taiwan)" and "NES LIMITED (Kaohsiung, Taiwan)" Both engineers confirmed that the mold shapes were correct and stated that in practical applications, it was crucial to avoid overflow during the simulation like in DEFORM-3D (Figure 7). The punch stroke must not be adjusted to the extent that the punch and the workpiece come into direct contact, as this might rupture and damage the punch and the mold. Therefore, the stopping position after each step was chosen to be about 0.1 mm before overflow, and attention must be paid to the geometric symmetry of the workpiece (Figure 8), as it affected the symmetry of the final product.

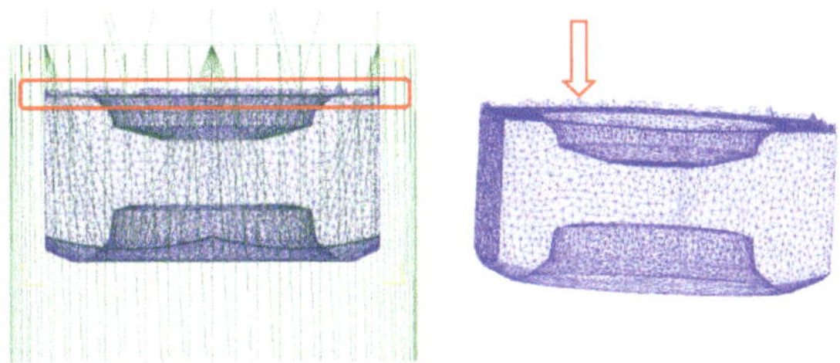

Figure 7. Excess material condition during the forming software simulation.

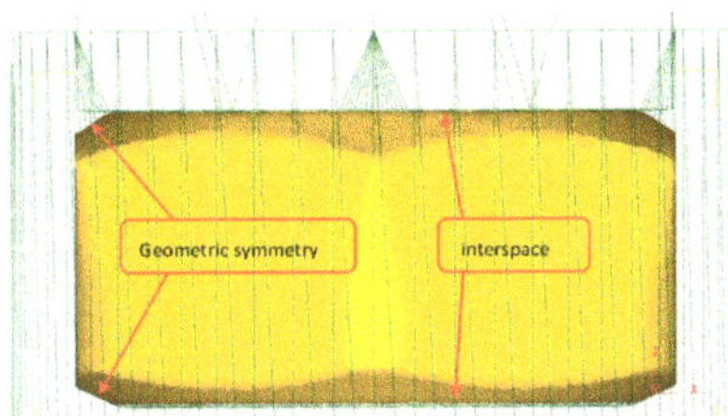

Figure 8. The gap between the mold and material, as well as the symmetry of geometric shapes during the analysis.

The corrected sketches were converted back into STL files and imported into DEFORM-3D for simulation. Since the mesh used was 90,000, a complete five-step simulation took approximately 18 h. After the simulation, the DEFORM-3D post-processor was used to obtain the relationship between the forging force and time for the punch. The data extracted from DEFORM-3D was then used to plot the relationship graphs for each step using EXCEL (Figure 9) and calculate the total forging force.

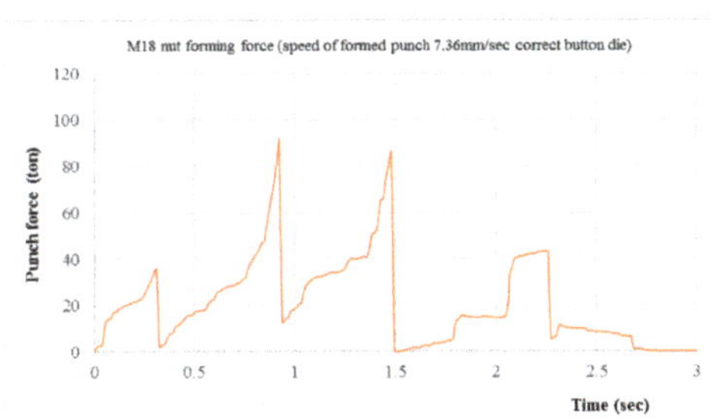

Figure 9. Distribution of the forging forces for five passes plotted in EXCEL using data extracted from the forming software.

From the extracted data, it was found that the sum of the peak forging forces for the M18 nut during the forming process was approximately 268 tons. On further inspection, an error was found in the original workpiece dimensions (height should be 14.43 mm but was incorrectly recorded as 14.13 mm). After making the necessary adjustments, the simulation was re-conducted, and the data were processed using EXCEL, resulting in a total forging force of about 265 tons (Figure 10), with minimal difference.

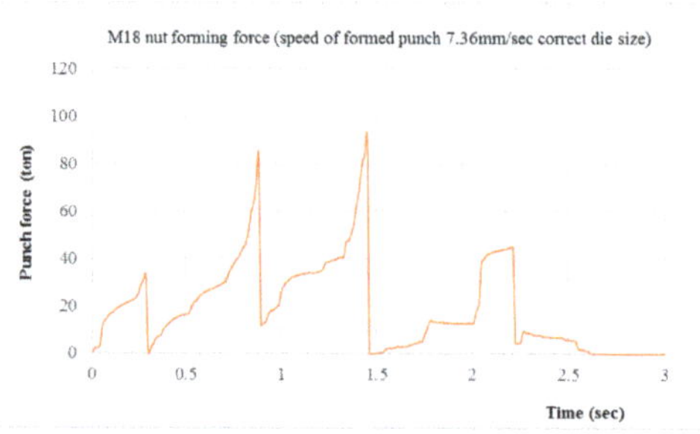

Figure 10. Calculation of the M18 nut simulation results extracted to EXCEL after correcting the original blank dimensions.

Similarly, the forging force data for the punch were extracted from the DEFORM-3D post-processor, and the sum of the forging forces was approximately 321 tons (Figure 11) and 329 tons (Figure 12) before and after correction, respectively. The same analysis process was repeated for M18 and M20, and errors in the original workpiece dimensions were identified. After correction, the data were extracted and used to plot the forging force relationship graphs for each step. The sum of the forging force peak values was approximately 343 tons (Figure 13) and 391 tons (Figure 14). The red box of the second pass in Figure 13 represented the corrected forging force, while the red box in Figure 14 represented the forging force before correction. Therefore, during simulation analysis, it is necessary to refer to practical experience in the field.

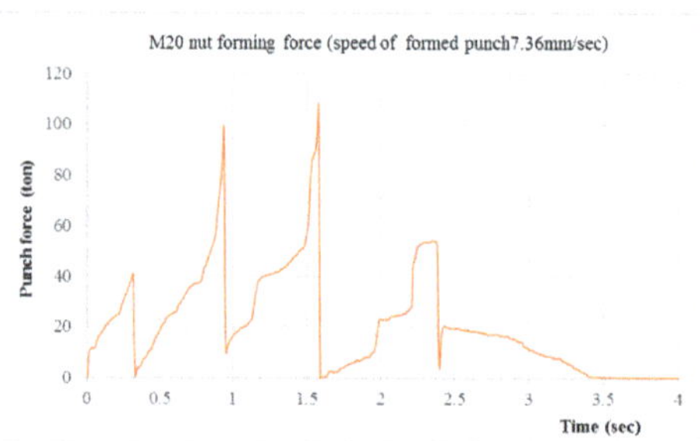

Figure 11. Distribution of the forging forces for the M20 nut at each pass.

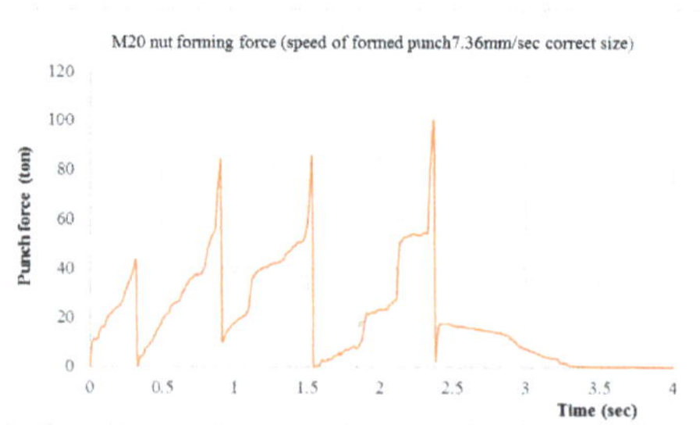

Figure 12. Distribution of the forging forces for the M20 nut at each pass after correcting the original blank dimensions.

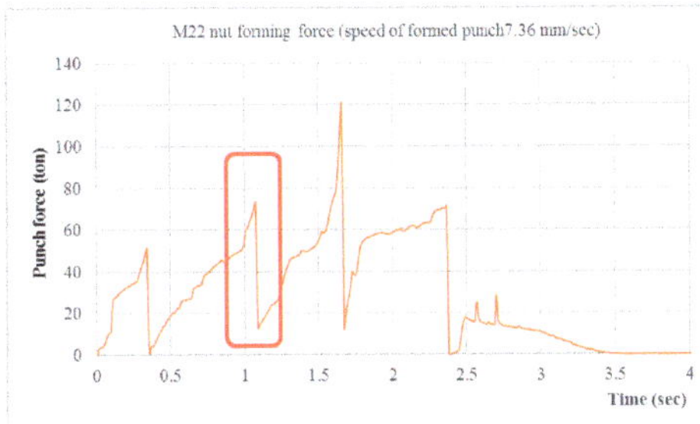

Figure 13. Distribution of the forging forces for the M22 nut at each pass; the area marked with the red box indicated the corrected forging force.

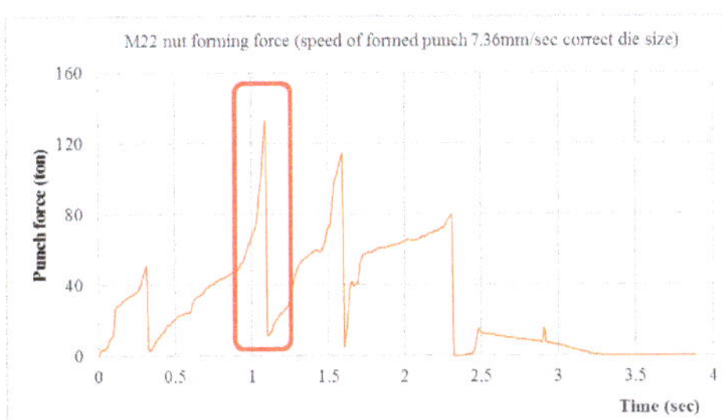

Figure 14. Distribution of the forging forces for the M22 nut at each pass after correcting the original blank dimensions; the area marked with the red box indicated the forging force before correction.

The results were obtained from the DEFORM-3D simulations of the M18, M20, and M22 nut-forming processes. Regression analysis was performed with the forging force set as the dependent variable (y) and the volume (or weight) of the original workpiece as the independent variable (x). From the data analysis, we obtained an equation y = 0.0057x + 253.10 (Figure 15). A comparison with the blank at 20 °C (●), using the manufacturer's catalog specifications (●), and the blank at 30 °C (●) revealed that the manufacturer's specifications were higher than the others. When the blank was set to a higher operating temperature, the forging force decreased significantly and uniformly.

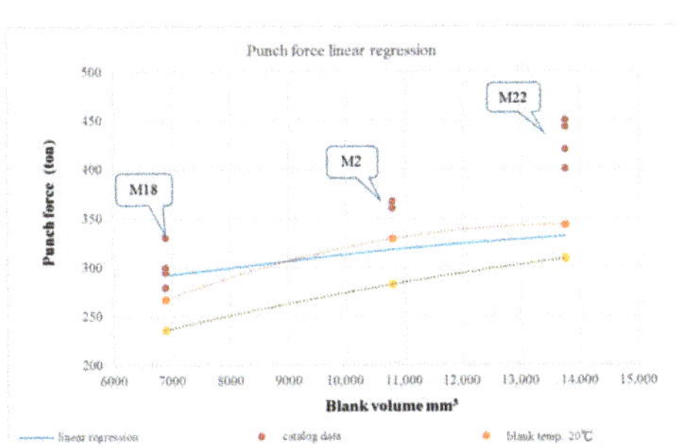

Figure 15. Comparison between the linear regression of forging forces from the simulation experiment and the manufacturer's catalog.

4. Conclusions

After a series of simulations and comparisons with the actual production process, the following conclusions were drawn from the results of this study. For the M18, M20, and

M22 nuts, the forging force specified for commercially available nut-forming machines was 8−20% higher than that from the initial simulation results. However, when considering the ambient temperature, the difference increased to 19−27%. Additionally, it was observed that the larger the size of the nut, the greater the disparity. The analysis result of abnormal simulation data showed that precise adjustment of the punch stroke was crucial as it directly affected the total forging force during the forming process. An experienced on-site engineer can maximize the efficiency of the machine and extend the lifespan of the mold based on the result of this study.

Author Contributions: Conceptualization, Y.-T.W. and S.-Y.H.; methodology, Y.-T.W. and S.-Y.H.; formal analysis, Y.-T.W.; writing—original draft preparation, Y.-T.W.; writing—review and editing, S.-Y.H. All authors have read and agreed to the published version of the manuscript.

Funding: This research received no external funding.

Institutional Review Board Statement: Not applicable.

Informed Consent Statement: Not applicable.

Data Availability Statement: Datasets related to these studies, findings, and results as reported are included in the manuscript itself.

Conflicts of Interest: The authors declare no conflict of interest.

References

1. O'Connell, M.; Painter, B.; Maul, G.; Altan, T. Flashless closed-die upset forging-load estimation for optimal cold header selection. *J. Mater. Process. Technol.* **1996**, *59*, 81–94. [CrossRef]
2. Kim, H.; Alfan, T. Cold forging of steel—Practical examples of computerized part and process design. *J. Mater. Process. Technol.* **1996**, *59*, 122–131. [CrossRef]
3. Vazquez, V.; Hannan, D.; Altan, T. Tool life in cold forging–example of design improvement to increase service life. *J. Mater. Process. Technol.* **2000**, *98*, 90–96. [CrossRef]
4. Falk, B.; Engel, U.; Geiger, M. Fundamental aspects for the evaluation of the fatigue behavior of cold forging tools. *J. Mater. Process. Technol.* **2001**, *119*, 158–164. [CrossRef]
5. Lee, Y.S.; Hwang, S.K.; Chang, Y.S.; Hwang, B.B. The forming characteristics of radial-forward extrusion. *J. Mater. Process. Technol.* **2001**, *113*, 136–140. [CrossRef]
6. Maccormack, C.; Monaghan, J. 2D and 3D finite element analysis of a three stage forging sequence. *J. Mater. Process. Technol.* **2002**, *127*, 48–56. [CrossRef]
7. Behrens, A.; Just, H. Extension of the forming limits in cold and warm forging by the FE based fracture analysis with the integrated damage model of effective stresses. *J. Mater. Process. Technol.* **2002**, *125–126*, 235–241. [CrossRef]
8. Landre, J.; Pertence, A.; Cetlin, P.R.; Rodrigues, J.M.C.; Martins, P.A.F. On the utilization of ductile fracture criteria in cold forging. *Finite Elem. Anal. Des.* **2003**, *39*, 175–186. [CrossRef]
9. Min, D.; Kim, M. A study on precision cold forging process improvements for the steering yoke of automobiles by the rigid–plastic finite-element method. *J. Mater. Process. Technol.* **2003**, *138*, 339–342. [CrossRef]
10. Hsia, S.Y. A numerical simulation of meshes mending used in the three-dimensional porthole die extrusion of thin wall profiles. *J. Eng. Res. Rep.* **2018**, *2*, 1–14. [CrossRef]
11. Hsia, S.Y.; Chou, Y.T.; Lu, G.F. Analysis of Sheet Metal Tapping Screw Fabrication Using a Finite Element Method. *Appl. Sci.* **2016**, *6*, 300. [CrossRef]

engineering proceedings

MDPI

Proceeding Paper

Timing Matters: Impact of Meal Timing on Daily Calorie Intake of Office Workers †

Xueyun Han

School of Transportation Science & Engineering, Beihang University, Beijing 100191, China; hanxy66@outlook.com

† Presented at the IEEE 5th Eurasia Conference on Biomedical Engineering, Healthcare and Sustainability, Tainan, Taiwan, 2–4 June 2023.

Abstract: This research aims to investigate the role of meal timing on calorie intake. A database of a company's cafeteria was used to track employees' food-purchasing behaviors. The findings show that starting meals early leads to an overall reduction in total daily calorie intake. However, the effect of meal timing differed between meals, with breakfast timing having the most significant impact followed by lunch and dinner timings. In terms of calorie intake per meal, breakfast timing was a decisive factor, followed by dinner, and lunch timing showed a weaker correlation to lunch intake. This result implied that early breakfast and dinner were important in managing calorie intake, while lunch timing had a lower impact. This research is a guide on how to lose weight or boost overall health through an appropriate dietary habit.

Keywords: over-weight; office workers; meal timing; calorie intake

Citation: Han, X. Timing Matters: Impact of Meal Timing on Daily Calorie Intake of Office Workers. *Eng. Proc.* **2023**, *55*, 69. https://doi.org/10.3390/engproc2023055069

Academic Editors: Teen-Hang Meen, Kuei-Shu Hsu and Cheng-Fu Yang

Published: 8 December 2023

1. Introduction

The global prevalence of obesity surpassed 1 billion individuals, comprising 650 million adults. "WHO estimates that by 2025, approximately 167 million people will become less healthy because they are overweight or obese" [1]. Being overweight and obese are major risks for non-communicable diseases (NCDs) such as type 2 diabetes and cardiovascular diseases as well as mental health issues. People with obesity are also three times more likely to be hospitalized for COVID-19 [2]. Maintaining an appropriate weight and keeping in good shape contributes to health and work performance and fosters a positive image. However, it is difficult to lose weight, particularly in terms of maintaining the loss. Maximal weight loss is usually achieved at 6 months, but most individuals regain weight thereafter [3]. Roughly 20% of overweight individuals were successful in long-term weight loss [4]. Office workers have challenges including high pressure from intense competition, limited time for exercise, and the burden of heavy workloads. Many weight loss programs do not appeal to men [5], as men prefer weight loss programs that include familiar participants and are convenient to visit frequently [6]. Thus, discovering an easy-to-stick-to-weight management approach in office settings is essential. Herein, to find an appropriate diet method, the role of meal timing needs to be investigated. Therefore, in this study, the impact of meal timing on daily caloric intake and the variations in different meal timings were explored.

2. Literature Review

Research on meal timing encompasses two main perspectives: the medical aspect of the correlation between meal timing and physiological indicators and the behavioral science aspect of how behavioral changes promote overall health.

Medical research on meal timing patterns focused on overnight fasting, breakfast skipping, and late-night eating which were linked to metabolic health parameters, including blood glucose and blood pressure. These associations were mediated through the circadian regulation of metabolic tissues, gut microbial characteristics, and health behaviors such as food intake, physical activity, and sleep [7]. However, these strategies were impractical for office workers compared to altering the time of each meal.

The role of meal timing in energy intake and weight management is not well understood as most relative studies focused on surveys associated with meal frequency or the variability of meal timing between certain days. There has been little evidence that reducing meal frequency was beneficial for reducing energy intake [8,9]. However, other research results showed the associations between the variability of meal timing on weekends versus weekdays and body mass index [10]. Few studies evaluated the association between the timing of breakfast, lunch, and dinner, and caloric intake. The only relevant research was carried out in 2014 in which 59 participants were asked to record their eating behavior for seven days. The results suggested that the timing of meals, particularly eating just before sleeping could lead to weight gain [11]. However, it is important to note that the conclusions of this study were based on a small sample size (N = 59) and a short duration (seven days). Test errors were unavoidable due to manual recording. Thus, further support from larger datasets is necessary to prove the validity of the results. Based on a large-scale dataset obtained from an actual company cafeteria, this study was carried out by using food purchase data to specifically examine the impact of breakfast, lunch, and dinner timing on daily calorie intake and meal-specific calorie consumption.

3. Background

A large archival dataset of individual-level food purchasing records of a large e-commercial firm in China was adopted in this study. The data was offered by the cafeteria located at the headquarters building. The transaction data included user ID, dish ID, calories per serving, serving numbers, transaction date and time, and weight of food. This dataset comprised 67,077 records over three years, during which each employee's calorie intake, food choice, and time were recorded. These records were categorized on a per-person per-day basis, resulting in 10,099 individual observations. Further analysis of the categorized data was conducted on cafeteria transactions on Saturdays (n = 1046), breakfast purchases (n = 7054), lunch purchases (n = 6867), and dinner purchases (n = 5922). Then, a highly detailed view of the user's entire eating habit and insights into the collective eating habits of office workers as a group were obtained.

4. Economical Model

The daily caloric intake for each employee was calculated by categorizing all transactions. The caloric intake in a meal was evaluated with a function of a predefined time for each meal, as can be seen in Equation (1).

$$\mathrm{DailyCal}_{it} = \alpha_0 + \alpha_1 \Delta \mathrm{Bkf}_{it} + \alpha_2 \Delta \mathrm{Lun}_{it} + \alpha_3 \Delta \mathrm{Din}_{it} + e_{it}, \quad (1)$$

where $\mathrm{DailyCal}_{it}$ refers to the overall calorie intake of employee i on a day. The cumulative sum of calories per serving was also calculated. The dataset was categorized into three distinct meal types: breakfast (6:00–10:59), lunch (11:00–15:59), and dinner (16:00–23:59). Four variables were defined ($\Delta\mathrm{Bkf}_{it}$, $\Delta\mathrm{Lun}_{it}$, $\Delta\mathrm{Din}_{it}$, and $\Delta\mathrm{AllMeal}$) which corresponded to the value of breakfast, lunch, dinner, and the sum of the previous three values. The descriptive statistics of all variables appear in Table 1.

These variables quantified the difference in minutes between the timing of each meal and the anchor time which were predetermined with standard values (6:00 for breakfast, 11:00 for lunch, and 16:00 for dinner). By examining these variables, the extent to which individuals deviated from the established meal timings was evaluated with the potential impact on calorie intake patterns. The analogous equations for specific meals were as follows.

$$\mathrm{BkfCal}_{it} = \beta_0 + \beta_1 \Delta \mathrm{Bkf}_{it} + \beta_2 \Delta \mathrm{Lun}_{it} + \beta_3 \Delta \mathrm{Din}_{it} + e_{it}, \quad (2)$$

$$\mathrm{LunCal}_{it} = \zeta_0 + \zeta_1\Delta\mathrm{Bkf}_{it} + \zeta_2\Delta\mathrm{Lun}_{it} + \zeta_3\Delta\mathrm{Din}_{it} + e_{it}, \quad (3)$$

$$\mathrm{DinCal}_{it} = \delta_0 + \delta_1\Delta\mathrm{Bkf}_{it} + \delta_2\Delta\mathrm{Lun}_{it} + \delta_3\Delta\mathrm{Din}_{it} + e_{it}, \quad (4)$$

Table 1. Variable definitions and descriptive statistics.

Variable	Obs	Mean	SD	Median
DailyCal	10,099	923.28	4666.21	730.00
BkfCal	10,099	231.94	331.56	157.00
LunCal	10,099	389.28	3786.02	289.00
DinCal	10,099	302.07	2713.82	141.00
ΔAllMeal	3977	342.97	104.51	330.00
ΔBkf	7073	132.38	139.34	90.00
ΔLun	6873	76.14	46.85	60.00
ΔDin	5935	170.29	74.91	150.00

Nine carry over (β_1, β_2, β_3; ζ_1, ζ_2, ζ_3; δ_1, δ_2, and δ_3) and three baseline coefficients (β_0, ζ_0, and δ_0) were determined for each variable in the data set. β_2, β_3, ζ_1, ζ_3, δ_1, and δ_2 were inter-meal carryover coefficients [12] to assess the influence of the timing of other meals on the caloric content of the current meal. β_1, ζ_2, and δ_3 were the intra-meal carryover coefficient to assess the impact of the timing of the same meal on calorie intake. Inter-meal coefficients showed how the effects between meals were balanced, while the intra-meal coefficient reflects the stability of intake concerning the biological clock. These coefficients were used to assess the impact of meal timing as their absolute sizes indicated the strength of carryover effects and their signs indicated the positive or negative valence of the effects.

5. Results

A positive correlation between meal timing and daily calorie intake was found which suggested that delaying the start of meals led to an overall increase in total daily calorie intake. The fixed effects (FE) model was used to examine the cumulative effects of meal timing represented by ΔAllmeal (the sum of ΔBreakfast, ΔLunch, and ΔDinner). A significant coefficient was 0.540 (t = 5.18). The impact of ΔBreakfast was the most pronounced with a significant coefficient of 0.739 (t = 5.42), while the effects of ΔLunch (t = 0.77, coefficient = 0.258) and ΔDinner (t = 1.00, coefficient = 0.213) were smaller and not statistically significant. These results are shown in Table 2.

Table 2. Regression results (daily calorie intake).

Variable	(1) OLS	(2) FE	(3) FE_robust	(4) RE	(5) RE_robust	(6) MLE	(7) PA	(8) BE
ΔAllMeal	0.433 [a] ** (2.19) [b]	0.540 *** (5.18)	0.540 * (1.88)	0.583 *** (6.03)	0.583 ** (2.13)	0.584 *** (6.03)	0.598 *** (5.70)	0.806 *** (3.31)
ΔBkf	0.998 ** (2.32)	0.739 *** (5.42)	0.739 * (1.91)	0.774 *** (6.08)	0.774 ** (1.98)	0.775 *** (6.08)	0.789 *** (5.65)	0.842 *** (2.59)
ΔLun	1.753 *** (3.56)	0.258 (0.77)	0.258 (0.32)	0.449 (1.40)	0.449 (0.63)	0.451 (1.40)	0.553 (1.57)	2.172 * (1.92)
ΔDin	−1.249 *** (−4.38)	0.213 (1.00)	0.213 (0.72)	0.195 (0.96)	0.195 (0.72)	0.195 (0.96)	0.178 (0.80)	0.014 (0.02)

[a] This number shows the coefficient. [b] This result represents t-value. The asterisks (*) indicate * $p < 0.1$; ** $p < 0.05$; *** $p < 0.01$.

In terms of breakfast calorie intake, only ΔBreakfast showed a significant effect (t = 5.96, coefficient = 0.407), while the deviations for lunch and dinner indicated non-significant effects. This suggested that people did not consider future meal timings when consuming breakfast (e.g., taking in more calories in breakfast when lunch was expected to be eaten late). Surprisingly, for lunch, none of the deviations for breakfast, lunch, or dinner showed significant effects. Regarding dinner, both breakfast (t = 3.40, coefficient = 0.238) and dinner (t = 2.76, coefficient = 0.303) timings showed significant effects, while lunch timing did show a significant influence. These results are shown in Tables 3–5.

Table 3. Regression results (breakfast).

Variable	(1) OLS	(2) FE	(3) FE robust	(4) RE	(5) RE robust	(6) MLE	(7) PA	(8) BE
ΔBkf	0.590 *** (2.79)	0.407 *** (5.96)	0.407 ** (2.05)	0.470 *** (7.46)	0.470 ** (2.31)	0.467 *** (7.39)	0.464 *** (7.45)	0.749 *** (5.17)
ΔLun	0.767 *** (3.38)	0.167 (0.99)	0.167 (0.55)	0.278 * (1.75)	0.278 (1.10)	0.268 * (1.68)	0.261 * (1.66)	0.629 (1.25)
ΔDin	−0.757 *** (−4.76)	0.093 (0.87)	0.093 (0.43)	0.022 (0.22)	0.022 (0.12)	0.027 (0.27)	0.030 (0.30)	−0.444 (−1.64)

The asterisks (*) indicate * $p < 0.1$; ** $p < 0.05$; *** $p < 0.01$.

Table 4. Regression results (lunch).

Variable	(1) OLS	(2) FE	(3) FE robust	(4) RE	(5) RE robust	(6) MLE	(7) PA	(8) BE
ΔBkf	0.093 (1.12)	0.094 (1.26)	0.094 (1.12)	0.055 (0.80)	0.055 (0.69)	0.051 (0.75)	0.043 (0.61)	−0.191 (−1.25)
ΔLun	0.661 *** (3.61)	−0.134 (−0.73)	−0.134 (−0.53)	−0.033 (−0.19)	−0.033 (−0.16)	−0.011 (−0.07)	0.044 (0.25)	0.701 (1.32)
ΔDin	−0.377 *** (−3.40)	−0.183 (−1.57)	−0.183 * (−1.72)	−0.149 (−1.38)	−0.149 (−1.40)	−0.148 (−1.38)	−0.150 (−1.34)	0.147 (0.52)

The asterisks (*) indicate * $p < 0.1$; *** $p < 0.01$.

Table 5. Regression results (dinner).

Variable	(1) OLS	(2) FE	(3) FE robust	(4) RE	(5) RE robust	(6) MLE	(7) PA	(8) BE
ΔBkf	0.316 (1.36)	0.238 *** (3.40)	0.238 (1.11)	0.256 *** (3.96)	0.256 (1.16)	0.259 *** (4.01)	0.263 *** (3.89)	0.284 * (1.82)
ΔLun	0.325 (1.22)	0.226 (1.31)	0.226 (0.54)	0.277 * (1.70)	0.277 (0.73)	0.286 * (1.75)	0.307 * (1.80)	0.842 (1.56)
ΔDin	−0.114 (−0.92)	0.303 *** (2.76)	0.303 ** (2.21)	0.312 *** (3.03)	0.312 ** (2.53)	0.313 *** (3.04)	0.313 *** (2.92)	0.311 (1.07)

The asterisks (*) indicate * $p < 0.1$; ** $p < 0.05$; *** $p < 0.01$.

6. Robustness of Model

To ensure the robustness of the model and its result, we conducted the robustness test using various econometric models. In addition to the fixed effects (FE) model, the results using seven alternative models including ordinary least squares (OLS), robust fixed effects (FE robust), random effects (RE), maximum likelihood estimation (MLE), population averaged (PA) and between effects (BE) were compared. The results are shown in Tables 3–5.

The results of these models showed that the seven models consistently supported the significant impact of meal timing on daily calorie intake. Similar consistency was observed for meal-specific calorie intake. The results of all models indicated a significant relationship between breakfast timing and calorie intake. Similarly, for lunch timing, seven to eight models yielded consistent results. For dinner timing, five to six models showed consistent findings. These consistent results across multiple models validated the robustness and reliability of the result of this study regarding the effects of meal timing on calorie intake.

7. Conclusions

The role of meal timing in overall calorie intake among office workers was investigated in this study. The effect of the timing of meals was validated using data from a large company cafeteria. The findings revealed that having meals early, particularly breakfast, played a significant role in reducing caloric intake. Instead of skipping or reducing dinner, having an early breakfast and an early dinner was effective in reducing calorie intake during dinner. Delaying lunchtime did not significantly affect calorie intake and daily calorie intake. This result showed healthier choices for individuals facing time constraints and having a busy work schedule. Evidence-based guidance was confirmed for individuals looking to lose weight or improve their overall health in this study. Importantly, the suggested meal timing can promote feasible and effective weight management in the daily routines of employees.

Though empirical data was analyzed to identify effective strategies for reducing calorie intake, the underlying medical mechanisms still need to be verified. Therefore, based on the findings of this study, it is necessary to explore how to decrease the body mass index (BMI) or improve overall health. The optimal dietary composition for achieving these outcomes also needs to be found out. For health management, the differential effects of meal timing on weekdays versus weekends are also required to be researched. Additionally, more data including health conditions and gender are necessary to provide a more comprehensive understanding of the relationship between meal timing and health management. Obese males intake twice as many calories between 10 PM and 4 AM compared to normal males [13]. Compared to obese females, normal-weight females eat dinner later on the weekends [14]. By incorporating these factors, additional insights can be obtained for personalized health management strategies for office workers.

The research findings of this study can be applied to the establishment of individual weight loss goals. Companies can adjust work schedules to encourage healthy diet habits and reduce the risk of chronic diseases associated with heavy work. By fostering a supportive environment that promotes healthy behaviors, companies can improve employees' working performance and enhance overall productivity and well-being in the workplace.

Funding: This research received no external funding.

Institutional Review Board Statement: Ethical review and approval were waived for this study due to the data containing only cafeteria order information in daily life scenario. It is not medical study.

Informed Consent Statement: Not applicable.

Data Availability Statement: Data is unavailable due to privacy or ethical restrictions.

Acknowledgments: The author thanks for the support from joint project (grant number 2023K20206) between Renmin University and Everest VC.

Conflicts of Interest: The author declares no conflict of interest.

References

1. Tedros. World Obesity Day 2022—Accelerating Action to Stop Obesity. WHO. 2022. Available online: https://www.who.int/news/item/04-03-2022-world-obesity-day-2022-accelerating-action-to-stop-obesity (accessed on 28 May 2023).
2. Stefan, N.; Birkenfeld, A.L.; Schulze, M.B. Global pandemics interconnected—Obesity, impaired metabolic health and COVID-19. *Nat. Rev. Endocrinol.* **2021**, *17*, 135–149. [CrossRef] [PubMed]
3. Garcia Ulen, C.; Huizinga, M.M.; Beech, B.; Elasy, T.A. Weight regains prevention. *Clin. Diabetes* **2008**, *26*, 100–113. [CrossRef]
4. Wing, R.R.; Phelan, S. Long-term weight loss maintenance. *Am. J. Clin. Nutr.* **2005**, *82*, 222S–225S. [CrossRef] [PubMed]
5. French, S.A.; Jeffery, R.W.; Forster, J.L.; McGovern, P.G.; Kelder, S.H.; Baxter, J.E. Predictors of weight change over two years among a population of working adults: The Healthy Worker Project. *Int. J. Obes. Relat. Metab. Disord. J. Int. Assoc. Study Obes.* **1994**, *18*, 145–154.
6. Sabinsky, M.S.; Toft, U.; Raben, A.; Holm, L. Overweight men's motivations and perceived barriers towards weight loss. *Eur. J. Clin. Nutr.* **2007**, *61*, 526–531. [CrossRef] [PubMed]
7. Gabel, K.; Varady, K.A. Current research: Effect of time restricted eating on weight and cardiometabolic health. *J. Physiol.* **2022**, *600*, 1313–1326. [CrossRef] [PubMed]

8. Mills, J.P.; Perry, C.D.; Reicks, M. Eating frequency is associated with energy intake but not obesity in midlife women. *Obesity* **2011**, *19*, 552–559. [CrossRef] [PubMed]
9. Schwingshackl, L.; Nitschke, K.; Zähringer, J.; Bischoff, K.; Lohner, S.; Torbahn, G.; Schlesinger, S.; Schmucker, C.; Meerpohl, J.J. Impact of meal frequency on anthropometric outcomes: A systematic review and network meta-analysis of randomized controlled trials. *Adv. Nutr.* **2020**, *11*, 1108–1122. [CrossRef] [PubMed]
10. Zerón-Rugerio, M.F.; Hernáez, Á.; Porras-Loaiza, A.P.; Cambras, T.; Izquierdo-Pulido, M. Eating jet lag: A marker of the variability in meal timing and its association with body mass index. *Nutrients* **2019**, *11*, 2980. [CrossRef] [PubMed]
11. Reid, K.J.; Baron, K.G.; Zee, P.C. Meal timing influences daily caloric intake in healthy adults. *Nutr. Res.* **2014**, *34*, 930–935. [CrossRef] [PubMed]
12. Khare, A.; Inman, J.J. Daily, week-part, and holiday patterns in consumers' caloric intake. *J. Public Policy Mark.* **2009**, *28*, 234–252. [CrossRef]
13. Andersson, I.; Rössner, S. Meal patterns in obese and normal weight men: The 'Gustaf' study. *Eur. J. Clin. Nutr.* **1996**, *50*, 639–646. [PubMed]
14. Corbalán-Tutau, M.D.; Madrid, J.A.; Garaulet, M. Timing and duration of sleep and meals in obese and normal weight women. Association with increase blood pressure. *Appetite* **2012**, *59*, 9–16. [CrossRef] [PubMed]

engineering proceedings

MDPI

Proceeding Paper

Procedure of Forming Power Law Functionally Graded Material (FGM) Plate Using ANSYS †

Thai Hien Nguyen [1], Nhat Tien Nguyen [1], Duy Anh Ly [1] and Trung Nghia Tran [2,*]

1 Department of Engineering Mechanics, Faculty of Applied Science, Ho Chi Minh City University of Technology (HCMUT), VNUHCM, 268 Ly Thuong Kiet Street, District 10, Ho Chi Minh City 72506, Vietnam; thnguyen@hcmut.edu.vn (T.H.N.); tien.nguyen_ssg@hcmut.edu.vn (N.T.N.); anh.lybku2019@hcmut.edu.vn (D.A.L.)
2 Laboratory of Laser Technology, Faculty of Applied Science, Ho Chi Minh City University of Technology (HCMUT), VNUHCM, 268 Ly Thuong Kiet Street, District 10, Ho Chi Minh City 72506, Vietnam
* Correspondence: ttnghia@hcmut.edu.vn
† Presented at the IEEE 5th Eurasia Conference on Biomedical Engineering, Healthcare and Sustainability, Tainan, Taiwan, 2–4 June 2023.

Abstract: The deflection, stress, and heat transfer of a plate made of functionally graded material (FGM) were investigated using the data of mechanical loads and temperatures using ANSYS Workbench. The 2D planar problem-solving method in ANSYS Workbench was used to simulate the FGM plate with the neutral surface for the mechanical problem. The problem of stable heat transfer for the plate was assumed for the solid block for the FGM plate. The results were compared with those of previous studies to validate the reliability of the newly fabricated FGM. The thin plate theory for materials with mechanical properties was considered with the power law distribution as each layer had identical thickness and temperature. Each layer showed material properties that changed with the thickness of the plate.

Keywords: functionally graded material; ANSYS; heat transfer; stress; deflection

Citation: Nguyen, T.H.; Nguyen, N.T.; Ly, D.A.; Tran, T.N. Procedure of Forming Power Law Functionally Graded Material (FGM) Plate Using ANSYS. *Eng. Proc.* **2023**, *55*, 70. https://doi.org/10.3390/engproc2023055070

Academic Editors: Teen-Hang Meen, Kuei-Shu Hsu and Cheng-Fu Yang

Published: 8 December 2023

1. Introduction

Materials with improved durability, plasticity, high bearing capacity, and temperature resistance have been researched in structural mechanics. As one of such materials, functional graded materials (FGM) are studied widely to apply them to aviation, space, construction, and mechanical engineering. FGM is a composite material characterized by continuously changing mechanical properties. Such properties help overcome the limitations of traditional composite materials and improve the ability to withstand mechanical and thermal loads in the structure. To expand the application of FGM, the behavior of materials with various methods and models is required. Research on the mechanical properties of FGM has been conducted by many researchers. Thai and Choi analyzed the FGM plate for bending and free vibration using the first-order shear theory [1]. Singha et al. analyzed it subjected to uniformly distributed loads by the finite element method [2]. Olley analyzed the application of FGM in aircraft structures [3,4]. The result of the previous research was compared to that of this research to provide reference and basic information for constructing appropriate models to understand the properties of FGM plates with muli-layers.

2. Materials and Methods

2.1. Structure of FGM

The ceramic and metal components are linearly distributed along the structural wall, which has either a ceramic-rich surface or a metal-rich surface in the FGM plate. The FGM plate with such properties is called a power law FGM plate (P-FGM). The structure of P-FGM is shown in Figure 1.

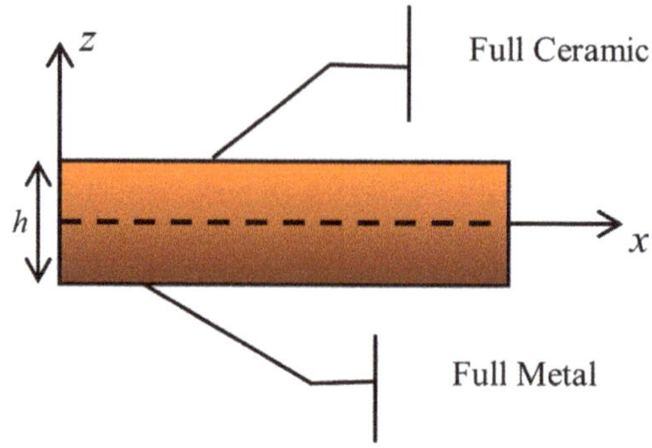

Figure 1. Structure of power-law FGM.

Material properties of P-FGM depend on the *p*-value and the position in the plate and vary according to a power law according to Equation (1).

$$\begin{aligned} V_c(z) &= \left(\frac{2z+h}{2h}\right)^p, \ -\frac{h}{2} \le z \le \frac{h}{2} \\ V_m(z) &= 1 - V_c(z) \end{aligned} \quad (1)$$

where p is a non-negative number called the volume fraction index, h is the thickness of the structure, and m and c indicate metal and ceramic components. According to the law of material distribution in Equation (1), when $p = 0$, the homogeneous plate is ceramic; when $p \ge 1$, the proportion of ceramic in the P-FGM plate decreases.

2.2. Heat Transfer in P-FGM

Assuming that a P-FGM plate has three layers with each layer's thickness of δ_1, δ_2, and δ_3 (m) and the thermal conductivity of λ_1, λ_2, and λ_3 (W/m·temp), and the surface temperatures are ($t_1 - t_4$), Figure 2 shows the layers affected by the temperature of the P-FGM plate.

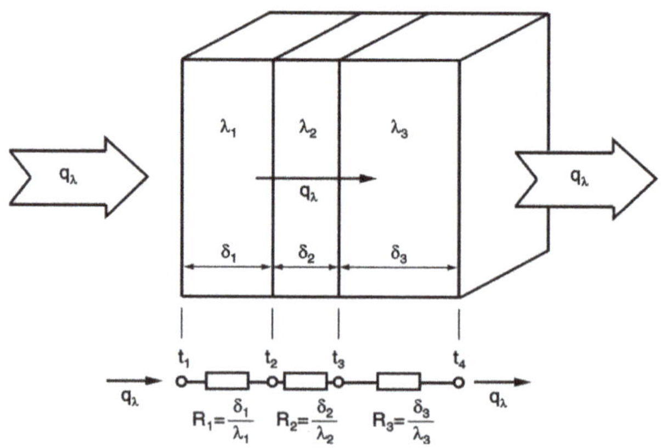

Figure 2. Heat transfer model through P-FGM plate consisting of three layers.

Then, heat flux through the layers is calculated using Equation (2).

$$\left.\begin{aligned} q &= \frac{\lambda_1}{\delta_1}(t_1 - t_2) \\ q &= \frac{\lambda_2}{\delta_2}(t_2 - t_3) \\ q &= \frac{\lambda_3}{\delta_3}(t_3 - t_4) \end{aligned}\right\} \quad (2)$$

From Equation (2), the following relationship is defined.

$$\left.\begin{array}{l} t_1 - t_2 = q\frac{\delta_1}{\lambda_1} \\ t_2 - t_3 = q\frac{\delta_2}{\lambda_2} \\ t_3 - t_4 = q\frac{\delta_3}{\lambda_3} \end{array}\right\} \tag{3}$$

$(t_1 - t_4)$ is calculated using Equation (4), and the heat flux is estimated using Equation (5).

$$(t_1 - t_4) = q\left(\frac{\delta_1}{\lambda_1} + \frac{\delta_2}{\lambda_2} + \frac{\delta_3}{\lambda_3}\right) \tag{4}$$

$$q = \frac{t_1 - t_4}{\frac{\delta_1}{\lambda_1} + \frac{\delta_2}{\lambda_2} + \frac{\delta_3}{\lambda_3}} = \frac{t_1 - t_4}{R_{\lambda_1} + R_{\lambda_2} + R_{\lambda_3}} \tag{5}$$

For the plate with n layers, the heat flux is estimated using Equation (6).

$$q = \frac{t_1 - t_{(n+1)}}{\sum\limits_{i=1}^{n} \frac{\delta_i}{\lambda_i}} = \frac{t_1 - t_{(n+1)}}{\sum\limits_{i=1}^{n} R_i} (\mathrm{W/m^2}) \tag{6}$$

where $\sum\limits_{i=1}^{n} \frac{\delta_i}{\lambda_i}$ is the thermal conductivity of a layer of the plate.

The heat flux of the contact layer of the plate is defined as

$$q = \frac{t_1 - t_{(n+1)}}{\sum\limits_{i=1}^{n} \frac{\delta_i}{\lambda_i}} = \frac{t_1 - t_{(n+1)}}{\sum\limits_{i=1}^{n} R_i} (W/m^2) \tag{7}$$

Therefore, the temperature of the layer is calculated as

$$t_{(k+1)} = t_1 - q\sum_{i=1}^{k} \frac{\delta_i}{\lambda_i} \tag{8}$$

3. Results

3.1. P-FGM Plate with Uniformly Distributed Load

For different FGM plates, bending under uniform load ($q_0 = 1\mathrm{MPa}$), and the responses of simple support with the SSSS boundary condition were estimated. The material properties of the plate are shown in Table 1. Numerical results are compared with those of previous studies to show the reliability of the proposed P-FGM model. Figure 3 shows the P-FGM plate with its dimensions.

Table 1. Material properties of P-FGM plate.

Material Properties	Metal Aluminum (Al)	Ceramic Alumina (Al_2O_3)
E (GPa)	70	380
ν	0.3	0.3

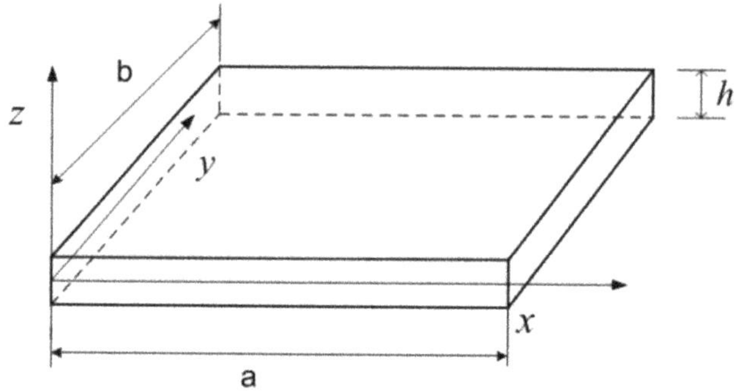

Figure 3. FGM plate with dimensions.

For an elastic square plate with a surface on the x-y plane and a thickness in the z-direction, tensile (compression) modulus E is defined as a function of E = E(z), and Poisson's ratio ν for smooth variation in thickness with a given rule is defined as a function of ν = ν(z). The square plate has a dimension of a × b × h = 100 × 100 × 10 or a/h = 10. As P-FGM has continuously varying mechanical properties, it was assumed that each layer of the P-FGM plate has different material parameters to model the plate in ANSYS Workbench. Using the power-law distribution, the details of 10 layers of the FGM plate were determined, as shown in Table 2.

Table 2. Layer's material properties of the P-FGM plate.

$\frac{z}{h}$	Young's Modulus (Pa)					
	$p = 0$	$p = 1$	$p = 2$	$p = 5$	$p = 10$	$p = \infty$
0.45	3.80×10^{11}	3.65×10^{11}	3.50×10^{11}	3.10×10^{11}	2.56×10^{11}	7.00×10^{10}
0.35	3.80×10^{11}	3.34×10^{11}	2.94×10^{11}	2.08×10^{11}	1.31×10^{11}	7.00×10^{10}
0.25	3.80×10^{11}	3.03×10^{11}	2.44×10^{11}	1.44×10^{11}	8.75×10^{10}	7.00×10^{10}
0.15	3.80×10^{11}	2.72×10^{11}	2.01×10^{11}	1.06×10^{11}	7.42×10^{10}	7.00×10^{10}
0.05	3.80×10^{11}	2.41×10^{11}	1.64×10^{11}	8.56×10^{10}	7.08×10^{10}	7.00×10^{10}
−0.05	3.80×10^{11}	2.10×10^{11}	1.33×10^{11}	7.57×10^{10}	7.01×10^{10}	7.00×10^{10}
−0.15	3.80×10^{11}	1.79×10^{11}	1.08×10^{11}	7.16×10^{10}	7.00×10^{10}	7.00×10^{10}
−0.25	3.80×10^{11}	1.48×10^{11}	8.94×10^{10}	7.03×10^{10}	7.00×10^{10}	7.00×10^{10}
−0.35	3.80×10^{11}	1.17×10^{11}	7.70×10^{10}	7.00×10^{10}	7.00×10^{10}	7.00×10^{10}
−0.45	3.80×10^{11}	8.55×10^{10}	7.08×10^{10}	7.00×10^{10}	7.00×10^{10}	7.00×10^{10}

For the material properties with $p = 2$ and load with the SSSS boundary condition, the deflection and the von Mises stress of the P-FGM plate under a uniform load are shown in Figures 4 and 5.

The non-dimensional result of the modeled P-FGM was similar to that of Ref. [1]. The following equations were used to convert numerical results to non-dimensional ones in ANSYS Workbench [1].

Non-dimensional deflection of FG plate under uniform load was defined as

$$\overline{w}_{\max} = \frac{10h^3 E_c}{p_0 b^4} w\left(\frac{a}{2}; \frac{b}{2}\right) \tag{9}$$

where $w\left(\frac{a}{2}; \frac{b}{2}\right)$ is deflection in the center of the plate.

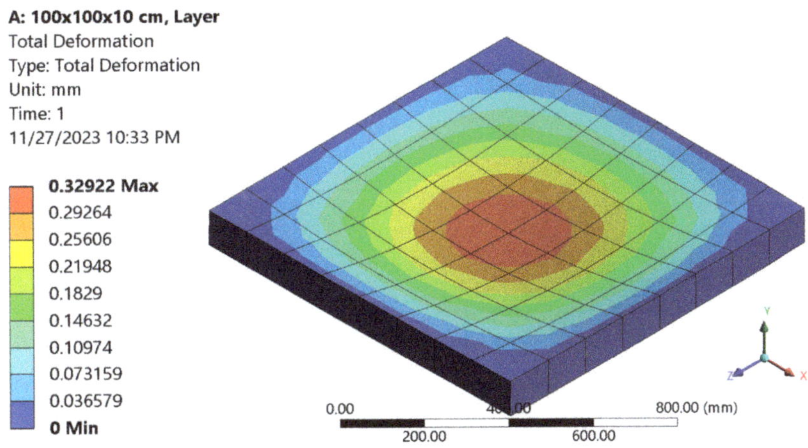

Figure 4. Deflection of P-FGM plate.

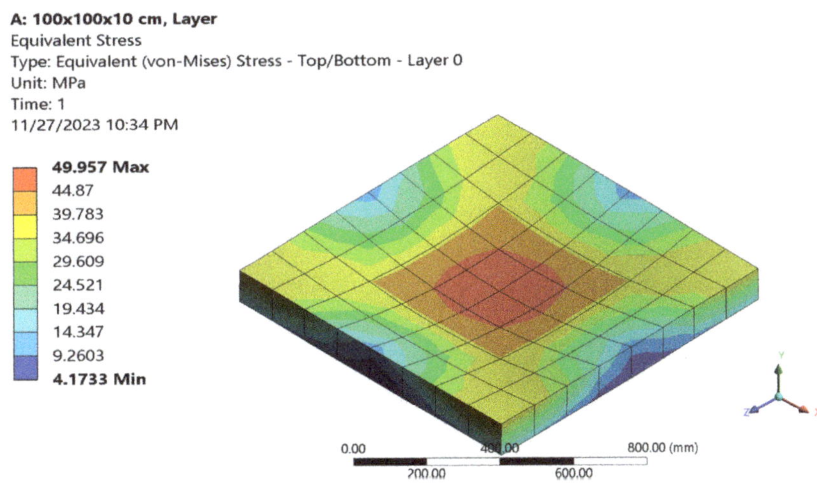

Figure 5. von Mises stress of P-FGM plate.

The non-dimensional von Mises stress of the P-FGM plate under uniform load was calculated as

$$\overline{\sigma}_{xx}(z) = \frac{h}{q_0 a}\sigma_x\left(\frac{a}{2}, \frac{b}{2}, z\right) \tag{10}$$

where $\sigma_x\left(\frac{a}{2}, \frac{b}{2}, z\right)$ is von Mises stress in the center of the plate.

For the P-FGM plate with 10 layers, the result of the modeled plate in this research and the other research was compared, as shown in Table 3.

Table 3. Non-dimensional result of P-FGM plate with 10 layers.

p	Reference	w_{max} (mm)	w_{max} (Non-Dimensional)	σ_{max} (MPa)	σ_{max} (Non-Dimensional)
2	This research	0.32922	1.251	49.957	4.9957
	[1]		1.1909		5.1852
	Error		4.8%		4%

For the P-FGM plate with 20 layers, the result of the modeled plate in this research and the other research was compared, as shown in Table 4.

Table 4. Non-dimensional result of P-FGM plate with 20 layers.

p	Reference	w_{max} (mm)	w_{max} (Non-Dimensional)	σ_{max} (MPa)	σ_{max} (Non-Dimensional)
2	This research	0.32858	1.2486	51.859	5.1859
	[1]		1.1909		5.1852
	Error		4.6%		4%

Remodeling the FG plate with 40 layers, the result of the modeled plate in this research and the other research was compared, as shown in Table 5.

Table 5. Non-dimensional result of FGM plate with 40 layers.

p	Reference	w_{max} (mm)	w_{max} (Non-Dimensional)	σ_{max} (MPa)	σ_{max} (Non-Dimensional)
2	This research	0.32858	1.2486	51.859	5.1859
	[1]		1.1909		5.1852
	Error		4.6%		4%

The results of dimensionless deflection and von Mises stress in different volume ratios were compared in Figures 6 and 7.

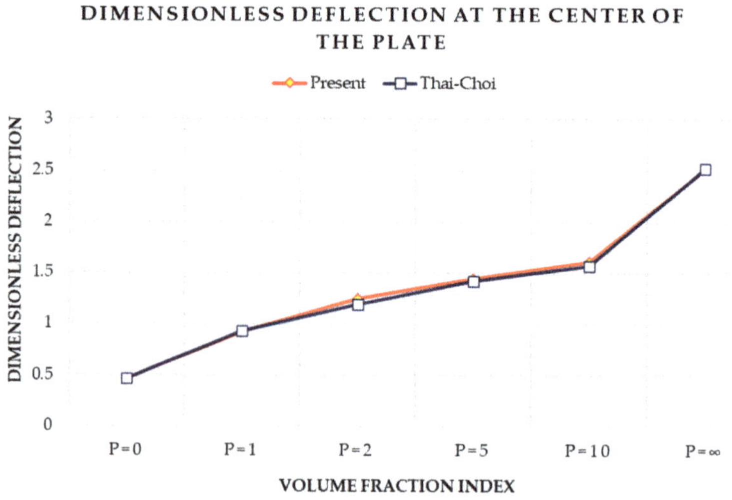

Figure 6. Dimensionless deflection at center of plate.

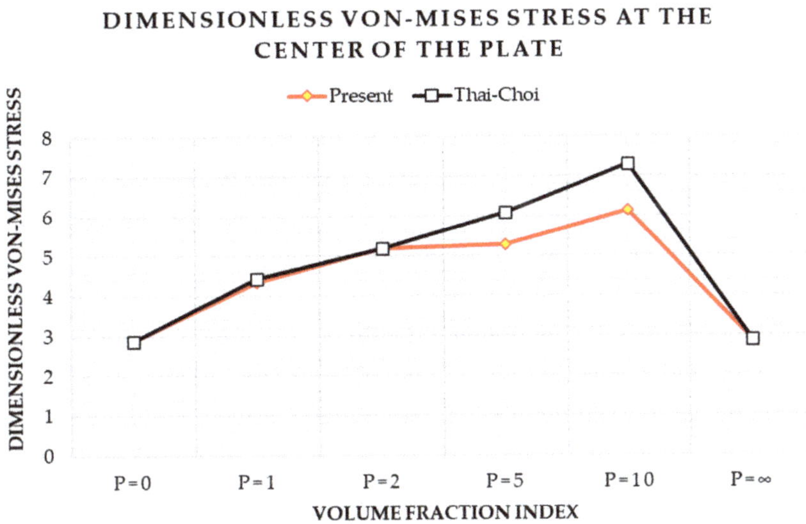

Figure 7. Dimensionless von Mises stress at center of plate.

3.2. *Influence of Temperature*

Assuming the elastic square plate of size 200 × 200 × 10 (mm), the square plate of P-FGM had a ceramic-rich top surface and metal-rich bottom surface, material parameters changed linearly according to the P-FGM model, and volume fraction index $p = 2$, it was found that the P-FGM plate was affected by temperature. The ceramic-rich surface endured heat up to 600 °C, and the metal-rich surface only afforded heat up to 20 °C. Details of the properties of ceramic and metal surfaces of the square plate are presented in Table 6.

Table 6. Properties of ceramic and metal surfaces of P-FGM plate.

Material Properties	Metal Aluminum (Al)	Ceramic Alumina (Al_2O_3)
E (GPa)	70	380
ν	0.3	0.3
ρ (kg/m^3)	2.707	3
Thermal Conductivity k (W/mK)	204	2.09

In this research, a steady heat transfer from one surface of the plate to the other was considered. By dividing the plate into many small layers and using their material properties, a solid plate model with dimensions of 200 × 200 × 10 (mm) was constructed, as shown in Figure 8.

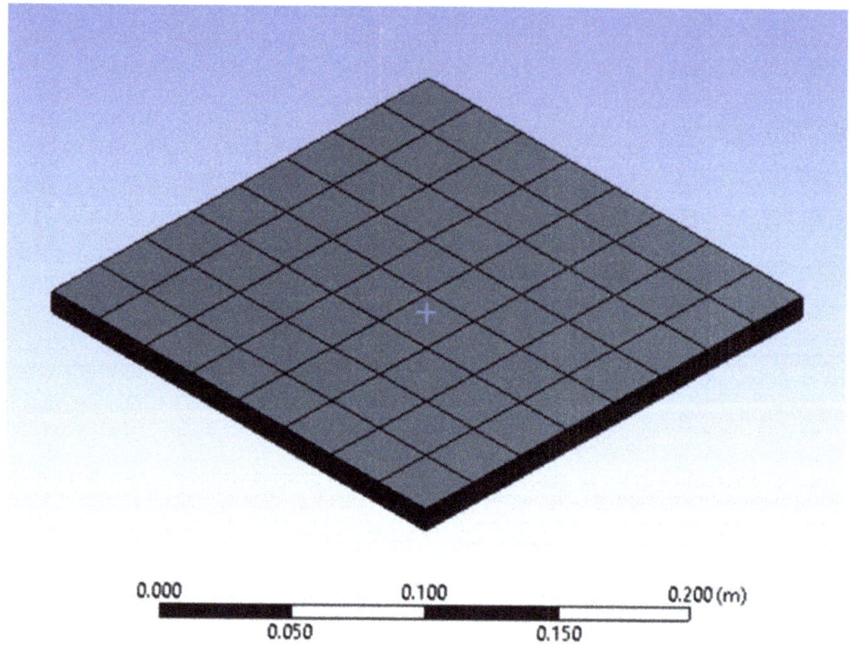

Figure 8. FGM plate model with multi-layers.

Table 7 presents the details of 10 layers of the P-FGM plate of the size of 200 × 200 × 10 (mm).

Table 7. Properties of 10 layers of P-FGM plate in 200 × 200 × 10 (mm).

$\frac{z}{h}$	Volume Faction Index (p = 2)			
	E (MPa)	**ρ (Kg/m^3)**	**ν**	**K (W/mK)**
0.45	1.47	2.985	0.3	12.19
0.35	1.39	2.956	0.3	32.38
0.25	1.31	2.927	0.3	52.57
0.15	1.23	2.897	0.3	72.76
0.05	1.15	2.868	0.3	92.95
−0.05	1.06	2.839	0.3	113.14
−0.15	9.84	2.810	0.3	133.33
−0.25	9.03	2.780	0.3	153.52
−0.35	8.22	2.751	0.3	173.71
−0.45	7.41	2.722	0.3	193.90

Using the steady-state analysis in Ansys Workbench, the heat transfer of the plate was simulated, as shown in Figure 9.

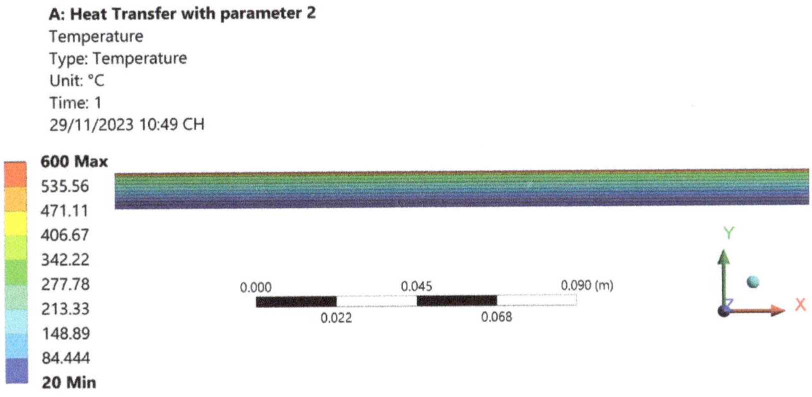

Figure 9. Heat transfer through plate.

The comparison of the results of the software simulation and the theory are shown in Table 8, which presents similar results for the two methods.

Table 8. Comparison of heat transfer results simulated using software and calculated with theory.

$\frac{z}{h}$	Volume Faction Index (p = 2)		
	T (°C) (Project)	**T (°C) (Theory)**	**Error**
0.45	600	600.00	0.00%
0.35	373.09	371.32	0.47%
0.25	287.63	285.64	0.69%
0.15	232.45	230.57	0.81%
0.05	190.26	188.61	0.87%
−0.05	155.09	153.77	0.85%
−0.15	124.20	123.24	0.77%
−0.25	96.031	95.46	0.60%
−0.35	69.60	69.44	0.23%
−0.45	44.21	44.47	0.59%

4. Conclusions

The mechanical behavior and heat transfer properties of the P-FGM plate were explored in this research. Specifically, a model was established according to the power-law distribution. The result showed the maximum deflection and stress at the center of the plate, as well as the heat transfer through the layers of the plate. A comparison of the results of other research proved the reliability of this research's model. It was found that the modeled P-FGM plate helped understand the properties and heat transfer of the plate and provided intuitive results. However, the model needs improvement as it was constructed based on the properties of the P-FGM plate estimated using software. Thus, variables might not be optimized for the model. This produced simple numerical results, which were not enough to explain the various problems in the plate's structure. The research method using ANSYS also needs to be improved for the construction of a better model.

Author Contributions: Conceptualization, T.H.N.; methodology, N.T.N.; software, D.A.L.; validation, T.H.N.; formal analysis, N.T.N.; investigation, N.T.N. and D.A.L.; resources, T.H.N.; data curation, N.T.N. and D.A.L.; writing—original draft preparation, N.T.N.; writing—review and editing, T.N.T.; visualization, T.H.N.; supervision, T.H.N.; project administration, T.H.N. and T.N.T. All authors have read and agreed to the published version of the manuscript.

Funding: This research received no external funding.

Institutional Review Board Statement: Not applicable.

Informed Consent Statement: Not applicable.

Data Availability Statement: Data underlying the results presented in this paper are not publicly available at this time but may be obtained from the authors upon reasonable request.

Acknowledgments: We acknowledge Ho Chi Minh City University of Technology (HCMUT), VNU-HCM for supporting this study.

Conflicts of Interest: The authors declare no conflict of interest.

References

1. Thai, H.-T.; Choi, D.-H. A simple first-order shear deformation theory for the bending and free vibration analysis of functionally graded plates. *Compos. Struct.* **2013**, *101*, 332–340. [CrossRef]
2. Singha, M.K.; Prakash, T.; Ganapathi, M. Finite element analysis functionally graded material plates under transverse load. *Finite Elem. Anal. Des.* **2011**, *47*, 453–460. [CrossRef]
3. Cooley, W.G. Application of Functionally Graded Materials in Aircraft Structures. Master's Thesis, Air Force Institute of Technology, Dayton, OH, USA, 2005.
4. Bich, D.H.; Do Long, V. Non-linear dynamical analysis of imperfect functionally graded material shallow shells. *Vietnam. J. Mech.* **2010**, *32*, 65–79. [CrossRef] [PubMed]

engineering proceedings

MDPI

Proceeding Paper

Bioinformatics of Serious Leisure in Playing Video Games and Learning English as a Sustainable Activity †

Li-Shiue Gau [1,*], Chung-Hsing Huang [2] and You-Ni Gau [3]

1 Department of Leisure and Recreation Management, Asia University, Taichung 413305, Taiwan
2 Department of Business Administration, Asia University, Taichung 413305, Taiwan; xing888@gmail.com
3 Chu-Jen Junior High School (CJJH), Taichung 403002, Taiwan; garygautw@gmail.com
* Correspondence: lsgau@asia.edu.tw; Tel.: +886-4-23323456 (ext. 20057)
† Presented at the IEEE 5th Eurasia Conference on Biomedical Engineering, Healthcare and Sustainability, Tainan, Taiwan, 2–4 June 2023.

Abstract: This study aimed to investigate why video gaming tends to become a serious leisure activity more easily than English learning, using a bioinformatics approach. Methods included interviews with 13 students aged 10–13 and an experimental research design with EEG (electroencephalograph) data from five students during rest, gaming, and English learning prior to and after a 4-week treatment. Interviews revealed that cram schools, school assignments, and future career aspirations influenced English learning habits. EEGs showed greater situational involvement in video games, but no significant improvement in English learning after the treatment. Encouraging English learning was challenging, perhaps due to participants' busy schedules.

Keywords: EEG; learning English; serious leisure; video game; elementary school

1. Introduction

In December 2019, several cases of coronavirus disease (COVID-19) were reported, raising global concerns about the threat of the new pandemic to public health [1]. In order to prevent the epidemic, governments imposed a series of strict prohibitions on gatherings [2], thereby increasing the chance for people to stay indoors or study from home. This has had a huge impact on changes in leisure patterns and likely led to a higher demand for digital leisure [3].

During the COVID-19 pandemic, pupils tend to be involved with digital leisure activities such as video games too seriously and hurt their eyesight. Given the development of digital technologies with unlimited access to the Internet, people's leisure lives have been changed greatly. Computers, video game consoles, smartphones, and other devices are adapted to daily leisure and entertainment activities [4]. However, the concern is that such video or online games are like a paradise to escape, causing pupils to become hooked and lose all notion of time.

Parents are worried about this type of unhealthy leisure habits and try to ask children to engage in other more positive leisure activities such as learning English. However, pupils wailed reluctantly at this request because seemingly learning English causes little pleasure, but rather induces pressure and is more like a job, a task, or a duty than leisure for children. This inspires the current research to examine why video gaming is an activity that more easily becomes a serious leisure activity than English learning from an aspect of bioinformatics. Biological data from electroencephalographs (EEGs) were collected while playing video games and learning English along with participant interviews.

Citation: Gau, L.-S.; Huang, C.-H.; Gau, Y.-N. Bioinformatics of Serious Leisure in Playing Video Games and Learning English as a Sustainable Activity. *Eng. Proc.* **2023**, *55*, 71. https://doi.org/10.3390/engproc2023055071

Academic Editors: Teen-Hang Meen, Kuei-Shu Hsu and Cheng-Fu Yang

Published: 8 December 2023

2. Literature Review

2.1. Sustainable Serious Leisure

Serious leisure is the systematic pursuit of a hobbyist core activity that people find so interesting and fulfilling that they launch themselves on a career centered on acquiring and expressing a combination of knowledge, affective involvement, and skills [5]. Previous studies indicated that serious leisure may generate flow experience [6], increase leisure satisfaction [7], and positively associate with eudemonic or subjective well-being [8]. However, those who are deeply involved in serious leisure may, on the other hand, face a risk of indulging in an excessive amount of such leisure participation [5]. Maladaptive patterns of excessive leisure activities might negatively influence health and ordinary lives.

If a strategy of balanced leisure patterns in various leisure activities can be adopted, multiple leisure activities may provide a healthy balance for distraction from some possible excessive leisure [9]. Since bilingualism is one of the major education policies to be implemented by the year 2030 by the Taiwanese government because of globalization, the cultivation of serious leisure in learning English for children or young teenagers seems a good choice. Comparison of these two different types of serious leisure in terms of playing video games and studying English is rare in prior studies. This study tries to fill the literature gap with two major objectives. First, it explores why cultivating a leisure habit of studying English is not an easy task. Second, it uses bioinformatics to compare these two leisure activities and derive useful information about how to design sustainable serious leisure.

Sustainable serious leisure refers to endurable leisure benefits of good health and well-being that are essential to sustainable development [10]. Earlier studies have shown that serious leisure could be practices of social sustainability [11,12]. For example, a group of old volunteers who participated in performing traditional Korean dance as their serious leisure activities reported positive experiential characteristics of the involvement including a perceived self-achievement of perseverance and enhancement of self-identity, leading to successful aging [12].

2.2. Learning English as a Serious Leisure Activity

Leisure experiences can be referred to four types of a 4-E model [13]: Educational, Entertainment, Esthetic, and Escapist experiences. While playing video games is more likely seen as a way of seeking refuge to escape from daily hassles, pupils learning English might experience educational epistemic satisfaction, as it is more like one of the liberal arts hobbies. The intrinsic goal of learning English is to acquire the knowledge and understanding of the language as an end in itself [6].

Within this type of cognitive liberal arts leisure, it is hard to generate a flow experience and therefore usually attracts little attention [6]. Without practices of perseverance, there would be little chance of cultivating a habit of learning English as serious leisure. In addition to the inherent value, however, some other possible practical reasons and instrumental purposes can become aspirations for learning English. For instance, fluency in English can facilitate social inclusion, allowing individuals to become a part of a preferred or envisioned community [14]. This would serve as a compelling incentive for an individual to learn English. Another possible example is to pursue a desired career where English competence is required.

2.3. Advantage and Disadvantage of Electroencephalograph

Little or lack of research thus far was conducted to investigate alterations in electroencephalogram (EEG) power spectra when playing video games and studying English. EEG power spectra are typically examined through frequency range consisting of delta (0.5–4 Hz), theta (4–8 Hz), alpha (8–13 Hz), and beta (13–30 Hz) [15–17]. An alpha frequency range is considered relevant to a feeling of relaxation without attention whereas beta power is related to the activation of the brain in active processing when on task [15,16,18,19]. The theta rhythm is brought on by early drowsiness or in a state of deep meditation [17,18] while the delta rhythm has been associated with the absence of consciousness or seen in the state of deep sleep [18,19].

The advantage of EEG is its noninvasive neuroimaging technique to investigate electrical activity in the brain [16,18,20], providing objective evidence. This is safe and straightforward to perform with the useful advantage that EEG can synchronously measure changes in brain activity in real time [18]. However, EEG's disadvantage is that the brain's activity at the individual neuronal level cannot be easily captured, causing failure to understand the nuanced and complex cognitive processes happening in the brain [20]. Another disadvantage is the sensitivity of EEG to the participant's facial movements, body shaking, state of consciousness, physical and mental activity, and the presence of different biological and environmental stimuli [17].

3. Research Method

3.1. Equipment and Software

A ProComp Infiniti encoder manufactured by Thought Technology Limited was utilized to acquire 1-channel EEG data from an electrode cup placed on the forehead and referenced to linked clip electrodes fixed on two earlobes. Before the electrodes were secured, a small amount of NuPrep gel was applied to abrade the skin, removing dead skin, sweat, and other impurities that might interfere with the EEG signal. Then, Ten20 conductive paste was filled in the electrode cup and put on the ear clips to help affix the electrodes on the forehead and ear lobes to improve the quality of the perceived EEG signal.

The EEG signal was transmitted by electrode cables and a DIN cable to the EEG-Z sensor, which was then connected to the encoder, TT-USB interface device, and ultimately to a personal computer [21] (Figure 1). Prior to recording the EEG signal, impedance checks were conducted for all electrodes. Data analysis was performed using Thought Technology Limited's BioGraph Infiniti software (Version 6.1).

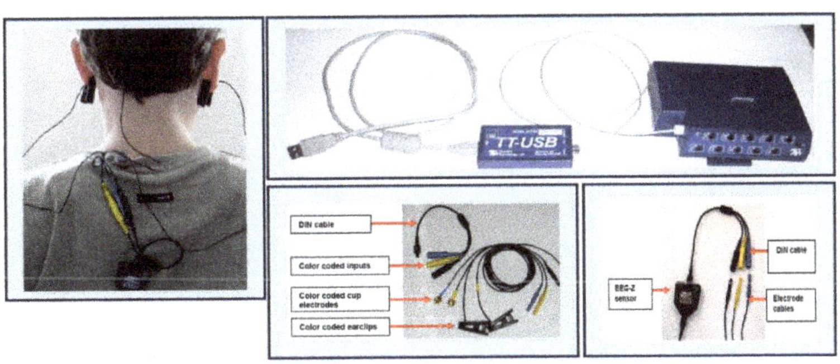

Figure 1. EEG data collection equipment reference [21], on pages 4, 12, 13, 18.

3.2. Data Collection

Two methods were used for data collection. First, 13 students between 10 and 13 years old were recruited for interviews. Second, among the 13 interviewees, 5 students participated in an experimental study with consent from their parents. In the interviews, outline questions (Table 1) attempted to determine the first thing that participants would like to do upon returning home from school and whether this was recognized as a positive or negative habit. The next question tried to understand participants' leisure habits and see whether the habits involved learning English. For answering the third question, participants recalled their memories to describe the process of how those habits were developed and to compare whether there were differences between good and bad habits. The final question focused on participants' interests, attitudes, and opinions toward the pursuit of learning English as a leisure habit and requested some explanations about the responses.

Table 1. Interview outline.

1. What is the first thing you would like to do when you come home after school? Are you aware of any good or bad habits?
2. What is your leisure habit? Is there any related to learning English?
3. How do you cultivate these habits? Is there any difference between good and bad habits?
4. How do you think of the cultivation of learning English as a leisure habit? Are you interested in this type of leisure habit? Why or why not?

For the experimental research design, EEGs were collected while resting, playing video games, and learning English. Each time, the duration was one and a half minutes, and, in total, there would be four collection times for every participant: resting with eyes closed, resting with eyes open, playing video games, and learning English. During the EEG recording, participants were instructed to be quiet, remain still, and maintain a comfortable posture without moving or shaking their body or facial muscles.

Prior to and after the four-week period of treatment, data collection was executed. For a period of four weeks, the treatment was provided with studying materials of English extracted from Studio Classroom Magazine in reading, speaking, listening, or writing five times a week. It was expected to take approximately between 20 and 25 min every time to study the materials. Participants were encouraged to cultivate a habit of learning English.

3.3. Data Analysis

A mode of auto-rejection setting in the software BioGraph Infiniti was used to keep or reject segments of sections for EEG analyses. This procedure primarily placed artifact rejection segments over sections of noisy signals for the purpose of generating reliable statistics [21]. A rejection threshold of 20 was entered initially as the upper limit for normal EEG microvolt levels. If the percentage of rejection durations were higher than 50% (included), the threshold was made more lenient to, for example, 25 or 30.

4. Results and Conclusions

4.1. Interviews

The results of interviews showed that most children started to do homework first thing when they arrived home after school. Bad habits included the overuse of smartphones. Some participants mentioned that a habit was developed day by day and gradually formed over time. It appeared that seemingly acquiring a negative habit was easier than establishing a positive one. The major factors influencing a habit of serious leisure in learning English included whether students went to a cram school, whether there were assignments required by their schools, and whether students had a vision of using English in their future careers.

4.2. Experimental Study

For the experiment, participants showed that it was hard to change their habits in learning English through encouragement from others because their schedule was tight. Two participants studied 75% of the treatment materials while the other three participants only studied approximately 25%. All participants mentioned that they went to English cram schools regularly.

The EEG results demonstrated higher theta and alpha wave activity when participants closed their eyes compared to when their eyes were open (Table 2), indicating that the equipment used was valid and possessed nomological validity. Alpha waves are most prominently observed when the eyes are closed, and their presence is typically attenuated through eye opening [17]. Paired sample t tests between eyes open and eyes closed in the ratio of alpha to beta showed statistical significance ($p < 0.05$) with higher ratio values when eyes were closed than those when eyes were open.

Table 2. EEG in resting.

Case	EEG	Resting (Prior)		Resting (After)	
		Eyes Open	Eyes Closed	Eyes Open	Eyes Closed
case 1	Delta	12.25	15.87	10.57	16.51
	Theta	9.17	11.71	7.74	11.26
	Alpha	6.91	18.64	5.67	18.97
	Beta	10.52	11.40	8.75	10.41
case 2	Delta	17.5	16.34	16.61	14.61
	Theta	9.36	10.93	9.89	10.75
	Alpha	6.98	9.64	6.92	7.95
	Beta	7.25	6.98	6.77	6.39
case 3	Delta	13.59	19.76	11.17	16.1
	Theta	16.29	27.22	15.26	21.54
	Alpha	14.84	22.08	14.18	19.91
	Beta	11.74	11.80	11.78	8.61
case 4	Delta	10.51	11.72	10.73	11.75
	Theta	8.90	12.04	9.48	12.70
	Alpha	6.70	13.41	7.62	12.93
	Beta	5.69	6.37	5.83	6.57
case 5	Delta	11.34	14.68	10.39	14.46
	Theta	9.68	15.35	10.39	15.39
	Alpha	7.91	13.82	8.10	14.31
	Beta	9.22	10.63	8.25	8.38

The comparison between playing video games and learning English in EEGs showed that participants tended to be more easily involved with playing games. The threshold and rejection percentage due to noise in data analysis were lower in playing video games than those learning English because it seemed unconsciously three male participants quickly concentrated on games. However, it seemed to be on a case-by-case basis whether the pupils would feel relaxed or excited when playing video games or learning English when checking the ratio of beta to alpha. One participant seemed to space out when learning English because the materials were difficult.

The prior and after comparison showed no significant difference, perhaps indicating it was not easy to cultivate a habit of learning English and have it become a serious leisure activity within four weeks. Further analyses showed that one participant had very high consistency between prior and after data (correlation coefficients between 0.971 and 0.994) and among the three conditions (for prior data between 0.992 and 0.997; for after data between 0.987 and 0.991). This seemed to indicate that this participant might have a stable personality, which might play the role of a disturbing factor in EEG research. Figure 2 was an example showing the relationship (0.997) between the EEG of prior data about learning English (prior_Eng) and EEG of prior data about playing a video game (prior_video) for this participant. Figure 3 was another example showing the relationship (0.971) between prior and after EEG data about learning English (prior_Eng, after_Eng) for this participant case.

4.3. Conclusions

Whether participants have already had regular habits in learning English or playing video games after school would influence the results of this research. Additionally, playing video games might more easily and quickly generate a state of flow experiences with high concentration than learning English. However, it seemed unclear that concentration in flow would cause relaxation or a feeling of stimulation [18]. As for learning English, it seemed also unclear if it caused stress or if students just spaced out. These various possible reasons would make it complex and difficult to interpret EEG data. A more strict research design is required to overcome these limitations.

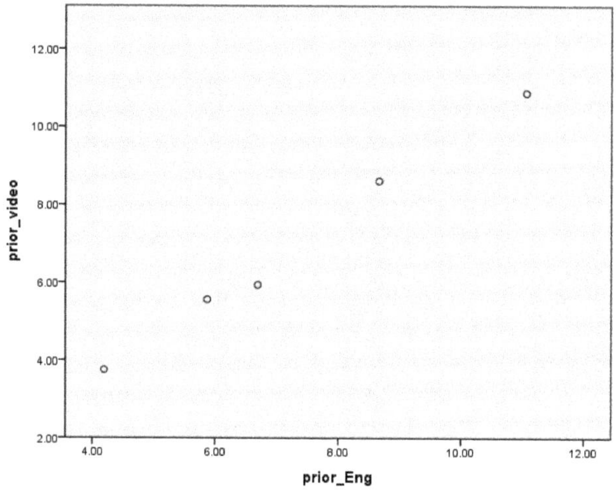

Figure 2. Correlation between prior EEG in learning English (prior_Eng) and playing a video game (prior_video) for one of the participants.

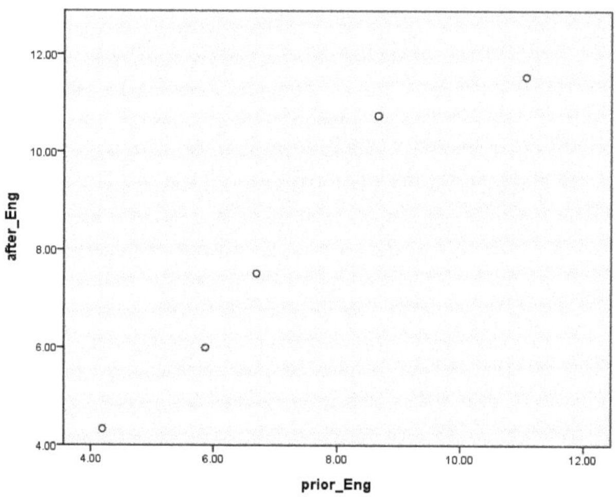

Figure 3. Correlation between prior and after EEG in learning English for one of the participants.

Author Contributions: Conceptualization, L.-S.G., C.-H.H. and Y.-N.G.; methodology, L.-S.G. and C.-H.H.; software, L.-S.G.; validation, L.-S.G., C.-H.H. and Y.-N.G.; formal analysis, L.-S.G.; investigation, L.-S.G., C.-H.H. and Y.-N.G.; resources, L.-S.G., C.-H.H. and Y.-N.G.; data curation, L.-S.G., C.-H.H. and Y.-N.G.; writing—original draft preparation, L.-S.G.; writing—review and editing, L.-S.G., C.-H.H. and Y.-N.G.; visualization, L.-S.G. and C.-H.H.; supervision, L.-S.G.; project administration, L.-S.G.; funding acquisition, L.-S.G. All authors have read and agreed to the published version of the manuscript.

Funding: This study appreciates the funding for the EEG equipment from Taichung Extinguishing Enterprise (TEE), Asia University (105-asia-02), and the National Science and Technology Council (MOST 106-2410-H-468-020) in Taiwan.

Institutional Review Board Statement: Ethical review and approval were waived for this study due to the lowest risk. Any risk suffered by the research subjects is not higher than those who do not participate in the study.

Informed Consent Statement: Informed consent was obtained from all subjects involved in the study.

Data Availability Statement: Data is unavailable due to privacy.

Conflicts of Interest: The authors declare no conflict of interest.

References

1. Zeydi, A.E.; Ghazanfari, M.J.; Sanandaj, F.S.; Panahi, R.; Mortazavi, H.; Karimifar, K.; Karkhah, S.; Osuji, J. Coronavirus disease 2019 (COVID-19): A literature review from a nursing perspective. *BioMedicine* **2021**, *11*, 5. [CrossRef]
2. The News Lens. "Taiwan's Level 3 COVID-19 Alert Expanded Nationwide. Available online: https://international.thenewslens.com/article/151210 (accessed on 13 March 2023).
3. López-Sintas, J.; de Francisco, L.R.; Álvarez, E.G. The nature of leisure revisited: An interpretation of digital leisure. *J. Leis. Res.* **2015**, *47*, 79–101. [CrossRef]
4. López-Sintas, J.; Rojas-DeFrancisco, L.; García-Álvarez, E. Home-based digital leisure: Doing the same leisure activities, but digital. *Cogent Soc. Sci.* **2017**, *3*, 1309741. [CrossRef]
5. Stebbins, R.A. Leisure reflections—Addiction to leisure activities: Is it possible? *LSA Newsl.* **2010**, *86*, 19–22.
6. Stebbins, R.A. Leisure reflections—Flow in serious leisure: Nature and prevalence. *LSA Newsl.* **2010**, *87*, 21–23.
7. Song, W.I.; Ahn, B.W. Effect of outdoor leisure activity participation on leisure attitude, serious leisure, recreational specialization, and leisure satisfaction. *Societies* **2023**, *13*, 155. [CrossRef]
8. Kono, S.; Ito, E.; Gui, J. Empirical investigation of the relationship between serious leisure and meaning in life among Japanese and Euro-Canadians. *Leis. Stud.* **2020**, *39*, 131–145. [CrossRef]
9. Gau, L.S.; Chen, M.I.; Chu, H.Y.; Chu, H.T.; Liu, M.S. EEG and ECG in reducing smartphone overuse by multiple leisure activities. In Proceedings of the 2020 IEEE 2nd Eurasia Conference on Biomedical Engineering, Healthcare and Sustainability (ECBIOS), Tainan, Taiwan, 29–31 May 2020. [CrossRef]
10. UNDP. Goal 3: Good Health and Well-Being, Sustainable Development Goals. Available online: https://www.undp.org/sustainable-development-goals (accessed on 13 March 2023).
11. Heley, J.; Jones, L. Growing older and social sustainability: Considering the 'serious leisure' practices of the over 60s in rural communities. *Soc. Cult. Geogr.* **2013**, *14*, 276–299. [CrossRef]
12. Lee, S.; Heo, J.; Chun, S.; Lee, J.H. Voluntary dance performance as serious leisure and successful ageing. *Leis. Stud.* **2020**, *39*, 1–15. [CrossRef]
13. Pine II, B.J.; Gilmore, J.H. *The Experience Economy*; Harvard Business School Press: Boston, MA, USA, 2011.
14. Kubota, R. Learning a foreign language as leisure and consumption: Enjoyment, desire, and the business of eikaiwa. *Int. J. Biling. Educ. Biling.* **2011**, *14*, 473–488. [CrossRef]
15. Ceccarelli, L.A.; Giuliano, R.J. Elevations in EEG power spectra during the recall of sport failure amongst University athletes. *Psyarxiv* **2020**. [CrossRef]
16. Jebelli, H.; Hwang, S.; Lee, S.H. An EEG signal processing framework to obtain high quality brain waves from an off-the-shelf wearable EEG device. *J. Comput. Civ. Eng.* **2018**, *32*, 4017070. [CrossRef]
17. Nayak, S.; Anilkumar, A.C. EEG Normal Waveforms. Available online: https://europepmc.org/article/nbk/nbk539805#free-full-text (accessed on 13 March 2023).
18. Lim, S.; Yeo, M.; Yoon, G. Comparison between Concentration and Immersion Based on EEG Analysis. *Sensors* **2021**, *19*, 1669. [CrossRef] [PubMed]
19. Frohlich, J.; Toker, D.; Monti, M.M. Consciousness among delta waves: A paradox? *Brain* **2021**, *144*, 2257–2277. [CrossRef] [PubMed]
20. Agarwal, S.; Dutta, T. Neuromarketing and consumer neuroscience: Current understanding and the way forward. *Decision* **2015**, *42*, 457–462. [CrossRef]
21. Thought Technology Limited. *EEG Suite: Reference Manual, for Use with BioGraph Version 6.1 or Later*; Thought Technology Ltd.: Montreal West, QC, Canada, 2019.

engineering proceedings

MDPI

Proceeding Paper

Application of Concurrent Design Strategy on the Design of Multifunction Hydrotherapy Bucket †

Shih-Wen Hsiao [1] and Chu-Hsuan Lee [2,*]

[1] Department of Industrial Design, National Cheng Kung University, Tainan 70101, Taiwan; swhsiao@mail.ncku.edu.tw
[2] Department of Culture Creativity and Digital Marketing, National United University, Miaoli 360001, Taiwan
* Correspondence: hsuan6389@gmail.com
† Presented at the IEEE 5th Eurasia Conference on Biomedical Engineering, Healthcare and Sustainability, Tainan, Taiwan, 2–4 June 2023.

Abstract: This study employs a systematic concurrent design approach to achieve product outcomes. It involves market research and analysis of products from various manufacturers. Through functional analysis and problem identification, TRIZ theory and a morphological diagram method are utilized for finding solutions. The best design scheme is then determined using the PUGH method. Additionally, a 3D model is created to present the concept that meets design goals, aiding the development of related products and boosting market share for home hydrotherapy buckets.

Keywords: concurrent design; morphological diagram method; PUGH method; TRIZ theory

Citation: Hsiao, S.-W.; Lee, C.-H. Application of Concurrent Design Strategy on the Design of Multifunction Hydrotherapy Bucket. *Eng. Proc.* **2023**, *55*, 72. https://doi.org/10.3390/engproc2023055072

Academic Editors: Teen-Hang Meen, Kuei-Shu Hsu and Cheng-Fu Yang

Published: 11 December 2023

1. Introduction

Rehabilitation is undoubtedly a difficult road for patients with physical disabilities or limb injuries. Regularly going to institutions for rehabilitation has become routine. Therefore, when operating cold rehabilitation equipment and the arduous rehabilitation process, they often feel uncomfortable. As Kim opinions [1], the fundamental principle of a human-centered design is premised on maintaining dignity. Therefore, when designing this functional physiotherapy rehabilitation equipment, the primary consideration is the user's psychological feelings. A successful product should be able to solve consumers' problems, or bring a more convenient life to the public, so designers and planners must identify problems and implement effective solutions when designing products and strive to develop solutions that can bring consumers high-quality products with different benefits [2]. At present, most of the multi-functional rehabilitation hydrotherapy pools provided by rehabilitation institutions are large-sized products used by many people, while personal rehabilitation hydrotherapy buckets take up little space but often have a single function. Considering that users may not be able to afford the huge cost of large-sized products and may not have sufficient space, this design case reduces the size of the multi-functional hydrotherapy bucket and modulates it for home use. In this way, the burden on the body, movement, and use of the rehabilitated person can be effectively reduced. At the same time, the developability and popularity of the product can be considered, and production costs and product size can be reduced as much as possible to benefit more users and enable them to perform personal rehabilitation with dignity. According to Cooper and Kleinschmidt [3], key factors for the success of a product to be launched are the design of the product itself, the characteristics of the product, the production methods, techniques aimed at producing a quality product, the implementation of collaborative production and design, and the benefits that the product brings to the user. In addition, a good product should carefully evaluate every detail during the development phase. Therefore, to study the competitiveness of products, we should start with the market performance, analyze the direct factors such as cost, price and quality, and then analyze the indirect factors such as resources, materials, capital and technology, and finally, find the specific reasons and ways to improve competitiveness. In this study, a concurrent design strategy was used as the main approach for the multifunctional hydrotherapy bucket, and different research methods were used to analyze and propose solutions for each problem in a clear box process, in the hope that the selected design solutions could effectively improve the problematic points of the product.

2. Background

2.1. Hydrotherapy

Lynda Huey, a Santa Monica, Calif.-based kinesiologist and sports trainer, considers that the 1970s were dominated by running and the 1980s by aerobics, but the 1990s belong to water exercise [4]. "Swimming is no longer the only form of activity for the water", says Huey. In fact, aquatic fitness has become popular because of the publication of "Deep Water Exercise for Health and Fitness". Author Glenn McWaters is a former Marine combat helicopter pilot who tried walking in water using a flotation belt while recovering from a thigh wound sustained in Vietnam [5]. He considers that deep-water running can improve exercise efficiency, because the resistance encountered in water movement is 12 times that of air movement. "A 20-min workout in the water is the equivalent of a 30- to 35-min workout on land", McWaters explains. Aquatic fitness was initially used in injured athletes but has since been applied to other patient users in orthopedics, rheumatology, and cardiology. This kind of program usually focuses on improving aerobic capacity, and the principles of traditional hydrotherapy are used to increase the difficulty of walking in water. Therefore, Aqua jogging has popped up, with programs covering many different aquatic fitness exercises, such as shallow water walking, aqua-aerobic, aqua-steps, aqua-nastics, hydrorobics, hydropower, and aqua-dynamics. The greatest advantage of rehabilitation in a hydrotherapy pool is that users are not afraid of falling, and reducing fear will help arouse their willingness to complete activities. Many experts in the field of rehabilitation believe that changes in the motor skills of the user can be seen when balance training in water, and if balance training in a hydrotherapy pool is effective, the chances of successful rehabilitation on land will increase; this argument is also confirmed in clinical research [6]. In addition, "Coaches found that athletes were coming out of rehabilitation in better shape than before they were injured", says Ruth Sova [7], president of the Aquatic Exercise Association in Port Washington, Wis., a research organization for professionals in aquatic fitness and therapy. Hence, we know that running in water and many other aquatic exercises trace their roots in therapy. In addition, there are several advantages of using hydrotherapy: improved movement ability by increasing joint range of motion and relaxation; use of an easy and simple method to strengthen weak muscles; movement can be performed with less force; low mechanical shock enhances cardiovascular endurance; stability, equilibrium, and adequate balance reactions can be facilitated; increased proprioceptive and exteroceptive input; reduced subjective complaints, such as shoulder pain, motivation and self-esteem can be increased; and body image can be improved [8]. To sum up, aquatic fitness in a hydrotherapy pool can provide a pleasurable and enjoyable situation for users and can encourage them to be active, reduce the distraction of rehabilitation on dry land, and at the same time, have substantial benefits for physical rehabilitation.

2.2. Lifestyle Change

Due to the large-scale spread of the Coronavirus Disease (COVID-19), many countries have created policies and legislation to reduce social interaction and curb the massive spread of this pandemic disease. Governments in most countries have imposed a blockade on all non-essential infrastructure. These rules include closing sports clubs, fitness centers, and community sports fields. Additionally, social gatherings of more than two people are banned, further limiting opportunities to exercise and exercise together. Therefore, many people have replaced organized physical activity with individual home-based workouts. Home-based workouts are definitely an easier and time-saving approach for those who already have the required equipment. They are able to continue exercising without assistance and instructions in this situation, but adequate space at home is necessary [9]. From this, it can be seen that if we want to design a home-type multifunctional hydrotherapy bucket, how to reduce storage space is also one of the key points that must be considered. Some experts even suggest people can take online exercise classes and use video- or app-guided aerobics training at home, because there are many useful workout videos available that can assist people in exercising on their own [10].

3. Materials and Methods

This research uses a concurrent design to solve the black box problem in design, and considers the practicality of the multi-functional hydrotherapy bucket, the storage capacity of household types, and safety during exercise. This research constructs a systematic, efficient, and fair design process.

3.1. Objectives Tree Method

The objectives tree method can effectively help designers to clarify the main objectives of the design and effectively organize consumer needs to make the design process more fluid. In addition, the ambiguity between consumers and the design team is minimized. The objectives tree method is presented in a graphical format so that the interrelationship between design goals and secondary goals is clearly presented [11].

3.2. TRIZ Theory

TRIZ is a combination of the initials of the Russian word "Teoriya Resheniya Izobreatatelskikh Zadatch", proposed by Soviet engineer Altshuller in 1946. TRIZ theory stands for "Theory of Inventive Problem Solving"; its English translation is "Theory of Inventive Problem Solving". The principle is that inventions should be universal and that certain principles inherent in creative inventions are usually the same or similar. Therefore, Altshuller uses these assumptions as the basis of TRIZ theory and integrates these principles. TRIZ theory compiles and summarizes 39 engineering parameters and 40 principles for solving inventive problems, and emphasizes that inventions or innovations can be made in accordance with certain procedures and steps [12]. As a result, TRIZ theory is regarded as a comprehensive, systematic solution to the problem of the invention and realization of technical innovation.

3.3. Morphological Chart Method

The morphological chart method is a method of analyzing and systematizing all generated concepts, and this method is based on the functional analysis method [13,14]. A new component is composed of abstract parameters built from concrete technical principles. The parameters and components are expanded and confirmed in parallel on the final morphology chart. Finally, the components are selected from each parameter, and each component that is combined together is considered a solution to the problem.

3.4. PUGH Method

The PUGH method is completed by comparing many problem options and is often used to select the best design solution. All question options have a weight that is used to compare with the datum [15,16]. If the question option is better than the datum, it is indicated by "+"; conversely, if it is worse than the datum, it is indicated by "−". If the question option is difficult to compare with the datum, it is indicated by "S". Finally, the relative scores of each option are used to obtain an optimal result.

4. Case Study

Most of the users of the multi-functional hydrotherapy bucket are physically disabled or disabled. Therefore, in addition to the free entry and exit of the bucket, it is more important to include multiple rehabilitation functions without taking up space. In this case, products of different brands on the market were selected for analysis and comparison, and their problem points were summarized using functional analysis. The solution to the problem was found using TRIZ theory and the morphological chart method. Finally, the best design scheme was obtained using the PUGH method.

4.1. Market Research

Generally speaking, market research is a good way to understand a consumer market. Through it, designers can better understand a market's needs. Therefore, this research starts with market research and current status analysis of related products by collecting data, so as to accurately understand the market demand for hydrotherapy pools and hydrotherapy buckets in the early stage of design. The market data reference sources in this research for the preliminary work are related product web pages, books, journals, as well as field visits to merchants, special education schools, and physiotherapy rehabilitation clinics to obtain the most authentic information. In addition to grasping the functional structure of the existing hydrotherapy buckets, the main design and improvement guidelines can also be obtained from the problem points. This study collects hydrotherapy pools and buckets launched in different countries. After collecting data and compiling the results, it was found that the small or individual hydrotherapy buckets on the market do not take up space, but most of them are single-use and single-function. Except for slightly different materials, the appearance has not changed much. The main function is rehabilitation, to relieve physical fatigue and soreness, and there are no other additional functions that can be used.

4.2. Design Specification

In order to clearly understand the hierarchical relationship between design goals, the results of market research, product analysis and field visits are drawn into a target tree diagram with a master–slave relationship (as shown in Figure 1).

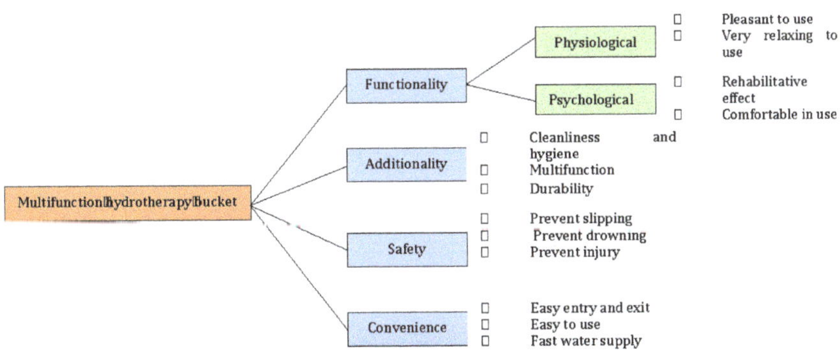

Figure 1. The total items of the objectives tree.

Then the TRIZ method was used to find out the parameters of improving features and worsening features. With the above methods, a set of design criteria was developed (Table 1), with D (Demand) indicating the design requirement as the "need" for essential functions and W (Wish) as the "expectation" for non-essential functions. Since most of the users who need hydrotherapy rehabilitation have physical disabilities or physical injuries, this product will consider both physical and psychological factors in terms of functionality, expecting that users can not only relieve physical pain, but that they also feel relaxed during the rehabilitation process. In this study, the focus was on the safety, functionality, and ease of use of the device, such as preventing slips, drowning, and injuries, which were considered as desired goals.

Table 1. Design specifications.

Functionality		Additionality		Safety(Prevent)		Convenience	
Pleasant	W	Hygiene	D	Slipping	D	Entry and exit	D
Relaxing	W	Multifunction	D	Drowning	D	Easy to use	D
Rehabilitation	D	Durability	D	Injury	D	Fast water supply	W
Comfortable	W						

4.3. Development Strategies

This product is still in the development stage, and there is still room for improvement in terms of acceptability, convenience, and ease of use, so this study uses different design strategies to improve the deficiencies.

4.3.1. TRIZ Theory

In this stage, TRIZ theory is used to explore the inadequate function of the hydrotherapy bucket, the problems accompanying its use, and the problem-solving model according to TRIZ theory. The steps are as follows.

- Analysis of problem points: this research draws up design specifications using functional analysis and use process, and also analyzes three problems, namely (1) the size of the multi-functional hydrotherapy pool is too large, (2) the function of the hydrotherapy buckets for home use is insufficient, (3) large hydrotherapy pools for the general public or medical use are inconvenient for people with physical disabilities to enter and exit.
- TRIZ inventive problem solving: after analyzing the problem, the improvement features were to enhance the equipment and functions with the least possible space, and to reduce the complexity of entering and leaving hydrotherapy buckets to avoid worsening features. These two features must be translated into the context of various engineering parameters of TRIZ. Then, in the contradictions matrix, the contradictions from the list of 39 that best fit the conflict statement must be selected. According to the TRIZ contradictions matrix, parameter 36 "Complexity of device" is introduced into the vertical axis (feature to change) and parameter 33 "Convenience of use" is introduced into the horizontal axis (undesired result). Finally, Principle 9 "Preliminary anti-action", Principle 24 "Intermediary", Principle 26 "Copying" and Principle 27 "Cheap short-living objects" should be expanded and solutions brainstormed. In this case, Principle 24 "Use an intermediary carrier article or intermediary process" and "Merge one object temporarily with another (which can be easily removed)" provided helpful ideas, and this study will use these concepts to solve the problems

4.3.2. Component Development

Generally speaking, regardless of the fact that the appearance and components of a product are always considered by many consumers as purchase factors, the product set up in this study is a multi-functional hydrotherapy bucket, so users' decisions are mostly based on the function, efficacy, practicality, ease of operation, durability, price, size, and other factors for purchase considerations rather than the appearance or the shape. The morphological diagram method is divided into four parts after disassembling the individual components and functions. First of all, Figure 2A shows the different ways of entering and leaving the bucket. Figure 2B is mainly for different massage and pressure relief functions. Figure 2C shows the handrails of different directions, sizes, and shapes. Figure 2D is the treadmill devices used for running or walking training.

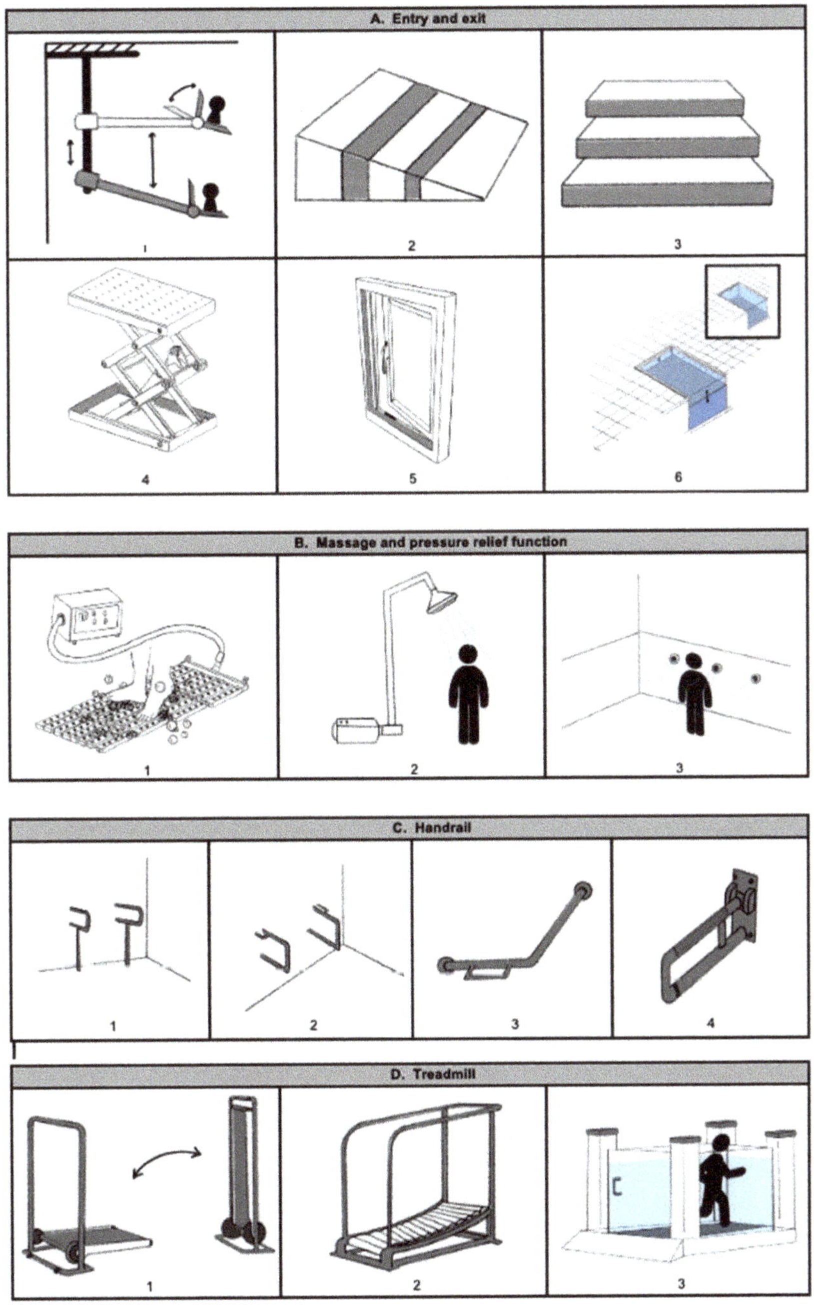

Figure 2. Morphological diagram of each component.

4.3.3. Concept Development

In this stage, the parts disassembled after the morphological diagram are combined to produce a total of four design concepts (Table 2), generated after the initial evaluation of this study for the subsequent screening of the best solution. The four concepts are (A) single function type for single person, (B) multifunction type for single person, (C) convenience type, and (D) combination storage type.

Table 2. Four design concepts.

A. Single function for single person	B. Multifunction type for single person
A6_B2_C3_D3	A5_B3_C1_D2
C. Convenience type	D. Combination storage type
A4_B3_C3_D2	A5_B1_C2_D1

4.3.4. Concept Selection and Evaluation

In this study, there are eight design criteria, which are: 1. does the hydrotherapy bucket take up space? 2. Is the hydrotherapy bucket economical? 3. Is the hydrotherapy bucket easy to access? 4. Is the hydrotherapy bucket easy to use? 5. Is the hydrotherapy bucket easy to assemble? 6. Is it too expensive to buy a hydrotherapy bucket? 7. Does the hydrotherapy bucket contain multiple functions? 8. Is the hydrotherapy bucket safe? Then, the weights of these eight design criteria were obtained using the tabular method (as in Table 3). Finally, the four design concepts (A–D) were compared to a commercially available personal hydrotherapy bucket that was considered a DATUM for this study. This personal hydrotherapy bucket is a movable lower body hydrotherapy bucket equipment set, which not only has wheels to push it but also includes a motor. The size of this hydrotherapy bucket is 107(L) × 51(W) × 71(H) centimeters and the capacity is 68 gallons. The hydrotherapy motor is 18 cm (diameter) × 96 cm (height), with a maximum water output of 55 gallons per minute and a weight of about 21 Kg. Then, the PUGH method (Table 4) was used to evaluate the most suitable decision scheme. After screening, Concept D received the highest score. In terms of cost, all design concepts are inferior to DATUM, but that is because DATUM is a single-function hydrotherapy bucket, so the cost must be much lower than a multi-function one. In addition, Concept D is superior to DATUM in terms of access, use, assembly, function, and even safety.

Table 3. The weight of each policy design criteria.

Criteria	**1**	**2**	**3**	**4**	**5**	**6**	**7**	**8**	**Total**	**Weight**
1		5	4	3	2	5	3	3	25	0.10
2	3		3	2	3	2	1	1	15	0.07
3	4	5		2	3	2	1	1	18	0.08
4	5	6	6		5	4	4	5	35	0.16
5	6	5	5	3		3	3	4	29	0.13
6	3	6	6	4	5		4	4	33	0.15
7	5	7	7	4	5	4		5	37	0.17
8	5	7	7	3	4	3	3		32	0.14
Total									224	1

Table 4. The weight of each design criteria.

Design Criteria		A	B	C	D	Weight
1	DATUM	S	S	S	S	0.10
2		S	S	S	S	0.07
3		S	+	+	+	0.08
4		−	S	+	+	0.16
5		−	S	S	+	0.13
6		−	−	−	−	0.15
7		+	+	+	+	0.17
8		−	S	+	+	0.14
+ Score		0.17	0.25	0.55	0.68	
− Score		0.58	0.15	0.15	0.15	
Total Score		−0.41	0.1	0.34	0.53	

5. Results

After generating the best design solution according to the above methodologies, the main design concept is D. Concept D enhances the function of combination and storage, because its massage and pressure relief functions and treadmill are detachable and storable, so it enhances the ideal of multiple functions of the bucket. Figure 3A–F show the detailed 3D drawing of each component. This product has a water-blocking door with a side anti-leakage plastic strip, so it can achieve the effect of preventing water leakage (Figure 3A). The two-stage closing function can enhance the tightness of the door plate and the hydrotherapy bucket (Figure 3B). Figure 3C combines the air foam cushion with the treadmill, so users can perform running and walking training at the same time as receiving the effect of a massage. In addition, this device can instantly eject small bubbles, a micro-vibration of about 10,000~15,000 times per second, to achieve the massage effect. Figure 3D shows a monitor, which is operated by the therapist and monitored externally, while Figure 3E shows a monitor that can be operated directly by the user from inside. Figure 3F is a schematic diagram of the treadmill and the air foam cushion that can be disassembled and stored.

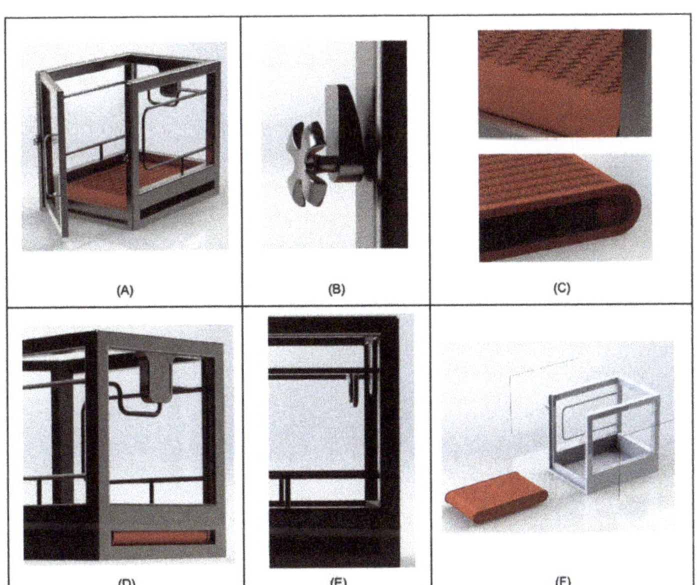

Figure 3. The detailed 3D drawing of each component (**A**–**F**).

6. Conclusions

In this study, the concurrent design strategy is applied to the development and re-design process of the multi-functional hydrotherapy bucket. Pre-design of products is carried out using the objectives tree method for decision making and design specifications. The TRIZ contradictions matrix and morphological diagram are used to find the problematic points and principles for solutions. Throughout the design process, the user experience and the PUGH method are used to select the best version of the design, and finally, the best one is presented in 3D with a computer-aided design.

This study adopts a systematic design process, which can effectively save development and design time, and is also closer to the real needs of users. Overall, this design proposal uses various methodologies and data to select the best solution and by compiling relevant data to build a user-oriented product as much as possible, but whether it can really achieve the effect of rehabilitation is still to be discovered after the test.

Author Contributions: Conceptualization, S.-W.H. and C.-H.L.; methodology, S.-W.H. and C.-H.L.; validation, S.-W.H. and C.-H.L.; investigation, C.-H.L.; resources, S.-W.H.; data curation, C.-H.L.; writing—original draft preparation, C.-H.L.; writing—review and editing, S.-W.H. and C.-H.L.; visualization, C.-H.L.; supervision, S.-W.H.; project administration, S.-W.H. and C.-H.L. All authors have read and agreed to the published version of the manuscript.

Funding: This research received no external funding.

Institutional Review Board Statement: Not applicable.

Informed Consent Statement: Not applicable.

Data Availability Statement: Data are contained within the article.

Conflicts of Interest: The authors declare no conflict of interest.

References

1. Kim, M. A study of dignity as a principle of service design. *Int. J. Des.* **2021**, *15*, 87–100.
2. Kwaku, A.G.; Wei, Y. The Vital Role of Problem-Solving Competence in New Product Success. *J. Prod. Innov. Manag.* **2011**, *28*, 81–98.
3. Cooper, R.G.; Kleinschmidt, E.J. *New Products: The Key Factors in Success*; Marketing Classics Press: Chicago, IL, USA, 2011.
4. Huey, L. The Waterpower Workout: The Stress-Free Way for Swimmers and Non-Swimmers Alike to Control Weight, Build Strength and Power, Develop Cardiovas. Available online: http://lyndahuey.com/ (accessed on 2 March 2021).
5. Glenn, J.M. *Deep Water Exercise for Health and Fitness*; Publitec Editions: Laguna Beach, CA, USA, 1988.
6. Poteat, S.A. Justifiable aquatic therapy: Scientific support for intervention, neurological and neurosurgical population. In *Congresboek: Hydrotherapie van Pracice Based Naar Evidence Based*; Lambeck, J., Bult, H., Eds.; Nederlands Paramedisch Instituut: Amersfoort, The Netherlands, 1999.
7. Ruth, S. RuthSova.com | Aquatic Therapy and Rehab Institute, Inc. Available online: http://ruthsova.com (accessed on 14 February 2022).
8. Lambeck, J. *Hydrotherapy in Adult Neurology*; EWAC Medical: Heerhugowaard, The Netherlands, 2002.
9. Mutz, M.; Gerke, M. Sport and exercise in times of self-quarantine: How Germans changed their behaviour at the beginning of the COVID-19 pandemic. *Int. Rev. Sociol. Sport* **2021**, *56*, 305–316. [CrossRef]
10. Mattioli, A.V.; Coppi, F.; Gallina, S. Importance of physical activity during and after the SARS-CoV-2/COVID-19 pandemic: A strategy for women to cope with stress. *Eur. J. Neurol.* **2021**, *28*, 78–79. [CrossRef] [PubMed]
11. Haik, Y.; Shahin, T.M.M. *Engineering Design Process*; Cengage Learning: Boston, MA, USA, 2010.
12. Altshuller, G. *40 Principles: TRIZ Keys to Technical Innovation*; Technical Innovation Center: Worcester, UK, 2002.
13. Roozenburg, N.F.M.; Eekels, J. *Product Design: Fundamentals and Methods*; Lemma: Utrecht, The Netherlands, 1995.
14. Cross, N. *Engineering Design Methods*; Wiley: Chichester, UK, 1989.
15. Ullman, D.G. A Taxonomy for Classifying Engineering Decision Problems and Support Systems. *Artif. Intell. Eng. Des. Anal. Manuf.* **1996**, *9*, 427–438. [CrossRef]
16. Pugh, S. *Total Design: Integrated Methods for Successful Product Engineering*; Addison Wesley: London, UK, 1991.

engineering proceedings

MDPI

Proceeding Paper

Research on Rescue Guiding Mechanism in Buildings: Illustrated by the Building Information Guiding System †

Chyi-Gang Kuo [1], Benson P. C. Liu [2], Chi-Wei Lee [3] and Chien-Wei Chiu [1,*]

1 Department of Architecture, Chaoyang University of Technology, Taichung 413310, Taiwan; chyigang@cyut.edu.tw

2 Department of Creative Design, National Yunlin University of Science and Technology, Yunlin 640301, Taiwan; bensonp.c.liu@gmail.com

3 Department of Visual Communication Design, Chaoyang University of Technology, Taichung 413310, Taiwan; cwl@cyut.edu.tw

* Correspondence: chiou1130@cyut.edu.tw

† Presented at the IEEE 5th Eurasia Conference on Biomedical Engineering, Healthcare and Sustainability, Tainan, Taiwan, 2–4 June 2023.

Abstract: When disasters such as fires or earthquakes occur, rescue personnel entering buildings are prone to spatial disorientation when the buildings have complex indoor compartments or compartments with high uniformity and similarity. This situation may expose both the people waiting for help and the rescue personnel to dangerous conditions. Therefore, this research study proposes that government search and rescue units can pre-establish a database of indoor spatial information for buildings. When a disaster occurs, the spatial information needed by rescue personnel can be transmitted in real-time and displayed on wearable 3D display systems. Given this proposal, we have developed a "Building Information Guiding System" (BIGS) to provide spatial information in real-time through the "Wearable Augmented-reality Seeking System" (WASS). BIGS uses QR codes presented at different information anchors in a building to display various embedded spatial information on horizontal and vertical escape paths, the locations of exits, firefighting equipment, and electrical power generator sets. According to the information provided by BIGS, 3D virtual arrows can be displayed in the air by the HoloLens augmented reality helmet worn by the rescue personnel in order to guide the direction toward the target. The BIGS system can read the Building Information Modeling (BIM) file submitted with new construction projects to the government construction management units and then input necessary spatial information to the file to establish an exclusive 3D spatial database for that new construction. BIGS can also construct an entire 3D spatial database for an old building by reading the BIM 3D point cloud model created through 3D laser scanning. This study explores the feasibility of using the BIM model to construct the BIGS system, the presentation mechanism of serial QR codes in real space, and the steps and modes of inputting spatial information into the 3D spatial database.

Keywords: rescue guiding system; augmented reality; spatial information; Building Information Modeling; 3D spatial database

Citation: Kuo, C.-G.; Liu, B.P.C.; Lee, C.-W.; Chiu, C.-W. Research on Rescue Guiding Mechanism in Buildings: Illustrated by the Building Information Guiding System. *Eng. Proc.* **2023**, *55*, 73. https://doi.org/10.3390/engproc2023055073

Academic Editors: Teen-Hang Meen, Kuei-Shu Hsu and Cheng-Fu Yang

Published: 11 December 2023

1. Introduction

When a construction disaster occurs, the first-line rescue personnel often enter the disaster site immediately, and every second counts when attempting to rescue the people who need help. However, the rescue personnel may not be familiar with the indoor layout of different buildings. If the indoor paths are complicated, or if smoke from a fire obstructs the line of sight, the rescue personnel are prone to spatial disorientation, which places the rescue personnel at increased risk of danger [1].

For evaluation purposes, we laser-scanned a daycare center to build a point cloud 3D digital model and added a dynamic fire smoke simulation effect into the virtual-reality scenery. Then, we invited the firefighters of the Fire Bureau of Taichung City Government for a test. The simulation revealed that even professional firefighters start to become nervous when they are in an unfamiliar and smoky space where they cannot see clearly. It takes them more time to find the emergency exit (Figure 1).

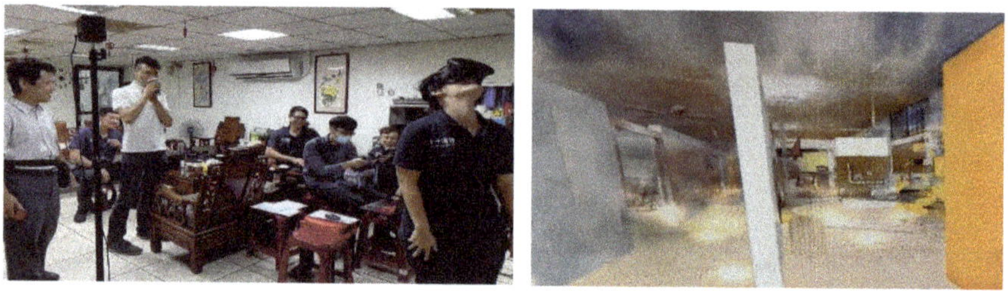

Figure 1. Firefighters of the Fire Bureau of Taichung City Government participating in a VR simulation test.

Therefore, it is necessary to create 3D spatial information for each building in advance and store the location information regarding firefighting facilities, escape equipment, and emergency exits in the database. Through an appropriate information presentation mechanism that provides timely 3D spatial information, the system can help to protect the lives of personnel even under extreme environmental conditions such as humidity, high temperature, dense smoke, power outage, and lack of indoor navigation signals.

The pre-established 3D spatial database of buildings can help rescue personnel quickly familiarize themselves with their surroundings when an accident occurs. The database can also be utilized in daily training or pre-mission training.

2. Literature and Case Study Review

2.1. Literature Review

Since interior spaces are often remodeled and furnished, they tend to be different from the original appearance of the building when they are initially constructed. This affects the rescue and escape path. Therefore, the database establishment for critical indoor facilities must first focus on making the 3D model of interior spaces in their current state.

With the advancement of technology, the most accurate way to build indoor digital 3D models is to use laser scanning techniques, and many products are already on the market. However, the price of laser scanning equipment is high, and the scanning speed is slow, so it is not suitable for operations requiring scanning a large number of buildings. Another method is 3D photo scanning. 3D photo scanning technology is not as accurate as laser scanning but has a much faster scanning speed. After continuously optimizing 3D photo scanning technology by researchers [2–5], it is possible to quickly complete spatial scanning modeling using tablets or even mobile phones. Since the 3D model used to display the location of critical indoor facilities does not need to be smooth and perfect, photo scanning technology is suitable for establishing a database of indoor spatial information for buildings.

In addition, we suggest that the scanned 3D format should be stored in a point cloud format (such as .e57 or .xyz) and integrated into the information model constructed by BIM software (such as ArchiCAD or Revit) to facilitate the integration and expansion of future architectural information.

2.2. Case Study

Companies in Taiwan have already built indoor spatial maps combined with mobile phone indoor positioning technology [6] and have applied them to department location guidance services between different hospital buildings. This technology focuses on indoor mobile phone positioning guidance, not constructing a 3D database. Thus, the user's mobile phone screen only displays a 2D map mode.

However, this indoor positioning method relies on electronic signal transmitters and receivers. It does not work under extreme environmental conditions such as power outages, high temperatures, humidity, dense smoke, and a lack of indoor navigation signals. Therefore, the simpler the indoor positioning method, the better. In this study, we conclude that relay QR code positioning combined with inertial navigation is the first choice for indoor positioning in extreme environments.

3. Establishing BIGS

This research study screens out the four most important building elements or facilities related to firefighting and rescue purposes for most buildings. The locations of these four items are set as information anchor points in a building. These items include a fire hose box, safety exit, fire extinguisher, and fire escape equipment.

The steps to construct a database of indoor spatial information for buildings are as follows.

3.1. Building Spatial Digital Information

- 3D laser scanning to create a point cloud model.
- Converting the created point cloud model into the BIM model.
- Adding "information anchor point" such as firefighting/escape equipment, emergency exit, and others. Text descriptions, image descriptions, and other information are added in each "Information Anchor Point" and stored in the database together.

The completed spatial information can be used by search and rescue units for daily training or pre-service education.

3.2. Building 3D Guiding Symbols on the Building Site

- Placing a QR Code at the actual location of each "Information Anchor Point"
- Building a list of "missions" in which every mission contains different series of "Information Anchor Points" in the space through the Dynamics 365 Guides software interface.
- Placing 3D guiding symbols or information tags floating in the air to connect a series of locations of information anchor points according to different missions.

3.3. Exploring Spatial Digital Information

When used in conjunction with a Wearable Augmented-reality Seeking System (WASS), the smart glasses can be used to read the QR Code embedded in each "Information Anchor Point" in order to read the set spatial information.

4. Experiment and Discussion

There are many paper books in libraries. If there is a fire, the books are at risk of producing a lot of thick smoke, which may hinder firefighters from finding the locations of fire hose boxes, safety exits, fire extinguishers, and/or fire escape equipment. Therefore, we used the 6F stack of the library of the Chaoyang University of Technology (Figure 2) as the location to conduct the spatial information embedding experiment of BIGS.

Figure 2. Panoramic view at the entrance of the sixth floor of Chaoyang University Library.

In Figures 3 and 4, there are four fire hose boxes, A, B, C, and D on the sixth floor. First, we started from fire hose box A at the entrance (Figure 5) and built mission options leading to the location of the other three fire hose boxes. In each task option, we used AR software (Dynamics 365 Guides) to place 3D guiding arrows floating in the air.

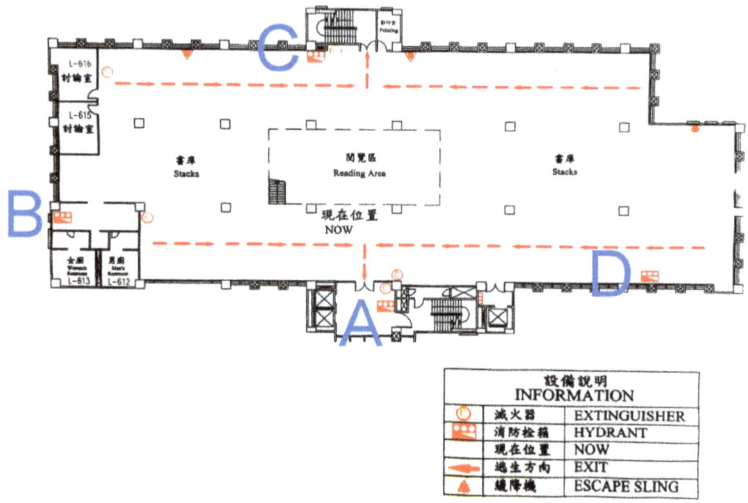

Figure 3. Sixth floor plan of Chaoyang University Library. The blue letters A, B, C, and D identify the location of four fire hose hydrant boxes.

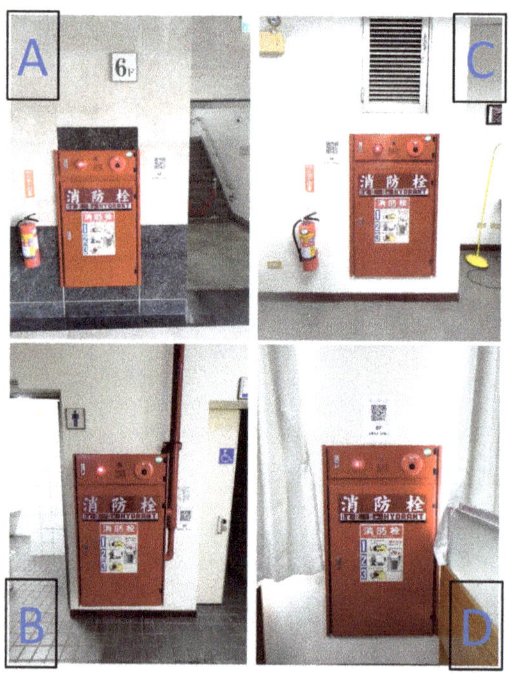

Figure 4. The photos of the four fire hose hydrant boxes labeled with A, B, C, and D in Figure 3.

Figure 5. Information anchor point #1 at the sixth floor elevator hall in Chaoyang University Library.

Since most buildings have a fire hose box at the main entrance to provide water in the event of a fire, an emergency power supply, and an light source, it is easier for search and rescue personnel to use the fire hose box at the main entrance as the starting point of spatial information to extend outward.

When the search and rescue personnel wear the smart glasses, they can choose the task they want to perform and see the corresponding 3D arrows in the smart glasses to guide them to the target (Figure 6). Since these 3D arrows are virtual objects that are floating in the air with inertial positioning, as long as they are positioned, the images do not disappear or shift and do not become invisible due to smoke blocking the line of sight (Figures 7 and 8).

Figure 6. The first two virtual arrows beside the first anchor point. The virtual arrows will not be blocked by objects or heavy smoke.

Figure 7. The third and fourth guiding arrows are positioned in the air.

Figure 8. The positioned arrows will not disappear or move much, even after returning from long distance movement from the first anchor point.

The Chaoyang University of Technology is establishing an innovative campus and has built many real-time monitoring systems for electric energy consumption, air quality, and indoor temperature (Figure 9). In the future, if BIGS can be pre-built in each campus corner, it can help firefighters or rescue personnel complete tasks quickly and accurately if/when a disaster occurs.

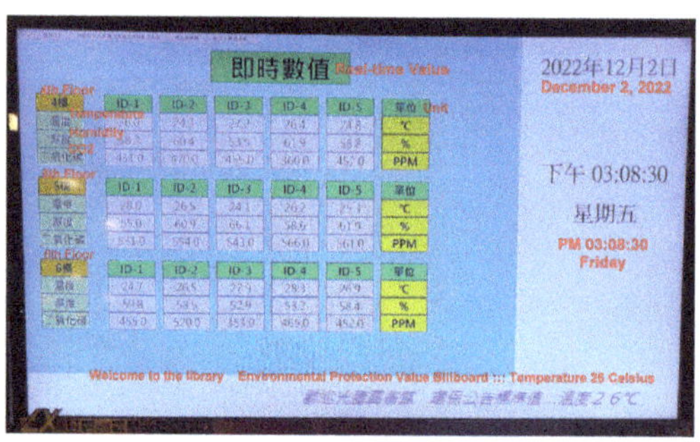

Figure 9. The real-time air quality monitoring system at the intelligent campus of Chaoyang University of Technology.

5. Conclusions

We have developed a model with the BIGS database to assist search and rescue personnel in completing tasks safely in unfamiliar building spaces. To keep functioning under extreme environmental conditions, such as power outages, high temperature, humidity, dense smoke, and lack of indoor navigation signals, we suggest that the greater accessibility of BIGS information, the better. Therefore, a QR code made with fireproof materials posted next to a fire hose box is the first choice for information anchor positioning. According to Point 6, Section 4, Chapter 1, "the Fire Bureau of Taichung City Government Guiding

Principles for Operation and Maintenance of Disaster Relief Equipment", at the scene of a fire, hot smoke accumulates in high places. Thus, the zone close to the floor can provide better visibility, and a low posture is advised for movement in these areas (Figure 10).

Figure 10. Firefighters are practicing moving forward in a low posture [7].

Therefore, it is recommended that the positioning anchor QR codes of BIGS should be posted around the lower area of a fire hose box. The first fire hydrant box can be used at the entrance as the first information anchor point to guide the fire search and rescue personnel to find the other information anchor points in sequence or return to the first information anchor point at the entrance. In this way, BIGS can help rescue personnel reach their destination without disorientation in buildings with complex internal partitions or high indoor feature consistency in an emergency situation. The database of BIGS can also be utilized in daily or pre-mission training. BIGS can combine Building Information Models (BIM) and Geometric Information Systems (GIS) to construct smart buildings and smart cities.

Author Contributions: Conceptualization, C.-G.K., B.P.C.L. and C.-W.L.; Methodology, C.-G.K. and C.-W.C.; Formal analysis, C.-G.K.; Resources, C.-G.K., B.P.C.L. and C.-W.C.; Data curation, C.-G.K.; writing—review and editing, C.-G.K.; project administration, B.P.C.L. All authors have read and agreed to the published version of the manuscript.

Funding: Ministry of Science and Technology of Taiwan (project number: MOST 108-2221-E-324-001-MY3).

Institutional Review Board Statement: Not applicable.

Informed Consent Statement: Not applicable.

Data Availability Statement: Data are contained within the article.

Acknowledgments: This research was supported by the Ministry of Science and Technology of Taiwan (project number: MOST 108-2221-E-324-001-MY3). The authors wish to extend their thanks to the Ministry of Science and Technology of Taiwan for the project support.

Conflicts of Interest: The authors declare no conflict of interest.

References

1. Ministry of the Interior, National Fire Agency. *Investigation Report of the Death of Firefighters on the Fire Accident in the Old Building of the Former Qiaoyou Department Store in Changhua County*; Ministry of the Interior: Taipei, Taiwan, 2022.
2. Suzuki, K.; Sakamoto, D.; Nishi, S.; Ono, T. SCAN: Indoor Navigation Interface on a User-Scanned Indoor Map. In Proceedings of the 21st International Conference on Human-Computer Interaction with Mobile Devices and Services, Taipei Taiwan, 1–4 October 2019; pp. 1–6. [CrossRef]
3. Pintore, G.; Mura, C.; Ganovelli, F.; Fuentes-Perez, L.; Pajarola, R.; Gobbetti, E. Automatic 3D reconstruction of structured indoor environments. In Proceedings of the ACM SIGGRAPH 2020 Courses, Virtual Event, 17 August 2020; pp. 1–218. [CrossRef]
4. Pintore, G.; Ganovelli, F.; Scopigno, R.; Gobbetti, E. Mobile metric capture and reconstruction in indoor environments. In Proceedings of the SIGGRAPH Asia 2017 Mobile Graphics & Interactive Applications on—SA '17, Bangkok, Thailand, 27–30 November 2017; pp. 1–5. [CrossRef]

5. Wang, J.; Wang, P.; Long, X.; Theobalt, C.; Komura, T.; Liu, L. NeuRIS: Neural Reconstruction of Indoor Scenes Using Normal Priors. In *Computer Vision—ECCV 2022*; Avidan, S., Brostow, G., Cissé, M., Farinella, G.M., Hassner, T., Eds.; Springer: Cham, Switzerland, 2022; Volume 13692, pp. 139–155. [CrossRef]
6. Patient-Centered Medical Indoor Navigation Service [Online]. Available online: https://tjcha.neocities.org/jct/product4.html (accessed on 1 December 2022).
7. Fire Bureau of Taichung City Government, Fire Bureau of Taichung City Government Guiding Principles for Operation and Maintenance of Disaster Relief Equipment. 2015. Available online: https://www.fire.taichung.gov.tw/df_ufiles/f/%E6%95%91%E7%81%BD%E5%99%A8%E6%9D%90%E6%93%8D%E4%BD%9C%E7%B6%AD%E8%AD%B7%E6%9A%A8%E4%BF%9D%E9%A4%8A%E6%8C%87%E5%B0%8E%E5%8E%9F%E5%89%87.pdf (accessed on 15 October 2022).

engineering proceedings

MDPI

Proceeding Paper

The Effects of Footbath Therapy Proved with Physiology Parameters, including Meridian Energy for Health Promotion †

Hsiao-Hui Chen [1] and Chih-Hung Lin [2,]*

1 Public Health Bureau, Tainan City Government, Tainan 730213, Taiwan; d00061@tncghb.gov.tw
2 Department of Pharmacy, Chia Nan University of Pharmacy & Science, Tainan 71710, Taiwan
* Correspondence: ochihung@mail.cnu.edu.tw
† Presented at the 5th IEEE Eurasia Conference on Biomedical Engineering, Healthcare and Sustainability 2023, Tainan, Taiwan, 2–4 June 2023.

Abstract: Footbath therapy is effective for health and wellness. We compared the effects of hot water footbaths (HFB), contrast water footbaths (CFB), and warm water footbaths (WFB) using physiological parameters, including blood pressure, calf volume, flexibility, and meridian energy for six weeks. Significant differences in diastolic blood pressure and calf volume were observed from the third to fifth week, and CFB showed the best effect. CFB and WFB demonstrated improvements in meridian energy, while HFB had no effect. These findings highlighted the effectiveness of CFB in managing stress and promoting health and well-being.

Keywords: foot bath; meridian energy; health promotion

Citation: Chen, H.-H.; Lin, C.-H. The Effects of Footbath Therapy Proved with Physiology Parameters, including Meridian Energy for Health Promotion. *Eng. Proc.* **2023**, *55*, 74. https://doi.org/10.3390/engproc2023055074

Academic Editors: Teen-Hang Meen, Kuei-Shu Hsu and Cheng-Fu Yang

Published: 12 December 2023

1. Introduction

Hydrotherapy has been used in medical fields for pain reduction, the improvement of joint mobility and muscle strength, enhancement of heart efficiency, promotion of blood and lymph circulation, and improvement of metabolic function, skin condition, and muscle control and coordination [1]. Although contrast bath therapy is beneficial for sports injuries, joint pain, muscle fatigue, inflammation, and poor blood circulation [2], its application is limited due to the need for specialized equipment and professional instructors. Kneipp's foot-treading method originated in Europe enhances blood circulation, reduces foot swelling and fatigue, and improves joint mobility [3]. Footbath therapy, especially with hot water, has also been proven to be effective in promoting health and wellness by alleviating symptoms of pain, fatigue, and insomnia, improving circulation, and enhancing overall well-being [4].

Foot health is essential to overall physical health, as foot blood circulation is closely linked to systemic blood circulation and metabolism, thereby promoting the regulation function of the nervous and endocrine systems [5]. Studies have shown that footbaths relieve fatigue and improve sleep quality in stroke and cancer patients, regulate heart rate variability [6], increase core body temperature, relieve menstrual pain and maternal anxiety, and restore low-meridian energy, among other health benefits [4–8]. A previous study revealed that occupational stress was a significant factor in high-stress industries such as high-tech and banking industries in Taiwan [9]. Strategies to overcome occupational stress include promoting work–life balance, creating a positive work environment, and providing necessary resources [10]. However, there is a dearth of simple and practical health promotion and guidance methods that can be easily used in daily life. Furthermore, limited research has been conducted to compare the effects of footbath in different temperatures on physiological parameters. Thus, this study was conducted to compare the effects of the hot water footbath and contrast water footbath on health promotion with warm water footbath as the control group. The study result provided evidence-based guidance on the use of footbath therapy as a health promotion strategy.

2. Materials and Methods

2.1. Subjects

This study was conducted with the approval of the Institutional Review Board (IRB: 103075). The participants were recruited from Chia Nan University of Pharmacy and Science, and were between the ages of 20 and 25. Volunteers were required to complete a health assessment form (Brief Symptom Rating Scale, BSRS-5) and a brief physical examination. Those who passed the screening then signed a consent form. A total of 33 participants were enrolled and allocated into three groups based on their decreasing body mass index (BMI) in a sequential order. Before the study, each group consisted of 11 participants. Five participants could not complete the experiment. Thus, the warm water footbath group (WFB, $n = 9$), hot water footbath group (HFB, $n = 9$), and contrast footbath group (CFB, $n = 10$) were included in the study. During the study, the participants' lifestyles were recorded, but they were subjected to no restrictions in their lives.

2.2. Methods

The study was conducted as a single-blind trial with participants undergoing a 15 min footbath once a week for six consecutive weeks at a room temperature of 25 ± 2 °C. Based on other research [2], water temperatures were set to 10 °C for cold water, 30 °C for warm water, and 40 °C for hot water. Skin electrical conductance was measured for psychological stress, while meridian energy was measured using the ARDK® on 24 special acupoints in the wrists and ankles. Most of the acupoints selected were the originating acupoints of the 12 meridians, namely the lung, large intestine, stomach, spleen, heart, small intestine, urinary bladder, kidney, pericardium, triple burner, gallbladder, and liver meridians [11]. Calf volume was measured using Archimedes' volume measurement bucket, and blood pressure and flexibility were measured using devices. All data were collected before and after a footbath for six weeks. The data were analyzed using inferential statistics such as ANOVA and paired t-tests with a significance level set at $p < 0.05$.

3. Results

3.1. BSRS-5 Scale

A total of 33 participants joined in a 3-month lifestyle assessment for the evaluation of their sleep, mood, disease, and medication conditions. The participants were 22.21 ± 1.36 years old, including nine males (27.3%) and 24 females (72.7%). The mean Body Mass Index (BMI) was 22.77 ± 4.38 kg/m^2. Among the participants, 30 (90.9%) had no smoking habit, one (3.0%) had an occasional smoking habit, and two (6.1%) had a social smoking habit. In total, 23 (69.7%) had no drinking habit, seven (21.2%) had an occasional drinking habit, and three (9.1%) had a social drinking habit. A total of 15 (45.5%) had no habit of drinking coffee, one (3.0%) drank coffee daily, and 17 (51.5%) drank coffee occasionally. A total of 12 (36.4%) had no habit of drinking tea, six (18.2%) drank tea daily, and 15 (45.5%) drank tea occasionally. Of the participants, 19 (57.6%) had a mostly sedentary lifestyle, while 14 (42.4%) did not. Before the experiment, participants underwent medical and foot-related condition assessments, and those with significant medical conditions were excluded. However, in the medical history assessment, one participant had a history of asthma, but since the condition had not occurred in the past 6 months, the participant was still eligible to participate in the study. The participants were assessed using the Brief Symptom Rating Scale 5 (BSRS-5), and all participants had BSRS-5 scores less than 10, indicating a good daily living status (Table 1). During the experiment, five participants terminated their participation due to personal reasons, resulting in a final sample size of 28.

Table 1. BSRS-5 assessment results.

Score \ Groups	Total (n = 33)	Footbath Groups		
		WFB (n = 11)	HFB (n = 11)	CFB (n = 11)
0–5	24 (72.7%)	9 (81.8%)	7 (63.6%)	8 (76.7%)
6–9	9 (27.3%)	2 (18.2%)	4 (36.4%)	3 (27.3%)
0–14	0	0	0	0
15–24	0	0	0	0

Note: BSRS (Brief Symptom Rating Scale), WFB (warm water footbath), HFB (hot water footbath), CFB (contrast water footbath), n = participants.

3.2. Evaluation of Physiological Parameters

We investigated the effect of footbath therapy on physiological parameters such as blood pressure, calf volume, and flexibility. The results of the three different foot bath groups were evaluated and compared.

The results indicated that the WFB group showed a statistically significant decrease in systolic blood pressure (SBP) by 8.33 mmHg before intervention (w0) to the end of the experiment (W7). The significant effect on SBP was observed from the third week of the experiment, while the effect on diastolic blood pressure (DBP) was minimal. However, the HFB and CFB groups did not show significant differences in SBP and DBP. ANOVA analysis was conducted to compare the values among the three groups and statistical differences in DBP were observed, but not SBP. The changes in DBP showed significant differences among the groups in the third week (p = 0.001), fourth week (p = 0.031), and fifth week (p = 0.007). The Tukey HSD test revealed that the CFB group had the most significant impact on DBP, followed by the WFB group and the HFB group. There were no long-term differences in SBP and DBP changes among the groups after the 6-week intervention (Table 2).

Table 2. Changes in blood pressure of three different groups after footbath three different groups.

	Parameters	Week	WFB (n = 9)	HFB (n = 9)	CFB (n = 10)	p-Value (Tukey HSD)
SBP	Original	Before (w0)	109.89 ± 10.914	104.56 ± 3.408	110.30 ± 13.549	
		After (w7)	101.56 ± 12.084 #	104.67 ± 3.575	103.70 ± 2.813	
	Net value	w1	−8.33 ± 5.657	0.11 ± 10.541	−3.90 ± 13.085	0.244
		w2	−1.00 ± 10.112	−4.11 ± 4.936	−4.00 ± 8.300	0.651
		w3	−2.33 ± 5.050 #	−4.11 ± 10.043	−2.40 ± 6.381	0.846
		w4	−4.56 ± 7.955	0.44 ± 5.981	−3.80 ± 6.596	0.267
		w5	0.56 ± 10.370 #	−1.56 ± 5.833	−4.20 ± 6.546	0.423
		w6	−1.67 ± 5.315 #	−3.20 ± 23.734	−3.30 ± 6.800	0.774
DBP	Original	Before (w0)	64.67 ± 6.856	59.78 ± 6.741	64.50 ± 8.059	
		After (w7)	60.89 ± 10.006	59.78 ± 6.037	63.10 ± 6.437	
	Net value	w1	−4.00 ± 5.979	0.00 ± 5.148	−0.90 ± 7.666	0.392
		w2	−0.78 ± 4.893	0.22 ± 4.086	−1.30 ± 6.165	0.812
		w3	0.78 ± 3.598	1.44 ± 3.468	−6.00 ± 3.801	0.001 * (a, c)
		w4	−3.78 ± 2.539	1.44 ± 4.503	−5.50 ± 7.764	0.031 * (c)
		w5	2.22 ± 6.320	1.78 ± 2.774	−4.90 ± 5.301	0.007 * (a, c)
		w6	−0.33 ± 5.099	−2.56 ± 4.447	−3.60 ± 5.317	0.365

Note: SBP (systolic blood pressure), DBP (diastolic blood pressure), WFB (warm water footbath), HFB (hot water footbath), CFB (contrast water footbath), w = weeks, n = participants, data presented M ± SD, significant: $p < 0.05$, * = ANOVA, [a] = WFB/CFB, [c] = HFB/CFB; # = pair t-test.

Table 3 shows significant differences between the groups in calf volume. Specifically, the CFB group had a significant reduction in calf volume from the first to the sixth week of intervention. The three groups revealed a significant decrease in calf volume. Post hoc tests showed that the decrease of the CFB group was larger than that of the WFB group and the HFB group. Although no statistically significant differences were found in flexibility among the three groups, significant differences were observed in the long-term follow-up results of the groups of CFB and HFB in paired t-tests between the beginning (W0) and end (W7) of the experiment.

Table 3. Changes in calf volume and flexibility changes in three different groups after footbath.

	Parameters	Week	WFB (n = 9)	HFB (n = 9)	CFB (n = 10)	p-Value (Tukey HSD)
Calf volume	Original	Before (w0)	24.40 ± 0.919	24.70 ± 0.750	24.68 ± 0.978	
		After (w7)	24.41 ± 0.929	24.81 ± 0.720 #	24.62 ± 1.009	
	Net value	w1	0.01 ± 0.105	0.11 ± 0.127	−0.06 ± 0.084	0.007 * (c)
		w2	0.04 ± 0.073	0.17 ± 0.068	−0.05 ± 0.071	0.000 * (a, b, c)
		w3	0.00 ± 0.071	0.16 ± 0.053	−0.01 ± 0.137	0.003 * (b, c)
		w4	0.03 ± 0.071	0.13 ± 0.087	−0.04 ± 0.070	0.000 * (b, c)
		w5	0.06 ± 0.073	0.11 ± 0.060	−0.06 ± 0.070	0.000 * (a, c)
		w6	−0.04 ± 0.088	0.12 ± 0.044	−0.09 ± 0.057	0.000 * (b, c)
Flexibility	Original	Before (w0)	28.39 ± 10.68	30.31 ± 9.726	29.06 ± 2.695	
		After (w7)	29.89 ± 11.75	31.83 ± 9.253 #	31.33 ± 2.409 #	
	Net value	w1	1.50 ± 2.194	1.52 ± 1.502	1.95 ± 2.692	0.881
		w2	1.61 ± 2.082	2.76 ± 2.017	0.57 ± 2.045	0.075
		w3	1.43 ± 1.097	2.56 ± 2.481	1.18 ± 2.124	0.304
		w4	1.48 ± 1.766	1.72 ± 1.161	1.72 ± 2.566	0.954
		w5	2.68 ± 1.663	2.46 ± 1.711	1.75 ± 0.951	0.365
		w6	1.72 ± 2.041	2.64 ± 1.816	1.19 ± 1.863	0.266

Note: SBP (Systolic blood pressure), DBP (Diastolic blood pressure), WFB (warm water footbath), HFB (hot water footbath), CFB (contrast water footbath), w = weeks, n = participants, data presented M ± SD, significant: $p < 0.05$, * = ANOVA, [a] = WFB/CFB, [b] = WFB/HFB, [c] = HFB/CFB; # = pair t-test.

3.3. Meridian Energy

Table 4 presents the results of meridian energy changes induced by footbaths in the three groups. The groups showed significant differences in the energy of the immune system (week 3, p = 0.022), skeletal muscle system (week 5, p = 0.070), liver function (week 4, p = 0.03), respiratory system (week 3, p = 0.013), and digestive system (week 4, p = 0.005). Specifically, the effects on the CFB group and WFB group on long-term meridian energy were larger than the HFb group.

Long-term energy changes were assessed by comparing the energy levels before (w0) and after (w7) intervention. The paired t-test analysis revealed that the CFB group had significant improvements in the long-term energy of the autonomic nervous system, skeletal muscle system, and respiratory system. Similarly, the WFB group showed significant improvements in the long-term energy of the total energy, endocrine system, and systemic report. However, the HFB group did not exhibit any significant changes in the energy levels of any physiological system. The WFB group showed a significant improvement in decreasing the energy levels of the cardiovascular system at the fourth (−10.89 ± 9.727), fifth (−17.22 ± 23.037), and sixth (−12.11 ± 15.964) weeks compared to those before intervention (w0). This finding was consistent with the experimental results presented in Table 2. Additionally, the WFB group showed a reduction in diastolic blood pressure, which supports and corroborates our results. Significant differences in immune, skeletal muscle, liver function, respiratory, and digestive system energy levels were also found. The CFB group improved the long-term energy of the autonomic nervous, skeletal muscle, and respiratory systems. The WFB group improved the total energy, endocrine, and systemic report energy levels and decreased cardiovascular system energy levels and diastolic blood pressure. However, the HFB group showed no effect of footbath on the energy levels of any physiological system.

Table 4. Changes in meridian energy between three different groups after footbath.

	Parameters	Week	WFB (n = 9)	HFB (n = 9)	CFB (n = 10)	p-Value (Tukey HSD)
Total energy	Original	Before (w0)	15.00 ± 14.697	21.78 ± 18.747	25.30 ± 21.618	
		After (w7)	33.44 ± 19.456 #	34.33 ± 20.512	38.50 ± 25.439	
	Net value	w1	16.67 ± 21.500	8.67 ± 28.688	13.20 ± 29.698	0.821
		w2	12.11 ± 14.146	−0.44 ± 22.973	15.60 ± 23.372	0.234
		w3	27.89 ± 18.210	−8.33 ± 14.009	4.20 ± 13.464	0.000 * (a, b)
		w4	2.11 ± 13.569	−3.67 ± 18.682	20.10 ± 26.780	0.048 * (c)
		w5	3.44 ± 26.302	−7.22 ± 21.022	16.30 ± 31.305	0.179
		w6	1.67 ± 15.588 #	−6.89 ± 24.670	−0.50 ± 31.529	0.755
systemic report	Original	Before (w0)	48.67 ± 12.845	50.78 ± 14.856	44.50 ± 13.385	
		After (w7)	57.33 ± 10.283	50.33 ± 8.426	52.30 ± 16.667	
	Net value	w1	6.11 ± 12.057	−1.33 ± 15.788	7.80 ± 10.075	0.279
		w2	4.44 ± 9.606	−3.22 ± 16.076	7.50 ± 16.009	0.267
		w3	13.89 ± 13.896	−3.89 ± 13.014	2.30 ± 18.172	0.061
		w4	5.89 ± 5.487	−4.78 ± 11.987	11.50 ± 12.572	0.009 * (c)
		w5	8.33 ± 13.153	−2.89 ± 8.054	5.00 ± 14.103	0.152
		w6	2.22 ± 8.913	−5.67 ± 9.138	5.80 ± 11.755	0.06

Note: WFB (warm water footbath), HFB (hot water footbath), CFB (contrast water footbath), w = weeks, n = participants, data presented M ± SD, significant: $p < 0.05$, * = ANOVA, [a] = WFB/CFB, [b] = WFB/HFB, [c] = HFB/CFB; # = pair t-test.

4. Discussion

Footbath therapy has been an ancient health practice for centuries. With the prevalence of stress in today's fast-paced work environment, finding effective health promotion methods to help employees learn and integrate stress management is crucial to achieving health self-management goals. To address this knowledge gap, we investigated the physiological effects of three footbath interventions: HFB, WFB, and CFB. The results showed that the CFB group had the best effects on reducing DBP, calf volume, and meridian energy. The ANOVA analysis of physiological parameters and meridian energy showed that the CFB group had the most significant adjusting effects, followed by the WFB and HFB groups.

Among the factors, CFB had the most significant impact, while the effect of HFB was the smallest. However, HFB provided better body-warming effects and health benefits for older adults and those who felt cold with poor circulation [4,12]. Although CFB is generally believed to have a stress-inducing effect on the cardiovascular system [13,14], the study results indicated that the impact on blood pressure was not enough for such an effect.

For the WFB, HFB, and CFB groups, total meridian energy changed significantly from the baseline at the end of the six-week intervention. However, no significant differences were observed after the fifth week. Based on the participants' lifestyle records, There were academic exams in the fifth week, which might cause stress and poor sleep, and interfered with their normal physiological life cycle. The impact of this interference was less pronounced on calf volume. The results of SBP and DBP also showed a significant rebound in the WFB group in the fifth week, but a continuous decreasing trend in the sixth week after the interference was relieved. Similar observations were made for the HFB group in the fourth and fifth weeks. Nevertheless, the results for the CFB group indicated that changes in their lifestyle patterns had a smaller impact on the improvement trend in their data. The results of this study also showed that the CFB group showed better effects in adjusting DBP, reducing calf volume, and improving ankle circumference, flexibility, and meridian energy. Overall, body composition and meridian energy demonstrated that their adjusting effects were large for the CFB group > WFB group > HFB group in order.

For the parameters, CFB had a more substantial impact, while the effect of HFB was the smallest. Nevertheless, HFB provided better body-warming effects and health benefits for older adults who felt cold with poor circulation [4,12]. Although CFB is generally believed to have a stress-inducing effect on the cardiovascular system, this study's results indicated that the impact on blood pressure was not large enough. These findings suggested that footbath therapy was an effective health promotion method for managing occupational stress and improving health self-management goals. Therefore, footbath therapy, specifically CFB, is recommended as a home health maintenance method.

5. Conclusions

We investigated the effects of three different footbath interventions on young adults. The WFB group showed a significant decrease in SBP from the third week, while the CFB group had the most significant impact on DBP in long-term follow-up. The CFB group had a significant reduction in calf volume, while the WFB group showed significant improvement in the energy levels of total energy and the cardiovascular system. These findings suggested that footbath therapy was effective in managing occupational stress, with CFB being particularly beneficial for health and stress management.

Author Contributions: Conceptualization, H.-H.C. and C.-H.L.; methodology, H.-H.C. and C.-H.L.; software, H.-H.C.; validation, H.-H.C.; formal analysis, H.-H.C.; investigation, H.-H.C. and C.-H.L.; resources, C.-H.L.; data curation, H.-H.C. and C.-H.L.; writing—original draft preparation, H.-H.C.; writing—review and editing, C.-H.L.; visualization, H.-H.C. and C.-H.L.; supervision, C.-H.L.; project administration, H.-H.C. and C.-H.L.; funding acquisition, C.-H.L. All authors have read and agreed to the published version of the manuscript.

Funding: This research received financial support of NT$169,240 from Chia Nan University of Pharmacy and Science, covering the period from 1 May 2015 to 31 December 2015.

Institutional Review Board Statement: The study was conducted in accordance with the Declaration of Helsinki, and approved by the Institutional Review Board of Chia-Yi Christian Hospital Ethics Committee (protocol code CYCH-IRB No. 103075 and 11/01/2015 of approval).

Informed Consent Statement: Informed consent was obtained from all subjects involved in the study. Written informed consent has been obtained from the all subjects to publish this paper.

Data Availability Statement: https://hdl.handle.net/11296/pa4xcr (accessed on 9 July 2015).

Acknowledgments: We acknowledge the financial support and experimental facilities provided by Chia Nan University of Pharmacy and Science.

Conflicts of Interest: The authors declare no conflict of interest.

References

1. Eidson, R. *Hydrotherapy for Health and Wellness: Theory, Programs & Treatments*; Erin O'Connor (Cengage Learning): Clifton Park, NY, USA, 2009.
2. Wilcock, I.M.; Cronin, J.B.; Hing, W.A. Physiological response to water immersion: A method for sport recovery? *Sports Med.* **2006**, *36*, 747–765. [CrossRef] [PubMed]
3. Uehleke, B.; Wöhling, H.; Stange, R. A Prospective "Study by Correspondence" on the Effects of Kneipp Hydrotherapy in Patients with Complaints due to Peripheral Neuropathy. *Ganzheits Med.* **2008**, *20*, 287–291.
4. Liao, W.C.; Wang, L.; Kuo, C.P.; Lo, C.; Chiu, M.J.; Ting, H. Effect of a warm footbath before bedtime on body temperature and sleep in older adults with good and poor sleep: An experimental crossover trial. *Int. J. Nurs. Stud.* **2013**, *50*, 1607–1616. [CrossRef] [PubMed]
5. Saeki, Y.; Nagai, N.; Hishinuma, M. Effects of footbathing on autonomic nerve and immune function. *Complement. Ther. Clin. Pract.* **2007**, *13*, 158–165. [CrossRef]
6. Vyas, S.C.; Mooventhan, A.; Manjunath, N.K. Effect of hot arm and foot bath on heart rate variability and blood pressure in healthy volunteers. *J. Complement. Integr. Med.* **2019**, *17*, 20180181. [CrossRef] [PubMed]
7. Miyazato, K.; Matsukawa, K. Decreased cardiac parasympathetic nerve activity of pregnant women during foot baths. *Jpn. J. Nurs. Sci.* **2010**, *7*, 65–75. [CrossRef] [PubMed]
8. Sung, E.J.; Tochihara, Y. Effects of Bathing and Hot Footbath on Sleep in Winter. *J. Physiol. Anthropol.* **2000**, *19*, 21–27. [CrossRef] [PubMed]
9. Hsu, Y.Y.; Bai, C.H.; Yang, C.M.; Huang, Y.C.; Lin, T.T.; Lin, C.H. Long Hours' Effects on Work-Life Balance and Satisfaction. *BioMed Res. Int.* **2019**, *2019*, 5046934. [CrossRef] [PubMed]
10. Quick, J.C.; Henderson, D.F. Occupational Stress: Preventing Suffering, Enhancing Wellbeing. *Int. J. Environ. Res. Public Health* **2016**, *13*, 459. [CrossRef] [PubMed]

11. Lin, C.H.; Chu, P.H.; Lin, C.C.; Yu, K.C.; Hsu, H.H. Improved Body Systemic Energy by the LOHAS Program on Spa Industry in Taiwan. In Proceedings of the 9th Asia Tourium Forum, Hualien, Taiwan, 7–9 May 2010; pp. 1041–1049.
12. Liao, W.C.; Chiu, M.J.; Landis, C.A. A warm footbath before bedtime and sleep in older Taiwanese with sleep disturbance. *Res. Nurs. Health* **2008**, *31*, 514–528. [CrossRef] [PubMed]
13. Fiscus, K.; Kaminski, T.; Powers, M. Changes in Lower-Leg Blood Flow During Warm-, Cold-, and Contrast-Water Therapy. *Arch. Phys. Med. Rehabil.* **2005**, *86*, 1404–1410. [CrossRef] [PubMed]
14. Ely, B.R.; Robinson, V.A.; Havens, C.W. The Impact of Acute Hot or Cold Water Immersion on Post-Exercise Blood Pressure Regulation. *FASEB J.* **2020**, *34*, 1. [CrossRef]

engineering proceedings

Proceeding Paper

A Study on Consumer Preferences for Incorporating Sustainable Development Goal Indicators in the Design Attributes of Music Imagery Products †

Lu Chen, Ying-Chieh Wu * and Jiann-Sheng Jiang

Institute of Cultural and Creative Design, Tung-Fang Design University, Kaohsiung 82941, Taiwan; chenlu@xujc.com (L.C.); jjs@mail.tf.edu.tw (J.-S.J.)
* Correspondence: tetsuyatw@me.com

† Presented at the IEEE 5th Eurasia Conference on Biomedical Engineering, Healthcare and Sustainability, Tainan, Taiwan, 2–4 June 2023.

Abstract: We explored consumer preferences for incorporating sustainable development goal (SDG) indicators into music imagery product design. Qualitative and quantitative investigations were conducted to survey and analyze consumer preferences and values. Through questionnaire surveys and focus group discussions, consumers' perceptions and values toward SDG indicators in music imagery products were investigated. The results showed that consumers' demands and preferences for music imagery products aligned with the SDGs. Consumer preferences and values were important in the design of music imagery products. By comparing the influence of different product design attributes on consumer attractiveness and purchase intention, this demonstrates the practical value of incorporating SDG indicators into music imagery products. The results serve as valuable references for the design and marketing of music imagery products to promote the application and advancement of the SDGs.

Keywords: SDG indicators; music imagery products; Miryoku engineering; importance–performance analysis

Citation: Chen, L.; Wu, Y.-C.; Jiang, J.-S. A Study on Consumer Preferences for Incorporating Sustainable Development Goal Indicators in the Design Attributes of Music Imagery Products. *Eng. Proc.* **2023**, *55*, 75. https://doi.org/10.3390/engproc2023055075

Academic Editors: Teen-Hang Meen, Kuei-Shu Hsu and Cheng-Fu Yang

Published: 12 December 2023

1. Introduction

The global promotion of sustainable development goals (SDGs) has gained significant attention, as product design and brand imaging play a crucial role in the modern consumer market. Consumers now prioritize not only practicality and functionality but also emotional and value aspects. Businesses are increasingly recognizing the importance of integrating social responsibility and sustainability into product design. Therefore, by focusing on consumer preferences for incorporating SDG indicators into the design attributes of music imagery products, their impact on consumer attitudes and purchase decisions was investigated in this study. Currently, consumer awareness and attention toward SDGs are not universal. Therefore, this research was carried out to deepen the understanding and concerns of consumers about sustainability and social responsibility by examining the incorporation of SDG indicators in music imagery product design. Integrating music imagery into product design evokes emotions and resonance through visual symbols and design elements, providing consumers with a unique and experiential interaction with the product. We also investigated how these attributes influenced consumer purchase intentions and brand loyalty. Recommendations to businesses were proposed to enhance the quality and social value of product design, promoting sustainable development practices and awareness.

2. Literature Review

2.1. Music Imagery in Visual Design

Numerous studies have focused on transforming abstract and imaginative ideas into tangible products or art designs [1–4]. Music serves as the imagery in relevant visual communication design. By harnessing the emotions and atmosphere conveyed by music, visual symbols, and design elements, they can be translated into design. The analysis of product design aims to explore how to visually represent the abstract characteristics and emotional resonance of music. The application of music imagery to visual design not only offers fresh creativity but also enriches the audience's understanding of the connection between music and visuals. By merging music and the visual arts, possibilities for heightened sensory experiences and creative exploration can be offered.

The interrelationship between music and visuals was investigated to creatively transform abstractive music into concrete visual symbols through design. A direct relationship between pitch and brightness was found in the context of music and color. Additionally, the content, tonality, and rhythmic sense of music influenced the visual form [5]. Music imagery was employed in visual communication design to visualize the perception of sound and music through color and shape and infuse new spiritual significance. With the elements of music, aesthetic forms, and the relationship between music and visual design, the shape and color of music imagery were analyzed, opting for a figurative approach to represent music, creating unique visual images, and exploring the abstract expressions of music [6]. The relationship between the shape elements of jazz drums and the imagery was analyzed to consider design elements. Through such imagery analysis, the correspondence between the structure of the imagery and the shape elements was analyzed, and the causal relationship between the imagery and overall evaluations of likability and beauty was examined [7].

Based on the previous result, we assumed that transforming music into visual symbols and design elements allowed products to uniquely evoke emotions and resonance in consumers. The use of music imagery not only infuses products with emotions and values but also provides consumers with an experiential interaction with the product. As music is an abstract expression, it evokes emotions and spiritual resonance in people. Through the rhythm, melody, and emotions of music, the design of the visual communication of products evokes consumers' deep emotional memories and experiences. By researching products with music as imagery in relevant visual communication design, emotionally resonant and differentiated products can be created, and innovation in design enriches fostering deeper interactions and connections between consumers and products. The use of music imagery elevates products to a higher artistic realm, creating richer and more valuable consumer experiences.

2.2. Importance–Performance Analysis (IPA)

IPA is a practical and feasible method for measuring the importance and performance of products or service attributes [8]. Martilla and James introduced IPA to prioritize attributes' importance and performance based on consumers' perceived importance and companies' performance related to quality attributes [9,10]. IPA is to assess consumers' perceived importance and satisfaction with quality attributes and provide a basis for service quality and marketing strategy decisions in the service industry [11]. It plots consumer satisfaction with product attributes on the horizontal axis and the importance of these attributes on the vertical axis. The central coordinate of the strategy matrix is determined by the average of satisfaction and importance values, dividing the matrix into four quadrants. Attributes are then categorized into quadrants based on their assessment values. The research methodology is used to conduct consumer surveys, using importance as the vertical axis and performance as the horizontal axis, and using the average values of each attribute to separate them on the x and y axes. The importance and performance of each attribute are marked by their positions in corresponding quadrants. Subsequently, specific strategies for quality improvement are determined and proposed based on the meanings represented by the attributes' locations in the quadrants (Figure 1).

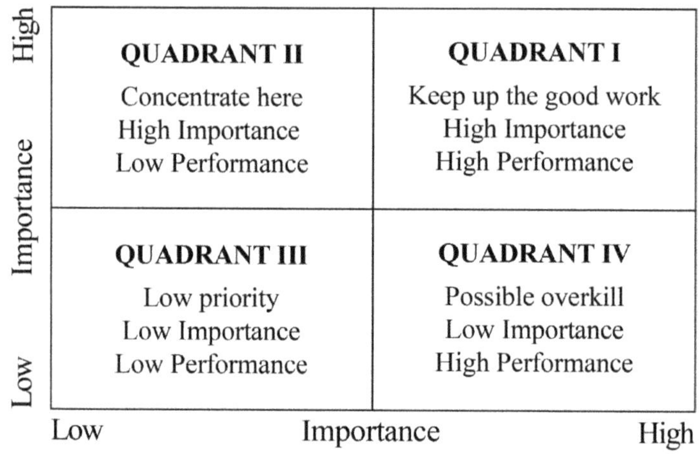

Figure 1. Chart of quadrants of IPA.

3. Research Methodology

We investigated consumer preferences for music imagery products with design attributes, considering SDG indicators. The Expert Group Meeting (EGM) method was used with six individuals, including three product design experts, two product marketing specialists, and one scholar. They identified and analyzed the attractiveness factors of music imagery products and clarified the relevance of incorporating SDG indicators into the design attributes. Based on the result, a correlation evaluation model for practical attractiveness was established for companies to integrate SDG indicators into music imagery product design and marketing strategy establishment. An IPA questionnaire was created to explore the design attributes of music imagery products and understand the strengths and weaknesses of product quality. The attractiveness factors incorporating SDG indicators into the design attributes were determined to investigate distinctive features to be maintained and improved for future music imagery product design. The data related to music imagery product design was obtained for importance and satisfaction ratings of how to incorporate SDG indicators into the design attributes of music imagery products. The values were calculated and plotted on a quadrant chart.

4. Results and Discussion

4.1. EGM Interview

Based on the simplified chart of attractiveness factors from the EGM (Figure 2), the following observations were obtained.

Visual design elements such as appearance, color selection, patterns, and material texture directly influenced the product's visual performance, and consumers held positive attitudes toward these design elements. The incorporation of music imagery in visual design, attractive colors, patterns, and textured materials attracted consumers and evoked their interest and resonance. Attractiveness factors related to sustainability, environmental protection, and economic integration indicated consumers' concern for products that integrated SDG indicators to promote sustainable development. This reflected the company's commitment to societal and environmental responsibility, resonating with consumers and enhancing the product's attractiveness. Attractiveness factors related to values, value, and emotional memories highlighted consumers' focus on the product's value and emotional connections. The use of music imagery infused products with more emotions and value, giving them special significance in consumers' minds.

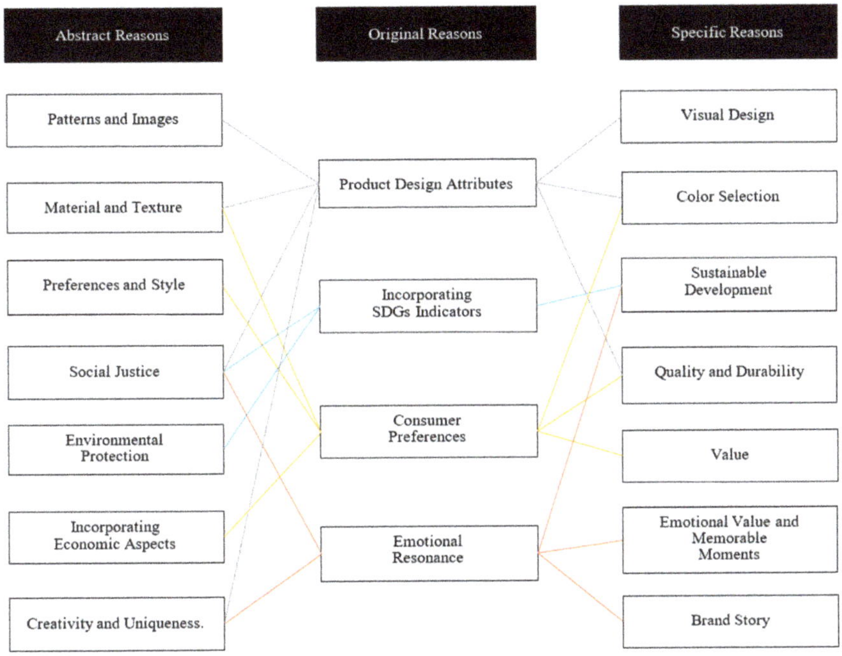

Figure 2. Summary of interview results of EGM.

The visual symbols and symbolic resonance of music imagery products significantly influenced consumer attitudes. By integrating SDG indicators and elements of sustainable development and emphasizing visual design, environmental protection, values, and brand storytelling, the visual appeal of products was enhanced to evoke emotional resonance with consumers. Additionally, by incorporating creative uniqueness and elements that aligned with individual preferences and styles, a stronger connection between the product and consumers was established, and its attractiveness and competitiveness were enhanced.

4.2. IPA Questionnaire Survey

In May 2023, the IPA questionnaire was conducted using Google Forms to collect responses from graduate students and professionals in the arts, design, and creative industries who were familiar with SDG concepts. A total of 110 valid questionnaires were collected. Table 1 presents the reliability and level of consistency of the IPA questionnaire survey using Cronbach's α. The importance and satisfaction of the IPA questionnaire showed Cronbach's α values of greater than 0.70, indicating a high level of internal consistency and stability of the items. The overall Cronbach's α for the entire questionnaire was 0.964. Within the range of $0.7 < \alpha \leq 0.9$, a high level of reliability was verified. Therefore, all items were retained for further analysis.

Table 1. Reliability of IPA questionnaire.

Questionnaire Item	Importance	Satisfaction
1. Visual Design	0.964	0.963
2. Color Selection	0.964	0.963
3. Patterns and Images	0.964	0.963
4. Material and Texture	0.964	0.963
5. Incorporating SDGs	0.964	0.963
6. Social Justice and Equity	0.964	0.963
7. Environmental Protection	0.964	0.963
8. Economic Sustainability	0.964	0.962
9. Creativity and Uniqueness	0.964	0.962
10. Personal Preferences and Style Alignment	0.964	0.963
11. Quality and Durability	0.964	0.963
12. Price and Value	0.964	0.963
13. Emotional Expression and Resonance	0.964	0.962
14. Brand Story and Background	0.963	0.962
15. Emotional Engagement and Connection	0.964	0.962
16. Emotional Value and Memorable Moments	0.964	0.962

4.3. Statistical Analysis

In Table 2 and Figure 3, the IPA results revealed three distinctive characteristics: visual design, patterns, and images. Material and texture were perceived as strengths and indicated consumers' positive attitudes toward the visual appearance and texture of music imagery product designs. These visual symbols enhanced the product appeal and resonated with consumers. Incorporating environmental protection and economic sustainability was emphasized for consumers' concern for products that integrate related indicators. These symbols conveyed social responsibility and encouraged consumers to choose sustainable products. Creativity and uniqueness, alignment with personal preferences and style, brand story and background, emotional engagement and connection, emotional value, and memorable moments were factors to be improved as consumers expected more creative and personalized elements in music imagery product designs for their preferences and emotions. These factors significantly influenced symbolic resonance, established deeper connections with consumers, and elicited emotional responses.

Table 2. Analysis results of IPA.

	Importance (M = 4.23)	Satisfaction (M = 3.29)	IPA Area
1. Visual Design	4.47	3.37	I.
2. Color Selection	4.35	3.28	II.

Table 2. *Cont.*

	Importance (M = 4.23)	Satisfaction (M = 3.29)	IPA Area
3. Patterns and Images	4.48	3.30	I.
4. Material and Texture	4.27	3.34	I.
5. Incorporating SDGs	4.00	3.36	IV.
6. Social Justice and Equity	3.53	2.92	III.
7. Environmental Protection	3.95	3.51	IV.
8. Economic Sustainability	3.94	3.43	IV.
9. Creativity and Uniqueness	4.54	3.24	II.
10. Personal Preferences and Style Alignment	4.26	3.23	II.
11. Quality and Durability	4.11	3.65	IV.
12. Price and Value	4.20	3.37	IV.
13. Emotional Expression and Resonance	4.45	3.29	I.
14. Brand Story and Background	4.30	3.17	II.
15. Emotional Engagement and Connection	4.51	3.10	II.
16. Emotional Value and Memorable Moments	4.36	3.05	II.

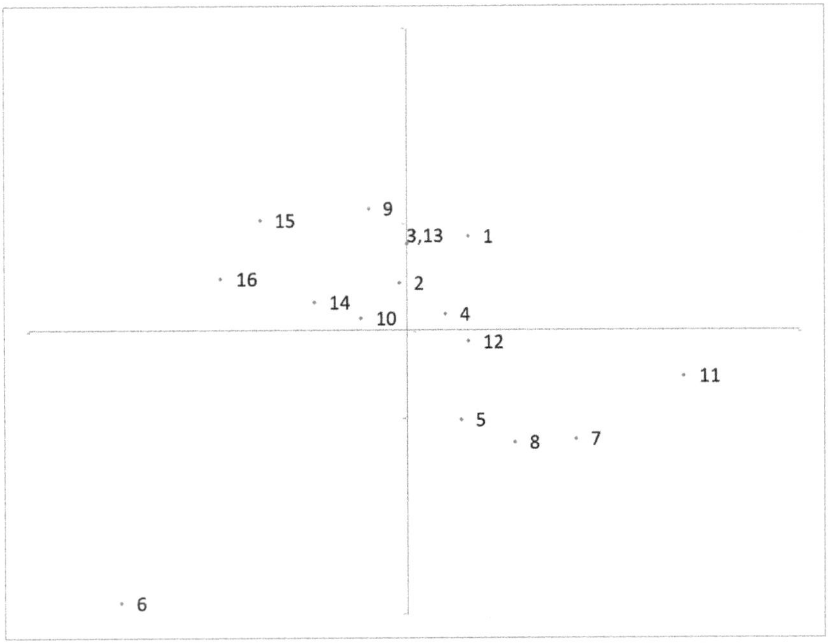

Figure 3. IPA matrix distribution.

Consumer preferences for incorporating SDG indicators into the design attributes of music imagery products emphasized the importance of visual symbols and symbolic resonance. Music imagery product design needs to emphasize creativity and uniqueness while aligning with consumers' personal preferences and style. Additionally, by integrating elements of environmental protection and economic sustainability, emotional connections and resonance are created. By adopting these design strategies, consumers' emotional expression and value reminiscence can be satisfied, enhancing product appeal and competitiveness.

5. Conclusions

We explored consumer preferences for incorporating SDG indicators into the design attributes of music imagery products. Using the importance–performance analysis (IPA) method, we assessed the correlation between consumers' perceived importance and satisfaction levels for different attributes. The results revealed diverse consumer preferences that provided crucial insights for businesses in product design. Consumers regarded visual design, patterns and images, and emotional expression and resonance as important for their satisfaction and emphasized the visual presentation and emotional connection of products. These attributes need to be optimized to increase consumer satisfaction and brand loyalty. Creativity and uniqueness, alignment with personal preferences, and style were perceived as more important than actual satisfaction, indicating they need to be enhanced. More creative and personalized designs are necessary to attract target consumers and enhance brand competitiveness. Consumers emphasized attributes related to SDG indicators, such as incorporating environmental protection and economic sustainability. The emphasis on these attributes reflected high expectations for businesses' social responsibility and environmental awareness. Thus, companies must actively integrate SDG indicators into product design while maintaining a balanced approach to quality elements. Consumers thought that price and value, quality, and durability were more important than actual satisfaction, expected cost-effectiveness, and product durability. Therefore, high-quality, durable, and reasonably priced products must be developed to meet consumer demands. Based on the findings, we propose the following recommendations for designing music imagery products for consumer preferences.

1. Strengthening visual design and emotional connection by optimizing appearance, patterns, and images and creating visually appealing effects that evoke emotional resonance, increasing product appeal and brand image
2. Offering creative and personalized choices by enhancing creativity and uniqueness, providing diverse choices that align with personal preferences and style, and enhancing product personalization and differentiation
3. Incorporating sustainable development goals by actively addressing SDG indicators and integrating sustainable development and social responsibility into product design to respond to consumers' interest in environmental protection and social justice.
4. Pursuing cost-effectiveness and product durability to aim for reasonably priced and durable product features by providing consumers with more valuable choices.

Author Contributions: The contribution of the author to the paper includes the provision of study materials or patients by L.C., critical revision of the article for important intellectual content, and final approval of the article by J.-S.J. Y.-C.W. provided administrative, technical, or logistic support. All authors have read and agreed to the published version of the manuscript.

Funding: This research did not receive external funding.

Institutional Review Board Statement: Not applicable.

Informed Consent Statement: Not applicable.

Data Availability Statement: The data supporting the findings of this study are not publicly available as the research is ongoing. However, they can be made available upon reasonable request. Interested researchers can access the data by contacting the author at tetsuyatw@me.com.

Conflicts of Interest: The authors declare no conflict of interest.

References

1. Ho, T.H. Symbolic Representation of Moon Imagery in Poetry: A Study on Dynamic Poster Creation. Master's Thesis, Department of Design, National Taiwan Normal University, Taipei City, Taiwan, 2022.
2. Lin, C.L. Application of Color Imagery in Flower Language to Color Suggestions and Analysis. Master's Thesis, In-Service Master's Program in Graphic Communication, National Taiwan Normal University, Taipei City, Taiwan, 2022.
3. Lee, Y.C. Construction of Creative Product Design Model: A Case Study of Analyzing and Applying Ocean Imagery. *J. Bus. Des.* **2023**, *26*, 85–99.
4. Chi-Hsiung, C.; Yeh-Hung, T. Research on Taiwan Cultural Imagery in Products: A Case Study of Tainan Lin Department Store's Cultural and Creative Products. *J. Appl. Des.* **2023**, *17*, 6–24.
5. Zhang, Z.X. Study on Form and Color Design of Music Imagery. Master's Thesis, Department of Design, National Taiwan Normal University, Taipei City, Taiwan, 2016.
6. Wang, L.F. Study on the Expression of Music Imagery in Visual Communication Design Works. Master's Thesis, Institute of Design, National Taiwan Normal University, Taipei City, Taiwan, 2001.
7. Wei, G.X. Study on the Formative Imagery of Musical Instrument Products: A Case Study of Jazz Drums. Master's Thesis, Department of Industrial Design, National Taipei University of Technology, Taipei City, Taiwan, 2004.
8. Sampson, S.E.; Showalter, M.J. The performance-importance response function: Observations and implications. *Serv. Ind. J.* **1999**, *19*, 1–25. [CrossRef]
9. Martilla, J.A.; James, J.C. Importance-performance analysis. *J. Mark.* **1997**, *41*, 77–79. [CrossRef]
10. Yang, W.H.; Liang, J.S. Application of Importance-Performance Analysis to Diagnose the Service Value of Taiwan International Port Logistics Center. *Marit. Q.* **2011**, *4*, 25–61.
11. Hansen, E.; Bush, R.J. I Understanding customer quality requirements: Model and application. *Ind. Mark. Manag.* **1999**, *28*, 119–130. [CrossRef]

engineering proceedings

MDPI

Proceeding Paper

Digitalization of the Quranic Reading Learning Program for Collage Students, Indonesia †

Mutimmatul Faidah [1,*], Siti Makrufah [2], Eliya Najma Muntazeri [3] and Safana Nejat Muntazeri [4]

1 Department of Cosmetology Education, Faculty of Engineering, Universitas Negeri Surabaya, Surabaya 60231, Indonesia
2 Faculty of Mathematics and Natural Sciences, Universitas Negeri Surabaya, Surabaya 60231, Indonesia; siti.23008@mhs.unesa.ac.id
3 Faculty of Economy and Islamic Business, Universitas Islam Negeri Sunan Ampel, Surabaya 6037, Indonesia; 08030421100@student.uinsby.ac.id
4 Faculty of Psychology and Health, Universitas Islam Negeri Sunan Ampel, Surabaya 6037, Indonesia; 11040123152@uinsa.ac.id
* Correspondence: mutimmatulfaidah@unesa.ac.id
† Presented at the IEEE 5th Eurasia Conference on Biomedical Engineering, Healthcare and Sustainability, Tainan, Taiwan, 2–4 June 2023.

Abstract: This study aimed to describe the planning of an online Quranic Reading Learning (QRL) program for students, explain the implementation of the online QRL program, and reveal the students' response on it. This study used a qualitative research approach. Data were collected using interviews, questionnaires, and observations that were conducted from July to October 2022. The results showed that the online QRL program was carried out as a response to the transition in lecturing mode from online to hybrid and fully offline. This program was attended by 4000 students programmed Islamic religious education courses and were guided by 50 teachers. Learning was carried out in eight online meetings. It progressed through a class grouping test, a grand opening, online learning, and a final evaluation. The identified problems were the large number of participants in a group, unequal assistance for individual participants, network constraints, noise that classical learning caused, and unclear articulation. The advantage of this program was the efficient cost, time, and place. All participants showed a positive response to the use of this program and suggested that it should be continued in the next semester with various improvements.

Keywords: online learning; articulation; reading; Quran

Citation: Faidah, M.; Makrufah, S.; Muntazeri, E.N.; Muntazeri, S.N. Digitalization of the Quranic Reading Learning Program for Collage Students, Indonesia. *Eng. Proc.* **2023**, *55*, 76. https://doi.org/10.3390/engproc2023055076

Academic Editors: Teen-Hang Meen, Kuei-Shu Hsu and Cheng-Fu Yang

Published: 13 December 2023

1. Introduction

The COVID-19 pandemic has had an impact on all sectors of life, including education. Education and learning should not be prevented in any situation. Adapting to the pandemic using online learning with various media platforms is an option [1]. Post-pandemic, online learning is believed to be an alternative and needs to be continued. The pandemic, on the one hand, has contributed to the acceleration of the digitization of education [2]. In the context of learning in universities in Indonesia, Islamic Religious Education courses are courses that must be taken by Muslim students. This course is in the family of Personality Development courses. The orientation of Islamic Religious Education is to study Islam and to find out how to pursue the right religion and source of knowledge. This course is expected to add religious insight and influence student attitudes and behaviors to be better, more religious, and more moderate [3]. Islamic lectures occupy two credits with 16 face-to-face meetings. In addition, there is one component of lectures outside of learning with lecturers, namely the program on learning to read the Quran.

Before the pandemic, the Quranic Reading Learning (QRL) program at Universitas Negeri Surabaya was carried out face to face. Each teacher teaches 20 students at the campus mosque. During the pandemic, the QRL program was abolished because there was no proper formula for this learning. Starting from the 2022/2023 academic year, this program will be implemented again with an online system. This program is interesting to study as the Quran has strict provisions of pronunciation of each letter and reading law [4]. This uniqueness needs to be studied further if learning to read the Quran is carried out online. Hence, this study aims to describe the planning of the online QRL program, explain the implementation of the online QRL program, and reveal the student's response to the implementation of the online QRL program to provide better improvements.

Research on learning to read the Quran online has been studied. Yahya researched an online learning mode of a Quran reading class during the pandemic. This study found that there was no significant difference between the student performance in the online and face-to-face learning modes. This study also looked at the saturation of online learning for longer periods of time [5]. In contrast to this research, this study explores the stages of implementing Quran reading education online and examine its problems in large class groups.

This research used a qualitative research approach. Data were collected using interviews, virtual observations, and questionnaires. The interviews were undertaken with the program management team on learning to read the Quran, teachers, and students as participants. The implementation of the Quran reading program was observed. And the questionnaire was given to the participants with a total of 1665 respondents. The research was conducted from July to October 2022 at Universitas Negeri Surabaya, Surabaya, Indonesia.

2. Discussion

2.1. Planning the Online Quranic Reading Learning

In Islam, the Quran is the guide for the life of a Muslim. For Indonesian Muslims, reading the Quran requires a gradual learning process because it is written in Arabic. It has different pronunciation rules from Latin letters. In ritual worship, such as prayer, Muslims use Arabic language Quranic sentences and verses, whether they can speak Arabic or not. The ability to read the Quran and read Arabic sentences is a necessity for every Muslim. Learning to read the Quran becomes an activity that is worthy of devotion and attention. This conception gives birth to the tradition of learning to read the Quran for Indonesian children from childhood to adulthood. Responding to these needs, various easier learning methods for reading the Quran have been developed, such as the *Iqra'*, *Qiroati*, *Tilawati*, and other methods. Each method emphasizes how to read the Quran according to the correct rules and proper pronunciation. A Quranic reader is required to follow the laws of pronunciation, the short length of reading, the articulation of pronunciation, and the nature of the letters correctly [6]. Reading the Quran is one of the mandatory and challenging tasks for most Muslim students in Indonesia.

Learning to read the Quran before the pandemic took place face to face in campus mosques in the form of study groups. The teacher classically gives an example of a correct reading and the student follows the reading by opening their textbook. For practice, the teacher calls the students' names in turn and asks them to read the Quran. The teacher corrects their readings in terms of pronunciation and reading law.

Universitas Negeri Surabaya is one of the state universities in East Java, Indonesia. The Islamic Religious Education course is compulsory. Meanwhile, students of religions other than Islam follow the schedule of religious lectures according to their religion. Based on interviews with the Academic Division of the campus, it is known that in 2022, 10,452 undergraduates were accepted and spread across eight faculties, namely the Faculty of Sports Science, Faculty of Social Sciences and Law, Faculty of Education, Faculty of Language and Arts, Faculty of Engineering, Faculty of Economics and Business, Faculty of Science and Mathematics, and Vocational Program. UNESA has two campuses, namely the Ketintang campus and the Lidah Wetan campus. Due to the large number of students, Islamic religious courses are issued in turn. Those who come from the Faculty of Education, Faculty of Language and Arts, Faculty of Sports Sciences, and Vocational Program participate at the same time. In even semesters, the course is conducted for those who are from Faculty of Engineering, Faculty of Economics and Business, Faculty of Mathematics and Science, and Faculty of Social Sciences and Law.

Based on an interview with the management team of the QRL program, the number of students programmed Islamic Religious Education in the current semester is 4000 students. The rest will take this course in the even semester from January to June. One of the obstacles in the implementation of face-to-face learning mode is the availability of space and learning facilities.

The planning stages of the QRL program are described as follows.

1. Program planning is prepared by the management team (Lecturer team and Student Activity Unit for Islamic Spirituality). In a meeting, the methods and textbooks used are discussed. There are various methods, but not all methods are easy to implement online and provide open access. Only the Iqra method provides open access for users and free downloads. The method consists of six Volumes of books (e.g., Volumes 1, 2, 3, 4, 5, and 6). The books are organized based on the order of Hijaiyah letters, arranged from concrete to abstract, ranging from easy to difficult, and ranging from simple to complex. The Iqra method has wider advantages and it has been applied throughout Indonesia and several Asian countries for its flexible, easy, and affordable mode of implementation [7].
2. Class mapping and coordination with the class leader are then conducted. Based on the incoming data, 4000 students take part in the program. They are divided into 100 study groups with 40 people and a teacher for each group.
3. To meet the needs of teachers on the required qualifications, open recruitment is carried out. This utilizes WA and Instagram with registration via Google Forms.
4. Tests and grouping of teachers are conducted. Teachers who have registered are tested online using the Google Meet platform. From the test results, they are grouped into four groups; Class A for Iqra 3, Class B for Iqra 4, Class C for Iqra 5, and Class D for Iqra 6.
5. Teacher training is undertaken to provide QRL methods and good forms of communicating.

2.2. Implementation of the Online QRL Program

The QRL implementation information is submitted via Instagram: tqq.unesa.official, which is related to:

1. The QRL registration link and contact person
2. The QRL implementation schedule
3. Management Team
4. Frequently Asked Questions
5. Implementation Guidelines
6. Class and Teacher Grouping
7. Activity Documentation

Based on the observation of the implementation of the program, it can be described as follows.

1. Socialization and registration. Registration through Google Forms can be accessed on the Committee's Instagram page.
2. Class grouping test. The test is carried out online using Google Meet. Participants are asked to read several verses from the Quran to assess the correctness of their readings.
3. Class and teacher determination. Based on the test results, students are grouped into four groups.
4. The grand opening conducted online via Zoom and YouTube is attended by all participants.

5. Implementation of learning with as many as six online meetings using Google Meet. The methods used include: (a) the teacher gives an example of the correct reading and the participants follow it; (b) the teacher calls participants' names in turn to read the Quran; (c) the teacher evaluates the participants' readings; and (d) 90 min of learning duration per meeting is undertaken.
6. Evaluation. The evaluation includes the psychomotor aspects of reading ability. Participants are asked to read according to their learning achievements and memorize several letters in the Quran. Evaluation is continued until the end of the program.

Based on this description, it is known that the digital patterns used are still limited, including: (1) recruitment utilizing social media, (2) registration using online forms, (3) a grand opening via Zoom and YouTube, (4) online implementation using the Google Meet platform, and (5) online evaluation. Social media plays a very significant role in the management of online learning [8].

The digitization of this program is limited to the use of digital media in the delivery of information and implementation. In fact, this program can be managed using digital optimization, such as a Learning Management System (LMS) that allows all processes to be on one digital network. An overview of the implementation of QRL activities is presented in Table 1.

Table 1. Learning schedule.

No.	Time	Activities
1	First 5 min	The introduction consists of greetings, praying together, motivation Classical
2	10 min	Classical repetition of material
4	50 min	Today's material (classical and individual)
3	20 min	Assignment (memorizing prayers and short letters from the Quran)
4	Last 5 min	Prayer

2.3. *Responses and Problems in Program Implementation*

Participants responded on the QRL program by providing answers to the following items:

1. I find it easy to follow QRL Online.
2. I find it difficult to interact with teachers when learning QRL online.
3. Online learning has more disadvantages than advantages.
4. I feel more time and effort efficient with online QRL.
5. The online platform determined by the committee is easy to run.
6. Online learning requires more preparation in learning.
7. I easily understand the correction of qur'anic readings delivered by the teacher in the online QRL.
8. I easily understand the correction of the Tajwid of the Qur'an that the teacher delivered in the online QRL.
9. I find classical learning easy to implement in online QRL.
10. I would recommend online QRL be implemented in next semester.
11. *Iqra* book is read clearly in online QRL.
12. I feel I can follow the teacher's Quranic reading in the online QRL.
13. I feel that learning is more effective in offline QRL than online QRL.
14. I feel that the set time duration is appropriate.
15. I feel that individual mentoring in online QRL is adequate.
16. Prayer and Memorization are easy to follow.

An overview of participant's responses of QRL activities is presented in Figures 1 and 2.

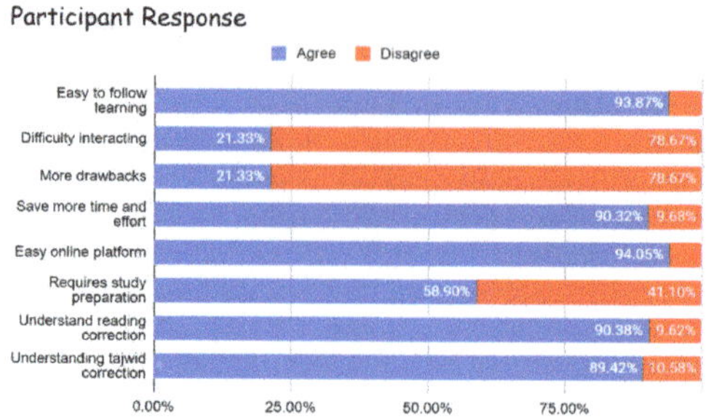

Figure 1. Participants' responses for items 1–8.

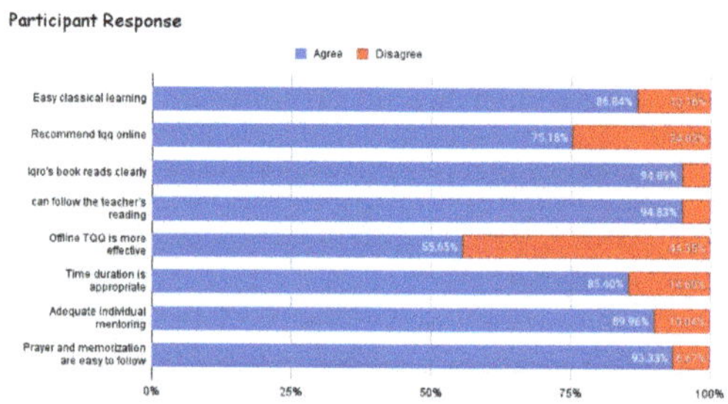

Figure 2. Participants' responses for items 9–16.

Positive responses that showed approval above 75 percent included the following: (1) easy to follow QRL online; (2) efficient use of time and effort; (3) easy platform; (4) understanding reading correction; (5) understanding *tajwid* correction; (6) easy classical learning; (7) recommending online QRL; (8) reading the *Iqra* book clearly; (9) easy to follow teachers' explanation; (10) appropriate duration; (11) sufficient for individual assistance; (12) and ease in following prayers and memorization.

The negative responses to items that received less than 25 percent agreement were these: (1) difficulty interacting with the teacher and (2) many disadvantages to online QRL. A negative response on online learning about requiring more preparation received approval from 59 percent of the respondents and offline QRL was suggested to be more effective by 54 percent.

In general, participants gave a positive response to the implementation of the online QRL. This is in accordance with the results of research by A. Odeh and I. Keshta that confirms that the pandemic has given new life to effective online learning [9].

Based on the responses and suggestions of the participants related to problems and suggestions for improving the program implementation, several findings can be mapped, namely: (1) the number of participants in one group is too large. This has an impact on mentoring participants individually less evenly and the time required becomes longer; (2) the network and sound are sometimes broken and unstable; (3) classical learning involves a lot of vocal noise, so that articulation is not clear; (4) in the learning of *makhraj* (articulation of Arabic letters), the participants experience difficulties because errors cannot be corrected directly. There is a time lag for corrections, so it cannot be implemented maximally. Most of the participants were off-camera on the grounds of network constraints, thus making it difficult to correct readings, and (5) participants struggled to conduct independent learning.

Suggestions for the program implementation encompass the following: (1) learning with small groups of 10–15 participants per group should be undertaken; (2) the *Iqra* book should be given to the participants; (3) it is necessary to structure the material of the *tajwid*; (4) the correction of readings is not optimal, so there need to be face-to-face meetings (a combination of online and offline learning); (5) the provision of time for rote memorization should be added since not all young people have the ability to "memorize quickly"; (6) the program planning has developed, so that there are not many changes needed to the concept that has been promoted; (7) the number of teachers should be considered so that individual mentoring is more evenly distributed; and (8) the timing of the implementation should be well managed.

Participants also give a positive response to online learning as it has its advantages. Both teachers and participants can meet on one platform from different regions. This minimizes distance, weather, and space/location constraints, and reduces costs. Participants with ill health can still take part in learning. The challenges include unstable network connection and bad weather making the voices of the participants or teachers indistinct. This is in accordance with previous research on the effectiveness of online learning [10].

3. Conclusions

Based on these findings, it is necessary to develop a more comprehensive online system. The system should allow all processes to be on one website, starting from participant and teacher registration, class grouping tests, teacher grouping tests, teacher graduation announcements, class divisions, learning schedules, and learning links, as well as textbooks. At the next stage, mobile learning should be developed with an application system that can be downloaded on Play store. Currently, mobile Quranic learning applications have begun to be developed. This research makes a valuable contribution to improving the effectiveness of Quran reading classes in situations that forces learning to continue to be conducted online.

Author Contributions: Conceptualization, M.F. and S.N.M.; methodology, M.F. and S.M.; software, S.M.; validation, M.F.; formal analysis, E.N.M. and S.N.M.; investigation, M.F. and E.N.M.; data curation, S.M.; writing—original draft preparation, M.F.; writing—review and editing, S.M and E.N.M.; visualization, S.M. and S.N.M. All authors have read and agreed to the published version of the manuscript.

Funding: This research received no external funding.

Institutional Review Board Statement: The study did not require ethical approval.

Informed Consent Statement: Informed consent was obtained from all subjects involved in the study.

Data Availability Statement: The data presented in this study are available on request. The data are not publicly available due to privacy reasons.

Conflicts of Interest: The Authors declare no conflict of interest.

References

1. Sundararasan, T.; Kalaiyarasan, G. Impact of Pandemic Period of COVID-19 on Online Learning and Teaching in Teacher Education. *Praksis* **2022**, *2*, 211–224. [CrossRef]
2. Agarwal, K.; Choudhury, S.; Tipirneni, S.; Mukherjee, P.; Ham, C.; Tamang, S.; Baker, M.; Tang, S.; Kocaman, V.; Gevaert, O.; et al. Preparing for the next pandemic via transfer learning from existing diseases with hierarchical multi-modal BERT: A study on COVID-19 outcome prediction. *Sci. Rep.* **2022**, *12*, 10748. [CrossRef] [PubMed]
3. Wijaya, C.; Abdurrahman, E.; Saputra, F. Management of Islamic Education Based on Interreligious Dialogue in The Learning Process in Schools as An Effort to Moderate Religion in Indonesia. *Rev. Int. Geogr. Educ. Online* **2021**, *11*, 4306–4314.
4. Kharisma, D.; Irzal, M.; Widyati, R. Rancang Bangun Aplikasi Makharijul Huruf dan Tajwid Berbasis Android Sebagai Penunjang Pembelajaran Tahsin Tilawah, (Design and Build an Android-Based Makharijul Surat and Tajwid Application to Support Tahsin Tilawah's Learning). *J-KOMA J. Ilmu Komput.* **2018**, *2*, 1–10.
5. Yahya, M.W.B.H.M.; Rahman, T.; Siddiq, A.A.; Parihat, P. Online learning in the quran reading class during COVID-19 pandemic. *Int. J. Learn. Teach. Educ. Res.* **2021**, *20*, 142–158. [CrossRef]
6. Adam, M.Z.; Shafie, N.; Abas, H.; Azizan, A. Analysis of Momentous Fragmentary Formants in Talaqi-like Neoteric Assessment of Quran Recitation using MFCC Miniature Features of Quranic Syllables. *Int. J. Adv. Comput. Sci. Appl.* **2021**, *12*, 533–540. [CrossRef]
7. Ulfah, T.T.; Assingkily, M.S.; Kamala, I. Implementasi Metode Iqro' Dalam Pembelajaran Membaca Al-Qur'an, (Implementation of the Iqro Method' In Learning to Read the Qur'an). *TA'DIBUNA J. Pendidik. Agama Islam.* **2019**, *2*, 44. [CrossRef]

8. Sobaih, A.E.E.; Moustafa, M.A.; Ghandforoush, P.; Khan, M. To use or not to use? Social media in higher education in developing countries. *Comput. Human Behav.* **2016**, *58*, 296–305. [CrossRef]
9. Odeh, A.; Keshta, I. Impact of COVID-19 pandemic on education: Moving towards e-learning paradigm. *Int. J. Eval. Res. Educ.* **2022**, *11*, 588–595. [CrossRef]
10. Omar, S.N.Z.; Musa, R.; Mohamad, M.; Cob, C.M.S.C.; Othman, A.Y.; Ramli, R. Efficiency of Online Learning during COVID-19 Pandemic. *WSEAS Trans. Bus. Econ.* **2023**, *20*, 30–39. [CrossRef]

engineering proceedings

MDPI

Proceeding Paper

Research on the Wearable Augmented Reality Seeking System for Rescue-Guidance in Buildings †

Chyi-Gang Kuo [1], Chi-Wei Lee [2], Benson P. C. Liu [3] and Chien-Wei Chiu [1,*]

1 Department of Architecture, Chaoyang University of Technology, Taichung 413310, Taiwan; chyigang@cyut.edu.tw
2 Department of Visual Communication Design, Chaoyang University of Technology, Taichung 413310, Taiwan; cwl@cyut.edu.tw
3 Department of Creative Design, National Yunlin University of Science and Technology, Yunlin 640301, Taiwan; bensonp.c.liu@gmail.com
* Correspondence: chiou1130@cyut.edu.tw

† Presented at the IEEE 5th Eurasia Conference on Biomedical Engineering, Healthcare and Sustainability, Tainan, Taiwan, 2–4 June 2023.

Abstract: When a construction disaster occurs, the first-line rescue personnel often enter the disaster site immediately, and every second counts in rescuing the people who need help. However, the rescue personnel may not be familiar with the indoor layouts of different buildings. If the indoor paths are complicated, or when the fire smoke obstructs the line of sight, the rescue personnel are prone to spatial disorientation, which usually causes the rescue personnel to fall into danger. Therefore, we have developed the "Wearable Augmented reality Seeking System" (WASS) to assist rescue personnel in reading the information provided by the "Building Information Guiding System". This system allows them to enter an unfamiliar space and reach the target rescue position, retreat to the entrance, or find an alternative escape route. The WASS is based on the HoloLens augmented reality system, which displays 3D digital information such as indoor layouts, one's current location, spatial images captured by an infrared camera and a depth camera, and 3D virtual guiding symbols or text. The WASS includes two modules: First, the augmented reality gesture interaction module allows one to read the positioning anchor information of the "Building Information Guiding System" (BIGS). The rescue personnel can communicate via gestures, select the task target, and follow the 3D virtual guidance symbols in the air to reach the relay anchor points and finally arrive at the target position. Second, the service support module, including a lighting source and backup power, ensures that the QR code recognition process and long-term operation of the WASS are successful.

Keywords: rescue-guiding system; augmented reality; building information; wearable devices; spatial information seeking system

Citation: Kuo, C.-G.; Lee, C.-W.; Liu, B.P.C.; Chiu, C.-W. Research on the Wearable Augmented Reality Seeking System for Rescue-Guidance in Buildings. *Eng. Proc.* **2023**, *55*, 77. https://doi.org/10.3390/engproc2023055077

Academic Editors: Teen-Hang Meen, Kuei-Shu Hsu and Cheng-Fu Yang

Published: 14 December 2023

1. Introduction

In 2021, a fire broke out in a building in Changhua County, Taiwan. A firefighter who was the first to enter the fire scene was left alone because his companion was injured and retreated. He was thought to be disoriented in the thick smoke and was eventually found to have exhausted his oxygen cylinder and died in a room without any windows.

According to the Ministry of Interior survey report in May 2022, the main cause of the firefighter's death was the inhalation of toxic gases, resulting in hypoxic shock and suffocation [1]. Therefore, we have established a Wearable Augmented reality Seeking System (WASS) to help firefighters and rescue personnel obtain 3D guidance information for the location of firefighting facilities or escape facilities immediately when entering an unfamiliar interior space and see through the thick smoke to figure out the objects and the heat sources in the surrounding area. We expect the WASS to improve rescue services' success rates and the safety of firefighters or rescue personnel.

2. Literature and Case Study Review

2.1. Literature Review

In the context of identifying the location of people indoors, the most frequently researched technology is indoor positioning technology [2–4]. Nowadays, the primary indoor positioning process first involves setting up auxiliary nodes with fixed positions in the indoor environment. The positions of these nodes are known. Position information such as radio frequency identification (RFID) tags is directly stored in the nodes, while other data, such as infrared and ultrasonic data, are stored in the databases of computer terminals [2].

The positioning system needs to measure the distance from the measured node to the auxiliary node to determine the relative position. Distance measurement usually requires transmitting and receiving equipment. According to the difference in the positions of the transmitter and receiver, the positioning techniques are divided into two types: In the first type, the transmitter is located at the measured node, and the receiver is located at the auxiliary node, and this involves techniques such as infrared, ultrasonic, and radio frequency identification (RFID). In the second type, the transmitter is located at the auxiliary node and the receiver is located at the measured node, involving techniques such as WiFi, ultra-wideband (UWB), and ZigBee [2].

The characteristics of the above positioning technologies are presented in the following Table 1.

Table 1. Characteristics of the positioning technologies [2–4].

Position Technology	Accuracy (5 Stars with the Highest Accuracy While 1 Star with the Lowest Accuracy)	Costs (5 Stars with the Highest Cost While 1 Star with the Lowest Cost)	Penetration Ability (5 Stars with the Highest Penetration Ability While 1 Star with the Lowest Cost Penetration Ability)	Disadvantage
Bluetooth	★★★★★ (Unit: cm)	★★ (USD 5~10 K)	★★★	software calibration
Infrared/Laser	★★★★★ (Unit: cm)	★★ (USD 5~10 K)	★	straight-line detection; easy to block
RFID	★★★ (Unit: m)	★★★ (USD 10~15 K)	★★	short transmission distance
Wi-Fi	★★ (Unit: m)	★★★ (USD 10~15 K)	★★★	complicated construction; high power consumption
ZigBee	★★ (Unit: m)	★★★ (USD 10~15 K)	★★★	susceptible to interference; high power consumption
UWB	★★★★★ (Unit: cm)	★★★★★ (USD > 20 K)	★★★★	susceptible to interference

However, none of the current mainstream indoor positioning systems mentioned above meet all of the relevant requirements (e.g., low cost and high accuracy). The auxiliary nodes "installed in the buildings" have to be without replacing batteries, and the system can still work even under extreme environmental conditions, such as during power outages, under high humidity, at high temperatures, and amidst dense smoke.

2.2. Case Study

In practical applications, thermal imaging cameras (TICs) are the best tools for firefighters to see through smoke and find the source of the fire in a fire scene obscured by dense smoke. In accordance with the "Fire Bureau of Taichung City Government Guiding Principles for Operation and Maintenance of Disaster Relief Equipment" [5], the appropriate circumstances in which one should use a thermal imaging camera include the following.

2.2.1. Fire Cases

TICs are used to detect and display the temperature around the fire site, to search for fire points and hidden fire sources, and determine the direction of fire spread.

2.2.2. Identifying the Environment

Under the influence of factors such as dense smoke, insufficient light, and a closed environment generated at the disaster site, visibility can be reduced, which is harmful to the safety of rescuers. The thermal images displayed by thermal imaging cameras can be used to preliminarily distinguish the terrain and features of the site.

2.2.3. Search and Rescue of Human Life

When the temperature of any object is above absolute zero, it emits different infrared rays due to the strength of internal molecular vibrations. Rescuers can use this feature to search for people who need to be rescued.

2.2.4. Chemical Tank Disasters

If the chemical tank body is impacted, overturned, and leaked, it may cause ignition or a fire, and the pressure accumulates as the temperature rises. Thermal imaging cameras can detect the temperature change of a tank body, allowing one to take appropriate protective measures to reduce the occurrence of more disasters (Figure 1).

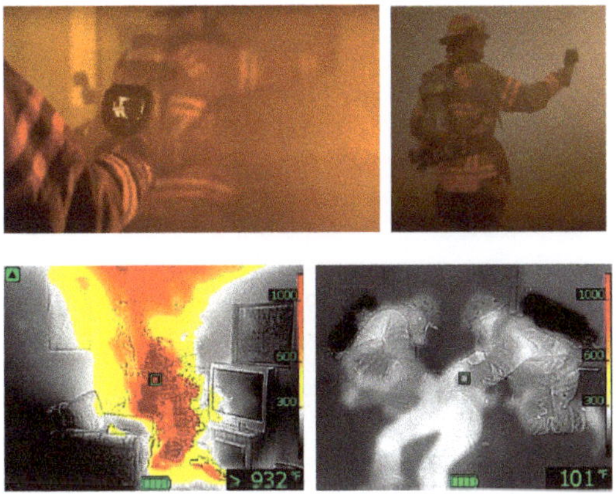

Figure 1. Images captured by thermal imaging cameras used by firefighters [6]. (From: https://www.flir.asia/discover/public-safety/no-excuse-for-firefighter-disorientation/ (accessed on 6 November 2022)).

Since the TICs are handheld devices, they hinders the ability of firefighters or rescue personnel to carry other things, extinguish fires, or assist rescuers. Moreover, TICs only have one single function. If their multiple functions were integrated into the design of a wearable device, the dexterity of the user's hands would be increased.

3. Design of the AR System

Based on the above analysis, we designed the Wearable Augmented reality Seeking System (WASS), which allows search and rescue personnel to directly see infrared images through smart glasses without using a handheld thermal camera. At the same time, through the depth sensor, the outline of surrounding objects can be seen clearly in the dark, even if these objects do not generate heat themselves. Additionally, through reading the information provided by the "Building Information Guiding System" (BIGS), the WASS lets rescue personnel view the 3D guiding arrows and information floating in the air to guide them in reaching the target rescue position, retreat to the entrance, or find an alternative escape route.

We used HoloLens as the basic device of the WASS. HoloLens has hardware equipment such as infrared cameras, depth sensors, and inertial sensors (Figure 2). It also has an open development environment "Research Mode" that allows researchers to develop software to control the hardware equipment on it.

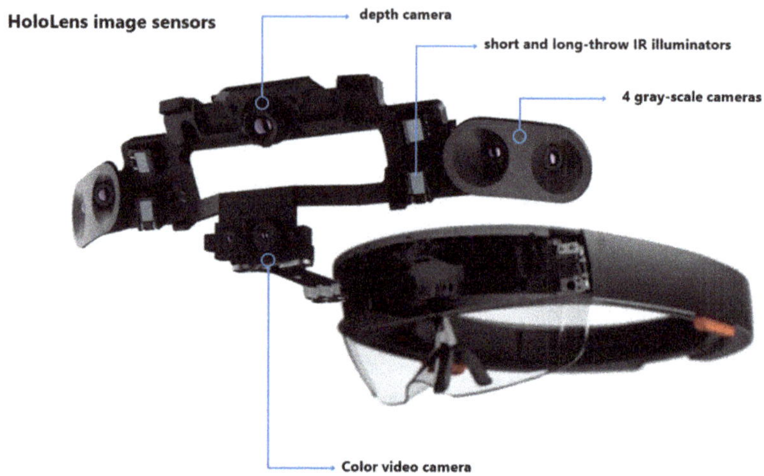

Figure 2. Hardware components of HoloLens 2 [7].

According to the introductory page of the website for HoloLens2, the Research Mode is for research applications and has access to the following streams:

1. Visible Light Environment Tracking Cameras—Grayscale cameras used by the system for head tracking and map building.
2. Depth Camera operating in two modes.

AHAT, high-frequency (45 FPS) near-depth sensing is used for hand tracking. Different from the first version's short-throw mode, AHAT gives a pseudo-depth with phase wrap beyond 1 m.

3. Long-throw, low-frequency (1–5 FPS) far-depth sensing used by spatial mapping.

Two versions of the IR-reflectivity stream are used by the HoloLens to compute depth. These images are illuminated by infrared and unaffected by ambient visible light.

The Research Mode is designed for academic and industrial researchers exploring new ideas in the fields of Computer Vision and Robotics. It is not intended for applications deployed in enterprise environments or available through the Microsoft Store or other distribution channels [8].

4. Experiment and Discussion

The WASS is based on the HoloLens augmented reality system. In the Research Mode, we built a Visual Studio environment to complete the programming so that, when one is wearing the HoloLens, they can see two visible light camera views on the left and right: an infrared camera view and a depth reader view (Figure 3).

As shown in the Figure 4, visible light cameras do not permit the viewing of images in a low-light environment. Still, infrared cameras allow one to see hot objects such as hands, and depth sensors can depict the contours and depths of surrounding objects. This helps search and rescue personnel see other rescuers and those who need rescuing and escape paths clearly in the dark and amidst thick smoke, also helping them to avoid hitting surrounding objects.

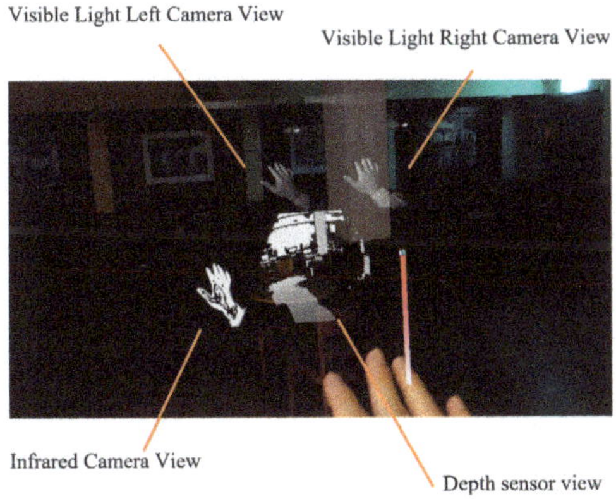

Figure 3. Four views captured by different cameras and sensors equipped on HoloLens launched by the program used for this study. The red-to-white gradient color bar on the screen is the coordinate showing the extent of the HoloLens tilt or rotation detected by the Inertial Measurement Unit (IMU).

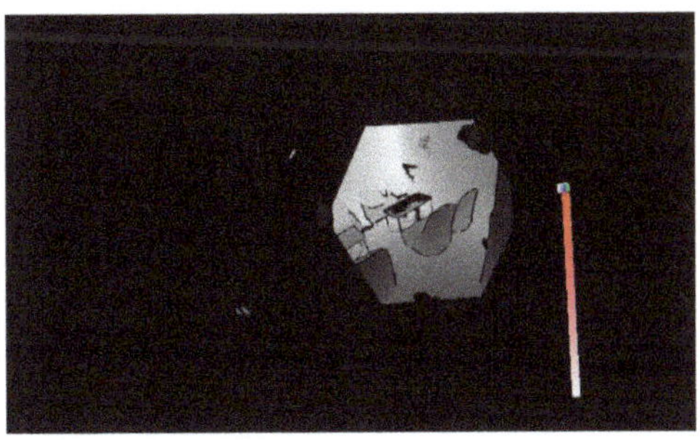

Figure 4. Visible light cameras and IR cameras cannot see objects clearly, but the depth sensor can allow one to see nearby things.

Compared with normal helmets used by firefighters (Figure 5), WASS displays 3D digital information in the HoloLens, including indoor layouts, one's current location, spatial images captured by the infrared camera and depth camera, and 3D virtual guiding symbols or text. Regarding the system's integration in firefighters' helmets, we used a fire helmet with a shallow brim so that the HoloLens glasses can be lifted when necessary. At the same time, there is an extended brim at the back of the fire helmet to protect the battery of the HoloLens from impact and other damage (Figure 6). The simulation of a firefighter wearing a WASS-based device is shown in Figure 7.

Figure 5. Helmets used by active-duty firefighters in Taichung City.

Figure 6. The WASS consists of a fire helmet and HoloLens 2, as well as other accessories.

Figure 7. Simulation of a firefighter wearing a WASS-based device.

The WASS includes the following modules:

1. The augmented reality gesture interaction module, which helps one to read the positioning anchor information of BIGS. The rescue personnel can communicate via gestures, select the task target, and follow the 3D virtual guidance symbols in the air to reach the relay anchor points and finally arrive at the target position.
2. The service support module, which includes a lighting source and backup power to ensure the QR code recognition process and the long-term successful operation of the WASS.

5. Conclusions

In this study, we applied HoloLens Research Mode in an open environment in combination with other creative programs to construct the WASS. The user of the WASS can see images provided by visible light cameras, infrared cameras, and depth cameras on the screen of the smart glasses. Then, using the WASS, rescue personnel can still work in dark or dense smoky environments and see their fellow rescuers, those that need to be rescued, surrounding objects, and escape routes. The WASS can also be used with the BIGS indoor space database to see 3D guidance arrows and information inertially positioned in space. The WASS can help firefighters overcome spatial disorientation in extreme environments.

Author Contributions: Conceptualization, C.-G.K., B.P.C.L. and C.-W.L.; Methodology, C.-G.K.; Formal analysis, C.-G.K.; Resources, C.-G.K., B.P.C.L. and C.-W.C.; Data curation, C.-G.K. and C.-W.C.; writing—review and editing, C.-G.K.; project administration, B.P.C.L. All authors have read and agreed to the published version of the manuscript.

Funding: This research was funded by Ministry of Science and Technology of Taiwan, project number: MOST 108-2221-E-324-001-MY3.

Institutional Review Board Statement: Not applicable.

Informed Consent Statement: Not applicable.

Data Availability Statement: Data are contained within the article.

Conflicts of Interest: The authors declare no conflict of interest.

References

1. National Fire Agency. *Investigation Report of the Death of Firefighters on the Fire Accident in the Old Building of the Former Qiaoyou Department Store in Changhua County*; Ministry of the Interior: Taipei City, Taiwan, 2 May 2022.
2. MocapLeader. Comparison of Several Indoor Positioning Methods for Intelligent Robots. 20 April 2022. Available online: https://blog.csdn.net/MocapLeader/article/details/124291440 (accessed on 6 November 2022).
3. Fudickar, S.; Amend, S.; Schnor, B. On the comparability of indoor localization systems' accuracy. In Proceedings of the Fifth ACM SIGSPATIAL International Workshop on Indoor Spatial Awareness—ISA'13, Orlando, FL, USA, 5 November 2013; pp. 21–28. [CrossRef]
4. Chen, Z. Centimeter-Level Accuracy Is Practical and Feasible; Indoor Positioning Technology Catches up with Demand. February 2018. Available online: https://www.2cm.com.tw/2cm/SpecialProductDetails.aspx?id=F731C54FD4B2431CBC3F1E01CE10ABFF&NodeID=1A3C9033A2204949B613EB84DBE18FE4&refID=6599E75087A24C0DBA99B12C18494CC0 (accessed on 15 November 2022).
5. Fire Bureau of Taichung City Government. Fire Bureau of Taichung City Government Guiding Principles for Operation and Maintenance of Disaster Relief Equipment. 2015. Available online: https://www.fire.taichung.gov.tw/df_ufiles/f/%E6%95%91%E7%81%BD%E5%99%A8%E6%9D%90%E6%93%8D%E4%BD%9C%E7%B6%AD%E8%AD%B7%E6%9A%A8%E4%BF%9D%E9%A4%8A%E6%8C%87%E5%B0%8E%E5%8E%9F%E5%89%87.pdf (accessed on 22 October 2022).
6. TELEDYNE FLIR. No Excuse for Firefighter Disorientation. 14 September 2020. Available online: https://www.flir.eu/discover/public-safety--transportation/no-excuse-for-firefighter-disorientation/ (accessed on 22 October 2022).
7. Pollefeys, M. Microsoft HoloLens Facilitates Computer Vision Research by Providing Access to Raw Image Sensor Streams with Research Mode. Microsoft Research Blog. 18 June 2018. Available online: https://www.microsoft.com/en-us/research/blog/microsoft-hololens-facilitates-computer-vision-research-by-providing-access-to-raw-image-sensor-streams-with-research-mode/ (accessed on 6 November 2022).
8. Vtieto, DhurataJ, and Tim Sherer with Aquent. HoloLens Research Mode. 22 April 2022. Available online: https://learn.microsoft.com/en-us/windows/mixed-reality/develop/advanced-concepts/research-mode (accessed on 15 November 2022).

engineering proceedings

MDPI

Proceeding Paper

Integrating Sustainable Development Goals into Project-Based Learning and Design Thinking for the Instructional Design of a Virtual Reality Course †

Pai-Hsun Chen

Department of Electronic Engineering, National Chin-Yi University of Technology, Taichung 411030, Taiwan; bhc@gm.ncut.edu.tw

† Presented at the IEEE 5th Eurasia Conference on Biomedical Engineering, Healthcare and Sustainability, Tainan, Taiwan, 2–4 June 2023.

Abstract: The integration of topics on Sustainable Development Goals (SDGs) into project-based learning (PBL) and design thinking (DT) was attempted for the instructional design of a virtual reality (VR) course in this study. It was investigated how to enhance the students' understanding of the SDGs and promote critical thinking and design thinking skills by integrating SDG topics into PBL and DT. A research design with mixed methods was created based on quantitative and qualitative data from the reflections and feedback of students and teachers. The findings suggested that SDG topics in PBL and DT provided students with a meaningful and engaging learning experience and enhanced their understanding of the interconnections between the SDGs. The importance of incorporating real-world challenges into the instructional design of VR courses and the potential of VR technology were highlighted to support SDG-related learning outcomes. Integrating SDG into PBL and DT was effective for the instructional design of VR courses by preparing students to become responsible global citizens and contribute to achieving the SDGs.

Keywords: sustainable development goals (SDGs); design thinking (DT); project-based learning (PBL); virtual reality (VR)

Citation: Chen, P.-H. Integrating Sustainable Development Goals into Project-Based Learning and Design Thinking for the Instructional Design of a Virtual Reality Course. *Eng. Proc.* **2023**, *55*, 78. https://doi.org/10.3390/engproc2023055078

Academic Editors: Teen-Hang Meen, Kuei-Shu Hsu and Cheng-Fu Yang

Published: 18 December 2023

1. Introduction

While teaching 3D and interactive design and supervising professional projects at the University, students pursuing game development-related degrees are required to collaborate and develop their graduate projects as the capstone requirement in their junior or senior year. However, several problems occurred in the students' final projects. (1) The students misunderstood the design and scope of the project which showed the gap with real-world needs. (2) They lacked practical experience, interdisciplinary application and teamwork skills. (3) They lacked higher-level cognitive thinking in planning the work. Hence, project-based learning (PBL) and design thinking (DT) methods were introduced in designing the virtual reality (VR) development. Orienting students towards the 17 Sustainable Development Goals (SDGs) [1] allowed them to focus on the current and future challenges of society and the environment and use the outcomes for their profession and skills. This approach relieved the students' feelings of disconnection from reality and increased social responsibility and global perspective in engineering courses.

Based on the result, it is necessary to demonstrate teaching design and the integration of SDG into PBL and DT into a VR technology course. Considering the quantitative and qualitative analysis of students' feedback and suggestions, the course was designed with the following objectives: (1) To understand the impact of SDGs on the students' graduate project, (2) To enhance their comprehension and utilization of DT tools; (3) To understand the impact of PBL on students' learning. The recommendations and feedback from course instructors were also considered for the establishment and implementation of teaching strategies.

2. Related Works

2.1. PBL

PBL [2,3] is a pedagogical approach that engages students in authentic and meaningful learning experiences. It involves students working on complex problems or challenges over an extended period, often collaboratively, to produce a tangible product or presentation. Project-based learning aims to develop the students' skills, such as critical thinking, creativity, communication, collaboration, and the understanding of content knowledge and world issues. Several key points in implementing project-based learning are described as follows.

- Defining the learning objectives and standards that the project addresses;
- Designing a driving question or challenge to guide the project and stimulate the students' curiosity;
- Planning the project activities and milestones, scaffolding, and formative assessments;
- Facilitating the project process and providing guidance and feedback to students as they work on their tasks and products;
- Evaluating project outcomes using rubrics or other tools to assess student performance and learning.

2.2. DT

Design thinking is an integrated approach in which participants think creatively and come up with innovative solutions [4]. This approach encompasses a variety of tools to think inductively and construct a problem in different ways [5]. DT proposes solutions to real-world problems, as it is user-centered, iterative, and structured. The key elements include goal setting for developing new solutions to real problems that can be solved by cross-disciplinary and collaborating teams. There are various frameworks for DT. One of them was described by the Design Council's Two Diamond 4D design thinking process [6].

- The exploration phase involves designers looking for new opportunities, new markets, new information, new trends, and new insights to understand users and empathize with their feelings.
- In the definition phase, initial insights are examined and either selected or discarded. This process includes the initial development of the project idea, during which the designer must engage with the wider context of the identified opportunity.
- In the development phase, the designed solution is developed, iterated, and tested internally by a multidisciplinary team using design tools such as brainstorming, sketches, scenes, renderings, or prototypes.
- In the delivery phase, the final concept undergoes final testing, production, and release.

3. SDGs-PBL-DT Pilot Pedagogy in VR Course

Figure 1 shows the SDGs-PBL-DT framework proposed in this study. The diagram was sectioned into three parts. The curriculum design is based on the core design of the PBL approach. The middle part of the diagram denotes the main four stages of the SDGs-PBL-DT VR course in the unified processing [7]: inception, elaboration, construction, and transition. The DT approach applied to the double-diamond model is integrated into the second phase of PBL, as shown in the left part of Figure 1. The right part of Figure 1 denotes the major activities in the four phases of the SDGs-PBL-DT VR course.

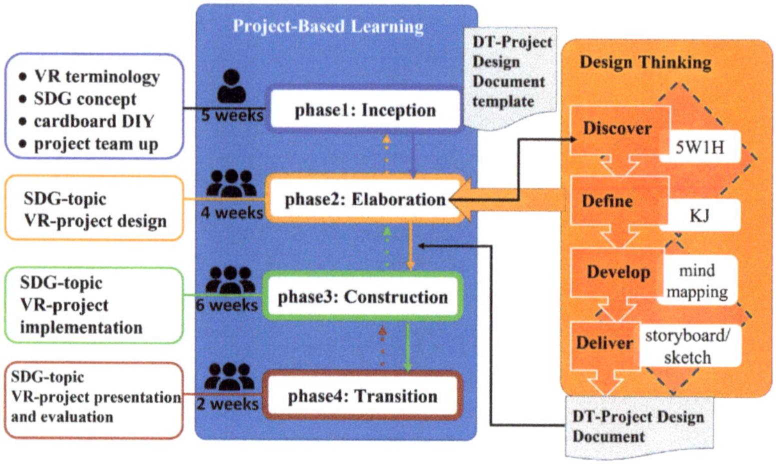

Figure 1. SDG-PBL-DT pilot pedagogy structure.

3.1. Phase 1: Inception

In this step, the project is launched based on the information on VR and SDGs provided before the project is planned and designed. The teacher explains that the course is based on a group project using VR technology with SDGs as the theme. The students must determine the SDG problems to be solved using VR. In the course, the teacher explains the terminology of VR technology so that the students can understand the application and key cores of VR. In addition, the teacher provides SDG-related internet resources for students to understand the 17 directions of SDG. The course also provides Google Cardboard VR glasses, paper models, and lenses for students to make their VR glasses at home. In this step, students are assessed on their conceptual learning and cardboard production, and are grouped to prepare for the subsequent group projects.

3.2. Phase 2: Elaboration

This step involves creating a project blueprint to complete a design brief for each project team. It introduces the design thinking approach, using the Double Diamond model and different DT tools at different stages. A Design Thinking Project Design Document (DT-PDD) template is developed and scored using a design scoring rubric that includes scoring and inter-group assessment by experts and teachers. Details are as follows.

- Discover by using the 5W1H tool to enable groups to think about the SDGs problem to be solved in terms of who, why, what, where, when, and how, including the target user and the cause and location of the problem
- Define by using the KT tool [7] to enable the sub-groups to integrate, cluster, organize, and clarify the design ideas of each group member.
- Develop by using the mind mapping tool to organize the ideas collated by the KT tool into a tree-style mind mapping to fit the VR design.
- Deliver by using storyboard or sketch tools to design system prototypes.

The diagrams generated by the above DT tools corresponding to the DT stages are collated in the DT-PDD of each project team.

3.3. Phase 3: Construction

In this phase, hands-on project development is carried out using each group's DT-PDD to start the technical implementation of VR projects. The instructor develops mobile-based teaching materials by implementing Google Cardboard VR. The Unity 3D game engine is used as the main implementation platform. In addition to providing basic hands-on teaching on the game engine, the VR interactive program must conform to the project blueprint design of the DT-PDD, while an iterative revision of the DT-PDD and the project must be carried out based on the technical evaluation of feasibility and usability.

3.4. Phase 4: Transition

The objective of this phase is to test and publish the project work, and present and evaluate it in various ways at the end of the period. The project presentation and oral report of the project with the video presentation on site are included. The final assessment is conducted with rubric assessment criteria, including teacher assessment, expert assessment, and inter-group assessment. Quantitative and qualitative assessments are performed to provide insightful suggestions for improvement and learnings from the project.

4. Pedagogy Implementation

The course was designed to use a VR design course and the Unity 3D game engine with Google Cardboard VR. The course was offered from September 2021 to January 2022. The students had already taken two semesters of 3D modeling and one semester of Unity 3D courses, and were expected to be able to build simple 3D scenes. The students were grouped in 7 groups of 3-4 students. Due to the pandemic at the beginning of the semester from September 2021 to February 2022, the first 3 weeks of the course were spent on online teaching. Lecture videos and Google quiz forms related to VR terminologies were offered for students to learn online. The course resumed later in the semester for on-site lessons.

5. Data Collection, Analysis, and Results

The students' feedback on the course was collected for quantitative and qualitative data. There were two feedback surveys, one at mid-term and one at the end of the course. The mid-term feedback survey was conducted to determine whether the SDGs impacted on the direction of students' graduated projects when students of the course had planned their DT-PDD projects using the SDG theme and integrating design thinking methods and DT tools.. The end-of-term feedback survey was conducted to determine the use of PBL in the students' decision making about developing a VR project.

5.1. Quantitative Analysis

5.1.1. Impact of SDG Topics

Before the class, 50% of the students had never heard of the term sustainable development or SDG. After Phase 2, each project team generated the SDG topics' DT-PDD document. Subsequently, it was surveyed whether the SDG topics project in the VR course influenced their project direction. Figure 2 denotes that 20% of the VR course students changed their graduate project to SDG-related, and 5% extended the SDGs-VR mechanism into their project, while 25% had already developed SDG-related graduate projects.

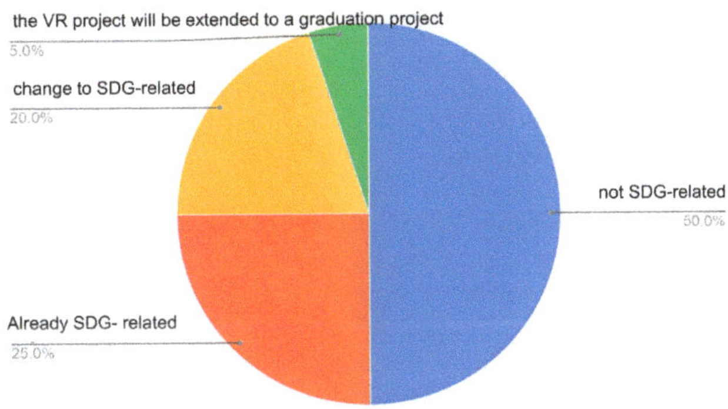

Figure 2. SDG integration into students' projects.

5.1.2. DT Tools

A total of 60% of the students in the course had studied the creative thinking course and understood relative tools for DT in the same semester. Figure 3 shows the proportion of students with prior knowledge of the DT tools before the DT lessons. A total of 65% of the students were familiar with and had employed the mind mapping tool, while 35% of the students were familiar with and had used the KJ method. Further, 20% of the students had never used any DT tools.

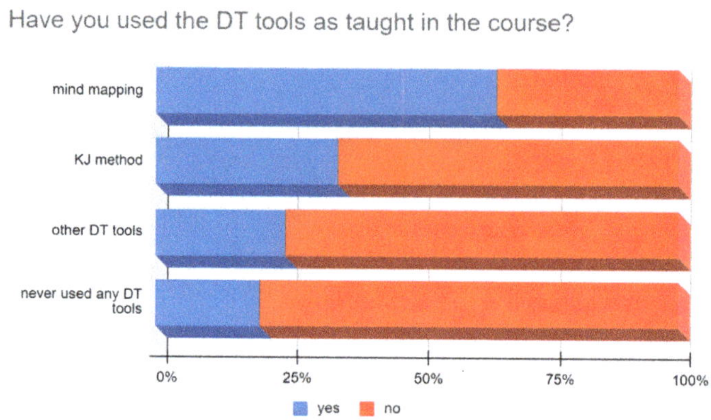

Figure 3. Proportion of students with prior knowledge of DT tools.

5.1.3. Survey Result

Table 1 presents the students' quantitative results from the survey. Though the students of the course had not heard about SDGs, the SGDs significantly contributed to their perceptions of DT.

Table 1. Students' feedback about SDGs, PBL, and DT.

Items	Mean
The SDG topics allowed me to think more about using practical project work to do something to make the world a better place or to solve its problems.	3.9
I understood that the SDG topics would be implemented in the teamwork of the course project. I think it would be useful for me to have project planning and implement practical works rather than merely knowing the concepts or practice techniques.	4.1
We had no direction and no idea what to do for the course project until the teacher suggested the SDG topics for the VR course project.	3.05
The SDG topics provided us with more possibilities for the VR course project.	3.75
I think various DT tools and different creative thinking methods are helpful in formulating project designs.	4.1
I understood the use of the KJ method after the lecturer discussed the topic and provided a template.	3.55
I understood the use of the mind mapping method after the lecturer explained the concept and provided a template.	3.9
I understood that working in groups is more conducive to a collaborative working environment, where consensus and results are achieved through coordination, discussion and division of labor, and where I can learn more about cross-disciplinary practices.	4.3
I think PBL is helpful in bridging the workplace because of problem-solving, team discussions, teamwork and practical work.	4.4

Table 1. *Cont.*

Items	Mean
PBL has helped me enhance my ability to think in multiple ways, as it is not just practical work.	4.5
Because PBL is a set application goal rather than a direct teaching of a bunch of program or function instructions, it stimulated my interest in learning.	4.25
In PBL, we had our own application goals, which helped me find solutions to achieve the goals I wanted to do.	4.31

5.2. Qualitative Result

In the same survey, we had one open-ended question to probe the students' commands about setting the SDGs topic for the VR course.

- Feedback from Student 1: "It is difficult to imagine how to use VR technology for the SDG topics for developing a VR project at first. Can you give more examples?"
- Feedback from Student 2: "The SDG topics were effective in helping us to find useful directions for our topics, and the SDGs project planning from the design thinking tools was interesting and informative, but the project results really depended on the group members' abilities and cooperations."
- Feedback from Student 3: "It is better to really talk to the target users to really understand their thoughts and needs."

5.3. Reflections on Curriculum Design

The lecturer discussed various points, described as follows.

- Although the first stage of the flipped classroom on VR terminology had to be carried out online because of COVID-19 and the video was given before the class, interaction and the reinforcement of concepts still occurred during the class.
- Based on classroom conversations with students, there was still a gap in translating the data generated from design thinking tools into a practical VR interactive project. Therefore, it was a challenge in curriculum design and research direction.
- It was worth noting that the DIY for Google Cardboard was still difficult. Although the students were provided with a Google Cardboard paper template and a live demonstration of the whole process, some still wanted to be provided with a video to watch again at home.
- The course lecturer had to have more training on applying DT in courses even if the lecturer attended more than five DT workshops. Continuous participation in DT workshops was essential to refine teaching.

6. Conclusions

In this study, the DT-PBL teaching method was successfully combined with SDGs to design a VR-based multimedia course. Unity 3D and Google Cardboard VR were useful as accessible and sustainable resources. The positive feedback from students highlighted the effectiveness of PBL focusing on SDGs. While improvements were made in the instructional method, the result of this study provides a reference for future SDGs-PBL-DT curriculum design and instruction in interactive multimedia courses.

Funding: This research received no external funding.

Informed Consent Statement: In this study, the informed consent form was provided to the students at the beginning of the course and each student signed the Chinese-version informed consent form for the course.

Data Availability Statement: No physical testing was conducted and hence no new data are available.

Conflicts of Interest: The author declares no conflict of interest.

References

1. General, A. *United Nations Transforming Our World: The 2030 Agenda for Sustainable Development*; Division for Sustainable Development Goals: New York, NY, USA, 2015.
2. Condliffe, B.; Quint, J.; Visher, M.G.; Bangser, M.R.; Drohojowska, S.; Saco, L.; Nelson, E. Project-Based Learning: A Literature Review. MDRC: Working Paper. 2017. Available online: https://www.mdrc.org/publication/project-based-learning (accessed on 31 May 2021).
3. Loyens, S.M.; Magda, J.; Rikers, R.M. Self-directed learning in problem-based learning and its relationships with self-regulated learning. *Educ. Psychol. Rev.* **2008**, *20*, 411–427. [CrossRef]
4. Dorst, K. The core of 'design thinking'and its application. *Des. Stud.* **2011**, *32*, 521–532. [CrossRef]
5. Plattner, H.; Meinel, C.; Leifer, L. (Eds.) Design Thinking Research: Making Distinctions: Collaboration versus Cooperation. Springer. The Design Process: What Is the Double Diamond. 2017. Available online: http://www.designcouncil.org.uk/news-opinion/design-process-what-double-diamond (accessed on 26 October 2019).
6. Kruchten, P. *The Rational Unified Process: An Introduction*; Addison-Wesley Professional: London, UK, 2004.
7. Scupin, R. The KJ method: A technique for analyzing data derived from Japanese ethnology. *Hum. Organ.* **1997**, *56*, 233–237. [CrossRef]

engineering proceedings

MDPI

Proceeding Paper

Study on the Impact of Public Attention Relative to Green Development Policies on the Return on Investment of Related Industries †

Jie He [1,*], Wen-Jia Zhao [1], Dong-Ni Jia [2,*] and Zheng-Yun Zhuang [3]

1 School of Computer & Computing Science, Hangzhou City University, Hangzhou 310015, China; hellozwj123@sina.com
2 Bentley University Graduate School of Business, Bentley University, Waltham, MA 02452, USA
3 Department of Civil Engineering, National Kaohsiung University of Science and Technology, Kaohsiung City 807, Taiwan
* Correspondence: hej@hzcu.edu.cn (J.H.); djia@falcon.bentley.edu (D.-N.J.); Tel.: +86-158-5820-6157 (J.H.)
† Presented at the IEEE 5th Eurasia Conference on Biomedical Engineering, Healthcare and Sustainability, Tainan, Taiwan, 2–4 June 2023.

Abstract: As an emerging development model, green development requires the participation of the government, the market, and the public. It needs the formation of a positive interaction between policy guidance, market demand, and public will in order to realize a win–win situation comprising sustainable economic development and environmental protection. Thus, we explored green-development-related policies in China and defined four areas: industrial energy preservation, petrochemicals, agriculture, and the integration of the Yangtze River Economic Belt. Then, we analyzed how public attention relative to green development policies impacted the return of stock indices in the green industry using the Baidu search index. The research study was conducted on all trading days of A-shares in China from 1 January 2013 to 1 January 2023. In total, 2430 valid datasets were observed and curated over these 10 years. The public attention to green development policies was classified into four categories: industrial energy efficiency, petrochemicals, agriculture, and the Yangtze River Economic Zone. The attention focused on a category was measured by the number of searches of keywords that were the category's names on the Baidu platform. The number of searches was considered the variable representing the public's attention relative to each area of green development policies. The results showed that the increase in public attention to green development policies was not correlated with the increase in the return on investment (ROI) of green and carbon-neutral industries. In addition, information on green development policies had no significant effect on encouraging individuals to participate in investments in the green industry and had little effect on the financing of the carbon-neutral industry. The lack of details for green development policies and effective guidelines for potential investors needs to be overcome in order to encourage investment and financing.

Keywords: green development; policy; green industry; stock index; public attention; longitudinal data introduction

Citation: He, J.; Zhao, W.-J.; Jia, D.-N.; Zhuang, Z.-Y. Study on the Impact of Public Attention Relative to Green Development Policies on the Return on Investment of Related Industries. *Eng. Proc.* **2023**, *55*, 79. https://doi.org/10.3390/engproc2023055079

Academic Editors: Teen-Hang Meen, Kuei-Shu Hsu and Cheng-Fu Yang

Published: 19 December 2023

1. Introduction

With the continuous development of the green industry, investors have paid attention to investment opportunities in the industry. Accordingly, the impact of the public's attention to green development policies on the return on investment (ROI) of the green industry has been explored extensively. Although many studies have shown a positive relationship between public attention and ROI, the increase in public attention did not necessarily have a positive effect on the ROI and even showed a negative impact. Therefore, it is necessary to elaborate on this topic systematically.

For the study, all government documents from 2008 to January 2023 related to "green development" were collected and summarized to understand the direction of policy development and sort the documents according to the frequency of each direction. The top three items with a frequency of more than 15 were "industrial energy conservation", "oil and petrochemical", and "Yangtze River Economic Belt". The policies in these three directions were selected as the research content in this study, and the corresponding Baidu search index was used to gauge public attention relative to the policies. Based on the collected content of green development policies, the stock prices of related companies in Tonghuashun market conditions were compiled to judge whether the stocks corresponded to the scope defined in the "Catalogue" in this study. "Carbon neutral", "oil and petrochemical", and "Yangtze River Economic Belt" were used as keywords to obtain the necessary stock information on the green industry.

As the unit of ROI and the Baidu search index were not identical, it was necessary to standardize their factors in order to build the model. Therefore, the changes in the ROI and the three-day moving average of changes in the Baidu search index were investigated for the model. In the model, the changes in the closing price of the Shenzhen Composite Index and the dummy variable of seasonal factors were selected as control variables, while the change in the ROI was used as the dependent variable. Variables were used to establish an ordinary least squares (OLS) regression model, and the effect on the model was analyzed based on the p-value and R^2.

2. Theories and Research Hypothesis

2.1. Definition of Concepts

The earliest document on green development policies was released in 2008. The Chinese government announced the 12th Five-Year Development Plan in 2009 [1]. China's first green development plan included a development plan that could strengthen energy conservation and emission reduction, develop a green economy and industries, build a green China, and promote green civilization. In the "green development" policy from 2008 to the present, the three aspects most frequently mentioned were industrial energy conservation, petrochemicals, and the Yangtze River Economic Belt. Subsequent research and reviews have been based on these three aspects, which were also used as the perspective of public attention in this study.

2.1.1. Public Attention

Public attention refers to the degree of public attention to a policy of a country. After a policy is announced, the response to this policy needs to be understood in order to propose timely adjustments accordingly. Thus, public attention needs to be straightforward and effective for such adjustments. It also affects the stock price of related industries. This study was conducted based on this concept to determine the impact of public attention relative to green development policies on the ROI of the green industry.

2.1.2. Industrial Energy Efficiency

The green development policy is related to industrial energy efficiency and has been developed over the years. Methods for evaluating energy efficiency have become an important indicator for analyzing the effect of China's industrial energy efficiency policies. Therefore, we included industrial energy efficiency as an indicator in this study.

2.1.3. Petrochemical Industry

The petrochemical industry is a pillar of China's economy and has contributed to its economic development significantly. However, the industry considerably harms the environment. Thus, the transformation of the petrochemical industry into a green industry is required. Since 2012, China has issued policies for the green development of the petrochemical industry, which have received attention from the public and have been researched. Recently, a multi-perspective investor attention index based on Google Trends was proposed, and it considers four perspectives, including the linkage of alternative energy prices, energy prices, macroeconomics, and geopolitical factors, to enhance the predictive ability of oil prices and their fluctuations [2]. It was claimed that investors' attention to macroeconomics and geopolitics affected oil prices.

2.1.4. Yangtze River Economic Belt

The Yangtze River Economic Belt is important in China's major development strategies. Thus, there are many reports on the environmental governance and protection of the belt's environment, showing significant improvements in environmental protection. We included public attention relative to the "Yangtze River Economic Belt" to study the impact of environmental protection on the ROI.

2.2. Theoretical Background

2.2.1. Ordinary Least Squares (OLS) Regression

OLS regression is a linear least squares method that selects unknown parameters in linear regression models by minimizing the sum of squared differences between the observed dependent variables. The least squares method shows the optimized and substantial effect observed in the Gauss–Markov theorem. The OLS model has been widely used. In Ref. [3], the relationship between parents and children's height was studied using the concept of "regression to the mean", which was the first research study that adopted the OLS model. OLS regression was used in this study for constructing the model to explain the impact of public attention on the fluctuation of closing stock prices.

2.2.2. Capital Asset Pricing Model (CAPM)

CAPM is based on the theory that the expected return of an asset is related to the market risk factor and irrelevant to other factors. However, the CAPM model cannot explain the fluctuation of the ROI of assets and related factors. To solve these problems, multi-factor models were proposed. In multi-factor models, it is assumed that the return rate of an asset is related to market risk factors, company size, book–market ratio, liquidity, and others. Risk factors are used to explain the volatility and differences in asset returns. A three-factor model [4] and a four-factor model (also known as the Fama–French three-factor model) [5] were proposed earlier. Since then, more multi-factor models, such as five- and six-factor models, have been proposed. Stock prices are influenced by multiple factors. Thus, Fama [6] proposed the efficient market hypothesis in which stock prices were explained with all available information, and the trend of stock prices was determined to be random and unpredictable. Malkiel [7] criticized the efficient market hypothesis, claiming that factors including fundamentals, technical factors, and psychology also influenced the trend of stock prices. Finally, Shiller [8] proposed using excessive volatility in stock prices based on the observation that the trend of stock prices exceeded the range of changes. Therefore, based on the summary of previous study results, public attention was used to explore the impact of public attention on stock prices.

2.3. Hypothesis for the Influence of Public Attention Relative to Green Policies

As environmental issues are increasingly recognized as serious, the public is paying more attention to environmental protection, and enterprises are taking responsibility for environmental protection. Therefore, the public's attention to green development may affect the stock price of the green industry, which needs more investigation. Improving energy efficiency can promote the green development of the industry. The government's practice of sustainable development and the improvement of the petroleum industry promote green development, especially in the Yangtze River Economic Belt. Therefore, green development is closely related to the green industry. We have selected two distinct methods for collecting data for public attention: one from PC users and the other from mobile phone users. Therefore, each aforementioned hypothesis will be analyzed with respect to the mutual impact on the stock price of the green industry based on public attention data from both

PCs and mobile platforms. Based on this, the following hypotheses were proposed in this study.

Hypothesis 1. *The public's attention to "industrial energy efficiency" based on pc platforms and the ROI of the industry related to "industrial energy efficiency" are mutually influential.*

Hypothesis 2. *The public's attention to "industrial energy efficiency" based on mobile platforms and the ROI of the industry related to "industrial energy efficiency" are mutually influential.*

Hypothesis 3. *The public's attention to "the petroleum industry" based on pc platforms and the ROI of the industry are mutually influential.*

Hypothesis 4. *The public's attention to "the petroleum industry" based on mobile platforms and the ROI of the industry are mutually influential.*

Hypothesis 5. *The public's attention to the "Yangtze River Economic Belt" based on pc platforms and the ROI of "environmental protection" are mutually influential.*

Hypothesis 6. *The public's attention to the "Yangtze River Economic Belt" based on the mobile platforms and the ROI of "environmental protection" are mutually influential.*

3. Research Design

3.1. Variable Selection

The three-day moving average of the Baidu search index was calculated using Equation (1).

$$Baidu\ search\ rate\ of\ change = \frac{\left\{\frac{\text{BD\_Index}(T-2)+\text{BD\_Index}(T-1)+\text{BD\_Index}}{3}\right\} - \text{BD\_Index}(T-2)}{\text{BD\_Index}(T-2)} \quad (1)$$

The three-day moving average (Baidu search change rate) was selected as a variable along with the change in the closing price and seasonal variables of the Shenzhen Composite Index. The change in the green industry stocks' closing price was chosen as a dependent variable. Control variables included seasonal changes and the changes in the Shenzhen Composite Index closing price.

3.2. Sample Data

Data from the Baidu search index were collected from 22 March 2013 to 31 December 2022, except for closed days of the market. The three-day moving average was calculated to obtain the Baidu search change rate, and the one-day forward shift was calculated from the PRE_Baidu search change rate. The change in the ROI and the Shenzhen Composite Index's closing price were obtained from Tonghuashun. The seasonal dummy variables were compiled as follows: spring (0) from March to March, summer (1) from April to June, autumn (2) from July to September, and winter (3) from October to December. The final sample time range is from 27 March 2013 to 30 December 2022.

3.3. OLS Regression Model Construction

The three sets of variables were summarized into the following three models.

$$TZH = \alpha + \beta_1 \text{pre}TZH + \beta_2 Index\_pc + \beta_3 preIndex\_pc + \beta_4 SZ + \beta_5 preSZ + \beta_6 Season + \varepsilon \quad (2)$$

$$TZH = \alpha + \beta_1 \text{pre}TZH + \beta_2 Index\_yd + \beta_3 preIndex\_yd + \beta_4 SZ + \beta_5 preSZ + \beta_6 Season + \varepsilon \quad (3)$$

$$SH = \alpha + \beta_1 \text{pre}SH + \beta_2 Index_{pc} + \beta_3 preIndex_{pc} + \beta_4 SZ + \beta_5 preSZ + \beta_6 Season + \varepsilon \quad (4)$$

$$SH = \alpha + \beta_1 \text{pre}SH + \beta_2 Index_{yd} + \beta_3 preIndex_{yd} + \beta_4 SZ + \beta_5 preSZ + \beta_6 Season + \varepsilon \quad (5)$$

$$CJ = \alpha + \beta_1 \text{preCJ} + \beta_2 Index\_pc + \beta_3 preIndex\_pc + \beta_4 SZ + \beta_5 preSZ + \beta 6\ Season + \varepsilon \quad (6)$$

$$CJ = \alpha + \beta_1 \text{preCJ} + \beta_2 Index\_yd + \beta_3 preIndex\_yd + \beta_4 SZ + \beta_5 preSZ + \beta 6\ Season + \varepsilon \quad (7)$$

TZH, SH, and CJ represent the stock board returns of "carbon neutrality", "petroleum and petrochemical", and "Yangtze River Economic Belt" industries, respectively. preTZH, preSH, and preCJ denote their respective previous day's stock board returns. Index signifies the Baidu search variation rate, while pre-index refers to its previous day's search variation rate. SZ stands for the Shenzhen Composite Index stock board returns, and preSZ represents the previous day's Shenzhen Composite Index stock board returns. Season indicates the seasonal index.

4. Empirical Analysis

4.1. Descriptive Analysis of Variables

"Industrial Energy Conservation" and "Carbon Neutrality"

On 21 March 2013, the China Nonferrous Metals Industry Association announced industrial energy conservation and green development. Provincial and municipal governments responded actively and issued their industrial green development plans. Since 2013, the national government has issued an industrial green development plan each year, and data from the Baidu search index with the keyword "industrial energy conservation" are shown in Figure 1.

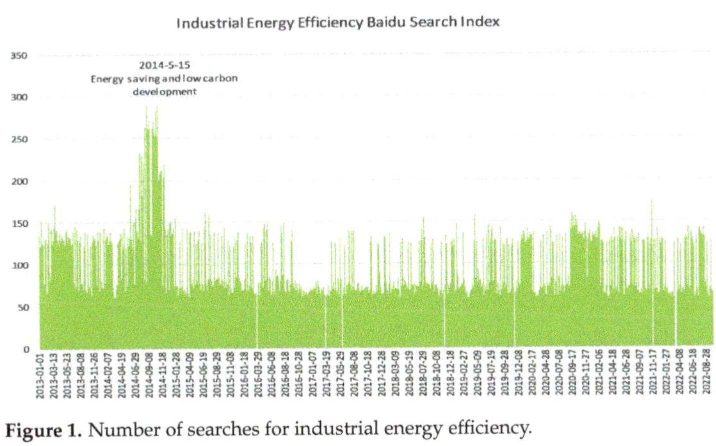

Figure 1. Number of searches for industrial energy efficiency.

The number of searches for "industrial energy conservation" reached its maximum between May and September 2014. As relevant policies during that period, the State Council issued the Action Plan on 15 May 2014. This plan aimed to achieve annual reductions of 3.9% in energy consumption per unit of GDP, 2% in chemical oxygen demand, 2% in sulfur dioxide, 2% in ammonia nitrogen, and 5% in nitrogen oxide emissions set in 2014 and 2015. Additionally, the plan targeted a 4 and 3.5% reduction in carbon dioxide emissions per unit of GDP in the two years. Since its introduction, this plan garnered significant public attention and response. In addition, the official website of the Chinese government provided an in-depth analysis of related policies. Consequently, there was a substantial increase in the number of searches.

Considering the emphasis on industrial energy conservation, the industrial chain related to carbon neutrality was represented with the stock price of the green industry. Furthermore, the trends in the closing prices of the stocks showed a lag compared to changes in public attention. This suggested the influence of public attention on the trends of the ROI.

The petrochemical industry is important and has high industrial relevance and a wide range of product coverage in China. However, as the economy develops, the demand for ecological and environmental protection increases, necessitating the green transformation of the petrochemical industry. In 2012, governments at various levels recognized this issue and began taking measures for the improvement of urban green development. The Yangzhou Municipal Government issued "Opinions" on 10 September 2012, outlining their plans. Subsequently, in 2016, the Hubei Province released the Petrochemical Industry Transformation Plan. The number of related searches with the keyword "petrochemical" is presented in Figure 2. The number of searches related to "petrochemical" remained relatively stable over time. However, there were periods when the number of searches increased. As the petrochemical industry has always been a subject of great interest and received attention from the public and government, changes in policies or significant events related to the industry led to a sudden surge in public attention.

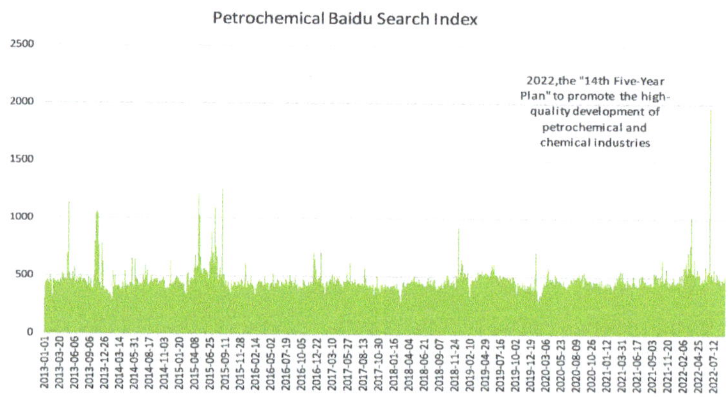

Figure 2. Number of searches for petrochemical industry.

In the stock market, there were 47 companies listed in the category of the petrochemical industry. In the period around June 2015, stock prices fluctuated when public attention to "petrochemicals" increased and declined. This showed a correlation between stock prices and the level of public attention relative to the petrochemical industry.

Since September 2013, the "Yangtze River Economic Belt" gradually gained attention from the public. The Belt covers 11 provinces and municipalities, including Shanghai, Jiangsu, Zhejiang, Anhui, Jiangxi, Hubei, Hunan, Chongqing, Sichuan, Yunnan, and Guizhou, spanning the eastern, central, and western regions. They have significantly contributed to China's economic growth. On 21 July 2013, during an inspection visit to Wuhan, General Secretary Xi Jinping emphasized the strengthening of cooperation in the Yangtze River basin for inland water transportation and transforming the entire basin into a "golden waterway". Figure 3 displays the number of searches with the keyword "Yangtze River Economic Belt".

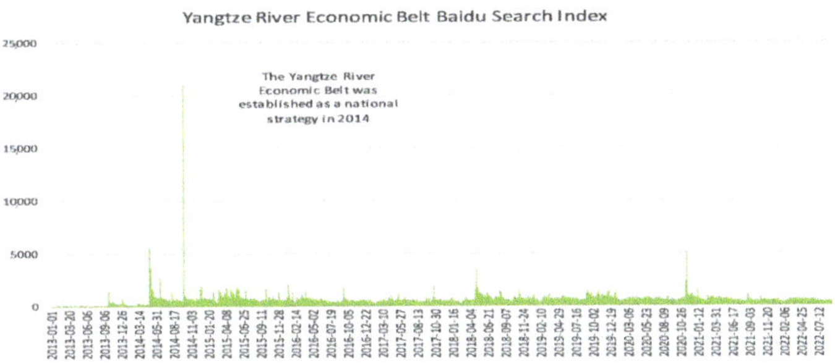

Figure 3. Number of searches for Yangtze River Economic Belt.

In September 2014, the Chinese government designated the Yangtze River Economic Belt as a national strategy for the construction of the economic belt based on the "golden waterway". Around this time, there was a peak in public attention and interest in the Yangtze River Economic Belt, as indicated by the number of searches, which reached an average of over 20,000 searches per day. This period had a significantly higher number of searches than other periods. Due to the focus on environmental protection and ecological conservation in the construction of the Yangtze River Economic Belt, the "environmental protection" industry was chosen to represent the corresponding industry. The trend of the stock price was influenced by a notable increase in public attention around October 2014. Several months later, the stock prices of companies for environmental protection also began to increase. Therefore, there was a correlation between the two, but further empirical verification was necessary to determine the specific impact.

4.2. Analysis of Regression Results

In this section, two distinct methods for collecting data for public attention were employed: one from PC users and the other from mobile phone users. This clear differentiation enhances the systematic presentation of our findings.

4.2.1. Industrial Energy Efficiency

The formula for the Baidu search index for "industrial energy efficiency" on the PC platform is as follows:

$$TZH = \alpha + \beta_1 \text{pre}TZH + \beta_2 Index\_pc + \beta_3 preIndex\_pc + \beta_4 SZ + \beta_5 preSZ + \beta_6 Season + \varepsilon \quad (8)$$

where Index_pc denotes the search variation rate for "industrial energy efficiency" on the PC platform.

The formula for the Baidu search index of "industrial energy efficiency" on the mobile platform is as follows:

$$TZH = \alpha + \beta_1 \text{pre}TZH + \beta_2 Index\_yd + \beta_3 preIndex\_yd + \beta_4 SZ + \beta_5 preSZ + \beta_6 Season + \varepsilon \quad (9)$$

where Index_yd represents the search variation rate for "industrial energy efficiency" on the mobile platform.

The Baidu search change rate was not statistically significant in the model, and there was no sufficient evidence to conclude that the rate impacted the ROI of the industry related to carbon neutrality. The OLS regression results for the Baidu search change rate for "Industrial Energy Efficiency" are presented in Table 1. The rate was not statistically significant, but the pre-search rate was significant at $p = 0.1$. This suggested public attention to "Industrial Energy Efficiency" and the ROI were correlated.

Table 1. Result of the OLS model of the Baidu search change rate and ROI of companies related to industrial energy efficiency.

	Coefficient	T	$p > \|T\|$
PRE_TZH	0.0718	3.616	0.000
ROC	-1.413×10^{-5}	−0.831	0.406
PRE_ROC	-2.948×10^{-5}	−1.736	0.083
SZ	0.0105	83.786	0.000
SZ_PRE	0.0009	3.504	0.000
SEASON	−0.0003	−2.519	0.012
R-SQUARED		0.751	
F-STAT		1195	
PROB (F)		0.000	

4.2.2. Petrochemical Industry

The formula for the Baidu search index of the "Petrochemical industry" on the PC platform is as follows.

$$SH = \alpha + \beta_1 \mathrm{pre}SH + \beta_2 Index\_pc + \beta_3 preIndex\_pc + \beta_4 SZ + \beta_5 preSZ + \beta_6 Season + \varepsilon \quad (10)$$

The formula for the Baidu search index of "Petrochemical industry" on the mobile platform is as follows.

$$SH = \alpha + \beta_1 \mathrm{pre}SH + \beta_2 Index\_yd + \beta_3 preIndex\_yd + \beta_4 SZ + \beta_5 preSZ + \beta_6 Season + \varepsilon \quad (11)$$

Based on the results of the OLS regression model (Equations (10) and (11)), the Baidu search change rate was not significant in the model, and no significant impact of the rate was observed on the ROI of the petrochemical industry. The OLS regression results for the Baidu ssearch change rate for "Petrochemical Industry" are presented in Table 2.

Table 2. Result of the OLS modeling of the Baidu search change rate and ROI of companies related to petrochemical industries.

	Coefficient	T	$p > \|T\|$
PRE_TZH	0.0724	3.644	0.000
ROC	-1.906×10^{-5}	−1.822	0.069
PRE_ROC	-6.11×10^{-6}	−0.584	0.559
SZ	0.0105	83.766	0.000
SZ_PRE	0.0008	3.432	0.001
SEASON	−0.0003	−2.527	0.012
R-SQUARED		0.751	
F-STAT		1194	
PROB (F)		0.000	

4.2.3. Yangtze River Economic Belt

The OLS model for the Yangtze River Economic Belt is as follows:

$$CJ = \alpha + \beta_1 \mathrm{preCJ} + \beta_2 Index\_pc + \beta_3 preIndex\_pc + \beta_4 SZ + \beta_5 preSZ + \beta 6\, Season + \varepsilon \quad (12)$$

where Index_pc denotes the search variation rate for the "Yangtze River Economic Belt" on the PC platform:

$$CJ = \alpha + \beta_1 \text{preCJ} + \beta_2 Index\_yd + \beta_3 preIndex\_yd + \beta_4 SZ + \beta_5 preSZ + \beta 6\, Season + \varepsilon \quad (13)$$

where Index_yd denotes the search variation rate for the "Yangtze River Economic Belt " on the mobile platform.

The three-day Baidu search change rate was not significant, while the rate before the three days was significant at $p = 0.1$. A significant correlation was observed between the public's attention to the Yangtze River Economic Belt and the ROI of the industries related to environmental protection. The OLS regression results for the Baidu ssearch change rate for "Petrochemical Industry" are presented in Table 3.

Table 3. Result of the OLS modeling of the Baidu search change rate and ROI of industries related to the Yangtze River Belt and environmental protection.

	Coefficient	T	$p > \|T\|$
PRE_TZH	0.0750	3.771	0.000
ROC	-2.149×10^{-5}	−0.560	0.576
PRE_ROC	-7.182×10^{-5}	−1.870	0.062
SZ	0.0089	69.265	0.000
SZ_PRE	0.0009	4.286	0.000
SEASON	−0.0003	−3.469	0.001
R-SQUARED		0.676	
F-STAT		823.7	
PROB (F)		0.00	

5. Discussions

In the constructed OLS regression models, the Baidu research change rate was not important for industrial energy efficiency. The overall explanatory power of the model was 70%, and the coefficient was negative, indicating that the rate influenced the ROI of the related industry. The rate affected the ROI of the petrochemical industry. The overall explanatory power of the model was 67.5%, suggesting that there was room for further improvement. The coefficients of the model were negative, indicating that public attention influenced the ROI. For the Yangtze River Belt and environmental protection, the model's explanatory power was 67.5%, and the coefficients were negative, suggesting that public attention had a negative impact.

As control variables in the models, the Shenzhen Composite Index and seasonal dummy variables mitigated the influence on the ROI. These two variables were significant at $p = 0.05$ in all models, demonstrating that the relationship between independent and dependent variables in the model was not ruined by the two variables. The Shenzhen Composite Index is a stock price index compiled by the Shenzhen Stock Exchange, reflecting the price trends of all A- and B-shares listed on the exchange, and it is significantly influenced by other markets. As stock market trends are correlated with seasons, seasonal factors and the Shenzhen Composite Index helped the model estimate the relationship between independent and dependent variables more accurately.

6. Conclusions

We explored the relationship between the public's attention to different green policies and the ROI of the green industry. The results showed that the higher the public's attention to green policies, the higher the ROI. Conversely, an increase in the ROI stimulated public attention relative to green policies.

Improving industrial energy efficiency is important and necessary to achieve carbon neutrality. By adopting highly efficient energy technologies, equipment, and management measures, energy consumption in the industry can be reduced with lowered carbon emissions. Then, carbon neutrality can be achieved. Empirical models proved a mutual relationship between the increase in public attention to industrial energy efficiency and the ROI of the industry related to carbon neutrality. Therefore, the government needs to improve industrial energy efficiency and disclose more details about implementing green development. The results of this study suggested that as public attention increased, investors paid more attention to investment opportunities in the petroleum industry, which increased the ROI. Increased public attention prompts the government to introduce more policies for developing the petroleum industry and increase the ROI. The government must actively practice sustainable development plans, strengthen corporate responsibility, and improve the green development of the petroleum industry.

As a national development strategy, the development of the Yangtze River Economic Belt has significant implications for sustainable development. On the one hand, with the continuous increase in public attention to environmental issues, the investment demand for the environmental protection industry increases accordingly. Implementing environmental protection policies and measures in the Yangtze River Economic Belt impacts the performance of the related industry, thereby affecting its RIO. Therefore, the government must exploit stakeholder theory. In conclusion, the government must improve the green development system, actively promote environmental greening, and facilitate green development.

While green policies are readily available, the selection of corresponding industry stocks is challenging. Most policies were not directly related to the industry. Thus, industries that corresponded the most to the policies were chosen in this study. Thus, the results of the models may vary depending on the methods for choosing the industries. The industry sectors for the policies of industrial energy efficiency and the petroleum industry were relatively well matched, while the sector for the "Yangtze River Economic Belt" was not satisfactorily matched. Therefore, in-depth research is needed to develop more accurate models.

Author Contributions: Conceptualization, J.H. and W.-J.Z.; methodology, J.H. and W.-J.Z.; software, D.-N.J. and W.-J.Z.; validation, J.H. and W.-J.Z.; formal analysis, J.H. and W.-J.Z.; investigation, W.-J.Z.; resources, J.H. and W.-J.Z.; data curation, J.H. and W.-J.Z.; writing—original draft preparation, W.-J.Z.; writing—review and editing, J.H., D.-N.J. and Z.-Y.Z.; visualization, D.-N.J.; supervision, J.H.; project administration, J.H.; funding acquisition, J.H. All authors have read and agreed to the published version of the manuscript.

Funding: This research was funded by 2019 Hangzhou Municipal New Professional Construction Project Plan Project "Internet+Data Intelligence"(Project No. Hangjiaogaojiao (2019) No. 5).

Institutional Review Board Statement: Not applicable.

Informed Consent Statement: Not applicable.

Data Availability Statement: Data are contained within the article.

Acknowledgments: This research was supported by the advanced computing resources provided by the Supercomputing Center of Hangzhou City University.

Conflicts of Interest: The authors declare no conflict of interest.

References

1. Zhu, Y. Research and reflection on the development of green stock index in China. *Financ. Vert.* **2018**, *7*, 59–64.
2. Qu, H.; Li, G. Multi-perspective investor attention and oil futures volatility forecasting. *Energy Econ.* **2023**, *119*, 106531. [CrossRef]
3. Francis, G. Regression Towards Mediocrity in Hereditary Stature. *J. Anthropol. Inst. Great Br. Irel.* **1886**, *15*, 246–263.
4. Fama, E.F.; French, K.R. The cross-section of expected stock returns. *J. Financ.* **1992**, *47*, 427–465.
5. Carhart, M.M. On persistence in mutual fund performance. *J. Financ.* **1997**, *52*, 57–82. [CrossRef]
6. Fama, E.F. Efficient capital markets: A review of theory and empirical work. *J. Financ.* **1970**, *25*, 383–417. [CrossRef]

7. Shiller, R.J. Do stock prices move too much to be justified by subsequent changes in dividends? *Am. Econ. Rev.* **1981**, *71*, 421–436.
8. Malkiel, B.G. The efficient market hypothesis and its critics. *J. Econ. Perspect.* **2003**, *17*, 59–82. [CrossRef]

engineering proceedings

MDPI

Proceeding Paper

Research on the Integration of the Bamboo Charcoal Industry as a Sustainable Development Goal for Creating Value Chains in Local Revitalization: A Case Study of the Longci District †

Fang-Ju Tsai [1], Chia-Hui Huang [1,*] and Chun-Chih Chen [2]

1 Institute of Cultural and Creative Design, Tung-Fang Design University, Kaohsiung 82941, Taiwan; iris9111100132@gmail.com
2 Department of Industrial Design, National Kaohsiung Normal University, Kaohsiung 80201, Taiwan; ccchen@nknu.edu.tw
* Correspondence: judy_huang@mail.tf.edu.tw

† Presented at the IEEE 5th Eurasia Conference on Biomedical Engineering, Healthcare and Sustainability, Tainan, Taiwan, 2–4 June 2023.

Abstract: In the past, research on regional revitalization has mainly focused on local culture and cultural industry. There is less research on the value composition of introducing Sustainable Development Goals (SDGs) into local industries. Thus, we analyzed existing local characteristic industries and the composition of the industry value chain from a consumer perspective. The main observation factors were determined for the introduction of products with sustainable development, value creation networks, and the formation of a sixth industrial structure under SDGs. The value creation configuration of local industries was also analyzed. By applying the two-dimensional Kano model of quality, consumers' attitudes towards products with SDG indicators were determined in local traditional industries more precisely. The correlation between the design of products with SDG indicators and satisfaction was also obtained. There was a two-dimensional correlation between the different qualities and satisfaction, rather than just a linear correlation. The research findings suggested that the value of the SDG indicators influenced consumers and societal trends significantly. Therefore, the value composition of the SDG indicators is the main determining factor for consumer expectations. The result provides a reference for stakeholders of local industries to construct a sustainable ecosystem in the regional revitalization system.

Keywords: egional revitalization; SDGs; industrial ecosystem construction; factor analysis

Citation: Tsai, F.-J.; Huang, C.-H.; Chen, C.-C. Research on the Integration of the Bamboo Charcoal Industry as a Sustainable Development Goal for Creating Value Chains in Local Revitalization: A Case Study of the Longci District. *Eng. Proc.* **2023**, *55*, 80. https://doi.org/10.3390/engproc2023055080

Academic Editors: Teen-Hang Meen, Kuei-Shu Hsu and Cheng-Fu Yang

Published: 20 December 2023

1. Introduction

In the past, research on egional evitalization was focused on the shaping of local culture and cultural industries, while there is relatively limited research on the value composition of Sustainable Development Goals (SDGs) in local industries. Therefore, we explored the composition of the industry value chain from a consumer perspective. By applying the two-dimensional Kano model, we determined a two-dimensional correlation of the factors rather than a linear relationship to understand consumer attitudes towards products from traditional local industries with SDG indicators and the correlation between the design of products with SDG indicators and satisfaction. The value of the SDG indicators is influenced by consumer and societal trends, determining consumer expectations. Regional evitalization and the industrial culture value of the indicatorsserve as a reference to incorporate the indicators into local industries and to establish the egional revitalization system as a sustainable ecosystem.

The Longci District has the lowest population and the most severe aging problem in Tainan City, Taiwan. Its natural and ecological features are characterized by the charcoal industry. In the sixth industrialization of agriculture, traditional agriculture was upgraded to integrate value chains in multiple dimensions of the industrial sectors. Comprehensive agricultural development was achieved using the resources of the natural environment, including bamboo forests, to promote diversified industrial development. This comprehensive approach helps the regional revitalization and the sustainable development of the rural area.

2. Literature Review

2.1. Regional Revitalization

"Regional Revitalization" is a policy initiated by the Japanese government in response to population decline, excessive population in Tokyo, and economic decline in local areas. The Japanese government enacted the "Act on the Promotion of City, People, and Jobs Creation" to create employment opportunities, promote migration, and support the younger generation in terms of marriage and child-rearing. The objective is to assist local areas to develop industries and improve living environments to attract young people back to these regions, and improve the demographic structure based on their industrial characteristics and cultural heritage [1]. Taiwan is facing similar problems. In 2019, the Executive Yuan declared it the "Year of Regional Revitalization" in Taiwan and approved the National Strategy Plan. This plan aims to promote comprehensive development in local communities, urban–rural areas, and regions, to achieve sustainable economic growth, social progress, and cultural prosperity [2]. The improvement of the quality of life for residents in rural areas and the resilience and competitiveness of local communities is also an objective. By protecting local characteristics and resources and promoting diversification and sustainable development in local areas, living environments in local communities are expected to improve to encourage residents to preserve local culture and traditions.

The National Strategy Plan for Regional Revitalization in Taiwan employs a "5-arrow" strategy to change population structures. These strategies include hometown investments by businesses, technology adoption, resource integration across ministries, fostering social participation, and establishing local brands. To achieve balanced development and attract young people back to their hometowns, it is vital to leverage and utilize local industry characteristics and creativity [3]. In rural communities, the production of eco-friendly and non-toxic handcrafted art pieces using local materials aligns with the principles of circular economy and traditional economic innovation. Through continuous evolution and innovation, the spirit of symbiosis and co-creation among community, craftsmanship, culture, industry, and environmental protection can be motivated [4]. According to Hsu [5], human force is inevitable in egional evitalization for local and sustainable development. In implementing egional evitalization, the participation and collaboration of stakeholders such as the government, businesses, communities, and residents are necessary. The government needs to consider the local characteristics and needs for the development of the local economy through investment and innovation, job opportunities, and social responsibility [5].

2.2. Consumer Value

The concept of consumer values was proposed by Hoffer [6]. Consumer behavior is influenced by social and psychological factors. Hoffer argued that people's consumption and their valuing are shaped by collective ideologies, culture, and social environments. Consumer values reflect individuals' evaluations, preferences, and goals regarding products and services, while expressing personal identity and value systems [6]. Sheth et al. proposed the Theory of Consumption Value in five dimensions of consumer value, including Functional Value, Emotional Value, Social Value, Conditional Value, and Novelty Value. These dimensions are used to explain the reasons behind purchasing decisions and elucidate consumer value perceptions and consumption behavior towards products or services [7]. The Theory of Consumption Value guides businesses and marketers to understand and predict consumer behavior. By understanding consumer needs and value perceptions, businesses can be designed to promote valuable products and services and meet consumer expectations to enhance market competitiveness. At the same time, the Theory of Consumption Value impacts consumer decision-making and inner satisfaction, aiding businesses and developers in gaining a deeper understanding of consumer choices and consumer psychology.

2.3. *Kano's Quality Attribute*

To improve the shortcomings of the linear hypothesis, Kano proposed the Kano Quality Model as a component of "Regression Analysis" to determine the relative reliability of "quality attributes." In the commonly employed method of "Regression Analysis", "quality" is considered as "sufficient quality" and "insufficient quality". "Regression coefficients" and "significance" are obtained to determine the "quality attributes" in Equation (1).

$$P = C + \beta 1 \times Kn + \beta 2 \times Kp \tag{1}$$

where P represents satisfaction, C is a constant term, Kn denotes the degree of insufficient quality, Kp represents the degree of sufficient quality, and β1 and β2 are their respective coefficients. In questionnaire surveys, Kn, Kp, and P are used to rate a scale ranging from −3 to +3. In the linear regression equation, individual "quality evaluations" are expressed with Kn and Kp. When the value of the quality evaluation is positive (e.g., 3), Kn = 0 and Kp = 3. When the value of the quality evaluation is negative (e.g., −2), Kn = −Kn = 2, and Kp = 0. By examining the relationship between β1 and β2, the quality attributes for each quality are determined. This relationship is shown in Table 1.

Table 1. Relationship between "Significance Of Regression Coefficients" and "Quality Attributes".

Quality Attributes	β1 (Backward) Significance	β2 (Forward) Significance	Remarks
Attractive	n.s.	*	$\beta 1 = 0$; $\beta 2 > 0$
One-Dimensional	*	*	$\beta 1 < 0$; $\beta 2 > 0$
Must-Have	*	n.s.	$\beta 1 < 0$; $\beta 2 = 0$
Indifferent	n.s.	n.s.	$\beta 1 = 0$; $\beta 2 = 0$
Reverse	*	*	$\beta 1 > 0$; $\beta 2 < 0$

(Sig. < 0.05 represents significance; n.s. represents non-significance; * represents significance). Source: Chen, C.J., and Wu, Y.Y. [8] (p. 45).

3. Methodology

We explored how integrating the bamboo charcoal industry with the concept of SDGs affected regional revitalization by co-creating the consumer value chain. Drawing on relevant research findings from scholars such as [1–3], we defined and explained the design attributes of co-creating consumer value of products with the SDG indicators. To clarify the level of consumer awareness of the SDG indicators and the relationship of the indicators with consumer value, consumer value theory, factor analysis, and the Kano model of two-dimensional quality were used. The results allowed for an understanding of consumer attitudes and behaviors toward co-creating consumer value goods design and SDGs [9–11].

4. Results and Discussions

4.1. *Factor Analysis*

In total, 268 questionnaires were distributed and returned. Bartlett's test of sphericity and the Kaiser–Meyer–Olkin (KMO) test were conducted to validate the questionnaire. As shown in Table 2, the *p*-value of Bartlett's test was 0.000. The KMO value was 0.963. A KMO statistic above 0.9 is considered "excellent", above 0.8 is "good", above 0.7 is "moderate", above 0.6 is "mediocre", above 0.5 is "poor", and below 0.5 is "unacceptable". Therefore, the questionnaire scale demonstrated validity. Thus, factor analysis was performed [12].

Table 2. KMO and Bartlett's test.

Kaiser–Meyer–Olkin Measure of Sampling Adequacy		0.963
Bartlett's Test of Sphericity	Approx. Chi-Square	6633.666
	df	276
	Sig.	0.000

After completing the factor analysis, the study advanced to an internal consistency evaluation to ascertain the reliability of the questionnaire. As elucidated in Table 3, Cronbach's Alpha values were employed to gauge the consistency across different dimensions of the consumer value chain. The overall item registered a Cronbach's Alpha value of 0.978, denoting an exceptionally high level of internal consistency. The dimension of environmental inclusiveness attained a Cronbach's Alpha of 0.974, the dimension of functional benefits stood at 0.917, and the sustainable value dimension reached 0.937. These findings further substantiate the questionnaire's reliability, suggesting that the individual items proficiently measure the respective concepts with commendable internal consistency.

Table 3. Consistency analysis of the consumer value chain dimensions.

	Cronbach's Alpha Value	**Item Number**
Overall item	0.978	
Environmental inclusiveness dimension	0.974	21, 25, 30, 26, 18, 24, 23, 20, 29, 19, 31, 22, 17
Functional benefits dimension	0.917	7, 6, 8, 5, 4, 2
Sustainable value dimension	0.937	13, 14, 11, 15, 10

Following the factor analysis, this study applied factor loadings to assess the items on the transformed axis of the consumer value chain. As delineated in Table 4, this table provides a summary of the composition of these factors. For example, the factor loading for confidence in environmental sustainability (Item 16) was 0.799, for a more profound understanding of eco-friendly products and sustainable lifestyles (Item 20) it was 0.781, and for the capability to be a distinctive topic or an appealing highlight (Item 23) it was 0.774. These results unveil the extent to which each item is associated with specific factors during the transformation of the consumer value chain, thereby enhancing our comprehension of the diverse dimensions of consumer value.

Three dimensions were identified as factors that influenced the integration of the bamboo charcoal industry into the value chain of SDGs. The first dimension, named "Environmental Coexistence Experience," consists of 13 items: questions 16, 20, 23, 21, 13, 19, 18, 15, 22, 14, 24, 17, and 12. This dimension represented consumers' experiences with bamboo charcoal products, including environmental sustainability, reduced negative emotions towards environmental issues, pride in environmental protection, expression of environmental sustainability values, sense of security in environmental protection, confidence in environmental sustainability, increased knowledge in the field of eco-friendly products, preference for innovation and innovative products, understanding of eco-friendly products and sustainable lifestyles, surprise and excitement from uniqueness, perceived convenience of use, alignment with social trends, creation of a fresh and comfortable environmental atmosphere, becoming a topic or highlight in special occasions or events, and displaying concern for environmental protection and sustainable living in specific situations. These items emphasized the benefits and value that consumers experienced when choosing and using bamboo charcoal products, highlighting the characteristics and advantages of bamboo charcoal products for sustainable consumption.

Table 4. Summary table of factor composition after the transformation of the consumer value chain axis.

Item	Factor 1	Factor 2	Factor 3
16. Confidence in environmental sustainability	0.799		
20. Greater understanding of eco-friendly products and sustainable lifestyles	0.781		
23. Ability to be a unique topic or an attractive highlight	0.774		
21. Surprise and excitement from uniqueness	0.762		
13. Pride in environmental protection	0.747		
19. Desire to try new things and innovative products	0.745		
18. Likability of innovation and willingness to use innovative products	0.737		
15. Sense of security in environmental protection	0.728		
22. Creation of a fresh and comfortable environment	0.725		
14. Expression of concern for environmental and sustainable values	0.716		
24. Display of concern for environmental protection and sustainable living in specific situations	0.701		
17. More knowledge in the field of eco-friendly products	0.693		
12. Lesser tendency to develop negative emotions towards environmental issues	0.664		
5. Removal of odors from pets		0.780	
4. Soil regulation and nutrient absorption		0.760	
6. Temperature regulation and moisture-wicking effects		0.703	
3. Prevention of mold issues caused by damp environments		0.691	
2. Increased value of home or office decoration		0.666	
1. Regulation of indoor humidity for fresher and more comfortable air		0.639	
9. Responsible use of bamboo resources for ecological preservation and biodiversity			0.808
10. Utilization of recycled resources in line with society's expectations for green production			0.769
8. Positive impact on local economy and employment through support for industry development			0.735
11. Reduction in demand for disposable products, minimizing waste and resource consumption			0.694
7. Decreased reliance on chemical products, friendlier to the environment and human health			0.657
Explained Variance	67.01	5.5	4.31
Cumulative Explained Variance	67.01	72.52	76.83

The second dimension, named "Functional Benefits", comprised six items: questions 5, 4, 6, 3, 2, and 1. This dimension emphasized the practical effects and benefits provided by bamboo charcoal products in various aspects, including air purification, humidity regulation, water purification, decorative value, moisture absorption, soil regulation, odor elimination, temperature regulation, and moisture-wicking properties. It was related to consumers' perceptions, preferences, and values regarding functional benefits, and the quality of bamboo charcoal products in meeting consumer needs and potential market demand.

The third dimension, named "Sustainable Value", consisted of five items: questions 9, 10, 8, 11, and 7. It represented the impact and contribution of bamboo charcoal products to environmental protection, sustainable development, and social values. Consumers' perceptions, importance, and preferences regarding the sustainable value of bamboo charcoal products were related to this dimension. The recognition of the contributions to environmental protection, reduced reliance on chemical products, support for bamboo industry development, emphasis on sustainable lifestyles, rational utilization of bamboo resources, environmental protection in the production process, utilization of renewable resources, and the benefits of reducing the demand for disposable products were involved in the dimension. Additionally, it highlighted the social value of bamboo charcoal products for environmental, social, and economic sustainability.

4.2. Kano Regression Analysis

One attribute, "reasonable utilization of bamboo resources for protecting the ecological environment and biodiversity", belonged to attractive quality, while the other 23 attributes to different qualities. This suggested that the current egional evitalization in Taiwan embraced a significant concept of sustainable development. Therefore, integrating the bamboo charcoal industry into SDGs provided additional value and attractiveness to consumers, making it more competitive in the market.

To further elucidate the attributes of bamboo charcoal products in terms of consumer perception, a detailed Kano quality classification was conducted. As presented in Table 5, the Kano model was applied to assess various quality attributes of bamboo charcoal products. This analysis categorized the attributes into different quality types based on their impact on consumer satisfaction and preference. For instance, the regulation of indoor humidity for fresher and more comfortable air (Attribute 1) and the increased value of home or office decoration (Attribute 2) were evaluated, among others. These findings provide a comprehensive understanding of how different attributes contribute to the overall appeal and utility of bamboo charcoal products from a consumer's perspective.

Table 5. Kano quality classification results.

Quality Attributes	β1	Significance Level	β2	Significance Level	R2	QC
1. Regulation of indoor humidity for fresher and more comfortable air	0.080	0.235	0.123	0.066	0.016	I
2. Increased value of home or office decoration	0.098	0.174	0.104	0.149	0.011	I
3. Prevention of mold issues caused by damp environments	0.021	0.753	0.038	0.572	0.001	I
4. Soil regulation and nutrient absorption	0.031	0.639	0.016	0.402	0.000	I
5. Removal of odors from pets	0.009	0.901	0.516	0.825	0.308	I
6. Temperature regulation and moisture-wicking effects	−0.031	0.645	0.016	0.809	0.002	I
7. Decreased reliance on chemical products, friendlier to the environment and human health	0.089	0.195	0.131	0.056	0.017	I
8. Positive impact on local economy and employment through support for industry development	0.084	0.217	0.135	0.048	0.017	I
9. Responsible use of bamboo resources for ecological preservation and biodiversity	0.075	0.272	0.168	0.015	0.025	A
10. Utilization of recycled resources in line with society's expectations for green production	−0.013	0.851	0.103	0.134	0.012	I
11. Reduction in demand for disposable products, minimizing waste and resource consumption	−0.025	0.716	0.087	0.201	0.010	I
12. Lesser tendency to develop negative emotions towards environmental issues	0.048	0.509	0.140	0.055	0.016	I
13. Pride in environmental protection	0.087	0.245	0.064	0.389	0.006	I
14. Expression of concern for environmental and sustainable values	0.085	0.231	0.056	0.427	0.006	I

Table 5. *Cont.*

Quality Attributes	β1	Significance Level	β2	Significance Level	R2	QC
15. Sense of security in environmental protection	0.083	0.257	0.115	0.116	0.011	I
16. Confidence in environmental sustainability	0.057	0.439	0.095	0.200	0.007	I
17. More knowledge in the field of eco-friendly products	0.009	0.901	0.012	0.863	0.000	I
18. Likability of innovation and willingness to use innovative products	0.061	0.391	0.024	0.738	0.003	I
19. Desire to try new things and innovative products	0.059	0.422	−0.041	0.575	0.007	I
20. Greater understanding of eco-friendly products and sustainable lifestyles	0.019	0.794	0.021	0.764	0.000	I
21. Surprise and excitement from uniqueness	0.115	0.130	0.029	0.707	0.011	I
22. Creation of a fresh and comfortable environment	0.064	0.360	0.011	0.873	0.004	I
23. Ability to be a unique topic or an attractive highlight	0.009	0.898	−0.004	0.958	0.000	I
24. Display of concern for environmental protection and sustainable living in specific situations	0.057	0.400	0.741	0.460	0.004	I

(QC: Quality Classification, A: Attractive, O: One-Dimensional, M: Must-Have, I: Indifferent, R: Reverse).

5. Conclusions

In the past, research on regional revitalization in Taiwan mainly focused on shaping local culture and cultural industries, with less attention on the integration of SDGs into local industries. Therefore, we analyzed the value chain composition of local industries from a consumer perspective to find an effective way to integrate SDGs. Based on the theory of consumer value, we identified the key factors in value creation for local industries and the importance of the SDG indicators. The study results suggested that consumers showed a positive attitude and satisfaction toward local traditional products with SDG attributes. Consumers were increasingly concerned about sustainable development and social responsibility, which influenced factors in their consumption behavior. Using the Kano regression analysis, the key factors in value creation were determined for local industries. The rational use of bamboo resources for protecting the ecological environment and biodiversity was expected by consumers. Stakeholders can use these findings as a reference to incorporate the SDG indicators into the industry value chain and improve consumer demand for sustainable products and services and the competitiveness and market position of local industries. The result can also be used to assist the development of regional revitalization systems and establish sustainable ecosystems.

The value composition of the SDG indicators was influenced by consumer needs and expectations and social trends and development. Therefore, stakeholders need to monitor social changes and updates in sustainable development goals to adjust products and services to correspond to these changes. The results of this study provided insights into the value creation of local industries in SDGs and regional revitalization. By applying consumer value theory, factor analysis, and the two-dimensional quality Kano model, consumers' attitudes towards local traditional products with SDG indicators were understood in terms of consumer attitudes and satisfaction towards the products. The results of this study are important for the industry and policymakers to make informed decisions and strategic arrangements for regional revitalization and sustainable development.

Author Contributions: The contribution of the author to the paper includes the provision of study materials or patients by F.-J.T., critical revision of the article for important intellectual content, and final approval of the article by C.-H.H. and C.-C.C. provided administrative, technical, or logistic support. All authors have read and agreed to the published version of the manuscript.

Funding: This research did not receive external funding.

Institutional Review Board Statement: The study did not involve human subjects or ethical concerns requiring approval by an Institutional Review Board or Ethics Committee, and was conducted in accordance with the Declaration of Helsinki.

Informed Consent Statement: Not applicable.

Data Availability Statement: The data that support the findings of this study are available from the corresponding author upon reasonable request.

Conflicts of Interest: The authors declare no conflict of interest.

References

1. Smile Taiwan. What is Regional Revitalization? Definition, Examples, and Actions in One go. Available online: https://smiletaiwan.cw.com.tw/article/5145 (accessed on 10 January 2022).
2. Taiwan Executive Yuan. Regional Revitalization 2.0. Available online: https://www.ey.gov.tw/Page/5A8A0CB5B41DA11E/de2c153a-2d59-4e35-95fa-00e89e23b116 (accessed on 10 May 2023).
3. Chen, M.-L. *National Strategy Plan for Local Revitalization in Taiwan*; National Development Council: Taipei City, China, 2017.
4. Li, Z.-Y. *2014 Record of Taiwanese Rural Crafts*; Water Conservation Bureau, Council of Agriculture: Nantou City, Taiwan, 2014.
5. Hsu, S.-M. The Role of University Education: Youth Cultivation and Local Revitalization. *Taiwan Econ. Rev.* **2021**, *19*, 36–46.
6. Hoffer, E. *The True Believer: Thoughts on the Nature of Mass Movements*; Harper Perennial Modern Classics: New York, NY, USA, 1955.
7. Sheth, J.N.; Newman, B.I.; Gross, B.L. Why we buy what we buy: A theory of consumption values. *J. Bus. Res.* **1991**, *22*, 159–170. [CrossRef]
8. Chen, C.J.; Wu, Y.Y. Exploring the Relationships between Consumers' Value Evaluations of Product Appearance and Purchase Intention--A Case of Digital Camera Designs. *J. Arts* **2012**, *8*, 39–62.
9. Chu, Y.Y. Exploring the Adoption Behavior of Consumers towards Tablet Computers Using the Theory of Consumption Values. Master's Thesis, EMBA Program, National Cheng Kung University, Tainan City, Taiwan, 2014.
10. Huang, X.W. The Relationship between Consumption Values, Offline Service Quality, and Intention to Use Food Delivery Platforms. Master's Thesis, Department of Business Administration, National Taipei University of Business, Taipei City, Taiwan, 2022.
11. Liao, S.R. A study on the consumption values and perceptions of Hakka cultural products. *J. Archit.* **2016**, *17*, 53–74.
12. Kaiser, H.F. An index of factorical simplicity. *Psychometrica* **1974**, *39*, 31–36. [CrossRef]

engineering proceedings

MDPI

Proceeding Paper

Dynamic Model of Andrographolide Therapy for COVID-19 †

Panittavee Yarnvitayalert and Teerapol Saleewong *

Department of Mathematics, Faculty of Science, King Mongkut's University of Technology Thonburi, Bangkok 10140, Thailand; panittavee.n@mail.kmutt.ac.th
* Correspondence: teerapol.sal@kmutt.ac.th
† Presented at the IEEE 5th Eurasia Conference on Biomedical Engineering, Healthcare and Sustainability, Tainan, Taiwan, 2–4 June 2023.

Abstract: *Andrographis paniculate* extract (APE) has been used as a Thailand traditional medicine owing to andrographolide which is effective for the viral clearance and prevention of disease progression. Several viral infections have been treated with APE, including SARS-CoV-2. The recommended dosage is 180 mg three times a day. We constructed a mathematical viral dynamic model of the SARS-CoV-2 model with andrographolide therapy by different doses. The pharmacokinetic/pharmacodynamic (PK/PD) model reduces the duration of viral clearance with a dose of 60 mg per day. Moreover, APE improved the therapeutic efficacy of COVID-19 therapy.

Keywords: andrographolide; COVID-19; PK/PD model; viral dynamic model

1. Introduction

Andrographolide is the major active component in Andrographis paniculate extract (APE) and has been used for the treatment of viral diseases such as COVID-19 due to its inhibited production of SARS-CoV-2 infectious virions [1]. Many researchers reported the clinical use of andrographolide for relieving symptoms of common colds and upper respiratory tract infections. In 2003, Kulichenko et al. tried the treatment of influenza with 30 mg and 45 mg of APE per day and decreased the duration of viral disease [2]. Saxena et al. found that 60 mg of APE per day relieved symptoms of uncomplicated upper respiratory tract infections (URTI) [3]. Thamlikitkul et al. researched the efficacy of APE (180–360 mg per day) for pharyngotonsillitis and observed the relief of fever and sore throat. Further, patients were satisfied with a higher dose of APE [4]. In 2021, Kulthanit et al. proposed COVID-19 treatment with 180 mg of APE per day for 5 days [5]. Nevertheless, there are limitations in the clinical use of APE. Thus, we investigated the potential mechanism and effect of andrographolide on the life cycle of SARS-CoV-2.

The pharmacokinetics (PK) of andrographolide on the life cycle of SARS-CoV-2 were explored in terms of in vivo mechanism that cannot be quantified without a blood test or reverse transcription-polymerase chain reaction (RT-PCR) tests to report the quantity of andrographolide in plasma and viral cells. Mathematical modeling is effective for the quantitative and qualitative study of COVID-19 because the model illustrates the amount of andrographolide, viral cell growth, and target cell infection. Goncalves et al. reported a viral dynamic model by using a standard target cell limit model with the eclipse phase to describe SARS-CoV-2 infection and determined viral growth parameters to predict the effects of antiviral treatments [6]. Dodds et al. discussed the potential drugs that influenced the SARS-CoV-2 viral cell cycle to reduce viral load and host cell infection [7].

For andrographolide treatment, we constructed a viral dynamics model which was a two-compartment model using pharmacokinetics and pharmacodynamics (PK/PD) in this study and investigated the efficacy of andrographolide on the SARS-CoV-2 infection. Andrographolide was measured in the plasma and tissue of subjects who took APE and its inhibitory potential was evaluated on viral production. The viral dynamic model

Citation: Yarnvitayalert, P.; Saleewong, T. Dynamic Model of Andrographolide Therapy for COVID-19. *Eng. Proc.* **2023**, *55*, 81. https://doi.org/10.3390/engproc2023055081

Academic Editors: Teen-Hang Meen, Kuei-Shu Hsu and Cheng-Fu Yang

Published: 21 December 2023

was refined with an effective andrographolide dose considering the behavior of target cell infection, virus-cell spread, and virus shedding duration. It was also explored if andrographolide dose affected the relief of symptoms such as fever, cough, and sore throat.

2. Mathematical Model

To investigate the effect of andrographolide dose on the viral life cycle, daily dosages of 30, 45, 60, 180, and 360 mg were given to the subjects, and PK/PD characteristics of andrographolide were explored [8]. PK was defined in in vivo distribution [8], whereas PK was determined to see if the drug's effect was related to its concentration. The relationship between PK and PD was investigated in the experiment, as seen in Figure 1.

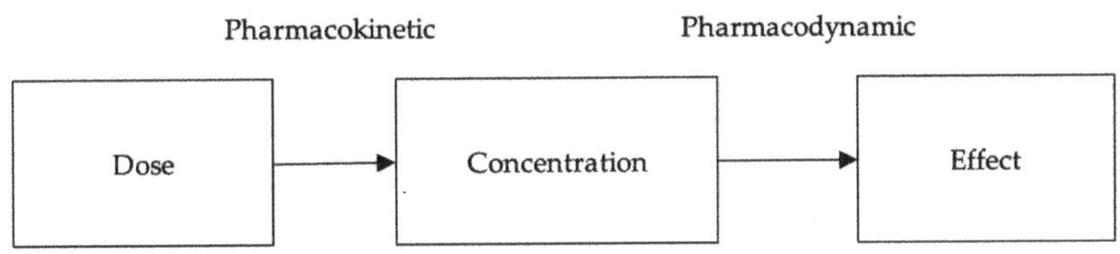

Figure 1. Relationship between pharmacokinetic and pharmacodynamic of PC.

2.1. PK Model

After patients took APE, andrographolide existed in plasma and tissue in the human body. Andrographolide was uptaken into plasma (C_p) at a certain rate of *Dose* and removed by a rate of k_{10}. Then, andrographolide was uptaken in peripheral tissue (C_t) at a rate of k_{12}, and transferred to plasma at a rate of k_{21}. PK of andrographolide is shown in Figure 2, which explains the concentration of andrographolide in the human body using the two-compartment model. The mathematical equation corresponding to the PK of andrographolide is expressed as

$$\begin{aligned} dC_p/dt &= Dose - (k_{10} + k_{12})C_p + k_{21}C_t, \\ dC_t/dt &= k_{12}C_p - k_{21}C_t, \end{aligned} \quad (1)$$

where C_p is the concentration of andrographolide in plasma, C_t is the concentration of andrographolide in peripheral tissue, and k_{10}, k_{12}, and k_{21} are rates.

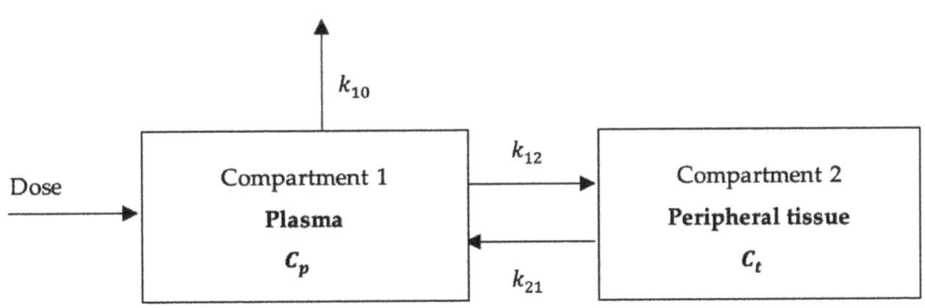

Figure 2. Pharmacokinetic of andrographolide.

2.2. PK/PD Model

The drug's effect is related to its concentration in plasma. Thus, the efficacy of andrographolide was assessed with the following equations.

$$\varepsilon\,(t) = C_p(t)/[C_p(t) + IC_{50}], \quad (2)$$

where IC_{50} is the half-maximal inhibitory concentration.

The average efficacy of andrographolide during the first 5 days of treatment was quantified as

$$\varepsilon_{mean} = 1/5 \int_0^5 [\, C_p(t)/(C_p(t) + IC_{50})]\, dt \quad (3)$$

As andrographolide inhibited the production of SARS-CoV-2 infectious virions, the inhibitory impact of andrographolide was adjusted in the proposed viral dynamic model.

2.3. Viral Dynamic Model

The coronavirus interacts with epithelial cells through the membrane by being bound between the receptor and the protein spike. Then, the uninfected target cell (T) is infected at a certain in the eclipse phase, β, and survives during the incubation period to productively infect cells at a rate of k. New viruses are RNA replicated at a productive rate of p. This process duplicates RNA to reproduce new viruses. The infected cell dies at a rate of δ. New viruses interact with and infect other epithelial cells. Virus cells die at a rate of c (Figure 3).

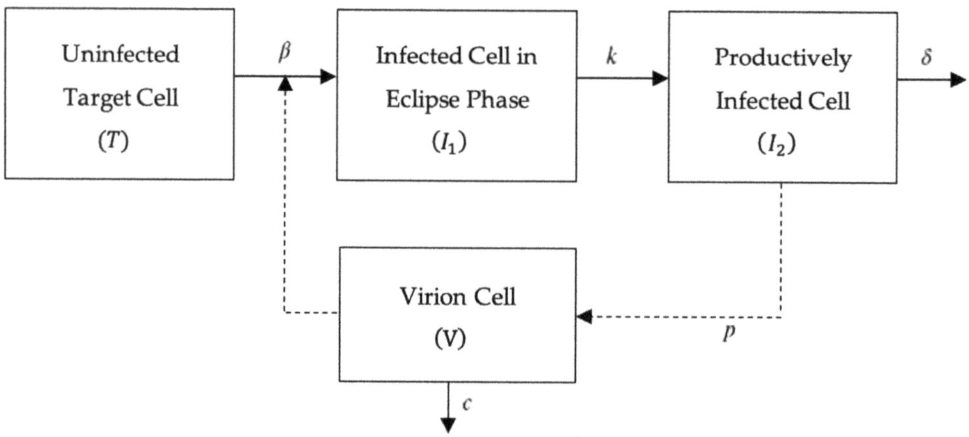

Figure 3. Viral dynamic model of SAR-CoV-2 in eclipse phase.

The mathematical equation corresponding to the viral dynamic model for SARS-CoV-2 treated with andrographolide is as follows.

$$\begin{aligned} dT/dt &= -\beta VT \\ dI_1/dt &= \beta VT - kI_1 \\ dI_2/dt &= kI_1 - \delta I_2 \\ dV/dt &= pI_2 - cV - \beta VT \end{aligned} \quad (4)$$

where T is target cells, I_1 is infected cells in the eclipse phase, I_2 is productively infected cells, V is virus cells, and β, k, δ, p, and c are constants.

2.4. Ordinary Differential Equation

Andrographolide inhibited RNA replication in the life cycle of viruses. RNA was replicated at a rate of p. Thus, the viral dynamic model was adjusted by the PK/PD of andrographolide, as shown in Figure 4.

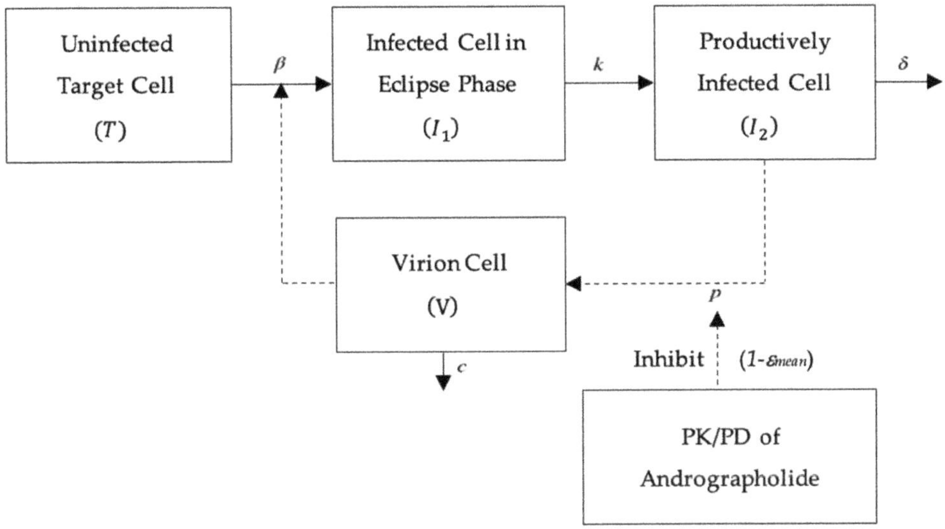

Figure 4. Viral dynamic model of SAR-CoV-2 treated with andrographolide.

Therefore, Equation (4) was rewritten as follows.

$$\begin{aligned} dC_p/dt &= Dose - (k_{10} + k_{12})C_p + k_{21}C_t, \\ dC_t/dt &= k_{12}C_p - k_{21}C_t, \\ dT/dt &= -\beta VT, \\ dI_1/dt &= \beta VT - kI_1, \\ dI_2/dt &= kI_1 - \delta I_2, \\ dV/dt &= (1 - \varepsilon_{mean})\, pI_2 - cV - \beta VT, \end{aligned} \quad (5)$$

where the efficacy is defined as $\varepsilon(t) = C_p(t)/[C_p(t) + IC_{50}]$, and the mean effectiveness of andrographolide in the first 5 days is given by $\varepsilon_{mean} = 1/5 \, {}_0\!\int^5 [C_p(t)/(C_p(t) + IC_{50})]$. The half-maximal inhibitory concentration, IC_{50}, was 9.54 μg/mL [5]. The variable of system equation as show in Table 1.

Table 1. Summary of variables of viral dynamics model.

	Variable	Unit
C_p	Andrographolide in plasma	ng/mL
C_t	Andrographolide in tissue	ng/mL
T	Target cells	cell/mL
I_1	Infected cells in eclipse phase	cell/mL
I_2	Productively infected cells	cell/mL
V	SARs-CoV-2	cell/mL

3. Results

We used the mathematical model based on Equation (5) to determine the viral load in the andrographolide therapy. Firstly, we chose the initial value $C_p(0)$ = Dose (ng/mL), $C_t(0)$ = 0 ng/mL, $T(0) = 1.33 \times 10^5$ cell/mL, $I_1(0)$ = 0 cell/mL, $I_2(0)$ = 0 cell/mL, $V(0) = 10^2$ cell/mL [6] and fixed the parameters as shown in Table 2. The rates k_{10}, k_{12}, and k_{21} were calculated by fitting the plasma concentration [9]. The chosen dosages for andrographolide were 30, 45, 60, 180, and 360 mg per day. The numerical simulation was carried out with MATLAB to illustrate how andrographolide diffusion in human plasma and tissue impacted the dynamic behavior of viruses (Figure 5). The plasma concentration of andrographolide increased substantially and remained steady for 10 h after beginning the experiment. The mean efficacy of andrographolide, ε_{mean}, is presented in Table 3. It was found that the behavior of target and viral cells was similar. The infection occurred on Day 5 and ended on Days 25–30.

Table 2. Related parameters of viral dynamics model.

Parameter	Values	Dimension	References
Dose	[30, 45, 60, 180, 360]	mg/day	-
k_{10}	3.6228	h^{-1}	[9]
k_{12}	4.2259	h^{-1}	[9]
k_{21}	1.4233	h^{-1}	[9]
β	2.21×10^{-5}	mL/cells/day	[6]
k	3.00	day^{-1}	[6]
δ	0.60	day^{-1}	[6]
p	22.71	day^{-1}	[6]
c	10.00	day^{-1}	[6]

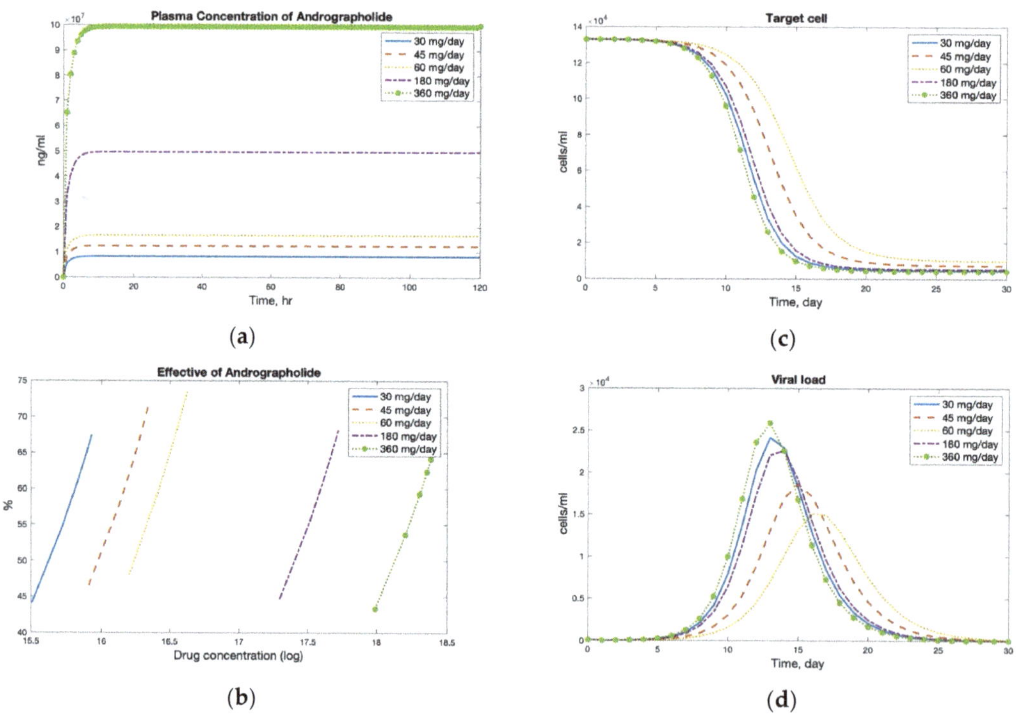

Figure 5. Numerical solution of PK/PD model and viral dynamic model of COVID-19 with andrographolide treatment in (**a**) the plasma concentration of drug dosage, (**b**) the effective of drug dosage, (**c**) the quantitative of target cell, and (**d**) the quantitative of viral cell.

In Thailand, the therapeutic dosage of andrographolide is 3 × 180 mg per day. The solution of the PK/PD model and the viral dynamics model with andrographolide treatment of the dosage is shown in Figure 6. In terms of PK, the concentration of andrographolide in plasma and tissues increased and remained steady for 10 h after beginning the dosage.

Considering the latent time of 3 days, target cell infection occurred on Day 5, and the virus load peaked on Day 15. The viral clearance occurred on Day 25.

Table 3. Mean efficacy of andrographolide to treat COVID-19.

Dose (mg/mL)	Mean Efficacy (ε_{mean})
30	66.30
45	69.90
60	72.11
180	67.13
360	65.34

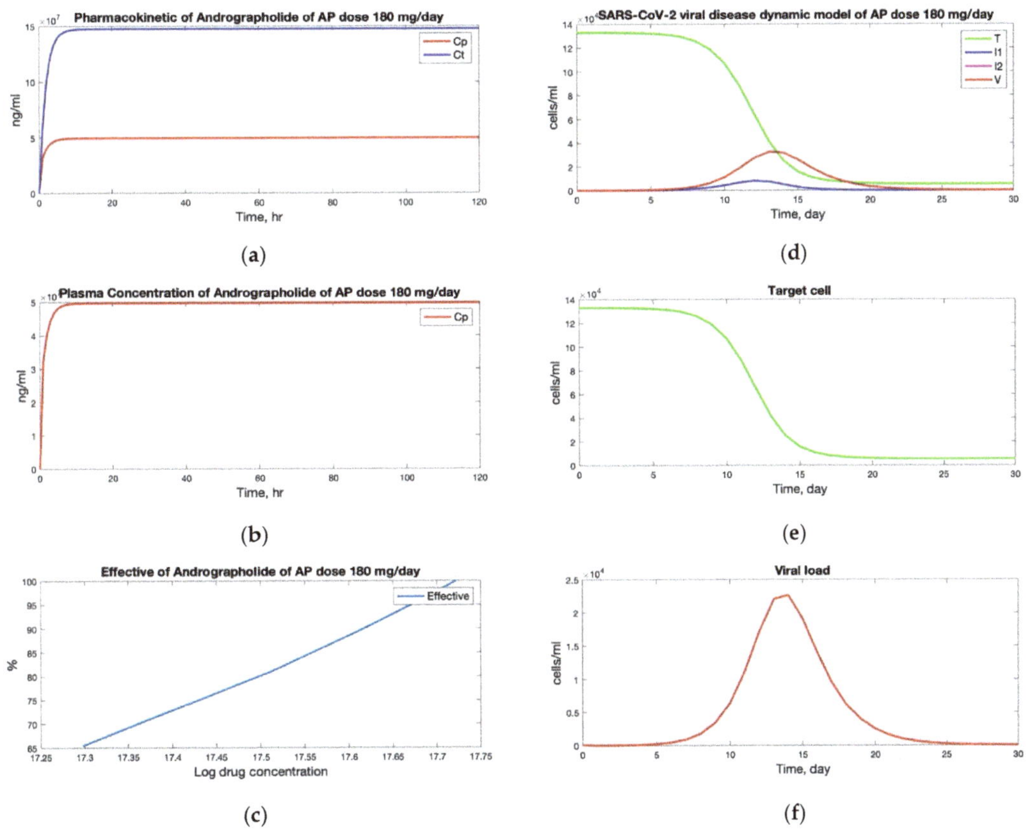

Figure 6. Numerical solution of PK/PD model and viral dynamic model of COVID-19 with a dose of 180 mg per day of andrographolide: (**a**) pharmacokinetics of daily dose of 180, (**b**) plasma concentration of andrographolide, (**c**) efficacy of andrographolide at IC_{50}, (**d**) number of COVID-19 cells, (**e**) number of uninfected target cell, and (**f**) viral load of viral dynamics model.

4. Conclusions

The PK/PD model and the viral dynamic model of andrographolide dose on COVID-19 were explored in this study. Andrographolide effectively reduced viral load and target cell infection, showing its efficacy. A dose of 60 mg per day of andrographolide dairy was the most efficient in preventing infection. Andrographolide can be used for the therapy of COVID-19. Due to possible toxicity, side effects, misuse, allergy symptoms, and interactions of APE with other medications, the use of andrographolide must be decided by physicians or pharmacists. As there is an increasing demand for APE with its benefits, the quality of APE in terms of safety and efficacy must be supervised and controlled to use it as medicine.

Author Contributions: Conceptualization, P.Y. and T.S.; methodology, P.Y. and T.S.; software, P.Y.; validation, P.Y. and T.S.; formal analysis, P.Y. and T.S.; investigation, P.Y. and T.S.; resources, P.Y.; data curation, P.Y. and T.S.; writing—original draft preparation, P.Y.; writing—review and editing, P.Y. and T.S.; visualization, P.Y. and T.S.; supervision, T.S. All authors have read and agreed to the published version of the manuscript.

Funding: This research received no external funding.

Institutional Review Board Statement: Not applicable.

Informed Consent Statement: Not applicable.

Data Availability Statement: Data are contained within the article.

Acknowledgments: This work was partially supported by the Science Achievement Scholarship of Thailand from the Department of Mathematics in the Faculty of Science and King Mongkut's University of Technology Thonburi.

Conflicts of Interest: The authors declare no conflict of interest.

References

1. Sa-Ngiamsuntorn, K.; Suksatu, A.; Pewkliang, Y.; Thongsri, P.; Kanjanasirirat, P.; Manopwisedjaroen, S.; Charoensutthivarakul, S.; Wongtrakoongate, P.; Pitiporn, S.; Chaopreecha, J.; et al. Anti-SARS-CoV-2 activity of andrographis paniculata extract and its major component andrographolide in human lung epithelial cells and cytotoxicity evaluation in major organ cell representatives. *J. Nat. Prod.* **2021**, *84*, 1261–1270. [CrossRef] [PubMed]
2. Kulichenko, L.L.; Kireyeva, L.V.; Malyshkina, E.N.; Wikman, G. A randomized, controlled study of Kan Jang versus amantadine in the treatment of influenza in Volgograd. *J. Herb. Pharmacother.* **2003**, *3*, 77–93. [CrossRef] [PubMed]
3. Saxena, R.C.; Singh, R.; Kumar, P.; Yadav, S.C.; Negi, M.P.; Saxena, V.S.; Joshua, A.J.; Vijayabalaji, V.; Goudar, K.S.; Venkateshwarlu, K.; et al. A randomized double blind placebo controlled clinical evaluation of extract of Andrographis paniculata (KalmCold) in patients with uncomplicated upper respiratory tract infection. *Phytomedicine* **2010**, *17*, 178–185. [CrossRef] [PubMed]
4. Thamlikitkul, V.; Dechatiwongse, T.; Theerapong, S.; Chantrakul, C.; Boonroj, P.; Punkrut, W.; Ekpalakorn, W.; Boontaeng, N.; Taechaiya, S.; Petcharoen, S. Efficacy of Andrographis paniculata, Nees for pharyngotonsillistis in adults. *J. Med. Assoc.* **1991**, *74*, 437–442.
5. Wanaratna, K.; Leethong, P.; Inchai, N.; Chueawiang, W.; Sriraksa, P.; Tabmee, A.; Sirinavin, S. Efficacy and safety of *Andrographis paniculata* extract in patients with mild COVID-19: A randomized controlled trial. *Arch. Intern. Med. Res.* **2022**, *5*, 423–427. [CrossRef]
6. Gonçalves, A.; Bertrand, J.; Ke, R.; Comets, E.; De Lamballerie, X.; Malvy, D.; Pizzorno, A.; Terrier, O.; Rosa Calatrava, M.; Mentré, F.; et al. Timing of antiviral treatment initiation is critical to reduce SARS-CoV-2 viral load. *CPT Pharmacomet. Syst. Pharmacol.* **2020**, *9*, 509–514. [CrossRef] [PubMed]
7. Dodds, M.G.; Krishna, R.; Goncalves, A.; Rayner, C. Model-informed drug repurposing: Viral kinetic modelling to prioritize rational drug combinations for COVID-19. *Br. J. Clin. Pharmacol.* **2021**, *87*, 3439–3450. [CrossRef] [PubMed]
8. Ducharme, M.P.; Braga, L. Introduction to pharmacokinetic and pharmacodynamic models and analyses. In *Shargel and Yu's Applied Biopharmaceutics and Pharmacokinetics*, 8th ed.; Ducharme, M.P., Shargel, L., Eds.; McGraw Hill: New York, NY, USA, 2022.
9. Panossian, A.; Hovhannisyan, A.; Mamikonyan, G.; Abrahamian, H.; Hambardzumyan, E.; Gabrielian, E.; Goukasova, G.; Wikman, G.; Wagner, H. Pharmacokinetic and oral bioavailability of andrographolide from Andrographis paniculata fixed combination Kan Jang in rats and human. *Phytomedicine* **2000**, *7*, 351–364. [CrossRef] [PubMed]

engineering proceedings

Proceeding Paper

Continuity versus Desertion in Pandemic Times: The Management of Engineering Schools in a Peruvian Private University via Big-Data Analysis †

Huber Nieto-Chaupis

Faculty of Engineering, Universidad Autónoma del Perú, Panamericana Sur. Km. 16.3, Lima 15842, Peru; hubernietochaupis@gmail.com

† Presented at the IEEE 5th Eurasia Conference on Biomedical Engineering, Healthcare and Sustainability, Tainan, Taiwan, 2–4 June 2023.

Abstract: The main consequence of the arrival of COVID-19 was the application of social distancing, which mainly had implications regarding the economies of affected countries. This affected students, whose economy was weakened because of social restrictions. Therefore, private universities that depend strongly on the continuity of students had to face some difficult periods, with an eventual increase in desertions. In this way, the general managers' offices had to find new and consistent directions for correct decision making without reducing the quality of the service. This paper presents a scheme that was applied in a private university in the south of Lima city during the period of 2020–2021. It is shown that the usage of a set of variables and their combinations to make a decision only appears to be relevant if a such decision is free of either external or internal noise. The results exhibit a minimization of 70% in the potential number of desertions in 2021 due to the gained experiences of 2020.

Keywords: engineering management; general topics for engineers; COVID-19

Citation: Nieto-Chaupis, H. Continuity versus Desertion in Pandemic Times: The Management of Engineering Schools in a Peruvian Private University via Big-Data Analysis. *Eng. Proc.* **2023**, *55*, 82. https://doi.org/10.3390/engproc2023055082

Academic Editors: Teen-Hang Meen, Kuei-Shu Hsu and Cheng-Fu Yang

Published: 27 December 2023

1. Introduction

In March of 2020, all countries were dramatically affected by the unexpected arrival of coronavirus disease [1–3] producing fatalities in its first pandemic wave. This triggered a chain of economic, political, and educational decisions in affected places that changed the day-to-day evolution of the whole of society in order to guarantee the health of the people [4,5].

Such decisions, such as to migrate to a virtual (or online) mode, were supported by technology, especially the Internet. In the age of 4G, the Internet emerged as a prominent exit from the problem of continuity in human development due to its interaction with a changing society dictated by the latest technologies.

Clearly, at the beginning of 2020, most countries adopted the so-called digitalization of their systems as an imminent action to counteract the rapid weakening of society due to the fast-increasing number of COVID-19 infections [6–10].

In this manner, the educational sector modified its classical style of blackboard teaching to one based entirely on video conferences [11–13]. This was undertaken as a protective action to avoid infections at schools and universities. Thus, after curfews were implemented, universities had to adjust their pedagogical methodologies towards online digital education.

Although advanced students, ranging from the second to the last years of their academic programs, might have interacted with video conferences prior to the beginning of the pandemic, this was not the case with freshman students. In effect, more of them had no experience with virtual education, so universities had to focus their educational machinery, in particular the mechanisms of virtual education, onto these freshman students as a critical action to keep their tuition and, therefore, to keep them enrolled over the coming years.

Clearly, universities faced a rise in desertion, which can be seen as an imminent threat to their institutional stability, particularly private ones [14,15].

Thus emerged the question: how interesting were the first online classes for freshman students? From this, one can then formulate a subsequent question: how stimulating were the first lectures for them whilst using software-based video conferences?

In this paper, attention has been paid to a private university in south Lima, with a population of around 7.5k students. Thus, from the beginning of the pandemic, this university continued with its evolution of whole-university dynamics through video-conference meetings and video-conference classes.

Thus, based on experience, it was noted that, although no strong desertion was reported, surveys underline the nonconformity of freshman students.

Although video technology most likely personalized students' education, one can see that the impact and nonconformity on the side of the students is not instantaneously felt. Thus, the success of online classes depends in on the one hand on technology, and on the other hand on the strong capability of instructors and how they can keep the attention of students throughout a class. In this manner, it is not clear if the quality of education was untouched [16,17]. It is merely a point to be studied in this paper.

The rest of this paper is structured as follows: In the second section, the main characteristics of online classes are viewed. In the third section, a statistical analysis is presented. In the fourth section, the results of the study are given. Finally, the conclusion regarding the results of paper is presented.

2. Basics of Online Classes

An online class, called also virtual teaching, can be defined as an interaction at a distance between a teacher and student through a software–hardware interface that aims to replace an ordinary class. In contrast to these ordinary classes, where a teacher employs a blackboard with chalk and there is a face-to-face interaction, a virtual class would present limitations (on the side of student), some of which can be listed below.

Table 1 above highlights up to five variables that were selected from applied surveys to freshman students. In the first column on the left, Internet connections are described as the most important variable to be met in a successful manner for virtual classes. It is clearly a crucial variable that demands to be unstoppable, as written in the third column. In the last column, one can see that this might not work for a class due to the following reasons: (i) the Internet provider might have increased costs in the pandemic epoch as a consequence of supply and demand; (ii) the fast-increasing number of new Internet clients had a noteworthy impact on the networks as well as on the quality of service. In this manner, connectivity to the Internet in South American countries can be seen as an intersection of social and technological problems.

Table 1. Description of variables studied at this paper.

Variables for Teachers and Students	Impact on the Classes	Noteworthy Comment about Variable	Works for All Students?
Internet connection	Very important	Needed to be unstoppable	Not
Software Video Conference	Adaptable	Students can be easily adapted	Yes
Expertise of VC usage	Not needed	Not neither the beginning nor the end	Not
Sociability	Adaptable	Not seen as necessary	Adopted
Confidence Examinations	Important	Highly relevant Assess the quality of education	Changed

Although, on the one hand, one can guarantee the stability of an Internet connection, on the other hand, one can wonder about the social role of a video conference. In fact, while an instructor carries out a class, based on either theory or exercises, it is noteworthy that this may not be comfortable for each of the attenders. A common mistake is likely made here in the sense that one assumes that virtual classes for a freshman student must be highly pleasant in all educational directions. A sensitive point in these virtual classes are the examinations. In 2020, various styles of examinations were undertaken, each carrying the risk that they may not be trusted as sincere qualifications. Thus, the application of virtual examinations might have not fulfilled the desire to know that one has truly learnt most parts of a course. Even with these subjective obstacles, courses in basic sciences such as physics and math implemented methodologies rather similar to the ordinary ones that kept up a face-to-face style, yielding interesting results.

3. Measuring the Satisfaction of Students

It is common that subjectivity cannot be measured in terms of numbers. Even in people with different levels of satisfaction or dissatisfaction, the measurements of a group turn out to be clouded by the presence of hidden variables that cannot be seen at a first glance.

Instead of that mentioned above, this study has opted to take probabilities as a firm basis to measure the levels of acceptance in freshman students and how, with these different levels of acceptance, one can build a mathematical methodology that allows us to identify the probability of desertion by also using the listed variables in Table 1.

A naïve manner to introduce mathematics in the construction of a phenomenological equation that yields reasonable probabilities of desertion [18] can be detailed as follows:

$$\mathcal{P} = \frac{u}{u+s}$$

where $\mathcal{P}$ is the probability of acceptance and u and s are the number of satisfied and unsatisfied students, respectively. A first equation can be arrived at below:

$$\mathcal{P} = \frac{u}{u(1+\frac{s}{u})} = \frac{1}{(1+\frac{s}{u})} \tag{1}$$

An unfortunate case is where $u \gg s$, which makes sense only if the number of unsatisfied students is large in comparison with the satisfied ones. In this manner, the following can be described:

$$\left(1+\frac{s}{u}\right) \approx 1 + \left(\frac{s}{u}\right) + \frac{1}{2!}\left(\frac{s}{u}\right)^2 + \frac{1}{3!}\left(\frac{s}{u}\right)^3 + \cdots \tag{2}$$

The following is the expansion in a series of the well-known exponential function:

$$\mathrm{Exp}\left(\frac{s}{u}\right) = \sum_{q=0}^{\infty} \frac{1}{q!}\left(\frac{s}{u}\right)^q. \tag{3}$$

With this, Equation (1), the dissatisfaction probability, can be written now as follows:

$$\mathcal{P}(u) = \mathrm{Exp}\left(-\frac{s}{u}\right) \tag{4}$$

Equation (4) might be seen as a simple mathematical expression. However, for the purpose of this study, the exact values of both "s" and "u" are required. Over the course of a semester of virtual classes, these quantities may have either oscillated or been constant. In this manner, one can state that these numbers depend on time. Thus emerges the question:

how to measure these quantities in an accurate way? In addition, Equation (4) can also be written as a satisfaction probability:

$$\mathcal{P}(s) = \text{Exp}\left(-\frac{u}{s}\right) \tag{5}$$

by which one can test that the sum of $\mathcal{P}(u) + \mathcal{P}(s) = 1$, which constitutes the starting point of the next section. Certainly, the involved probabilities should be dependent upon time. This point is boarded below.

Probability of Desertion

Equations (4) and (5) constitute the basic definitions towards a solid relationship that can communicate to us, to some extent, the feasibility of having a quantitative idea of desertion.

In virtue of the above equations, one can define the following:

$$\mathcal{P}_D = \frac{\beta_1 \text{Exp}\left(-\frac{s}{u}\right)}{\beta_1 \text{Exp}\left(-\frac{s}{u}\right) + \beta_2 \text{ Exp}\left(-\frac{u}{s}\right)} \tag{6}$$

where the constant is a kind of weight in the formulation. In order to illustrate Equation (6), one can assume the limit case where $u = s$. It yields the following:

$$\mathcal{P}_D = \frac{\beta_1}{(\beta_1 + \beta_2)} \tag{7}$$

so that, if $\beta_1 \approx \beta_2$, then one has a probability of desertion of 50%. Clearly, one has ignored tangible details that denote the true causes of possible desertion. Then, one can wonder about all these details and how they can all be entered into a pure mathematical description, as is given in all of the above equations.

Consequently, one can see the entire responsibility of the constants β_1 and β_2. Again, from experience, one can state that the disagreement and nonconformity with the virtual sessions might not be constant throughout the academic semester, but is also under variation based on time.

Then, one finds that the time as an independent variable cannot be incorporated in a straightforward manner inside the arguments of exponentials in Equation (6).

Alternatively, one can opt for the variables of Table 1: (i) the type of video conference and (ii) confidence in examinations. In virtue of Equation (7), a liner model of desertion can be written below:

$$\mathcal{P}_D(t) = \frac{a + \beta_1 t}{a + \beta_1 t + b + \beta_2 t^{n+\Delta}} \tag{8}$$

where, while dissatisfaction increases linearly with time, satisfaction depends directly on t^n, with n being an integer number that, in Equation (8), is assumed to be negative, and Δ is a kind of error with the identification of the true number of students that are satisfied with virtual classes.

Consider the case that no error is perceived and the probability of desertion is free of errors. Thus, one may obtain below the normalized probability of desertion:

$$\mathcal{P}_D(t) = \frac{(1/0.8)(1 + (\beta_1 = 1)t)}{2 + (\beta_1 = 1)t + \ (\beta_2 = 1)\ t^{-2+q}} \tag{9}$$

where $\beta_1, \beta_2 = 1$ and $a = b = 1.0$. On the other hand, the factor $1/0.8$ is required to normalize to unity. In Figure 1, one can see the different distributions of the probability of desertion based on Equation (9) for q = 1, 2, 3, and 4. While the green color distribution clearly increases with $q = 1$, the orange color decreases with $q = 4$.

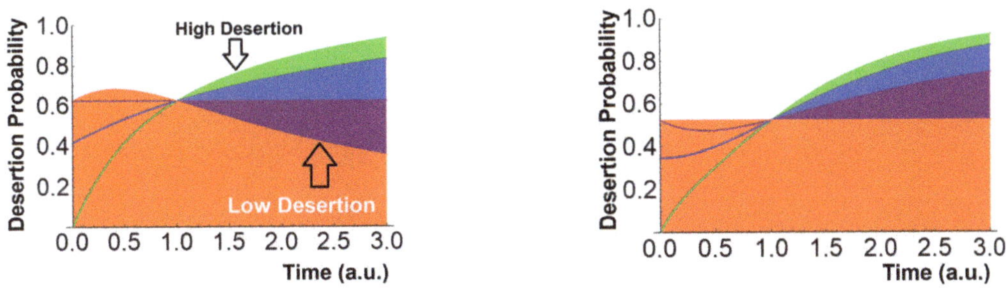

Figure 1. (**left**) The different distributions of probability for 4 values of integer "q" according to Equation (9). Here q = 1, 2, 3 and 4 for colors green, blue, magenta and orange, respectively. The arrows indicate high and low desertion. (**right**) Plotting of Equation (9) with the quadratic terms in "t". The orange color denotes the scenario where desertion is not well defined. Green, blue and magenta are denoting scenarios of high desertion.

In Figure 2, Equation (9) has been plotted, but with the case where $a + \beta_1 t \implies a + \beta_1 t^2$. The quadratic case is of interest because it exhibits a rapid increase in desertion despite the fact that the term t^{-2+q} increases with $q > 2$. Thus, in Figure 2, the orange color appears to be constant at around 50%. Then, one can wonder which parameter from Equation (8) can be changed in order to decrease the probability of desertion. In fact, the parameter "a" as well as β_1 can be slightly varied according to the variables listed in Table 1. The results of an applied survey for 78 physics students are detailed below.

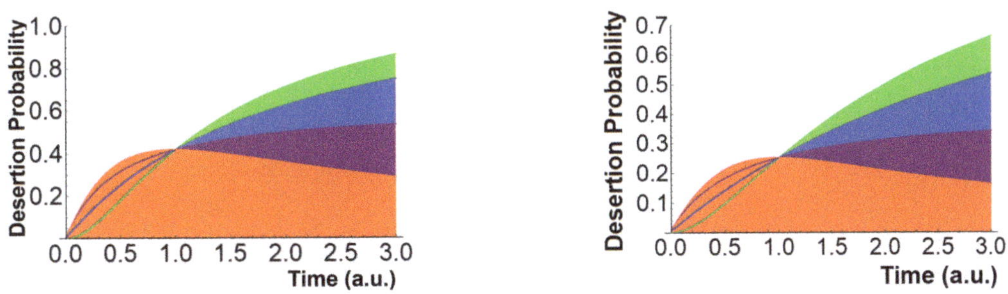

Figure 2. (**left**) Distributions of probability of desertion, where 1 is the entire acceptance of students on a physics course (2020). The influence of constant "a" equal to 0.001 is seen as a kind of alleviation. Here q = 1, 2, 3 and 4 for colors green, blue, magenta and orange, respectively. (**right**) The case when $0.01 + 0.5\,t$ where the desertion appears to be lower than previous cases. The role of the university is assumed to manage the academic level in order to maintain the attention of students and avoid their desertion.

4. Alternatives Scenarios of Desertion

As stated above, one can define the free parameters "a" and β_1 to acquire a set of values that decrease the probability of desertion [19]. In this way, one can write below that

$$a = \frac{N_{VC}}{N_{CC} + N_{NA}} \tag{10}$$

where N_{VC}, N_{CC}, and N_{NA} are the number of students that are not satisfied with the video-conference software (or do not like it), the number of students that approve of the content of the course (see Table 2), and the ones that do not agree with virtual classes, respectively. For example, the case where $N_{CC} + N_{NA}$ represents a high percent of order of 80% up to 90% tells us that N_{CC} is a large number, whereas N_{VC} would represent a minimum number;

then, "a" acquires a small size. With this in mind, the resultant probabilities can be seen in Figure 2. In fact, from Equation (10), one can obtain a value for "a" = 0.01; the distributions of Figure 2 based on the equation are written below:

$$\mathcal{P}_D(t) = \frac{\left(\frac{1}{0.8}\right)(0.01 + t)}{0.01 + t + \ 1 + t^{-2+q}}. \tag{11}$$

Table 2. Variables for freshman students.

Variable	Score	Confidence
Video-Conference Software (zoom V.5.6.6)	I don't like	>75%
Style of instructor	Standard	>75%
Content of course	Standard	>75%
Manner of examination	I don't like	>50%
Do not agree with online classes	Yes	>65%

In Figure 2 (**left**), one can see a kind of acceptance of the virtual classes in the freshman students. Only the green color, that is, the case of "q = 1", turns out to be above 50%. The blue and other colors range between 35% and 65%, expressing the fact that an excellent presentation of the course content might guarantee, to some extent, the minimization of the number of desertions.

On the other hand, the case that is proportional to "t", the time, is of interest for this study. In this manner, the term "t" becomes "0.5 t", acquiring the following form:

$$\mathcal{P}_D(t) = \frac{\left(\frac{1}{0.8}\right)(0.01 + 0.5\,t)}{0.01 + 0.5\,t + \ 1 + t^{-2+q}}. \tag{12}$$

In Figure 2 (**right**), the resulting probabilities from Equation (12) are plotted. In comparison to previous plots, this case turns out to be optimal in the sense that the number of desertions might be at the mid-range of the probabilities, around 50%. However, the orange color displays the range of less than 25%, indicating that one of four students is likely willing to abandon their university studies. In respect of the shape of the distributions of probability, one can see that the mathematical function modeling the willingness to desert of freshman students can be somewhat given as

$$\mathcal{P}_D(t) = \alpha[\, Tan(\beta_1 t) + Exp(-\beta_2 t)] \tag{13}$$

with "α" being a constant of normalization, and β_1 and β_2 are defined above. While the tanh distribution denotes the deserters, the negative exponential explains the expected behavior to decrease desertion through the parameter "β_2".

5. Conclusions

The present study consisted of a quantitative analysis of a set of mathematical models that may shed light on the problem of desertion amongst freshman students at a private university located in south Lima. Essentially, a sample of 78 students was considered over the course of the 2020 academic year, encompassing the very beginning of the COVID-19 pandemic.

The main variables used within the mathematical model are (i) satisfaction and (ii) dissatisfaction. It is assumed and is the argument of this paper that if this information is possible to identify through well-designed surveys aimed at students throughout a course, then by using the information that there exists a number of unsatisfied students, a rapid response on the side of the instructor or teacher should be undertaken. As seen in Table 2 above the content of a course and how it is presented would be a valuable to be considered in order to avoid desertion. Furthermore, the applied survey also determined that the style in which a teacher or instructor carries out their class would be an interesting point to be considered in the future. Nevertheless, the pandemic blocked all available ways to avoid the imminent, undesired fact of the desertion of freshman students [20].

Funding: This research received no external funding.

Institutional Review Board Statement: Not applicable.

Informed Consent Statement: Not applicable.

Data Availability Statement: Data are contained within the article.

Conflicts of Interest: The author declares no conflict of interest.

References

1. Bahl, P.; Doolan, C.; De Silva, C.; Chughtai, A.A.; Bourouiba, L.; MacIntyre, C.R. Airborne or droplet precautions for health workers treating coronavirus disease 2019? *J. Infect. Dis.* **2020**, *225*, 1561–1568. [CrossRef] [PubMed]
2. Bourouiba, L. Turbulent gas clouds and respiratory pathogen emissions: Potential implications for reducing transmission of COVID-19. *JAMA* **2020**, *323*, 1837–1838. [CrossRef] [PubMed]
3. Mittal, R.; Ni, R.; Seo, J. The flow physics of COVID-19. *J. Fluid Mech.* **2020**, *894*, F2. [CrossRef]
4. Lewis, D. Mounting evidence suggests coronavirus is airborne—But health advice has not caught up. *Nature* **2020**, *583*, 510–513. [CrossRef] [PubMed]
5. Seminara, G.; Carli, B.; Forni, G.; Fuzzi, S.; Mazzino, A.; Rinaldo, A. Biological fluid dynamics of airborne COVID-19 infection. *Rend. Fis. Acc. Lincei* **2020**, *31*, 505–537. [CrossRef] [PubMed]
6. Dietz, L.; Horve, P.F.; Coil, D.A.; Fretz, M.; Eisen, J.A.; Van Den Wymelenberg, K. 2019 novel coronavirus (COVID-19) pandemic: Built environment considerations to reduce transmission. *mSystems* **2020**, *5*, e00245-20. [CrossRef] [PubMed]
7. The World Health Organisation. Coronavirus Disease (COVID-19)_Advice_for_the_Public. 2020. Available online: https://www.who.int/emergencies/diseases/novel-coronavirus-2019/advice-for-public (accessed on 1 July 2022).
8. Hoehl, S.; Rabenau, H.; Berger, A.; Kortenbusch, M.; Cinatl, J.; Bojkova, D.; Behrens, P.; Böddinghaus, B.; Götsch, U.; Naujoks, F.; et al. Evidence of SARS-CoV-2 infection in returning travelers from Wuhan, China. *N. Engl. J. Med.* **2020**, *382*, 1278–1280. [CrossRef] [PubMed]
9. Lai, C.C.; Liu, Y.H.; Wang, C.-Y.; Wang, Y.-H.; Hsueh, S.-C.; Yen, M.-Y.; Ko, W.-C.; Hsueh, P.-R. Asymptomatic carrier state, acute respiratory disease, and pneumonia due to severe acute respiratory syndrome coronavirus 2 (SARS-CoV-2): Facts and myths. *J. Microbiol. Immunol. Infect.* **2020**, *53*, 404–412. [CrossRef] [PubMed]
10. Sohrabi, C.; Alsafi, Z.; O'Neill, N.; Khan, M.; Kerwan, A.; Al-Jabir, A.; Iosifidis, C.; Agha, R. World Health Organization declares global emergency: A review of the 2019 novel coronavirus (COVID-19). *Int. J. Surg.* **2020**, *76*, 71–76. [CrossRef] [PubMed]
11. Kita, Y.; Yasuda, S.; Gherghel, C. Online education and the mental health of faculty during the COVID-19 pandemic in Japan. *Sci. Rep.* **2022**, *12*, 8990. [CrossRef] [PubMed]
12. Wang, Z.H.; Wang, Z.-H.; Yang, H.-L.; Yang, Y.-Q.; Liu, D.; Li, Z.-H.; Zhang, X.-R.; Zhang, Y.-J.; Shen, D.; Chen, P.-L.; et al. Prevalence of anxiety and depression symptom, and the demands for psychological knowledge and interventions in college students during COVID-19 epidemic: A large cross-sectional study. *J. Affect. Disord.* **2020**, *275*, 188–193. [CrossRef] [PubMed]
13. Aguilera-Hermida, A.P. College students' use and acceptance of emergency online learning due to COVID-19. *Int. J. Educ. Res. Open* **2020**, *1*, 100011. [CrossRef] [PubMed]
14. Rogowska, A.M.; Kuśnierz, C.; Bokszczanin, A. Examining anxiety, life satisfaction, general health, stress and coping styles during COVID-19 pandemic in Polish sample of university students. *Psychol. Res. Behav. Manag.* **2020**, *13*, 797–811. [CrossRef] [PubMed]
15. Son, C.; Hegde, S.; Smith, A.; Wang, X.; Sasangohar, F. Effects of COVID-19 on college students' mental health in the United States: Interview survey study. *J. Med. Internet Res.* **2020**, *22*, e21279. [CrossRef] [PubMed]
16. Nieto-Chaupis, H. Quantifying and Modeling the Measurement of the Quality of Education in Electrical Engineering Programs: A Heuristic Approach. In Proceedings of the 2018 IEEE XXV International Conference on Electronics, Electrical Engineering and Computing (INTERCON), Lima, Peru, 8–10 August 2018; pp. 1–4. [CrossRef]
17. Nieto-Chaupis, H. Experiences on the quality of education assessment in a network engineering program in Lima city. In Proceedings of the 2017 IEEE URUCON, Montevideo, Uruguay, 23–25 October 2017; pp. 1–4. [CrossRef]
18. Merchán Rubiano, S.M.; López-Cruz, O.; Gómez Soto, E. Teaching computer programming: Practices, difficulties and opportunities. In Proceedings of the 2015 IEEE Frontiers in Education Conference (FIE), El Paso, TX, USA, 21–24 October 2015; pp. 1–9. [CrossRef]

19. Da Costa, F.J.; de Souza Bispo, M.; de Faria Pereira, R.d.C. Dropout and retention of undergraduate students in management: A study at a Brazilian Federal University. *RAUSP Manag. J.* **2018**, *53*, 74–85. [CrossRef]
20. Palacios, C.A.; Reyes-Suárez, J.A.; Bearzotti, L.A.; Leiva, V.; Marchant, C. Knowledge Discovery for Higher Education Student Retention Based on Data Mining: Machine Learning Algorithms and Case Study in Chile. *Entropy* **2021**, *23*, 485. [CrossRef] [PubMed]

engineering proceedings

MDPI

Proceeding Paper

Computer-Aided Simulation on the Impact of the Combination of High-Rise Building Wall and Roof Green Coverage Ratio on Urban Microclimate †

Ying-Ming Su and Yu-Ting Hsu *

Department of Architecture, National Taipei University of Technology, Taipei 106, Taiwan; ymsu@ntut.edu.tw
* Correspondence: syuting634@gmail.com

† Presented at the IEEE 5th Eurasia Conference on Biomedical Engineering, Healthcare and Sustainability, Tainan, Taiwan, 2–4 June 2023.

Abstract: Environmental issues related to global warming and urbanization are becoming more serious. Many studies have shown that urban vertical planting can effectively reduce ambient temperature. However, the impact of different vertical planting combinations on urban microclimate has rarely been studied in Taiwan. Thus, in this study, the impact of different proportions of green walls and green roofs on the environment is explored. Referring to 6 times 6 high-rise buildings of 90 m in the ideal city. FLUENT was used to simulate the average climatic conditions of the Taipei Station in the past ten years' summer. Since the actual building has openings that cannot reach 100% vertical plant coverage, the coverage is calculated based on the proportion of the green coverage area to the area of bare walls and roof decks. We had four options, including case 1 without greening, case 2 (green wall 25% + green roof 75%), case 3 (50% green wall + 50% green roof), and case 4 (75% green wall + 25% green roof). The research results show that at the height of the pedestrian layer (1.5 m), the wind speed of urban streets is reduced due to the obstruction of surrounding buildings. The installation of wall greening slows down the wind speed and reduces the ambient temperature, which is better than roof greening. In the urban canopy (90.5 m), as the Z-axis height increases, the higher the green roof ratio, the higher the wind speed. To improve the overall urban wind below 100% of the total greening balance of walls and roofs, it is recommended that wall greening be 50–75% and roof greening be 25–50%.

Keywords: urban microclimate; high-rise building; vertical greening; green coverage ratio

Citation: Su, Y.-M.; Hsu, Y.-T. Computer-Aided Simulation on the Impact of the Combination of High-Rise Building Wall and Roof Green Coverage Ratio on Urban Microclimate. *Eng. Proc.* **2023**, *55*, 83. https://doi.org/10.3390/engproc2023055083

Academic Editors: Teen-Hang Meen, Kuei-Shu Hsu and Cheng-Fu Yang

Published: 27 December 2023

1. Introduction

As environmental issues related to global warming and urbanization become more serious, high-density and high-rise buildings cause the temperature in urban areas to be higher than those in rural areas, resulting in the urban heat island effect. Summarized the factors that affect the heat island effect at home and abroad. The urban scale includes the total population, topography, green coverage, and use intensity, as well as factors such as the number of plants, the aspect ratio of the street, the reflectivity of the pavement material, and the sky visibility. Gromke et al. [1] pointed out that the global average temperature will increase by 1.5 to 4.5 °C by 2100. Improving the urban green cover rate and achieving a natural ventilation environment can be adjusting targets to slow down the urban heat island effect. Urban air corridors generate different wind directions and speeds due to urban density, building height, and street canyon width. To actively shape an excellent urban environment and microclimate, many countries have vigorously promoted and rewarded policies with the primary goal of developing three-dimensional greening of buildings Yeh [2]. On 24 February 2011, the Taipei City Government announced the revision of the "Taipei City Urban Renewal Building Volume Incentive Measures", adding the "Building Roof Platform and Vertical Greening of Buildings" volume reward project

[Appendix A]. Moreover, the Taichung Government promulgated in 2019, "Taichung City Encourages Migrants". Building Facilities Setup and Feedback Measures [Appendix B].

The three-dimensional greening of the building increases ventilation and reduces the ambient temperature via the shading and evapotranspiration capacity of the plants. Due to the different types of plants, the porosity, canopy size, and thickness are different. Various studies have pointed out that façade greening methods ranging from single to street-level buildings benefit environmental cooling and improve comfort. Urban greening is considered the most suitable and effective strategy for mitigating climate change. Perini et al. [3], Lu [4], and Huang [5] explored the differences between planting types, greening methods, and air wall thicknesses on building surface temperature or air temperature at a certain distance via actual measurement. Numerical simulations also reveal the influence of different three-dimensional greening factors: wall, street tree, shelterbelt, a combination of green coverage, green coverage, and height.

The above literature discussed the design of building wall green coverage rate or simulating the relationship between greening and wind and heat environment in the natural base configuration but lacks discussion of the three-dimensional (wall and roof) green coverage combination of buildings at the street scale. Therefore, we simulated a high-rise and high-density ideal city to explore the combination of greening ratios on building walls and roofs using computational fluid dynamics (CFD). Four schemes, including no greening and greening (25% of the wall + 75% of the roof; 50% of the wall + 50% of the roof; and 75% of the wall + 25% of the roof), are configured for changes in the urban environment at pedestrian height (1.5 m).

2. Literature Review

2.1. Urban Heat Island Effect and Warm Environments

Due to the lack of green space, over-concentration of the population, sizeable artificial heat dissipation, and the influence of building materials, the urban environment makes the city a hot island. Liu [6] made three cases of adjusting the urban form and the width of the street profile and analyzed them with ACH and PET. The research results showed that the airflow can be increased when the street profile configuration is parallel to the wind direction. The ventilation effect can be improved. In addition to the lateral airflow, the longitudinal airflow also increases the airflow, and ACH is proportional to the volume flow of the inflow space. The high-rise building area has a considerable distance between buildings, which is conducive to air circulation, and the heat island effect tends to weaken. Wen [7] studied the factors that affect the thermal environment due to the high-temperature phenomenon of the walking environment in the campus open space via field measurement, a questionnaire survey, thermal environmental factor investigation, and CFD software simulation. The research results showed that the sky visibility rate (SVF) in the campus open space walking environment during the hot season is the ratio of the degree of shading affecting the walking environment to the direct sunlight walking environment. It is necessary to reduce the visibility of the walking environment, such as green plants, pavement materials, water bodies, artificial heat dissipation, wind environment, sky visibility factors, and urban density and height. Ting [8] used the computational fluid dynamics of FLUENT to make different road widths, building setback scales, and building setback heights. The research showed that the type of building setback volume with large and small building setback height scales was helpful for urban ventilation and ventilation efficiency. To improve the overall urban environment, the mass design of the building must have a large scale back toward the building. Zhang et al. [9] analyzed the frontal density of building arrays in terms of thermal comfort and air quality for pedestrian layers above four floors, considering realistic solar radiation (north, south, east, and west). They showed that the local solar position (0800LST) and (1400LST) wind speed and air change were different from (1200LST) as the frontal density increased.

2.2. Urban Canopy and Pedestrian Wind Field

The urban microclimate of Hong Kong. They believed that high-density buildings affected the air convection and exchange in the inner city, increasing the heat load in summer. Mei et al. [10] proposed that lower building density improves ventilation efficiency. Better ventilation performance can be obtained in compact urban development by reducing frontal area density or building number. yang et al. [11] set up experimental models with an average height of 30 m and a staggering degree of 0, 5, 10, 15, 20, 40, and 60 m. The results showed that the staggering buildings were beneficial in guiding the airflow from the upper floor to the ground and promoting the air on the leeward side. Lin [12] discussed the actual measurement and simulation of the wind environment in different areas of the city by adjusting the building type and the width of the street profile. For ventilation efficiency and area ventilation, Luis [13] analyzed the influence of different urban types on the thermal effect. He proposed that the building group coverage rate of 30 and 40% provided a good balance between the environmental impact in the city and the external trade-off. As the height of the building increased, the air velocity on the outer surface of the building gradually increased. Ku [14] showed that the urban form indicators related to building height significantly correlated with other types of urban form indicators, indicating that building height played a vital role in the construction of the urban environment. Lin [15] used the removal flow rate (PFR) and air exchange rate (ACH) to evaluate the impact of different building heights, building coverage, and ambient wind direction on urban canopy ventilation. Studies have shown that changes in building height increase airflow around high-rise buildings but reduce ventilation in low-rise buildings.

After summarizing the relevant literature, the factors affecting urban microclimate include building types (building coverage, average height, and plot ratio), and it was found that the average building height significantly impacts the wind environment.

2.3. Urban Vertical Planting and Warm Environments

The effect of plant cutting to improve the urban microclimate depends on the climatic environment (season, temperature, speed, and wind direction), silk planting wind, and plant cutting type (roof greening, three-dimensional, and street greening). Hsieh [16] studied that when the green coverage rate of the balcony was 50 or 100%, the average temperature on the leeward side significantly reduced. As the green coverage rate of the balcony increases, the overall average temperature is lower. Xi et al. [17] showed that the average temperature of green roofs dropped by 1 °C in summer and winter, the average temperature of street trees dropped by 2 °C, and the average temperature of shelter forests dropped by less than 1 °C. The three greening combinations have better cooling benefits than single greening Ziaul et al. [18]. The most effective strategy is to reduce the temperature by 2.6 °C when the roof and wall greening cover 100% of the open mid-rise configuration. Herath et al. [19] showed that a 100% green roof reduced the temperature slightly more than 50%, and the combination of the street tree with a 50% green roof and 50% green wall reduced the temperature by up to 1.9 °C compared with the solid base. The building facade in all directions. The results showed that the average temperature of the wall surface decreased by 0.7 °C, the maximum cooling in the east was about 12.6 °C, the air temperature was decreased on average by about 0.8 to 2.1 °C, and the relative humidity was higher than that around the bare wall. Aflaki et al. [20] Compared with the glass building surface, the three-dimensional greening system with 100% coverage can reduce the ambient temperature by about 1 °C. Ouyang et al. [21] experimented with green coverage in different urban densities, ranging from 2 to 30%, and whole area greening (56% greening outside buildings). Setting tree height (9 m), trunk height (3 m), and crown width (7 m), the study found that a low density of 20–30% of planting coverage is the most effective cooling value. Peng et al. [22] simulated the typical urban form of Nanjing with a green coverage rate of 35–100%. The results showed that high-density greening of 61–81% achieved 3% energy savings during the day. Morakinyo et al. [23] used ENVI-met to study the effect of three-dimensional green coverage and orientation on cooling in a typical high-density urban configuration in Hong Kong. The wind direction was set with the southwesterly wind in summer. A reduction of 12 to 14 °C, followed by a reduction of 7 to 1 °C in the south direction, produced better thermal benefits when the same green coverage was used in the east–west and north–south directions. Zhao [24] found the influence of the most popular human height wind field related to the high and warm trend. Kong et al. [25] measured the impact of different tree species on the high-density spatial microclimate. They found that trees with large, short trunks and dense canopies had the best effect on reducing the mean radiant temperature. Djedjig et al. [26] used a scale model to conduct actual measurements. Compared to the differences between green roofs and facade greening, both greening types improved environmental comfort via evapotranspiration, but the facade greening effect was better.

3. Parameter Setting

A 6 times 6 conceptual city model with a total of 36 buildings Yang et al. [9] was used in this study. The dimensions of a single building are 30 m long (A) times 30 m wide (A), the X-axis channel and Y-axis channel width (W) is 20 m (B), and the configuration range of the research area is 280 times 280 m. As shown in Figures 1 and 2, the urban wind and temperature field distribution was analyzed via CFD numerical simulation, and the combination of greening ratios on building walls and roofs was discussed. The green coverage area accounted for 100% of the area of bare walls and roof platforms. Table 1 shows the different cases in this study.

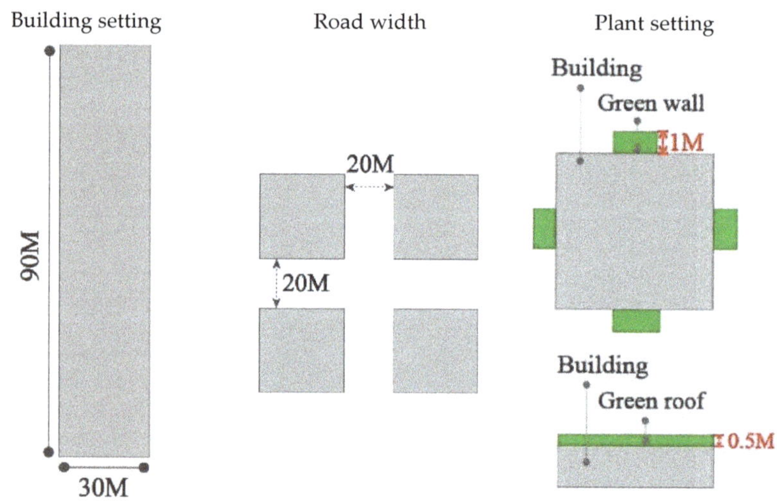

Figure 1. Architectural model and planting layer settings.

ANSYS FLUENT V18, the numerical simulation software, was used to carry out a three-dimensional simulation analysis of high-speed end flow and incompressible flow, steady state, thermal conductivity, and heat transfer. According to Yang et al. [9], the boundary model must be set for an outlet, inlet, lateral, and top. The air inlet was 5 H. The 5 H high, exit boundary and model were kept at 15 H (H is the length of the long side of the overall building model). The height of the highest building model from the upper boundary was 6 B (B is the height of the highest building) to achieve a complete wake, as shown in Figure 3.

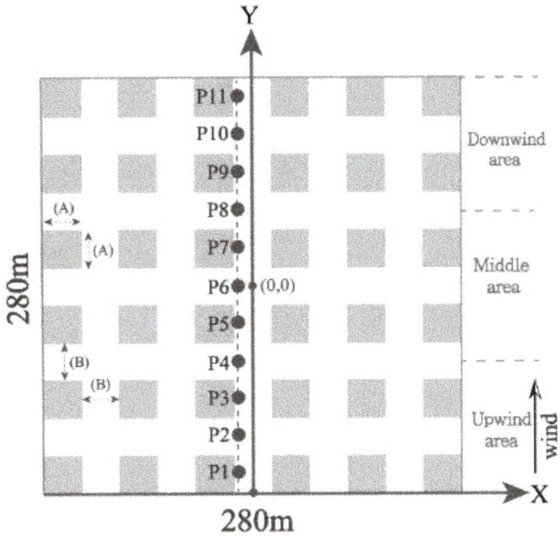

Figure 2. Area setting and measurement points.

Table 1. Different cases for simulation.

Wind		Green Wall	Green Roof
Case1	↗	0%	0%
		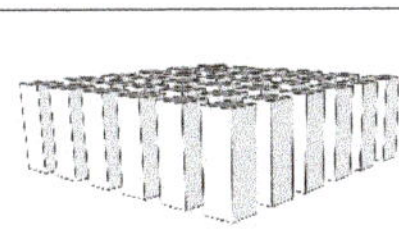	
Case2	↗	25%	75%

Case3	↗	50%	50%

Case4	↗	75%	25%
			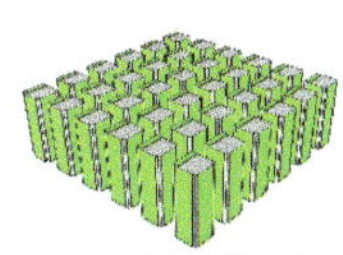

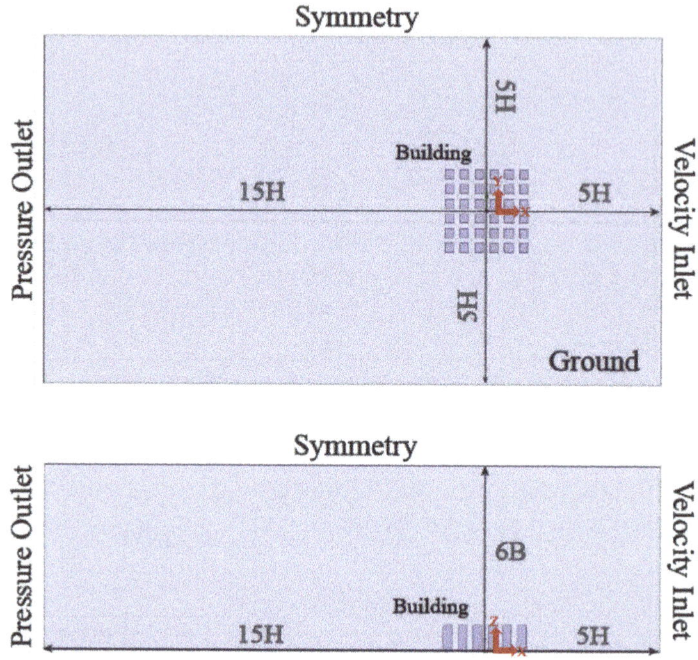

Figure 3. Boundary condition.

Material conditions and the building wall were set to concrete; the ground was set to asphalt; and the planting material and parameters were set based on Buccolieri et al. [27], Baxevanou et al. [28], and Koch et al. [29] It is a porous material with uniform filling and distribution and inelasticity. The planting porosity is 0.9, the thickness of the green wall is 1 M, and the thickness of the green roof is 0.5 M, as shown in Tables 1 and 2.

Table 2. Planting parameter settings.

Materials	**Density (kg/m^3)**	**Specific Heat (J/kg·k)**	**Thermal Conductivity**
Porous Materials	700	2310	0.173
Porosity (ε)	**Permeability (α) (m^2)**	**Particle diameter (m)**	**Emissivity (ε)**
0.9	4.86×10^{-3}	0.1	0.46

Eleven measuring points were set in the Y-axis (parallel wind direction channel) as 1 m away from the outer wall of the building, which were P1–P11. In order to analyze the changes in different street valleys, the Z-axis was for the pedestrian height of 1.5 m, the middle height of 45 m, and the urban canopy of 90.5 m to monitor the changes in wind speed and temperature thoroughly, as the changes in wind speed and temperature thoroughly as shown in Table 2.

The meteorological parameters were set based on the average wind speed, wind direction, and the average temperature at the Taipei Station (466920) from 2012 to 2021 essential reference meteorological data. The inflow wind direction was set as the east wind, the average wind speed was 1.92 m/s, and the average temperature was 29.67 °C, as shown in Table 3.

Table 3. Meteorological data of Taipei Station in the summer of 2012–2021 (June–August).

Station	Wind Speed	Wind Direction	Average Temperature
466920_Taipei	1.92 m/s	East wind	29.67 °C
IMAGE	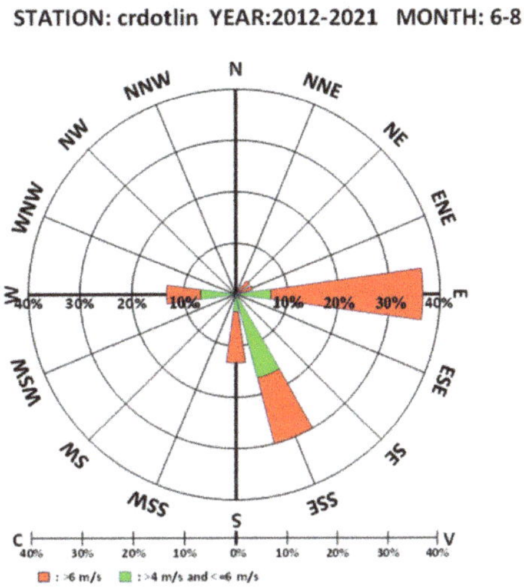		

4. Simulation Analysis

4.1. Wind Field Simulation Analysis

Table 4 shows the changes in the wind flow field caused by the pedestrian height of 1.5 m due to the four three-dimensional green coverage ratio combinations. The analysis results show that the average wind speed in the upwind area of case 1 is 1.3 m/s, case 2 is 0.69 m/s, and case 3 is 0.85 m/s, case 4 is 0.79 m/s. Compared with case 1 without the greening configuration, the wind speed of other schemes is slightly lower due to the influence of planting porosity and surface friction, while among the three greening configurations, the upper wind of case 2 has the largest average. The average wind speeds of each scheme in the middle area are 0.89, 0.47, 0.67, and 0.55 m/s. The downwind area is the lowest average wind speed, which is 0.83, 0.21, 0.45, and 0.27 m/s in sequence. The overall average wind speed of cases 1 to 4 is 1.01, 0.41, 0.6, and 0.48 m/s. In the greening configuration scheme, the average wind speed of case 3 is 0.24 m/s. It is higher than the average wind speed of case 2, which increases the ambient wind speed by 31%. The average wind speed of case 3 is 0.12 m/s higher than the average wind speed of case 4. M. The ambient wind speed is increased by 20%.

In case 2, the average wind speed is lower than in other schemes. It is inferred that the proportion of the bare surface area of the building is too large compared to the green wall area, affecting the wind speed performance in each area. In the greening scheme, case 3 achieves better wind field performance.

Table 5 shows the effects of pedestrian height (1.5), middle layer (45), and urban canopy (90.5) on different flow fields and wind speeds for the four schemes. The building height of each scheme is 1.5 to 90 m, and the upwind place is between 1.3 and 2.1 m/s, 0.69 and 1.46 m/s, and 0.85 and 1.39 m/s. The result shows that the average wind speed at the lower part of the greening scheme (1.5 m) becomes weaker in the order of case 1 > case 3 > case 4 > case 2, and the average wind speed at the upper part (90.5 m) becomes weaker in the order of case 1 > case 2 > case 3 > case 4. The greening design parallel to the wind direction increases the wind speed compared with the vertical design.

Table 4. Z-axis plane (Z = 1.5, 45, 90.5) wind velocity simulation results.

	1.5 m	45 m	90.5 m
	Case 1 (No green) 		
WIND ←	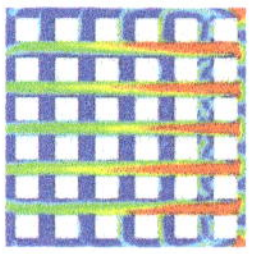	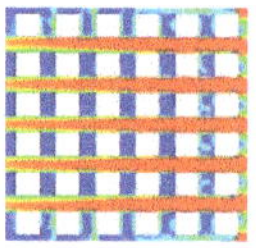	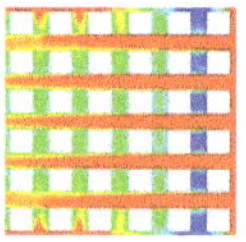
	Case 2 (GW25% + GR75%) 		
WIND ←		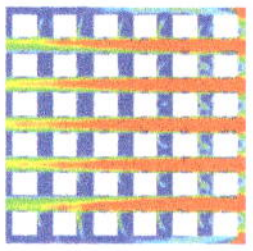	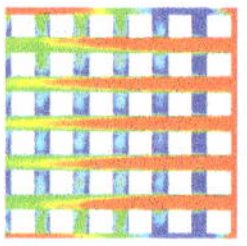
	Case 3 (GW50% + GR50%) 		
WIND ←	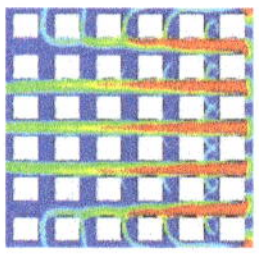	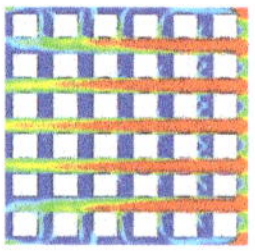	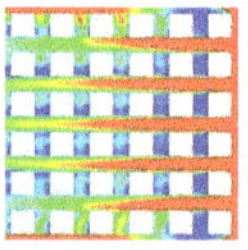
	Case 4 (GW75% + GR25%) 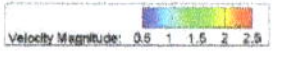		
WIND ←			

4.2. Temperature Field Simulation Analysis

Table 5 shows that the four three-dimensional green coverage rate combination schemes result in the temperature performance of the pedestrian height of 1.5 m. With greening, the temperature of each scheme gradually increases between the windward area and the downwind area. Compared with case 1 without the greening scheme, the average temperature of the greening scheme dropped slightly. The average temperature of each scheme in the upwind area is 35 °C for case 1, 34.1 °C for case 2, 32.8 °C for case 3, and 33.7 °C for case 4. Compared with the average temperature of case 1, the average temperature of each greening scheme decreases by 2 to 6%. Case 3 has the best temperature and the highest wind speed. Cases 3 and 4 have the most apparent temperature increase from the middle area to the downwind. In the middle zone, the average temperature of cases 1 to 4 is 36.6, 36.2, 34.8, and 35.2 °C, which are 1, 4, and 3% lower than case 1. The overall average temperature of each scheme is 36.5, 35.9, 34.6, and 35.3 °C, which are 1, 5, and 3% lower than case 1.

Table 5. Z-axis plane (Z = 1.5, 45, 90.5) temperature simulation result.

	1.5 m	45 m	90.5 m
	Case 1 (No green) 		
WIND ↑			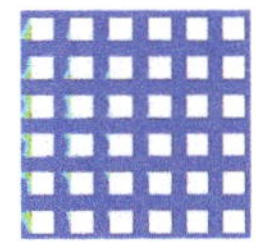
	Case 2 (GW25% + GR75%) 		
WIND ↑			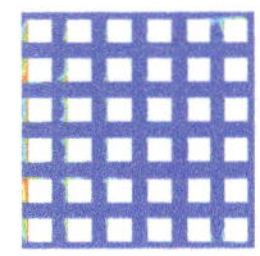
	Case 3 (GW50% + GR50%) 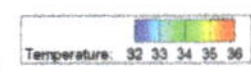		
WIND ↑	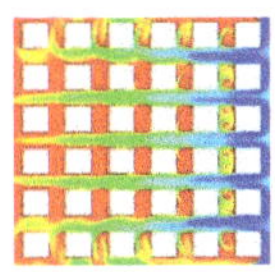	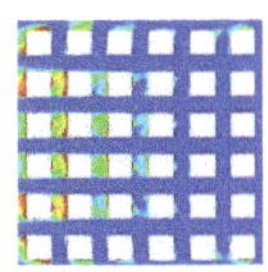	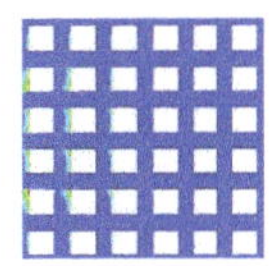
	Case 4 (GW75% + GR25%) 		
WIND ↑		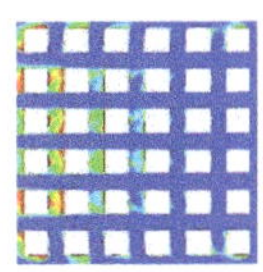	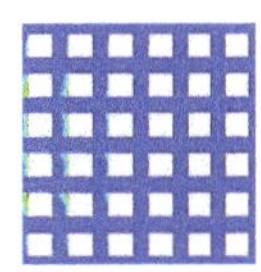

In the greening scheme, the wind flow upwind of Case 3 is more complicated, so the overall average wind speed is faster, and the average temperature is also lower. In pedestrian height (1.5 m), middle layer (45 m), and urban canopy (90.5 m) to different thermal fields in the four schemes. The average temperature of each scheme is slightly lower. The overall average temperature of each scheme ranges from 36.5 to 30.9 °C, 35.9 to 30.9 °C, 34.6 to 31.1 °C, and 35.3 to 31.3 °C. The ambient temperature decreases by 15% for case 1, 13% for case 2, 10% for case 3, and 11% for case 4. The greening scheme shows a decreasing optimal temperature in case 2. At the pedestrian height, the wind speed of urban streets is reduced due to obstruction of the surrounding buildings, so the temperature is higher, and the greening scheme of the wall is set to reduce the ambient temperature by 1, 5, and 3%, respectively. Since there is no greening scheme at the urban canopy's height, the wind flow is unobstructed, and the average temperature is also lower. The configuration of roof greening increases the ambient temperature by 1, 0.3, and 1%. The results show that urban wall greening has a better reduction effect than roof greening. Herath et al. (2018) [10] showed that green walls reduce ambient temperature better than green roofs.

5. Conclusions

The simulation results of the environmental wind field and thermal field of the four different schemes are summarized as follows.

1. At the height of the pedestrian layer (1.5 m), the greening scheme slows down the wind speed and reduces the ambient temperature due to the influence of plant porosity and surface friction. The wind speed of the greening scheme from large to small is "case 3 (0.6 m/s) > case 4 (0.41 m/s) > case 2 (0.53 m/s). Case 3 has the best wind performance. In the urban canopy (90.5 m), the average wind speed of the greening scheme shows case 2 (1.46 m/s) > case 3 (1.39 m/s) > case 4 (1.22 m/s). As the Z-axis height increases, the higher the green roof ratio, the higher the wind speed.
2. The results show that urban wall greening has a better cooling effect than roof greening. Compared with the urban canopy, at the pedestrian height, the wind speed of the urban street is reduced due to the obstruction of the surrounding buildings, and the temperature is higher. The greening on the wall has a more significant cooling effect. The setting of the greening scheme reduces the ambient temperature by 1% (case 2), 5% (case 3), and 3% (case 4). At the height of the urban canopy, the no-greening scheme has a lower average temperature due to unobstructed wind flow, while the greening scheme increases the ambient temperature by 0.1% (case 2), 0.3% (case 3), and 1% (case 4), respectively.
3. From the pedestrian height (1.5 m) to the urban canopy (90.5 m), case 1 reduces the ambient temperature by 12%, case 2 by 13%, case 3 by 11%, and case 4 by 11%. Case 2 decreases the optimal temperature. The average temperature of each scheme decreases from low to high.
4. Based on the above analysis, if the total green area of walls and roofs is less than 100%, it is recommended to improve the overall urban wind farm by 50–75% on the walls and 25–50% on the roofs.
5. The green coverage ratio combination of a single building height with different three-dimensional greening is discussed. In the future, more urban factors can be added to compare variables, such as different building heights and street widths and green coverage ratio ratios, and to find the possibility of lowering the ambient temperature.

Author Contributions: Writing—original draft, Y.-T.H. Validation, Y.-T.H.; Writing—review & editing, Y.-M.S.; Supervision, Y.-M.S. All authors have read and agreed to the published version of the manuscript.

Funding: This research received no external funding.

Institutional Review Board Statement: Not applicable.

Informed Consent Statement: Not applicable.

Data Availability Statement: Data are contained within the article.

Acknowledgments: Thanks to the Ministry of Science and Technology Program (Program Code 110-2221-E-027-013-MY2) for the sponsorship of this research, but this research received no external funding.

Conflicts of Interest: The authors declare no conflict of interest.

Appendix A

On 24 February 2011, the Taipei City Government announced amendments to the "Taipei City Urban Renewal Building Volume Rewards Measures", adding the volume reward program of "Building Roof Platforms and Vertical Greening of Facades". Unless otherwise stipulated by urban planning or other laws and regulations, the greening area of the roof deck of a new building shall be 50% of the roof deck area, and the greening area shall be calculated based on the actual covered area.

Appendix B

In 2019, the Taichung government promulgated the "Taichung City's Measures for the Installation and Feedback of Building Facilities to Encourage Migrants". Among them, paragraph 2 of Article 5 stipulates that the exterior wall of the building shall be set with a planted wall and shall not protrude from the exterior wall by more than 2 m.

References

1. Gromke, C.; Blocken, B.; Janssen, W.; Merema, B.; van Hooff, T.; Timmermans, H. CFD analysis of transpirational cooling by vegetation: Case study for specific meteorological conditions during a heat wave in Arnhem, Netherlands. *Build. Environ.* **2015**, *83*, 11–26. [CrossRef]
2. Yeh, T.-Y. The Evaluation and Promotion Strategies of Building Greenery. Master's Thesis, National Taipei University of Technology, Taipei, Taiwan, 2018.
3. Perini, K.; Ottelé, M.; Fraaij AL, A.; Haas, E.M.; Raiteri, R. Vertical greening systems and the effect on air flow and temperature on the building envelope. *Build. Environ.* **2011**, *46*, 2287–2294. [CrossRef]
4. Lu, Y.-H. Influences of Plants on Wall Cooling Effect. Master's Thesis, Feng Chia University, Taichung, Taiwan, 2012.
5. Huang, S.-S. Influences of Different Airspace Depth of Green Wall on Building Cooling Effect. Master's Thesis, National Cheng Kung University, Tainan, Taiwan, 2013.
6. Liu, P.-J. CFD Analysis of Outdoor Ventilation with Different Urban Geometry and Unequal Block Width in Taiwan. Master's Thesis, National Taipei University of Technology, Taipei, Taiwan, 2016.
7. Wen, C.-J. The Thermal Comfort of Walking Environment in Campus Open Space—A Case Study of National ChengChi University. Master's Thesis, National ChengChi University, Taipei, Taiwan, 2018.
8. Ting, T.-Y. The Influence of High-Rise Building Setback Type of Wind Environment. Master's Thesis, National Taipei University of Technology, Taipei, Taiwan, 2017.
9. Li, Z.; Zhang, H.; Wen, C.Y.; Yang, A.S.; Juan, Y.H. Effects of frontal area density on outdoor thermal comfort and air quality. *Build. Environ.* **2020**, *180*, 107028. [CrossRef]
10. Mei, S.J.; Hu, J.T.; Liu, D.; Zhao, F.Y.; Li, Y.; Wang, Y.; Wang, H.Q. Wind driven natural ventilation in the idealized building block arrays with multiple urban morphologies and unique package building density. *Energy Build.* **2017**, *155*, 324–338. [CrossRef]
11. Yang, J.; Zhang, T.; Tan, Y. Technological evolution, and evaluation system integration of urban wind environment research. *South. Archit.* **2014**, *3*, 31–38.
12. Lin, C.-Y. The Study of the Impact on Environment Comfort of Mass Layout in Waterfront Building. Master's Thesis, National Taipei University of Technology, Taipei, Taiwan, 2017.
13. Santos, L.G.; Nevat, I.; Pignatta, G.; Norford, L.K. Climate-informed decision-making for urban design: Assessing the impact of urban morphology on urban heat island. *Urban Clim.* **2021**, *36*, 100776. [CrossRef]
14. Ku, C.-A.; Liu, J.-T. Spatial Relationship between Land Use and Flood Potential-A Case Study of Taichung City. *Urban Plan.* **2020**, *47*, 27–52.
15. Lin, M.; Hang, J.; Li, Y.; Luo, Z.; Sandberg, M. Quantitative ventilation assessments of idealized urban canopy layers with various urban layouts and the same building packing density. *Build. Environ.* **2014**, *79*, 152–167. [CrossRef]
16. Hsieh, C.-J. A Study on the Influence of Balcony Greening in High-Rise Buildings on Urban Microclimate. Master's Thesis, National Taipei University of Technology, Taipei, Taiwan, 2019.
17. Xi, C.; Ding, J.; Wang, J.; Feng, Z.; Cao, S.J. Nature-based solution of greenery configuration design by comprehensive benefit evaluation of microclimate environment and carbon sequestration. *Energy Build.* **2022**, *270*, 112264. [CrossRef]
18. Ziaul, S.; Pal, S. Modeling the effects of green alternative on heat island mitigation of a meso level town, West Bengal, India. *Adv. Space Res.* **2020**, *65*, 1789–1802. [CrossRef]
19. Herath, H.M.P.I.K.; Halwatura, R.U.; Jayasinghe, G.Y. Evaluation of green infrastructure effects on tropical Sri Lankan urban context as an urban heat island adaptation strategy. *Urban For. Urban Green.* **2018**, *29*, 212–222. [CrossRef]
20. Aflaki, A.; Mirnezhad, M.; Ghaffarianhoseini, A.; Ghaffarianhoseini, A.; Omrany, H.; Wang, Z.H.; Akbari, H. Urban heat island mitigation strategies: A state-of-the-art review on Kuala Lumpur, Singapore, and Hong Kong. *Cities* **2017**, *62*, 131–145. [CrossRef]
21. Ouyang, W.; Morakinyo, T.E.; Ren, C.; Ng, E. The cooling efficiency of variable greenery coverage ratios in different urban densities: A study in a subtropical climate. *Build. Environ.* **2020**, *174*, 106772. [CrossRef]
22. Peng, L.L.; Jiang, Z.; Yang, X.; Wang, Q.; He, Y.; Chen, S.S. Energy savings of block-scale facade greening for different urban forms. *Appl. Energy* **2020**, *279*, 115844. [CrossRef]
23. Morakinyo, T.E.; Lai, A.; Lau, K.K.L.; Ng, E. Thermal benefits of vertical greening in a high-density city: Case study of Hong Kong. *Urban For. Urban Green.* **2019**, *37*, 42–55. [CrossRef]
24. Zhao, X.; Li, G.; Gao, T. The adjustment mechanism of thermal comfort effect and morphological characteristics of typical street trees in Harbin in summer. *Landsc. Archit.* **2016**, *12*, 74–80.
25. Morakinyo, T.E.; Kong, L.; Lau KK, L.; Yuan, C.; Ng, E. A study on the impact of shadow-cast and tree species on in-canyon and neighborhood's thermal comfort. *Build. Environ.* **2017**, *115*, 1–17. [CrossRef]

26. Djedjig, R.; Bozonnet, E.; Belarbi, R. Experimental study of the urban microclimate mitigation potential of green roofs and green walls in street canyons. *Int. J. Low Carbon Technol.* **2015**, *10*, 34–44. [CrossRef]
27. Buccolieri, R.; Gromke, C.; Di Sabatino, S.; Ruck, B. Aerodynamic effects of trees on pollutant concentration in street canyons. *Sci. Total Environ.* **2009**, *407*, 5247–5256. [CrossRef]
28. Baxevanou, C.; Fidaros, D.; Bartzanas, T.; Kittas, C. Numerical simulation of solar radiation, air flow and temperature distribution in a naturally ventilated tunnel greenhouse. *Agric. Eng. Int. CIGR J.* **2010**, *12*, 48–67.
29. Koch, K.; Samson, R.; Denys, S. Aerodynamic characterisation of green wall vegetation based on plant morphology: An experimental and computational fluid dynamics approach. *Biosyst. Eng.* **2019**, *178*, 34–51. [CrossRef]

engineering proceedings

MDPI

Proceeding Paper

Investigation of Behavioral Sciences for Survival in the Food Industry during the COVID-19 Crisis [†]

Yung-Fu Huang [1], Ming-Wei Weng [1,*], Kuang-Mao Deng [2], Hung-Jen Tsai [3] and Kai-Fu Yang [3]

[1] Department of Marketing and Logistics Management, Chaoyang University of Technology, Taichung 413310, Taiwan; huf@cyut.edu.tw
[2] Language Center, Chaoyang University of Technology, Taichung 413310, Taiwan; kmdeng@cyut.edu.tw
[3] Department of Business Administration, Chaoyang University of Technology, Taichung 413310, Taiwan; vpben@vsprite.com.tw (H.-J.T.); skynyu@gmail.com (K.-F.Y.)
* Correspondence: mwweng@cyut.edu.tw
† Presented at the IEEE 5th Eurasia Conference on Biomedical Engineering, Healthcare and Sustainability, Tainan, Taiwan, 2–4 June 2023.

Abstract: The COVID-19 pandemic has dramatically impacted the beverage industry. It directly causes job losses, reduced income, and changes in customer preferences, and it influences the relationship between franchisors and franchisees. COVID-19 has spread rapidly around the world in the past two years. Meanwhile, the impact has reverberated from abroad to the headquarters of companies in Taiwan. Eight Taiwanese food and beverage brands saw steep drops in sales as the pandemic grew in severity. Thus, it is important to identify the critical elements of running a franchise store. Collected with Analytic Hierarchy Process (AHP) technology, the results show that the five key factors in franchises are going direct-to-consumer, establishing a B2B portal for distributors, assessing supply chain elastic limit, optimizing inventory, and streamlining e-commerce to meet changing customer needs. Guidelines and directions are provided for decision-makers through this study to design mobile applications in the simplest platform.

Keywords: franchisees; AHP technology; COVID-19; beverage industry

Citation: Huang, Y.-F.; Weng, M.-W.; Deng, K.-M.; Tsai, H.-J.; Yang, K.-F. Investigation of Behavioral Sciences for Survival in the Food Industry during the COVID-19 Crisis. *Eng. Proc.* **2023**, *55*, 84. https://doi.org/10.3390/engproc2023055084

Academic Editors: Teen-Hang Meen, Kuei-Shu Hsu and Cheng-Fu Yang

Published: 27 December 2023

1. Introduction

Taiwan's unique food culture has developed the food service industry that never seems to hit its maximum capacity. According to research [1] on "Sales and Annual Growth Rate of Trade and Food Services" published in 2022 by the Department of Statistics, the Taiwanese food chain accounted for 7280 NTD in 2021. Over the past decade, Taiwanese foods, such as bubble tea and the XiaoLongBao (broth-filled steamed pork dumplings) of the Din Tai Fung restaurant chain and Gua Bao, have become world-renowned. How a food or beverage chain standardizes its management and production process is the key factor in deciding whether it is competitive. For instance, when more "human factors" are involved in preparing Chinese cuisine, it is difficult to set up a standardized process so that every branch offers the same food quality and taste. However, foreign franchisees must also overcome problems specific to the Chinese market. Franchisees in China often have difficulties finding local managers who can understand how to run a business. Previous research on customer expectations and perception in the food industry has shown valued attributes, e.g., service, location, image, brand name, low price, food quality (food tastes and nutrition properties), and value for money [2,3]. The early theorization of the nature of the franchise contract can be traced back to Rubin [4]. Falbe and Welsh [5] explained the impact of successes and failures by analyzing franchise executives' perceptions based on franchisors in Canada, Mexico, and the United States. Various approaches to franchisee issues are discussed in Refs. [6–8]. Chow et al. [9] presented a conceptual framework for linking quality and satisfaction in catering place operations. Cheng et al. [10] provided insights into successfully implementing the international franchise system. Afni et al. [11] analyzed the development strategies and franchise-based business models of J.R. Tea Makassar. Studies [12,13] have stressed that resource scarcity, agency theory, risk spreading, and life cycle are fundamental for company ownership versus franchisee ownership. In terms of resource scarcity, franchising is presented as a means of rapid marketing. Scarce capital resources, managerial expertise, and knowledge of local markets are provided by franchisees. The related franchisees are classified into four major categories: license chain, franchise, voluntary chain, and cooperation chain. The limits of authority issues included unit franchise, multiple franchises, and regional franchise. In 1960, Penrose's theory offered a sounder theoretical basis for the resource-based view (R.B.V.). Wernerfelt [14] provided extensive discussions on the applications of R.B.V. Barney [15] examined the links among firms' resources, sustained competitive advantage-value, rareness, imitability, and substitutability. Many others, including Gallon et al. [16], Hunt [17], Ho and Tsai [18], and Lin et al. [19], have pointed out that the theoretical models of R.B.V. firm-based resources may be tangible (physical assets, e.g., financial resources and human resources, including cash, plants, machinery, real estate, and raw materials, etc.) or intangible (an organization's culture, reputation, know-how, accumulated experience, and relationships with customers, suppliers or key stakeholders). Lacking supplies needed to maintain life or a certain quality of life is resource scarcity. It is one of the basic ideas in the study of economics. Oxenfeldt and Kelly [20] proposed that firms should franchise access to scarce resources, particularly capital and managerial resources. Curran and Stanworth [21] explored the emergence and role of franchised economic activities at three levels: social, organizational, and motivational. Barney [15] examined the implications of this firm-resource model of sustained competitive advantage for other business disciplines. Several studies [22–24] have provided extensive discussions on the applications of resource scarcity. Resource dependency theory examines the relationship between external resources and organizational behavior. In the literature, this conflict has been analyzed within two quite disparate perspectives: agency theory and resource dependency theory. The authors provide an empirical assessment of various agency-theory explanations for capital to explain both franchisors' decisions about the terms of their contracts (royalty rates and up-front franchise fees). Lieberman and Montgomery [25] investigated the role of mechanisms conferring advantages and disadvantages on first-mover firms. Justis et al. [26], Mohr and Spekman [27], and Kumar et al. [28] described the fields of application in resource dependency theory. Rubin [4] observed that the nature of the franchise is examined using agency theory. Shane [29] proposed that hybrid organizational forms provide a way to overcome the agency problems of adverse selection and moral hazard in selecting, assimilating, and monitoring new managers. Eisenhardt [30] explained that agency theory is the behavior of a firm from the perspectives of various contracts between different parties. Many studies have interrogated this question: how can owners maximize the value of a franchise system before the sale process? [12,22]. Garg and Rasheed [31] examined several agency problems inherent in multi-unit franchising: bonding, adverse selection, information flow, shirking, inefficient risk-bearing, free-riding, and quasi-rent appropriation. Researchers have investigated the classification of stream research, especially franchising, by pointing out the contents and main contributions [32,33]. However, social exchange theory is based on a relationship created through a cost–benefit analysis between two people, especially a business-to-business relational exchange. Several studies have also investigated the impact of the franchisors' role performance and cultural sensitivity on franchisees' trust in and satisfaction with franchise partnerships [34,35]. When small-business owners differ in terms of motives to start a business, a key success factor is to help owners become responsible for managing the daily activities of an enterprise. Owning a business franchise offers an alternative to starting a business from scratch that appeals to many entrepreneurs.

In this study, marketing strategies in the service franchise industry such as tobacco and liquor stores, coffee chains, convenience store chains, language schools, restaurants, and beverage chains were investigated to show how the strategies affected the franchisor–

franchisee relationships. The structure of this research is described below. Section 1 provides a brief introduction to impact of COVID-19 in food industries. Section 2 describes the general design of methodology. Section 3 provides an example of data analysis and results. Section 4 results are then presented, with a thorough description of the case study. Finally results are discussed and conclusions are drawn.

2. Methodology

2.1. AHP Technology

The AHP is a mathematical tool for problem-solving to predict weighting items of a measurement scale. Table 1 shows a set of pairwise comparisons for each of the lower levels. The AHP consists of 5 primary steps to evaluate business performance in the food and beverage industry.

- Step 1: Identifying the decision, options, and criteria.
- Step 2: Constructing pairwise comparison matrices [36].

$$A = \begin{bmatrix} 1 & A_{12} & \ldots A_{1n} \\ 1/A_{12} & 1 & \ldots A_{2n} \\ \vdots & \vdots & \cdots \vdots \\ 1/A_{1n} & 1/A_{2n} & \cdots 1 \end{bmatrix}, \tag{1}$$

- In matrix A, all the diagonal elements are self-compared with survival strategies; thus $a_{ij} = 1$, where $i =$ j, $i, j = 1, 2 \cdots, $n. Let $Ia_{ij} = 1/a_{ji}$, where $a_{ij} > 0, i \neq$ j;
- Step 3: Determining the important weight of each criterion. Let w_i represent the degree of importance of the ith attribute of store operations; then,

$$w_i = \frac{1}{n}\sum_{j=1}^{n} \frac{a_{ij}}{\sum\limits_{i=1}^{n} a_{ij}} \quad i, j = 1, 2, \cdots, n, \tag{2}$$

$$w_i = \frac{\left(\prod\limits_{j=1}^{n} a_{ij}\right)^{\frac{1}{n}}}{\sum\limits_{i=1}^{n}\left(\prod\limits_{j=1}^{n} a_{ij}\right)^{\frac{1}{n}}} \quad i, j = 1, 2, \cdots, n, \tag{3}$$

$$w_i = \frac{\left(\frac{1}{\sum\limits_{i=1}^{n} a_{ij}}\right)}{\sum\limits_{j=1}^{n}\left(\frac{1}{\sum\limits_{i=1}^{n} a_{ij}}\right)} \quad i, j = 1, 2, \cdots, n, \tag{4}$$

$$\lambda_{\max} = \frac{1}{n}\left(\frac{W_1}{W_2} + \frac{W_2}{W_2} + \cdots + \frac{W_n}{W_n}\right), \tag{5}$$

- Step 4: Achieving matrix consistency. According to Saaty, the consistency index is CR to measure the consistency ratio, which is expected to be less than 0.10. RI values can be calculated as matrices.

$$C.I. = \frac{\lambda_{\max} - n}{n - 1}\begin{cases} = 0 \\ > 0.1, \\ \leq 0.1 \end{cases} \tag{6}$$

- Step 5: Identifying the best option by calculating utility functions.

If CR is greater than 0.1, the comparison matrix is not consistent. The comparison matrix needs revising in this situation, and consistency arrangements can be conducted [36].

Table 1. Strategies for minimizing the impacts of COVID-19 pandemic on the food and beverage industry.

Dimensions	Weight	Sub-Dimensions	Weight	Rank	Overall Weight	Overall Rank
Manpower	0.067	Personality traits	0.138	4	0.0135	16
		Before job training	0.232	3	0.0155	14
		On job training	0.246	2	0.016	13
		Management ability	0.384	1	0.025	10
Environment	0.143	Government subsidy	0.122	4	0.017	12
		Food festival	0.237	2	0.033	7
		Streamlining e-commerce to meet customer needs	0.461	1	0.065	5
Equipment	0.056	Competitor	0.180	3	0.0249	11
		POS data analysis	0.499	1	0.028	8
		Uber Eats/Foodpanda	0.272	2	0.015	15
		Store design	0.229	3	0.013	17
Product	0.379	Going direct-to-consumer	0.556	1	0.211	1
		Assessing supply chain resiliency	0.373	2	0.141	3
		Production packaging	0.071	3	0.027	9
Customer	0.355	Establishing a B2B customer portal for distribution networks	0.519	1	0.184	2
		Consumer group	0.137	3	0.049	6
		Optimizing inventory	0.344	2	0.122	4

2.2. Participants

Nineteen subjects including ten consultants and nine university professors were asked to complete a questionnaire that elicited information on their attitudes and professional knowledge of running a franchise store. The subjects, aged fifty to sixty years old, came from different provinces in China. For the present research, Figure 1 describes the framework for survival opportunities with five main evaluation dimensions: manpower, environment, equipment, product, and customer strategies related to beverage service to headquarters in Taiwan. The framework aims to test service quality and performance.

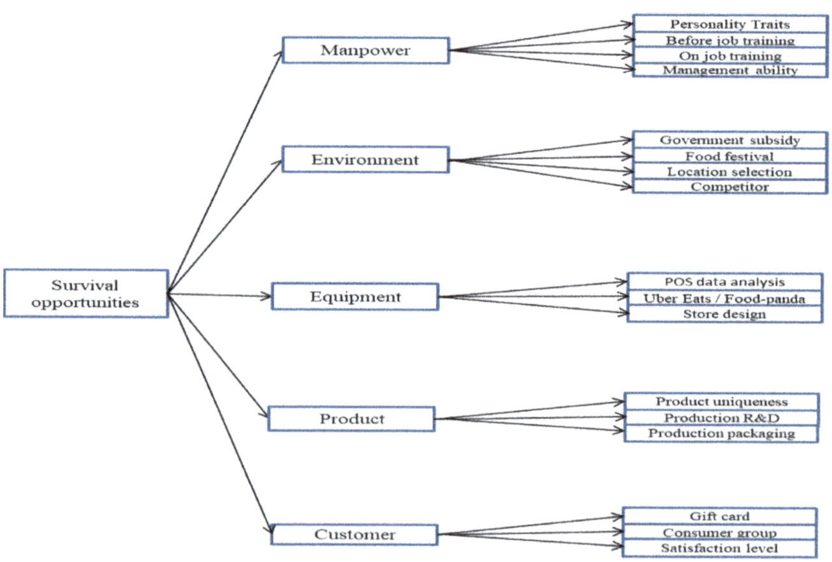

Figure 1. Impact of COVID-19 on plans to implement IT to reduce cost.

3. Data Analysis and Results

The data analysis for all small and medium-sized enterprises was conducted after two in-depth interviews averaging 2 h each. A two-phase study was designed to explore four critical success factors from the literature, and the experts judged these factors. The subjects indicated the level of agreement with each of the fundamental scales of (1) equal importance, (3) moderate importance, (5) essential or strong importance, (7) very strong importance, and (9) extreme importance. The fundamental scales of (2), (4), (6), and (8) showed intermediate values between the two adjacent judgments. Table 1 shows the result of the weight analysis of general indicators on the website in similar contexts.

The global rank of KFS was identified. The KFS summary table shows 15 behavioral indicators (Table 1): (1) a going direct-to-consumer, (2) establishing a B2B customer portal for distribution networks, (3) assessing supply chain resiliency, (4) optimizing inventory, (5) streamlining e-commerce to meet changing customer needs, (6) diversified transactions, (7) delivery time, (8) system accessibility, (9) market share, (10) browsing speeds, (11) return, (12) economize workforce, (13) maintenance, (14) website traffic, and (15) cost. Table 2 shows the results of the sub-dimensional analysis in the workforce within dimension weight. The sub-dimensions are management ability (38.4%), on-the-job training (24.6%), before-job training (23.2%,) and personality traits (13.8%). Further, the sub-dimensional environment was characterized by location selection (46.1%), food festival (23.7%), competitor (18%), and government subsidy (12.2%). Sub-dimensional equipment included POS data analysis (49.9%), Uber Eats/Food (27.2%), and store design (22.9%). The sub-dimensional products were product uniqueness (21.1%), production R&D (14.1%), and product packaging (2.7%). The sub-dimensions of customers were gift cards (51.9%), satisfaction level (12.2%), and consumer group (4.9%). Similarly, by comparing the attribute values of all the alternatives, they were ranked accordingly.

Table 2. Analysis of matrix and weight of product, customer, environment dimension, manpower, and equipment dimensions.

Dimensions	Sub-Dimensions							
			I_{11}	I_{12}	I_{13}	I_{14}	Weight	Rank
Manpower (CR = 0.014 < 0.1)	Management ability (I_{11})	I_{11}	1	7	1/6	5	0.384	1
	On-the-job training (I_{12})	I_{12}	1/7	1	8	4	0.246	2
	Before-job training (I_{13})	I_{13}	6	1/8	1	9	0.232	3
	Personality traits (I_{14})	I_{14}	1/5	1/4	1/9	1	0.138	4
			I_{21}	I_{22}	I_{23}	I_{24}	Weight	Rank
Environment (CR = 0.025 < 0.1)	Streamlining e-commerce to meet customer needs (I_{21})	I_{21}	1	1/6	1/8	5	0.461	1
	Food festival (I_{22})	I_{22}	6	1	1/7	1/8	0.237	2
	Competitor (I_{23})	I_{23}	8	7	1	7	0.180	3
	Government subsidy (I_{24})	I_{24}	1/5	8	1/7	1	0.122	4
			I_{31}	I_{32}	I_{33}		Weight	Rank
Equipment (CR = 0.021 < 0.1)	POS data analysis (I_{31})	I_{31}	1	9	7		0.499	1
	Uber Eats/Foodpanda (I_{32})	I_{32}	1/9	1	6		0.272	2
	Store design (I_{33})	I_{33}	1/7	1/6	1		0.229	3
			I_{41}	I_{42}	I_{43}		Weight	Rank
Product (CR = 0.031 < 0.1)	Going direct-to-consumer (I_{41})	I_{41}	1	6	9		0.211	1
	Assessing supply chain resiliency (I_{42})	I_{42}	1/6	1	8		0.141	2
	Production packaging (I_{43})	I_{43}	1/9	1/8	1		0.027	3
Customer (CR = 0.034 < 0.1)			I_{51}	I_{52}	I_{53}		Weight	Rank
	Establishing a B2B portal for distribution networks (I_{51})	I_{51}	1	5	8		0.519	1
	Optimizing inventory (I_{52})	I_{52}	1/5	1	7		0.122	2
	Consumer group (I_{53})	I_{53}	1/8	1/7	1		0.049	3

4. Case Study

In Chinese, "Gong cha" means offering superlative tea to the emperor from all possessions, symbolizing the highest quality and self-expectation in Taiwan. Gong cha has become one of the world's most renowned quality tea providers. Nowadays, Gong cha has blossomed in over 20 regions worldwide, such as Korea, Japan, Hong Kong, Macau, Taiwan, Myanmar, Vietnam, Cambodia, and the Philippines. During the COVID-19 pandemic, customers stayed at home. However, social distancing did not stop them from going out to enjoy their favorite bubble tea combo. Alternatively, customers ordered their favorite fruit and milk tea through a food delivery service.

Although most of our stores were closed, several stores opened. As the COVID-19 crisis continued, many employers faced the difficulty of achieving significant cost reductions due to sudden revenue reduction. For the survival of beverage companies, the following were the top five success factors: going direct-to-consumers, establishing a B2B customer portal for the distribution network, assessing supply chain resiliency, optimizing inventory, and streamlining e-commerce to adapt to changing customer needs. During the crisis, the CEOs were forced to consider cost-saving plans, for example, free-cost placement, cloud computing, subscription service, and minimum maintenance charge. Figure 2 summarizes the cost-cutting strategy for Gong Cha during the COVID-19 period.

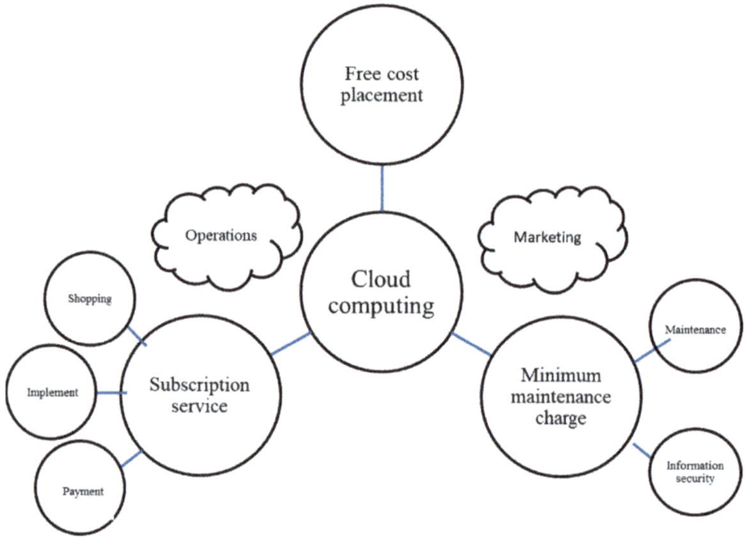

Figure 2. Framework for survival opportunities.

5. Conclusions

The Chinese economy is the world's second-largest in terms of GDP. Traditional business franchising, such as food and beverage (F and B) and retail, enters the Chinese market. Best practices for doing business in China have been established by major international franchises. For instance, without changing the core product, they localize their products when necessary. To achieve rapid expansion and mass acceptance, they are willing to minimize the price of the final product and the franchising fee. However, the COVID-19 outbreak impacted the franchising sector, with consequences for the business activities and integrity of the franchise system. Also, the COVID-19 crisis created complex challenges in the dynamic between franchisees and franchisors. The five major KFS for implementation were going direct-to-consumer (0.211), establishing a B2B customer portal for distribution networks (0.184), assessing supply chain resiliency (0.141), optimizing inventory (0.122), and streamlining e-commerce to meet changing customer needs (0.065). It implied that CEOs could provide made-to-order products to improve brand awareness during the COVID-19 period, paying for R&D costs associated with a new product/service. Gift cards can give cash-strapped franchises immediate cash flow.

Author Contributions: Conceptualization, Y.-F.H. and M.-W.W.; methodology, M.-W.W.; software, M.-W.W.; validation, M.-W.W. and K.-M.D.; formal analysis, M.-W.W.; investigation, H.-J.T.; resources, K.-F.Y.; data curation, K.-F.Y.; writing—original draft preparation, M.-W.W.; writing—review and editing, M.-W.W.; visualization, K.-M.D.; supervision, Y.-F.H.; project administration, Y.-F.H.; funding acquisition, Y.-F.H. All authors have read and agreed to the published version of the manuscript.

Funding: This research received no external funding.

Institutional Review Board Statement: Not applicable.

Informed Consent Statement: Not applicable.

Data Availability Statement: Data sharing is not applicable.

Acknowledgments: The researchers would like to thank Chaoyang University of Technology for funding the publication of this project no. TCJ-111J940.

Conflicts of Interest: The authors declare no conflict of interest.

References

1. Sales and Annual Growth Rate of Trade and Food Services. 2022. Available online: https://www.mckinsey.com/capabilities/growth-marketing-and-sales/our-insights/revenue-growth-management-unlocking-value-in-foodservice (accessed on 1 June 2023).
2. Johns, N.; Howard, A. Customer expectations versus perceptions of service performance in the food service industry. *Int. J. Serv. Ind. Manag.* **1998**, *9*, 248. [CrossRef]
3. Tam, W.Y.; Yung, N.L.A. Managing Customers for Value in the Catering Industry (Fast Food) in Hong Kong. Master's Thesis, The Chinese University of Hong Kong, Hong Kong, China, 2003.
4. Rubin, P. The Theory of the Firm and the Structure of the Franchise Contract. *J. Law Econ.* **1978**, *21*, 222–233. [CrossRef]
5. Falbe, C.M.; Welsh, D.H.B. Nafta and Franchising: A comparison of franchisor perceptions of characteristics associated with franchisee success and failure in Canada, Mexico, and the United States. *J. Bus. Ventur.* **1998**, *13*, 151–171. [CrossRef]
6. Schewe, C.D. *Marketing Concepts and Applications*, 2nd ed.; N.Y. McGraw-Hill: New York, NY, USA, 1983; p. 415.
7. Kotler, P.; Clarke, R.N. *Marketing for Health Organization*; Prentice-Hall: Hoboken, NJ, USA, 1987.
8. Justis, R.T.; Judd, R.J. *Franchising*, 2nd ed.; Dame Publications: Houston, TX, USA, 2002.
9. Chow, I.H.; Lau, V.P.; Lo, T.W.C.; Sha, Z.; Yun, H. Service quality in restaurant operations in China: Decision- and experiential-oriented perspectives. *Int. J. Hosp. Manag.* **2007**, *26*, 698–710. [CrossRef]
10. Cheng, J.M.S.; Lin, J.Y.C.; Tu, H.H.J.; Wu, N.S.H. Toward a stage model of the international franchise system development. *J. Mark. Channels* **2007**, *14*, 64–83. [CrossRef]
11. Afni, R.; Supatminingsih, T.; Muhammad, H.; Dinar, M. Development strategy and franchise-based business model at beverage shop. *Hum. South Asean J. Soc. Stud.* **2021**, *1*, 11–21.
12. Combs, J.G.; Castrogiovanni, G.J. Franchisor strategy: A proposed model and empirical test of franchise versus company ownership. *J. Small Bus. Manag.* **1994**, *17*, 37–48.
13. LaFontaine, F.; Kaufmann, P. The evolution of ownership patterns in franchise systems. *J. Retail.* **1994**, *70*, 97–113. [CrossRef]
14. Wernerfelt, B. A special case of dynamic pricing policy. *Mark. Sci.* **1986**, *32*, 1562–1566. [CrossRef]
15. Barney, J. Firm resources and sustained competitive advantage. *J. Manag.* **1991**, *17*, 99–120. [CrossRef]
16. Gallon, M.R. Putting core competency thinking into Practice. *Res. Technol. Manag.* **1995**, *38*, 20–28. [CrossRef]
17. Hunt, S.D. Resource-advantage theory: An evolutionary theory of competitive firm behavior. *J. Econ. Issues* **1997**, *31*, 59–77. [CrossRef]
18. Ho, Y.C.; Tsai, T.H. The Impact of Dynamic Capabilities with Market Orientation and Resource-Based Approaches on NPD Project Performance. *J. Am. Acad. Bus.* **2006**, *8*, 215–228.
19. Lin, B.W.; Chen, C.J.; Wu, H.L. Patent portfolio synergy, technology strategy and firm performance. *IEEE Trans. Eng. Manag.* **2006**, *53*, 17–26.
20. Oxenfeldt, A.; Kelly, A. Will Successful Franchise Systems Ultimately Become Wholly-Owned Chains? *J. Retail.* **1969**, *44*, 69–83.
21. Curran, J.; Stanworth, J. Small business research in Britain. In *Small Bus*; Routledge: London, UK, 2020; pp. 127–152.
22. Elango, B.; Fried, V.H. Franchising Research: A literature review and synthesis. *J. Small Bus. Manag.* **1997**, *35*, 68–81.
23. Litz, R.A.; Stewart, A.C. Franchising for sustainable advantage? Comparing the performance of independent retailers and trade-name franchisees. *J. Bus. Ventur. Insights* **1998**, *13*, 131–150. [CrossRef]
24. Roddick, A. *Body and Soul: Profits with Principles—The Amazing Story of Anita Roddick and the Body Shop*; Crown Publishers: New York, NY, USA, 1994.
25. Lieberman, M.B.; Montgomery, D.B. First-mover advantages. *Strateg. Manag. J.* **1988**, *9*, 41–58. [CrossRef]
26. Justis, R.T.; Olsen, J.E.; Chan, P. Using marketing research to enhance franchisee/franchisor relationships. *J. Small Bus. Manag.* **1993**, *31*, 121–127.

27. Mohr, J.; Spekman, R. Characteristics of Partnership Success: Partnership Attributes, Communication Behavior, and conflict resolution Techniques. *Strategic* **1994**, *15*, 135–152. [CrossRef]
28. Kumar, N.; Scheer, L.; Steenkamp, J.E.M. The Effects of Perceived Interdependence on Dealer Attitude. *J. Mark. Res.* **1995**, *32*, 348–356. [CrossRef]
29. Shane, S. Hybrid organizational arrangements and their implications for firm growth and survival: A study of new franchisors. *Acad. Manag. J.* **1996**, *39*, 216–234. [CrossRef]
30. Eisenhardt, K. Agency theory: An assessment and review. *Acad. Manag. Rev.* **1989**, *14*, 57–74. [CrossRef]
31. Garg, V.K.; Rasheed, A.A. International multi-unit franchising: An agency theoretic explanation. *Int. Bus. Rev.* **2003**, *12*, 329–348. [CrossRef]
32. Phan, P.H.; Butler, J.E.; Lee, S.H. Crossing Mother: Entrepreneur-Franchisees' Attempts to Reduce Franchisor Influence. *J. Bus. Ventur. Insights* **1996**, *11*, 379–402. [CrossRef]
33. Münch, R. *Sociological Theory: From the 1850s to the Present*; Nelson-Hall Publishers: Chicago, IL, USA, 1993.
34. Drucker, P.F. The Network Society. *The Wall Street Journal*, 29 March 1995; A14.
35. White, G.P. A survey of recent management science applications in higher education. *Interfaces* **1987**, *17*, 97–108. [CrossRef]
36. Sariçali, G.; Kundakçi, N. AHP ve COPRAS yöntemleri ile otel alternatiflerinin değerlendirilmesi. *Int. Rev. Econ. Manag.* **2016**, *4*, 45–66. [CrossRef]

engineering proceedings

Proceeding Paper

Delivering Sustainable G2B Messages to Ensure Food Safety: The Case of Taiwan's Online Food Business †

Kuan Chuan Tao [1] and Abbott Po Shun Chen [2,*]

1 Department of Leisure Services Management, Chaoyang University of Technology, Taichung 413310, Taiwan; willytao0218@cyut.edu.tw
2 Department of Marketing and Logistics Management, Chaoyang University of Technology, Taichung 413310, Taiwan
* Correspondence: chprosen@gm.cyut.edu.tw
† Presented at the IEEE 5th Eurasia Conference on Biomedical Engineering, Healthcare and Sustainability, Tainan, Taiwan, 2–4 June 2023.

Abstract: The Internet brings convenience and happiness to consumers. The government must guide food safety standards. The government must provide implementation details, schedule planning, description-of-use documents, relevant regulations on food safety, and address matters needing attention and critical points of on-site counseling and inspection, the formulation of pertinent form documents, the evaluation of the feasibility of the form, and the time required. The Taichung City Government mainly conducts this research to investigate and guide online food sale companies. The industries are food manufacturing, online sales, and catering. The information was surveyed from major website platforms, comprising 640 incidents. These sites include Shopee, PC Home, MOMO, PChome, eBay, and Gomaji. This study uses corporate interview surveys and narrative statistics. The research results show that there are ten food safety issues. The most common ones are the items that should and should not be recorded in the stereotypical food or catering service contract stipulated by communication transactions. That is the most common food safety violation by businesses. This research provides academic and practical references for the positive message transmission of online sales.

Keywords: message deliver; food safety; G2B; online business; website service

Citation: Tao, K.C.; Chen, A.P.S. Delivering Sustainable G2B Messages to Ensure Food Safety: The Case of Taiwan's Online Food Business. *Eng. Proc.* **2023**, *55*, 85. https://doi.org/10.3390/engproc2023055085

Academic Editors: Teen-Hang Meen, Kuei-Shu Hsu and Cheng-Fu Yang

Published: 28 December 2023

1. Introduction

Food safety is essential for all consumers. When we search for the return and volatility spillover effects on Asian Dragons, Yahoo Finance provided the monthly statistics as an example [1]. Still, food safety can't easily find how to make stable rules. Making sense of the food safety qualification is also essential [2]. We usually inform advisory committee members to check food safety. For example, we need a full-time lecturer or someone of greater authority in a food-related department of a college or university teaching food safety courses, those who are currently or have worked in health agencies and have experience in food hygiene management audits, and specialists in the field of food science with audit experience. Secondly, we should evaluate the advisory committee. After the meeting, one must take the assessment personnel test, and the score must reach 80 points or more, and one must report to the Food Safety Department for approval before one can participate in the assessment and counseling work. Finally, we manage the mechanism. To protect the rights and interests of the industry, the counseling committee members must sign a non-disclosure agreement on the spot for each session, and they are not allowed to take photos or videos without permission. If a report is reported and verified to be true, the membership of the counseling committee member will be disqualified immediately.

Randomly, we select live auction operators. The on-site inspection of GHP and inspection of "items that should be recorded and not recorded in the finalized contract of food or catering services established by means of communication transactions", "food import inspection registration", and "food packaging labeling" are in compliance with the regulations. At least two live broadcast operators are selected. We also search for illegal food advertisements (executed according to the inspection and counseling form)

For online platforms such as Shopee, open-air, and Yahoo stores, use search keywords for illegal food advertisements, such as detoxification, anti-oxidation, and hangover. Private message operators implement online food advertising content laws and regulations, provide guidance on changing illegal content or taking down illegal webpages, complete at least 640 cases of Internet operators in Taichung City (the place of shipment is Taichung City), and track the results of follow-up counseling (more than 50% must be completed to remove illegal webpages or change forbidden words and sentences).

2. Literature

We handle consensus meetings of the expert committee on the implementation details of the plan, scheduling planning, document use description, and food safety-related laws'. Regulations or on-site guidance and inspection should pay attention to critical points, formulate relevant form documents, and evaluate the feasibility and needs of the form [3]. After a certain period, it can be used in the implementation of this plan only after obtaining the approval of the Ministry of Food and Drug Safety.

The counseling committee should be a consensus meeting and educational training. Usually, we give consensus meetings and education and training lectures through a group of expert committee members. We also handle the guidance committee consensus meeting and education training: hold one session, 3 h per session. On-site audit counseling committee members are invited to audit counseling committee members who specialize in food science, have audit experience, and have received training to serve as audit counselors [4]. The list of check counselors must be reviewed and approved by the Food Safety Department. After the assessment is passed, 20 guidance committee members will be checked and approved by Taichung City for food and drug safety review [5]. Finally, we create a Line community for review counselors: Scholars and experts in the plan and review counselors join in and give immediate feedback on relevant questions and consultations during the review and counseling process [6].

3. Methods

3.1. GHP Assessment Guidance for the Online Food Retailing Industry

There are too many challenges to need that we should adopt initiatives to supply chain management for manufacturing industries [7], such as follows

1. Based on the list of operators provided by the Department of Food Safety, at least 80 or more counselors will be provided. On-site counseling will encourage operators to implement online information inspections and hygiene counseling at physical locations to comply with relevant food safety and hygiene regulations.
2. On-the-spot counseling for online food sellers to handle "Food Business Registration," "Webpage content must conform to the stereotyped contract," "Imported food should be legally registered for import inspection," and "Complete Chinese labeling is required," etc. project.
3. Personnel who have passed the review of the Food Safety Department will be in charge, one person at a time, and two inspections are required each time, and the Taichung City Food and Drug Safety Department will send personnel to check jointly.
4. Data collection: In the end-of-period report, statistical analysis should be carried out on the reinspection results. Suggestions for improved methods and management models should be proposed for common deficiencies in the industry.
5. To protect the rights and interests of the industry, the counseling committee members must sign a non-disclosure agreement on the spot for each session. They are not allowed to take photos or record videos without permission. The counseling committee members will be disqualified immediately if the reports are verified.

6. If it is impossible to confirm whether a business has ceased operation and cannot be contacted, it should still be recorded accurately. Still, it will not be included in calculating the number of results.

3.2. Implementation of Online Food Sales Inspection Project

3.2.1. Counseling Objects and Content

Randomly select live auction operators to check GHP on the spot and check whether the "items that should be recorded and should not be recorded in the finalized contract of food or catering services established using communication transactions", "food import inspection registration," and "food outer packaging labeling" comply with the regulations. This year, at least two live broadcast operators are expected to be selected. For online platforms such as Shopee, open-air, and Yahoo stores, use search keywords for illegal food advertisements, such as detoxification, anti-oxidation, and hangover. Private message operators implement online food advertising content laws and regulations and provide guidance on changing illegal content or taking down illegal webpages, complete at least 640 cases of Internet operators in Taichung City (the place of shipment is Taichung City), and track the results of follow-up counseling (more than 50% must be completed to remove illegal webpages or change forbidden words and sentences).

3.2.2. Qualifications of the Advisory Committee

This will be implemented by counselors approved by the Department of Food and Drug Safety. The counseling committee will implement the counseling items according to the formulated form and follow the Food and Drug Administration's announcement, "Food and related product labeling advertisements involve false exaggeration, misunderstanding or medical efficacy identification criteria," to guide the industry to improve it or remove it from the shelves.

3.2.3. Data Collection

In the end-of-term report, the roster of operators, the date of counseling, the execution review and counseling committee members, and the counseling results should be compiled. In addition, the common deficiencies of the employees should be included, and improvement methods and management countermeasures should be developed for the policy planning reference of the Food and Drug Safety Department.

4. Results

The implementation of this case began on 8 October 2011, and on 13 January 2011, 80 online food entity stores were counseled. On 13 March 2011, 640 food violations and guidance were completed, and 350 items were improved.

There are 30 entities in the general catering industry, 17 in the food manufacturing industry, and 33 in online sales.

Table 1 shows the top 10 deficiencies in the counseling of entity operators. That would build how to check the standards for fitting the industry. Table 2 shows the top 10 deficiencies in the guidance of food manufacturers. Ref. [7] suggests that those also face challenges in adopting green initiatives in supply chain management. Table 3 shows the top 10 deficiencies in the counseling of online sellers. That lets buyer-seller connect for value-in-use [8]. Table 4 shows the general top ten deficiencies in available catering business counseling. Table 5 shows the counseling of illegal projects on various platforms. Table 6 shows the counseling of each platform to improve the situation.

Table 1. Top 10 deficiencies in the counseling of entity operators.

Items	a	b	c
Comply with the items that should be recorded and should not be recorded in the food or catering service stereotyped contract established through communication transactions.	18	57	5
Items should be sorted and stored on pallets, shelves, or other effective measures should be taken and kept tidy, and should not be placed directly on the ground.	31	37	12
Work and business premises are equipped with hand washing equipment, running tap water, hand sanitizer, paper towels or electric hand dryers, trash cans with lids, and nail brushes. When necessary, appropriate disinfection facilities should be installed. The location of hand washing and hand drying equipment should be appropriate, and the number should be sufficient. Hang a simple and easy-to-understand method of washing hands in an obvious position (six steps of hand washing).	27	26	27
The business owner manages the records independently.	43	26	11
The industry should assign sanitation personnel to fill in the sanitation management records daily, including the sanitation work specified by GHP.	31	25	24
Food is distributed in retail, and food labels are marked according to regulations and in compliance with relevant laws and regulations.	50	22	8
The walls, pillars, ground, floors, or ceilings should be kept clean, and the floors or ceilings directly above the food exposure must not have mildew, peeling, dust accumulation, dirt, or condensation.	49	21	10
Raw materials, semi-finished products, and packaging supplies should be properly stored and clearly marked with date records so the industry can manage them independently.	44	18	18
Food handlers shall accept hygiene lectures or training conducted by the competent health authority or relevant agencies (institutions), schools, and legal persons approved or entrusted during their working period and keep records.	31	16	33
On food, food additives, food detergents, food utensils, food containers, or packaging, the name of the product or words, pictures, symbols, or attached instructions are recorded. Nutrition labeling refers to the food container or package that records the nutritional content, content, and nutrition claims of the food.	60	16	4

a. fit; b. no fit; c. no suitable.

Table 2. Top 10 deficiencies in the guidance of food manufacturers.

Items	a	b	c
Comply with the items that should be recorded and should not be recorded in the food or catering service stereotyped contract concluded through communication transactions.	7	9	1
The walls, pillars, ground, and floors or ceilings should be kept clean, and the floors or ceilings directly above the food exposure must not have mildew, peeling, dust accumulation, dirt, or condensation.	10	7	0
The ventilation is good, and there is no bad smell. Entrances, doors, windows, vents, and drainage systems should be kept clean, and facilities should be set up to prevent the intrusion of vectors. The appearance of piping should be kept clean.	10	7	0

Table 2. *Cont.*

Items	a	b	c
Work and business premises are equipped with hand washing equipment, running tap water, hand sanitizer, paper towels or electric hand dryers, trash cans with lids, and nail brushes. When necessary, appropriate disinfection facilities should be installed. The location of hand-washing and hand-drying equipment should be appropriate, and the number should be sufficient. Hang a simple and easy-to-understand method of washing hands in an obvious position (six steps of hand washing).	9	7	1
Items should be classified and stored on pallets, shelves, or other effective measures should be taken and kept tidy and should not be placed directly on the ground.	10	7	0
Raw materials, semi-finished products, and packaging supplies should be properly stored and marked with date records so the industry can manage them independently.	11	6	0
Food additives that have not been added without approval shall be managed by special personnel, special counters, and special books according to regulations and shall be properly recorded.	8	5	4
The indoor air is ventilated, and there are facilities to prevent the intrusion of vectors (such as automatic doors, screen doors, and air curtains).	13	4	0
The industry should assign sanitation personnel to fill in the sanitation management records daily, including the sanitation work specified by GHP.	12	4	1
The business owner manages the records independently.	13	4	0

a. fit; b. no fit; c. no suitable.

Table 3. The top 10 deficiencies in the counseling of online sellers.

Items	a	b	c
Comply with the items that should be recorded and should not be recorded in the stereotypical food or catering service contract concluded using communication transactions.	4	28	1
Items should be classified and stored on pallets or shelves; otherwise, other effective measures should be taken to keep them tidy and not placed directly on the ground.	5	17	11
The business owner manages the records independently.	13	13	7
Recording the product's name or words, pictures, symbols, or attached instructions on food, food additives, detergents, utensils, containers, or packaging. Nutrition labeling refers to the food container or package that records the food's nutritional content, content, and nutrition claims.	21	10	2
Food is in retail circulation, and food labels shall be marked by regulations and compliant with relevant laws and regulations.	21	10	2
The industry should assign sanitation personnel to fill in the sanitation management records daily, including the sanitation work specified by GHP.	6	9	18
Complete food import inspection and registration according to the law	15	7	11
Freezers (cabinets) should be kept below minus 18 degrees Celsius, cold storage (cabinets) should be kept below 7 degrees Celsius and above the freezing point, and frozen storage equipment should be equipped with temperature indicators in apparent places and recorded regularly. Freezer (storage) cabinets and refrigerators (storage) should be routinely defrosted and kept clean.	15	6	12

Table 3. *Cont.*

Items	a	b	c
Work and business premises are equipped with hand washing equipment, running tap water, hand sanitizer, paper towels or electric hand dryers, trash cans with lids, and nail brushes. When necessary, appropriate disinfection facilities should be installed. The location of hand washing and hand drying equipment should be appropriate, and the number of machines should be sufficient. Hang a simple and easy-to-understand method of washing hands in a prominent position (six steps of hand washing).	8	4	21
Toilets should be kept clean and free from odors. "Wash your hands after using the toilet," and the washing steps should be marked.	9	4	20

a. fit; b. no fit; c. no suitable.

Table 4. General top ten deficiencies in available catering business counseling.

Items	Fit	No Fit	No Suitable
Comply with the items that should be recorded and should not be recorded in the stereotypical food or catering service contract concluded using communication transactions.	7	20	3
Work and business premises have hand washing equipment, running tap water, hand sanitizer, paper towels or electric hand dryers, trash cans with lids, and nail brushes. When necessary, appropriate disinfection facilities should be installed. The location of hand washing and hand drying equipment should be correct, and the number should be sufficient. Hang a simple and easy-to-understand method of washing hands in a prominent position (six steps of hand washing).	10	15	5
The walls, pillars, ground, and floors or ceilings should be kept clean, and the floors or ceilings directly above the food exposure must not have mildew, peeling, dust accumulation, dirt, or condensation.	15	14	1
Items should be classified and stored on pallets or shelves; otherwise, other effective measures should be taken, and items should be kept tidy and should not be placed directly on the ground.	16	13	1
The industry should assign sanitation personnel to fill in the sanitation management records daily, including the sanitation work specified by GHP.	13	12	5
Food handlers shall accept hygiene lectures or training conducted by the competent health authority or relevant agencies (institutions), schools, and legal persons approved or entrusted during their working period and keep records.	13	10	7
The business owner manages the records independently.	17	9	4
New employees should pass the medical examination before being employed; the employer should take the initiative to go through the health examination at least once a year.	19	8	3
The personal clothes of food workers should be placed in the changing place and not be brought into the food workplace; the entry and exit of non-food workers should be properly controlled.	18	8	4
Raw materials, semi-finished products, and packaging supplies should be stored appropriately and marked with date records so the industry can manage them independently.	18	8	4

a. fit; b. no fit; c. no suitable.

Table 5. Counseling illegal projects on various platforms.

Types	a	b	c	d	e	f	g
Shopee	354	69%	157	30%	5	1%	516
Ruten (PChome)	58	60%	39	40%	0	0%	97
Carousell	21	91%	2	9%	0	0%	23
Firm website	1	50%	1	50%	0	0%	2
Rakuten	1	50%	1	50%	0	0%	2
Total	435	68%	200	31%	5	1%	640

a. Exaggeration; b. Percentage; c. Involved in medical treatment efficacy; d. Percentage; e. Joint violation; f. Percentage; g. Violation Subtotal.

Table 6. Counseling each platform to improve the situation.

Types	a	b	c	d	e
Shopee	320	126	39%	194	61%
Ruten (PChome)	15	6	40%	9	60%
Carousell	13	12	92%	1	8%
Firm website	1	0	0%	1	100%
Rakuten	1	0	0%	1	100%
Total	350	144	41%	206	59%

a. Completed improved; b. Deleted; c. Percentage; d. Partial improved; Percentage.

5. Conclusions

The information was surveyed from major website platforms, comprising 640 incidents. These sites include Shopee, PC Home, MOMO, PChome, eBay, and Gomaji. This study uses corporate interview surveys and narrative statistics. The research results show that there are ten food safety issues. The most common items should and should not be recorded in the food or catering service stereotyped contract stipulated by communication transactions. That is the most common food safety violation by businesses. This research provides academic and practical references for the positive message transmission of online sales.

Author Contributions: Conceptualization, K.C.T. and A.P.S.C.; methodology, A.P.S.C.; validation, A.P.S.C.; formal analysis, A.P.S.C.; investigation, K.C.T.; resources, K.C.T.; data curation, A.P.S.C.; writing—original draft preparation, A.P.S.C.; writing—review and editing, K.C.T. and A.P.S.C.; visualization, K.C.T.; supervision, K.C.T.; project administration, K.C.T.; funding acquisition, K.C.T. All authors have read and agreed to the published version of the manuscript.

Funding: This research received no external funding.

Institutional Review Board Statement: Not applicable.

Informed Consent Statement: Not applicable.

Data Availability Statement: Data sharing is not applicable.

Conflicts of Interest: The authors declare no conflict of interest.

References

1. Lee, C.W.; Chen, S.H.; Huruta, A.D.; Dewi, C.; Chen, A.P.S. Net Transmitter of Stock Market Volatility and Safe Haven for Portfolio Investors in the Asian Dragons. *Economies* **2022**, *10*, 273. [CrossRef]
2. Suhartanto, D.; Ismail, T.A.T.; Leo, G.; Triyuni, N.N.; Suhaeni, T. Behavioral intention toward online food purchasing: An analysis at different purchase levels. *Int. J. E-Bus. Res. (IJEBR)* **2020**, *16*, 34–50. [CrossRef]
3. Sun, J.; Buijs, J. Online food regulation in China: The role of online platforms as a critical issue. *Eur. Food Feed Law Rev.* **2018**, *13*, 503–513.
4. Limon, M.R. Food safety practices of food handlers at home engaged in online food businesses during COVID-19 pandemic in the Philippines. *Curr. Res. Food Sci.* **2021**, *4*, 63–73. [CrossRef] [PubMed]

5. Osaili, T.M.; Al-Nabulsi, A.A.; Al-Jaberi, T.M. Food safety knowledge, attitudes, and practices among female food handlers engaged in home-based online food businesses in Jordan during the COVID-19 pandemic. *Heliyon* **2022**, *8*, e10427. [CrossRef] [PubMed]
6. Suhartanto, D.; Kartikasari, A.; Najib, M.; Leo, G. COVID-19: Pre-purchase trust and health risk impact on M-commerce experience–young customers experience on food purchasing. *J. Int. Food Agribus. Mark.* **2022**, *34*, 269–288. [CrossRef]
7. Chen, A.P.S.; Huang, Y.F.; Do, M.H. Exploring the Challenges to Adopt Green Initiatives to Supply Chain Management for Manufacturing Industries. *Sustainability* **2022**, *14*, 13516. [CrossRef]
8. Chen, A.P.S.; Wang, J.H. Is B2B marketing enable in value-in-use? The co-creation perspective. *Int. J. Adv. Sci. Technol.* **2019**, *28*, 577–581.

engineering proceedings

Proceeding Paper

Urban Traffic Flow Prediction Using LSTM and GRU †

Hung-Chin Jang * and Che-An Chen

Department of Computer Science, National Chengchi University, Taipei City 116302, Taiwan; an5566x@gmail.com
* Correspondence: jang@cs.nccu.edu.tw
† Presented at the IEEE 5th Eurasia Conference on Biomedical Engineering, Healthcare and Sustainability, Tainan, Taiwan, 2–4 June 2023.

Abstract: For smart cities, the issue of how to solve traffic chaos has always attracted public attention. Many studies have proposed various solutions for traffic flow prediction, such as ARIMA, ANN, and SVM. With the breakthrough of deep learning technology, the evolutionary models of RNN, such as LSTM (Long Short-Term Memory) and GRU (Gated Recurrent Units) models, have been proven to have excellent performance in traffic flow prediction. By using LSTM and GRU models, we explore more features and multi-layer models to increase the accuracy of traffic flow prediction. We compare the prediction accuracy of LSTM and GRU models in urban traffic flow prediction. The data collected in this study are divided into three categories, namely "regular traffic flow data", "predictable episodic event data", and "meteorological data". The regular traffic flow data source is the "Vehicle Detector (VD) data of Taipei Open Data Platform". Predictable episodic event data are predictable as non-routine events such as concerts and parades. We use a crawler program to collect this information through ticketing systems, tourism websites, news media, social media, and government websites and the meteorological data from the Central Meteorological Bureau. Through these three types of data, the accuracy in predicting traffic flow is enhanced to predict the degree of traffic congestion that may be affected.

Keywords: traffic flow prediction; deep learning; LSTM (Long Short-Term Memory); GRU (Gated Recurrent Units)

Citation: Jang, H.-C.; Chen, C.-A. Urban Traffic Flow Prediction Using LSTM and GRU. *Eng. Proc.* **2023**, *55*, 86. https://doi.org/10.3390/engproc2023055086

Academic Editors: Teen-Hang Meen, Kuei-Shu Hsu and Cheng-Fu Yang

Published: 2 January 2024

1. Motivation

There are many factors causing traffic congestion. When the number of vehicles exceeds the area's load capacity and the traffic system's design is poor, unsolvable traffic jams occur. Most of the solutions include adjusting the time phase of traffic lights, widening the road, restricting the entry and exit of vehicles, charging, and scheduling driving lanes. However, these solutions often encounter many obstacles or require high costs in implementation. As an extension of the work of Jang [1], in which a "smart traffic control platform" is proposed, we proposed the time-phased control of traffic lights that is adaptively adjusted according to the current traffic conditions for effectively relieving congestion and reducing driving time. Reference [1] proposed a symptomatic solution to resolve the traffic jam, while this study proposes a fundamental solution through the prediction of the possible traffic flow in the future. Then, combined with a smart traffic control platform, it becomes feasible to relieve traffic congestion. An accurate and instant traffic flow prediction is essential to the Intelligent Transportation System (ITS). ITS uses various computer science and communication technologies to manage the entire transportation system effectively. With the Internet of Things, existing data can be effectively collected and analyzed through many IoT devices to provide feedback to the transportation infrastructure.

Since the 20th century, many researchers have proposed various solutions for traffic flow prediction, such as the Autoregressive Integrated Moving Average model (ARIMA), Artificial Neural Networks (ANNs), and Support Vector Machines (SVMs). With the successful application of deep learning in ImageNet classification [2], many researchers have attempted to use deep neural networks for traffic-related predictions instead of shallower neural networks such as SVM. Researchers have confirmed that the LSTM model has good performance for the time-series-based traffic flow data by using the data from Caltrans Performance Measurement System (PeMS) in California, USA. Few studies have used the data on urban street traffic flow. Thus, based on the vehicle detector (VD) data from the Songshan District, 2017, provided by the Taipei Municipal Transportation Bureau, we propose a prediction model suitable for urban road sections with relatively low error through the deep learning method, combining it with the weather data in Songshan District and the relevant data of large-scale events. The evolutionary model of RNN in deep learning systems such as LSTM and GRU is used to obtain more features and multi-layer models to increase the accuracy. Finally, we compare the performance of LSTM and GRU models in urban traffic flow prediction.

2. Methodology

We used the data on regular traffic flow, predictable episodic events, and weather and the LSTM model for module A and model B, and the GRU model for module C to predict. The effectiveness of different models for traffic prediction in urban areas is then compared with other results.

- Module A: the LSTM module that does not include the data of "predictable episodic event" (dataset: VD data);
- Module B: the LSTM module that includes the data of "predictable episodic events" (datasets: VD data, weather, and events held by the Taipei Arena);
- Module C: the GRU module that includes the data of "predictable episodic events" (data set: VD data, weather, and events held by the Taipei Arena).

Regarding the role of module A, we demonstrated in a previous study [3] that the data of "predictable episodic event" are beneficial for predicting traffic flow. Therefore, we focused on the role of modules B and C.

2.1. Data Collection

The datasets for predicting urban traffic flow are divided into three categories: regular traffic flow data (VD data), predictable episodic event data, and weather data.

2.1.1. Regular Traffic Flow Data

The regular traffic flow data source is "Vehicle Detector (VD) data of Taipei City Government Data Open Platform". Five vehicle detectors are deployed on the main roads around Taipei Arena: VMCN800, VKRM820, VKWNV20, VKWN800, and VL7PX00. The data are provided by the Traffic Control Center of Taipei City for the entire year of 2017.

2.1.2. Predictable Episodic Events Data

Predictable and non-routine events include concerts and parades. Such information can be found through ticketing systems, tourism websites, news media, social media, and government websites. In Ref. [4], we implemented a crawler program with Python 3.6.5 to collect the event data of the Taipei Arena to predict the degree of traffic congestion.

2.1.3. Weather Data

On the subscription system of the Central Meteorological Administration, the hourly observation data of all observation stations in Taiwan since 2010 have been published. The data include temperature, humidity, rainfall, and other weather data. We wrote a crawler to download a CSV file containing all the required weather data.

2.2. Data Preprocessing

2.2.1. Filter Data Field

The vehicle detector (VD) data, weather data, and predictable episodic event data contain several fields that are not used, so these fields must be removed first. The remaining data fields after filtering are as follows.

- Vehicle detector (VD) data: fields of traffic flow, vehicle speed, and vehicle occupancy (%);
- Weather data: fields of temperature, relative humidity, rainfall, week, and hour;
- Episodic event data: field of event.

2.2.2. Feature Scaling

After the data field is normalized, the convergence speed of learning can be sped up. Depending on the type of data, there are two methods of normalization.

- For general data fields such as traffic flow and rainfall, we use the following Z-score formula to perform feature scaling and shift the data range to a smaller range.

$$X_{normalization} = \frac{(x - \mu)}{\sigma} \quad (1)$$

where x is the value of the data field, μ is the mean of the parent data, and σ is the standard deviation.

- For the time periodic data field (hour, week), x, we make the scaled value have circularity. The range of numerical data can be scaled through normalization to $[-1, 1]$ (Equation (2)).

$$X_{normalization} = \sin(x \times 2 \times \pi/24) \quad (2)$$

2.3. Deep Learning

2.3.1. Build Up Training Data

Training data consist of these three data types, and we input different lengths of data, as shown in Figure 1. The 24 h input data are an example. The VD data from the 0th hour to the 24th hour are used as the input data (X_train [5]), and the output data are the next time point (Y_train [5]) of the last input data. That is, the traffic flow data after 30 min are the traffic flow data of the 24.5th hour; the VD data from the 0.5th hour to the 24.5th hour are used as the input data (X_train [4]), and the output data are the 25th-hour traffic flow data (Y_train [4]), and so on. The last input data are the VD data from the first hour to the I+24th hour, and the output data are the traffic flow data of the I+24.5th hour.

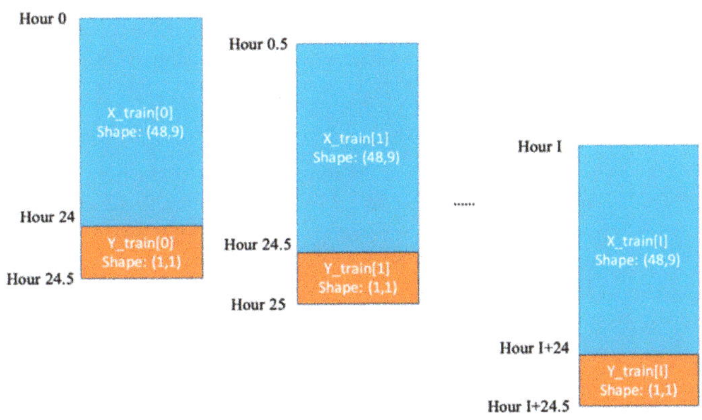

Figure 1. Training data arrangement.

2.3.2. Build Up Models

We propose module B (LSTM model) and module C (GRU model) in this study. Both modules adopt three architectures: the basic Vanilla architecture, the stacked architecture [4,5], and the encoder–decoder architecture [6].

- Vanilla architecture: an LSTM or GRU model with only a single layer;
- Stacked architecture: an LSTM or GRU model with multiple layers;
- Encoder–decoder architecture: convert the input sequence into a fixed-length vector through the encoder, then convert the fixed-length vector into the output sequence through the decoder.

(a) LSTM model

In practice, LSTM is divided into four types: one-to-one, one-to-many, many-to-one, and many-to-many, according to the number of input and output data. The input for this simulation is VD data for the 24 h prior to the target time, plus episodic event data and weather data. The output is the traffic flow for 30 min from the target time. It means that 48 pieces of data are input and one piece of data is output, which is a many-to-one LSTM.

- Vanilla LSTM: single-layer LSTM with a Dense layer to output traffic flow;
- Stacked LSTM: a four-layer LSTM with a Dense layer to output the traffic flow. However, simply using this module may encounter overfitting. We adjust the parameters of the dropout layer or increase the number of dropout layers to reduce the overfitting, depending on the situation;
- Encoder–decoder LSTM is used to customize how many layers of LSTM the encoder and decoder have.

(b) GRU model

In implementing the GRU model, we adopt the same design concept as the LSTM model, divided into Vanilla, stacked, and Encoder–decoder architectures.

- Vanilla GRU: a single-layer GRU with a Dense layer to output traffic flow;
- Stacked GRU: a three-layer GRU with a Dense layer to output traffic flow. Moreover, increase the number of dropout layers to prevent overfitting;
- Encoder–decoder GRU: a layer of GRU is used as encoder and decoder, respectively.

2.3.3. Model Training

In two different modules, we used 80% of the data for the entire year as the training data and 20% of the data as the validation data. Since the model trained from the training data expressed the behavior of the training data, it did not express other behaviors shown by the test data. Therefore, we used the validation data to find the best model, and at the same time, we used Adam Optimizer to speed up the training process and set the learning rate to 0.001.

2.4. Performance Evaluation

In the simulation, the remaining 20% of the data in the entire year was used as test data. We used MAPE, MSE, and MAE to evaluate the difference between the actual and predicted values and compare Modules B and C's performance. MAPE and MAE considered the magnitude of the error between the predicted value and the actual value, regardless of a positive or negative error. MSE took the square of the error between the predicted and actual values, so the MSE value was prone to being affected when there were extreme values. Therefore, to avoid extreme error conditions, we observed them through the MAPE and MAE metrics. Among the three metrics of MAPE, MAE, and MSE, the smaller the value, the smaller the error between the predicted value and the actual value, and the model has a better predictive ability. The MAE and MSE were relative evaluation metrics. The smaller the value, the better the performance. Regarding the MAPE evaluation metric, we referred to the classification of the effectiveness of the evaluation metric proposed by Lewis [7].

3. Experimental Simulation

The experiment consisted of three simulations: simulation 1 was the Vanilla architecture, simulation 2 was the stacked architecture, and simulation 3 was the Encoder–decoder architecture. Each simulation used LSTM and GRU models and MAPE, MSE, and MAE metrics to evaluate the prediction accuracy. We collected data from the five VD stations deployed around Taipei Arena according to different driving directions, including VKRM820 in the east lane, VKRM820 in the west lane, VKWN800 in the north lane, VKWN800 in the south lane, and VKWNV20 in the east lane, VKWNV20 in the west lane, VMCN800 in the south lane, and VL7PX00 in the north lane, as shown in Figure 2. The parameter settings of the three simulations are listed in Table 1.

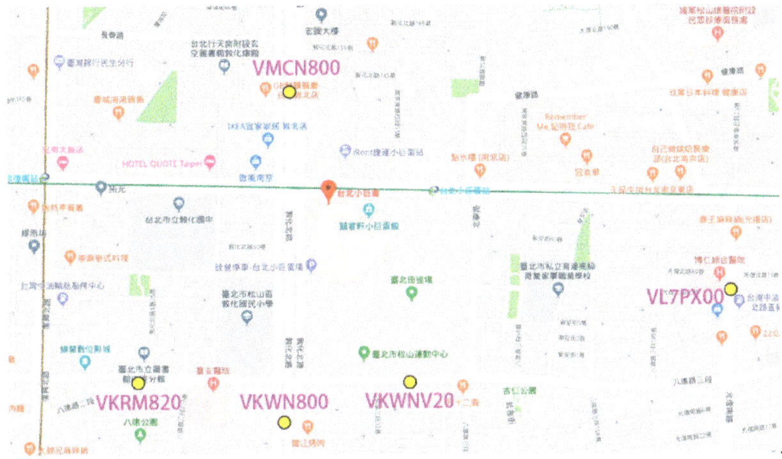

Figure 2. Five VD stations deployed around Taipei Arena.

Table 1. Simulation parameter setting.

Parameter	Value	Parameter	Value
Learning rate	0.001	Optimizer	Adam
Dropout rate	0.5	Batch size	128
Recurrent dropout rate	0.3	Timestep	96
Loss function	MAPE		

3.1. Simulation 1: Vanilla Architecture

In the simulation, we used the basic Vanilla architecture, i.e., LSTM or GRU, with only a single layer. Tables 2 and 3 show the evaluation of the Vanilla LSTM and Vanilla GRU prediction in the east lane of VKRM820 under various units. The overall performance of the LSTM model was better than that of the GRU model, but the training time was longer. When there were only a single layer of LSTM and GRU, and the unit parameters were 4, 8, 16, 32, and 64. It was prone to encountering the problem of learning bottlenecks (training loss and validation loss are almost unchanged) or low-level fitting. When the unit parameters were 128 and 256, more epoch training reduced the loss value with its limit. According to Lewis' MAPE accuracy classification, its prediction was ranked "inaccurate".

Table 2. Prediction of VANILLA LSTM in the east lane of VKRM820.

	4 Units	8 Units	16 Units	32 Units	64 Units	128 Units	256 Units
MAPE	68.563	60.414	56.783	54.975	48.175	51.994	50.421
MSE	0.303	0.198	0.188	0.146	0.138	0.116	0.118
MAE	0.428	0.299	0.279	0.218	0.198	0.159	0.157
Time	29 min 48 s	19 min 33 s	28 min 39 s	28 min 10 s	20 min 57 s	59 min 40 s	1 h 40 s

Table 3. Prediction of VANILLA GRU in the east lane of VKRM820.

	4 Units	8 Units	16 Units	32 Units	64 Units	128 Units	256 Units
MAPE	70.198	79.199	72.030	66.403	68.731	57.508	50.012
MSE	0.336	0.229	0.197	0.175	0.174	0.124	0.132
MAE	0.454	0.345	0.297	0.257	0.255	0.177	0.185
Time	19 min 46 s	16 min 5 s	15 min 36 s	15 min 23 s	15 min 40 s	46 min 8 s	48 min 40 s

Tables 4 and 5 present the result of the Vanilla LSTM and Vanilla GRU prediction in the west lane of VKRM820 under various units. The overall performance of the LSTM model was better than that of the GRU model, but the training time was longer. The evaluations of the MAPE, MSE, and MAE were significantly worse than that of the VKRM820 eastward lane, and according to Lewis' MAPE accuracy classification, it is ranked "inaccurate" prediction.

Table 4. Prediction of VANILLA LSTM in the west lane of VKRM820.

	4 Units	8 Units	16 Units	32 Units	64 Units	128 Units	256 Units
MAPE	78.214	70.654	64.325	60.794	65.512	61.249	63.350
MSE	0.703	0.466	0.371	0.283	0.239	0.236	0.224
MAE	0.592	0.473	0.402	0.330	0.298	0.292	0.281
Time	29 min 2 s	28 min 46 s	27 min 59 s	27 min 55 s	27 min 58 s	57 min 57 s	1 h 23 s

Table 5. Prediction of VANILLA GRU in the west lane of VKRM820.

	4 Units	8 Units	16 Units	32 Units	64 Units	128 Units	256 Units
MAPE	89.821	83.798	80.621	75.071	70.644	68.073	69.716
MSE	0.715	0.602	0.595	0.566	0.463	0.259	0.244
MAE	0.645	0.561	0.531	0.500	0.433	0.309	0.297
Time	15 min 57 s	15 min 44 s	14 min 59 s	15 min 14 s	14 min 56 s	45 min 40 s	48 min 36 s

Tables 6 and 7 reveal the Vanilla LSTM and Vanilla GRU prediction in the north lane of VKWN800 under various units. The overall performance of the LSTM model was better than that of the GRU model, but the training time was longer. The evaluations of the MAPE, MSE, and MAE were significantly worse than that of the VKRM820 eastward lane. Even if increasing the number of units can effectively reduce the loss value, it has its limit. According to Lewis' MAPE accuracy classification, it is ranked "inaccurate" prediction.

Table 6. Prediction of VANILLA LSTM in the north lane of VKWN800.

	4 Units	8 Units	16 Units	32 Units	64 Units	128 Units	256 Units
MAPE	77.510	80.423	75.476	69.035	64.242	54.430	55.740
MSE	0.779	0.657	0.592	0.511	0.446	0.310	0.324
MAE	0.565	0.471	0.420	0.368	0.317	0.189	0.199
Time	29 min 41 s	28 min 42 s	28 min 6 s	28 min 34 s	29 min 3 s	57 min 42 s	1 h 13 s

Table 7. Prediction results of VANILLA GRU in the north lane test set of VKWN800.

	4 Units	8 Units	16 Units	32 Units	64 Units	128 Units	256 Units
MAPE	100.787	98.284	98.909	94.709	91.722	73.398	92.598
MSE	1.168	0.760	1.109	0.792	0.667	0.511	0.620
MAE	0.831	0.590	0.802	0.609	0.473	0.350	0.521
Time	15 min 57 s	15 min 44 s	15 min 13 s	15 min 9 s	15 min 13 s	45 min 50 s	48 min 26 s

Tables 8 and 9 show the Vanilla LSTM and Vanilla GRU prediction in the south lane of VKWN800 under various units. The overall performance of the LSTM model was better than that of the GRU model, but the training time was longer. In the Vanilla LSTM of 8 units, the three evaluations of MAPE, MSE, and MAE were better than that of 16, 32, and 64 units. This phenomenon is opposite to the previous VKRM820 in the east and west lanes. The evaluations of the MAPE, MSE, and MAE were significantly worse than that of the VKRM820 eastward lane. Even if increasing the number of units effectively reduced the loss value, it showed its limit. According to Lewis' MAPE accuracy classification, it is an "inaccurate" prediction.

Table 8. Prediction of VANILLA LSTM in the south lane of VKWN800.

	4 Units	8 Units	16 Units	32 Units	64 Units	128 Units	256 Units
MAPE	101.627	97.693	132.424	144.599	120.241	89.442	129.854
MSE	0.958	0.354	0.757	0.719	0.672	0.254	0.209
MAE	0.848	0.447	0.720	0.687	0.649	0.315	0.284
Time	28 min 36 s	28 min 46 s	28 min 19 s	28 min 19 s	27 min 48 s	57 min 33 s	1 h 1 min 23 s

Table 9. Prediction of VANILLA GRU in the south lane of VKWN800.

	4 Units	8 Units	16 Units	32 Units	64 Units	128 Units	256 Units
MAPE	101.278	107.800	103.922	108.222	105.009	114.851	237.453
MSE	0.964	0.938	0.962	0.938	0.957	0.789	0.886
MAE	0.851	0.838	0.849	0.838	0.847	0.752	0.803
Time	15 min 57 s	15 min 30 s	15 min 10 s	15 min 5 s	15 min 27 s	46 min 23 s	48 min 19 s

Tables 10 and 11 show the evaluation of the Vanilla LSTM and Vanilla GRU prediction in the east lane of VKWNV20 under various units. The simulation showed that the overall performance of the LSTM model was better than that of the GRU model, but the gap was not significant. The evaluations of the MAPE, MSE, and MAE were significantly worse than that of the VKRM820 eastward lane. Even if increasing the number of units effectively reduced the loss value, the effect was not as good as the VKRM820 in the east lane and VKWN800 in the north lane. According to Lewis' MAPE accuracy classification, it is an "inaccurate" prediction.

Table 10. Prediction of VANILLA LSTM in the east lane of VKWNV20.

	4 Units	8 Units	16 Units	32 Units	64 Units	128 Units	256 Units
MAPE	137.058	126.985	113.882	113.211	113.736	120.333	135.323
MSE	3.724	3.703	3.717	3.689	3.675	3.505	3.410
MAE	0.514	0.465	0.508	0.471	0.463	0.366	0.343
Time	33 min 9 s	29 min 21 s	30 min 45 s	32 min 10 s	28 min 59 s	59 min 28 s	59 min 37 s

Table 11. Prediction of VANILLA GRU in the east lane of VKWNV20.

	4 Units	8 Units	16 Units	32 Units	64 Units	128 Units	256 Units
MAPE	105.374	147.958	112.256	112.211	122.547	137.412	140.778
MSE	3.777	3.663	3.756	3.741	3.690	3.668	3.657
MAE	0.595	0.481	0.575	0.570	0.514	0.449	0.449
Time	15 min 34 s	15 min 35 s	15 min 34 s	15 min 30 s	15 min 10 s	47 min 6 s	48 min 50 s

Tables 12 and 13 present the Vanilla LSTM and Vanilla GRU prediction in the west lane of VKWNV20 under various units. The simulation shows that the overall performance of the LSTM model was better than that of the GRU model, but the gap was not significant. The evaluations of the MAPE, MSE, and MAE were significantly worse than that of the VKRM820 westward lane. Even if increasing the number of units effectively reduced the loss value, it had its limit. According to Lewis' MAPE accuracy classification, it is an "inaccurate" prediction.

Table 12. Prediction of VANILLA LSTM in the west lane of VKWNV20.

	4 Units	8 Units	16 Units	32 Units	64 Units	128 Units	256 Units
MAPE	89.634	100.658	105.359	107.994	108.121	140.375	146.721
MSE	1.697	1.481	1.471	1.453	1.444	1.410	1.370
MAE	0.567	0.464	0.445	0.398	0.392	0.358	0.343
Time	29 min 44 s	30 min 5 s	28 min 7 s	28 min 28 s	29 min 14 s	58 min 49 s	58 min 35 s

Table 13. Prediction of VANILLA GRU in the west lane of VKWNV20.

	4 Units	8 Units	16 Units	32 Units	64 Units	128 Units	256 Units
MAPE	100.886	111.012	104.821	104.347	107.894	114.547	117.554
MSE	1.731	1.431	1.487	1.425	1.418	1.378	1.362
MAE	0.651	0.452	0.468	0.415	0.391	0.344	0.332
Time	15 min 48 s	15 min 57 s	15 min 34 s	15 min 32 s	15 min 49 s	47 min 14 s	49 min 29 s

Tables 14 and 15 present the Vanilla LSTM and Vanilla GRU prediction in the south lane of VMCN800 under various units. The simulation shows that the overall performance of the LSTM model was still better than that of the GRU model. The evaluations of the MAPE, MSE, and MAE were significantly better than that of the VKRM820 eastward lane, and according to Lewis' MAPE accuracy classification, it is a "reasonable" prediction.

Table 14. Prediction of VANILLA LSTM in the south lane of VMCN800.

	4 Units	8 Units	16 Units	32 Units	64 Units	128 Units	256 Units
MAPE	39.129	28.168	34.247	28.623	28.921	24.401	22.992
MSE	0.119	0.096	0.122	0.110	0.115	0.094	0.080
MAE	0.223	0.184	0.188	0.178	0.177	0.157	0.135
Time	28 min 37 s	28 min 55 s	27 min 59 s	27 min 58 s	27 min 53 s	56 min 53 s	1 h 1 min 16 s

Table 15. Prediction of VANILLA GRU on the south lane of VMCN800.

	4 Units	8 Units	16 Units	32 Units	64 Units	128 Units	256 Units
MAPE	41.525	30.118	30.101	20.017	25.460	23.634	28.544
MSE	0.128	0.123	0.161	0.148	0.105	0.094	0.106
MAE	0.256	0.233	0.242	0.222	0.175	0.158	0.182
Time	19 min 12 s	15 min 30 s	14 min 58 s	15 min 18 s	15 min 23 s	45 min 56 s	48 min 17 s

Tables 16 and 17 reveal the Vanilla LSTM and Vanilla GRU prediction in the north lane of VL7PX00 under various units. The simulation shows that the overall performance of the LSTM model was still better than that of the GRU model. The MAPE, MSE, and MAE evaluations were better than that of the VKRM820 eastward lane. Overall, increasing the number of units effectively reduced the loss of value. According to Lewis' MAPE accuracy classification, Vanilla LSTM was a "reasonable" prediction when units are 8, 16, 32, 64, 128, and 256, while Vanilla GRU is a "reasonable" prediction when units are 64, 128, and 256.

Table 16. Prediction of VANILLA LSTM in the north lane of VL7PX00.

	4 Units	8 Units	16 Units	32 Units	64 Units	128 Units	256 Units
MAPE	65.353	49.848	47.432	44.103	40.602	40.502	41.524
MSE	0.436	0.249	0.221	0.190	0.159	0.154	0.149
MAE	0.507	0.325	0.286	0.243	0.206	0.175	0.174
Time	29 min 21 s	29 min 14 s	28 min 49 s	28 min 26 s	28 min 53 s	58 min 28 s	1 h 1 min 12 s

Table 17. Prediction of VANILLA GRU in the north lane of VL7PX00.

	4 Units	8 Units	16 Units	32 Units	64 Units	128 Units	256 Units
MAPE	68.616	57.033	55.862	50.743	45.745	39.649	43.304
MSE	0.473	0.290	0.285	0.237	0.208	0.176	0.185
MAE	0.541	0.366	0.349	0.301	0.258	0.200	0.210
Time	19 min 35 s	15 min 39 s	15 min 17 s	15 min 29 s	15 min 37 s	45 min 43 s	48 min 33 s

3.2. Simulation 2: Stacked Architecture

In simulation 2, a stacked architecture was used in an LSTM or GRU that stacks multiple layers. In the architectures of stacked LSTM and stacked GRU, the LSTM model generally outperformed the GRU model, except for the VL7PX00. However, even if the stacked architecture was used for training, the training results were different from those obtained in Refs. [6,8]. Therefore, we suspect that a dataset problem caused the training results to be inaccurate. Although the dataset used by Ref. [6] is different from ours, the MAPE of the final prediction is the lowest, so we adopt their architecture to verify whether it is a problem with the architecture or the dataset.

3.3. Simulation 3: Encoder–Decoder Architecture

In simulation 3, the encoder–decoder architecture was used to verify whether our dataset has problems. Although Ref. [6] only uses the LSTM model for simulations, we tested the effectiveness of the GRU model. Due to the space limit, we skip the details of simulation data. In the architecture of the Encoder–decoder LSTM and Encoder–decoder GRU, the LSTM model was no longer better than the GRU model. In the south lane of VKWN800, the performance of the GRU model was even better, and the training time was short. However, even referring to the architecture of [6], we were still unable to achieve as good as their MAPE below 10, and the predictions made by many stations were still ranked "inaccurate".

4. Conclusions

Three types of architecture were simulated in this study. Simulation 1 was conducted for traffic flow prediction for five stations through the basic Vanilla architecture. The simulation results showed that only a single layer of LSTM or GRU was not complex enough to make accurate predictions. Even if the number of neurons in LSTM or GRU increased or the number of epochs increased, the improvement in predictions was limited. In simulation 2, multiple layers of LSTM or GRU models were stacked through the stacked architecture. In the stacked architecture, the predictions of either the LSTM model or the GRU model were better than that of the LSTM model and GRU model of the Vanilla architecture, but the training time was long. However, even with the stacked architecture, the MAPE was still far from the MAPE of [6], so we verified whether it was a problem with the dataset by simulating the encoder–decoder architecture. In the encoder–decoder architecture of simulation 3, the predictions obtained using LSTM were generally not different from the stacked architecture, but the training time was short. The GRU model performed better than the stacked architecture, and the training time was short. We could not obtain an excellent performance like the MAPE of [6], which was below 10%, so there were data missing in the dataset. The LSTM outperformed the GRU in the three simulations, but the training time was long. In the encoder–decoder architecture of simulation 3, the GRU model occasionally outperformed the LSTM model.

Author Contributions: Conceptualization, H.-C.J. and C.-A.C.; methodology, H.-C.J. and C.-A.C.; software, C.-A.C.; validation, H.-C.J. and C.-A.C.; formal analysis, H.-C.J. and C.-A.C.; investigation, H.-C.J. and C.-A.C.; resources, H.-C.J. and C.-A.C.; data curation, C.-A.C.; writing—original draft preparation, C.-A.C.; writing—review and editing, H.-C.J.; visualization, C.-A.C.; supervision, H.-C.J.; project administration, H.-C.J.; funding acquisition, H.-C.J. All authors have read and agreed to the published version of the manuscript.

Funding: This study is sponsored by the Ministry of Science and Technology, Taiwan (Grant No. MOST 109-2221-E-004-010-).

Institutional Review Board Statement: Not applicable.

Informed Consent Statement: Not applicable.

Data Availability Statement: The regular traffic flow data source is "Vehicle Detector (VD) data of Taipei City Government Data Open Platform". Five vehicle detectors are deployed on the main roads around Taipei Arena: VMCN800, VKRM820, VKWNV20, VKWN800, and VL7PX00. The data are provided by the Traffic Control Center of Taipei City for the entire year of 2017.

Acknowledgments: The authors gratefully thank the reviewers for their precise and constructive remarks, which significantly helped improve the manuscript.

Conflicts of Interest: The authors declare no conflicts of interest.

References

1. Jang, H.C.; Lin, T.K. Traffic-aware traffic signal control framework based on SDN and cloud-fog computing. In Proceedings of the 2018 IEEE 88th Vehicular Technology Conference (VTC 2018-Fall), Chicago, IL, USA, 27–30 August 2018.
2. Krizhevsky, A.; Sutskever, I.; Hinton, G.E. ImageNet classification with deep convolutional neural networks. In Proceedings of the NIPS'12, 25th International Conference on Neural Information Processing Systems, Red Hook, NY, USA, 3–6 December 2012; Volume 1, pp. 1097–1105.
3. Jang, H.C.; Chang, Y.H. Traffic flow forecast for traffic with forecastable sporadic events. In Proceedings of the 12th International Conference on Ubi-Media Computing (Ubi-Media 2019), Bali, Indonesia, 6–9 August 2019.
4. Du, X.; Zhang, H.; Nguyen, H.V.; Han, Z. Stacked LSTM deep learning model for traffic prediction in vehicle-to-vehicle communication. In Proceedings of the 2017 IEEE 86th Vehicular Technology Conference (VTC-Fall), Toronto, ON, Canada, 24–27 September 2017.
5. Chen, Y.Y.; Lv, Y.; Li, Z.; Wang, F.Y. Long short-term memory model for traffic congestion prediction with online open data. In Proceedings of the IEEE 19th International Conference on Intelligent Transportation System, Rio de Janeiro, Brazil, 1–4 November 2016.
6. Shao, H.; Soong, B.H. Traffic flow prediction with long short-term memory networks (LSTMs). In Proceedings of the 2016 IEEE Region 10 Conference (TENCON), Singapore, 22–25 November 2016.
7. Lewis, C.D. *Industrial and Business Forecasting Methods: A Practical Guide to Exponential Smoothing and Curve Fitting*; Butterworth Scientific: London, UK, 1982.
8. Kang, D.; Lv, Y.; Chen, Y.Y. Short-term traffic flow prediction with LSTM recurrent neural network. In Proceedings of the 2017 IEEE 20th International Conference on Intelligent Transportation Systems (ITSC), Yokohama, Japan, 16–19 October 2017.

engineering proceedings

MDPI

Proceeding Paper

Age-of-Information-Based Transmission Protocol in Vehicular Network †

Hung-Chin Jang * and Chung-Yen Huang

Department of Computer Science, National Chengchi University, Taipei 11605, Taiwan; diabloyenyen@gmail.com
* Correspondence: jang@cs.nccu.edu.tw

† Presented at the IEEE 5th Eurasia Conference on Biomedical Engineering, Healthcare and Sustainability, Tainan, Taiwan, 2–4 June 2023.

Abstract: With the development of the Internet of Things (IoT) and communication networks, the concept of a smart city emerged spontaneously. In the traffic control of a smart city, the vehicular network plays an important role. If the driving information of vehicles can be collected from the vehicular network and then aggregated into a smart city system, traffic control facilities can be adjusted in real time to improve traffic or increase traffic safety. This study is based on WAVE/DSRC under the IEEE 802.11p and IEEE 1609 standards. When the vehicle is moving, the On-Board Unit (OBU) on the vehicle and the Roadside Unit (RSU) transmit data through the 5.9 GHz (5.85–5.925 GHZ) frequency band to establish vehicle-to-vehicle communication (V2V) and vehicle-to-infrastructure communication (V2I). The immediacy of the transmission of beacon messages is also discussed over a vehicular network. A beacon message is essential for communication in a vehicular network, including various safety messages, such as driving directions, driving speed, and location. We introduce the Age-of-Information (AoI) indicator to reflect the immediacy of information. AoI is used for the elapsed time after the sender samples the message until the receiver receives the message. We propose a centralized AoI-based protocol and a decentralized AoI-based protocol. By using Random-Walk and SUMO, we simulate driving dynamics and various RSU setting scenarios. Finally, we verify the performance of the proposed AoI-based protocol through experimental simulations.

Keywords: age of information; transmission protocol; vehicular network

Citation: Jang, H.-C.; Huang, C.-Y. Age-of-Information-Based Transmission Protocol in Vehicular Network. *Eng. Proc.* **2023**, *55*, 87. https://doi.org/10.3390/engproc2023055087

Academic Editors: Teen-Hang Meen, Kuei-Shu Hsu and Cheng-Fu Yang

Published: 2 January 2024

1. Motivation

This research aims to design a transmission protocol that enables the vehicle and the RSU to receive the most real-time driving information. When the received information has the minimum average AoI, it is the most real-time driving information. If the vehicle information can be collected from the vehicular network and then aggregated into a smart city system, traffic control facilities can be adjusted in real time to improve traffic and increase traffic safety. Traffic control is associated with many real-time applications, and the quality of applications is based on whether the information is collected in real time. For applications that need to be predicted or monitored, outdated information does not match reality, resulting in poor quality. In order to continuously monitor and control the traffic system, it is necessary to constantly collect real-time information from the vehicular network, not simply the minimum delay or maximum throughput. The transmission period proposed by Baiocchi [1] has a functional relationship with the average AoI of the vehicular network in a known network topology, and an optimal value exists. We thus design a decentralized vehicular network transmission protocol that dynamically adjusts the transmission period. Ni [2] proposed a centralized vehicular network, and its experimental design, context setting, and control group are all references for this study. The difference in the channel connection status between this study and Ni's research is that we assume that the vehicular network is variable, and the channel connection status is unstable. In Kaul's work [3] the vehicular network employed the CSMA model and adjusted the contention window and broadcast period. It was experimentally proven that AoI is mainly affected by the broadcast period instead of the contention window. Kadota [4] proposed a design strategy for wireless communication networks, and we refer to the max-weight strategy of Kadota's work and design a centralized vehicular network transmission protocol.

2. Methodology

2.1. Network Architecture

The network topology of the vehicular network proposed in this study is divided into two types: centralized and decentralized. The centralized vehicular network includes the RSU, which receives the vehicles' messages, transmits the collected information to the server for subsequent applications, or sends warning messages to the vehicles to maintain driving safety. However, deploying many RSUs in a wide area is expensive. The centralized vehicular network is only suitable for urban areas. The decentralized vehicular network consists only of vehicles, and the transmission is carried out through the vehicles among the network. This network architecture can be applied to most areas, but it lacks centralized equipment to collect driving information, making it challenging to develop more advanced applications. In addition, those areas with sparse vehicle density cannot support a decentralized vehicular network due to intermittent connection.

The vehicular network includes the OBU installed in the vehicle and the RSU installed on the roadside. The OBU generates beacon messages, including the vehicle's speed, direction, and location. In addition, the OBU can transmit beacon messages with RSUs and other OBUs. The OBU records when the beacon message is transmitted. The RSU is responsible for managing the connection status between its channel and each OBU and it transmits beacon messages to the OBU. If the OBU obtains and exchanges the safety information of its vehicle with other vehicles, other vehicles and the RSU know the current state of the vehicle to prevent accidents and improve driving safety. The OBU records when the beacon message is received and subtracted from the current time to calculate the AoI. The connection state between the RSU's channel and each OBU may change due to obstacles in the driving environment or other factors. The RSU constantly communicates with the OBU to obtain the vehicle's driving information and keep the connection states updated for upcoming transmissions.

2.2. Traffic Model

The binomial equation [5,6] is used for the workload modeling of the vehicular network. The binomial represents the probability of the successful transmission of messages by applications in individual nodes, and the arrival function, $\lambda(x)$, represents the distribution function of the successful transmission of messages by applications, as shown in Equation (1).

$$\mathrm{P}[T_i(x) > t] = (1 - p(x))^{\frac{t}{t_c}} = e^{-\lambda(x)t} \tag{1}$$

where $p(x)$ represents the probability that a node successfully transmits a message in each slot, $T_i(x)$ is the size of a single slot, t is the time required to transmit a message, and t_c is the maximum time needed to transmit a message, as shown in Equation (2).

$$t_c = \frac{maximum\ length\ of\ packets}{transmission\ rate} \tag{2}$$

Equation (3) can be obtained by taking the logarithmic operation on $(1 - p(x))^{\frac{t}{t_c}} \cong e^{-\lambda(x)t}$ in Equation (1).

$$\lambda(x) = \frac{1}{t_c}\ln\left(\frac{1}{1 - p(x)}\right) \tag{3}$$

The main application in the vehicular network is the transmission of safety messages. We only consider the transmission based on the immediacy of messages. When a node transmits a message, it has its exclusive slot in each time frame, so the value of $p(x)$ is determined by the number of slots in the frame. The slot, $T_i(x)$, needs to be greater than t_c to ensure that each message is successfully delivered. In Equation (4), there are 20 slots in each frame, and the $p(x)$ can be expressed as follows:

$$p(x) = \frac{length\ of\ a\ slot}{length\ of\ a\ frame} = \frac{1}{20} \quad (4)$$

By substituting $p(x)$ in Equation (4) into Equation (3), we can derive Equation (5).

$$\lambda(x) = \frac{0.051}{t_c} > \frac{0.051}{T_i(x)} = \frac{0.051}{slot} \quad (5)$$

The slot, $T_i(x)$, must be greater than t_c and make the arrival rate $\lambda(x) < 1$. If $\lambda(x) > 1$, the slot size does not allow the application to transmit messages. Therefore, the slot size must be set appropriately so that $\lambda(x) < 1$.

2.3. *Age-of-Information-Based Transmission Protocol*

2.3.1. Centralized AoI-Based Transmission Protocol

In the centralized AoI-based transmission protocol, the time allocated to the dedicated channel is based on the frame/slot structure. When the vehicle receives the update message, the AoI is updated, as shown in Figure 1.

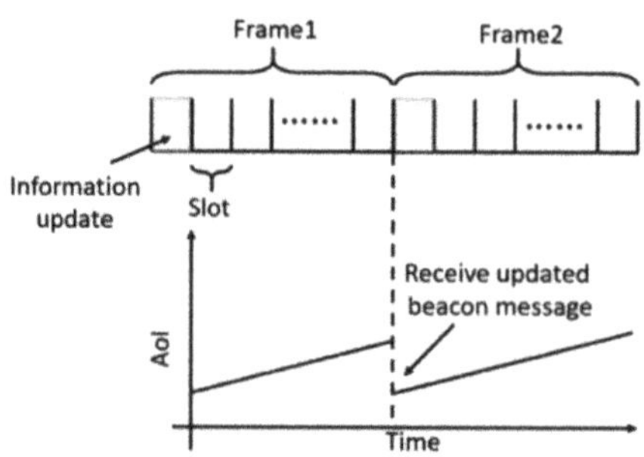

Figure 1. Variation of AoI in frame/slot structure.

In Figure 1, the frame is the period during which the vehicle generates the beacon message. Generally, a frame is about hundreds of milliseconds, and the length of the frame is fixed and does not change no matter how many vehicles there are in the RSU coverage. When the RSU or vehicle receives the beacon message during the frame, the vehicle's driving information is updated, and the AoI reduces. The maximum time required for each slot to transmit the beacon message depends on the packet size of each beacon message and the transmission speed of the channel. In this study, the size of the beacon message is fixed. Therefore, after setting the transmission speed of the channel, the slot length is a fixed value. Because a frame consists of multiple slots, vehicles in a given frame broadcast beacon messages in a single RSU area by using the channel and the time allocated by the RSU to avoid collisions. At the beginning of each frame, there is a short time for the RSU to update the information of the vehicles in the coverage so that the RSU knows which vehicles exist in the coverage. The AoI-based protocol updates the RSU for the vehicles under the coverage, the combination of channels used, and the selected slot. Then, it publishes a message to notify those vehicles in the next frame that the beacon message can be broadcast. The actual length of the frame and slot is determined by the experiment. We assume that the vehicle sends the beacon message to all vehicles in the area and the centralized RSU through broadcasting. The relationship between the AoI of the beacon message received by the RSU and the AoI of the overall network is as follows: Let there be n vehicles, $v_1, v_2, \ldots, v_n$. At time, t, the AoIs of the beacon messages of other vehicles stored by each vehicle can be represented by an $n \times n$ matrix, V_t.

$$V_t = \begin{bmatrix} v_{11(t)} & \cdots & v_{1n(t)} \\ \vdots & \ddots & \vdots \\ v_{n1(t)} & \cdots & v_{nn(t)} \end{bmatrix} \quad (6)$$

where v_{ij} represents the AoI obtained by vehicle v_i observing vehicle v_j, which must satisfy the following properties.

$$\begin{gathered} v_{ij(t)} = 0, if i = j \\ v_{ij(t+1)} = \left(v_{ij(t)} + 1\right) \times u_i \\ u_i = \begin{cases} 0\,,\ if\ v_i\ updates \\ 1\,,\ if\ v_i\ doesn't\ update \end{cases} \end{gathered} \quad (7)$$

v_{ii} = 0 means that each vehicle always knows its information instantly, and $v_{ij(t+1)}$ means v_{ij} at the next time point. The AoI changes according to whether the vehicle transmits the beacon message. If v_i is updated, the AoI is set to zero. Otherwise, the AoI increases with time. Next, we calculate the average AoI matrix M stored by each vehicle.

$$M = [m_1\,,\ m_2,\ \ldots m_n] \text{ where } m_i = \frac{v_{1i} + v_{2i} + \ldots + v_{ni}}{n},\ 1 \leq \text{i} \leq \text{n} \quad (8)$$

The RSU receives the driving information of each vehicle and records the AoI, but the RSU does not have the driving information to be transmitted. Therefore, the AoI of the RSU itself does not exist. Let the RSU record the $1 \times n$ matrix X of the AoI of each vehicle at time t as

$$X_t = \left[x_{1(t)}\,,\ x_{2(t)},\ \ldots x_{n(t)}\right] \quad (9)$$

where $x_{i(t)}$ represents the AoI of the vehicle v_i observed by the RSU.

$$x_{i(t+1)} = \left(x_{i(t)} + 1\right) \times u_i, 1 \leq \text{i} \leq \text{n} \quad (10)$$

The AoI observed by the RSU is the same as the AoI observed by the vehicles, so the AoI obtained by the vehicles other than vehicle i and the RSU satisfy the following property:

$$\begin{gathered} x_i = v_{1i} = v_{2i} = \ldots = v_{ji},\ 1 \leq j \leq \text{n}, \text{j} \neq \text{i} \\ x_i = \frac{v_{1i}+v_{2i}+\ldots+v_{ji}+\ldots+v_{ni}}{n-1} \end{gathered} \quad (11)$$

Next, comparing m_i and x_i, we obtain

$$x_i = \frac{n}{n-1} \times m_i \quad (12)$$

$$X = \frac{n}{n-1} \times M \quad (13)$$

From the above equations, we find the coefficient relationship between the AoI of the vehicle observed by the RSU and the average AoI of the overall vehicular network. In other words, we need to pay attention to the AoI observed by the RSU, which indicates the AoI of the whole network. When considering the centralized vehicular network, it is necessary to avoid message collision due to competing channels and transmission failure. When the information continues to fail to be transmitted and it is not updated for a long time, the AoI increases, and the immediacy of the vehicle information is lost. In this study, the channel assignment is performed by the RSU to avoid channel contention and packet collision. Due to the characteristics of vehicle movement, the connection status of each vehicle and channel changes dynamically, so we need to continuously update the channel connection information.

At the beginning of each frame, the RSU selects specific vehicles to update the driving information and thus knows the AoI of the vehicles and the channel–vehicle connection status. We denote the specific channel as c_i and the vehicle as v_i, as shown in Figure 2. We make a graph of the connection between the channel and the vehicle. The points on the left and right represent the channel and the vehicle, respectively, and the individual AoI of the vehicle weights the edge connected between them. We use the maximum matching algorithm to select the channel–vehicle combination of the AoI with the maximum sum to assign the channel to a specific vehicle. As shown in Figure 3, the solid line represents the selected channel–vehicle combination this time.

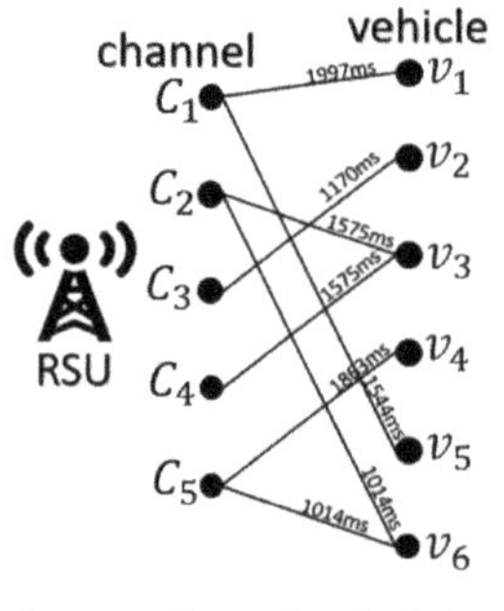

Figure 2. Channel–vehicle bipartite graph.

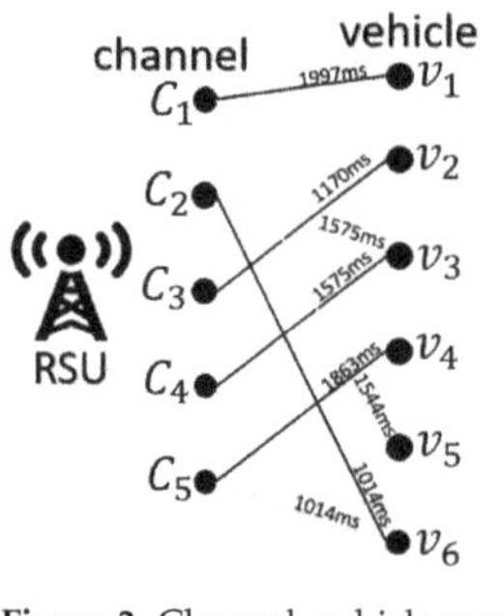

Figure 3. Channel–vehicle maximum matching.

The example in Figure 3 shows that the channel–vehicle combination selected for driving information update is (c_1,v_1), (c_2,v_6), (c_3,v_2), (c_4,v_3), (c_5,v_4). The updated AoI sum is 1997 + 1170 + 1575 + 1863 + 1014 = 7619 ms, which is the sum of the maximum matching of this channel–vehicle combination. The AoI of v_5 is 1544 ms, which is not the lowest, so it cannot be updated. We follow the above steps each time, picking the maximum matching channel–vehicle combination to obtain the maximum sum of the AoIs. The vehicles selected in the combination broadcast the beacon message and update the AoI that has not been updated for the longest time in the overall network. Based on this principle, we propose an AoI-based protocol that is suitable for centralized vehicular networks. The AoI-based protocol must be installed on the RSU and vehicle's OBU. At the beginning of each frame, the RSU obtains the connection statuses of vehicles and then calculates the AoI of each vehicle as follows:

1. For the vehicles that existed in the previous frame, use the current time minus the time of the previous vehicle's beacon message update as the vehicle's AoI.

2. For the new vehicles that appear in the current frame, give the initial value as the vehicle's AoI.

After calculating the AoI of the vehicles in each area, the RSU creates a channel–vehicle bipartite graph based on the previously obtained vehicle connection status. The calculated AoI selects the max matching to determine which vehicles broadcast the beacon message in the subsequent slots of the frame.

2.3.2. Decentralized AoI-Based Transmission Protocol

The proposed AoI-based protocol for decentralized vehicular networks is based on the concept of the work of Baiocchi [1]. Baiocchi improved the IEEE 802.11p (https://standards.ieee.org/ieee/802.11p/3953/ (accessed on 1 December 2023)) CSMA model and proposed a protocol to change the transmission period dynamically. Their work considers only periodic one-hop message exchange. In IEEE 802.11p CSMA, T_{msg} is defined as the period for the vehicle to transmit messages periodically. B_1 and B_2 in Figures 4 and 5 are the first and second message transmission times, respectively. B_1 and B_2 include the time spent transmitting the message, the time T spent in the DIFS (Distributed Inter-frame Spacing) process, and the time C spent returning the back-off counter to zero. Y represents the AoI of each message interval, which is defined as the current message arrival time minus the last message arrival time. t_1 and t_2 represent the arrival times of the first and second messages, respectively. Figure 4 shows that the node can successfully transmit the message in the transmission period (T_{msg}), and Figure 5 shows that the node cannot successfully transmit the message in the transmission period (T_{msg}).

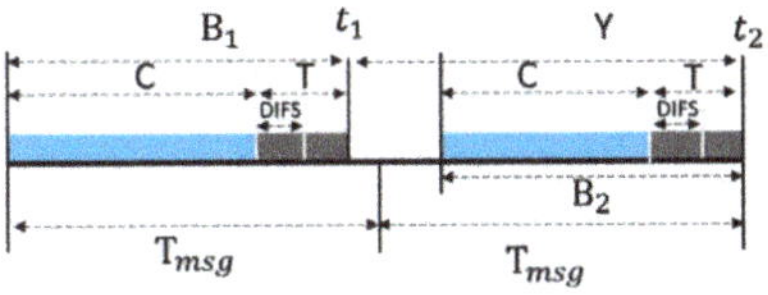

Figure 4. Decentralized model: $T_{msg} > B_1$.

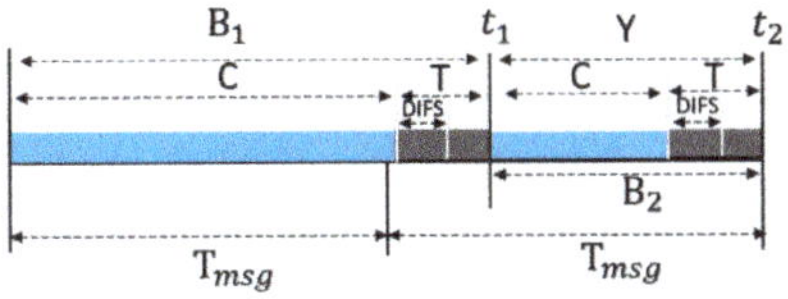

Figure 5. Decentralized model: $T_{msg} < B_1$.

The decentralized model uses updated processing to model the behaviors of generic nodes to transmit packets. We assume that the node sends beacon messages at times t_1 and t_2. At time t_2, the AoI is the current time minus the time t_1 when the last message was received, so AoI Y = $t_2 - t_1$. Because the competition for the transmission channel keeps the channel busy, if the waiting time C for the back-off counter to return to zero is increased, the transmission time B increases. According to the length of the node transmission time B, we consider two cases: the regular transmission time (T_{msg}) is greater than the message transmission time (B_1), and the transmission is successful within the specified time (Figure 4), or the regular transmission time (T_{msg}) is less than the message transmission time (B_1), and it cannot be successfully transmitted within the specified time (Figure 5). When the periodic message transmission time (T_{msg}) is greater than the message transmission time (B_1), Y = B_2 + ($T_{msg} - B_1$), and when the periodic message transmission time is less than the message transmission time (B_1), Y = B_2. Therefore, the general equation for the relationship between AoI Y and B is as follows:

$$Y = B_2 + \max\{0, T_{msg} - B_1\} \tag{14}$$

Equation (14) implies that the minimum AoI Y $= B_2 + 0$; that is, T_{msg} and B_1 have the same value. Our goal is to minimize the AoI Y, and the AoI Y can be reduced by adjusting the values of C and back-off counter to make T_{msg} close to B_1. The decentralized vehicular network is established under the IEEE 802.11p standard. When the nodes compete for the transmission channel and the node senses that the channel is not busy, it waits for a slot and subtracts it from the back-off counter. The slot's length, δ, is the IEEE 802.11p back-off slot duration. When the node perceives that the channel is busy, it freezes the back-off counter and waits for a time, T, so that other nodes can complete the transmission. T is the time it takes to complete the transmission and the subsequent DIFS process, and we use ε to represent the length of time T. C is the sum of δ and ε and it is represented by Equation (15):

$$C = \sum_{j=1}^{N} X(j) \tag{15}$$

where N is a discrete random variable with uniform distribution with values in the range $[0, W_0 - 1]$, and W_0 is the contention window size of IEEE 802.11p, where $W_0 = 16$. *X(j)* is generally a random variable of an independently identically distribution (iid), and it is defined as having the same distribution as X.

$$X = \begin{cases} \delta \text{ with probabilty } 1 - b \\ \varepsilon \text{ with probabilty } b \end{cases} \tag{16}$$

where b is the probability that the node perceives that the channel is busy. Baiocchi proved the functional relationship between T_{msg} and the AoI of adjacent vehicles by using the CSMA model mentioned above and Equations (10) and (11) and by using the adjacency matrix topology of the vehicular network. If the adjacency matrix of the vehicular network does not change, that is, when the relative locations of the vehicles do not change, there is a specific value T_{msg}, so that the average AoI of the entire vehicular network is minimum [1]. The function consists of a decreasing function and an increasing function, as shown in Figure 6.

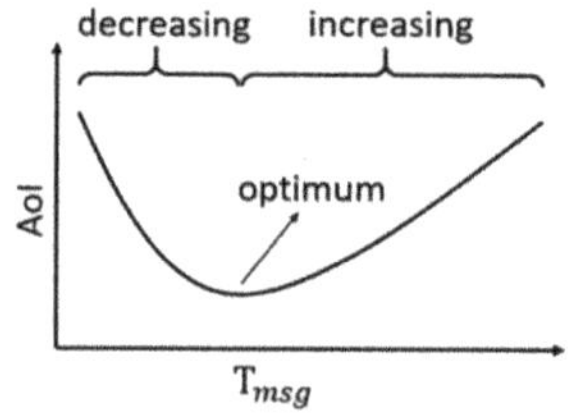

Figure 6. Relationship between average AoI and T_{msg} in a fixed network topology.

In driving, the status regarding whether the vehicles are in the same area or connected with other vehicles changes randomly over time, which causes the adjacent matrix to change accordingly. With the proof of ref. [1], we use the protocol with a variable transmission period to increase or decrease the length of T_{msg} to make the AoI minimal. If we increase T_{msg} and find that the AoI is higher than the value obtained last time, then T_{msg} decreases, and vice versa.

3. Experimental Simulations and Results

The experiment is simulated by NS-3 (network simulator-3). The vehicle density of the simulated environment is set to be between 50 and 400 vehicles/km^2, which means that the driving environments range from sparse to crowded. The frame size and slot sizes are 120 ms and 6 ms, respectively. The propagation loss model uses Friis, the vehicle speed uses the moving speed transmission module provided by NS-3 and is set to 40 km/h, and the time of each simulation is 40 s. The default value of the transmission period of the decentralized AoI-based protocol is set to 100 ms. The increase/decrease time to adjust the transmission period is set to 50 ms. We have two simulations to verify the proposed

AoI-based transmission protocols. The first simulation uses Random Walk to simulate the mobility model of vehicles, and the second uses SUMO (Simulation of Urban Mobility) to simulate the urban vehicle mobility model. Because the simulation is limited by the number of vehicles and vehicle density that a single subnet can accommodate, we set the vehicle to roam in an area of 799 × 799 m. The vehicle density ranges from 50 to 400 vehicles/km^2. The number of RSUs is set to 0, 1, 2, and 4, respectively. By setting different numbers of RSUs, we simulate the performance of centralized, decentralized, and hybrid vehicular networks. Each experiment is carried out for 40 s, the AoI is sampled for all vehicles in the network every 0.5 s, and the average AoI is calculated. The control group includes Round Robin (in the centralized vehicular network only) and IEEE 802.11p default transmission (in both centralized and decentralized vehicular networks).

3.1. Simulation 1: Random Walk Mobility Model

The first experiment simulation uses Random Walk as the mobility model, assuming that the vehicles move randomly in the area. It is also assumed that each vehicle and RSU in the area install the proposed AoI-based protocol so that the vehicle roams freely and transmits beacon messages to other vehicles and RSUs. We have several scenarios in simulation 1. The first scenario has one RSU in the area as the centralized AoI-based protocol. In the second scenario, we set up four RSUs as the centralized AoI-based protocol applied in a multiple-RSU scenario, simulating the impact of handoff with vehicles passing through different RSUs. In the third scenario, we do not set up any RSUs as the decentralized AoI-based protocol. The last scenario sets up two RSUs as a hybrid vehicular network, enabling vehicles to switch between centralized and decentralized AoI-based protocols. Regarding the control group, Round-Robin is conducted in a single RSU and four RSU scenarios since it requires a centralized RSU for scheduling. The IEEE 802.11p default is conducted with the number of RSUs being 0, 1, 2, and 4. The simulation results are shown in Table 1.

Table 1. Random Walk: AoI-based protocol improvement.

No. of RSU.	Avg. Improvement Compared with Round-Robin	Avg. Improvement Compared with 802.11p
RSU = 0	N/A	33.3%
RSU = 1	40.4%	56.9%
RSU = 2	N/A	38.5%
RSU = 4	30.0%	41.3%

3.2. Simulation 2: SUMO Simulates Vehicle Dynamics

SUMO is a traffic simulation software for microscopic traffic flow, which is highly portable and simulates continuous traffic flow and discrete-time traffic events for large-scale road networks. Simulation 2 uses SUMO and OSM (Open Street Map) to generate real driving dynamics. The OSM is used to extract "Sec. 2, Zhinan Rd., Wenshan Dist., Taipei City" in front of National Chengchi University as the central area for simulation, as shown in Figure 7. The same area corresponding to the Google map is shown in Figure 8.

In addition to basic street routes, SUMO also includes traffic control systems for traffic lights and intersections and has an actual driving dynamic simulation.

In simulation 2, we compare the performance of the proposed AoI-based protocol, Round-Robin, IEEE 802.11p default, and the AoI-based protocol [7]. Table 2 shows that the AoI-based protocol improves the AoI of the IEEE 802.11p default, Round-Robin, and AoI-based protocol [7] in various RSU scenarios.

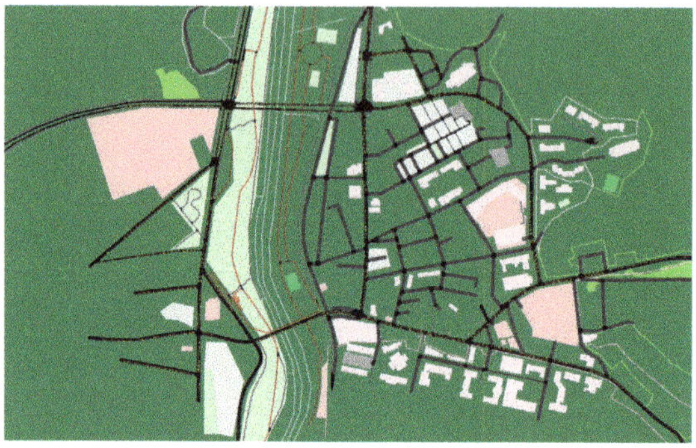

Figure 7. An example of real map around "Section 2, Zhinan Rd." extracted by sumo-gui.

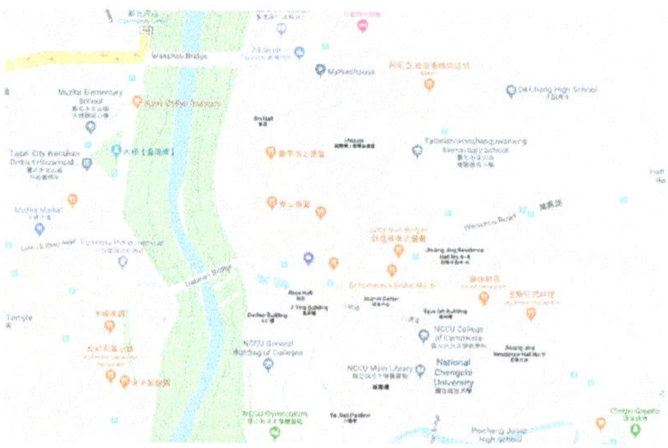

Figure 8. Map around "Section 2, Zhinan Rd." extracted from Google Maps.

Table 2. SUMO: AoI-based protocol improvement.

No. of RSU.	Avg. Improvement Compared with Round-Robin	Avg. Improvement Compared with 802.11p	Avg. Improvement Compared with AoI-Based Protocol [7]
RSU = 0	N/A	24.5%	N/A
RSU = 1	40.1%	57.8%	N/A
RSU = 2	N/A	34.85	N/A
RSU = 4	32.7%	51.1	19.7%

4. Conclusions

The experimental simulations show that under the four RSU setting scenarios, the AoI-based protocol outperforms the average AoI of Round-Robin and IEEE 802.11p default. We also simulate the actual driving environment through SUMO and OSM to verify the practicability of the AoI-based protocol. We found that when using the AoI-based protocol in four simulation scenarios with RSU = 0, 1, 2, and 4, when the vehicle density is 50–400 vehicle/km^2, regardless of the Random Walk or SUMO simulations, there is a different degree of improvement, as shown in Table 3.

Table 3. Vehicles' overall AoI-based protocol improvement.

No. of RSU	Random Walk	SUMO	Average Improvement
RSU = 0	33.3%	24.5%	28.9%
RSU = 1	48.7%	41.0%	44.9%
RSU = 2	38.5%	34.9%	36.7%
RSU = 4	35.6%	41.9%	38.8%

Author Contributions: Conceptualization, H.-C.J. and C.-Y.H.; methodology, H.-C.J. and C.-Y.H.; software, C.-Y.H.; validation, H.-C.J. and C.-Y.H.; formal analysis, H.-C.J. and C.-Y.H.; investigation, H.-C.J. and C.-Y.H.; resources, H.-C.J. and C.-Y.H.; data curation, C.-Y.H.; writing—original draft preparation, C.-Y.H.; writing—review and editing, H.-C.J.; visualization, C.-Y.H.; supervision, H.-C.J.; project administration, H.-C.J.; funding acquisition, H.-C.J. All authors have read and agreed to the published version of the manuscript.

Funding: This study is sponsored by the Ministry of Science and Technology, Taiwan (Grant No. MOST 109-2221-E-004-010-).

Institutional Review Board Statement: Not applicable.

Informed Consent Statement: Not applicable.

Data Availability Statement: Data are contained within the article.

Acknowledgments: The authors gratefully thank the reviewers for their precise and constructive remarks, which significantly helped improve the manuscript.

Conflicts of Interest: The authors declare no conflicts of interest.

References

1. Baiocchi, A.; Turcanu, I. A model for the optimization of beacon message age-of-Information in a VANET. In Proceedings of the 2017 29th International Teletraffic Congress (ITC 29), Genoa, Italy, 4–8 September 2017.
2. Ni, Y.; Cai, L.; Bo, Y. Vehicular beacon broadcast scheduling based on age of information (AoI). *China Commun.* **2018**, *15*, 67–76. [CrossRef]
3. Kaul, S.; Gruteser, M.; Rai, V.; Kenney, J. Minimizing age of information in vehicular networks. In Proceedings of the 2011 8th Annual IEEE Communications Society Conference on Sensor, Mesh and Ad Hoc Communications and Networks, Salt Lake City, UT, USA, 27–30 June 2011.
4. Kadota, I.; Uysal-Biyikoglu, E.; Singh, R.; Modiano, E. Scheduling policies for minimizing age of information in broadcast wireless networks. *IEEE/ACM Trans. Netw.* **2016**, *26*, 2637–2650. [CrossRef]
5. Asgari, M.; Yousefi, S. Traffic modeling of safety applications in vehicular networks. In Proceedings of the 2013 5th Conference on Information and Knowledge Technology, Shiraz, Iran, 28–30 May 2013.
6. Khekare, G.S.; Sakhare, A.V. A smart city framework for intelligent traffic system using VANET. In Proceedings of the 2013 International Mutli-Conference on Automation, Computing, Communication, Control and Compressed Sensing (iMac4s), Kottayam, India, 22–23 March 2013.
7. Li, Y.; Chen, W.; Peeta, S.; Wang, Y. Platoon control of connected multi-vehicle systems under V2X communications: Design and experiments. *IEEE Trans. Intell. Transp. Syst.* **2019**, *21*, 1891–1902. [CrossRef]

engineering proceedings

MDPI

Proceeding Paper

Design of a Singing Evaluation System of Heyuan Hua Chao Opera Based on Mel-Frequency Cepstral Coefficients †

Shuping Sun *, Yulei Zhu and Yanhui Wang

School of Art, Xiangtan University, Xiangtan 411100, China; zhuyulei0@163.com (Y.Z.); wyh23yan@163.com (Y.W.)
* Correspondence: 18670901605@163.com

† Presented at the IEEE 5th Eurasia Conference on Biomedical Engineering, Healthcare and Sustainability, Tainan, Taiwan, 2–4 June 2023.

Abstract: Heyuan Hua Chao opera is affected by modern culture and is facing a limited number of teaching staff, little time available for teachers to be dispatched, a small scope of popularity, the inability for continuous tracking in teaching, unsystematic and coherent learning among students, and difficulty of independent learning. Therefore, the voice feature parameters required for the evaluation model were extracted through MFCC coefficient features and later input into a convolutional neural network to generate data sets and training sets. These feature-labeled sets are segmented to output singing feedback. With a comprehensive overview of the artistic characteristics of Hua Chao opera singing and the research and interviews conducted by the Hua Chao opera heritage development center and local people, a system is designed to evaluate the singing voice of Hua Chao opera singers based on the MFCC. The aim is to effectively help students participate in learning and intelligently evaluate their learning effects. The model can be applied to other opera repertoires to promote the preservation and dissemination of traditional opera culture.

Keywords: Mel-Frequency Cepstral Coefficients; Hua Chao opera; vocal evaluation; opera into the campus

Citation: Sun, S.; Zhu, Y.; Wang, Y. Design of a Singing Evaluation System of Heyuan Hua Chao Opera Based on Mel-Frequency Cepstral Coefficients. *Eng. Proc.* **2023**, *55*, 88. https://doi.org/10.3390/engproc2023055088

Academic Editors: Teen-Hang Meen, Kuei-Shu Hsu and Cheng-Fu Yang

Published: 3 January 2024

1. Introduction

Hua Chao opera is a local traditional drama in Zijin County, Heyuan City, Guangdong Province, and is one of the forms of national intangible cultural heritage. The current opera in schools is active. On this basis, Huachao Opera has built a talent training system of "elementary school–junior high school–high school–university" with a close connection and coordinated development. However, Hua Chao opera on campus still faces the problem of a limited number of teaching staff, small teaching coverage, the inability of teachers to effectively detect students' learning effects in time, and students' lack of systematic and coherent learning and difficulty in independent learning. Considering the current problem of introducing Hua Chao opera to the campus, we decided to explore new ways and ideas of Hua Chao opera culture education and inheritance using new media and design a singing-assessment system based on Mel-Frequency Cepstral Coefficients. The result is expected to improve the initiative of students' participation in learning Hua Chao opera and intelligently assess the learning effect. At the same time, we propose a method of speech classification by fusing MFCC coefficient features with convolutional neural networks, which effectively improves the recognition accuracy and noise resistance of the model and provides a reference for the study of speech processing methods to apply the model to other opera repertoires.

2. Design of the Singing Evaluation System

The design of the Heyuan Hua Chao opera singing-voice-evaluation system mainly realizes the functions of learning–grading–singing voice opinion feedback. The system is designed to use the singing voice in Huayuan Hua Chao opera as the object of study and establish the evaluation criteria of the vocal character, clear and accurate singing words, and the stable pitch of the tune. By building a database and adopting quantitative methods, i.e., pre-emphasis, and using the fast Fourier transform method, the musical characteristics and speech spectrum information of Hua Chao opera singing are extracted [1]. The inverse spectrum-analysis method is also used to analyze the semantic features of speech signals. The large volume of singing speech signals is standardized and analyzed for objective and scientific evaluation results to achieve the noise resistance, robustness, and accuracy of the evaluation system.

2.1. Core Design

2.1.1. Design Flow of Mel-Frequency Cepstral Coefficients

1. Sample data and labels

The participants of the Hua Chao opera singing -evaluation system were students in grades 3 to 5, and the collected sample data were divided into training and testing audio sets. A total of 20,316 audio items with a chanting style were collected from the training set, while there were 4060 audio items from the training set with a fast singing style and 10,084 audio items from the training set with a traditional singing style. After the audio storage paths of the files were obtained, the labels of the speech signals, i.e., the specific feedback of the three singing voices, were imported.

2. Specific implementation of MFCC coefficient generation speech signal model

After the training set of audio sample data of Hua Chao opera was obtained, the continuous speech signals with a 16 KHz sampling frequency in 1 s were needed to improve the high-frequency part to reduce the noise output of the pre-emphasized overlapped sampling points to reduce the change in frame number in the sub-frame and increase the continuity of the left and right ends of the frame. Then, the spectral leakage of the plus window was reduced, and the time domain signal was converted into frequency domain energy to better observe the signal characteristics of the fast Fourier transform. The speech signal envelope and spectral details were extracted to obtain the sound properties of the cepstrum analysis using five processing methods. Finally, the speech signal model was generated to be used in the Heyuan Hua Chao opera singing-evaluation system (Figure 1).

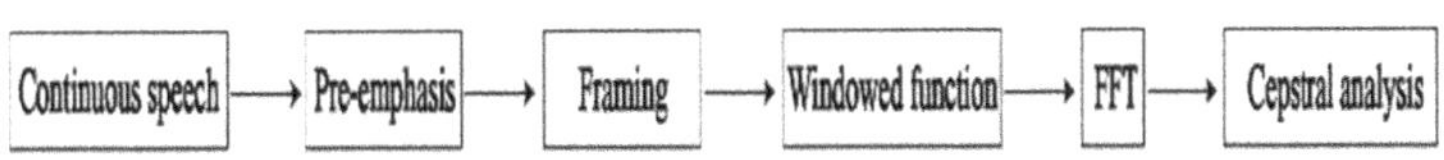

Figure 1. Specific implementation steps of the speech signal model.

The sound interval of Hua Chao opera, such as more step-in, less jump-in, regular rhythm, and high-frequency signal transmission, can easily weaken [2]. Based on the audio sample collection of vocal skills, volume, environment, and radio effect, the signal frequency varies, and a large number of speech signals are piled up in the low-frequency and high-frequency parts. The low-frequency part of the value gap is the large spectrum tilt (Spectral Tilt). The speech signal was taken through a high-pass filter method, a differential first-order signal value to reduce the high-frequency part (high-frequency change position) differential value and increase the low-frequency part (low-frequency change position) differential value so that the speech spectrum maintained a stable state (Figure 2, Equation (1)).

$$y(t) = x(t) - \alpha x(t-1),\ 0.95 < \alpha < 0.99 \tag{1}$$

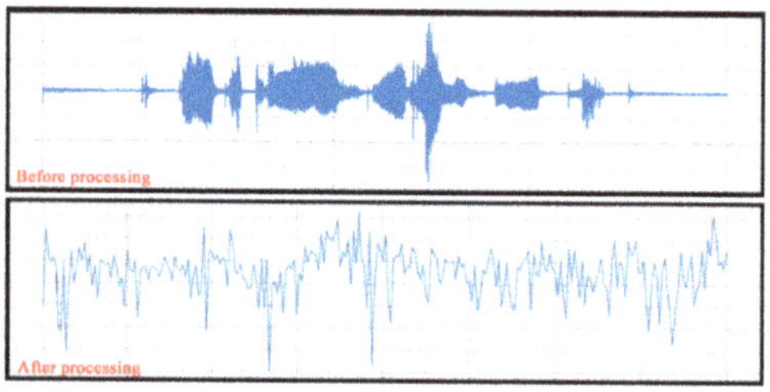

Figure 2. Spectrogram of changes before and after the pre-emphasis method.

The voice in Hua Chao opera is simple and healthy, the lyrics are catchy, and the voice signal changes slowly. The frequency in the voice signal after pre-emphasis becomes unstable with the change in time, so it needs to be processed in frames. However, the length of frames must not be too long or short. An excessive length reduces the time resolution, while a short length aggravates the cost of operation. As the speech signal sampling frequency was 16 kHz, the standard resolution was 25 ms, and the frame length of 16 kHz signal had $0.025 \times 16{,}000 = 400$ samples (the chart unit is N). As the frameshift in the speech signal is usually 10 ms, there were $0.01 \times 16{,}000 = 160$ samples in the frame [3]. The difference between frame length and frameshift in the overlapping area in the speech signal, the so-called "frame iteration", accounted for about 1/3 of each frame, that is, $400 - 160 = 240$ samples (the chart unit is M). The main role of the 240 samples was to maintain the frame number difference. The starting value of the speech frame with 400 sample points was 0, the starting value of the second speech frame was 160, and the starting value of the third speech frame was 320 (Figure 3). According to this rule, the frames were divided until the end of the speech signal.

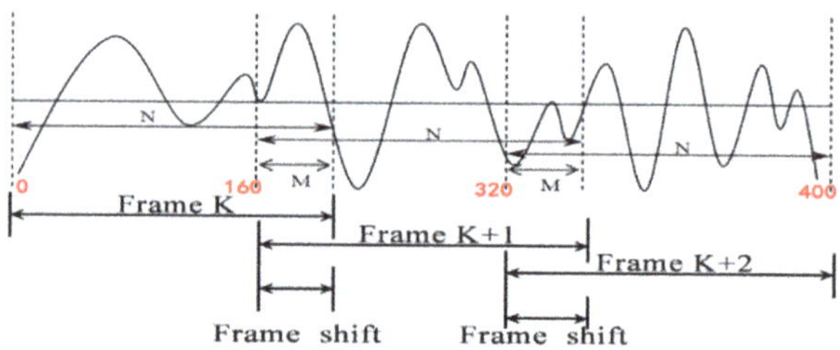

Figure 3. Schematic diagram of frame splitting.

The speech signal of Hua Chao opera is a time series, but the middle domain values are disconnected when carrying out the frame splitting process, which causes the spectral leakage phenomenon. Considering the long-time fluctuation of the speech signal and the disconnection of the endpoints, it is necessary to process its fixed characteristics. Each frame number is substituted into the Hamming Windows function (Hamming Windows), which truncates the original speech sample signal with amplitude–frequency to achieve the finite speech signal and then supplements the continuity of the left end and right end of the frame. By formulating the out-of-window value of the Hamming Windows function as 0 (based on Equation (2)), one takes a generic 0.46 and applies the function to each frame number, the sharp angle in the truncated finite speech signal is blunted, and the waveform amplitude slowly tapers to 0, thus reducing the truncation effect of the speech frames (Figure 4).

$$w(n,\alpha) = (1-\alpha) - \alpha cos\frac{2\pi n}{N-1}, (0 \le n \le N-1) \tag{2}$$

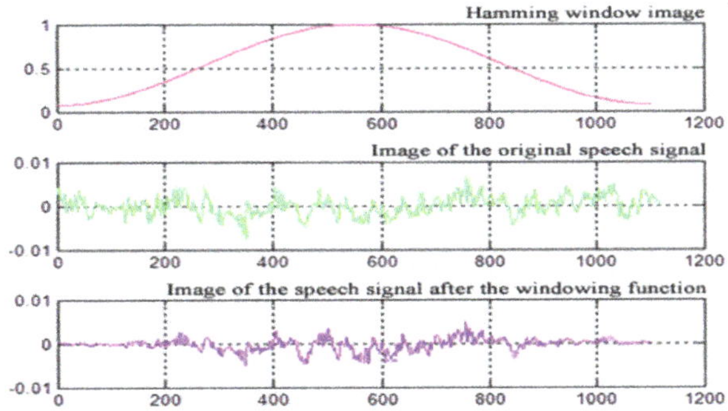

Figure 4. Schematic diagram of the Hamming Window function.

After the speech signal is substituted into the Hamming window function, the individual frames in the signal need to be fast-Fourier-transformed, because the melodic direction of the Hua Dynasty opera tune is rich and variable. It is more difficult to capture its dynamic information when the speech signal belongs to the one-dimensional signal. Thus, the time-domain information cannot be seen as frequency-domain information. Therefore, the speech signal must be processed and output into a 500 × 20 two-dimensional matrix model group as the MFCC matrix model generation code (Figure 5). After that, the speech signal is arranged according to frame, and each frame of speech corresponds to a spectrum using coordinates. In the horizontal coordinate, the duration and vertical coordinate become the decibels of the sound, indicating the relationship between frequency and energy. The amplitude of the speech signal is mapped to a gray-level embodiment, and the amplitude value is proportional to the corresponding gray area. The larger the amplitude value, the darker the corresponding area. The display duration of the speech spectrum is increased to obtain a change over time depicting the sound spectrum of the speech signal (Figures 6 and 7). By doing this, its energy distribution on the spectrum is obtained, and the influence of frequency points higher than the sampled signal is removed to make static and dynamic information parameters more intuitive for the improved robustness of speech signal data acquisition.

```
def wav2mfcc(path, max_pad_size=11):
    y, sr = librosa.load(path=path, sr=None, mono=False)
    if len(y) == 2:
        y = y[0, :]
    audio_mac = librosa.feature.mfcc(y=y, sr=16000)
    audio_mac = cv2.resize(audio_mac, (500,20))
    y_shape = audio_mac.shape[1]
    if y_shape < max_pad_size:
        pad_size = max_pad_size - y_shape
        audio_mac = np.pad(audio_mac, ((0, 0), (0, pad_size)), mode='constant')
    else:
        audio_mac = audio_mac[:, :max_pad_size]
    return audio_mac
```

Figure 5. MFCC matrix model generation code.

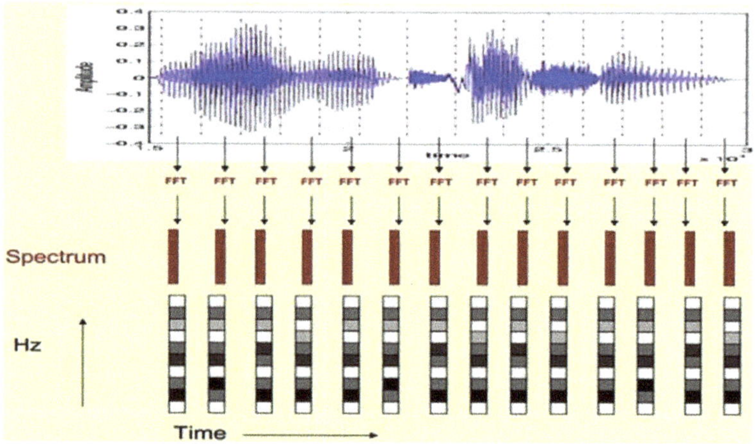

Figure 6. Fast-Fourier-transform process.

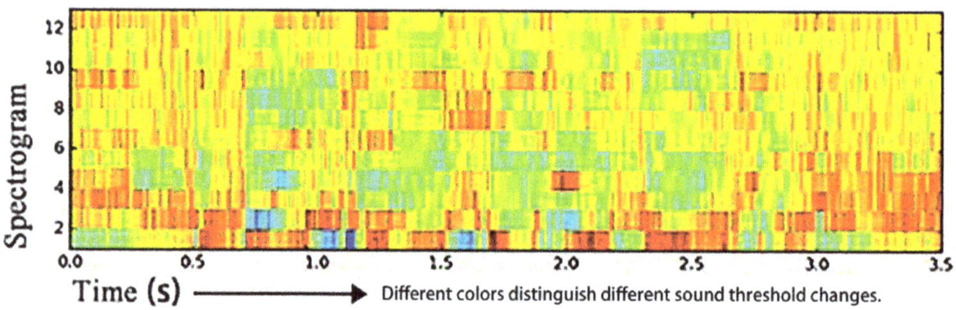

Figure 7. Sound spectrum after conversion.

Since Hua Chao opera is sung in the Zijin dialect and its transcription recognition is poor, it is important to perform cepstral analysis on speech signals to extract speech components. The peak resonance peaks on the speech spectrogram carry important sound-recognition properties, while the smooth curve connecting the resonance peaks is called the envelope [4]. To better identify the acoustic properties of the speech signal, the envelope needs to be separated from the peaks using inverse spectral analysis. Considering the peak as a detail of the spectrum, the horizontal axis is considered as the envelope E of the low-frequency component, which is considered as a sinusoidal signal with four cycles per second, giving a peak at 4 Hz on the coordinate axis. The vertical axis is the spectral detail H[k] of the high-frequency component, which is considered as a sinusoidal signal with 100 cycles. A peak is assigned to the position of 4 Hz on the pseudo-coordinate axis. The vertical axis is the spectral detail of the high-frequency component, H[k], viewed as a sinusoidal signal with 100 cycles per second, and a peak is assigned at 100 Hz on the coordinate axis (Figure 8).

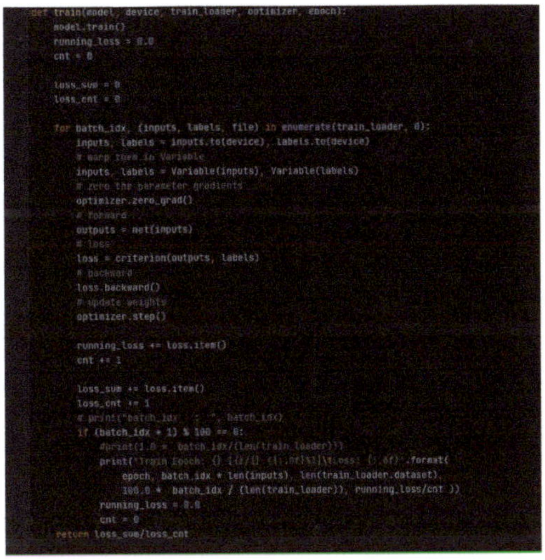

```
def train(model, device, train_loader, optimizer, epoch):
    model.train()
    running_loss = 0.0
    cnt = 0

    loss_sum = 0
    loss_cnt = 0

    for batch_idx, (inputs, labels, file) in enumerate(train_loader, 0):
        inputs, labels = inputs.to(device), labels.to(device)
        # warp them in Variable
        inputs, labels = Variable(inputs), Variable(labels)
        # zero the parameter gradients
        optimizer.zero_grad()
        # forward
        outputs = net(inputs)
        # loss
        loss = criterion(outputs, labels)
        # backward
        loss.backward()
        # update weights
        optimizer.step()

        running_loss += loss.item()
        cnt += 1

        loss_sum += loss.item()
        loss_cnt += 1
        # print("batch_idx : ", batch_idx)
        if (batch_idx + 1) % 100 == 0:
            #print(1.0 * batch_idx/(len(train_loader)))
            print('Train Epoch: {} [{}/{} ({:.0f}%)]\tLoss: {:.6f}'.format(
                epoch, batch_idx * len(inputs), len(train_loader.dataset),
                100.0 * batch_idx / (len(train_loader)), running_loss/cnt ))
            running_loss = 0.0
            cnt = 0
    return loss_sum/loss_cnt
```

Figure 8. MFCC training function process.

Firstly, the multiplicative signal (x[k]) is turned into a multiplicative signal via the convolution of the two superpositions; then, the multiplicative signal is transformed into an additive signal by taking the logarithm. Finally, an inverse transformation is performed, and the additive signal is restored to a convolutive signal, which is returned to the matrix model to generate a new speech signal model (Figure 9).

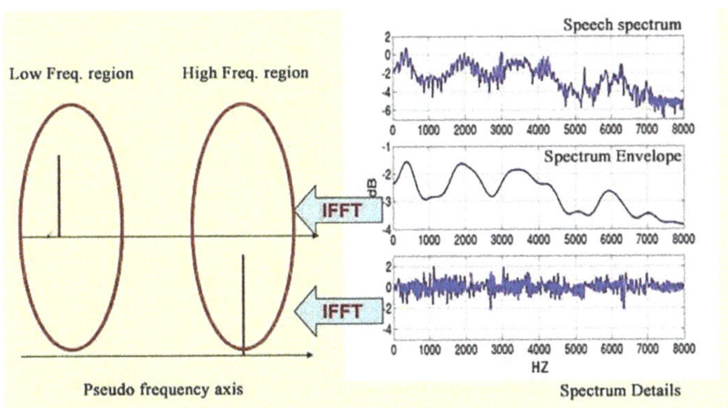

Figure 9. Inverse spectrum analysis process.

3. Accuracy testing of speech signal training models

The training model is built on python, and the training set of audio sample data of the three singing voices collected in the early stage is imported into the model. There are 458 test audio samples of chanting-style singing, 124 test audio samples of fast-singing-style singing, and 301 test audio samples of traditional-singing-style singing. Then, the processed training set audio signals are imported into the test model together with the relevant speech parameters, and the accuracy of a test result is generated in the model through constant comparison of the two parameters. If the deviation between the two data is large, the model reverses the transmission to adjust the parameters of the speech layer until the accuracy rate rises, and the accuracy of the tested speech signal model is 65.1% (Figure 10 shows the test function and the resulting accuracy results).

```
def test(model, device, test_loader):
    model.eval()
    test_loss = 0
    correct = 0
    with torch.no_grad():
        for inputs, labels, files in test_loader:
            inputs, labels = inputs.to(device), labels.to(device)
            output = model(inputs)
            test_loss += criterion(output, labels)  #
            pred = output.max(1, keepdim=True)[1]  #
            correct += pred.eq(labels.view_as(pred)).sum().item()

            # for i in range(len(pred)):
            #     print(output[i]," : ",pred[i]," : ", labels[i])
            # break

    test_loss /= len(test_loader.dataset)
    print("\nTest set: Average loss: {:.4f}, Accuracy: {}/{} ({:.4f}%) \n".format(
        test_loss, correct, len(test_loader.dataset),
        100. * correct / len(test_loader.dataset)
    ))
    temp = 100. * correct / len(test_loader.dataset)
    return temp, test_loss
```

Figure 10. Test function and accuracy results.

2.1.2. Convolutional Neural Network for Classification Processing

In the two-dimensional matrix processing of the speech signal in the training model and the frequency-domain information processing of the speech signal, the classification method of the convolutional neural network is integrated by using convolutional fixity to suppress the multiplicity of the speech signal [5]. The main principle is to input the corresponding speech signal parameters, read the parameters through a layer of convolution, process the convolved information using pooling, and then perform the same operation as in the previous step. The secondary processed information is passed into the two neural layers (fully connected), which form an unusual two-layer neural network layer. Finally, a classifier is used to connect the layers for classification.

The convolutional neural network is not limited by having to process the input information of each pixel. Based on its characteristic of having a batch filter that continuously rolls over the two-dimensional graph, the information is collected in the graph. The fusion of MFCC coefficients is used to extract feature parameters and implement classification processing for each small block of the pixel area in the two-dimensional matrix model. The fusion enhances the continuity of vocal information in the speech signal such that the neural network can acquire graphical information, not a single pixel point, and deepens the neural network's knowledge of graphical information, allowing the classification and collation to have an actual presentation.

2.2. Back-End Data Acquisition and Processing Design

2.2.1. Overall Construction of the Back-End Framework

4. Implementation of algorithm scoring function and overall algorithm deployment

The singing evaluation system is based on the testing algorithm of speech recognition to complete the communication connection between the scoring system and the back-end port. The standardized speech and the test speech signal are compared and identified based on melody and number of words. The difference in degree between the two is obtained by calculating the ratio of the two. The smaller the value of the different degree, the higher the matching degree. The test data shows that nearly 70% of the test speech signals are higher than 60 points, of which the average score is 67 and the highest score is 90.

The choice of back-end server is a Linux system, with the use of FTP terminal simulation software (v5.0.1221). With nohup background running services, the unit is used in the form of numbers instead of singing feedback comments. Different values represent different singing feedback comments. These parameters are requested for the audio path of the server to access the relevant score and numerical parameters, and finally, the prediction results are returned to the WeChat applet via HTTP to achieve data communication.

5. Overall back-end deployment

The core part of the back-end development is to open the communication with the algorithm related to singing assessment, the processing of the audio reception path, and the collection of different types of singing learning data and answering data from local users to link the system-affiliated functions. The specific implementation process includes three core modules: the data-collection module, the data-analysis module, and the data feedback module. The data-collection module collects users' audio data and historical behavior record information, including scoring parameters of singing voice learning, feedback parameters, question quantity, and accuracy rate for questions. The data-analysis module starts the MFCC algorithm deployed on the server to analyze the audio and answer data set output by the data-collection module. The data feedback module analyzes the information dataset traversed by the module, stores it in the MySQL database twice, and returns the response results to the application-side display on the WeChat applet through the path. At the same time, the API interface is called to obtain the user login name, school, and other identifying information to build the user experience record dataset.

2.2.2. Front-End and Back-End Data Communication Implementation

The whole front-end and back-end communication-implementation method is mainly divided into three steps. The front-end logic is to build the official API in the applet: wx. request () interface, direct calls js logic function. Then, the back-end server receives and processes the request passed by the front end. In the process of returning the response data, it runs the web framework program derived from Flask under python to handle network reception and sending and other behaviors. The background interface accesses the database and the database-related data, which are returned to the front-end through JSON format and the final front-end HTML for the display of feedback results.

3. Singing Voice Evaluation System Test

The carrier of the Heyuan Hua Chao opera singing-evaluation system is the WeChat applet, and the hardware environment to test the system is Android and iOS. The test included functional and performance tests of the applet. The functional test of each module meets the business requirements and the expected results. The system's performance test is carried out for the applet's response time, concurrent users, throughput rate, business success rate, error rate, thinking time, and system resources. The performance test report shows that the maximum concurrent access results in 1 s include an access-success rate of 100% for 80 users, and the system is running stably. CPU memory occupation is not more than 50%, and the number of errors reported in a 1 s response time is 0. This shows that the Heyuan Hua Chao opera singing-assessment system runs normally and stably.

4. Conclusions

At this stage, the Heyuan Hua Chao opera singing assessment system has been put into use normally, named "Zhaohua opera", from the harmonic sound "Dawn Blossoms Plucked at Dusk". The main interface comprises a practice, competition, list, and four functional modules (Figure 11). Through the interaction logic of the video-learning–user-recording–audio-playback–system-operation–feedback, it aims to convey the core educational functions of system singing-voice selection, singing-voice experience, and singing-voice feedback. It has the advantageous features of noise resistance, robustness, and accuracy, and users conduct independent learning and receive real-time feedback, which is practical and agile. Traditional opera is a content provider and a content disseminator. By using new media such as WeChat applets, we can expand the scope of the audience and the group base of Hua Chao opera for its survival. At the same time, support for future generations with a database of audio samples of Hua Chao opera is required to inspire more scholars to think about the path for the preservation and dissemination of traditional opera culture.

Figure 11. Core UI interface of "Zhaohua opera" (the language is set to Chinese).

Author Contributions: Conceptualization, Y.Z. and S.S.; methodology, Y.Z.; software, Y.W.; validation, Y.Z., Y.W. and S.S.; formal analysis, S.S.; investigation, Y.W.; resources, Y.Z.; data curation, Y.Z.; writing—original draft preparation, Y.Z.; writing—review and editing, S.S.; visualization, Y.Z.; supervision, S.S.; project administration, S.S.; funding acquisition, S.S. All authors have read and agreed to the published version of the manuscript.

Funding: This research received no external funding.

Institutional Review Board Statement: There are no ethical issues involved in this article, as we did not study any human or animal subjects, nor did we collect any personal information or sensitive data.

Informed Consent Statement: Informed consent was obtained from all participants prior to registration for publication.

Data Availability Statement: The datasets generated or analyzed during this study are available from the corresponding author upon reasonable request.

Conflicts of Interest: The authors declare no conflict of interest.

References

1. Chen, K.; Zhang, T. A preliminary study on the feature engineering of Mei Lanfang's singing voice using artificial intelligence technology. *Mei Lanfang J.* **2020**, 137.
2. Wu, G. *Selected Essays on Hakka Gupi Studies*; Propaganda Department of the CPC Heyuan Municipal Committee: Heyuan, China, 2010; Volume 191.
3. Li, S.; Wang, X.; Zhang, Y.; Li, H.; Xiang, H. Road rage emotion recognition based on improved MFCC fusion features and FA-PNN. *Comput. Eng. Appl.* **2021**, *9*, 306–313.
4. Wang, X.G.; Zhu, J.W.; Zhang, A.X. A vocal pattern identity identification method based on MFCC features. *Comput. Sci.* **2021**, *12*, 343–348.
5. Long, H.; Zhang, L.P.; Shao, Y.B.; Du, Q.Z. Speaker feature-constrained speech enhancement for multi-task convolutional networks. *Small Microcomput. Syst.* **2021**, *10*, 2178–2183.

engineering proceedings

Proceeding Paper

Cloud-Based Payment Systems in Australia: How Security Affects Consumer Satisfaction †

Domingos Mondego * and Ergun Gide

School of Engineering and Technology, Central Queensland University, Sydney, NSW 2000, Australia; e.gide1@cqu.edu.au
* Correspondence: d.yamagutimondego@cqumail.com

† Presented at the IEEE 5th Eurasia Conference on Biomedical Engineering, Healthcare and Sustainability, Tainan, Taiwan, 2–4 June 2023.

Abstract: Over the past years, online payments or cloud-based payments have significantly increased around the globe. Cloud-based payment systems (CBPS) are more involved in the payment process due to their convenience and flexibility. Although CBPS offers obvious benefits, their adoption rates among Australian users are comparatively lower compared to those in other countries. People are dissatisfied with current payment methods or are unaware of the advantages of CBPS. Using the technology acceptance model (TAM) with perceived ease of use and perceived usefulness, a qualitative research method was applied through semi-structured interviews to collect data from 20 consumers in Australia. The findings pointed out an appreciation for security features such as two-factor authentication and cutting-edge technologies. The banks were trusted by Australians, but a lack of education and additional fees on digital payment platforms were sources of concern. In the context of CBPS, it was observed that electronic devices were easy to use and proved to be useful. Service providers need to improve security measures and implement innovative technologies to enhance user privacy and prevent fraudulent activities. Overall, Australians expressed satisfaction with their banks; however, there are opportunities for enhancement, particularly in bolstering security measures and providing education on emerging payments options.

Keywords: CBPS; cloud-based payment systems; consumer satisfaction; security; price value; perceived ease of use; perceived usefulness; trust; TAM; technology acceptance model; Australia

Citation: Mondego, D.; Gide, E. Cloud-Based Payment Systems in Australia: How Security Affects Consumer Satisfaction. *Eng. Proc.* **2023**, *55*, 89. https://doi.org/10.3390/engproc2023055089

Academic Editors: Teen-Hang Meen, Kuei-Shu Hsu and Cheng-Fu Yang

Published: 5 January 2024

1. Introduction

The development of technology has significantly impacted traditional payment systems, leading to the advent of innovative methods such as contactless payments and online banking [1]. This has resulted in a reduction in dependence on cash transactions, moving towards a cashless system. Several studies [2–4] have shown a notable decline in cash transactions and an increased preference for electronic payment methods. Cash is only used for a small portion for daily transactions, mostly for payments of small amounts [5]. The decline in cash use can be attributed to the growth of e-commerce, which provides consumers with more electronic payment options [6,7]. Technologies such as cloud computing and mobile technology have transformed the modes of commerce transactions, making it easier for businesses and consumers to conduct transactions [8]. By leveraging these technologies, banks increase their profits and gain a competitive advantage [9].

The widespread use of the Internet and mobile technology has also contributed to the development of online banking and digital payment systems [10]. In this sense, cloud-based payment systems (CBPS) are gaining extensive popularity as they enable the acceptance and processing of payments through the Internet instead of physical devices [1]. CBPS offers customers ubiquitous and diverse payment options, allowing individuals to use any mobile device at any time and location [11,12]. However, despite the advantages of

CBPS, Australians use it at a lower adoption rate compared to other countries [13]. This indicates that users are either dissatisfied with existing payment systems or are unaware of the benefits of CBPS [12].

This study was carried out to examine the impact of security on user satisfaction with CBPS. We explored the technology acceptance model (TAM), integrating security as an external variable alongside the key factors of perceived ease of use and perceived usefulness. Additionally, trust and price value were added to the model as constructs given their recognition in prior research as significant factors influencing the acceptance of CBPS [1].

2. Literature Review

Satisfaction plays a pivotal role in assessing the overall experience of shopping or using a service or product [12]. It is a critical construct influencing long-term consumer responses [12]. User satisfaction encompasses the emotional responses, expectations, and attitudes that individuals have regarding the high-quality service delivered by providers through various interactions [1,14]. It essentially reflects the level of pleasure derived from perceiving exceptional service [15]. Given that user satisfaction significantly impacts the continued usage of information technology [16], banks and financial institutions are meticulously monitoring key areas such as security and privacy. As financial institutions offer comparable products and services, it becomes crucial to differentiate and consider customer preferences to enhance user satisfaction with the provided services [17].

2.1. Constructs

The primary focus of this study was placed on TAM, a widely employed framework for forecasting the acceptance of technology and its determinants [1,12]. TAM comprises four components: attitude towards use, actual system use, perceived ease of use, and perceived usefulness [1,18]. Nonetheless, in this investigation, only two primary constructs (perceived usefulness and perceived ease of use) were utilized, alongside other factors such as security, trust, and price value. Through this methodology, the study explored the influence of these additional variables on users' intention to embrace new payment systems. While prior studies delved into diverse aspects involving users, merchants, banks, and providers [1] affected by technological evolution and study duration, our focus predominantly centered on consumers.

2.1.1. Security

The concept of security is to protect personal information and ensure its confidentiality, integrity [19], authenticity, and non-replicability [20]. User perception of security is crucial in e-payment service providers' ability to safeguard sensitive data [21]. With the increase in e-commerce and, as a result, the growth of CBPS interactions [12], technology is being used to detect and prevent fraud [22], which has become more prevalent due to the lack of face-to-face interactions [23]. In this context, technology has become an essential tool in combating fraud and ensuring the security of e-commerce transactions [24,25].

2.1.2. Trust

Trust is crucial in payment systems, serving to authenticate monetary systems and ensuring confidence in the fulfillment of obligations by involved parties [1,7]. Trust plays a significant role in risky financial transactions, particularly when users are vulnerable to financial loss [26]. Research in Singapore [27] and China [28] highlighted the importance of trust in mobile payment adoption. The process of trust is dynamic, evolving from the initial establishment of trust to the ongoing development of trust over time [29]. Continuous trust is based on the performance of artificial intelligence and its ability to serve its objective [29].

2.1.3. Perceived Usefulness

The term perceived usefulness (PU) is characterized by an individual's belief that their performance at work will be improved through the utilization of a particular system [17]. This belief is an important construct in TAM, explaining the acceptance and use of technology by users [17]. Extensive research has delved into PU, establishing it as a pivotal element in TAM, with multiple studies showcasing its beneficial influence on the acceptance and adoption of technology by users [1]. Its relevance extends to online environments, illustrating how a particular technology can assist individuals in attaining their desired outcomes [30]. Furthermore, PU positively impacts acceptance and adoption by enabling individuals to discern the prospective advantages of employing a financial transaction platform [31].

2.1.4. Perceived Ease of Use

Perceived ease of use (PEOU) refers to an individual's perception of the minimal effort required when utilizing a specific system [18]. It holds significance in embracing emerging technologies [1], playing a crucial role in influencing consumers' acceptance of technology [32]. The assessment of PEOU revolves around evaluating how systems contribute to accomplishing tasks, enhance output, improve overall execution, and optimize effectiveness [33]. PEOU is influenced by both the concrete benefits of using technology and the act of utilizing it [6].

2.1.5. Price Value

Price value is the exchange value that users make between the perceived advantages of an application and its associated economic expenditures [26]. This exchange reflects the concession users make between the costs and benefits of using the application [34]. Besides, when development does not consider values, each demand is viewed as equally important, leading to a product with numerous important features that may not offer substantial benefits to the overall consumer base [35].

2.2. Proposed Conceptual Research Framework

We used TAM by incorporating three new constructs including security, price value, and trust to investigate the impact of security constructs on other variables (trust, PU, PEOU, and price value) and assess the influence on user satisfaction. We analyzed the security construct from an organizational perspective. Security constructs are widely utilized by organizations globally for anti-fraud measures, risk management, anti-money laundering initiatives, and identity verification [36]. Figure 1 illustrates the proposed conceptual research framework:

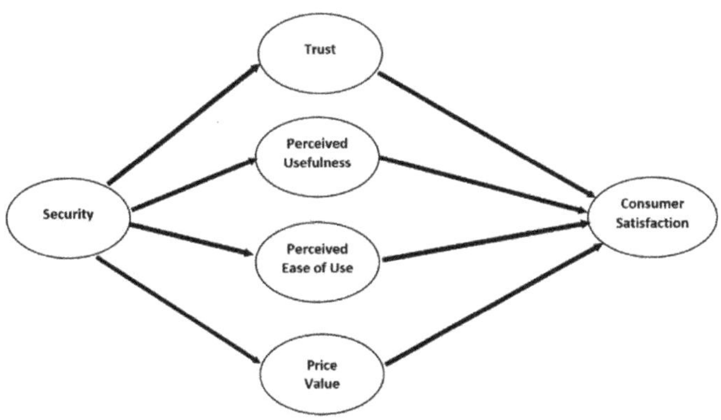

Figure 1. Proposed Conceptual Research Framework.

3. Research Method

Semi-structured interviews were employed as a qualitative approach to collect information from users of the CBPS in Australia. The data were collected from 20 Australian participants, 18 years old or older, who fully responded to the semi-structured interviews. We conducted interviews, recorded the content, and transcribed it. The transcripts were meticulously examined and categorized. Open coding was employed to analyze the combined constructs within the theoretical framework.

Following the collection of qualitative data through semi-structured interviews, content analysis was employed to scrutinize the acquired information. Content analysis is a systematic and impartial method that condenses words into a smaller number of categories based on their content to delineate and measure research issues [37]. The interpretation of the findings took into account the social context of every interview environment [6]. The primary aim of content analysis is to objectively and systematically classify and analyze information in order to derive valuable insights and draw conclusions [1,38]. Hence, the manual coding process was employed to analyze the qualitative data gathered from each interview. This involved considering the participant's exact words to comprehend the importance of each research question. This approach was chosen to prevent any potential oversight of data patterns that automated coding systems might encounter.

4. Data Analysis

Most consumers prefer using debit cards for in-store payments. When it comes to online payments, debit cards are still popular, followed by credit cards and PayPal. Customers also use their debit cards on mobile phones. While a considerable number of consumers opt for desktop computers when making online payments, there is a segment that favors the use of mobile phones. Computers are considered practical, safe, and time-saving by most consumers. However, several individuals felt nervous about making payments over the Internet due to potential issues it may present. C6 hesitates to employ electronic devices for payments, citing concerns about potential errors and a lack of fully grasping the procedure. Security concerns were common among consumers, including C17. Nevertheless, consumers such as C1 expressed a willingness to engage in online payments after gaining greater familiarity with the process.

4.1. Security

Participants were presented with statements and a security question. The first statement asked about the safety of electronic payment methods in Australia, with most participants expressing their willingness to use them if assured of security. Several believed security had to be the responsibility of both companies and the government, while others acknowledged the inherent risk in all payment methods. The second question inquired about security measures provided by Australian financial providers, with the majority expressing confidence in their banks' measures, such as two-construct authentication. Several participants had positive experiences with additional security layers, while others suggested that banks should consider improvements such as facial recognition. When asked to prioritize security and cost, participants prioritized security over cost, being willing to pay more for a secure payment method.

4.2. Trust

The following set of questions pertained to consumers' trust in CBPS (Appendix A). The majority of participants (18) trusted their banks, while two participants (C4 and C17) did not. C4 had issues with extra fees charged by banks, while C17 felt compelled to trust financial service providers due to government and organizational influence. Regarding trust in mobile devices for payments, the majority of participants expressed a favorable attitude towards their use, but reported infrequent utilization. Factors contributing to this included a lack of awareness (C6) and understanding of device features (C1), concerns about the loss of control (C3), and gadget malfunctions. However, several participants appreciated features such as such as two-factor authentication, facial recognition, and fingerprint recognition. The third question focused on the impact of family and friends on trust in emerging digital payment methods. The majority of participants (13) disagreed, citing it being the first opportunity in their family to experiment with new technologies (C12 and C13), convenience (C20), and a preference for making their own decisions (C20). On the other hand, seven participants concurred, expressing a hesitancy to be early adopters in trying out new services (C10) and preferring to trust methods endorsed by family or friends (C3).

4.3. PU

When questioned about the usefulness of electronic devices for paying bills (Appendix A), all respondents expressed agreement that computers contribute to the convenience of their daily routines. One respondent (C1) emphasized that digital gadgets empower them to complete various tasks from the comfort of their homes, rendering them particularly valuable. Furthermore, C20 asserted that digital gadgets prove particularly beneficial for individuals who are frequently on the move or have hectic schedules, as they streamline the process. C8 highlighted the discomfort associated with conventional banking strategies, pointing out that the time and effort involved in physically visiting financial institutions and waiting in line can be extensive. Finally, C7 emphasized the expediency of digital devices as a quicker and more efficient alternative, resulting in substantial time savings.

4.4. PEOU

When asked about the ease of using electronic devices (Appendix A), all participants responded that they were user-friendly. Three participants (C13, C15, and C16) acknowledged that, despite not being tech-friendly, managing payments using digital gadgets was achievable. C14 noted that individuals might make mistakes when first learning to use electronic devices, but the majority of financial institutions assisted with any issues. One participant, C10, mentioned that electronic devices were usually designed to be user-friendly and had self-help or click-on buttons to assist people, while C15 emphasized that the only obstacle involved adjusting to the digital tools and being receptive to its usage. Finally, C18 proposed that users facing challenges with technology could encounter difficulty in operating electronic devices, whereas those with greater technological familiarity likely found it effortless.

4.5. Price Value

When asked about the possibility of customers being open to paying higher fees for financial services incorporating cutting-edge technologies such as cloud computing and artificial intelligence to protect their privacy and detect fraud, just over half of the participants (C8, C9, C10, C11, C12, C13, C14, C15, C16, C17, and C18) disagreed with the statement. Several cited reasons such as believing it was just an excuse to charge more money, that these companies already make a lot of money, and that protecting customer information is just part of doing business. On the other hand, 9 participants stated that they would pay more for better data privacy protection, as long as the cost is not extreme and only a small percentage of their transactions. One participant was unsure about the cost, and another noted that the services were currently free.

4.6. User Satisfaction

The final section of the study included a set of five questions (Appendix A) regarding user satisfaction. In the first question (security), all participants expressed contentment with their bank's security protocols, as they had not experienced any difficulties. Several reasons were provided, such as the highly regulated nature of the financial services sector in the country (C15), the use of secure browsers by banks (C2), and the financial incentive for banks to maintain strong security measures to prevent financial losses (C4). However, participant C18 indicated that there is still potential for enhancement. They specifically mentioned implementing two-construct authentication and biometric measures such as facial recognition and fingerprint identification to enhance customer data security further.

For the second question (trust), most participants (16) expressed agreement with the satisfactory performance of their banks. However, C2, C3, C4, C5, and C8 expressed uncertainty about this claim. One participant (C1) emphasized the importance of financial service providers making every effort to protect customers. C3 stated that, although they are not completely certain, they must trust that their financial institutions deploy the best technology to secure their accounts. On the other hand, C2 argued that there is no definite way to know. C11 disagreed and mentioned that determining if banks are using the latest technology is challenging but they still use the services regardless.

In the third question (usefulness), all participants agreed that CBPS is convenient and useful. Responses ranged from 'really useful' to 'extremely useful' and 'convenient'. One participant mentioned that they would not want to go back to the previous method of payment as it was much more convenient now. Several participants pointed out that they do not need to physically go to the bank anymore, and everything can be conducted through their computer or mobile phone. While CBPS is considered safe and easy to use, one participant still felt hesitant and preferred handling things over the phone due to concerns about information security due to the lack of security and the potential for someone to misuse their information at any time.

In the fourth question (easy to use), the majority of participants (19) agreed with the question and gave reasons such as it being fairly easy to use (C1 and C14), relatively user-friendly (C5, C18, and C20), and relatively self-explanatory (C20), stating that with the apps, they could use services while they were outside (C2). The services can be accessed from almost anywhere, making it much easier. However, C3 responded that it was complicated, but admitted to not being tech-friendly and usually requiring assistance in understanding how to use it.

For the last question (price value), the majority of participants (16) believed that the amount of money they paid for the financial services offered by their banks was reasonable. C14 said that, while they would like to pay less, they believed that the services provided were fair considering the costs incurred by banks. C15 stated that they were unaware of the cost but assumed that it must be fair since they did not notice it. However, C3 and C6 disagreed with the statement, with one participant saying that banks make a lot of money while providing the same services at a higher cost. C5 and C15 did not know how much they paid for the services, while others (C8, C11, and C12) believed that they paid little for them. C8 stated that it was almost free. One participant (C11) believed that the cost was competitive but not necessarily fair.

5. Findings and Discussion

Most Australian consumers prefer using debit cards as their primary mode of payment, with credit cards and Paypal ranking next in preference. Although cash continues to be employed for transactions, desktop computers take precedence as the preferred device for online payments, followed by mobile phones. However, consumers avoid using electronic devices due to concerns about potential issues and financial costs. In terms of security, participants felt secure with their bank's security procedures, especially with two-factor authentication. Several participants did not mind paying more for added security. However, it was suggested that banks improve security by using biometrics such as face recognition or fingerprint identification for online purchases instead of relying on passwords.

Regarding trust, most participants had confidence in their banks and did not complain. However, they were concerned about the additional charges imposed by financial institutions and governmental pressures advocating for opening a bank account. The participants generally trusted mobile devices for payments, but their use was limited due to a lack of education. Better demonstrations of features could increase usage, but they were concerned about losing control and device failures. Critical security protocols, such as two-step verification, fingerprint identification, and facial recognition were required. Family and friends significantly influenced the adoption of innovative digital payment methods.

Concerning PU, all participants agreed that electronic devices proved advantageous for bill payments due to their convenience, enabling payments to be made from home or while traveling. Besides, electronic devices eliminated the need for daily visits to financial institutions and to endure long queues, thereby saving time and facilitating faster decision-making. For PEOU, participants unanimously agreed that the digital gadgets were simple to learn (meaning easy to use), even for individuals not highly proficient in technology, such as bank support customers. Electronic devices were user-friendly, with self-help options and intuitive buttons. However, the necessity of technology in payments and the ease of using these devices was found to be affected by proficiency in technology use. Those who were comfortable with technology found it simple to use, while those who struggled with it found it challenging.

In terms of price value, financial service providers utilized advanced technologies, including cloud computing and artificial intelligence, to detect fraud and enhance customer privacy without imposing additional charges. Safeguarding customer data was a crucial aspect of banking, and a significant number of customers were willing to contribute a small amount for enhanced privacy protection.

Regarding consumer satisfaction with financial service providers in Australia, participants generally expressed contentment with the security features implemented by their financial institutions. However, they recommended additional improvements, such as the incorporation of biometric measures and two-step verification. Most survey respondents expressed trust in their financial institutions' utilization of advanced technologies to safeguard their information and deter fraudulent activities. Although many respondents pointed out CBPS as convenient and beneficial, concerns emerged regarding data security and the complexity of the system. In general, the majority of participants felt that the fees associated with financial services were fair, although there were dissenting opinions among some individuals.

6. Managerial Implications

TAM was revised in this study to evaluate the satisfaction levels of consumers who used CBPS. The updated model included four new elements that measure how satisfied the consumers are with the system. The outcomes of this investigation provided useful information for businesses and organisations operating in the payment system, offering potential practical guidance and strategic direction. The findings of the study suggested that financial institutions improve electronic payment systems, offering biometric measures for security, providing better education and transparency to increase usage and trust, investing in user-friendly technology, advanced security measures, and exploring the impact of friends and family and information security concerns on consumer satisfaction. Additionally, the importance of improving electronic payment systems was emphasized to keep up with the increasing demand for cashless transactions. Furthermore, potential benefits were highlighted for both consumers and financial institutions, such as increased convenience, cost-effectiveness, and security. We also acknowledged the challenges associated with implementing and maintaining these systems, such as ensuring user privacy, addressing technology literacy gaps, and mitigating fraud risks.

7. Conclusions

In this study, we provided consumers' insights into payment preferences, security, trust, PU, PEOU, price value, and satisfaction with Australians banks. The results indicated a preference for debit cards as the primary mode of payment, with credit cards and PayPal coming afterward. Cash is still in use for minor transactions. Personal computers were the favored choice for online transactions, and participants emphasized the importance of security features like authentication and biometrics using advanced technologies. While participants expressed confidence in their banks, concerns were raised regarding additional fees and insufficient knowledge regarding the utilization of mobile payments. Nevertheless, electronic devices were regarded as easy to use, convenient, and useful. Financial service providers are urged to enhance security measures and adopt advanced technologies to prevent fraudulent activities and safeguard user privacy, all while minimizing additional costs for consumers. Overall, the majority of participants expressed satisfaction with their financial institutions, although there is potential for improvement, especially in bolstering security measures and offering guidance on emerging payment methods.

There were some constraints to consider in this study. In terms of geolocation, the participants were recruited only in one region (NSW) of Australia. The queries were formulated to capture precise information, leading to a restricted scope. The possibility of sampling bias arises due to the challenge of interviewing every representative across demographic categories such as gender, income, age, and education. Additionally, social desirability bias may be present, as some participants might be reluctant to disclose information, leading to socially desirable responses. Finally, these biases can constrain the accuracy of the collected data, influenced by the body language, beliefs, or assumptions of the participants.

To address the gaps in the current literature, further studies are required to explore the personal characteristics of consumers. Additionally, it is essential to conduct comparative studies spanning various regions and countries to explore both the similarities and differences related to geographical location and culture. Furthermore, a more comprehensive investigation into constructs such as trust, education, and influence from social networks that affect the adoption of electronic payment methods must be conducted. Additionally, a comparative study is needed to compare the efficacy of biometric measures such as facial recognition or fingerprint identification for online purchases with traditional security measures such as PIN codes and passwords. Moreover, it is necessary to explore the effects of service quality improvements on customer satisfaction and willingness to pay. Finally, further investigation is required to understand how technology affects pricing in financial services to identify the elements that influence the expenses linked with the adoption of new technological advancements.

Author Contributions: Conceptualisation, methodology, software, validation formal analysis, investigation, D.M.; resources, data curation D.M. and E.G.; writing—original draft preparation, D.M.; writing—review and editing D.M. and E.G.; visualization, D.M.; supervision, E.G.; project administration, D.M. and E.G. All authors have read and agreed to the published version of the manuscript.

Funding: This research received no external funding.

Institutional Review Board Statement: Ethical review and approval were not required for this study as no participant information was collected.

Informed Consent Statement: Verbal informed consent has been obtained from the participants to publish this paper.

Data Availability Statement: The datasets used and/or analyzed during the current study are available from the corresponding author.

Conflicts of Interest: The authors declare no conflicts of interest.

Appendix A

Questions for the semi-structured interviews—Consumers

(1) What's your preferred method to make payments (cash, cheque, debit cards etc.)?
(2) What do you think about making payments using electronic devices (desktop computer, mobile phones etc.)?

Constructs

- **Security**

(1) I would pay my bills from consumers' electronic devices (desktops, mobile phones, etc.) if I feel that it is safe (meaning secure) to do so. What do you think about this statement? Why? [1,39]
(2) Do you think that the Australian financial service providers offer good security measures (e.g., PIN, passwords, etc.) to protect your privacy and data confidentiality? Why? [7]
(3) Security rather than cost is more important to me when it comes to adopting a new payment method. What do you think about this statement? Why? [40]

- **Trust**

(1) Do you believe that the service provided by my financial service provider is trustworthy? Why? [1,6]
(2) Do you trust in mobile devices to make payments? Why? [1,7]
(3) If none of my friends or relatives had used any of the cloud-based services (Internet banking, mobile apps, etc.) provided by Australian financial service providers, I would not trust in it. What do you think about this statement? Why? [1,41]

- **Perceived usefulness**

(1) I think that paying bills using electronic devices (e.g., desktops, mobile phones, etc.) is really useful. What do you think about this statement? Why? [1,42]

- **Perceived ease of use**

(1) Do you think that the use of any of the cloud-based payment methods (electronic payments, mobile payments, mobile banking, etc.) is ease of use and simple to learn? Why? [1,6]

- **Price value**

(1) I would pay more for services provided by my financial service provider if I knew that it is using the latest technology (e.g., artificial intelligence, cloud computing, etc.) to protect my privacy and detect fraud activities. What do you think about this statement? Why? [1,43]

- **User satisfaction**

(1) I am satisfied with the security measures that financial service providers offer in Australia. What do you think about this statement? Why? [1,15]
(2) I trust that financial service providers have been using the best technology available to detect and protect my privacy, as well as avoid fraud activities. What do you think about this statement? Why? [1,44]
(3) Do you think that payments using the CBPS, offered by my financial service provider, is really useful and convenient? Why? [1,15]
(4) Do you think that payment using the CBPS, offered by my financial service provider, is user-friendly (meaning: ease of use)? Why? [1,15]
(5) I believe that the price I have been paying for the services provided by my financial service provider is fair. What do you think about this statement? Why? [1,15]

References

1. Mondego, D.; Gide, E. Examining the influence of service quality on the factors that affect consumer satisfaction in cloud-based payment systems in Australia. In Proceedings of the 2023 11th International Conference on Information and Communication Technology (ICoICT), Melaka, Malaysia, 23–24 August 2023; pp. 80–85. [CrossRef]
2. Mulqueeney, J.; Livermore, T. Cash Use and Attitudes in Australia. Available online: https://www.rba.gov.au/publications/bulletin/2023/jun/pdf/cash-use-and-attitudes-in-australia.pdf (accessed on 16 June 2023).
3. Guttmann, R.; Livermore, T.; Zhang, Z. The Cash-Use Cycle in Australia. Available online: https://www.rba.gov.au/publications/bulletin/2023/mar/pdf/the-cash-use-cycle-in-australia.pdf (accessed on 25 June 2023).
4. Bank for International Settlements (BIS). Digital Payments Make Gains but Cash Remains. Available online: https://www.bis.org/statistics/payment_stats/commentary2301.pdf (accessed on 16 February 2023).

5. Lowe, P. A Payments System for the Digital Economy, RBA—Address to the 2019 Australian Payments Network Summit, 1–12. Available online: https://www.rba.gov.au/speeches/2019/sp-gov-2019-12-10.html (accessed on 10 December 2019).
6. Hampshire, C. Exploring UK Consumer Perceptions of Mobile Payments Using Smart Phones and Contactless Consumer Devices through an Extended Technology Adoption Model. Ph.D. Thesis, University of Chester, Chester, UK, 2016.
7. Yamaguti Mondego, D. A Framework to Build Trust in Mobile Payment Systems for Australian Consumers. Master's Thesis, Central Queensland University, Sydney, Australia, 2019. [CrossRef]
8. Mangruwa, R.D.; Nuraeni, R.; Ferdinand, N. Mobile Commerce Model with Cloud Technology to Support Smes Sales Performance. In Proceedings of the International Conference on Industrial Engineering and Operations Management (IEOM), Istanbul, Turkey, 7–10 March 2022. Available online: https://ieomsociety.org/proceedings/2022istanbul/703.pdf (accessed on 25 October 2022).
9. Elhag, H.M. Enhancing Online Banking Transaction Authentication by Using Tamper Proof and Cloud Computing. Ph.D. Thesis, University of Surrey, Surrey, UK, 2015.
10. Alkhowaiter, W.A. Digital payment and banking adoption research in Gulf countries: A systematic literature review. *Int. J. Inf. Manag.* **2020**, *53*, 102102. [CrossRef]
11. Donoghue, M. Six Reasons You Need to Consider a Cloud-Based Payment System. Available online: https://ipsi.com.au/six-reasons-need-consider-cloud-based-payment-system/# (accessed on 1 February 2018).
12. Mondego, D.; Gide, E. The impact of demographics on user satisfaction in cloud-based payment systems in Australia. In Proceedings of the 2021 IEEE Asia-Pacific Conference on Computer Science and Data Engineering (CSDE), Brisbane, Australia, 8–10 December 2021; pp. 1–6. [CrossRef]
13. Fisher, B. Contactless Competition and Digital Limitations Stifle Proximity Mobile Payment Growth in the UK. Available online: https://www.insiderintelligence.com/content/contactless-competition-and-digital-limitations-stifle-proximity-mobile-payment-growth-in-the-uk (accessed on 30 January 2020).
14. Sharma, G.; Lijuan, W. The effects of online service quality of e-commerce websites on user satisfaction. *Electron. Libr.* **2015**, *33*, 468–485. [CrossRef]
15. Patel, R.K. Examining Predictors of Satisfaction with Mobile Payment Systems among Small Business Users. Ph.D. Thesis, Northcentral University, Scottsdale, AZ, USA, 2016.
16. Albashrawi, M.A. Understanding Mobile Banking Usage Behavior: A Multi-Model Perspective. Ph.D. Thesis, University of Massachusetts, Lowell, MA, USA, 2017.
17. Liébana-Cabanillas, F.; Munoz-Leiva, F.; Rejón-Guardia, F. The determinants of satisfaction with e-banking. *Ind. Manag. Data Syst.* **2013**, *113*, 750–767. [CrossRef]
18. Davis, F.D. PU, PEOU, and user acceptance of information technology. *MIS Q.* **1989**, *13*, 319–340. [CrossRef]
19. Teoh, W.M.Y.; Chong, S.C.; Lin, B.; Chua, J.W. Constructs affecting consumers' perception of electronic payment: An empirical analysis. *Internet Res.* **2013**, *3*, 465–485. [CrossRef]
20. Zhang, W.K.; Kang, M.J. Constructs affecting the use of facial-recognition payment: An example of Chinese consumers. *IEEE Access* **2019**, *7*, 154360–154374. [CrossRef]
21. Yaokumah, W.; Kumah, P.; Okai, E.S.A. Demographic influences on e-payment services. *Int. J. E-Bus. Res.* **2017**, *13*, 44–65. [CrossRef]
22. Chen, Z.; Khoa, L.D.; Teoh, E.N.; Nazir, A.; Karuppiah, E.K.; Lam, K.S. Machine learning techniques for anti-money laundering (AML) solutions in suspicious transaction detection: A review. *Knowl. Inf. Syst.* **2018**, *57*, 245–285. [CrossRef]
23. Hossain, A.; Udin, M.N. A differentiate analysis for credit card fraud detection. In Proceedings of the 2nd International Conference on Innovations in Science, Engineering and Technology (ICISET), Chittagong, Bangladesh, 27–28 October 2018. Available online: https://ieeexplore.ieee.org/abstract/document/8745592 (accessed on 16 September 2019).
24. Capgemini. World Payments Report 2019, Capgemini Research Institute. Available online: https://www.capgemini.com/news/world-payments-report-2019/ (accessed on 17 September 2019).
25. Zirkle, A. The Critical Role of Artificial Intelligence in Payments Tech. Available online: https://www.fintechnews.org/the-cririal-role-of-artificial-inteliigence-in-payments-tech/ (accessed on 27 May 2019).
26. Slade, E.; Williams, M.; Dwivedi, Y.; Piercy, N. Exploring consumer adoption of proximity mobile payments. *J. Strateg. Mark.* **2015**, *23*, 209–223. [CrossRef]
27. Chandra, S.; Srivastava, S.C.; Theng, Y.L. Evaluating the role of trust in consumer adoption of mobile payment systems: An empirical analysis. *Commun. Assoc. Inf. Syst.* **2010**, *27*, 561–588. [CrossRef]
28. Yan, H.; Yang, Z. Examining mobile payment user adoption from the perspective of trust. *Int. J. u-e-Serv. Sci. Technol.* **2015**, *8*, 117–130. [CrossRef]
29. Siau, K.; Wang, W. Building trust in artificial intelligence, machine learning, and robotics. *Cut. Bus. Technol. J.* **2018**, *31*, 47–53.
30. Liébana-Cabanillas, F.; Sánchez-Fernández, J.; Muñoz-Leiva, F. Antecedents of the adoption of the new mobile payment systems: The moderating effect of age. *Comput. Hum. Behav.* **2014**, *35*, 464–478. [CrossRef]
31. Malaquias, R.F.; Silva, A.F. Understanding the use of mobile banking in rural areas of Brazil. *Technol. Soc.* **2020**, *62*, 101260. [CrossRef]
32. Liébana-Cabanillas, F.; Sánchez-Fernández, J.; Muñoz-Leiva, F. Role of gender on acceptance of mobile payment. *Ind. Manag. Data Syst.* **2014**, *114*, 220–240. [CrossRef]

33. Munoz-Leiva, F.; Climent-Climent, S.; Liébana-Cabanillas, F. Determinants of intention to use the mobile banking apps: An extension of the classic TAM model. *Span. J. Mark.—ESIC* **2017**, *21*, 25–38. [CrossRef]
34. Killian, D.; Kabanda, S. Mobile payments In South Africa: Middle income earners' perspective. In Proceedings of the Twenty First Pacific Asia Conference on Information Systems (PACIS), Langkawi Island, 16–20 July 2017. Available online: https://aisel.aisnet.org/pacis2017/53/ (accessed on 17 September 2019).
35. Jenkins, P.; Ophoff, J. Constructs influencing the intention to adopt NFC mobile payments—A South African perspective. In Proceedings of the International Conference on Information Resources Management (CONF-IRM), Cape Town, South Africa, 18–20 May 2016; pp. 1–12. Available online: https://aisel.aisnet.org/cgi/viewcontent.cgi?referer=&httpsredir=1&article=1031&context=confirm2016 (accessed on 17 September 2019).
36. Digalaki, E. The Impact of Artificial Intelligence in the Banking Sector and How AI Is Being Used in 2020. Available online: https://www.businessinsider.com/ai-in-banking-report?r=AUandIR=T (accessed on 3 February 2022).
37. Duan, X. An Integrated Solution to the Adoption of Electronic Market in Australian Small-and-Medium Sized Enterprises. Ph.D. Thesis, RMIT University, Melbourne, Australia, 2012.
38. Bengtsson, M. How to plan and perform a qualitative study using content analysis. *Nurs. Plus Open* **2016**, *2*, 8–14. [CrossRef]
39. Tounekti, O.; Ruiz-Martínez, A.; Gomez, A.F.S. Users supporting multiple (mobile) electronic payment systems in online purchases: An empirical study of their payment transaction preferences. *IEEE Access* **2019**, *8*, 735–766. [CrossRef]
40. Hebie, A.P. Improving Mobile Phone Banking Usefulness, Usability, Risk, Cost, and Intention to Adopt. Ph.D. Thesis, Walden University, Minneapolis, MN, USA, 2017.
41. Mehrad, D.; Mohammadi, S. Word of Mouth impact on the adoption of mobile banking in Iran. *Telemat. Inform.* **2016**, *34*, 1351–1363. [CrossRef]
42. Lee, V.H.; Hewa, J.J.; Leong, L.Y.; Tan, G.W.H.; Ooi, K.B. Wearable payment: A deep learning-based dual-stage SEM-ANN analysis. *Expert Syst. Appl.* **2020**, *157*, 113477. [CrossRef]
43. Apanasevic, T.; Markendahl, J.; Arvidsson, N. Stakeholders' expectations of mobile payment in retail: Lessons from Sweden. *Int. J. Bank Mark.* **2016**, *34*, 37–61. [CrossRef]
44. Mbama, C.I. Digital Banking Services, Customer Experience and Financial Performance in UK Banks. Ph.D. Thesis, Sheffield Hallam University, Sheffield, UK, 2018.

engineering proceedings

MDPI

Proceeding Paper

Computer-Aided Simulation Analysis on the Impact of Various Opening Patterns in High-Rise Opening Building towards Pollutants Dispersion †

Chih-Hsien Huang * and Ying-Ming Su

Department of Architecture and Urban Design, National Taipei University of Technology, No. 1, Sec. 3, Zhongxiao E. Rd., Da'an Dist., Taipei City 106344, Taiwan, China; ymsu@ntut.edu.tw

* Correspondence: jemmy880318@gmail.com

† Presented at the IEEE 5th Eurasia Conference on Biomedical Engineering, Healthcare and Sustainability, Tainan, Taiwan, 2–4 June 2023.

Abstract: Taiwan is the fourth most urbanized country in Asia, where the urban spatial structure of high-rise and density hinders urban ventilation. Studies have proven that opening buildings reduce the area of windward surfaces, which can effectively mitigate the urban heat island effect and disperse pollutant accumulation. Until now, most researchers have discussed the differences in heights and sizes of openings in the opening buildings, but few discussed the influence of opening patterns on urban ventilation. Thus, we set the building unit to 30 × 30 m with 160-m height with the opening height as tall as 0.45 times the building height and a 9% opening rate, distributed in 6 × 6 ideal city configuration Four cases (case A: no opening, case B: middle square, case C: right square, and case D: middle rectangular) with different arrays of opening buildings were compared with ANSYS Fluent v18 to simulate the wind environment and NO_2 pollutants. The results showed that the opening building improved the permeability of street ventilation and air circulation, which greatly increased the wind speed at a height of 72 m. The distribution of pollutants was affected by the distance from the pollution source and the width of the street. Pollutants were gradually dispersed as the height increased. Case D of a long-narrow rectangular opening (adjacent to the pedestrian floor) and the venture effects formed eddy currents above and below the opening, which effectively improved the ventilation in the street canyon. Therefore, it had the best wind speed on the pedestrian level among the cases. The wind speed of the 72 m-high floor was much higher than that of case A, and the vortex generated by the airflow flowing through the opening in the street canyon increased the diffusion effect of pollutants. Overall, the opening building with a rectangular opening was the optimum solution in terms of wind speed improvement and pollutant removal. In addition to the opening design in the building facade, it is recommended to provide sufficient open space to improve air circulation in the building block and disperse pollutants.

Keywords: high-rise opening building; urban street canyon; traffic pollutant (NO_2); CFD analysis

Citation: Huang, C.-H.; Su, Y.-M. Computer-Aided Simulation Analysis on the Impact of Various Opening Patterns in High-Rise Opening Building towards Pollutants Dispersion. *Eng. Proc.* **2023**, *55*, 90. https://doi.org/10.3390/engproc2023055090

Academic Editors: Teen-Hang Meen, Kuei-Shu Hsu and Cheng-Fu Yang

Published: 8 January 2024

1. Introduction

Major cities are gradually developing into compact cities to respond to urbanization, increase population carrying capacity, and maximize land use. High-density development in compact cities challenges urban growth, planning, and design. For example, the wall effect formed by high-rise buildings easily hinders urban ventilation and makes it difficult to discharge pollutants. Long-term exposure to pollutants can be harmful to health. Nitrogen dioxide can easily cause diseases such as macular degeneration, diabetes, asthma, and fetal heart disease, and its concentration is positively correlated with the incidence of COVID-19 [1]. Therefore, we explore the impact of NO_2 on the urban environment.

Ventilation problems and pollutant accumulation are the inevitable results of high-density living and urbanization in developing countries [2]. High-rise buildings block wind flow, and areas with no wind and weak wind are prone to heat and pollutant accumulation. Therefore, the wind corridor design has been implemented in many countries recently. However, it is not easy to re-plan the urban air corridors in densely built cities, so the opening buildings conducive to urban ventilation have gradually received attention [3]. In the past, the research on opening buildings mainly focused on the structure of building mass and the reduction of wind resistance [4–6]. The related research still focuses on the simulation of the opening building's opening rate, height, and street aspect ratio [3,7,8]. There is a lack of relevant research on the impact of different opening types of opening buildings on urban microclimates.

In this study, ANSYS Fluent v18 was used to simulate the wind environment and pollutants and to discuss the influence of four openings on the diffusion of NO_2 pollutants in street canyons. We referred to the ideal city 6 × 6 configuration of Li et al. [9], and the building units were set to be 30 × 30 × 160 m. The research is carried out as follows:

(1) Comparing the wind speed distribution at the pedestrian floor (1.5 m) and the upper floor (72 m) of the four buildings with different openings

(2) The pollutant diffusion distribution of four different opening types of high-rise buildings at different heights (1.5, 30, and 61.5 m) was compared.

2. Literature Review

2.1. Relationship between Cities of Different Densities and Urban Wind Environment

Due to the limited land resources in urban areas, the buildings gradually become high-rise and high-density. High-rise buildings hinder urban ventilation, quickly generate strong winds, or affect air circulation. Strong winds have an impact on pedestrian wind fields and safety [10], and areas with no wind and weak wind are prone to heat accumulation and pollutant accumulation. This affects the pedestrian wind field. For compact urban areas, urban design planning can effectively enhance urban wind energy potentials changing urban wind conditions by adjusting urban density, street width, building geometry, and layout can affect [11]. The joint research report of the Institute of Architecture of the Ministry of the Interior [12] pointed out that the air corridor can effectively use the wind's ability to regulate heat to cool down the urban high temperature, improve the quality of the living environment, and reduce the accumulation of harmful pollutants. Owing to the importance of wind corridor construction, many countries have successfully implemented urban ventilation corridors such as Vauban, an ecological community in Freiburg, Germany, the draft axis planned for Tokyo Bay, and the construction of five urban ventilation corridors in Munich, Germany. According to the research, the wall effect formed as a result of high-rise buildings affects the urban wind environment. At the same time, trees and objects cause secondary interference [13]. Tsai [14] selected street profile samples in Kaohsiung's city center and pointed out that the composition of the urban wind field is closely related to the street profile form.

The increase and decrease in wind speed have different effects. However, the building height has the most significant impact on improving the road ventilation effect which creates the urban heat island effect and the impact on ventilation efficiency. Research by the Institute of Architecture of the Ministry of the Interior [15] pointed out that high-rise buildings create obstacles to the wind field, which changes the state and speed of the airflow around the building, forms a massive obstacle to the airflow, and results in an undercut, shrinkage, channelization, vortex, and angle. Flow, wake, shading, and cross-window effects of the strong wind generated by this effect affect the safety and comfort of pedestrians. Lin [16] studied the effects of different building heights, building densities, and ambient wind directions on the urban canopy (UCL) ventilation and used the removal flow rate (PFR) and air exchange rate (ACH) to evaluate the building. The height change increases the airflow around high-rise buildings but reduces the ventilation effect near low-rise buildings. Mei et al. [17] proposed that lower building density can improve ventilation efficiency. Better ventilation performance can be obtained in compact urban development by reducing frontal area density or building number. Luis [18] analyzed the influence of different urban types on the thermal effect. He pointed out that when the building group coverage rate is 30−40%, the city's environmental impact and external trade-offs can be well balanced.

2.2. Impact of Air Pollutants in Urban Areas on Wind Environment

Various types of waste discharge make the amount, concentration, and duration of smoke and harmful gases in the atmosphere reach a certain level and cause a significant burden on the environment, known as air pollution [19,20]. The main factors affecting the diffusion of pollutant $PM_{2.5}$ in cities are microclimate, including temperature, wind speed, wind direction, and air pressure [21]. The frequent industrial and commercial activities and the development of high-rise and high-density buildings caused by the urban high population density change the city's original environmental wind field. Studies have pointed out that improving air quality is not only to suppress the emission of pollution sources but also to effectively improve the urban ventilation environment through urban planning and architectural design—influence [3,7,16]. The ventilation effect also becomes different under the urban development patterns of different densities (street pattern, building height, and building density). If the street pattern becomes more mixed-used and the density increases, the polluted area becomes enlarged, and the permeability reduces. The higher street network space is conducive to the diffusion of pollutants due to its better ventilation effect. When the street's height-to-width ratio (H/W) increases, the street's closedness is enhanced with the relative increase of the street-side buildings' height and seriously hinders the diffusion of pollutants. In the case of poor ventilation efficiency, it is easier to cause pollutants to accumulate, and the increase in building density reduces air circulation and makes it difficult for pollutants to diffuse [21].

2.3. High-Rise Opening Building

In recent years, in response to the dense population generated by rapid urbanization in Taiwan, environmental factors such as the landscape, building mass, configuration, and street layout in the city have gradually developed towards high-rise and high density, which hurts the original wind field of the city. Obstacles disturb the airflow and generate strong winds or no wind areas between the streets, which aggravate environmental problems such as the heat island effect and pollutant accumulation and seriously affect the pedestrian wind field and human comfort. They also influence the urban wind field and excavate its potential wind energy in the urban street profile to prevent achieving the goal of sustainable development.

For the study of the relationship between opening buildings and urban pollutant emissions, Hang et al. [22] confirmed that in high-rise compact urban areas, changing the proportion of buildings and open spaces and the height of building volumes increases the air permeability and affects the wind speed of the urban canopy. Research on high-rise opening buildings is divided into two parts.

One of them is the research and analysis of the building structure. For example, Li et al. [21] found that after openings were installed in high-rise buildings, the distribution law of floor wind became different and reduced the wind load of the building mass. Zhang et al. [4] pointed out that setting openings reduced the overall average wind load on buildings. However, it does not mean that the larger the opening, the more significant the reduction in wind load. Opening in the upper part of the building is beneficial to reduce the bending moment of the foundation while opening in the middle and upper part is more effective in reducing the average wind load. Xia et al. [5] studied the influence of wind characteristics on high-rise buildings with openings of different heights. They proposed that the wind speed must be the highest for the slit effect to be formed in the opening. The wind direction and the base bending moment have an influence, too. When the opening is located at 0.65 h, the lateral wind pressure coefficient is smaller, and the base bending moment decreases the most. Thus, opening at 0.65 h is the most favorable. Chen et al. [5] conducted a rigorous model force test on a high-rise opening building (0.5 and 0.85 h).

The results showed that the upper opening is better than the lower opening in reducing the average bending moment of the substrate, and the larger the opening ratio, the more obvious the effect.

The second is a study on the impact of high-rise opening buildings on urban microclimate. Fourteen groups of plans were analyzed for the improvement benefits of high-rise opening buildings and the distance between adjacent buildings in the urban environment. For example, Chen [3] discussed the relationship between open high-rise buildings and the distance between different adjacent buildings. Li [7] discussed the effects of different openings (13.5, 9, and 4.5%) and opening heights (68 and 36 m) on the urban wind, temperature, and pollutant concentration. Yeh [8] designed a total of twenty simulation cases based on different building and street space forms (height of openings, street aspect ratio, and building orientation) and planting configuration (green coverage). Furthermore, he discussed the impact of different architectural design cases on the urban microclimate. Voordeckers et al. [23] aggregated more than 200 studies covering different configuration variables (street canyons, buildings, and in-canyon configurations). Nineteen urban planning strategies were formulated to adjust for different building forms (Table 1).

Table 1. High-rise Opening Building cases.

Legend				
name	Hysan Place	South Korea Amore Pacific	OUE Twin Peaks	Baohui Qiuhonggu
year	2012	2017	2015	2015
high	36F.4B/204 m	22F.7B	35F	41F.4B/160.98 m
opening	Multiple rectangular	middle square	right square	middle rectangular
address	500 Hennessy Road, Causeway Bay, Wanchai District, Hong Kong Island	Yongsan District, Seoul, South Korea	33 Lianni Hill Road, Singapore	No.166, Shicheng North 6th Road, Xitun District, Taichung City, Taiwan

From the above literature review, it is known that high-rise opening buildings have a significant effect on urban ventilation. However, in the past, research mainly focused on the building structure, bending moment, and surface wind pressure. The relevant research on the impact of urban microclimate still carries out numerical simulation analysis on the air penetration height, air penetration opening ratio, and the same building height. Thus, it is necessary to explore the impact of different opening patterns in high-rise buildings on urban wind fields and NOx emission study as shown in Table 2.

Table 2. Different patterns high-rise open-building Simulation program.

Single Building	Multiple Building	10*9Array Building
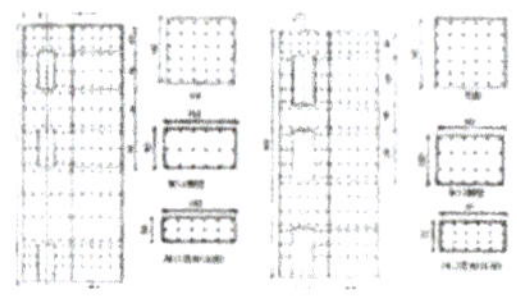	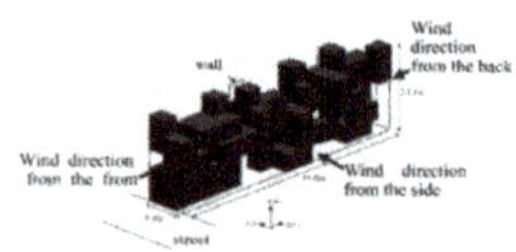	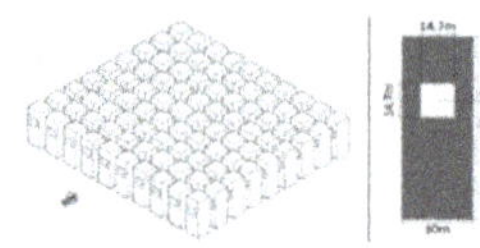
Three Openings [4]	Multiple Openings [24]	Single Opening [3]
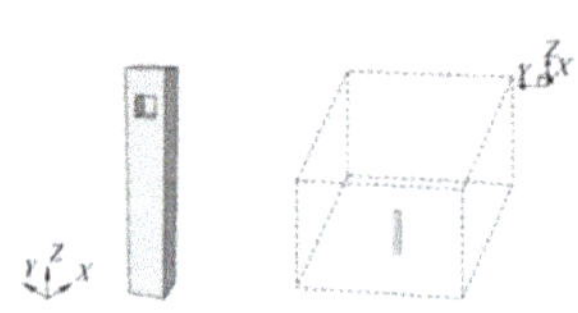	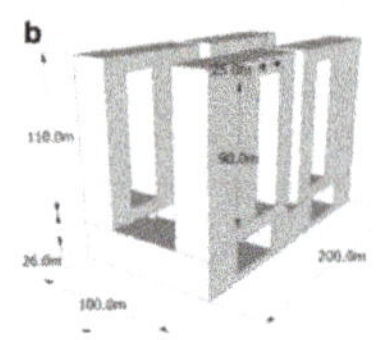	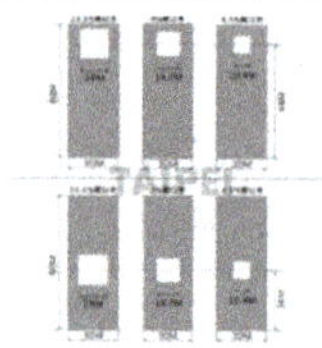
Single Opening [5]	Multiple Openings [25]	Single Opening [7]

3. Research Design

In this study, ANSYS Fluent 18 is used to simulate operations with CFD. The required spatial boundary range needs to maintain an appropriate distance from the building volume to ensure the accuracy of the numerical calculation results with the size of the model calculation domain (Computational Domain) set. The entrance and lateral boundaries must be at least 5 H away from the model (H is the length of the long side of the overall building model). The exit boundary must keep a distance of more than 10 h, and the height of the highest building model from the top boundary must be at least 5 h (h is the highest building height) to achieve a complete flow field as shown in Table 3. The meteorological parameters are set based on the 10-year summer meteorological data from the Taipei Station (466920) from 2012 to 2021. The annual average wind speed is 1.92 m/s, the wind direction is east, and the annual average temperature is 29.67 °C. The inlet is the wind speed gradient ABL Profile, the outlet is atmospheric pressure, and the ground roughness is 0.016 m.

Table 3. Boundary Condition.

Wind	Velocity	Temperature	NO_2	Mesh	Ground	Element Size
East	1.92 m/s	29.67 °C	649 ppb	32 million	0.016 m	1 m

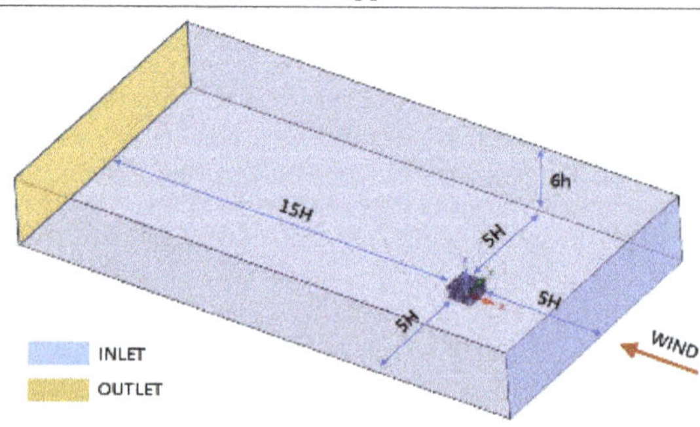

The opening rate of the building facade is 9% according to the research of Zhang et al. [4]. This study refers to the 6 × 6 ideal city case of Li et al. [9] as the setting basis, and the research range is 330 × 330 m as shown in Figure 1. Referring to domestic and foreign cases of opening buildings, each building unit is 30 × 30 × 160 m, and four types of openings are designed as shown in Figure 2 and Table 3. The opening height is 0.45 h according to the research of Xia et al. [5] and Yeh [8] (h is the height of the building). The square and rectangular openings' dimensions are 21 × 21 m and 10 × 44 m as shown in Figure 2.

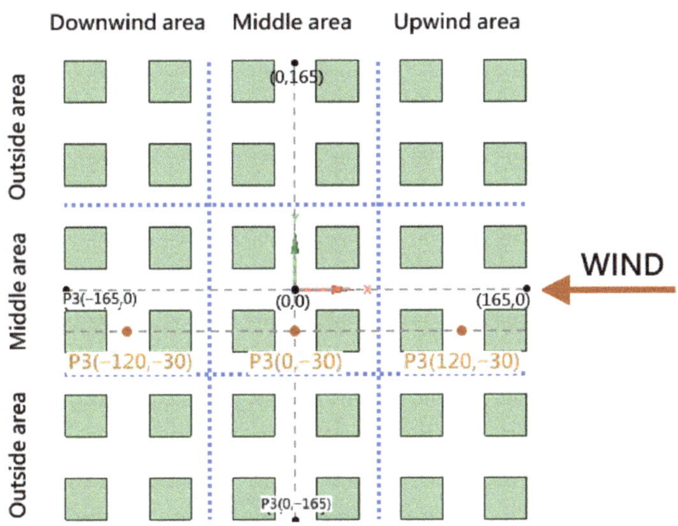

Figure 1. Ideal city and measuring point.

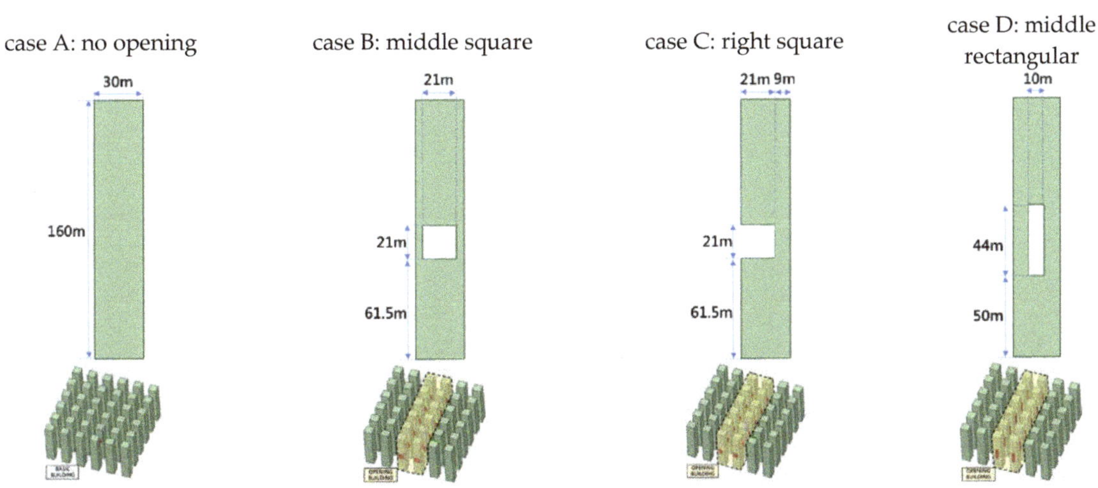

Figure 2. High-rise opening building cases.

4. Simulation Analysis

4.1. Wind Field Simulation Analysis

For simulation, a 1.5 m pedestrian floor and a 72 m high floor (at the center of the opening) were used. The analysis of wind speed measurement points is carried out as follows:

- At a high floor of 72 m, case A does not have an opening. When the building facade is perpendicular to the flow of people, the windward area of the building is more extensive than in other cases of the opening building. Thus, the obstruction to air circulation is relatively large, resulting in a high degree of wind speed attenuation. The average wind speed at the leeward and downstream buildings of the affected buildings is low. At the 72-m high floor, the wind speed of P1 is 1.3 m/s, the wind speed of P2 is 0.82 m/s, and the wind speed of P3 is 0.12 m/s. In contrast, cases B, C, and D with openings reduce the obstruction of air circulation due to the increased ventilation cross-section and the Venture Effect of the opening, and the overall wind speed has a significant effect. Case B (middle square opening) is at a high floor of 72 m, the wind speed of P1 is 4.37 m/s, and the wind speed of P2 is 2.72 m/s. The wind speed of P3 is 1.59 m/s, and the opening is 72 m; the upper and lower street canyons generate a wake, which is beneficial to urban ventilation. For case C (the right side of the square is open), the wind speed of P1 is 3.93 m/s, the wind speed of P2 is 2.17 m/s, and the wind speed of P3 is 1.57 m/s. Because the opening is on the right side of the building, the windward side is the smallest, which effectively allows the wind to flow in, so the wind speed attenuation at the end of P3 is the least among all the cases. In case D (the center rectangle is open), the wind speed of P1 is 3.8 m/s and that of P2 is 2.03 m/s. The wind speed of P3 is 0.59 m/s. Since the opening length is the longest and is close to the ground, the air flows through the opening. It hinders the downstream building facade and generates a windward vortex, which increases the street canyon. The vertical airflow flows, thus reducing the flow velocity at the upper floors and increasing the flow velocity at the pedestrian level. Hence, the wind velocity at the end of P3 is lower than that of cases A and B. The above analysis result shows that at a high floor of 72 m, the ventilation benefit is case B > case C > case D > case A.
- At 1.5 m from the pedestrian level, since case A does not have an opening, the wind bounces in the street canyon between the buildings to generate windward eddies when the building facade is perpendicular to the flow of people. However, in the urban canopy, the wind speed at the layer is significantly attenuated, but it can make the airflow into the pedestrian layer. The wind speed of P1 is 0.86 m/s, the wind speed of P2 is 0.62 m/s, and the wind speed of P3 is 0.45 m/s. For case B, the wind speed at P1 is 0.52 m/s, the wind speed at P2 is 0.34 m/s, and the wind speed at P3 is 0.27 m/s. For case C, the wind speed at P1 is 0.55 m/s, the wind speed at P2 is 0.25 m/s, and the P3 wind speed is 0.23 m/s. There is little difference in wind speed between the two cases. The openings on the facade significantly increase the wind speed of the upper floors, and the wake forms at the upper and lower parts of the building openings, which is beneficial to the ventilation of the urban canopy. It flows into the pedestrian layer and causes an apparent weak wind area. For case D, the wind speed of P1 is 0.49 m/s, the wind speed of P2 is 0.44 m/s, and the wind speed of P3 is 0.39 m/s. The opening length is long and close to the pedestrian layer. Therefore, the wind speed on the upper floors is higher than that of case A, and the vortex formed under the building opening is closer to the ground so that the wind speed of the pedestrian floor is better than that of cases B and C. The above analysis result shows that at the pedestrian level of 1.5 m, ventilation benefits are in the order of case A > case D > case B > case C (Figure 3).

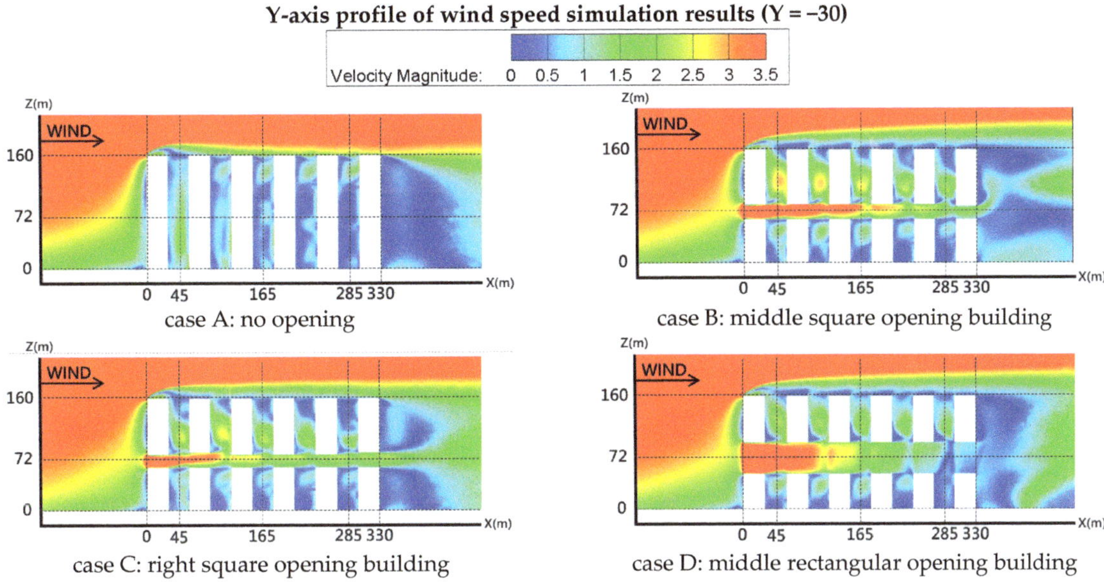

Figure 3. Profile of wind speed simulation results.

4.2. NO_2 Field Simulation Analysis

The 1.5 m pedestrian floor, the 30 m middle floors (the average height of the pedestrian floor and high floors), and the 61.5 m high floor (the height of the lower edge of the opening) were used as pollutant measurement points for analysis (Figure 4 and Table 4).

4.2.1. Case A: Without Opening

In the absence of air-permeable openings, with the building facade perpendicular to the flow of people, the X-axis street parallel to the wind direction is the place with the highest overall wind speed, which carries pollutants away from the street area. The areas with a higher concentration of pollutants were mainly concentrated in the lee of the building. At a 1.5 m pedestrian level, the concentration of pollutants in the lee of the front end of the city was the lowest. As a result, the pollutant concentration gradually increased. In contrast, the pollutants at the height of 30 m on the middle floor and 61.5 m on the upper floor were mainly concentrated in the urban end area. However, the distribution of pollutant concentration decreased with the increase of the Z-axis height.

4.2.2. Case B: The Middle Square Opening

In case B with air openings, because of the increase of the ventilation section, the obstruction of air circulation was small, and the windward vortex was formed above and below the opening so that the heights of 30 and 61.5 m benefited from the increased wind speed of the air openings. Effective diffusion of pollutants and pollutant removal were better than that of case A without openings. Nevertheless, because the building floor was too high, a weak wind area was generated at the 1.5 m pedestrian level, causing pollutants to accumulate at the end of the city—the lee of the building along the X-axis street in the area.

4.2.3. Case C: Right Side Square Opening

For case C with an air-permeable opening, the obstruction of air circulation is small due to the increase of the ventilation section. A windward vortex was formed above and below the opening. The heights of 30 and 61.5 m benefited from the increased wind speed of the air-permeable opening. Pollutants were effectively diffused. However, because the opening was located on the right side, the airflow in the street flew out directly, and the Venturi effect was weak. The pollutant removal effect was worse than that of the middle opening case B, and the building floor was too high. The height was far from the ground, a weak wind area was generated at the pedestrian level of 1.5 m, and pollutants were not effectively diffused. Hence, pollutants accumulated in the leeward of buildings and X-axis streets in the urban end area. Regardless of different heights (1.5, 30, and 61.5 m), the accumulation of pollutants was higher than in the other open cases (cases B and D).

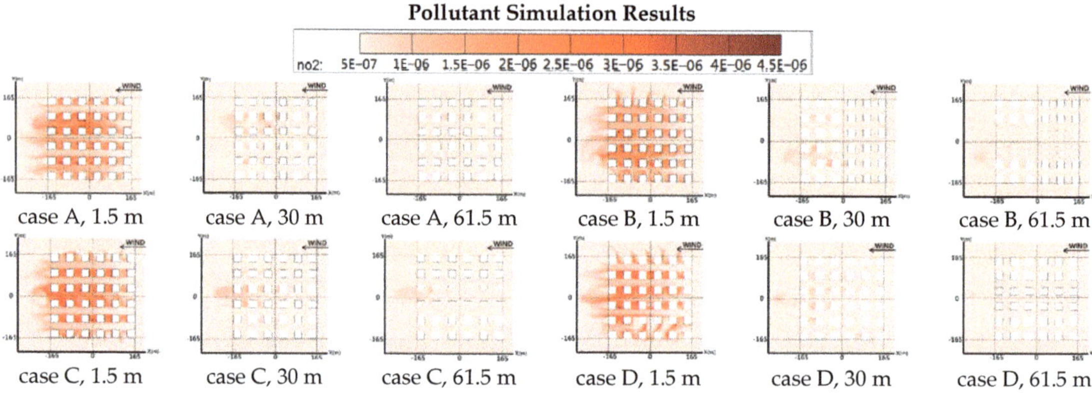

Figure 4. Floor plan of pollutant simulation results.

4.2.4. Case D: Rectangular Opening

In case D with a rectangular air-permeable opening, because the opening was long and narrow and adjacent to the pedestrian floor, the airflow entered the street canyon more effectively under the Venturi effect and was affected by the downstream building facade. The area where pollutants were gathered, so the removal effect of pollutants at different heights was better than the other opening solutions (cases B and C).

Table 4. Distribution characteristics of pollutants in street Canyons with different heights.

Height	Pedestrian Level 1.5 m	Middle Floor 30 m	High Floor 61.5 m
Street Canyon Space	continuous and closed	continuous and closed	0.45 h square opening's lower edge
Pollutant Distribution	continuous linear pollution	Small blocks gather at the end of the city	Dotted and scattered at the end of the city
Pollutant Concentration	Wide range and high concentration	Low concentration in local area	Low concentration in sporadic areas

5. Conclusions

This study simulates urban wind and pollutant changes using ideal cities. Future research should consider real urban factors like climate, design laws, and opening heights. The focus is on how various high-rise building openings affect the urban environment.

5.1. Wind Speed Distribution

Case A, without opening, had the largest area on the windward side, and the wind speed hindered the air circulation on the upper floors, so the wind speed was the lowest among the cases at 72 m. Overall, there was little difference in wind speed in case B with a square in the middle, and, in case C, with a square on the right. However, the wind speed of case C was slightly lower than that of case B at 1.5 m from the pedestrian level. Case D had the most extended opening length so that the eddy current formed under the building opening became closer to the ground, making it easier for airflow to enter the pedestrian floor from the high floor. Therefore, the wind speed of case D was better than that of the opening cases B and C for the pedestrian floor. The wind speed was much faster than the no-opening case A. Overall, case D was the optimal ventilation case.

5.2. Concentration Distribution of Pollutants

Pollutant distribution was influenced by wind speed, causing pollutants in case A to concentrate downwind of the building. Conversely, cases B and C saw pollutants concentrated in the middle and lower city layers, with cases B and C's middle and upper floors particularly affected. Effective airflow through vents limited ground-level contaminant spread. Case D, with the most extensive opening near the pedestrian floor, created a vortex evacuating pollutants, presenting an optimized solution for pollutant removal.The conclusions and suggestions of the four simulation cases for no-opening and different types of opening buildings are as follows.

1. The Opening Building Improved the Permeability of the Street and Air Circulation and Increased the Pollutant Removal Effect

The air-permeable of the opening building facade increased the ventilation section. It improved the permeability of the overall street in the city, significantly increasing the wind speed on high floors. The wind speed reached the maximum value at the opening. However, because the building height was too high (160 m), it was difficult for the eddy current under the opening to enter the pedestrian layer, resulting in the weak wind area affecting the pedestrian wind field.

2. Different Opening Types Affected the Diffusion of Pollutants

The rectangular opening was the best optimization case for removing pollutants at different heights (1.5, 30, and 61.5 m). There was little difference in wind speed and pollutant concentration field between cases B and C which had square openings. However, the wind speed benefit of rectangular openings at 1.5 m of the pedestrian level was better than others. This case benefited from its narrow and long opening adjacent to the street traffic pollution source. The vortex generated by the airflow flowing through the opening in the street canyon improved the diffusion of pollutants.

3. Affected by the Distance from the Pollution Source and the Space of the Street Canyon, the Distribution of Pollutants Changed with the Height of the Z-Axis

Pollutants gradually dispersed and decreased in concentration as the height increased. The pollutant distribution was continuous, linear, and high in concentration at 1.5, 30, and 61.5 m. The wind speed increased due to the adjacent openings. The pollutants diffused into several small blocks at 30 m, and the pollutants diffused at 61.5 m in sporadic distribution.

4. The Wind Speed of Street Canyons at the End of the City Decreased with the Increase in the Number of Buildings

The continuous and long streets hindered air circulation, making it difficult for the wind to penetrate the end of the city and resulting in the inability of pollutants to diffuse naturally through air convection. Therefore, in addition to penetrating the building facade, it is necessary to avoid excessively long streets and leave open space appropriately to achieve the ideal ventilation effect and reduce the accumulation of pollutants in the street canyon. It is necessary to develop a city sustainably under the high-rise and high-density development mode.

Author Contributions: Conceptualization, Methodology: Y.-M.S. Software, Validation, Formal Analysis: C.-H.H. All authors have read and agreed to the published version of the manuscript.

Funding: This research was funded by Ministry of Science and Technology Program (Program Code 110-2221-E-027-013-MY2).

Institutional Review Board Statement: Not applicable.

Informed Consent Statement: This article does not involve human subjects in the research.

Data Availability Statement: Due to privacy concerns, we are unable to provide the data.

Conflicts of Interest: The authors declare no conflict of interest.

References

1. Codo, A.C.; Davanzo, G.G.; de Brito Monteiro, L.; de Souza, G.F.; Muraro, S.P.; Virgilio-da-Silva, J.V.; Prodonoff, J.S.; Carregari, V.C.; de Biagi Junior, C.A.O.; Crunfli, F.; et al. Elevated glucose levels favor SARS-CoV-2 infection and monocyte response through a HIF-1α/glycolysis-dependent axis. *Cell Metab.* **2020**, *32*, 437–446. [CrossRef] [PubMed]
2. Manning, W.J. Urban environment: Defining its nature and problems and developing strategies to overcome obstacles to sustainability and quality of life. *Environ. Pollut.* **2011**, *159*, 1963–1964. [CrossRef] [PubMed]
3. Chen, T.-F. Research on the Relation between Opening Buildings and Adjacent Buildings to the Influence on Urban Environment. Master's Thesis, National Taipei University of Technology, Taipei Tech, Taipei, Taiwan, China, 2019.
4. Wang, C.-G.; Zhang, Y.-C.; Qin, Y. *Wind Tunnel Tests Study on Mega Tall Buildings with Opening Rigidity Models;* School of Civil Engineering, Harbin Institute of Technology: Harbin, China, 2004.
5. Xia, X.; Chen, R.; Ren, H.; Hu, D.; Deng, X.; Zhang, C. Numerical simulation of the influence of openings at different heights on the wind characteristics of tall buildings. *J. Anhui Univ. Technol. (Nat. Sci. Ed.)* **2016**, *33*, 354–359. [CrossRef]
6. Hang, J.; Li, Y.; Sandberg, M.; Buccolieri, R.; Di Sabatino, S. The influence of building height variability on pollutant dispersion and pedestrian ventilation in idealized high-rise urban areas. *Build. Environ.* **2012**, *56*, 346–360. [CrossRef]
7. Li, C.-Y. The Impact of High-Density High-Rise Opening Buildings with Different Opening Size and Position on Microclimate. Master's Thesis, National Taipei University of Technology, Taipei Tech, Taipei, Taiwan, 2019.
8. Yeh, P.-T. Evaluating the Impact of Three-Dimensional Greening in Opening Building on Microclimate. Master's Thesis, National Taipei University of Technology, Taipei Tech, Taipei, Taiwan, 2020.
9. Li, Z.; Zhang, H.; Wen, C.Y.; Yang, A.S.; Juan, Y.H. Effects of frontal area density on outdoor thermal comfort and air quality. *Build. Environ.* **2020**, *180*, 107028. [CrossRef]
10. He, M.; Fang, F.; Li, Y.; Cai, Y.; Liu, W.; Zhong, Z.; Xu, J.; Lai, G.; Wu, Y. *Analysis and Analysis of Wind Environment Circulation Effects in Urban Areas;* Collaborative Research Report of the Institute of Architecture of the Ministry of the Interior; The Institute of Architecture of the Ministry of the Interior: Taipei, Taiwan, China, 2015.
11. Juan, Y.H. Analysis of Urban Wind Energy Potential around High-Rise Buildings in Close Proximity Using Computational Fluid Dynamics. Ph.D. Thesis, Eindhoven University of Technology, Eindhoven, The Netherlands, 2021.
12. Wang, A.; Lin, Z.; Liang, W.; Guo, J.; Zeng, S.; Chen, Y.; Hou, K.; Lan, S. *Analysis of the Construction of Wind Corridors across Different Land Conditions and Assessment of Urban Ventilation Environment;* Collaborative Research Report of the Institute of Architecture, Ministry of the Interior; The Institute of Architecture, Ministry of the Interior: Taipei, Taiwan, China, 2018.
13. Shi, X.; Zhu, Y.; Duan, J.; Shao, R.; Wang, J. Assessment of pedestrian wind environment in urban planning design. *Landsc. Urban Plan.* **2015**, *140*, 17–28. [CrossRef]
14. Tsai, K.-J. Study on the Shadowing Effects of Wind Distributon Pattern by the Building Mass in the Street Blocks-Preliminary Case Study in Kaohsiung City. Master's Thesis, Shu-Te University, Kaohsiung, Taiwan, 2013.
15. Guo, J. *Research on the Application of Wind Tunnel Tests to the Environmental Impact Assessment of Pedestrian Wind Fields;* Architectural Research Institute of the Ministry of the Interior: Taipei, Taiwan, China, 2009.
16. Lin, M.; Hang, J.; Li, Y.; Luo, Z.; Sandberg, M. Quantitative ventilation assessments of idealized urban canopy layers with various urban layouts and the same building packing density. *Build. Environ.* **2014**, *79*, 152–167. [CrossRef]
17. Mei, S.J.; Hu, J.T.; Liu, D.; Zhao, F.Y.; Li, Y.; Wang, Y.; Wang, H.Q. Wind driven natural ventilation in the idealized building block arrays with multiple urban morphologies and unique package building density. *Energy Build.* **2017**, *155*, 324–338. [CrossRef]
18. Santos, L.G.; Nevat, I.; Pignatta, G.; Norford, L.K. Climate-informed decision-making for urban design: Assessing the impact of urban morphology on urban heat island. *Urban Clim.* **2021**, *36*, 100776. [CrossRef]
19. Liu, J. *Urban Environmental Physics;* China Architecture and Building Press: Beijing, China, 2010.
20. He, M. *Experimental Study on Urban Street Outer Air Pollution Diffusion and Street-Ground Wind Environmental Assessment;* Research Report of the Institute of Architecture, Ministry of the Interior; The Institute of Architecture, Ministry of the Interior: Taipei, Taiwan, China, 2013.
21. Borck, R.; Schrauth, P. Population density and urban air quality. *Reg. Sci. Urban Econ.* **2021**, *86*, 103596. [CrossRef]
22. Hang, J.; Li, Y. Ventilation strategy and air change rates in idealized high-rise compact urban areas. *Build. Environ.* **2010**, *45*, 2754–2767. [CrossRef]
23. Voordeckers, D.; Lauriks, T.; Denys, S.; Billen, P.; Tytgat, T.; Van Acker, M. Guidelines for passive control of traffic-related air pollution in street canyons: An overview for urban planning. *Landsc. Urban Plan.* **2021**, *207*, 103980. [CrossRef]

24. Murakami, S.; Kato, S.; Ooka, R.; Shiraishi, Y. Design of a porous-type residential building model with low environmental load in hot and humid Asia. *Energy Build.* **2004**, *36*, 1181–1189. [CrossRef]
25. Yuan, C.; Ng, E. Building porosity for better urban ventilation in high-density cities—A computational parametric study. *Build. Environ.* **2012**, *50*, 176–189. [CrossRef] [PubMed]

engineering proceedings

MDPI

Proceeding Paper

The Effectiveness of Lighting Design for Improved Patient Care Considering Energy Conservation †

Samaneh Aghajari [1,*] and Cheng-Chen Chen [2]

[1] Department of Design, National Taipei University of Technology, Taipei 10608, Taiwan
[2] Department of Architecture and Urban Design, National Taipei University of Technology, Taipei 10608, Taiwan; coolhas@ntut.edu.tw
* Correspondence: maaaahsa.aj@gmail.com

† Presented at the IEEE 5th Eurasia Conference on Biomedical Engineering, Healthcare and Sustainability, Tainan, Taiwan, 2–4 June 2023.

Abstract: Unquestionably, hospital patient rooms require a proper lighting design. Dissimilar to cultural and artistic settings, where artistic discourse on light has significant importance, in medical settings, the most crucial conversation refers to standards. Research indicates that light in hospital settings has an impact on a patient's physical and mental health. Effective lighting in medical settings can enhance the hospital's positive experience and the speed at which patients recover from their diseases. It can also increase staff attentiveness and productivity. It is also critical to consider reducing electricity consumption in hospital settings that require lighting 24/7. Due to the high cost of lighting, access to natural light in combination with time-of-day controls minimizes energy consumption when daylight is available and impacts the hospital's bottom line. The effect of light on hospital users was investigated in this article; therefore, it is important to understand both natural and artificial light sources in this regard. Natural light has many benefits for humans, and when it comes to electricity consumption, it is the best method because it is a free source; but, since natural light is not always available and cannot be used throughout the day, there is a need to have an artificial light source that gives the best lighting effect in terms of visual comfort and visual performance for users. Secondly, proper artificial light sources can reduce electricity consumption; hence, these two critical aspects were underlined in this study.

Keywords: patient care lighting; LED; energy savings; daylight; electric light

Citation: Aghajari, S.; Chen, C.-C. The Effectiveness of Lighting Design for Improved Patient Care Considering Energy Conservation. *Eng. Proc.* **2023**, *55*, 91. https://doi.org/10.3390/engproc2023055091

Academic Editors: Teen-Hang Meen, Kuei-Shu Hsu and Cheng-Fu Yang

Published: 9 January 2024

1. Introduction

Among the human senses, sight has always been the most powerful for obtaining information about the surrounding world. On the other hand, the visual quality of the environment has a positive effect on the users' feeling of the space. This issue becomes more critical in the case of hospitals and other healthcare centres whose goal is to accelerate the recovery process of patients. In this regard, lighting is one of the aesthetic and practical elements that can provide attractive and pleasant conditions for patients, visitors, and staff. Light plays an essential role in human visual activities; on the other hand, it is very effective in mental and physical health. Multiple research investigations have demonstrated the importance of light in healing disorders such as jaundice in newborns, lowering depression and exhaustion, enhancing alertness, and altering the circadian rhythm [1].

Unlike historical, cultural, etc., areas, where in their lighting design, an artistic view is critical, in hospital areas, the main goal is the implementation of standards and to meet the needs of each hospital area, from office areas to reception and patient rooms, etc. Therefore, lighting should be such that it meets the needs of each healthcare area and has an effect on the improvement of the staff work process and the recovery of the patients.

It should be added that an average hospital in the United States uses about 31.0 kWh of electricity per square foot, according to research conducted by the Business Energy

Advisor. Here, 65% of the total energy consumption is dedicated to lighting, space, and water heating, which shows that the use of suitable and low-consumption lamps in the lighting sector can be a great way to reduce energy consumption [2].

2. Importance of Lighting in a Hospital

- For patients: The environment in which patients are cared for in the hospital is related to the patient's satisfaction, patient safety, and the patient's recovery process, and patients need an environment with standard lighting.
- For staff: Facilitating the work of the hospital staff is another goal of lighting the hospital; light with appropriate intensity and standard colour increases the energy of the personnel. Therefore, it also increases the performance and efficiency of these staff. Moreover, the personnel can better attend to the hospital's different departments in sufficient light.
- For visitors: The requirements are different from hospital's staff; they might need to relax and rest at night rather than remain awake [3].

Proper lighting fulfils human needs, puts the mind at ease, and creates a sense of comfort and security. In contrast to this situation, inappropriate lighting can fill the place with an uncomfortable atmosphere or make the residents nervous, worried, and anxious; the person constantly feels that something is not correct or regular. It is even possible that low and inappropriate light can cause headaches, eye fatigue, anxiety, or even collisions due to low vision [4].

Important Factors for Evaluating the Quality of the Lighting System

- Visual Comfort: Visual comfort means not feeling tired by being in the environment. Factors that are effective in achieving visual comfort include good colour rendering and uniform brightness distribution.
- Visual Performance: The environment's lighting should be such that it meets the needs of the human eye to see objects. This means that the environment should be bright enough, and the factors that cause glare should be limited.
- Visual Ambience: One of our goals in placing a lighting system is to obtain a visual ambience. Visual ambience means that by choosing the right direction of light and choosing the right light colour, we can recognize objects in three dimensions.
- Glare: Glare is one of the aggravating factors in lighting that limits the field of vision and causes fatigue for people, which should be limited as much as possible. Factors that can cause glare include use of inappropriate lights, placing lights or windows in an inappropriate position, and the high reflection of different surfaces.
- Uniformity: If there is a significant difference between the luminance of the environment that the person's eyes are constantly dealing with and the surrounding environment, the person becomes tired. In general, it is recommended that the brightness of a person's surrounding environment should be at least one-third of the brightness of the work surface. On the other hand, if there is no significant difference between the brightness of different points in an environment, the environment looks uniform and causes a person to feel tired [5].

3. Lighting System Design

3.1. Different Aspects Taken into Consideration When Designing Hospital Lighting Systems

3.1.1. Natural Illumination

Increasing the quality of natural light in a space, access to windows, and creating suitable conditions for seeing objects have obvious effects on people's emotions and moods. Research results show that light affects people in two ways, direct and indirect; its immediate effect is through changes in the quality of vision, and its indirect impact is on emotions, mood, and even body hormones. Natural light radiation and visual communication with the outside environment reduce anxiety, improve behaviour and personality, and maintain and increase health and comfort [6] (Table 1).

Table 1. The Influence of Natural Light on Humans.

Physical Impact		Psychological Impact	
Increase	**Decrease**	**Increase**	**Decrease**
Vitamin D	Carcinogenic Probability	Temperament	Dejection
Vision	Bone Deformation	Mental Ability	Strain
Sleeping Quality		Attention	Depression
Circadian Entrainment		Intellect	Violence

By passing visible light through a glass prism, Sir Isaac Newton famously demonstrated the division of visible light into its constituent colors. It is a well-known fact that sunlight contains a spectrum of electromagnetic energy known as photons, which was discovered by Sir Albert Einstein. A photon's energy level is inversely related to its wavelength, in addition to colour. Numerous investigations have demonstrated that highly energetic photons with a wavelength of 290 nm can trigger the secretion of hormones from deep inside the brain's endocrine glands and induce a Vitamin D reaction in the epidermis. The sun's vacuum UV spectrum reaches the Earth's surroundings and can cause a variety of pathological illnesses. Although UV wavelengths are required for humans to create Vitamin D3, excessive exposure, even at moderate levels, can raise the risk of heart failure, stroke, and the development of cardiac abnormalities. Researchers in photobiology have shown that exposure to daylight can considerably lower blood pressure and serum cholesterol levels. Surprisingly, daylighting has a natural healing effect on the surrounding environment [7].

Much research demonstrates that not only does access to daylight improve patient outcomes, but it can also bring restorative advantages to other users in their particular surroundings, such as medical staff. As a result, daylight should be incorporated in hospital lighting design not only because it is helpful to patients and workers, but also because it is free. As a result, it can take the lead in energy conservation, contributing to sustainability. We should also mention that the penetration of the sun should be limited so as not to cause thermal and visual discomfort.

3.1.2. Artificial Illumination

With all the advantages of natural light, it is impossible to use natural light 24 h a day. Even when there are certain weather conditions, such as cloudy and rainy days, the possibility of using daylight is lost, and even in situations in different parts of the hospital, it may be necessary to move the light; for example, light is needed in one position and not in another region. With the advancement of technology, it is possible to be inspired by day and night circulation and create similar natural light through appropriate artificial lights. Since the tops of the opposing walls and the ceiling are the patients' common lines of sight in the hospital, the design should minimize glare for patients while maintaining good visibility for medical staff. For every application, a limited glare index is advised [8].

3.2. Three Main Parameters to Be Considered in an Artificial Source of Lighting Design for the Hospital

3.2.1. Lighting Level or Illuminance

Illuminance is the total luminous flux incident on a surface per unit area or the amount of light measured on a planar surface. Lighting intensity is expressed in either footcandles (Lumens per square foot) or lux (Lumens per square meter). Lighting levels have been issued by numerous international standards organizations and standardized for a range of applications and occupancies, including The British and European Standard BS-EN12464 and the Illuminations Engineers Society of North America (IESNA) [9] (Table 2).

Table 2. List of lighting levels in some hospital areas.

Location	Lighting Level
Lobby Area	50 lx (5 fc)
Waiting Area	100 lx (10 fc)
Patient Ward Rooms	300 lx (30 fc)
Medical Laboratory	500 lx (50 fc)
Operating Room	3000–10,000 lx (300–1000 fc)
Critical Care Areas (Examination)	500 lx (50 fc)

3.2.2. Colour Rendering Index (CRI)

The CRI represents the lighting source's capability to realistically and naturally depict the colors of objects, which has a scale between 0 and 100 percent; the excellent light sources are those with a CRI above 90 (Tables 3 and 4).

Table 3. Different CRI.

CRI + 90	CRI + 80	CRI < 65
Excellent	Good	Reasonable

Table 4. Various artificial light sources.

Lamp Type	Lumens/Watt	Avg. Lumens/Watt	CRI	Life (h)
Incandescent	8–18	14	100	1000
T12 Fluorescent	40–70	55	92	8000
T8 Fluorescent	60–80	70	85	6000
T5 Fluorescent	100–105	102.5	85	9000
Mercury	44–57	50	50	24,000
HPS (High Pressure Sodium)	66–121	90	21	50,000+
LPS (Low Pressure)	101–175	150	10	60,000+
LED	75–200	137.5	98	50,000+

3.2.3. Colour Temperature

Measured in Kelvin units, the colour temperature expresses the brightness and colour of light. All lamps emit light with a certain colour. The higher the degree of Kelvin, the cooler and brighter the emitted light will be, and in the same way, the lower degree of Kelvin indicates warm lights such as yellow and red (Table 5).

Table 5. Different ranges of colour temperature.

Temperature (K)	Colour	Description
2000–3500	Orange/Yellow	Ultra-Warm or Warm White
3500–5000	Paper White	Natural/Neutral White
5100–6500	Bluish White	Cool White

A CRI above 90 shows objects very similar to what we see under the natural light of the sun. The suggested colour temperature range is between 3000 Kand 6500 K, according to the IEC standard 60601-2-41 [10], which provides certain specifications for surgical and diagnostic luminaires. The guideline also stipulates that the Colour Rendering Index (CRI) must be between 90 and 100 percent.

Generally, a 4000 K colour temperature or more is needed in most hospital areas, but for areas where high accuracy is important, such as treatment areas and operating areas, the colour temperature should be 5000 K, with a CRI of 90 to 100 percent, and in areas where patient comfort is important, a colour temperature of 3000 K, which is warmer, and a CRI of 85 to 100 percent are appropriate [11].

4. Results

Two important capabilities in light output, as well as for the requirements of health care environments, are the light colour temperature and CRI; colour temperature creates a balance between comfort and biological effect. According to the mentioned standards, we need a minimum colour temperature range of 3000 K to 6500 K in healthcare environments. The knowledge shows that LEDs with a colour temperature range of 6000 K to 8000 K have the best biological effect, and 3000 K to 5000 K colour temperature ranges are the closest ranges to Daylight White. Also, a high CRI is needed for the comfort of patients and the staff and different environments of health centres. According to Table 3, LED lamps have the highest CRI compared to other lights. For hospital environments that need lighting 24 h a day, 365 days a year, apart from light colour temperature and CRI, it is also a requirement to consider energy savings. The efficiency of a light source is measured in lumens per watt, often known as "luminous efficacy" or "efficacy" (watts converted into lumens). As Table 3 shows, the efficacy of old incandescent lamps is between 8 and 18 lumens/watt, depending on the type of lamp and their manufacture, while LEDs are 75 to 200 lumens/watt. Therefore, LEDs have a longer lifespan of up to 50,000 h and can reduce energy usage by up to 70%.

5. Conclusions

Studies show that proper lighting in hospitals has an unavoidable effect on improving patients and the work process of employees and doctors. In an environment like a hospital, where the main goal is to improve the patient's health, we must achieve a high lighting standard and avoid any visual discomfort caused by lighting for patients and employees, such as glare, the negative effects of which can cause headaches, burning eyes, and fatigue. In hospital environments, we need light sources that are both low consumption to save energy and cost and give us light similar to daylight. LED lamps with a CRI of 98% and a colour temperature range between 3000 and 6800 are the best options for hospital and treatment environments, and at the same time, they are also the most energy-efficient lamps. Natural light also should be used as much as possible, because in addition, patients respond better to lighting that creates a restful atmosphere and best supports their circadian cycle and has other positive effects on staff and patients; also, they can use this light at no cost.

Author Contributions: S.A.; writing original draft preparation, C.-C.C.; writing—review and editing. All authors have read and agreed to the published version of the manuscript.

Funding: This research received no external funding.

Institutional Review Board Statement: Ethical approval is not applicable for this systematic review since the data are from previous published studies in which informed consent was obtained by the primary investigators.

Informed Consent Statement: Not applicable.

Data Availability Statement: Data supporting this systematic review are available in the reference section. In addition, the analysed data that were used during the current systematic review are available from the author on reasonable request.

Conflicts of Interest: The authors declare no conflicts of interest.

References

1. Ulrich, R.S.; Zimring, C.; Joseph, A.; Quan, X.; Choudhary, R. *The Role of the Physical Environment in the Hospital of the 21st Century: A Once-in-a-Lifetime Opportunity*; The Center for Health Design: Concord, Australia, 2014.
2. Alimoglu, M.K.; Donmez, L. Daylight Exposure and the Other Predictors of Burnout among Nurses in a University Hospital. *Int. J. Nurs. Stud.* **2005**, *42*, 549–555. [CrossRef] [PubMed]
3. Baehr, E.; Fogg, L.F.; Eastman, C.I. Intermittent Bright Light and Exercise to Entrain Human Circadian Rhythms to Night Work. *Am. J. Physiol.* **1999**, *277*, 1598–1604. [CrossRef]
4. Beauchemin, K.M.; Hays, P. Sunny Hospital Rooms Expedite Recovery from Severe and Refractory Depressions. *J. Affect. Disord.* **1996**, *40*, 49–51. [CrossRef] [PubMed]
5. Beauchemin, K.M.; Hays, P. Dying in the Dark: Sunshine, Gender and Outcomes in Myocardial Infarction. *J. R. Soc. Med.* **1998**, *91*, 352–354. [CrossRef] [PubMed]
6. Begemann, S.H.A.; van den Beld, G.J.; Tenner, A.D. Daylight, Artificial Light and People in an Office Environment: Overview of Visual and Biological Responses. *Int. J. Ind. Ergon.* **1997**, *20*, 231–239. [CrossRef]
7. Mehrotra, S.; Basukala, S.; Devarakonda, S. Effective Lighting Design Standards Impacting Patient Care: A Systems Approach. *J. Biosci. Med.* **2015**, *3*, 54–61. [CrossRef]
8. Van Den Wymelenberg, K.; Inanici, M. A critical investigation of common lighting design metrics for predicting human visual comfort in offices with daylight. *Leukos* **2014**, *10*, 145–164. [CrossRef]
9. Alzubaidi, S. Energy Efficient Lighting System Design for Hospitals Diagnostic and Treatment Room—A Case Study, April 2012. *J. Light Vis. Environ.* **2012**, *36*, 23–31. [CrossRef]
10. Dalke, H.; Little, J.; Niemann, E.; Camgoz, N.; Steadman, G.; Hill, S.; Stott, L. Colour and lighting in hospital design. *Opt. Laser Technol.* **2006**, *38*, 343–365. [CrossRef]
11. Thorn Lighting. *Thorn Technical Handbook*; Thorn Lighting: Bellerive, Australia, 2017; pp. 69–77.

engineering proceedings MDPI

Proceeding Paper

Gamepad Design for Touch Generation: Evaluation of First-Person Shooter/Third-Person Shooter Game Control and Possibility of Touched-Based Control as Norm †

Chia-En Chen [1,*] and Fang-Wu Tung [2]

[1] Department of Design, National Taiwan University of Science and Technology, Taipei 106335, Taiwan
[2] Department of Arts and Design, National Tsing Hua University, Hsinchu 300044, Taiwan; fwtung@mail.ntust.edu.tw
* Correspondence: xplusn@gmail.com

† Presented at the IEEE 5th Eurasia Conference on Biomedical Engineering, Healthcare and Sustainability, Tainan, Taiwan, 2–4 June 2023.

Abstract: This research compares a Steam controller, a DualSense controller, two gestures on a prototype touchpad-implemented gamepad design, an iPad Pro, a Logitech G304 mouse with latency matching the iPad, and a Razer 8K gaming mouse using Fitts Task 2, 3D Aim Trainer for a performance and gameplay experience test in a Death Stranding firing range and gathers feedback on the devices. With the participants categorized by their gaming experience, the differences between the experienced groups are recorded in terms of performance and preference. The average result shows that the touch-based input has the potential of substituting a mouse when the latency condition is equal, and the players that already have touch-based FPS/TPS gaming experience tends to favor touch over a mouse input. However, the prototype controller designed to implement a larger touchpad did not meet the expectations in terms of performance and preference, but the knowledge, data and feedback gathered in this study will aid future touch-based gamepad designs in the emerging handheld console market.

Keywords: HCI; Fitts' law; game controller evaluation; touchscreen; mouse; gyroscope; mobile game

Citation: Chen, C.-E.; Tung, F.-W. Gamepad Design for Touch Generation: Evaluation of First-Person Shooter/Third-Person Shooter Game Control and Possibility of Touched-Based Control as Norm. *Eng. Proc.* **2023**, *55*, 92. https://doi.org/10.3390/engproc2023055092

Academic Editors: Teen-Hang Meen, Kuei-Shu Hsu and Cheng-Fu Yang

Published: 9 January 2024

1. Introduction

Shooter games have become an iconic game genre and the most popular choice across different gaming platforms [1]. They have various input methods based on the platform's hardware limitations. The mouse and keyboard inputs of a PC are known to be the most accurate and efficient input methods [2]. However, console games are played with gamepads to adapt to the living room environment for diverse game genres. As they are different from a mouse providing direct coordinate information, the analog sticks on a gamepad provide vector angle data that are translated into the direction and force inputs for aiming in a game. The gamepad is inferior to the mouse in shooter game environments. Thus, game developers need to add an assist function for the players. Adding a gyro input to the gamepad is used to improve the gamepad's functionality [3] and its efficiency [4].

Mobile shooter games require touch screen as a substitute for the mouse, which provides coordinated data by using a gyroscope. With the success of the PUBG Mobile on the platform [5], many developers are porting their IPs such as Apex Legends [6] for mobile shooter players. Companies such as Valve have attempted to transform more PC games into "mobile" ones with devices such as Steam Deck, a handheld game console for playing most of the titles developed for the PC platform. In Steam Deck, two touchpads and a gyroscope on the device replace the mouse and keyboard controls [7]. With the advancement of the mobile gaming platform and the increasing number of players familiar with the touch and gyro input method, it is necessary to measure the players' performance and preference for the control methods and explore the potential of touch and gyro as the norm in future gamepad designs. The participants' gaming experience also needs to be documented to understand their experience and preference for game genres.

2. Literature Review

2.1. Fitts' Law Testing

Fitts' law is used to evaluate the performance of human–computer input devices. Although the pointer–background relationship is opposite to that of 3D shooter games, in games, the control moves the background, and the crosshair stays in place. The Fitts test and normal GUI interface make the pointer move with the input on the stationary background. Fitts tasks are seen as a validified method to test the performance of peripherals for shooter games [8].

The original Fitts task experiment (now referred to as a 1D task) requires two parallel target areas, with their width determined and a certain distance. Modern tests often use a 2D version of the circular targets arranged in a circular array. As shown in Figure 1, with the target size identified as W (width), and the distance between the pointer and target being A (amplitude), the MT movement time is measured for each task. With Equation (1), the throughput is calculated as the efficiency of the tested device (the error rate is also recorded).

$$TP = \frac{log_2\left(\frac{Ae}{We} + 1\right)}{MT} \tag{1}$$

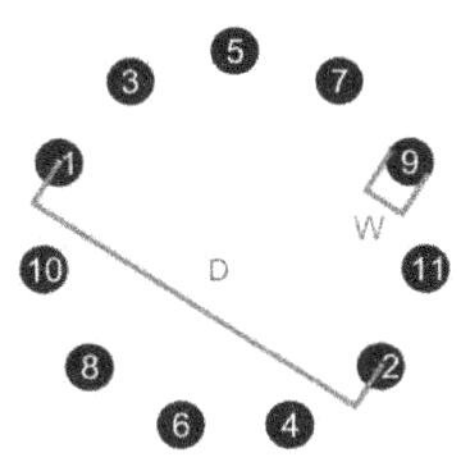

Figure 1. Two-dimensional Fitts task example.

2.2. Testing with Games

Testing with existing games by using in-game scores or time as data is a method to evaluate the different input methods [9]. Opinions for playing actual games are collected too [10] for understanding the gaming experience and performance.

2.3. History of PUBG Mobile

PUBG Mobile is one of the first battle royal games on the mobile platform that inherits the full experience from its original PC platform version [11]. This was traced back to a game called ARMA 2, a realistic battlefield first-person shooter simulator that was used by militaries to train their soldiers. A player named Brendan Greene created a modification called DayZ inspired by titles like The Hunger Games and started the survival shooter genre. He later worked in Bluehole Studio to finally turn his vision into the game PlayerUnknown's Battleground. Lightspeed & Quantum (a subsidiary of Tencent) created the mobile port, which later became the highest-grossing mobile game across iOS and Android [5]. The success led Bluehole to create a mobile port of PUBG New State. As EA games cooperated with Lightspeed & Quantum to create the mobile version of Apex Legends, the number of mobile shooter games and players increased with COD M and Free Fire in the market.

A mobile phone has different hardware inputs compared to a PC or console, as these games require the touch screen for the players to customize the button location, sizes, and opacity. This results in diverse control methods between the players with phones and tablets of different sizes. However, such as PC mice, the gestures of mobile shooter players are categorized by the number of fingers they are using [12].

Some players invest money in devices to add physical buttons to their mobile devices for better ergonomics and advantages in the game.

2.4. Devices with Similar Input Methods

Valve, the company behind the Steam platform and games, such as Half-Life, Counter Strike, Dota, Team Fortress, and Portal, can stream the PC experience to living rooms with Steam Link. They need to design a substitute for the mouse to be used on the couch. Their solution develops from replacing the mouse with a trackball and results in a gyro-included gamepad with two multifunction circular concave haptic touchpads to replace the d-pad and the right thumb stick. Having Steam Input as the backbone software, this controller is customized to replace the keyboard and mouse on the steam game launcher. However, without the traditional d-pad and right thumb stick, many users find it time-consuming to learn. A Steam controller was released in 2019, while the Steam Input software allows any controller to be fully customized. In 2022, Valve introduced the Steam Deck, a handheld, Linux-based gaming console that can run most of the games on the Steam platform. The controls are a successor to the Steam controller, with all the traditional gamepad buttons resized to fit on the device.

In 2020, the Youtuber Tech Yesterday, who was used to the KB+M (keyboard and mice) controls, realized that it was impossible to switch back to the traditional gamepad for shooter games, and with an FPS gamepad with a mousepad area and a miniaturized mouse sensor for the right thumb, they achieved similar gaming performance to that with the KB+M. The later developments added more buttons and a mouse wheel, while adjusting sensor positions to improve the ergonomics and performance [13].

2.5. Latency-Related Research

While developing a touched-based prototype for this research, the importance of latency became clear. Figure 2 is based on Nvidia's explanation of system latency and the included definition. Gaming monitors often show their a latency lower than 1 ms (display latency), PC latency often takes the most part in total system latency. While mouse input latency is often ignored in the modern era, there is no touchpad on the market with a latency low enough for shooter games. A latency as low as 41 ms impacts the input efficiency [14], while the peripheral latency of the touchpads on the market ranges from 80 to 150 ms.

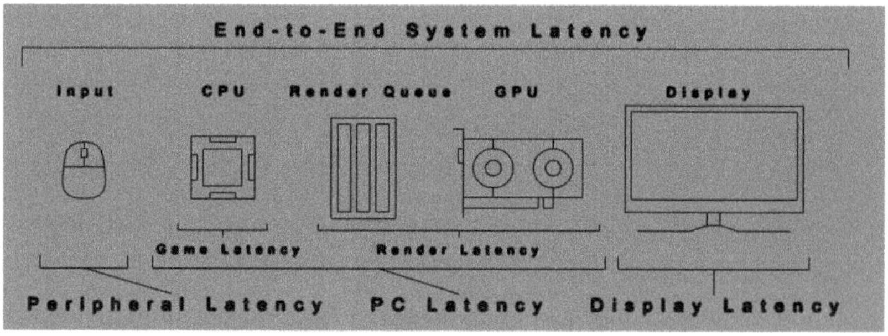

Figure 2. Simplified illustration of system latency terminology from Nvidia [15].

2.6. Other Related Studies

Apart from custom button preferences and touchpad sizes, there are many factors in the ergonomics design of gamepads. The research designs demand tasks to test and receive quantifiable results [16] by using questionnaires, interviews, or reviews from Amazon as references [17].

3. Approach

3.1. Designing Prototype Gamepad Model

The gestures and buttons used by players are obtained from YouTube with the keyword "Handcam". The following (Figures 3–5) are illustrations based on the video observation to demonstrate hand gesture, finger placement and their controlling functions.

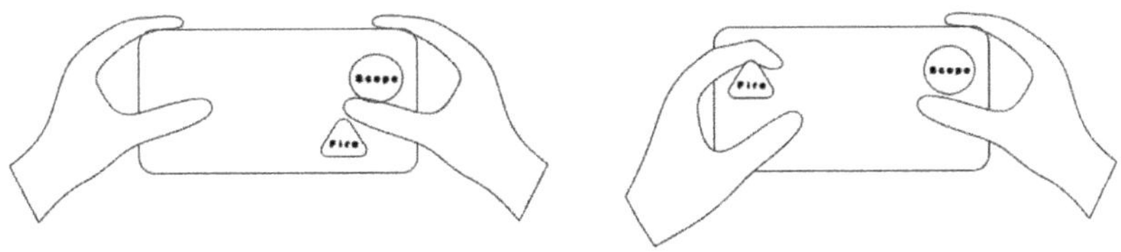

Figure 3. Two (**left**) [18] and three (**right**) [19] fingers + gyro layout.

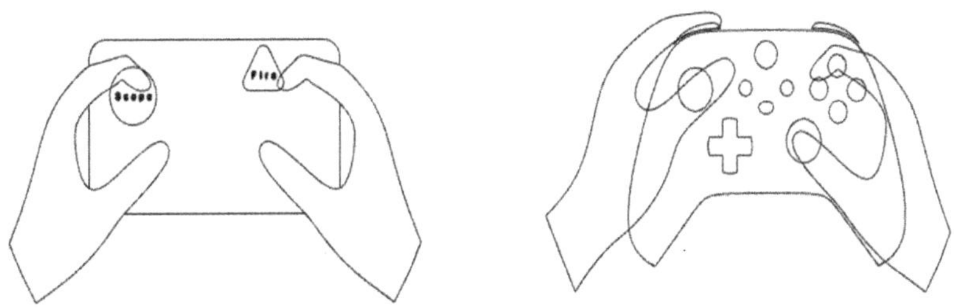

Figure 4. Four fingers with gyro layout [20]. Xbox Elite 2 Controller Apex gameplay handcam [21].

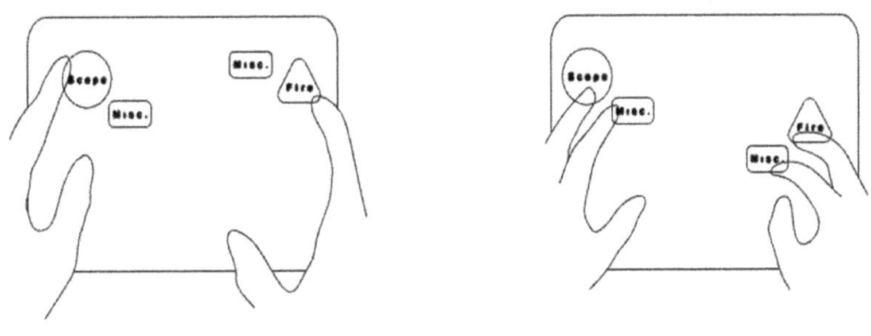

Figure 5. Four (**left**) [22] and six (**right**) [23] fingers with no gyro layout.

The prototype is based on non-gyro ergonomics (Figure 6) with physical buttons to replace the mouse. However, the latency of the T650 wireless touchpad is too high for shooter games. For testing the trackpads of an iPad Pro 10.5's 120 Hz screen as a frame counter, a PC with Win11, i7-8700K, RTX 2060 super, a BenQ GW2480 Full HD 60 Hz screen, and Samsung Note 9's 960fps camera is used to estimate the total system latency (Figure 6).

The Steam Link from the steam controller era has been improved to make phones and tablets that have the lowest-latency touchpad. While the steam controller is used to represent one of the designs a controller product for a touch-based gamer, the prototype design continues with the major iterations shown below (Figure 7).

Device tested with desktop	Frames(+ - 2 frames)	Rough Latency Cal (x8.3)
Apple magic trackpad wired	13	108ms
No Brand industrial usb touchpad	18	150ms
Logitech T650 wireless touchpad	12+ teleports	100ms+
MANO-703UB touchpad wired	12	100ms
Bosto usb touchpad	10	83ms
iPad Pro 10.5 with Steam Link	6~8	50~67ms
Kensington Trackball	3~4	25~33ms
MSI Mouse	1~2	8~16ms
Steam touch Bluetooth	9	75ms
Steam touch USB Dongle	8	67ms
Steam touch USB wired	7+ teleport?	58ms+
Steam Stick USB wired	12	100ms
Steam Gyro USB wired	5	42ms
Xbox dongle + Controller Companion	10	83ms
Xbox dongle stick tilt / wired	10 / 6	83 / 50ms
Ipad pubg touch latency	6~7	50~58ms

Figure 6. A prototype made with Logitech T650 and Arduino (**left**) and total system latency estimates for several devices (**right**).

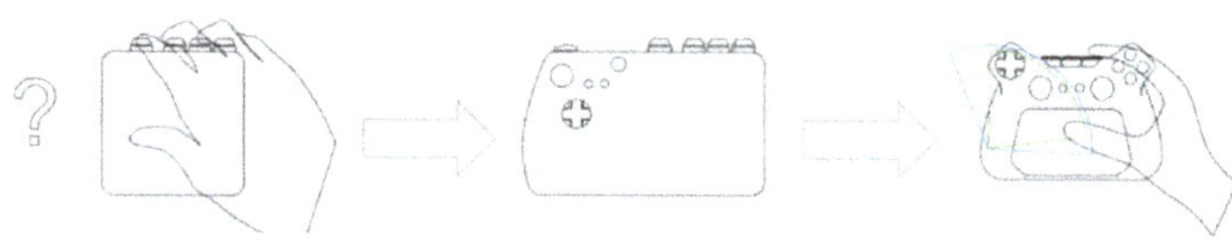

Figure 7. Design iterations of the touch gamepad prototype.

The second design iteration aims to replace the mouse for basic testing. The gamepad inputs for the left side replace the keyboard controls. However, the absence of traditional gamepad inputs may be radical for players to adapt to and set up for unsupported games. Thus, a large touchpad must be added to the traditional control scheme as a new design objective. The traditional controller has an analog thumb stick and buttons (or d-pad) close to the center line of the thumb's moveable angle (the larger blue pan shape in Figure 8). The angles are adjusted so that the thumb can comfortably reach the smaller green sector. Using the arrangement of the PlayStation, the thinness of the buttons (and d-pad) mechanisms are used to move on top of the index finger. More room can be arranged for the touchpad under the thumbstick.

Figure 8. Existing and prototype controller ergonomic design.

A prototype is modeled in a VR environment by using Gravity-Sketch and printed for testing and feedback. It is also designed to be usable with several grip styles in mind (Figure 9). There are two-finger-style physical triggers (which aligns with most console gamers' ergonomics), a three-to-four-finger claw grip (several console gamer also uses this grip style for the right index finger to access the four face buttons), and lastly, the style I call the "tripod" (with the ring and middle fingers as anchors and the thumb resting on the touchpad, while the index finger hovers on the firing button). With the grip style of the prototype (Figure 6), mobile gamers are used to 4+ finger controls.

Figure 9. Ergonomics of the prototype gamepad design.

Compared to the traditional gamepads that allow the thumb to rest in between the thumb stick and buttons (or d-pad), this prototype controller allows the thumb to rest on the analog sticks.

The next prototype (Figure 10) has buttons and analog sticks wired to an Arduino, emulating Xinput gamepad and connecting to the type-c port of a Pixel 3 running Steam Link.

Figure 10. Final design of the functional prototype gamepad.

3.2. Designing the Remote Testing Unit

All the equipment for testing is packaged into a pelican 1510 case. It comprises 4G internet and a Wi-Fi router, a surface book 2 15″ as the test computer, and a Logitech Stream Cam for recording (Figure 11).

Figure 11. Portable remote testing unit.

4. Methodology

The test was randomly carried out with devices. After understanding the test subject's gaming history, each device was given to perform the Tile frenzy test for the participant to be familiar with the control. Fitts Task 2 2D task was then performed, followed by the completion a questionnaire for feedback. After the firing range feature of the game Death Stranding Director's Cut was played with the devices, a final questionnaire was given, and an interview was performed. There were 7 participants from universities (3 males and 4 females). Four were used to play the game with a mouse. Three had experienced with FPS games before the test, while two had no FPS gaming experience.

4.1. Devices Setup

The following devices were selected for testing (Figure 12).

Figure 12. List of input devices in this experiment.

- The Pixel 3 running Steam was linked with 5 Ghz Wi-Fi. It is a phone-size controller with touch and gyro features. Its system latency is 50–75 ms. It had a touch button on the screen for a trigger, and, later, the functional prototype gamepad with traditional (P_Traditional) and "Tripod" grip styles (P_Tripod) was adopted as the input. Both styles use a physical button for the trigger. An analog stick was not used in the performance tests.
- The iPad Pro 10.5 running steam was linked with 5 Ghz Wi-Fi. It is a tablet-sized touch controller. The system latency is 50–75 ms. With a 3D-printed stand placing the tablet at 15 degrees on the surface, the touch button was used as the trigger, and the gyro input was disabled for this device.
- The Logitech G305 was wirelessly connected to an intel NUC (BOXNUC7CJYSAMN1) running Win10. The surface book 2 can be connected via Steam Remote Play, while running Clumsy 0.2. Fifty ms lag occurs via the internet connection, making this mouse have a similar input latency to those of the phone and tablet running Steam Link. The system latency is 50–67 ms. The participants were informed of the artificial latency after all the tests were completed.
- The Razer Viper 8K runs on a 2K pooling rate and has a USB connection. It is represented by a pinnacle shape. Since several apps and games have problems with 8k, 2k was chosen for the test. The system latency is 8–16 ms.
- The Steam Controller runs on a USB connection. Gyro is included for mobile shooter players. The system latency is 42–58 ms for the touch feature and 8–16 ms for the gyro feature. The participants chose to use the right touchpad or the gyro for aiming and selecting a target.
- The Dual Sense Controller runs on a USB connection. It has a gyro and a pinnacle shape. The system latency is 50–83 ms for the touch feature and 8–16 ms for the gyro feature. The participants chose to use the right analog stick or the gyro for aiming and selecting a target with the right bumper.

4.2. 3D Aim Trainer Tile Frenzy

This web-based game is intended for players to train and improve their aiming in shooter games with many different challenges. Three yellow square appeared on the screen randomly, and the players aimed with the crosshair on the center of the screen. Hitting and destroying the target were performed with one click. The targets reappeared in a different locations on the screen. The score was determined by the number of targets hit in one minute. This was chosen to be used for the test subjects to gain familiarity with the testing devices before performing the Fitts task. With its 3D environment, the game is used as a test for player performance in a real gaming scenario. Tile frenzy level was chosen for the players. The participants took 1–3 tries to be familiar with the input device. The best score was recorded.

4.3. Fitts Task 2

The testing setup with Fitts Task 2 was used in a previous study [4], with a 2D task type, 15 targets, and amplitudes of 128, 256, and 512. The target widths were 20 and 35 pixels. The error threshold was set at 50%. Sixteen targets were arranged in a circular array, with the next "task" highlighted in blue (Figure 13). Audio feedback was given to notify if the task succeeded or failed. The movement time, error rate, throughput, target re-entry count, and movement path were logged with the task settings.

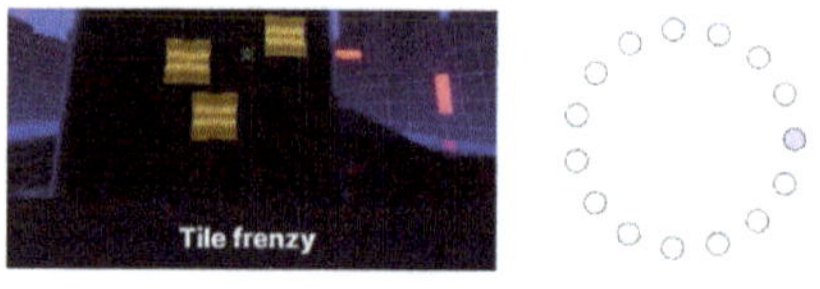

Figure 13. Tile frenzy level in the 3D Aim Trainer (**left**) and Fitts Task 2 target example (**right**).

4.4. In-Game Experience: Death Stranding Director's Cut Firing Range—Drill13

This level (Figure 14) was selected for the participants to experience the input devices in a real game scenario with 13 human-shaped targets and 13 hexagon-shaped targets. Several targets were stationary, while the others moved in a predictable linear path. The Dual Sense, Razer 8k, iPad Pro, and prototype with both grip styles were tested in this order. Since the game was designed for a controller with a KB+M, it was used first for the testers to gain familiarity with the game.

Figure 14. Screenshot of the in-game experience testing.

4.5. Questionnaire/Interview

A questionnaire was used to survey each device understanding the participant's preference for the input methods. The Likert-scale-like survey questions included the following aspects: ergonomics, likeness, handiness, usage scenario imagination, and the desire for adjustments. A brief interview was also conducted for the task of arranging the game platforms and input methods according to preference and the possibility of purchase.

4.6. Participants

Eleven participants were recruited from the university and social media groups of related mobile games, with four males and seven females aged 16~31. The participants were later grouped into five groups: non-FPS (not used to 3D FPS/TPS games), MsW/ME (used to operate FPS/TPS game with a mouse and has mobile FPS/TPS experience), Msw/oME (used to operate FPS/TPS games with a mouse, but had no FPS/TPS experience), MS/Stylus (had experience using the Steam controller or drawing tablet for gaming), MobileGm (experienced mobile FPS/TPS players).

5. Results

In the preliminary findings, the Razer 8K emerges as the frontrunner, exhibiting superior scores, throughput, and the lowest error rate, as illustrated in Figures 15 and 16. Notably, specific participant groups, such as MsW/ME and MobileGM, demonstrate higher performance on the iPadPro when compared to the latency-matched mouse G304. The questionnaire underscores distinctions in usage scenarios and likability perceptions across various input devices, as depicted in Figure 17.

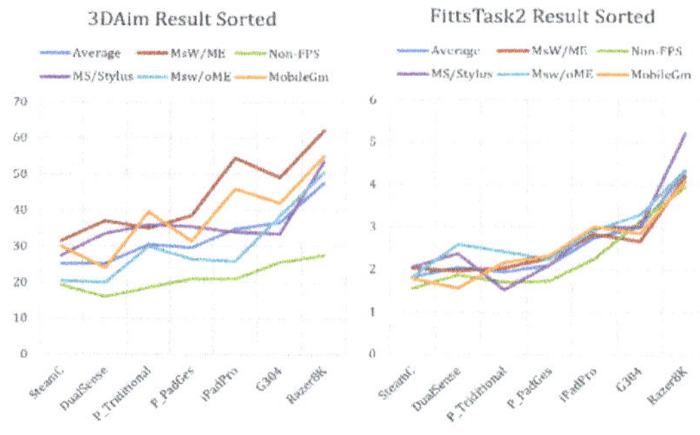

Figure 15. Three-dimensional Aim Trainer score (**Left**) and Fitts Task 2 throughput (**Right**).

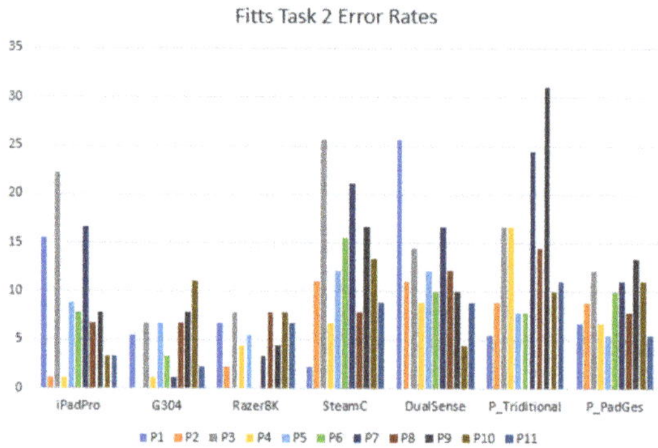

Figure 16. Error rate of participants across devices in Fitts Task 2.

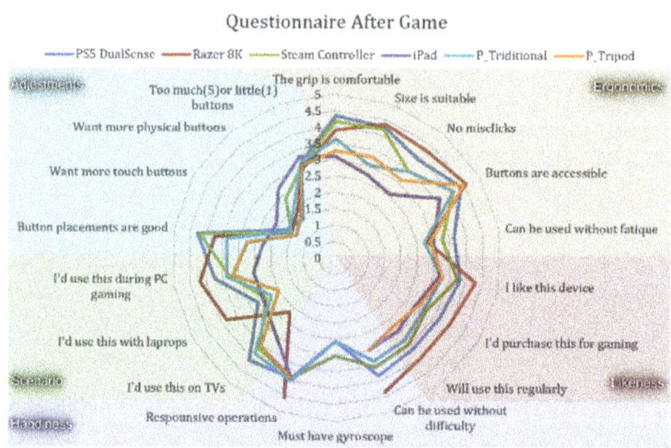

Figure 17. Likert scale result between devices.

6. Discussion of Results

The average score for the 3D Aim Trainer and throughput in the Fitts Task 2 shows similar results to those of previous studies, where the mouse has the highest input efficiency, followed by touch-based inputs, with a gap in between. However, touch-based inputs seems to be heavily affected by the touchpad size and gyroscope, where the Steam controller scored the lowest with its small touchpad and enabled gyroscope. Interestingly, while the iPad achieved a score close to the similar latency mouse G304, the mouse error rate seems to be not very affected by the increased latency. The gyro-enabled devices also show the highest error rate on average. The amounts of direction change in movement with the mouse, the iPad Pro, and the prototype with the tripod grip were relatively low and similar. The higher direction change with other input devices was caused by unfamiliarity with the gyro-aided controls. The results from the 3D Aim Trainer showed that the touch input devices and the latency-matched G304 mouse had similar throughputs to those of the Fitts test. This was caused by the larger target for the touchpad and gyro.

The 3D Aim Trainer shows more diverse results between the experience groups compared to those in the Fitts Task 2. The non-FPS participants scored the worst, and all the participants achieved their highest score with the Razer 8K mouse. Apart from the non-FPS and Msw/oME participants, the iPad performs similar or often better than the G304 that has a similar latency (artificially added). The Steam controller, Dual Sense, and both grip styles of the prototype controller performed relatively poorly and similarly to the other devices. The prototype controller performed worse than expected, with the P_PadGes intended to be a replacement for the iPad, offering a more ergonomic grip style and physical fire button. This may be caused by the types of switches used for the button, the touchpad size (phone screen) that is not big enough or shaped correctly for the task, the grip style not being used by the participants, or the overall size of the device being a bit too big for the test participants. Interestingly, the P_PadGes has a much lower error rate compared to that of the P_Triditional and is more similar to the iPad's error rate. This may be the result of similar grip style and lack of a gyro input, which distracted them more than aiding them.

The favorability of the devices has a relationship with their performance. However, devices such as the Dual Sense controller obtained a higher score on the likeness scale, but this may be the result of its familiarity and versatility among many game genres, as well as the great ergonomic design and interesting force feedback mechanism.

7. Relationship with Previous Work

The test shows similar results to those of previous studies, where mouse control was the best, most efficient input method, followed by having a gap between the touchpad and gamepad. By having a mouse with an artificial latency similar to the touchpad device, this research shows the possibility of a touchpad having a similar performance to the mouse. The attempt of designing a gamepad with a large touchpad was also pursued, and although the result did not match the expectations, the knowledge gained will help with future designs. Using the standard Fitts Task throughput measurement means the results can be easily compared and analyzed, with more detailed participation labeling and grouping and testing with 3D Aim Trainer, which is an easier setup that has produced a more significant result; this study may aid future FPS/TPS input method studies.

8. Limitations

Without a proper touchpad that can match the latency of a gaming mouse, proving that the touch efficiency in this research (iPad with Steam Link) is similar to that of the latency-matched mouse (artificially latency added Logitech G304) only increases the possibility for touchpads to have a similar performance to that of a mouse in an FPS gaming environment; a touchpad that actually has similar low latency to that of a gaming mouse is needed for further testing to answer if the touchpad input can replace the mouse input. The surface of the touchpad also needs to be improved since many participants complained that the glossy surface (due to the age of the device and using many alcohol wipes for the "cleaning" of the surface possibly removing the coating) of the glass caused friction that reduced the smoothness of the operation (while natural grease on the finger helps reducing sudden friction, it is inconsistent and not pleasant). If possible, haptic feedback similar to that of the Steam controller is recommended, along with a sanded/etched surface. The prototype controller also received negative feedback on its size and weight since it is limited to the shape and weight of the phone used.

9. Future Work

This research shows that touchpads have a higher throughput compared to those of the traditional gamepads and have a similar performance to that of gaming mice with a similar latency. To understand if a touchpad can achieve a mouse-like performance and push the industry to adapt more touch-based controls, a prototype touchpad with latency and a poll rate matching the current gaming mouse specs is required. With this low-latency touchpad prototype that has been built, a test with fewer devices and scenarios can be conducted with more test subjects to understand the potential of the touch-based aiming input. The touchpad may also need to be designed for such use, but it is recommended to have a good sliding surface with haptic feedback; more research on the different types of gestures/grip styles used for touch-based controls is also needed to design a controller for such players. The surface may also be concave like the Steam controller touchpad, or have a complex surface, suiting thumb ergonomics. The buttons and triggers on the prototype also need improvement, with an actual low-latency touchpad instead of a mobile phone for a touchpad; the size, weight and shape of the prototype should be more flexible and likable.

10. Conclusions

This research shows that a touchpad-based input can have a similar performance to a mouse with similar latency. Participant groups with different gaming experiences can have very diverse preferences and performance outcomes. While the prototype touchpad implemented controller design did not achieve the expected performance or preference results, the knowledge and experiences gained will aid future studies and designs of such devices. With mobile FPS/TPS titles continuing to become more popular and more portable gaming devices, such as the Steam Deck, Logitech G Cloud, and GPD Win debut, gaining footing in the gaming market, using a touchpad as a mouse substitute input may become more common. However, more research is needed on the design of such devices to maximize the touch-based aiming performance; a prototype with a gaming-mouse-like latency and more test subjects from the mobile FPS/TPS community are also needed to conclude if a touchpad can become the new standard input method in the future.

This test was randomly carried out with the devices. After understanding the test subject's gaming history, each device was given to perform the Tile frenzy test for the participant to become familiar with the control. A Fitts Task 2 2D task was then performed, followed by the presentation of a questionnaire for feedback. After the firing range feature in the game Death Stranding Director's Cut was played with the devices, a final questionnaire was given, and an interview was performed. There were seven participants from the universities (three males and four females). Four were used to playing games with a mouse. Three had experienced with FPS games before the test, while two had no FPS gaming experience.

Author Contributions: Conceptualization, C.-E.C. and F.-W.T.; methodology, C.-E.C.; validation, C.-E.C. and F.-W.T.; formal analysis, C.-E.C.; investigation, C.-E.C.; resources, C.-E.C.; data curation, C.-E.C.; writing—original draft preparation, C.-E.C.; writing—review and editing, C.-E.C. and F.-W.T.; visualization, C.-E.C.; supervision, F.-W.T.; project administration, F.-W.T. All authors have read and agreed to the published version of the manuscript.

Funding: This research received no external funding.

Informed Consent Statement: Informed consent was obtained from all subjects involved in the study.

Data Availability Statement: Link to test data https://docs.google.com/spreadsheets/d/1og_NJEdUCP_idyjTnADWZ3iATj0s2ou_sroRypMrcC0/edit?usp=sharing.

Conflicts of Interest: The authors declare no conflict of interest.

References

1. Samdo. 2021 Essential Facts about the Video Game Industry. Entertainment Software Association, 7 February 2022. Available online: https://www.theesa.com/resource/2021-essential-facts-about-the-video-game-industry/ (accessed on 1 January 2022).
2. Isokoski, P.; Martin, B. Performance of Input Devices in FPS Target Acquisition. In Proceedings of the International Conference on Advances in Computer Entertainment Technology (ACE '07), Salzburg, Austria, 13–15 June 2007; Association for Computing Machinery: New York, NY, USA, 2007; pp. 240–241. [CrossRef]
3. Splatoonus. Live from Squid Research Lab. Tumblr, 30 May 2015. Available online: https://splatoonus.tumblr.com/post/120246428234/one-of-our-scientists-has-made-a-rather-stunning#notes (accessed on 1 January 2022).
4. Ramcharitar, A.; Teather, R.J. A Fitts' Law Evaluation of Video Game Controllers: Thumbstick, Touchpad and Gyrosensor. In Proceedings of the 2017 CHI Conference Extended Abstracts on Human Factors in Computing Systems (CHI EA '17), Denver, CO, USA, 6–11 May 2017; Association for Computing Machinery: New York, NY, USA, 2017; pp. 2860–2866. [CrossRef]
5. Global Consumer Spending in Mobile Apps Reached $133 Billion in 2021, up Nearly 20% from 2020. Sensor Tower—Market-Leading Digital & Mobile Intelligence. Available online: https://sensortower.com/blog/app-revenue-and-downloads-2021 (accessed on 1 January 2022).
6. Apex Legends Mobile Regional Betas Start Soon. Available online: https://www.ea.com/games/apex-legends/news/apex-legends-mobile-regional-betas (accessed on 1 January 2022).
7. Steam Deck™. Welcome to Steam, 27 May 2022. Available online: https://store.steampowered.com/steamdeck (accessed on 1 January 2023).
8. Looser, J.; Cockburn, A.; Savage, J. On the Validity of Using First-Person Shooters for Fitts' Law Studies. *People Comput. XIX* **2012**, 2, 33–36.
9. Baldauf, M.; Fröhlich, P.; Adegeye, F.; Suette, S. Investigating On-Screen Gamepad Designs for Smartphone-Controlled Video Games. *ACM Trans. Multimed. Comput. Commun. Appl.* **2015**, *12*, 21. [CrossRef]
10. Toktaş, A.O.; Serif, T. Evaluation of Crosshair-Aided Gyroscope Gamepad Controller. In *Lecture Notes in Computer Science*; Springer: Cham, Switzerland, 2019. [CrossRef]
11. Haider, M.W. PUBG Mobile: Five Reasons Why Pubg the Most Popular Battle Royale Game in India. News18, 22 May 2019. Available online: https://www.news18.com/news/tech/pubg-mobile-five-reasons-why-pubg-the-most-popular-battle-royale-game-in-india-2152567.html (accessed on 1 May 2022).
12. PUBG Mobile Controls: Two Fingers vs. Four Finger Claw vs. Air Triggers—Technology News, Firstpost. Tech2, 29 March 2019. Available online: https://www.firstpost.com/tech/gaming/pubg-mobile-controls-two-fingers-vs-four-finger-claw-vs-air-triggers-6343821.html (accessed on 1 March 2022).
13. I Built the *Most Op* Controller | Optical Mouse Controller | Fortnite Warzone. YouTube, 28 October 2020. Available online: https://www.youtube.com/watch?v=KC9tUrWcFmw (accessed on 1 March 2022).
14. Ivkovic, Z.; Stavness, I.; Gutwin, C.; Sutcliffe, S. Quantifying and Mitigating the Negative Effects of Local Latencies on Aiming in 3D Shooter Games. In Proceedings of the 33rd Annual ACM Conference on Human Factors in Computing Systems—CHI '15, Seoul, Republic of Korea, 18–23 April 2015. [CrossRef]
15. How to Reduce LAG—A Guide to Better System Latency. NVIDIA. Available online: https://www.nvidia.com/en-us/geforce/guides/system-latency-optimization-guide/ (accessed on 1 March 2022).
16. Lander, R.; Reid, A.; Khasawneh, M.T. Impact of ergonomics on game-pad design: Qualitative and quantitative analysis. In Proceedings of the IIE Annual Conference, Orlando, FL, USA, 20–24 May 2006; pp. 1–6. Available online: https://login.ezproxy.lib.ntust.edu.tw/login?url=https://www.proquest.com/scholarly-journals/impact-ergonomics-on-game-pad-design-qualitative/docview/192457965/se-2?accountid=8014 (accessed on 1 March 2022).
17. Merdenyan, B.; Petrie, H. User Reviews of Gamepad Controllers: A Source of User Requirements and User Experience. In Proceedings of the 2015 Annual Symposium on Computer-Human Interaction in Play, London, UK, 5–7 October 2015; pp. 643–648. [CrossRef]
18. 2 Finger + Gyroscope HANDCAM | Sensitivity | PUBG Mobile. YouTube, 10 September 2019. Available online: https://www.youtube.com/watch?v=SthNRvg469A (accessed on 1 May 2022).
19. 3 Finger Claw Guide/Tutorial (PUBG Mobile/BGMI) Tips and Tricks Setup/Sensitivity Settings (Handcam). YouTube, 10 July 2021. Available online: https://www.youtube.com/watch?v=gx5o3z0et1k (accessed on 1 May 2022).
20. 4 Fingers + Gyroscope Handcam. YouTube, 16 July 2021. Available online: https://www.youtube.com/watch?v=wbS-tmVtkgQ (accessed on 1 May 2022).
21. What 5000 Hours of Apex on Controller Looks Like... (7000 Damage). YouTube, 5 February 2022. Available online: https://www.youtube.com/watch?v=HFoHtsSHPRw&list=PLay2U7qcvV8Pa8w_vAKn4WMpJlCXCJnHP&index=3&t=696s (accessed on 1 May 2022).

22. IPAD 8 Generation | PUBG Mobile | HANDCAM | 4 Finger Claw + No Gyro | Montage | #1. YouTube, 27 December 2020. Available online: https://www.youtube.com/watch?v=VDqyKKFVqK0&list=PLay2U7qcvV8Pa8w_vAKn4WMpJlCXCJnHP&index=6&t=53s (accessed on 1 May 2022).
23. [Pubg Mobile] Six Fingers Claw Handcam L Non-Gyro IPAD Sensitivity #22. YouTube, 8 November 2021. Available online: https://www.youtube.com/watch?v=AIki_zkIYpY&list=PLay2U7qcvV8Pa8w_vAKn4WMpJlCXCJnHP&index=5&t=118s (accessed on 1 May 2022).

engineering proceedings

MDPI

Proceeding Paper

Study on Building Information Modeling Application for Building Space Design Conflict Effects †

Ren-Jwo Tsay

Department of Internal Design and Civil Engineering, Vanung University, Taoyuan 320313, Taiwan; trj@vnu.edu.tw

† Presented at the IEEE 5th Eurasia Conference on Biomedical Engineering, Healthcare and Sustainability, Tainan, Taiwan, 2–4 June 2023.

Abstract: In building structure design, refinements during construction for practical use or space conflicts induce construction delays and budget increases. Building information modeling (BIM) provides a 3D building model which includes the structure's size, material, and construction information so designers clearly understand the whole construction process. For an irregular shaped building or elements such as a ramp or a curved roof, space conflict problems occur. The Autodesk's Navisworks program provides a visual analysis tool for users to roam in a 3D model to simulate possible space conflicts and avoid them before the final design. In this study, we used the AutoDesk Revit program to set up a building system with a Navisworks space distance measurement tool to confirm the design's rationality. An optimal structure system design process is proposed for practical applications to reduce the possibility of design changes.

Keywords: space conflicts; BIM; Revit; Navisworks

Citation: Tsay, R.-J. Study on Building Information Modeling Application for Building Space Design Conflict Effects. *Eng. Proc.* **2023**, *55*, 93. https://doi.org/10.3390/engproc2023055093

Academic Editors: Teen-Hang Meen, Kuei-Shu Hsu and Cheng-Fu Yang

Published: 18 January 2024

1. Introduction

For building systems, the establishment of a spatial structure system needs to consider the user's requirements so that a discontinuous design structure system or insufficient space and height can be avoided. Changes in design will delay construction time and impact the budget. Scholars have discussed the issue. Fazil et al. [1] used BIM combined with Navisworks to set up a construction management (CM) system. Latiffi et al. [2] discussed the application of BIM in the Malaysian industry system and its benefit. Xu et al. [3] integrated Revit and Tekla into Navisworks to set up a steel structure building system and compared the difference. In real building structure design, engineers define the element sizes to confirm the building's safety and ensure its use function fits the space need. Thus, we combine BIM and Navisworks programs to develop a useful design process for structural designers to reduce design change problems.

2. BIM for Irregular Structure System Design and Spatial Impact Analysis

2.1. *Irregular Structure System Design via BIM (Revit)*

For an irregular structure system such as a ramp or a non-rectangular building, the designer uses a 2D concept layout structure system. When an irregular 3D shape causes a space to be blocked by beams and columns, the Revit BIM 2021 program is used to set up building models. Its most important function is its ramp making tool, as shown in Figure 1. Users can use the function to model a building-floor-height ramp system. Using this function, the ramp floor support beam system locations are defined.

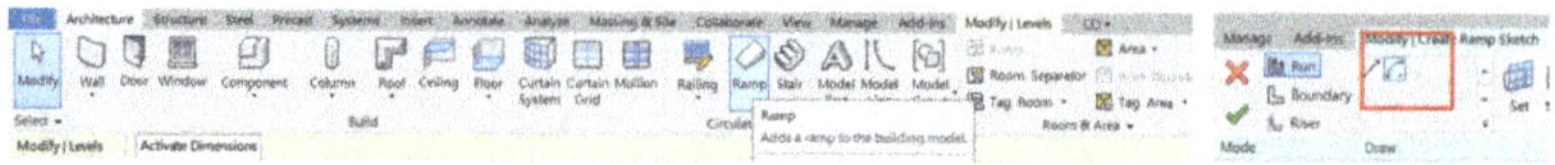

Figure 1. Revit's ramp-making function.

2.2. *Code for Car Parking Lot Structures*

The building code of Taiwan includes the following regulations.

2.2.1. Car Height Limit

The height of small cars shall not exceed 1.5 times their full width, and their maximum height shall not exceed 2.85 m.

2.2.2. Lane Slope Limit

The slope of the lane shall not exceed one to six, and the surface shall be made of a rough surface or other non-slippery materials. The radius of the inner curve of the lane shall be more than five meters.

2.2.3. Clear Height Restrictions for Fire Pipeline Facilities Limit

When an airtight sprinkler head is installed under a beam, the distance between the return plate and the bottom of the beam shall be less than 10 centimeters, and the distance from the floor or ceiling shall be less than 50 cm.

Thus, we simulated a ramp slope space from bottom to top, as shown in Figure 2.

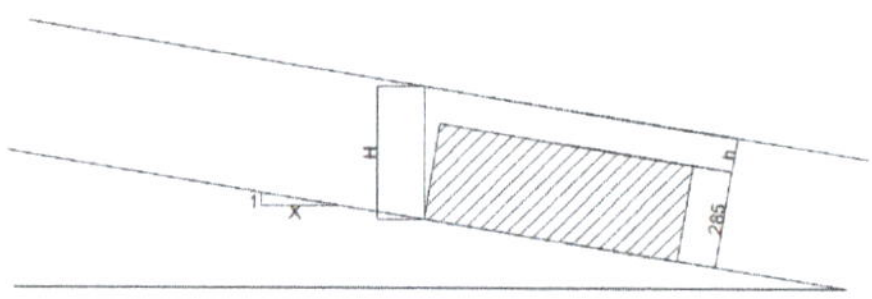

Figure 2. Net height for parking lot ramp.

$$tan\theta = 1/X \quad (1)$$

$$H_{min} = (2.85 + h)sec\theta \quad (2)$$

θ: the slope angle;
X: the unit's height increase in the horizontal dimension;
h: the net height from the car top.

By using Equation (2), we simulated the space from a car top of h = 0.15 m, h = 0.5 m, and h = 1 m to obtain the min. space to avoid impact, as shown in Tables 1 and 2. The different colors mean the clear height fit the design required.

Table 1. Straight ramp slope: net height calculation.

H (Rise) (m)	X (m)	θ (deg)	Hmin = 3.85/cos(θ)	Hmin = 3.35/cos(θ)	Hmin = 3/cos(θa)
1	6	9.4623	3.9031	3.3962	3.0414
1	7	8.1301	3.8891	3.3840	3.0305
1	8	7.1250	3.8800	3.3761	3.0233
1	9	6.3402	3.8737	3.3706	3.0185
1	10	5.7106	3.8692	3.3667	3.0150

Table 2. Circular ramp slope: net height calculation.

H (m)	R (m)	θ (deg)	Hmin = 3.85/cos(θ)	Hmin = 3.35/cos(θ)	Hmin = 3/cos(θ)
3.5	12	2.657777633	3.8541	3.3536	3.0032
3.5	13	2.453593441	3.8535	3.3531	3.0027
3.5	14	2.278528576	3.8530	3.3526	3.0024
3.5	15	2.126771136	3.8527	3.3523	3.0021
4	12	3.036793779	3.8554	3.3547	3.0042
4	13	2.803582438	3.8546	3.3540	3.0036
4	14	2.603612686	3.8540	3.3535	3.0031
4	15	2.430254036	3.8535	3.3530	3.0027
5	12	3.793995785	3.8584	3.3573	3.0066
5	13	3.502906792	3.8572	3.3563	3.0056
5	14	3.253257186	3.8562	3.3554	3.0048
5	15	3.036793779	3.8554	3.3547	3.0042

2.2.4. For a Circular Ramp

$$X = 2\pi R \tag{3}$$

$$tan\theta = H/X \tag{4}$$

X: the circumference of a circle with a radius R;
H: the building's height.

2.3. Navisworks Spatial Conflict Analysis

In AutoDesk Navisworks 2021 software, the model created by the Revit program is exported in an "nwc" format file and then imported into Navisworks to create a spatial model. Navisworks space distance measurement tool is used to measure the x-, y-, and z-axis distances of two spatial points in the design model to obtain the real space distance. Then, users can confirm the design results using codes. In addition, Navisworks also provides a space cruise function. Users can understand the feasibility of the design results through animation and find problems before real construction process.

3. Case Application

The above method was used to analyze a five floor parking lot space as an example. Its plan view is shown in Figure 3. The ramp design coefficients are shown in Figure 4. The building's north elevation is shown in Figure 5. The N-S and E-W section profiles are shown in Figures 6 and 7. From the profile map, we define the support beam's real location and set up the support frame system as shown in Figure 8.

In order to define the ramp and plane floor connection and improve the car passing joint comfortableness, we set up a profile map (Figure 9) to discuss the joint function detail sections. To understand the effect of sunlight, we used Revit to simulate the effect to discuss an energy-reducing design (Figure 10).

Then, we transferred the model to AutoDesk Navisworks 2021 software, and the whole building model is shown in Figure 11. We simulated the structure system design as shown in Figure 12.

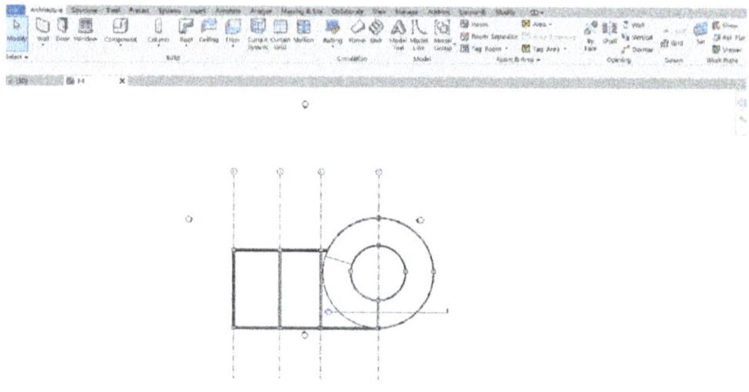

Figure 3. Revit's structure system.

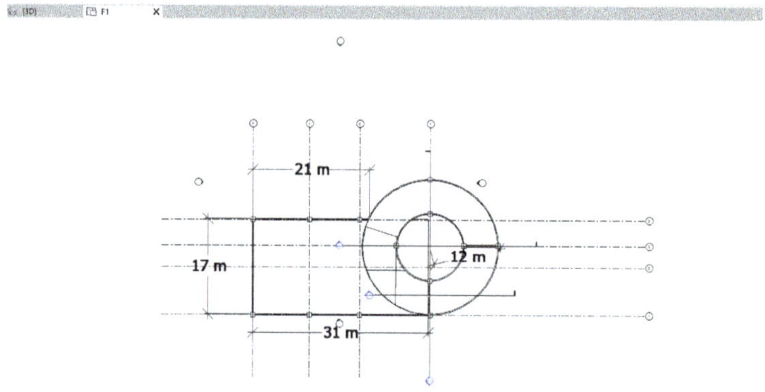

Figure 4. Parking lot ramp design coefficient.

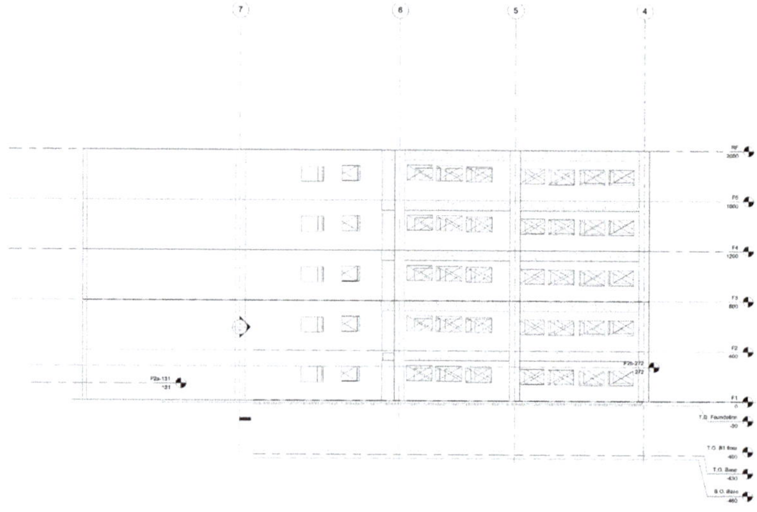

Figure 5. North vertical section map.

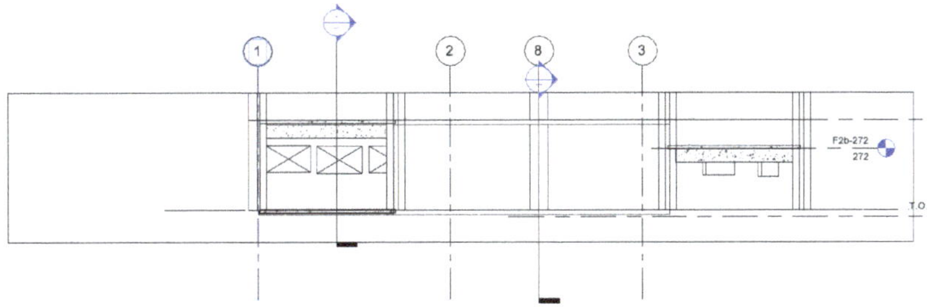

Figure 6. Profile for N-S section.

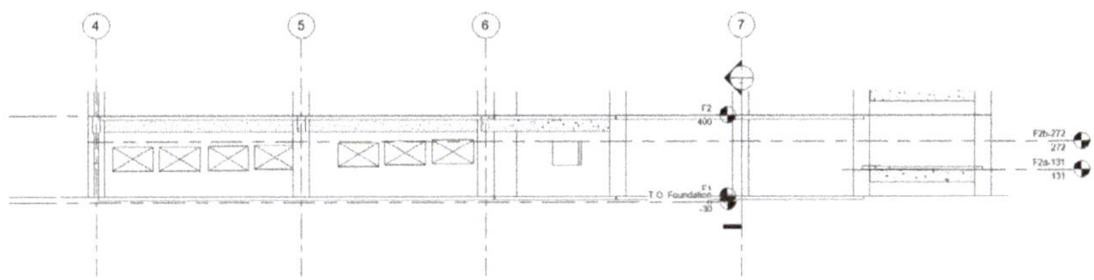

Figure 7. Profile for E-W section.

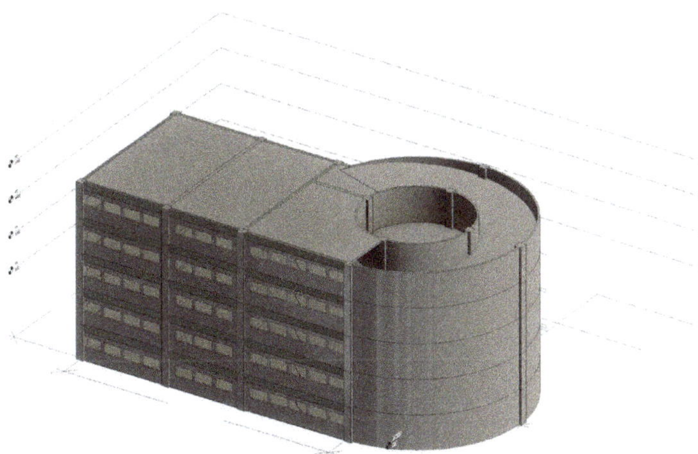

Figure 8. 3D Structure system render effect for parking lot in Revit.

Then, we used a roaming function to traverse the building model and measured the spatial two point distance to obtain the net height space in the structure system (Figure 13). The ramp width was also obtained, as shown in Figure 14. In Figures 13 and 14 red green and blue box mean the measure results for x, y, z axis distance.

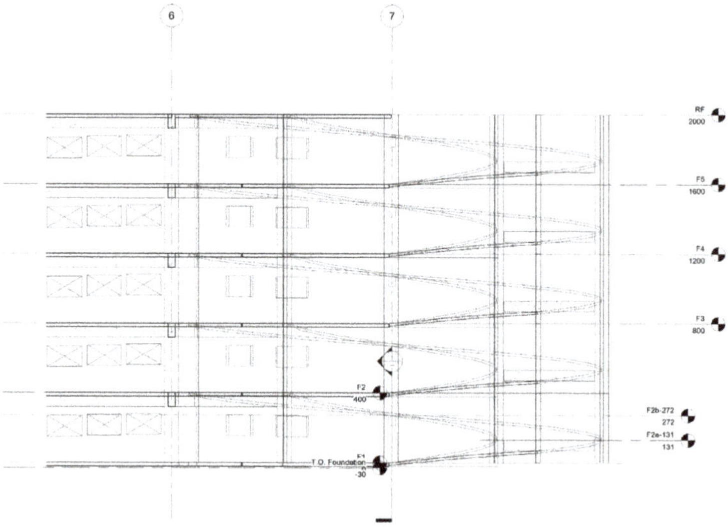

Figure 9. Details of the ramp section interaction of the building floor.

Figure 10. Map for sunlight rendering effect in building.

Figure 11. Navisworks 3D building view.

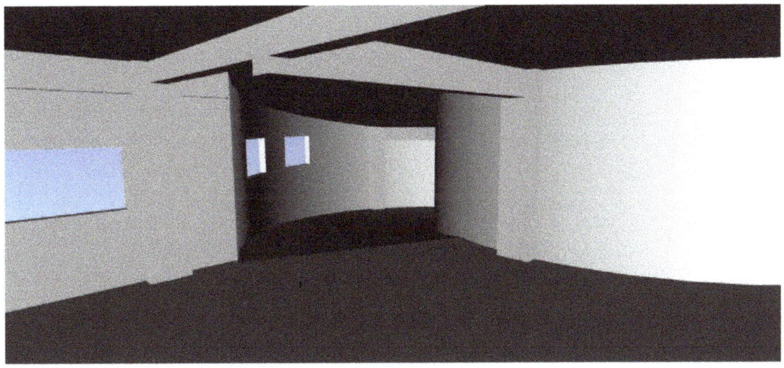

Figure 12. Internal space view result in Navisworks.

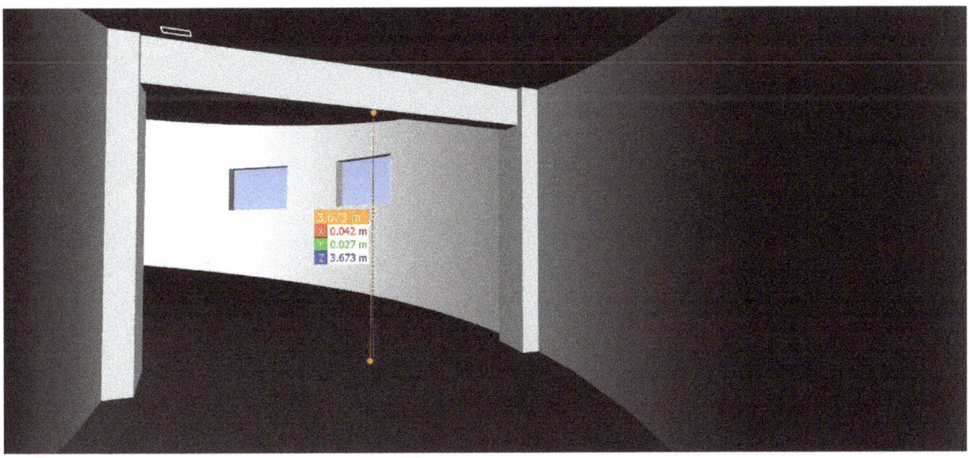

Figure 13. Space net height measurement in Navisworks.

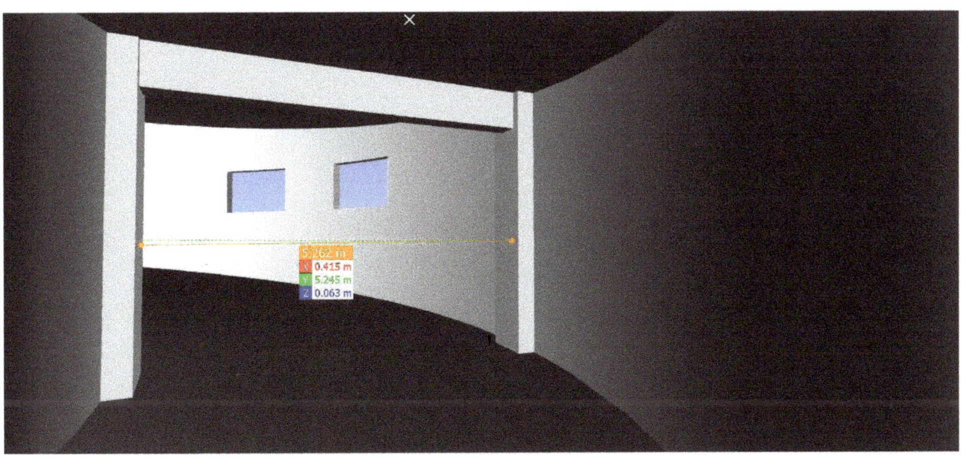

Figure 14. Ramp width measurement in Navisworks.

4. Conclusions

From the above analysis, the following conclusions were made from a real-case application.

The combination of the Autodesk Revit and Autodesk Navisworks programs provides a real simulation for the actual structure system's spatial relationships. Through a spatial analysis of the structural system, it can avoid unreasonable design results. A design combined with BIM and a Navisworks simulation provides the a reasonable structure system and improves architectural lighting and the building's energy savings.

Funding: This research received no external funding.

Institutional Review Board Statement: Not applicable.

Informed Consent Statement: Informed consent was obtained from all subjects involved in the study.

Data Availability Statement: The data that support the findings of this study are available on request from the corresponding author, upon reasonable request.

Conflicts of Interest: The author declares no conflict of interest.

References

1. Fazil, S.M.; Sultan, C.R.; Kumar, P.N. Planning and scheduling of residential building using MS project and BIM. *Int. Res. J. Eng. Technol.* **2021**, *8*, 1473–1479.
2. Latiffi, A.A.; Mohd, S.; Kasim, N.; Fathi, M.S. Building Information Modeling (BIM) Application in Malaysian Construction Industry. *Int. J. Constr. Eng. Manag.* **2013**, *2*, 1–6.
3. Xu, Y.; Zhang, J.; Li, D.; Ao, C. BIM Model Integration of Concrete and Steel Structures in Assembled Substations. *Adv. Comput. Sci. Res.* **2019**, *91*, 65–70.

engineering proceedings

MDPI

Proceeding Paper

La^{3+}-Induced Band-Gap Modifications in Barium Hexaferrite: An Investigation of the Structural, Optical, and Dielectric Properties †

Gowravi Subramanyam, Nishkala Kadambu Rao and Mamatha Durgadas Daivajna *

Department of Physics, Manipal Institute of Technology, Manipal Academy of Higher Education, Manipal 576104, India; sgowravi1@gmail.com (G.S.); nishkalakr@gmail.com (N.K.R.)

* Correspondence: mamata.shet@yahoo.com

† Presented at the IEEE 5th Eurasia Conference on Biomedical Engineering, Healthcare and Sustainability, Tainan, Taiwan, 2–4 June 2023.

Abstract: M-type barium hexaferrites, $BaFe_{(12-x)}La_xO_{19}$ (x = 0.0, 0.05, 0.1, 0.15, and 0.2), were prepared by a low-cost solid-state reaction method. The specimens crystallized in a non-centrosymmetric hexagonal magnetoplumbite structure that belonged to the $P6_3/mmc$ space group. Morphologically, the samples were dense with hexagonal plate-like grains and size variations of around 1.19–1.70 µm. The optical band gap of the system was reduced in the wide band-gap region from 1.78 to 1.74 eV. The band-gap values would be useful in photocatalysis and photovoltaics. The AC conductivity was enhanced with La^{3+} substitution, following Jonscher's power law. Maxwell–Wagner-type polarization was observed in the specimen, and tangent loss decreased with La^{3+} substitution. The values of the tangent loss were in the appropriate range for electromagnetic shielding applications.

Keywords: M-type hexaferrite; water splitting; dielectric properties; M–W-type relaxation

Citation: Subramanyam, G.; Rao, N.K.; Daivajna, M.D. La^{3+}-Induced Band-Gap Modifications in Barium Hexaferrite: An Investigation of the Structural, Optical, and Dielectric Properties. *Eng. Proc.* **2023**, *55*, 94. https://doi.org/10.3390/engproc2023055094

Academic Editors: Teen-Hang Meen, Kuei-Shu Hsu and Cheng-Fu Yang

Published: 29 January 2024

1. Introduction

Concerns have been growing about the energy crisis and environmental pollution, especially pollutants in water bodies. Various methods have been employed to reduce the pollutants in water and produce 'clean' energy by alternative methods. Photocatalysis is used to produce hydrogen, which is regarded as a new energy source. In using photocatalysis, tuning the band gap is a challenging process. Metal oxide semiconductors with energies in the narrow band-gap region are highly stable and show high absorption in the visible range. Also, they have good charge and separation properties [1,2]. Ferrite-based materials are widely known for their excellent photocatalysis properties, relatively low band gaps, thermal and chemical stability, and optical and magnetic properties [3,4].

M-type barium hexaferrite ($BaFe_{12}O_{19}$) belongs to the ferrite family and shows a hexagonal crystal structure. It is the most common and simple hexaferrite. It shows a high Curie temperature, high saturation magnetization, large magnetocrystalline anisotropy, high chemical stability, high electrical resistance, and a good optical band gap. Each unit cell of barium hexaferrite needs two formula units and comprises ten layers of oxygen anions with four oxygen ions in each layer. Two Ba ions replace one of the oxygen ions in the middle layers. Fe^{3+} ions occupy the interstitial sites created by the oxygen ions. However, the Fe^{3+} ions in the interstitial sites are in five different crystallographic environments, namely, *12k*, $4f_1$, $4f_2$, *2a*, and *2b* [5,6].

Many research works have been conducted on the optical, electrical, and photocatalytic properties of ferrites [7–11]. The main reason for choosing hexagonal ferrite is the number of oxygens in its lattice (19 per formula unit), as the redox activity depends on the oxygen storage capability of the molecule in its lattice [12]. Also, during sintering, ferrites tend to create oxygen vacancies, which enhances the reduction by fixing oxygen in the existing vacancies and tuning the band gap. Its electrical and optical properties, such as its band gap, are tuned using doping. In summary, these materials are promising for photocatalytic applications. In the present research, La-substituted barium hexaferrites were synthesized using a solid-state reaction method and their structural, morphological, optical, and dielectric properties were studied.

2. Materials and Methods

The La^{3+}-substituted barium hexaferrites, $BaFe_{(12-x)}La_xO_{19}$ (x = 0−0.20, Δx = 0.5), were prepared using a solid-state reaction. The precursors were weighed stoichiometrically according to Equation (1).

$$BaCO_3 + \frac{x}{2}La_2O_3 + (6 - \frac{x}{2})Fe_2O_3 \rightarrow BaLa_xFe_{12-x}O_{19} + CO_2 \quad (1)$$

The weighed precursors were mixed and ground using an agate mortar and pestle for 4 h. These powders were calcined at 1200 °C for 6 h. The calcination was repeated 3 times. The powders were mixed with 1% PVA solution, then pressed using a hydraulic press and sintered at 1200 °C for 6 h. X-ray diffraction measurements were carried out to study the structural properties and phase formation of the prepared $BaLa_xFe_{12-x}O_{19}$ using a Brooker (D2 Phaser) instrument in the 2Θ range of 20–80°. The morphological properties were studied with field emission scanning electron microscopy (FE-SEM) measurements using a CARL ZEISS SIGMA 03-81, and elemental analysis was performed using EDS measurements (OXFORD Instruments, made in Germany). A UV-Vis-NIR spectroscope in diffused reflectance mode was used to study the optical properties, with wavelengths ranging from 180 to 2500 nm.

3. Results and Discussions

3.1. Structure

Powder X-ray diffraction measurements were carried out to study the structures and phases of the prepared samples. Rietveld analysis of the XRD patterns shown in Figure 1 confirmed the single-phase magnetoplumbite structure of barium hexaferrite that crystallized in a hexagonal crystal structure of the *P6₃/mmc* space group. The values of the lattice parameters, unit cell volumes, average crystallite sizes, and refinement parameters are presented in Table 1. The lattice parameters, 'a' and 'c', varied slightly, and the unit cell volume, 'V', increased after doping of La^{3+} as compared to the pristine sample. These variations, along with the variation in unit cell volume, depended on the ionic radii of the Fe^{3+} and La^{3+} ions. Sherrer's formula [13] was used to calculate the average crystallite size, 'D', given in Equation (2).

$$D = \frac{k\lambda}{\beta \cos} \quad (2)$$

where 'D' is the average particle size, 'λ' is the wavelength of the incident beam, 'k' is the shape constant, 'β' is the full width at half maximum, and 'θ' is the angle of diffraction. The average crystal size increased up to x = 0.10 and then slightly decreased.

Table 1. List of lattice parameters, unit cell volumes, average crystallite sizes, and refinement parameters.

La Content (x)	Crystallite Size D (nm)	a = b (Å)	c (Å)	c/a Ratio	Unit Cell Volume $(Å)^3$	Rp	Rwp	Rexp	χ^2
0.0	34.224	5.890	23.215	3.940	697.671	12.1	15.1	10.44	2.10
0.05	44.256	5.896	23.213	3.936	699.019	23.7	23.2	21.35	1.19
0.10	54.947	5.895	23.208	3.936	698.637	14.3	17.6	15.95	1.22
0.15	45.029	5.894	23.202	3.936	698.201	18.3	20.6	15.42	1.79
0.20	46.858	5.895	23.209	3.936	698.687	19.2	21.4	16.13	1.77

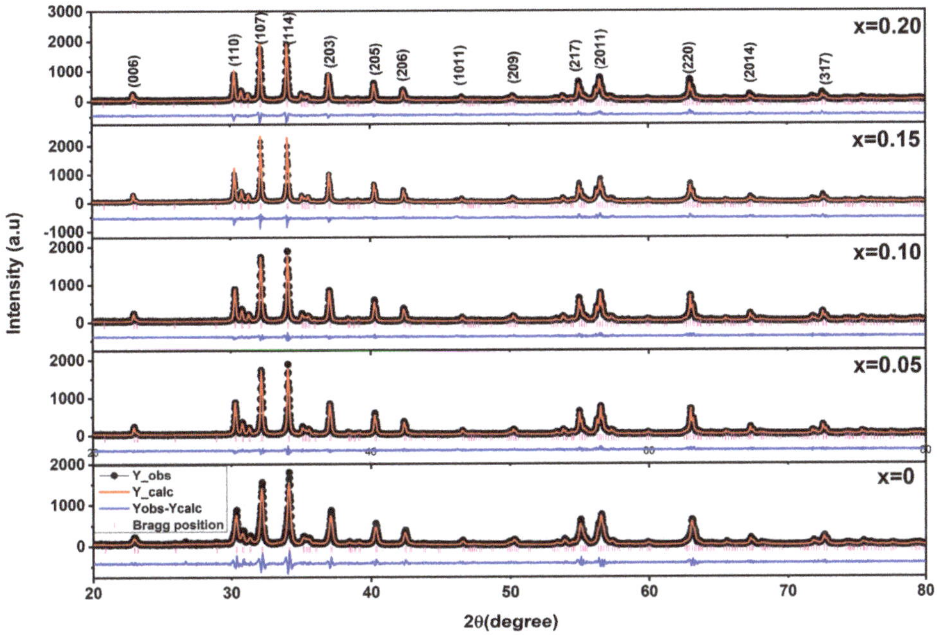

Figure 1. Rietveld refinement of the X-ray diffraction pattern of the $BaLa_{(x)}Fe_{12-x}O_{19}$, (x = 0–0.20) samples.

3.2. Morphological Study Using FE-SEM

FE-SEM was used to study the surface morphology of the prepared samples, as shown in Figure 2. The images were captured at a magnification of 40 kX. Close-packed, hexagonal plate-like grains of different sizes were observed in the images. The average grain size was measured using 'ImageJ' software (Figure 3) and found to be in the range of 1.19 to 1.70 μm.

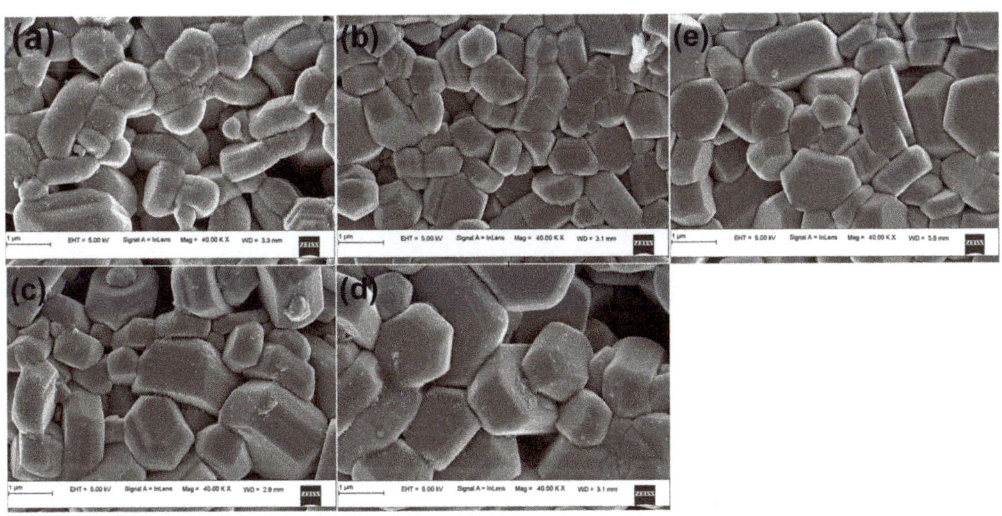

Figure 2. FE-SEM images of (**a**) x = 0, (**b**) x = 0.05, (**c**) x = 0.10, (**d**) x = 0.15, and (**e**) x = 0.20.

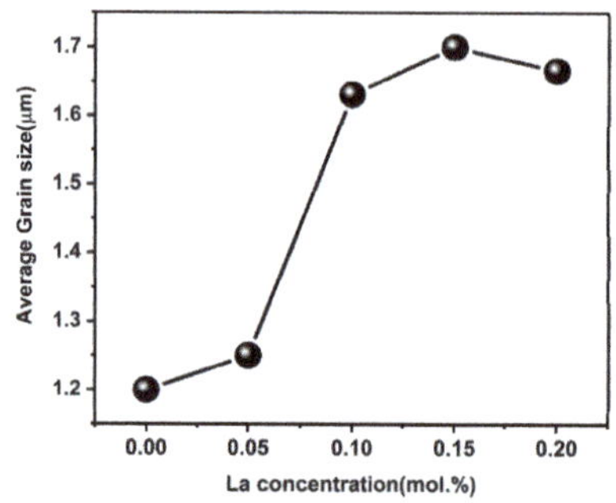

Figure 3. Variation in average grain size with La concentration.

3.3. Optical Properties

The band gaps of the prepared samples were measured to understand the optical properties. A UV-VIS-NIR spectrometer was used in diffused reflectance mode to measure the gaps [14]. The percentage of diffused reflectance (R %) spectra of the $BaLa_xFe_{12-x}O_{19}$ ($0 \leq x \leq 0.2$) are shown in Figure 4. The samples absorbed the light in the range of 200–550 nm, and the reflection increased above 550 nm. The reflectance of each sample decreased with the increase in La concentration. The optical absorption coefficient, α, was computed using the Kubelka–Munk function [15].

$$\alpha = {(1-R)^2}/{2R} \tag{3}$$

where α is the absorption coefficient and 'R' is the diffused reflectance. The optical band-gap energy was estimated by the Tauc relation [15].

$$\alpha h\nu = A(h\nu - E_g) \tag{4}$$

where '$h\nu$' is the photon energy, 'A' is a constant, and the absorption characterized index is given by 'n'. As $BaLa_xFe_{12-x}O_{19}$ is considered a direct band-gap material, n was $\frac{1}{2}$. A graph of $(\alpha h\nu)^2$ v/s Energy '$h\nu$' was plotted. The linear portion of $(\alpha h\nu)^2$ was extrapolated to calculate the band gap, as shown in Figure 5. The variation in band gap values with La concentration is shown in Figure 6. The band-gap values fell between 1.79 and 1.74 eV in the visible region. The maximum band-gap value was 1.79 eV for x = 0, and it slightly decreased with the concentration of La (x = 0.05, 0.10, 0.15, and 0.20).

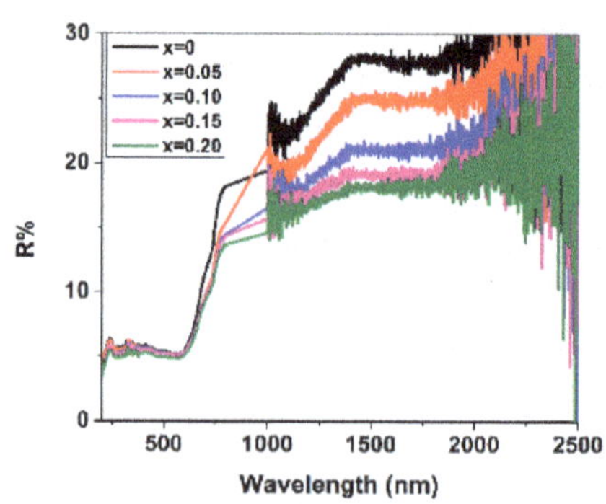

Figure 4. Variation in diffused reflectance with wavelength.

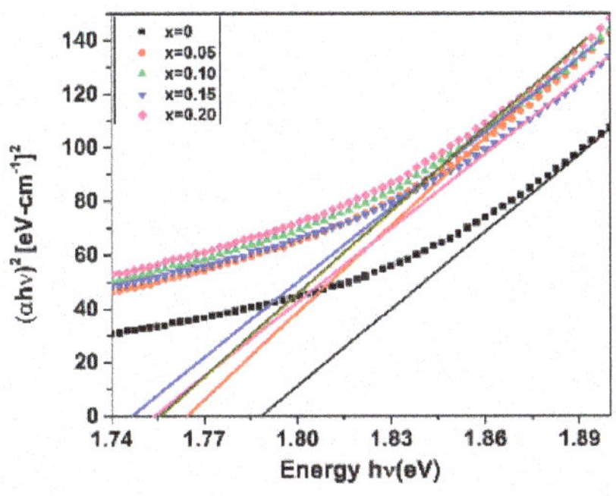

Figure 5. Tauc plot of $Ba_{1-x}La_xFe_{12}O_{19}$.

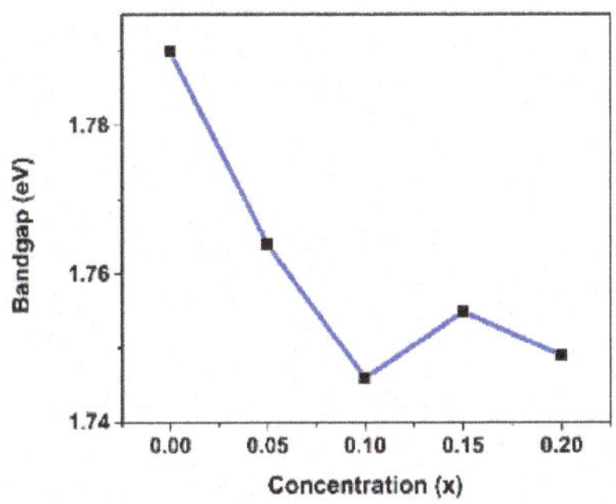

Figure 6. Variation in band gap with La concentration.

As the concentration of La^{3+} increased, a decrease in the bandwidth of the conduction band and valence band was observed, which reduced the band gap significantly. The smaller crystallite size was a reason for the lower-energy band gaps, which were different from the higher band-gap energy values reported around 3.6 and 1.9 eV [16] and 3.92 eV for undoped samples prepared by the sol–gel method [17]. However, band gap tuning of barium hexaferrite with Mn and Mg doping shows 2.94 eV for undoped sample and decreases to 1.90 eV with Mn and Mg doping [9]. The band gap of Sr-Ba hexaferrite with Tm-Tb doping [18] increased from 1.73 to 2.65 eV. $Ba_{1-x}La_xFe_{12}O_{19}$ (x = 0.10–0.04) ferrite nanoparticles [19] showed a band-gap increase from 1.87 to 2.53 eV. In this research, the values were observed in the visible band-gap region, which is promising for photocatalytic applications. The decrease in the band gap with La doping may have been due to the generation of sub-bands. That is, the E_g value decreased with La concentration. A continuous band was developed with the combination of conduction and sub-bands [20]. The decrease in the band gap was caused by the decrease in energy levels within the bands such that the electrons needed greater energy to travel from valence to conduction bands [21].

3.4. Dielectric Spectroscopy

Dielectric spectroscopy was used to explore the frequency of dielectric constant, *tanδ*, and conductivity. The plots are shown in Figures 7–10. The dielectric constant and *tanδ* decreased with frequency and were independent of frequency at higher frequencies. Such behavior can be explained using Maxwell–Wagner polarization [22–24]. According to this theory, heterogeneous materials such as hexaferrites consist of conducting grains separated by insulating grain boundaries. These grain boundaries oppose the electrical conduction and aid in polarization. When an electric field is applied, the charge carriers align at the grain boundaries due to high resistance. This builds up charge carriers near the grain boundaries, which induces large space-charge polarization and a large dielectric constant. In conduction, according to Koops [25], grain boundaries dominate at low frequencies and grains dominate at high frequencies. Thus, as the frequency increases, the grain boundaries lose control and a small space charge builds up, which results in a low value of polarization and decreases the value of the dielectric constant. At low frequencies, grain boundaries become more effective, and charge carriers need more energy to cross the boundaries, causing high loss. Grains are more effective at high frequencies and conduct more charge carriers. This results in loss tangent at high frequencies.

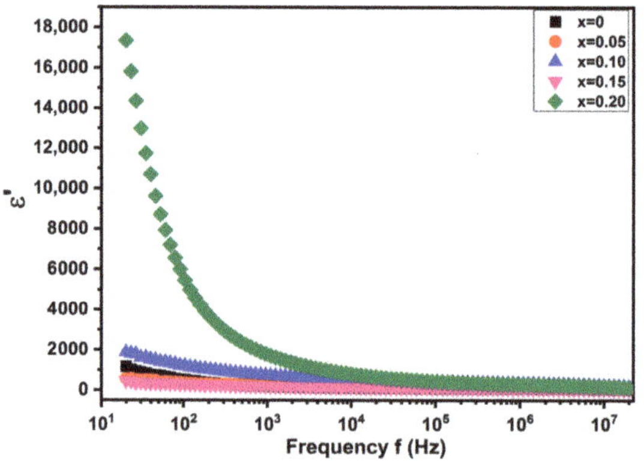

Figure 7. Variation in ε′ with frequency.

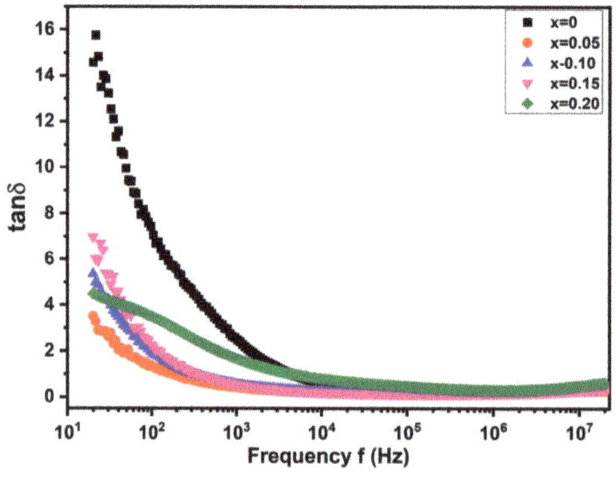

Figure 8. Variation in tanδ with frequency.

The AC conductivity increased with the log frequency, as the AC conductivity plots show (Figures 9 and 10). AC conductivity was independent of frequency at low frequencies but increased sharply at high frequencies. This behavior followed Jonscher's power law. When compared with the La concentration, the AC conductivity remained constant at x = 0.0, 0.05, and 0.15, but it increased rapidly for La concentrations of 0.1 and 0.2. This may have been due to electron hopping between Fe^{3+} and Fe^{2+} ions. The hopping of electrons increased as the frequency of the applied field increased. The frequency responses of the dielectric constant and the loss tangent to La concentration are shown in Figures 11 and 12. Electron hopping was caused by the higher electronegativity of La^{3+} (1.1) compared to Ba^{2+} (0.89) but which is lesser than Fe^{3+}. La^{3+} donate electrons easily. With the increase in La content, the number of charge carriers increased, which, again, increased the polarization and the dielectric constant. The dielectric constant increased as the frequencies increased, being highest for the sample with x = 0.2. On the other hand, the loss tangent decreased with an increase in frequency, and the loss was high for the pure barium hexaferrite. The loss decreased with an increase in the doping concentration of La (x = 0.05, 0.1, 0.15, and 0.2). The lowest concentration was found for the sample with the La concentration at x = 0.05 because of the smallest grain size of La at x = 0.

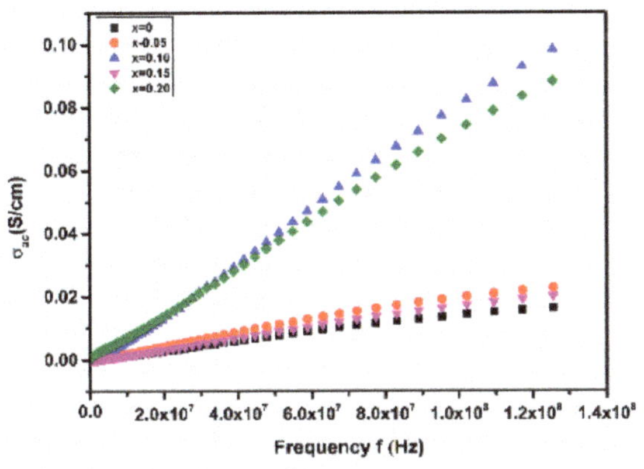

Figure 9. Variation in AC conductivity with frequency.

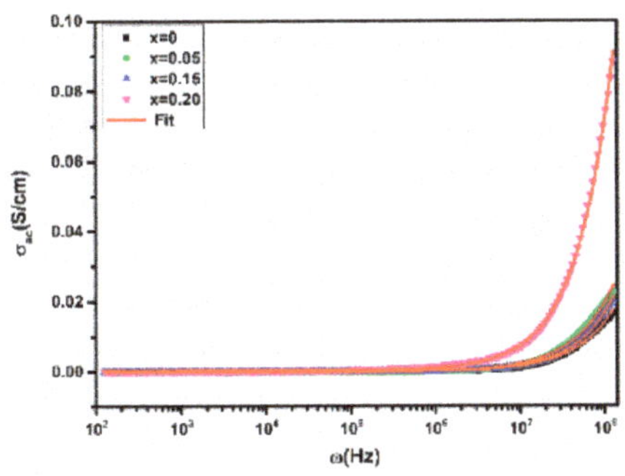

Figure 10. Jonscher's power law plot.

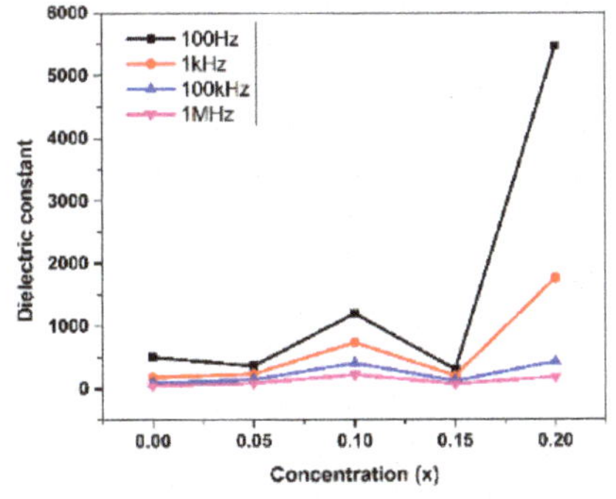

Figure 11. Variation in ε′ with La content.

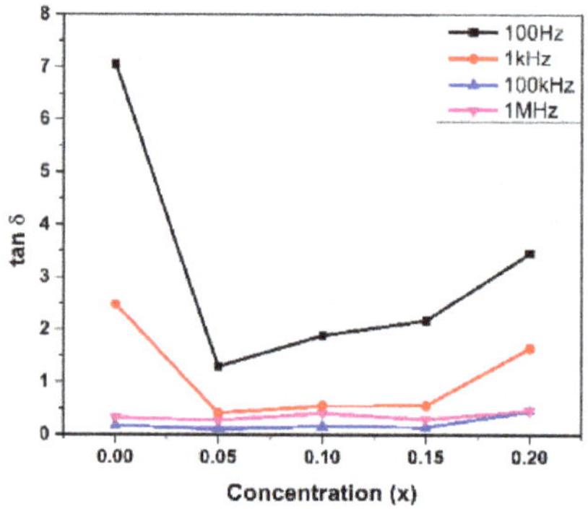

Figure 12. Variation in tanδ with La content.

4. Conclusions

La-doped barium hexaferrite, $BaLa_xFe_{12-x}O_{19}$ ($0 \leq x \leq 0.2$), was prepared using the solid-state reaction method. The X-ray diffraction patterns of the Rietveld refinement confirmed the formation of a single-phase hexagonal magnetoplumbite structure, which belonged to the *P6₃/mmc* space group. Its crystal size increased with the increase in La concentration because of the lattice distortion caused by the substitution of La^{3+} ions in the Fe^{3+} sites. FE-SEM images showed hexagonal plate-like grains of different sizes. The diffused reflectance spectra revealed a slight decrease in band gap with La content due to a decrease in energy levels within the valence and conduction bands. The dielectric studies showed a decrease in the loss tangent and an increase in the AC conductivity at high frequencies as the concentration of La^{3+} ions in the barium hexaferrite increased. The dielectric constant was found to be inversely proportional to the frequency.

Author Contributions: Synthesis and data analysis, G.S.; Writing—original draft preparation and editing, N.K.R.; Supervision, M.D.D. All authors have read and agreed to the published version of the manuscript.

Funding: The present research did not receive any external funding.

Institutional Review Board Statement: Not applicable.

Informed Consent Statement: Not applicable.

Data Availability Statement: (Data is confidential, is not available in public). To access any of the data presented in the present article, one can write to the corresponding author.

Acknowledgments: The authors were very grateful to the Manipal Academy of Higher Education, Manipal, for financial assistance; S. Rayaprol, UGC-DAE CSR Centre, Indore, for X-ray diffraction measurements; and Murari M.S., DST-PURSE Program, Mangalore University, Mangala Gangotri, for FESEM measurements.

Conflicts of Interest: The authors declare no conflicts of interest. The funders had no role in the design of the study; in the collection, analyses, or interpretation of the data; in the writing of the manuscript; or in the decision to publish the results.

References

1. Horta-Piñeres, S.; Britto Hurtado, R.; Avila-Padilla, D.; Cortez-Valadez, M.; Flores-López, N.S.; Flores-Acosta, M. Silver nanoparticle-decorated silver nanowires: A nanocomposite via green synthesis. *Appl. Phys. A Mater. Sci.* **2020**, *126*, 15. [CrossRef]
2. Ata, S.; Shaheen, I.; ul-Ayne, Q.; Ghafoor, S.; Sultan, M.; Majid, F.; Bibi, I.; Iqbal, M. Graphene and silver decorated ZnO composite synthesis, characterization and photocatalytic activity evaluation. *Diam. Relat. Mater.* **2018**, *90*, 26–31. [CrossRef]
3. Shaikh, S.F.; Ubaidullah, M.; Mane, R.S.; Al-Enizi, A.M. Types, Synthesis methods and applications of ferrites. In *Spinel Ferrite Nanostructures for Energy Storage Devices*; Mane, R.S., Jadhav, V.V., Eds.; Elsevier: Amsterdam, The Netherlands, 2020; pp. 51–82.

4. Sugimoto, M. The past, present, and future of ferrites. *J. Am. Ceram. Soc.* **2004**, *82*, 269–280. [CrossRef]
5. Pullar, R.C. Hexagonal ferrites: A review of the synthesis, properties and applications of hexaferrite ceramics. *Prog. Mater. Sci.* **2012**, *57*, 1191–1334. [CrossRef]
6. Goldman, A. *Modern Ferrite Technology*, 2nd ed.; Springer Science & Business Media: Pittsburgh, PA, USA, 2006; pp. 35–106.
7. Ismael, M. Ferrites as solar photocatalytic materials and their activities in solar energy conversion and environmental protection: A review. *Sol. Energy Mater. Sol. Cells* **2021**, *219*, 110786. [CrossRef]
8. Kaur, P.; Kaushik, A.; Singhal, S. Comparative evaluation of photocatalytic performance of hexagonal ferrites procured via sol-gel route. *Mater. Today Proc.* **2019**, *14*, 426–434. [CrossRef]
9. Bibi, I.; Naz, T.; Majid, F.; Kamal, S.; Ata, S.; Almoneef, M.M.; Iqbal, S.; Iqbal, M. Band gap tuning of $BaFe_{12}O_{19}$ with Mg and Mn doping to enhance solar light absorption for photocatalytic application. *Int. J. Energy Res.* **2021**, *7*, 11193–11205. [CrossRef]
10. Koutavarapu, R.; Reddy, C.V.; Syed, K.; Reddy, K.R.; Saleh, T.A.; Lee, D.Y.; Shim, J.; Aminabhavi, T.M. Novel Z-scheme binary zinc tungsten oxide/nickel ferrite nanohybrids for photocatalytic reduction of chromium (Cr (VI)), photoelectrochemical water splitting and degradation of toxic organic pollutants. *J. Hazard. Mater.* **2022**, *423*, 127044. [CrossRef] [PubMed]
11. Dillert, R.; Taffa, D.H.; Wark, M.; Bredow, T.; Bahnemann, D.W. Research Update: Photoelectrochemical water splitting and photocatalytic hydrogen production using ferrites (MFe_2O_4) under visible light irradiation. *J. Hazard. Mater.* **2015**, *3*, 104001. [CrossRef]
12. López, Y.O.; Vázquez, H.M.; Gutiérrez, J.S.; Velderrain, V.G.; Ortiz, A.L.; Martínez, V.C. Synthesis method effect of $CoFe_2O_4$ on its photocatalytic properties for H_2 production from water and visible light. *J. Nanomater.* **2015**, *16*, 76–85. [CrossRef]
13. Brandon, D.; Kaplan, W.D. *Microstructural Characterisation of Materials*, 2nd ed.; John Wiley & Sons, Ltd.: West Sussex, UK, 2008.
14. Morales, A.E. Use of diffuse reflectance spectroscopy for optical characterization of un-supported nanostructures. *Rev. Mex. Física* **2007**, *53*, 18–22.
15. Tholkappiyan, R.; Vishista, K. Synthesis and characterization of barium zinc ferrite nanoparticles: Working electrode for dye sensitized solar cell applications. *Sol. Energy* **2014**, *106*, 118–128. [CrossRef]
16. Anjum, S.; Sattar, M.; Mustafa, Z. Structural, optical and multiferroic properties of La^{+3}-substituted M-type barium hexaferrite properties $BaLa_xFe_{12-x}O_{19}$. *J. Mater. Sci. Mater. Electron.* **2021**, *32*, 232–245. [CrossRef]
17. Kaur, T.; Kaur, B.; Bhat, B.H.; Kumar, S.; Srivastava, A.K. Effect of calcination temperature on microstructure, dielectric, magnetic and optical properties of $Ba_{0.7}La_{0.3}Fe_{11.7}Co_{0.3}O_{19}$ hexaferrites. *Phys. B Condens. Matter.* **2015**, *456*, 206–212. [CrossRef]
18. Slimani, Y.; Almessiere, M.A.; Guner, S.; Alahmari, F.S.; Yasin, G.; Trukhanov, A.V.; Baykal, A. Influence of Tm–Tb substitution on magnetic and optical properties of Ba–Sr hexaferrites prepared by ultrasonic assisted citrate sol-gel approach. *Mater. Chem. Phys.* **2020**, *253*, 123324. [CrossRef]
19. Sajjad, M.; Qarnain, Z.; Chi, S.; Bi, J.; Hou, Z.; Li, W.; Qian, Z. Tuning Structural and Optical Properties of M-Type Barium Hexaferrites Under Different Substitutions of Lanthanum by Thermal Heating. Available online: https://ssrn.com/abstract=4182759 (accessed on 5 August 2022).
20. Almessiere, M.A.; Slimani, Y.; Sertkol, M.; Nawaz, M.; Baykal, A.; Ercan, I. The impact of Zr substituted Sr hexaferrite: Investigation on structure, optic and magnetic properties. *Results Phys.* **2019**, *13*, 102244. [CrossRef]
21. Alzaid, M. Enhancement in Optical Properties of Lanthanum-Doped Manganese Barium Hexaferrites under Different Substitutions. *Adv. Condens. Matter Phys.* **2021**, *2021*, 8849595. [CrossRef]
22. Kremer, F.; Schönhals, A. (Eds.) *Broadband Dielectric Spectroscopy*; Springer Science & Business Media: Pittsburgh, PA, USA, 2002; pp. 1–98.
23. Maxwell, J.C. *A Treatise on Electricity and Magnetism*; Dover: New York, NY, USA, 1954; Volume 2.
24. Wagner, K.W. The theory of incomplete dielectricity. *Ann. Phys.* **1913**, *40*, 817. [CrossRef]
25. Koops, C.G. On the Dispersion of Resistivity and Dielectric Constant of Some Semiconductors at Audiofrequencies. *Phys. Rev.* **1951**, *83*, 121–124. [CrossRef]

engineering proceedings

MDPI

Proceeding Paper

Discussion on Satisfaction and Loyalty of Women with Pelvic Treatment in Postpartum Period †

Wann-Jyi Horng [1,*] and Ming-Chia Yeh [2]

1 Department of Social Work, Chia Nan University of Pharmacy & Science, Tainan City 717301, Taiwan
2 Department of Hospital and Health Care Administration, Chia Nan University of Pharmacy & Science, Tainan City 717301, Taiwan; larry11yeh37@gmail.com
* Correspondence: hwj7902@mail.cnu.edu.tw
† Presented at the IEEE 5th Eurasia Conference on Biomedical Engineering, Healthcare and Sustainability, Tainan, Taiwan, 2–4 June 2023.

Abstract: The satisfaction and loyalty of women in the postpartum period with their pelvic health and related treatment were investigated to provide a reference for the decision-making of operators. The study results showed that women's age, education level, occupation, postpartum time, and mode of delivery influenced the cognition of and satisfaction with pelvic health recovery. The results also provided a reference for postpartum care centers or related practitioners to offer better choices in pelvic health care to secure the loyalty of women in the postpartum period. Based on these results, women in the postpartum period can improve their quality of life.

Keywords: pelvic conditioning; rehabilitation; satisfaction

1. Introduction

Women experience tremendous physical and mental changes during pregnancy and childbirth. If the loose pelvic floor ligaments and muscles do not recover within three months after delivery, it may cause a change in body shape, lower back pain, walking problems, and even severe pelvic organ prolapse and urinary incontinence. Such symptoms limit women's physical, psychological, and social functions and affect their quality of life. Most postpartum education focuses on breastfeeding, baby bathing, jaundice precautions, emotional distress, nutrition, and wound care after giving birth [1]. However, postpartum pelvic recovery and pelvic health are crucial for women in the postpartum period. Institutions educate women in the postpartum period about the importance of postpartum pelvic recovery and provide related services to improve women's postures, promote pelvic health, and regain previous self-confidence and style.

Researchers have also pointed out that dissatisfaction and poor self-image in women in the postpartum period are related to depression or other maternal psychological distress [2]. Therefore, this research was carried out to understand how much women in the postpartum period are satisfied with pelvic health recovery and the relationship between their satisfaction and the effectiveness of treatment. In this research, significant differences in satisfaction with pelvic health recovery and treatment were investigated to suggest how to treat and serve women in the postpartum period.

Citation: Horng, W.-J.; Yeh, M.-C. Discussion on Satisfaction and Loyalty of Women with Pelvic Treatment in Postpartum Period. *Eng. Proc.* **2023**, *55*, 95. https://doi.org/10.3390/engproc2023055095

Academic Editors: Teen-Hang Meen, Kuei-Shu Hsu and Cheng-Fu Yang

Published: 1 February 2024

2. Pelvic Health

The pelvis is a skeletal structure with muscles at the bottom for important functions of support and contraction [3] to hold the organs such as the bladder, intestine, and uterus. The sphincter muscles control the opening and closing of the urethra, vagina, and anus on the pelvic floor. Childbirth is conducted by the proper contraction and relaxation of the pelvic floor muscles [4,5]. Professional techniques are used to maintain the normal function of the pelvis by the muscles, tendons, and membranes. For its functions, the movements of various parts including the air joints and nodules are corporated. In Chinese medicine, the qi of the meridians and viscera is important to allow the smooth flow of qi and blood and coordination between the outside and the inside of the body. By improving the functions of the viscera, tissues, and organs, the physical balance in the pelvis and the mental well-being of women in the postpartum period can be obtained.

3. Satisfaction and Loyalty

3.1. Satisfaction

Cardozo [6] pointed out that satisfaction affected the behavior and willingness of customers to buy and maintain their loyalty to a product or service. Many researchers have different explanations for the definition of customer satisfaction. Boshoff and Gray [7] believed that customer satisfaction could be measured through the evaluation of their experience and reaction after using the product. Joewono and Kubota [8] pointed out that customer satisfaction was the evaluation of products and services based on customers' experience. According to Baker and Crompton [9], customer satisfaction refers to the psychological and emotional state after an experience. If the expectation before use exceeded that after use, customers were satisfied.

3.2. Loyalty

Customer loyalty refers to the customer's feelings after using a product or service. For short contacts to develop into loyalty, experience is always required. Therefore, the key to creating customer loyalty is to deliver more and higher customer value. The competition in the industry affects customers' decisions on whether to continue to use the products and services and relies on the premise that the products are convenient to use. Depending on the decision and comparison, customers choose products or services and decide to maintain a relationship with the service provider [10]. To maintain a long-term relationship, providers must provide education on the advantages of their products and services so that customers can visit or buy the product and service repeatedly. This is called repeated purchase or re-consumption [11]. Providers must offer information and promise to customers for mutual trust. Therefore, Stank et al. [12] defined loyalty as a long-term commitment to repurchase, including the cognition of the provider for repeated purchases.

4. Methods

A questionnaire survey was conducted for women in the postpartum period after receiving pelvic treatment. The responses were analyzed to obtain descriptive statistics and the results of variance, correlation, and regression analyses using SPSS21. The results were investigated to understand the satisfaction with and loyalty to the postpartum treatment. The results of this research provided postpartum care homes or related practitioners with a reference for better care and improvement of their treatments and services. The data were collected from 1 March to 30 April 2022, from 381 subjects who were in the postpartum period. In this research, the following hypotheses were proposed.

Hypothesis 1. *There are significant differences in satisfaction with pelvic treatment and service according to different demographic variables.*

Hypothesis 2. *Women in the postpartum period have different loyalty to the treatment and service depending on different demographic variables.*

Hypothesis 3. *Satisfaction with pelvic treatment and service of women in the postpartum period has a positive impact on their loyalty.*

5. Result and Discussion

5.1. Reliability of Questionnaire Survey

The four dimensions of satisfaction and loyalty are shown in Table 1. The Cronbach α values were 0.952, 0.935, 0.950, 0.950, and 0.921 for each dimension. The results indicated satisfactory reliability of the questionnaire survey.

Table 1. Reliability test results of questionnaire survey.

Dimension of Satisfaction and Loyalty	Cronbach's α	Number of Questions
Space and environment	0.952	6
Service attitude	0.935	5
Professionalism	0.95	5
Cost	0.95	4
Loyalty	0.921	6

5.2. Survey Result

Table 2 presents the descriptive statistics of the questionnaire survey results. Respondents aged 31−35 accounted for 40.4% of the total number of respondents. Those who were 36−40 years old and over 41 years old accounted for 37.8%. Those who were younger than 30 years old accounted for 21.8%. For the education level, 80.8% of respondents graduated from universities or higher. In total, 29.7% were students or unemployed while the remaining 70.3% had occupations. A total of 48.3% gave birth 1 year before the survey, 20.7% delivered their babies 1–3 years before the survey, and 31% did so 4 years before it. A total of 38.0% had experienced a Caesarean section. Table 3 shows the sum and the average of scores for the questions in each dimension. The average score of each item was above 4, which showed that the respondents were satisfied with the dimensions and had loyalty to them.

Table 2. Results of descriptive analysis of background data.

Variable	Item	Number of Respondents	Ratio (%)
Age	Under 25 years old (inclusive)	18	4.7
	26–30 years old	65	17.1
	31–35 years old	154	40.4
	36–40 years old	72	18.9
	41 years old (inclusive) and above	72	18.9
Education	High school/vocational (inclusive) or below	73	19.2
	University/college	247	64.8
	Graduate school (inclusive) or above	61	16.0
Occupation	None (student/housekeeping)	113	29.7
	service industry	100	26.2
	Military education	17	4.5
	Business	29	7.6
	Technology industry	19	5.0
	Manufacturing industry	32	8.4
	Others	71	18.6
Years after delivery	Within 1 year	184	48.3
	1–3 years	79	20.7
	4 years	118	31.0
Mode of delivery	Spontaneous delivery	236	61.9
	Caesarean section	119	31.2
	Both	26	6.8

Table 3. Descriptive statistical results of each dimension.

Dimension	Sum of Scores (Number of Questions)	Average Score
Space and environment	27.86 (6)	4.64
Service attitude	24.09 (5)	4.82
Professionalism	24.11 (5)	4.82
Cost	18.56 (4)	4.64
Loyalty	27.63 (6)	4.61

5.3. Variance Analysis

There was no significant difference between the groups of different ages for the four dimensions at the significance level of 0.05. The education level of the respondents showed a significant impact on satisfaction with the space and environment of the institutions for treatment and services. However, it did not influence the service attitude, professionalism, cost, and loyalty of the institutions at a significance level of 0.05. The occupations of respondents also did not impact their satisfaction with and loyalty to the institutions. The mode of delivery influenced the satisfaction with professionalism significantly but did not impact the space and environment, service attitude, cost, and loyalty. The results of multiple regression analysis of the four dimensions (Table 4) present that the adjusted R-squared (coefficient of determination) reached 0.80. This indicated a significant relationship between the four dimensions. In particular, a significant relationship between cost and loyalty was observed in the results of the multiple regression analysis.

Table 4. Multiple regression analysis results of four dimensions of satisfaction and loyalty.

Dimension	Coefficient	Standard Errors	t Value (p Value)
Space and environment	1.371	1.001	1.369 (0.172)
Service attitude	−0.127	0.034	−3.740 (0.000)
Professionalism	−0.444	0.122	−3.638 (0.000)
Cost	0.768	0.123	6.241 (0.000)
Loyalty	1.183	0.054	21.982 (0.000)
	0.80		
			F value (p value)
			381.781 (0.000)

6. Conclusions

The satisfaction and loyalty of women in the postpartum period were explored in this study. The results provide the required information for institutions of pelvic treatment and maternal services including beauty spas, beauty parlors, hospitals, and maternal care centers. It also can be used to establish a mother–infant-friendly institution with better treatments and services. Women in the postpartum period showed satisfaction and loyalty, the degrees of which differed with education levels and the delivery method they experienced. Those who were at different ages showed significant differences in loyalty, and so did their different occupations. The preliminary research results need to be complemented with future research.

Author Contributions: Writing—original draft, W.-J.H. and M.-C.Y.; Validation, W.-J.H.; Writing—review & editing, W.-J.H.; Supervision, W.-J.H.; Project Administration, M.-C.Y. All authors have read and agreed to the published version of the manuscript.

Funding: This research received no external funding.

Institutional Review Board Statement: Not applicable.

Informed Consent Statement: Not applicable.

Data Availability Statement: No new data was analyzed in this study. Data is not applicable to share in this article.

Acknowledgments: Thanks to Chia Nan University of Pharmacy & Science for providing administrative support.

Conflicts of Interest: The authors declare no conflict of interest.

References

1. Crowley, C. Addressing a gap in postpartum care. *Canadia* **2014**, *110*, 12–13.
2. Gjerdingen, D.; Fontaine, P.; Crow, S.; McGovern, P.; Center, B.; Miner, M. Predictors of Mothers' Postpartum Body Dissatisfaction. *Women Health* **2010**, *49*, 491–504. [CrossRef] [PubMed]
3. Ibrahim, I.K.; Hameed, M.M.A.; Taher, E.M.; Shaheen, E.M.; Elsawy, M.S.A.G. Efficacy of biofeedback-assisted pelvic floor muscle training in females with pelvic floor dysfunction. *Alex. J. Med.* **2015**, *51*, 137–142. [CrossRef]
4. Bartling, S.J.; Zito, P.M. Overview of Pelvic Floor Dysfunction Associated with Pregnancy. *Int. J. Childbirth Educ.* **2016**, *31*, 18–20.
5. Li, H.; Wu, R.F.; Qi, F.; Xiao, A.M.; Ma, Z.; Hu, Y.; Wang, Z.C. Postpartum pelvic floor function performance after two different modes of delivery. *Genet. Mol. Res. GMR* **2015**, *14*, 2994–3001. [CrossRef] [PubMed]
6. Cardozo, N.R. An experimenetal study of customer effort, expectation, and satisfaction. *J. Martketing Res.* **1965**, *2*, 244–249. [CrossRef]
7. Boshoff, C.; Gray, B. The Relationships between Service Quality, Customer Satisfaction and Buying Intentions in the Private Hospital Industry. *S. Afr. J. Bus. Manag.* **2004**, *35*, 27–37. [CrossRef]
8. Joewono, T.B.; Kubota, H. User Satisfaction with Paratransit in Competition with Motorization in Indonesia: Anticipation of Future Implications. *Transportation* **2007**, *33*, 337–355. [CrossRef]
9. Baker, D.A.; Crompton, J.L. Quality, satisfaction and behavioral Intentions. *Ann. Tour. Res.* **2000**, *27*, 785–804. [CrossRef]
10. Singh, J.; Sirdeshm, D. Agency and trust mechanisms in consumer satisfaction and loyalty judgments. *J. Acad. Mark. Sci.* **2000**, *28*, 150–167. [CrossRef]
11. Oliver, R.; Rust, R.; Varki, S. Customer Delight. *J. Retail.* **1997**, *73*, 311–336. [CrossRef]
12. Stank, T.P.; Goldsby, T.J.; Vickery, S.K.; Savitski, K. Logistics service performance: Estimating its influence on market share. *J. Bus. Logist.* **2003**, *24*, 27–55. [CrossRef]

engineering proceedings

MDPI

Proceeding Paper

Construction Safety Risk Assessment and Cause Analysis for High-Cable Tower Cranes †

Zhensong Ni [1], Shuri Cai [2,*] and Cairong Ni [2]

1 School of Big Data and Artificial Intelligence, Fujian Polytechnic Normal University, Fuzhou 350300, China; nizhensong@sina.com
2 Institute of Highway Science, Ministry of Transport, Beijing 100005, China; cairongni@gmail.com
* Correspondence: caishuri@126.com

† Presented at the IEEE 5th Eurasia Conference on Biomedical Engineering, Healthcare and Sustainability, Tainan, Taiwan, 2–4 June 2023.

Abstract: In the construction of modern bridges, tower cranes are used for vertical transportation and hoisting. In erecting and removing tower cranes, a high degree of risk occurs. Thus, we evaluated the risk of using tower cranes in construction and proposed preventive measures.

Keywords: high-cable tower; tower crane; risk events; truss

Citation: Ni, Z.; Cai, S.; Ni, C. Construction Safety Risk Assessment and Cause Analysis for High-Cable Tower Cranes. *Eng. Proc.* **2023**, *55*, 96. https://doi.org/10.3390/engproc2023055096

Academic Editors: Teen-Hang Meen, Kuei-Shu Hsu and Cheng-Fu Yang

Published: 26 February 2024

1. Introduction

Risks exist in life with the uncertainty of their occurrence. Haynes defined risk as an economic factor. Until the 1950s, risk management was not an independent discipline. Yet, the main research focuses on risks related to hazardous events and their probability of occurring. In this study, we proposed monitoring and early warning technology for preventing risks in bridge construction. The major risks in bridge construction are caused by the structural form of the bridge. The risk of occurrence is affected by various factors, such as changes in load and material, the inaccuracy of the calculation model, and human errors. Through risk assessment, a loss of investment, accidents, and social impacts can be reduced [1]. High-cable tower cranes are often used in construction, so the risk assessment of the cranes is important. Therefore, we analyzed the cause of risky events of the tower cranes to prevent accidents and propose preventive measures [2].

2. Risk of High-Cable Tower Crane

The safety risks of the high-cable tower crane include the toppling of the tower crane, the impact of buildings, falling off of heavy objects, broken arms of the tower crane, and falling of laborers. For those risks, the potential of the risk needs to be evaluated [3]. The risk assessment method of bridge construction and the mathematical model for uncertainty have been proposed using probability theory and statistics. We used random variables to indicate the times of the risk event and to evaluate the total amount of loss caused by risk events as follows:

$$L_{i.j} = \sum_{k=1}^{N_i} L_{i,j,k} \tag{1}$$

Referring to the previous research on the risk assessment of the construction safety of high-cable tower cranes, we classified the losses into three types: personnel fall, time delay, and monetary loss. A personnel fall refers to accidents in which workers fall on the site [4]. Time delay refers to a reduction in construction time due to the occurrence of risk events, and monetary loss is the value of material loss due to accidents. We weighted the three types of loss differently to obtain the combined effect of loss as follows:

$$L = L_h \times w_h + L_t \times w_t + L_m \times w_m \tag{2}$$

where w_h is the casualty weight, 0.45 according to the reference, w_t is the time delay weight, 0.25, and w_m is the currency loss weight, 0.3.

3. Probability Model of Risk Event

The probability model of the risk of the high-cable tower crane was defined as the probability and statistics related to the occurrence of the loss [5]. The state of personnel loss is expressed by the state equation z = r − s < 0, where *s* is the generalized effect under the action of a risk event *H*, and *r* is the generalized resistance. Then, the probability of loss is expressed as follows [6]:

$$P = p(R \leq S) = \int_R^{\infty} f(S)\mathrm{dS} \tag{3}$$

The generalized effect *S* is related to the risk event *H*, so the probability density function for (*S*, *H*) is

$$f(S,H) = f(S|H)f(H) \tag{4}$$

where *f*(*S* | *H*) is the conditional probability density function for the failure of the limit state in a given risk *H*, and *f*(*H*) is the probability density function for the risk state *H*. The law of total probability is expressed with the following equations:

$$P = p(R \leq S) = \int_R^{\infty} \left[\int_{-\infty}^{+\infty} f(S|H)f(H)\mathrm{dH} \right] \mathrm{dS} \tag{5}$$

$$\begin{aligned} P = p(R \leq S) &= \int_R^{\infty} \left[\int_{-\infty}^{+\infty} f(S|H)f(H)\mathrm{dH} \right] \mathrm{dS} \\ &= \int_0^{\infty} \left[\int_R^{\infty} f(S|H)\mathrm{dS} \right] f(H)\mathrm{dH} \end{aligned} \tag{6}$$

$$F_s(H) = \int_R^{+\infty} f(S|H)\mathrm{dS}$$

where

$$P = \int_0^{+\infty} F_s(H)f(H)\mathrm{dH}$$

Considering the upper and lower limits H_1 and H_2 of the level at which a risk event *H* causes a loss, the above formula is transformed into the following:

$$P = \int_{H_1}^{H_2} F_s(H)f(H)\mathrm{dH} \tag{7}$$

where $F_s(\overline{H}_i)$ is the limit state failure probability for the H_i level of risk events in *i* interval, $\Delta F_0(\overline{H}_i)$ is the interval probability for the *i* level of risk events, and *N* is the number of the interval for the divided risk event level.

4. As Low as Reasonably Possible (ALARP) Principle

We used the ALARP criterion to determine the safety risk criterion in the construction of the high-cable tower crane [7]. First, we defined param to measure the construction risk and the basic characteristics of its mathematical functions according to param according to special values of the function to determine the level of and response to risk. The representative value of the function for decision-making was obtained through the questionnaire survey with the divided level of the risk grade as shown in Table 1.

Table 1. Level of risk classification of decision-makers.

Level of Risk	Negligible	Acceptable (ALARP)
Division level	$0 \leq R \leq 3$	$3 \leq R \leq 5$
Reasonable Control (ALARP)	**Strictly Controlled (ALARP)**	**Unacceptable**
$5 \leq R \leq 6$	$6 \leq R \leq 7$	$7 \leq R \leq 10$

The determined risk level from the risk level interval division table is presented in Figure 1. The division of risk level intervals was determined by the attitude of the risk decision-maker, and the attitude depended on the risk effect function. The risk level interval determines the names of each division area according to the ALARP risk decision-making criteria. The entire risk area was divided into unacceptable risk areas, negligible risk areas, and ALARP areas. Risk situations that fall into unacceptable areas must be reduced using mandatory measures. The risk situation in the negligible area can be ignored as the risk probability is far lower than the construction safety threshold.

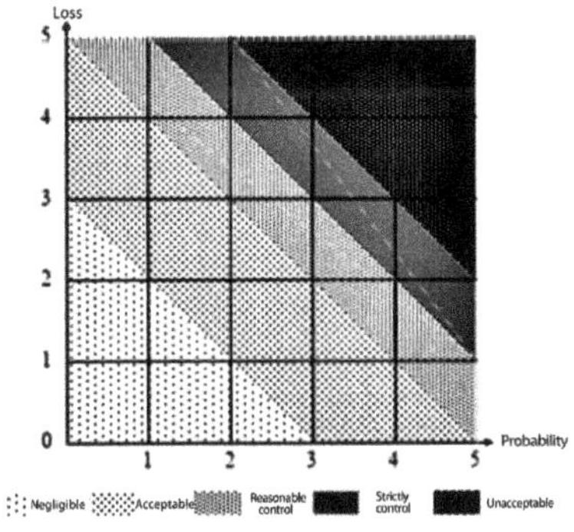

Figure 1. Risk level of decision-making.

The ALARP criteria are used for continuous functions and discrete functions. The risk matrix presents the embodiment of ALARP criteria in the form of a discrete function. The ALARP area, based on the different loss assessments of risk events, is divided into acceptable risk areas, reasonable risk control areas, and strict risk control areas.

5. Risk Event of High-Cable Tower Crane

- Tower crane overturning (GSTTD01):
  Due to the failure to meet the requirements for bearing capacity and ground flatness, it is easy for the tower body to tilt and the tower crane to overturn. During tower lifting and dismantling, tower overturning accidents may occur due to weak anchoring, insufficient strength or connection of steel wire ropes, control system failure, and improper use of traction capacity.
- Impact on tower body or other buildings (GSTTD02):
  Due to errors in cooperation between signal workers and operators, accidents sometimes occur when heavy objects collide with tower bodies or other buildings during the lifting process.
- Heavy objects falling off or tower crane arm breakage (GSTTD03):
  In the use of the tower crane, due to illegal lifting (the weight of the lifting object is not separated from the steel hook of the object below, the weight of the lifting object is unknown, and the operation action of the tower crane is suddenly changed), forced lifting, the hook falling off, the weight falling, or the tower crane limiter malfunctioning. These cause accidents of the tower crane overturning or arm breakage [8].

- Personnel fall (GSTTD04):
  In the process of lifting the tower crane, due to illegal operations, hidden dangers are not promptly eliminated, resulting in electric shock and personnel falling.

For the above four risk events, the losses of the Maanshan Tower crane with a high-cable tower were assessed. The assessment results are shown in Table 2. The loss assessment was calculated using Equation (2). The proportion of each weight is shown in Figure 2. The abscissa of Figure 2 shows the size of the loss assessment [9].

Table 2. Linear ratio of voltage to current.

Risk Event	Probability Level of Occurrence	Personnel Casualties	Time Delays	Currency Losses	Combined Effect	Loss Assessment
Tower crane overturning (GSTTD01)	2	3	1	2	2.4	4.4
Impact on tower body or other buildings (GSTTD02)	2	2	1	2	1.95	3.95
Heavy objects falling off or tower crane arm breakage (GSTTD03)	3	2	2	2	2.3	5.3
Personnel falling (GSTTD04)	2	2	1	2	1.95	3.95

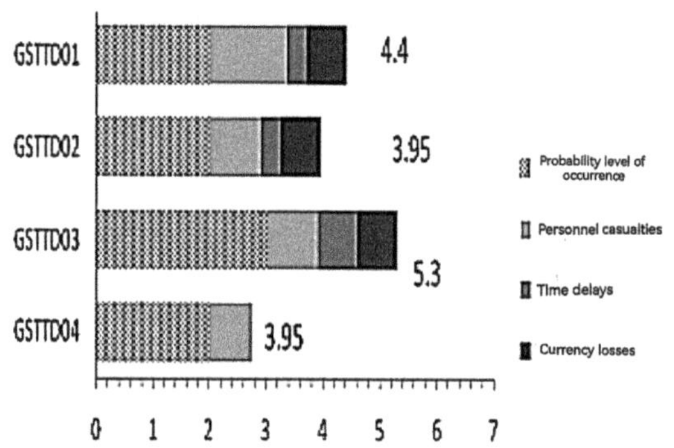

Figure 2. Proportion of each weight in risk assessment of steel tower column installation and loss assessment.

The loss assessment results of the risk situations during the construction of the high-cable tower crane were plotted in the risk level interval division table, as shown in Figure 2. The risk events GSTTD01, GSTTD02, GSTTD03, and GSTTD04 are located in the ALARP area. For the risks, reasonable safety precautions were required to reduce the occurrences. GSTTD01, GSTTD02, and GSTTD04 are located in the reasonable risk area and the acceptable risk area. General management measures were needed to reduce their construction risks without the need for further measures for the risks. For GSTTD03, prevention and caution were required to prevent the occurrence. In addition to general risk management, it was also necessary to consider the comparison result of the value of risk reduction and effect. Reasonable measures had to be taken for prevention and control to reduce its risks [10].

Possible risks need to be prevented and controlled. In Figure 3, in GSTTD03, heavy weight falling off or tower crane arm breaking are the risk events we focus on. Risk prevention and control measures must be formulated to reduce the risk of tower crane operation and avoid the occurrence of accidents [11].

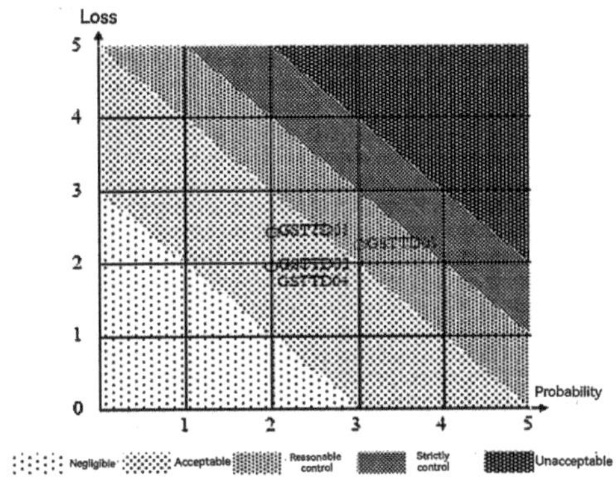

Figure 3. Risk event assessment.

6. Prevention and Control Technology

6.1. Tower Crane Safety Calculation

- Underframe: The upper part of the underframe is connected to the tower body, and the lower part is directly installed on a dedicated concrete foundation. It is composed of a base, a foundation section, and a diagonal brace [12].
- Tower body: The standard section of the tower body has two specifications according to its design strength, namely the lower tower body and the standard section. Each section is connected with 12-M39 special high-strength bolts, and the standard section of the same specification has interchangeability. The vertical members inside and outside the standard section are square cross-section members with a side length of 15 cm, while the horizontal members are square cross-section members with a side length of 10 cm. Each standard section has a length of 1.5 m, and every two standard sections are grouped. According to the construction requirements of the main tower, the tower crane used in this project has a height of 210 m.
- Lifting arm: The lifting arm is divided into 11 sections. There is a pull rod lifting point set on the upper chord of the third, seventh, and eighth sections of the arm. Each section of the arm is connected by a pin shaft, and a trolley traction mechanism is installed on the first section of the arm. The arm's end is equipped with a steel wire rope with an anti-torsion device. The maximum arm length is 70 m, and it can also be assembled into six types of arm lengths, including 65, 60, 52.5, 45, and 30 m. In this study, a 70 m boom was used for the analysis. The three main members on the outer side are square cross-section members with a side length of 15 cm. The remaining diagonal rods are steel pipes with an outer diameter of 7 cm and a wall thickness of 1 cm.
- Balancing arm: The balancing arm is divided into three sections, which are two sections of 7.5 m and one section of 4 m, connected by a pin. When the length of the lifting arm is 75, 65, or 60 m, the length of the balancing arm is 19 m. When the length of the lifting arm is 52.5 m, 45 m, or 30 m, the length of the balancing arm is 15 m. With the help of a pin, the balance arm and balance arm pull rod are connected to the rotating tower body and top as a whole, and there are railings and walkways on both sides of the balance arm. The two outer members of the balance arm are square cross-section members with a side length of 15 cm. The rest are steel pipes with an outer diameter of 10cm and a wall thickness of 1 cm.
- Upper and lower supports: The upper support is installed on top of the slewing bearing and connected to the inner ring of the slewing bearing. The lower support is a box-shaped support for the non-rotating part of the crane, which is equipped with an outer gear ring of the slewing support on its upper plane. The lower support is connected to the outer ring of the slewing support through bolts, and the relative rotational motion between the upper and lower supports is achieved through the use

of a slewing mechanism. The tower is connected to the upper support and turn-around, and the tower to the standard section below the lower support.

- Rotating tower body: The lower end of the rotating tower body is connected to the upper support with 16 high-strength bolts, the upper part is connected to the tower top pin, and the front and rear ear plates are connected to the balance arm and lifting arm, respectively. The upper part is equipped with a lifting weight limiter.
- Tower top: The tower top is an inclined cone, with the upper end connected to the lifting arm and balance arm through a pull rod to keep the two arms horizontal. The lower end is connected to the rotating tower body with four pins. To install the lifting arm pull rod and balance arm pull rod, a working platform and pulley block are installed on the upper part of the tower top.

6.2. Parameter Description: Q235 Type Steel

According to the "Code for Design of Steel Structures" (GB50017-2003), design tensile and compressive and flexural strength are f = 215 N/mm^2. Design shear strength is fv = 125 N/mm^2. The elastic modulus is E = 2.05 × 10^3 N/mm^2.

6.3. Model Establishment

The model-simplified tower crane consisted of a main support truss and a boom truss. The boom was connected by four cables to bear the load of counterweights and goods. The base was fixed [7]. A 3D beam element beam44 was used for the model of the truss of the tower crane. LINK10 was used to model the cable in Figure 4. Because the structural layout of the tower crane has repetitive characteristics, the basic structural cell of the truss is constructed and then copied. The counterweight of the tower crane and the lifted goods were treated as load equivalents. After constructing the tower crane structure, the strength analysis of the components in the structure and the layout analysis of the cable positions were carried out.

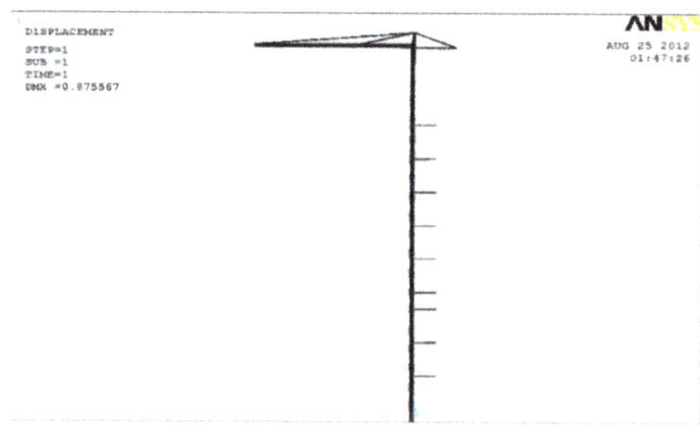

Figure 4. Truss of tower crane.

The only deformation of the tower crane without a load added at the boom end is shown in Figure 5. One end of the lifting arm was slightly raised by 0.8 m, and the entire arm end remained horizontal. Under normal use, the displacement and deformation of the structure met the requirements. When the lifting arm was 70 m away and the ultimate load was 3.4 tons, the entire structure moved down along one side of the lifting arm at a distance of 2 m. However, due to the long lifting arm, the entire structure remained approximately horizontal.

For a 3.4-ton weight applied at the end of the lifting arm, the internal force distribution of the entire tower crane is shown in Figure 6.

The maximum values of tension and pressure borne by the members were observed at the diagonal cables and lifting goods. In addition, the stress at the connection between the lifting arm and the tower body made the entire tower crane structure weak. According to the calculation results, the maximum stress borne by the cable was 907,180 Pa. Therefore, the stress met the strength requirement of less than 215 MPa.

Figure 5. Truss of tower crane with load.

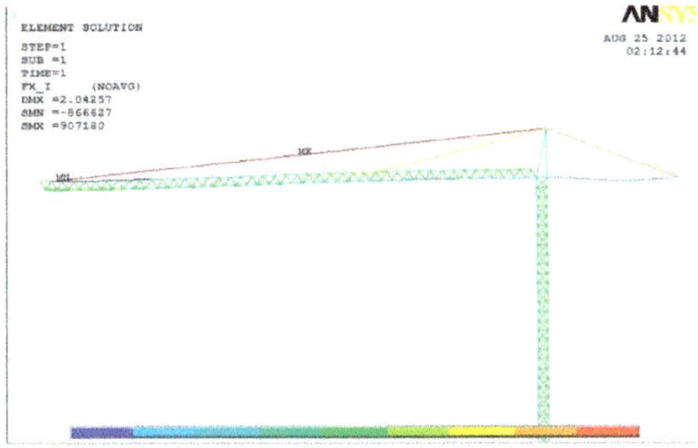

Figure 6. Internal force distribution diagram of tower crane truss.

Figures 7 and 8 show tower crane safety monitoring.

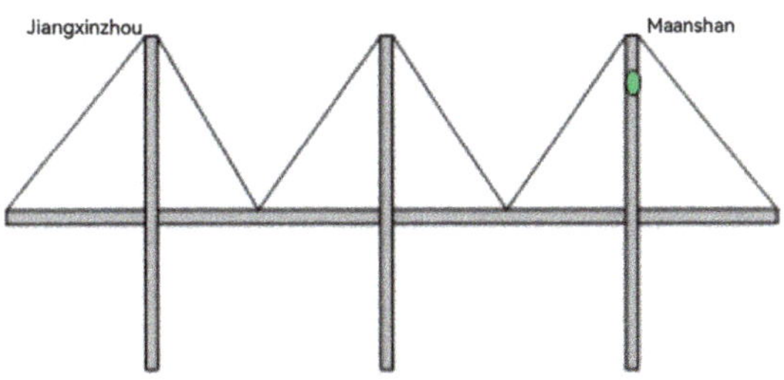

Figure 7. Overall schematic diagram of tower crane safety monitoring.

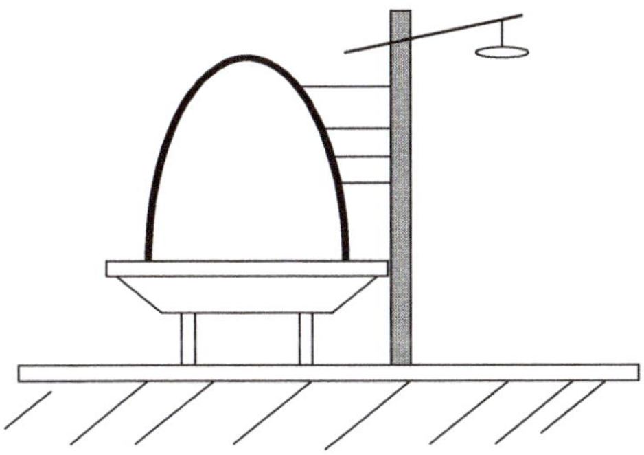

Figure 8. Cross-section diagram of tower crane safety monitoring.

A wireless cable force sensor was installed at the position of the diagonal cable to measure the changes in cable force and install an alarm on-site. Once the cable force sensor detected that the cable force value exceeded the calculated value, an alarm was issued immediately to stop the tower crane work and check the condition of the tower crane. In addition, the stress at the connection between the boom and the tower body made the entire tower crane structure weak. The stress change at the connection between the boom and the tower body was measured using wireless dynamic strain sensors, and an alarm was issued when the threshold was exceeded.

7. Conclusions

In bridge construction, bridge piers and cable towers are installed at a high altitude with small working spaces. The selection and layout of tower cranes are the keys to construction, as the time for the total construction of the entire bridge is affected by the performance of the cranes. The selection and layout of tower cranes are conducted with a comprehensive analysis process, and the relevant parameters of tower cranes must be determined based on the actual situation and construction requirements of each bridge's structural form, scale, and terrain conditions at the bridge location. On this basis, the tower crane must be reasonably matched and arranged according to its performance indicators. Based on the different sources of risk, the structure and construction characteristics of tower cranes can be analyzed. We determined safety prevention and control technology for tower crane construction and proposed corresponding preventive measures for risk events. A 3D beam element beam44 was used to model the truss of the tower crane, and LINK10 was used to model the cable. The maximum stress borne by the cable was calculated to be less than the threshold value. Using the proposed model of the truss of the crane, the changes in cable force and installed alarms on-site were measured. Once the cable force sensor detected the cable force exceeding the calculated value, an alarm was issued immediately to stop the tower crane operation and check its condition to avoid accidents.

Author Contributions: Conceptualization, Z.N., S.C. and C.N.; methodology, Z.N.; software, S.C.; validation, Z.N., S.C. and C.N.; formal analysis, Z.N.; investigation, Z.N.; resources, Z.N.; data curation, Z.N.; writing—original draft preparation, C.N.; writing—review and editing, C.N.; visualization, C.N.; supervision, S.C.; project administration, S.C.; funding acquisition, S.C. All authors have read and agreed to the published version of the manuscript.

Funding: This study was supported by the Natural Science Foundation of Fujian Province (2018J0106) and the phased research result of the provincial major research project on education and teaching reform of undergraduate colleges and universities in Fujian Province (project name: Research on Three Innovation Education Projects of Internet of Things Engineering (project no. fbjg202101018)).

Institutional Review Board Statement: Not applicable.

Informed Consent Statement: Not applicable.

Data Availability Statement: The data that support the findings of this study are available from the corresponding author upon reasonable request.

Acknowledgments: This work was supported by Fujian Polytechnic Normal University Grant in 2022~2023.

Conflicts of Interest: The authors declare no conflicts of interest.

References

1. Chen, B.; Su, J.L.; Lin, S.; Chen, G.; Zhuang, Y.; Tabatabai, H. Development and application of concrete arch bridges in China. *J. Asian Concr. Fed.* **2017**, *3*, 12–19. [CrossRef]
2. Chen, B.C.; Wang, T.L. Overview of concrete filled steel tube arch bridges in China. *Pract. Period. Struct. Des. Constr.* **2009**, *14*, 70–80. [CrossRef]
3. Wei, J.; Chen, B. Application and research advancement of long span concrete arch bridges abroad. *World Bridge* **2009**, *2*, 4–8.
4. Xie, X.; Qin, R.; Peng, W.; Deng, Z. Theoretical analysis of creep and shrinkage effects in SRC arch bridge. *Eng. Sci.* **2001**, *3*, 80–84.
5. Ministry of Employment and Labor (MOEL). *Safety Inspection Notice, 2020*; No. 2020-43; MOEL: Seoul, Republic of Korea, 2020.

6. Shepherd, G.; Kahler, R.; Cross, J. Crane fatalities—A taxonomic analysis. *Saf. Sci.* **2000**, *36*, 83–93. [CrossRef]
7. Olearczyk, J.; Al-Hussein, M.; Bouferguène, A. Evolution of the crane selection and on-site utili-zation process for modular construction multilifts. *Autom. Constr.* **2014**, *43*, 59–72. [CrossRef]
8. Riga, K.; Jahr, K.; Thielen, C.; Borrmann, A. Mixed integer programming for dynamic tower crane and storage area optimization on construction sites. *Autom. Constr.* **2020**, *120*, 103259. [CrossRef]
9. Chen, B.; Huang, Q. Study on the design of 600 m span concrete arch bridge. *J. China Foreign Highw.* **2006**, *26*, 80–82.
10. Salonga, J.; Gauvreau, P. Comparative study of the proportions, form, and efficiency of concrete arch bridges. *J. Bridge Eng.* **2014**, *19*, 04013010. [CrossRef]
11. Li, W.H. Suitable Middle-Pylon Stiffness and Mechanical Transmission Effecting Kilometer Level Multi-Pylon Continuous Suspension Bridge. Ph.D. Thesis, Beijing Jiaotong University, Beijing, China, 2017.
12. Guo, H. Wind Stability of Three-Tower Suspension Bridges under Construction. Master's Thesis, Zhejiang University of Technology, Hangzhou, China, 2010.

engineering proceedings

MDPI

Proceeding Paper

Impact of Gd_2O_3 Incorporation in Structural, Optical, Thermal, Mechanical, and Radiation Blocking Nature in HMO Boro-Tellurite Glasses †

Ashwitha Nancy D'Souza [1], M. I. Sayyed [2] and Sudha D. Kamath [1,*]

1 Department of Physics, Manipal Institute of Technology, Manipal Academy of Higher Education, Manipal 576104, India; nancyash983@gmail.com
2 Department of Physics, Faculty of Science, Isra University, Amman 11622, Jordan; drmabualssayed@gmail.com
* Correspondence: sudha.kamath@manipal.edu
† Presented at the IEEE 5th Eurasia Conference on Biomedical Engineering, Healthcare and Sustainability, Tainan, Taiwan, 2–4 June 2023.

Abstract: The glass system of B_2O_3-SiO_2-TeO_2-Bi_2O_3-ZnO-BaO doped with Gd_2O_3 (x = 0, 1, 2, 3, and 4 mol%) (BiTeGd-x) was prepared by using the melt-quench technique. The density of glasses increased from 5.323–5.579 g cm^{-3} for 0–4 mol% with an increase in Gd_2O_3 concentration. The simulation results obtained using Photon Shielding and Dosimetry (PSD) software (Phy-X version) produced the maximum mass attenuation coefficient (MAC) and minimum half-value layer (HVL) in the entire photon energy spectrum 0.015–15 MeV, suggesting the highest potential of BiTeGd-4 glass to act as a shield against low and high-energy radiation photons.

Keywords: Gd^{3+} ions; tellurite glass; radiation shielding; mass attenuation coefficient

Citation: D'Souza, A.N.; Sayyed, M.I.; Kamath, S.D. Impact of Gd_2O_3 Incorporation in Structural, Optical, Thermal, Mechanical, and Radiation Blocking Nature in HMO Boro-Tellurite Glasses. *Eng. Proc.* **2023**, *55*, 97. https://doi.org/10.3390/engproc2023055097

Academic Editors: Teen-Hang Meen, Kuei-Shu Hsu and Cheng-Fu Yang

Published: 26 March 2024

1. Introduction

The usefulness of glass materials for protection against radiation has been extensively explored by researchers. Simple fabrication techniques, non-toxicity, transparency, chemical durability, and chemical flexibility are the main contributing factors of glasses to enhance radiation shielding proficiency and achieve high density [1]. Certain glasses designed have transcended conventional shielding materials such as concretes and bricks with appropriate attenuation coefficients and half-value layers (HVL). Tellurite glasses show promising results for thermal and chemical stability with a low melting temperature and density [2].

Rare-earth oxide such as Gd_2O_3 has been reported to improve the physical, optical, and mechanical properties of tellurite glasses by increasing their density, refractive index, and hardness values [3]. Kaewjaeng et al. [4] suggested that Gd^{3+} doping reduced HVT significantly, allowing the glass to perform better than commercial X-ray windows, concrete, and bricks [5]. There are limited investigations on the effect of Gd^{3+} ions and the overall improvement of the stability of glasses and radiation shielding properties. The present study was carried out to explore the potential of this glass system in radiation field application by analyzing the physical, optical, and radiation-blocking development of B_2O_3-SiO_2-Gd_2O_3-TeO_2-Bi_2O_3-ZnO-BaO glass system (BiTeGd-x).

2. Materials and Methods

Melt quenching was conducted to produce the BiTeGd-x system where Gd_2O_3 mol% varied as 0, 1, 2, 3, and 4. A furnace temperature of 1100–1120 °C was used for its synthesis [6]. The density of prepared glasses was determined using Archimedes theory and distilled water. Carl Zeiss FESEM recorder was used to study its surface morphology through EDAX measurement. Theoretical values of terms to evaluate the gamma-ray shielding property such as mass attenuation coefficient (MAC), and HVL of the synthesized glasses were obtained by using Photon Shielding and Dosimetry (PSD)(Phy-X version) software in an energy region of 0.015–15 MeV [7].

3. Results and Discussion

3.1. Physical Properties

The synthesized Gd^{3+} tellurite glasses are shown in Figure 1, where Gd_2O_3 incorporation improved the transparency and changed the color of the glass from reddish-orange to light yellow. Archimedes' principle was used to determine the density of the sample (Table 1). Physical parameters such as molecular weight and molar volume were calculated for the fabricated glasses with other parameters using the following relations.

Figure 1. Pictures of synthesized BiTeGd glasses.

Table 1. Different physical parameters calculated for Gd^{3+}-doped glasses.

Sample Code	BiTeHost	BiTeGd-1	BiTeGd-2	BiTeGd-3	BiTeGd-4
Average molecular weight, M (g/mol)	159.76	161.78	163.81	165.84	167.87
Density, ρ (g/cc) (±0.01)	5.2844	5.4229	5.4429	5.4935	5.5793
Molar volume, V_m (cm^3)	30.231	29.8335	30.0967	30.1888	30.0882
Number density of Gd^{3+} ions in host glass, N_{Gd} ($\times 10^{23}$ ions/mol)	0	0.202	0.400	0.598	0.801
Inter-ionic separation between Gd^{3+}, r_i (nm)	0	36.727	29.236	25.566	23.202

For the number density of Gd^{3+} ions,

$$N_{Gd} = \frac{xN_{A\rho}}{M} \quad (1)$$

The interionic separation between Gd^{3+} ions was obtained with

$$r_i = \left(\frac{1}{N_{Gd}}\right)^{\frac{1}{3}} \quad (2)$$

All calculated parameters are summarized in Table 1. Gd^{3+} increased the density of the glass from 5.323 to 5.5793 gcm^{-3} from 0 to 4 mol% of Gd_2O_3. The high molecular weight of Gd_2O_3 compared to TeO_2 caused a density increase in the glass. The tendency of Gd^{3+} ions to form a closed-packed network by filling the interstitial spaces has also increased the density. Moreover, the Gd^{3+} ions have an ionic radius of 1.19 Å which is greater than that of Te^{4+} and Bi^{3+} ions (0.99 and 1.03 Å), which also increased the density. Molar volume (V_m) initially decreased but increased depending on the concentration of Gd_2O_3. BiTeGd-1 glass with the minimum molar volume confirmed polymerization in this glass network [8]. The number density of Gd^{3+} ions increased with the increase in Gd_2O_3 concentration. The interionic radius (r_i) decreased with successive addition of Gd_2O_3 molecules, indicating the shrinkage of ionic clouds due to Gd^{3+} ions.

The XRD images of the Gd^{3+} glasses are shown in Figure 2a. Sharp crystalline peaks were absent in the XRD images, assuring the amorphous nature of the current glasses. In addition, the broad hump observed in Bragg's angle of 20–30° also reflected the non-crystallinity of the glasses. The surface morphology of Gd^{3+}-doped tellurium borosilicate glass BiTeGd-2 was examined by using SEM images (Figure 2b).

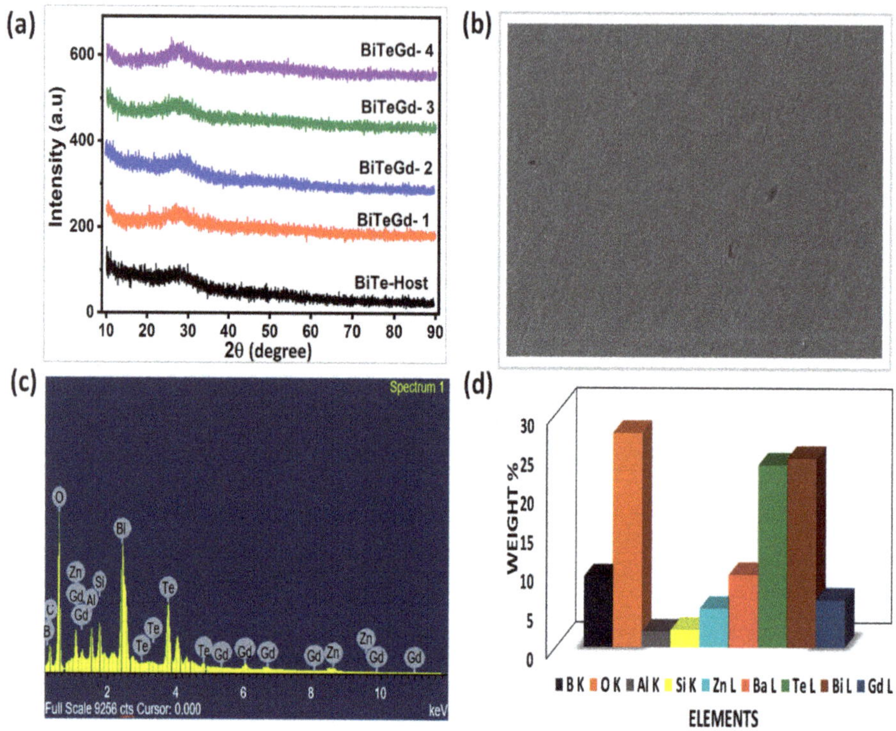

Figure 2. (**a**) XRD profiles, (**b**) SEM micrograph, (**c**) EDAX record, and (**d**) chart representing weight percentage of all the constituent elements in BiTeGd-2 glass.

The occurrence of smooth and homogenous texture in the SEM images without any cluster of unresolved particles proved the amorphous character of the synthesized glasses. The compositional analysis of the BiTeEu-2 glass was performed using EDAX measurement. The EDAX spectrum shows properly distributed elements such as boron (B), oxygen (O), silicon (Si), europium (Eu), tellurium (Te), bismuth (Bi), barium (Ba), and zinc (Zn) (Figure 2c). Aluminium (Al) was detected in the glass composition because of the alumina crucible used in the glass melting process. A bar chart representing the weights of all constituent elements (Figure 2d) exhibited the highest weight of Bismuth (Bi) as Bi has the heaviest atomic weight.

3.2. Optical Properties

The function of Gd^{3+} ions in enhancing the optical properties of the glass was studied by calculating parameters such as refractive index (n), dielectric constant (ε), molar refractivity (R_m), reflectance loss (R in %), and molar electron polarizability (α_m) [9] using the following set of equations.

$$\varepsilon = n^2 \quad (3)$$

$$R_m = \left(\frac{n^2 - 1}{n^2 + 2}\right) V_m \quad (4)$$

$$R = \left(\frac{n^2 - 1}{n^2 + 2}\right)\% \quad (5)$$

$$\alpha_m = \frac{3}{4\pi N_A} R_m \quad (6)$$

The values are presented in Table 2. The refractive index continuously increased with doping Gd_2O_3. The refractive index was influenced by a larger atomic radius (1.79 Å) of Gd [10] which is greater than that of tellurium (1.6 Å) and boron (0.98 Å). The higher polarization ability of cations resulting from higher cation radius of Gd^{3+} ions induces high *n* values, providing a platform for current Gd^{3+}-doped glasses in the non-linear optical application. An enhanced refractive index was also associated with a high dielectric constant, molar refractivity, reflectance loss, and molar electron values as shown in Table 2.

Table 2. Optical parameters of Gd^{3+}-doped glasses.

Sample Codes	R. I n	Dielectric Constant ε	R_m (cm^3)	R (%)	Molar Electron Polarizability ε_m ($Å^3$)	Band Gap E_g (eV)
BiTeHost	1.975	3.901	14.7544	0.9508	5.854	3.153
BiTeGd-1	1.989	3.955	14.8053	0.9778	5.875	2.957
BiTeGd-2	1.991	3.9663	14.9635	0.9832	5.938	3.121
BiTeGd-3	1.998	3.9939	15.0792	0.9969	5.984	3.196
BiTeGd-4	2.01	4.0407	15.1455	1.0204	6.01	2.791

The absorption spectra recorded in the UV-visible region for the Gd-doped glasses are shown in Figure 3a. The synthesized glasses including the undoped glass showed maximum absorption in the UV region (300–400 nm). In addition, all the glasses showed a broad low intense absorption peak around 500 nm which corresponds to the absorption of Bi^{3+} ions. Higher transmittance observed in the visible of all samples was proof of improved transparency of the Gd-doped glass.

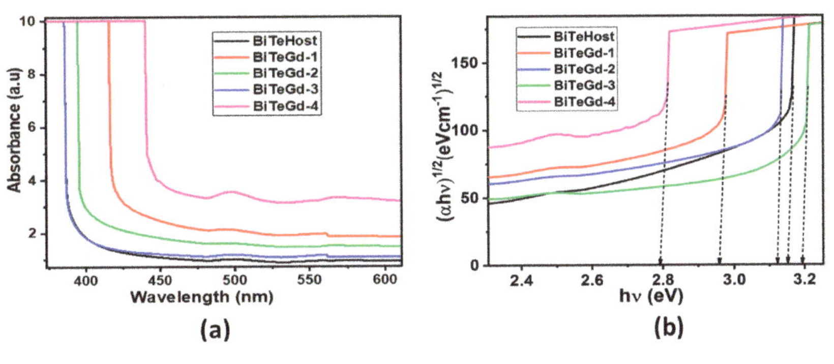

Figure 3. (**a**) UV-visible absorption spectra and (**b**) Tauc's plots drawn for BiTeGd glasses.

The relationship between absorbance and optical band gap E_g was provided by Tauc's relation as follows.

$$\alpha = \frac{B(h\nu - E_g)^{\gamma}}{h\nu} \quad (7)$$

where α is the absorption coefficient, also given by α = 2.303 A/t, A is the absorbance, t is the thickness of the sample material, B is the band tailing parameter, and the exponent γ depends on the kind of electronic transition mechanism. Because glasses are amorphous, we took γ = 2 as the transitions are indirect in nature. The resulting Tauc plot is represented in Figure 3b, where the extrapolation of the linear part of the curve is taken as E_g (Table 2).

The E_g values showed that the band gap of Gd^{3+}-doped glass was less than that of the undoped glass except for the BiTeGd-3 sample.

3.3. Radiation Shielding Parameters

MAC and HVL data simulated by Phy-X/PSD software in the 0.015–15 MeV photon energy spectrum are represented in Figure 4 [7,10]. The influence of energy on the MAC values of the glass was evident from the rapidly falling trend in the lower energy range, constancy in the intermediate range, and an increase at the higher end of the spectrum. This showed the occurrence of photoelectric absorption, Compton scattering, and electron–positron pair formation in the three energy regions. The two sharp peaks at 0.035 and 0.1 MeV were the K-edges associated with Te and Bi. Furthermore, the effect of varying the Gd_2O_3 content in MAC graphs on the continuous increase in MAC with the Gd^{3+} content was studied. Such an effect was observed because of the increasing density values of Gd_2O_3 from 5.323 to 5.579 gcm^{-3} from 0 to 4 mol% in concentration. Therefore, BiTeGd-4 glass exhibited the maximum attenuation compared to other Gd glasses.

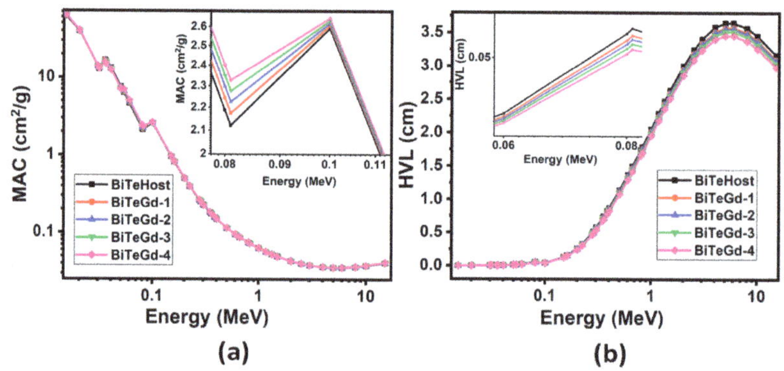

Figure 4. (**a**) MAC and (**b**) HVL of BiTeGd glasses simulated for energies in 0.015–15 MeV range using Phy-X/PSD software.

In Figure 4b, low-energy photons generated low HVL values due to the high photon absorption mechanism. HVL increased continuously in the energy range of 0.1–5 MeV, showing the maximum value at 5 MeV and a decrease to the energy of 15 MeV. This implied that when the energy increased, the radiation penetrated deeper into the glass, and therefore it was essential to increase the thickness of the material to shield it from the high radiation photons. The maximum HVL was found at 5 MeV for BiTeGd-4 glass as 3.4619 cm, which was thick enough to protect the glass from high-energy radiation. Additionally, the HVL decreased with successive doping of Gd_2O_3 in the range of 0.015–15 MeV suggesting that BiTeGd-4 glass had the lowest HVL.

4. Conclusions

Improved transparency with enhanced density values was proved with tellurite glasses doped with Gd_2O_3. Non-crystalline nature and smooth glass surface morphology were verified by XRD and FESEM results. Optical parameters including the refractive index continuously increased with an increase in Gd_2O_3. MAC and HVL parameters computed by PSD software in the energy range of 0.015–15 MeV showed that BiTeGd-4 was the optimum glass for gamma radiation shielding.

Author Contributions: Writing and data collection: A.N.D.; review and interpretation: M.I.S.; conceptualization and supervision: S.D.K. All authors have read and agreed to the published version of the manuscript.

Funding: The authors have not received any funding support for the research.

Institutional Review Board Statement: Not applicable.

Informed Consent Statement: Not applicable.

Data Availability Statement: All data generated or analyzed during this study are included in this published article.

Acknowledgments: One of the authors, Ashwitha Nancy D'Souza, expressed their gratitude to the Manipal Academy of Higher Education for providing a Ph.D. research fellowship (TMA Pai Fellowship).

Conflicts of Interest: The authors declare that there are no conflicts of interest.

References

1. Al-Hadeethi, Y.; Sayyed, M.I. Effect of Gd_2O_3 on the radiation shielding characteristics of Sb_2O_3–PbO–B_2O_3–Gd_2O_3 glass system. *Ceram Int.* **2020**, *46*, 13768–13773. [CrossRef]
2. Azianty, S.; Yahya, A.K. Enhancement of elastic properties by WO_3 partial replacement of TeO_2 in ternary (80 − X)TeO_2-20PbO-xWO_3 glass system. *J. Non-Cryst. Solids* **2013**, *378*, 234–240. [CrossRef]
3. Azlan, M.N.; Eevon, C.; Halimah, M.K.; El-Mallawany, R.; Hii, S.L. Effect of Gd^{3+} on optical and thermal properties of tellurite glass. *J. Theor. Appl. Phys.* **2020**, *14*, 137–147. [CrossRef]
4. Kaewjaeng, S.; Wantana, N.; Kothan, S.; Rajaramakrishna, R.; Kim, H.J.; Limsuwan, P.; Kaewkhao, J. Effect of Gd_2O_3 on the radiation shielding, physical, optical and luminescence behaviors of Gd_2O_3–La_2O_3–ZnO–B_2O_3–Dy_2O_3 glasses. *Radiat. Phys. Chem.* **2021**, *185*, 109500. [CrossRef]
5. Gaylan, Y.; Bozkurt, A.; Avar, B. Investigating Thermal and Fast Neutron Shielding Properties of B_4C-, B_2O_3-, Sm_2O_3-AND Gd_2O_3-doped Polymer Matrix Composites using Monte Carlo Simulations. *Süleyman Demirel Üniversitesi Fen Edebiyat Fakültesi Fen Dergisi* **2021**, *16*, 490–499. [CrossRef]
6. D'Souza, N.A.; Padasale, B.; Murari, M.S.; Karunakara, N.; Sayyed, M.I.; Elsafi, M.; Al-Ghamdi, H.; Almuqrin, A.H.; Kamath, S.D. TeO_2 for enhancing structural, mechanical, optical, gamma and neutron radiation shielding performance of bismuth borosilicate glasses. *Mater. Chem. Phys.* **2023**, *293*, 126657. [CrossRef]
7. Şakar, E.; Özpolat, Ö.F.; Alım, B.; Sayyed, M.I.; Kurudirek, M. Phy-X/PSD: Development of a user friendly online software for calculation of parameters relevant to radiation shielding and dosimetry. *Radiat. Phys. Chem.* **2020**, *166*, 108496. [CrossRef]
8. Gökçe, M.; Koçyiğit, D. Structural and optical properties of Gd^{+3} doped Bi_2O_3-GeO_2 glasses and glass-ceramics. *Mater. Res. Express* **2018**, *36*, 4620–4625. [CrossRef]
9. Wagh, A.; Raviprakash, Y.; Upadhyaya, V.; Kamath, S.D. Composition dependent structural and optical properties of PbF_2-TeO_2-B_2O_3-Eu_2O_3 glasses. *Spectrochim. Acta Part A Mol. Biomol. Spectrosc.* **2015**, *151*, 696–706. [CrossRef] [PubMed]
10. Samanta, B.; Dutta, D.; Ghosh, S. Synthesis and different optical properties of Gd_2O_3 doped sodium zinc tellurite glasses. *Phys. B Condens. Matter* **2017**, *515*, 82–88. [CrossRef]

MDPI AG
Grosspeteranlage 5
4052 Basel
Switzerland
Tel.: +41 61 683 77 34

Engineering Proceedings Editorial Office
E-mail: engproc@mdpi.com
www.mdpi.com/journal/engproc